SPACE SCIENCE SERIES
Tom Gehrels, General Editor

Planets, Stars and Nebulae, Studied with Photopolarimetry
Tom Gehrels, editor, 1974, 1133 pages

Jupiter
Tom Gehrels, editor, 1976, 1254 pages

Planetary Satellites
Joseph A. Burns, editor, 1977, 598 pages

Protostars and Planets
Tom Gehrels, editor, 1978, 756 pages

Asteroids
Tom Gehrels, editor, 1979, 1181 pages

Comets
Laurel L. Wilkening, editor, 1982, 766 pages

Satellites of Jupiter
David Morrison, editor, 1982, 972 pages

Venus
Donald M. Hunten et al., editors, 1983, 1143 pages

Saturn
Tom Gehrels and Mildred S. Matthews,
editors, 1984, 968 pages

Planetary Rings
Richard Greenberg and André Brahic,
editors, 1984, 784 pages

Protostars & Planets II
David C. Black and Mildred S. Matthews,
editors, 1985, 1293 pages

Satellites
Joseph A. Burns and Mildred S. Matthews,
editors, 1986, 1021 pages

The Galaxy and the Solar System
Roman Smoluchowski, John N. Bahcall,
and Mildred S. Matthews,
editors, 1986, 485 pages

Meteorites and the Early Solar System
John F. Kerridge and Mildred S. Matthews,
editors, 1988, 1269 pages

Mercury
Faith Vilas, Clark R. Chapman,
and Mildred S. Matthews,
editors, 1988

METEORITES AND THE EARLY SOLAR SYSTEM

METEORITES AND THE EARLY SOLAR SYSTEM

Editors

John F. Kerridge
Mildred Shapley Matthews

With 69 collaborating authors

THE UNIVERSITY OF ARIZONA PRESS
TUCSON

About the cover:
Painting by William K. Hartmann. The cover painting illustrates the contemporary view of the origin of meteorites in the collision of two asteroidal bodies. This view shows the collision of a dark C-class object with a higher-albedo S-class object. Fragments and vaporized material are being dislodged. The collision will unfold over a matter of a minute or more, with much of the mass of both bodies being fragmented, with the resultant possible production of monomict breccias, polymict breccias, a Hirayma family, and other phenomena familiar to meteorite and asteroid observers.

The back cover shows a chondrule in the Felix CO carbonaceous chondrite. The width is 200 microns.

THE UNIVERSITY OF ARIZONA PRESS

Copyright © 1988
The Arizona Board of Regents
All Rights Reserved

This book was set in 10/12 Times Roman.
Manufactured in the U.S.A.

Library of Congress Cataloging-in-Publication Data

Meteorites and the early solar system/editors, John F. Kerridge,
  Mildred Shapley Matthews; with 69 collaborating authors.
      p.   cm. — (Space science series)
    Bibliography: p.
    Includes index.
    ISBN 0-8165-1063-6 (alk. paper)
    1. Meteorites.   2. Solar System.   I. Kerridge, John F., 1937–
  II. Matthews, Mildred Shapley.   III. Series.
  QB755.M485 1988
  523.5′1 — dc19   88-21820
                                                               CIP

British Library Cataloguing in Publication data are available.

To
Harvey H. Nininger
and
Harold C. Urey

CONTENTS

	COLLABORATING AUTHORS	xii
	FOREWORD	xiii
	PREFACE	xv

Part 1—INTRODUCTION

1.1.	OVERVIEW AND CLASSIFICATION OF METEORITES *D. W. G. Sears and R. T. Dodd*	3

Part 2—SOURCE REGIONS

2.1.	ASTEROIDS AND METEORITES *G. W. Wetherill and C. R. Chapman*	35

Part 3—SECONDARY PROCESSING

3.1.	SIGNIFICANCE OF SECONDARY PROCESSING *J. F. Kerridge*	71
3.2.	IGNEOUS ACTIVITY IN THE EARLY SOLAR SYSTEM *R. H. Hewins and H. E. Newsom*	73
3.3.	THERMAL METAMORPHISM *H. Y. McSween, D. W. G. Sears and R. T. Dodd*	102
3.4.	AQUEOUS ALTERATION *M. Zolensky and H. Y. McSween, Jr.*	114
3.5.	METEORITE REGOLITHIC BRECCIAS *T. E. Bunch and R. S. Rajan*	144
3.6.	SHOCK EFFECTS IN METEORITES *D. Stöffler, A. Bischoff, V. Buchwald and A. E. Rubin*	165

Part 4—IRRADIATION EFFECTS

4.1.	IRRADIATION RECORDS IN METEORITES *M. W. Caffee, J. N. Goswami, C. M. Hohenberg, K. Marti and R. C. Reedy*	205

CONTENTS

Part 5—SOLAR-SYSTEM CHRONOLOGY

5.1. PRINCIPLES OF RADIOMETRIC DATING — 249
G. R. Tilton

5.2. AGE OF THE SOLAR SYSTEM — 259
G. R. Tilton

5.3. DATING OF SECONDARY EVENTS — 276
G. Turner

5.4. COMPACTION AGES — 289
M. W. Caffee and J. D. Macdougall

Part 6—THE EARLY SOLAR SYSTEM

6.1. CHONDRITES AND THE EARLY SOLAR SYSTEM: INTRODUCTION — 301
J. F. Kerridge

6.2. PROTOSTELLAR COLLAPSE, DUST GRAINS AND
P. Cassen and A. P. Boss

6.3. A REVIEW OF SOLAR NEBULA MODELS — 329
J. A. Wood and G. E. Morfill

6.4. FORMATION PROCESSES AND TIME SCALES FOR METEORITE PARENT BODIES — 348
S. J. Weidenschilling

Part 7—CHEMISTRY OF CHONDRITES AND THE EARLY SOLAR SYSTEM

7.1. THE COSMOCHEMICAL CLASSIFICATION OF THE ELEMENTS — 375
J. W. Larimer

7.2. ELEMENTAL VARIATIONS IN BULK CHONDRITES: A BRIEF REVIEW — 390
G. W. Kallemeyn

7.3. REFRACTORY LITHOPHILE ELEMENTS — 394
J. W. Larimer and J. T. Wasson

7.4. SIDEROPHILE-ELEMENT FRACTIONATION — 416
J. W. Larimer and J. T. Wasson

7.5. MODERATELY VOLATILE ELEMENTS — 436
H. Palme, J. W. Larimer and M. E. Lipschutz

7.6. HIGHLY LABILE ELEMENTS — 462
M. E. Lipschutz and D. S. Woolum

7.7.	OXIDATION STATE IN CHONDRITES A. E. Rubin, B. Fegley and R. Brett	488
7.8.	PLANETARY COMPOSITIONS S. R. Taylor	512
7.9.	TRAPPED NOBLE GASES IN METEORITES T. D. Swindle	535
7.10.	PLANETARY ATMOSPHERES D. M. Hunten, R. O. Pepin and T. C. Owen	565

Part 8—MAGNETIC FIELDS IN THE EARLY SOLAR SYSTEM

8.1.	MAGNETIC STUDIES OF METEORITES N. Sugiura and D. W. Strangway	595

Part 9—CHONDRULES

9.1.	PROPERTIES OF CHONDRULES J. N. Grossman, A. E. Rubin, H. Nagahara and E. A. King	619
9.2.	EXPERIMENTAL STUDIES OF CHONDRULES R. H. Hewins	660
9.3.	FORMATION OF CHONDRULES J. N. Grossman	680
9.4.	ENERGETICS OF CHONDRULE FORMATION E. H. Levy	697

Part 10—PRIMITIVE MATERIAL SURVIVING IN CHONDRITES

10.1.	POTENTIAL SIGNIFICANCE OF PRISTINE MATERIAL J. F. Kerridge	715
10.2.	PRIMITIVE MATERIAL SURVIVING IN CHONDRITES: MATRIX E. R. D. Scott, D. J. Barber, C. M. Alexander, R. Hutchison and J. A. Peck	718
10.3.	PRIMITIVE MATERIAL SURVIVING IN CHONDRITES: REFRACTORY INCLUSIONS G. J. MacPherson, D. A. Wark and J. T. Armstrong	746
10.4.	PRIMITIVE MATERIAL SURVIVING IN CHONDRITES: MINERAL GRAINS I. M. Steele	808

CONTENTS

10.5. ORGANIC MATTER IN CARBONACEOUS
CHONDRITES, PLANETARY SATELLITES,
ASTEROIDS AND COMETS 819
J. R. Cronin, S. Pizzarello and D. P. Cruikshank

Part 11—MICROMETEORITES

11.1. INTERPLANETARY DUST PARTICLES 861
J. P. Bradley, S. A. Sandford and R. M. Walker

Part 12—INHOMOGENEITY OF THE NEBULA

12.1. HETEROGENEITY IN THE NEBULA: EVIDENCE
FROM STABLE ISOTOPES 899
M. H. Thiemens

Part 13—SURVIVAL OF PRESOLAR MATERIAL IN METEORITES

13.1. CIRCUMSTELLAR MATERIAL IN METEORITES:
NOBLE GASES, CARBON AND NITROGEN 927
E. Anders

13.2. INTERSTELLAR CLOUD MATERIAL IN
METEORITES 956
E. Zinner

13.3. ASTROPHYSICAL IMPLICATIONS OF PRESOLAR
GRAINS 984
J. A. Nuth, III

Part 14—NUCLEOSYNTHESIS

14.1. SOLAR-SYSTEM ABUNDANCES AND PROCESSES
OF NUCLEOSYNTHESIS 995
D. S. Woolum

14.2. STELLAR NUCLEOSYNTHESIS AND CHEMICAL
EVOLUTION OF THE SOLAR NEIGHBORHOOD 1021
D. D. Clayton

14.3. IMPLICATIONS OF ISOTOPIC ANOMALIES FOR
NUCLEOSYNTHESIS 1063
T. Lee

Part 15—EXTINCT RADIONUCLIDES AND NUCLEOCOSMOCHRONOLOGY

15.1. EXTINCT RADIONUCLIDES 1093
F. A. Podosek and T. D. Swindle

15.2. NUCLEOCOSMOCHRONOLOGY 1114
F. A. Podosek and T. D. Swindle

15.3. IODINE-XENON DATING *T. D. Swindle and F. A. Podosek*	1127

Part 16—SUMMARY

16.1. BOUNDARY CONDITIONS FOR THE ORIGIN OF THE SOLAR SYSTEM *J. F. Kerridge and E. Anders*	1149
16.2. FUTURE DIRECTIONS IN METEORITE RESEARCH *E. Anders and J. F. Kerridge*	1155
APPENDIX 1: MINERAL NAMES	1187
APPENDIX 2: IMPACT BRECCIAS	1191
APPENDIX 3: CHONDRITE BULK ELEMENTAL ANALYSES	1195
APPENDIX 4: SOLAR-SYSTEM ABUNDANCES OF THE NUCLIDES	1199
APPENDIX 5: IODINE-XENON AGES	1209
GLOSSARY	1219
ACKNOWLEDGMENTS	1249
INDEX	1253

COLLABORATING AUTHORS

C. Alexander, *718*
E. Anders, *xiii, 927, 1155*
J. T. Armstrong, *746*
D. Barber, *718*
A. Bischoff, *165*
A. P. Boss, *304*
J. P. Bradley, *861*
R. Brett, *488*
V. Buchwald, *165*
T. E. Bunch, *144*
M. W. Caffee, *205, 289*
P. Cassen, *304*
C. R. Chapman, *35*
D. D. Clayton, *1021*
J. R. Cronin, *819*
D. P. Cruikshank, *819*
R. T. Dodd, *3, 102*
B. Fegley, *488*
J. N. Goswami, *205*
J. N. Grossmann, *619, 680*
R. H. Hewins, *73, 660*
C. M. Hohenberg, *205*
D. M. Hunten, *565*
R. Hutchison, *718*
G. W. Kallemeyn, *390*
J. F. Kerridge, *xv, 71, 301, 715, 1149, 1155*
E. A. King, *619*
J. W. Larimer, *375, 394, 416, 436*
T. Lee, *1063*
E. H. Levy, *697*
M. E. Lipschutz, *436, 462*
J. D. Macdougall, *289*
G. J. MacPherson, *746*
K. Marti, *205*

H. Y. McSween, *102, 114*
G. E. Morfill, *329*
H. Nagahara, *619*
H. E. Newsom, *73*
J. A. Nuth, *984*
T. C. Owen, *565*
H. Palme, *436*
J. A. Peck, *718*
R. O. Pepin, *565*
S. Pizzarello, *819*
F. A. Podosek, *1093, 1114, 1127*
R. S. Rajan, *144*
R. C. Reedy, *205*
A. E. Rubin, *165, 488, 619*
S. A. Sandford, *861*
E. R. D. Scott, *718*
D. W. G. Sears, *3, 102*
I. M. Steele, *808*
D. Stöffler, *165*
D. W. Strangway, *595*
N. Sugiura, *595*
T. D. Swindle, *535, 1093, 1114, 1127*
S. R. Taylor, *512*
M. H. Thiemens, *899*
G. R. Tilton, *249, 259*
G. Turner, *276*
R. M. Walker, *861*
D. A. Wark, *746*
J. T. Wasson, *394, 416*
S. J. Weidenschilling, *348*
G. W. Wetherill, *35*
J. A. Wood, *329*
D. S. Woolum, *462, 995*
E. Zinner, *956*
M. Zolensky, *114*

FOREWORD

This book is dedicated to the memory of Harvey N. Nininger (1887–1986) and Harold C. Urey (1893–1981). Each man, in his own way, has profoundly influenced meteorite research in the 20th century.

Nininger, Professor of Biology at McPherson College in Kansas, was drawn to meteorites in 1923, by an article in the *Scientific Monthly*. He soon gave up his faculty position in order to devote himself fully to the collection and study of meteorites. As a collector, he was incredibly successful, recovering no fewer than 218 finds and 8 falls. Indeed, on a map of meteorite finds, one can tell from the density of marks where Nininger lived and traveled in the 1930s to the 1950s (Kansas, Colorado, Texas, New Mexico, Nebraska). A single lecture in each county usually sufficed to turn most farmers into avid meteorite hunters, and one is left wondering what might have happened if Nininger had chosen to go into politics.

Through eight popular books, Nininger exposed a large public to meteorites, thus leading to the disappearance of geologists "who . . . seriously questioned the existence [of meteorites]." Another of his lasting influences was the observation that all diamond-bearing meteorite fragments at the Barringer Crater were reheated, which led to the realization that the diamonds were made by the impact shock, not by high static pressures in a planetary body.

Urey consumed meteorites rather than recovered them, but in the process he transmuted them into data and ideas that have shaped the field to this day. His first paper on meteorites, "Measurement of O^{16}/O^{18} in 6 Chondrites," dates from 1934, the year he won the Nobel Prize for the discovery of deuterium. It has an uncannily modern ring, being an attempt to check whether these meteorites—alleged to have hyperbolic orbits—had oxygen-isotope anomalies indicative of an interstellar origin.

Urey's full-time entry into meteoritics came in 1950, when he was Professor of Chemistry at the University of Chicago. In an incredible spurt of creativity during the next six years, Urey and his coworkers produced a number of landmark papers, as well as the classic book "The Planets." Some of the research topics of this era are: the oxygen-isotope paleotemperature scale (with S. Epstein and others); the origin of the Earth and Moon; the origin of life (with S. L. Miller); composition and classification of chondrites (with H. Craig); chemical fractionations in meteorites; radioactive elements in the

Earth and Moon; abundance of the elements (with H. E. Suess); classification of C-chondrites (with H. B. Wiik); K-Ar, Rb-Sr, and cosmic-ray exposure ages of meteorites (with G. J. Wasserburg, J. Geiss, and E. Schumacher); and composition of comets (with B. Donn).

Many of the above topics appear in the present book, but even when Urey's papers are not cited, his influence is clearly traceable: if not to his final answers, then to some discarded ideas; if not to his data, then to his methodology. It is a book he would have loved to write.

Edward Anders

PREFACE

With so many books about meteorites already on the market, why produce another one? The answer lies in an experience I had a few years ago when preparing a graduate course on the Origin of the Solar System. Although several of the standard texts on meteorites provided good introductions to the subject, none of them interpreted the meteoritic record, at a level appropriate for graduate students, in terms of constraints on theories of the origin and early evolution of the solar system. Because those constraints represent both some of the most direct tests of such theories and also some of the most signal achievements of meteorite research, it seemed worthwhile to pull together into a single volume those pertinent observations and interpretations of the meteoritic record that were previously available only in review articles or research papers. Hence this book.

Although the project was conceived initially with the graduate-student population in mind, it quickly became apparent that such a book would also prove valuable to established scientists in other disciplines who needed to draw upon the results of meteorite research in connection with their own studies. Astrophysicists, with interests ranging from nucleosynthesis through interstellar-cloud chemistry to star formation, as well as those working on the origin of the solar system, obviously represented one such potential readership, but others included exobiologists, planetologists, and those studying planetary atmospheres, as well as those geologists whose interests go back beyond 3.8 Gyr. We believe that meeting the needs of all of these different groups has not necessitated any compromise of the original aims of this book.

The book was planned from the start as a coherent narrative rather than as a collection of self-contained articles. For that reason, authors were given quite specific guidelines about which topics should be covered in each chapter, and they invariably responded nobly to the challenge. Also, in the interests of broad and balanced coverage, many individual chapters were assigned to teams of authors, and we have been most impressed by the high degree of constructive interaction that resulted from such arrangements.

What has proved more elusive has been a consistent level of treatment of the material. Though deemed desirable, and urged upon the authors, achieving such consistency would have necessitated subjecting the manuscripts to a reiterative process that both authors and editors would have found unacceptably burdensome. Similarly, any hopes we may have had of imposing a "house style" on the book have been only partly realized: the transition in style from one chapter to the next is, in many cases, more marked than we would have wished. We have been more successful at achieving consistent use of units, symbols and notations, but caution the reader that there are nonetheless a few cases where a symbol has different meanings in different chapters. (More than one person has suggested that such inconsistencies make the book more like the real world.)

For the benefit of readers unfamiliar with meteorite research, a few words about nomenclature are in order. Traditionally, the meteorites themselves have been named after the nearest post office to their point of fall or discovery. Recently, however, the discovery in Antarctica of meteorites in numbers exceeding those of post offices by at least three orders of magnitude has resulted in a new identification system for those meteorites: letters denote the collection site and numbers the year and order of discovery. Thus, EETA79001 was the first meteorite to be found during the 1979 collecting season at the Elephant Morraine site.

Interplanetary dust particles are also assigned numbers, identifying the stratospheric flight on which they were collected, but in some cases the investigators studying them have informally given them more fanciful names that have passed into general circulation. IDP nomenclature is therefore currently something of a hodgepodge. A similar blend of formal and informal nomenclature is also found in the study of meteorite inclusions, particularly of CAIs from carbonaceous chondrites such as Allende. The resulting mixture may be somewhat confusing to the neophyte, but one quickly learns to recognize the key players, such as HAL or EK 1-4-1.

Jargon is inseparable from science, and meteoritics is no exception. We have attempted to keep it to a minimum but have included a glossary of such specialized terms as have slipped through our net. Similarly, Appendix 1 lists the chemical formulas of the minerals named in the text. Four other appendices include tabulations of data of more or less general relevance.

Underlying much of the discussion in this book is a concept manifested by the term "solar nebula": a few words of clarification are in order. Although originally applied specifically to the Laplacean construct of an equatorial disk of material torn out of the Sun to fill the space now occupied by the planets, the term is now used more generally to describe the material around the Sun from which the planets, comets and asteroids (meteorites) formed, regardless of its mode of emplacement there. In fact, the possibility of analogous nebulae, or subnebulae, around the major planets, leading to formation of their regular satellite systems, has been recently debated. Because a term is needed when referring to the preplanetary distribution of matter in the solar system, we use "solar nebula" in that general sense throughout the book, particularly since it is no longer needed in its original Laplacean sense.

While organizing the material of this book, it became apparent that good reasons could be found for placing every chapter before every other chapter.

In practice, our structure reflects both our emphasis on the record of the early solar system and our recognition of the fact that meteorites are not handpicked samples of nebular matter but are, to put it simply, rocks. First, therefore, we establish the meteorites' credentials as probes of the early solar system. We note that most meteorites are extremely old, and that for much of their existence they have been parts of larger objects: asteroids. In such a setting, their record of nebular, or earlier, processes could be perturbed by secondary alteration. For those meteorites known as chondrites, however, we show that that alteration was modest. Second, we sketch an outline of the theoretical background to early-solar-system studies, as presently understood. Third, we assess the chemical, petrographic, magnetic and isotopic properties of chondrites in terms of conditions and processes in the early solar system. We conclude with a brief summary of boundary conditions that can currently be applied to models of the early solar system, and a discussion of future directions in meteorite research that may help refine those boundary conditions further.

We hope that organizing the material in this way will result in readers approaching the meteoritic record in a properly cautious frame of mind, recognizing that nature does not yield her secrets easily, and that meteorites come with no guarantee that they will answer all our questions about the early solar system. Meteorites are often described as Rosetta Stones from the solar nebula. We should be so lucky; the Rosetta Stone came with one inscription in a script already known. For the deciphering of meteorite properties in nebular terms, Linear B is, alas, the appropriate analogue.

The enthusiastic support for this project by the meteorite community as a whole has been extremely welcome, but two individuals in particular gave of their time and energy to a degree that went way beyond the call of duty. They are Edward Anders and Alan Rubin, and I am immeasurably grateful to them both.

Finally, it has been both a privilege and a great pleasure to work alongside my co-editor Mildred Shapley Matthews and her assistant Melanie Magisos. Their unfailing dedication, expertise and energy have been a source of inspiration and encouragement. Without them, the idea would never have become reality.

John F. Kerridge

PART 1
Introduction

1.1. OVERVIEW AND CLASSIFICATION OF METEORITES

DEREK W. G. SEARS
University of Arkansas

and

ROBERT T. DODD
State University of New York at Stony Brook

Before interpreting properties of meteorites in terms of putative processes and conditions in the early solar system, it is necessary to understand just what sort of objects meteorites are. Such understanding begins with classification. In this chapter, we summarize the current taxonomy of meteorites, and show how certain stone meteorites, the chondrites, possess chemical and petrographic features that make them potentially attractive as probes of the early solar system. The prevalence of secondary alteration effects, often capable of perturbing the primitive record even in chondrites, is also emphasized.

1.1.1. METEORITE CLASSIFICATION

The purpose of classification is to sort the meteorites into broadly similar types of object, so that their origins and relationships can be better understood. Meteorites are a unique source of information about the materials present and conditions prevailing in the solar system during the earliest phases of its history: that is the central tenet of this book. It is clear that not all meteorites are the same; there are various compositional and physical differences between one meteorite and another, sometimes quantitatively minor, and sometimes major. One has to distinguish fundamental, primary properties, attributable to the starting material from which the meteorite was made, from

TABLE 1.1.1
Meteorite Classes and Numbers[a]

Class	Falls[b]	Fall frequency (%)[c]	Finds[d] Non-Antarctic[b]	Finds[d] Antarctic[e]
Chondrites				
CI	5	0.60	0	0
CM	18	2.2	5	34
CO	5	0.60	2	6
CV	7	0.84	4	5
H	276	33.2	347	671
L	319	38.3	286	224
LL	66	7.9	21	42
EH	7	0.84	3	6
EL	6	0.72	4	1
Other	3	0.36	3	3
Achondrites				
Eucrites	25	3.0	8	13
Howardites	18	2.2	3	4
Diogenites	9	1.1	0	9
Ureilites	4	0.48	6	9
Aubrites	9	1.1	1	17
Shergottites	2	0.24	0	2
Nakhlites	1	0.12	2	0
Chassignites	1	0.12	0	0
Anorthositic breccias	0	0	0	1
Stony-irons				
Mesosiderites	6	0.72	22	2
Pallasites	3	0.36	34	1
Irons				
IAB	6	0.73	97	4
IC	0	0.08	11	0
IIAB	5	0.45	60	6
IIC	0	0.05	7	0
IID	3	0.09	12	0
IIE	1	0.10	13	0
IIF	1	0.03	4	0
IIIAB	8	1.42	189	0
IIICD	2	0.14	19	0
IIIE	0	0.10	13	0
IIIF	0	0.05	6	0
IVA	3	0.39	52	1
IVB	0	0.09	12	0
Other irons	13	1.32	175	0

[a] Groups within a classification scheme conventionally contain 5 or more members; meteorites not falling into a group are generally termed "anomalous." In cases where it is useful to identify associations of less than 5 meteorites, the term "grouplet" is occasionally used; for example, Al Rais and Renazzo are normally considered CM chondrites, but they are more reduced than the others and may be referred to as the CR grouplet.
[b] Graham et al. (1985), with minor modification.
[c] Iron-meteorite fall statistics calculated from finds, scaled to percentage of total iron meteorite falls.

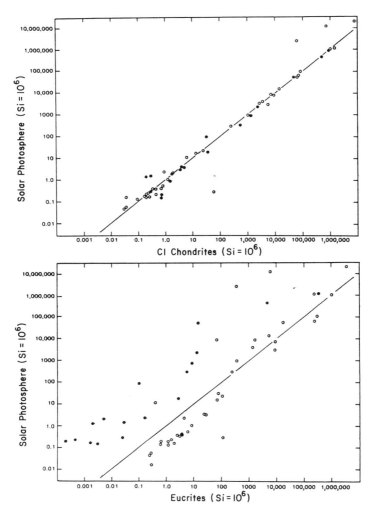

Fig. 1.1.1. Plots of element abundance in the solar photosphere against element abundance in the CI chondrites (a) and the eucrite achondrites (b). Siderophile and chalcophile elements are represented by filled symbols, others by open symbols. With the exception of several highly volatile elements and Li (which is underabundant in the Sun as a result of destructive nuclear reactions), the elements are present in similar proportions in the CI chondrites and the solar photosphere. On the other hand, in the eucrites (b), siderophiles and chalcophiles are 100-fold depleted and lithophiles are 10-fold enriched, relative to the solar photosphere (Sears 1987).

---

[d] Data for finds are given to provide an indication of available material. The unusual conditions in the Antarctic favor the recovery of large numbers of meteorites without the selection biases of non-Antarctic regions (e.g., in non-Antarctic regions, stony meteorites, especially achondrites, are more easily confused with terrestrial rocks than iron meteorites). The statistics for Antarctic finds, therefore, more closely resemble those of falls than non-Antarctic finds. In fact, several rarer classes are overrepresented in the Antarctic collections.

[e] US finds in the Antarctic (Antarctic Meteorite Working Group 1986); probable pairings have been taken into account, but more may still be present. In addition, >6000 meteorites have been recovered from the Antarctic by Japanese teams.

TABLE 1.1.2
The Anders and Ebihara Table of Cosmic Abundances of the Elements Based on Analyses for CI Chondrites[a]

Element	Atomic Abundance (atoms/10^6 Si)	Orgueil[c] Concentration
1 H	2.72×10^{10}	20.2 mg/g
2 He	2.18×10^9	56 nL/g
3 Li	59.7	1.59 µg/g
4 Be	0.78	26.7 ng/g
5 B	24	1.25 µg/g
6 C	1.21×10^7	34.5 mg/g
7 N	2.48×10^6	3180 µg/g
8 O	2.01×10^7	464 mg/g
9 F	843	58.2 µg/g
10 Ne	3.76×10^6	203 pL/g
11 Na	5.70×10^4	4830 µg/g
12 Mg	1.075×10^6 [b]	95.5 mg/g
13 Al	8.49×10^4	8620 µg/g
14 Si	1.00×10^6	106.7 mg/g
15 P	1.04×10^4	1180 µg/g
16 S	5.15×10^5 [b]	52.5 mg/g
17 Cl	5240	698 µg/g
18 Ar	1.04×10^5	751 pL/g
19 K	3770	569 µg/g
20 Ca	6.11×10^4	9020 µg/g
21 Sc	33.8	5.76 µg/g
22 Ti	2400	436 µg/g
23 V	295	56.7 µg/g
24 Cr	1.34×10^4	2650 µg/g
25 Mn	9510	1960 µg/g
26 Fe	9.00×10^5 [b]	185.1 mg/g
27 Co	2250	509 µg/g
28 Ni	4.93×10^4	11.0 mg/g
29 Cu	514	112 µg/g
30 Zn	1260	308 µg/g
31 Ga	37.8	10.1 µg/g
32 Ge	118	32.2 µg/g
33 As	6.79	1.91 µg/g
34 Se	62.1	18.2 µg/g
35 Br	11.8	3.56 µg/g
36 Kr	45.3	8.7 pL/g
37 Rb	7.09	2.30 µg/g
38 Sr	23.8	7.91 µg/g
39 Y	4.64	1.50 µg/g
40 Zr	10.7	3.69 µg/g
41 Nb	0.71	250 ng/g
42 Mo	2.52	920 ng/g
44 Ru	1.86	714 ng/g
45 Rh	0.344	134 ng/g
46 Pd	1.39	557 ng/g
47 Ag	0.529	220 ng/g
48 Cd	1.69	673 ng/g

TABLE 1.1.2 (continued)

Element	Atomic Abundance (atoms/10^6 Si)	Orgueil[c] Concentration
49 In	0.184	77.8 ng/g
50 Sn	3.82	1680 ng/g
51 Sb	0.352	155 ng/g
52 Te	4.91	2280 ng/g
53 I	0.90	430 ng/g
54 Xe	4.35	8.6 pL/g
55 Cs	0.372	186 ng/g
56 Ba	4.36	2270 ng/g
57 La	0.448	236 ng/g
58 Ce	1.16	619 ng/g
59 Pr	0.174	90 ng/g
60 Nd	0.836	462 ng/g
62 Sm	0.261	142 ng/g
63 Eu	0.0972	54.3 ng/g
64 Gd	0.331	196 ng/g
65 Tb	0.0589	35.3 ng/g
66 Dy	0.398	242 ng/g
67 Ho	0.0875	54 ng/g
68 Er	0.253	160 ng/g
69 Tm	0.0386	22 ng/g
70 Yb	0.243	166 ng/g
71 Lu	0.0369	24.3 ng/g
72 Hf	0.176	119 ng/g
73 Ta	0.0226	17 ng/g
74 W	0.137	89 ng/g
75 Re	0.0507	36.9 ng/g
76 Os	0.717	590 ng/g[d]
77 Ir	0.660	473 ng/g
78 Pt	1.37	953 ng/g
79 Au	0.186	145 ng/g
80 Hg	0.52	390 ng/g
81 Tl	0.184	143 ng/g
82 Pb	3.15	2430 ng/g
83 Bi	0.144	111 ng/g
90 Th	0.0335	28.6 ng/g
92 U	0.0090	8.1 ng/g

[a] See Anders and Ebihara (1982) for details and sources.
[b] Average of mean values for individual meteorites. For the remaining elements, a straight average of all acceptable analyses was used.
[c] Abundances in Orgueil, the "typical," i.e., most widely studied, CI chondrite, are adequately close to the CI chondrite mean, except for REE, where the values given in Table 6 of Anders and Ebihara (1982) are preferable. For maximum accuracy, a CI chondrite mean, for elements other than H, C, N, O and the noble gases, may be calculated from the atomic abundances in column 2, using the relation $C = 3.788 \times 10^{-3} HA$, where C = weight concentration (μg/g), H = atomic abundance, and A = atomic weight.
[d] The Os value is a correction to that of Anders and Ebihara (1982), based on a personal communication from E. Anders.

properties which were produced by later events and which may overlay the original properties; sometimes later events may totally destroy primary properties. Several of the details of classification and interpretation are therefore intertwined, and while some interpretations are simple and well understood, and classifications based on them are secure, others may be more controversial. The route to the present meteorite classification schemes was therefore long and tortuous, and some abortive early efforts still permeate the literature (see Sears 1978, p. 60, for a review). The meteorite classes commonly recognized today are briefly described in this chapter and are listed in Table 1.1.1, along with some statistics of falls and finds. Brief definitions of the classes and other terms commonly used in meteorite studies, which might not be familiar to readers, are listed in the Glossary at the back of the book.

The simplest classification of meteorites is into stones, irons and stony-irons. However, further subdivision is essential because there is considerable diversity within these divisions; in fact, these three divisions unite objects of dissimilar character and history, and separate objects that we now know to be closely related. The irons are pieces of metal (essentially Fe-Ni alloys) while the stony-irons are mixtures of stony material and metal. Among the stones there is enormous diversity in chemical and physical properties. The most fundamental distinction is that some, termed chondrites (since they usually contain distinctive features known as chondrules), have a composition which is remarkably similar to that of the solar photosphere for all but the most volatile elements (Fig. 1.1.1a), while others, the achondrites (and other differentiated meteorites) differ considerably from the Sun in composition (Fig. 1.1.1b). While the details may differ from one class of meteorite to another, chondrites are approximately solar in composition while achondrites are enriched in lithophile elements, and depleted in siderophile and chalcophile elements, by about an order of magnitude or more.

The compositional similarity between chondrites and the photosphere, two very different types of extraterrestrial material, has profound implications for the study of both. The Sun is a fairly normal star, and astrophysical observations indicate that much of the cosmos has a composition similar to it. The concept of "cosmic abundances" is not without its flaws, but the idea that there is a mean starting composition from which solar-system objects were formed, has become important in modeling meteorite and solar histories. Table 1.1.2 is a compilation of current best estimates of cosmic abundances.

On the basis of subtle differences in the proportions of major, nonvolatile elements, and differences in mineralogy and mineral composition, the chondrites are divided into 9 classes: the CI, CM, CO and CV chondrites, many of which are characterized by relatively high C contents and which are collectively termed the "carbonaceous chondrites"; the H, L and LL classes, which are qualitatively similar to each other, constitute the most abundant meteorite type and are collectively termed the "ordinary chondrites" and the EH and

Fig. 1.1.2. The internal structure of (a) a chondrite and (b) an achondrite are clearly apparent when the samples are prepared as thin (30 μm) sections and viewed through a microscope using transmitted light. In the Semarkona chondrite, the fine-grained opaque matrix and chondrules, with a variety of internal structures, are readily apparent. The section is about 1.2 mm across. In the achondrite (Serra de Magé), large well-formed crystals of feldspar with occasional small grains of pyroxene and troilite are visible. The width of field is about 2.6 mm. (Both photographs kindly provided by R. Hutchison.)

EL chondrites, which are collectively termed the "enstatite chondrites" after their dominant mineral. Significantly, isotopic properties can also be used to identify most of these classes. The details of the classification of chondrites are discussed further below. There are also considerable petrological differences between the chondrites and the others; the chondrites are aggregates of different sorts of material ("cosmic sediments") while the achondrites are either igneous rocks or derived from igneous rocks. The major lithic components in chondrites are the chondrules, matrix, metal and sulfide and, in certain classes, inclusions rich in Ca, Al and similar refractory elements. Chondrules are an especially important component, both in terms of their high abundance and their scarcity in other rocks. They are essentially silicate mineral assemblages, typically 0.1 to 1.0 mm in size, which had an independent existence prior to the formation of the meteorite. They have a wide variety of internal structures, but generally consist of silicate grains enclosed either in glass or in crystals which formed from a glass (Gooding and Keil 1981; see also Chapter 9.1). Chondrules have been partially or totally melted (Chapter 9.2) and many owe their droplet form to their liquid origin. The origin of chondrules is discussed in Chapters 9.3 and 9.4. Figure 1.1.2 compares the overall structure of a chondrite and an achondrite, as seen in transmitted light under a low-powered microscope. Figure 1.1.3 shows some hand specimens of various meteorites.

The achondrites and stony-iron meteorites are relatively few in number but very diverse in their properties and histories. Brief descriptions of their major mineralogy are given in the Glossary. All have experienced major planetary processing, typically including melting, which has virtually obscured the nebular phases of their history, but they sometimes provide an

Fig. 1.1.3. Some hand specimens of meteorites. (a) The Fayetteville regolith breccia showing the light-dark structure of the interior and the black fusion crust produced by the heat of atmospheric passage. Scale cube is 1 cm. (b) The Allan Hills A81005 achondrite which is from the Moon. Scale cube is 1 cm. (c) The Crab Orchard stony-iron meteorite (mesosiderite). (d) The Edmonton, Kentucky, III C,D iron meteorite showing an etched, polished face exhibiting the Widmanstätten structure. (Photographs were kindly provided by (a) C. Schwarz, (b) J. Annexstad, (c) R. Hutchison, (d) J. T. Wasson.)

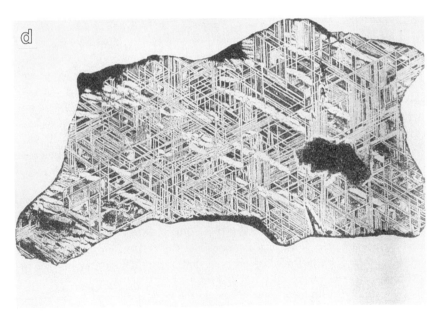

11

indication of the scale and type of igneous processes that occurred in the early solar system (Chapter 3.2). These, in turn, sometimes provide insights into available heat sources, sizes of parent objects and time of accretion.

Pallasites consist of networks of Fe-Ni alloy containing angular or rounded nodules of olivine (typically 4 to 5 mm in size). They are clearly of igneous origin, and probably formed at the interface between a large molten metal body (such as a core or large metal nugget) and a large magma chamber in which olivine could form and sink to the bottom.

Three groups of achondrite (eucrites, diogenites and howardites) and the pyroxene-plagioclase stony-irons, or mesosiderites, are closely related. The eucrites and diogenites are magmatic rocks, while the others are breccias of these two classes. In the case of the mesosiderites, a large metallic component is also present, while howardites may contain fragments of more extreme composition than normally seen in eucrites and diogenites. Most authors agree that the parent material for these classes was chondritic, and that their varied textures and compositions reflect differences in cooling history, complicated considerably in some cases by metamorphism and shock.

The shergottites, nakhlites and Chassigny, collectively termed the SNC meteorites, are also closely related achondrites. Their bulk compositions, trapped noble gases and relatively young formation ages (around 1.25 Gyr) suggest that they came from Mars (Chapters 5.3, 7.8 and 7.10). The four anorthositic breccia achondrites found in the Antarctic have no established class name, though a wide variety of compositional, chronological and petrographic data indicate that they are of lunar origin. The other major achondrite classes are aubrites and the ureilites. The aubrites may be partially melted EL chondrites, but the matter is disputed. The ureilites are magmatic rocks to which significant amounts of C and volatile-rich material have been added.

There are many unusual achondrites and stony-irons, some of which may be related to established classes. A unique achondrite is Angra dos Reis that was formed by magmatic processes 4.55 Gyr ago, very close to the time of formation of the solar system at 4.56 Gyr (Chapter 5.2). Lodran is a unique stony-iron meteorite of uncertain origin, possibly related to the IAB irons, see below, and several unusual chondrites like Winona. It has been suggested that these meteorites may constitute an additional class of meteorite. Bencubbin and Weatherford are unclassified breccias that contain chondritic and achondritic clasts.

The study of iron meteorites illustrates the strong influence of taxonomy on interpretation, and the effect of advancing technology on both. Iron meteorites are essentially Fe-Ni alloys with minor amounts of C, S and P. The alloy segregates into two phases, a low-Ni body-centered-cubic phase known as the α phase or "kamacite," and a high-Ni face-centered-cubic phase known as the γ phase or "taenite." Iron-meteorite structures may range from pure kamacite, with shock-twins (Chapter 3.6) as their only structural feature (these are structurally termed "hexahedrites"), to meteorites which are structureless to the naked eye but consist of taenite with microscopic grains of kamacite ("ataxites"). In between these extremes are the majority of irons, which consist of plates of kamacite, in an octahedral arrangement, with taenite filling the interstices; an example appears in Fig. 1.1.3d. Irons with this structure are termed "octahedrites," and their characteristic structure is called the "Widmanstätten structure" after one of its discoverers. The structure of an iron meteorite, and its coarseness, depend on the bulk Ni content and the cooling rate of the meteorite. Until recently, irons were classified almost en-

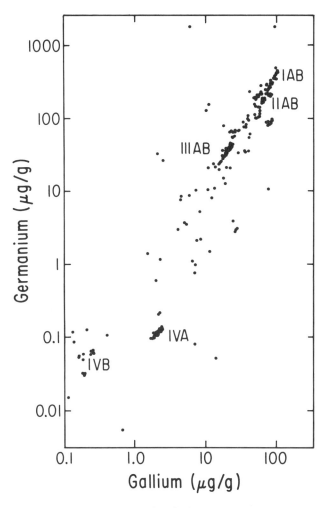

Fig. 1.1.4. Plot of Ge content against that of Ga for iron meteorites. Some of the larger classes are indicated (figure from Sears 1978).

tirely in terms of their structure and, although this scheme has largely given way to one based on composition, structure is clearly relevant to major questions concerning starting composition and subsequent history.

The structures and Ni contents of iron meteorites do not produce well-defined groups. The structural types grade into each other and, while there are some peaks in the histogram of Ni contents, there are no gaps. This is not true for the volatile trace elements Ga and Ge (Fig. 1.1.4), which show a 10^4- and 10^6-fold range in abundance, respectively, throughout the irons as a whole, but produce many tight clusters within which these elements vary by less than a factor of 2. It seems that these elements are unique because (1) their volatility made them especially sensitive to small differences in the pressure and temperature in the region of the solar nebula in which the irons formed, and (2) their distributions were relatively insensitive to subsequent igneous processes. Irons are therefore commonly classified on the basis of Ga, Ge and Ni content, qualified occasionally by structural and mineralogical factors. The most populous classes are known as IAB, IIAB, IIIAB, IVA and IVB, and there are a great many smaller classes (the nomenclature originally reflected Ga and Ge abundance, but has become somewhat arbitrary). The classification of iron meteorites has been reviewed by Scott and Wasson (1975).

It should be stressed that while classification sorts meteorites which differ from each other in a manner which can help us identify origins and formation mechanisms, it does not preclude genetic relationships between classes. In Fig. 1.1.5, an attempt has been made to indicate the relationships, or lack thereof, between the various classes. For example, the two types of

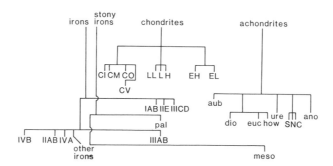

Fig. 1.1.5. A "family tree" of meteorite classes and an indication of possible interclass relationships; for example, the IAB, IIE and IIIAB iron meteorites and the pallasites (pal) may be related to the ordinary chondrites, although it is clear that the relationship is closest for the IAB and IIE irons (which contain silicates of approximately chondritic composition); the mesosiderites (meso) may be related to the howardite and eucrite achondrites, and the aubrites have been related to the enstatite chondrites. Abbreviations for the achondrites are as follows: aubrites, aub; diogenites, dio; eucrites, euc; howardites, how; ureilites, ure; shergottites, nahklites and chassignites, SNC; the anorthositic breccias, ano.

stony-iron meteorites discussed above share the relatively trivial characteristic of being about half metal and half silicate, but the pallasites are much more closely related to iron meteorites than they are to the mesosiderites. The mesosiderites are, in turn, more closely related to eucrites and howardites than they are to the pallasites. The CI, CM and CO chondrites are, likewise, more closely related to each other than any are to the CV chondrites.

1.1.2. THE CHONDRITES: PRIMARY CLASSIFICATION

After many decades of having been considered to be almost identical in composition, small systematic differences in composition among the chondrites were discovered in the early 1950s. These differences led, ultimately, to the definition of the 9 chondrite classes (Table 1.1.3). There appear to have been a number of processes which caused elemental abundances to depart from their mean cosmic values. In terms of the kinds of elements involved, and how they are used in classification, the taxonomic discriminants are as follows:

1. Nonvolatile lithophile elements cause the cleanest separations; for example, Mg/Si is 1.05 for carbonaceous chondrites, 0.95 for ordinary chondrites and 0.83 for enstatite chondrites (Fig. 1.1.6) (Ahrens et al. 1968). If refractory elements (e.g., Ca/Si), are taken into account, the CV class separates from the others (Van Schmus and Hayes 1974);
2. The siderophile elements lead to further subdivision of these classes; the

TABLE 1.1.3
Chondrite Classes and Mean Properties[a]

Group	Mg/Si (a/a)	Ca/Si (a/a)	Fe/Si (a/a)	Fa (mol%)	Co in Kamacite (mg/g)	Fe_{met}/Fe_{tot}	$\delta^{18}O$ (‰)[c]	$\delta^{17}O$ (‰)[c]
CI	1.05	0.064	0.86	—	—	0	~16.4	~8.8
CM	1.05	0.068	0.80	—	—	0	~12.2	~4.0
CO	1.05	0.067	0.77	—	—	0–0.2	~−1.1	~−5.1
CV	1.07	0.084	0.76	—	—	0–0.3	~0	~−4.0
H	0.96	0.050	0.81	16–20	5.2	0.58	4.1	2.9
L	0.93	0.046	0.57	23–26	10.6	0.29	4.6	3.5
LL	0.94	0.049	0.52	27–32	22.9[b]	0.11	4.9	3.9
EH	0.77	0.035	0.95	—	—	0.76	5.6	3.0
EL	0.83	0.038	0.62	—	—	0.83	5.3	2.7

[a] Data from Von Michaelis et al. (1969); Anders and Ebihara (1982); Sears and Axon (1976); Sears and Weeks (1986); Wiik (1969); Mason (1966); Van Schmus and Hayes (1974); R. N. Clayton 1981 and other sources.
[b] While many of the LL chondrites cluster around the value indicated, they show a "tail" up to 110.0 mg/g (see Fig. 1.1.9).
[c] Relative to SMOW.

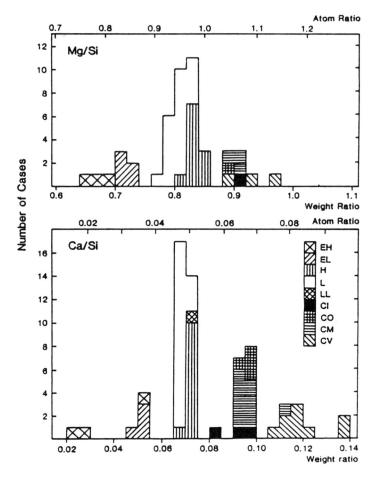

Fig. 1.1.6. Histograms of the Mg/Si and Ca/Si contents of chondrites. These elemental ratios are the basis of classification into enstatite (EH and EL), ordinary (H, L and LL) and carbonaceous (CI, CM, CO and CV) chondrites. Further resolution into individual classes is sometimes possible; for example, the CV chondrites are clearly resolved from the other carbonaceous chondrites on the basis of Ca/Si (figure after Von Michaglis et al. 1969; Van Schmus and Hayes 1974).

ordinary chondrites into the H, L and LL classes (Urey and Craig 1953; Craig 1964) (for high Fe, low Fe and low Fe-low metal) and the enstatite chondrites into the EH and EL classes (Sears et al. 1982). Carbonaceous chondrites can also be subdivided to some degree on the basis of Fe/Si, but the same subdivisions are more readily made on the basis of other criteria;

3. Oxidation state, or, more precisely, the proportion of Fe in the silicates and sulfides as compared to that in metal, varies throughout the classes defined by lithophile and siderophile elements (Keil 1968; Fredriksson

and Keil 1964; Fredriksson et al. 1969; McSween 1979). To discuss oxidation-state-related parameters further, it is necessary to look briefly into the relevant mineralogy.

As of 1975, more than 100 minerals had been identified in meteorites, many of them unknown elsewhere (a list of the better documented minerals may be found in Appendix 1). However, most of them are present in trace amounts. The important minerals for the classification of ordinary chondrites are olivine, Ca-poor pyroxene (commonly orthopyroxene), feldspar, sulfide and metal. Olivine and pyroxene are solid solutions of Fe-, Mg- and Ca-silicates. Olivine is a solid solution of two main components, fayalite (Fe_2SiO_4, symbol Fa) and forsterite (Mg_2SiO_4, Fo). Pyroxenes are solid solutions of ferrosilite ($FeSiO_3$, Fs), enstatite ($MgSiO_3$, En) and Wollastonite ($CaSiO_3$, Wo). Other components are more abundant in pyroxene than olivine, and the structure of pyroxene allows for large amounts of "impurities." Feldspar is a solid solution of three Al-silicate end members, albite ($NaAlSi_3O_8$, symbol Ab), anorthite ($CaAl_2Si_2O_8$, An) and orthoclase ($KAlSi_3O_8$, Or). Various names are given to feldspars of intermediate compositions; for example, feldspar in ordinary chondrites is typically oligoclase ($Ab_{84}An_{10}Or_6$). The sulfide in ordinary chondrites is almost always pure FeS, known as troilite after an eighteenth century monk who observed the mineral with the naked eye in a newly fallen meteorite. The metal occurs as two phases (kamacite and taenite), as in iron meteorites; they exist as small, widely dispersed grains surrounded by silicates.

The enstatite and the carbonaceous chondrites generally have a distinctly different mineralogy from that of the ordinary chondrites. The enstatite chondrites are predominantly metal, enstatite and various sulfides (Mason 1966; Keil 1968). In addition to troilite, which is unusual for its large content of Ti, the enstatite chondrites contain a number of remarkable sulfide, phosphide and carbide minerals (see Appendix 1). There is also an oxynitride of Si, not observed elsewhere, called sinoite (Si_2N_2O).

The carbonaceous chondrites, likewise, have varied mineralogy (McSween 1979). The CI and CM chondrites consist mainly of a variety of clay-like hydrous sheet silicates often identifiable with terrestrial phyllosilicates such as serpentine (see Chapter 3.4). They also contain hydrous Mg- and Ca-sulphates (sometimes in veins that suggest deposition from aqueous solutions), magnetite in a variety of morphologies, and considerable quantities of C, as carbonate or complex organic compounds (Chapter 10.5). Unlike the CM chondrites, the CI chondrites do not contain chondrules. The CO and CV chondrites are similar to the ordinary chondrites in their major mineralogy, but contain olivines with a mean composition that is richer in Fe. The CO chondrites are noteworthy for the smallness of their chondrules.

The CV chondrites are especially known for the presence of mm-sized white or pinkish inclusions of Ca- and Al-rich minerals (Ca-Al-rich inclusions or CAI, see Chapter 10.3). CAIs are also frequently observed in other carbonaceous classes, and occasionally in ordinary chondrites. They may be either fine grained (< 1 μm) and composed essentially of spinel and pyroxene, with sodalite, nepheline and grossularite, or coarse grained (100 μm to 1 mm) and composed predominantly of spinel and melilite (type A) or spinel, anorthite and pyroxene (type B). Both types of CAI occasionally contain sulfide-metal assemblages, referred to as "Fremdlinge," whose metal is rich in rare refractory elements like Re, Os and Ir. High-temperature mineral assemblages similar to those observed in the CAIs sometimes occur as chondrules (Grossman 1980).

We now return to the question of the classification of chondrites and, in particular, the relevance of parameters that involve the degree of oxidation of Fe. Figure 1.1.7 compares the amount of Fe in the "reduced" form, i.e., as metal, with the amount in the "oxidized" and "sulfurized" form, usually as

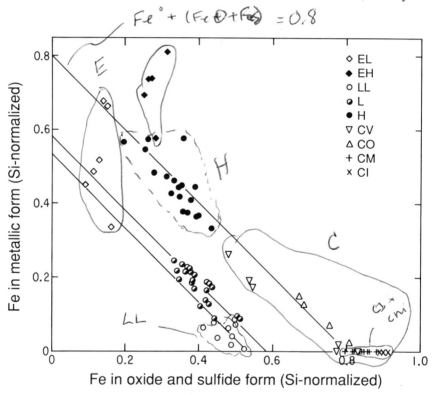

Fig. 1.1.7. Plot of the Fe/Si ratio for the sulfides and metal against Fe/Si in the oxidized phases (mainly silicates) for the chondritic meteorites. The bulk Fe/Si for the meteorites determines the position of the diagonal, while the oxidation state of the Fe determines the location of a given data point on the diagonal. The enstatite, ordinary and carbonaceous chondrites are clearly resolved, but resolution on a finer scale is also possible, especially for EH and EL chondrites and H, L and LL chondrites. See also Fig. 7.4.1.

silicates and sulfides. A meteorite with a specific Fe/Si ratio, but with all the Fe in the reduced form, would plot at the top-left end of one of the diagonal lines. A meteorite with the same bulk Fe/Si ratio, but with a high level of oxidation, would plot on the same diagonal but at the lower-right end of the line. Intermediate degrees of oxidation or reduction would cause meteorites to plot at intermediate locations on the diagonal. Different bulk Fe/Si ratios would correspond to different diagonals. Thus the plot reveals two trends in the data, (1) variations in Fe oxidation (see also Chapter 7.7) and (2) variations in bulk Fe/Si (see also Chapter 7.4). On the basis of such data, the enstatite, ordinary and carbonaceous chondrites are resolved, since the enstatite chondrites are highly reduced, the carbonaceous chondrites are highly oxidized and the ordinary chondrites are intermediate. These data enable the ordinary chondrites to be resolved further into the H, L and LL classes (Urey and Craig 1953; Craig 1964).

These differences in Fe oxidation within the ordinary chondrites are also reflected in their mineralogy. A plot of the fayalite content of olivine against the ferrosilite content of pyroxene, for instance, clearly resolves the H, L and LL classes (Fig. 1.1.8). Although other factors are involved, the Co content of the kamacite is also in large part determined by the degree of Fe oxidation

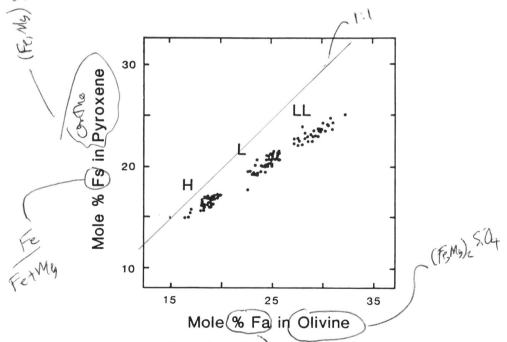

Fig. 1.1.8. Plot of the ferrosilite in the low-Ca pyroxene against the fayalite content of the olivine for the equilibrated (types 4–6) ordinary chondrites. This measure of the oxidation state of the Fe sorts the ordinary chondrites into the H, L and LL classes (figure after Fredricksson and Keil 1964; Fredricksson et al. 1969).

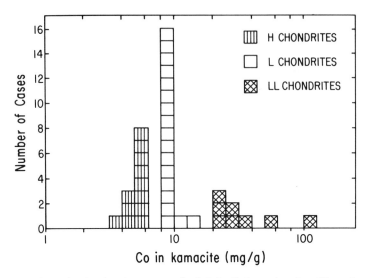

Fig. 1.1.9. Histogram showing the amount of cobalt in the kamacite of equilibrated ordinary chondrites. This measurement is acutely sensitive to the extent to which Fe has been oxidized, the H and L chondrites having 4–6 and 8–14 mg/g of Co in the kamacite, respectively. The LL chondrites show much larger amounts of Co in the kamacite and show a wide range of values (Sears and Weeks 1986).

and can be used to resolve the groups (Fig. 1.1.9). These data indicate that the LL chondrites are not only more oxidized than the others, but they also show a much greater spread in the degree of Fe oxidation.

It should be stressed that although certain elemental ratios (e.g., Mg/Si, Ca/Si) were singled out for discussion above, the elements chosen are representative rather than unique; for example, one can generally substitute any refractory lithophile element and produce a histogram like that of Ca/Si (see Chapter 7.3). Nevertheless, elements that are geochemically similar may vary systematically in volatility and this may also help with classification. Thus, plots of elemental abundance, with elements ranked according to their volatility in a gas of solar composition, can also be meaningful for distinguishing the classes from each other as they constitute a "finger-print." Figure 1.1.10 shows schematic diagrams for the element abundances as a function of volatility for the 9 chondrite classes (see also Chapters 7.5 and 7.6). When both lithophile and siderophile/chalcophile element trends are taken into account, each chondrite class has a unique abundance pattern.

The relative abundances of the 3 isotopes of O have provided an independent means of classification (Clayton et al. 1976; Fig. 1.1.11). A convenient way to examine the data is to plot differences in the $^{17}O/^{16}O$ ratio between the sample and a standard, such as standard mean ocean water (SMOW), against the corresponding differences in $^{18}O/^{16}O$. The differences

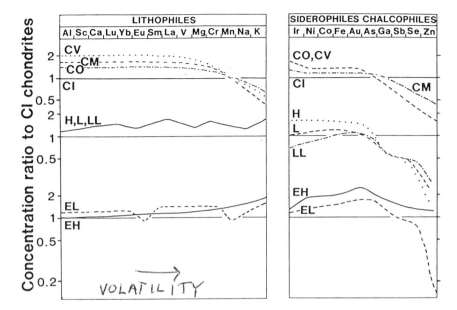

Fig. 1.1.10. Schematic diagrams showing elemental abundance patterns in the chondritic meteorite classes. The concentration of each element in each class is expressed as a ratio to its concentration in CI chondrites. Taking this ratio removes the effects of differences in elemental abundances produced during nucleosynthesis and enables us to examine small differences from CI patterns. Each class has a unique elemental abundance pattern (see also Chapters 7.3, 7.4, 7.5 and 7.6).

are expressed as parts per thousand, or per mil (‰). Two processes seem to have determined the isotopic proportions of O in a given meteorite or part of a meteorite: (1) mixing of two or more components with different isotopic proportions and, (2) mass fractionation. An unusual component of O is enriched in ^{16}O, and mixtures of this with more "normal" O (i.e., containing greater amounts of ^{17}O and ^{18}O) result in lines of slope very near unity in a plot such as Fig. 1.1.11. Mass fractionation is the adjustment of isotopic proportions by physical and chemical processes which are mass dependent (see also Chapter 12.1). Most processes (e.g., volatilization, crystallization, chemical reactions) are of this kind and tend to distribute samples along a line of slope close to 0.5 on a 3-isotope plot. The H, L and LL classes of ordinary chondrites plot in similar regions of the plot, with the H chondrites clearly resolved from the others and the L chondrites weakly resolved from the LL chondrites. The carbonaceous-chondrite classes plot in very widely dispersed regions of the diagram, and it was the distribution of data for minerals in CAIs from the Allende CV chondrite along a slope-1 line which first indicated the presence of exotic ^{16}O-rich material, probably of extrasolar-system origin, in meteorites. The enstatite chondrites plot on the line occupied by terrestrial

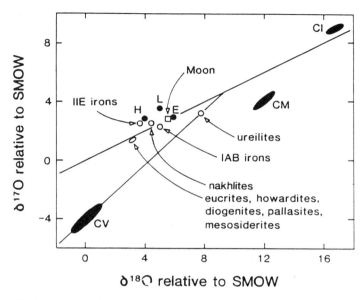

Fig. 1.1.11. Plot showing the distribution of the three isotopes of O in meteorites. The δ parameter refers to the difference in the ratio ^{17}O or ^{18}O to ^{16}O (in parts per thousand, or per mil) between the sample and standard mean ocean water (SMOW). Terrestrial and lunar samples plot along the line with slope ~ 0.5 which is referred to as the "terrestrial mass-fractionation" line. The line with slope ~ 1.0 is produced by components from Allende and other meteorites and reflects mixing of an ^{16}O-enriched component and a more "normal" component (figure after Clayton 1981).

samples and, while the EH and EL chondrites differ slightly in their O-isotope systematics, the two groups are not well resolved.

1.1.3. THE CHONDRITES: SECONDARY PROPERTIES

During or after the accretion of chondritic material in the solar nebula, this material underwent several processes that modified its original nebular properties. As discussed above, the classification of meteorites often involves judgment concerning these primary (nebular) and secondary properties. Over the last several decades, schemes have been devised which attempt to describe the nature and extent of the secondary processes and they are often regarded as part of the meteorite's classification in as much as they can be reported, in a routine way, along with the meteorite's class. The processes which have received most attention are metamorphism, shock, aqueous alteration and brecciation. These are described in some detail in Part 3 of this book and are dealt with only briefly here as they relate to classification.

The few chondrites that escaped significant levels of metamorphism (up to 1200°C) on their parent bodies are particularly helpful for identifying pro-

TABLE 1.1.4
Definitions of Petrographic Types[a]

	Petrographic Types					
	1	2	3	4	5	6
(i) Homogeneity of olivine and pyroxene compositions	—	>5% mean deviations		<5% mean deviations to uniform	Uniform	
(ii) Structural state of low-Ca pyroxene	—	Predominantly monoclinic		monoclinic		orthorhombic
				>20%	<20%	
(iii) Degree of development of secondary feldspar	—	Absent		<2 μm grains	<50 μm grains	>50 μm grains
(iv) Igneous glass	—	Clear and isotropic primary glass; variable abundance		Turbid if present	Absent	

(v) Metallic minerals (maximum Ni content)	—	(<20%) Taenite absent or very minor	kamacite and taenite present (>20%)			
(vi) Sulfide minerals (average Ni content)	—	>0.5%	<0.5%			
(vii) Overall texture	no chondrules	very sharply defined chondrules	well-defined chondrules	chondrules readily delineated	poorly defined chondrules	
(viii) Texture of matrix	all fine-grained, opaque	much opaque matrix	opaque matrix	transparent microcrystalline matrix	recrystallized matrix	
(ix) Bulk carbon content	~3.5%	1.5–2.8%	0.1–1.1%	<0.2%		
(x) Bulk water content	~6%	3–11%	<2%			

[a] Table modified after Van Schmus and Wood (1967). The strength of the vertical line is intended to reflect the sharpness of the type boundaries. A few ordinary chondrites which show signs of partial melting, in response to higher metamorphic temperatures than those associated with type 6, are described as petrographic type 7. Water contents do not include loosely bound, i.e., terrestrial water.

cesses that occurred before accretion, especially at the level of individual grains. Van Schmus and Wood (1967) have devised a widely used scheme in which chondrites are assigned to one of six petrographic types on the basis of a variety of petrographic and mineralogic properties (Table 1.1.4). [Editor's Note: Traditionally, the term "petrologic" has been used in connection with these types, but in fact "petrographic" is the correct term for use within a taxonomic context. The latter is therefore used throughout this book, even in chapters where the authors' versions used "petrologic." Of course, the relevant observations have also a petrologic significance.] As petrographic type increases, minerals that were initially inhomogeneous become more homogeneous, the dominant form of Ca-poor pyroxene changes from monoclinic to orthorhombic, volatile inventories decrease and primary textures become obliterated and chondrules less well defined. The glassy material in the chondrules, which is characteristic of the lower petrographic types, becomes turbid in type 4 and disappears in the higher types, to be replaced by feldspar which increases in grain size at the highest types. In recent years, meteorites and meteorite clasts have been identified in which metamorphism seems to have caused partial melting; these are sometimes referred to as type 7.

The type 3 ordinary chondrites have experienced a considerable range of metamorphic intensity which is evident in several of the parameters discussed above (Dodd et al. 1967), but especially in thermoluminescence (TL) sensitivity (Sears et al. 1980). TL sensitivity is acutely dependent on the extent of crystallization of the chondrule glass and, while it displays a 10^5-fold range

TABLE 1.1.5
Definition of Subtypes of Type 3 Ordinary Chondrites[a]

Type	TL Sensitivity (Dhajala = 1)	C.V.(Fa) in Olivine[b]	C.V.(Co) in Kamacite[c]	Matrix Recrystallization (%)	F/FM Ratio[d]
3.9	2.2–4.6	5–10	<1.5–1.9	>60	<1.0
3.8	1.0–2.2	11–20	1.9–3.1	>60	1.0–1.1
3.7	0.46–1.0	21–30	3.1–6.3	>60	1.1–1.2
3.6	0.22–0.46	31–40	6.3–7.5	>60	1.2–1.3
3.5	0.10–0.22	41–50	7.5–10.0	~50	1.3–1.4
3.4	0.046–0.10	>50	10.0–12.5	~20	1.4–1.5
3.3	0.022–0.046	>50	12.5–16.3	10–20	1.5–1.6
3.2	0.010–0.022	>50	16.3–21.3	10–20	1.6–1.7
3.1	0.0046–0.010	>50	21.3–26.3	10–20	1.7–1.9
3.0	<0.0046	>50	>26.3	<10	>1.9

[a] Table from Sears et al. (1980).
[b] Coefficient of variation of Fa in olivine. Standard deviation expressed as a percentage of the mean.
[c] Coefficient of variation of Co in kamacite.
[d] FeO/(FeO+MgO) in the matrix normalized to the same quantity for the whole rock.

over the ordinary chondrites as a whole, within the type 3 ordinary chondrites there is a 10^3-fold range. Thus the type 3 ordinary chondrites can be subdivided into types 3.0–3.9 (Table 1.1.5). There are also significant changes in the way in which TL is emitted as petrographic type increases; these reflect the state of the feldspar and can be used to provide unique insights into palaeotemperatures (Chapter 3.3). CO chondrites share many mineralogical properties with type 3 ordinary chondrites, and like them constitute a metamorphic sequence. Using petrographic criteria similar to those used for type 3 ordinary chondrites, McSween (1979) identified three subdivisions which he labeled I, II and III. One of the parameters identified by McSween was the disappearance of glass and the formation of feldspar; as expected, TL sensitivity also increases along the metamorphic sequence (Keck and Sears 1987). However, the emission characteristics differ from the ordinary chondrite case, implying different cooling histories for the two classes.

As noted above, and discussed in Chapter 3.4, CI1 and CM2 chondrites show evidence of considerable aqueous alteration, so that type 3 chondrites may better represent original nebular material. Recently, hydrous silicates, calcite and magnetite have been found in two type 3.0–3.1 ordinary chondrites, so it seems possible that several of their unusual properties, once thought due to weak thermal metamorphism, are actually aqueous-alteration effects (Hutchison et al. 1987). A cartoon, which summarizes the probable relative roles of metamorphism and aqueous alteration in carbonaceous and ordinary chondrites, is shown in Fig. 1.1.12.

Brecciation and shock are common features of meteorites of all classes; however, the effects vary more randomly from group to group than the sec-

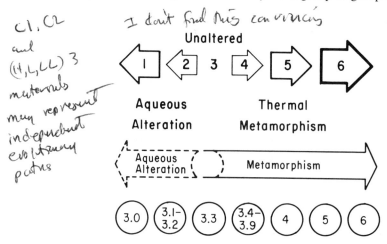

Fig. 1.1.12. Sketches suggesting possible relationships between aqueous alteration, metamorphism and petrographic type for all chondrites (*above*, from McSween [1979]) and for ordinary chondrites (*below*, based on Guimon et al. [1987]). These meteorites have apparently experienced aqueous alteration, but its importance relative to metamorphism is unclear.

TABLE 1.1.6
Classification of Chondritic Breccias[a]

Breccia Type	Description
Lithic fragments in breccias	— xenolithic—clasts of different chemical class to host
	— cognate—impact-melt fragments
Regolith breccias	— fragmental debris on surface of body (contain solar-wind gases, solar-flare tracks, agglutinates, etc.), usually have light-dark structure
Fragmental breccias	— fragmental debris with no regolith properties
Impact melt breccias	— unmelted debris in igneous matrices
Granulitic breccias	— metamorphosed fragmental breccias
Primitive breccias	— type 3 ordinary chondrite breccias

[a] Table after Keil (1982).

ondary properties summarized by Van Schmus and Wood (1967), and lesser significance is usually given to them in modern classification schemes. The terms polymict, monomict and genomict are often used to describe breccias. Polymict breccias are those in which the fragments differ from each other compositionally, while monomict means that the fragments of the breccia are similar compositionally. Genomict means that the fragments differ from each other petrographically but not compositionally. However, this scheme is less precise and comprehensive than that recently suggested by Keil (1982), which is described in Table 1.1.6; but, there is again a complex interplay between classification and interpretation. Brecciation is discussed in detail in Chapter 3.5.

Shock played a major part in disrupting the parent objects of the meteorites and perhaps in sending the fragments to Earth. Systematic studies of shock effects have resulted in several classification schemes, some based on metal structures and others based on silicate properties. The widely used scheme of Dodd and Jarosewich (1979) is summarized in Table 1.1.7, with some additional data. Identified are six shock "facies" labeled a–f which represent increasing levels of shock. Laboratory calibration experiments provide rough estimates of the shock level described by each facies and these are given in Table 1.1.7. However, it should be stressed that chondrites, which are complex mixtures of minerals with highly diverse physical properties, behave very differently from pure minerals during the passage of shock waves so that the classification is essentially qualitative. Shock effects on meteorites, and minerals which are relevant to meteorite studies, are discussed in Chapter 3.6.

TABLE 1.1.7
Definition and Properties of 6 Shock Facies in Ordinary Chondrites[a]

Shock Facies	Pressure[b] (GPa)	Olivine	Plagioclase	Melt Pockets	^{40}Ar	TL Sensitivity	Trace Elements
a	<5	fractured	<50% deformed	None	normal	normal	normal
b	5–20	fractured & undulose extinction	<50% deformed	None	normal	normal	normal
c	20–22	fractured & undulose extinction	>50% deformed	None	~0.7 × normal	0.15–0.4	Bi,Ag,Zn <0.8
d	22–35	fractured & mosaic extinction	100% deformed or maskelynite	Present	~0.7 × normal	0.0025–0.15	In,Tl,Cd <0.5
e	35–57	fractured, mosaicked & marginal granulation	>50% maskelynite	Present	~0.1 × normal	0.025–0.15	Bi,Ag,Zn <0.5
f	>57	recrystallized	100% maskelynite	Present	~0.1 × normal	0.025–0.15	In,Tl,Cd <0.1

[a] Table after Dodd and Jarosewich (1979); Walsh and Lipschutz (1982); Sears et al. (1984); Dennison and Lipschutz (1986). A recalibrated scale of shock effects can be found in Table 3.6.2.
[b] 1 Gigapascal (GPa) equals 10 kbar.

1.1.4. SIGNIFICANCE OF THE CHONDRITE CLASSES

It is clear from four lines of evidence that the 9 chondrite classes offer unique insights into conditions in the early solar system. First, as described in detail above, they are compositionally very similar to the Sun and quite unlike other known samples of solid material in the solar system. This suggests a relatively simple origin from a common pool of starting material. Second, they have formation ages similar to those inferred for the Earth, Moon and Sun, and presumably that of the solar system as a whole, namely about 4.55 Gyr (Chapter 5.2). Radiometric systems which can readily be disturbed, such as those dependent on the retention of gases, sometimes yield lower ages, but in these cases there is usually evidence that the ages refer to a subsequent event such as shock heating. Third, virtually all chondrites, especially some of the low type 3 ordinary chondrites, contain highly non-equilibrium assemblages which have been altered very little, if at all, since the components came together. An example is the coexistence of Si-bearing metal and silicates (see Chapter 7.7). Fourth, chondrites contain isotopic evidence for accumulation quite soon after the end of element synthesis (Chapter 15.1). Also some elements are present with isotopic proportions that are not encountered in any other known solar-system material; apparently this material escaped the isotopic homogenization experienced by most solids (Chapter 12.1). In some instances, there is a resemblance between these unusual isotopic ratios and those in interstellar molecules (Chapters 13.1 and 13.2).

Given this background, the existence of 9 distinct classes of chondrite is significant because it leads to the conclusion that those classes were established in the solar nebula. This means that the properties that distinguish one class from another—the variations in Mg/Si, Fe/Si, oxidation state and O-isotope ratios, for instance—in some way reflect nebular processes. For example, there was apparently a nebular process for separating metal and silicate to produce the Fe/Si variations (Chapter 7.4). Another process caused mixing of reservoirs with different O-isotope properties. Some of these properties seem to be paralleled in the different planets to some extent (Fe/Si and oxidation state variations, for example) which may mean that the chondrite classes sampled different regions of the nebula. Thus each of the classes, and any unusual property each displays, may be providing a snap-shot view of conditions ~ 4.55 Gyr ago at a particular nebular location. The major challenge of modern meteorite studies is to take these complex, ancient samples from the earliest days of solar-system history, disentangle the diverse properties which have resulted from their great age and the various processes that they have experienced, and thus learn something about a unique time and place in the history of our planetary system.

Acknowledgments. We are grateful to C. Schwarz, A. Davis, R. Hewins, A. Rubin and, especially, R. Hutchison for providing figures and comments on various parts of the manuscript, and F. Hasan and M. Mabie for technical assistance, and U. Marvin and R. Hutchison for providing reviews.

REFERENCES

Ahrens, L. H., Von Michaelis, H., Erlank, A. J., and Willis, J. P. 1968. Fractionation of some abundant lithophile element ratios in chondrites. In *Meteorite Research,* ed. P. M. Millman (Dordrecht: D. Reidel), pp. 166–173.

Anders, E., and Ebihara, M. 1982. Solar-system abundances of the elements. *Geochim. Cosmochim. Acta* 46:2363–2380.

Antarctic Meteorite Working Group. 1986. Index of classified meteorites from the 1976–1985 Antarctic collection (as of September 1986). *Antarctic Meteorite Newsletter* 9 (4).

Clayton, R. N. 1981. Isotopic variations in primitive meteorites. *Phil. Trans. Roy. Soc.* A303:339–349.

Clayton, R. N., Onuma, N., and Mayeda, T. K. 1976. A classification of meteorites based on oxygen isotopes. *Earth Planet. Sci. Lett.* 30:10–18.

Craig, H. 1964. Petrological and compositional relationships in meteorites. In *Isotopic and Cosmic Chemistry,* eds. H. Craig, S. L. Miller, and G. J. Wasserburg (Amsterdam: North-Holland), pp. 401–451.

Dennison, J. E., and Lipschutz, M. E. 1986. Chemical studies of H chondrites—II. Weathering effects in the Victoria Land, Antarctic population and comparison of two Antarctic populations with non-Antarctic falls. *Geochim. Cosmochim. Acta* 51:741–754.

Dodd, R. T., Van Schmus, W. R., and Koffman, D. M. 1967. A survey of the unequilibrated ordinary chondrites. *Geochim. Cosmochim. Acta* 31:921–951.

Dodd, R. T., and Jarosewich, E. 1979. Incipient melting in and shock classification of L-group chondrites. *Earth Planet. Sci. Lett.* 44:335–340.

Fredriksson, K., and Keil, K. 1964. The iron, magnesium, and calcium distribution in coexisting olivines and rhombic pyroxenes in chondrites. *J. Geophys. Res.* 69:3487–3515.

Fredriksson, K., Nelen, J., and Fredriksson, B. J. 1968. The LL-group chondrites. In *Origin and Distribution of the Elements,* ed. L. H. Ahrens (New York: Pergamon Press), pp. 457–466.

Gooding, J. L., and Keil, K. 1981. Relative abundances of chondrule primary textural types in ordinary chondrites and their bearing on conditions of chondrule formation. *Meteoritics* 16:17–43.

Graham, A. L., Bevan, A. W. R., and Hutchison, R. 1985. *Catalogue of Meteorites,* fourth edition (Tucson: Univ. of Arizona Press).

Grossman, L. 1980. Refractory inclusions in the Allende meteorite. *Ann. Rev. Earth Planet. Sci.* 8:559–608.

Guimon, R. K., Lofgren, G. E., and Sears, D. W. G. 1987. Chemical and physical studies of type 3 chondrites—IX: Thermoluminescence and hydrothermal annealing experiments and their relationship to metamorphism and aqueous alteration in type < 3.3 ordinary chondrites. *Geochim. Cosmochim. Acta,* in press.

Hutchison, R., Alexander, C. M. O., and Barber, D. J. 1987. The Semarkona meteorite: First recorded occurrence in an ordinary chondrite, and its implications. *Geochim. Cosmochim. Acta* 51:1875–1882.

Keck, B. D., and Sears, D. W. G. 1987. Chemical and physical properties of type 3 chondrites—VIII: Thermoluminescence and metamorphism in CO chondrites. *Geochim. Cosmochim. Acta,* in press.

Keil, K. 1968. Mineralogical and chemical relationships among enstatite chondrites. *J. Geophys. Res.* 73:6945–6976.

Keil, K. 1982. Composition and origin of chondritic breccias. *Workshop on Lunar Breccias and Their Meteoritic Analogs,* eds. G. J. Taylor and L. L. Wilkening, LPI Tech. Rept. 82-02 (Houston: Lunar and Planetary Inst.), pp. 65–83.

Mason, B. 1966. The enstatite chondrites. *Geochim. Cosmochim. Acta* 30:23–39.

McSween, H. Y. 1979. Are carbonaceous chondrites primitive or processed? A review. *Rev. Geophys. Space Phys.* 17:1059–1078.

Scott, E. R. D., and Wasson, J. T. 1975. Classification and properties of iron meteorites. *Rev. Geophys. Space Phys.* 13:527–546.

Sears, D. W. 1978. *The Nature and Origin of Meteorites* (New York: Oxford Univ. Press).
Sears, D. W. G. 1987. *Thunderstones: The Meteorites of Arkansas* (Fayetteville: Univ. of Arkansas Museum/Univ. of Arkansas Press), in press.
Sears, D. W. G., and Axon, H. J. 1976. Nickel and cobalt contents of chondritic meteorites. *Nature* 260:34–35.
Sears, D. W. G., Bahktiar, N., Keck, B. D., and Weeks, K. S. 1984. Thermoluminescence and the shock and reheating history of meteorites: II. Annealing studies of the Kernouve meteorite. *Geochim. Cosmochim. Acta* 48:2265–2272.
Sears, D. W., Grossman, J. N., Melcher, C. L., Ross, L. M., and Mills, A. A. 1980. Measuring metamorphic history of unequilibrated ordinary chondrites. *Nature* 287:791–795.
Sears, D. W., Kallemeyn, G. W., and Wasson, J. T. 1982. The compositional classification of chondrites: II. The enstatite chondrite groups. *Geochim. Cosmochim. Acta* 46:597–608.
Sears, D. W. G., and Weeks, K. S. 1986. Chemical and physical studies of type 3 chondrites—VI: Siderophile elements in ordinary chondrites. *Geochim. Cosmochim. Acta* 50:2815–2832.
Urey, H. C., and Craig, H. 1953. The composition of the stone meteorites and the origin of the meteorites. *Geochim. Cosmochim. Acta* 4:36–82.
Van Schmus, W. R., and Hayes, J. M. 1974. Chemical and petrographic correlations among carbonaceous chondrites. *Geochim. Cosmochim. Acta* 38:47–64.
Van Schmus, W. R., and Wood, J. A. 1967. A chemical-petrologic classification for the chondritic meteorites. *Geochim. Cosmochim. Acta* 31:747–765.
Von Michaelis, H., Ahrens, L. H., and Willis, J. P. 1969. The composition of stony meteorites—II. The analytical data and an assessment of their quality. *Earth Planet. Sci. Lett.* 5:387–394.
Walsh, T. M., and Lipschutz, M. E. 1982. Chemical studies of L chondrites—II. Shock induced trace element mobilization. *Geochim. Cosmochim. Acta* 46:2491–2500.
Wiik, H. B. 1969. On regular discontinuities in the composition of meteorites. *Commentat. Phys. Math.* 34:135–145.

PART 2
Source Regions

2.1. ASTEROIDS AND METEORITES

GEORGE W. WETHERILL
Carnegie Institution of Washington

and

CLARK R. CHAPMAN
Planetary Science Institute

There is general agreement among meteoriticists that most meteorites are fragments of asteroids, produced by relatively recent collisions in the asteroid belt. Two obstacles have stood in the way of simple acceptance of this consensus: (1) known dynamical mechanisms did not appear able to provide the observed mass flux and orbits of meteorites, and (2) spectrophotometric data indicate that asteroids with the composition of the most abundant meteorite class are rare or absent in the asteroid belt. Recent work has shown that resonance mechanisms are adequate to solve the dynamic problem, but the spectrophotometric problem remains. Hypotheses are suggested to resolve this paradox, but it is not at all clear that our understanding is adequate to accept any of them at this time. If it is assumed that most meteorites are fragments of asteroids, then meteoritics and the planetary geology of the asteroid belt are two names for the same scientific subject. An understanding of the origin and evolution of the asteroid belt is essential to an adequate basis for interpretation of meteoritic data. Conversely, the detailed record of early asteroidal history preserved in the meteorites provides evidence of a kind unavailable for the earliest history of any other planetary bodies, and will be of great value in testing alternative hypotheses regarding asteroidal origin.

2.1.1. INTRODUCTION

The study of meteorites represents one aspect of the long-standing and continuing human inquiry into the origin of our Earth and solar system. From the perspective of planetary science, meteorites may be regarded as rocks broken from outcrops in the present solar system. Evidence from the meteorites themselves tells us that almost all their sources are relics of primordial, relatively pristine and presumably small solar-system bodies.[a] Without much loss of certainty the great majority of meteorites can be thought of as fragments of objects in the present-day asteroid belt. The uncertainty refers only to the possibility that some meteorites may be of cometary origin.

That most meteorites come from asteroids has been the prevailing opinion for several decades, and for good reasons. However, it may be noted that this consensus is fairly recent. As discussed by Olivier (1925), at that time the limited observational evidence available provided support for a cometary or even an interstellar origin of meteorites. This included observation of fireballs, apparently similar to those that accompany meteorite falls, in meteor streams associated with known comets. Furthermore, the heliocentric orbits of meteorites appeared to be parabolic or hyperbolic, based on visual velocity determinations, now known to be inaccurate. It is often argued that because "there has never been an observed fall from a shower meteor" an association between the cometary sources of shower meteors and meteorites is very unlikely. Although it is now regarded simply as a coincidence, in fact the Mazapil iron meteorite fell during one of the greatest meteor showers ever observed, the Bielids of 27 November 1885. This was formerly regarded by many authorities as evidence for a cometary origin of at least iron meteorites.

Abundant geochemical and petrological evidence implies that meteorites are derived from subplanetary-size bodies. After the Apollo missions taught us what lunar meteorites should look like, the only remaining plausible choices for the source of at least the most abundant meteorites were the comets and the asteroids. Cometary spectra and *in situ* analyses provide strong support for the belief that comets are volatile-rich bodies, consistent with their orbits and their presumed origin in the outermost solar system. In contrast, the most abundant meteorites, the ordinary chondrites, are depleted in volatiles to about the same extent as the highly depleted planet Earth. There are also indications that live comet nuclei resemble D- and P-type asteroids, very unlike the ordinary chondrites (Hartmann et al. 1985; Tholen et al. 1986;

---

[a]The principal exceptions are a group of rare achondrites—shergottites, nakhlites, chassignites (SNC)—that have the appearance of being samples of a larger and more geologically active planet (Papanastassiou and Wasserburg 1974; Walker et al. 1979). If so, the most plausible candidate is the planet Mars (Wasson and Wetherill 1979; Nyquist et al. 1979). These SNC meteorites have been reviewed by McSween (1985). In addition, a few small achondrites are of lunar origin.

Campins et al. 1987; Millis et al. 1987). Furthermore, there is increasingly clear evidence for late collisional events, recorded in the isotopic systems of ordinary chondritic meteorites that are difficult to understand if the bodies from which they were derived were in the Oort cometary cloud ~ 500 Myr ago (Anders 1964; Bogard 1979; Turner 1979).

A serious difficulty that lay in the way of simply accepting the asteroidal origin of meteorites was pointed out by Öpik (1963). He showed that the known mechanisms for bringing meteorite material from the asteroid belt to Earth were quantitatively inadequate and in disagreement with the observed ratio of purely Mars-crossing to Earth-crossing bodies of asteroidal appearance. This led to his suggestion that, despite conventional opinion, the ordinary chondrites were derived from Earth-crossing remnants of comets that had lost their volatile portions by solar heating. This somewhat outrageous proposal provided the necessary challenge to find a quantitatively adequate mechanism for delivering asteroidal collision fragments to Earth on the appropriate time scale and in the required quantity.

An adequate response to this challenge required extension of our understanding of the heliocentric orbits of meteorites and of the dynamical mechanisms that control the evolution of these orbits. The observational evidence for ordinary-chondrite orbits is summarized in Sec. 2.1.2. Section 2.1.3 outlines the theoretical basis for identifying asteroids rather than comets with the source region of ordinary chondrites, and in a somewhat less direct way with most carbonaceous meteorites and achondrites. The recent dynamical work not only strongly supports an asteroidal origin of ordinary chondrites, but also shows that a large fraction of the Apollo and Amor objects are probably asteroidal-collision fragments.

Spectrophotometric data, discussed in Sec. 2.1.4, provide a way to compare the mineralogical composition of asteroids and meteorites. Qualitatively, considerable resemblance is found between the mineralogies of meteorites and those of their presumed asteroidal sources. In more detail, however, important unanswered questions remain. The most serious of these is the quantitative disagreement between the contents of olivine, pyroxene and metal of the most abundant meteorite class, the ordinary chondrites, and those inferred for the S asteroids, otherwise the most promising asteroidal source for these meteorites.

The origin of the asteroidal sources of meteorites is discussed in Sec. 2.1.5. Perhaps the greatest impediment to theoretical studies of the origin of the solar system becoming a more "respectable" science is the great difficulty of finding observational evidence with which to test theoretical predictions. It may be expected that the well-preserved record in the meteorites will provide the means of testing hypotheses regarding the origin of asteroids, evidence of a nature unavailable for the larger planets. Conversely, full use of the detailed evidence for events in the solar system preserved in the meteorites requires understanding the planetological context in which these events took place.

Little attention, however, has been given to detailed exploitation of this important opportunity. Only an outline of possible courses of events can be provided at this time, together with some suggestions for future work.

2.1.2. METEORITE ORBITS

Before colliding with the Earth, meteorites are moving in orbits about the Sun that intersect the orbit of the Earth. These Earth-intersecting orbits are a subset of Earth-crossing orbits, i.e., with perihelia closer to the Sun than the Earth and aphelia beyond the orbit of the Earth. If the solar system were two-dimensional, all Earth-crossing orbits would be Earth-intersecting orbits. In the actual three-dimensional solar system, a body in an Earth-crossing orbit will usually be above or below the plane of the orbit of the Earth at the time it is at the Earth's distance from the Sun and therefore will not actually intersect the orbit of the Earth. It is found, however, that gravitational perturbations by the planets will cause Earth-crossing orbits to "precess" in such a way that on a time scale of about 10^4 yr, most Earth-crossing orbits will repeatedly be temporarily Earth-intersecting (see, e.g., Öpik 1951; Wetherill 1974; Wetherill and Shoemaker 1982). Because 10^4 yr is short compared to the 10^6 to 10^8 yr collisional and dynamical lifetimes of bodies in Earth-crossing orbits, Earth-crossing bodies may be thought of as potential Earth impactors.

The orbits of three meteorites, all ordinary chondrites, have been determined by astrometric and photometric photography by bright meteor "fireball" networks (Fig. 2.1.1). All of these orbits have perihelia fairly near Earth's orbit, penetrate into the asteroid belt, but remain well inside the orbit of Jupiter. By use of phenomena associated with the observed atmospheric trajectory of these meteorites (i.e., deceleration, fragmentation, light curve and depth of atmospheric penetration), it has proven possible to identify about 30 photographed, but unrecovered, very bright meteors (Ceplecha and McCrosky 1976; Wetherill and ReVelle 1981) that appear to have physical properties indistinguishable from those of the photographed ordinary chondrites. In contrast, the majority of the fireballs do not appear to share the properties of ordinary chondrites, and must be associated with bodies of other kinds, probably carbonaceous meteorites, as well as cometary fragments that are not capable of surviving their passage through the atmosphere. Because the recovered ordinary chondritic fireballs are among the very brightest photographed meteors, it may be expected that they should be accompanied by a much larger number of less massive ordinary chondrites. These must be well represented among the fireballs, and it seems most reasonable to identify them with those fireballs with atmospheric trajectories and lightcurves that resemble those of the ordinary chondrites, rather than with those that do not.

If this plausible identification is made, the statistical significance of the data set is enlarged considerably, and the number of meteorite orbits becomes

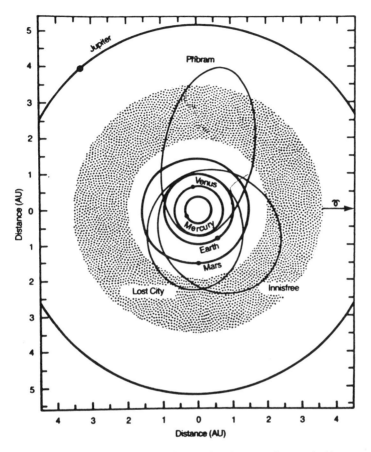

Fig. 2.1.1. The orbits of the three recovered meteorites that were photographed by astrometric and photometric fireball networks (figure from Wasson 1985).

sufficient to be of diagnostic value with regard to their potential sources. An important feature of these observed orbits is that of their perihelion distribution (Fig. 2.1.2a). These are strongly concentrated between 0.9 and 1.0 AU. This in agreement with perihelia determined from visual radiants of ordinary chondrites, calculated by Simonenko (1975) (Fig. 2.1.2b). Significant, but somewhat less concentration of perihelia toward the Earth's orbit is exhibited by the data of Halliday et al. (1984) of Canadian MORP fireballs with terminal masses >50 g (Fig. 2.1.2c). It is possible that the bodies in Fig. 2.1.2c are not all ordinary chondrites, but contain an admixture of possibly cometary meteors that are in the process of disintegrating into small and probably unrecoverably small fragments. As will be mentioned in the next section, it should not be expected that such cometary material will have perihelia con-

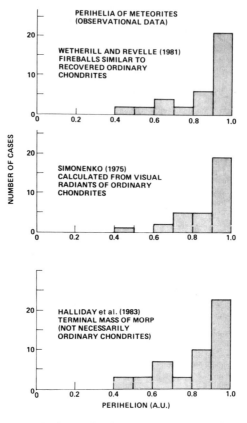

Fig. 2.1.2. Observed perihelia of meteorites that at least primarily consist of ordinary chondrites (see text for discussion).

centrated toward 1 AU, and inclusion of some objects of this kind may be expected to blur the concentration of orbits near 1 AU.

These completely determined orbits are in excellent agreement with earlier inferences based on visual radiants and time-of-fall statistics (Wood 1961; Wetherill 1968). The time of fall (daylight afternoon/daylight total falls) is a particularly simple, sensitive, and therefore useful parameter for testing whether or not a proposed dynamical mechanism of meteorite origin is supported by observations. Ordinary chondrites are observed to fall about twice as often in the daylight afternoon hours, as in the daylight morning hours. In contrast, on a 24-hour basis, probably biased against midnight to 6 AM falls, the basaltic achondrites exhibit no significant PM excess (Table 2.1.1). Although the source of these meteorites is also almost certainly asteroidal, quite likely it is different from those of the ordinary chondrites.

TABLE 2.1.1
A. Observed Fall Times of Chondrites Between 6 AM and 6 PM: Total AM and PM Falls, Including Nighttime, in Parentheses

	L	LL	H	C
Morning Falls	54(68)	10(12)	48(64)	2(3)
Afternoon Falls	100(146)	22(36)	75(120)	7(13)
Daytime PM Falls/ Total Daytime Falls	0.65	0.69	0.61	

B. Observed Fall Times of Differentiated Meteorites: Both Day and Night Falls

	Basaltic Achondrites	Other Achondrites (not SNC)	Irons	Stony Irons
AM	16	6	9	2
PM	17	9	15	6

2.1.3. DYNAMICAL MECHANISMS FOR DELIVERY OF METEORITES TO EARTH

The results of dynamical studies of Williams (1969) provided the opportunity for the first serious response to Öpik's demonstration that the known mechanisms for deriving meteorites from the asteroid belt were inadequate, and to his view that for this reason a cometary source is more likely. Williams showed that the ν_6 secular resonance just inside the asteroid belt (2.04 AU) was capable of accelerating material in its vicinity into Earth-crossing orbit by gravitational mechanisms (Williams 1973; Wetherill and Williams 1979). An additional resonance mechanism was proposed at about the same time, based on nonlinear interaction of the effect of encounters within ~ 0.5 AU of Jupiter and the 2:1 Jovian commensurability resonance at 3.27 AU, responsible for the observed Kirkwood gap at that heliocentric distance (Zimmerman and Wetherill 1973).

The possibility of a more quantitatively satisfactory resonance mechanism was provided by Wisdom (1983, 1985) who showed that there is a chaotic zone in the asteroid belt, associated with the 3:1 commensurability resonance and Kirkwood gap at 2.50 AU. This is capable of accelerating asteroidal material into Earth-crossing orbit on a ~ 1 Myr time scale. It had been known for some time (Wetherill 1968) that an initial Earth-crossing orbit with a semimajor axis of about 2.5 AU was required to explain the chondritic orbital data, but prior to Wisdom's work, it was not clear that a dynamical mechanism adequate to provide enough material in these orbits existed. Further elaboration of this problem showed that Wisdom's mechanism quantitatively provided both the orbital distribution and Earth-impacting flux of chondritic meteorites (Wetherill 1985), provided of course that a large fraction of the asteroids near the 3:1 resonance are of ordinary chondritic composition. Furthermore, it also provided a way to "normalize" other potential resonance sources, particularly the $\dot{\nu}_6$ resonance, to the Earth-impacting yield of the 3:1 commensurability resonance, and thereby provide a quantitatively adequate determination of the yield of meteorites from the entire asteroid belt (Wetherill 1987). If Wisdom's conclusions regarding the 3:1 resonance can be extended to the more distant Kirkwood gaps, it can be concluded that few meteorites are derived from asteroids beyond 2.6 AU, because these resonances would cause material to be lost into Jupiter-approaching orbits before they become Earth crossing.

These calculations show that the $\dot{\nu}_6$ resonance should contribute about 10% as much meteorite-size fragments as the 3:1 resonance, and that no excess of afternoon falls should be expected for meteorites from this source. This source has been proposed as a possible, but by no means certain, source of some types of achondrites (Wetherill 1987). This yield is sufficient to provide the observed flux of basaltic achondrites as well as that of several other types of achondrites. Because no afternoon excess is found for these achondrites, the predicted absence of asymmetry poses no problem for this hypothesis.

A further consequence of discovery of these mechanisms is a quantitative understanding of the delivery of ~ 1 to 20 km diameter Apollo-Amor objects into Earth-approaching orbits. Within this general theoretical framework those Apollo-Amor objects of asteroidal origin are seen to be equivalent to big meteorites, the largest members of a continuous spectrum of Earth-approaching asteroidal debris, that also includes intermediate size (10 m–100 m) bodies, as well as meteorite size objects. Most of these Earth-approaching bodies will continue to have asteroidal aphelia. While passing through the asteroid belt they will continue to collide with small debris in the asteroid belt and therefore will continue to fragment into meteorite-size bodies at a rate comparable to that of their counterparts of similar size in the main asteroid belt. Earth-crossing meteorite-size fragments will be generated by this entire complex of bodies, in the asteroid belt, in purely Mars-crossing orbits, and in Earth-crossing orbits, and all these fragments will contribute to the Earth-impacting flux of meteorites. In these calculations it has been assumed that the physical strength of these bodies is independent of mass and that they follow a size distribution like that proposed by Dohnanyi (1969) as the theoretical steady-state distribution for a self-fragmenting asteroidal swarm. In the size range of sources found to be most important for production of meteorite-size fragments (< 20 km diameter), this theoretical distribution is in agreement with the observed distribution of asteroids (Kuiper et al. 1958; Van Houten et al. 1970) if a uniform average albedo is assumed. Bias-corrected size distributions for larger asteroids (Zellner 1979; Chapman 1987) exhibit interesting deviations from this simple theory. An understanding of the causes of these deviations would be very valuable. Nevertheless, in the small-size range, this approximation should be roughly valid.

Under these assumptions, it is calculated that the "immediate parent bodies" of most meteorites are quite small asteroidal bodies (10 m–100 m diameter), and that most of the meteorite-size fragments are produced by collision of these immediate parent bodies with even smaller main-belt bodies while in Earth-crossing orbits (Wetherill 1985). Under the assumptions made in these calculations it is found that only about 10% of the meteorite flux is provided by bodies that were reduced to a meteorite size while their orbits were still entirely in the asteroid belt. Therefore, conceptually this mechanism may be regarded as one in which meteorites are secondary-collision products of Apollo objects, with the qualification that the "Apollo objects" are mostly smaller than those observed.

If for some reason (see, e.g., Greenberg and Chapman 1983), fragments of larger asteroids are preferentially ejected at the necessary velocities (50–200 m/s) into the resonance regions, the proportion of meteorites actually generated entirely in the asteroid belt will increase, but the distribution of meteorite orbits will not change very much. An adequate mass yield of meteorites and Apollo objects does appear to require a major contribution from the fragmentation of smaller bodies, not much larger than the meteorites themselves.

The results of these same calculations argue, on dynamical grounds, against a cometary origin of ordinary chondrites. Neither active comets nor those Apollo objects most likely to be volatile-depleted comets exhibit the concentration of perihelia near 1 AU required to satisfy the orbital requirements of ordinary chondrites. In fact, a principal argument for a cometary component among the Apollo objects is the need to match the observed frequency of Apollo objects with perihelia $\lesssim 0.8$ AU and with semimajor axes $\gtrsim 2.1$ AU. A steady-state distribution of orbits of this kind is predicted to result from evolution of short-period comet orbits into those resembling the orbit of the observed comet Encke. This class of orbits, however, will provide an Earth-impacting flux of bodies with no morning vs afternoon asymmetry in fall times, nor will it match any of the other observed characteristics of ordinary chondrites. Therefore, even if some of the collision fragments of this population represent recoverable meteoritic material, they are unlikely to be ordinary chondrites. Data bearing on the orbits of carbonaceous meteorites are more scant, but are consistent with the expected origin of most of these meteorites from the vicinity of the 3:1 Kirkwood gap. There is also abundant evidence that most cometary material is very weak and does not survive atmospheric entry, except in the form of interplanetary dust and micrometeorites (see Chapter 11.1). Nevertheless, a cometary contribution of a small number of carbonaceous meteorites cannot be excluded, and may even be suggested by fireball observations (Wetherill and ReVelle 1982).

In all these calculations it has been assumed that asteroidal collisions and ejection into Earth-crossing orbits is a continuous process. In fact, there is a significant, but probably not dominant stochastic component associated with an asteroidal size distribution in which most of the mass is in the largest bodies. A quantitative evaluation of this stochastic effect is required if we are to understand phenomena such as the 7 Myr H-chondrite peak in exposure ages (Crabb and Schultz 1981; see also Chapter 4.1).

The conclusions of these dynamical investigations (Wetherill 1985, 1987a,b) are the following:

1. There are two major asteroidal source regions for meteorites and Apollo-Amor objects. The first is adjacent to the 3:1 Kirkwood gap that extends from 2.48 to 2.52 AU. It includes main-belt asteroids between about 2.45 and 2.48 AU and 2.52 to 2.55 AU, with smaller contributions from asteroids with somewhat greater and smaller semimajor axes. The second is in the innermost belt, in the vicinity of the Flora asteroidal group. The $\dot{\nu}_6$ resonance at 2.04 AU plays an important role in accelerating fragments of these asteroids into Earth-crossing orbits.
2. Most ordinary chondrites should be derived from the vicinity of the 3:1 gap. It is also expected that a comparable number of carbonaceous meteorites originate in this region.
3. The innermost belt is a good candidate source region for several kinds of achondrites, including the basaltic achondrites. Both the calculated mass yield and the time-of-fall statistics of these meteorites can be understood if this is the case (Wetherill 1987a).
4. It is unlikely that any significant number of meteorites originate from orbits beyond about 2.6 AU, although further study of the 5:2 resonance at 2.82 AU is warranted.

2.1.4. SPECTROPHOTOMETRIC AND RADAR EVIDENCE LINKING ASTEROIDS AND METEORITES

Remote-sensing Techniques

In the previous section, we have considered the dynamical and collisional processes that can deliver meteorites to the Earth, both directly and through the intermediary of Apollo objects, from the main asteroid belt (primarily from the inner third of the belt). An entirely different approach to the problem is to examine the asteroids, and other potential source bodies, using remote-sensing techniques to see if these bodies are, in fact, composed of minerals like those of the meteorites. Ideally, the techniques would be so definitive and uniquely diagnostic that we could identify specific parent bodies for the different meteorite types. In reality, there are complications to remote sensing that will be elaborated in this section.

Remote-sensing techniques seek to measure characteristics of distant objects for comparison with laboratory measurements. The chief technique that has been applied to the asteroids is visible and near-infrared reflection spectroscopy. Recently, useful results have also been obtained from groundbased radar-reflection measurements. In neither case are the techniques as diagnostic as spectroscopy of gases, but useful results have been obtained. Some of the more recent reviews of asteroidal spectral-reflectance studies are those of Gaffey and McCord (1978) and Chapman (1979, 1981).

The possibility of identifying asteroid mineralogies and linking them to meteoritic mineralogies was first demonstrated by McCord et al. (1970) for Vesta, which was identified as being very similar to basaltic achondritic meteorites. This identification has stood the test of time, although for dynamical reasons, it is difficult to understand how Vesta itself can actually be the basaltic achondrite parent body.

In essence, sunlight, in the course of being reflected from an asteroidal surface, is transmitted through mineral crystals in the surface layer and acquires a spectral signature caused by the composition and particular lattice structure of the minerals. These are characterized by having "slopes" (colors) and several broad absorption features, whose positions, widths and shapes may be diagnostic of even fairly minor differences in mineralogy. These data are combined with an understanding of "plausible" assemblages, based on cosmic abundances. Particular multimineral assemblages are more difficult to model, but a wide body of laboratory calibrations exist. Results of such measurements on asteroids can then be compared with those on a nearly complete sample of meteorite types, measured by Gaffey (1976). The best matches to asteroid spectra (e.g., Fig. 2.1.3) are found when the meteorite samples are pulverized to simulate an asteroidal regolith.

The use of groundbased radar for asteroid compositional studies is in its infancy; a recent paper that reviews this topic is by Ostro et al. (1985). Radar echoes are strongly affected by the dielectric constant of the reflective surface, so the technique is exceptionally sensitive to metal, which has been its chief application to asteroid compositions so far.

The technique that was used by McCord et al. (1970) for Vesta has been applied in more recent years to other taxonomic types of asteroids. Of particular interest has been the effort to identify in the asteroid belt the sources of the most abundant type of stony meteorite, the ordinary chondrites. The dynamical studies described in the previous section suggest that the asteroids in the vicinity of the 3:1 Kirkwood gap at 2.5 AU may be expected to provide a yield comparable to that observed and may be regarded as *prima facie* prime candidates for this identification.

The Composition of S Asteroids and the Spectrophotometric Paradox

About a decade ago there apeared to be two major paradoxes that had to be faced in understanding the relationship between asteroids and meteorites.

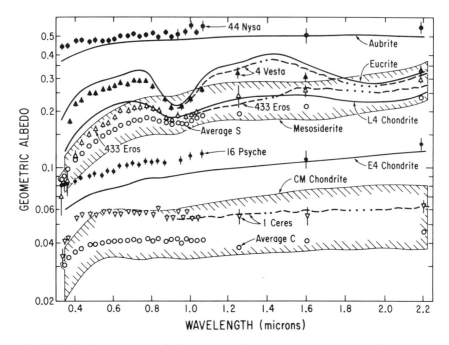

Fig. 2.1.3. Reflectance spectra (visible through infrared) of representative asteroids and meteorites. The two shaded bands indicate the ranges of spectral albedo curves for typical mesosiderites and typical CM chondrites. It is evident that the L4 chondrite spectrum bears some similarities to S types, but most ordinary chondrites resemble S types even less. In general, however, detailed analysis of reflectance spectra show that the range of diverse mineral assemblages found in meteorite collections are also represented in the asteroid belt.

Although on cosmochemical grounds essentially all meteoriticists believed that meteorites were of asteroidal origin, it also seemed that:

1. There was no adequate mechanism for transferring material from the main asteroid belt into Earth-crossing orbit. As discussed in Secs. 2.1.2 and 2.1.3, important progress has been made in resolving this "dynamical paradox."
2. Even if a dynamically adequate mechanism were found, the problem would still remain that there were virtually no asteroids in the main belt that had the reflectance spectra expected for ordinary chondrites. This "spectrophotometric paradox" remains unresolved.

The most abundant asteroids in the inner belt are classified as S and C asteroids (Chapman et al. 1975). The C asteroids are most plausibly associated with meteorites rich in opaque minerals with low albedos, i.e., the carbonaceous meteorites, and the S asteroids are therefore the most promising ordinary chondrite sources. Qualitatively, the mineralogy inferred from reflection spectra of S-type asteroids is the same as found in ordinary chondrites: olivine, pyroxene and metallic Fe. Careful quantitative consideration of the spectrophotometric data leads to the conclusion, however, that the proportions and in some cases the compositions of these mineral constituents differ from those found in the ordinary chondrites (Gaffey 1984b,1986).

Chapman (1979) summarized the essential problem in interpreting the reflectance spectrum of S-type asteroids. These spectra, which would seem likely to be as diagnostic as the Vesta spectrum, show two near-infrared absorption bands—with variable bandwidths and band centers—and other prominent spectral features, in particular the long-wavelength shoulder of a deep ultraviolet charge-transfer absorption band (Fig. 2.1.4). Indeed, there

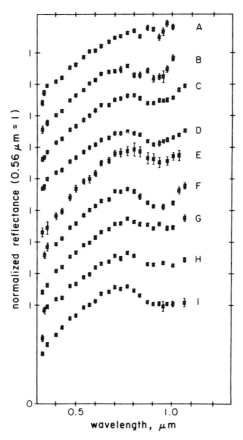

Fig. 2.1.4. Forty-five visible and near-infrared spectra of S types by Chapman and Gaffey (1979) are placed into nine groups and averaged, illustrating the range of spectral-curve shapes. The differences reflect different proportions of metal, olivine and pyroxene as well as different chemistry of the silicates.

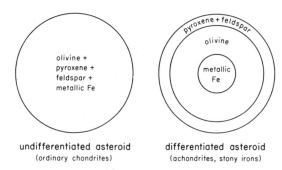

Fig. 2.1.5. Schematic cross sections of two meteorite parent bodies. If the differentiated parent body becomes thoroughly "jumbled" by collisional fragmentation and reaccumulation processes, its surficial mineralogical assemblages will crudely resemble the undifferentiated parent body.

has been little controversy over the interpretation of the S-type spectra as a combination of pyroxene, olivine and metallic Ni-Fe. However, the reflectance spectra of two very different meteorite types are primarily dominated by these three minerals: the undifferentiated ordinary chondrites, and highly differentiated stony-iron meteorites, the pallasites and mesosiderites (Fig. 2.1.5). The proportions of these minerals differ, however, in these two classes of meteorites.

Stony-iron meteorites have comparable amounts of metal and silicates (see Chapter 1.1). The ferro-magnesian silicates may be chiefly olivine (in the case of pallasites) or pyroxene (in the case of mesosiderites). Still other metal/silicate compositions are represented among known meteorites. Ordinary chondrites, however, have relatively minor amounts of reduced metal (as much as 20 wt.% in the case of H types, but < 5% in the case of LL types); the silicates in ordinary chondrites are a mixture (in somewhat different proportions) of olivines and pyroxenes. Thus, in summary, the major problem of sorting out the implied composition of S types in terms of well-known meteorite types relates to determining the relative proportions of the three major components to within factors of better than two or three.

The most direct comparison of asteroid spectra and meteorites is done by crushing meteorites and measuring their spectra in the laboratory. By this test, S types fail to match ordinary chondrites. (It is more difficult to perform the test on stony-iron meteorites because they cannot be crushed due to the ductility of their metallic matrix.) S types have a larger signature from metal than is shown by laboratory measurements of ordinary chondrites. Another important difference is that most S types are very olivine rich compared with the olivine/pyroxene ratios in even the olivine-rich LL types, based on laboratory calibration of these ratios. Feierberg et al. (1982) argued that within the large errors of the astronomical data, some S types fell within the range of ordinary chondrites. Gaffey (1984a) has studied several S types in great detail, including one (8 Flora) that Feierberg et al. (1982) identified as least olivine rich in their sample of asteroids for which mid-infrared spectra are available. He finds that Flora cannot be an ordinary chondrite, primarily because the manner in which the spectra vary with rotation imply spatially heterogenous mineral-phase and composition distributions that are not consistent with those expected for ordinary chondrites. The technique Gaffey has used can be applied only to the brightest asteroids, but he believes that Flora typifies large S-type asteroids and concludes that they are of achondritic (in this case stony-iron) composition.

The nonspectrophotometric evidence for an asteroidal origin of ordinary chondrites described earlier in this chapter provides a strong expectation for there being a large mass of ordinary chondrite sources near the 3:1 Kirkwood gap or at least elsewhere in the asteroid belt. Even if the immediate sources of ordinary chondrites are Earth-crossing Apollo objects, there must be an adequate source for these objects. At present, no very promising resolution of this paradox is known. In view of the central nature of this problem to the most fundamental questions of the asteroid-meteorite connection, an effort to reconcile these conflicting data is a matter of highest scientific priority.

For this reason, we will discuss several conceivable ways of resolving this paradox. No claim is being made that present evidence supports the conclusion that any of these hypotheses is likely to be correct. In fact, it is not hard to make the case that all of them are wrong. However if, as seems very likely, ordinary chondrites are actually of asteroidal origin after all, such a negatively oriented approach would fail to discover this truth. For this reason, the difficulties that will be pointed out in these hypotheses should not be thought of as refutations of them, but rather as explanations of why we cannot at this time simply accept them and thereby dispose of the spectrophotometric paradox.

Ordinary chondrites are fragments of S asteroids, but the present sampling is incomplete. There are about three dozen S-type asteroids larger than 100 km diameter. Hundreds of smaller asteroids have been observed that are classifiable as S types on the basis of available astronomical data (Zellner 1979; Tholen 1984; Zellner et al. 1985; Chapman 1987). Although all S-type asteroids contain silicates and exhibit the reddening ascribed to Ni-Fe metal, only a fraction of the available data constrain the pyroxene/olivine ratio. Numerous asteroids not yet well observed may have appropriate pyroxene/olivine ratios. While Flora is perhaps the most pyroxene-rich asteroid of eleven in the Feierberg et al. (1982) sample, many other S types are known that are more pyroxene rich (e.g., 11 Parthenope, which is near the 3:1 resonance and is among the asteroids discussed by Feierberg et al. [1982], for which mid-infrared spectra were not available). Other examples are given by Bell et al. (1987).

It is not plausible to carry this incomplete sampling hypothesis too far, however. It is sometimes suggested that the paradox can be resolved by recognition that the ordinary-chondritic S types are an extremely special sample of the asteroid belt, and that the dynamical mechanisms by which asteroidal debris is selected for transfer to Earth-crossing orbits are so highly selective that an almost unobservably rare quantity of material dominates the meteoritic impact flux on Earth. A variant of this hypothesis is that the stochastic nature of asteroid-disrupting impacts is responsible for the predominance of ordinary chondrites, and that if the Earth's impact flux were averaged over a sufficiently long interval of time, the observed differences would disappear.

It would be excessively dogmatic to rule out such hypotheses completely. The sampling process is certainly selective. As mentioned earlier, the 2.5 AU Jovian commensurability resonance greatly enhances the probability that asteroidal debris produced in this region will ultimately hit the Earth. The nature of the asteroidal size distribution, in which most of the mass is concentrated in the largest objects, requires serious consideration of stochastic effects. On the other hand, it must be emphasized that, at present, there is no theoretical basis at all for believing that either very special selection, or stochastic fluctuations, can explain how most meteorites can come from rare asteroids. The observed terrestrial mass flux in itself places a severe burden on the efficiency of any proposed orbital-transfer mechanism. Selection of rare asteroids would require extremely high efficiencies. Furthermore, there is a natural "mixing length" in the asteroid belt of about 0.03 AU imposed by the variation in semimajor axis of asteroidal fragments collisionally ejected at velocities of 0.05 to 0.2 km/s (Gault et al. 1963). For this reason, it can be understood how the mechanism associated with the 2.48 to 2.52 AU Kirkwood gap can preferentially sample regions between about 2.45 and 2.48 AU and between 2.52 and 2.55 AU. But these are not extremely small regions; rather they include about 5% of the entire asteroid belt. Moreover, the mass distribution of asteroids, although concentrated in the larger bodies, still contains a sufficiently large number of smaller bodies to cause the contribution of their fragmentation debris to be comparable to that of the larger bodies. Stochastic effects, associated with episodic large impacts on large bodies, will thereby be smoothed out by a comparable background flux of material from smaller bodies. For this reason, stochastic effects of the order of about a factor of two might not be too surprising, but these effects should not be expected to dominate completely.

Comparison of Antarctic and non-Antarctic H5 chondrites provides some evidence for stochastic variations in the compositions of ordinary chondrites on a very short time scale ($\sim 3 \times 10^5$ yr) (Dennison and Lipschutz 1986,1987). Because the cosmic-ray exposure ages of both Antarctic and non-Antarctic meteorites are much longer (~ 10 Myr) than the $\sim 3 \times 10^5$ yr difference in their terrestrial residence times, these differences cannot be explained as the result of a single stochastic event that occurred during the last

3×10^5 yr. Rather, it would seem to require that following their fragmentation from a larger object, meteorites remain in compact streams, analogous to cometary meteor streams for times ~ 10 Myr, and the Earth intersects different streams at different times. As pointed out by Wetherill (1986), the existence of such long-lived meteorite streams would be unexpected because it is not only contrary to present understanding of the dynamics of bodies in Earth-approaching orbits, and is not in general supported by analysis of Prairie Network fireball data (Wetherill and ReVelle 1982). Halliday (1987) has noted the similarity of the orbit of an unrecovered Canadian fireball to that of Innisfree, and McCrosky (personal communication 1987) pointed out that these orbits are also similar to that of the "chondritic" unrecovered fireball PN40996. Although there is thus some limited observational support for meteorite streams, in any case only a small fraction of the chondrites belong to such streams. If more compelling evidence can be obtained for the preterrestrial origin of the differences reported for Antarctic meteorites *and* the reality of such short-term variations in the meteorite flux (e.g., reproducibility of the results), the consequences would be more far-reaching than simply a demonstration of stochastic fluctuations in the fragmentation events responsible for the formation of meteorites.

Ordinary chondrites come from S asteroids, but the quantitative mineralogical interpretation of the S-type spectra is in error, e.g., because of surficial effects that mask the actual mineral composition of the asteroids. Spectrophotometry samples only a very thin layer of the asteroid surface (< 1 mm) even for relatively transparent minerals. It is conceivable that the relative proportions of olivine, pyroxene and metal on the outermost surface of the asteroid is not representative of the asteroid as a whole, or that phenomena peculiar to the surface, e.g., impact of micrometeorites or solar irradiation, have altered the optical properties of the surficial material. A significant amount of consideration has been given to this problem, and no specific hypothesis has yet been proposed that appears capable of resolving the paradox. This does not necessarily mean that no satisfactory phenomena exist, however. It would be premature to give up the attempt to identify such processes, considering the rudimentary state of knowledge concerning the properties of the regolith of asteroidal bodies in the space environment. Even gas-rich meteorites, presumably exposed to a surficial environment early in the history of the asteroid region, are well-indurated rocks and may not represent well the dusty coatings on the surface grains of the present asteroids. It is often said that this question can only be resolved by sufficiently sophisticated spacecraft missions to asteroids. There is no doubt that such missions will play an invaluable role in furthering our understanding of asteroids and meteorites, but this should not preclude doing what can be done in the laboratory in the meantime. Such work could be more timely even if less definitive, and also will help to provide the scientific basis required for successful spacecraft experiments.

One such question that has received attention concerns the apparent metal richness of S types relative to ordinary chondrites. It has been noted (see, e.g., Pieters et al. 1976; Anders, 1978) that the physical differences between metal and silicate (e.g., density, strength and ductility) might result in a bias in the inferred metal/silicate ratio or even a physical fractionation in the regolith. For example: (1) the little lumps of Ni-Fe present in ordinary chondrites do not crush as readily as silicate grains in laboratory preparation for spectral analysis, which could result in underrepresentation of the metal; (2) there have been no laboratory experiments that adequately model the regolith-generation process for materials of ordinary chondritic composition subject to hypervelocity impact; and (3) at the low temperatures on asteroids (~ 160 K), the metal may respond in a more brittle manner than is common in the laboratory. Gaffey (1986) has addressed these issues, and concludes that it remains difficult to reconcile the different spectra. In particular, he finds that the actual color of metallic grains in ordinary chondrites is neutral, rather than reddish (as observed for iron meteorites in the laboratory, or observed for M-type asteroids). He believes this is a condition present in space, rather than one caused by, for example, terrestrial oxidation. However, the nature of the coatings on metallic grains in ordinary chondrites has not been determined definitively and, as mentioned earlier, our understanding of microscale processes occurring on the surfaces of asteroids is very incomplete.

Ordinary chondrites come from S asteroids, and represent a portion of a differentiated asteroid that somehow escaped differentiation (Gaffey 1984b). Perhaps after fragmentation, the undifferentiated objects are concentrated in the meteorite-size fraction or the small-size (e.g., ~ 1 km) end of the mass spectrum from which meteorites are likely to be derived. Difficulties with this hypothesis are the rarity of achondritic clasts in chondritic meteorites, and the differences in O-isotope composition between achondritic and chondritic meteorites.

Ordinary chondrites do not come from S asteroids, but for example, come from some other asteroid class that has not been sampled well, possibly because its members are concentrated in the smallest-size range. Bell (1986) has proposed that the Q asteroids (Tholen 1984) are good candidates for being of ordinary chondrite composition. Three asteroids of this class have been identified (1862 Apollo, 1980 WF and 1981 QA). These are all small (1–2 km diameter) Apollo-Amor objects. Although it is not possible at present to make reliable individual distinctions between asteroidal and cometary Apollo-Amor objects, these bodies have no orbital or physical attributes that suggest they are of cometary origin. If they are asteroidal, then there must be a fairly large population of such bodies in the asteroid belt as well, presum-

TABLE 2.1.2
Classified Asteroids Near 3:1 Gap

No.	a	Bias Fac.	D (km)	Alb.	Type[a]
11	2.452	1.0	155	0.134	S
17	2.468	1.0	93	0.139	S
46	2.524	1.0	128	0.043	P
178	2.460	1.1	38	0.211	S?
198	2.459	1.0	59	0.194	S
248	2.471	1.1	53	0.048	C?
329	2.476	1.0	81	0.037	C
335	2.472	1.0	94	0.053	F?
421	2.542	11.9	12	0.000	S
472	2.544	1.0	48	0.243	S
556	2.466	1.0	40	0.210	S
619	2.520	1.3	30	0.000	S?
623	2.461	2.5	46	0.037	C?, P?
650	2.459	2.4	11	0.000	unusual
660	2.536	1.0	46	0.142	S
695	2.538	1.0	51	0.164	S
714	2.535	1.0	41	0.242	S
797	2.538	2.7	23	0.000	S?
877	2.486	2.0	40	0.047	F
887	2.493	3.3	4	0.198	S
897	2.542	3.0	24	0.212	S
914	2.457	1.0	79	0.084	C?
969	2.463	2.4	21	0.038	F?
974	2.534	1.8	25	0.197	S
1012	2.482	2.4	23	0.039	P?, F?
1076	2.476	2.0	25	0.029	F
1391	2.548	11.9	12	0.000	S
1644	2.548	3.0	18	0.000	S
1740	2.467	24.0	12	0.000	F
1768	2.451	6.0	13	0.000	F
1915	2.533	3.3	1	0.000	unusual
2089	2.534	3.0	16	0.000	S?
2139	2.459	2.4	18	0.000	F
2278	2.451	6.0	14	0.000	F?
2279	2.457	6.0	17	0.040	F

[a] Types shown are as defined by Tholen (see Table 2.1.3); types shown with question marks are due to Chapman's unpublished analysis.

ably in the small-size range, because the spectral class has not been found among the nearly completely studied large asteroids (> 50 km diameter) with semimajor axis < 2.60 AU. This corresponds well to the conclusion of Wetherill (1985) that ordinary chondrites are primarily supplied by the small-mass end of the asteroidal size spectrum (< 20 km diameter).

Evidence relevant to this hypothesis can be obtained by use of bias-

TABLE 2.1.3
Description of Asteroid Classes[a]

Low-Albedo (≤ 0.1) Classes

C a very common asteroid type in the outer part of the main belt; they typically have a flat spectrum longward of 0.4 μm and an absorption feature of varying strength at shorter wavelengths; presumably similar in surface composition to some carbonaceous chondrites; the strength of the ultraviolet absorption feature appears to be correlated with the presence of water of hydration; subclasses include:
        B higher albedo than the average C-type
        F weak to non-existent ultraviolet absorption feature
        G strong ultraviolet absorption feature

D although rare in the main belt, this asteroid type becomes progressively more dominant beyond the 2:1 resonance with Jupiter at 3.25 AU; their spectra are neutral to moderately reddish at shorter wavelengths, with rapidly increasing reflectivity beyond 0.6 μm; coloring may be due to kerogen-like materials

P a fairly common asteroid type near the outer edge of the main belt; they have flat to slightly reddish but otherwise featureless spectra (identical to the M and E classes); presumably C rich, but surface composition is otherwise unknown

T a rare asteroid type of unknown composition; their spectra are slightly redder than for a typical D-type at shorter wavelengths, but tend to level off longward of 0.8 μm

Moderate-Albedo Classes

A a rare asteroid type with extremely reddish spectra shortward of 0.7 μm and a very strong infrared absorption feature due to olivine (the spectral signature of pyroxene is very weak to undetectable)

M a fairly common asteroid type in the main belt; they have flat to slightly reddish but otherwise featureless spectra (identical to P and E classes); presumably of metallic (Ni-Fe) composition, but with varying metal content

Q a spectrum unique to 1862 Apollo and possibly a couple of other Earth-approaching asteroids; shortward of 0.7 μm, the spectrum is similar to that of 4 Vesta, but the infrared absorption feature is not as pronounced and is centered at a longer wavelength; possibly similar to ordinary chondrites

R a spectrum unique to 349 Dembowska; the spectrum is very red shortward of 0.7 μm; a fairly strong infrared absorption feature is also present, probably due to a combination of olivine and pyroxene, but the relative proportions of these two silicates and the amount of metal present are uncertain

S a very common asteroid type in the inner part of the main belt and also among Earth-approaching asteroids; their spectra are moderately to very reddish shortward of 0.7 μm; infrared absorption features are usually but not always present, with varying proportions of olivine and pyroxene indicated

V a spectrum unique to 4 Vesta and the Amor-type object 3551 1983 RD; the spectrum is very red shortward of 0.5 μm, moderately red from 0.5 to 0.7 μm, with strong pyroxene absorption features at longer wavelengths

High-Albedo (≥ 0.3) Class

E an uncommon asteroid type; these objects have flat to slightly reddish but otherwise featureless spectra (identical to P and M classes); surface composition is possibly similar to enstatite achondrites

[a] This table was especially prepared by D. J. Tholen for this chapter.

corrected asteroid statistics and taxonomy (Chapman 1987). Table 2.1.2 lists the 35 asteroids near the 3:1 resonance (between 2.45 and 2.55 AU) for which good classification data are available. (Types shown are as defined by Tholen [Table 2.1.3]; types shown with question marks are due to Chapman's unpublished analysis.) The bias factors shown indicate that the observed asteroid is expected to be representative (in type and size) of (F-1) additional unobserved asteroids where F is the bias factor. In general, asteroids near the 3:1 resonance with S-like albedos have been completely sampled down to 50 km diameter, about 40% sampled down to 25 km, and significantly ($\sim$ 10%) sampled down to 10 km. No Q types exist in the observed sample. Expanding the sample to include asteroids from 2.4 to 2.6 AU and less-well-characterized asteroids, including those for which the only data are albedos measured by IRAS (Matson 1986), there are approximately 85 S-type asteroids among the 212 observed asteroids. Only a single asteroid (2151 Hadwiger) shows some similarity to a Q asteroid, but is classified as a U rather than a Q (see Table 2.1.3). This study, therefore, gives no encouragement that there could be a dominant population of Qs hidden among the smaller asteroids in this, or any other, part of the main asteroid belt. Therefore this hypothesis requires that concentration of Q types into the small-size end be quite extreme, i.e., be predominantly in bodies < 10 km diameter.

The more general idea that there may be preferential sizes for asteroids of different compositions is supported by the bias-corrected statistics of Zellner (1979) and especially by the recent work of Chapman (1987). Unfortunately for the hope of finding small ordinary-chondritic asteroids, the admittedly limited data suggest that the smaller S types appear to be preferentially olivine rich and, hence, even less likely to be of ordinary-chondritic composition.

A further problem, of a more theoretical nature, is that present collisional models argue against existence of a major population of such small bodies unless their loss by collisional destruction is balanced by fragmentation of larger bodies. If, as a result of a surface density in the asteroid belt varying smoothly between 1 AU and 5 AU, the early asteroid belt contained $\sim$ 1000 times as many bodies as it does today, then even size ranges as large as 100 km diameter are unlikely to contain compositional classes that differ from those represented by the very largest surviving asteroids.

A conceivable way out of the above dilemma would be some mechanism that causes the spectral reflectance to change as fragmentation transfers material from the higher-mass end to the low-mass end of the asteroidal-size spectrum. For example, if S asteroids are igneously differentiated bodies, it could be speculated that bodies in the 50 km size range are strong olivine-plus-metal cores of $\sim$ 100 km bodies, whereas the more pyroxene-rich larger bodies have not experienced fragmentation quite that severe. This particular speculation obviously is not relevant to the formation of undifferentiated bodies like ordinary chondrites, however.

Ordinary chondrites do not come from S asteroids, but instead come from comets, e.g., from Apollo objects of cometary origin. Despite the geochemical arguments against a cometary origin (see, e.g., Anders 1978), it has been suggested from time to time (see, e.g., Öpik 1963; Wetherill 1971; Wilkening 1979) that there is in the interior of comets some rocky material that has been devolatilized during early solar-system history, e.g., heating by decay of ^{26}Al.

As discussed in Sec. 2.1.3, the more we learn about cometary composition and the expected orbital distribution of cometary Apollo objects, the less plausible this hypothesis appears to be. Also, as mentioned earlier, the continuing collisional history of the ordinary-chondrite sources, revealed by resetting of ^{39}Ar–^{40}Ar ages, points strongly to their having been in the asteroid belt throughout most of solar-system history, rather than in the Oort cloud.

All of the ways the authors believe have any chance of being helpful have been included among the above hypotheses. We challenge and invite readers to show that this statement is incorrect, and hope that they will thereby contribute to understanding this puzzling situation.

Asteroidal Sources of Differentiated Meteorites

There are other spectrophotometric types of main-belt asteroids that may be sources of differentiated meteorites: achondrites, iron meteorites and stony-irons.

One problem with identification of differentiated meteorites with the spectral reflectance of asteroids is that there is no *a priori* reason to believe why an entire asteroid should have the same composition as a particular sample of an igneous differentiation sequence, particularly if any original igneous layering was disrupted by a fragmentation history sufficiently intense to expose metallic asteroidal cores. This may be the reason that several types of achondrites, e.g., the ureilites, have not yet been clearly identified with an asteroid type. Nevertheless, some principally monomineralic asteroids are observed, and collisional fragments of these bodies could be sampled as meteorites. For example, a rare type of main-belt asteroid, the A type, has been identified as having reflectance spectra of nearly pure olivine (Cruikshank and Hartmann 1984). They have been associated with pallasite meteorites, although metallic-Fe-free olivine meteorites such as Brachina are another possibility, depending on the quantity of metal.

It was pointed out by Chapman et al. (1975) that the M asteroids could have the mineralogy of two very different meteorite types: the enstatite chondrites or the metallic meteorites. Enstatite chondrites consist of metal grains embedded in an enstatite matrix; since enstatite is colorless and transparent, only the color of the metal shows through, so they look spectrally identical to metallic meteorites. The ambiguity appears to have been resolved recently by groundbased radar. In 1985, Ostro et al. reported measurements of the radar reflectivity of the large M-type asteroid 16 Psyche, which they interpreted as almost certainly requiring a nearly pure metallic composition. More recently, two Earth-approaching M-type asteroids (Gradie and Tedesco 1986) have also been found to have high radar reflectivities interpreted as pure metal (Ostro et al. 1986). The quantity of metal contained in enstatite chondrites is too low to explain the radar results, so it seems that the M-type asteroids, as a group, are of metallic composition.

Several types of asteroid spectra exhibit an inadequate variety of spectral features to permit unique mineralogical identification. These include the rare E types, which have the highest known albedos among the asteroids, but quite featureless reflection spectra. Among known meteorite types, only the enstatite achondrites (distinguished from enstatite chondrites primarily by lacking metal) have such characteristics, so the identification is plausible despite the lack of spectral signature. The low-albedo P types and the D types have featureless spectra—one somewhat "redder" than the E types (similar to M types) and the other much redder—that do not match laboratory measurements of known meteorite types. This is perhaps not unexpected, because (despite the existence of at least one P type near the 3:1 resonance: see Table 2.1.2) they are found mainly at semimajor axes greater than 2.6 AU, distances at which resonance mechanisms will primarily result in ejection from the solar system by Jupiter perturbations, rather than storage in relatively long-lived Earth-crossing orbits. The mineralogy is unknown, although the low albedos of both types probably imply a significant C content and several suggestions have been made about the probable nature of the reddening.

A particularly important question is that of the source of the relatively abundant basaltic achondrites, the eucrites and the howardites. Together, they constitute about 5% of the observed stony-meteorite falls. Their O-isotopic composition is distinctly different from that of the chondrites. This, together with the symmetry of their fall-time distribution suggests they are not derived from the same source region as chondrites of any kind.

McCord et al. (1970) demonstrated the similarity of the reflectance spectrum of the asteroid 4 Vesta to that of the basaltic achondrites. This conclusion is supported by subsequent work (Larson and Fink 1975; Feierberg et al. 1980; Gaffey 1984a). For this reason, a number of workers have concluded that Vesta is the parent body of at least the eucrites (see, e.g., Drake 1979).

Although this identification is very attractive in some ways, it also involves serious difficulties. The orbital elements of Vesta are about as distant as possible from either of the resonances found to provide an efficient mechanism for transfer of material from the asteroid belt to Earth-crossing orbit. Quantitative calculations (Wetherill 1987a) lead to the result that no more than ~ 1 kg/yr of Vesta ejecta should impact the Earth. Even if as much as 10% of the smaller asteroidal population between 2.30 and 2.40 AU were Vesta-fragmentation debris, and therefore "second-hand" sources of Vesta-derived meteorites, the Earth impact rate would still be no more than $\sim 10^5$ g/

yr, about 1% of that required. Spectrophotometric data do not support this abundance of Vesta-like small bodies. Another problem is that the principal route by which Vesta ejecta are found to reach Earth is via the 3:1 resonance. This leads to the expectation of an orbital distribution similar to that of the ordinary chondrites whereas the basaltic achondrites exhibit no clear AM/PM fall-time asymmetry at all. Identification of M asteroids with metallic cores, A asteroids with pallasites, and at least some S asteroids with mesosiderites suggests that very extensive fragmentation took place on other bodies in the same region of the solar system, probably during an early epoch when the density of asteroidal bodies was much greater than it is today. If this is the case, it is difficult to understand how Vesta could still have on its surface a thick layer of basaltic meteorites, radiometrically dated at 4.5 Gyr (see Chapter 5.2), unmixed with larger quantities of its mantle rocks. Perhaps the basaltic mineralogy observed on the surface of Vesta was acquired from magmatism significantly later than 4.5 Gyr and thereby escaped this intense bombardment. Although it is possible to speculate how this may have happened, no quantitative thermal history has been proposed to explain it, however.

An alternative source region for these basaltic meteorites is the innermost asteroid belt adjacent to the ν_6 resonance. The calculated yield of meteorites from this region, primarily through the intermediary of Apollo and sub-Apollo size bodies is in good agreement with observation, and the agreement between the calculated and observed fall-time distributions is also excellent.

The largest asteroid in this region is 8 Flora (160 km diameter). As a result of his detailed spectrophotometric study of Flora, Gaffey (1984a) concluded that Flora is probably the residual metal-plus-Fe-silicate core of a magmatically differentiated planetesimal. He suggested that the metal-poor smaller asteroids of the Flora family may represent the original crust and mantle of this planetesimal. This is in one way consistent with the proposal of Wetherill (1987a) that the basaltic achondrites could be fragments of the smaller bodies in this region. On the other hand, no convincing way has been suggested to explain the great abundance of "crustal" basaltic meteorites relative to those with the expected peridotitic composition of the residual mantle. A similar deficiency of peridotitic material relative to material of basaltic composition is observed in the stony-iron mesosiderites. A quantitatively satisfactory explanation of these puzzles would constitute a valuable contribution to our understanding of the relationship between asteroids and meteorites.

Zoning of the Asteroid Belt

There are a number of other observational data that are relevant to the origin and earliest history of the asteroid belt. Perhaps foremost among these is the zoning of asteroidal compositions inferred from spectrophotometric observations (Zellner 1979; Gradie and Tedesco 1982; Chapman 1987). This zoning is the tendency for asteroids of different types to predominate (although with a spread of about 1 AU) at different heliocentric distances, with the S types in the inner belt, the Cs in the outer belt, the Ps dominating in the Hilda region and the Ds among the Trojans. Even if the explanation of the spectrophotometric paradox discussed earlier is found to require revision of present identification of spectra with mineral composition, it is quite likely that the classification will provide a means of "fingerprinting" the original composition of the asteroid belt. If so, this zoning argues, although not in a compelling way, for a relatively unmixed asteroid belt. In spite of the presumed depletion by a factor of $\sim 10^3$ in mass, the process by which this was achieved was apparently not so violent as to homogenize completely the composition of the asteroid belt. A meteoritic expression of this could be the rarity of exotic clasts in ordinary chondrites. Although a large fraction of these rocks are breccias, they are almost entirely monomict or "genomict" breccias belonging to the same class (H, L or LL) of ordinary chondrites, rather than a jumble of all kinds of meteorites (Wasson 1974; see also Chapter 3.5).

2.1.5. ORIGIN OF ASTEROIDS

Probably because meteorites bear such striking evidence of close association with the earliest events in solar-system history, and even with presolar events, there has been a tendency to think of them as directly derived from hypothetical parent bodies in the primordial solar nebula and to forget that they are really recently derived fragments of asteroids that exist today, or have existed until recently. Most meteorites were naturally hammered out not long ago from asteroidal outcrops in the present real-life solar system. When considered in this way, we realize that in addition to their being solar-nebula relicts, meteorites also have been processed by events, primarily irradiation, impacts and fragmentation that took place on the asteroids since the solar system became "normal," perhaps ~ 4.0 Gyr ago (see also Chapters in Parts 3 and 4).

Such secondary processing is undoubtedly important, particularly with regard to a contribution to regolith generation (Housen et al. 1979; Housen and Wilkening 1982; Chapter 3.5) and resetting of 4.5 Gyr ages (Bogard 1979; Chapter 5.3). In fact, the prevalence of these resetting events resulting from the relatively high frequency of asteroidal collisions, constitutes a major reason for believing that meteorites, and especially ordinary chondrites are of asteroidal origin rather than cometary, despite the spectrophotometric problems discussed in the previous section.

In addition to this rather mundane processing in the present asteroid belt, there are good reasons to expect there was much more extensive processing of asteroids earlier in solar-system history. It at least seems reasonable to suppose that the density of matter in the solar nebula varied smoothly with distance from the Sun. At present, this matter is clumped into planets. Between the planets there are gaps, either because the orbits in these positions are unstable at present, or presumably because matter in these regions was swept up by the planets as they grew. These explanations do not explain the low density of matter in the asteroid belt. Except in the vicinity of resonances, at least the inner (<2.7 AU) asteroidal belt seems to be stable on a time scale comparable to the age of the solar system. Otherwise, the lunar cratering record would show a decaying impact rate that is not observed. There is no theoretical reason to believe that Jupiter was able to sweep up material so far from its orbit without the assistance of bodies large enough to perturb this matter into Jupiter-approaching orbits. For these reasons, it is often, and quite likely correctly, hypothesized that the original number of asteroidal bodies was $\sim 10^3$ to 10^4 times greater than is observed today. If so, the asteroids from which the meteorites are being derived today are but the surviving remnant of a much greater early population. As survivors, they must have experienced processing events of much greater frequency than those occurring in the relatively tranquil present-day asteroidal belt. The meteoritic record would therefore reflect not only the "normal" processing that is taking place today, but this ancient and much more extensive processing as well. In addition, the physical and chemical nature of the meteorites must have been affected by the processes that accompanied the formation of the asteroids from smaller objects. Only after the effects of all these events are subtracted can the meteoritic record simply portray the nature of the solar nebula and of the presolar history of meteoritic matter. Until we accept the challenge of acquiring a much better understanding of the actual evolutionary history of the asteroidal bodies of which the meteorites are fragments, it can be expected that attempts to interpret detailed petrographic and geochemical signatures in the meteoritic record will be frustrated by uncertainty. For example, if a particular mineral assemblage can be shown to represent a gas-solid equilibrium, and thereby permit inferences concerning the composition, temperature and pressure of the "nebula," the problem of what "nebula" we are talking about will still remain. Is it the nebula from which the Sun formed, a residual planet-forming nebula of dust and gas left over after the formation of the Sun, a dust-enriched central layer of that nebula, or something else we have not thought of yet?

At present, we are a long way from being able to untangle all of this. It is generally thought that in some way, the formation of the giant planets, especially Jupiter, is responsible for removal of almost all the material originally in the asteroid belt. If so, then we are led directly to the need to solve a major unsolved problem, the origin of Jupiter and the other giant planets.

If the clearing of the asteroid region is to be accomplished by Jupiter, this planet must be formed quickly enough to forestall the formation of Earth-size (or even larger) planets in the asteroid belt, and as far in toward the Sun as the orbit of Mars. The problem of the relative rate of growth of bodies in Jupiter vs Mars orbit does not depend on absolute time scales. Specifying an approximate time scale, however, does help in thinking about the processes in terms of events that are often discussed in the framework of absolute time scales, i.e., the $\sim 10^5$ to 10^6 yr time scale of solar formation (Chapter 6.2), the $\sim 10^7$ yr characteristic time for $\sim 1/2$ the growth of the terrestrial planets (Chapter 6.4), the 10^6 to 10^7 yr estimates for outflow of solar gas during an early active stage of solar evolution, and the radiometric time scales obtained from laboratory studies of meteorites (Chapters 5.2 and 15.3). In terms of these time scales, a quantitative model that would lead to the formation of Jupiter in $\sim 10^6$ yr would appear to be quite satisfactory.

At the present time, only the barest outlines of such a quantitative model for the formation of Jupiter exists. The fractionated composition of Jupiter, as well as is internal differentiation, is thought to require that Jupiter's > 10 Earth-mass core formed first (Stevenson 1982). This was then followed by the rapid gravitational capture of its massive gaseous mantle and atmosphere after the core reached a critical size (Mizuno et al. 1978; Mizuno 1980; Bodenheimer 1985; Bodenheimer and Pollack 1986).

By appropriate choice of otherwise uncertain parameters, it is possible to show that runaway growth of Jupiter cores of ~ 2 Earth masses could possibly take place in about 5×10^5 yr. Choice of somewhat larger nebular densities (Lissauer 1987) could produce even larger (~ 10 Earth-mass) cores. Figure 2.1.6 shows the results of such a calculation by use of the approach described by Wetherill and Stewart (1986,1987). Here it is assumed that

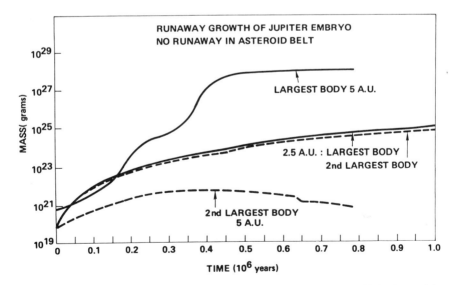

Fig. 2.1.6. Calculated growth of planetesimals in the asteroid belt and at 5 AU. Under the special assumptions described in the text, within ~ 0.5 Myr a body twice the mass of the Earth formed at 5 AU, as a result of runaway growth. During this same time period, a number of bodies comparable in mass to the largest present-day asteroids formed at 2.5 AU.

planetesimal surface density decreased as the reciprocal of the heliocentric distance at least as far as Jupiter and Saturn. This is a less rapid decrease than that inferred from the distribution of matter in the present solar system, but is not inconsistent with current nebular models. It is assumed that C, N and O are primarily in condensed phases beyond 5 AU, and that the largest body in the initial mass distribution constitutes a "seed," having a radius of about twice that of the second largest body. This runaway is enhanced by the increase in gravitational-capture cross section found as a result of failure of the two-body approximation at low relative velocities (Wetherill and Cox 1985). Greenzweig and Lissauer (1986) have found that this effect is greater at 5 AU.

To our knowledge, no quantitative work has been done to determine whether or not the Jupiter cores found to form in $\sim 5 \times 10^5$ yr under these conditions can then accumulate into the full-sized Jupiter and Saturn cores required to capture the massive gaseous envelopes of these planets, and if so, whether or not this is possible on a short, i.e., $\sim 10^6$ yr time scale. It is likely that a whole host of unstudied phenomena will be involved in the evolution of these bodies in a gaseous environment. These include the possible role of massive H_2O-H_2 atmospheres even before the runaway bodies become larger than 1 Earth mass (Stevenson 1984), and the interaction of these very massive bodies with the nebular gas, e.g., the formation of spiral density waves and the opening of gaps in the nebula that may prevent further gas accretion (Ward 1986).

The result of another calculation at the asteroidal distance of 2.5 AU is also shown in Fig. 2.1.6. It is assumed that no initial seed is present, and that the CNO ices are not condensed. It is also assumed that the relative velocity of the asteroidal planetesimals is not simply determined by the balance between their mutual gravitational perturbations, collisions and gas drag. In addition, in this calculation these velocities are externally driven by perturbations of moderate magnitude, corresponding to a "background" eccentricity of 0.001 for the first 2×10^5 yr, and after that increasing linearly at a rate of $0.001/10^5$ yr. These external perturbations may be thought of as caused by the more rapidly growing larger bodies in the region of Jupiter, as a consequence of the presence of seeds at that heliocentric distance, and because of the increase in density associated with the condensed state of solid ices. After 5×10^5 yr, the largest bodies of the swarm have gone to ~ 600 km radius, and the largest body is only slightly more massive than the second-largest body.

As a result of this conceivable, but by no means convincing chain of events, one can imagine that it may be possible to form the Jupiter core on a time scale of $\lesssim 10^6$ yr, at which time the asteroids could be no larger than those observed today and thereby vulnerable to collisional destruction at eccentricities comparable to the modest present-day asteroidal eccentricities. Following the rapid collapse of nebular gas onto Jupiter, it can then be imagined that sweeping resonances resulting from all the commotion in the giant-planet region (Ward 1980), as well as perturbations by residual bodies of ~ 2 Earth masses scattered into the asteroid belt from the Jupiter-Saturn region, could increase asteroidal eccentricities to their present values, and possibly even higher. Then, along the general lines of earlier calculations (see, e.g., Davis et al. 1979), the asteroid belt could grind itself down by mutual collisions. The largest bodies would tend to be preserved, whereas the smaller bodies would relax toward a collisional steady-state distribution in which most of the mass is located in the residual large asteroids. As pointed out by Davis et al. (1985), the principal problem with a collisional history of this kind is retention of a basaltic crust on Vesta, as discussed in Sec. 2.1.4.

It must be reiterated that in the absence of quantitatively working out the myriad physical (and chemical) problems involved, the foregoing falls far short of qualifying as a theory. In addition to working out all these problems, it will be necessary to face the observational consequences that will be predicted by such a model. These include such questions as "where did the asteroidal material go?" and the extent to which the answer to this question is consistent with the chemical composition of Earth, Moon and Mars, and with their cratering history. If it turns out that Jupiter-scattered planetesimals played a major role in driving material from the asteroid belt by gravitational scattering, one must understand why orbits in the present asteroid belt appear to be more than marginally stable with respect to loss to Jupiter. It would be excessively optimistic to expect that when all this theoretical work is carried out and compared with observation, that everything will become clear. More likely, serious problems with our conventional ideas will be exposed, new concepts will need to be generated, and the work will have to begin anew. The author of Moby Dick wrote that small works "may be finished by their first architects; grand ones, true ones, ever leave the copestone to posterity. God keep me from ever completing anything. This whole book is but a draft—nay but the draft of a draft. Oh Time, Strength, Cash, and Patience."

In our more lucid moments, it is clear to those of us working on the formation of the solar system that we are in the same boat with Melville. Only a few engaged in this quest are likely to be present when the goal is reached. This need not lead to despair. There are many specific problems of the kind discussed here that must be addressed before a well-ordered understanding of planetary formation can be realized. Knowledge that there are these well-defined steps to be completed can give substance to our individual efforts, and offset the desire to climb our personal Mount Pisgah and proclaim too strongly our individual vision of the distant Promised Land.

Acknowledgments. We thank J. Dunlop for her work in preparing the manuscript. This work was supported in part by a NASA grant and is part of a larger program at DTM also supported by a grant from NASA.

REFERENCES

Anders, E. 1964. Origin, age, and composition of meteorites. *Space Science Rev.* 3:583–714.
Anders, E. 1978. Most stony meteorites come from the asteroid belt. In *Asteroids: An Exploration Assessment*, eds. D. Morrison and W. C. Wells, NASA CP-2053, pp. 57–75.
Bell, J. F. 1986. Mineralogical evolution of meteorite parent bodies. *Lunar Planet. Sci.* XVII:985–986 (abstract).
Bell, J. F., Hawke, B. R., Owensby, P. D., and Gaffey, M. J. 1987. Atlas of asteroid infrared reflection spectra (0.08–2.5 microns). Preprint.
Bodenheimer, P. 1985. Evolution of the giant planets. In *Protostars & Planets II*, eds. D. C. Black and M. S. Matthews (Tucson: Univ. of Arizona Press), pp. 873–894.
Bodenheimer, P., and Pollack, J. B. 1986. Calculations of the accretion and evolution of giant planets: The effects of solid cores. *Icarus* 67:391–498.
Bogard, D. D. 1979. Chronology of asteroid collisions as recorded in meteorites. In *Asteroids*, ed. T. Gehrels (Tucson: Univ. of Arizona Press), pp. 558–578.
Campins, H., A'Hearn, M. F., and McFadden, L. A. 1987. The bare nucleus of Comet P/Neujmin 1. *Astrophys. J.* 316:847–857.
Ceplecha, Z., and McCrosky, R. E. 1976. Fireball end heights: A diagnostic for the structure of meteoric material. *J. Geophys. Res.* 81:6257–6275.
Chapman, C. R. 1979. The asteroids: Nature, interrelations, origin, and evolution. In *Asteroids*, ed. T. Gehrels (Tucson: Univ. of Arizona Press), pp. 25–60.
Chapman, C. R. 1981. Remote sensing of the asteroids. In *Proc. 15th Internatl. Symposium on Remote Sensing of the Environment*.
Chapman, C. R. 1987. Compositional structure of the asteroid belt and its families. *Icarus*, submitted.
Chapman, C. R., Morrison, D., and Zellner, B. 1975. Surface properties of asteroids: A synthesis of polarimetry, radiometry, and spectrophotometry. *Icarus* 25:104–130.
Crabb, J., and Schultz, L. 1981. Cosmic ray exposure ages of the ordinary chondrites and their significance for parent body stratigraphy. *Geochim. Cosmochim. Acta* 45:2151–2160.
Cruikshank, D. P., and Hartmann, W. K. 1984. The meteorite-asteroid connection: Two olivine-rich asteroids. *Science* 223:281–283.
Davis, D. R., Chapman, C. R., Greenberg, R., Weidenschilling, S. J., and Harris, A. W. 1979. Collisional evolution of asteroids: Populations, rotations, and velocities. In *Asteroids*, ed. T. Gehrels (Tucson: Univ. of Arizona Press), pp. 528–557.
Davis, D. R., Chapman, C. R., Weidenschilling, S. J., and Greenberg, R. 1985. Collisional history of asteroids: Evidence from Vesta and the Hirayama families. *Icarus* 62:30–53.
Dennison, J. E., and Lipschutz, M. E. 1987. Chemical studies of H chondrites. II. Weathering effects in the Victoria Land, Antarctic population and comparison of two Antarctic populations with non-Antarctic falls. *Geochim. Cosmochim. Acta* 51:741–754.
Dennison, J. E., Lingner, D. W., and Lipschutz, M. E. 1986. Antarctic and non-Antarctic meteorites in different populations. *Nature* 319:390–393.
Dohnanyi, J. W. 1969. Collisional model of asteroids and their debris. *J. Geophys. Res.* 74:2531–2554.
Drake, M. J. 1979. Geochemical evolution of the eucrite parent body: Possible nature and evolution of asteroid 4 Vesta? In *Asteroids*, ed. T. Gehrels (Tucson: Univ. of Arizona Press), pp. 765–782.
Feierberg, M. A., Larson, H. P., Fink, U., and Smith, H. 1980. Spectroscopic evidence for two achondrite parent bodies: Asteroids 349 Dembowska and 4 Vesta. *Geochim. Cosmochim. Acta* 44:513–524.
Feierberg, M. A., Larson, H. P., and Chapman, C. R. 1982. Spectroscopic evidence for undifferentiated S-type asteroids. *Astrophys J.* 257:361–372.
Gaffey, M. J. 1976. Spectral reflectance characteristics of the meteorite classes. *J. Geophys. Res.* 81:905–920.
Gaffey, M. J. 1984a. Rotational spectral variations of asteroid (8) Flora: Implications for the nature of the S-type asteroids and for the parent bodies of the ordinary chondrites. *Icarus* 60:83–114.
Gaffey, M. J. 1984b. The asteroid (4) Vesta: Rotational spectral variations, surface material het-

erogeneity, and implications for the origin of the basaltic achondrites. *Lunar Planet. Sci.* XIV:231–232 (abstract).
Gaffey, M. J. 1986. The spectral and physical properties of metal in meteorite assemblages: Implications for asteroid surface materials. *Icarus* 66:468–486.
Gaffey, M. J., and McCord, T. B. 1978. Asteroid surface materials: Mineralogical characterizations from reflectance spectra. *Space Sci. Rev.* 21:555–628.
Gault, D. E., Shoemaker, E. M., and Moore, H. J. 1963. Spray ejected from the lunar surface by meteoroid impact. NASA TN-1767.
Gradie, J., and Tedesco, E. 1982. Compositional structure of the asteroid belt. *Science* 216:1405–1407.
Gradie, J., and Tedesco, E. F. 1986. M-class asteroids among the near-Earth asteroid population. *Bull. Amer. Astron. Soc.* 18:797 (abstract).
Greenberg, R., and Chapman, C. R. 1983. Asteroids and meteorites: Parent bodies and delivered samples. *Icarus* 55:455–481.
Greenzweig, V., and Lissauer, J. J. 1986. A scaling law for accretion zone sizes. *Bull. Amer. Astron. Soc.* 18:817 (abstract).
Halliday, I. 1987. Detection of a meteorite "stream": Observations of a second meteorite fall from the orbit of the Innisfree chondrite. *Icarus* 69:550–556.
Halliday, I., Griffin, A. A., and Blackwell, A. T. 1983. Meteorite orbits from observations by camera networks. In *Highlights of Astronomy,* vol. 6, ed. R. N. West (Dordrecht: D. Reidel), pp. 399–404.
Halliday, I., Blackwell, A. T., and Griffin, A. A. 1984. The frequency of meteorite falls on the Earth. *Science* 223:1405–1407.
Hartmann, W. K., Cruikshank, D. P., and Tholen, D. J. 1985. Outer solar system materials: Ices and color systematics. In *Ices in the Solar System,* eds. J. Klinger, D. Benest, A. Dollfus, and R. Smoluchowski (Dordrecht: D. Reidel), pp. 169–181.
Housen, K. R., and Wilkening, L. L. 1982. Regoliths on small bodies in the solar system. *Ann. Rev. Earth Planet. Sci.* 10:355–376.
Housen, K. R., Wilkening, L. L., Chapman, C. R., and Greenberg, R. 1979. Asteroidal regoliths. *Icarus* 39:317–351.
Kuiper, G. P., Fujita, Y., Gehrels, T., Groeneveld, I., Kent, J., Van Biesbroeck, G., and Van Houten, C. J. 1958. Survey of asteroids. *Astrophys. J. Suppl.* 3:289–428.
Larson, H. A., and Fink, U. 1975. Infrared and spectral observations of asteroid 4 Vesta. *Icarus* 26:420–427.
Lissauer, J. J. 1987. Timescales for planetary accretion and the structure of the protoplanetary disk. *Icarus* 69:249–265.
Matson, D. 1986. Infrared Astronomical Satellite Asteroid and Comet Survey. Preprint 1, JPL/IPAC, California Inst. of Technology.
McCord, T. B., Adams, J. B., and Johnson, T. V. 1970. Asteroid Vesta: Spectral reflectivity and compositional implications. *Science* 168:1445–1447.
McSween, H. Y. 1985. SNC meteorites: Clues to martian petrologic evolution? *Rev. Geophys.* 23:391–416.
Millis, R. L., A'Hearn, M. F., and Campins, H. 1987. An investigation of the nucleus and coma of Comet P/Arend-Rigaux. *Astrophys. J.* 324:1194–1209.
Mizuno, H. 1980. Formation of the giant planets. *Prog. Theor. Phys.* 64:544–557.
Mizuno, H., Nakazawa, K., and Hayashi, C. 1978. Instability of gaseous envelope surrounding planetary core and formation of giant planets. *Prog. Theor. Phys.* 60:699–710.
Nyquist, L. E., Wooden, J., Bansal, B., Wiesmann, H., McKay, G., and Bogard, D. D. 1979. Rb-Sr age of the Shergotty achondrite and implication for metamorphic resetting of isochron ages. *Geochim. Cosmochim. Acta* 43:1057–1074.
Olivier, C. P. 1925. *Meteors.* (Baltimore: Williams and Wilkins), Chapter 24.
Öpik, E. J. 1951. Collision probabilities with the planets and the distribution of interplanetary matter. *Proc. Roy. Irish Acad.* 54A:165–199.
Öpik, E. J. 1963. Survival of comet nuclei and the asteroids. *Adv. Astron. Astrophys.* 2:219–262.
Ostro, S. J., Campbell, D. B., and Shapiro, I. I. 1985. Mainbelt asteroids: Dual polarization radar observations. *Science* 229:442–446.

Ostro, S. J., Campbell, D. B., and Shapiro, I. I. 1986. Radar detection of 12 asteroids from Arecibo. *Bull. Amer. Astron. Soc.* 18:796 (abstract).
Papanastassiou, D. A., and Wasserburg, G. J. 1974. Evidence for late formation and young metamorphism in the achondrite Nakhla. *Geophys. Res. Lett.* 1:23–26.
Pieters, C., Gaffey, M. J., Chapman, C. R., and McCord, T. B. 1976. Spectrophotometry (0.33 to 1.07 μm) of 433 Eros and compositional implications. *Icarus* 28:105–115.
Simonenko, A. N. 1975. Orbital elements of 45 meteorites. Atlas. *Nauka*, Moscow.
Stevenson, D. J. 1982. Structure of the giant planets: Evidence for nucleated instabilities and post-formational accretion. *Lunar Planet. Sci.* XIII:770–771 (abstract).
Stevenson, D. J. 1984. On forming the giant planets quickly (Superganymedean puffballs). *Lunar Planet Sci.* XV:822–823 (abstract).
Tholen, D. J. 1984. Asteroid Taxonomy from Cluster Analysis of Photometry. Ph.D. thesis, Univ. of Arizona.
Tholen, D. J., Cruikshank, D. P., Hartmann, W. K., Lark, N., Hammel, H. B., and Piscatelli, J. R. 1986. A comparison of the continuum colors of P/Halley, other comets, and asteroids. In *Proc. 20th ESLAB Symposium on the Exploration of Halley's Comet*, vol. 3, ESA SP-250, pp. 503–507.
Turner, G. 1979. A Monte Carlo fragmentation model for the production of meteorites: Implications for gas retention ages. *Proc. Lunar Planet. Sci. Conf.* 10:1917–1941.
Van Houten, C. J., Van Houten-Groeneveld, I., Herget, P., and Gehrels, T. 1970. The Palomar-Leiden survey of faint minor planets. *Astron. Astrophys. Suppl.* 2:339–348.
Walker, D., Stolper, E. M., and Hays, J. F. 1979. Basaltic volcanism: The importance of planet size. *Proc. Lunar Planet. Sci. Conf.* 10:1995–2015.
Ward, W. R. 1980. Scanning secular resonances: A cosmogonical broom? *Lunar Planet. Sci.* XI:1199–1201.
Ward, W. R. 1986. Density waves in the solar nebula: Differential Lindblad Torque. *Icarus* 67:164–180.
Wasson, J. T. 1974. *Meteorites* (New York: Springer-Verlag).
Wasson, J. T., and Wetherill, G. W. 1979. Dynamical, chemical, and isotopic evidence regarding the formation locations of asteroids and meteorites. In *Asteroids*, ed. T. Gehrels (Tucson: Univ. of Arizona Press), pp. 926–974.
Wasson, J. T. 1985. *Meteorites* (New York: Freeman).
Wetherill, G. W. 1968. Time of fall and origin of stone meteorites. *Science* 159:79–82.
Wetherill, G. W. 1971. Cometary vs. asteroidal origin of chondritic meteorites. In *Physical Studies of Minor Planets*, ed. T. Gehrels, NASA SP-267, pp. 447–460.
Wetherill, G. W. 1974. Solar system sources of meteorites and large meteoroids. *Ann. Rev. Earth Planet. Sci.* 2:303–331.
Wetherill, G. W. 1985. Asteroidal source of ordinary chondrites. *Meteoritics.* 20:1–22.
Wetherill, G. W. 1986. Unexpected Antarctic chemistry. *Nature* 319:357–358.
Wetherill, G. W. 1987a. Dynamical relationship between asteroids, meteorites, and Apollo-Amor objects. *Phil. Trans. Roy. Soc. London.* A323:323–337.
Wetherill, G. W. 1987b. Solar system sources of Apollo-Amor objects. *Icarus*, submitted.
Wetherill, G. W., and Cox, L. P. 1985. The range of validity of the two-body approximation in models of terrestrial planet accumulation, II. Gravitational cross-sections and runaway accretion. *Icarus* 63:290–303.
Wetherill, G. W., and ReVelle, D. O. 1981. Which fireballs are meteorites? A study of the Prairie Network photographic meteor data. *Icarus* 48:308–328.
Wetherill, G. W., and ReVelle, D. O. 1982. Relationship between comets, large meteors, and meteorites. In *Comets* ed. L. L. Wilkening (Tucson: Univ. of Arizona Press), pp. 297–319.
Wetherill, G. W., and Shoemaker, E. M. 1982. Collision of astronomically observable bodies with the Earth. In *Large Body Impacts and Terrestrial Evolution*, ed. L. T. Silver, Geol. Soc. Amer. Special Paper 190, pp. 1–13.
Wetherill, G. W., and Stewart, G. R. 1986. The early stages of planetesimal accumulation. *Lunar Planet. Sci.* XVII:939 (abstract).
Wetherill, G. W., and Stewart, G. R. 1987. Factors controlling early runaway growth of planetesimals. *Lunar Planet. Sci.* XVIII:1077 (abstract).
Wetherill, G. W., and Williams, J. G. 1979. Origin of differentiated meteorites. In *Origin and Distribution of the Elements*, ed. L. H. Ahrens (Oxford: Pergamon Press) pp. 19–31.

Wilkening, L. L. 1979. The asteroids: Accretion, differentiation, fragmentation, and irradiation. In *Asteroids*, ed. T. Gehrels (Tucson: Univ. of Arizona Press), pp. 61–74.
Williams, J. G. 1969. Secular perturbations in the solar system. Ph.D. thesis, Univ. of California, Los Angeles.
Williams, J. G. 1973. Meteorites from the asteroid belt? *Eos, Trans. AGU* 54:233 (abstract).
Wisdom, J. 1983. Chaotic behavior and the origin of the 3/1 Kirkwood gap. *Icarus* 56:51–74.
Wisdom, J. 1985. Meteorites may follow a chaotic route to Earth. *Nature* 315:731–733.
Wood, J. A. 1961. Stony meteorite orbits. *Mon. Not. Roy. Astron. Soc.* 122:79–88.
Zellner, B. 1979. Asteroid taxonomy and the distribution of the compositional types. In *Asteroids*, ed. T. Gehrels (Tucson: Univ. of Arizona Press), pp. 783–806.
Zellner, B., Tholen, D. J., and Tedesco, E. F. 1985. The eight-color asteroid survey: Results for 589 minor planets. *Icarus* 61:355–416.
Zimmerman, P. D., and Wetherill, G. W. 1973. Asteroidal source of meteorites. *Science* 182:51–53.

PART 3
Secondary Processing

3.1. SIGNIFICANCE OF SECONDARY PROCESSING

JOHN F. KERRIDGE
University of California at Los Angeles

Having dealt with the classification and source regions of meteorites, we now consider ways in which secondary processes may have perturbed the primary record that is of principal interest to this book. In this context, "primary" means that the observational record reflects nebular or prenebular conditions, whereas "secondary" means any process that postdated accumulation of planetesimals in the early solar system. Note that certain processes (e.g., chondrule formation) are primary by this definition, but are described elsewhere as secondary in that they operated upon preexisting material.

3.1.1. WHY IS SECONDARY PROCESSING IMPORTANT?

As pointed out succinctly and accurately in Chapter 2.1, virtually all meteorites have been broken off outcrops on or in asteroids. That observation is valuable both for establishing the provenance of the meteorites that we study and also for reminding us that those meteorites are not simply providentially solidified chunks of nebular gas but are in fact rocks. This is significant because we think of rocks as having been shaped by planetary processes whereas there is often a tendency to overlook the fact that even chondrites have spent most of their existence as parts of subplanetary objects. Such objects, though in most cases far less geologically active than the Earth, or even the Moon, nonetheless have not been devoid of activity: that modest level of activity has been enough in many cases to perturb the record of preaccretionary, i.e., nebular or prenebular, processes.

In the following five chapters, we explore the effects that processes on

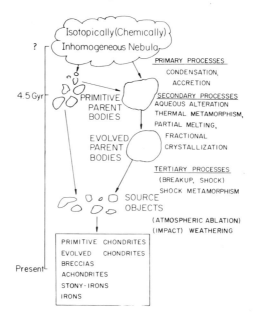

Fig. 3.1.1. Schematic representation of episodes in the history of meteoritic material in the solar system. Compositions of meteorites today reflect the action of one or more of the processes listed on the right. Some meteorites may conceivably be derived from unaltered, sub-km-sized source objects formed directly from the nebula (left-hand path). However, most if not all of those described in this book have been derived from larger, asteroidal objects (right-hand path) in which post-accretionary, parent-body processes could take place. Insofar as a meteorite, or a meteoritic component, escaped secondary (or tertiary) processing, or if the effects of such processing can be compensated for, conditions during primary nebula processes can be inferred. (Figure by courtesy of M. E. Lipschutz.)

asteroids, either endogenic, such as internal heating, or exogenic, like meteoroid bombardment, have had on meteorites. Our emphasis will be twofold: first to assess the impact that secondary processes have had on the primary record, and second to discuss what can be learned from the secondary record itself. Processes such as igneous differentiation, thermal metamorphism, aqueous alteration, brecciation and shock metamorphism represent manifestations of conditions that need to be included in any comprehensive picture of the early solar system.

A schematic representation of the principal events in the evolution of a meteorite is given in Fig. 3.1.1. Events termed "tertiary" in the figure, are subsumed under the heading of secondary in the following chapters.

3.2. IGNEOUS ACTIVITY IN THE EARLY SOLAR SYSTEM

R. H. HEWINS
Rutgers University

and

H. E. NEWSOM
University of New Mexico

Although the main emphasis of this book is on what can be learned about the early solar system from material that has escaped secondary processing, the study of differentiated meteorites can provide unique insights into the processes of basalt generation and core formation. Such processes would have been of fundamental importance during the evolution of planetary objects, including the terrestrial planets, early in solar-system history. In this chapter, we summarize the properties of igneous meteorites, focusing in particular on the howardite-eucrite-diogenite (HED) suite. Geochemical and petrologic trends in those meteorites are discussed with the objective of defining the thermal and chemical evolution of the HED parent body. A major issue is the nature of the dominant source of heat in the early solar system. Heating by decay of ^{26}Al is a possibility, but a major role for electromagnetic induction heating by a putative T Tauri phase of the early Sun appears likely.

3.2.1. INTRODUCTION

A large number of meteorites—stony, stony-iron and iron—record melting events in at least seventy small parent bodies (Scott 1979) essentially at the time of formation of the solar system, and remote sensing demonstrates that a large number of asteroids are differentiated, i.e., were partly or entirely melted. Several mechanisms have been proposed to explain such heating, including decay of radioactive ^{26}Al and electromagnetic induction during a hypothetical T-Tauri phase of the Sun. The compositional relationships among different achondrite meteorites from a single parent body can reveal the extent of melting of the parent body, e.g., magma ocean or serial magmatism (Walker 1983), and the time scale of the heating. This knowledge would clarify the thermal evolution of small bodies in the early solar system with implications for the nature of the early Sun and the distribution of extinct radionuclides. Understanding the processes of core formation and igneous differentiation on asteroid-sized bodies may also provide clues to the operation of these processes on the larger terrestrial planets.

3.2.2. HEATING IN THE EARLY SOLAR SYSTEM

Although only one large asteroid with basaltic rocks on its surface is known (Gaffey 1983; Consolmagno and Drake 1977), there are many other asteroids inferred to be differentiated, including A-type (olivine-rich), S-type (metal-olivine-pyroxene) and M-type (metal-rich) (Tholen 1984; Bell 1986). The lack of currently detectable basaltic asteroids is undoubtedly due to the collision history of asteroids, since silicates are brittle and can be stripped off the tougher metallic cores (Bell 1986). The differentiated asteroids and asteroids with chondrite-like surfaces have distinctly different distributions in the solar system, with the more strongly heated bodies concentrated at the inner edge of the main asteroid belt (Gradie and Tedesco 1982; Bell 1986). This distribution is probably a primordial feature of the asteroid belt (Patterson 1987); the heating mechanism was apparently more effective closer to the Sun.

Many processes which efficiently heat planets are unable to melt smaller bodies such as asteroids. The accretion process itself causes heating and impact melting, but the velocities and thus the kinetic energy of planetesimals impacting on an asteroid must necessarily be small for the asteroid to accrete and grow rather than be disrupted. Also, the heat deposited during the accretion of small bodies tends to escape rapidly relative to the accretion time scale. In general, the heat released by changes in the gravitational potential energy during accretion is inadequate to melt bodies smaller than the Earth's Moon (Basaltic Volcanism Study Project 1981). Long-lived radioactive isotopes, energy of differentiation (including core formation), adiabatic compression and tidal heating were also insignificant in asteroidal heating (Basaltic Volcanism Study Project 1981).

Among all the short-lived radioactive isotopes, ^{26}Al has the greatest potential for asteroidal heating (Basaltic Volcanism Study Project 1981) and there is evidence for the former presence of ^{26}Al in some components in chondritic meteorites (Lee et al. 1976). Herndon and Herndon (1977) calculated thermal histories for asteroids assuming ^{26}Al as the heat source and found that the maximum central temperatures were very sensitive to ^{26}Al/^{27}Al ratio for a given estimate of total Al concentration. An asteroid of radius 200 km would have reached near-solidus temperatures with a value of 10^{-5} for ^{26}Al/^{27}Al but would have been totally melted with a value of 2×10^{-5}. Miyamoto et al. (1981) achieved realistic thermal metamorphism in chondrite parent bodies with a ratio of 5×10^{-6} and their prediction of a higher temperature for L compared to H chondrites is consistent with pyroxene thermometry (Colucci and Hewins 1984). The abundance ratio of 6×10^{-5} estimated by Lee et al. (1976) would have required assembly of the asteroid within a few half-lives, 1 or 2 Myr, if it were melted by ^{26}Al. More complex modeling was performed by Minear et al. (1979) who considered incipient melting throughout an asteroid followed by gravitational separation of basaltic and Fe-rich liquids from the residuum. Efficient removal of the basalt, into which almost all of the Al would partition, to the surface would have prevented the total melting of the body. However, concentration of ^{26}Al in the crust could have led to remelting of early igneous rocks (Smith 1982). Thus live ^{26}Al has the potential to melt asteroids and create complex geochemical relationships, although it is far from certain that it was in fact present in asteroids in sufficient abundance to cause differentiation.

The observation that asteroids closer to the Sun are more heated suggests that the mechanism was related to the Sun via, for example, electromagnetic induction. Under the conditions of strong solar wind and strong but variable solar magnetic field, electric currents could have been induced in a planetary body either electrically or magnetically or both. Passage of the current through the body would have caused heating. Herbert and Sonett (1978) have shown that induction heating would have been adequate for asteroidal melting, requiring only reasonable assumptions about initial temperature and conductivity of the asteroid. What is far from certain is the state of the Sun at that time, specifically whether the very strong solar winds necessary to the model, and commonly associated with a T-Tauri stage, were in fact available. However, irradiation effects in several meteorite regolith breccias (see Chapter 4.1) are consistent with an early active Sun. The fact that asteroids close to the Sun were melted, and those farther out were not, is consistent with a T-Tauri stage and induction heating. Induction heating is also more effective for larger bodies, permitting small ordinary-chondrite bodies (although not yet identified; see Chapter 2.1) to occur alongside large achondrite bodies (Herbert and Sonett 1978). Another possible solar effect is heating in a superluminous Hayashi stage, but even if this did occur, only a very thin surface layer would have been heated (Sonett and Reynolds 1979).

Both ^{26}Al and induction-heating models are able to provide basalts within 1 Myr, satisfying the constraint of very old ages for eucrites. The ability of ^{26}Al to partition into basaltic melt and migrate makes it an attractive heat source, if most eucrites are primary magmas formed at low degrees of partial melting from a chondritic source (Basaltic Volcanism Study Project 1981), or if differences between eucrites and diogenite parent magmas could be explained by remelting of early plutons (Mittlefehldt 1979; Smith 1982). The distribution of asteroids, with differentiated bodies closer to the Sun, argues for an important role for electromagnetic induction heating.

Possibly both ^{26}Al and induction heating operated in tandem; it should not be overlooked that certain complex relationships between achondrites suggest multiple heating (Fukuoka et al. 1977; Mittlefehldt 1979; Consolmagno 1979; Smith 1982; Delaney 1986). Early estimates placed the time of induction heating at about 10 Myr after any heating by ^{26}Al (Ezer and Cameron 1962; Hostetler and Drake 1980). Recently, however, the lifetimes of T-Tauri stars have been taken at most as a few Myr (Mercer-Smith et al. 1984), so that the two heat sources might have operated simultaneously or sequentially. Establishment of the degree of complexity in igneous activity, and especially the nature of the source regions, will clarify the nature of the heat source or sources. This problem is discussed in detail for one group of meteorites in Sec. 3.2.4.

3.2.3. THE METEORITIC RECORD

Iron Meteorites

A very wide extent of melting is implied for small bodies in the early solar system, particularly from iron rather than silicate meteorites. There are over sixty groups and grouplets of iron meteorites, the majority of which formed by fractional crystallization of molten metal (Scott 1979), most probably in the cores of small asteroids. The structure and geochemistry of irons are well documented, yet conflicting interpretations of the nature, size and history of their parent bodies have arisen. The sheer number of parent bodies makes resolution of the problems with the melting and cooling of irons critical for understanding the melting of asteroids.

The groups and grouplets of irons are recognized from the clustering of abundances of Ni and trace elements, especially Ga and Ge (Scott and Wasson 1975; see also Chapter 1.1). This implies different nebular formation locations (or times), with the different compositions arising from solid-vapor equilibrium under different P-T conditions (Kelly and Larimer 1977; Scott 1979). Within most groups fractionation trends of the compatible and incompatible trace elements demonstrate convincingly that the irons resulted from solidification of a liquid. Exceptions to this generalization include the IAB irons which contain abundant silicate inclusions that are discussed below in a separate section.

The origin of some or most iron meteorites from fractionally crystallized melts in asteroidal cores could be confirmed if the solidification of these cores could be modeled. This requires a knowledge of the partition coefficients of the trace elements between solid and liquid metal. Sellamuthu and Goldstein

(1985) showed that partition coefficients depend on the concentrations of minor elements such as P and S. The most recent measurements of partition coefficients were obtained in reversed isothermal experiments by Malvin et al. (1986).

Understanding the chemical data is necessary for determining the actual mode of core solidification, e.g., by settling of crystals to the center or by crystallization from the outside inwards. Narayan and Goldstein (1982) proposed freezing of cores from the mantle interface by dendritic solidification rather than by plane-front solidification. Esbensen et al. (1982) suggested an alternative model to dendritic solidification for group IIIAB irons, in which blocks of metal dislodged from the roof of the core (e.g., by impact) sank into late liquid and were partly homogenized. The compositional data do not, however, indicate the size of the cores or magma chambers and the planetesimals or asteroids in which irons formed: for this, structural and mineralogical properties must be considered.

Iron meteorites can be classified on the basis of their structure or texture. Many irons with intermediate Ni contents cooled through the two-phase region of the Fe-Ni phase diagram so as to produce exsolution lamellae of the low-temperature α phase (kamacite) in an octahedral pattern. They are "octahedrites" and are subdivided according to the width of the kamacite lamellae (Goldstein and Axon 1973; Scott and Wasson 1975). With a knowledge of diffusion coefficients at low temperatures, the cooling conditions to match the Ni-concentration gradient in the residual high-temperature γ phase (taenite) can be calculated. Because reactions at low temperatures are slow and affect small volumes, diffusion measurements are difficult and have been frequently revised. The current cooling rates are about five times higher than the earliest estimates and are consistent with asteroid-sized parent bodies (Saikumar and Goldstein 1988).

Because of the high thermal conductivity, one would expect samples of a given Fe-Ni core to have virtually identical cooling rates. However, Narayan and Goldstein (1985) found several groups with a range of cooling rates greater than that expected from experimental error and they concluded such irons formed as isolated pods in a silicate matrix. Metal pods might be formed by reaccretion of original core and mantle fragments after asteroidal disruption, assuming small metal cores can be shattered, or by late accretion of differentiated planetesimals, or by partial differentiation.

There is a very strong presumption that most iron meteorites come from asteroidal cores, excluding unusual groups like IAB irons. The data are not yet totally consistent with this picture, but given the frequency of their revision, they may yet turn out to indicate differentiated cores. The presence of radiogenic ^{107}Ag, from decay of ^{107}Pd (halflife = 6.5 Myr), in many iron meteorites (see, e.g., Kaiser and Wasserburg 1983) shows that core differentiation in asteroids occurred very early in solar-system history (see also Chapter 15.1).

Silicate-Rich Differentiated Meteorites

SNC Group. Among silicate-rich meteorites the SNC group (shergottites, nakhlites and Chassigny) do not appear relevant to this discussion because their young ages, noble-gas composition, oxidation state, and presence of water suggest an origin on Mars (Reid and Bunch 1975; Floran et al. 1978; Wasson and Wetherill 1979; Bogard et al. 1984; see also Chapters 5.3, 7.8 and 7.10). This cannot be confirmed, however, until the return of known Mars samples.

Bencubbin and Weatherford. The Bencubbin and Weatherford meteorites are identical highly shocked polymict breccias containing metal and silicate (mostly enstatite) clasts (Newsom and Drake 1979). These meteorites are probably not the products of extensive melting on a parent body because of the primitive and variable metal compositions (Newsom and Drake 1979), the unfractionated rare earths in the so-called "aubritic" or "host" silicates (Kallemeyn et al. 1978), and an extremely anomalous N-isotopic composition (Prombo and Clayton 1985).

Winonaite/IAB Iron Meteorites. There exists a suite of modified primitive meteorites (rich in magnesian olivine and orthopyroxene), including winonaites and IAB irons with silicate inclusions, which may have experienced minor fractionation due to partial melting of silicate and segregation of a metal-sulfide eutectic (Prinz et al. 1983; Kracher 1985). Wasson et al. (1980), however, interpreted both IAB and IIICD irons as impact-melt rocks (not to be confused with shock-melted chondrites such as Shaw; Taylor et al. 1979). Other groups of meteorites that are similar to the winonaite/IAB meteorites, but which come from different parent bodies, on the basis of O isotopes and chemical data, include the meteorite Kakangari and another group including Acapulco, Lodran and probably Allan Hills A77081 (Palme et al. 1981). Brachina, formerly regarded as an SNC, may represent the least-reduced winonaite-like material (Prinz et al. 1983). Although more than one parent body may be represented by the winonaite/IAB meteorites, their properties are very similar and are consistent with formation during a single heating event within a moderately large parent body (Kracher 1985). During heating of the parent body, liquids close to FeS in composition probably began segregating from the silicates, particularly near the silicate solidus of about 1100°C, in this model. Nevertheless, the impact-melt model cannot be ruled out.

Angra dos Reis (ADOR). The ADOR meteorite is a unique ultramafic pyroxenite containing about 93% fassaite, a Ti-rich pyroxene (Prinz et al. 1977). Fassaite occurs in both metamorphic and igneous terrestrial rocks. Incompatible trace elements are enriched to roughly 20 times the chondritic level in whole-rock and pyroxene separates, with a convex rare earth element (REE) pattern. Textural and mineralogical data (Prinz et al. 1977) suggest that the ADOR assemblage crystallized at low pressures from a highly undersaturated basaltic melt, and formed as a pyroxene (olivine) cumulate which has been mildly recrystallized or annealed. The composition, especially the presence of a negative Eu anomaly (Ma et al. 1977), and mineralogy suggest a complicated history for the ADOR parent body that requires either removal of melilite by fractional crystallization from a parent magma before crystallization of the ADOR assemblage, or crystallization of a magma produced by partial melting of a previously fractionated aluminous-pyroxene source rock (Prinz et al. 1977).

Aubrites. The aubrites (enstatite achondrites) are mostly breccias, and consist largely of cumulus enstatite crystals up to 10 cm in length. The aubrites apparently formed by melting of a highly reduced parent body (very low-O fugacity) similar to, but distinct from the parent body of the enstatite chondrites (Brett and Keil 1986). A complicated igneous history for the aubrites is indicated by the negative Eu anomalies in most of the analyzed samples, suggesting possible segregation of a plagioclase fraction (Boynton and Schmitt 1972; Wolf et al. 1983). The aubrites are depleted in siderophile and chalcophile elements, but the nature and timing of a possible core-formation event have not been worked out. The highly siderophile elements are less depleted than in diogenites (part of the HED parent body discussed in detail below), possibly indicating less complete removal of a metal phase (Wolf et al. 1983). The study of the igneous history of the aubrites has been hampered by their large grain size, the brecciated nature of almost all of the samples, and the partitioning of many of the diagnostic trace elements, such as the rare earths, into minor phases such as oldhamite (CaS) that are formed only under extremely reducing conditions (Newsom et al. 1986).

Ureilites. The ureilites are carbonaceous olivine-pyroxene achondrites whose igneous history is complex (Berkley and Jones 1982; Goodrich et al. 1987). A two-stage model for the formation of the ureilites has been proposed (Goodrich et al. 1987): chondritic material underwent a large degree of partial melting (10–25%) leaving olivine-pigeonite-plagioclase(?) residues. Later melting (<10%) of mafic cumulates from the earlier magmas produced the ureilite parent magmas with high Ca/Al ratios and negative Eu anomalies. Carbon becomes reducing at low pressure and, when the C-bearing ureilite magmas were intruded into the crust, they equilibrated at varying depths and were reduced to different extents, losing different amounts of Fe metal ranging from 20% to 27%. The cumulus ureilite minerals then crystallized from the reduced magmas. In this scenario, the FeO and siderophile-element contents imply that the source regions for the ureilite magmas did not experience a core formation event. The ureilites were finally intruded by less than

1% of an interstitial liquid, unrelated to the original magmas from which the cumulates formed. Goodrich et al. (1987) suggest that a large parent body (>470 km diameter) is required to explain the ureilites. However, new O isotope data show that ureilites are igneous but not related to each other by differentiation of a common source, which means they cannot have been formed by large-scale igneous activity on a single parent body (Clayton et al. 1987). The model of Takeda (1987) involving partial melting of carbonaceous-chondritic material therefore deserves serious consideration.

Howardite, eucrite, diogenite, mesosiderite, main group pallasite and IIIAB iron O-isotope group. Howardites, eucrites and diogenites probably formed on one parent body, since howardites contain clasts very similar to eucrites and diogenites, though some (Jones 1984) would disagree. Remote sensing has shown basaltic rocks on the surface of only one large asteroid, Vesta, which has therefore been considered as a possible parent body for HED meteorites (Gaffey 1983; Consolmagno and Drake 1977). However, there are no dynamical mechanisms to deliver large numbers of meteorites from Vesta (Wasson and Wetherill 1979; see also Chapter 2.1).

Oxygen-isotope data indicate derivation of HED meteorites plus mesosiderites, main group pallasites and IIIAB iron meteorites from a common nebular reservoir (Clayton et al. 1986). However, the Eagle Station and two similar pallasites require a different O-isotopic reservoir, and hence also parent body (Clayton et al. 1976). The IIIAB irons cooled more rapidly than the mesosiderites which contain much more of the low-temperature ordered phase tetrataenite (Hewins 1983), suggesting different parent bodies. Basalts were present, therefore, probably on at least three meteorite parent bodies: the mesosiderite, IIIAB-main-group pallasite and the Eagle Station pallasite bodies. They are also currently observed on the asteroid Vesta but there are dynamical difficulties in concentrating Vesta ejecta into Earth-approaching orbits. The HED meteorites may be derived from collision fragments from a small asteroid (Wasson and Wetherill 1979).

3.2.4. BASALT FORMATION IN THE HOWARDITE-EUCRITE-DIOGENITE PARENT BODY

Our most detailed information about the differentiation and geochemical evolution of an asteroid comes from work on the HED meteorites. In this section and the next we describe the evidence that suggests a complicated multiple stage igneous history for the HED parent body, involving core formation, formation of primary magmas and fractionation and possibly remelting of early igneous rocks. The complexity of this igneous activity has implications for the nature of the heat source.

The simplest models (Table 3.2.1) for the origin of the eucrites involve melting of a chondritic parent body with three alternatives: magma-ocean

TABLE 3.2.1
Possible Origins for Eucrites that are Either Poor or Rich in Incompatible Elements (IE)

	IE-Poor (Trend A)	IE-Rich (Trend B)
I Primary magmas		
(i) chondritic source (Mg basalts later event)	Higher % melting	Lower % melting
(ii) differentiated sources (Mg basalts cogenetic)	Higher % melting	Lower % melting
II Fractionated magmas (polybaric?)		
(i) from magma ocean:	At top (volatile loss)	
(ii) from partial melt:		Parent lower % melting
(iii) common parent:		Magma mixing too
III Combinations of I and II		

fractionation, fractionation of high-degree partial melt(s), and survival of primary magmas, the first two being hard to distinguish. More complex models consider the remelting of early igneous rocks. Mason (1962) noted the sequential relationship in mineralogy and continuity of Fe/Fe + Mg ratios in the HED suite, with very little diversity of rock type at any one Fe/Fe + Mg ratio. He discussed the equilibrium crystallization of liquids in the system forsterite-anorthite-silica as a guide to the origin of meteorites and implied that the eucrites formed as fractionated liquids. This approach was extended by Takeda (1979) by considering the thermal history revealed by exsolution textures so as to place various diogenites and eucrites at different depths in a layered crust. The Mason model became explicitly a magma-ocean model, as developed by Ikeda and Takeda (1985), but a complex one in which the magma was stratified with volatile loss near the surface.

Stolper (1977) proposed that most eucrites represent primary or nearly primary magmas, based on both phase diagrams and geochemical considerations. A liquidus phase diagram for a given chemical system illustrates the solid phase(s) in equilibrium with a melt and geometrically defines crystallization and melting paths. Stolper performed melting experiments on eucrites so that components not in the simple system used by Mason (1962) were considered and interpreted the resulting phase diagram rigorously. He argued against the fractional crystallization model for eucrites, because crystallization could not lead to the eucrites at point X in Fig. 3.2.1a. Accumulation of olivine from a highly magnesian parent liquid A would drive the liquid composition to the olivine-pyroxene field boundary B where, olivine no longer being stable, crystallization of pyroxene alone would push the residual liquid composition across the projection so as to miss the position of eucrites (path A-B-C in Fig. 3.2.1). However, the eucrite compositions project near the

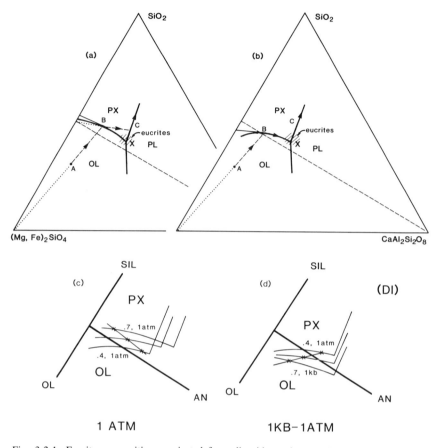

Fig. 3.2.1. Eucrite compositions projected from diopside to the pseudoternary plane silica-olivine-anorthite. Fields for liquids initially crystallizing olivine, pyroxene and plagioclase are shown. (a) The 1 atm. olivine-pyroxene boundary is shown as a reaction curve with Fe/Mg increasing from B to X. Liquid A crystallizes olivine and is pushed to the boundary at B. Since the tangent to the boundary curve at B falls outside the line joining the olivine and orthopyroxene compositions, no olivine can form. Fractional crystallization of orthopyroxene pushes liquid compositions across the pyroxene field to C, so as to miss peritectic point X where common eucrite compositions cluster (Stolper 1977). With this geometry, eucrites must arise as primary magmas. (b) The presence of olivine and orthopyroxene in diogenites suggests that the olivine-pyroxene boundary is a cotectic curve, shown with the tangent at B hitting the sideline between olivine and orthopyroxene compositions, possibly due to pressure effects. A liquid evolved to B by olivine subtraction crystallizes orthopyroxene and minor olivine to pass down the curve to point X. With this geometry eucrites can be generated by fractionation. (c) The boundary curve with fractional crystallization at 1 atm, as in (a) above, arises by crossing the equilibrium (constant Mg/Fe + Mg of 0.7–0.4) boundaries, as successive liquids become more Fe-rich. (d) Schematic fractionation path crossing liquidus boundaries with Mg/Fe + Mg at 0.7–0.4: the cotectic geometry in (b) may arise by polybaric fractionation. Note that the cotectic effect is possible with only a small pressure drop.

peritectic point X. This is where initial melts from a plagioclase peridotite source, liquids saturated with olivine, pyroxene and plagioclase, would plot and thus eucrites could be partial melts produced during the beginning of melting of a chondritic mantle.

The primary-magma model depends critically on the nature of the olivine-pyroxene field boundary, taken to be a reaction curve (Fig. 3.2.1a) by Stolper (1977). Lipin (1978), Morse (1980), Jones (1984), Ikeda and Takeda (1985), Hewins (1986) and Delaney (1986) have argued that this boundary is, in fact a cotectic, with a geometry like that in Fig. 3.2.1b. In this case, the liquid would descend the cotectic curve by crystallizing both olivine and pyroxene to generate eucrites by fractionation. Morse (1980) in particular gives a very complete discussion of the transition from reaction curve toward cotectic curve as Fe/Mg increases. The picture is complicated by the displacement of the boundary curve towards the olivine corner as Fe/Mg increases. Because Fe/Mg increases with fractionation, the liquid can still step across equilibrium (constant Fe/Mg) cotectic curves (Fig. 3.2.1c) to yield a reaction relationship (Stolper 1977; Delaney 1986). Warren (1985) suggested that the boundary, though possibly a reaction curve, is at least a virtual cotectic, and so close to the pyroxene-plagioclase join that orthopyroxene crystallization produces eucritic liquids anyway. Longhi and Pan (1987) and Beckett and Stolper (1987) have reaffirmed that the boundary is a reaction curve; however, the boundary curve becomes a cotectic with a modest increase in pressure (Longhi and Pan 1987; Warren and Wasson 1979). In the forsterite-silica system, the reaction relationship disappears at 1.3 kbar (Chen and Presnall 1975). Stolper (1977) recognized that polybaric fractionation of olivine and orthopyroxene could produce eucritic liquids at the peritectic point, but discounted this process principally because of the small size of the parent body. Regardless of the exact positions of boundary curves at different pressures, fractionation paths will probably be effectively cotectic with a small pressure drop, i.e., if the magma crystallizes as it rises (Fig. 3.2.1d).

The nature of the applicable boundary curve is not certain from experimental studies, but the meteorite mineral data are consistent with a cotectic. Diogenites are cumulates containing both olivine (Fo_{76-62}) and orthopyroxene (En_{79-69}); olivine and pyroxene clasts in howardite breccias range up to Fo_{92} and En_{86} (Fredriksson and Keil 1963; Dymek et al. 1976; Fuhrman and Papike 1981; Desnoyers 1982). Because the rocks contain equilibrium mineral pairs, except for the most magnesian olivine, olivine and pyroxene crystallized together in natural liquids along a cotectic boundary curve, so that eucrites could have formed by fractionation. This suggests crystallization during uprise through the parent body, followed by extrusion of late liquids.

Stolper (1977) used concentrations of incompatible trace elements (IE) as a second major argument in favor of a primary-magma origin for eucrites. He showed that the most Fe-rich eucrites (e.g., Nuevo Laredo; Fig. 3.2.2) fall on the same enrichment trends as found in liquids crystallized in his labo-

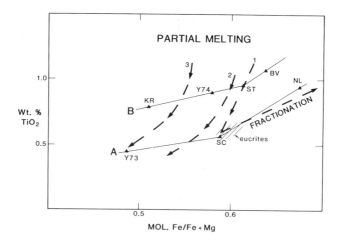

Fig. 3.2.2. Relation between a highly incompatible element (IE) like Ti and a weakly compatible element like Fe during basalt evolution. Crystal fractionation moves liquids from left to right, rising slightly, whereas increased degrees of partial melting drop liquids from top right to the bottom of the diagram. Trend A and B basalts (Ikeda and Takeda 1985) are represented by Yamato 7308 clast PE1(Y73), Sioux County(SC), Nuevo Laredo(NL), Kapoeta clast Rho(KR), Yamato 74450(Y74), Stannern(ST) and Bouvante(BV). Partial melting track 1 and the observed experimental fractionation trend (essentially leg SC-NL of trend A) are taken from Stolper (1977). Partial melting tracks 2 and 3, taken from Smith (1982), were calculated for La and source regions with $Fe/Fe+Mg$ of 0.35 and 0.29. If all IE-rich basalts are primary, very different source regions are required. Partial melting similar to track 3 must have operated if trends A and B result from fractionation of magnesian primary magmas.

ratory (e.g., line labeled "fractionation" in Fig. 3.2.2). These eucrites are sometimes designated as the "Nuevo Laredo trend" (see, e.g., Basaltic Volcanism Study Project 1981). Other eucrites, e.g., Stannern, are too rich in IE such as Ti (and also La; Smith 1982; W: Palme and Rammensee 1981; Hf: Warren and Jerde 1987; and Na: see below) to plot on this trend (Fig. 3.2.2). Stolper (1977) inferred increased melting with decreasing incompatible elements (trend 1 of Fig. 3.2.2) and subsequently Consolmagno and Drake (1977) calculated that Stannern would correspond to 4% partial melting, in contrast to the common eucrites which could require 10 to 15% melting. The chondrite/basalt trace-element ratio increasing down Line 1 yields increased estimates of parent melting. Stannern and similar eucrites have been designated the "Stannern trend" (see, e.g., Basaltic Volcanism Study Project 1981). Eucrites that show no obvious affinity with either Stannern or Nuevo Laredo are, in this nomenclature, "Main Group" eucrites.

The partial-melting trend of Stolper, shown as Line 1 in Fig. 3.2.2, was based mainly on the composition of Stannern, since the common eucrites cluster. (Note that Ibitira is not a good indicator for a partial-melting track because, although the Ti concentration is elevated, REE, W and Hf are not, and Na is depleted (Stolper 1977; Palme and Rammensee 1981; Smith 1982;

Warren and Jerde 1987). Wasson (1985) calculated that Yamato 74450 corresponds to 6% partial melting, and similarly Christophe Michel-Lévy et al. (1987) suggested that Bouvante is primary. However, the other two meteorites do not fall on the trend from Stannern to common eucrites (Fig. 3.2.2). Smith (1982) derived curves for partial melting of sources with different Fe/Mg ratios, using La as the trace element, shown in Fig. 3.2.2. IE-rich basalts might represent primary magmas from source regions with Fe/Fe + Mg of 0.29 to 0.35 or fractionated liquids from magnesian primary magmas, alternatives listed in Table 3.2.1. Mittlefehldt (1979), Smith (1982) and Delaney (1986) argued that heterogeneous source regions might arise from failure of melt to migrate to the surface and/or from remelting of early plutons. Such complex scenarios are very difficult to model (Consolmagno 1979; Smith 1982). However, Bouvante, Stannern and Y 74450 (Fig. 3.2.2) all plot on fractionation trend B of Ikeda and Takeda (1985). Since partial melting to produce primary eucrites is not conclusively proven, the evidence for fractionation, particularly as revealed by recently recovered polymict breccias that contain basalt clasts more magnesian than common eucrites, must be evaluated.

Delaney et al. (1981) found two groups of basaltic clasts in howardites and polymict eucrites. One, with similar mineral compositions to eucrites, was termed peritectic basalt; the other, with a wide range of mineral compositions, was called evolved basalt (Fig. 3.2.3). Ikeda and Takeda (1985) confirmed the existence of evolved basalts, which they called trend B, and found

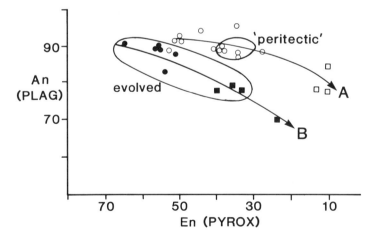

Fig. 3.2.3. Mineral compositions defining two groups of basalts in howardites: "peritectic" (like common eucrites) and evolved (Delaney et al. 1981). These groups are confirmed and extended in the Yamato 7308 howardite, as trend A (open circles) and B (solid circles) basalts (Ikeda and Takeda 1985). Filled squares are calculated compositions for liquids in equilibrium with cumulates. Extremely Fe-rich trend A rocks (open squares) contain olivine because low-Ca pyroxene is no longer stable. Trend B basalts are richer in IE (e.g., Na, Ti) than trend A basalts.

that the "peritectic" basalts are part of another fractionated group which they called trend A. Trends A and B are IE-poor and IE-rich basalts, respectively, and a geochemical classification is desirable since both equilibrated and unequilibrated eucrites are found. The two trends are represented in figures other than Fig. 3.2.3, by three or four samples each but it should be remembered that other analyses scatter about the trend lines shown. Both Delaney et al. (1981,1984) and Ikeda and Takeda (1985) explained the basalts, including eucrites, by fractionation but whereas Delaney et al. suggested the two trends are derived from two high-degree partial melts, Ikeda and Takeda assumed a magma ocean. One of the main differences between the two groups is that trend-A basalts are Na-poor and might represent liquids from the top of the magma ocean which experienced volatile loss to space (Ikeda and Takeda 1985). However, since trend-A basalts are also poor in nonvolatile elements like Ti and La, the difference is at least partly due to the incompatible nature of Na and can be explained in terms of different degrees of partial melting (on trend 3 of Fig. 3.2.2) with subsequent fractionation. In other words, a serial-magmatism model is better than a magma-ocean model. However, Mittlefehldt (1987) has shown that eucrites have experienced some volatile loss relative to the parent magmas of diogenites.

The specific fractionations involved in trends A and B can be determined by calculations which mix crystals and daughter liquids to match assumed parental liquid compositions. Analyses of liquids plotted in Fig. 3.2.2, taken from Duke and Silver (1967), Wänke et al. (1977), Ikeda and Takeda (1985), Dymek et al. (1976) and Christophe Michel-Lévy et al. (1987), were used. Yamato 7308 clast PE1 (trend A) and Kapoeta clast Rho (trend B) are taken as parent liquids because they are the two most magnesian samples available with the appropriate IE-poor and IE-rich chemistries. Trend B can be reproduced by crystallizing magnesian pigeonite from Kapoeta Rho, plus minor chromite, followed by plagioclase plus more ferroan pigeonite to move from Stannern to Bouvante. The mathematical fit is improved a little if some olivine is formed in generating Stannern, although little or no olivine crystallization is expected (Fig. 3.2.4).

Trend A requires crystallization of 10% olivine along with ferroan orthopyroxene to reach the common eucrites, although Y7308PE1 itself does not contain olivine because of reaction at the peritectic point, and then, as widely accepted (see, e.g., Stolper 1977), plagioclase plus pigeonite to generate ferroan eucrites like Nuevo Laredo. Note that the low-Ca orthopyroxene rather than magnesian pigeonite is needed for IE-poor trend A because, where Na is low, available Ca must match Al to make plagioclase: the relatively constant plagioclase composition in eucrites is due to low initial Na coupled with orthopyroxene crystallization failing to reduce the Ca content of daughter liquids. The lavas modeled may not represent perfect fractional crystallization because other processes, possibly accumulation of the phenocrysts (Dymek et al. 1976; Warren and Jerde 1987) or magma mixing (Hewins 1987), cause

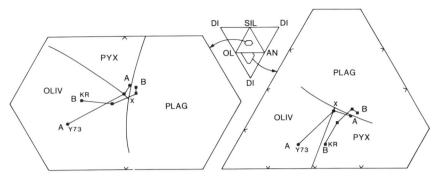

Fig. 3.2.4. Basalts in the quaternary system diopside-silica-olivine-anorthite projected onto the silica-olivine-anorthite (left) and anorthite-olivine-diopside pseudoternary planes. Fractionation trends A and B are represented by the same basalts as in Fig. 3.2.2. Common eucrites plot very close to Sioux County, the datum closest to peritectic point X. Divisions correspond to 10 mol %. Crystallization of olivine and pyroxene from parental liquids (Y7308PE1 and Kapoeta Rho) appears to drive the daughter liquids near peritectic point X before plagioclase-pyroxene crystallization begins. Note that Bouvante (trend B) is displaced from the peritectic in the sense explained by plagioclase-pyroxene fractionation, just as Nuevo Laredo (trend A) is.

dispersion of individual analyses about the general trend line. However, the low sums of squares of residuals for major elements show that fractionation within trends A and B is plausible, whereas derivation of a trend B liquid from trend A, or vice versa, is highly improbable.

The role of fractionation in the two basalt trends can be examined by projecting liquid compositions onto pseudo-ternary phase diagrams. The liquid compositions are expressed in terms of silica, olivine, plagioclase and diopside components and shown projected from quaternary space onto two of the bounding triangles (Fig. 3.2.4). The liquidus-field boundaries are based on the 1 atm melting experiments of Stolper (1977) and Longhi and Pan (1988). Liquids produced by high degrees of partial melting beyond exhaustion of plagioclase would stretch from peritectic point X up the olivine-pyroxene boundary curve. Liquids plotting in the olivine field should evolve by crystallizing olivine until they reach the olivine-pyroxene boundary and eventually they will descend the pyroxene-plagioclase boundary curve. Note that these diagrams can be misleading because, in projecting from the tetrahedron to the triangles, liquids can appear to be in the olivine field (Fig. 3.2.4a) or in the pyroxene field (Fig. 3.2.4b). Figure 3.2.4 also suggests that only olivine should have crystallized from Y7308PE1 on the first leg of trend A, indicating that to permit some pyroxene crystallization, the actual pressure must have been higher than 1 atm, shrinking the olivine field. This is consistent with the arguments above, that cotectic crystallization of olivine and orthopyroxene is related to magma uprise (Fig. 3.2.1d). Since the basalts are

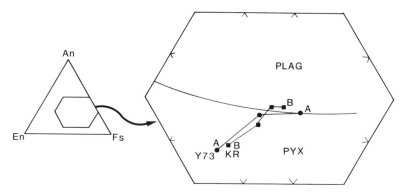

Fig. 3.2.5. Projection of trend A and B basalts onto the CaAl$_2$Si$_2$O$_8$-MgSiO$_3$ - FeSiO$_3$ plane from augite (olivine and silica projections are similar). The change of direction at the onset of plagioclase fractionation, after crystallization of olivine and pyroxene from Y7308PE1 and Kapoeta Rho parent liquids, and the extent of Fe-Mg fractionation are clearly shown.

plagioclase-pyroxene rocks, a convenient projection is onto the plane An-En-Fs (Fig. 3.2.5), which clearly shows the fractionation in terms of a wide range of Mg/Fe ratios. After any olivine crystallization from the parental liquids, pyroxene subtraction drives the liquids to the boundary curve where plagioclase begins to crystallize with pyroxene. The sharp bends in trends A and B in Figs. 3.2.4b and 3.2.5 clearly show descent of the liquids down the plagioclase-pyroxene cotectic to reach Nuevo Laredo and Bouvante, as required by the mixing calculations discussed above.

The eucrites could possibly be fractionated from more magnesian basalts rather than primary. The high concentrations of IE in some eucrites indicate low-degree partial melting, but it is not necessary that they are primary magmas: they can be derived by fractionation of a magnesian primary magma caused by a low degree of partial melting (trend 3 of Fig. 3.2.2). Bouvante, Stannern and Y74450 have negative Eu anomalies suggestive of solid plagioclase remaining in the source region; a test of the fractionation model is that Kapoeta Rho should also have a negative Eu anomaly. It is therefore possible that the entire trend B was generated before the exhaustion of plagioclase in a rather plagioclase-rich parent asteroid with a lower Fe/Mg ratio (Dreibus and Wänke 1980; Smith 1982) than that estimated by Consolmagno and Drake (1977). Trend-A liquids are poorer in IE and have no Eu anomalies (unless they are extensively fractionated) because melting continued after all plagioclase was exhausted. Thus, the arguments of Stolper (1977) and Consolmagno and Drake (1977) applied only to eucrites could be reapplied to all the HED basalts, as suggested by Fig. 3.2.2. One problem with this is that diogenites contain a trapped basaltic liquid with a negative Eu anomaly, but trend-B liquids are not magnesian enough to produce all the diogenites. Yet another alternative (Table 3.2.1) is to take a high-degree trend-A partial melt as parent magma for both trends A and B. Trend A could be produced by simple fractional crystallization and trend B by contamination or magma-mixing of parent with IE-rich daughter liquids (Hewins 1987).

Because there are both IE-rich and IE-poor eucrites, a chondritic magma-ocean origin is unlikely. The geochemical diversity of HED basalts could be attributed mainly to fractionation, mainly to partial melting, even mainly to magma mixing (Table 3.2.1), or to some combination. The fractionation alternative involves a two-stage melting history, IE-rich basalts followed by IE-poor, with a magnesian chondritic source (Dreibus and Wänke 1980). However, the IE-rich basalts could perhaps be generated by extensively melting a chondritic source and magma-mixing primitive IE-poor liquids with IE-rich differentiates. If all the IE-rich basalts are instead primary magmas, source regions varying in Fe/Mg ratio are required (Smith 1982). The range of Fe/Mg ratios could have been generated by intruding early magmas and remelting the plutons, rather than by heterogeneous accretion (Mittlefehldt 1979; Delaney 1986). Although such situations may be impossible to model uniquely (Consolmagno 1979), Smith (1982) showed that eucrite REE patterns could have been derived by melting differentiated crust.

There is a clear need to pursue coordinated petrologic-geochemical studies especially of new Antarctic meteorites so as to define relationships in HED basalts and cumulates. Documentation of remelting of early igneous rocks would suggest extended or multiple heating, perhaps a combination of ^{26}Al and induction, or migrations of ^{26}Al with basalt (analogous to the role of KREEP on the moon). If remelting were not needed, a major role for ^{26}Al heating would not be indicated, as asteroid data implicate a solar mechanism, such as electromagnetic induction heating.

3.2.5. BASALTS AND CORE FORMATION IN THE HED PARENT BODY

In studying eucritic basalts, we have tried to distinguish between primary melts that survived from the lowest fraction of partial melting, when the body was below 1150°C and largely solid, and those like Nuevo Laredo that formed by fractionation from more magnesian basalts. Core formation, or separation of metal from an originally chondritic mantle, must have taken place at some point between the extremes of partial melting required. Metal segregation or core formation causes a decrease or depletion of siderophile elements in the residual silicate mantle and silicate melts. The concentrations of siderophile elements can show whether a eucrite melt was in contact with metal (e.g., earliest basaltic partial melts) or whether the melts formed after the separation of a core (e.g., by fractionation in the serial-melting model). The evidence that core formation occurred before the igneous fractionation events is an important clue to the thermal history of the HED parent body. Recent analytical work on siderophile-element abundances in samples from the HED and experimental determinations of partition coefficients can be used to calculate the metal content of the HED parent body (Palme and Rammensee 1981; Newsom and Drake 1982, 1983; Newsom 1985).

The depletion of siderophile elements relative to their initial chondritic abundances cannot be calculated directly from their absolute abundances if the siderophile elements are fractionated by igneous processes. The depletions can be obtained, however, from correlations with refractory nonsiderophile elements whose chondritic initial abundance in the parent body can be assumed. The siderophile element and the nonsiderophile element should have a similar geochemical behavior during igneous fractionation *in the absence of metal,* such that an enrichment of the siderophile element due to fractionation is mirrored by an enrichment of the nonsiderophile elements resulting in a constant ratio of siderophile to nonsiderophile. For example, assuming that the siderophile-element-to-La ratios were originally chondritic, the depletion of W is determined from the W/La ratio in eucrites (Palme and Rammensee 1981) compared to the W/La ratio in CI chondrites, and the depletion of Mo is determined from the Mo/La ratio (Newsom 1985).

The existence in HED meteorite samples of constant ratios of incompatible siderophile and nonsiderophile elements, over a wide range in absolute concentrations suggests that the HED, like the Earth, experienced a core-formation event *before* the igneous fractionation events. This early core-formation event is supported by the constant ratio of Mo to W in the eucrites (Fig. 3.2.6), even though the absolute concentrations of these elements are variable, because Mo is much more siderophile than W. Similar data for the incompatible siderophile element P also support this idea (Newsom and Drake 1983). If metal had been present in the mantle during formation of the different eucrite magmas by partial melting, the metal would have buffered the absolute concentrations of the otherwise incompatible siderophile elements and we would not observe the variations (Fig. 3.2.6) that actually exist. The only way around this conclusion is the unlikely possibility that the highly fractionated eucrites, such as Stannern, Nuevo Laredo and Bouvante, were derived from source regions with less metal than the source regions for the normal eucrites (Newsom and Drake 1982).

Determining the depletion of compatible siderophile elements, such as Co and Ni, is not as easy as for those that are incompatible, such as Mo and W, because much of the Co and Ni content of the silicate portion of the HED parent body is retained in the unsampled olivine-rich mantle. The Co and Ni depletions used by Newsom (1985) were those reported by Dreibus and Wänke (1980). The depletions of Co and Ni used in this work are less than the depletions derived for the parent body by Dreibus and Wänke (1980) for reasons illustrated in Fig. 3.2.7. Recently, Delano (1986) has shown that the depletions of Co and Ni in the Earth and Moon can be calculated using correlations with Mg. Figure 3.2.7 illustrates the depletion of the Co/Mg ratio compared to the trend for terrestrial samples and the Co/Mg ratio in CI me-

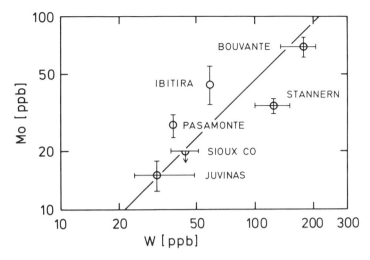

Fig. 3.2.6. Molybdenum and tungsten concentrations in eucrites (Newsom 1985). The constant ratio of the highly siderophile and chalcophile element Mo to the moderately siderophile element W, independent of absolute concentrations, is consistent with igneous fractionation in the absence of metal.

teorites. The special composition, termed PR* by Dreibus and Wänke (1980) is derived from a mixing calculation based on howardites, assuming that howardites are a mixture of eucrites and diogenites. The PR* composition is also characterized by chondritic ratios of all of the incompatible refractory elements, essentially representing the composition of the melt fraction produced by a large degree of partial melting of the HED. The composition labeled EPB from Dreibus and Wänke (1980) was obtained by adding 43% of the pallasite Marjalahti olivine composition to 57% PR* composition, to get back to the bulk composition of the HED. The Marjalahti olivine, however, has very low siderophile-element abundances (7 ppm Co) compared to the PR* composition (17 ppm Co) and this lowered the Co/Mg ratio in the EPB composition resulting in too large a depletion for Co. The Co content of olivine in equilibrium with the HED meteorites, however, should be about 4 times the Co content of the HED meteorites (Table 3.2.2), suggesting that the depletion of Co was not as great as indicated by Dreibus and Wänke (1980). The Co/Mg ratio of the HED meteorites and the PR* composition, compared to the Co/Mg ratio of CI meteorites and terrestrial samples with the same MgO contents represents a better indication of the depletion of Co in the HED and is the value used in Figs. 3.2.8 and 3.2.9 (Table 3.2.2). A similar situation occurs for Ni, but the uncertainties are much greater because of poorer analytical data and the larger and more uncertain value of the olivine/melt partition coefficients, which are strongly dependent on temperature, ranging from 20 at 1200° C to 8 at 1300° C (Irving 1978).

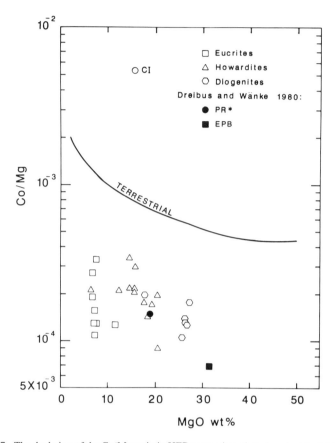

Fig. 3.2.7. The depletion of the Co/Mg ratio in HED meteorites plotted against the MgO content of the meteorites. The figure illustrates the depletion of the Co/Mg ratio compared to the trend for terrestrial samples (J. W. Delano, personal communication, 1986) and the Co/Mg ratio in CI meteorites. The PR* composition of Dreibus and Wänke (1980) is derived from a mixing calculation based on howardites, which confirms that howardites are probably a mixture of eucrites and diogenites. The composition labeled EPB from Dreibus and Wänke (1980) was obtained by adding 43% of a Marjalahti olivine composition to 57% PR* composition, to get back to the bulk composition of the HED. The very low siderophile-element abundances of the Marjalahti olivine, however, results in too large a depletion for Co. The Co/Mg ratio of the HED meteorites and the PR* composition, compared to the Co/Mg ratio of CI meteorites and terrestrial samples with the same MgO contents, represents a better indication of the depletion of Co in the HED. Data are from: Duke and Silver (1967); McCarthy et al. (1972); Fukuoka et al. (1977); Laul et al. (1972); and the Mainz Laboratory (Palme et al. 1978, and references therein).

TABLE 3.2.2
Eucrite Parent Body Depletion Factors and Partition Coefficients.[a]

	Depletion Factor	D(metal/silicate)	D(solid/liquid-silicate)
W	31 ± 8 (a)	$25 ^{+8}_{-5}$ (b)	< 0.01 (c)
Mo	570 ± 230 (d)	$600 ^{+300}_{-200}$ (e)	< 0.01 (c)
Co	$36 ^{+40}_{-16}$ (f)	$250 ^{+75}_{-50}$ (g,h)	4 ± 0.5 (c)
Ni	$220 ^{+280}_{-170}$ (f)	4000 ± 1000 (g,h)	15 ± 5 (c)
P	40 ± 16 (i)	10 ± 8 (h,j)	< 0.01 (c)

[a] Data from: (a) Palme and Rammensee (1981); (b) Newsom and Drake (1982); (c) Irving (1978); (d) Newsom (1985); (e) Rammensee (1978); (f) Dreibus and Wänke (1980) PR* composition; (g) Jones and Drake (1985); (h) Schmitt (1984); (i) Weckwerth et al. (1983); (j) Newsom and Drake (1983).

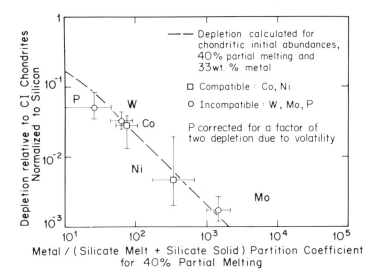

Fig. 3.2.8. The depletion relative to chondritic abundances of the siderophile elements Co, W, Ni, Mo and P. The depletion factors are normalized to Si, such that the assumed initial abundances in the HED fall at the top of the diagram. The depletion of P indicated on the figure has been corrected to show the depletion due to the siderophile nature of P, by assuming that the initial abundance of P in the HED was depleted by a factor of 2 due to volatility. The metal/(silicate melt + silicate solid) partition coefficients for each element are calculated for 40% partial melting using the data from Table 3.2.2, which allows for the different partitioning behavior between solid silicates and liquid silicates of the compatible elements Co and Ni, compared to the incompatible elements W, Mo and P. The calculated depletion trend assumes chondritic initial siderophile-element abundances, metal-silicate equilibrium at 40% partial melting of the silicates, and 33 wt% metal in the HED.

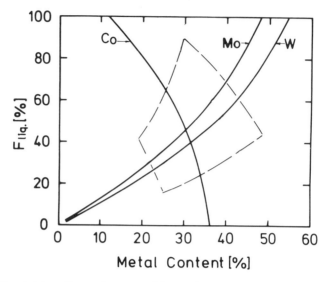

Fig. 3.2.9. Constraints on the metal content and degree of partial melting of the silicates required to obtain the observed depletions of Mo, W and Co in the EPB. The dashed lines represent the maximum uncertainties for the depletion factors and partition coefficients in the area of overlap. The illustrated data are consistent with a metal content between 20 wt% and 40 wt% and metal segregation between 20% and 70% partial melting of the silicates.

The depletions of siderophile elements in the HED are consistent with their appropriate metal/silicate partition coefficients (Fig. 3.2.8). Fortunately, in the HED parent body, segregation of a sulfide melt was apparently of minor importance because the HED silicates are not depleted in chalcophile elements, in contrast to the depletions of elements such as V, observed in the winonaite/IAB silicates (Bild 1977; Newsom and Drake 1987). The very highly siderophile elements, such as Ir, are consistent with the trend in Fig. 3.2.8, but their metal-silicate partition coefficients have much larger uncertainties so they have not been plotted. The volatile siderophile elements such as P, Ga and Ge are depleted below the trend in Fig. 3.2.8, suggesting that volatility as well as a core formation event has contributed to their depletion (Newsom 1985).

Calculating the metal content of the HED from the depletion of siderophile elements depends critically on understanding the partitioning behavior of the siderophile elements between solid silicates and liquid silicate. Assuming that metal segregation occurs at some relatively low degree of partial melting, the bulk-metal/total-silicate partition coefficient $D^{M/BS}$ can be calculated from the relationship

$$D^{M/BS} = \frac{D^{M/LS}}{F^{LS} + (D^{SS/LS})(1 - F^{LS})} \quad (1)$$

where $D^{M/LS}$ is the solid-metal/liquid-silicate partition coefficient and $D^{SS/LS}$ is the solid-silicate/liquid-silicate partition coefficient. F^{LS} is the fraction of silicate melt divided by the total fraction of silicates such that

$$BS = F^{LS} + F^{SS}. \qquad (2)$$

For compatible trace elements, such as Co and Ni, with $D^{SS/LS} > 1$, $D^{M/BS}$ is smaller than $D^{M/LS}$ because of retention of the trace element in silicate minerals. For incompatible trace elements, such as W and Mo, with $D^{SS/LS} < 1$, $D^{M/BS}$ is larger than $D^{M/LS}$ because the trace element is excluded from the silicate minerals.

Given the metal/total-silicate partition coefficient, the equation for the weight fraction of metal X required to achieve a certain depletion a in a core-formation event has been derived by Rammensee and Wänke (1977), assuming equilibrium:

$$X = \frac{a - 1}{D^{M/BS} + a - 1}. \qquad (3)$$

In Fig. 3.2.8, the bulk-metal/silicate partition coefficients ($D^{M/BS}$) have been calculated for each siderophile element according to Eq. 1, assuming 40% partial melting and the values in Table 3.2.2. At higher or lower degrees of partial melting the correlation of the incompatible and compatible siderophile elements is not as good. The dashed line in Fig. 3.2.8 is the amount of depletion expected for a given amount of metal calculated from Eq. (3). The assumption of 40% partial melting results in a very good correlation of the siderophile elements within the uncertainties of the depletions and partition coefficients. The uncertainties for Ni are much larger because the analytical data for Ni in samples from the HED are limited, and because of large uncertainties in both the metal/silicate partition coefficients and the solid-silicate/liquid-silicate partition coefficients. A more complete discussion of the uncertainties in these calculations is found in Newsom (1985).

A different way of illustrating the limits on the metal content of the HED from the available data is shown in Fig. 3.2.9. In this figure the amount of metal required to achieve the observed depletions of the siderophile elements is calculated as a function of the degree of partial melting. Figure 3.2.9 graphically illustrates the opposite behavior of the incompatible elements W and Mo compared to the compatible element Co. By themselves, the depletions of the elements W and Mo are consistent with a wide range of metal contents, depending on the degree of partial melting (Newsom and Drake 1982). The opposite dependence of Co constrains the metal content to between 20% and 40%, with metal segregation at a degree of partial melting between 20% and 70%. The Ni data are not shown, although the large uncertainties of the Ni data completely overlap the outlined area where Co overlaps Mo and W.

The metal content of the HED from the results of these calculations (20%–40%) is slightly less than the results of Newsom (1985) (30%–50%), and overlaps the results of Palme and Rammensee (1981), but it is still significantly larger than found in most chondrites. While the metal content may actually be large, another possibility is that the HED may be enriched in refractory lithophiles (Jones 1984). Thus, elements such as Co and Ni may not have been originally present in CI chondritic abundances relative to the refractory lithophile and siderophile elements as assumed in the calculations. The degree of partial melting (20% to 70%) required for metal segregation also covers a large range, although experimental and theoretical data on metal segregation are almost nonexistent.

Another important conclusion of the siderophile-element data is the evidence (Fig. 3.2.7) that metal segregation occurred before the fractionation events that produced the magmas represented in the samples available from the HED. Because complete metal segregation probably requires a relatively high degree of partial melting ($>10\%$), consistent with the results described above (20% to 70%), formation of the highly fractionated eucrites at very low degrees of partial melting ($\sim 5\%$; Consolmagno and Drake 1977) *before* metal segregation is unlikely. The siderophile-element data, therefore, support a fractional crystallization origin for the eucrite magmas (Mason 1962; Delaney et al. 1984; Ikeda and Takeda 1985), in contrast to the partial-melting model (Stolper 1977). The earliest event recorded in eucrites, after isolation from the solar nebula, is depletion of siderophile elements due to core formation, prior to the separation of the magnesian magmas parental to the eucrites themselves.

3.2.6. CONCLUSIONS

Even though the chondrites are numerically more abundant in the meteorite collections, the number of parent bodies, approximately 70, represented by the igneous meteorites is much larger than the number of parent bodies represented by the primitive meteorites. This conclusion is consistent with the observation that the parent bodies of the igneous meteorites, the differentiated asteroids, are relatively close to us in the asteroid belt. The howardite-eucrite-diogenite parent body, though not conclusively identified with a known asteroid, is the one we know most about. The data for HED basalts and cumulates indicate that their parent body was very extensively but not totally melted. Establishment of remelting of igneous crust for the origin of any HED magmas (see, e.g., Smith 1982) would suggest electromagnetic-induction heating combined with ^{26}Al decay or ^{26}Al alone, because the latter heat source could have migrated to the planetary surface. Complicated igneous histories are also observed in samples from the aubrite and ureilite parent bodies and in the

Angra dos Reis meteorite. Induction heating probably played a major role in differentiating asteroids in view of the location of such asteroids relatively close to the Sun.

Acknowledgments. Mixing calculations and projections were performed using the program IGPET by M. J. Carr. Detailed reviews by J. H. Jones and P. H. Warren were invaluable and comments from K. Keil improved the final draft. The work was partially supported by the National Aeronautics and Space Administration.

REFERENCES

Basaltic Volcanism Study Project. 1981. *Basaltic Volcanism on the Terrestrial Planets* (New York: Pergamon Press), pp. 1144–1151, 1212–1217.

Beckett, J. R., and Stolper, E. 1987. Constraints on the origin of eucritic melts: An experimental study. *Lunar Planet. Sci.* XVIII: 54–55 (abstract).

Bell, J. F. 1986. Mineralogical evolution of the asteroid belt. *Meteoritics* 21: 333–334 (abstract).

Berkley, J. L., and Jones, J. H. 1982. Primary igneous carbon in ureilites: Petrological implications. *Proc. Lunar Planet. Sci. Conf.* 13, *J. Geophys. Res. Suppl.* 87: A353–A364.

Bild, R. W. 1977. Silicate inclusions in group IA irons and a relation to the anomalous stones Winona and Mt. Morris (Wis). *Geochim. Cosmochim. Acta* 41: 1439–1456.

Bogard, D. D., Nyquist, L. E., and Johnson, P. 1984. Noble gas contents of shergottites and implications for the Martian origin of SNC meteorites. *Geochim. Cosmochim. Acta* 48: 1723–1739.

Boynton, W. V., and Schmitt, R. A. 1972. The europium anomaly and rare earth and other abundances in calcium-poor achondrites. Internatl. Assn. Geochem. Cosmochem. Symp. on Cosmochemistry, August 14–18 (abstract).

Brett, R., and Keil, K. 1986. Enstatite chondrites and enstatite achondrites (aubrites) were not derived from the same parent body. *Earth Planet. Sci. Lett.* 81: 1–6.

Chen, C.-H., and Presnall, D. C. 1975. The system Mg_2SiO_4–SiO_2 at pressures up to 25 kb. *Amer. Mineral.* 60: 398–406.

Christophe Michel-Lévy, M., Bourot-Denise, M., Palme, H., Spettel, B., and Wänke, H. 1987. L'eucrite de Bouvante chimie, petrologie et minéralogie. *Bull. Soc. Fr. Mineral. Crist.* 110: 449–458.

Clayton, R. N., Mayeda, T. K., and Davis, A. M. 1976. Parent bodies of stony-iron meteorites. *Lunar Science* VII: 160–162 (abstract).

Clayton, R. N., Mayeda, T. K., Prinz, M., Nehru, C. E., and Delaney, J. S. 1986. Oxygen isotope confirmation of a genetic association between achondrites and IIIAB iron meteorites. *Lunar Planet. Sci.* XVII: 141 (abstract).

Clayton, R. N., Mayeda, T. K., and Yanai, K. 1987. Oxygen isotopes in ureilites. *Meteoritics* 22, in press.

Colucci, M. T., and Hewins, R. H. 1984. Ordinary chondrite pyroxene thermometry—One more time. *Lunar Planet. Sci.* XV: 180–181 (abstract).

Consolmagno, G. J. 1979. REE patterns versus the origin of the basaltic achondrites. *Icarus* 40: 522–530.

Consolmagno, G. J., and Drake, M. J. 1977. Composition and evolution of the eucrite parent body: Evidence from rare earth elements. *Geochim. Cosmochim. Acta* 41: 1271–1282.

Delaney, J. S. 1986. Phase equilibria for basaltic achondrites and the basaltic achondrite planetoid. *Lunar Planet. Sci.* XVIII: 164–167 (abstract).

Delaney, J. S., Prinz, M., Nehru, C. E., and Harlow, G. E. 1981. A new basalt group from howardites: Mineral chemistry and relationships with basaltic achondrites. *Lunar Planet. Sci.* XII: 211–213 (abstract).

Delaney, J. S., Prinz, M., and Takeda, H. 1984. The polymict eucrites. *Proc. Lunar Planet. Sci. Conf.* 15, *J. Geophys. Res. Suppl.* 89: C251–C288.

Delano, J. W. 1986. Abundances of cobalt, nickel and volatiles in the silicate portion of the

Moon. In *Origin of the Moon*, eds. W. K. Hartmann, R. J. Phillips, and G. J. Taylor (Houston: Lunar and Planetary Inst.), pp. 231–248.

Desnoyers, C. 1982. L'olivine dans les howardites: Origine, et implications pour le corps parent de ces météorites achondritiques. *Geochim. Cosmochim. Acta* 46:667–680.

Dreibus, G., and Wänke, H. 1980. The bulk composition of the eucrite parent asteroid and its bearing on planetary evolution. *Z. Naturforsch.* 35a:204–216.

Duke, M. B., and Silver, L. T. 1967. Petrology of eucrites, howardites and mesosiderites. *Geochim. Cosmochim. Acta* 31:1637–1665.

Dymek, R. F., Albee, A. L., Chodos, A. A., and Wasserburg, G. J. 1976. Petrography of isotopically-dated clasts in the Kapoeta howardite and petrologic constraints on the evolution of its parent body. *Geochim. Cosmochim. Acta* 40:1115–1130.

Esbensen, K. H., Buchwald, V. F., Malvin, D. J., and Wasson, J. T. 1982. Systematic compositional variations in the Cape York meteorite. *Geochim. Cosmochim. Acta* 46:1913–1920.

Ezer, D., and Cameron, A. G. W. 1962. A study of solar evolution. *Canadian J. Phys.* 43:1497–1517.

Floran, R. J., Prinz, M., Hlava, P. F., Keil, K., Nehru, C. E., and Hinthorne, J. R. 1978. The Chassigny meteorite: A cumulate dunite with hydrous amphibole-bearing melt inclusions. *Geochim. Cosmochim. Acta* 42:1213–1229.

Fredriksson, K., and Keil, K. 1963. The light-dark structure in the Pantar and Kapoeta stone meteorites. *Geochim. Cosmochim. Acta* 27:717–739.

Fuhrman, M., and Papike, J. J. 1981. Howardites and polymict eucrites: Regolith samples from the eucrite parent body. Petrology of Bholgati, Bununu, Kapoeta and ALHA 76005. *Proc. Lunar Planet. Sci. Conf.* 12:1257–1279.

Fukuoka, T., Boynton, W. V., Ma, M.-S., and Schmitt, R. A. 1977. Genesis of howardites, diogenites and eucrites. *Proc. Lunar Sci. Conf.* 8:187–210.

Gaffey, M. J. 1983. The asteroid (4) Vesta: Rotational spectral variations, surface material heterogeneity and implications for the origin of the basaltic achondrites. *Lunar Planet. Sci.* XIV:231–232 (abstract).

Goldstein, J. I., and Axon, H. J. 1973. The Widmanstätten figure in iron meteorites. *Naturwissenschaften* 60:313–321.

Goodrich, C. A., Jones, J. H., and Berkley, J. L. 1987. Origin and evolution of the ureilite parent magmas: Multi-stage igneous activity on a large parent body. *Geochim. Cosmochim. Acta* 51:2255–2273.

Gradie, J. C., and Tedesco, E. 1982. Compositional structure of the asteroid belt. *Science* 216:1405–1407.

Herbert, F., and Sonett, C. P. 1978. Primordial metamorphism of asteroids via electrical induction in a T-Tauri-like solar wind. *Astrophys. Space Sci.* 55:227–239.

Herndon, J. M., and Herndon, M. A. 1977. Aluminum-26 as a planetoid heat source in the early solar system. *Meteoritics* 12:459–465.

Hewins, R. H. 1983. Impact versus internal origins for mesosiderites. *Proc. Lunar Planet. Sci. Conf.* 14, *J. Geophys. Res. Suppl.* 88:B257–B266.

Hewins, R. H. 1986. Serial melting or magma ocean for the HED achondrites? *Meteoritics* 21:396–397 (abstract).

Hewins, R. H. 1987. Partial melting, fractionation and magma mixing in HED basalts. *Meteoritics* 22, in press.

Hostetler, C. J., and Drake, M. J. 1980. On the early global melting of the terrestrial planets. *Proc. Lunar Planet. Sci. Conf.* 11:1915–1929.

Irving, A. J. 1978. A review of experimental studies of crystal/liquid trace element partitioning. *Geochim. Cosmochim. Acta* 42:743–770.

Ikeda, Y., and Takeda, H. 1985. A model for the origin of basaltic achondrites based on the Yamato 7308 howardite. *Proc. Lunar Planet. Sci. Conf.* 15, *J. Geophys. Res. Suppl.* 90:C649–C663.

Jones, J. H. 1984. The composition of the mantle of the eucrite parent body and the origin of eucrites. *Geochim. Cosmochim. Acta* 48:641–648.

Jones, J. H., and Drake, M. J. 1985. Experiments bearing on the formation and primordial differentiation of the Earth. *Lunar Planet. Sci.* XVI:412–413 (abstract).

Kaiser, T., and Wasserburg, G. J. 1983. The isotopic composition and concentration of Ag in iron meteorites and the origin of exotic silver. *Geochim. Cosmochim. Acta* 47:43–58.

Kallemeyn, G. W., Boynton, W. V., Willis, J., and Wasson, J. T. 1978. Formation of the Bencubbin polymict meteoritic breccia. *Geochim. Cosmochim. Acta* 42:507–515.
Kelly, W. R., and Larimer, J. W. 1977. Chemical fractionations in meteorites—VIII. Iron meteorites and the cosmochemical history of the metal phase. *Geochim. Cosmochim. Acta* 41:93–111.
Kracher, A. 1985. The evolution of partially differentiated planetesimals: Evidence from iron meteorite groups IAB and IIICD. *Proc. Lunar Planet Sci. Conf.* 15, *J. Geophys. Res. Suppl.* 90:C689–C698.
Laul, J. C., Keays, R. R., Ganapathy, R., Anders, E., and Morgan, J. W. 1972. Chemical fractionations in meteorites—V. Volatile and siderophile elements in achondrites and ocean ridge basalts. *Geochim. Cosmochim. Acta* 36:329–345.
Lee, T., Papanastassiou, D. A., and Wasserburg, G. J. 1976. Demonstration of ^{26}Mg excess in Allende and evidence for ^{26}Al. *Geophys. Res. Lett.* 3:109–112.
Lipin, B. R. 1978. The system Mg_2SiO_4-Fe_2SiO_4-$CaAl_2Si_2O_8$-SiO_2 and the origin of Fra Mauro basalts. *Amer. Mineral.* 63:350–364.
Longhi, J., and Pan, V. 1987. Olivine/low-Ca pyroxene liquidus relations and their bearing on eucrite petrogenesis. *Lunar Planet. Sci.* XVIII:570–571 (abstract).
Longhi, J., and Pan, V. 1988. A reconnaissance study of phase boundaries in low-alkali basaltic liquids. *J. Petrol.,* in press.
Ma, M.-S., Murali, A. V., and Schmitt, R. A. 1977. Genesis of the Angra dos Reis and other achondrite meteorites. *Earth Planet. Sci. Lett.* 35:331–346.
Malvin, D. J., Jones, J. H., and Drake, M. J. 1986. Experimental investigations of trace element fractionation in iron meteorites. III: Elemental partitioning in the system Fe-Ni-S-P. *Geochim. Cosmochim. Acta* 50:1221–1231.
Mason, B. 1962. *Meteorites* (New York: Wiley), pp. 116–119.
McCarthy, T. S., Ahrens, L. K., and Erlank, A. J. 1972. Further evidence in support of the mixing model for howardite origin. *Earth Planet. Sci. Lett.* 15:86–93.
Mercer-Smith, J. A., Cameron, A. G. W., and Epstein, R. I. 1984. On the formation of stars from disk accretion. *Astrophys. J.* 279:363–366.
Minear, J. W., Clow, G., and Fletcher, C. R. 1979. Thermal models of asteroids. *Lunar Planet. Sci.* X:842–843 (abstract).
Mittlefehldt, D. W. 1979. The nature of asteroidal differentiation processes: Implications for primordial heat sources. *Proc. Lunar Planet. Sci. Conf.* 10:1975–1993.
Mittlefehldt, D. W. 1987. Volatile degassing of basaltic achondrite parent bodies: Evidence from alkali elements and phosphorus. *Geochim. Cosmochim. Acta* 51:267–278.
Miyamoto, M., Fujii, N., and Takeda, H. 1981. Ordinary chondrite parent body: An internal heating model. *Proc. Lunar Planet. Sci.* 12:1145–1152.
Morse, S. A. 1980. *Basalts and Phase Diagrams* (New York: Springer-Verlag), pp. 328–331.
Narayan, C., and Goldstein, J. I. 1982. A dendritic solidification model to explain Ge-Ni variations in iron meteorite chemical groups. *Geochim. Cosmochim. Acta* 46:259–268.
Narayan, C., and Goldstein, J. I. 1985. A major revision of iron meteorite cooling rates—An experimental study of the growth of the Widmanstätten pattern. *Geochim. Cosmochim. Acta* 49:397–410.
Newsom, H. E. 1985. Molybdenum in eucrites: Evidence for a metal core in the eucrite parent body. *Proc. Lunar. Planet. Sci. Conf.* 15, *J. Geophys. Res. Suppl.* 90:C613–C617.
Newsom, H. E., and Drake, M. J. 1979. The origin of metal clasts in the Bencubbin meteoritic breccia. *Geochim. Cosmochim. Acta* 43:689–707.
Newsom, H. E., and Drake, M. J. 1982. The metal content of the eucrite parent body: Constraints from the partitioning behavior of tungsten. *Geochim. Cosmochim. Acta* 46:2483–2489.
Newsom, H. E., and Drake, M. J. 1983. Experimental investigation of the partitioning of phosphorus between metal and silicate phases: Implications for the Earth, Moon and eucrite parent body. *Geochim. Cosmochim. Acta* 47:93–100.
Newsom, H. E., and Drake, M. J. 1987. Formation of the Moon and terrestrial planets: Constraints from V, Cr, and Mn abundances in planetary mantles and from new partitioning experiments. *Lunar Planet. Sci.* XVIII:716–717 (abstract).
Newsom, H. E., Keil, K., and Scott, E. R. D. 1986. Dark clasts with variable REE contents in

the Khor Temiki aubrite: Origin by impact blackening of heterogeneous target material. *Meteoritics* 21:469–470.
Palme, H., and Rammensee, W. 1981. The significance of W in planetary differentiation processes: Evidence from new data on eucrites. *Proc. Lunar Planet. Sci.* 12:949–964.
Palme, H., Baddenhausen, H., Blum, K., Cendales, M., Dreibus, G., Hofmeister, H., Kruse, H., Palme, C., Spettel, B., Vilcsek, E., and Wänke, H. 1978. New data on lunar samples and achondrites and a comparison of the least fractionated samples from the earth, moon and eucrite parent body. *Proc. Lunar Planet. Sci. Conf.* 9:25–57.
Palme, H., Schultz, L., Spettel, B., Weber, H. W., Wänke, H., Christophe Michel-Lévy, M., and Lorin, J. C. 1981. The Acapulco meteorite: Chemistry, mineralogy and irradiation effects. *Geochim. Cosmochim. Acta* 45:727–752.
Patterson, C. W. 1987. Three-body resonance trapping and the asteroid belt. *Lunar Planet. Sci.* XVIII:766–767 (abstract).
Prinz, M., Keil, K., Hlava, P. D., Berkley, J. L., Gomes, C. B., and Curvello, W. S. 1977. Studies of Brazilian meteorites, III. Origin and history of the Angra dos Reis achondrite. *Earth Planet. Sci. Lett.* 35:317–330.
Prinz, M., Nehru, C. E., Delaney, J. S., and Weisberg, M. 1983. Silicates in IAB and IIICD irons, winonaites, lodranites and Brachina: A primitive and modified-primitive group. *Lunar Planet Sci.* XIV:616–617 (abstract).
Prombo, C. A., and Clayton, R. N. 1985. A striking nitrogen isotope anomaly in the Bencubbin and Weatherford meteorites. *Science* 230:935–937.
Rammensee, W. 1978. Verteilungsgleichgewichte von Spurenelementen zwischen Metallen und Silikaten. Ph.D. thesis, Mainz Univ., Federal Republic of Germany.
Rammensee, W., and Wänke, H. 1977. On the partition coefficient of tungsten between metal and silicate, and its bearing on the origin of the Moon. *Proc. Lunar Sci. Conf.* 8:399–409.
Reid, A. M., and Bunch, T. E. 1975. The nakhlites part II: Where, when and how. *Meteoritics* 10:317–324.
Saikumar, V., and Goldstein, J. I. 1988. An evaluation of the methods to determine the cooling rates of iron meteorites. *Geochim. Cosmochim. Acta,* in press.
Schmitt, W. 1984. Experimentelle Bestimmung von Metall/Sulfid/Silikat-Verteilungskoeffizienten Geochemisch Relevanter Spurenelemente. Ph.D. thesis, Mainz Univ., Federal Republic of Germany.
Scott, E. R. D. 1979. Origin of anomalous iron meteorites. *Mineral. Mag.* 43:415–421.
Scott, E. R. D., and Wasson, J. T. 1975. Classification and properties of iron meteorites. *Rev. Geophys. Space Phys.* 13:527–546.
Sellamuthu, R., and Goldstein, J. I. 1985. Analysis of segregation trends observed in iron meteorites using measured distribution coefficients. *Proc. Lunar Planet. Sci. Conf.* 15, *J. Geophys. Res. Suppl.* 90:C677–C688.
Smith, M. R. 1982. A chemical and petrologic study of igneous lithic clasts from the Kapoeta howardite. Ph.D. thesis, Oregon State Univ.
Sonett, C. P., and Reynolds, R. T. 1979. Primordial heating of asteroidal parent bodies. In *Asteroids,* ed. T. Gehrels (Tucson: Univ. of Arizona Press), pp. 822–848.
Stolper, E. 1977. Experimental petrology of eucritic meteorites. *Geochim. Cosmochim. Acta* 41:587–611.
Takeda, H. 1979. A layered-crust model of a howardite parent body. *Icarus* 40:455–470.
Takeda, H. 1987. Mineralogy of Antarctic ureilites and a working hypothesis for their origin and evolution. *Earth Planet. Sci. Lett.* 81:358–370.
Taylor, G. J., Keil, K., Berkley, J. L., Lange, D. E., Fodor, R. V., and Fruland, R. M. 1979. The Shaw meteorite: History of a chondrite consisting of impact-melted and metamorphic lithologies. *Geochim. Cosmochim. Acta* 43:323–337.
Tholen, D. J. 1984. Asteroid taxonomy from cluster analysis of photometry. Ph.D. thesis, Univ. of Arizona.
Walker, D. 1983. Lunar and terrestrial crust formation. *Proc. Lunar Planet. Sci. Conf.* 14, *J. Geophys. Res.* 88:B17–B25.
Wänke, H., Baddenhausen, H., Blum, K., Cendales, M., Dreibus, G., Hofmeister, H., Kruse, H., Jagoutz, E., Palme, C., Spettel, B., Thacker, R., and Vilcsek, E. 1977. On the chemistry of lunar samples and achondrites. Primary matter in the lunar highlands a re-evaluation. *Proc. Lunar Sci. Conf.* 8:2191–2213.

Warren, P. H. 1985. Origin of howardites, diogenites and eucrites: A mass balance constraint. *Geochim. Cosmochim. Acta* 49:577–586.

Warren, P. H. and Jerde, E. A. 1987. Composition and origin of Nuevo Laredo Trend eucrites. *Geochim. Cosmochim. Acta* 51:713–725.

Warren, P. H., and Wasson, J. T. 1979. Effects of pressure on the crystallization of a "chondritic" magma ocean and implications for the bulk composition of the moon. *Proc. Lunar Planet. Sci. Conf.* 10:2051–2083.

Wasson, J. T. 1985. *Meteorites: Their Record of Early Solar-System History* (New York: Freeman), pp. 102–103.

Wasson, J. T., and Wetherill, G. W. 1979. Dynamical, chemical and isotopic evidence regarding the formation locations of asteroids and meteorites. In *Asteroids,* ed. T. Gehrels (Tucson: Univ. of Arizona Press), pp. 926–974.

Wasson, J. T., Willis, J., Wai, C. M., and Kracher, A. 1980. Origin of iron meteorite groups IAB and IIICD. *Z. Naturforsch.* 35a:781–795.

Weckwerth, G., Spettel, B., and Wänke, H. 1983. Phosphorus in the mantle of planetary bodies. *Terra Cognita* 3:79–80.

Wolf, R., Ebihara, M., Richter, G. R., and Anders, E. 1983. Aubrites and diogenites: Trace element clues to their origin. *Geochim. Cosmochim. Acta* 47:2257–2270.

3.3 THERMAL METAMORPHISM

HARRY Y. McSWEEN, JR.
University of Tennessee

DEREK W. G. SEARS
University of Arkansas

and

ROBERT T. DODD
State University of New York, at Stony Brook

Most chondrites have experienced thermal metamorphism, resulting in changes in texture, mineralogy and possibly chemical composition. The physical conditions for metamorphism range from approximately 400 to 1000° C at low lithostatic pressure. Metamorphism may have resulted from decay of short-lived radionuclides, electromagnetic induction or accretion of hot materials. Several thermal models for chondrite parent bodies have been proposed. The least metamorphosed type 3 chondrites probably carry the most information about the early solar system, but even these have been affected to some degree by thermal processing.

3.3.1. INTRODUCTION

The Van Schmus and Wood (1967) classification scheme for chondrites, already presented in Chapter 1.1 divides meteorites by chemical group and petrographic type (Van Schmus and Wood actually used the term "petrologic type"; "petrographic type" is substituted here to conform with its usage in this book). Most meteoriticists now accept that the textural and mineralogical variations summarized by petrographic types for the most part reflect the effects of thermal metamorphism, that is, the adjustment of rocks to a new set of physical conditions. However, the idea that chondrites are metamorphosed has been contested by some workers since it was first proposed by Merrill (1921). As we shall see, there may be some additional complexities that affect the metamorphic interpretation. Before chondrites can be used to infer conditions in the early solar system, the kinds of changes caused by thermal processing must be thoroughly understood.

3.3.2. TEXTURAL EVIDENCE FOR METAMORPHISM

Recrystallization during metamorphism resulted in progressive blurring of the distinctive chondritic texture, that is, integration of chondrules with matrix, as illustrated in Fig. 3.3.1. This textural change is the most obvious but certainly not the only modification that these meteorites experienced. Increasing metamorphism is also indicated by devitrification of glass in chondrules, growth of secondary feldspar, increase in the proportion of orthopyroxene relative to low-calcium clinopyroxene, and replacement of fine-grained opaque matrix by more transparent recrystallized matrix. Most metamorphic processes appear to have been nearly isochemical in terms of the major-element composition of the bulk sample. However, the compositions of major phases like olivine and pyroxene changed systematically. In general, olivines and pyroxenes tended to become more iron-rich as they underwent chemical exchange with matrix during metamorphism, and their compositional variabilities decreased. Olivines achieved homogeneity before pyroxenes because Fe-Mg diffusion is slower in the latter. Statistical measures of compositional variability for olivine and pyroxene (standard deviation or percent mean deviation) are commonly used to rank chondrites according to their metamorphic grades. This tendency toward homogenization of silicate phases is the basis for the terms "unequilibrated" and "equilibrated", used to describe petrographic types 3 and types 4 through 7, respectively.

Many chondrites are breccias, commonly containing lithic clasts of more highly equilibrated material within hosts of the same chemical group but lower petrographic type (Binns 1968; see also Chapter 3.5). This association, and the lack of mixing between groups, indicate that chondrites of various petrographic types formed within the same parent bodies and were later mixed together by impact processes. Dodd et al. (1967) and Scott et al. (1985) recognized individual chondrules with aberrant olivine compositions in some chondrites without obvious clasts; from this observation, they argued that most chondrites may be unrecognized breccias and that the Van Schmus-Wood classification system does not identify samples with more complex thermal histories. This is probably an extreme viewpoint, although recognition of events in the thermal history is certainly difficult in many cases.

Despite the problem of brecciation, it is possible to assess how wide-

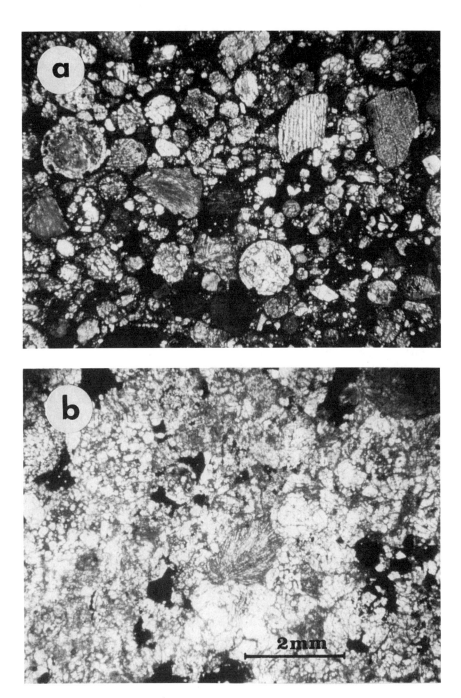

Fig. 3.3.1. Photomicrographs illustrating typical textures of (a) type 3, Tieschitz and (b) type 6, Dhurmsala ordinary chondrites. Recrystallization of type 6 chondrites has blurred the outlines of chondrules. The scale for both photographs is as shown in (b).

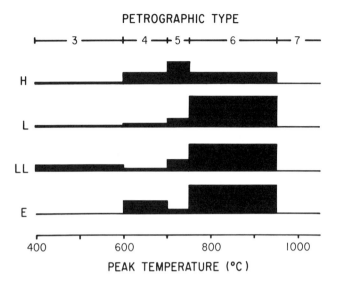

Fig. 3.3.2. The relative proportions of different petrographic types of the various chondrite groups illustrated by histograms of well-classified non-Antarctic chondrites in each category. Classification data are from Motylewski (1978); a few type 7 ordinary chondrites and E3 chondrites are now recognized, but were not included in this compilation. The width of each petrographic type is based on its estimated temperature range from Dodd (1981).

spread the effects of thermal metamorphism are among chondrites of the various chemical groups. Although some recent studies identify the range of petrographic types in chondritic breccias, earlier classifications of chondrites were based on the petrographic types of the most abundant materials. Nevertheless, these data provide at least a rough index of the proportions of unequilibrated and equilibrated chondrites. Petrographic type assignments for well-classified chondrites are summarized in Fig. 3.3.2; Antarctic chondrites are omitted from this tabulation because of uncertainties in pairings. It is clear that equilibrated samples dominate most chondrite classes. For these data to have any significance for parent-body thermal models, we must assume that falls have sampled chondrite parent bodies randomly, yielding the various petrographic types in their correct volumetric proportions, although it is uncertain that this was indeed the case.

3.3.3. TEMPERATURE OF METAMORPHISM

The physical conditions under which chondrite metamorphism occurred are difficult to quantify. Although the metamorphic trend for a group of chondrites forms a continuum, the divisions between petrographic types are arbitrarily based on certain critical petrographic observations, summarized in Chapter 1.1; see Table 1.1.4. These divisions therefore do not necessarily correspond to equal temperature intervals. Chondrite metamorphic temperatures can be bracketed from observations of metal and troilite. The absence of eutectic melting in type 6 chondrites suggests temperatures of less than 950 to 1000° C for such meteorites. Depletion of Fe and S from some type 7 chondrites, from which it has been inferred that a melt fraction was drained off, suggests temperatures at or above this range. Compositions of metal-grain interiors in the least metamorphosed type 3 chondrites suggest maximum temperatures near 400° C. The two-pyroxene geothermometer, based on the temperature-dependent partitioning of Fe and Mg between orthopyroxene and diopside, provides some of the best available estimates of metamorphic temperatures between these limits, but its application is limited by the absence of metamorphic diopside from most meteorites of low petrographic type. The partitioning of Ca between coexisting pyroxenes also varies with temperature. Analyses of equilibrated chondrites using both these thermometers have been summarized by Olsen and Bunch (1984). Phase transformations in feldspar, which can be monitored by certain characteristics of thermoluminescent emission, provide another indication of paleotemperatures (Guimon et al. 1985). While some type 3 ordinary chondrites contain feldspar predominantly in the high-temperature form, others contain predominantly the low-temperature form. Guimon et al. suggest that this boundary within the type 3 group (to which they assign a value of 3.5) coincides with the order/disorder transformation temperature (500–600° C). The distribution of ^{18}O between coexisting minerals is a function of, and has also been used to constrain estimates of temperature (Onuma et al. 1972). Other mineral-exchange geothermometers, specific to one chondrite group or another, have been formulated with mixed success. The approximate ranges for peak metamorphic temperatures for chondrites of types 3 to 7 are summarized in Fig. 3.3.2. Dodd (1981) provides detailed information on the application of geothermometers to chondrites and their uncertainties.

3.3.4. PRESSURES OF METAMORPHISM

Lithostatic pressures during thermal metamorphism, inferred from mineralogy, were invariably low. Although some chondrites show deformation due to stress (see, e.g., Cain et al. 1986), such effects do not appear to correlate with petrographic type (Dodd 1965). Pressures even at the centers of asteroids are less than 2 kbar, and very limited pressure-sensitive chemical substitutions in chondritic pyroxenes are consistent with this interpretation. Pyroxene components measured by Heyse (1978) are very close to analytical uncertainties and more work of this kind is needed. Heyse argued nonetheless that systematic variations in such substitutions with increasing temperature in LL-group chondrites indicate that metamorphic intensity increased with depth in the parent body. In this model, petrographic type should be inversely correlated with cooling rate, because deeply buried rocks would cool more slowly than those near the surface. Meteorite cooling rates can be inferred from the retention of fission tracks in different minerals (see, e.g., Pellas and Storzer 1977) or the zoning profiles of Ni in metal grains (Wood 1967). In the case of LL-group chondrites, cooling rates based on fission tracks agree with Heyse's conclusion, but metallographic cooling rates do not. From a synthesis of cooling rates based on metallographic data, Scott and Rajan (1981) suggested that metamorphic intensities are not related in any straightforward way to burial depths within parent bodies. The issue is clouded further by the problem that cooling rates determined by the metallographic and fission track techniques do not agree with those that appear to be required by K-Ar and ^{40}Ar-^{39}Ar isotopic dating (Wood 1979). The relationship, if any, between temperature and pressure in chondrite parent bodies remains an unsolved problem, though an important one because of its implications for asteroid thermal models.

3.3.5. LACK OF METAMORPHIC FLUIDS

The persistence of many of the same minerals in chondrites throughout the metamorphic sequence is very different from the terrestrial metamorphic experience and can be explained by the virtual absence of fluids (H_2O and CO_2). This in turn may explain why type 6 chondrites contain some relict textures. Equilibrated chondrites contain no hydrous or carbonated phases, and earlier accounts of fluid inclusions contained in them have now been discounted (Rudnick et al. 1985). The apparent absence of fluids from ordinary and enstatite chondrites is consistent with their very low measured or calculated oxygen fugacities. Carbonaceous chondrites might be expected to have contained a fluid phase during metamorphism because some of them contain hydrated phyllosilicates and carbonates at low grade, but the properties of type 4 carbonaceous chondrites are very similar to those of the anhydrous classes (Scott and Taylor 1985). Fluids undoubtedly played a role in altering some chondrites at very low temperatures (see Chapter 3.4), but they must have been largely driven off at or below the onset of thermal metamorphism.

3.3.6. METAMORPHIC HEAT SOURCES

The source of heat for chondrite parent body metamorphism is still controversial. The most commonly held view is that initially cool material accreted to form parent bodies and was subsequently heated. Peak temperatures were obtained within a few tens of millions of years. Similar Rb-Sr ages for equilibrated and unequilibrated chondrites constrain metamorphism to have occurred soon after chondrite formation, because these ages are reset by thermal effects. Decay of short-lived radionuclides such as ^{26}Al (Lee et al. 1976) is a plausible heating mechanism, although external heat sources such as electromagnetic induction by a massive solar wind (Sonett et al. 1970) have also been suggested. Note that a body heated internally by radioactivity should show an increase in metamorphic intensity with depth, whereas an object externally heated by induction may show the opposite trend. An alternative view to these heating mechanisms for asteroidal parent bodies after they are assembled is that small, rapidly accreted clumps of chondritic material were metamorphosed by residual heat from condensation; Larimer and Anders (1967) suggested that accretion at varying temperatures would explain the different petrographic types. This kind of model can also be adapted to larger parent bodies. For example, Heyse (1978) favored continuous accretion of hot material onto a progressively cooling body of asteroidal size. One test that might distinguish among some of these models would be to establish whether oxygen fugacity varied systematically with petrographic type. If metamorphism was caused by accretion during condensation, more oxidizing conditions would have obtained as the temperature fell. Another test involves determining the relationship, if any, between temperature and pressure during metamorphism (Heyse 1978).

3.3.7. PARENT-BODY THERMAL STRUCTURES

Thermal models for several chondrite parent bodies have now been formulated based on information concerning heat sources, peak metamorphic temperatures, and the relative proportions of various petrographic types. From the discussion above, it should be clear that these models are highly speculative because of gross uncertainties in all of these input parameters. The orthodox view of chondrite parent bodies is the "onion-shell" model, in which internal heating has produced a succession of concentric layers of different petrographic type grading from equilibrated material in the interior to unequilibrated at or near the surface. Miyamoto et al. (1981) constructed onion-shell models for ordinary chondrite parent bodies, one of which is illustrated in Fig. 3.3.3a. A feature of this model is that the volume fraction of each petrographic type matches very closely the distribution of petrographic types among classified falls; this, of course, assumes random sampling of chondrite parent bodies.

In order to explain the absence of a correlation between petrographic type and metallographic cooling rate, Scott and Rajan (1981) formulated an entirely different kind of parent body model, called the "rubble-pile". They hypothesized that peak metamorphic temperatures were reached during, rather than after, accretion, so that metamorphism occurred in small planetesimals. These were subsequently assembled into composite asteroid-sized bodies in which the metallographic cooling rates were controlled by burial depth. A sketch of a rubble-pile parent body is given in Fig. 3.3.3b. Wood (1979) attempted to circumvent the problem of lack of correlation between the metallographic cooling rate and that seemingly required by ^{40}Ar-based age

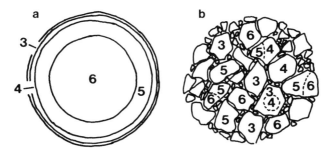

Fig. 3.3.3. Thermal models for ordinary chondrite parent bodies: (a) an onion-shell model for the H chondrite parent body calculated by Miyamoto et al. (1981), with concentric shells of petrographic types; (b) sketch of a rubble-pile parent body advocated by Scott and Rajan (1981).

dating by invoking parent bodies coated with thermally insulating particulate matter. In this model, cooling to the temperature for Ar isotopic diffusion to cease occurred rapidly in small planetesimals; these were then accreted into larger, insulated bodies for further cooling through the temperature at which metallographic cooling rates were set. Grimm (1985) concluded that rubble-pile models are only applicable to highly insulating, ^{26}Al-rich planetesimals that can accrete without being shattered by impact. He advocated a hybrid model in which onion-shell parent bodies were collisionally fragmented during metamorphism and then gravitationally reassembled on a short time scale.

3.3.8. METAMORPHIC REDISTRIBUTION OF VOLATILES

Recognition of the effects of thermal metamorphism in chondrites has some important implications for our ability to decipher the information carried in these meteorites. As one example, the concentrations of trace elements with different volatilities can, in principle, provide important constraints on the condensation process in the early solar system; however, many of these elements are thermally mobile so that their abundances may have been changed during metamorphism (see Chapter 7.6). Gas permeabilities of some chondrite classes are high enough that volatile elements within them may have exhibited open-system behavior (Sugiura et al. 1984). Inverse correlations between abundances of the noble gases and petrographic type are now firmly established. Larimer and Anders (1967) argued that the volatile-element distributions were established during condensation and are unrelated to metamorphism. However, Dreibus and Wänke (1980) suggested that such elements may have been mobilized from parent-body interiors and ultimately deposited in the cooler exterior regions.

Differences in the carbon contents and oxygen isotopic compositions of unequilibrated and equilibrated ordinary chondrites have been explained by oxidation and subsequent loss of C in the form of CO during metamorphism (Clayton et al. 1981). Calculations suggest that it may also be possible to transport other elements like bismuth, thallium and indium, known to be inversely correlated with petrographic type (Sugiura et al. 1985), but uncertainties in the thermodynamic data used in such calculations limit their usefulness. Ikramuddin et al. (1977) attempted to distinguish between volatile (in the sense of condensation) and mobile (in the sense of metamorphism) trace elements experimentally. These kinds of behavior are different, in that volatility is calculated from thermodynamics and mobility is a kinetic process that is measured empirically. The order of some element mobilities suggested by heating experiments (summarized more fully in Chapter 7.6) is Te < Bi < In < Zn < Tl < Cd; the comparable volatility sequence is Te < Zn < Cd < Bi < Tl < In (Chapter 7.6). Apart from Cd, the most mobile (but not most volatile) elements appear to be correlated with petrographic type (Lipschutz et al. 1983), and temperatures in the uppermost parent body regions may simply have been too high for Cd redeposition. Supercosmic abundances of some highly mobile elements in type 3 chondrites (Lipschutz et al. 1983) may also be evidence that at least some elements have been redistributed during parent-body heating. No definitive conclusions can be made concerning the mobilization of elements during metamorphism, but it remains an intriguing possibility.

3.3.9. METAMORPHISM IN UNEQUILIBRATED CHONDRITES

Because of potential changes in the petrography, chemical composition and isotopic systems of chondrites during thermal metamorphism, it is of obvious importance to identify and concentrate our efforts on the least altered meteorites. However, even type 3 chondrites have experienced various degrees of mild thermal processing. Dodd et al. (1967) used the dispersion of olivine compositions in unequilibrated ordinary chondrites as a criterion for metamorphic ranking of these meteorites. Huss et al. (1981) proposed subdivisions for these chondrites based on changes in matrix, and Afiattalab and Wasson (1980) offered a subclassification based on metal compositions. McSween (1977) identified a metamorphic sequence within one class of type 3 carbonaceous chondrites (CO3), based on systematic changes in metal and matrix compositions and subtle textural blurring. At present, the most useful scale for assessing the degree of metamorphism within type 3 ordinary chondrites is based on changes in thermoluminescence (TL) discovered by Sears et al. (1980). TL sensitivity for ordinary chondrites increases markedly with petrographic type, and type 3 chondrites exhibit an especially large range, as shown in Fig. 3.3.4. This behavior reflects the formation of secondary feldspar, which acts as a TL phosphor, during metamorphism at modest temperatures. Based on the excellent correlations between TL sensitivity and olivine compositions, metal compositions, degree of matrix recrystallization, and

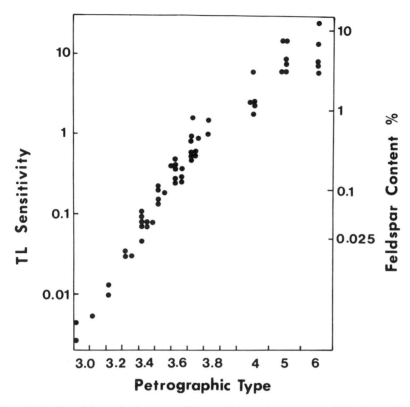

Fig. 3.3.4. Plot of thermoluminescence (TL) sensitivity (relative to that of Dhajala = 1) vs petrographic type for ordinary chondrites, after Sears et al. (1980). The large TL range exhibited by type 3 chondrites provides a means of subdividing them according to their thermal histories. TL sensitivity is a function of devitrification of glass to form feldspar, as illustrated by the right-hand axis of this diagram.

carbon and noble gas contents, Sears et al. (1980) assigned type 3 ordinary chondrites to ranks of 3.0 to 3.9 with increasing metamorphism. Complications in TL patterns, introduced by aqueous alteration or shock, make the classification of some chondrites ambiguous. The TL sensitivity is more complex in carbonaceous chondrites than in ordinary chondrites, but Keck and Sears (1985) also found systematic changes within the CO3 group that correlate with McSween's (1977) proposed sequence. Enstatite chondrites of petrographic type 3 are so uncommon that an analogous study has not yet been done for this group. It would be desirable to be able to subdivide equilibrated chondrites into finer subdivisions as well (using, for example, feldspar grain size), but little work of this type has been done.

3.3.10. SUMMARY

The metamorphic process in chondrite parent bodies is not well understood, but it is of crucial importance that its effects be recognized and quantified. Important problems yet to be solved include determining whether chondrite metamorphism occurred in a closed or open system, and quantifying pressure effects and possible correlations with temperature. The present classification system based on discrete petrographic types is serviceable, but it should be replaced by one that recognizes the gradational nature of the metamorphic process, such as a continuous scale based on maximum temperature attained. The apparent lack of correlation between cooling rates and petrographic type must be investigated further, and its implications understood. Finally, more refined thermal models for asteroids are needed. Without an appreciation of chondrite metamorphism, the usefulness of these meteorites as probes of early solar system processes and events is very limited.

REFERENCES

Afiattalab, F., and Wasson, J. T. 1980. Composition of the metal phases in ordinary chondrites: Implications regarding classification and metamorphism. *Geochim. Cosmochim. Acta* 44:431–446.

Binns, R. A. 1968. Cognate xenoliths in chondritic meteorites: Examples in Mezö-Madaras and Ghubara. *Geochim. Cosmochim. Acta* 32:299–317.

Cain, P.M., McSween, H. Y., Jr., and Woodward, N. B. 1986. Structural deformation of the Leoville chondrite. *Earth Planet. Sci. Lett.* 77:165–175.

Clayton, R. N., Mayeda, T. K., Gooding, J. L., Keil, K., and Olsen, E. J. 1981. Redox processes in chondrules and chondrites. *Lunar Planet. Sci.* XII:154–156 (abstract).

Dodd, R. T. 1965. Preferred orientation of chondrules in chondrites. *Icarus* 4:308–316.

Dodd, R. T. 1969. Metamorphism of the ordinary chondrites: A review. *Geochim. Cosmochim. Acta* 33:161–203.

Dodd, R. T. 1981. *Meteorites: A Petrologic-Chemical Synthesis* (London: Cambridge Univ. Press).

Dodd, R. T., Van Schmus, W. R., and Koffman, D. M. 1967. A survey of the unequilibrated ordinary chondrites. *Geochim. Cosmochim. Acta* 31:921–951.

Dreibus, G., and Wänke, H. 1980. On the origin of the excess volatile trace elements in the dark portion of gas-rich chondrites. *Meteoritics* 15:284–285 (abstract).

Grimm, R. E. 1985. Penecontemporaneous metamorphism, fragmentation, and reassembly of ordinary chondrite parent bodies. *J. Geophys. Res.* 90:2022–2028.

Guimon, R. K., Keck, B. D., Weeks, K. S., DeHart, J., and Sears, D. W. G. 1985. Chemical and physical studies of type 3 chondrites—IV: Annealing studies of a type 3.4 ordinary chondrite and the metamorphic history of meteorites. *Geochim. Cosmochim. Acta* 49:1515–1524.

Heyse, J. V. 1978. The metamorphic history of LL-group ordinary chondrites. *Earth Planet. Sci. Lett.* 40:365–381.

Huss, G. R., Keil, K., and Taylor, G. J. 1981. The matrices of unequilibrated ordinary chondrites: Implications for the origin and history of chondrites. *Geochim. Cosmochim. Acta* 45:33–51.

Ikramuddin, M., Matza, S. D., and Lipschutz, M. E. 1977. Thermal metamorphism of primitive meteorites—V. Ten trace elements in Tieschitz H3 chondrite heated at 400–1000°C. *Geochim. Cosmochim. Acta* 41:1247–1256.

Keck, B. D., and Sears, D. W. G. 1985. Thermoluminescence and the metamorphic history of the CO chondrites. *Lunar Planet. Sci.* XVII:412–413 (abstract).

Larimer, J. W., and Anders, E. 1967. Chemical fractionations in meteorites—II. Abundance patterns and their interpretation. *Geochim. Cosmochim. Acta* 31:1239–1270.
Lee, T., Papanastassiou, D. A., and Wasserburg, G. J. 1976. Demonstration of excess ^{26}Mg in Allende and evidence for ^{26}Al. *Geophys. Res. Lett.* 3:109–112.
Lipschutz, M. E., Biswas, S., and McSween, H. Y., Jr. 1983. Chemical characteristics and origin of H chondrite regolith breccias. *Geochim. Cosmochim. Acta* 47:169–179.
McSween, H. Y., Jr. 1977. Carbonaceous chondrites of the Ornans type: A metamorphic sequence. *Geochim. Cosmochim. Acta* 41:477–491.
Merrill, G. P. 1921. On metamorphism in meteorites. *Geol. Soc. Amer. Bull.* 32:395–416.
Miyamoto, M., Fujii, N., and Takeda, H. 1981. Ordinary chondrite parent body: An internal heating model. *Lunar Planet. Sci. Conf.* 12:1145–1152.
Motylewski, K. 1978. *The Revised Cambridge Chondrite Compendium* (Cambridge: Smithsonian Astrophys. Obs.).
Olsen, E. J., and Bunch, T. E. 1984. Equilibration temperatures of the ordinary chondrites: A new evaluation. *Geochim. Cosmochim. Acta* 48:1363–1365.
Onuma, N., Clayton, R. N., and Mayeda, T. K. 1972. Oxygen isotope temperatures of "equilibrated" ordinary chondrites. *Geochim. Cosmochim. Acta* 36:157–168.
Pellas, P., and Storzer, D. 1979. Differences in the early cooling histories of the chondritic asteroids. *Meteoritics* 14:513–515 (abstract).
Rudnick, R. L., Ashwal, L. D., Henry, D. J., Gibson, E. K., Roedder, E., Belkin, H. E., and Colucci, M. T. 1985. Fluid inclusions in stony meteorites—a cautionary note. *Proc. Lunar Planet. Sci. Conf.* 15, *J. Geophys. Res. Suppl.* 90:C669–C675.
Scott, E. R. D., and Rajan, R. S. 1981. Metallic minerals, thermal histories, and parent bodies of some xenolithic, ordinary chondrites. *Geochim. Cosmochim. Acta* 45:53–67.
Scott, E. R. D., and Taylor, G. J. 1985. Petrology of types 4–6 carbonaceous chondrites. *Proc. Lunar Planet. Sci. Conf.* 15, *J. Geophys. Res. Suppl.* 90:C699–C709.
Scott, E. R. D., Lusby, D., and Keil, K. 1985. Ubiquitous brecciation after metamorphism in equilibrated ordinary chondrites. *Proc. Lunar Planet. Sci. Conf.* 16, *J. Geophys. Res. Suppl.* 90:D137–D148.
Sears, D. W., Grossman, J. N., Melcher, C. L., Ross, L. M., and Mills, A. A. 1980. Measuring metamorphic history of unequilibrated ordinary chondrites. *Nature* 287:791–795.
Sonett, C. P., Colburn, D. S., Schwartz, K., and Keil, K. 1970. The melting of asteroidal-sized bodies by unipolar dynamo induction from a primordial T Tauri sun. *Astrophys. Space Sci.* 7:446–488.
Sugiura, N., Brar, N. S., and Strangway, D. W. 1984. Degassing of meteorite parent bodies. *Proc. Lunar Planet. Sci. Conf.* 14, *J. Geophys. Res. Suppl.* 89:B641–B644.
Sugiura, N., Arkani-Hamed, J., and Strangway, D. W. 1985. On the possible transport of volatile trace elements in meteorite parent bodies. *Lunar Planet. Sci.* XVII:843–844.
Van Schmus, W. R., and Wood, J. A. 1967. A chemical-petrologic classification for the chondritic meteorites. *Geochim. Cosmochim. Acta* 31:747–765.
Wood, J. A. 1967. Chondrites: Their metallic minerals, thermal histories, and parent bodies. *Icarus* 6:1–49.
Wood, J. A. 1979. Review of the metallographic cooling rates of meteorites and a new model for the planetesimals in which they formed. In *Asteroids*, ed. T. Gehrels (Tucson: Univ. of Arizona Press), pp. 849–891.

3.4. AQUEOUS ALTERATION

MICHAEL ZOLENSKY
NASA Johnson Space Center

and

HARRY Y. McSWEEN, JR.
University of Tennessee

In the two preceding chapters, we have considered the effects that postaccretional heating can have on the record of nebular processes contained in meteorites. Heat was not the only agent of alteration, however; of even greater importance in the evolution of some of the most compositionally primitive meteorites was the chemical reactivity of water at relatively low temperatures. Our knowledge of the nature of aqueous-alteration products in meteorites has improved to the point where we can begin to model the course of such alteration processes. These models are still at a primitive level, but even at that level it is clear that in several cases mineralization on the parent body has replaced the preaccretionary lithology, though apparently without significantly perturbing the bulk chemistry. Isotopic studies show that at least some of this alteration took place very early in solar-system history.

3.4.1. INTRODUCTION

An assumption underlying much of this book is that we can recognize and study primitive materials in meteorites. This chapter explores the low-temperature chemical-alteration processes that in many cases have perturbed those earliest-formed samples. The primary reactions involved in chemical alteration are oxidation, hydrolysis and carbonation; however, in this chapter we are principally concerned with the action of the second process, i.e., aqueous alteration. The unequaled ability of water to act as a solvent reflects its dipolar nature and strong molecular polarizability, and in acidic or alkaline solutions the capacity of H^+ or OH^- ions to form complexes with solute species. Since most minerals have a high degree of polarity in their chemical bonds, water is a common solvent effecting material transport. The origin and nature of aqueous activity in primitive extraterrestrial materials is a fascinating topic in its own right, but for the purposes of this chapter we are more concerned with the products of those reactions and their effect on the cosmochemical record. Aqueous reactions promote chemical redistribution, changes in petrologic relationships and isotopic fractionation. In addition, aqueous alteration can complicate thermoluminescence (TL) patterns obtained from chondrites (Sears et al. 1980), rendering uncertain the metamorphic grade of those materials, as revealed by TL. Thus, it is important to recognize the signs of aqueous alteration in primitive materials, so that allowance may be made for its effects.

The recognition and interpretation of the processes and products of aqueous alteration are largely based upon our understanding of terrestrial alteration processes (over a wide range of conditions and time scales). Thus, the study of extraterrestrial alteration processes relies heavily upon the use of terrestrial analogues. With this concept in mind, however, one must nonetheless understand that most terrestrial weathering has operated upon rocks of different initial composition from the presumed starting material of the chondrites. Thus, an important task is to understand how preterrestrial alteration processes and products may differ from proposed terrestrial analogues.

Aqueous reactions may occur under any *P-T* conditions where liquid water is stable. The unusually wide range of thermal stability for liquid water reflects the strength of H-bonding between water molecules. Solutes also affect the stability of liquid water. Solids are generally more soluble in liquid water than in water ice or vapor, and therefore tend to increase liquid-water stability, shifting the ice-liquid stability curve to lower temperatures and pressures, and the liquid-vapor curve to higher temperatures and pressures. Because gases are more soluble in water vapor than in liquid water, and in liquid water than in ice, gases in solution tend to shift both the liquid-vapor and ice-liquid curves to lower temperatures and pressures.

Water may also be stable below the equilibrium freezing point of 0° C. Ugolini and Anderson (1973) and Gooding (1984) have described the process of hydrocryogenic alteration, due to the action of "unfrozen" water. Thin (<100 Å) films of liquid pore water are stabilized, at temperatures down to at least $-20°$C, by the surface energy of solid materials. Gooding (1984) and Rietmeijer (1985) have applied this concept to alteration on a meteorite parent body, and shown that the greatest stability for unfrozen water is achieved adjacent to phyllosilicates. The amount of unfrozen water increases, for any solid, with increasing pressure (overburden).

Thus, aqueous alteration may occur over a wide range of temperatures and pressures. Our task is to define, as accurately as we can, the P-T-X-H_2O conditions attending aqueous alteration of extraterrestrial samples, so that we may discover how and where this alteration occurred, and how it has affected the meteoritic record of earlier events.

Our strategy here is first to describe the products of aqueous alteration observed within primitive extraterrestrial materials: carbonaceous and unequilibrated ordinary chondrites, and interplanetary dust particles (IDPs). (Further descriptions of meteorite matrices and IDPs are given in Chapters 10.2 and 11.1, respectively). We then discuss the theories that have been formulated to account for those aqueous-alteration parageneses and products. Such theories should attempt to explain the source(s) of the aqueous solutions as well as the physico-chemical conditions attending aqueous alteration. As we shall see, these endeavors have not yet been entirely successful, so that this is currently a fruitful area for research.

3.4.2. THE MINERALOGICAL PRODUCTS OF AQUEOUS ALTERATION

The minerals produced by aqueous alteration that are most commonly encountered within primitive extraterrestrial materials include phyllosilicates, tochilinites (coherently interstratified sulfide-hydroxide phases), sulfates, oxides, carbonates and hydroxides. Most of these alteration products are matrix phases, although alteration of larger components (e.g., chondrules, aggregates and inclusions) figures prominently in the earlier literature, prior to the application of fine-scale microstructural and microchemical techniques to meteorites.

There has been a great deal of confusion in the literature regarding the nomenclature and correct chemical formulae of minerals encountered within the matrices of altered meteorites. Thus, we present the currently accepted mineral formulae and other pertinent data for the phyllosilicates, hydroxides and tochilinites found in primitive extraterrestrial materials in Table 3.4.1, so that an element of commonality may be brought to these discussions. Formulae for additional minerals in these meteorites are given in Table 3.4.2. The information in these tables comes from Brindley and Brown (1981), Fleischer (1987), Zolensky (1987), and references cited therein.

All of the compositional formulae given in Table 3.4.1, and many in Table 3.4.2, are subject to chemical substitutions to varying degrees. For example, most of the phyllosilicates exhibit Mg-Fe isomorphous series, which are of particular interest to meteoriticists. In addition, many of these mineral species have overlapping chemical compositions. Thus, for most of the alteration species discussed here, gross chemical compositions are not sufficient to identify unambiguously the mineral species under consideration. It is also necessary to collect morphological and structural data (X-ray or

TABLE 3.4.1
Characteristics of Hydrated Minerals in Chondrites

Group Name	Mg-Bearing Members	Fe-Bearing Members	Morphology	Characteristic Diffraction Spacing(s), Å
Phyllosilicates				
Serpentine	Chrysotile $Mg_3Si_2O_5(OH)_4$ Lizardite $Mg_3Si_2O_5(OH)_4$ Antigorite $Mg_3Si_2O_5(OH)_4$	Cronstedtite $Fe_2^{+2}Fe^{+3}(SiFe^{+3})O_5(OH)_4$ Greenalite $(Fe^{+2},Fe^{+3})_{2-3}Si_2O_5(OH)_4$ Ferroan Antigorite $(Mg,Fe,Mn)_3(Si,Al)_2O_5(OH)_4$	Tubes, Fibers, Flat and Corrugated Sheets, Cones	7.0–7.3
	Berthierine $(Fe^{+2},Fe^{+3},Mg)_{2-3}(Si,Al)_2O_5(OH)_4$		Sheets and Fibers	7.0–7.1
	Amesite $Mg_2Al(SiAl)O_5(OH)_4$		Sheets and Fibers	14
Cholorite	Clinochlore $(Mg,Fe^{+2})_5Al(Si_3Al)O_{10}(OH)_8$	Chamosite $(Fe^{+2},Mg,Fe^{+3})_5Al(Si_3Al)O_{10}(OH,O)_8$	Sheets	14
Smectite	Montmorillonite $(Na,Ca)_{0.3}(Al,Mg)_2Si_4O_{10}(OH)_2 \cdot nH_2O$ Saponite $(Ca/2,Na)_{0.3}(Mg,Fe^{+2})_3(Si,Al)_4O_{10}(OH)_2 \cdot 4H_2O$ Sobokite $(K,Ca/2)_{0.3}(Mg_2Al)(Si_3Al)O_{10}(OH)_2 \cdot 5H_2O$	Nontronite $Na_{0.3}Fe_2^{+3}(Si,Al)_4O_{10}(OH)_2 \cdot nH_2O$	Sheets	14–15 (Fully Hydrated) ~10 (Fully Dehydrated)
Vermiculite	$(Mg,Fe^{+2},Al)_3(Al,Si)_4O_{10}(OH)_2 \cdot 4H_2O$		Sheets	14–15 (Fully Hydrated), ~9 (Fully Dehydrated)
Hydroxides				
	Brucite $Mg(OH)_2$	Amakinite $(Fe^{+2},Mg)(OH)_2$	Sheets and Filled Tubes	4.7–4.9
Tochilinites				
	Tochilinite $2[(Fe,Mg,Cu,Ni[])S] \cdot 1.57–1.85[(Mg,Fe,Ni,Al,Ca)(OH)_2]$ Haapalaite $4(Fe,Ni)S \cdot 3(Mg,Fe^{+2})(OH)_2$		Massive, Tubes, Fibers, Flat and Corrugated Sheets	10.7–11.3, 5.4
Pumpellyite	$Ca_2(Mg,Fe^{+2})Al_2(SiO_4)(Si_2O_7)(OH)_2 \cdot H_2O$		Fibers and Plates	9.5

TABLE 3.4.2
Compositions of Selected Minerals in CI and CM Chondrites

Sulfates	Carbonates
Epsomite $MgSO_4 \cdot 7H_2O$	Ferroan Magnesite $(Mg,Fe^{+2})CO_3$
Hexahydrite $MgSO_4 \cdot 6H_2O$	Calcite $CaCO_3$
Blodite $Na_2Mg(SO_4)_2 \cdot 4H_2O$	Dolomite $Ca(Mg,Fe^{+2},Mn)(CO_3)_2$
Gypsum $CaSO_4 \cdot 2H_2O$	Ankerite $Ca(Fe^{+2},Mg,Mn)(CO_3)_2$
	Kutnohorite $Ca(Mn,Mg,Fe^{+2})(CO_3)_2$
Sulfides	
	Oxides
Cubanite $CuFe_2S_3$	
Pentlandite $(Fe,Ni)_9S_8$	Maghemite gamma-Fe_2O_3
Troilite FeS	Magnetite $Fe^{+2}Fe_2^{+3}O_4$
Pyrrhotite $Fe_{1-x}S$ ($x = 0-0.17$)	Spinel $MgAl_2O_4$
	Chromite $Fe^{+2}Cr_2O_4$
	Hercynite $Fe^{+2}Al_2O_4$

electron diffraction, or high-resolution TEM imaging: HRTEM). Likewise, HRTEM imaging alone cannot completely characterize these mineral species, as many of them display nearly identical diffraction spacings (see Table 3.4.1). Much of the confusion in the literature springs from these experimental difficulties in mineral identification. Finally, many of these minerals are hydrated, which means that they are subject to dehydration during electron-beam analyses.

It is obvious that for studies of aqueous alteration of extraterrestrial materials to have credibility, the alteration must be demonstrated to be preterrestrial. Thus, unsubstantiated reports of aqueous-alteration products within meteoritic "finds" (as opposed to "falls") (including the Antarctic meteorites [see Gooding 1986]) should always be viewed with suspicion.

3.4.3. OBSERVATIONAL EVIDENCE FOR AQUEOUS ALTERATION

CV and CO Chondrites

Possible products of aqueous alteration within CV and CO chondrites have received limited attention because, although present in these meteorite classes, they are not as pervasive as in the CIs and CMs. Incompletely described Mg-Fe layer-silicates have been reported from the matrices of Vigarano (CV) (Bunch and Chang 1980) and Lancé (CO) (Christophe Michel-Lévy 1969; Kurat 1975). The identities of these minerals have not been determined, although they are reported as being enriched in Ca and Al, and depleted in Na, with respect to phyllosilicates within CMs. In addition, chemically uncharacterized layer silicates and clino-chrysotile, a hydrated Mg-silicate, have been identified in Mokoia (CV), Lancé and Ornans (CO) (Kerridge 1964).

Fortunately, phyllosilicates found decorating inclusions in Allende and Mokoia (CV) have been more comprehensively characterized (Cohen et al. 1983; Tomeoka and Buseck 1982b, 1985). These phases have entered the literature under the acronyms "LAP" and "HAP," for "low-" and "high-aluminum phyllosilicate," respectively. Compositional and structural data for these phases vary, but appear to indicate that LAP is saponite, while HAP seems to be an Al-rich smectite, possibly sobotkite (although Gooding [1985] has suggested pumpellyite). These minerals appear under identical circumstances in both Allende and Mokoia, although they are more common in Mokoia. Both minerals are poorly crystallized, forming crystals no larger than 1000 Å across. "Saponite" is pervasive in these meteorites, but "sobotkite" has only been observed in CAIs and chondrules, where it appears to represent the alteration product of Ca-Al rich glass and Ca-clinopyroxene (see Figs. 3.4.1 and 3.4.2). There are two explanations for this mineralogical distribution. Either (1) "sobotkite" formed before incorporation of the CAIs and chondrules into the meteorite, and "saponite" afterwards, or (2) high Ca-Al glass and pyroxene present only in the CAIs and chondrules have localized the nucleation and growth of "sobotkite," in which case the timing of alteration is uncertain.

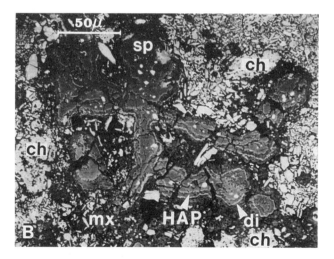

Fig. 3.4.1. Backscattered-electron image of a Mokoia (CV) CAI. "Sobotkite" (HAP) and diopside (di) rim spinel (sp) grains. Inclusion matrix (mx) is a porous intergrowth of olivine laths (light) and saponite (dark). Reprinted by permission from Cohen et al. (1983), *Geochimica et Cosmochimica Acta,* 47, Copyright 1983, Pergamon Journals, Ltd.

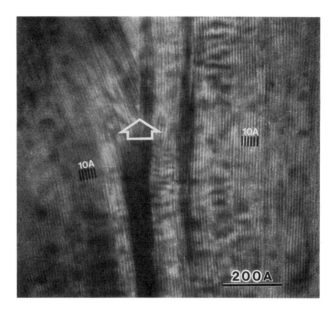

Fig. 3.4.2. A HRTEM image of Al-rich smectite (possibly intergrown with a mica) rimming feldspathic and spinel grains in an Allende CAI. Overlapping fringes are indicated by the arrow. This view is near the periphery of the CAI, suggesting a relationship to adjacent matrix material. Reprinted by permission from Tomeoka and Buseck (1982b), *Nature*, 299, 326, Copyright © 1982 Macmillan Magazines Limited.

Meeker et al. (1983) describe embayed pyroxenes within five Allende CAIs which they feel have experienced an alteration episode involving Ca-rich high-temperature metasomatic fluids. In this instance the pyroxenes are mantled with secondary melilite, indicating that this process might more properly be classed as a metamorphic reaction.

CM Chondrites

The products of aqueous alteration are ubiquitously present in CM chondrites. These minerals include serpentines, montmorillonites, sulfates, tochilinite, tochilinite-serpentine intergrowths and carbonates. The layer-structured materials that predominate within CM chondrites have been thoroughly reviewed by Barber (1985).

There is an inherent difficulty in separating the alteration history of CM matrix materials from the larger components (e.g., chondrules, aggregates, inclusions and phenocrysts) contained therein. Although there may commonly be clear indications of alteration within a chondrule (for example), and different alteration mineralogy/textures outside the chondrule, the early literature includes many examples where the textural context of possible alteration phases is not noted. Such information is critical to the interpretation of mineralogy. This problem does not in general extend to the CIs, where such large components are rare, or to COs and CVs, where alteration products are rare and generally nonpervasive.

Many early studies of CMs reported the presence of phyllosilicates with a principal diffraction spacing of ~7 Å. More comprehensive later studies have concluded that these phases are probably various serpentines (Barber 1981). Less clear is the identity of the specific serpentines, although these are often Fe rich, and are therefore probably largely cronstedtite, ferroan antigorite and greenalite (Tomeoka and Buseck 1985; Zolensky et al. 1987). In fact, it seems that virtually all of the serpentines are present within the matrices of CMs, with at least 3 different serpentine phases commonly being reported from a single meteorite. It has become obvious that serpentines (Fe-rich serpentines in particular) are the dominant matrix phases in CM chondrites.

Other phyllosilicates have also been reported from CM chondrites, although they are volumetrically less important than the serpentines. Montmorillonite has been reliably characterized within Cold Bokkeveld (Caillère and Rautureau 1974) and Nawapali matrix (Barber 1981).

The crystal size of these materials typically ranges from 100 to 1000 Å only, and so are best observed using HRTEM techniques. Morphologies most commonly observed include: (1) platelike or flaky; (2) rounded forms, often with spherulitic or sectored appearance; (3) sinuous individual fibers and bundles; and (4) poorly crystalline fernlike and spongelike forms (Barber 1981). Most of these forms are shown in Fig. 3.4.3.

Of considerable interest is the green, fibrous phyllosilicate material observed within chondrules and anhydrous silicate aggregates in all CMs, called "spinach" by Fuchs et al. (1973). An example of a chondrule containing "spinach" is shown in Fig. 3.4.4. Fuchs et al. (1973) and later workers (Olsen and Grossman 1978; Richardson 1981) have suggested that "spinach" represents an alteration product of anhydrous silicates (forsterite and clinopyroxene) and glass, which necessitates enrichment of the alteration products in Fe relative to Mg, with progressive alteration. This observation could place interesting constraints on the alteration process, which require that the exact nature of the alteration products be determined. Bunch and Chang (1980) and Tomeoka and Buseck (1985) concluded that this material is an intergrowth of Fe-rich serpentines; however, Gooding (1985) noted that the chemical and structural information available on this material (for Nogoya, at least) fit a ferroan chlorite equally well. Clearly, additional work on this material is needed.

In an early study of Murchison, Fuchs et al. (1973) described a group of Fe-S-Ni-O and Fe-S-Ni-Si-O rich phases which they termed "poorly characterized phases" (or PCPs). These phases are found throughout CM matrices, as well as in embayed chondrules and aggregates, and within fine-grained rims. Forms exhibited by these phases include massive, hollow and filled tubes, fibers, flat and corrugated sheets. The grain size of this material is

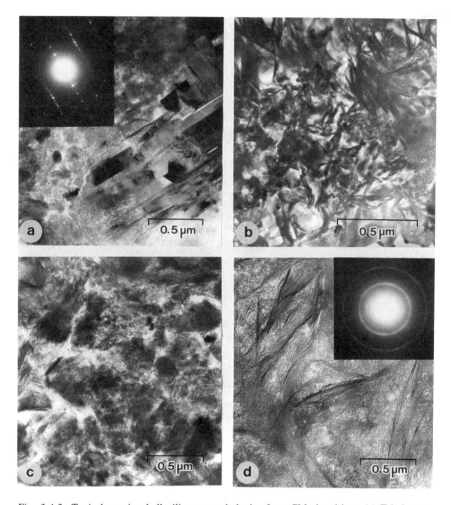

Fig. 3.4.3. Typical matrix phyllosilicate morphologies from CM chondrites. (a) Tabular and platelike crystals in Nawapali. (b) Fibers and curved flakes intergrown with hollow fibers in Cold Bokkeveld. (c) Patches of flaky crystals and spongy material in Murchison. (d) Fernlike poorly crystalline material in Nawapali. Figure from Barber (1981), reprinted with permission.

usually below 1000 Å. Building on the earlier work, Mackinnon and Zolensky (1984) demonstrated that PCPs are composed, dominantly, of two phases. These are (1) tochilinite (an unusual coherently interstratified Fe-Ni sulfide and Fe-Mg hydroxide), with principle diffraction spacings of 10.8 and 5.4 Å (Fig. 3.4.5), and (2) a still unnamed mineral consisting of coherently interstratified serpentine and tochilinite, with a diagnostic interlayer spacing of 17 to 18 Å (Fig. 3.4.6). Structural models for these two phases are shown in

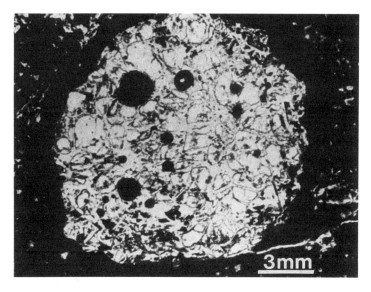

Fig. 3.4.4. A plane-polarized light view of a Nogoya chondrule originally consisting of three large forsterite crystals, now largely altered to massive and fibrous phyllosilicates ("spinach"). Reprinted by permission from Bunch and Chang (1980), *Geochimica et Cosmochimica Acta*, 44, Copyright 1980, Pergamon Journals, Ltd.

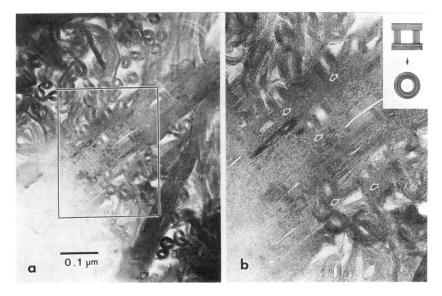

Fig. 3.4.5. (a) Cylindrical growth of alternating tochilinite and cronstedtite from the matrix of Murchison. (b) An enlargement of the framed area. "Proto-tubes," indicated by arrows, seem to be an intermediate morphology between platy and cylindrical types shown in the inset. Figure from Tomeoka and Buseck (1985), reprinted with permission.

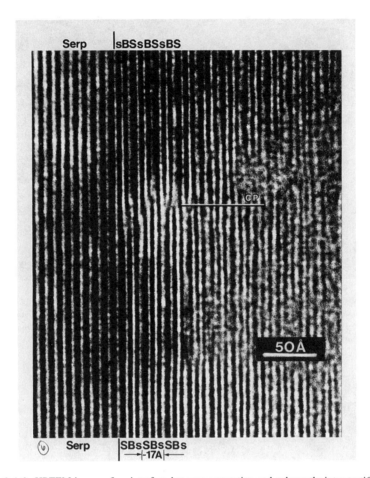

Fig. 3.4.6. HRTEM image of an interface between serpentine and coherently interstratified tochilinite-serpentine, from Mighei matrix. "SBs" notation indicates serpentine-brucite-sulfide (the latter two phases constituting the tochilinite). The discontinuity of lattice fringes is the result of a stacking error in which the c* directions of SBs-type material are reversed on either side of a common plane (CP). Figure from Mackinnon (1982), reprinted by permission.

Fig. 3.4.7. These phases have since been located within most CM chondrites, and the mineralogy of extraterrestrial tochilinites has been considerably developed (Zolensky 1984, 1987; Zolensky and Mackinnon 1986). Tochilinite and the tochilinite-serpentine phase are of sufficient abundance that the stability of these phases can constrain the conditions under which the aqueous alteration of the CM chondrites occurred (Zolensky 1984; Tomeoka and Buseck 1985; Zolensky et al. 1987). The probable nature of these conditions is discussed later in this chapter.

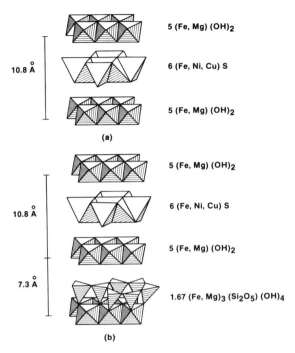

Fig. 3.4.7. Models for tochilinite phases. (a) Model of tochilinite, indicating sulfide and hydroxide layers. (b) Model of coherently interstratified tochilinite-serpentine, indicating the relative stacking of the sulfide, hydroxide and serpentine layers. Figure from Mackinnon and Zolensky (1984), reprinted by permission.

Carbonates are ubiquitously present in the matrices of CMs, as both calcite and aragonite (Bunch and Chang 1980; Barber 1981). Carbonates intimately intergrown with matrix phyllosilicates and tochilinite have been notably reported in Murchison, Nogoya, Cold Bokkeveld, Nawapali and Cochabamba (Fuchs et al. 1973; Müller et al. 1979; Bunch and Chang 1980; Barber 1981). An example of this relationship is shown in Fig. 3.4.8. Several generations of carbonates may be present in CM matrix, and examples of aragonite apparently nucleating upon earlier-formed calcite have been reported (Fig. 3.4.9) (Barber 1981). Armstrong et al. (1982) report calcite intimately intergrown with hibonite in the core of a large CAI, and argue that its presence is best explained by aqueous alteration on (in?) a planetary body.

Magnesium and Ca sulfates have been reported from CM chondrites (for example, Murchison; Fuchs et al. 1973; Barber 1981), but these are minor phases. However, as these are important phases in the CI chondrites they may deserve closer examination in the CMs as well.

Studies of CM chondrites have revealed the presence of a complex suite of soluble and insoluble organic compounds, at least some of which may have

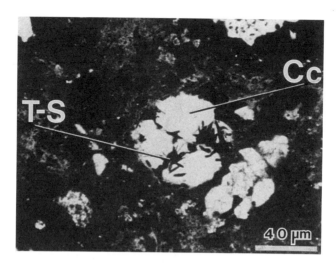

Fig. 3.4.8. Transmitted light (crossed polars) view of crystals of coherently interstratified tochilinite-serpentine (T-S) enclosed within grains of calcite (Cc), in Murray matrix. Figure reprinted and adapted by permission from Bunch and Chang (1980), *Geochimica et Cosmochimica Acta*, 44, Copyright 1980, Pergamon Journals, Ltd.

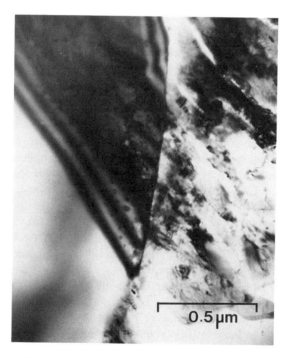

Fig. 3.4.9. Contact between coarse-grained calcite and polycrystalline aragonite in Cold Bokkeveld matrix. Figure from Barber (1981), reprinted by permission.

been altered by, or even synthesized during, the epoch of aqueous activity (e.g., possible amino-acid production via the Strecker-cyanohydrin synthesis; see Chapter 10.5).

CI Chondrites

The CI chondrites are composed entirely of what would, in the other carbonaceous chondrites, be considered matrix phases. These are primarily phyllosilicates, oxides (primarily magnetite), sulfides, carbonates and sulfates. McSween and Richardson (1977) found that values of S/Si, Ni/Si and Fe/Si are lower in bulk CI matrix material than in bulk CM matrix, and noted that these differences are minimized if a phase equivalent to tochilinite is eliminated from bulk CM matrix. This suggests that CI matrix could have been derived from CM matrix material, if the latter somehow lost tochilinite. On the other hand, Kerridge (1977) noted compositional and spatial correlations between Ni and S in Orgueil phyllosilicates, and suggested that this indicated a chemical fossil reflecting the former presence of the phase now known as tochilinite.

Serpentine has been reported in various CIs (see, Bass 1971; Barber 1985), although some of this material could be chamosite (Zaikowski 1979). Montmorillonite has also been fairly reliably reported in Orgueil (Bass 1971; Caillère and Rautureau 1974) in agreement with inferences drawn from calcination studies (Fanale and Cannon 1974). Other phyllosilicates are possibly present as well (Mackinnon 1985).

Carbonates are important constituents of CIs, comprising up to 5 vol. % of Orgueil, for example (Boström and Fredriksson 1966). The species which have been reported are dolomite, ankerite, ferroan magnesite (formerly called breunnerite) and a pure Ca-carbonate commonly assumed to be calcite but actually either aragonite or vaterite (Kerridge et al. 1980). DuFresne and Anders (1962) noted the high perfection of dolomite crystals from Ivuna, and surmised that such crystals required on the order of 10^3 yr to form. Macdougall et al. (1984) calculated the age of carbonates in Orgueil, on the basis of Rb-Sr systematics, and concluded that they had crystallized within 100 Myr of the formation of the parent body, with the oldest carbonates being, within uncertainties, contemporaneous with the oldest meteoritic material yet dated (excluding presolar material). On the other hand, they concluded that at least some of the Ca-sulfate present had formed (or recrystallized) at some later date.

Much has been made of the Na-, Ca- and Mg-sulfates in CI chondrites, Orgueil in particular. The common sulfate-filled veining displayed by Orgueil presents some of the strongest evidence for the action of an aqueous fluid on the CI parent body(ies). Such veining in Orgueil is shown in Fig. 3.4.10, the principal sulfates in CIs being epsomite and hexahydrite, with the former predominating (Boström and Fredriksson 1966; Richardson 1978). Minor amounts of gypsum and blödite are also present.

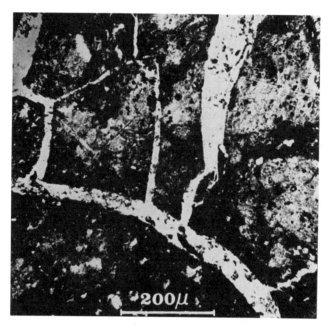

Fig. 3.4.10. Transmitted light image of cross-cutting sulfate-filled cracks in Orgueil. Figure reprinted by permission from Richardson (1978).

Elemental S has been reported within CIs (Mason 1962; DuFresne and Anders 1962), associated with pyrrhotite, commonly misidentified in early work as troilite. The presence of elemental S in CIs is potentially important for the elucidation of the physico-chemical conditions attending aqueous alteration, because its stability is sensitive to the overall concentrations of aqueous species, and to solution Eh and pH. The source of CI elemental S is almost certainly the pyrrhotite with which it is associated and whose crystals often appear to be heavily corroded.

Magnetite is another mineral whose presence in CIs is a potentially important indicator of the conditions of aqueous alteration. The morphologies of CI magnetite include framboids (Fig. 3.4.11), spherulites and platelets (Jedwab 1971), unusual structures apparently caused by crystallization from gels or aqueous solutions (Kerridge et al. 1979a). The I-Xe age of this magnetite is among the oldest ever measured (Lewis and Anders 1975), indicating that these magnetites crystallized very early in solar-system history (see also Chapter 15.3).

Unequilibrated Ordinary Chondrites

A detailed description of the matrix mineralogy of unequilibrated ordinary chondrites is given in chapter 10.2. Only the mineralogy pertinent to possible aqueous alteration of these materials is summarized here.

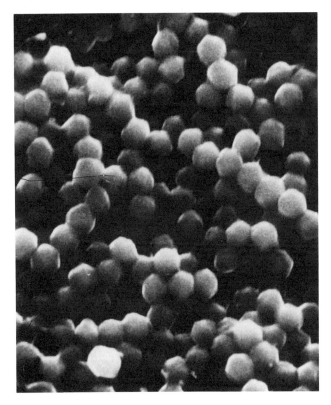

Fig. 3.4.11. A SEM image of framboidal magnetite in Orgueil. The field of view measures 8 μm across.

Recent work has uncovered probable aqueous alteration products of limited extent within the matrices of four unequilibrated ordinary chondrites: Semarkona (LL3), Bishunpur (LL3), Tieschitz (H3) and Krymka (L3) (Alexander et al. 1986; Hutchison et al. 1987). In addition, Hutchison et al. (1987) report that the mesostases of some chondrules and clasts in Semarkona display devitrification textures, suggesting that these objects also experienced aqueous alteration to a limited extent. All four meteorites are "falls" which were recovered within short periods of time, and so the alteration phases present are assumed to be preterrestrial in origin.

Na, Fe- and Ca-smectites are reported from Semarkona and Bishunpur, respectively. Hutchison et al. (1987) report that the structure and chemistry of the Semarkona phyllosilicate is best matched by nontronite; however, Gooding (1985) suggests that the composition of the phyllosilicate in Semarkona is equally well fit by vermiculite, chlorite or a tri-octahedral mica. In Bishunpur the possible Ca-smectite (saponite or sobotkite?) is reported to be replacing amorphous, fine-grained matrix material (glass?). No phyllosilicates have been located within Tieschitz or Krymka; however, a Na-zeolite has been reported from the former (Alexander et al. 1986).

In these meteorites, matrix olivines and pyroxenes show incipient dissolution features (embayment), and are intimately associated with maghemite and other poorly described Fe-oxide phases. Calcite is also reported (Alexander et al. 1986).

Interplanetary Dust Particles

The mineralogy of interplanetary dust particles (IDPs) is described in chapter 11.1. Here we are concerned with evidence that some of these particles may have experienced aqueous alteration. This evidence consists, entirely, of the presence of phyllosilicates within the serpentine and smectite classes of chondritic IDPs (Mackinnon and Rietmeijer 1987; Bradley 1987). These two chondritic IDP classes will be discussed separately.

The smectite class is the most common type of chondritic IDP, and is characterized by the presence of finely intergrown Fe-, Mg-rich phyllosilicate and an Fe-rich glass which is described as having the composition of an Fe-pyroxene (Bradley 1987). An example of this relationship is shown in Fig. 3.4.12. This phyllosilicate, which has a basal spacing of 10 to 12 Å, may

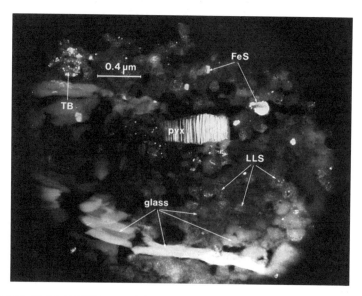

Fig. 3.4.12. Darkfield TEM image of the matrix of an IDP of the smectite class. The dominant phases in this view are a poorly-crystalline phyllosilicate (LLS) in intimate contact with a ferromagnesian glass. Also present are balls of finer-grained material (TB) and FeS. Figure reprinted by permission from Bradley (1987), *Geochimica et Cosmochimica Acta*, 51, Copyright 1987, Pergamon Journals, Ltd.

indeed be a smectite, but within analytical uncertainty could also be chlorite or vermiculite, or a combination of these phases. The associated glass has a lower silica and higher Fe content than this phyllosilicate, but the intimate relationship between these phases suggests a genetic relationship. Associated minerals most commonly encountered within this class of IDPs include olivine, pentlandite, pyrrhotite and magnetite. Mg- and Fe-carbonates are common constituents of some of these particles.

The least common type of chondritic IDP, particles representative of the serpentine IDP class, are composed predominently of a phyllosilicate with a characteristic 7 Å basal spacing. This phase(s) is probably a serpentine-group mineral, although fine intergrowths with pentlandite, pyrrhotite and magnetite make accurate compositional analysis currently impossible (Bradley 1987). No other silicate phases have been found within particles of this IDP class. No solar-flare tracks have been found in any representative of the serpentine IDP class, and thus there is as yet no hard evidence that these particles existed as separate entities in space (although the presence of solar-flare tracks is certainly not a sufficient requirement for a particle to be an IDP). In other words, while these particular particles could be the collision products of CI or CM chondrite parent bodies in space, they might also be the products of the atmospheric ablation of CI or CM carbonaceous chondrite meteorites.

3.4.4. AQUEOUS ALTERATION CONDITIONS

Location of the Aqueous Alteration Events in the Early Solar System: Nebula vs Parent Body

There are several possible major sources of water available for aqueous alteration reactions in the early solar system. Water vapor would have been available for reactions in the primitive solar nebula and, in the absence of evidence for the direct condensation of hydrous minerals in the nebula, it is plausible that these phases could have formed via the hydration (and reconstructive transformation) of preexisting anhydrous silicates. Such a scenario has, in fact, been commonly assumed (see, e.g., Grossman and Larimer 1974; Barshay and Lewis 1976); however, more recent theoretical work (Prinn and Fegley 1987) has indicated that silicate hydration in the nebula would probably have been kinetically inhibited. A consequence would have been that water would not have accreted until the temperature had dropped to a point where water-ice could have formed.

The petrographic vidence for aqueous alteration in chondrites is inconsistent with that alteration having taken place by hydration in the nebula, though that does not in itself rule out the prior occurrence of such nebular reactions. The meteoritic record is, however, entirely consistent with the conclusions of Prinn and Fegley (1987). Aqueous alteration within meteorite parent bodies is required by the following observations: (1) alteration minerals bridging chondrules, aggregates and phenocrysts with matrix (for example, tochilinite appearing at the interface between fractured chondrule grains and matrix in Mighei [Tomeoka and Buseck 1985]); (2) CI magnetite morphologies that apparently involved crystallization from a gel or fluid phase (Kerridge et al. 1979b); (3) relict chemical zoning or correlations within matrix and altered chondrules and aggregates (e.g., the bulk compositional trends reported for CMs by McSween [1979a]; and (4) alteration minerals lining fractures and forming veins in a meteorite (e.g., sulfate veining in Orgueil described by Richardson [1978] and calcite filling cracks in a Murchison CAI, reported by Armstrong et al. [1982]). In addition, the ubiquitous presence of carbonates in CI and CM matrices and in at least one CAI demands aqueous alteration on a planetary body, since the CO_2 partial pressure necessary to stabilize carbonates was unlikely to be attained in the solar nebula (Armstrong et al. 1982)

This secondary mineralization required the transport of major elements (e.g., Ca, Fe, Mg and Mn) over distances of mm, at least. In addition, most chondrites are breccias, and some, notably the CIs, display abundant fracturing, implying the operation of impact brecciation at, or near, the surface of parent bodies. Some chondrites (e.g., Murchison and Yamato 74662) also contain slightly flattened xenoliths, or oriented phyllosilicates, implying gentle compaction (Fujimura et al. 1982; Fujimura et al. 1983). These effects are also best described by regolith processes on a parent body, probably dynamic compaction (due to impacts) or static compaction (due to burial), the latter being preferred by Fujimura et al. (1983). Thus, we may conclude that the present evidence requires that the aqueous alteration of primitive solar-system material occurred upon parent bodies, and not in the solar nebula.

Sources of Water on Meteorite Parent Bodies

Water, presumably in the form of ice (Prinn and Fegley 1987), would have accreted along with lithic material to form the original meteorite parent bodies. Once contained within these planetary bodies this ice would have been potentially available for the aqueous alteration of adjacent lithic material, which probably consisted mainly of anhydrous oxides, silicates and metal. The next step is the tricky one. How was the liquid water required for aqueous-alteration processes derived from the oiginally accreted ice? Several possible solutions to this problem have been proposed.

Water-ice could have promoted hydrocryogenic alteration of anhydrous material directly through the action of "unfrozen water" as described earlier (Gooding 1984; Rietmeijer 1985). In controlled experiments this phenomenon is effective down to a temperature of approximately $-11°C$. However, given the much greater time scales available for reaction on primitive meteorite parent bodies than in the laboratory, it is conceivable that aqueous alteration through the action of thinner layers of unfrozen water could have proceeded at lower temperatures (Rietmeijer 1985). As a potential test of the effectiveness of hydrocryogenic alteration it would be instructive to determine whether unfrozen water could produce sulfate veins or carbonate grains of the dimensions observed in Orgueil.

True aqueous alteration on the meteorite parent body would have been possible if liquid water were available. Several mechanisms for this have been advanced. The first process involves melting of water ice and chemically bound water by heat generated within the parent body (DuFresne and Anders 1962) by decay of short-lived radionuclides. Through this process, a narrow zone of liquid water would have been produced within the parent body, presumably migrating towards the surface as heating continued. Eventually a zone of liquid water would have underlain a surficial icy layer, and subsequent loss of water from the body would have been then largely controlled by sublimation of surface ice, as modified by impact processes. DuFresne and Anders (1962) showed that a body 100 km in radius in a main-belt asteroidal orbit would have retained liquid water for about 200 Myr, which is certainly long enough to produce any of the alteration products observed. The lifetime of liquid water would have been decreased by a smaller parent-body radius and/or closer proximity to the Sun. This model runs into trouble if the necessary radionuclides were present in quantities too small to generate sufficient heat, or if there had been appreciable decay of these radionuclides before the parent bodies had accreted to the point where heat could be stored (Sonett 1971), though Rietmeijer (1985) argues that this was probably not the case. There are also various theories wherein heat was provided to a rapidly accreting body by residual heat from the condensation process (Larimer and Anders 1967).

Sonett (1971) developed a theory that parent bodies located as far from the Sun as the present main asteroid belt would have been heated by magnetic induction, through interaction with a strong solar plasma, if the early Sun had passed through a T-Tauri stage. Heating by this process may have proceeded in the opposite direction to that ensuing from the internally-heated model, i.e., from the outside inwards. Of course, it is not certain that our Sun passed through such a T-Tauri stage (but see Chapter 4.1).

Another mechanism for the production of liquid water in a parent body is by release of water ice and structural water in phyllosilicates during impacts onto the surface of the parent body (Lange et al. 1985). Water vapor released during shock events accompanying impacts could be buried by redistributed regolith, and subsequently condensed as liquid. This mechanism would have operated most effectively during the accretion of the parent body, when impacts were most frequent. It remains for this process to be rigorously modeled, to establish just how long impact-generated water would remain liquid, before either freezing or evaporating into space.

Evidence for the activity of liquid or unfrozen water in CM chondrites is provided by an O-isotopic study of matrix phyllosilicates and calcite in Murchison (Clayton and Mayeda 1984). By making the assumption that these minerals formed simultaneously, from the same fluid reservoir, Clayton and

Mayeda (1984) determined that the temperature of the aqueous fluid was < 20° C, and that a water/rock ratio of approximately 1/1 was required. This water/rock ratio may be a conservative estimate, as it assumes that all water interacted with the rock. A temperature just below 0° C was not precluded by the experimental results. Extrapolation of this work (by Clayton and Mayeda) to other CMs shows similar results, while extension of these results to the CI Orgueil suggests somewhat higher temperatures during alteration (for example 140° C for matrix carbonates) and larger volumes of water, which is consistent with the more pervasive aqueous alteration displayed by these meteorites. Note, however, that these extrapolations are model dependent, requiring verification by more comprehensive work on these primitive meteorites.

Location of Primitive Meteorite Parent Bodies

In a review of the possible locations of primitive meteorite parent bodies within which aqueous alteration could occur, Kerridge and Bunch (1979) favored an asteroidal source. It is generally accepted that asteroids have regoliths, and there is convincing spectroscopic evidence for the presence of hydrated minerals or ice on the surfaces of main-belt C-type asteroids (spectroscopically similar to the carbonaceous chondrites) measuring over 100 km in diameter (Feierberg et al. 1985). Asteroids provide some fraction of IDPs, but the percentage is controversial (Bradley 1987; Mackinnon and Rietmeijer 1987; see also Chapter 11.1).

On the other hand, there are problems with the selection of cometary surfaces as the source of altered meteorites. Although the surface of a comet could exceed 0° C during passage through the inner solar system (measured surface temperatures for Comet Halley varied up to approximately 130° C [Emerich et al. 1986]), any material so heated would be quickly lost by ablation (dust) or evaporation (ice). Thus, although unfrozen water or a regolith could form on a cometary surface it would eventually be blown off. These difficulties may not apply to either possible protocometary cores or dead cometary nuclei. Comets were once presumed to be the major source of IDPs, but this assumption is now hotly debated (see above).

3.4.5. AQUEOUS ALTERATION SCENARIOS

Workers are beginning to attempt the construction of plausible scenarios for aqueous alteration of primitive meteoritic material. This development has been delayed by poor understanding of the mineralogy of the fine-grained material within matrices, altered chondrules and aggregates of anhydrous silicates within carbonaceous chondrites. This fundamental problem persists, although recent studies have begun to bring the mineralogy of these fine-grained minerals into sharper focus. Unfortunately, although some alteration products are well (or at least partly) characterized, all too often the relationships between these phases within the meteorites are not known. Thus, mineral parageneses derived for these materials are typically tentative. In the future, alteration products and their surrounding phases should be characterized together, not piecemeal. Only thorough studies of these phases within a textural context will result in accurate descriptions of alteration parageneses. All of the alteration models described below are subject to some extent to this shortcoming.

An entirely new dimension would be added to the study of the aqueous alteration of primitive meteorites if fluid inclusions could be located within matrix materials (carbonates, for example) or adjacent olivine crystals. Such fluids would be very sensitive indicators of the aqueous fluid responsible for alteration, and of incipient diagenetic changes in the solids. Fluid inclusions have periodically been reported from meteorites, but subsequent studies have indicated that the majority (at least) of these inclusions are due to terrestrial contamination (mostly during thin-section preparation) (Rudnick et al. 1985).

It is important to keep in mind that multiple environments and complex time scales are required by the apparent disequilibrium assemblages often present within meteorites, and the fact that many (most?) of these materials are breccias. Many workers have reported finding highly altered meteoritic materials in direct contact with material that appears to have escaped alteration completely. Thus, one simple model will probably never be formulated which successfully accounts for all of the features displayed by, for example, the CM chondrites.

CV, CO and Unequilibrated Ordinary Chondrites

We have only just begun to characterize the products of aqueous alteration processes in the CV, CO and unequilibrated ordinary chondrites, so that only a few tentative suggestions regarding the nature of these processes is possible. In these meteorites, Ca-pyroxenes in the CAIs and Ca,Al-rich glasses appear to have been the materials most susceptible to alteration. These components alter to what have been interpreted as Ca,Al-, and sometimes Na-smectites. These minerals are probably montmorillonite, saponite and sobotkite, although those identifications remain tentative. Unfortunately, since the mineralogical relationships between these and surrounding phases are not well known, suitable models cannot be derived to describe the aqueous fluid conditions. For example, not enough observations have been made for us to know whether these phyllosilicates are transformation (simple replacement alteration) or neoformation (new growth from a solution) materials. Without this knowledge, we cannot begin to construct adequate models for the growth of these materials.

CM and CI Chondrites

Much more work has been performed upon the products of aqueous alteration in the CM and CI chondrites than in the other chondrites, as befits the more pervasive nature of alteration in these meteorites. In addition, the alteration mineralogies of these chondritic types are similar enough that they are often treated together, and there have been suggestions that they are genetically related (McSween 1979a). In an early review of the mineralogy and phase relations of the CM and CI chondrites, DuFresne and Anders (1962) concluded that these meteorites had undergone alteration by liquid water, at approximately 20° C. This aqueous fluid was reported to have been characterized by an Eh of -0.2V, pH of 8 to 10 for the CIs and 6 to 8 fr the CMs, and contained at various times NH_3, SO_2, CO_2 and some H_2S. DuFresne and Anders (1962) also concluded that the aqueous-alteration process for these meteorites operated at, or very near, equilibrium. They envisioned this alteration as occurring near the surface of an internally heated main-belt asteroid, although other water-producing mechanisms are possible, as described earlier. The temperature regime of these models is roughly consistent with that proposed by Clayton and Mayeda (1984) on the basis of O-isotopic systematics.

As a useful comparison, it should be noted that Zolensky (1984) proposed similar aqueous conditions for the alteration which produced tochilinite within CM chondrites. Noting that tochilinite structurally consists of coherently interstratified amakinite (Fe-brucite) and mackinawite, and is paragenetically associated with serpentinites (in terrestrial rocks), he argued that tochilinite should be stable where the stability fields of amakinite, mackinawite and serpentines overlap. He concluded that the presence of tochilinite therefore constrained the aqueous fluid from which it precipitated to very low fO_2 (Eh well below 0V), high fS_2, pH 10 to 12, and a temperature below 170° C. Since these pH values are probably too high for Fe to be mobile, Fe may have initially precipitated as some other phase at a more moderate pH, and a change in pH and/or fS_2 made production of tochilinite thermodynamically possible.

Other attempts at cosmothermometry for the aqueous alteration on CI and CM parent bodies yield peak temperatures which lie within the range suggested by Zolensky (1984), approximately 85° C for CIs and 105° to 125° C for CMs (see Hayatsu and Anders 1981), but which are all higher than the temperatures derived by Clayton and Mayeda (1984). Bunch and Chang (1980) performed a detailed survey of the evidence for aqueous alteration of carbonaceous chondrites, and also concluded that the temperature of the aqueous fluid was below 125° C. Since these studies typically focused on different components of these meteorites, which are breccias with complex histories, it is interesting that these proposed temperatures are as similar as they are.

McSween (1979a), noting that CM chondrites are characterized by a wide range of the relative proportion of matrix (from 57 to > 90 vol. %), and assuming that aqueous alteration produced at least some of the matrix in these meteorites from chondrules and xenoliths, concluded that a scale for alteration of CM chondrites could be estimated by measuring the relative proportion of matrix. Using this scale, he then found that the Fe/Si ratio in CM matrices decreased with progressive alteration, due probably to the formation of matrix phyllosilicates characterized by higher Mg/Fe ratios (at the expense of olivine and pyroxene in chondrules and aggregates), and of opaque phases which were no longer counted as matrix material. He concluded that the alteration process in CM chondrites appears to have been isochemical for all but the most highly mobile components (water was added, certain organic constituents were lost). He suggested that as CI chondrites are essentially completely "matrix," relative to the CMs, similar aqueous processes which affected the CMs were active in processing the CIs, but were more pervasive in the latter. Based on these results, McSween (1979b) proposed a reinterpretation of petrographic types 1 through 3 in the chondrite classification of Van Schmus and Wood (1967) (see Fig. 1.1.12). On the basis of bulk elemental compositions, Ebihara et al. (1982) concluded that alteration in CIs was also isochemical.

Tomeoka and Buseck (1985) have proposed specific mineralogical reactions to account for the compositional variations noted by McSween (1979a), and, subsequently by Richardson (1981), and their own petrographic and HRTEM observations of CM chondrites. Tomeoka and Buseck (1985) proposed the following multistage aqueous-alteration sequence for these meteorites:

1. Tochilinite (called FESON by them) was produced from the dissolution of kamacite and sulfides in chondrules and aggregates in the parent body regolith by the addition of water. Mesostasis glasses in chondrules and aggregates became hydrated and altered to "spinach," which they feel is most similar to berthierine, but which could also be another Fe-serpentine or chlorite as well.
2. Magnesian cronstedtite (an Fe-end member serpentine) formed from fine-grained matrix olivine and pyroxene. Tochilinite and spinach became separated from the chondrules and aggregates by regolith gardening. Cronstedtite and tochilinite incompletely reacted to form a quantity of the coherently interstratified tochilinite-serpentine phases. Presumably the spinach reacted with matrix phyllosilicates to form ferroan serpentine and additional magnesian cronstedtite.
3. Tochilinite in the matrix reacted with Mg and Si provided by the dissolution of olivine and pyroxene, becoming an intergrowth of tochilinite and ferroan antigorite (called Type II PCP by Tomeoka and Buseck). In a similar manner, magnesian cronstedtite became increasingly enriched in Si and Mg, and altered to ferroan antigorite as well.
4. Ultimately, as revealed by the most heavily altered CMs, all of the remaining tochilinite broke down to form magnetite and sulfides.

McSween (1987) used mass-balance calculations to estimate the relative proportions of tochilinite, cronstedtite, and Fe-, and Mg-serpentines in CM

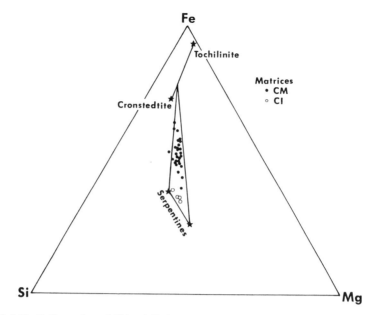

Fig. 3.4.13. Bulk matrices of CM and CI chondrites, plotted onto an Fe-Si-Mg ternary diagram, along with reference points defined by the compositions of serpentines, cronstedtite and tochilinite. Using this figure the lever rule may be employed to determine the percentage of each of these three minerals in the chondrite matrices. Figure reprinted by permission from McSween (1987), *Geochim. Cosmochim. Acta* 51, Copyright 1987, Pergamon Journals, Ltd.

chondrite matrices (Fig. 3.4.13). This matrix modeling is completely dependent upon which phases are chosen as dominant, and the assumption must be made that no other phases are present in significant amounts. McSween (1987) concluded that progressive loss of Fe from matrix phyllosilicates and tochilinite during alteration reflected breakdown of tochilinite and cronstedtite, and formation of more Mg-rich serpentine (in agreement with Tomeoka and Buseck [1985]). CI chondrites apparently contain little, if any, cronstedtite and tochilinite. His results generally supported the model of Tomeoka and Buseck (1985).

However, the true alteration sequence for the CM chondrites is probably more complex still than described by these models. Zolensky et al. (1987) have performed complementary mass-balance calculations, modeling the matrix of CM chondrites by a combination of tochilinite, magnesian cronstedtite and ferroan antigorite, and have obtained different results. They find that the cronstedtite/antigorite ratio correlates poorly with modal matrix amount, and that the amount of tochilinite does not vary significantly among CM chondrites. Nogoya is one of the most completely altered CM chondrites available (Bunch and Chang 1980), so that the model of Tomeoka and Buseck (1985) would predict that this meteorite should contain little remaining tochilinite.

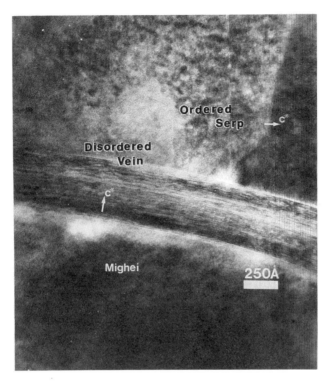

Fig. 3.4.14. A HRTEM image of Mighei matrix, showing a vein of mixed-layered tochilinite and serpentine (disordered) cutting a well-ordered serpentine crystal. Figure reprinted by permission from Mackinnon et al. (1982), *Micron* 13, Copyright 1982, Pergamon Journals, Ltd.

Nevertheless, Nogoya is one of the most tochilinite-rich CMs known (Bunch and Chang 1980). Additional evidence for alteration sequence complexities can be seen in Fig. 3.4.14. In this HRTEM image of Mighei matrix, a vein of mixed-layer tochilinite and serpentine (disordered) cuts a well-ordered serpentine grain. The model of Tomeoka and Buseck (1985) would predict the opposite to be the observed.

Carbonates, most commonly calcite, are ubiquitously present within the matrices of CMs, as well as within at least one CAI. It is common for other phases, for example tochilinite, to be observed in an intimate intergrowth with calcite (see Fig. 3.4.8), suggesting that these phases crystallized simultaneously. This observation requires that some of the fluid responsible for alteration was characterized by a CO_2 partial pressure of as much as 10^{-6} bar, depending upon temperature (Armstrong et al. 1982).

The same alteration processes proposed for CM chondrites may be extended to explain the aqueous alteration products observed in the CIs. The CI chondrites contain predominantly phyllosilicates, carbonates, sulfides and oxides, with cross-cutting fractures filled with carbonates and sulfates. Richardson (1978) has unraveled the sequence of mineral precipitation within the veins in Orgueil, from vein cross-cutting relationships. The earliest stage of this alteration was characterized by intense fracturing of the meteorite parent material, and contemporaneous filling of these fractures by carbonates. Following the cessation of carbonate deposition, Ca-sulfates were precipitated, followed, ultimately, by the precipitation of Mg-sulfates. In this model, the vein-filling minerals were provided by the dissolution of olivine, sulfides and phyllosilicates, with progressive local loss of CO_2 and H_2O with time. Apparently, Fe and Mn liberated in the early stages of this dissolution process were partially oxidized and reprecipitated early on as magnetite and carbonates, because later vein-filling minerals are generally low in those elements (Fredriksson et al. 1980).

Interplanetary Dust Particles

Even compared to all other types of primitive extraterrestrial material, our understanding of IDPs is at a very low level. However, from a consideration of the mineralogy of the two classes of chondritic IDPs which contain hydrated phases, several suggestions can be advanced. Particles of the so-called smectite class (which may not contain any smectites) show similarities to CV, CO and unequilibrated ordinary chondrites, these being the presence of phyllosilicates in intimate contact with glass. No altered pyroxenes are, however, observed in the smectite class IDPs, so this analogy is not complete. It seems probable that the phyllosilicate phase(s) within these particles was derived from the glass by aqueous alteration, perhaps yielding the observed finely admixed Fe sulfides and magnetite in the process. However, it would be premature to attempt further characterization of this alteration process at this time.

There is currently no evidence to indicate that the serpentine class of chondritic IDPs represents anything other than ablation fragments of CI or CM chondrites. At the very least, they seem closely related to those meteorites by mineralogy.

3.4.6. CONCLUSIONS

Our knowledge of the nature of products of aqueous alteration in primitive meteoritic materials has improved to the point where we can begin to model the course of the alteration processes. These models, however, are still at a primitive level. Much important work remains to be performed in this field, and we can safely conclude that many exciting discoveries have yet to be made. It is clear from isotopic studies that aqueous solutions were present within parent bodies during the very earliest stages of solar-system history. Thus, the study of those aqueous processes represents an important cosmochemical topic in its own right. In addition, unfortunately, those aqueous processes destroyed a portion of the record of the early solar system. Glassy and fine-grained materials were very subject to such alteration. Pyroxenes, olivines, metal, refractory oxides and sulfides were also altered or destroyed. There is evidence that later generations of oxides and sulfides were generated during or following the peak of aqueous alteration, adding possible confusion to the cosmochemical record.

Acknowledgments. The authors thank I. D. R. Mackinnon, J. L. Gooding, J. R. Cronin and F. Vilas for valuable comments and suggestions. MEZ wishes to acknowledge support from the NASA Planetary Materials/Geochemistry Program, and HYM from a National Aeronautics and Space Administration grant.

REFERENCES

Alexander, C., Barber, D. J., and Hutchison, R. 1986. Hydrous phases and hydrous alteration in U.O.C.s. *Meteoritics* 21:328 (abstract).
Armstrong, J. T., Meeker, G. P., Huneke, J. C., and Wasserburg, G. J. 1982. The Blue Angel: I. The mineralogy and petrogenesis of a hibonite inclusion from the Murchison meteorite. *Geochim. Cosmochim. Acta* 46:575–596.
Barber, D. J. 1981. Matrix phyllosilicates and associated minerals in C2M carbonaceous chondrites. *Geochim. Cosmochim. Acta* 45:945–970.
Barber, D. J. 1985. Phyllosilicates and other layer-structured materials in stony meteorites. *Clay Minerals* 20:415–454.
Barshay, S. S., and Lewis, J. S. 1976. Chemistry of primitive solar material. *Ann. Rev. Astron. Astrophys.* 14:81–94.
Bass, M. N. 1971. Montmorillonite and serpentine in Orgueil meteorite. *Geochim. Cosmochim. Acta* 35:139–147.
Bostrom, K., and Fredriksson, K. 1966. Surface conditions of the Orgueil meteorite parent body as indicated by mineral associations. *Smithsonian Misc. Coll.* 151:1–39.
Bradley, J. P. 1987. Chondritic interplanetary dust. *Geochim. Cosmochim. Acta* 51, in press.
Bunch, T. E., and Chang, S. 1980. Carbonaceous chondrites—II. Carbonaceous chondrite phyllosilicates and light element geochemistry as indicators of parent body processes and surface conditions. *Geochim. Cosmochim. Acta* 44:1543–1577.
Caillere, S., and Rautureau, M. 1974. Determination des silicates phylliteux des meteorites carbonees par microscopie et microdiffraction electroniques. *Compte Rendus Acad. Sci.* D279:539–542.
Christophe Michel-Lévy, M. 1969. Étude minéralogique de la chondrite CIII de Lancé. In *Meteorite Research,* ed. P. Millman (New York: Springer-Verlag), pp. 492–499.
Clayton, R. N., and Mayeda, T. K. 1984. The oxygen isotope record in Murchison and other carbonaceous chondrites. *Earth Planet. Sci. Lett.* 67:151–161.
Cohen, R. E., Kornacki, A. S., and Wood, J. A. 1983. Mineralogy and petrology of chondrules and inclusions in the Mokoia CV3 chondrite. *Geochim. Cosmochim. Acta* 47:1739–1757.
DuFresne, E. R., and Anders, E. 1962. On the chemical evolution of the carbonaceous chondrites. *Geochim. Cosmochim. Acta* 26:1085–1114.
Ebihara, M., Wolf, R., and Anders, E. 1982. Are Cl chondrites chemically fractionated? A trace element study. *Geochim. Cosmochim. Acta* 46:1849–1862.
Emerich, C., Lamarre, J. M., Moroz, V. I., Combes, M., Sanko, N. F., Nikolsky, Yu. V., Rocard, F., Gispert, R., Coron, N., Bibring, J. P., Encrenaz, T., and Crovisier, J. 1986. Temperature and size of the nucleus of Halley's comet deduced from the I.K.S. infrared Vega 1 measurements. *Proc. 20th ESLAB Symp. on the Exploration of Halley's Comet,* vol. 1, *ESA* SP-250, pp. 381–383.

Feierberg, M. A., Lebofsky, L. A., and Tholen, D. J. 1985. The nature of C-class asteroids from 3-μm spectrophotometry. *Icarus* 63:183–191.

Fleischer, M. 1987. *Glossary of Mineral Species* (Tucson: Mineralogical Record Press).

Fredriksson, K., Jarosewich, E., Beauchamp, R., and Kerridge, J. 1980. Sulphate veins, carbonates, limonite and magnetite: Evidence on the late geochemistry of the C-1 regoliths. *Meteoritics* 15:291–292 (abstract).

Fuchs, L. H., Olsen, E., and Jensen, K. J. 1973. Mineralogy, mineral-chemistry, and composition of the Murchison (C2) meteorite, *Smithsonian Contrib. Earth Sci.* 10.

Fujimura, A., Kato, M., and Kumazawa, M. 1982. Preferred orientation of phyllosilicates in Yamato-74642 and -74662 in relation to the deformation of C2 chondrites. In *Proc. 7th Symp. on Antarctic Meteorites*, Memoirs of the Natl. Inst. of Polar Research, Special Issue 25, ed. T. Nagata (Tokyo: Natl. Inst. of Polar Research), pp. 207-215.

Fujimura, A., Kato, M., and Kumazawa, M. 1983. Preferred orientation of phyllosilicate [001] in matrix of Murchison meteorite and possible mechanisms of generating the oriented texture in chondrules. *Earth Planet. Sci. Lett.* 66:25–32.

Gooding, J. L. 1984. Aqueous alteration on meteorite parent bodies: Possible role of "unfrozen" water and the Antarctic meteorite analogy. *Meteoritics* 19:228–229 (abstract).

Gooding, J. L. 1985. Clay minerals in meteorites: Preliminary identification by analysis of goodness-of-fit to calculated structural formulas. *Lunar Planet. Sci.* XVI:278–279 (abstract).

Gooding, J. L. 1986. Clay-mineraloid weathering products in Antarctic meteorites. *Geochim. Cosmochim. Acta* 50:2215–2223.

Grossman, L., and Larimer, J. W. 1974. Early chemical history of the solar system. *Rev. Geophys. Space Phys.* 12:71–101.

Hayatsu, R., and Anders, E. 1981. Organic compounds in meteorites and their origins. In *Cosmo- and Geochemistry*, vol. 99, *Topics in Current Chemistry*, (Berlin: Springer-Verlag), pp. 1–37.

Hutchison, R., Alexander, C. M. O., and Barber, D. J. 1987. The Semarkona meteorite: First recorded occurrence of smectite in an ordinary chondrite, and its implications. *Geochim. Cosmochim. Acta* 51:1875–1882.

Jedwab, J. 1971. La magnetite de la meteorite d'Orgueil vue au microscope electronique a balayage. *Icarus* 15:319–340.

Kerridge, J. F. 1964. Low-temperature minerals from the fine-grained matrix of some carbonaceous meteorites. *Ann. N.Y. Acad. Sci.* 119:41–53.

Kerridge, J. F. 1977. Correlation between nickel and sulfur abundances in Orgueil phyllosilicates. *Geochim. Cosmochim. Acta* 41:1163–1164.

Kerridge, J. F., and Bunch, T. E. 1979. Aqueous alteration on asteroids: Evidence from carbonaceous meteorites. In *Asteroids*, ed. T. Gehrels (Tucson: Univ. of Arizona Press), pp. 745–764.

Kerridge, J. F., Macdougall, J. D., and Marti, K. 1979a. Clues to the origin of sulfide minerals in CI chondrites. *Earth Planet. Sci. Lett.* 43:359–367.

Kerridge, J. F., Mackay, A. L., and Boynton, W. V. 1979b. Magnetite in CI carbonaceous chondrites: Origin by aqueous activity on a planetesimal surface. *Science* 205:395–397.

Kerridge, J. F., Fredricksson, K., Jarosewich, E., Nelen, J., and Macdougall, J. D. 1980. Carbonates in CI chondrites. *Meteoritics* 15:313–314 (abstract).

Kurat, G. 1975. Der köhlige chondrit Lancé: Eine petrologische Analyse der komplexen Genese eines Chondriten. *Tschermaks Mineralogische und Petrographische Mitteilungen* 22:38–78.

Lange, M. A., Lambert, P., and Ahrens, T. J. 1985. Shock effects on hydrous minerals and implications for carbonaceous meteorites. *Geochim. Cosmochim. Acta* 49:1715–1726.

Larimer, J. W., and Anders, E. 1967. Chemical fractionations in meteorites—II. Abundance patterns and their interpretation. *Geochim. Cosmochim. Acta* 31:1239–1270.

Lewis, R. S., and Anders, E. 1975. Condensation time of the solar nebula from extinct ^{129}I in primitive meteorites. *Proc. Natl. Acad. Sci. USA* 72:268–273.

Macdougall, J. D., Lugmair, G. W., and Kerridge, J. F. 1984. Early solar system aqueous activity: Sr isotope evidence from the Orgueil CI meteorite. *Nature* 307:249–251.

Mackinnon, I. D. R. 1982. Ordered mixed-layer structures in the Mighei carbonaceous chondrite matrix. *Geochim. Cosmochim. Acta* 46:479–489.

Mackinnon, I. D. R. 1985. Fine-grained phases in carbonaceous chondrites: Alais and Leoville. *Meteoritics* 20:702–703 (abstract).

Mackinnon, I. D. R., and Rietmeijer, F. J. M. 1987. Mineralogy of chondritic interplanetary dust particles. *Rev. Geophys. Space Phys.* 25, in press.
Mackinnon, I. D. R., and Zolensky, M. E. 1984. Proposed structures for poorly characterized phases in C2M carbonaceous chondrite meteorites. *Nature* 309:240–242.
Mason, B. 1962. The carbonaceous chondrites. *Space Science Rev.* 1:621–646.
McSween, H. Y., Jr. 1979a. Alteration in CM carbonaceous chondrites inferred from modal and chemical variations in matrix. *Geochim. Cosmochim. Acta* 43:1761–1770.
McSween, H. Y., Jr. 1979b. Are carbonaceous chondrites primitive or processed? A review. *Rev. Geophys. Space Phys.* 17:1059–1078.
McSween, H. Y., Jr. 1987. Aqueous alteration in carbonaceous chondrites: Mass balance constraints on matrix mineralogy. *Geochim. Cosmochim. Acta* 51, in press.
McSween, H. Y., Jr., and Richardson, S. M. 1977. The composition of carbonaceous chondrite matrix. *Geochim. Cosmochim. Acta* 41:1145–1161.
Meeker, G. P., Wasserburg, G. J., and Armstrong, J. T. 1983. Replacement textures in CAI and implications regarding planetary metamorphism. *Geochim. Cosmochim. Acta* 47:707–722.
Olsen, E., and Grossman, L. 1978. On the origin of isolated olivine grains in type 2 carbonaceous chondrites. *Earth Planet. Sci. Lett.* 41:111–127.
Prinn, R. G., and Fegley, B. J. 1987. The atmospheres of Venus, Earth and Mars: A critical comparison. *Ann. Rev. Earth Planet. Sci.* 15:171–212.
Richardson, S. M. 1978. Vein formation in the Cl carbonaceous chondrites. *Meteoritics* 13:141–159.
Richardson, S. M. 1981. Alteration of mesostasis in chondrules and aggregates from three C2 carbonaceous chondrites. *Earth Planet. Sci. Lett.* 52:67–75.
Rietmeijer, F. J. M. 1985. A model for diagenesis in proto-planetary bodies. *Nature* 313:293–294.
Rudnick, R. L., Ashwal, L. D., Henry, D. J., Roedder, E., Belkin, H. E., and Colucci, M. T. 1985. Fluid inclusions in stony meteorites—A cautionary note. *Proc. Lunar Planet. Sci. Conf.* 15, *J. Geophys. Res. Suppl.* 90:C669–C675.
Sears, D. W., Grossman, J. N., Melcher, C. L., Ross, L. M., and Mills, A. A. 1980. Measuring metamorphic history of unequilibrated ordinary chondrites. *Nature* 287:791–795.
Sonett, C. P., 1971. The relationship of meteoritic parent body thermal histories and electromagnetic heating by a pre-main sequence T Tauri sun. In *Physical Studies of Minor Planets,* ed. T. Gehrels, NASA SP-267, pp. 239–245.
Tomeoka, K., and Buseck, P. R. 1982. Intergrown mica and montmorillonite in the Allende carbonaceous chondrite. *Nature* 299:326–327.
Tomeoka, K., and Buseck, P. R., 1985. Indicators of aqueous alteration in CM carbonaceous chondrites: Microtextures of a layered mineral containing Fe, S, O and Ni. *Geochim. Cosmochim. Acta* 49:2149–2163.
Ugolini, F. C., and Anderson, D. M. 1973. *Ionic Migration and Weathering in Frozen Antarctic Soils,* NASA Contractor Report 2283.
Van Schmus, W. R., and Wood, J. A. 1967. A chemical-petrologic classification for the chondritic meteorites. *Geochim. Cosmochim. Acta* 31:747–765.
Zaikowski, A. 1979. Infrared spectra of the Orgueil (C-1) chondrite and serpentine minerals. *Geochim. Cosmochim. Acta* 43:943–945.
Zolensky, M. E. 1984. Hydrothermal alteration of CM carbonaceous chondrites: Implications of the identification of tochilinite as one type of meteoritic PCP. *Meteoritics* 19:346–347 (abstract).
Zolensky, M. E. 1987. Tochilinite in C2 carbonaceous chondrites: A review with suggestions. *Lunar Planet. Sci.* XVIII:1132–1133 (abstract).
Zolensky, M. E., and Mackinnon, I. D. R. 1986. Microstructures of cylindrical tochilinites. *Amer. Mineral.* 71:1201–1209.
Zolensky, M. E., Barrett, R. A., and Gooding, J. L. 1987. Mineralogical variations within the matrices of CM carbonaceous chondrites. *Meteoritics* 22, in press.

3.5. METEORITE REGOLITHIC BRECCIAS

T. E. BUNCH
NASA Ames Research Center

and

R. S. RAJAN
Jet Propulsion Laboratory

In addition to endogenic processes such as heating and aqueous activity, meteorite parent bodies were subjected also to exogenic processing brought about by the impact of other solar-system objects. Such impacts can produce a variety of effects, ranging from shock metamorphism of individual mineral grains, discussed in Chapter 3.6, to production of breccias, i.e., rocks consisting of mixtures of disparate lithic units. This chapter focuses on the study of such breccias, which has generated significant information about accretional growth of parent bodies, as well as their evolution, composition, stratigraphy and geological processing.

3.5.1. INTRODUCTION

The Apollo Missions showed that nearly the entire lunar surface is covered by a veneer, several to tens of meters thick, of loose "soil" that is referred to as regolith. Regolith consists of noncoherent rock material that ranges in grain size from submicrometer to several meters in diameter and overlies more coherent rock. The lunar regolith is a product of repeated impacts of varied magnitude, which result in a mixture of broken and crushed lithic and mineral fragments together with impact-produced glasses and clasts, and igneous materials. The regolith is permeated with solar-wind-implanted elements and solar-flare tracks. Regolith components are derived from local rocks and material transported from more distant regions by highly energetic primary and secondary impacts. A special type of impact-produced glass, termed agglutinates, is typically vesicular and contains inclusions of other regolithic material. Agglutinates are formed by very small, local impacts and are an important indication of regolith maturity, i.e., as the regolith matures by repeated impacts and turnover, the agglutinate content increases along with other exposure indicators (e.g., content of solar-wind implanted elements and density of solar-flare particle tracks) while the mean grain size decreases. Induration processes transform regolithic materials into abundant breccias and microbreccias. Much information on the lunar regolith and the lithologies that comprise it has been obtained from analysis of the Apollo cores; observations include grain-size variations, grain morphology, chemical and mineralogical variations, layering and fabric, gardening turnover rates, porosity and solar activity through time (Papike and Simon 1982; McKay et al. 1974). These surface products of on-going events serve as a guide to the interpretation of surfaces on other planetary objects subjected to meteoritic bombardment. Our interest in this chapter is the understanding of meteorite parent-body origin and evolution gained by the study of meteorite regolithic samples and other types of breccia.

Meteorite breccias (e.g., those found among howardites, eucrites, diogenites and ordinary chondrites) have been recognized since the early days of meteorite science (Partsch 1843) and were brought into perspective by Wahl (1952). They have received considerable attention in the post-Apollo period because of similarities with lunar breccias and the knowledge we have gathered of breccia-formational processes. Moreover, a single meteorite breccia sample can provide information from an extensive region of a parent body. The numerous studies of meteoritic breccias are reviewed by Prinz et al. (1977), Bunch (1975), Dymek et al. (1976), Rajan (1974), Stolper (1977) and Taylor and Wilkening (1982). Our terminology throughout this chapter is consistent with Keil (1982) unless otherwise indicated.

Unmanned missions to Mercury and fly-by imagery of Phobos, Deimos and satellites of the outer planets, have shown heavily cratered terrains on all of these objects; the Martian satellites, despite their small size, show thick ejecta-regolith blankets. The study of meteorite regolith samples can provide insight into surface and subsurface activities during accretionary stages, information on surface mixing histories, stratigraphy, thermal histories of regolith blankets, early solar-wind and flare activities, lithification processes, and otherwise unknown rock types that are present as clasts in breccias. This information greatly expands the data base with which we can assess spectral surface compositional measurements of asteroids, impact energetics, depths of excavation, burial rates and surface maturation, in addition to other theoretical, imaging and spectral studies.

Since the study of parent-body surface samples transects time boundaries, this presentation overlaps other primary and secondary processing mechanisms that are described in other chapters of this book.

3.5.2. METEORITE REGOLITH AND ACCRETIONARY AGGLOMERATES: REPRESENTATIVES AND CLASSIFICATION

In many ways, direct comparison of meteoritic breccias with lunar surface samples invites the "apples and oranges" criticism. One important point to be borne in mind is that most lunar breccias sampled by the Apollo program were made more recently than the asteroidal breccias available to us as meteorites. In addition, the formational and evolutionary histories of the Moon and asteroids are different and we cannot expect many direct similarities. For example, the turnover rate of surface materials calculated for asteroid surfaces is different from that inferred for the Moon. Modeling calculations by Housen et al. (1979) have shown that, independent of physical characteristics or size, blanketing rates exceed excavation rates on asteroids. Consequently, on a cm-to-m scale, asteroidal regoliths are poorly gardened, which could explain the relatively short exposure times to cosmic rays and other immature characteristics of gas-rich meteorites, most of which are regolith breccias, compared to the lunar soil. However, on a μm-to-mm scale, the grains on asteroidal surfaces had to be turned over much more often than typical lunar soil grains. This constraint is required to explain the much smaller anisotropy in the irradiation dosages on different surfaces of grains from gas-rich meteorites compared to grains from the lunar soil. Therefore, in order to provide an adequate representation of surface and near-surface activities of these parent bodies, we include all types of breccias and agglomerates that may have sustained at least short surface residency times or were involved in surface mixing events.

These limitations do not imply that meteorite parent bodies did not or do not now possess lunar-like soil or regolith breccias more numerous than present sampling implies and more consistent with the lunar model. Loose soil ejected from its parent body does not make a coherent meteorite; it simply augments the amount of interplanetary dust.

The prevelance of breccias in the major categories of stony meteorites is summarized in Table 3.5.1, together with the names of representative examples. Not all of these breccias are true *regolith breccias,* formed by shock lithification of unconsolidated fragmental debris. The different types of breccia are listed in Table 3.5.2. Stöffler et al. (1979) and Taylor (1982) have speculated on the geologic setting for formation of these breccia types, based on lunar and terrestrial field examples. These settings may also apply to meteorite breccias that have similarities with their respective lunar analogue.

Fragmental breccias consisting of clastic material of various lithologies formed around craters larger than a few hundred meters and as breccia layers within craters below or intermingled with impact melt sheets. *Cataclastic*

TABLE 3.5.1
Distribution of Breccias Among Types of Stony Meteorites[a]

	Class	Breccia Abundance	Examples
CHONDRITES	Enstatite	common	Abee, Hvittis, Blithfield
	H group	common	Plainview, Tysnes Island
	L group	common	Mafra, Mezö-Madaras
	LL group	abundant	Kelly, Bhola
	CV group	common	Allende, Leoville
	CO group	rare	Colony, Isna
	CM group	common	Murchison, Murray
	CI group	all[b]	Orgueil, Alais
ACHONDRITES			
Ca-poor	Aubrites	abundant	Norton County, Cumberland Falls
	Diogenites	abundant	Johnstown
	Chassignites	none	
	Ureilites	common	North Haig, Nilpena, ALHA81101
Ca-rich	Nakhlites	none	
	Howardites	all	Kapoeta, Bununu
	Eucrites	abundant	Pasamonte, ALHA76005

[a] Table modified from Keil (1982).
[b] Obscured by aqueous alteration.

breccias occur in continuous ejecta blankets around craters and as clasts in fallback breccias. They are monomict breccias with extensive intergranular brecciation. *Impact-melt breccias* have igneous matrices containing clastic debris; the rocks probably formed in crater floors at depths >1 km, below breccia piles, and in melt pods beyond crater rims. *Granulitic breccias* are metamorphosed, polymict fragmental breccias; they formed either in impact melts inside large craters (held at $1000°$ C for long periods of time) or during extended periods of regional granulite metamorphism.

In addition to lunar-like *regolith breccias* consisting of shock-lithified regolithic debris (Taylor 1982), other types of meteorite breccia occur that shed light on accretionary processes. *Primitive breccias* consist predominantly of type 3 chondritic material. In some cases, fragments of more-metamorphosed chondrites are also present. These breccias may have formed inside accreting parent bodies by mixing of rock fragments with unconsolidated, primitive material during disruption and reassembly of asteroid-sized objects (Keil 1982; Scott and Taylor 1982).

Accretionary agglomerates, represented by carbonaceous chondrites, were probably formed by low-velocity infall of accreting materials and lack the dominant cataclastic character of regolith breccias, although the matrices may have been generated from gentle comminution of brittle material during

TABLE 3.5.2
Meteorite Breccias and Agglomerates

PRIMITIVE BRECCIAS
  Contain well-preserved chondrules, glass, and opaque silicate matrix. They are common to *type 3 ordinary chondrites*. They may have formed on planetesimals or previous to regolith breccias on parent bodies.

REGOLITH BRECCIAS
  Exhibit lithification of unconsolidated fragmental debris, possibly with accretionary material. *Ordinary chondrite* examples have characteristic light-dark structure some of which are shocked. They also contain solar-wind gases and solar-flare tracks. *HED achondrite* examples may contain surface-exposure indicators, in addition to carbonaceous-chondrite xenoliths, impact-derived glasses and agglutinates.

FRAGMENTAL/CATACLASTIC BRECCIAS
  Consist of rock and mineral fragments that represent many lithologies. They are devoid of surface-exposure indicators. This may contain cogenetic melted material. Such breccias are found among *ordinary carbonaceous* and *enstatite chondrites* and brecciated *eucrites*.

IMPACT-MELT BRECCIAS
  Contain melted and unmelted clasts or mineral grains set in igneous-textured matrices. They have low abundance of quench textures. Several *ordinary chondrites* are impact-melt breccias; melt-breccia clasts also occur in *ordinary chondrites* and *howardites*.

GRANULITIC BRECCIAS
  Consist of metamorphosed fragmental breccias with metamorphic textured matrices (granoblastic to poikiloblastic). They are represented by some *LL chondrites* and clasts in *howardites*. They are buried deeply after brecciation or, alternatively, disrupted by large impacts, reassembled and buried deeply.

REGOLITH AGGLOMERATES
  Consist of bedded sedimentary textures in *CV3 meteorites*, with less than 50 vol. % fragments unless matrix is considered as clastic. *CM meteorites* could be altered equivalents. They contain mostly nonangular inclusions. Xenoliths of "dark inclusions" could represent fragments transported by impact from other source areas on the same parent body.

---

burial and compaction. All of these samples (or at least some components thereof) resided on or close to the surface of a parent body at some point in time. They have similarities to lunar breccias, although some have unique characteristics. Surprisingly few samples contain solar-flare tracks, agglutinates, glassy objects, shocked materials and micrometeorite craters. These characteristics strongly suggests that meteorite breccias are less mature than lunar regolith samples. In addition, the size distribution of grains in a breccia can reflect its relative maturity. In lunar breccias, the number of grains rapidly increases at small sizes. In contrast, gas-rich meteorites tend to have flat size distributions: meteorite breccias tend to be coarse by lunar standards (Bhattacharya et al. 1975). The essence of meteorite surface samples is not how similar they are to lunar surface rocks, but what unique information they contain, which can ultimately aid in understanding the formation and growth of planetary objects in the early stages of the solar system. For example, the manner in which a parent regolith evolves in response to its local environment of bombardment and irradiation determines the physical (and some compositional) properties of a breccia.

TABLE 3.5.3
Clast Populations in Brecciated Stony Meteorites

ORDINARY CHONDRITES

H group	*L group*	*LL group*
CM clasts	CM clasts	Carbonaceous
CI-CM transitional	C3 clasts	(new types)
LL clast	Cognate clasts	H clasts
UOC (new types)	melt	Cognate clasts
Devolatilized	L3-L6 clasts	melt
Cognate clasts		skeletal
melt		aphanitic
poikilitic		microporphyritic
microporphyritic		shock-darkened
spinifex		LL clasts (other)
shock-darkened		
H3-H5 clasts		

CARBONACEOUS   ENSTATITE

CM group	*CO,CV groups*	*EH group*	*EL group*
CM clasts	CM clasts	EH clasts	Sulfide-rich clasts
CI-CM transitional	C3 (anhydrous)	Dark inclusions	Cognate clasts
CO inclusions	C3 (hydrous)	Silica-rich clasts	melt
C3 clasts		Sulfide-rich clasts	
		Cognate clasts	
		melt	

ACHONDRITES

Ca-rich		*Ca-poor*
CM clasts		Other achondrite
C3 clasts		clasts
Other achondrite		
clasts		
Cognate clasts		
aphanitic	spinifex	
microbreccias	glass spherules	
microporphyritic	shock-darkened	
glass-laden	recrystallized	
melt	agglutinates	
poikilitic		

3.5.3. CHARACTERISTICS OF BRECCIAS

The many textures, compositionally different clasts, and other features of these complex meteorites would appear, on first approximation, to be of endless proportions. Assessment of the copious data available to us indicates that general trends emerge for each class of surface sample, which allow us to draw conclusions about the physical and chemical constraints under which these samples formed. In contrast, informational gaps also emerge, which leave many questions unanswered. Some of these will become apparent as we relate here a synthesis of pertinent information.

An outline of major breccia characteristics by class is given in Table 3.5.3. Ordinary chondrites, being the most abundant meteorite, also have the largest number of breccias. All three ordinary chondrite groups include members that contain carbonaceous-chondrite fragments as xenoliths, and many examples of cognate clasts (genetically related to the host rock). The dominant clasts in each group are lithic fragments of the same compositional group. Apart from carbonaceous-chondrite fragments, the occurrence of clasts of different groups is rare and occurs only in a few LL and H chondrites.

Fig. 3.5.1. (a) The Cangas de Onis chondrite regolith breccia showing light clasts in a dark, equilibrated matrix (scale in mm; photograph courtesy of K. Keil). (b) Slice of Kelly regolith breccia showing a variety of multicolored and multishaped components; base of slice is 9 cm.

Cognate clasts are more varied in H and LL groups. A slice of an H-chondrite regolith breccia (Fig. 3.5.1a) exemplifes the typical appearance of light-colored clasts in a dark clastic matrix, which is intrinsic to this type of breccia (Williams et al. 1985). A more diverse example (LL4 Kelly chondrite) is shown in Fig. 3.5.1b. In addition to the light clast-dark matrix signature typical of regolith breccias, Kelly has dark clasts, large quenched melt droplets, and recrystallized clasts of extremely varied metamorphic textures but similar compositions (Bunch and Stöffler 1974). Metamorphism of ordinary chondrites is discussed in Chapter 3.3. Additional examples of cognate clasts in chondritic and achondritic breccias are shown in Figs. 3.5.2. and 3.5.3.

Achondritic breccias have closer textural affinities to lunar and terrestrial impact breccias than are found in chondritic breccias. Large terrestrial single impacts have produced, in addition to less abundant types, breccias that are monomict (consisting of fragments and matrix of identical composition and origin), polymict (containing some foreign rock fragments) and polymict with

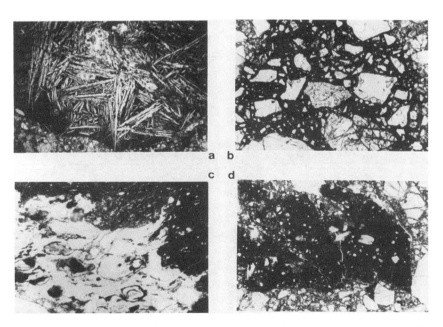

Fig. 3.5.2. Photomicrographs of thermally processed materials that are commonly observed in chondritic and HED regolith breccias (agglutinates are rare in any occurrence). (a) Rapidly cooled melt-droplet showing dendritic growth of pyroxene (Kelly chondrite; basal width is 2.5 mm). (b) Dark clast of fragmented rocks and minerals set in a sulfide matrix (Washougal howardite; basal width is 2.5 mm). (c) Agglutinate mass in Jodzie howardite (basal width is 400 μm). (d) Fine-grained to aphanitic matrix clast in the Jodzie howardite (basal width is 2.5 mm; see Fig. 3.5.3 for details).

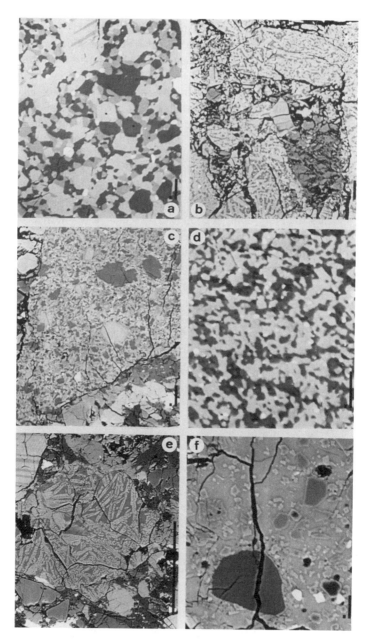

Fig. 3.5.3. (a,b,c,d) SEM-BSE (Back Scattered Electron) images of matrices in recrystallized eucritic clasts of the type shown in Fig. 3.5.2d (scale bars = 10 μm). Darkest phase is anorthite; intermediate is ferroaugite; light is ferrohypersthene, free of exsolution lamellae. (e) SEM-BSE image of quench crystals (ferrohypersthene) in plagioclase-normative-rich glass (scale bar = 100 μm). (f) SEM-BSE image of aphanitic matrix in shock-darkened clast (scale bar = 10 μm). All specimens in the Jodzie howardite.

melted material. These correspond to deep subsurface strata, ejecta and ejecta overtaken by melt, respectively. HED achondrites (howardites, eucrites and diogenites) tend to resemble this simple picture and may reflect a similar geological setting, with lower gravity scaling. Diogenites are almost exclusively monomict; eucrites and howardites contain clasts of both types in addition to clasts of unknown meteorite types. There is general acceptance that the HED achondrites came from the same parent body (see Chapter 3.2). A typical example of one of these meteorites is shown in Fig. 3.5.4.

Petrographic examination of HED achondrites indicates many secondary textures that are common to ordinary chondrites (Table 3.5.3). Extreme care must be exercised in distinguishing between primary igneous and impact-produced igneous-like textures (Figs. 3.5.5 and 3.5.6).

In addition to clasts of recognizable meteorite types, howardites contain rock types not sampled as distinct meteorites. Examples of these include basaltic clasts enriched in FeO (ferrobasalts), quartz-normative clasts (pyroxene-rich), anorthositic and one example of a hypersthene cumulative clast (Bunch 1975). Bunch et al. (1979) reported on carbonaceous-chondrite-like material present in Jodzie as xenolithic fragments. A square cm of Jodzie

Fig. 3.5.4. A sawn slice of Kapoeta showing typical texturally features, on a hand-specimen scale, of the howardite Kapoeta (long diameter is 3.25 cm).

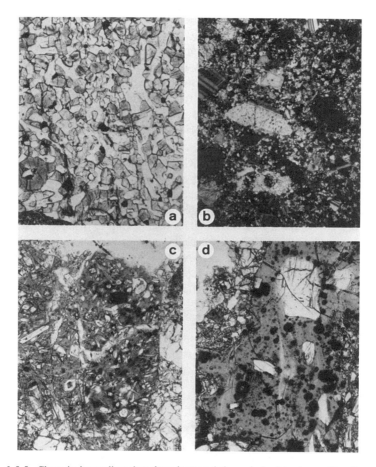

Fig. 3.5.5. Clasts in howardites that show increased thermal shock/metamorphic effects. (a) Partially recrystallized microgabbro clast in Jodzie. Unknown meteorite type although bulk and mineral compositions are eucritic (basal width is 800 μm). (b) Poikilitic breccia clast in Yuturk howardite (basal width is 350 μm). (c) Melt with quench crystals in Bununu howardite (basal width is 1.2 mm). (d) Shock-melted clast with flow, resembling suevite, in Bununu (basal width is 350 μm).

contained hundreds of such fragments that range in size from mm to μm (Fig. 3.5.7). Scanning-electron-microscope (SEM) examination of these fragments together with microprobe analyses, strongly suggested that more than one type of carbonaceous chondrite is present. Textural studies indicate that many such fragments are mechanically (and possible thermally) altered, whereas other fragments show no obvious processing other than aqueous alteration. Apparently Jodzie sampled several compositionally distinct carbonaceous-chondritic objects characterized by a range of impact energetics.

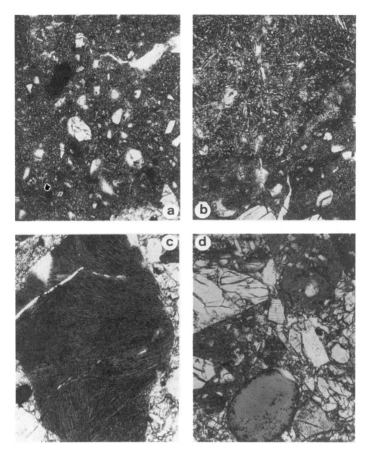

Fig. 3.5.6. Photomicrographs of recrystallized glassy-melt clasts (a,b,c) and glass spherules in Bununu (d). Basal widths are 350 μm.

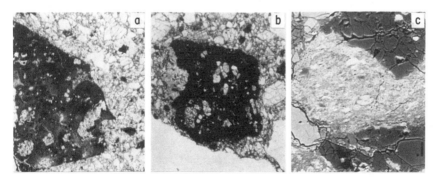

Fig. 3.5.7. Examples of carbonaceous-chondrite clasts (normal and processed) in Jodzie. (a) Photomicrograph of normal clast. (b) Photomicrograph of a different normal clast. (c) Processed clast; scale bar = 10 μm. Note granulation and foliation of material.

As discussed in Chapter 3.4, carbonaceous chondrites have been subjected to aqueous alteration on their parent bodies, ranging from extensive for CI and CM2 to minor for CO3 and CV3 (see also Chapter 10.2). The Allende meteorite (CV3) contains fine-grained, dark inclusions (Fruland et al. 1978) that have been interpreted as xenolithic remnants of earlier-formed parent-body regolithic units (Bunch et al. 1979). Allende is itself probably a sample of a regolithic unit. A decrease in volatile-element contents together with petrographic and SEM observations strongly suggest an association with metamorphic temperature/heating-time regimes that acted within various regolith units of the parent body. That dark inclusions are representative of regolith materials is shown by their apparent lithification as a result of dynamic metamorphism, their record of multiple brecciation events, veining and grain-size distributions. Typical examples of type I (least metamorphosed) and type IV (most metamorphosed) inclusions are shown in Figs. 3.5.8a and b. Multiple brecciation (3 generations) and multiple veining (2 time periods) are graphically shown in Fig. 3.5.9. Dark-inclusion regolithic units were probably formed during the phase of asteroidal accretion in which the accumulation of material exceeded ejection. This period has been referred to as the building of "accretional megaregoliths'" (Housen 1982). The asteroid grew by accumulation of broken, weakly indurated material that sustained cycles of redistribution by weak-to-moderate impacts. As the layers grew,

Fig. 3.5.8. (a) Photomicrograph of an Allende type I dark inclusion that shows only fine-grained dust and no components larger than 30 μm. (diameter = 4 mm). (b) Photomicrograph of an Allende type IV dark inclusion that shows flattened and lineated inclusions of highly processed ferromagnesian materials. Normal Allende in top and left of the section (width = 4 mm). The simplest explanation for the differences between the two dark inclusions is that they represent two different regolithic units.

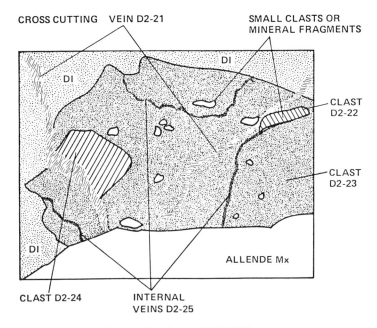

Fig. 3.5.9. Sketch of multiple-event clasts in type II DI (X38).

they were modified by the action of heating events, more energetic impacts, and the action of internal lithostatic pressures (Bunch et al. 1986).

Other clasts in Allende include *exogenic* or exotic clasts (derived from processes external to the parent body) of CO and CV material and endogenic clasts (originating by processes from within or on the parent body) that may represent examples of earlier regolithic periods. All components in these clasts exhibit incomplete thermal alterations, possibly caused by a rapid heating event(s) (Cassen et al. 1985), which may also have acted on most components in Allende. Other notable xenolithic inclusions in CV, CO and CM meteorites include various forms of CAIs (with high contents of refractory elements such as Ca, Al and Ti; see also Chapter 10.3), ferromagnesian inclusions and chondrules.

3.5.4. SOLAR-FLARE RECORD IN GAS-RICH METEORITES

As noted in the previous section, gas-rich meteorites generally have a characteristic light-dark structure. The first such meteorite was discovered by Gerling and Levski (1956), who found that it contained high concentrations of He, Ne and Ar. In each of these meteorites, the large excess of noble gases is found in the dark portion. The light part has much lower levels of noble gases, which can be accounted for easily in terms of radioactivity and exposure to cosmic rays in interplanetary space (see Chapter 4.1). Suess et al. (1964) suggested that the main fraction of the noble gases was probably incorporated by solar-wind implantation into the surfaces of grains, followed by diffusion into somewhat deeper layers. Ingenious experiments by Wänke (1965) and Eberhardt et al. (1965) confirmed this by showing that almost all the gas resided in the top 1000 Å or less, which is roughly the range of 1 keV/nucleon solar-wind particles. The epoch of this irradiation, i.e., the compaction age of the gas-rich breccias is difficult to determine precisely, but is believed to be around 4.3 Gyr, at least for chondrites (see Chapter 5.4).

The irradiation of individual grains by ancient solar wind, suggests the possibility of contemporaneous irradiation by ancient solar flares. Such irradiation should manifest itself in a population of tracks identifiable on the basis of either (1) track densities that are orders of magnitude higher than the background tracks from galactic cosmic rays, or (2) a track density gradient over several tens of μm, or (3) both. (Tracks are linear regions of lattice damage produced by the energy released during deceleration of the solar-flare or cosmic-ray particle.) Lal and Rajan (1969) and Pellas et al. (1969) discovered such grains with huge track densities and characteristic gradients (hereafter called track-rich grains) in addition to a host of normal grains in the dark portions of several gas-rich meteorites. The discovery of such grains established that the grains had indeed been bombarded by low-energy solar-flare particles (0–20 MeV/nucleon) in addition to the solar-wind (~1 keV/nucleon) particles, which implanted the gases. The observation that even in the dark portion, only a small fraction (between 2 and 20%) of grains had solar-flare tracks, demonstrated unequivocally that irradiated and unirradiated materials were intimately mixed on a 10-to-100 μm scale.

One of the crucial questions which can be effectively answered from a detailed study of solar-flare track records, pertains to the identification of the location in which irradiation of these grains took place, which clearly preceded the brecciation of the meteorites themselves. The two probable locales for the prebrecciation irradiation of these grains are: (1) irradiation in free space, and (2) irradiation in a regolith. Detailed study of the irradiation geometry of individual track-rich grains showed that most of them exhibited strong anisotropy, as would be expected for irradiation in a regolith (cf. Rajan 1974). In contrast, irradiation in free space, which would lead to isotropic irradiation, can be ruled out on the basis of such studies. The formation of gas-rich meteorites on a parent body has been discussed by a number of authors (Fredriksson and Keil 1963; Anders 1964; Eberhardt et al. 1965; Wilkening 1971; Poupeau and Berdot 1972; Pellas 1972), based on evidence obtained from petrographic and rare-gas studies.

3.5.5. MODELS OF ASTEROIDAL REGOLITH FORMATION AND LITHIFICATION PROCESSES

Not having had the opportunity to do field work on asteroidal bodies, we are relegated to the position of piecing together the geology of such bodies by laboratory investigations on meteorite samples and by constructing models that describe cratering events and regolith evolution. Regolith modeling for asteroids of different sizes has been performed by many investigators and are summarized by Housen and Wilkening (1982). Housen et al. (1979), found that the deepest regoliths can be expected to occur on rocky asteroids ~ 300 km in diameter. A body of this size could have accumulated 3.5 km of regolith. The difference in regolith depth between the Moon and the largest asteroids is mostly the result of differences in cratering flux and body size. The cratering flux is higher for asteroids at 3 AU than it is for the Moon at 1 AU. If a 300 km diameter asteroid were moved from 3 AU to 1 AU, it would accumulate only 30 m of regolith instead of 3.5 km over a period of 3.5 Gyr, due to the reduced flux rate. Moreover, smaller asteroids form thinner regoliths because of ejecta loss resulting from the weak gravity of small bodies. Regolith gardening is minimal compared with the Moon because the formation rate of ejecta blankets exceeds the excavation rate by cratering. Thus, grains are excavated few times or not at all compared with the lunar regolith where gardening is a dominant process. The net result is that lunar regolith is much more mature than the asteroidal regoliths from which brecciated meteorites are derived. As noted earlier, lunar breccias commonly exhibit greater exposure times to solar-wind and solar-flare irradiation and higher contents of glasses, agglutinates and microcraters, compared with meteoritic breccias.

The calculations discussed above pertain to the present-day solar system; conditions were presumably very different 4.5 Gyr ago. Housen (1982) defined two distinct periods of regolith evolution. (1) Early in the growth period of a small body, accreted material was fractured and comminuted. These bodies grew by building successive layers of rocky debris (regolith). The collection of weak coherent debris is referred to as "accretional megaregolith." (2) When the relative velocities of impacting material and asteroid reached the presently observed value of 5 km/s, the net effect was to lose material. In the event of large, nearly catastrophic collisions on large asteroids, deeply buried rocks may be excavated and in turn surficial layers may be mixed to great depths (a "megaregolith"). Seismic waves may also have played an important role in mixing surface material (Cintala et al. 1979; Hörz and Schall 1981).

On a strong, rocky asteroid such as one covered with basalt, ejecta from large craters are spread over large regions. In-flight comminution and secondary cratering also play important roles in the reduction of grain size and distribution of ejecta. Langevin (1982) concluded that shock comminution on medium-sized basaltic asteroids should result in a mean grain size larger than for the Moon, because of a very different size distribution for the impacting objects, although the finer grain size of meteoritic breccias compared with theoretical calculations is unexplained. Langevin also postulated that carbonaceous regolith, such as is believed to be present on C-type asteroids (Chapter 2.1) should be very fine grained. The comminution shock required to break up carbonaceous material is much less than for basaltic rock. This is consistent with observations of carbonaceous chondrites, although the very early accretionary history of these meteorites should also be considered.

We have seen that impact events form regoliths and that they also may be responsible for the lithification of regolith breccias. Ashworth and Barber (1976) showed that chondritic breccias were affected by shock which resulted in grain boundary melting and recrystallization. They distinguished breccias, that show limited grain-boundary melting from those that contain greater amounts of interstitial recrystallized material, presumably as a result of having experienced higher shock pressures. Furthermore, they suggested that this process was responsible for the lithification of gas-rich chondritic regolith breccias. Bischoff et al. (1983) identified an additional class of breccia, which showed an even greater magnitude of shock melting and lithification that was inversely correlated with noble-gas contents. They also concluded that melt formation was contemporaneous with the consolidation process and was responsible for the regolith-to-rock conversion.

Little similar work has been done with brecciated achondrites, such as the HED group. We have examined matrix in four howardites, (Yurtuk, Noblesboro, Jodzie and Petersburg) and one monomict eucrite (Sioux County) and observed minor interstitial melt at grain boundaries in Noblesboro (although heterogeneously distributed) and Sioux County. Our initial conclusions are that HED meteorites formed in a manner similar to that advocated by Ashworth and Barber (1976) and Bischoff et al. (1983). Fig. 3.5.10 shows SEM images of different HED meteorite matrices that appear to have a range of textures similar to those described by these authors. Possibly, much may be learned about breccia formation by applying analytical and observational techniques to all types of breccia matrix.

CI and CM2 chondrites are aqueously altered and little or no information from the epoch prior to alteration is available. The well-studied CV, Allende, is foliated with lineation of components in the plane of foliation, which is consistent with a compaction origin. Supportive evidence of this concept is shown in Fig. 3.5.11 in which a series of SEM images of a "pinched" area of Allende matrix are shown. This kind of feature is common in Allende, especially in matrix between very large objects. No evidence of shock lithification has been found.

Also in Allende, certain dark inclusions may have been subjected to medium-to-high temperatures for hours or days. These inclusions have been shown to have lost volatiles such as S, Cl, Zn and C, and noble gases (Ne and Ar). In addition, they are mostly recrystallized with a distinct lineated fabric contemporaneous with recrystallization (Bunch et al. 1986). The data

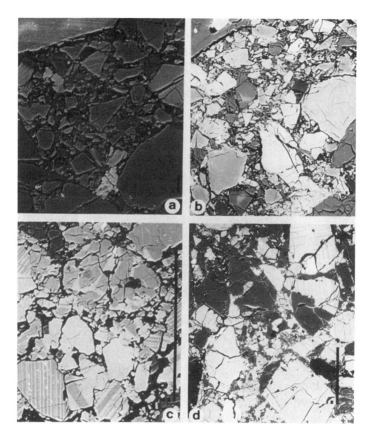

Fig. 3.5.10. (a) Angular grains in Washougal (Class A breccia) (SEM-BSE; scale bar = 10 μm). (b) Matrix of Frankfort (scale bar = 100 μm). (c) Matrix of the eucrite Sioux County (scale bar = 100 μm). (d) Crushed and "welded" appearance of Noblesboro; darkest grains are plagioclase, intermediate are ferroaugite, light are ferrohypersthene (scale bar = 100 μm).

strongly suggest that dark inclusions underwent lithification, within a thick hot ejecta blanket, in a manner analogous to Apollo 14 breccias (Williams 1972).

3.5.6. SUMMARY AND PROBLEM AREAS FOR FUTURE CONSIDERATION

In summary, meteorite regolith breccias have yielded significant information about parent-body growth, evolution, composition, stratigraphy, igneous and metamorphic processes, in addition to unique components (e.g., refractory inclusions in CM, CV and CO meteorites [Chapter 10.3]). A few problem areas for future consideration are offered below:

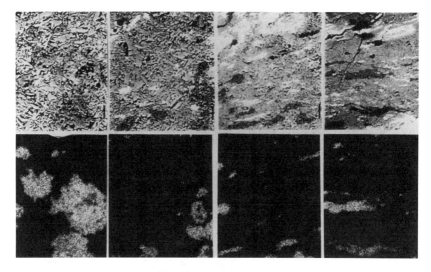

Fig. 3.5.11. Series of SEM-BSE images (upper) of "pinched" area in Allende matrix between two large inclusions. Note decrease in porosity and increased elongation of inclusions going from left to right. Lower series are Ca-X-ray images showing Ca-rich inclusions in upper BSE images.

1. Far too little comprehensive information on key regolith samples is available. Meteoritic breccias are logical candidates for study by consortia employing a broad range of analytical and observational techniques. Systematic investigations should be performed on the same components of these meteorites. Results could provide a better data base for investigators including theoreticians and modelers.
2. There is growing evidence that medium- to large-sized parent bodies (300–1000 km) may have had transient atmospheres during their accretionary-megaregolith stages. How could they have been generated and what were the lifetimes? What were the effects on surface materials and material inputs to the parent body?
3. Refined regolith modeling has made successful inroads towards better understanding of small-body regolith characteristics. Experimentation with multiple-layered target materials coupled with better theoretical understanding of how ejecta velocities should scale with impact energy, momentum, gravitational acceleration, etc. are needed. Ballistic experiments can also aid in understanding the distributions and mixing of variable component materials.
4. Heating and dynamic lithification experiments on simulated regolith soil samples could give better insight into lithification processes, compaction and metamorphic conditions of meteorite parent bodies. In addition,

volatile-element transport, enrichment and fractionation patterns could also be studied under these experimental conditions.

REFERENCES

Anders, E. 1964. Origin, age and composition of meteorites. *Space Sci. Rev.* 3:583–714.
Ashworth, J. R., and Barber, D. J. 1976. Lithification of gas-rich meteorites. *Earth Planet. Sci. Lett.* 30:222–233.
Bhattacharya, S. K., Goswami, J. N., Lal, D., Patel, P., and Rao, M. N. 1975. Lunar regolith and gas-rich meteorites: Characterization based on particle tracks and grain-size distributions. *Proc. Lunar Sci. Conf.* 6:3509–3526.
Bischoff, A., Rubin, A. E., Keil, K., and Stöffler, D. 1983. Lithification of gas-rich chondrite regolith breccias by grain boundary and localized shock melting. *Earth Planet. Sci. Lett.* 66:1–10.
Bunch, T. E. 1975. Petrography and petrology of basaltic achondrite polymict breccias (howardites). *Proc. Lunar Sci. Conf.* 6:469–492.
Bunch, T. E., and Stöffler, D. 1974. The Kelly chondrite: A parent body surface metabreccia. *Contrib. Mineral. Petrol.* 44:157–171.
Bunch, T. E., Chang, S., Frick, U., Neil, J., and Moreland, G. 1979. Carbonaceous chondrites—I. Characterization and significance of carbonaceous chondrite (CM) xenoliths in the Jodzie howardite. *Geochim. Cosmochim. Acta* 43:1727–1742.
Bunch, T. E., Chang, S., Cassen, P., and Reynolds, R. 1986. Allende: Profile of parent body growth. *Lunar Planet. Sci.* XVII:89–90 (abstract).
Cassen, P., Reynolds, R., Lissauer, J., Bunch, T., and Chang, S. 1985. Heating of CAIs in the environment of an accreting body. *Lunar Planet. Sci.* XVI:117–118 (abstract).
Cintala, M. J., Head, J. W., and Wilson, L. 1979. The nature and effect of impact cratering on small bodies. In *Asteroids*, ed. T. Gehrels (Tucson: Univ. of Arizona Press), pp. 579–600.
Dymek, F. R., Albee, A. L., Chodos, A. A., and Wasserburg, G. J. 1976. Petrology of isotopically-dated clasts in the Kapoeta howardite and petrologic constraints on the evolution of its parent planet. *Geochim. Cosmochim. Acta* 40:1115–1130.
Eberhardt, P., Geiss, J., and Grogler, N. 1965. Uber die Verteilung der Uredelgaso in Meteoriten Khor Temiki. *Tschermaks Mineral. Petrog. Mitt.* 10:535.
Fredriksson, K., and Keil, K. 1963. The light-dark structure in the Pantar and Kapoeta stone meteorites. *Geochim. Cosmochim. Acta* 27:717–739.
Fruland, R. M., King, A. E., and MacKay, D. S. 1978. Allende dark inclusions. *Proc. Lunar Planet. Sci. Conf.* 9:1305–1329.
Gerling, E. K., and Levskii, L. K. 1956. On the origin of the rare gases in stony meteorites. *Dokl. Akad. Nauk SSSR* 110:750–753.
Horz, F., and Schaal, R. B. 1981. Asteroidal agglutinate formation and implications for asteroidal surfaces. *Icarus* 46:337–353.
Housen, K. R. 1982. Modeling the evolution of asteroidal regoliths. In *Workshop on Lunar Breccias and Soils and Their Meteoritic Analogs*, eds. G. J. Taylor and L. L. Wilkening, Tech. Rept. 82–02 (Houston: Lunar and Planetary Inst.), pp. 38–46.
Housen, K. R., and Wilkening, L. L. 1982. Regoliths on small bodies in the solar system. *Ann. Rev. Earth Planet. Sci.* 10:355–376.
Housen, K. R., Wilkening, L. L., Chapman, C. R., and Greenberg, R. J. 1979. Asteroidal regoliths. *Icarus* 39:317–351.
Keil, K. 1982. Composition and origin of chondritic breccia. In *Workshop on Lunar Breccias and Soils and Their Meteoritic Analogs*, eds. G. J. Taylor and L. L. Wilkening, Tech. Rept. 82–02 (Houston: Lunar and Planetary Inst.), pp. 65–83.
Lal, D., and Rajan, R. S. 1969. Observations on space irradiation of individual crystals of gas-rich meteorites. *Nature* 223:269–271.
Langevin, Y. 1982. Evolution of an asteroid regolith: Granulometry, mixing, and maturity. In *Workshop on Lunar Breccias and Soils and Their Meteoritic Analogs*, eds. G. J. Taylor and L. L. Wilkening, Tech. Rept. 82–02 (Houston: Lunar and Planetary Inst.), pp. 97–109.
McKay, D. S., Fruland, R. M., and Heiken, G. H. 1974. Grain size and the evolution of lunar soils. *Proc. Lunar Sci. Conf.* 5:887–906.

Partsch, P. 1843. Die meteoriten oder Von Himmel gefallenen Stein-und Eisenmassen. In *K. K. Hofmineralienkabinette zu Wien*, Catalog of Vienna Collection, Vienna.

Papike, J. J., and Simon, S. B. 1982. The lunar regolith: Chemistry, mineralogy, and petrology. *Rev. Geophys. Space Phys.* 20:761–826.

Pellas, P. 1972. Irradiation history of grain aggregates in ordinary chondrites. Possible clues to advanced stages of aggregation. In *From Plasma to Planet, Proc. Nobel Symposium 21*, ed. A. Elvins (New York: Wiley), pp. 65–92.

Poupeau, G., and Berdot, J. L. 1972. Irradiations ancienne et récent des aubrites. *Earth Planet. Sci. Lett.* 14:381–396.

Prinz, M., Foder, R. V., and Keil, K. 1977. Comparison of lunar rocks and meteorites. In *The Soviet-American Conference on Cosmochemistry of the Moon and Planets*, eds. J. H. Pomeroy and N. J. Hubbard, NASA SP-370, pp. 183–199.

Rajan, R. S. 1974. On the irradiation history and origin of gas-rich meteorites. *Geochim. Cosmochim. Acta* 38:777–788.

Scott, E. R. D., and Taylor, G. J. 1982. Primitive breccias among type 3 ordinary chondrites—Origin and relation to regolith breccias. In *Workshop on Lunar Breccias and Soils and Their Meteoritic Analogs*, eds. G. J. Taylor and L. L. Wilkening, Tech. Rept. 82-02 (Houston: Lunar and Planetary Inst.), pp. 131–142.

Stöffler, D., Knoll, H. D., and Maerz, U. 1979. Terrestrial and lunar impact breccias and the classification of lunar highland rocks. *Proc. Lunar Planet. Sci. Conf.* 10:639–675.

Stolper, E. 1977. Experimental petrology of eucritic meteorites. *Geochim. Cosmochim. Acta* 41:587–611.

Suess, H. E., Wänke, H., and Wlotzka, F. 1964. On the origin of gas-rich meteorites. *Geochim. Cosmochim. Acta* 28:595–607.

Taylor, G. J. 1982. Composition and origin of chondritic breccias. In *Workshop on Lunar Breccias and Soils and Their Meteoritic Analogs*, eds. G. J. Taylor and L. L. Wilkening, Tech. Rept. 82-02 (Houston: Lunar and Planetary Inst.), pp. 143–151.

Taylor, G. J., and Wilkening, L. L., eds. 1982. *Workshop on Lunar Breccias and Soils and Their Meteoritic Analogs*, Tech. Rept. 82-02 (Houston: Lunar and Planetary Inst.).

Wahl, W. 1952. The brecciated stony meteorites and meteorites containing foreign fragments. *Geochim. Cosmochim. Acta* 2:91–117.

Wilkening, L. L. 1971. Particle track studies and the origin of gas-rich meteorites. Nininger meteorite award paper, pp. 1–36.

Williams, C. V., Rubin, A. E., Keil, K., and San Miguel, A. 1985. Petrology of the Cangas de Onis and Nulles regolith breccia: Implication for parent body history. *Meteoritics* 20:331–345.

3.6. SHOCK EFFECTS IN METEORITES

D. STÖFFLER, A. BISCHOFF
University of Münster

V. BUCHWALD
The Technical University of Denmark

and

A. E. RUBIN
University of California at Los Angeles

Impacts between objects on intersecting orbits in the solar system can generate a wide variety of effects in their constituent materials. These phenomena range from regolith production and breccia formation, introduced in the preceding chapter, to shock-induced deformations and transformations that can create new mineral phases or melt old ones. Such shock-metamorphic effects have been observed in all major groups of meteorites. They affect not only the petrographic characteristics but also the chemical and isotopic properties and the ages of primordial meteoritic material. It appears that collision-induced hypervelocity impacts took place prior to, simultaneously with and subsequent to the end of accretion and early differentiation of the parent bodies to which ordinary chondrites and differentiated meteorites are related. A full understanding of shock metamorphism and breccia formation in meteorites is necessary in order to resolve more clearly not only the early processes of accretion, differentiation and regolith evolution of the parent bodies, but also the primordial composition of the accreted material itself.

3.6.1. INTRODUCTION AND HISTORY OF RESEARCH

Investigators in the 19th century described certain textural features in meteorites that we now believe to be produced by shock waves during parent-body collisions. Partsch (1843) and von Reichenbach (1860) recognized "polymict breccias" and Tschermak (1872) identified "maskelynite" as an isotropic form of plagioclase whose shock origin was not revealed until 1963 (De Carli and Milton 1963). Brecciation and the occurrence of veins, interpreted later as shock effects by Fredriksson et al. (1963), were used as basic criteria of the Rose-Tschermak-Brezina classification of meteorites (Brezina 1904). In later classification schemes (see, e.g., Prior 1920; Mason 1962), the shock-related properties of meteorites were considered nonfundamental. Their genetic importance emerged only recently by the obvious analogies of shocked and brecciated meteorites to lunar and terrestrial impact breccias (see, e.g., Prinz et al. 1977; Stöffler et al. 1980; Taylor and Wilkening 1982).

It should be emphasized that our current understanding of impact cratering and collisional fragmentation processes suggests a strong genetic relationship between the two major effects of shock in meteorites: shock metamorphism and brecciation (see, e.g., Roddy et al. 1977; Grieve et al. 1977; Stöffler et al. 1979). Although a fundamental review of meteoritic breccias was given by Wahl (1952), it was not until the post-Apollo epoch that meteorites were seriously dealt with as parts of impact formations (see, e.g., Keil et al. [1969]; Fodor et al. [1972]; and later papers reviewed in Keil [1982]; Taylor [1982]).

Important chemical and mineralogical properties of meteorites as well as "young" ages were not attributed to shock metamorphism until the early 1960s (see, e.g., Anders [1964]; papers in French and Short [1968] and references therein) with the exception of shock effects in irons (see, e.g., Uhlig 1955). Although shock effects in meteoritic minerals were included in some general publications on shock metamorphism (French and Short 1968; Stöffler 1972,1974) and short sections on shock and brecciation are contained in some textbooks (Wasson 1974,1985; Buchwald 1975; Dodd 1981), a comprehensive review of shock metamorphism *and* brecciation in the field of meteoritic research has not been available until now.

A summary of the types of meteoritic impact breccias and their possible geologic settings is given in Appendix 2.

3.6.2. BASIC PHYSICS OF SHOCK COMPRESSION OF MINERALS AND ROCKS

General Aspects

Sudden acceleration of the free surface of a solid by hypervelocity impact of a projectile or by a chemical or nuclear explosion results in a *shock wave*, also called shock front, propagating with supersonic velocity. In contrast to normal compression waves (e.g., seismic waves) the material engulfed by a shock wave moves behind the shock front with a high *particle velocity* (which is lower than the *shock velocity*). Theoretically, the shock front represents a discontinuity in pressure P, specific volume V or density, and internal energy E. Assuming a plane wave geometry and hydrodynamic conditions (absence of material strength) the parameters characterizing the compressed material are related to the corresponding parameters P_o, V_o, E_o of the unshocked material by the Rankine-Hugoniot equation

$$E - E_o = (P + P_o)(V_o - V). \qquad (1)$$

Details of the physics of shock waves in homogeneous solids may be found in the following reviews: Duvall and Fowles (1963); McQueen et al. (1970); Davison and Graham (1979) and papers in Akimoto and Manghnani (1982).

The graphical representation of the Rankine-Hugoniot equation in the P, V plane is termed the *Hugoniot curve*. It defines the locus of all shock states achievable in any material by shock waves of variable intensity, e.g., by various impact velocities of a projectile (Fig. 3.6.1). In practice, a shock front has a finite extent or thickness at which the peak pressure is maintained until the shock front is overtaken by a rarefaction wave originating from a free surface such as the rear side of the projectile. In the case of a projectile impact on an infinite target, the pressure decay with increasing distance from the point of impact is enhanced for geometrical reasons because of the spherical geometry of the propagating shock front (see, e.g., papers in Roddy et al. [1977]).

As seen from the Hugoniot curve (Fig. 3.6.1), irreversible work is done in a shock process resulting in a certain amount of post-shock heat (waste heat) which increases with increasing peak shock pressure. The material shocked to state A is compressed along the Rayleigh line and decompressed along an adiabat (Fig. 3.6.1). The waste heat which is proportional to the area between the two curves may lead to melting or even vaporization if the shock pressure is high enough.

It is important to emphasize at this point that three types of thermal states or temperatures must be distinguished in characterizing the thermal history of the constituents of meteorite breccias:

1. The shock temperature achieved in a single rock fragment by the shock compression prior to the breccia formation;
2. The post-shock temperature of this fragment resulting from the shock-induced waste heat;
3. The equilibrium temperature achieved after deposition of a polymict breccia formation which consists of rock fragments of variable post-shock temperature, e.g., melt particles, shocked and unshocked rock fragments, all mixed together.

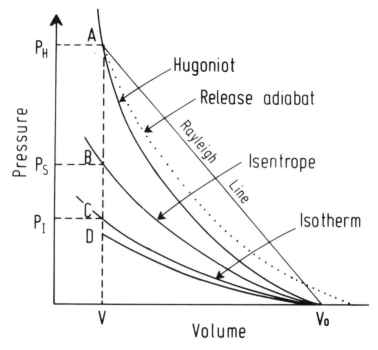

Fig. 3.6.1. Generalized Hugoniot curve A of a solid for shock compression under ideal hydrodynamic conditions. D is hydrostatic compression. Irreversible heat produced by a shock wave with peak pressure P_H is represented by the area between the release adiabat and the Rayleigh line (figure from Stöffler 1984).

Special Conditions for Polycrystalline and Porous Materials

Regarding the shock compression of polycrystalline multiphase and porous solids, additional factors not mentioned above are important: crystallographic orientation of the mineral constituents with respect to the shock front, differences in shock impedance (shock wave velocity times density) of the constituent phases (minerals, liquid, gas, vacuum), grain size, grain morphology and porosity. These heterogeneities of the target material essentially cause locally multiple shock-wave reverberations and therefore stress and temperature concentrations at grain boundaries and free surfaces. These effects are most severe in porous rocks or sediments (Kieffer 1971,1975). In addition, the equilibrium peak pressure in a heterogeneous solid is reached via multiple shock-wave reverberations consisting of shock and rarefaction wavelets produced at grain boundaries. This results in a distinctly different stress-time profile for such materials compared to single crystals. Furthermore, the amount of post-shock heat deposited in the material by a shock wave of a certain intensity strongly increases with increasing porosity.

3.6.3. PHYSICAL AND GEOLOGICAL CONTEXT OF SHOCK METAMORPHISM AND BRECCIA FORMATION RECORDED IN METEORITES

Shock metamorphism and breccia formation as recorded in the present-day meteorites are the result of collisions of their parent bodies during their history. Since the time when the Rose-Tschermak-Brezina classification of meteorites was abandoned by the cosmochemists, shock metamorphism and brecciation have been considered as secondary or even tertiary processes of alteration obliterating the "important" genetically relevant features. We believe that this is a misconception which in fact obliterates the recognition of important processes in the very early history of the meteorites' parental material.

It appears useful to introduce a working hypothesis about the time, type and intensity of collision-induced shock wave processes which may have occurred during the evolution of meteorites and parent bodies.

Evolutionary Phases of Small Bodies and Related Types of Collisions

Theoretically, collision-induced shock waves may have been produced in different ways during five major phases of the evolution of meteorite parent bodies:

a. Collision of dust particles or of aggregates of particles during the formation of an accretion disk in the solar nebula;
b. Collision of particles during the accretion of planetesimals;
c. Collision of planetesimals during the accretion of the meteorite parent bodies;
d. Interplanetary collision of the meteorite parent bodies immediately after accretion;
e. Late collisions of the meteorite parent bodies and formation of Earth-crossing meteoroids.

Collision-induced Shock Metamorphism and Brecciation

The relative collision or impact velocities and the size of the colliding bodies are greatly variable in going from phases a to e and also within a particular phase. The effects of hypervelocity collisions depend on the relative and absolute size (masses) of the colliding bodies (Fig. 3.6.2). Other critical parameters are the relative impact velocity and the specific impact energy (projectile kinetic energy/initial target mass). These parameters have been studied experimentally and theoretically (see, e.g., Gault and Wedekind 1969; Fujiwara et al. 1977; Hartmann 1979; Fujiwara 1980; Cintala and Hörz 1984) and can only be mentioned here briefly.

The above mentioned phases a and b are characterized by low-velocity grain collisions in general. True shock effects can hardly be expected except for fracturing of mineral grains as observed rarely in refractory inclusions of

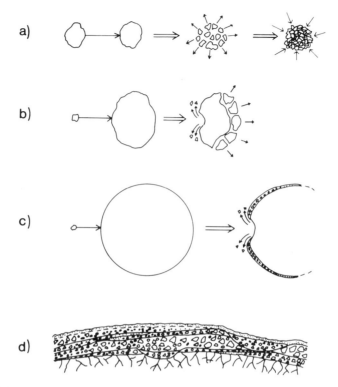

Fig. 3.6.2. Effects of interplanetary hypervelocity collisions for variable size ratios of impactor and target: a = total disruption of both bodies and possibility of reaccretion; b = cratering and spallation; c = cratering and regolith formation; d = cratering and formation of regolith and megaregolith (superposition of individual ejecta blankets; projectile not shown).

primitive chondrites. Low-velocity collisions are still predominant in phases b and c. However, "accretionary breccias" (without shock effects) as defined by Scott and Taylor (1982), Kracher et al. (1985) and Grimm (1985) may have been formed during these phases. Phase d contains three new elements: impact cratering, shock metamorphism with intense shock waves (melting) and impact-induced breccia formation. There is a limiting shock pressure and hence impact velocity below which shock metamorphism can no longer be recognized petrographically (see, e.g., Stöffler 1982a,1984): 1 to 5 GPa for particulate rock material and iron and 5 to 10 GPa for nonporous crystalline rocks corresponding to impact velocities of 0.5 to 1.5 km/s and 0.7 to 1.5 km/s, respectively (the ranges relate to the density range of the impactors). With these facts in mind, we define *shock metamorphism* as the mechanical deformation and transformation (subsolidus and solid-liquid-vapor) of rocks by shock-wave compression without substantial relative movement of the rock constituents. *Breccia formation* involves mass transport, either ballistic or

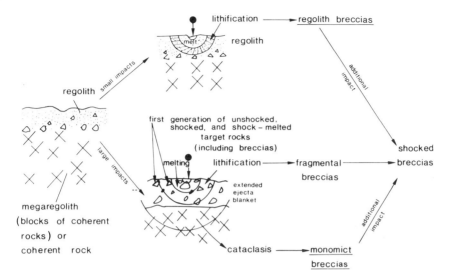

Fig. 3.6.3. Geological scenarios for the formation of various types of meteoritic impact breccias and lithic clasts of polymict meteorites. Half-spherical lines indicate transient shock-pressure isobars before crater excavation; filled circle is the projectile.

nonballistic, and therefore, the relative movement of rock fragments and their displacement from the primary location in the source material(s). Breccias commonly contain unshocked and shock-metamorphosed (shocked) clasts which were aggregated during crater excavation and ejecta deposition. These rocks have been termed "impact breccias" (Fig. 3.6.3; Stöffler et al. 1979; Keil 1982; Taylor 1982) in contrast to "accretionary breccias."

Collision Modes Bearing on Shock Metamorphism and Breccia Formation

Figure 3.6.2 depicts the basic modes of collision of meteorite parent bodies with the relative size of the bodies as the primary variable. It is important to keep in mind that the impact velocities which depend on the bodies' masses due to the gravity-induced acceleration, and the specific impact energy play an important role for the final geologic effect of any collision.

Case *a* of Fig. 3.6.2 which holds for nearly equally sized bodies (diameter ratio: 0.01–1) is relatively rare but leads to very specific effects. In the low-velocity/low-energy regime, catastrophic fragmentation and subsequent reaccretion may occur if the impact strength of the bodies is exceeded (6×10^4 to 3×10^7 erg/cm^3; Davis et al. 1979). Thereby *accretionary breccias* can be formed. More details of case *a* are discussed in Sec. 3.6.9. In cases *b* to *d* of Fig. 3.6.2, impact cratering and formation of regolith and megaregolith (impact breccias) are the dominant processes whereby the efficiency of fragmenting and displacing target material decreases in this sequence at any given mass and velocity of the projectile because the specific impact energy decreases. Moreover, the relative proportion of highly shocked to weakly or unshocked (but displaced) material increases with increasing mass (gravity) of the target body at constant impact parameters. The processes typical of cases *c* and *d* of Fig. 3.6.2 are more specifically characterized in Fig. 3.6.3.

3.6.4. SHOCK DEFORMATION AND TRANSFORMATION OF METEORITIC MINERALS

The principal types of residual shock effects in rock-forming minerals have been described in several review articles and monographs (see, e.g., French and Short 1968; Lipschutz 1968; Jain and Lipschutz 1971; Stöffler 1972,1974,1984; Buchwald 1975). They are related to specific pressure-volume regimes of the Hugoniot curve (see Stöffler 1972) and can be classified into the following categories with increasing peak pressure (Fig. 3.6.4; see also Sec. 3.6.2): (a) fracturing; (b) plastic deformations; (c) phase transformations; (d) melting or thermal dissociation; and (e) vaporization and condensation.

The characteristic deformation and transformation features observed in minerals relevant for meteorites are graphically represented on the basis of a shock pressure scale which has been obtained by shock-recovery experiments of various authors (Fig. 3.6.4). The terminology used in this presentation is taken from Stöffler (1972,1984).

For certain minerals like plagioclase and quartz, the pressure calibration is more accurate in the 25 to 50 GPa range than shown in Fig. 3.6.4. This fine calibration is based primarily on the refractive index, birefringence, and on recrystallization behavior (for details see Stöffler 1974,1984; Ostertag and Stöffler 1982; Ostertag 1983; Stöffler et al. 1986). It should be noted that the commonly used shock-pressure scale for chondrites (Dodd and Jarosewich 1979) is based on a pressure scale for shocked plagioclase (Van Schmus and Ribbe 1968) which is outdated by the present data (Fig. 3.6.4).

3.6.5. THE PRESSURE AND TEMPERATURE REGIMES OF SHOCKED METEORITES

The residual shock effects which can be produced in polycrystalline rocks by shock waves having particular peak pressures depend strongly on the initial porosity of the rock material under shock compression and on other factors (see Sec. 3.6.2.). Because of the dominant role of porosity, the pressure and temperature scales for the progressive stages of shock metamorphism will be discussed separately for nonporous and porous rocks.

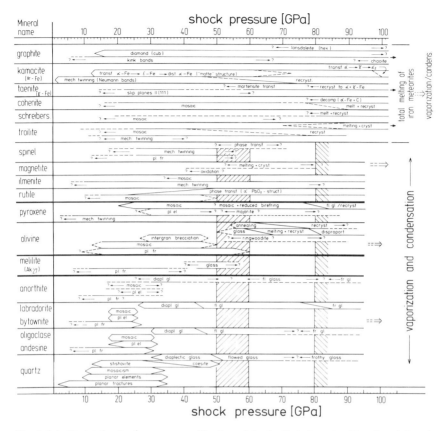

Fig. 3.6.4. Best estimates for pressure calibration of shock effects in meteoritic minerals based on observations from shock-recovery experiments, natural shocked rocks and Hugoniot data. Data sources are reviewed in Stöffler (1972, 1974, 1984); additional data sources not reviewed therein are: Lambert and Grieve (1984); Syono et al. (1981). Abbreviations are as follows: pl.fr. = planar fractures; pl.el. = planar elements; diapl. = diaplectic; mosaic. = mosaicism; fl. = flowed; fr. = frothy; gl. = glass; recryst. = recrystallization; disproport. = disproportionation; birefring. = birefringence; trans. = transformation; mech. = mechanical; cryst. = crystallization; decomp. = decomposition; hex. = hexagonal; cub. = cubic; schreibers. = schreibersite. The vertical hatched columns indicate the onset of complete melting in acidic (50–60 GPa) and basic (> 80 GPa) rocks.

Pressure-Temperature Scales for Shocked Nonporous and Nonbrecciated Meteorites

Results of controlled shock-recovery experiments on terrestrial basalts and on dunite combined with experiments on particular mineral phases such as feldspar, olivine, pyroxene and ore minerals serve most appropriately to establish a P-T scale of progressive shock metamorphism of nonporous and nonbrecciated (monomict) meteorites (Kieffer et al. 1976a; James 1969;

TABLE 3.6.1
Progressive Stages of Shock Metamorphism of Basaltic Achondrites[a]

Shock Stage	Equilibrium Peak Pressure (GPa)	Post-shock Temperature[b] (°C)	Texture and Characteristic Shock Effects	Accompanying Disequilibrium Shock Effects
0			unshocked (no unequivocal shock effects)	none
	~1 – 5	~0		
1			fractured silicates; mechanical twinning in pyroxene and ilmenite kink bands in mica if present; rock texture preserved	none
	~20 – 22.5	~50 – 150		
2 a b	~28 – 34	~200 – 250	plagioclase with planar elements (isotropic lamellae) below 28–34 GPa; above: completely isotropic plagioclase; mechanical twinning in pyroxene and ilmenite mosaicism in olivine and other silicates	incipient formation of localized "mixed melt", and glassy veins (above ~30 GPa)

3	~42 – 45 ——— ~900 (?)	plagioclase glass with incipient flow structure and vesicles; mafics and ore as in stage 2
	——— ~1100 (?)	
4	~60	normal plagioclase glass with vesicles and flow structure; incipient edge-melting of pyroxene; incipient re-crystallisation of olivine
	——— ~1500 (?)	
5	~80	plagioclase increasingly mixed with melt of mafic and ore minerals olivine re-crystallized; whole rock melts at pressures of ~>100 GPa

melt pockets and glassy veins

[a] Table based on data of Kieffer et al. (1976b); Schaal and Hörz (1977); James (1969) Ostertag (1983); Stöffler (1984) and Stöffler et al. (1986).
[b] Post-shock temperatures are relative to 0°C ambient temperature and in part based on Raikes and Ahrens (1979).

Schaal and Hörz 1977; Reimold and Stöffler 1978; Bauer et al. 1979; Ashworth 1985). Other data sources are reviewed in Stöffler (1972,1974,1984). Two major schemes of progressive shock metamorphism are proposed, one for basaltic-gabbroic rocks (Table 3.6.1) and one for olivine-rich rocks (Table 3.6.2) as analogues for various groups of achondrites and chondrites of very low porosity (recrystallized petrographic types 4–6).

In most cases, the degree of shock deformation or transformation of the individual mineral constituents of a monomict meteorite (e.g., the degree of mosaicism in all olivine grains) appears to be rather homogeneous throughout the whole meteorite so that analysis of a thin section or measurements on randomly selected grains are sufficient for establishing the shock stage of a specimen. However, important local deviations from this homogeneity occur which result from stress and temperature concentrations at grain boundaries (see Sec. 3.6.2). This mechanism is most probably responsible for localized melting and hence for the formation of *glassy veins* (similar to pseudotachylite veins in terrestrial and lunar shocked rocks; see Stöffler et al. 1979; Stöffler 1984) and of *melt pockets*. These *disequilibrium shock effects* have not been produced in small-scale shock-recovery experiments. However, through observations on meteorites, lunar rocks and shocked rocks of terrestrial impact craters (Dodd and Jarosewich 1979; Stöffler et al. 1986), it was possible to correlate them with the equilibrium shock pressure experienced by the meteorite as a whole (Tables 3.6.1 and 3.6.2). The pressure-temperature scales and the main characteristics for progressive stages of shock of basaltic achondrites and of recrystallized chondritic meteorites are summarized in Tables 3.6.1 and 3.6.2, respectively.

Pressure-Temperature Scales for Shocked Porous Meteorites and Shocked Regolith

The shock-pressure calibration for porous and particulate meteoritic materials which may include polymict breccias, can be based on shock experiments and Hugoniot data of analogue materials such as terrestrial alluvial soil (Kieffer 1975), quartz sand and sandstone (Kieffer et al. 1976*b*; Stöffler et al. 1975), lunar soil (Schaal and Hörz 1980; Kieffer 1975; Christie et al. 1973; Simon et al. 1986), artificial mixtures of mineral detritus and particulate basalt (Gibbons et al. 1975; Schaal et al. 1979) and chondrite powder (Bischoff and Lange 1984). The calibration is given in Table 3.6.3.

In contrast to shocked nonporous rocks, the most characteristic feature of shocked porous materials is the extreme heterogeneity of the shock pressure on the microscale of a sample. In other words, individual grains vary spatially from unshocked to shock melted (Kieffer et al. 1976*a,b;* Schaal and Hörz 1980).

Another important aspect of shock metamorphism of detrital material is the lithification by intergranular melt (matrix glass) which serves as welding material (Ashworth and Barber 1976; Bischoff et al. 1983). The process in-

TABLE 3.6.2
Pressure Calibration of Shock Effects in Olivine, Plagioclase and Ordinary Chondrites[a]

Equilibrium Shock Pressure: 0, 10, 20, 30, 40, 50, 60, 80 GPa

Shock Recovery experiments

Olivine
- Irregular fractures
- Planar fractures
- Undulatory extinction
- Mosaicism
- Intergranular brecciation
- Recrystallization
- Intergr. melt. and recryst.
- Total melting

Plagioclase (Oligoclase)
- Fractures
- Planar elements
- Mosaicism
- Diaplectic glass (maskelynite)
- Normal glass (vesiculated)
- Refractive index (average ñ for An$_{24}$): 1.54, 1.53, 1.52, 1.51

	Observations on chondrites						
Plagioclase U = undeformed D = deformed M = diaplectic glass (maskelynite) N = normal glass	100% 60% 20%	U		D	M		N
Disequilibrium shock effects glassy veins and blackening melt pockets							whole rock melting above about 80–100 GPa
Recalibrated Dodd and Jarosewich (1979) scale[b]		a	b+c	d	e	f	g
Alternative shock stages		O	I	II	III	IV	V
Main petrographic characteristics		irr. fr. in all minerals	olivine pl.fr., u.ext., mos. pyroxene m.tw.	olivine pl. fr., mos., inc.br. plagioclase pl.el., mos. pyroxene m.tw. inc.gl.ve. + m.p.	olivine pl.fr., mos., br. plagioclase diapl gl. pyroxene m.tw.	olivine recr., inc.m + recr. plagioclase glass	whole rock melting
						glass veins and melt pockets	

[a] Table based on data from shock recovery experiments with olivine (Reimold and Stöffler 1978; Bauer et al. 1979; Snee and Ahrens 1975) and plagioclase (Stöffler 1972,1974; Ostertag 1983) and on the classification of shocked chondrites by Dodd and Jarosewich (1979).

[b] Note the revision of the pressure scale for the Dodd and Jarosewich classification. Irr. fr. = irregular fractures; m. tw. = mechanical twinning; mos. = mosaicism; u. ext. = undulatory extinction; inc. br. = incipient brecciation; pl. el. = planar elements; inc. gl. ve. = incipient glass veins; br. = brecciation; diapl. gl. = diaplectic glass; recr. = recrystallization; inc. m. = incipient melting; m.p. = melt pockets.

TABLE 3.6.3
Progressive Shock Metamorphism of Particulate Materials[a]

Shock Pressure (GPa)	Particulate Basalt 75035[b]	Lunar Soils 15101[c] 45–150 μm	Lunar Soils model soil[d]	Lunar Soils 65010[e]	H5-Chondrite Powder[f] ~16% poros.[g] <150 μm	H5-Chondrite Powder[f] <5% poros.[g] <150 μm	Quartz Sand[h] 63–250 μm
40	vesiculated glass	vesiculated glass	vesiculated glass	glass bonding	?	?	vesiculated glass
30	lithification by glass cement	lithification by glass cement			lithification by glass cement	lithification by glass cement	
20	minor intergranular glass	minor intergranular glass	lithification	lithification	minor intergranular glass / lithification and compaction	minor intergranular glass / lithification and compaction	minor intergranular glass / lithification without glass
10	lithification and compaction	lithification and compaction	compaction		?	?	?

← decreasing plagioclase content of sample

[a] Table based on data from shock recovery experiments and theoretical models (see, e.g., Kieffer 1975); [b] Schaal et al. (1979); [c] Schaal and Hörz (1980); [d] Kieffer (1975); [e] Christie et al. (1973); [f] Bischoff and Lange (1984); [g] Poros. = porosity; [h] Stöffler et al. (1975).

volves melting at grain boundaries and jetting of melt into pores (Kieffer 1975). It is important to note that incipient melting in detrital material is already apparent at pressures as low as ~ 5 GPa (Table 3.6.3), whereas disequilibrium melting in shocked nonporous basaltic material starts only at equilibrium shock pressures above ~ 30 GPa (Table 3.6.1; Stöffler et al. 1986). In feldspar-bearing detritus, the average intergranular melt is always feldspar-normative, indicating the preferred melting of feldspar compared to mafic minerals. This may explain the observed negative correlation between the abundance of feldspar in the detritus and the shock pressure required for the onset of intergranular melting (Table 3.6.3).

The main characteristics of progressive shock metamorphism, as explained in Table 3.6.3, are similar in both porous rocks and detrital material in the absence of gas or liquid in the pore space.

However, shock-compressed gas or liquid in the pores of the target material may lead to comminution upon pressure release in the lower shock stages due to the explosive expansion of the highly compressed fluid (Kieffer 1975). In the absence of these effects, shock compression of planetary surface detritus leads to *regolith breccias* or *fragmental breccias* or, if the target is a preconsolidated porous rock, to *shocked regolith breccias or shocked fragmental breccias,* or more generally to a shocked (nonporous) sedimentary rock (see also Fig. 3.6.3).

3.6.6. SHOCK METAMORPHISM AND BRECCIATION IN CHONDRITES

Shock effects have been found in all chondrite groups. Since about 90% of all chondrites belong to the group of ordinary chondrites, the following more detailed descriptions will be largely restricted to shock effects in ordinary chondrites. Shock-induced features in enstatite chondrites and carbonaceous chondrites have not been studied in any systematic detail. Therefore, they will be discussed only very briefly at the end of this chapter.

Ordinary Chondrites

As a large fraction of ordinary chondrites are polymict breccias, it is necessary to deal with shock metamorphism and breccia formation separately in the following sections. The first section relates to progressive shock metamorphism of monomict, nonbrecciated chondrites as coherent crystalline rocks. The second and third sections are devoted to the constituents of polymict breccias and to the lithification process leading to coherent polymict breccias, respectively.

Equilibrium Shock Effects in Nonbrecciated Ordinary Chondrites. Virtually all chondrites are shocked to some degree ranging from nearly unshocked to impact-melted. At present there is only one systematic petrographic study on shock effects in ordinary chondrites (L group chondrites; Dodd and Jarosewich 1979). It is known that on average L chondrites are more heavily shocked than H chondrites. Strongly shocked H chondrites are rare. The characteristic shock effects observed in chondrites are tabulated in Table 3.6.2 and correlated with an experimentally established shock-pressure scale. The effects observed by Dodd and Jarosewich (1979)—deformations and transformations in olivine and plagioclase, shock veins, blackening and melt pockets—have been recalibrated in Table 3.6.2 based on more recent data. We propose 6 stages of shock which correspond to the Dodd and Jarosewich scale $a-f$ except for stage I which comprises $b + c$. Dodd and Jarosewich (1979) examined 52 type L4–L6 ordinary chondrites and did not find any unshocked meteorite. Even if the plagioclase was undeformed (< 5 GPa), the olivine was fractured or showed undulose extinction. These meteorites (6%) belong to shock stage a of Table 3.6.2; 29% and 19% of all L chondrites studied by Dodd and Jarosewich (1979) fall in shock stages $b + c$ and d, respectively. 46% of the L chondrites were exposed to even higher shock pressures, shock stages e (40%) and f (6%).

Disequilibrium Shock Effects in Nonbrecciated Ordinary Chondrites. Besides the typical features of shocked mineral grains, many ordinary chondrites contain shock veins, metal- and troilite-rich assemblages with quenched textures, vugs and melt pockets (see Rubin 1985, and references therein) that might be classified as disequilibrium shock effects because their formation is believed to result from local deviations of the peak shock pressure from the equilibrium pressure experienced by the whole rock (see Sec. 3.6.5).

Pseudotachylite-like *shock veins* are dark, fine-grained to glassy shock-induced filaments that vary from a few μm to > 1 cm in width (Fig. 3.6.5a). Probably they were formed by shock-induced local melting, injection through a rock along pre-existing fractures and, in some cases, simultaneous incorporation of fragments from the target rocks. Shock veins, basically chondritic in composition, are restricted to chondrites shocked in excess of ~ 20 GPa (Dodd 1981; Dodd and Jarosewich 1979). In the veins of seven L6 chondrites, the high-pressure phases ringwoodite and majorite have been found (Rubin 1985) attesting to a peak shock pressure of at least 50 GPa (Fig. 3.6.4).

Many ordinary chondrites contain fine-grained *assemblages of metal and troilite* that, in many cases, have dendritic or cellular textures. The assemblages appear to have been melted and then rapidly cooled (1–300 K/s; Scott 1982). The physical association of some assemblages with melt-rock clasts indicate that they were formed by shock melting of metal and troilite grains. Most chondrites that contain these mixtures, however, are regolith or fragmental breccias.

Melt pockets in ordinary chondrites (< 100 to 700 μm in size; Fig. 3.6.5b) are of subspherical to sinuous shape. They are cryptocrystalline to glassy and have a composition of a silicate melt formed by the mixed melting

 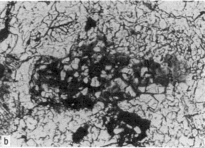

Fig. 3.6.5. (a) Dark and light silicate veins in the Chantonnay L6 chondrite. Horizontal dimension of field of view is 4 mm. Sample photographed in plane polarized transmitted light (photograph from Dodd et al. 1982). (b) Melt pocket in the Aumale L6 chondrite. The melt pocket is 300 μm long and consists of brown glass with xenocrysts of olivine and pyroxene and both dropshaped and irregular masses of metal and troilite. Sample photographed in plane polarized transmitted light (photograph from Dodd and Jarosewich 1979).

of the main mineral constituents of the parent meteorite. Melt pockets often contain irregular or droplet-shaped masses of metal and troilite. Many melt pockets are composed of olivine and low-Ca-pyroxene identical in composition to the corresponding phases in the host (Dodd and Jarosewich 1979,1980). Melt pockets have been found only in chondrites that have been severely shocked in excess of 20 to 25 GPa (Dodd and Jarosewich 1979; Table 3.6.2). Obviously, a disequilibrium shock pressure of 45 to 80 GPa is necessary to produce localized in-situ melting of a chondrite (see Sec. 3.6.5). Melt pockets are usually enriched in a plagioclase component; this suggests preferential melting of plagioclase due to its low impedance to shock compression (Schaal et al. 1979; Dodd and Jarosewich 1982).

Vugs, quite common features in ordinary chondrites (Olsen 1981), have been interpreted as shock effects. Based on a study by Kieffer (1975) the suitable environment required for the formation of vugs by shock-induced vapor expansion (Olson 1981) is unlikely to occur on meteorite parent bodies. We believe that a shock-related origin of vugs remains questionable.

Shock Effects in Chondritic Breccias. Keil (1982) and Scott et al. (1985) have noted that a large fraction of the ordinary chondrites are breccias. They suggested that further studies would show that most chondrites were brecciated and lithified after metamorphism.

Shock veins, metal-troilite mixtures and *melt pockets* as described above were also found in chondritic breccias. These shock features as well as equilibrium shock features such as mosaicism in olivine and maskelynite formation may have been already present in the lithic clasts (coherent rock, regolith or megaregolith of the parent body) prior to breccia aggregation and lithification. However, in some cases they may have been formed in the same event that caused the formation of breccias from displaced target rocks, the lithification of fragmental debris or the ejection of material from the parent body. These two contrasting possibilities demonstrate the difficulties in the interpretation of the time sequence of shock features in chondritic breccias. Besides the shock effects described above, which affect the whole breccia, shock features related to individual constituents such as lithic clasts will be described in the following three paragraphs.

Impact-melt rock clasts are light-colored objects (< 50 μm to several cm in size) that are depleted in metal and sulfide relative to the host chondrites. Such clasts display nonchondritic, poikilitic, porphyritic or skeletal textures. They were formed by *total* melting of parent-body material and loss of an immiscible metal and sulfide liquid (Fodor and Keil 1978; Rubin 1985, and references therein). Essentially, impact-melt rock clasts have the same composition as their parent rocks, except for the loss of metal and troilite. Impact-melt rock clasts, often accompanied by metal-troilite mixtures, must have experienced extremely high peak shock pressures. Whole-rock melting is only possible in excess of about 80 to 100 GPa resulting in post-shock temperatures of $> 1500°$ C (Table 3.6.2).

Impact-melt breccias are melted rocks that contain variable amounts of clastic, unmelted debris. They occur as light-colored lithic fragments in chondritic breccias and as individual meteorites (Keil 1982; Rubin 1985). The coexistence of melted and unmelted materials in impact-melt breccias is a common phenomenon in melt sheets of terrestrial impact craters and lunar impact-melt breccias (Grieve et al. 1977; Stöffler et al. 1979). The formation of such melt breccias obviously needs a relatively large impact site. Based on the slow post-shock cooling rates ($3K/10^3$ yr), Taylor et al. (1979) suggested that these meteorites were derived from melt breccia sheets at the floor of impact craters several kilometers in diameter. The fragmental debris could have resulted from incomplete melting of local target material or from the incorporation of "cold" debris from the country rock during crater excavation. In the former case, such fragmental debris could show severe shock effects—supported by impact-melt breccias containing ringwoodite and majorite (Keil 1982, and references therein); in the latter case incorporation of weakly shocked materials is possible as found in many terrestrial impact-melt breccias (see, e.g., Grieve et al. 1977; Bischoff and Stöffler 1984). The melt fraction of the impact-melt breccias corresponds to the highest stage of shock metamorphism (stage 5 of Table 3.6.2) which is achieved in a transient melt zone near to the impacting projectile during cratering.

Agglutinates are abundant in lunar soils and extremely rare in ordinary chondrite regolith breccias (Rubin 1985, and references therein). Agglutinates are irregularly shaped, vesicular, glass-bonded aggregates of rock, glass and mineral fragments formed by the impact of micrometeorites into regolith (McKay et al. 1972). Typical *impact-melt spherules,* as found in lunar regolith breccias, are essentially absent in ordinary-chondrite regolith breccias.

Enstatite and Carbonaceous Chondrites

Enstatite chondrites occur as nonbrecciated meteorites (~70%) and fragmental breccias (~25%) (Keil 1982; Rubin 1983a,b; Rubin and Keil 1983; Rubin 1984; Nehru et al. 1984; Brett and Keil 1986). Only one solar-wind-bearing regolith breccia is known (McKinley and Keil 1984; Wieler et al. 1985). Although a systematic study on shock features in enstatite chondrites is missing, the observations are similar to those made in ordinary chondrites and include shocked minerals, veins and metal- and silicate-rich melt-rock clasts (Rubin 1985, and references therein).

Also carbonaceous chondrites have not been thoroughly studied regarding shock effects. Many carbonaceous chondrites contain materials of diverse petrogenetic history (Kracher et al. 1985, and references therein) indicating that they are breccias (Keil et al. 1969; McSween 1977; Kracher et al. 1982; Metzler and Bischoff 1987). Kracher et al. (1985) suggested that Leoville belongs to a special type of breccia termed "accretionary breccia," whose parent body accreted after xenolithic fragments had formed and undergone alteration (see also Chapter 3.5). Individual mineral constituents of carbonaceous chondrites are affected by shock. In most carbonaceous chondrites olivines and plagioclase (if present) frequently show fracturing and undulatory extinction. Shocked minerals (melilite, fassaite) have been found also in CAIs (Grossman 1975; McSween 1977; Bischoff et al. 1987). Obviously, many constituents were shocked prior to the accretion of the carbonaceous chondrite material.

Lithification of Chondritic Breccias

Shock-induced lithification of clastic rock and mineral debris is an important process in the formation of coherent fragmental and regolith breccias of the chondrite group. Kieffer (1975) suggested that the welding process results from impact-induced shock melting at grain contacts. This local grain-boundary melting in chondritic breccias was first observed by Ashworth and Barber (1976). Bischoff et al. (1983) demonstrated that an interstitial, feldspar-normative, shock-melted material is responsible for the lithification of formerly loose clastic debris. They found a continuous gradation in matrix textures from nearly completely clastic to highly cemented breccias in which the clasts are completely surrounded by shock-melted material. The progressive stages of shock in particulate meteoritic regolith are summarized in Table 3.6.3.

Chemical and Physical Effects of Shock Metamorphism

Besides the microscopically visible shock effects in chondrites (see previous sections), shock metamorphism has also caused changes of the abundances of noble gases and highly volatile trace elements, and of the thermoluminescence (TL) properties.

Kirsten et al. (1963) and Heymann (1967) found shock-induced loss of radiogenic noble gases such as ^{40}Ar and ^{4}He in certain chondrites resulting in low K-Ar and U-He ages (gas-retention ages). Low gas-retention ages associated with petrographic evidence of shock are now known for all groups of chondrites, but they are most frequent in the L Group (Anders 1964; Heymann 1967; Taylor and Heymann 1969; Dodd and Jarosewich 1979). Bogard et al. (1976) and Turner and Enright (1977) demonstrated that highly shocked meteorites show complex time-temperature spectra.

No noticeable ^{40}Ar loss was reported from an experimentally shocked L6 chondrite (29 GPa; Bogard et al. 1987) and terrestrial basalt (27 GPa; Davis 1977). Even at a shock pressure of 45 GPa, Jessberger and Ostertag (1982) did not find Ar loss in samples of a terrestrial labradorite single crystal. This indicates that the gas loss in chondrites is not directly caused by the shock-induced heating. Shock metamorphism has also disturbed the Rb-Sr system of ordinary chondrites (Gopalan and Wetherill 1971; Minster and Allègre 1978,1979). The partial or total resetting of the Rb-Sr system is obviously caused by thermal effects following the shock compression. Deutsch et al. (1986) found that artificial shock pressures of 47.5 GPa alone do not affect the Rb-Sr systematics of geologic samples.

Chemical studies on L chondrites also demonstrate the mobilization of trace elements due to shock. As pointed out by Walsh and Lipschutz (1982), volatile trace elements that are mobilized from chondrites during experimental heating progressively decrease in abundance in L chondrites with increasing degree of shock. Huston and Lipschutz (1984) obtained significantly lower mean concentrations of volatile trace elements in 26 strongly shocked L4–6 chondrites than in 14 mildly shocked ones. They suggest an elemental loss in shock-formed FeS-Fe eutectic and/or by vaporization during cooling of shock-heated collisional debris. Shock-induced melting is also considered to be responsible for anomalously low S/Mg and Fe/Mg ratios (Cripe and Moore 1975; Dodd 1981).

Sears et al. (1984b) studied the thermoluminescence (TL) sensitivities of shocked meteorites and found that unshocked meteorites have much higher TL sensitivities than the heavily shocked chondrites. TL studies on an artificially shocked H chondrite confirm this finding (Sears et al. 1984a). Shock-induced lithification processes of chondritic breccias caused degassing of solar-wind-implanted gases (Bischoff et al. 1983). The abundances of solar-wind-implanted ^{4}He and ^{20}Ne are inversely correlated with the abundance of shock-induced interstitial melts that cemented the formerly loose regolith materials to form a tough chondritic regolith breccia.

It should be emphasized that the present experimental data clearly indicate that the short thermal pulse induced by the shock wave cannot be the prime cause for gas losses, mobilization of volatile elements and perhaps also not for changes in thermoluminescence of meteorites. We believe that an additional heat source is required such as the one of a hot-impact formation or any other heat source through which the shocked constituents are annealed over extended periods of time.

3.6.7. SHOCK METAMORPHISM AND BRECCIATION IN ACHONDRITES

Different classes of achondrites (Table 3.6.4) have been processed differently by impact-induced shock and brecciation (Scott et al. 1985) depending on their geological setting (e.g., size of their parent bodies). Table 3.6.4 lists the frequency of the various types of unshocked, shocked and brecciated achondrites in comparison with SNC meteorites (probably of Martian origin) and lunar meteorites.

Since the latter two groups are not directly relevant for the processes in the early solar system the following paragraph will be restricted to a description of shock features in the HED meteorite class (howardites, eucrites, diogenites) and in mesosiderites, ureilites and aubrites.

The Howardite-Eucrite-Diogenite-Mesosiderite Meteorite Group

Although there has been no systematic study on shock metamorphism of HED meteorites and meteorite clasts, it is indisputable that all stages of progressive shock as defined in Table 3.6.1 have been observed. Unshocked or weakly shocked meteorites or clasts prevail by far but basalts of shock stages 2 to 5 (Table 3.6.1) are present among lithic clasts in howardites and polymict eucrites (see, e.g., Fuhrmann and Papike 1981; Delaney et al. 1984; Metzler 1985). The scarcity of highly shocked rocks is consistent with the fact that the volume of target rock shocked above ~ 20 GPa in any hypervelocity impact is very small (probably much less than 10% of the displaced rocks). In addition, this volume may be even less in the asteroid environment because of the lower impact velocities compared with the Earth-Moon system and because of the low gravity which increases the ratio of weakly shocked to highly shocked material in the displaced rocks.

Impact-melt lithologies represent about 20 to 40 vol.% of all lithic clasts in howardites (Fuhrmann and Papike 1981) and less than 10% in polymict eucrites (Metzler 1985). The lunar analogues (fragmental breccias from Apollo 16) contain an average fraction of 44% melt lithologies (Stöffler et al. 1985). In the case of mesosiderites, some meteorites as a whole and specific clasts in the recrystallized types of mesosiderites are interpreted as impact-melt breccias (Floran 1978; Hewins 1984).

It is conspicuous that only about 16% of the HED meteorites are non-brecciated and that nearly 50% of the HEDs and 100% of the mesosiderites are polymict breccias (Table 3.6.4). Impact-melt breccias are not found as individual HED meteorites in contrast to the mesosiderites. These data clearly point to an effective large-scale impact-cratering episode in the early evolution of the parent body not unlike the one which produced the lunar highland

TABLE 3.6.4
Frequency of the Various Types of Unshocked and Brecciated Achondrites,
Lunar Meteorites and Lunar Highland Rocks[a]

	Crystalline Rocks	Monomict Breccias	Regolith Breccias	Polymict Breccias			Number of Samples
				Polym. Fragm. Breccias	Impact-Melt Breccias		
Eucrites	18	46	5	31	—		39
Cum. eucr.	67	16.5	—	16.5	—		6
Howardites	—	—	15	85	—		19
Diogenites	13	67	—	20	—		15
Total HED	16.5	37	5	41.5	—		79
Mesosider.	—	—	68		32		22
Aubrites	10	40	30	20	—		10
Ureilites	90	—	—	10	—		21
SNC-meteorites	63* 37**	—	—	—	—		8
Lunar meteorites	—	—	67	33	—		3
Apollo 16 highland rocks (>20g)	—	19.1	7.8	15.7	57.4		115
Lithic clasts in Apollo 16 fragmental breccias (0.04–1 mm)	—	44	—	—	56		2213

[a] Figures are given in number frequency percent. Cum. eucr. = cumulate eucrites; mesosider. = mesosiderites; polym. fragm. = polymict fragmental; SNC = Shergottites, Nakhlites, Chassigny. Data for achondrites, SNCs and Apollo 16 rocks from Scott et al. (1985), McSween (1985), and Ryder (1981), respectively; mesosiderite data from Hewins (1984); data on lithic clasts in Apollo 16 breccias (Stations 11 and 13, North Ray) from Stöffler et al. (1985).
* Unshocked
** Shocked

breccias. Although a number of models for the early evolution of the parent bodies have been devised (Takeda 1979; Delaney et al. 1984; Hewins 1983; Wasson and Rubin 1985), much more detailed analysis of the polymict breccias is required to reveal the early processes.

Several attempts have been made to date shock and breccia-forming events by radiometric methods (see, e.g., Bogard et al. 1985, and references therein; Dickinson et al. 1985). There is evidence for a wide range of dates of shock-induced resetting of the ca. 4.5 Gyr old meteorites. It covers the time from about 4.2 Gyr to < 1 Gyr.

Ureilites and Aubrites

According to the mineralogy of ureilites (see, e.g., Berkley et al. 1980) and aubrites (Watters and Prinz 1979), the appropriate shock criteria can be found in Fig. 3.6.4 and Table 3.6.2. Data of Table 3.6.1 could also be relevant for aubrites. Observations on shock effects in both types of meteorites are rarely reported in any detail and are mostly qualitative and sometimes unclear.

Vdovykin (1970) classified the ureilites according to the degree of shock (low or high) into two groups. More recently Berkley et al. (1980) and Berkley (1986) have added "low-moderate" and "very high" degree of shock. Two meteorites out of 24 are polymict breccias. The criteria for the shock classes do not make use of the available pressure calibrations (see, e.g., Bauer et al. 1979; Reimold and Stöffler 1978). Diamond and lonsdaleite found in ureilites are obviously due to shock transformation of graphite requiring a minimum peak pressure of about 13 GPa. The upper limit to the shock pressure experienced by ureilites is not clear from the descriptions in the literature. It may be as high as 60 GPa (see also Vdovykin 1970).

In aubrites, disorder in enstatite has been reported (Watters and Prinz 1979) which might be due to shock (Reid and Cohen 1967). Because plagioclase appears to be birefringent in all aubrites (Watters and Prinz 1979), the maximum shock pressure experienced by aubrites cannot exceed 20 to 25 GPa. It is of interest for the early evolution of the parent body that among aubrites, all types of breccias (monomict, polymict) are represented (Table 3.6.4). The polymict aubrite breccia Cumberland Falls contains chondritic clasts (Binns 1969).

3.6.8. SHOCK EFFECTS IN IRON METEORITES AND IN METALLIC PHASES OF METEORITES

Since most iron meteorites are well-processed bodies with a long and complex prehistory, the imprints of early solar-system events will usually have become thoroughly eliminated. Since, however, the observations on iron meteorites also pertain to Fe-Ni grains in chondrites and mesosiderites, it may be of some interest to present a brief overview.

The major phases of iron meteorites are the metallic phases kamacite and taenite and their various duplex intergrowths. Minor phases are troilite, daubreelite, schreibersite, cohenite, graphite, diamond, phosphates and silicates. For shock studies the ubiquitous kamacite and troilite phases are the most interesting, while the other phases may, in certain meteorites, add supporting information. Shock effects in various constituents of iron meteorites are listed in Fig. 3.6.4. Data were taken from Lipschutz (1968), Jain and Lipschutz (1971) and Buchwald (1975) and references therein.

Neumann bands are shock-induced mechanical twins attesting to very low shock pressures on the order of 1 GPa (Jain and Lipschutz 1971, and references therein). If a pressure of ~ 13 GPa is reached or exceeded, kamacite will transform to a stable close-packed ε-iron of high density (Smith 1958; Maringer and Manning 1962; Takahashi and Bassett 1964). Upon pressure release, the ε-phase reverts to the α-phase. The result of a very rapid α-ε transformation is a distorted α-structure visible as a densely hatched martensite; upon annealing, this shock-hatched phase will recover and recrystallize (Buchwald 1977). Some iron-meteorite groups appear to have been formed from impact-melt pools in chondritic bodies (Wasson et al. 1980); their silicates show chondritic major-element abundances (Rubin 1985, and references therein).

From a technical standpoint, the shock processing of metals has been studied since the 1950s (Leslie et al. 1962; Meyers and Murr 1981; Murr et al. 1986). Many observations from these fields are applicable to the interpretation of shock phenomena in meteorites.

The metal phase is usually characterized by the combined assessment of hardness, e.g., Vickers hardness at 100 g load, and microstructure (Buchwald 1975,1977). In some cases, electron-microscopic work has been presented (Jago 1974; Nielsen and Buchwald 1979).

The unique feature of shock loading of metals is the large increase in hardness and the associated small dimensional changes, provided that no additional shearing takes place. The major difficulty in meteoritics is the problem of estimating, from hardness and structure, how much damage is due to shock, how much to heating associated with shock and/or shearing, and how much to any later reheating without a shock or a shearing component.

The complex microstructures of shocked kamacite may be described qualitatively in a very simplified picture (Fig. 3.6.6). A sample mass of Ni-Fe is exposed to various shock events, and the shock paths are superimposed on a schematic *P-T* diagram for Fe-6%Ni. A shock of 20 GPa, obtained, e.g., by the flying plate technique, will transform massive homogeneous kamacite to the ε-phase, which on shock relaxation reverts to the distorted α-phase (case 1). The sample is at $< 40°$ C immediately after the shock. The α-phase is characterized by high hardness, 325–350 Vickers; the microstructure is twinned and of a martensitic hatched appearance. On the submicroscopic level, there are dense tangles of dislocations and a profu-

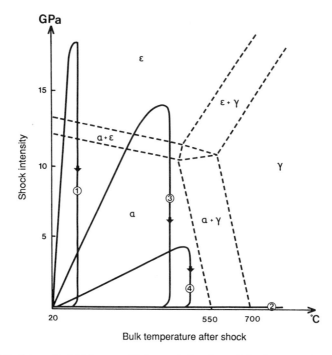

Fig. 3.6.6. Schematic *P-T* diagram with four typical regimes of shock compression and heating of iron meteorites as deduced from metallography (see text for details).

sion of microtwins. The dislocation density is very high, on the order of 10^{11} to $10^{12}/cm^2$.

Case 3 suggests a situation where the shock wave is attenuated by internal cracks and inclusions (troilite, silicates), so that strong post-shock heating results and significant plastic deformation accompanies the shock event. In this case, the phase-transformed, extremely fine-grained kamacite may, due to residual heat, start recrystallizing and ultimately attain a low hardness of 160. The kamacite will appear as 1 to 5 μm equiaxed units, and on the electronmicroscopic scale the dislocation density will be reduced a factor of 10^4 to 10^5, relative to case 1.

Case 4 represents a situation where the shock intensity is low, but where major additional adiabatic heating occurs, associated with plastic deformation. The kamacite hardness is 200 to 250, and there are mechanical twins and heavy shear bands, often 10 to 50 μm wide. The dislocation density is intermediate between case 1 and 3, and the twins are much larger than in case 1.

Case 2 suggests a situation where the shock part of the event is negligible, while heating is essential. The case is comparable to the brief surface heating of any meteorite that penetrates the atmosphere. The kamacite briefly transforms to γ, then reverts to disordered α_2. This has a hardness of 180 to 200, and displays a ragged microstructure, very different from that of case 1. The dislocation density is intermediate; it is rich in tangles and with many Fe-Ni martensite laths.

When troilite inclusions happen to be located in the shock-exposed material, they will suffer a local temperature rise, even to the point of melting, about 1200° C. If troilite melts, it will dissolve part of the adjacent metal (Fe-6%Ni), and since the troilite is rather pure FeS, the resulting, rapidly solidified material will appear as a fine-grained eutectic of Fe-Ni-S. If troilite and graphite are present, the shock event may lead to diamond formation (Nininger 1956; Lipschutz and Anders 1961). In many meteorites, the troilite is associated with schreibersite, daubreelite and chromite. While the latter minerals rarely melt under the shock event, they usually fragment and become dispersed as discrete splinters in the eutectic.

One and the same meteorite may display shock and thermally induced features, produced at very different pressure and temperature conditions (Buchwald 1975; Bevan et al. 1981). A recent case is Verkhne Dnieprovsk (Buchwald and Clarke 1987) where kamacite of the hatched and the Neumann-band variety coexists with recrystallized kamacite. The schreibersite is micromelted and forms local pockets which largely have solidified in situ. The more mobile troilite melts have apparently been ejected.

3.6.9. QUANTITATIVE RELATIONS BETWEEN IMPACT VELOCITIES, TYPES OF COLLIDING PARENT BODIES, AND THE DEGREE OF SHOCK OR TYPE OF BRECCIATION IN METEORITES

The peak shock pressure P is related to the particle velocity u of the compressed material moving behind the shock front through the equation

$$P - P_o = \rho_o U u \qquad (2)$$

where P_o and ρ_o are the ambient pressure and density and U is the shock wave velocity.

By applying the impedance-match method (see, e.g., Duvall and Fowles 1963), the peak shock pressure P achieved by a certain velocity v of the impactor can be determined by means of the so-called "free surface approximation" ($v = 2u$) if the Hugoniot curves of impactor and target are known (Fig. 3.6.7). The P-u state common to both colliding bodies (P_1, u_1) is given in Fig. 3.6.7 by the intersection of the mirror image of the impactor's Hugoniot curve with the target's Hugoniot curve. This mirror-image curve intersects the abscissa at the velocity v of the impactor (see inset of Fig. 3.6.7). Hugoniot curves for 4 materials (Stöffler 1982b) which may serve as analogues for 4 major types of meteorites or parent bodies are plotted in Fig.

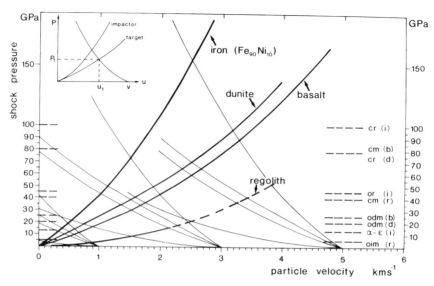

Fig. 3.6.7. Hugoniot curves of various meteorite analogue materials in the shock pressure-particle velocity plane (thick solid curves). Various sets of mirror curves for impact velocities of 1, 3 and 5 km/s are given to explain the method of impedance matching, from which the peak shock pressure achieved, in a given target material by a given impactor, can be obtained (see inset): e.g., an iron projectile impacting basalt at 3 km/s produces a peak shock pressure of ~ 42 GPa but only 21 GPa if regolith is the target material. Data for iron ($Fe_{90}Ni_{10}$), Twin Sisters dunite, lunar basalt 12063, and lunar regolith 70051 (42% porosity) are from McQueen et al. (1970), McQueen and Marsh (1966), Ahrens and Watt (1980) and Ahrens and Cole (1974), respectively. Dashed horizontal lines indicate the required peak pressure for the onset of certain shock effects cr, cm, or, etc. (as explained in Table 3.6.5) in iron (i), basalt (b), dunite (d) and regolith (r).

3.6.7. Unfortunately, there are no measured Hugoniot data for chondrites in the literature. We believe that the Hugoniot curve of recrystallized ordinary chondrites and E chondrites (petrographic types 4–6) will be intermediate between those of dunite and basalt (Fig. 3.6.7). Hugoniots of C chondrites and of porous chondrites are most likely intermediate between basalt and regolith. The dunite and basalt curves are representative of ureilites and basaltic achondrites, respectively.

From Fig. 3.6.7 minimum collision velocities for the formation of critical shock effects observed petrographically in meteorites can be derived graphically for all types of impactors and targets. The data are given in Table 3.6.5. Some of the effects selected for this analysis are also critical for the formation of certain types of breccias. The onset of intergranular melting at 5 GPa is required for the production of regolith breccias. The complete melting of all types of coherent stony meteorites above 80 GPa is a prerequisite for the formation of polymict fragmental and regolith breccias which

TABLE 3.6.5
Minimum Impact Velocities (v_{min}) in km/s for Certain Shock Effects in Analogue Materials for Various Meteorites[a,b]

	regolith → regolith	basalt → regolith	dunite → regolith	iron → regolith
v_{min} oim (5 GPa)	2.2	1.45	1.4	1.3
v_{min} cm (40 GPa)	~7	5.3	4.9	4.35
	regolith → basalt	basalt → basalt	dunite → basalt	iron → basalt
v_{min} odm (25 GPa)	4.0	2.6	2.3	2.0
v_{min} cm (80 GPa)	>7	6.1	5.8	4.65
	regolith → dunite	basalt → dunite	dunite → dunite	iron → dunite
v_{min} odm (20 GPa)	3.2	1.9	1.6	1.35
v_{min} cr (80 GPa)	>8	5.8	5.5	4.35
	regolith → iron	basalt → iron	dunite → iron	iron → iron
v_{min} $\alpha-\varepsilon$ (13 GPa)	2.25	1.1	0.9	0.7
v_{min} or (45 GPa)	~4.5	3.1	2.75	2.1
v_{min} cr (100 GPa)	>7.5	~5.4	~5.2	3.6

[a] Data from Fig. 3.6.8; Impactor → target.
[b] oim = onset of intergranular melting = lithification;
   cm = complete melting;
   odm = onset of disequilibrium melting (veins, pockets);
   cr = complete recrystallization (upon melting);
   $\alpha-\varepsilon$ = phase transition of kamacite;
   or = onset of recrystallization.

contain impact-melt rock clasts. The data of Fig. 3.6.7 and Table 3.6.5 have been summarized for the more general case of small projectiles impacting large target bodies (Fig. 3.6.8) and for the special case (Fig. 3.6.9) of colliding equal-sized asteroids established by Hartmann (1979, his Fig. 1).

Collisions considered in Fig. 3.6.8 lead to impact cratering and the associated shock effects above a certain limit of the impact velocity if the limiting case of total disruption of the target body is excluded. The shocked material resides in ejecta formations and crater basements. Below the limiting impact velocity, accretion will be the dominant process where shock effects

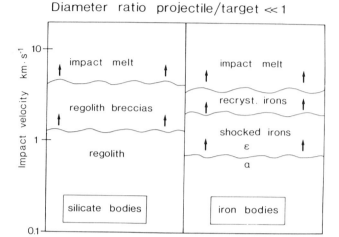

Fig. 3.6.8. Minimum impact velocities required for the onset of certain shock effects in regolith-bearing silicate and iron target bodies (if iron projectiles are assumed (see Table 3.6.5). These relations characteristic for cratering on asteroidal surfaces can also be applied to the case of colliding equal-size bodies since impact velocities and not approach velocities are considered in contrast to Fig. 3.6.9. In this case the shock-metamorphosed materials will escape (see text).

will be absent. Collisions of equal-sized bodies (Fig. 3.6.9) cause different mechanical effects in both bodies which are velocity and size dependent. Note that above a certain asteroid size, the actual impact velocity of the bodies drastically exceeds the approach velocity (plotted in Fig. 3.6.9) due to gravitational acceleration. All observable, unequivocally shocked material resides in escaping fragments produced in disruption events. Accretionary breccias which may be formed in the regime "fragmentation and reaccretion" (Fig. 3.6.9) will hardly display shock effects.

It is clear from Figs. 3.6.8 and 3.6.9 and Table 3.6.5 that genuine, petrographically recognizable shock metamorphism as well as impact-induced breccia formation requires collision velocities in excess of at least 0.7 km/s (iron, M asteroids) or 1.3 km/s (porous regolith or C chondrites). Regolith breccias or fragmental breccias with melt-rock clasts must originate from parent bodies which suffered collisions at velocities of at least 4.5 to 5 km/s. This velocity may be slightly less than 4 km/s for polymict breccias with metal and silicate rock material (e.g., mesosiderites). These velocities are all well within the expected range for the impact velocities in the main asteroid belt (Hartmann 1983).

In conclusion, an attempt can be made to derive, to a first approximation, the main characteristics of the early collisional history of the major types of meteorite parent bodies on the basis of known shock and brecciation features. The C chondrites reveal evidence for breccia formation by accretion

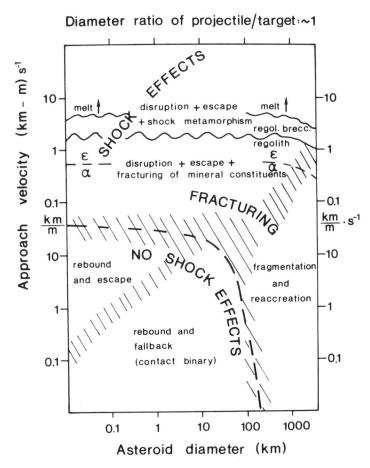

Fig. 3.6.9. Regimes for collision-induced effects in the meteorites' parent bodies as a function of approach velocity (velocity at infinity) and size of the colliding bodies which are assumed to be of similar size (figure is a modified and supplemented diagram of Hartmann, [1979]). Minimum impact velocities for the onset of certain shock effects are given by dashed and wavy lines (dashed line for the $\alpha-\varepsilon$ phase transition in iron meteorites). Data for these minimum impact velocities are based on Table 3.6.5. Note that the impact velocity exceeds the approach velocity for bodies larger than about 100 km because of gravity effects ($v^2_{impact} = v^2_{approach} + v^2_{escape}$) which leads to a decrease of the approach velocity required for the onset of certain shock effects.

only (Kracher et al. 1985) but no clear evidence for hypervelocity shock on their parent bodies. An accretionary-breccia model has also been proposed for certain ordinary chondrites (Scott and Taylor 1982; Grimm 1985). In general, the ordinary chondrites and E chondrites indicate post-accretional/premetamorphic as well as post-metamorphic (Scott et al. 1985) hypervelocity impact cratering with regolith formation and formation of melt rocks very early in their history. The same holds for the basaltic achondrite and mesosiderite parent body or bodies. The picture is less clear for ureilites and aubrites because of insufficient reliable data. Polymict breccias are known in both meteorite groups but any indication for impact melting is lacking. The involvement of these meteorites in regolith processes is improbable although the relatively high-shock level of ureilites requires collision velocities of at least ~ 3 km/s. The time of the impact processing of these meteorites' parent bodies is not known. This is a general problem for all types of meteoritic breccias. Therefore, it is one of the major tasks for future research to determine the maximum shock, the composition, and age of breccia clasts for all meteorite groups which might belong to a common parent body, or bodies of the same type, in order to reveal not only the early dynamical evolution but also the internal structure and composition of small planetary bodies at the time of their formation.

Acknowledgments. The authors wish to thank G. A. McCormack, F. Bartschat, M. Figgemeier, R.-M. Swietlik, V. Müller-Mohr, M. Endreß and H.-J. Schneider for their excellent technical assistance in preparing the manuscript. We gratefully acknowledge the constructive reviews by F. Hörz and M. E. Lipschutz that substantially improved the manuscript. The work of the Münster group was supported by grants from the Deutsche Forschungsgemeinschaft.

REFERENCES

Ahrens, T. J., and Cole, D. M. 1974. Shock composition and adiabatic release of lunar fines from Apollo 17. *Proc. Lunar Sci. Conf.* 5:2333–2345.

Ahrens, T. J., and Watt, J. P. 1980. Dynamic properties of mare basalts. Relation of equations of state to petrology. *Proc. Lunar Planet. Sci. Conf.* 11:2059–2074.

Akimoto, J., and Manghani, M. H. 1982. *High Pressure Research in Geophysics, Adv. Earth Planet. Sci.* 12 (Dordrecht: D. Reidel), p. 632.

Anders, E. 1964. Origin, age, and composition of meteorites. *Space Sci. Rev.* 3:583–714.

Ashworth, J. R. 1985. Transmission electron microscopy of L-group chondrites, 1. Natural shock effects. *Earth Planet. Sci. Lett.* 73:17–32.

Ashworth, J. R., and Barber, D. J. 1976. Lithification of gas-rich meteorites. *Earth Planet. Sci. Lett.* 30:222–233.

Bauer, J. F., Hörz, F., and Schaal, R. B. 1979. Shock metamorphism of granulated lunar basalts. *Proc. Lunar Planet. Sci. Conf.* 10:2547–2571.

Berkley, J. L. 1986. Four Antarctic ureilites: Petrology and observations on ureilite petrogenesis. *Meteoritics* 21:169–189.

Berkley, J. L., Taylor, G. J., Keil, K., Harlow, G. E., and Prinz, M. 1980. The nature and origin of ureilites. *Geochim. Cosmochim. Acta* 44:1579–1597.

Bevan, A. W. R., Kinder, J., and Axon, H. J. 1981. Complex shock-induced Fe-Ni-S-Cr-C melts in the Haig (IIIA) iron meteorite. *Meteoritics* 16:261–267.

Binns, R. A. 1969. A chondritic inclusion of unique type in the Cumberland Falls meteorite. In *Meteorite Research,* ed. P. M. Millman (Dordrecht: D. Reidel), pp. 696–704.

Bischoff, A., and Lange, M. A. 1984. Experimental shock-lithification of chondritic powder: Implications for ordinary chondrite regolith breccias. *Lunar Planet. Sci.* XV:60–61 (abstract).

Bischoff, A., and Stöffler, D. 1984. Chemical and structural changes induced by thermal annealing of shocked feldspar inclusions in impact melt rocks from Lappajärvi Crater, Finland. *Proc. Lunar Planet. Sci. Conf.* 14, *J. Geophys. Res. Suppl.* 89:B645–B656.

Bischoff, A., Rubin, A. E., Keil, K., and Stöffler, D. 1983. Lithification of gas-rich chondrite regolith breccias by grain boundary and localized shock melting. *Earth Planet. Sci. Lett.* 66:1–10.

Bischoff, A., Palme, H., and Spettel, B. 1987. A37—A coarse-grained, volatile element-poor Ca,Al-rich inclusion with huge Fremdlinge. *Lunar Planet. Sci.* XVIII:81–82 (abstract).

Bogard, D. D., Husain, L., and Wright, R. J. 1976. ^{40}Ar-^{39}Ar dating of collisional events in chondrite parent bodies. *J. Geophys. Res.* 81:5664–5678.

Bogard, D. D., Taylor, G. J., Keil, K., Smith, M. R., and Schmitt, R. A. 1985. Impact melting of the Cachari eucrite 3.0 Gy ago. *Geochim. Cosmochim. Acta* 49:941–946.

Bogard, D. D., Hörz, F., and Johnson, P. 1987. Shock effects and argon loss in samples of the Leedey L6 chondrite experimentally shocked to 29–70 GPa pressure. *Geochim. Cosmochim. Acta* 51:2035–2044.

Brett, R., and Keil, K. 1986. Enstatite chondrites and enstatite achondrites (aubrites) were not derived from the same parent body. *Earth Planet. Sci. Lett.* 81:1–6.

Brezina, A. 1904. The arrangement of collections of meteorites. *Trans. Amer. Phil. Soc.* 43:211–247.

Buchwald, V. F. 1975. *Handbook of Iron Meteorites: Their History, Distribution, Composition, and Structure* (Berkeley: Univ. of California Press).

Buchwald, V. F. 1977. The mineralogy of iron meteorites. *Phil. Trans. Roy. Soc. London* 286A:453–491.

Buchwald, V. F., and Clarke, R. S. 1987. The Verkhne Dnieprovsk iron meteorite specimens in the Vienna collection and the confusion of Verkhne Dnieprovsk with Augustinovka. *Meteoritics* 22:121–135.

Christie, J. M., Griggs, D. T., Heuer, A. H., Nord, G. L., Radcliff, S. V., Lally, J. S., and Fisher, R. M. 1973. Electron petrography of Apollo 14 and 15 breccias and shock-produced analogs. *Proc. Lunar Sci. Conf.* 4:365–382.

Cripe, J. D., and Moore, C. B. 1975. Total sulphur content of ordinary chondrites. *Meteoritics* 10:387–388 (abstract).

Davis, P. K. 1977. Effects of shock pressure on ^{40}Ar/^{39}Ar radiometric age determination. *Geochim. Cosmochim. Acta* 41:195–205.

Davis, D. R., Chapman, C. R., Greenberg, R., Weidenschilling, S. J., and Harris, A. W. 1979. Collisional evolution of asteroids: Populations, rotations, and velocities. In *Asteroids*, ed. T. Gehrels (Tucson: Univ. of Arizona Press), pp. 528–557.

Davison, L., and Graham, R. A. 1979. Shock compression of solids. *Phys. Rept.* 55:255–379.

De Carli, P. S., and Milton, D. J. 1965. Stishovite, synthesis by shock wave. *Science* 147:144–145.

Delaney, J. S., Prinz, M., and Takeda, H. 1984. The polymict eucrites. *Proc. Lunar Planet. Sci. Conf.* 15, *J. Geophys. Res. Suppl.* 89:C251–C288.

Deutsch, A., Quandt, B., and Hornemann, U. 1986. The response of the Sr isotopic system in geological samples to artificial shock pressure. *Meteoritics* 21:354–355.

Dickinson, T., Keil, K., Lapaz, L., Bogard, D. D., Schmitt, R. A., and Smith, M. R. 1985. Petrology and shock age of the Palo Blanco Creek eucrite. *Chem. Erde* 44:245–257.

Dodd, R. T. 1981. *Meteorites: A Petrologic-Chemical Synthesis* (New York: Cambridge Univ. Press).

Dodd, R. T., and Jarosewich, E. 1979. Incipient melting in and shock classification of L-group chondrites. *Earth Planet. Sci. Lett.* 44:335–340.

Dodd, R. T., and Jarosewich, E. 1980. Chemical variations among L-group chondrites, I: The Air, Apt and Tourinnes-la-Grosse (L6) chondrites. *Meteoritics* 15:69–83.

Dodd, R. T., and Jarosewich, E. 1982. The compositions of incipient shock melts in L6 chondrites. *Earth Planet. Sci. Lett.* 59:355–363.

Dodd, R. T., Jarosewich, E., and Hill, B. 1982. Petrogenesis of complex veins in the Chantonnay (L6f) chondrite. *Earth Planet. Sci. Lett.* 59:364–374.

Duvall, G. E., and Fowles, G. R. 1963. Shock waves. In *High Pressure and Chemistry*, vol. 2, ed. R. S. Bradley (London: Academic Press), pp. 209–291.

Floran, R. J. 1978. Silicate petrography, classification, and origin of the mesosiderites: Review and new observations. *Proc. Lunar Planet. Sci. Conf.* 9:1053–1081.

Fodor, R. V., and Keil, K. 1978. Catalog of lithic fragments in LL-group chondrites. *Spec. Publ. No. 19,* Univ. of New Mexico Inst. of Meteoritics.

Fodor, R. V., Keil, K., and Jarosewich, E. 1972. The Oro Grande, New Mexico, chondrite and its lithic inclusion. *Meteoritics* 7:495–507.

Fredriksson, K., De Carli, P. S., and Aaramae, A. 1963. Shock-induced veins in chondrites. *Proc. Third Internatl. Space Sci. Symp.,* Washington, D.C. (Amsterdam: North Holland Publ. Co.), pp. 974–983.

French, B. M., and Short, N. M., eds. 1968. *Shock Metamorphism of Natural Materials* (Baltimore: Mono Book Corp.).

Fuhrman, M., and Papike, J. J. 1981. Howardites and polymict eucrites: Regolith samples from the eucrite parent body. Petrology of Bholgati, Bununu, Kapoeta and ALHA 76005. *Proc. Lunar Planet. Sci. Conf.,* 12:1257–1279.

Fujiwara, A. 1980. On the mechanism of catastrophic destruction of minor planets by high-velocity impact. *Icarus* 41:356–364.

Fujiwara, A., Kamimoto, G., and Tsukamoto, A. 1977. Destruction of basaltic bodies by high-velocity impact. *Icarus* 31:277–288.

Gault, D. E., and Wedekind, J. A. 1969. The destruction of tektites by micrometeoroid impact. *J. Geophys. Res.* 74:6780–6794.

Gibbons, R. V., Morris, R. V., Hörz, F., and Thompson, T. D. 1975. Petrographic and ferromagnetic resonance studies of experimental shocked regolith analogs. *Proc. Lunar Planet. Sci. Conf.* 6:3143–3171.

Gopalan, K., and Wetherill, G. W. 1971. Rubidium-strontium studies on the black hyperthene chondrites: Effects of shock and reheating. *J. Geophys. Res.* 76:8484–8492.

Grieve, R. A. F., Dence, M. R., and Robertson, P. B. 1977. Cratering processes: As interpreted from the occurrence of impact melts. In *Impact and Explosion Cratering,* eds. D. J. Roddy, P. O. Pepin, and R. B. Merrill (New York: Pergamon), pp. 791–814.

Grimm, R. E. 1985. Penecontemporaneous metamorphism, fragmentation, and reassembly of ordinary chondrite parent bodies. *J. Geophys. Res.* 90:2022–2028.

Grossman, L. 1975. Petrography and mineral chemistry of Ca-rich inclusions in the Allende meteorite. *Geochim. Cosmochim. Acta.* 39:433–454.

Hartmann, W. K. 1979. Diverse puzzling asteroids and a possible unified explanation. In *Asteroids,* ed. T. Gehrels (Tucson: Univ. of Arizona Press), pp. 466–479.

Hartmann, W. K. 1983. *Moon and Planets* (Belmont: Wadsworth Publ. Co.).

Hewins, R. H. 1983. Impact versus internal origins for mesosiderites. *Proc. Lunar Planet. Sci. Conf.* 14, *J. Geophys. Res. Suppl.* 88:B257–B266.

Hewins, R. H. 1984. The case for a melt matrix in plagioclase-POIK mesosiderites. *Proc. Lunar Planet. Sci. Conf.* 15, *J. Geophys. Res. Suppl.* 89:C289–C297.

Heymann, D. 1967. On the origin of hypersthene chondrites: Ages and shock effects of black chondrites. *Icarus* 6:189–221.

Huston, T. J., and Lipschutz, M. E. 1984. Chemical studies of chondrites—III. Mobile elements and $^{40}Ar/^{39}Ar$ ages. *Geochim. Cosmochim. Acta* 48:1319–1329.

Jago, R. A. 1974. A structural investigation of the Cape York meteorite by transmission electron microscopy. *J. Materials Sci.* 9:564–568.

Jain, A. V., and Lipschutz, M. E. 1971. Shock history of iron meteorites and their parent bodies: A review, 1967–1971. *Chem. Erde* 30:199–215.

James, O. B. 1969. Shock and thermal metamorphism of basalt by nuclear explosion, Nevada Test Site. *Science* 116:1615–1620.

Jessberger, E. K., and Ostertag, R. 1982. Shock effects on the K-Ar-system of plagioclase feldspar and the age of anorthosite inclusions from North-eastern Minnesota. *Geochim. Cosmochim. Acta* 46:1465–1471.

Keil, K. 1982. Composition and origin of chondritic breccias. In *Workshop on Lunar Breccias and Soils and Their Meteoritic Analogs,* eds. G. J. Taylor and L. L. Wilkening, LPI Tech. Rept. 82-02 (Houston: Lunar and Planetary Inst.), pp. 65–83.

Keil, K., Huss, G. I., and Wiik, H. B. 1969. The Leoville, Kansas, meteorite: A polymict breccia of carbonaceous chondrites and achondrite. In *Meteorite Research,* ed. P. M. Millman (Dordrecht: D. Reidel), p. 217.

Kieffer, S. W. 1971. Shock metamorphism of the Coconino sandstone at Meteor crater, Arizona. *J. Geophys. Res.* 76:5449–5473.

Kieffer, S. W. 1975. From regolith to rock by shock. *The Moon* 13:301–320.
Kieffer, S. W., Phakey, P. P., and Christie, J. M. 1976*a*. Shock processes in porous quarzite: Transmission electron microscope observations and theory. *Contrib. Mineral. Petrol.* 59:41–93.
Kieffer, S. W., Schaal, R. B., Gibbons, R., Hörz, F., Milton, D. J., and Dube, A. 1976*b*. Shocked basalts from Lonar impact crater, India, and experimental analogues. *Proc. Lunar Sci. Conf.* 7:1391–1412.
Kirsten, T., Krankowsky, D., and Zähringer, J. 1963. Edelgas- und Kaliumbestimmungen an einer größeren Zahl von Steinmeteoriten. *Geochim. Cosmochim. Acta* 27:13–42.
Kracher, A., Keil, K., and Scott, E. R. D. 1982. Leoville (CV3)—An accretionary breccia? *Meteoritics* 17:239 (abstract).
Kracher, A., Keil, K., Kallemeyn, G. W., Wasson, J. T., Clayton, R. N., and Huss, G. I. 1985. The Leoville (CV3) accretionary breccia. *Proc. Lunar Planet. Sci. Conf.* 16, *J. Geophys. Res. Suppl.* 90:D123–D135.
Lambert, P., and Grieve, R. A. F. 1984. Shock experiments on maskelynite-bearing anorthosite. *Earth Planet. Sci. Lett.* 68:159–171.
Leslie, W. C., Hornbogen, E., and Dieter, G. E. 1962. The structure of shock-hardened iron before and after annealing. *J. Iron and Steel Inst.* 200:622–633.
Lipschutz, M. E. 1968. Shock effects in iron meteorites: A review. In *Shock Metamorphism of Natural Materials*, eds. B. M. French and J. M. Short (Baltimore: Mono Book Corp.), pp. 571–599.
Lipschutz, M. E., and Anders, E. 1961. Origin of diamonds in iron meteorites. *Geochim. Cosmochim. Acta* 24:83–105.
Maringer, R. E., and Manning, G. K. 1962. Some observations on deformation and thermal alteration in meteoritic iron. In *Researches on Meteorites*, ed. C. B. Moore (New York: John Wiley), pp. 123–144.
Mason, B. 1962. *Meteorites* (New York: John Wiley).
McKay, D. S., Heiken, G. H., Taylor, R. M., Clanton, D. A., Morrison, D. A., and Ladle, G. H. 1972. Apollo 14 soils: Size distribution and particle types. *Proc. Lunar Sci. Conf.* 3:983–994.
McKinley, S. G., and Keil, K. 1984. Petrology and classification of 145 small meteorites from the 1977 Alan Hills collections. *Smithsonian Contrib. Earth Sci.* 26:55–71.
McQueen, R. G., and Marsh, S. P. 1966. In *Handbook of Physical Constants*, ed. S. P. Clark, *Geol. Soc. Amer. Mem.* 87 (New York: Geol. Soc. Amer.), pp. 87–147.
McQueen, R. G., Marsh, S. P., Taylor, J. W., Fritz, J. N., and Carter, W. J. 1970. The equation-of-state of solids from shock waves studies. In *High Velocity Impact Phenomena*, ed. R. Kinslow (New York: Academic Press), pp. 294–419.
McSween, H. Y., Jr. 1977. Petrographic variations among carbonaceous chondrites of the Vigarano type. *Geochim. Cosmochim. Acta* 41:1777–1790.
McSween, H. Y., Jr. 1985. SNC meteorites: Clues to Martian petrologic evolution. *Rev. Geophys.* 23:391–416.
Metzler, K. 1985. Gefüge und Zusammensetzung von Gesteinsfragmenten in polymikten achondritischen Breccien. Diplomarbeit, Universität Münster.
Metzler, K., and Bischoff, A. 1987. Accretionary dark rims in CM-chondrites. *Meteoritics*, in press.
Meyers, M. A., and Murr, L. E., eds. 1981. *Shock Waves and High-Strain-Rate Phenomena in Metals* (New York: Plenum Publ. Corp.).
Minster, P. M., and Allègre, C. J. 1978. ^{87}Rb-^{87}Sr dating of L and LL chondrites: Effects of shock and brecciation. *Meteoritics* 13:563–566 (abstract).
Minster, P. M., and Allègre, C. J. 1979. ^{87}Rb-^{87}Sr dating of L chondrites: Effects of shock and brecciation. *Meteoritics* 14:235–248.
Murr, L. E., Staudhammer, K. P., and Meyers, M. A., eds. 1986. *Metallurgical Applications of Shock Wave and High-Strain-Rate Phenomena* (New York: Marcel Dekker Inc.).
Nehru, C. E., Prinz, M., Weisberg, M. K., and Delaney, J. S. 1984. Parsa: An unequilibrated enstatite chondrite (UEC) with an aubrite-like impact melt clast. *Lunar Planet. Sci.* 15:597–598 (abstract).
Nielsen, H. P., and Buchwald, V. F. 1979. Fe-Ni alloys after shock-loading, cold-rolling and annealing. *Meteoritics* 14:495–497.

Nininger, H. H. 1956. *Arizona Meteorite Crater* (Denver: World Press).
Olsen, E. 1981. Vugs in ordinary chondrites. *Meteoritics* 16:45–59.
Ostertag, R. 1983. Shock experiments on feldspar crystals. *Proc. Lunar Planet. Sci. Conf.* 14, *J. Geophys. Res.* 88:B364–B376.
Ostertag, R., and Stöffler, D. 1982. Thermal annealing of experimentally shocked feldspar crystals. *Proc. Lunar Planet. Sci. Conf.* 13, *J. Geophys. Res.* 87:A457–463.
Partsch, P. 1843. Die Meteoriten oder vom Himmel gefallene Stein- und Eisenmassen. *Im K. K. Hofmineralienkabinette zu Wien,* Catalog of Vienna Collection, Vienna.
Prinz, M., Fodor, R. V., and Keil, K. 1977. Comparison of lunar rocks and meteorites: Implication to histories of the moon and parent meteorite bodies. *Sov. Amer. Conf. Cosmochem. Moon and Planets,* NASA SP-370, pt. 1, pp. 183–199.
Prior, G. T. 1920. The classification of meteorites. *Mineral. Mag.* 10:51–63.
Raikes, S. A., and Ahrens, T. J. 1979. Post-shock temperatures in minerals. *J. Roy. Astron. Soc.* 58:717–747.
Reichenbach, V. 1860. *Poggendorffs Annal. 1860* 3:353–386.
Reid, A. M., and Cohen, A. J. 1967. Some characteristics on enstatite from enstatite chondrites. *Geochim. Cosmochim. Acta* 31:661–672.
Reimold, W. U., and Stöffler, D. 1978. Experimental shock metamorphism of dunite. *Proc. Lunar Planet. Sci. Conf.* 9:2805–2824.
Roddy, D. J., Pepin, R. O., and Merrill, R. B., eds. 1977. *Impact and Explosion Cratering* (New York: Pergamon Press).
Rubin, A. E. 1983a. The Adhi Kot breccia and implications for the origin of chondrules and silica-rich clasts in enstatite chondrites. *Earth Planet. Sci. Lett.* 64:201–212.
Rubin, A. E. 1983b. Impact melt-rock clasts in the Hvittis enstatite chondrite breccia: Implications for a genetic relationship between EL chondrites and aubrites. *Proc. Lunar Planet. Sci. Conf.,* 14, *J. Geophys. Res.* 88:B293–B300.
Rubin, A. E. 1984. The Blithfield meteorite and the origin of sulfide-rich, metal-poor clasts and inclusions in brecciated enstatite chondrites. *Earth Planet. Sci. Lett.* 67:273–283.
Rubin, A. E. 1985. Impact melt products of chondritic material. *Rev. Geophys. Space Phys.* 23:277–300.
Rubin, A. E., and Keil, K. 1983. Mineralogy and petrology of the Abee enstatite chondrite breccia and its dark inclusions. *Earth Planet. Sci. Lett.* 62:118–131.
Ryder, G. 1981. Distribution of rocks at the Apollo 16 site. In *Workshop on Apollo 16,* eds. O. B. James and F. Hörz, LPI Tech. Rept. 81-01 (Houston: Lunar and Planetary Institute), pp. 112–119.
Schaal, R. B., and Hörz, F. 1977. Shock metamorphism of lunar and terrestrial basalts. *Proc. Lunar Sci. Conf.* 8:1697–1729.
Schaal, R. B., and Hörz, F. 1980. Experimental shock metamorphism of lunar soils. *Proc. Lunar Planet. Sci. Conf.* 11:1679–1695.
Schaal, R. B., Hörz, F., Thompson, T. D., and Bauer, J. F. 1979. Shock metamorphism of granulated lunar basalts. *Proc. Lunar Planet. Sci. Conf.* 10:2547–2571.
Scott, E. R. D. 1982. Origin of rapidly solidified metal-troilite grains in chondrites and iron meteorites. *Geochim. Cosmochim. Acta* 46:813–823.
Scott, E. R. D., and Taylor, G. J. 1982. Primitive breccias among type 3 ordinary chondrites: Origin and relation to regolith breccias. In *Workshop on Lunar Breccias and Soils and Their Meteoritic Analogs,* eds. G. J. Taylor and L. L. Wilkening, LPI Tech. Rept. 82-02 (Houston: Lunar and Planetary Inst.), pp. 130–134.
Scott, E. R. D., Lusby, D., and Keil, K. 1985. Ubiquitous brecciation after metamorphism in equilibrated ordinary chondrites. *Proc. Lunar Planet. Sci. Conf.* 16, *J. Geophys. Res. Suppl.* 90:D137–D148.
Sears, D. W. G., Ashworth, J. R., and Broadbent, C. P. 1984a. Studies of an artificially shock-loaded H group chondrite. *Geochim. Cosmochim. Acta* 48:343–360.
Sears, D. W. G., Bakhtiar, N., Keck, B. D., and Weeks, K. S. 1984b. Thermoluminescence and the shock and reheating history of meteorites—II: Annealing studies of the Kernouve meteorite. *Geochim. Cosmochim. Acta* 48:2265–2272.
Simon, S. B., Papike, J. J., Hörz, F., and See, T. H. 1986. An experimental investigation of agglutinate melting mechanics: Shocked mixtures of Apollo 11 and 16 soils. *Proc. Lunar Planet. Sci. Conf.* 17, *J. Geophys. Res.* 91:E64–E74.

Smith, C. S. 1958. Metallographic study of metals after explosive shock. *Trans. Amer. Inst. Min. Metal. Eng.* 212:574–589.
Snee, L. W., and Ahrens, T. J. 1975. Shock-induced deformation features in terrestrial peridot and lunar dunite. *Proc. Lunar Sci. Conf.* 6:833–842.
Stöffler, D. 1972. Deformation and transformation of rock-forming minerals by natural and experimental shock processes. I. Behavior of minerals under shock compression, *Fortschr. Mineral.* 49:50–113.
Stöffler, D. 1974. Deformation and transformation of rock-forming minerals by natural and experimental shock processes. II. Physical properties of shocked minerals. *Fortschr. Mineral.* 51:256–289.
Stöffler, D. 1982a. Terrestrial impact breccias. In *Workshop on Lunar Breccias and Soils and Their Meteoritic Analogs*, eds. G. J. Taylor and L. L. Wilkening, LPI Tech. Rept. 82-02 (Houston: Lunar and Planetary Inst.), pp. 139–146.
Stöffler, D. 1982b. Density of minerals and rocks under shock compression. In *Landolt-Börnstein-Numerical Data and Functional Relationships in Science and Technology*, New Series, Group V. Geophysics and Space Research, Vol. 1, Subvol. a, ed. K.-H. Hellwege (Berlin: Springer Verlag), pp. 120–183.
Stöffler, D. 1984. Glasses formed by hypervelocity impact. *J. Non-Crystalline Solids* 67:465–502.
Stöffler, D., Gault, D. E., Wedekind, J., and Polkowski, G. 1975. Experimental hypervelocity impact into quartz sand: Distribution and shock metamorphism of ejecta. *J. Geophys. Res.* 80:4062–4077.
Stöffler, D., Knöll, H.-D., and Maerz, U. 1979. Terrestrial and lunar impact breccias and the classification of lunar highland rocks. *Proc. Lunar Planet. Sci. Conf.* 10:639–675.
Stöffler, D., Knöll, H. D., Marvin, U. B., Simonds, C. H., and Warren, P. H. 1980. Recommended classification and nomenclature of lunar highland rocks—a committee report. *Proc. Conf. Lunar Highlands Crusts*, LPI Contrib. 394, pp. 51–70.
Stöffler, D., Bischoff, A., Borchardt, R., Burghele, A., Deutsch, A., Jessberger, E. K., Ostertag, R., Palme, H., Spettel, B., Reimold, W. U., Wacker, K., and Wänke, H. 1985. Composition and evolution of the lunar crust in the Descartes Highlands, Apollo 16. *Proc. Lunar Planet. Sci. Conf.* 15, *J. Geophys. Res.* 90:C449–C506.
Stöffler, D., Ostertag, R., Jammes, C., Pfannschmidt, G., Sen Gupta, P. R., Simon, S. M., Papike, J. J., and Beauchamp, R. M. 1986. Shock metamorphism and petrography of the Shergotty achondrite. *Geochim. Cosmochim. Acta* 50:889–903.
Syono, Y., Goto, T., Takei, H., Tokonami, M., and Nobugai, K. 1981. Dissociation reaction in forsterite under shock compression. *Science* 214:177–179.
Takahashi, T., and Bassett, W. A. 1964. High-pressure polymorph of iron. *Science* 145:483–486.
Takeda, H. 1979. A layered crust model of a howardite parent body. *Icarus* 40:445–470.
Taylor, G. J. 1982. Petrologic comparison of lunar and meteoritic breccias. In *Workshop on Lunar Breccias and Soils and Their Meteoritic Analogs*, eds. G. J. Taylor and L. L. Wilkening, LPI Tech. Rept. 82-02 (Houston: Lunar and Planetary Inst.), pp. 153–167.
Taylor, G. J., and Heymann, D. 1969. Shock reheating, and the gas retention ages of chondrites. *Earth Planet. Sci. Lett.* 7:151–161.
Taylor, G. J., and Wilkening, L. L., eds. 1982. *Workshop on Lunar Breccias and Soils and Their Meteoritic Analogs*, LPI Tech. Rept. 82-02 (Houston: Lunar and Planetary Inst.).
Taylor, G. J., Keil, K., Berkley, J. L., and Lange, D. E. 1979. The Shaw meteorite: History of a chondrite consisting of impact-melted and metamorphic lithologies. *Geochim. Cosmochim. Acta* 43:323–337.
Tschermak, G. 1872. Die Meteoriten von Shergotty und Gopalpur. *Sitzungsber. Akad. Wiss. Wien, Math.-Naturwiss.*, Kl. 65, Teil 1:122–145.
Turner, G., and Enright, M. C. 1977. Meteorite ages and $^{40}Ar/^{39}Ar$ release patterns. *Meteoritics* 12:372–373 (abstract).
Uhlig, H. H. 1955. Contributions of metallurgy to the origin of meteorites, Pt. II, Neumann bands. *Geochim. Cosmochim. Acta* 7:34–42.
Van Schmus, W. R., and Ribbe, P. H. 1968. The composition and structural state of feldspar from chondritic meteorites. *Geochim. Cosmochim. Acta* 32:1327–1347.
Vdovykin, G. P. 1970. Ureilites. *Space Sci. Rev.* 10:483–510.

Wahl, W. 1952. The brecciated stony meteorites and meteorites containing foreign fragments. *Geochim. Cosmochim. Acta* 2:91–117.

Walsh, T. M., and Lipschutz, M. E. 1982. Chemical studies of L chondrites—II. Shock-induced trace element mobilization. *Geochim. Cosmochim. Acta* 46:2491–2500.

Wasson, J. T. 1974. *Meteorites* (Heidelberg: Springer-Verlag).

Wasson, J. T. 1985. *Meteorites* (New York: W. H. Freeman & Co.).

Wasson, J. T., and Rubin, A. E. 1985. Formation of mesosiderites by low-velocity impacts as a natural consequence of planet formation. *Nature* 318:168–169.

Wasson, J. T., Willis, J., Wai, C. M., and Kracher, A. 1980. Origin of iron meteorite groups IAB and IIICD. *Z. Naturforsch.* 35a:781–795.

Watters, T. R., and Prinz, M. 1979. Aubrites: Their origin and relationship to enstatite chondrites. *Proc. Lunar Planet. Sci. Conf.* 10:1073–1093.

Wieler, R., Baur, H., Graf, T., and Signer, P. 1985. He, Ne and Ar in Antarctic meteorites: Solar noble gases in an enstatite chondrite. *Lunar Planet. Sci.* XVI:902–903 (abstract).

PART 4
Irradiation Effects

4.1. IRRADIATION RECORDS IN METEORITES

M. W. CAFFEE
Lawrence Livermore National Laboratory

J. N. GOSWAMI
Physical Research Laboratory

C. M. HOHENBERG
Washington University

K. MARTI
University of California at San Diego

and

R. C. REEDY
Los Alamos National Laboratory

During the course of a meteorite's 4.6 Gyr history it will experience at least one, and perhaps more, exposures to energetic particles. From a meteorite's exposure to galactic cosmic rays it is possible to determine how long a meteorite existed as a m-sized body before its collision with the Earth. It is also possible in some cases to determine the preatmospheric size of the meteorite. This information plays a role in determining the dynamics of transporting meteorites from their source region in the asteroid belt to Earth-crossing orbits. It also seems likely that some meteorites were exposed to energetic particles before ejection from their parent bodies. Exposure to energetic particles at that time requires that the meteorite reside in the upper several m of the parent-body regolith. Irradiation records obtained during that time may contain infor-

mation about regolith processes on the parent bodies. Finally, it is possible that meteorites contain records from yet an earlier epoch, perhaps as early as 4.6 Gyr ago. Conditions in the early solar system were considerably different from the contemporary solar system. Our Sun was a young star and it may have been more active. Also, the meteorite parent bodies were in the early stages of formation and it is not known what type of shielding conditions prevailed at that time. It is possible, for example, that there was still considerable gas and dust in the solar cavity then. Also unknown is the size of the bodies exposed to energetic particles during that epoch. Thus, irradiation records obtained during the early stage of solar-system development are an important piece of information in reconstructing events in the early solar system.

4.1.1. INTRODUCTION

The apparent emptiness of the interplanetary space belies its complex nature. The interplanetary medium is populated by a variety of energetic particles originating both within and outside our solar system. Earth-based ionization measurements established the presence of galactic cosmic rays and energetic solar-flare particles in 1912 and 1942, respectively (cf. Pomerantz and Duggal 1974). Based on the observation of comet plasma tails, Biermann (1951,1953) first deduced the presence of solar-wind particles in the interplanetary medium. More recently, satellite probes of the contemporary energetic-particle environment around Earth have enabled us to determine the flux of these particles, their energy spectra and elemental and isotopic abundances. In addition to these man-made probes, extraterrestrial materials, such as meteorites and lunar surface material, have been used to extend our knowledge of the interplanetary energetic-particle environment into the distant past.

Considerable work has been done in this field and there are still many active areas of research but our primary concern is meteorites and the early solar system. Accordingly we will focus on the meteorites' exposure to energetic particles during their 4.5 Gyr history. Understanding the details of this exposure can be difficult because of the many possible histories experienced by a particular bit of meteorite material. Figure 4.1.1 shows a time line illustrating the possible exposure times of a meteorite over its 4.5 Gyr history. All meteorites underwent a recent exposure to energetic particles which began when collisions produced m-sized objects from larger bodies in the asteroid belt or in Earth-crossing orbits and ended when the meteorite landed on Earth. The duration of this recent exposure to galactic cosmic rays is the "galactic-cosmic-ray exposure age." This age can be calculated if the meteorite did not undergo further changes in its exposure geometry (its size or shape) after excavation from its parent body. If additional fragmentation of the m-sized body did occur, then several galactic-cosmic-ray exposure ages can be determined and the history is described as "complex."

Before its most recent exposure in space, a meteorite was buried, by at least several m of material, inside a larger parent body. In the simplest case,

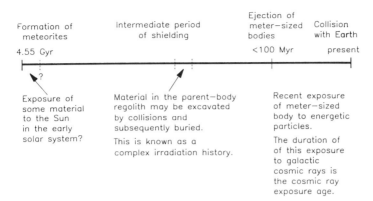

Fig. 4.1.1. Time line showing a possible evolutionary history of a meteorite. Irradiation by energetic particles could plausibly occur during several different stages.

the meteorite remained shielded over the entire time interval from formation of the parent body until collisions fragmented the parent body. However, it is possible that exposure occurred while the meteorite was still on the parent body but located within several m of the surface. Churning of the surface layers by collisions can result in repeated, complex sequences of exposure and burial on the parent body.

Finally, it is possible that some solid material was exposed to energetic particles 4.5 Gyr ago, prior to its compaction into the meteorite-sized object. Irradiation records from this early period are useful for constraining various models of meteorite formation and in studying the activity of the early Sun. However, more recent exposures of the meteorite to energetic particles make finding unambiguous evidence of an irradiation occurring 4.5 Gyr ago difficult.

4.1.2. NATURE OF THE ENERGETIC-PARTICLE ENVIRONMENT

Since all meteorites have undergone a recent exposure, the first step in interpreting the total irradiation record in meteorites is to understand the contemporary particle environment in our solar system. The energetic-particle environment in our solar system consists primarily of four classes of particles (Table 4.1.1): solar-wind particles, solar energetic particles, galactic cosmic rays and particles accelerated by interplanetary shock waves. Effects from this last source have not been identified in meteorites or lunar samples so they will not be discussed here.

The main source of energetic particles in our solar system is the Sun. The continual expansion of the solar corona in conjunction with the acceleration provided by the thermal energy of the solar atmosphere (Parker

TABLE 4.1.1
Nuclear Particle Effects in Extraterrestrial Materials[a]

Radiation Source	Energy and Characteristic Penetration Distance	Major Observable Effects
Solar wind	1 keV/nuc several hundred ångstroms	Direct implantation (e.g., surface correlated rare gases) Re-implantation of lunar atmospheric species (e.g., ^{40}Ar excess in lunar soils) Radiation damage (e.g., amorphous layers on lunar dust grains)
Solar flares	< 1 MeV/nuc to ≳ 100 MeV/nuc mm to cm many more low-energy than high-energy particles	Radionuclide production (e.g., ^{26}Al, ^{53}Mn) Track production (principally tracks produced by slowing down VH nuclei) Electronic defects (e.g., thermoluminescence)
Galactic cosmic rays	≥ 100 MeV/nuc typically ~ 3 geV/nuc cm to m	Radionuclide production Stable-isotope production (e.g., ^{21}Ne, ^{15}N) Nuclear effects due to buildup of nuclear cascades with depth (e.g., N-capture in Gd) Tracks (spallation recoils in addition to slowing down heavy nuclei)

[a] Walker (1980).

1958, 1969; Dessler 1967) supplies most of the energetic particles observed in the solar system. This outward flow of plasma (ionized nuclei and electrons), known as the solar wind, results in a particle flux at 1 AU of about 3×10^8 protons/cm^2-s. Solar-wind ions all have about the same velocity, typically 400 km/s at 1 AU (~ 1 keV/nuc), but the average value is somewhat variable in time. The solar wind also affects the number of particles entering our solar system from outside by "modulating" energetic charged particles entering the solar system from sources elsewhere in the Galaxy. Solar modulation is caused by a magnetic field which is effectively "frozen" into the expanding solar-wind plasma. Since the magnetic field lines are rooted to the Sun, the rotation of the Sun makes the field pattern in interplanetary space, at least in the plane of the ecliptic resemble an Archimedes spiral (Parker 1958). Current theories (Axford 1972; Holzer 1979) predict that this expansion continues for at least 50 AU from the Sun. Spacecraft observations have verified this to be the case at least out to 32 AU (Van Allen and Randall 1985).

The Sun is also the source of a number of other solar energetic particles, particularly those associated with solar-flare events. Solar-flare events release energy in a variety of ways, the emission of energetic ionized nuclei accounting for few percent of the total released energy. Unlike solar-wind particles, solar-flare particles have a range of energies with the flux decreasing rapidly with increasing energy. This can be expressed either as a power law in kinetic energy, E, or as an exponential in rigidity, R (momentum per unit charge):

$$dN/dE = constant \times E^{-\gamma} \qquad (1)$$

or

$$dN/dR = constant \times e^{-R/R_o} \qquad (2)$$

where γ, the power-law exponent, and R_o, the characteristic rigidity, define the spectral hardness. For example, a high value of γ or a small value for R_o will represent a steeply falling energy spectrum for the solar-flare particles. Typical values for γ and R_o in contemporary solar flares range from 2 to 4, and 50 to 200 MeV (million electron volts), respectively (Reedy and Arnold 1972; Lal 1972). The average long-term flux of solar-flare particles at 1 AU is approximately 100 nuclei/cm^2-s for particles with kinetic energy $>$ 10 MeV/nuc. The typical energy of solar-flare ions is more than three orders of magnitude higher than the solar wind as is evident in Fig. 4.1.2.

The source of the solar-flare particles is most likely the region just above the visibly active area of the Sun. However, the local region of the flare cannot supply the total energy observed in large flares. The most likely pipeline for this energy is the solar magnetic field. This field is linked to the convective motion in the near-surface and sub-surface regions of the Sun, so the ultimate source of energy is in the solar interior. The acceleration mechanism, however, for solar-flare particles is not well understood (see Svestka 1976; Sturrock 1980 for a variety of acceleration models).

Galactic cosmic rays, the most energetic particles in the interplanetary medium, come from outside our solar system. The flux of the galactic cosmic-ray nuclei is about 3 nuclei/cm^2-s for particles having kinetic energy $>$1 GeV/nuc (Table 4.1.1). Like solar-flare particles, their energy spectrum obeys a power law in energy (Fig. 4.1.2). Above $\sim$ 10 GeV, the number of particles is proportional to $E^{-2.65}$, where E is the energy of the energetic particle. The solar modulation influences the number of galactic cosmic-ray

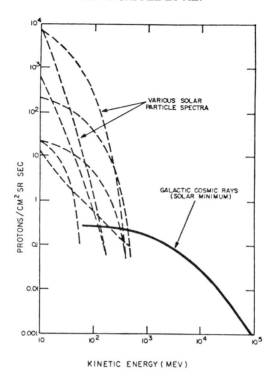

Fig. 4.1.2. Energy spectra from several moderate-size solar-flare proton events compared with the galactic-cosmic-ray proton spectrum. In both cases the number of particles decreases as the energy increases; however, for solar-flare particles the decrease is much steeper.

nuclei penetrating the solar system, this effect being most pronounced for the low-energy component (< 1 GeV/nuc) of the galactic cosmic rays.

The elemental and isotopic abundances in the solar system, the local interstellar matter and our Galaxy are fundamental to theories pertaining to the origin of the elements, origin of the solar system and processes occurring in the early solar system. Since the Sun constitutes most of the material in our solar system, its elemental and isotopic abundances are naturally of interest. Direct spectroscopic observations of the outer layers of the Sun have yielded abundances of a number of elements in the Sun. The solar abundances of several elements have also been determined from direct measurements of solar-wind and solar-flare particles. These are, after all, samples of the outer atmosphere of the Sun, although it is likely that they have been altered by the processes that accelerated them. Likewise, galactic cosmic rays represent the most accessible measure of the elemental and isotopic composition of the interstellar matter of our own Galaxy, although, undoubtedly, their composition has been altered by acceleration and transit in interstellar space for millions of years before reaching our solar system.

Solar Wind

Elemental Composition. Spacecraft instruments have measured the He/H, O/H, Si/H and Fe/H ratios in the modern solar wind (Bame 1972; Bame et al. 1983). Mass-spectrometric studies of aluminum foils exposed to solar wind on the lunar surface have allowed a determination of the solar-wind Ne/He and Ar/He ratios. Within the uncertainties of the techniques, these elemental ratios agree with those obtained from studies of the surface layers of the Sun (Geiss 1973; Geiss and Bochsler 1979; Geiss et al. 1979; for a review of these see Pepin [1980]). The similarity between the elemental abundances of the solar wind as measured on the lunar surface and spectroscopic measurements of the Sun is a strong argument that the solar wind is not altered by mass-dependent fractionation during acceleration.

Isotopic Ratios. The present-day solar-wind isotopic composition of the light noble gases was determined from foils exposed on the lunar surface (Geiss 1973). Eberhardt et al. (1972) measured the solar-wind-implanted noble gases in lunar soils. There is not a similar direct measurement from foils for the isotopic composition of N, although Becker and Clayton (1977) have measured the N-isotopic composition of several recently exposed lunar soils.

Solar Energetic Particles

Elemental Composition: $Z \leq 4$. The He/H ratio of solar-flare events varies considerably from flare to flare (Freir and Webber 1963; Biswas and Fichtel 1963) and with time within a flare. To understand the long-term behavior of solar flares, several investigators have integrated proton and alpha fluxes over several events, although these events have different energy ranges. Values for the He/H ratio are 0.001 to 0.04 (Lanzerotti 1973; Hsieh and Simpson 1970; Goswami et al. 1988), with an average value of ~0.02 for the last two solar cycles. The comparable chromospheric ratio is 0.06 and the average solar-wind value is 0.045 (Hirshberg 1973; Hundhausen 1972).

Elemental Composition: $Z > 4$. Direct measurements from spacecraft and lunar samples generally show that solar-flare particles possessing >20 MeV/nuc have approximately the same elemental abundance as those in the surface layers of the Sun (Biswas and Fichtel 1963; McDonald et al. 1974; McGuire et al. 1986), presumably their site of production. Low-energy solar-flare particles (< 20 MeV/nuc) are harder to characterize. Experiments based on track studies indicate that at sufficiently low energies, heavy elements from solar flares are enriched relative to their abundance at higher energies (Price et al. 1971; Shirck 1974; Goswami et al. 1980). However, more recent satellite measurements of solar energetic particles (Breneman and Stone 1985) do not substantiate the track data. Based on the data of Breneman and Stone, it appears that the average solar energetic-particle-derived Fe/Si ratio agrees with photospheric values.

Isotopic Ratios. The isotopic ratios of several elements in solar-flare events have been determined. Using satellite-borne detectors, Mewaldt et al. (1984) measured the isotopic composition, of He, C, N, O, Ne and Mg in the energy range ~ 5 to ~ 50 MeV/nuc. Wieler et al. (1986) and Nautiyal et al. (1986) measured the directly implanted solar-flare noble gases from lunar grains. Wieler et al. (1986) obtained a ^{20}Ne/^{22}Ne ratio of 11.7 ± 0.3 while Mewaldt et al. (1984) obtained a comparable value of $9.17(^{+2.17}_{-1.60})$ from satellite data. Both of these values are well below the accepted value of 13.7 ± 0.3 for the solar wind (Geiss et al. 1979). That the isotopic ratios of solar-wind Ne and solar energetic-particle Ne differ is perplexing since both presumably come from the same region of the Sun. Perhaps the most likely explanation for this discrepancy is mass-dependent isotopic fractionation of either the solar-energetic or solar-wind particles (Mewaldt et al. 1984, and references therein). Since several lines of evidence argue against fractionation of the solar wind, this leaves fractionation of the solar energetic particles the most likely possibility.

Galactic Cosmic Rays

Elemental Abundances. Observed elemental abundances of galactic cosmic rays exhibit both similarities and differences when compared to solar-system abundances (Anders and Ebihara 1982). The similarities suggest that cosmic rays have been synthesized by the same or similar processes that were responsible for the production of the nuclei now comprising the Sun. The high energy of the cosmic radiation links it to high-energy phenomena in the Galaxy. The differences are presumably the result of different histories, most likely the acceleration of cosmic rays to high energies and their transport through the interstellar medium at high velocities. For instance, this propagation through the interstellar medium results in the production of the light elements Li, Be and B, via nuclear interactions, causing them to be overabundant when compared to solar-system values.

Isotopic Ratios. Recently it has become possible to measure the isotopic ratios of several elements in galactic cosmic rays. The isotopic composition of these elements usually agrees with the solar-system abundances. An outstanding exception to this is the ^{20}Ne/^{22}Ne ratio. In galactic cosmic rays, ^{22}Ne is enriched relative to either solar-wind-implanted Ne or the trapped Ne component (see Chapter 7.9) observed in many meteorites (see Mewaldt [1982] for a review of galactic-cosmic-ray isotopic abundances). This observation is interesting in light of the occurrence of essentially pure ^{22}Ne in several meteorites (Eberhardt 1974; Eberhardt et al. 1979; see also Chapter

13.1). There are a variety of explanations for the isotopically anomalous Ne in cosmic rays but at this time this observation remains unresolved.

4.1.3. MODES AND PRODUCTS OF INTERACTIONS

Energetic particles interact with solid matter in a variety of ways, and to decipher the exposure of a meteorite it is necessary to understand these. The dominant modes of interaction are determined by the energy and mass of the incident particle (see Table 4.1.1). At one end of the spectrum are the solar-wind nuclei. Their low energies only enable them to be directly implanted in material. At the other end of the spectrum are protons and α-particles in the galactic cosmic rays. Their high energies allow them to induce nuclear reactions in material. Intermediate in this spectrum are the solar energetic particles, capable of interactions ranging from direct implantation (especially heavy nuclei) to nuclear reactions (for some protons).

Direct Implantation

Energetic particles in the solar wind having very low energy (~ 1 keV/nuc) can only penetrate naturally occurring solids to depths of ~ 50 nm, although subsequent diffusion of the implanted nuclei within the solid may result in an altered depth distribution. For the low-energy solar-wind ions, direct implantation is their primary mode of interaction, although they are also capable of causing radiation damage on the surface of solids, as is the case with mineral grains exposed to solar wind on the lunar surface (Borg et al. 1970). Owing to the high flux of solar-wind ions, mineral grains exposed to the solar wind can become saturated, especially for lighter elements. For instance, at 1 AU, 200 yr suffice to reach He saturation. Not surprisingly, many lunar grains are saturated with solar-wind-implanted He. Solar-wind-implanted ions are far more abundant than ions implanted from other sources; however, Wieler et al. (1986) and Nautiyal et al. (1986) have detected directly implanted solar-flare Ne and Ar in lunar samples.

Energy Loss and Lattice Damage

Energetic very heavy particles with $Z > 20$ (*VH* nuclei) nearing the end of their range in many silicate materials (typically the last 10 to 20 μm for Fe nuclei) produce enough lattice damage to create permanent latent damage trails. These damage trails are observed directly by transmission electron microscopy or chemically etched and enlarged to produce a conical hole that is visible by optical microscopy (Fig. 4.1.3). These damage trails or etch holes are termed nuclear tracks (see Fleischer et al. [1975] for a review of this technique). The tracks in silicate grains are produced by *VH* ($20 < Z < 28$) and *VVH* ($Z > 30$) nuclei from both solar flares and galactic cosmic rays. These two sources of tracks can be distinguished from each other. Galactic cosmic-ray *VH* nuclei can penetrate solid matter to depths of several cm

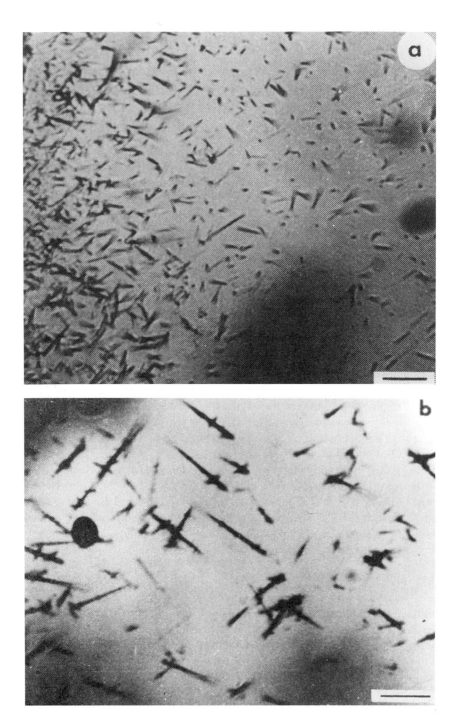

Fig. 4.1.3. (a) Tracks produced by solar-flare heavy ions. (b) Tracks produced by galactic-cosmic-ray heavy ions. These tracks become visible by optical microscopy after chemical etching. Scale bars = 1 μm.

whereas solar-flare *VH* nuclei have a range of <0.1 cm in solid matter. In this shorter range, solar-flare-produced tracks are far more prevalent than galactic cosmic-ray-produced tracks. Thus solar-flare-produced tracks are usually characterized by high track densities and track-density gradients, reflecting the steeply falling energy spectra of the solar-flare nuclei. Figure 4.1.4 shows track-density profiles from several solar-flare-irradiated lunar and meteorite grains along with the track-density profile from a piece of glass carried on the Surveyor spacecraft, that was exposed to contemporary solar-flare particles on the lunar surface. The slope of these profiles allows us to determine the spectral shape of the incident track-forming heavy nuclei in flares. The relationship between the track density and the depth can be described by the power law

$$\rho(x) \propto x^{-\eta} \tag{3}$$

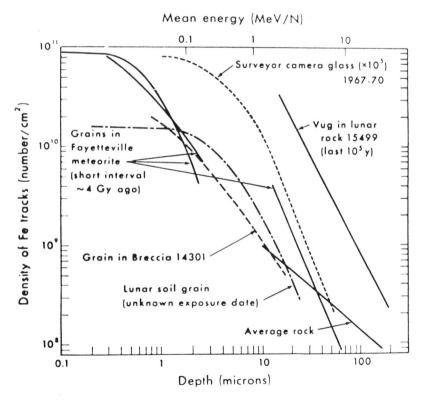

Fig. 4.1.4. Track-density profiles in grains from meteoritic and lunar samples along with the measured track-density profile in Surveyor glass. The number of tracks decreases sharply with increasing depth. The epoch of solar-flare irradiation is noted in some cases.

where η, the power-law exponent, is the slope of the track-density profile. The relationship between the energy of the incident particle and its range S is given by

$$S \propto E^\beta. \tag{4}$$

Since tracks are produced only towards the end of the range of a particle, and the track density at a given shielding depth x is proportional to the number of track-forming particles stopping at depth x, we have

$$\rho \propto \left(\frac{dN}{dS}\right)_x = \left(\frac{dN}{dE} \cdot \frac{dE}{dS}\right)_x. \tag{5}$$

The quantity dN/dE is given by Eq. (1) and dE/dS can be derived from Eq. (4). Substituting these derivatives into Eq. (5) and equating it to Eq. (3) produces

$$\eta = (1/\beta)(\gamma + \beta - 1). \tag{6}$$

This equation describes the dependence of the spectral shape of the track-forming heavy nuclei on the power-law exponent η of the incident primary nuclei. For a more detailed discussion on the production of tracks in extraterrestrial samples see Fleischer et al. (1967, 1975) and Battacharya et al. (1973).

Most tracks in meteorites are produced by stopping Fe nuclei whose primary rate of ionization exceeds that necessary for track production for about the last 15 μm of their range. Heavier nuclei have larger rates of ionization and give rise to longer tracks. The existence of very heavy nuclei ($Z \geq 30$) in cosmic radiation, and the first measurements of their relative abundance were established by track measurements in meteorites (Fleischer et al. 1967).

It is also possible to determine the duration of exposure of the material to the *VH* nuclei. Since the number of tracks is a function of shielding, it is essential to determine the shielding depth of the analyzed sample.

Nuclear Reactions

Charged particles lose energy in traversing material at a rate that depends on their velocity and charge. Whatever the initial energy, a particle may ultimately slow down and come to rest. However, if the energy exceeds the Coulomb barrier (typically ~ 10 MeV for an incident proton), the charged particle can suffer inelastic nuclear collisions and may be lost in the process. For a relativistic particle, whose kinetic energy exceeds its rest-mass energy, nuclear collisions dominate and the number of surviving particles drops off

$$^{56}Fe + p \rightarrow {}^{56}Co + n$$

exponentially with depth. The cross-over point where the probability of loss by nuclear collision is about equal to the probability of survival until stopping by ionization loss is ~ 300 MeV/nuc for a proton. Since the ionization loss is proportional to Z^2, the crossover energy is higher for heavily charged particles. However, at relativistic energies, nuclear collisions always dominate and this is why, for example, the density of nuclear particle tracks (which are produced by heavy nuclei) drops off exponentially with depth in a meteorite.

To induce a nuclear reaction, the bombarding particles must have at least several MeV of energy. Of the energetic particles in our solar system, protons and α-particles from solar flares and galactic cosmic rays are capable of producing nuclear reactions. For example, low-energy solar-cosmic-ray primaries can produce nuclei very close in mass to the targets, via reactions such as ^{56}Fe$(p,n)^{56}$Co. This notation represents a reaction between an energetic proton and an ^{56}Fe target nucleus. The residual, or spallogenic nucleus is ^{56}Co. A secondary neutron is also produced. Galactic cosmic rays have a much broader energy spectrum than solar cosmic rays (solar energetic particles) and induce a large variety of low- and high-energy nuclear reactions. Galactic cosmic-ray primaries are responsible for most high-energy reactions; however, the secondary particle cascade, especially of the electrically neutral neutrons, produces many spallogenic nuclides in lower-energy reactions (< 100 MeV).

Owing to the low flux of energetic cosmic-ray particles, nuclear reactions in meteorites are rare. Over a ~ 10 Myr galactic-cosmic-ray exposure age of a meteorite, only about one in every 10^8 nuclei undergoes a nuclear transformation. In order to detect solar or galactic cosmic-ray-produced (cosmogenic) nuclides we must look for those that are either radioactive or very rare in initial abundance, such as noble gases (Reedy et al. 1983). Table 4.1.2 shows a compilation of the measured cosmogenic nuclides in meteorites.

The accumulation of cosmogenic nuclides in a meteorite can serve as a "clock," allowing us to determine the duration of exposure of the meteorite to cosmic rays. To obtain cosmic-ray exposure ages, the clock must be calibrated, i.e., we need the production rates of the various cosmogenic nuclides. However, these production rates are sometimes difficult to determine accurately. For instance, the production rate is a function of shielding and it is not always known where a particular sample was within the meteorite. Since the production rate is proportional to the particle flux, it is necessary to determine whether the particle flux has changed over the course of the exposure.

One approach for determining the production rate of a particular nucleus is to model the production rate $P(R,d)$ as

$$P(R,d) = \sum N_j \sum \int \sigma_{ij}(E) F_i(E,R,d) dE \tag{7}$$

where N_j is the abundance of the jth target element, $\sigma_{ij}(E)$ is the cross section at energy E for making the nuclide from particle i reacting with element j,

TABLE 4.1.2
Cosmogenic Nuclides Frequently Measured in Meteorites

Nuclide	Half-life[a] (yr)	Main Targets
^{3}H	12.3	O, Mg, Si, Fe
^{3}He, ^{4}He	S	O, Mg, Si, Fe
^{10}Be	1.6×10^{6}	O, Mg, Si, Fe
^{14}C	5730	O, Mg, Si, Fe
^{20}Ne, ^{21}Ne, ^{22}Ne	S	Mg, Al, Si, Fe
^{22}Na	2.6	Mg, Al, Si, Fe
^{26}Al	7.1×10^{5}	Si, Al, Fe
^{36}Cl	3.0×10^{5}	Fe, Ca, K, Cl
^{36}Ar, ^{38}Ar	S	Fe, Ca, K
^{37}Ar	35 days	Fe, Ca, K
^{39}Ar	269	Fe, Ca, K
^{40}K	1.3×10^{9}	Fe
^{39}K, ^{41}K	S	Fe
^{41}Ca	1.0×10^{5}	Ca, Fe
^{46}Sc	84 days	Fe
^{48}V	16 days	Fe
^{53}Mn	3.7×10^{6}	Fe
^{54}Mn	312 days	Fe
^{55}Fe	2.7	Fe
^{59}Ni	7.6×10^{4}	Ni
^{60}Co	5.27	Co, Ni
^{81}Kr	2.1×10^{5}	Rb, Sr, Zr
^{78}Kr, ^{80}Kr, ^{82}Kr, ^{83}Kr	S	Rb, Sr, Zr
^{129}I	1.6×10^{7}	Te, Ba, La, Ce
$^{124-132}$Xe	S	Te, Ba, La, Ce, (I)

[a] S denotes that the nuclide is stable.

and $F_i(E,R,d)$ is the flux of particle i at energy E in a meteoroid of radius R and at a depth d (Reedy 1985, 1987a). The basic shapes for $F_i(E,R,d)$ are fairly well known, especially for high-energy galactic cosmic rays. Factors influencing $F_i(E,R,d)$ are the size and shape of a meteoroid (Reedy 1985). If the radius of a meteoroid is less than the interaction length of the galatic cosmic rays (~ 100 g/cm^2), particles entering from any direction can reach most of the meteorite. In this case the production rates do not decrease much near the center. In much larger meteoroids, galactic cosmic rays interact before they reach the center, causing the production rates to decrease from their

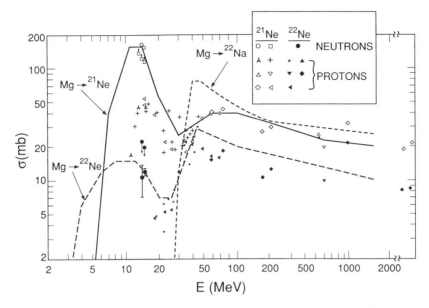

Fig. 4.1.5. Cross sections for the production of ^{21}Ne and ^{22}Ne from natural Mg (Reedy et al. 1979; Bieri and Rutsch 1962; Goebel et al. 1964; Herzog 1975; Walton et al. 1976).

peak values near the surface. The energy of the incident particles also influences the production profile. For instance, low-energy solar-flare energetic particles have very steep production profiles, only inducing reactions in the upper several cm of material. Conversely, production profiles from galactic cosmic rays vary less than those of solar energetic particles.

The other parameter governing the production rate is the cross section for the production of a particular nuclide. Most cross sections are based upon experimental data or extrapolations from experimental data, although some are calculated from theoretical considerations. Figure 4.1.5 shows the cross sections as a function of incident particle energy (excitation functions) for making ^{21}Ne and ^{22}Ne from Mg with both protons and neutrons. Combining the particle flux (both primary and secondary), excitation functions and target chemistry according to Eq. (7) yields the production profile for a particular nuclide. Figure 4.1.6 shows the production profile for ^{21}Ne, a low-energy reaction. Figure 4.1.7 shows the production profile for ^{3}He, produced primarily in high-energy reactions.

Other approaches have been employed to predict the production profiles of cosmogenic nuclides in meteorites. Honda and Arnold (1964) observed that in high-energy spallation reactions the yield of the product obeys a power law with respect to the product mass. Signer and Nier (1960) applied a semi-empirical model to the spallogenic ^{3}He, ^{4}He, ^{21}Ne and ^{38}Ar measured in a slab

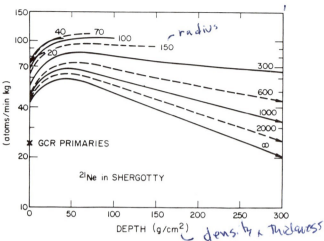

Fig. 4.1.6. Calculated production rates of ^{21}Ne as a function of meteorite radius and depth.

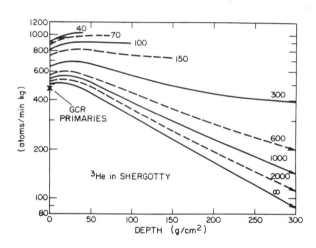

Fig. 4.1.7. Calculated production rates of ^{3}He as a function of meteorite radius and depth.

of the Grant iron meteorite. These results were used to predict spallogenic noble-gas profiles in spherical iron meteorites of any radius.

It is also possible to simulate galactic-cosmic-ray bombardment of meteorites using accelerator techniques. In these experiments, thick targets are irradiated with energetic protons (Honda 1962). The thick-target production profiles are then translated into production profiles for a hemispherical object (Kohman and Bender 1967). This approach has been successful in predicting the production of cosmogenic nuclides in large iron meteorites. More recently, Michel et al. (1986) have simulated the production of cosmogenic nuclides in small meteorites by irradiating a spherical target with 600 MeV protons. This particular experiment has the advantage of being more realistic since a 4π-exposure geometry is obtained by continuous rotation of the sphere. In conjunction with the experimental work, theoretical calculations using Monte Carlo techniques have been performed. The production profiles obtained from the theoretical calculations are generally in agreement with those obtained experimentally. It is not practical to conduct simulation experiments for all the possible exposure geometries of meteorites, so theoretically derived production profiles in agreement with simulated irradiations give us confidence that the underlying theory can reliably predict production profiles for meteorites of different sizes. More experiments of this type at different energies are being planned in the near future.

In lieu of reliable absolute production rates, it is sometimes necessary to use pairs of cosmogenic nuclides. If the production rate of a particular nuclide and the ratio of the production rates are known, it is possible to infer the production rate of the other nuclide. Examples of this are ^{22}Na/^{22}Ne, ^{36}Cl/^{36}Ar, ^{39}Ar/^{38}Ar, ^{40}K/^{41}K and ^{81}Kr/^{83}Kr. The ^{81}Kr/^{83}Kr ratio is especially useful since the measurement of both isotopes is done at the same time mass-spectrometrically, making this technique capable of yielding extremely precise measurements. The ^{81}Kr/^{83}Kr production-rate ratio is a function of the amount of shielding and can usually be obtained from the other cosmogenic Kr isotopes (Marti 1967; Eugster et al. 1967).

Exposure Ages

The most common method of determining exposure ages uses spallogenic ^{21}Ne (cf. Bogard et al. 1971; Cressey and Bogard 1976; Herzog 1975; Crabb and Schultz 1981). Since ^{21}Ne is stable, this technique can be applied to meteorites having very long exposure ages. Unfortunately, the ages obtained from this method do not always agree with those obtained by other methods, most notably ^{26}Al (Nishiizumi et al. 1980). This inconsistency, presumably the result of inaccurate production rates, makes caution necessary when comparing exposure ages of meteorites that have been dated using different nuclides. However, as long as ages from one technique are used, *relative* exposure ages can be accurately determined. In other words, if a consistent set of ^{21}Ne production rates is used for determining the exposure age of many meteorites, the *differences* in exposure ages between those meteorites can be accurately determined. However, because of possible uncertainties in the production rates, the *absolute* exposure ages themselves may be systematically wrong by as much as 50%.

Figure 4.1.8 shows a histogram of exposure ages for H- and L-group chondrites based on ^{21}Ne concentrations. Almost half of the H chondrites have a major peak near 6 to 7 Myr that contains about half of this type of meteorite. The distribution of exposure ages in Fig. 4.1.8 for L chondrites,

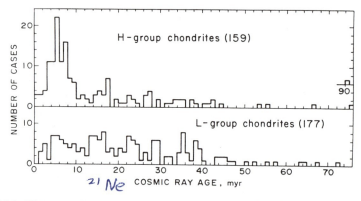

Fig. 4.1.8. Histogram of galactic-cosmic-ray exposure ages for H and L chondrites. The exposure ages for H chondrites show a major peak around 6 to 7 Myr.

on the other hand, shows no statistically significant peaks. The LL chondrites have a very broad cluster near 15 Myr. The carbonaceous chondrites tend to have shorter exposure ages than the other types of chondrites, with ages above 20 Myr being fairly rare. Achondrites are more likely to have exposure ages above 30 Myr than are the chondrites, most of them clustering between 20 to 35 Myr. Iron meteorites have much longer exposure ages than stony meteorites (Fig. 4.1.9) (Voshage and Hintenberger 1963), most likely indicating that greater physical strength helps a meteorite survive longer in space. Alternatively, exposure-age differences could be attributed to the orbits of the meteoroids or their parent bodies. Objects in orbits where collisions are less frequent would naturally survive longer.

From exposure-age distributions the frequency of events and perhaps the structure of the parent body can be deduced. The large peak in H-chondrite exposure ages suggests that a single large collision produced all these meteorites. Contained in this cluster are representatives of all the petrographic types, including gas-rich members. If at the time of the collision the parent body had a simple stratified structure with each layer consisting of a single petrographic type, those types in the upper layers would be likely to populate the peak in the exposure-age frequency distribution. Since we observe members from all the petrographic types in the peak, it may be inferred that at the time of the collision the parent body did not have a simple layered structure. If the assumptions used in the above line of reasoning are correct, then the picture we have of the parent body is one where the various petrographic types are somehow randomly mixed and reside side by side with each other. In other words, the surface material has been mixed to a depth of $\lesssim 1$ km prior to the collision that produced m-sized bodies.

There are several indications that some meteorites have received some portion of their cosmogenic products before their exposure as m-sized bodies.

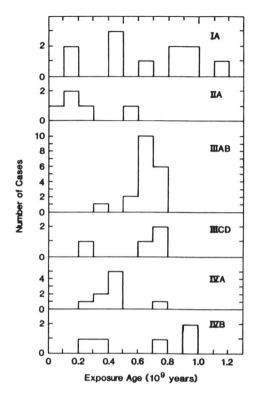

Fig. 4.1.9. Histogram of galactic-cosmic-ray exposure ages for six types of iron meteorites as calculated from measured ratios of $^{40}K/^{41}K$ and $^{4}He/^{21}Ne$.

Among iron meteorites, for example, about 30% appear to have complex histories (Schaeffer et al. 1981). Complex irradiation histories have also been observed in chondrites. Jilin (H-5) for example, evidently had a two-stage exposure (Honda et al. 1982; see also Begemann [1985] for a consortium report on this meteorite). From documented sample positions and ^{60}Co activities, the pre-atmospheric radius of Jilin was determined to be ~ 85 cm (Heusser et al. 1985). While this size, Jilin was exposed to galactic cosmic rays between 0.4 and 0.6 Myr (Pal et al. 1985; Heusser et al. 1985). This 4π exposure was immediately preceded by a 2π exposure in a much larger body lasting between 6.2 and 10 Myr (Begeman et al. 1985; Pal et al. 1985; Heusser et al. 1985). These exposure ages were based on a combination of ^{21}Ne, ^{53}Mn, ^{26}Al and ^{10}Be.

A slightly more complicated irradiation history is seen in some gas-rich chondrites. Schultz et al. (1972) studied seven different light inclusions from Weston (H-5). In this study, noble-gas and track-counting methods were combined. Of the samples studied, one inclusion had a 20% excess of ^{21}Ne, cor-

TABLE 4.1.3
Noble Gas and Radionuclide Exposure Ages of Gas-Rich Meteorites[a]

| Meteorite | Type | Exposure Age (Myr) | | | | T^*_{pc} (yr) |
| | | (Stable nuclides) | | (Radionuclides) | | |
		^{21}Ne	^{38}Ar	^{26}Al	^{53}Mn	
Orgueil	CI	4.5, 10.7†	10†	>2**	≥10**	<10⁶
Alais	CI	8.0	—	—	—	?
Murchison	CM	0.87, 1, 1.3†	—	1.6	>2	<10⁵
Murray	CM	3.5	—	>2**	5.7	<10⁵(?)
Nogoya	CM	0.21	—	0.15	0.21	<10⁵
Cold Bokkeveld	CM	0.36, 0.35†	—	0.34	—	<10⁵
Kilbourne	H	5.9	—	—	4.7	<10⁶
Leighton	H	6.5	—	—	5.1	≤10⁶
Pantar	H	4.0, 5.5	—	—	4.8	<10⁶
Kapoeta	How	3–6	3.0	>2	3.0	<10⁶

*The difference between stable nuclide and radionuclide ages is considered to represent the precompaction irradiation duration (T_{pc}).
**Exposure age estimates based on assumption of saturated activity.
†Based on analyses of mineral separates.
[a] Source of data: Bogard et al. (1971), Goswami and Nishiizumi (1982), Heymann and Anders (1967), Jeffery and Anders (1970), Kerridge et al. (1979), Macdougall and Phinney (1977), Mazor et al. (1970), Rajan et al. (1979), Rowe and Clarke (1971), Goswami and Nishiizumi (1987).

responding to a 4 Myr exposure. By using the ^{3}He/^{21}Ne ratio as an indicator of shielding (^{3}He is primarily produced in high-energy reaction and ^{21}Ne is more likely to be produced by secondary neutrons [Reedy 1985]), a shielding depth of 40 cm was obtained. This particular inclusion also has more galactic cosmic-ray-produced tracks than can be accounted for by the most recent exposure. Since 40 cm is too deep for any significant track production, Schultz et al. (1972) proposed that this inclusion underwent a very short irradiation at a shallower depth after the irradiation at 40 cm. Similar observations have been made in St. Mesmin (Schultz and Signer 1977) and Djermaia (Lorin et al. 1970). On the basis of these studies, it seems reasonable to conclude that collisions were responsible for the excavation of previously shielded material. As a result of further cratering, this material subsequently moved around in the upper several m of the regolith and at some time was close enough to the surface to record tracks from exposure to galactic cosmic-ray VH nuclei. Finally, ejecta from other cratering events buried the material, totally shielding it from further exposure. The length of this precompaction exposure is based on the stable nuclide ^{21}Ne so it is difficult to determine when the exposure occurred. Schultz and Signer (1977) suggest, on the basis of K-Ar ages, that at least for St. Mesmin, the regolith was not compacted to its present structure until as late as 1.36 Gyr ago.

Other direct methods for detecting precompaction exposures are measurements of neutron detectors, such as certain isotopes of Br, Kr and Gd (Clark and Thode 1964; Marti et al. 1966; Göbel et al. 1982; Eugster et al. 1970), which indicate exposure in a large body.

Another way of detecting a complex history is to compare exposure ages based on different cosmogenic nuclides. Since radioactive cosmogenic nuclides decay, any records from an ancient exposure to energetic particles would be lost. Stable nuclides, on the other hand, integrate the entire exposure history. In principle, the difference between a radionuclide exposure age and a stable-nuclide exposure age is a measure of any prior exposure. In practice, this technique is limited because of uncertainties in production rates. Nonetheless, it is possible to set some limits on the duration of an early exposure for *bulk* material. Table 4.1.3 shows a comparison of exposure ages from radioactive and stable nuclides. For most meteorites the duration of an early exposure is ≤ 1 Myr; however, these numbers are only upper limits. If more reliable production parameters were available for more cosmogenic nuclides, it would be possible to obtain more certain differences between stable- and radioactive-nuclide-based exposure ages.

4.1.4. TIME VARIATIONS OF THE ENERGETIC-PARTICLE ENVIRONMENT

General Properties and Time Scales

Variations in the energetic-particle environment can occur in a variety of ways. Over short time scales (years), solar flares are known to vary both in energy and particle flux from event to event. Over longer time scales (1 Gyr) it is possible that the galactic cosmic-ray flux has changed. To understand long-term variations we must examine material exposed for long durations, such as lunar and meteoritic material. More difficult though, is evaluating the possibility that early in the history of the solar system, during the formation of the meteorites, the energetic-particle environment was significantly different from contemporary conditions. Since the particle flux inside the solar system is determined by the Sun and its magnetic field, any unusual solar behavior would produce a different particle environment.

Variations in the Solar Cosmic Rays

Satellite-based detectors have measured particles emitted by large flares since about 1960 although indirect measurements go back a few more decades (Reedy 1980; Pomerantz and Duggal 1974). Meteorites are exposed to solar flares as m-sized bodies but lose these records upon entry into the atmosphere (see Sec. 4.1.6). Superbly suited for studying the record of solar-flare particles over the last 10 Myr are lunar rocks. Measurements of radionuclides such as ^{14}C ($t_{1/2}$ = 5730 yr), ^{81}Kr ($t_{1/2}$ = 2.1 × 10^5 yr), ^{26}Al ($t_{1/2}$ = 7 × 10^5 yr) and ^{53}Mn ($t_{1/2}$ = 3.7 Myr) allow a determination of the average solar-cosmic-ray flux for two half-lives before the present. Based on the available observations, it appears that the average fluxes and spectral shapes of solar protons during the last few Myr are not perceptibly different from contemporary solar flares (Goswami et al. 1988). Based on track records in lunar grains it appears that neither the *average* energy spectra of the *VH* nuclei nor the *VH/VVH* ratio have varied. These are averages and less is known about past individual solar-flare events (Goswami et al. 1988), although tree-ring records of ^{14}C indicate that no flares having 10 times or more greater particle fluxes than those observed since 1956 have occurred since ~ 5000 B.C. Also, based on studies of lunar samples, it seems unlikely that many solar flares during the last 10 Myr were orders of magnitude larger than contemporary solar flares (Reedy et al. 1983).

Solar-Activity-Induced Variations in Galactic-Cosmic-Ray Fluxes (1 to 1000 yr)

Evans et al. (1982) measured gamma-ray-emitting radionuclides in a number of recent meteorite falls. These results indicate that the shielding-corrected variations in activity were larger for the shorter-lived radionuclides ^{46}Sc and ^{54}Mn than for longer-lived nuclides. This is consistent with the observed anticorrelation between galactic cosmic-ray protons and solar activity over the 11-yr solar cycle.

In the period from 1645 to 1715, known as the Maunder Minimum, sunspots and aurorae were rare. During this time there was enhanced production of ^{14}C in the Earth's atmosphere (Eddy 1976). Forman and Schaeffer (1980) observed that the mean activity in some meteorites of ^{37}Ar ($t_{1/2}$ = 35 days) is below that for ^{39}Ar ($t_{1/2}$ = 269 yr). This excess of ^{39}Ar (~ 18%) was attributed to enhanced fluxes of galactic cosmic rays during the Maunder Minimum. A more definitive experiment would be measurements of ^{39}Ar in meteorites that fell just after the Maunder Minimum.

Meteorites (0.1 to 10 Myr)

The ratio of a measured radionuclide activity to its calculated production rate can be used as an indicator of cosmic-ray variability. In iron meteorites, this ratio varies by a factor of two, most likely due to uncertainties in both the calculations and measurements (Arnold et al. 1961). Based on meteorite, lunar and terrestrial samples, it appears that the galactic-cosmic-ray flux variations over the last 10^5 yr are < 30% (Reedy et al. 1983).

Iron Meteorites (> 0.1 to 1 Gyr)

Galactic-Cosmic-Ray Variations. Studies of cosmic-ray variations for longer time periods are based upon measurements from iron meteorites, which have much longer exposure ages than stone meteorites. Exposure ages of iron meteorites based on different long-lived radionuclide/stable-nuclide pairs generally agree within experimental uncertainties implying no major variations in the cosmic-ray flux over the last few Myr. However, exposure ages determined with ^{40}K ($t_{1/2} = 1.26$ Gyr) are 45% greater than ages inferred with the shorter-lived radionuclides (Voshage and Hintenberger 1963). These results have been confirmed by Hampel and Schaeffer (1979). The higher exposure ages based on ^{40}K could indicate either a change in the cosmic-ray flux or a change in the shielding conditions. For example, an erosion rate of $\sim 2 \times 10^{-8}$ cm/yr could produce enough of a shielding change to account for the observed differences in the exposure ages. Schaeffer et al. (1981) simulated the erosion rates of iron meteorites in laboratory experiments, obtaining a value of 2.2×10^{-9} cm/yr, and concluded that erosion could not account for the exposure-age ratios. However, it is possible that the physical properties of Fe at room temperature are different from those under space conditions or that the flux of micrometeoroids responsible for the erosion was higher in the past or in other regions of the solar system.

Nevertheless, the generally accepted conclusion is that the high exposure ages determined with ^{40}K were caused by a $\sim 50\%$ higher flux of cosmic-ray particles during the last ~ 10 Myr when compared to the last 100 Myr to 1 Gyr. (The recent higher galactic cosmic-ray flux increased the activities of the shorter-lived radionuclides, thus increasing the inferred production rate for the stable nuclide and lowering the exposure age.) A $\sim 50\%$ variation in the flux of galactic cosmic-ray particles is relatively small, much less than the variation seen over a typical 11-year solar cycle due to modulation; however, it is evidently a long-term phenomenon. This variation could be attributed to either an increase in the cosmic-ray flux during the last ~ 10 Myr or increased modulation of galactic cosmic rays by the solar wind > 10 Myr ago.

Solar-Wind Variations

Most of our knowledge about solar-wind activity during the last several Gyr has come from the study of the lunar regolith. Volatiles such as the noble gases and N are depleted in the Moon so solar-wind-derived noble gases and N are readily observed in lunar materials (see, e.g., Frick et al. 1988). Since both the isotopic composition and abundance of He in the solar wind can vary over time scales from hours to years, it is difficult to determine systematic long-term variations. The situation is further complicated by the ease with which He diffuses out of most minerals. Nevertheless, Geiss (1973) noted a systematic difference in ^{3}He/^{4}He ratios between old breccias and relatively young soils. These observations have been confirmed by other similar studies (cf. Kerridge 1980). Furthermore, Eberhardt et al. (1972) observed that gas-rich aubrites have ^{3}He/^{4}He ratios $\sim 33\%$ lower than modern-day values.

Geiss (1973) has argued that the abundance of the solar-wind-implanted Xe in lunar soils is greater than can be accounted for by integrating the present-day flux over the age of the regolith. Similar reasoning leads to the conclusion that N is also overabundant in lunar soils (Clayton and Thiemens 1980). Taken together, these two observations may indicate that the present-day solar-wind flux is lower by a factor of two or three than the average over the last few Gyr. It has also been noted that the ^{15}N/^{14}N ratio has increased by $\sim 33\%$ over the last 2 to 4 Gyr in lunar soil samples (Kerridge 1980). Murty and Marti (1986) have studied the N-isotopic composition of the gas-rich meteorite Pesyanoe to see if meteorites show similar effects. In their initial study, a "light" lunar-like N component was not found. There have been a number of explanations proposed to explain the secular variation in the N-isotopic composition observed in lunar materials. At this time, these observations have no simple and accepted explanation (see Kerridge [1980] for a review of these models). However, despite the lack of an acceptable model explaining the variation in isotopic composition or flux, it seems that over time scales of a few Gyr the nature of the solar wind has changed. This observation is particularly interesting in light of the observation in iron meteorites that the flux of galactic cosmic rays in the solar system was evidently less in the distant past.

4.1.5. PARTICLE ENVIRONMENT OF THE EARLY SOLAR SYSTEM

Conditions unique to the early solar system could have resulted in different particle fluxes compared with the present day. Differences could reflect either an enhanced particle flux coming from a more active Sun or a decrease in particle fluxes due to residual gas and dust.

Astronomical observations have established that young solar-mass stars go through a phase of increased activity (T Tauri) while en route to the main sequence (see Chapter 6.2). This period is characterized by strong stellar winds, increased solar-flare activity, and increased ultraviolet luminosity. These observations constitute circumstantial evidence that our own Sun may have gone through a similar phase during its early evolution. During such a phase, particle fluxes would be considerably different from contemporary particle fluxes. For example, increased solar-flare activity would have resulted in increased solar-cosmic-ray production rates. Changes in the solar wind would have altered both the galactic-cosmic-ray-production rate and the length of time a meteorite grain would have to be exposed to the Sun before becoming saturated with ^{4}He. There is considerable uncertainty about the duration of this phase but it seems unlikely that it would have lasted longer than 10 Myr, although there is some evidence that young stars might have increased ultraviolet luminosity for a considerably longer period (Zahnle and Walker 1982). However, it is not precisely known when this phase occurred with respect to the formation of the meteorites. If the extremely active phase of the Sun ended prior to the formation of meteorites, then the meteorites themselves would contain no direct evidence for the existence of increased stellar activity

IRRADIATION RECORDS 229

(exceptions to this might be spallation reactions in gaseous material later incorporated into solid material). On the other hand, considerable solid material may have existed during this phase but other circumstances may have prevented this material, which may be presently incorporated in meteorites, from recording their exposure to an enhanced particle flux. For example, the presence of sufficient amounts of gas and/or dust could have shielded them from exposure to an active early Sun. This is especially true for solar wind and solar flares. Solar-wind ions cannot penetrate more than 2×10^{-4} g/cm^2 of material so even a minor amount of gas and dust would fully shield bodies forming at >1 AU from solar-wind exposure. Solar-flare ions have more energy than solar-wind ions but nebular gas and dust could still keep solar-flare ions from penetrating to the region of meteorite formation (Housen and Wilkening 1980).

Further complicating the picture is our lack of knowledge about the early history of the meteorites themselves. A meteorite's most recent exposure to the energetic-particle environment occurred as a m-sized body. Based on our knowledge of the contemporary energetic-particle environment, we can determine the duration of this exposure. However, we are not allowed the same luxury when we study meteorites exposed to energetic particles in the early solar system. Were the meteorites exposed as small bodies during accretion or was their exposure dominated by regolith processes on larger bodies? Was the ancient particle environment similar to the contemporary particle environment or did increased solar activity change the particle flux? In short, we have more unknowns than equations. Exposure ages cannot be obtained without absolute production rates. Conversely, without some independent chronometer that allows us to place at least an upper limit on the exposure duration, it is difficult to determine production rates.

4.1.6. ALTERATION OF RECORDS

Before considering the implications for the early solar-system chronology of irradiation records, we will briefly discuss the possibility that isotopic records were altered by exposure to energetic particles and that the irradiation records themselves were altered by processes such as space erosion and ablation during atmospheric entry. Irradiation records can also be altered by redistribution and diffusive loss (see Wieler et al. [1983] for a discussion of this in lunar samples).

In the study of spallation-produced stable nuclides, it is necessary to distinguish the cosmogenic component from those nuclides of other origins. Conversely, to study the noncosmogenic components present in the meteorite, the cosmogenic component must be subtracted. In noble-gas studies, it is common practice to separate these components. As an example, consider the various components of Ne shown in Fig. 4.1.10. The ^{20}Ne/^{22}Ne and ^{21}Ne/^{22}Ne ratios of spallogenic Ne are roughly unity, whereas the other compo-

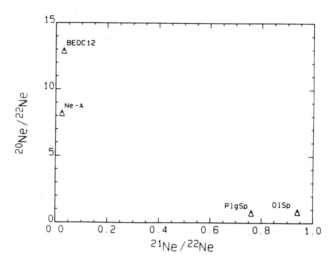

Fig. 4.1.10. Neon three-isotope plot. The expected galactic-cosmic-ray spallogenic Ne components from olivine and plagioclase target material are denoted by OLSP and PLGSP, respectively. The solar-wind Ne isotopic composition is represented by BEOC-12. Ne-A represents a trapped component found in many meteorites (see Chapter 7.9).

nents, such as solar-wind-implanted Ne and Ne-A (see Chapters 7.9 and 13.1) are deficient in ^{21}Ne and thus lie close to the ordinate. If the Ne extracted from a sample has a composition that lies on a line between a spallogenic component and some other component, Ne-A for example, then the relative contribution from each component can be uniquely determined. This idea can be generalized to n components provided there are n isotopes. For Ne it is possible to determine the contributions from up to three components.

Erosion on the lunar surface has removed at least 0.3 to 0.6 mm during the past 10 Myr. The rate varies a little depending on the type of material (i.e., mechanical strength) and the micrometeorite bombardment rate (Hörz et al. 1974). When galactic-cosmic-ray VH ion tracks are used to determine the shielding depth and exposure geometry of a meteorite, the erosion rate may be important. Fortunately, unless the space exposure times are very long (> 20 Myr), the loss of surficial material in space is not as important as the loss of material during atmospheric entry. This loss amounts to several cm of surface material. Additionally, the heating pulse created upon atmospheric entry causes the formation of a fusion crust several mm thick.

Since the penetration depth of the solar-wind ions is of the order of several hundred Å, these records are lost upon atmospheric entry. Likewise, with only a few exceptions (Nishiizumi et al. 1986), solar-flare records are also lost. Generally, the only records of the meteorite's recent exposure to energetic particles that survive atmospheric entry are galactic cosmic-ray-produced effects. However, there are many meteorites that do contain large quantities of solar-wind-implanted noble gases. Individual components of these gas-rich meteorites must have been exposed to the Sun at some earlier time, following which they were shielded by at least several meters of material, and remained buried within their parent body until being re-excavated by collisions that ultimately led to their being placed in an Earth-crossing orbit. The irradiation records in gas-rich meteorites may provide information about both the energetic-particle environment in the early solar system and the formation of gas-rich meteorites.

4.1.7. IMPLICATIONS FOR THE EARLY SOLAR SYSTEM

Irradiation Records in Gas-rich Meteorites

Solar-Wind Records. Gerling and Levskii (1956) reported the discovery of an extremely high concentration of the noble gases He, Ne and Ar in the meteorite Pesyanoe. Later, other laboratories discovered more gas-rich meteorites. These studies demonstrated that gas-richness was neither rare nor confined to a particular class of meteorites (Table 4.1.4). Suess et al. (1964) hypothesized that the excess gas in these meteorites was implanted solar wind. Experiments by Eberhardt et al. (1965) and Wänke (1965) conclusively demonstrated that the excess gas was sited in the surface layers of the mineral grains, making solar-wind-implanted ions the most plausible source of the excess noble gases (see Goswami et al. [1984] for a review of gas-rich meteorites). Similar effects in lunar grains and direct measurements of the solar-wind isotopic and elemental composition demonstrated conclusively that the solar wind is indeed the source of this excess noble gas (Geiss 1973). Poupeau et al. (1974) observed in a detailed study that 1 grain in 500 from gas-rich aubrites had an amorphous coating, characteristic of exposure to the solar wind. They also found that meteorite grains were not saturated with solar-wind-implanted ^{4}He. It is significant that these observations are in contrast to lunar grains, where amorphous coatings and saturated ^{4}He are common.

The elemental and isotopic composition of the solar wind has been measured in gas-rich achondrites. Marti et al. (1972) demonstrated that the noble gases comprising the solar-type component in gas-rich meteorites and lunar soils had elemental abundance ratios that agree with those predicted by interpolation in the cosmic-abundance table using nuclear abundance rules. Marti (1969) also determined the Xe-isotopic composition of the solar-type Xe from the gas-rich achondrite Pesyanoe. Its composition is compatible with the isotopic composition of modern-day solar-wind Xe in lunar samples. This seems to indicate that the isotopic composition of the solar-wind Xe has not changed for 3 to 4 Gyr.

Solar-Flare VH Ion Tracks. Since some constituents of the meteorites had been exposed to the solar wind it was reasoned that they might also contain solar-flare records. These were subsequently discovered by Lal and Rajan

TABLE 4.1.4
Abundance of Gas-Rich Meteorites in Different Meteorite Classes[a]

Meteorite Type	No. of Gas-Rich Meteorites Known	Percent Abundance[†]
Chondrites*		
CI chondrites	4	100
CM chondrites	8	57
CV3 chondrites	5	55
H chondrites	31	15
L chondrites	6	3
LL chondrites	3	9
Anomalous chondrites	1	25
Achondrites*		
Howardites (Ho)	4	22
Aubrites (Au)	3	37

*Among the chondrites, there are no gas-rich members in the CO3 class and enstatites. Only howardites and aubrites among achondrites have gas-rich members.
†This is based on the total number of meteorites in each class analyzed for their noble gas records.
[a]Goswami et al. 1984.

(1969) and Pellas et al. (1969), who detected solar-flare VH ion tracks in mineral grains from several gas-rich meteorites. The fraction of grains containing solar-flare VH ion tracks in these meteorites is small, usually less than a few percent of the total grains. Additionally, most of these grains do not have tracks evenly distributed over the entire grain (Rajan 1974). The lack of grains having a 4π track distribution eliminates the possibility of irradiation of these grains in free space. The most likely alternative is that the irradiation took place on the surface of their parent body. Poupeau et al. (1974) established a qualitative correlation between solar-wind-implanted gases and solar-flare tracks. They concluded that the solar-flare irradiated grains were simultaneously exposed to the solar wind and solar flares in the regolith of their parent body. They also reasoned that the exposure to the solar wind was short, < 100 yr, assuming modern-day particle fluxes. If we assume that the ancient solar-flare VH ion intensity was similar to contemporary values, then the typical track exposure age of these grains is of the order of 10^4 yr. This exposure duration is considerably shorter than track exposure ages of lunar grains, which are typically $> 10^5$ yr.

The energy spectrum of the VH solar-flare ions was determined from measured track-density depth profiles in grains from gas-rich meteorites. For Fayetteville (Rajan 1974), the slope of the curve is essentially the same as that of the Surveyor glass, indicating that the energy spectra of the ancient solar-flare heavy ions that produced the irradiation effects in Fayetteville are the same as for contemporary flares. Similar studies on carbonaceous chondrites (Goswami et al. 1980) find in some cases a smaller slope. One possible explanation for this is the presence of a small amount of shielding material during the irradiation. Lunar grains, for example, often have flatter slopes than the Surveyor glass, indicating shielding or erosion effects. The alternative possibility is that the solar-flare energy spectra of the ancient Sun were different from contemporary flare spectra.

Spallation Records. It follows that since individual grains from gas-rich meteorites were exposed to the Sun during an earlier irradiation those meteorites should also contain cosmogenic nuclides. The cosmogenic nuclides could be produced from either solar or galactic cosmic rays. Usually, spallation reactions occurring in meteorites are dominated by those caused by galactic cosmic rays. Of considerable interest is the duration of this exposure. We must first separate the spallation records of the early exposure from those of the most recent exposure. By comparing ages determined by radionuclides to those based on stable nuclides, we conclude that the difference in exposure ages is $\lesssim 1$ Myr for most meteorites. More precisely, the *average* duration of an early irradiation is $\lesssim 1$ Myr. It is, however, possible that a small portion of the grains comprising the bulk sample underwent a longer exposure.

Specifically, we can compare the amount of a stable spallogenic nuclide (e.g., ^{21}Ne) in grains known to have been exposed to the Sun, as evidenced by the presence of solar-flare tracks, to the amount in grains not containing tracks (Caffee et al. 1983,1987). Differences between these two represent an early exposure to energetic particles. Table 4.1.5 shows the results from such an experiment. In all cases, the solar-flare-irradiated grains contain large excesses of spallation-produced ^{21}Ne. If we at first assume the source of the excess spallogenic ^{21}Ne to be galactic-cosmic-ray bombardment and that the irradiation occurred in a regolith, the grains must have been exposed for 30 to 200 Myr, depending on the meteorite.

Let us first qualitatively consider the exposure of grains containing solar-flare tracks. Based on the solar-wind implanted ions, the grains' surface exposure is ~ 100 yr. From the track records we estimate their exposure in the upper mm of regolith to be 10^3 to 10^5 yr. If galactic cosmic rays are the source of the spallogenic ^{21}Ne, then these same grains were resident within the upper several meters for at least 30 and perhaps several hundred Myr. As we have previously mentioned, several gas-rich meteorites (most notably Weston) appear to have inclusions that were irradiated before compaction of the parent-body regolith. Is it possible that the excess ^{21}Ne observed in solar-flare track-rich grains is a result of the same type of regolith process? At this time there is not a definitive answer to this question. On the one hand, it is tempting to ascribe both irradiations to the same type of processes (Occam's razor). On the other hand, there are very significant differences. The first difference is the very magnitude of the excess in some of the samples, especially Kapoeta, where the regolith exposure age could be as much as several 10^2 Myr. This

TABLE 4.1.5
Spallogenic Ne and Apparent Ne Exposure Ages

irr = solar flare tracks

Sample	Spallation-produced ^{21}Ne ($\times 10^{-9}$ cm^3 STP/g)	Ne Cosmic* Ray Exposure "Ages" (Myr)	Apparent† Pre-Compaction Exposure	Regolith‡ Exposure Age (Myr)
Murchison				
non-irr. olivines-II	5.0	0.78	—	—
irr. olivines-I	141	28	27	108
irr. olivines-II	33	7.6	7	28
Kapoeta				
non-irr. pyroxenes	1.6	0.47	—	—
non-irr. feldspars	6.1	3.5	—	—
irr. pyroxene	178	57	54	216
irr. feldspars	56	33	30	120
Weston				
non-irr. olivines	148	28	—	—
irr. olivines	203	38	10	40
irr. chondrules	192	47	19	76

*Production rates are from Reedy et al. (1979).
†Represents difference in apparent exposure to galactic cosmic rays between irradiated and non-irradiated grains.
‡Estimated exposure duration of irradiated grains if the irradiation occurred in a regolith (see text).

observation is hard to understand in light of our knowledge of the lunar regolith. On the Moon, in older regoliths (~ 100 Myr), the solar-wind-implanted ^{4}He is at saturation level and the track densities are much higher than those observed in gas-rich meteorites. Furthermore, in an old regolith, a high percentage ($> 50\%$) of the grains have been exposed to the Sun, as evidenced by the presence of solar-flare tracks, unlike the gas-rich meteorites, where only a small percentage of the grains have solar-flare tracks. Another difference is the observation that the grains in gas-rich meteorites that do not contain tracks also do not contain any measurable excess spallogenic ^{21}Ne. In the upper several meters of a parent-body regolith, there should exist many grains that did not reside in the upper mm, hence contain no solar-flare tracks, but which were nevertheless exposed to galactic cosmic rays and should have spallogenic products from cosmic-ray interactions. Such effects are indeed observed in lunar grains, where the spallation is due to galactic cosmic rays (Caffee et al. 1986). These effects have not been observed in the non solar-flare irradiated grains from gas-rich meteorites. Based on comparison with the lunar surface, excepting the large amounts of cosmogenic Ne in some gas-rich meteorite grains, the gas-rich breccias appear to be very "immature" (see Chapter 3.5). The difficulty in explaining the large excesses of cosmogenic Ne in gas-rich meteorites despite their immaturity leads us to speculate that either the source of the energetic particles is not the galactic cosmic rays or our understanding of parent-body regolith formation is entirely inadequate, or perhaps both.

As noted earlier, energetic particles in solar flares are also capable of inducing spallation reactions in the top cm of regolithic material. At 1 AU, the production of cosmogenic Ne as a result of solar energetic particles (assuming a modern-day particle flux) is roughly 10 times greater (at 1 gm/cm^2) than the production of Ne from galactic cosmic rays. However, at 3 to 4 AU, the most likely region for formation of the gas-rich meteorites (Anders 1975,1978), the production of Ne by solar energetic particles would be lower by an order of magnitude, assuming an inverse-square dependence of flux with distance from the Sun. This makes the production of Ne by solar energetic particles essentially the same as that from galactic cosmic rays at 3 to 4 AU. Thus, exposure to solar cosmic rays cannot reduce the required exposure duration. In fact, if we assume modern-day particle fluxes, the problem is aggravated. If the grains spent up to several Myr in the upper cm, this would require the underlying several m of material to experience even longer exposure times resulting in the accumulation of galactic-cosmic-ray effects even greater than those observed in lunar material. This is clearly not observed. The short range of solar protons does hold some promise of explaining the correlation between solar-flare tracks and excess spallogenic ^{21}Ne. However, given the difficulties outlined above, it is unlikely that solar flares having contemporary particle fluxes could account for these results, since the required exposure durations would be too long.

Alternatively, it is possible that these grains were exposed to the Sun very early in the history of the solar system and that at that time the particle flux was higher than the contemporary particle flux. For example, increasing the galactic-cosmic-ray flux lowers the required exposure time of the solar-flare-irradiated grains. However, the correlation between tracks and excess ^{21}Ne would still be unexplained. Apparently what is needed are more low-energy protons, capable of penetrating only ~ 1 cm of material, thereby concentrating the spallogenic ^{21}Ne in the track-bearing grains. One possibility is that outside our solar system (and the effects of solar modulation), the galactic cosmic rays have an increased flux of low-energy protons and that in the early solar system this low-energy component of the galactic cosmic rays was not modulated as effectively as it is now. This would require a young Sun having less solar-wind activity than the current Sun rather than more intense stellar winds as is generally observed in young Sun-like stars. Reedy (1987b) has estimated that the flux of energetic particles in the contemporary demodulated galactic rays would be no higher than a factor of 5 above current fluxes at 1 AU, so even demodulated galactic cosmic rays are not sufficiently intense to explain the ^{21}Ne excesses in the preirradiated meteorite grains.

The next alternative is increased ^{21}Ne production from solar-flare protons. We have already discussed the astronomical observations that lead us to believe our Sun may have gone through an active phase. Such a model for the ^{21}Ne production is attractive for several reasons. The increased production rate would reduce the required exposure durations. The smaller penetration depth of solar protons qualitatively explains the correlation between solar-flare tracks and spallogenic ^{21}Ne. Thus, exposure to an active early Sun might explain the presence of these irradiation effects (solar wind, solar-flare tracks, and spallogenic Ne) in at least three very different types of meteorite parent body; carbonaceous and ordinary chondrites and howardites.

However, there are several shortcomings that need to be addressed. The first involves the energy spectrum of the protons. The ^{21}Ne/^{22}Ne ratio can serve as an indicator of the energy of the incident protons. In high-energy galactic-cosmic-ray reactions, ^{21}Ne and ^{22}Ne are produced in about the same proportion, whereas in low-energy reactions, ^{21}Ne is not as easily produced in some minerals. Preliminary calculations indicate a hard power-law spectrum for solar-flare protons (i.e., flatter energy spectra) does increase the ^{21}Ne/^{22}Ne ratio (Caffee et al. 1987), as observed in grains from gas-rich meteorites. This implies that the solar-flare proton spectra were flatter for the early Sun. On the other hand, the depth profiles in solar-flare-irradiated grains from gas-rich meteorites do not require flatter solar-flare spectra, although for several carbonaceous chondrites flatter spectra would not be inconsistent with the observed track results. More work needs to be done in this area. Studies that establish the compaction age (see Chapter 5.4) of gas-rich meteorites would be extremely beneficial. For example, the regolith exposure ages (assuming a galactic cosmic-ray or solar cosmic-ray source) are several hundred Myr. If it can be unambiguously shown that these grains were buried and shielded from subsequent irradiation 4.5 Gyr ago, then this would be strong evidence that there was indeed an enhanced particle flux in the early solar system. Conversely, if many of these grains were buried deep within a parent body (see Chapter 5.4), as has been proposed by Pellas and Storzer (1981) and Pellas (1972) based on fission-track cooling rates, and were not uncovered until much later, then a galactic-cosmic-ray irradiation is most plausible.

4.1.8. FORMATION MODELS OF GAS-RICH METEORITES

One of the major goals in studying irradiation records in gas-rich meteorites is to place constraints on various models for the formation of meteorite parent bodies, and of regoliths in them, during the early history of the solar system. Unfortunately, it is not yet possible to determine uniquely how the gas-rich meteorites were formed; however, there are several models. The irradiation records play an important role in these models.

Regolith Model. Some of the early attempts to develop models for gas-rich meteorites produced in a parent-body regolith environment were largely influenced by the study of lunar samples (Anders 1975,1978; Price et al. 1975). The important conclusions drawn from these studies are:

1. Gas-rich meteorites were formed within 10 AU of the Sun, most likely within 5 AU, making the asteroid belt the prime candidate as the source region;
2. These asteroidal regoliths were formed by bombardment of impacting objects. The dominant process of this regolith growth was addition of new material rather than recycling of existing material.

These observations are consistent with many of the irradiation features seen in gas-rich meteorites. For example, a fast-growing regolith would probably result in fewer irradiated grains, a lack of glassy agglutinates, etc. Building on these ideas, Housen et al. (1979a,b) and Langevin and Maurette (1980) developed more detailed models of asteroidal regolith development (see also Chapter 3.5). The Housen et al. model starts by assuming a mass-frequency distribution of impactors in the asteroid belt. This is combined with relationships between the kinetic energy of the impactor and the resulting crater diameter. Using these it is then possible to estimate the depth of the regolith and the gardening rate for a variety of different compositions (basalt and sand being the extreme cases). However, there are parameters needed in this model that are difficult to obtain. The mass distributions of both the contemporary and ancient asteroid belt are unknown. The physics of cratering in a low-gravity-field environment is also not well understood. In this model, it is possible for some asteroidal regoliths to become several km deep. This is much deeper than the lunar regolith, which is only about 10 m or less deep. A deep regolith is needed to explain the high fraction of gas-rich meteorites in some chemical classes of meteorites. The deeper regolith occurs for two reasons; a higher impactor flux at 3 to 4 AU than at 1 AU, and the small size of the bodies, which allows more widespread ejecta. Many of the irradiation features in gas-rich meteorites are qualitatively consistent with this model. Most notably, the small doses of solar-wind-implanted ^{4}He and short track exposure ages point to rapid burial of material. Likewise, precompaction exposure ages ≤ 1 Myr for bulk material reinforce this idea. Not as easy to interpret is the excess ^{21}Ne in grains containing solar-flare tracks. In a fast-growing regolith, such excesses of cosmogenic Ne should not occur (Caffee et al. 1986). If the spallogenic ^{21}Ne results from exposure to an active early Sun, then the idea of a fast-growing regolith is salvaged; however this implies that during parent-body regolith processes the Sun was still active. Indeed, for the basaltic achondrite Kapoeta, differentiation must have preceded its exposure to an active Sun. If we assume this period of increased activity is ≤ 100 Myr then this represents an upper limit on the time-scale for regolith formation.

Small-Body Model. Goswami and Lal (1979) and Goswami and Macdougall (1983) have preformed detailed studies of the track distributions in gas-rich carbonaceous chondrites. The salient observations of their studies are:

1. Only about 2 to 3% of the grains in CM chondrites and ~ 10% of the grains in CI chondrites contain precompaction irradiation records;
2. 30 to 50% of the irradiated grains exhibit a gradient in track density;
3. Very few of the grains have a 4π exposure geometry;
4. Most of the grains were irradiated in a simple single-stage geometry.

The authors contend that a regolith model similar to that outlined above cannot simultaneously account for the irradiation features in gas-rich carbonaceous chondrites and the fact that ~ 70% of the CI and CM chondrites are gas-rich. The high percentage of gas-rich members among the gas-rich carbonaceous meteorites is a problem for the regolith models of Housen et al. (1979a,b) and Langevin and Maurette (1980), that predict thin (cm to m) regoliths for the small, weak carbonaceous asteroids. Goswami and Lal (1979) proposed a model in which the grains were exposed to the solar wind and solar flares during residence in or on the surface of cm- to m-sized bodies (planetesimals). This swarm of planetisimals accreted to form larger parent bodies within a time scale of 10^5 yr. This time scale is consistent with believed precompaction time scales of carbonaceous chondrites, based on studies of bulk material or mineral separates (see Table 4.1.3). The high percentage of gas-rich meteorites among CM and CI chondrites is the result of the entire meteorite parent body being built from irradiated components.

The most troubling aspect of this model is that it proposes dramatically different irradiation settings for different classes of meteorites. This is especially so in light of the observation of excess spallogenic ^{21}Ne in at least three distinct classes of meteorites. If the spallogenic ^{21}Ne in carbonaceous chondrites is due to galactic cosmic rays, then the exposure duration (hence accretion time scale) is > 10 Myr, roughly two orders of magnitude more than originally anticipated in this model. On the other hand, if the spallogenic ^{21}Ne is the result of an active early Sun, and if this phase is short, then it follows that as the carbonaceous chondrites were being irradiated as small bodies, the remainder of the chondrites and achondrites were already km-sized bodies, capable of maintaining thick regoliths.

Megaregolith. Yet another possible origin for the gas-rich meteorites involves the collisional break up and reassembly of the meteorite parent bodies as suggested by Davis et al. (1979), Hartmann (1979) and Chapman et al. (1982). While the frequency or even plausibility of collisional fragmentation and reassembly is a subject of debate (Chapman et al. 1982), if this process can occur, it might provide a reasonable explanation for the high abundances of gas-rich meteorites of certain types (e.g., H chondrites). Note that even if the regolith of the H-chondrite parent body is 1 km thick, one large excavation would not generate the proper proportion of gas-rich and nongas-rich meteorites; in particular, there would not be enough gas-rich members (Housen et al. 1979a,b). In fact Housen et al. (1979a) also proposed that collisional break up and reassembly of meteorite parent bodies could lead to the formation of a megaregolith.

In the megaregolith model, an asteroidal body with an already developed regolith is almost catastrophically fragmented by a low-velocity collision. However, the resulting orbits of the fragments are such that they can approach each other at relatively low velocities and perhaps reaccumulate. This will lead to incorporation of previously generated regolith material inside the reassembled asteroid and also provide fresh surfaces for development of new regolith. If the above process is repeated several times, a megaregolith is produced. Recently, Goswami and Nishiizumi (1988) have proposed that even in the megaregolith model one can imagine that solar irradiation of the gas-rich components took place not during regolith growth, but as a result of the exposure of individual components to the Sun following fragmentation of the asteroid but before its reaccumulation. It is worth noting that the Goswami and Lal (1979) and Goswami and Nishiizumi (1988) models are similar in that the irradiation features in individual grains are produced during the grain's residence on the surface of or inside small bodies, rather than in the regolith of an asteroidal body. While these models do not rule out irradiation during regolith formation on meteorite parent bodies, the contribution from such a process to the total irradiation record is considered to be small.

Since the dynamics of collisional fragmentation and reaccumulation processes are not well understood, this model does not seem to have any readily testable features. If the time scales for reaccumulation are long then m-sized (or smaller) bodies could be exposed to galactic cosmic rays for long periods of time. This could provide a possible explanation for the excess spallogenic ^{21}Ne observed in gas-rich meteorites except that it would still not explain the correlation between the presence of tracks and spallogenic Ne. Conversely, if the spallogenic Ne does result from irradiation by an active early Sun, then this model would not be required to explain the irradiation features in gas-rich meteorites. Whether acquisition of the precompaction irradiation record in small bodies or megaregoliths is consistent with petrographic features of gas-rich meteorites, such as the evidence for aqueous alteration in carbonaceous chondrites (Chapter 3.4), is also not yet clear.

4.1.9. SUMMARY

Our aim is to use the irradiation records in meteorites as probes into the early solar system. From these records we can potentially learn about early solar activity, formation of the meteorite parent bodies and perhaps other events unique to the early solar system. However, to demonstrate unambiguously that a meteorite was exposed to energetic particles in the early solar system, it is necessary to subtract the effects of irradiations obtained more recently. All meteorites have been recently exposed to galactic cosmic rays. The irradiation records obtained from this exposure to galactic cosmic rays can be used to learn about the meteorites' recent irradiation history. Prior to this exposure, some meteorites were exposed to galactic cosmic rays and solar irradiation during an intermediate period. Finally, there is some evidence that some gas-rich meteorites received their irradiation records 4.5 Gyr ago. The existence of such ancient irradiation records in meteorites is an exciting possibility since much could be learned about the early solar system from their study. However, before this evidence can be accepted without question, it will be necessary to devise a reliable independent chronometer capable of telling when these irradiation records were obtained.

REFERENCES

Anders, E. 1975. Do stony meteorites come from comets? *Icarus* 24:363–371.
Anders, E. 1978. Most stony meteorites come from the asteroid belt. In *Asteroids: An Exploratory Assessment*, eds. D. Morrison and W. C. Wells, NASA CP-2053, pp. 57–76.
Anders, E., and Ebihara, M. 1982. Solar-system abundances of the elements. *Geochim. Cosmochim. Acta* 46:2363–2380.
Arnold, J. R., Honda, M., and Lal, D. 1961. Record of cosmic-ray intensity in the meteorites. *J. Geophys. Res.* 66:3519–3531.
Axford, W. I. 1972. The interaction of solar wind with the interstellar medium. In *Solar Wind*, eds. C. P. Sonett, P. S. Coleman, and J. M. Wilcox, NASA SP-308, pp. 609–660.
Bame, S. J. 1972. Spacecraft observations of solar wind. In *Solar Wind*, eds. C. P. Sonett, P. S. Coleman, and J. M. Wilcox, NASA SP-308, pp. 535–558.
Bame, S. J., Feldman, W. C., Gosling, J. T., Young, D. T., and Zwickl, R. D. 1983. What magnetospheric workers should know about solar wind composition. In *Energetic Ion Composition in the Earth's Magnetosphere*, ed. R. G. Johnson (Tokyo: Terra Scientific Publ.), pp. 73–98.
Battacharya, S. K., Goswami, J. N., Gupta, S. K., and Lal, D. 1973. Cosmic ray effects induced in a rock exposed on the Moon or in free space: Contrast in patterns for tracks and isotopes. *The Moon* 8:253–286.
Becker, R. H., and Clayton, R. N. 1977. Nitrogen isotopes in lunar soils as a measure of cosmic-ray exposure and regolith history. *Proc. Lunar Sci. Conf.* 8:3685–3704.
Begemann, F. 1985. Jilin Consortium Study I. *Earth Planet. Sci. Lett.* 72:246.
Begemann, F., Li, Z., Schmitt-Strecker, S., Weber, H. W., and Xu, Z. 1985. Noble gases and the history of Jilin meteorite. *Earth Planet. Sci. Lett.* 72:247–262.
Bieri, R. H., and Rutsch, W. 1962. Erzeugungsquerschnitte fur Edelgase aus Mg, Al, Fe, Ni, Cu, and Ag bei Bestrahlung mit 540 MeV protonen. *Helv. Phys. Acta* 35:553–554.
Biermann, L. 1951. Kometenschweife und solare korpuskularstrahlung. *Z. Astrophys.* 29:274–286.
Biermann, L. 1953. Physical processes in comet tails and their relation to solar activity. *Extrait des Mem. Soc. Roy. Sci. Liege Quatr. Ser.*, pp. 291–302.
Biswas, S., and Fichtel, C. E. 1963. Nuclear composition and rigidity spectra of solar cosmic rays. *Astrophys. J.* 139:941–950.
Bogard, D. D., Clark, R. S., Keith, J. E., and Reynolds, M. A. 1971. Noble gases and radionuclides in Lost City and other recently fallen meteorites. *J. Geophys. Res.* 76:4076–4083.
Borg, J., Dran, J. C., Durrieu, L., Jouret, C., and Maurette, M. 1970. High voltage electron microscope studies of fossil nuclear particle tracks in extraterrestrial matter. *Earth Planet. Sci. Lett.* 8:379–386.
Breneman, H. H., and Stone, E. C. 1985. Solar coronal and photospheric abundances from solar energetic particle measurements. *Astrophys. J.* 299:L57–L61.
Caffee, M. W., Goswami, J. N., Hohenberg, C. M., and Swindle, T. D. 1983. Cosmogenic neon

from precompaction irradiation of Kapoeta and Murchison. *Proc. Lunar Planet. Sci. Conf.* 14, *J. Geophys. Res.* 88:B267–B273.

Caffee, M. W., Hohenberg, C. M., Swindle, T. D., and Goswami, J. N. 1986. Pre-compaction irradiation of meteorite grains. *Lunar Planet. Sci.* XVII:99–100.

Caffee, M. W., Hohenberg, C. M., Swindle, T. D., and Goswami, J. N. 1987. Evidence in meteorites for an active early Sun. *Astrophys. J.* 313:L31–L35.

Chapman, C. R., Davis, D. R., and Greenberg, R. 1982. Apollo asteroids: Relationships to main-belt asteroids and meteorites. *Meteoritics* 17:193 (abstract).

Clark, W. B., and Thode, H. G. 1964. The isotopic composition of Krypton in meteorites. *J. Geophys. Res.* 69:3673–3679.

Clayton, R. N., and Thiemens, M. H. 1980. Lunar nitrogen: Evidence for secular change in the solar wind. In *Proc. Conf. Ancient Sun*, eds. R. O. Pepin, J. A. Eddy, and R. B. Merrill (New York: Pergamon Press), pp. 463–473.

Crabb, J., and Schultz, L. 1981. Cosmic-ray exposure ages of the ordinary chondrites and their significance for parent body stratigraphy. *Geochim. Cosmochim. Acta* 45:2151–2160.

Cressey, P. J., Jr., and Bogard, D. D. 1976. On the calculation of cosmic-ray exposure ages of stone meteorites. *Geochim. Cosmochim. Acta* 40:749–762.

Davis, D. R., Chapman, C. R., Greenberg, R., Weidenschilling, S. J., and Harris, A. W. 1979. Collisional evolution of asteroids: Populations, rotations, and velocities. In *Asteroids*, ed. T. Gehrels (Tucson: Univ. of Arizona Press), pp. 528–557.

Dessler, A. J. 1967. Solar wind and interplanetary magnetic field. *Rev. Geophys.* 5:1–41.

Eberhardt, P. 1974. A neon-E-rich phase in the Orgueil carbonaceous chondrite. *Earth Planet. Sci. Lett.* 24:182–187.

Eberhardt, P., Geiss, J., and Grogler, N. 1965. Further evidence on the origin of trapped gases in the meteorite Khor Temiki. *J. Geophys. Res.* 70:4375–4378.

Eberhardt, P., Geiss, J., Graf, H., Mendia, M. D., Morgeli, M., Schwaller, H., Stettler, A., Krähenbühl, U., and von Gunten, H. R. 1972. Trapped solar wind noble gases in Apollo 12 lunar fines 12001 and Apollo 11 breccia 10046. *Proc. Lunar Planet. Sci. Conf.* 3:1821–1856.

Eberhardt, P., Jungck, M. H. A., Meier, F. O., and Niederer, F. 1979. Presolar grains in Orgueil: Evidence from neon-E. *Astrophys. J.* 234:L169–L171.

Eddy, J. A. 1976. The Maunder Minimum. *Science* 192:1189–1202.

Eugster, O., Eberhardt, P., and Geiss, J. 1967. ^{81}Kr in meteorites and ^{81}Kr radiation ages. *Earth Planet. Sci. Lett.* 2:77–82.

Eugster, O., Tera, F., Burnett, D. S., and Wasserburg, G. J. 1970. Neutron capture of ^{157}Gd in Norton County. *Earth Planet. Sci. Lett.* 7:436–440.

Evans, J. C., Reeves, J. H., Rancitelli, L. A., and Bogard, D. D. 1982. Cosmogenic nuclides in recently fallen meteorites: Evidence for galactic cosmic ray variations during the period 1967–1978. *J. Geophys. Res.* 87:5577–5591.

Fleischer, R. L., Price, P. B., Walker, R. M., and Maurette, M. 1967. Origin of fossil charged particle tracks in meteorites. *J. Geophys. Res.* 72:333–353.

Fleischer, R. L., Price, P. B., and Walker, R. M. 1975. *Nuclear Tracks in Solids* (Berkeley: Univ. California Press).

Forman, M. A., and Schaeffer, O. A. 1980. ^{37}Ar and ^{39}Ar in meteorites and solar modulation of cosmic rays during the last thousand years. In *Proc. Conf. Ancient Sun*, eds. R. O. Pepin, J. A. Eddy, and R. B. Merrill (New York: Pergamon Press), pp. 279–292.

Freier, P. S., and Webber, W. R. 1963. Exponential rigidity spectrums for solar-flare cosmic rays. *J. Geophys. Res.* 68:1605–1629.

Frick, U., Becker, R. H., and Pepin, R. O. 1988. Solar wind record in the lunar regolith: Nitrogen and noble gases. In *Proc. Lunar Planet. Sci. Conf.* 18:87–120.

Geiss, J. 1973. Solar wind composition and implications about the history of the solar system. *13th Intl. Cosmic Ray Conf.* 5:3375–3398.

Geiss, J., and Boschler, P. 1979. On the abundances of rare ions in the solar wind. In *Proc. Solar Wind Conf. 4th* (Burghausen: Springer-Verlag).

Geiss, J., Buhler, F., Cerutti, P., Eberhardt, P., and Filleux, C. 1979. Solar wind composition experiment. *Apollo 16 Prelim. Sci. Rept.* NASA SP-314, 14-1, to 14-10.

Gerling, E. K., and Levskii, L. K. 1956. On the origin of the rare gases in stony meteorites. *Dokl. Akad. Nauk. SSR* 110:750–755.

Göbel, R., Begemann, F., and Ott, U. 1982. On neutron-induced and other noble gases in Allende inclusions. *Geochim. Cosmochim. Acta* 46:1777–1792.

Goebel, K., Schultes, H., and Zähringer, J. 1964. Production Cross Sections of Tritium and Rare Gases in Various Target Elements. C.E.R.N. Rept., CERN-64-12.

Goswami, J. N., and Lal, D. 1979. Formation of the parent bodies of the carbonaceous chondrites. *Icarus* 10:510–521.

Goswami, J. N., and Macdougall, J. D. 1983. Nuclear track and compositional studies of olivines in CI and CM chondrites. *Proc. Lunar Planet. Sci. Conf.* 13, *J. Geophys. Res. Suppl.* 88:A755–A764.

Goswami, J. N., and Nishiizumi, K. 1982. Constraints on the irradiation history of the gas-rich meteorites. In *Workshop on Lunar Breccias and Soils and Their Meteoritic Analogs*, LPI Tech. Rept. 82-02 (Houston: Lunar and Planetary Inst.), pp. 44–48.

Goswami, J. N., and Nishiizumi, K. 1988. Constraints on the irradiation history of gas-rich chondrites. *Geochim. Cosmochim. Acta*, submitted.

Goswami, J. N., Lal, D., and Macdougall, J. D. 1980. Charge composition and energy spectra of ancient solar flare heavy nuclei. In *Proc. Conf. Ancient Sun*, eds. R. O. Pepin, J. A. Eddy, and R. B. Merrill (New York: Pergamon Press), pp. 347–364.

Goswami, J. N., Lal, D., and Wilkening, L. L. 1984. Gas-rich meteorites: Probes for the particle environment and dynamical processes in the inner solar system. *Space Sci. Rev.* 37:111–159.

Goswami, J. N., McGuire, R. E., Reedy, R. C., Lal, D., and Jha, R. 1988. Solar flare proton and alpha particles during the last three solar cycles. *J. Geophys. Res.*, in press.

Hampel, W., and Schaeffer, O. A. 1979. ^{26}Al in iron meteorites and the constancy of cosmic ray intensity in the past. *Earth Planet. Sci. Lett.* 42:348–358.

Hartmann, W. K. 1979. Diverse puzzling asteroids and a possible unified explanation. In *Asteroids*, ed. T. Gehrels (Tucson: Univ. of Arizona Press), pp. 466–479.

Herzog, G. F. 1975. The production of ^{21}Na by low-energy protons in meteorites. *J. Geophys. Res.* 80:1109–1112.

Heusser, G., Ouyang, T., Kirsten, T., Herpers, U., and Englert, P. 1985. Conditions of the cosmic ray exposure of the Jilin chondrite. *Earth Planet. Sci. Lett.* 72:263–272.

Heymann, D., and Anders, E. 1967. Meteorites with short cosmic ray exposure ages, as determined from their ^{26}Al content. *Geochim. Cosmochim. Acta* 31:1793–1809.

Hirshberg, J. 1973. Helium abundance of the Sun. *Rev. Geophys. Space Phys.* 11:115–131.

Holzer, T. E. 1979. Solar wind and related astrophysical phenomena. In *Solar System Plasma Physics*, eds. E. N. Parker, C. F. Kennel, and L. J. Lanzerotti (Amsterdam: North-Holland Publ. Co.), pp. 101–176.

Honda, M. 1962. Spallation products distributed in a thick iron target bombarder by 3-Bev protons. *J. Geophys. Res.* 67:4847–4858.

Honda, M., Nishiizumi, M., Imamura, M., Takaoka, N., Nitoh, O., Horei, K., and Komura, K. 1982. Cosmogenic nuclides in the Kirin chondrite. *Earth Planet. Sci. Lett.* 57:101–109.

Honda, M., and Arnold, J. R. 1964. Effects of cosmic rays on meteorites. *Science* 143:203–212.

Hörz, F., Schneider, E., and Hill, R. E. 1974. Micrometeoroid abrasion of lunar rocks: A Monte-Carlo simulation. *Proc. Lunar Sci. Conf.* 5:2397–2412.

Housen, K. R., and Wilkening, L. L. 1980. Solar-ion penetration in the early solar nebula. *Proc. Lunar. Planet. Sci.* 11:1251–1269.

Housen, K. R., Wilkening, L. L., Chapman, C. R., and Greenberg, R. 1979a. Asteroidal regoliths. *Icarus* 39:317–351.

Housen, K. R., Wilkening, L. L., Chapman, C. R., and Greenberg, R. 1979b. Regolith development and evolution on asteroids and the moon. In *Asteroids*, ed. T. Gehrels (Tucson: Univ. of Arizona Press), pp. 601–627.

Hsieh, K. C., and Simpson, J. A. 1970. The relative abundances and energy spectra of ^{3}He and ^{4}He from solar flares. *Astrophys. J.* 162:L191–L196.

Hundhausen, A. J. 1972. *Coronal Expansion and Solar Wind* (New York: Springer-Verlag).

Jeffery, P. M., and Anders, E. 1970. Primordial noble gases in separated meteorite minerals—I. *Geochim. Cosmochim. Acta* 34:1175–1198.

Kerridge, J. F. 1980. Secular variations in composition of the solar wind: Evidence and causes.

In *Proc. Conf. Ancient Sun,* eds. R. O. Pepin, J. A. Eddy, and R. B. Merrill (New York: Pergamon Press), pp. 475–489.
Kerridge, J. F., Macdougall, J. D., and Marti, K. 1979. Clues to the origin of sulfide minerals in CI chondrites. *Earth Planet. Sci. Lett.* 43:359–367.
Kohman, T. P., and Bender, M. L. 1967. Nuclide production by cosmic rays in meteorites and on the moon. In *High-Energy Reactions in Astrophysics,* ed. B. S. P. Shen (New York: Benjamin), pp. 169–245.
Lal, D. 1972. Hard rock cosmic ray archaeology. *Space Sci. Rev.* 14:93–102.
Lal, D., and Rajan, R. S. 1969. Observations on space irradiation of individual crystals of gas-rich meteorites. *Nature* 233:269–271.
Langevin, Y., and Maurette, M. 1980. A model for small body regolith evolution: The critical parameters. *Lunar Planet. Sci.* XI:602–604.
Lanzerotti, L. J. 1973. *High Energy Phenomena on the Sun Symposium Proceedings,* eds. R. Ramaty and R. G. Stone, NASA Goddard X-693-73-193, 427, 1973.
Lorin, J. C., Pellas, P., Schultz, L., and Signer, P. 1970. Evidence for different irradiation histories of xenoliths from gas-rich Djermaja chondrite. *Trans. Amer. Geophys. Union* 51:340.
Macdougall, J. D., and Phinney, D. 1977. Olivine separates from Murchison and Cold Bokkeveld: Particle tracks and noble gases. *Proc. Lunar Sci. Conf.* 8:293–311.
Marti, K. 1967. Mass-spectrometric detection of cosmic-ray-produced ^{81}Kr in meteorites and the possibility of Kr-Kr dating. *Phys. Rev. Lett.* 18:264–266.
Marti, K. 1969. Solar-type xenon: A new isotopic composition of xenon in the Pesyanoe meteorite. *Science* 166:1263–1265.
Marti, K., Eberhardt, P., and Geiss, J. 1966. Spallation, fission and neutron capture anomalies in meteoritic krypton and xenon. *Z. Naturforsch* 21a:398–413.
Marti, K., Wilkening, L. L., and Suess, H. E. 1972. Solar rare gases and the abundances of the elements. *Astrophys. J.* 173:445–450.
Mazor, E., Heymann, D., and Anders, E. 1970. Noble gases in carbonaceous chondrites. *Geochim. Cosmochim. Acta* 34:781–824.
McDonald, F. B., Fichtel, C. E., and Fisk, L. A. 1974. Solar particles (observations, relationship to the Sun, acceleration, interplanetary medium). In *High Energy Particle and Quanta in Astrophysics,* eds. F. B. McDonald and C. E. Fichtell (Cambridge: MIT Press), pp. 212–272.
McGuire, R. E., von Rosenvinge, T. T., and McDonald, F. B. 1986. The composition of solar energetic particles. *Astrophys. J.* 301:939–961.
Mewaldt, R. A. 1982. The elemental and isotopic composition of galactic cosmic ray nuclei. *Rev. Geophys. Space Phys.* 21:295.
Mewaldt, R. A., Spalding, J. D., and Stone, E. C. 1984. A high-resolution study of the isotopes of solar flare nuclei. *Astrophys. J.* 280:892–901.
Michel, R., Dragovitsch, P., Englert, P., Peiffer, F., Stuck, R., Theis, S., Begemann, F., Weber, H., Signer, P., Wieler, R., Filges, D., and Cloth, P. 1986. On the depth dependence of spallation reactions in a spherical thick diorite target homogeneously irradiated by 600 MeV protons: Simulation of production of cosmogenic nuclides in small meteorites. *Nucl. Instrum. Methods* B16:61–82.
Murty, S. V. S., and Marti, K. 1986. Pursuit of solar nitrogen: Nitrogen components in the Pesyanoe meteorite. *Lunar Planet. Sci.* XVII:591–592.
Nautiyal, C. M., Padia, J. T., Rao, M. N., and Venkatesan, T. R. 1986. Solar flare neon composition and solar cosmic-ray exposure ages based on lunar mineral separates. *Astrophys. J.* 301:465–470.
Nishiizumi, K., Regnier, S., and Marti, K. 1980. Cosmic ray exposure ages of chondrites, preirradiation and constancy of cosmic ray flux in the past. *Earth Planet. Sci. Lett.* 50:156–170.
Nishiizumi, K., Arnold, J. R., Goswami, J. N., Klein, J., and Middleton, R. 1986. Solar cosmic ray effects in Allan Hills 77005. *Meteoritics* 21:472–473.
Pal, D. K., Moniot, R. K., Kruse, T. H., Tuniz, C., and Herzog, G. F. 1985. Spallogenic ^{10}Be in the Jilin chondrite. *Earth Planet. Sci. Lett.* 72:273–275.
Parker, E. N. 1958. Interaction of the solar wind with the geomagnetic field. *Phys. Fluids* 1:171–187.

Parker, E. N. 1969. Theoretical studies of the solar wind phenomena. *Space Sci. Rev.* 9:325–360.

Pellas, P. 1972. Irradiation history of grain aggregates in ordinary chondrites. Possible clues to the advanced stages of accretion. In *From Plasma to Planet,* Proc. Nobel Symp. 21, ed. A. Elvius (New York: Wiley), pp. 65–92.

Pellas, P., and Storzer, D. 1981. ^{244}Pu fission track thermometry and its application to stony meteorites. *Phil. Trans. Roy. Soc. London* A347:253–270.

Pellas, P., Poupeau, G., Lorin, J. C., Reeves, H., and Audouze, J. 1969. Primitive low-energy particle irradiation of meteoritic crystals. *Nature* 223:272–274.

Pepin, R. O. 1980. Rare gases in the past and present solar wind. In *Proc. Conf. Ancient Sun,* eds. R. O. Pepin, J. A. Eddy, and R. B. Merrill (New York: Pergamon Press), pp. 411–421.

Pomerantz, M. A., and Duggal, S. P. 1974. The Sun and cosmic rays. *Rev. Geophys. Space Phys.* 12:343–361.

Poupeau, G., Kirsten, T., Steinbrunn, F., and Storzer, D. 1974. The record of solar wind and solar flares in aubrites. *Earth Planet. Sci. Lett.* 24:229–241.

Price, P. B., Hutcheon, I. D., Cowsik, R., and Barber, D. J. 1971. Enhanced emission of iron nuclei in solar flares. *Phys. Rev. Lett.* 26:916–919.

Price, P. B., Hutcheon, I. D., Braddy, D., and Macdougall, J. D. 1975. Track studies bearing on solar system regoliths. *Proc. Lunar Sci. Conf.* 6:3449–3469.

Rajan, R. S. 1974. On the irradiation history and origin of gas-rich meteorites. *Geochim. Cosmochim. Acta* 38:777–788.

Rajan, R. S., Huneke, J. C., Smith, S. P., and Wasserburg, G. J. 1979. Argon 40-argon 39 chronology of lithic clasts from the Kapoeta howardite. *Geochim. Cosmochim. Acta.* 43:957–971.

Reedy, R. C. 1980. Lunar radionuclide records of average solar-cosmic-ray fluxes over the last ten million years. *Proc. Conf. Ancient Sun,* eds. R. O. Pepin, J. A. Eddy, and R. B. Merrill (New York: Pergamon Press), pp. 365–386.

Reedy, R. C. 1985. A model for GCR-particle fluxes in stony meteorites and production rates of cosmogenic nuclides. *Proc. Lunar Planet. Sci. Conf.* 15, *J. Geophys. Res. Suppl.* 90:C722–C728.

Reedy, R. C. 1987*a*. Predicting the production rates of cosmogenic nuclides in extraterrestrial matter. *Nucl. Instrum. Meth.* B29:251–261.

Reedy, R. C. 1987*b*. Cosmogenic nuclide production in the early solar system. *Meteoritics* 21:489–490 (abstract).

Reedy, R. C., and Arnold, J. R. 1972. Interaction of solar and galactic cosmic-ray particles with the Moon. *J. Geophys. Res.* 77:537–555.

Reedy, R. C., Arnold, J. R., and Lal, D. 1983. Cosmic-ray records in solar system matter. *Ann. Rev. Nucl. Part. Sci.* 33:505–537.

Reedy, R. C., Herzog, G. F., and Jessberger, E. K. 1979. The reaction Mg(n,α)Ne at 14.1 and 14.7 MeV: Cross sections and implications for meteorites. *Earth Planet. Sci. Lett.* 44:341–348.

Rowe, M. W., and Clark, R. S. 1971. Estimation of error in the determination of ^{26}Al in stone meteorites by indirect X-ray spectrometry. *Geochim. Cosmochim. Acta* 35:727–730.

Schaeffer, O. A., Nagel, K., Fechtig, H., and Neukum, G. 1981. Space erosion of meteorites and the secular variation of cosmic rays (over 10^9 years). *Planet. Space Sci.* 29:1109–1118.

Schultz, L., and Signer, P. 1977. Noble gases in the St. Mesmin chondrite: Implications to the irradiation history of a brecciated meteorite. *Earth Planet. Sci. Lett.* 36:363–371.

Schultz, L., Signer, P., Lorin, J. C., and Pellas, P. 1972. Complex irradiation history of the Weston chondrite. *Earth Planet. Sci. Lett.* 15:403–410.

Shirck, E. K. 1974. Observations of trans-iron solar flare nuclei in an Apollo 16 command module window. *Astrophys. J.* 190:695–702.

Signer, P., and Nier, A. O. 1960. The distribution of cosmic-ray-produced rare gases in iron meteorites. *J. Geophys. Res.* 65:2947–2964.

Sturrock, P. A. 1980. Flare models. In *Solar Flares,* ed. P. A. Sturrock (Boulder: Colorado Associated Univ. Press), pp. 411–448.

Suess, H. E., Wänke and Wlotzka, F. 1964. On the origin of gas-rich meteorites. *Geochim. Cosmochim. Acta* 28:595–607.

Svestka, Z. 1976. *Solar Flares* (Dordrecht: D. Reidel).

Van Allen, J. A., and Randall, B. A. 1985. Interplanetary cosmic ray intensity: 1972–1984 and out to 32 AU. *J. Geophys. Res.* 90:1399–1412.

Voshage, H., and Hintenberger, H. 1963. The cosmic-ray exposure ages of iron meteorites as derived from the isotopic composition of potassium and the production rates of cosmogenic nuclides in the past. In *Radioactive Dating* (Vienna: IAEA), pp. 367–379.

Walker, R. M. 1980. Nature of the fossil evidence: Moon and meteorites. In *Proc. Conf. Ancient Sun*, eds. R. O. Pepin, J. A. Eddy, and R. B. Merrill (New York: Pergamon Press), pp. 11–28.

Walton, J. R., Heymann, D., Yaniv, A., Edgerley, D., and Rowe, M. W. 1976. Cross sections for He and Ne isotopes in natural Mg, Al, and Si, He isotopes in CaF_2, Ar isotopes in natural Ca, and radionuclides in natural Al, Si, Ti, Cr, and stainless steel induced by 12- to 45-MeV protons. *J. Geophys. Res.* 81:5689–5699.

Wänke, H. 1965. Der sonnenwind as quelle der uredelgase in Steinmeteoriten. *Z. Naturf.* B20a:946–949.

Wieler, R., Etique, P., Signer, P., and Poupeau, G. 1983. Decrease of the solar flare/solar wind flux ratio in the past several aeons deduced from solar neon and tracks in lunar soil plagioclases. *Proc. Lunar Planet. Sci. Conf.* 13, *J. Geophys. Res. Suppl.* 88:A713–A724.

Wieler, R., Baur, H., and Signer, P. 1986. Noble gases from solar energetic particles revealed by closed system stepwise etching of lunar soil minerals. *Geochim. Cosmochim. Acta* 50:1997–2017.

Zahnle, K. J., and Walker, J. C. G. 1982. The evolution of solar ultraviolet luminosity. *Rev. Geophys. Space Phys.* 20:280–292.

PART 5
Solar-System Chronology

5.1. PRINCIPLES OF RADIOMETRIC DATING

G. R. TILTON
University of California at Santa Barbara

One of the major contributions that the study of meteorites has made to our understanding of the origin of the solar system is in defining when that event took place. In addition, several other important events in early solar-system history have been dated using radiochronological techniques applied to meteorites. These topics are discussed in a number of chapters throughout the book. In this chapter, we outline the principles on which those applications of radiometric dating are based.

5.1.1. INTRODUCTION

Six decay systems have been utilized in most meteorite dating, namely Rb-Sr, Sm-Nd, K-Ar, Th-Pb and the two U-Pb systems. The methods date chemical differentiation of radioactive parent elements from the daughter elements due to physical processes. As in terrestrial dating, different systems may date different physical processes, even when applied to the same sample material. Here, however, we will simply review the dating techniques, since interpretations are discussed in later chapters.

In any kind of radiometric dating, several assumptions and conditions must be satisfied in order to obtain meaningful results. These are:

1. The decay constant of the radioactive parent is accurately known;
2. The radiogenic component of the daughter nuclide can be distinguished from the initial, or nonradiogenic, component;
3. The isotopic composition of the daughter element was homogeneous at the time of the chemical differentiation;

4. The dated material has been a closed chemical system with respect to parent and daughter nuclides since the differentiation event.

The following sections discuss methods of evaluating the above points, which in turn control the reliability of the ages.

5.1.2. DECAY CONSTANTS

The decay constants are known with varying degrees of accuracy. The constants for alpha emitters are generally the most accurately known because the particles are monoenergetic and create high ion densities in traversing matter, thereby yielding good counting data. The greatest problems are encountered with low-energy beta emitters. Table 5.1.1 lists currently accepted values of the decay constants for various systems that have been applied to meteorite dating. Brief discussions of the current status of each constant follow.

^{238}U, ^{235}U, ^{232}Th. The values summarized in Steiger and Jäger (1977) are in general use. Those are based on absolute alpha counting, and have stated accuracies in the range between 0.2 and 0.3%.

^{147}Sm. The value obtained by Lugmair and Marti (1978), based on a weighted mean of measurements compiled from the literature, is in general use. The standard error for the measurements is 0.8%. Nunes (1981) compared accurate U-Pb data on zircon with whole-rock Sm-Nd age measure-

TABLE 5.1.1
Decay Constants

Nuclide	Decay Constant (per year)	Reference[a]
^{238}U	1.55125×10^{-10}	(1)
^{235}U	9.8485×10^{-10}	(1)
^{232}Th	4.9475×10^{-11}	(1)
^{147}Sm	6.54×10^{-12}	(2)
^{87}Rb	1.42×10^{-11}	(1)
^{40}K	4.962×10^{-10} (beta)	(1)
	0.581×10^{-10} (K-electron capture)	(1)
^{187}Re	1.52×10^{-11}	(3)
^{176}Lu	1.94×10^{-11}	(4)

[a] (1): Steiger and Jäger (1977); (2): Lugmair and Marti (1978); (3): Luck and Allègre (1983); (4) Patchett et al. (1981).

ments from the Stillwater complex (age = 2.71 Gyr), obtaining results indicating that the Sm and U decay constants are consistent within ca. 0.5% when 6.54 × 10^{-12}/yr is used for ^{147}Sm.

^{87}Rb. This constant has been the most difficult to evaluate because the large fraction of very low-energy beta particles in the energy spectrum make absolute beta counting difficult (Neumann and Huster 1976). The report of Steiger and Jäger (1977) adopted 1.42 × 10^{-11}/yr for the decay constant after a review of various values in the literature. This was weighted heavily on the result of Davis et al. (1977), 1.415 ± 0.012 (2σ) × 10^{-11}/yr, based on aging of Rb salt in a laboratory experiment, which has the advantage of circumventing the low-energy beta problem (Neumann and Huster 1976). Most Rb-Sr ages are currently reported using 1.42 × 10^{-11}/yr for the decay constant of ^{87}Rb. The uncertainty in the decay constant is probably greater than cited above, perhaps on the order of 1%. For example, Minster et al. (1982) show that a value of 1.402 ± 0.008 × 10^{-11}/yr would bring the Rb-Sr ages of chondrites into agreement with the U-Th-Pb ages. Nyquist et al. (1986) recommend the same value based on their study of eucrite Y75011 and lunar samples.

^{40}K. A discussion of the constants adopted for the branched decay of ^{40}K to ^{40}Ar by K-electron capture and to ^{40}Ca by beta emission can be found in Steiger and Jäger (1977).

In addition to the above, two other decay constants have been evaluated by measuring the ratios of radiogenic daughter nuclides to parent nuclides in meteorite samples, assuming closed-system behavior over 4.55 Gyr. These are ^{187}Re: 1.52 ± 0.04 × 10^{-11}/yr (Luck and Allegre 1983) and ^{176}Lu: 1.94 ± 0.07 × 10^{-11}/yr (Patchett et al. 1981).

5.1.3. NORMALIZATION CONVENTIONS

The long half-lives of the parent elements cause Rb-Sr and Sm-Nd ages to be quite sensitive to corrections for the non-radiogenic daughter isotopes. For most meteorite samples, the component of radiogenic ^{87}Sr or ^{143}Nd is in the range of a few percent or less of the total ^{87}Sr or ^{143}Nd. In many cases, ratio measurements with accuracies at the 0.002% level are required in order to obtain useful age data; however, measured ratios in thermal-ionization mass spectrometry change by many times that proportion due to fractional distillation of isotopes from the filament in the course of a run. Normalization procedures in which the ratios of two nonradiogenic isotopes are assigned constant values, with adjustment of all other ratios according to mass differences, can correct for such variations. In the case of Sr, the universal convention has been to normalize ^{86}Sr/^{88}Sr to 0.1194 (Steiger and Jäger 1977) and assume that fractionation is a linear function of mass difference. With this approach normalized, ^{87}Sr/^{86}Sr ratios for replicate runs on a reference standard typically yield ^{87}Sr/^{86}Sr ratios that agree within ca. 0.005% or better (2 σ).

For Nd, several normalization conventions are widely used in the literature. One is based on ^{146}Nd/^{142}Nd = 0.636151, or its equivalent, ^{150}Nd/^{142}Nd = 0.2096 (Wasserburg et al. 1981); another on ^{148}Nd/^{144}Nd = 0.241572 (Lugmair et al. 1976). The Lugmair et al. (1976) normalization yields 0.721895 ± 15 for ^{146}Nd/^{144}Nd, which many laboratories have rounded to 0.721900 for normalization of data. The CalTech and UCSD scales are not equivalent, requiring conversion factors to be applied when intercomparing data normalized by the two methods. To convert ^{143}Nd/^{144}Nd ratios based on ^{148}Nd/^{144}Nd = 0.241572 to ratios based on ^{146}Nd/^{142}Nd = 0.636151, multiply by 0.998454; for ^{146}Nd/^{144}Nd = 0.721900, multiply by 0.998457. (These factors assume that fractionation is a linear function of mass difference.) Providing consistent normalization schemes are followed throughout all calculations, age results will be virtually identical, although for meteorites the two conventions lead to differences in the initial, or primordial, Nd isotope ratios, as noted below. In using Nd data it is therefore important to know which normalization scheme is utilized, and to be certain that all data are normalized in the same manner.

Since Pb has only one nonradiogenic isotope, ^{204}Pb, it has not been possible to use internal normalization to correct for mass fractionation. Recently, however, Todt et al. (1984) have developed a ^{202}Pb-^{205}Pb double-spike procedure that allows internal normalization of Pb-isotope ratios as well as determination of absolute isotope abundances. Fractionation for meteorite Pb data currently in the literature has been somewhat controlled by interspersing measurements on known standard references samples with unknowns. Fortunately, the variations in Pb-isotope ratios are for the most part quite large compared to those for Nd and Sr, reducing the dependence of results on fractionation corrections.

5.1.4. INITIAL RATIOS

Calculation of ages requires corrections for initial, or nonradiogenic, isotopes. In meteorite dating, the isotopic composition of the initial daughter element may either be measured directly, or inferred. When the initial ratios are inferred, the ages are designated as "model ages" because they use assumed corrections for nonradiogenic daughter nuclides. For meteorite model ages, the isotopic composition of initial Pb, Sr or Nd is assumed to be that of the primordial element as determined in other meteorite samples. For Pb, the ratios are taken from Canyon Diablo troilite, while Sr calculations utilize the initial ratios from basaltic achondrites or Allende CAIs. For Nd the most used primordial ratios are based either on the Juvinas achondrite value (Lugmair et al. 1976) or on average data from chondrites and achondrites (Jacobsen and Wasserburg 1984). A number of complexities arise in assessing the Nd ratio,

TABLE 5.1.2
Primordial Ratios

Element	Ratio	Value	Reference[a]
Pb	^{206}Pb/^{204}Pb	9.307 ± 3	(1)
Pb	^{207}Pb/^{204}Pb	10.294 ± 3	(1)
Pb	^{208}Pb/^{204}Pb	29.476 ± 9	(1)
Nd[b]	^{143}Nd/^{144}Nd	0.50677 ± 10	(2)
Nd[c]	^{143}Nd/^{144}Nd	0.505893 ± 20	(3)
Sr[d]	^{87}Sr/^{86}Sr	0.69877 ± 2	(4)
Sr[e]	^{87}Sr/^{86}Sr	0.69898 ± 4	(5,6)

[a] (1): Tatsumoto et al. (1973); (2): Lugmair et al. (1976); (3): Jacobsen and Wasserburg (1984); (4): Gray et al. (1973); (5): Papanastassiou and Wasserburg (1969); (6): Allègre et al. (1975).
[b] Internal isochron for Juvinas achondrite, based on ^{148}Nd/^{144}Nd = 0.241572 (^{146}Nd/^{144}Nd = 0.721895).
[c] From whole-rock analyses of 9 chondrites and 3 achondrites, with an assumed age of 4.56 Gyr, ^{147}Sm/^{144}Nd = 0.1967 (present time), ^{146}Nd/^{142}Nd = 0.636151. The authors do not include an uncertainty estimate; that shown is estimated by G.R.T. For ^{146}Nd/^{142}Nd = 0.721900, the initial ratio is 0.50668.
[d] Ratio measured on Rb-poor inclusions in Allende carbonaceous chondrite.
[e] Ratio for basaltic achondrites.

and the reader is referred to the excellent discussion of interlaboratory results in Jacobsen and Wasserburg (1980) for details. A summary of primordial ratios is listed in Table 5.1.2 and further discussions of initial Sr ratios are given in Chapter 5.2.

5.1.5. ISOCHRON DIAGRAMS

If a sufficient number of samples are available, it is often possible to measure the initial ratios directly by isochron methods, which provide age as well as information on the initial isotopic composition of the daughter, providing four conditions are met: (a) the samples are coeval; (b) the initial isotopic compositions of the daughter element were identical for all samples; (c) all samples have been closed chemical systems for parent and daughter elements since crystallization; and (d) there is sufficient variation in parent/daughter ratios to generate a spread in daughter isotope ratios. When those conditions are met, a plot of parent vs daughter nuclide, each normalized to a nonradiogenic daughter-element nuclide, yields a straight line called an isochron. The slope of the isochron is a function of the age of the samples, while the initial isotopic composition of the daughter element is given by the intercept of the isochron with an axis for which parent/daughter ratio = zero. A schematic example for the Rb-Sr system is shown in Fig. 5.1.1.

Still another isochron method is available for the U-Pb system owing to

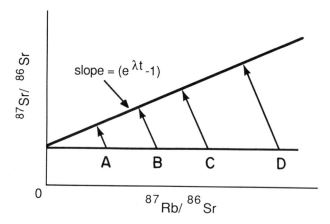

Fig. 5.1.1. Schematic isochron diagram for the ^{87}Rb-^{87}Sr system. Phases A, B, C and D all have identical initial ^{87}Sr/^{86}Sr ratios at $t = 0$, with varying ^{87}Rb/^{86}Sr ratios. With closed-system behavior for Rb and Sr, the ratios evolve as shown by the arrows to define an isochron for which t is the age of the system, and λ is the decay constant for ^{87}Rb.

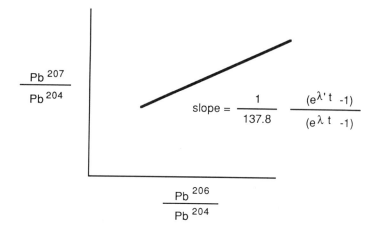

Fig. 5.1.2. Isochron in the ^{207}Pb-^{206}Pb system, drawn for samples with varying U/Pb ratios, all having the same age and same initial ^{207}Pb/^{204}Pb and ^{206}Pb/^{204}Pb ratios. Closed-system chemical behavior for U and Pb is assumed. λ' is the decay constant for ^{235}U, λ the decay constant for ^{238}U.

the coupled decay of the two U isotopes to two isotopes of Pb. For this method, sample suites must meet the four requirements described above for single decay systems, with the exception that recent loss of Pb, or gain or loss of U will not change the age values. Figure 5.1.2 illustrates this method. Although the age is given by the isochron, note that it does not yield information on the initial Pb isotope ratios, which require U and Pb concentration data. For example, a plot of ^{206}Pb/^{204}Pb against ^{238}U/^{204}Pb would yield an initial ^{206}Pb/^{204}Pb ratio at the y-axis intercept. However, meteorite data frequently show disturbances that preclude such calculations (see, e.g., Unruh et al. 1977). If Pb-Pb isochrons through groups of meteorite samples also fit the Pb-isotopic composition of Canyon Diablo troilite, the initial Pb ratios are probably accurately determined by the troilite data, even if the U and Pb concentrations are unknown. Qualitatively, this follows because of the short half-life of ^{235}U (0.7 Gyr) compared to that of ^{238}U (4.5 Gyr). For example, in the time range 4.56 to 4.4 Gyr, the ^{207}Pb/^{204}Pb ratio has changed at a rate twice that of ^{206}Pb/^{204}Pb for any given U/Pb ratio.

5.1.6. CONCORDIA DIAGRAMS

When disturbances have occurred in the U-Pb system, concordia diagrams (Wetherill 1956; Tera and Wasserburg 1972) can still sometimes permit evaluation of the crystallization age and a time of disturbance. Assume a suite of samples for which the initial (nonradiogenic) Pb components are known, allowing accurate determination of the radiogenic components. Then, further

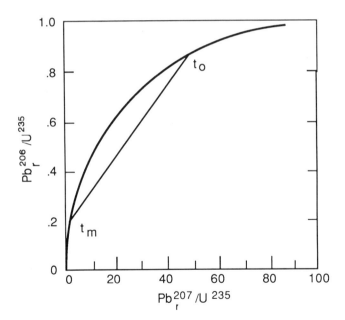

Fig. 5.1.3. Schematic concordia diagram. The curve is the locus of samples with equal ^{207}Pb/^{235}U and ^{206}Pb/^{238}U ages. Subscript r denotes the radiogenic components of ^{207}Pb and ^{206}Pb. Samples with crystallization age of t_o that lost varying proportions of radiogenic Pb at time t_m, but were closed chemical systems at all other times will plot along the chord connecting t_o and t_m.

assume that a single disturbance has occurred, with closed-system behavior at all other times. When radiogenic $^{206}Pb/^{238}U$ is plotted against radiogenic $^{207}Pb/^{235}U$ (Wetherill 1956), or radiogenic $^{207}Pb/^{206}Pb$ against ^{238}U/radiogenic ^{206}Pb (Tera and Wasserburg 1972), a chord results whose upper intersection with the concordia (equal age) curve corresponds to an original (crystallization) age, while the lower intersection dates the disturbance. This method requires accurate knowledge of any nonradiogenic Pb incorporated in the samples at the time of crystallization. A schematic example is shown in Fig. 5.1.3. Although relatively few meteorites have been analyzed in a manner that permits plotting of data in concordia diagrams, an example of such a solution is cited in Chapter 5.2, and the application is discussed there in more detail. Additional discussions of isochron and concordia diagrams are given in Faure (1986).

5.1.7. EXTINCT-NUCLIDE DATING

This method uses nuclides with half-lives in the range of 10 Myr to measure the time interval between the cessation of nucleosynthesis and the condensation of matter in the solar system. The most widely used chronometer employs the beta decay of ^{129}I to ^{129}Xe with a half-life of 17 Myr. A number of requirements and assumptions must be met in order to calculate a meaningful decay interval. One must (a) know the isotopic composition of initial Xe in order to determine any radiogenic component present; (b) show that the radiogenic ^{129}Xe is correlated with ^{127}I rather than inherited from an external source; (c) specify the production rates of ^{129}I and ^{127}I as a function of time in the nucleosynthetic process; and (d) assume that the ratio $^{129}I/^{127}I$ in initial solar-system matter was uniform. In addition to the above requirements, the calculated interval is correct only if the sample has been a closed system for I and Xe since formation. Further possible complications are discussed in Chapter 15.3, and will not be pursued here. The general approach is to attempt to determine initial $^{129}I/^{127}I$ ratios in various meteorites, thereby establishing a relative time scale for the formation interval. Because the half-life of ^{129}I is so short, the presence of any ^{129}Xe correlated with ^{129}I signifies a relatively short ^{129}I-decay interval, regardless of the accuracy of the assumptions.

5.1.8. ANALYTICAL METHODS AND INSTRUMENTATION

Analytical methods have steadily improved over the past several decades, both in the chemical processing of samples and in instrumentation. This has been most important in U-Pb dating, for which Pb blanks are difficult to control compared to other elements owing to the ubiquity of Pb in the environment, and the generally low levels of Pb in meteorite samples. Lead-processing blanks have steadily decreased from ca 0.1 μg around 1960 to 0.1 ng at present for processing of 50 to 100 mg samples.

Improvements in mass spectrometry also promise to play a major role in future isotope measurements. Multiple-collector machines that allow simultaneous collection of 5 or more isotopes are now available from two manufacturers. Simultaneous collection offers the advantage of increased sensitivity since a larger fraction of time is spent measuring individual peaks, together with elimination of filament noise. Examples of the performance of such machines have been cited for Sr and Nd (Blenkinsop 1983) and for Pb (Todt et al. 1984). Todt et al. note that with multiple collection and the ^{202}Pb-^{205}Pb double spike it should be possible to measure ^{206}Pb/^{204}Pb ratios within 2σ absolute errors of $\pm$ 0.002 (0.01%) on 50 ng of common Pb. This can be compared with a measured ^{206}Pb/^{204}Pb ratio of 13.757 $\pm$ 0.008 on 20 ng of Pb from the Tennasilm chondrite with a single-collector machine and a ^{205}Pb spike (Unruh 1982). The result includes a fractionation correction of 0.10 $\pm$ 0.03% per mass unit, whose uncertainty must be added to the absolute uncertainty.

It is apparent that with the new instrumentation and analytical techniques now available, future measurements with improved precision and accuracy are possible in many cases. The improvement should be especially marked for Pb-isotope data. In fact, techniques for Pb analysis are now such that reduction in blank levels, both preanalysis and chemical processing contamination, may be the controlling factor on the accuracies of age measurements. For Sr and Nd, a large advantage of the newer instrumentation will be the saving in time for multiple-collection mass spectrometry compared to single-collector modes because multiple-collector machines can attain precisions in runs of ca. 2 hr that require 6 to 8 hr of running on single-collector machines. Moreover, multiple collection should in many cases yield improved precision over that attainable by single collection. Static multiple collection will allow isotope measurements to be carried out on smaller samples, an important advantage for samples where material is limited. Clearly, much of the past meteorite work can profitably be repeated with the newer methodologies.

A brief discussion of principles of mass spectrometry, together with a list of general references can be found in Faure (1986).

Acknowledgment. I thank L. E. Nyquist for constructive criticism of the manuscript.

REFERENCES

Allègre, C. J., Birck, J. L., Fourcade, S., and Semet, M. P. 1975. Rubidium-87/strontium-87 age of Juvinas basaltic achondrite and early igneous activity in the solar system. *Science* 187:436–438.

Blenkinsop, J. 1983. A preliminary evaluation of multiple collection in solid-source mass spectrometry. *Geol. Assoc. of Canada Program with Abstracts* 8:A6.

Davis, D. W., Gray, J., Cumming, G. L., and Baadsgaard, H. 1977. Determination of the ^{87}Rb decay constant. *Geochim. Cosmochim. Acta* 41:1745–1749.

Faure, G. 1986. *Principles of Isotope Geology* (New York: John Wiley and Sons).

Gray, C. M., Papanastassiou, D. A., and Wasserburg, G. J. 1973. The identification of early condensates from the solar nebula. *Icarus* 20:213–219.

Jacobsen, S. B., and Wasserburg, G. J. 1980. Sm-Nd isotopic evolution of chondrites. *Earth Planet. Sci. Lett.* 50:139–155.

Jacobsen, S. B., and Wasserburg, G. J. 1984. Sm-Nd isotopic evolution of chondrites and achondrites, II. *Earth Planet. Sci. Lett.* 67:137–150.

Luck, J.-M., and Allègre, C. J. 1983. ^{187}Re-^{187}Os systematics in meteorites and cosmological consequences. *Nature* 302:130–132.

Lugmair, G. W., and Marti, K. 1978. ^{143}Nd/^{144}Nd: Differential evolution of the lunar crust and mantle. *Earth Planet. Sci. Lett.* 39:349–358.

Lugmair, G. W., Marti, K., Kurtz, J. P., and Scheinin, N. B. 1976. History and genesis of lunar troctolite 76535 or: How old is old? *Proc. Lunar Sci. Conf.* 7:2009–2033.

Minster, J. F., Birck, J.-L., and Allègre, C. J. 1982. Absolute age of formation of chondrites studied by the ^{87}Rb-^{87}Sr method. *Nature* 300:414–419.

Neumann, W., and Huster, H. 1976. Discussion of the ^{87}Rb half-life determined by absolute beta counting. *Earth Planet. Sci. Lett.* 33:277–288.

Nunes, P. D. 1981. The age of the Stillwater complex—A comparison of U-Pb and Sm-Nd isochron systematics. *Geochim. Cosmochim. Acta* 45:1961–1963.

Nyquist, L. E., Takeda, H., Bansal, B. M., Shih, C.-Y., Wiesmann, H., and Wooden, J. L. 1986. Rb-Sr and Sm-Nd internal isochron ages of a subophitic basalt clast and a matrix sample from the Y75011 eucrite. *J. Geophys. Res.* 91:8137–8150.

Papanastassiou, D. A., and Wasserburg, G. J. 1969. Initial strontium isotopic abundances and the resolution of small time differences in the formation of planetary objects. *Earth Planet. Sci. Lett.* 5:361–376.

Patchett, P. J., Kouvo, O., Hedge, C. E., and Tatsumoto, M. 1981. Evolution of continental crust and mantle heterogeneity: Evidence from Hf isotopes. *Contrib. Mineral. Petrol.* 78:279–297.

Steiger, R. H., and Jäger, C. 1977. Subcommission on geochronology: Convention on the use of decay constants in geo- and cosmochronology. *Earth Planet. Sci. Lett.* 36:359–362.

Tatsumoto, M., Knight, R. J., and Allègre, C. J. 1973. Time differences in the formation of meteorites as determined from the ratio of lead-207 to lead-206. *Science* 180:1279–1283.

Tera, F., and Wasserburg, G. J. 1972. U-Th-Pb systematics in three Apollo 14 basalts and the problem of initial lead in lunar rocks. *Earth Planet. Sci. Lett.* 14:281–304.

Todt, W., Cliff, R. A., Hanser, A., and Hofmann, A. W. 1984. ^{202}Pb + ^{205}Pb spike for lead isotopic analysis. *Terra Cognita* 4:209 (abstract).

Unruh, D. M. 1982. The U-Th-Pb age of equilibrated chondrites and a solution to the excess radiogenic Pb problem in chondrites. *Earth Planet. Sci. Lett.* 58:75–94.

Unruh, D. M., Nakamura, N., and Tatsumoto, M. 1977. History of the Pasamonte achondrite: Relative susceptibility of the Sm-Nd, Rb-Sr and U-Pb systems to metamorphic events. *Earth Planet. Sci. Lett.* 37:1–12.

Wasserburg, G. J., Jacobsen, S. B., DePaolo, D. J., McCulloch, M. T., and Wen, T. 1981. Precise determination of Sm/Nd ratios, Sm and Nd isotopic abundances in standard solutions. *Geochim. Cosmochim. Acta* 45:2311–2323.

Wetherill, G. W. 1956. Discordant uranium-lead ages. *Trans. Amer. Geophys. Union* 37:320–326.

5.2. AGE OF THE SOLAR SYSTEM

G. R. TILTON
University of California at Santa Barbara

Calcium-aluminum-rich inclusions of the Allende CV chondrite, with a model $^{207}Pb/^{206}Pb$ age of 4.559 ± 0.004 Gyr, yield the oldest high-precision meteorite date, thus providing the best estimate of the first condensation of matter in the solar system. The achondrites, Angra dos Reis and Juvinas are dated at 4.551 ± 0.002 and 4.539 ± 0.004 Gyr, respectively. The initial $^{87}Sr/^{86}Sr$ ratios for the three meteorites correlate with age in a manner indicating isotopic evolution of Sr of the achondrites in parent matter that had Rb/Sr equal to the solar abundance ratio between 4.559 Gyr and the times of magma crystallization. The magmas then formed with very low Rb/Sr ratios. The observations best fit a model in which the achondrite parent bodies condensed ca. 4.56 Gyr ago, at the same time as the Allende CAIs, then were remelted in a major impact event(s) in a manner similar to that recently postulated for formation of the Earth's Moon. With one exception, ages for ordinary chondrites have larger uncertainties that encompass the Allende-basaltic achondrite range. Whitlockite from St. Séverin has been dated at 4.552 ± 0.003 Gyr by $^{207}Pb/^{206}Pb$. Evidence for metamorphic disturbances between 4.3 and 4.48 Gyr is reviewed.

5.2.1. INTRODUCTION

In order to discuss the "age of the solar system" we must first define what is meant by the term, and what is measured by various chronometers. In one sense, the age of the solar system is equal to that of the elements of which it is composed, for which nucleosynthesis models place the upper limit at around 20 Gyr. However, here we are more interested in the time(s) of formation of the various bodies in the solar system. Since the chronometers date chemical differentiation rather than physical processes, it is necessary to identify the ages with specific events. Those processes can in principle be associated with any of (a) first condensation of solid particles from a nebular gas; (b) differentiation associated with accretion of particles to form planetary and planetesimal-sized objects; (c) post-accretion differentiation associated with impacts, melting or thermal metamorphism. The following discussions review the most precise meteorite ages found in the literature and their model interpretations.

In practice, the generally accepted age of the solar system has changed very little over the past 30 years since Patterson (1956) reported a ^{207}Pb/^{206}Pb age of 4.57 ± 0.07 Gyr for 3 stone meteorities paired with the Canyon Diablo troilite Pb, which represented primordial Pb. That age was controlled mainly by the isotopic compositions of the troilite and the Nuevo Laredo eucrite. It was assumed that the age represented chemical differentiation that occurred within a few Myr of the formation of the solar system. When Patterson's data are recalculated using newer values for primordial Pb and decay constants, the age is reduced to 4.50 Gyr. As noted below, such an age agrees rather well with the present ^{207}Pb/^{206}Pb ages for eucrites, while the oldest known Pb/Pb ages are still 4.56 Gyr. We will consider several meteorite age groups in the following sections, then discuss physical models that explain the ages.

5.2.2. ALLENDE CV3 CHONDRITE: A CASE HISTORY

Petrographic studies of chondrules and inclusions in Allende (see, e.g., Marvin et al. 1970; Grossman 1975; see also Chapter 10.3) reveal highly refractory minerals rich in Ca, Ti and Al similar in composition to those predicted to have condensed early in a gas of solar composition (Larimer 1967; Grossman 1972). This observation led to the suggestion that such Ca, Al-rich inclusions (CAI) might be some of the oldest material to be found in meteorites. Other evidence is seen in the relatively short ^{129}I/^{129}Xe decay interval age for certain Allende CAIs (Podosek and Lewis 1972; see also Chapter 15.3) compared to most other meteorites, and evidence within some CAIs for the presence of excess ^{26}Mg from the decay of extinct ^{26}Al (Lee et al. 1976; see also Chapter 15.1). These observations indicate that accurate age measurements on Allende CAIs are of great importance to early solar-system cosmochronology.

In one of the first measurements, Fireman et al. (1970) reported a K-Ar age of 4.44 Gyr for a chondrule from Allende. Gray et al. (1973) attempted to measure a Rb-Sr age using separated fractions of Allende, but were unable to obtain an accurate value due to disturbance of the system. They defined a model age of 4.6 Gyr from several chondrules, plus bulk samples and inclusions from Allende. They further noted that Rb-Sr ratios in some of the phases have been disturbed more recently than 3.6 Gyr. Tatsumoto et al. (1976) found a similar spread in the Rb-Sr data in an isochron diagram, which prevented precise age determination. In a low-Rb inclusion Gray et al. (1973) made the important discovery of the lowest ^{87}Sr/^{86}Sr ratio yet found in any solar system matter, 0.69877 ± 2. This is an observed value, which corrects to 0.69876 when corrected for *in-situ* Rb decay over 4.6 Gyr. The value of 0.69877 has been adopted in the literature and named ALL. Extrapolated values from studies of Wetherill et al. (1973) and Tatsumoto et al. (1976) agree with the Gray et al. value within larger error limits.

The first precise isotopic Pb ages of Allende inclusions were given by Tatsumoto et al. (1976) and by Chen and Tilton (1976). Both groups separated inclusions and chondrules from matrix for age studies. Tatsumoto et al. (1976) reported a well-defined model $^{207}Pb/^{206}Pb$ age of 4.553 ± 0.004 Gyr, Chen and Tilton 4.565 ± 0.004 Gyr. The difference between the two studies is outside of experimental errors. There is some indication that it may be due to differences in weighting of chondrules and inclusions in the sampling, but the matter needs further study. The data from Chen and Tilton (1976) are reproduced in Fig. 5.2.1.

Both groups found that the U-Pb ages, like the Rb-Sr ages, indicate postcrystallization disturbances. When plotted in concordia diagrams (see Fig. 5.1.3) the approximate time of disturbance is 0.105 ± 0.07 Gyr for the data of Tatsumoto et al. and 0.28 ± 0.07 Gyr for those of Chen and Tilton, indicating that the disturbance(s) occurred relatively recently. Presumably the Rb-Sr system was also disturbed at the same time. A more recent study by Chen and Wasserburg (1981), based entirely on coarse-grained inclusions (type B of Grossman [1975]), gives a Pb/Pb age of 4.559 ± 0.004 Gyr. Their data

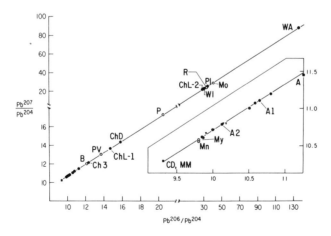

Fig. 5.2.1. $^{207}Pb/^{204}Pb-^{206}Pb/^{204}Pb$ isotope correlation diagram for Allende and other chondrite samples. Unlabeled solid circles: Allende total rocks; solid triangles: Allende matrix; Ch, ChL: Allende chondrules; WA, WI and Pl: Allende inclusions. Other symbols: CD, A, B, PV, My, R, Mo: Canyon Diablo troilite, Allende, Beardsley, Plainview, Murray, Richardton, and Modoc whole-rock samples from Tatsumoto et al. (1973). MM, Mn, P and A1–A2: Mezö Madaras, Murchison, Pultusk and Allende whole rocks from Tilton (1973). Figure reproduced from Chen and Tilton (1976).

likewise indicate disturbances in the U-Pb ratios, but at a negative time (i.e., the lower intersection in a concordia diagram plots below 0 Gyr). In concordia diagrams, the upper intersections are within error limits of 4.56 Gyr for all three studies. Arden and Cressey (1984) also analyzed chondrules and inclusions from Allende, finding a model Pb/Pb age of 4.567 ± 0.004 Gyr, a result close to that of Chen and Tilton (1976).

The result of Chen and Wasserburg (1981), 4.559 ± 0.004 Gyr, is adopted here as the best estimate of the Allende CAI age. It is close to the average of all of the measurements (4.561 Gyr), used only inclusions, and is based on over twice as many inclusions as any of the other three investigations. Since the Pb-Pb age is a model age, based upon an assumed initial or nonradiogenic isotopic composition, the question naturally arises as to what effect that assumption has on the reliability of the results. Several factors argue that the assumption is valid within error limits. (a) Although the observed $^{206}Pb/^{204}Pb$ ratios vary from 49 to 263, the calculated $^{207}Pb/^{206}Pb$ model ages of each individual inclusion, using Canyon Diablo troilite Pb as an index of initial Pb, show no correlation with the $^{206}Pb/^{204}Pb$ ratios. (b) A regression line fitted to the inclusions and the Canyon Diablo Pb yields a Pb/Pb age of 4.559 ± 0.004 Gyr, while if the Canyon Diablo Pb is omitted, the Pb/Pb age is 4.568 ± 0.005 Gyr. (c) Although, as noted above, the U/Pb ages are discordant, the disturbance (or disturbances) has occurred within the past few hundred Myr at most, and appears not to have affected the U/Pb ratios in most of the samples by more than ±10%. Such a disturbance would not seriously affect the Pb/Pb ages. A model calculation in which a 4.56 Gyr-old sample lost 10% of its radiogenic Pb 0.2 Gyr ago and retained Pb quantitatively from 0.2 Gyr up to the present indicates that the derived Pb/Pb age is reduced by only 1 Myr. For those reasons, the Allende CAIs appear to be reliably dated with high precision, and are the oldest meteorite ages presently known.

Jessberger (1984) has found $^{40}Ar/^{39}Ar$ ages of 4.85 Gyr in two CAI fragments from Allende. The significance of this result is not yet understood. Possible explanations are critically reviewed in Jessberger (1984) and in references cited therein.

Tatsumoto et al. (1976) and Chen and Tilton (1976) also measured the isotopic composition of Pb in Allende matrix material. Although the Pbs are not very radiogenic ($^{206}Pb/^{204}Pb$ = 9.73 to 10.17; cf. Table 5.2.1), the resulting Pb/Pb model ages calculated using Canyon Diablo troilite for initial isotopic composition range from 4.481 to 4.549, averaging 4.50 Gyr, all uniformly lower than 4.56 Gyr. Arden and Cressey (1984), in a more detailed study, reported Pb-isotope data for 15 matrix samples of Allende, finding $^{206}Pb/^{204}Pb$ ratios in the range of 9.65 to 11.99. Their data yield a Pb/Pb age of 4.502 Gyr ± 0.030. Although all of these experiments indicate that the model Pb ages of the matrix material are distinctly younger than that of the inclusions and chondrules of Allende, no physical significance can

TABLE 5.2.1
Basaltic Achondrite Pb Isotope Data and Ages

Sample	Atom Ratios $\frac{^{206}Pb}{^{204}Pb}$	$\frac{^{207}Pb}{^{204}Pb}$	Ages (Gyr)[a] $\frac{^{207}Pb}{^{206}Pb}$	$\frac{^{206}Pb}{^{238}U}$	$\frac{^{208}Pb}{^{232}Th}$	Reference[b]
Nuevo Laredo	222.4	140.1	4.529 (0.005)	4.63	4.65	(1)
Sioux County	195.3	123.3	4.526 (0.004)	4.54 (0.06)	4.29	(1)
Passamonte	138	91.0	4.573 (0.011)	4.87 (0.04)	4.68 (0.09)	(2)
Ibitira	148.9	96.8	4.554	4.68	4.71	(3)
Ibitira	599	376	4.556 (0.006)	4.567 (0.013)	4.658 (0.022)	(4)
Kapoeta	99.9	65.2	4.524	4.77	5.35	(3)
Béréba	254.0	148.0	4.415	4.08	3.93	(3)
Stannern	286.0	157.1	4.329	4.40	4.20	(3)
Juvinas	37.9	27.9	4.543	4.65	4.14	(3)
Juvinas			4.539[c] (0.004)	—	—	(5)
CD troilite	9.307	10.294				(1)

[a] Ages in parentheses are 2σ uncertainties.
[b] (1): Tatsumoto et al. 1973; (2): Unruh et al. 1977; (3): Mahnes et al. 1975; (4): Chen and Wasserburg 1985; (5): Manhes et al. 1984.
[c] Age based on intercept of internal chord with the concordia curve. Lower intercept = 1.92 AE.

be attached to the age difference owing to the possible effects of post-crystallization metamorphic disturbances. We next consider some additional high-precision ages that are associated with early planetary volcanism and metamorphism.

5.2.3. ANGRA DOS REIS ACHONDRITE

Angra dos Reis is a Ca, Al, Ti-rich achondrite (angrite) that appears on the basis of petrographic criteria to have crystallized from a magma, thus relating it to an early planetary igneous process. It is distinguished also by marked depletions in volatile elements (Larimer 1967), i.e., elements that are volatile in vacuum at temperatures of ca. 1000° C. These include the alkali elements and Pb. Angra dos Reis is an especially favorable sample for dating because the bulk sample has a high U/Pb ratio and especially because the calcium phosphate, whitlockite, is present in sufficient quantities to allow separation for analyses. Both fractions have been dated, as shown in Table 5.2.2. The average of the model Pb/Pb ages is 4.551 ± 1.8 $(2\sigma/\sqrt{n-1})$ Gyr

TABLE 5.2.2
Angra dos Reis U-Pb Ages in Gyr[a]

Sample	$\dfrac{^{207}Pb}{^{206}Pb}$	$\dfrac{^{206}Pb}{^{238}U}$	$\dfrac{^{208}Pb}{^{232}Th}$	Reference[b]
Whole rock	4.555 (0.005)	4.57	4.67	(1)
Whole rock	4.551 (0.004)	4.55 (0.06)	—	(2)
Whitlockite	4.553 (0.008)	4.50 (0.20)	—	(2)
Whole rock	4.552 (0.001)	4.63 (0.07)	4.53 (0.10)	(3)
Whitlockite A	4.549 (0.001)	4.61 (0.07)	4.60 (0.11)	(3)
Whitlockite B	4.550 (0.001)	4.546 (0.038)	4.540 (0.038)	(3)

[a] Ages in parentheses are 2σ uncertainties.
[b] (1): Tatsumoto et al. 1973; (2): Chen and Wasserburg 1981; (3): Wasserburg et al. 1977.

for all determinations, and 4.551 ± 1.4 Gyr for the CalTech data alone. The U/Pb and Pb/Pb ages agree closely, in contrast to the Allende CAI data. Angra dos Reis can apparently be dated with high precision by the U-Pb system.

The low concentration of Rb, presumably due to loss by volatilization, causes the Rb/Sr ratio in Angra dos Reis to be exceptionally low, thereby precluding accurate age determination by the Rb-Sr system. Wasserburg et al. (1977) reported an age of 4.4 ± 0.2 Gyr for Angra dos Reis based on an internal isochron. On the other hand, an accurate measurement of initial $^{87}Sr/^{86}Sr$ was possible using whitlockite, which contains large concentrations of Sr and negligible Rb. Wasserburg et al. (1977) reported a measured ratio of 0.69883 ± 2 for $^{87}Sr/^{86}Sr$ in Angra dos Reis whitlockite, with no correction for *in-situ* Rb decay required. This ratio, which the authors identified by the acronym, ADOR, is statistically greater than that for Allende (ALL).

Sm-Nd data (Lugmair and Marti 1977) yield an age of 4.55 ± 0.04 Gyr, a value in agreement with the Pb/Pb age, but with considerably larger uncertainty. The initial $^{143}Nd/^{144}Nd$ ratio is 0.50682 ± 5 (normalized to $^{148}Nd/^{144}Nd = 0.241572$). Jacobsen and Wasserburg (1984) report a Sm-Nd age of 4.564 ± 0.037 Gyr with an initial ratio of 0.505895 ± 44 (normalized to $^{146}Nd/^{142}Nd = 0.636151$), which converts to 0.506678 for $^{148}Nd/^{144}Nd = 0.241572$. (See Chapter 5.1 for a discussion of Nd normalization procedures.) As noted in Jacobsen and Wasserburg (1980), a systematic difference that involves more than normalization schemes seems to exist between the two laboratories that still awaits resolution. Jacobsen and Wasserburg (1984) further note that for any Nd data set to be consistent, all results should be obtained in the same laboratory. Much of the problem may be due to nonlinearity in amplifier systems, which would affect the normalization results. Newer instrumentation (Chapter 5.1) may help to resolve some of these difficulties.

In summary, the chemical differentiation age for Angra dos Reis is ca. 10 Myr younger than that for Allende, while the initial Sr ratio is higher. It is significant that a magmatic meteorite gives an age so close to that of Allende, which shows no evidence for such a process. An attempt to measure an ^{129}I decay interval for Angra dos Reis failed, presumably due to a very low I content (Hohenberg 1970), although Hohenberg did find evidence for now-extinct ^{244}Pu.

5.2.4 BASALTIC ACHONDRITES

Basaltic achondrites include the eucrite and howardite classes, which, like Angra dos Reis, have crystallized from a magma. Howardites are polymict breccias; eucrites are, with a few exceptions, monomict breccias (cf. Dodd 1981). Both groups are depleted in those elements that can be volatilized at ca. 1000° C in vacuum, causing the whole-rock samples to contain highly radiogenic Pb and nonradiogenic Sr. From an analytical standpoint these achondrites are favorable samples for U-Pb-method dating and initial Sr-isotope ratios, although brecciation and metamorphic history complicate the interpretations in many cases.

Papanastassiou and Wasserburg (1969) first established the initial ^{87}Sr/^{86}Sr ratio of 7 basaltic achondrites (designated BACHs hereafter) at 0.69898 ± 6 and coined the acronym BABI for "basaltic achondrite best initial." Although accurate information was obtained for the initial ratio, the age of the samples was poorly defined at 4.30 ± 0.26 Gyr, owing to the nonradiogenic nature of the Sr. In a second study of 10 eucrites Birck and Allègre (1978) found very similar values of 0.69899 ± 4 and 4.57 ± 0.13 Gyr. More recently, Nyquist et al. (1986) report values of 0.69896 ± 3 and 0.69894 ± 2 for initial ^{87}Sr/^{86}Sr of matrix and a clast, respectively, from Y75011, a pristine Antarctic eucrite. It is clear that the initial ^{87}Sr/^{86}Sr ratio for achondrites is measurably higher than those for Angra dos Reis and Allende CAI.

Sm-Nd ages from internal isochrons for 6 eucrites are summarized in Van Schmus (1981), and newer measurements on Y75011 are reported in Nyquist et al. (1986). Most ages are compatible with a value of 4.56 Gyr, with uncertainties of ±0.02 to 0.10 Gyr. Nyquist et al. (1986) find no measureable difference between initial ^{143}Nd/^{144}Nd in Y75011 and the Jacobsen and Wasserburg (1984) primordial value based on chondrites. Surprisingly, two cumulate (unbrecciated) eucrites have distinctly younger ages. Jacobsen and Wasserburg (1984) report a Sm-Nd age 4.46 ± 0.03 Gyr for Moama, with an initial ^{143}Nd/^{144}Nd ratio of 0.506027 ± 52 (Cal Tech normalization) that fits the CHUR evolution curve for that age. Lugmair et al. (1977) found a Sm-Nd age of 4.41 ± 0.02 Gyr for another cumulate eucrite, Serra de Magé.

U-Pb dating offers the possibility of higher precision dating, as noted above. Patterson (1956) published the first achondrite U-Pb date for the howardite, Nuevo Laredo, obtaining a model Pb/Pb age of 4.50 Gyr (recalculated). We now examine newer BACH data. Tatsumoto et al. (1973) reanalyzed Nuevo Laredo using improved techniques over those available to Patterson, and found a model ^{207}Pb/^{206}Pb age of 4.529 ± 0.005 (2σ) Gyr. For the Sioux County eucrite they reported 4.526 ± 0.010 Gyr. A number of U/Pb and Pb/Pb ages for BACHs, taken largely from the summary by Van Schmus (1981), are reproduced in Table 5.2.1. The Pb/Pb ages show a range of values from 4.57 ± 0.01 (Passamonte) to 4.329 (Stannern). Both of these ages are somewhat suspect since the Rb-Sr isochrons for these achondrites are disturbed (Birck and Allègre 1978). A ^{207}Pb-^{206}Pb isochron for the data in Table 5.2.1, omitting the Béréba and Stannern results which seem anomalously low, and adding Canyon Diablo troilite Pb as initial Pb-isotopic composition yields a Pb/Pb age of 4.543 ± 0.016 Gyr (2σ). The U/Pb ages indicate that most of the isotopic-Pb ages are somewhat discordant.

Since the model Pb/Pb ages assume that Canyon Diablo troilite Pb gives the isotopic composition of initial Pb for the samples, their accuracy needs discussion. As a plausible perturbation (Birck and Allègre 1978), we may assume that the achondrite source had a U/Pb ratio similar to that of average H- or L-chondritic matter, for which the higher present-day ^{206}Pb/^{204}Pb ratios average ca. 70 (Unruh 1982), and for which ^{238}U/^{204}Pb (or μ) = 60. Assume that the source evolved with such a μ from 4.56 to 4.54 Gyr, and was then fractionated to yield the achondrite material. If the Pb/Pb age of Nuevo Laredo (for which μ = 233, a typical value for BACHs) is recalculated using the source Pb at 4.54 Gyr instead of primordial Pb, the Pb/Pb age is 4.545 Gyr. For such a history, the model Pb/Pb age is an upper limit on the true age. Many chondrites, especially carbonaceous chondrites, have still lower ^{238}U/^{204}Pb ratios, which would affect the model age to an even smaller extent. It thus appears that the BACH Pb/Pb ages that are slightly younger than 4.56 Gyr are not strongly influenced by inherited radiogenic Pb that evolved in a parent-body precursor during the first tens of Myr of solar system history.

The most thoroughly studied eucrite is Juvinas, for which Rb-Sr, Sm-Nd and U-Pb internal-isochron data exist. Allègre et al. (1975) determined an internal Rb-Sr isochron age of 4.50 ± 0.07 Gyr, which is still one of the two most accurately known eucrite Rb-Sr isochrons, the other being Y75011 (Nyquist et al. 1986). The initial Sr ratio for Juvinas, calculated from plagioclase, is 0.69898 ± 4.

Lugmair ct al. (1976) measured a Sm/Nd age of 4.56 ± 0.08 Gyr, with an initial ^{143}Nd/^{144}Nd ratio of 0.50677 ± 10 (^{148}Nd/^{144}Nd = 0.241572); Jacobsen and Wasserburg (1980) derived an initial ^{143}Nd/^{144}Nd ratio of

0.505888 ± 22 (^{146}Nd/^{142}Nd = 0.636121) for a whole-rock analysis of Juvinas, recalculated for an age of 4.56 Gyr in place of the 4.6 Gyr age used in the paper. For the Lugmair et al. (1976) normalization, the result converts to 0.506671.

Manhes et al. (1984) studied the internal U-Th/Pb isotope systematics of Juvinas. They showed that the system has been disturbed since initial crystallization and that the Juvinas Pb/Pb model ages based on whole-rock isotopic compositions using Canyon Diablo troilite initial ratios appear to be erroneous. The model Pb/Pb ages of 8 fragments so calculated range between 4.567 and 4.553 Gyr. However, plagioclase separates yield Pb/Pb ages of 4.796 to 4.821 Gyr and ^{206}Pb/^{238}U ages of 7.0 Gyr, while pyroxene separates yield model ^{206}Pb/^{238}U ages of ca. 4.76 Gyr, showing that the minerals in Juvinas have not behaved as simple closed systems over 4.56 Gyr. Whole-rock fragments gave ^{206}Pb/^{238}U ages of 4.62 to 4.69 Gyr, also indicating disturbance. The authors next solved for the "nonradiogenic" Pb component by plotting ^{204}Pb/^{206}Pb against ^{238}U/^{206}Pb and obtained 13.026 for ^{206}Pb/^{204}Pb at ^{238}U/^{206}Pb = 0. A similar plot for ^{235}U and ^{207}Pb yielded 13.510 for the inherited nonradiogenic ^{207}Pb/^{204}Pb. In a concordia diagram using those ratios to correct for the nonradiogenic Pb component, the data accurately defined a chord intersecting the equal-age curve at 4.539 ± 0.004 and 1.92 ± 0.06 Gyr (2 σ). 4.539 Gyr is thus a time of complete Pb-isotope equilibration between mineral phases—presumably due to melting; the younger age is attributed to an impact disturbance that added Pb with the above composition to the meteorite. The authors take the close fit of the solution to the data to indicate that any nonradiogenic Pb contained in the meteorite prior to 1.92 Gyr ago was swamped by the Pb added at the time of impact. Note that this method did not assume primordial Pb ratios for the nonradiogenic component, and in that sense differs from the usual model-age approach. This is an extremely important study that cautions against reaching conclusions about Pb isotope ages solely on the basis of the model Pb/Pb ages of whole rocks. The Rb-Sr, Sm-Nd and U-Pb ages agree within error limits, but the U-Pb age is the most precise owing to the relatively short half-life of ^{235}U, and is the probable crystallization age of the meteorite.

Investigations on the pristine eucrite Y75011 (Nyquist et al. 1986) yield a Rb-Sr age of 4.50 ± 0.05 Gyr with initial ^{87}Sr/^{86}Sr 0.69896 ± 2 for a clast, and 0.69893 for the matrix. The initial ^{143}Nd/^{144}Nd ratio is compatible with the CHUR value of Jacobsen and Wasserburg (1984). It is unfortunate that no Pb data are available for this eucrite since it is less altered than Juvinas, and could be expected to yield more reliable data for that reason.

Another eucrite, Ibitera, has received much attention because of its unbrecciated nature (Wilkening and Anders 1975). It does, however, exhibit both static and shock metamorphism (Steele and Smith 1976). Strontium-isotope data (Birck and Allègre 1978) plot within error limits along the same isochron as the highly brecciated eucrites and the howardite, Nuevo Laredo

(Papanastassiou and Wasserburg 1969; Birck and Allègre 1978) and yield an initial ratio indistinguishable from BABI. No Nd data are presently available; however, several groups have reported U-Pb ages (Manhes et al. 1975; Wasserburg et al. 1977; Chen and Wasserburg 1985). The data of Chen and Wasserburg (1985) are the most precise (Table 5.2.1) and give a concordant model age of 4.56 Gyr, within analytical uncertainty. The relationship between the initial $^{87}Sr/^{86}Sr$ ratio and the apparent 4.556 ± 0.006 model Pb/Pb age is puzzling. Perhaps an internal U-Pb isochron study such as was done for Juvinas would reveal disturbances not evident in the whole-rock data. Perhaps the isotopic composition of Sr was not uniform in solar matter 4.56 Gyr ago. Other possibilities are easily imagined, but further speculation is hardly profitable with presently available data.

Many basaltic achondrites appear to have crystallized between perhaps 4.56 and 4.53 Gyr, with an average age of ca. 4.54. Some show later disturbances (Birck and Allègre 1978), presumably due to impacts. With emphasis on the Juvinas data, and average model Pb/Pb ages of all BACH data, and the nearly concordant Nuevo Laredo and Sioux County ages (Table 5.2.1), an age 4.539 ± 0.004 Gyr is tentatively adopted here as the best age for basaltic achondrites corresponding to the BABI initial Sr ratio. The correlation must be viewed with caution since it is known that not all BACHs are coeval, as noted above for Moama and Serra de Magé. Furthermore, the apparent older age of Ibitira must be evaluated.

5.2.5. CHONDRITES

A considerable body of data exist for the Sr, Nd and Pb geochronometers from chondrites. Most values are consistent with ages of 4.56 Gyr, but within rather coarse error limits that more than encompass the age range covered by Allende CAIs and the achondrites. One exception is the LL chondrite, St. Séverin, for which Chen and Wasserburg (1981) report model Pb/Pb ages of 4.551 ± 0.003 and 4.552 ± 0.004 Gyr for two whitlockite samples, with $^{206}Pb/^{238}U$ ages = 4.55 and 4.52 ± 0.02 Gyr. Manhes et al. (1978) reported similar data for St. Séverin, and studied the Rb-Sr system as well. Unfortunately, the whitlockite contained radiogenic Sr ($^{87}Sr/^{86}Sr$ = 0.70237), leading to only an approximate value of 0.6990 ± 2 for the initial Sr ratio. The authors believe that the Rb-Sr system in the chondrite was disturbed subsequent to initial crystallization. The Pb/Pb age of the St. Séverin whitlockite agrees with that for the Angra dos Reis whitlockite within narrow limits. In newer high-precision work Manhes et al. (1987) report Pb/Pb ages of 4.552 ± 0.003 Gyr for the L5 chondrites, Knyahinya and Homestead.

The most comprehensive evaluation of Rb-Sr data is the summary of Minster et al. (1982), based upon combined H, E and LL chondrite whole-rock analyses, which gives an age of 4.498 ± 0.015 Gyr ($\lambda^{87}Rb$ = 1.42 × 10^{-11}/yr) with initial $^{87}Sr/^{86}Sr$ = 0.69885 ± 10 (Fig. 5.2.2). Minster et al.

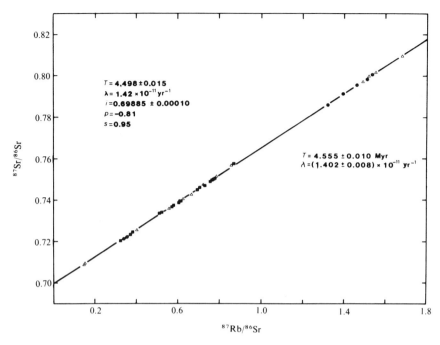

Fig. 5.2.2. $^{87}Sr/^{86}Sr-^{87}Rb/^{86}Sr$ isochron diagram for whole-rock chondrites. Classifications are ■: H; △: LL; ●: E. Figure reproduced from Minster et al. (1982).

point out that the age would become 4.555 Gyr, in close agreement with U-Pb ages, if the decay constant of ^{87}Rb is changed from 1.420 to 1.402 × 10^{-11}/yr. Such an adjustment approaches the error limits for the value quoted by Davis et al. (1977), 1.419 ± 0.012 × 10^{-11}/yr. Nyquist et al. (1986) also argue for a value of 1.402 × 10^{-11}/yr for the decay constant on the basis of their lunar and meteorite studies. It is not possible to exclude that value with present data.

Disturbed Systems

There is clear evidence from internal Rb-Sr isochrons in E chondrites for disturbances in the age range of 4.3 to 4.45 Gyr (Minster et al. 1979), although the whole rock isochron is indistinguishable from that for H and LL chondrites as noted above (Minster et al. 1982).

Hanan and Tilton (1985) cited further evidence for ages substantially younger than 4.56 Gyr from internal isochrons of the Mezö-Madaras (L3) and Sharps (H3) unequilibrated chondrites. They measured Pb/Pb ages of 4.472 ± 0.005 Gyr for Sharps and 4.480 ± 0.010 for Mezö-Madaras from internal isochrons. Although preanalysis terrestrial Pb contamination could cause the low Pb/Pb ages, the authors showed that acid washes and residual samples yielded identical Pb/Pb ages within statistical error limits. This suggests that the ages are real, and due to post-crystallization metamorphism, perhaps impact metamorphism as Manhes et al. (1984) postulate for the discordance in Juvinas U-Pb ages.

Unruh (1982), in a study of 7 L4-L6 chondrites, has made a good case for terrestrial Pb contamination in the range of 5 to 120 ng/g in the samples, compared to total Pb concentrations of 33 to 340 ng/g. With corrections for the calculated amounts of terrestrial Pb, the chondrites yield a model Pb/Pb age of 4.550 ± 0.005 Gyr.

In summary, it is presently somewhat difficult to place the ordinary chondrites within a highly precise time framework due to uncertainties in the analytical data, lack of precise knowledge of the ^{87}Rb decay constant, and the influence of terrestrial Pb contamination on the model Pb/Pb ages. The model Pb/Pb ages mostly fall within the range of 4.53 to 4.56 Gyr. Three chondrites yield high-precision model Pb/Pb ages of ca. 4.552 ± 0.003 Gyr. That is probably the best estimate from U-Pb data, but the age is obviously quite tentative. The best estimate for initial ^{87}Sr/^{86}Sr is probably that given in Minster et al. (1982), 0.69885 ± 10. The initial ^{143}Nd/^{144}Nd ratio is accurately defined by the CHUR parameters of Jacobsen and Wasserburg (1984). There is considerable evidence for early metamorphic processes that affected some chondrite ages.

5.2.6. DISCUSSION

The oldest reliably and precisely dated meteoritic material consists of certain CAIs from Allende, which yield an age of 4.559 ± 0.004 Gyr. Since the radiometric ages date chemical differentiation processes *sensu stricto*, we need to define the process responsible for the age. Starting in a nebula of solar composition as given by Anders and Ebihara (1982), the isotopic evolution of Pb is effectively arrested due to the low U/Pb and Th/Pb ratios, while the Rb/Sr and Sm/Nd ratios cause Sr and Nd isotopic compositions to evolve. The isotopic-Pb ages therefore date a process that fractionated volatile Pb from refractory U and Th, allowing isotopic evolution of Pb to commence. This process is reasonably associated with the condensation of matter from the nebula (see, e.g., Dodd 1981), which for the Allende CAIs requires cooling through a temperature of approximately 1600 K. There is good evidence from the presence of extinct ^{129}I in meteoritic matter that cooling through the range of 1600 to ~400 K was completed in a few Myr. Thus Lewis and Anders (1975) found I-correlated radiogenic ^{129}Xe in magnetite separated from the carbonaceous chondrites Murchison (CM2), Orgueil (CI) and Karoonda (CV4) indicating that Xe retention in these began ca. 6.5 Myr after retention of Xe in Allende CAIs (see also Chapter 15.3). This observation is even more remarkable considering that there is good evidence (Kerridge et al.

TABLE 5.2.3
Comparison of Initial Sr Ratios with Model Pb/Pb Ages in Gyr[a]

Meteorite	^{207}Pb/^{206}Pb Age	Initial ^{87}Sr/^{86}Sr
Allende CAI	4.559 ± 0.004	0.69877 ± 0.00002
Angra dos Reis	4.551 ± 0.002	0.69883 ± 0.00002
Eucrites	4.539 ± 0.004	0.69898 ± 0.00004
Ibitira	4.556 ± 0.006	0.69898 ± 0.00004

[a] See text for data selection.

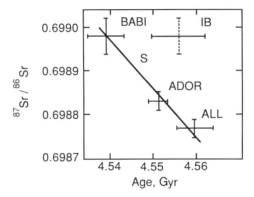

Fig. 5.2.3. Initial ^{87}Sr/^{86}Sr ratios plotted against Pb/Pb model ages of meteorites. S: calculated ^{87}Sr/^{86}Sr evolution for systems with solar abundance ratio for Rb/Sr from Anders and Ebihara (1982). Initial ratios are: ALL: Allende CAI; ADOR: Angra dos Reis; BABI: basaltic achondrite best initial. IB: data for Ibitira. The initial ratio for Ibitira is based on the agreement of the whole-rock data with other eucrite whole-rock data that define the BABI initial ratio (Birck and Allègre 1978). See text for discussion of Ibitira Pb/Pb age.

1979) that the Orgueil magnetite is the result of an aqueous alteration, rather than a nebular condensation, process.

We turn next to the achondrite data. Two key factors are the younger isotopic-model Pb ages for most BACHs and the higher initial Sr ratios compared to the Allende inclusions. A comparison of the initial ^{87}Sr/^{86}Sr ratios for Allende, Angra dos Reis and the basaltic achondrites with the best estimates of isotopic-Pb ages is shown in Table 5.2.3, and plotted in Fig. 5.2.3. Except for Ibitira, the correlation corresponds within error limits to evolution of Sr in environments having solar, or CI chondrite (Anders and Ebihara 1982), Rb/Sr ratios of ca. 0.25. The average Rb/Sr ratio of ordinary chondrites also agrees closely with the solar abundance value; however, O-isotope data (Clayton et al. 1976) preclude derivation of basaltic achondrites from identically the same sources that produced H, L or LL chondrites.

Assuming that volatilization of Rb had occurred prior to differentiation, the sources of Angra dos Reis and the basaltic achondrites must have formed with initial Sr ratios equal to ADOR and BABI, respectively, since the Rb/Sr ratios of the parent materials would have been equal to or lower than those observed in the meteorites today. Calculations assuming 10% partial melting to produce the magmas (Consolmagno and Drake 1977), show that $^{87}Sr/^{86}Sr$ would have evolved by less than 0.00001 in 50 Myr within the parent body. Even if total melting occurred at the time of magma production, the result is unchanged for all achondrites except Stannern and Jonzac (Birck and Allègre 1978), for which the value becomes 0.00002. The basaltic achondrite sources must then have evolved from 4.56 to 4.54 Gyr either in a nebula of solar composition or in an object/objects with approximately solar Rb/Sr ratios. Since Pb data suggest that evolution of Pb isotopes commenced around 4.54 Gyr, the U/Pb ratios in the parent material must also have been near the solar abundance ratio up to that time. Again Ibitira may be an exception.

Gray et al. (1973), using a value for the Rb/Sr ratio based on solar spectroscopic data that is higher than the chondritic value, proposed that the basaltic achondrite material condensed 3 to 8 Myr later than the Allende inclusions. The data in Fig. 5.2.3 suggest that Sr evolution occurred in source materials having Rb/Sr ratios that are lower than the Gray et al. value by about a factor of 2.5. If the achondrite parent body condensed from nebular gas at or near 4.54 Gyr, the depletion of volatile elements must have occurred at the time of condensation. This model requires either that some volume of the solar nebula was preserved as a gas over a 20 Myr interval, or that it was reheated at 4.54 Gyr. The first hypothesis requires a longer-time interval than is usually assumed for condensation of the nebular gas (Wetherill 1986). The second hypothesis requires a special history for achondrite parent material that did not affect the Allende inclusions, and possibly other chondritic parent material as well.

An alternate model consistent with all data is one in which Allende CAIs and the parent bodies of the basaltic achondrites crystallized approximately simultaneously, with later disturbance of the achondrite parents by thermal pulses. The pulses could be associated with collisional processes between large-sized planetesimals in a manner analogous to the one postulated to have formed the Moon (Hartman and Davis 1975; Wetherill 1985). Depending on the collision energy, this could cause complete melting or vaporization of parent materials, which would account for the low abundance of volatile lithophile elements and the normal or enriched abundances of nonvolatile lithophile elements that characterize the basaltic achondrites as well as the Moon. Strontium- and Pb-isotopic evolution between 4.56 and 4.54 Gyr would have to have occurred in sources with solar abundances of volatile, as well as nonvolatile, elements. In fact, Ikeda and Takeda (1985) suggest a C- or LL-chondrite source for the howardite, Y7308 on the basis of chemical and petrologic data. Such a source would satisfy the Nd data as well. Wetherill (1977) has shown in model calculations how ca. 10^{23} g bodies, perhaps formed from collisions between still larger (up to 10^{26} g) bodies, from the Earth's planetesimal swarm might be transferred to orbits in the inner asteroidal belt and stored for times up to 4.5 Gyr. Wetherill (1977) also speculated that some of these objects are the source of the differentiated meteorites.

The above model does not explain the Pb data for Ibitira, assuming that the concordant model U-Pb age of 4.556 ± 0.006 Gyr is correct. There is, in any case, difficulty at present in reconciling that age with initial $^{87}Sr/^{86}Sr$ = BABI.

The Moon appears to have followed a similar history to the model outlined above, although the lunar data are less precisely known. The best estimates for the lunar initial $^{87}Sr/^{86}Sr$ ratio are obtained from anorthosites. Papanastassiou and Wasserburg (1972), measured a ratio of 0.69896 ± 3 for the total-rock sample of anorthosite 60025, which yields an initial ratio of 0.69894 when corrected for ^{87}Rb decay over 4.5 Gyr. Nyquist et al. (1973) reported 0.69891 ± 3 (adjusted to the CalTech scale to allow for a small systematic bias between the two laboratories) for the anorthosite portion of breccia 61016, with no correction required for radiogenic Sr. These estimates are probably close to the true value since the Moon, like the basaltic chondrites, is known to have a very low Rb/Sr ratio. The age of the Moon is not precisely known, but the Sr data may indicate that it has a formation age intermediate between the Angra dos Reis and basaltic achondrite values.

Concluding Remark

This chapter has attempted to speculate on some future directions for research in meteorite geochronology as well as to review a few noteworthy past accomplishments. The review is obviously by no means comprehensive. Much work remains to be done on the dating and deciphering of events that took place early in the history of the solar system. There has been a tendency to lump most meteorite ages together in a "4.56 Gyr time bin," but careful, detailed work with newer techniques has shown this to be an oversimplification. We can anticipate that future high-precision measurements with improved analytical techniques and instrumentation will provide tests of the time scale suggested in this review, and discover still other events as well.

Acknowledgments. Numerous comments by L. E. Nyquist have aided greatly in improving and clarifying the manuscript. I thank D. Crouch for preparation of the figures.

REFERENCES

Allègre, C. J., Birck, J. L., Fourcade, S., and Semet, M. P. 1975. Rubidium-87/strontium-87 age of Juvinas basaltic achondrite and early igneous activity in the solar system. *Science* 187:436–438.

Anders, E., and Ebihara, M. 1982. Solar-system abundances of the elements. *Geochim. Cosmochim. Acta* 46:2363–2380.
Arden, J. W., and Cressey, G. 1984. Thallium and lead in the Allende C3V carbonaceous chondrite. A study of the matrix phase. *Geochim. Cosmochim. Acta* 48:1899–1912.
Birck, J. L., and Allègre, C. J. 1978. Chronology and chemical history of the parent body of basaltic achondrites studied by the ^{87}Rb-^{87}Sr method. *Earth Planet. Sci. Lett.* 39:37–51.
Chen, J. H., and Tilton, G. R. 1976. Isotopic lead investigations on the Allende carbonaceous chondrite. *Geochim. Cosmochim. Acta* 40:635–643.
Chen, J. H., and Wasserburg, G. J. 1981. The isotopic composition of uranium and lead in Allende inclusions and meteoritic phosphates. *Earth Planet. Sci. Lett.* 52:1–15.
Chen, J. H., and Wasserburg, G. J. 1985. *Lunar Planet. Sci.* XVI:119–120 (abstract).
Clayton, R. N., Onuma, N., and Mayeda, T. K. 1976. A classification of meteorites based on oxygen isotopes. *Earth Planet. Sci. Lett.* 30:10–18.
Consolmagno, G. J., and Drake, M. J. 1977. Composition and evolution of the eucrite parent body: Evidence from rare earth elements. *Geochim. Cosmochim. Acta* 41:1271–1282.
Davis, D. W., Gray, J., Cumming, G. L., and Baadsgaard, H. 1977. Determination of the ^{87}Rb decay constant. *Geochim. Cosmochim. Acta* 41:1745–1749.
Dodd, R. T. 1981. *Meteorites* (Cambridge: Cambridge Univ. Press).
Fireman, E. L., DeFelice, J., and Norton, E. 1970. Ages of the Allende meteorite. *Geochim. Cosmochim. Acta* 34:873–881.
Gray, C. M., Papanastassiou, D. A., and Wasserburg, G. J. 1973. The identification of early condensates from the solar nebula. *Icarus* 20:213–219.
Grossman, L. 1972. Condensation in the primitive solar nebula. *Geochim. Cosmochim. Acta* 36:597–619.
Grossman, L. 1975. Petrography and mineral chemistry of Ca-rich inclusions in the Allende meteorite. *Geochim. Cosmochim. Acta* 39:433–454.
Hanan, B. B., and Tilton, G. R. 1985. Early planetary metamorphism in chondritic meteorites. *Earth Planet. Sci. Lett.* 74:209–219.
Hartmann, W. K., and Davis, D. R. 1975. Satellite-sized planetesimals and lunar origin. *Icarus* 24:504–515.
Hohenberg, C. M. 1970. Xenon from the Angra dos Reis meteorite. *Geochim. Cosmochim. Acta* 34:185–191.
Ikeda, Y., and Takeda, H. 1985. A model for the origin of basaltic achondrites based on the Yamato 7308 howardite. *Proc. Lunar Sci. Conf.* 15, *J. Geophys. Res.* 90:C649–C663.
Jacobsen, S. B., and Wasserburg, G. J. 1980. Sm-Nd isotopic evolution of chondrites. *Earth Planet. Sci. Lett.* 50:139–155.
Jacobsen, S. B., and Wasserburg, G. J. 1984. Sm-Nd evolution of chondrites and achondrites, II. *Earth Planet. Sci. Lett.* 67:137–150.
Jessberger, E. K. 1984. ^{39}Ar recoil and the apparent persistence of the presolar age of an Allende inclusion. *Earth Planet. Sci. Lett.* 69:1–12.
Kerridge, J. F., MacKay, A. L., and Boynton, W. V. 1979. Magnetite in CI carbonaceous meteorites: Origin by aqueous activity on a planetesimal surface. *Science* 205:395–397.
Larimer, J. W. 1967. Chemical fractionation in meteorites—I. Condensation of the elements. *Geochim. Cosmochim. Acta* 31:1215–1238.
Lee, T., Papanastassiou, D. A., and Wasserburg, G. J. 1976. Demonstration of ^{26}Mg excess in Allende and evidence for ^{26}Al. *Geophys. Res. Lett.* 3:109–112.
Lewis, Roy S., and Anders, E. 1975. Condensation time of the solar nebula from extinct ^{129}I in primitive meteorites. *Proc. Natl. Acad. Sci. U.S.* 72:268–273.
Lugmair, G. W., and Marti, K. 1977. Sm-Nd-Pu timepieces in the Angra dos Reis meteorite. *Earth Planet. Sci. Lett.* 35:273–284.
Lugmair, G. W., Marti, K., Kurtz, J. P., and Scheinen, N. B. 1976. History and genesis of lunar troctolite 76535 or: How old is old? *Proc. Lunar Sci. Conf.* 7:2009–2033.
Lugmair, G. W., Scheinen, N. B., and Carlson, R. W. 1977. Sm-Nd systematics of the Serra de Mage eucrite. *Meteoritics* 10:300–301 (abstract).
Manhes, G., Tatsumoto, M., Unruh, D., Birck, J. L., and Allègre, C. J. 1975. Comparative ages of basaltic achondrites and early evolution of the solar system. *Lunar Planet. Sci.* VI:546–547 (abstract).
Manhes, G., Minster, J. F., and Allègre, C. J. 1978. Comparative U-Pb and Rb-Sr study of the

St. Severin amphoterite: Consequences for early solar system chronology. *Earth Planet. Sci. Lett.* 39:14–24.
Manhes, G., Allègre, C. J., and Provost, A. 1984. U-Th-Pb systematics of the eucrite "Juvinas": Precise age determination and evidence for exotic lead. *Geochim. Cosmochim. Acta* 48: 2247–2264.
Manhes, G., Gopel, C., and Allègre, C. J. 1987. High resolution chronology of the early solar system given by lead isotopes. *Terra Cognita* 7:377 (abstract).
Marvin, U. B., Wood, J. A., and Dickey, J. S., Jr. 1970. Ca-Al-rich phases in the Allende meteorite. *Earth Planet. Sci. Lett.* 7:346–350.
Minster, J.-F., Ricard, L.-P., and Allègre, C. J. 1979. ^{87}Rb-^{87}Sr chronology of enstatite chondrites. *Earth Planet. Sci. Lett.* 44:420–440.
Minster, J.-F., Birck, J.-L., and Allègre, C. J. 1982. Absolute age of formation of chondrites by the ^{87}Rb-^{87}Sr method. *Nature* 300:414–419.
Nyquist, L. E., Hubbard, N. J., and Gast, P. W. 1973. Rb-Sr systematics for chemically defined Apollo 15 and 16 materials. *Proc. Lunar. Sci. Conf.* 4:1823–1846.
Nyquist, L. E., Takeda, H., Bansal, B. M., Shih, C.-Y., Wiesmann, H., and Wooden, J. L. 1986. Rb-Sr and Sm-Nd internal isochron ages of a subophitic basalt clast and a matrix sample from the Y75011 eucrite. *J. Geophys. Res.* 91:8137–8150.
Papanastassiou, D. A., and Wasserburg, G. J. 1969. Initial strontium isotopic abundances and the resolution of small time differences in the formation of planetary objects. *Earth Planet. Sci. Lett.* 5:361–376.
Papanastassiou, D. A., and Wasserburg, G. J. 1972. Rb-Sr systematics of Luna 20 and Apollo 16 samples. *Earth Planet. Sci. Lett.* 17:52–63.
Patterson, C. C. 1956. Age of meteorites and the earth. *Geochim. Cosmochim. Acta* 10: 230–237.
Podosek, F. A., and Lewis, R. S. 1972. ^{129}I and ^{244}Pu abundances in white inclusions of the Allende meteorite. *Earth Planet. Sci. Lett.* 15:101–109.
Steele, I. M., and Smith, J. V. 1976. Mineralogy of the Ibitira eucrite and comparison with other eucrites and lunar samples. *Earth Planet. Sci. Lett.* 33:67–78.
Tatsumoto, M., Knight, R. J., and Allègre, C. J. 1973. Time differences in the formation of meteorites as determined from the ratio of lead-207 to lead-206. *Science* 180:1279–1283.
Tatsumoto, M., Unruh, D. M., and Desborough, G. A. 1976. U-Th-Pb and Rb-Sr systematics of Allende and U-Th-Pb systematics of Orgueil. *Geochim. Cosmochim. Acta* 40:617–634.
Tilton, G. R. 1973. Isotopic lead ages of chondritic meteorites. *Earth Planet. Sci. Lett.* 19:321–329.
Unruh, D. M. 1982. The U-Th-Pb age of equilibrated L chondrites and a solution to the excess radiogenic Pb problem in chondrites. *Earth Planet. Sci. Lett.* 58:75–94.
Unruh, D. M., Nakamura, N., and Tatsumoto, M. 1977. History of the Pasamonte achondrite: Relative susceptibility of the Sm-Nd, Rb-Sr and U-Pb systems to metamorphic events. *Earth Planet. Sci. Lett.* 37:1–12.
Van Schmus, W. R. 1981. Chronologic and isotopic studies on basaltic meteorites. In *Basaltic Volcanism on the Terrestrial Planets* (New York: Pergamon Press), pp. 935–947.
Wasserburg, G. J., Tera, F., Papanastassiou, D. A., and Hunecke, J. C. 1977. Isotopic and chemical investigations on Angra dos Reis. *Earth Planet. Sci. Lett.* 35:294–316.
Wetherill, G. W. 1977. Evolution of the earth's planetesimal swarm subsequent to the formation of the earth and moon. *Proc. Lunar Sci. Conf.* 8:1–16.
Wetherill, G. W. 1985. Occurrence of giant impacts during the growth of the terrestrial planets. *Science* 228:877–879.
Wetherill, G. W. 1986. Accumulation of the terrestrial planets and implications concerning lunar origin. In *Origin of the Moon,* eds. W. K. Hartmann, R. J. Phillips, and G. J. Taylor (Houston: Lunar and Planetary Inst.), pp. 519–550.
Wetherill, G. W., Mark, R., and Lee-Hu, C. 1973. Chondrites: Initial strontium-87/strontium-86 ratios and the early history of the solar system. *Science* 182:281–283.
Wilkening, L. L., and Anders, E. 1975. Some studies of the unusual eucrite: Ibitira. *Geochim. Cosmochim. Acta* 39:1205–1210.

5.3. DATING OF SECONDARY EVENTS

GRENVILLE TURNER
University of Sheffield

Secondary processes, i.e., those occurring after accretion of planetesimals in the early solar system, are capable not only of effecting petrographic and chemical changes in meteorites, as shown earlier, but also of perturbing isotopic systems used for radiometric dating. Such resetting of radiochronometers can interfere with dating of the earliest events in solar-system history but can also yield useful information about the evolution of solar-system objects. This chapter discusses the acquisition of such information, focusing on the $^{40}Ar/^{39}Ar$ technique and its record of thermal evolution and impact-induced shock processing of meteorite parent bodies.

5.3.1. INTRODUCTION

Any event which brings about, or brings to an end, the redistribution of a radiometric daughter isotope has the potential to be dated. There are a number of significant processes and events in the history of meteorites which produce detectable isotopic changes but which, in the context of the early history of the solar system, must be regarded as secondary. Thermal metamorphism in the interiors of the parent bodies, shock reheating during major impacts, late-stage igneous activity and aqueous alteration or mineralization are the main events which fall into this category (see the chapters in Part 3 of this book). For the first three of these, gas-retention ages, and in particular ^{40}Ar-^{39}Ar ages, have been especially useful in sorting out the time scale.

5.3.2. GAS-RETENTION AGES AND ^{40}AR-^{39}AR DATING

Gas-retention ages are based on the production of ^{4}He from the decay of U and Th, and ^{40}Ar from the decay of ^{40}K. As the name suggests, the age obtained is, at its simplest, a measure of the length of time for which the gaseous daughter products have been accumulating in the minerals concerned. Ages based on the production of Xe isotopes from the decay of now extinct ^{129}I and ^{244}Pu are also gas retention ages, but because of their particular relevance to the early solar system have been considered elsewhere (see Chapters 15.1 and 15.3).

Gas-retention ages on bulk samples are commonly obtained as a matter of course during most noble-gas studies. However, unless parallel measurements of U, Th and K abundances are carried out, the ages calculated are based on nominal values of these quantities and are therefore inaccurate. More important, even when an accurate age determination has been carried out, the actual significance of the numerical result may be ill defined due to the fact that partial loss of He and Ar frequently occurs during shock-induced reheating. For this reason, U,Th-He and K-Ar ages are not used other than as a general indication of the presence or absence of shock reheating. Precise age information, when obtainable, is based on the ^{40}Ar-^{39}Ar variant of K-Ar dating (Merrihue and Turner 1966; Turner et al. 1966).

^{40}Ar-^{39}Ar dating is based on the use of fast-neutron activation and the production of ^{39}Ar by way of the reaction ^{39}K(n,p)^{39}Ar. The ^{40}Ar/K ratio from which the K-Ar age is calculated is obtained from a measurement of the ^{40}Ar/^{39}Ar ratio of Ar released from the sample by heating. The conversion factor between ^{39}Ar and K is determined by the irradiation of a standard in parallel with the sample. The fact that the age determination is based on a single measurement of an isotope ratio in the same sample, rather than on an elemental ratio measured on aliquots of the sample, adds greatly to the power of the technique. Measurements can be made on small inhomogeneous samples which may be rare and either difficult or impossible to separate into representative aliquots. This ability is now being exploited to advantage by the use of pulsed lasers to melt locally μ-sized regions of meteorite sections. These "laser probe" measurements make it possible to look for and interpret small-scale age variations, as might arise, for example, as a result of diffusive loss.

The most widely used approach to ^{40}Ar-^{39}Ar dating is that involving stepped-heating experiments. In these, Ar is released from the sample at a series of increasing temperatures and the ^{40}Ar/^{39}Ar ratio in each step converted to an apparent age. Constancy of apparent age over most of the release is often taken as indicating the time of a significant event, e.g., cooling down of the parent body, and the absence of major secondary reheating events. Variations in apparent age with (typically) low ages in the low-temperature release increasing to high ages at high temperatures has been taken as evidence for recent reheating and partial Ar loss. In cases where the degree of loss is very high, an indication of the time of the reheating can sometimes be obtained from the low-temperature ages. Alternatively, when the loss is relatively small, the existence of a high-temperature "age plateau" may be used to infer an "original" age.

In some instances, variations are observed which have been shown to be an artifact of the stepped-heating technique. An example of such an effect is that arising from recoil of ^{39}Ar, from the mineral in which it is produced, into an adjacent mineral. This is most serious in very fine-grained materials (e.g., petrographic type 3 chondrites) since the recoil distance of the ^{39}Ar following neutron activation is typically 0.1 μm. In a fine-grained sample, recoil leads to a separation of ^{39}Ar and ^{40}Ar into different mineral phases and can lead to variations in ^{40}Ar/^{39}Ar during stepped heating which have no chronological significance. The interpretation of ^{40}Ar-^{39}Ar release patterns is based on the accumulated experience of a large number of studies of terrestrial and extraterrestrial rocks. In many cases, the interpretations can be substantiated by comparison with other techniques. In other cases, interpretations are less certain and limited to providing an indication of the disturbance and of the level of uncertainty associated with the chronology of the sample concerned. Even this limited information may be preferable to a K-Ar age of unknown significance. It is not the purpose of this chapter to provide a comprehensive review of ^{40}Ar-^{39}Ar dating; for more detailed discussion the reader is referred to a number of reviews (see, e.g., Turner 1977; Faure 1986; McDougal and Harrison 1987).

Figure 5.3.1 summarizes most of the more definitive ^{40}Ar/^{39}Ar data in the form of an age histogram. There is clear evidence for discrete peaks in the histogram. The highest of these, around 4.5 Gyr, corresponds approximately to the age of the solar system and is discussed in the next section in terms of the history of parent-body cooling. The younger age peaks are thought to arise mainly from shock heating brought about as a result of parent-body collisions or cratering events in the regoliths of parent bodies. Shock reheating events are discussed in Sec. 5.3.4. In some of the SNC meteorites, low ages result from late igneous activity which is discussed in Sec. 5.3.5.

5.3.3. THERMAL METAMORPHISM AND COOLING AGES

The meteorites represented in Fig. 5.3.1 have been analyzed for a variety of reasons by different analysts. In a statistical sense they are not a strictly representative selection. Nevertheless, from what is known of the distribution of bulk K-Ar ages, it is probable that the histogram is broadly representative of the overall picture. It is clear that a significant proportion of chondrites began to accumulate radiogenic Ar around 4.5 Gyr ago and have remained sufficiently cold for the whole of the period since then to retain it.

On examining in detail the meteorites represented in the high-age group,

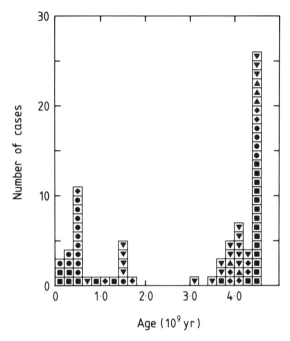

Fig. 5.3.1. Histogram of ^{40}Ar-^{39}Ar ages of meteorites. Symbols are as follows: squares: H chondrites; circles: L chondrites; diamonds: LL chondrites; upward triangles: other chondrites; downward triangles: achondrites and irons.

a number of features emerge. All chondrite classes and all petrographic grades are represented. Both brecciated and unbrecciated, shocked and unshocked examples are seen. Achondrites and silicate inclusions from iron meteorites are also represented. Small but significant differences in age, on the order of 100 Myr, are seen between the oldest and the youngest. For example, in one survey, Turner et al. (1978) observed a range of ages of 15 chondrites, from 4.52 ± 0.03 Gyr to 4.42 ± 0.05 Gyr with a mean of 4.48 ± 0.03 Gyr. Although most of the ages were indistinguishable from each other, the difference between the oldest ages (Forest Vale and Pontlyfni) and the youngest (Shaw and Guareña) was significant.

The Ar data are consistent with evidence of thermal metamorphism based on Rb-Sr and Pb-Pb measurements. A Rb-Sr internal isochron on Guareña (H6) by Wasserburg et al. (1969) was interpreted in terms of a metamorphic event 74 Myr after formation, on account of the evolved initial ^{87}Sr/^{86}Sr ratio. Similarly evolved initial ratios have been observed in other chondrites by Minster et al. (1982). Pb-Pb age determinations on Mezö-Madaras (L3; 4.480 ± 0.011 Gyr) and Sharps (H3; 4.472 ± 0.005 Gyr) have been interpreted by Hanan and Tilton (1985) in terms of an early planetary metamorphism (see Chapter 5.2).

The ^{40}Ar/^{39}Ar ages are generally taken to indicate the times at which the region of the parent body from which the meteorite originated, cooled below what is sometimes referred to as the closure temperature for Ar diffusion. This is the temperature below which further diffusive loss of Ar becomes insignificant in a steadily cooling object. Regarding the age as a "parent-body cooling age" may be something of an oversimplification given the observation that many of the high-age meteorites show evidence of brecciation and/or shock. Both processes can cause heating and Ar loss. The old ages in these meteorites may therefore be affected by a very early period of brecciation and shock, for example, as the parent bodies were accreting. Alternatively, in some instances, the ages may be genuine parent-body cooling ages and the shock features may result from more recent events too mild to cause Ar loss. What seems most likely is that in some cases the old ages are indicative of the time of parent-body cooling while in others they represent the time of early collisional heating (and subsequent cooling) following some localized impact on the parent body. One interpretation of the petrology of the Shaw L7 chondrite, for example, is that it represents a recrystallized surface impact breccia.

In a steadily cooling system, the temperature at which Ar retention effectively begins, the closure temperature, is a function of the cooling rate, being higher in a rapidly cooling object than in a slowly cooling object. As a system cools, diffusion coefficients decrease, and the mean time required for the radiogenic Ar being produced, to be lost by diffusion, increases. When this time becomes comparable to the time during which the diffusion coefficient itself decreases by an appreciable factor, the system becomes closed. The closure temperature T is given by the expression:

$$\frac{D}{a^2} = \frac{D_0}{a^2} \exp\left(-\frac{E}{RT}\right) = \frac{ACE}{RT^2} \quad (1)$$

where D is the diffusion coefficient, a the effective grain radius, E the activation energy for the diffusion process, R the gas constant and C the cooling rate. A is a numerical coefficient dependent on the grain geometry: it has the value 0.018 for spherical grains, 0.049 for cylindrical grains and 0.16 for planar grains of thickness a (Dodson 1973).

Diffusion coefficients and activation energies can be estimated from stepped-heating data. Measured activation energies for chondrites are typically in the range 28 to 62 Kcal/mole, and estimated closure temperatures in the range 120 to 360°C for an assumed cooling rate of 10°C/Myr. Because of the exponential dependence of diffusion coefficient on temperature, increasing the cooling rate by a factor of 10 increases the closure temperature by only 30°C on average.

As explained above, closure is a dynamic process. However, in order to remain closed to diffusive loss over the 4.5 Gyr lifetime of the solar system, cooling must continue to lower temperatures until $D/a^2 \simeq 10^{-20}$/s (Anders 1964). Based on measured activation energies for the meteorites with 4.5 Gyr ages, this long-term storage must have involved temperatures in the range $-30°$ C to $290°$ C, or lower.

The major inferences to be drawn from the narrow range of ^{40}Ar-^{39}Ar ages in the high-age peak of the histogram and the narrow interval between the gas-retention ages and Pb-Pb estimates of the age of the solar system is that the chondrite parent bodies were of such a size as to permit cooling from a range of metamorphism temperatures to the Ar closure temperatures of 240 ± 120° C in times of 100 Myr or less. Estimates of parent-body sizes, based on these observations of cooling times are model dependent but typically lead to radius estimates (or depth estimates) of 100 km or less. The estimates of long-term storage temperatures lead to similar conclusions.

5.3.4. SHOCK REHEATING DURING MAJOR IMPACTS

Approximately three quarters of all meteorites analyzed by the ^{40}Ar-^{39}Ar technique show some evidence of Ar loss at times more recent than 4.4 Gyr. Of these the loss has been sufficiently extensive in about a quarter that they exhibit "low-temperature age plateaux," from which a rough estimate can be made of the time of the loss. While these events do not relate directly to the early history of the solar system, their effects must be considered and understood, if only as representing a significant disturbance to the meteorite, which potentially obscures earlier features and events. Volatile-trace-element distributions are particularly sensitive to disturbance by events capable of causing radiogenic-Ar loss. As can be seen from Fig. 5.3.1, the distribution of ages is by no means uniform but appears to be biased to both very early and very recent events. There is also a suggestion from the discrete nature of the age histogram that the present-day flux of meteorites may be biased by the debris from a few recent major collisions.

Several features of the outgassing-age distribution are worth noting. LL chondrites appear to be strongly represented in the age grouping around 4.0 Gyr (Mangwendi, Appley Bridge, Ensisheim, Soko Banja, Alta-Ameen), as do the howardites (Kapoeta, Bununu, Malvern) and eucrites (Pasamonte, Yamato 74159, Yamato 74450, Cachari). Rb-Sr mineral isochrons close to 4.1 Gyr have also been obtained for two eucrites, Sioux County and Béréba (Birck and Allègre, 1978; cf. Chapter 5.2), and have been included in Fig. 5.3.1. This period is, of course, close to the period of intense bombardment of the lunar surface. There is clearly a suggestion in the observations that the meteorite parent bodies were involved in their own version of the lunar cataclysm or early bombardment.

It is perhaps worth contrasting at this stage the meteorite ^{40}Ar-^{39}Ar age histogram with that of the lunar highlands, which shows a sharp peak of ages at 3.9 Gyr, a few ages (~ 10%) up to 4.2 Gyr and essentially none at 4.5 Gyr (Turner 1977). Since both histograms have been affected by impacts, why are they so different? Three major differences need explaining. The meteorites have retained a memory stretching back to 4.5 Gyr while the Moon's was largely reset by the impacts which occurred between 4.2 and 3.9 Gyr ago. Secondly, the meteorite age histogram contains a much higher proportion of recent impact events. A third distinction is a tentative observation that the higher part of the meteorite age distribution does not appear to cut off so sharply at 3.9 Gyr as does that of the lunar highland samples.

The trimodal age distribution of the meteorites and the contrast with the lunar situation arises from basic differences in the way meteorites and the lunar surface have sampled the impact history of the solar system. To a first order, the surface of the Moon can be regarded as a closed system which carries an integrated record of bombardment since the crust formed. Meteorites, on the other hand, are samples of m-sized fragments of interiors and surfaces of a range of different-sized bodies. Interior samples are protected from the effects of surface bombardment, which accounts for the survival of a high proportion of 4.5 Gyr ages in meteorites. Moreover, since the meteorite population is not a closed system but is subjected to continual fragmentation and comminution, it is likely that a high proportion of those objects which were asteroidal surface material at the time of the lunar cataclysm, have been preferentially destroyed and so lost from the system. Conversely those meteorites, principally LL chondrites and some achondrites, which have preserved a memory of an early-bombardment phase must at that time have been near-surface material on large asteroids. The sharp cut-off in the lunar bombardment rate is well established from the comparatively low crater density of old (3.8 Gyr) mare surface. If the early bombardment of meteorite parent bodies does not show the sharp cut-off (and this remains to be convincingly established), it may imply important differences in the two populations of impacting bodies.

The existence of a low-age peak in the meteorite age histogram is a reflection of the production mechanisms of meteorites whereby asteroidal bodies are either fragmented or have surface material ejected in energetic impacts prior to being transported to the Earth. Once small objects have been produced by impact, their lifetime is strictly limited by erosion and further fragmentation. The size distribution of fragments is such that the lifetime of an object in space against being hit by another object of sufficient size to fragment it still further is proportional to the square root of its radius. Thus, while 100 km-sized objects in suitable orbits in the asteroid belt may survive for 4.5 Gyr, m-sized bodies may have lifetimes of the order of only 10 Myr. At epochs in the past (and the future), the meteorite age distribution would therefore have had (will have) a near-zero-age peak or peaks, together with

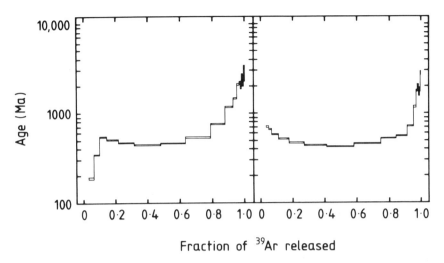

Fig. 5.3.2. ^{40}Ar-^{39}Ar stepped heating measurements of the Peace River shocked L-group chondrite showing a well-defined low-temperature plateau age of 450 Myr.

peaks corresponding to the formation of the solar system and the early bombardment.

The earliest applications of ^{40}Ar-^{39}Ar dating involved the study of the meteorite Bruderheim, an L6 chondrite which was shown to have experienced major (> 95%) Ar loss an estimated 550 Myr ago. Since then, a substantial number of L chondrites have been shown to have been involved in a major heating event, or events, within the last 400 to 500 Myr (Bogard et al. 1976). One of the clearest examples is Peace River (L6), which shows an extremely well-developed low-temperature age plateau corresponding to an outgassing age of 450 ± 30 Myr (Fig. 5.3.2). Most or all of the L chondrites showing evidence of recent outgassing of radiogenic Ar show even clearer evidence of having lost radiogenic He. A clustering of U, Th-He ages around 400 Myr led Anders (1964) to cite this as evidence for a major parent-body collision around this time.

The interpretation of the low-temperature plateau ages as indicating a real event has not been universally accepted, largely for dynamical reasons. If the L-chondrite parent body was disrupted in a major collision 400 Myr ago, where have the fragments been stored since that time? Had they been in Earth-crossing orbits 400 Myr ago, they would have long since disappeared. The conclusion must be that they are stored in a region of the asteroid belt from which they are relatively easily perturbed by Jupiter into Earth-crossing orbits on a several hundred Myr time scale. An alternative view is that the L-chondrite parent body is still in existence, the L chondrites representing surface ejecta continuously being produced by impacts. The ejecta are assumed to be placed into orbits from which they can be perturbed into Earth-crossing orbits. In this view, the 400 Myr ages may either represent an event which produced heating on the parent body, but which is unrelated to the cratering events that are the immediate source of the meteorites, or alternatively it is argued that the 400 Myr ages are simply wrong for technical reasons. The technical reasons cited are that the ^{40}Ar required to give rise to an age of 400 Myr is only 3% of that produced over 4.5 Gyr and that this amount might be retained in a recent heating event such as that which ejected the meteorite from its parent body and commenced its exposure to cosmic rays.

Artificial annealing experiments on the 4.5 Gyr-old Barwell chondrite have shown that zero-age release patterns can be produced provided annealing occurs at solidus temperatures. Quenching from temperatures where partial melting has occurred can, however, lead to false age plateaux if the sample is not subsequently annealed. Recent laser-probe ^{40}Ar-^{39}Ar dating experiments have provided the clearest evidence in support of the validity of the 400 Myr ages (Fig. 5.3.3). Argon extracted from μg regions of Peace River (L6) show large variations in K concentration and correlated variations in radiogenic ^{40}Ar corresponding to the stepped-heating age of 450 Myr. In no instance was

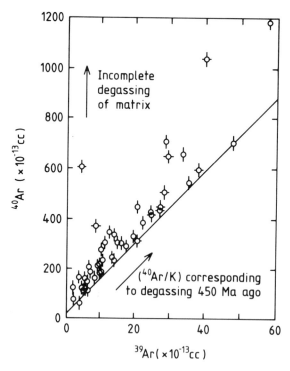

Fig. 5.3.3. Laser probe ^{40}Ar-^{39}Ar measurements of Peace River. Each point represents Ar released by melting a few μg of sample. At this resolution there is no evidence of Ar loss later than 450 Myr ago.

there any evidence on the μg scale for regions outgassed more recently than 450 Myr ago.

A significant number of H chondrites also show evidence of recent outgassing, though as a group they appear to have been less affected by recent heating events than the L chondrites. Although there is no *a priori* reason to expect that the two classes should have experienced similar impact histories, much of the difference in the degrees of outgassing observed can be understood in terms of differences in diffusion properties. Argon diffusion in the H chondrites is generally characterized by higher activation energies (and higher closure temperatures) than for the L chondrites. In stepped-heating experiments in the laboratory, this is manifested by the major gas release from L chondrites occurring 100 to 200°C lower in temperature than for the H chondrites.

In addition to the chronological information obtained from ^{40}Ar-^{39}Ar experiments, stepped degassing combined with simple diffusion calculations can be used to provide thermal information about the events which produced the gas loss. A simple parameter which can be measured directly is the temperature required in the laboratory heating experiment to reproduce the fractional radiogenic Ar loss which occurred in the heating event. Activation energies can also be estimated. Both parameters can be used to constrain plausible thermal histories for the impact events.

Perhaps the most surprising feature of these calculations is the indication that relatively high temperatures have been experienced by a high proportion of meteorites (Turner 1981), corresponding for a typical chondrite to a mean energy deposition of 2×10^5 J/kg. This is two orders of magnitude higher than the energy required to fragment moderate-sized bodies in space. It is also an order of magnitude greater than the mean energy deposited in the ejecta of a moderate-sized lunar crater, which accounts for the well-known observation that the K-Ar ages of most lunar rocks are unaffected by involvement in recent cratering events.

It is not too difficult to account for a significant amount of reheating of meteorites which were part of an asteroidal regolith during the early bombardment period. Although only a small part of the ejecta close to a particular impact is heated, repeated impacts increase the probability of a given sample being in such a favored position.

The reheating associated with the young-age peak cannot be accounted for in this way. It is also difficult to account for the young ages as resulting from a substantial proportion of meteorites being outgassed in cratering events which started them on their journey to Earth. The ejecta would contain a very high proportion of unheated material. One possibility is that a selection mechanism operates preferentially on the ejecta which have been most strongly heated. In this context, it is worth noting that ejecta with the highest velocity are also expected on theoretical grounds to be most strongly heated in the crater-forming impacts.

An alternative explanation for the extensive reheating of a high proportion of the L chondrites is that a single major event, involving the break up of the parent body 400 Myr ago, was responsible for the high proportion of these meteorites which have been outgassed. An attractive feature of this hypothesis is the relative absence of "gas-rich" L chondrites. Gas-rich meteorites are those with a high proportion of implanted solar wind, a feature characteristic of regolith breccias (see Chapter 3.5). Eucrites, howardites and H and LL chondrites all have gas-rich members consistent with their derivation from cratering of asteroidal regoliths. The absence of gas-rich L chondrites is consistent with deeper sampling as would arise from disruption of the parent body.

5.3.5. THE CHRONOLOGY OF SNC METEORITES

The SNC meteorites (shergottites, nakhlites and chassignites) comprise three groups of nine achondrites characterized by cumulate textures and high ratios of volatile to refractory elements. A comparison of the isotopic composition of noble gases and N in the shergottite EETA 79001 with the measurements by Viking of the Martian atmosphere provided strong evidence supporting the suggestion that the SNC meteorites are impact ejecta from the surface of Mars (Bogard and Johnson 1983; Becker and Pepin 1984). Prior to these observations, the chronology of the SNC meteorites appeared to require recent igneous activity on the parent body (or bodies), which in turn implied that the parent body was of planetary size. The alternative possibilities that impact-heating events on smaller bodies were responsible either for producing magma pools from which the SNCs crystallized, or alternatively for resetting the isotopic clocks by heating at subsolidus temperatures, face a number of difficulties.

The chronology of the nakhlites (Nakhla, Lafayette, Governador Valdares) has been thoroughly studied by ^{40}Ar-^{39}Ar, Rb-Sr, U,Th-Pb and Sm-Nd techniques. All methods provide clear evidence for ages close to 1.3 Gyr. The textures of the nakhlites show preferred mineral orientations, and an absence of post-crystallization shock effects or of recrystallization. This evidence indicates that 1.3 Gyr is the crystallization age and not a metamorphism age. Only on a large parent body of planetary size could the mechanism of radioactive heating generate magma at this late stage in solar-system history. Age information on Chassigny, which shows moderate shock and annealing, is more restricted but indicates a similar chronology.

The shergottites (Shergotty, Zagami, ALHA 77005, EETA 79001) are essentially basaltic in composition with a high content of feldspar ($\sim 23\%$) which has been almost totally converted to maskelynite by shock pressures on the order of 30 GPa. Pockets of impact melt in EETA 79001 are the hosts of the putative Martian atmospheric gases. There has been an extensive study of the chronology of the Shergottites and though the results are somewhat confusing, they also appear to indicate an igneous crystallization age younger than 4.6 Gyr (Shih et al. 1982).

Sm-Nd model ages indicate upper limits for the crystallization ages in the range 2.8 to 3.6 Gyr. A whole-rock Sm-Nd isochron for the group as a whole suggests a crystallization age around 1.34 Gyr. However, both REE measurements and a 4.6 Gyr whole-rock Rb-Sr isochron indicate that the meteorites were not comagmatic 1.34 Gyr ago (at that time the $^{87}Sr/^{86}Sr$ ratios would have been very different, implying that they were not parts of the same isotopically mixed reservoir). Although similar to the age of the nakhlites this age is of uncertain significance. For it to be the crystallization age would require the (isolated) source regions of the shergottites to have had similar time-averaged (sub-chondritic) Sm-Nd ratios for the period 4.6 Gyr to 1.34 Gyr.

Based on internal Rb-Sr isochrons for these samples, the shock event is tentatively dated at around 180 Myr. The lowest ^{40}Ar-^{39}Ar ages, around 250 Myr, indicated by the maskelynite, is thought to represent partial resetting by the shock event. Limits on the post-shock annealing temperature (based on the survival of the maskelynite) combined with the inferred diffusion properties of Ar (based on ^{40}Ar-^{39}Ar temperature release data) imply post-shock annealing in a large (~ km-sized) body. If this body were indeed ejecta from Mars, there remains the problem of storing it in space for 180 Myr. The cosmic-ray exposure age of 2 Myr indicates that Shergotty itself was broken from the larger object at that later time.

5.3.6. THE CHRONOLOGY OF AQUEOUS ALTERATION

There is clear evidence, in the form of veins of water-soluble minerals, for the action of liquid water on the CI parent body (see Chapter 3.4). The chronology of this aqueous activity is not well established, though Sr-isotope studies in Orgueil suggest that carbonates in this meteorite formed early. $^{87}Sr/^{86}Sr$ ratios in the carbonate are in the range 0.699 to 0.702. On the assumption that these Sr compositions evolved in an environment with a high Rb/Sr ratio, corresponding to bulk Orgueil, the Rb-poor, Sr-rich carbonate must have been formed within the first 50 to 170 Myr of solar-system history (Macdougall et al. 1984).

REFERENCES

Anders, E. 1964. Origin, age and composition of meteorites. *Space Sci. Rev.* 3:583–714.

Becker, R. H., and Pepin, R. O. 1984. The case for a martian origin of the shergottites: Nitrogen and noble gases in EETA 79001. *Earth Planet. Sci. Lett.* 69:225–242.

Birck, J. L., and Allègre, C. J. 1978. Chronology and chemical history of the parent body of basaltic achondrites studied by the ^{87}Rb-^{87}Sr method. *Earth Planet. Sci. Lett.* 39:37–51.

Bogard, D. D., and Johnson, P. 1983. Martian gases in an Antarctic meteorite? *Science* 221:651–654.

Bogard, D. D., Husain, L., and Wright, R. J. 1976. ^{40}Ar-^{39}Ar dating of collisional events in chondritic parent bodies. *J. Geophys. Res.* 81:5664–5678.

Dodson, M. H. 1973. Closure temperatures in cooling geochronological and petrological systems. *Contrib. Mineral. Petrol.* 40:259–274.

Faure, G. 1986. *Principles of Isotope Geology* (New York: John Wiley).

Hanan, B. B., and Tilton, G. R. 1985. Early planetary metamorphism in chondritic meteorites. *Earth Planet. Sci. Lett.* 74:209–219.

Macdougall, J. D., Lugmair, G. W., and Kerridge, J. F. 1984. Early solar system aqueous activity: Sr isotopic evidence from the Orgueil CI meteorite. *Nature* 307:249–251.

McDougall, I., and Harrison, T. M. 1987. *Geochronology and Thermochronology by the ^{40}Ar-^{39}Ar Method* (Cambridge: Cambridge Univ. Press).

Merrihue, C. M., and Turner, G. 1966. Potassium-argon dating by activation with fast neutrons. *J. Geophys. Res.* 71:2852–2857.

Minster, J.-F., Birck, J.-L., and Allègre, C. J. 1982. Absolute age of formation of chondrites by the ^{87}Rb-^{87}Sr method. *Nature* 300:414–419.

Shih, C.-Y., Nyquist, L. E., Bogard, D. D., McKay, G. A., Wooden, J. L., Bansal, B. M., and Wiesman, H. 1982. Chronology and petrogenesis of young achondrites, Shergotty, Zagami, and ALHA77005: Late magmatism on a geologically active planet. *Geochim. Cosmochim. Acta* 46:2323–2344.

Turner, G. 1977. Potassium-argon chronology of the Moon. *Phys. Chem. Earth* 10:145–195.

Turner, G. 1981. Argon-argon age measurements and calculations of temperatures resulting from asteroidal break-up. *Proc. Roy. Soc. London* 374:281–198.

Turner, G., Miller, J. A., and Grasty, R. L. 1966. The thermal history of the Bruderheim meteorite. *Earth Planet. Sci. Lett.* 1:155–157.

Turner, G., Enright, M. C., and Cadogan, P. H. 1978. The early history of chondrite parent bodies inferred from ^{40}Ar-^{39}Ar ages. *Proc. Lunar Planet. Sci. Conf.* 9:989–1025.

Wasserburg, G. J., Papanastassiou, D. A., and Sanz, H. G. 1969. Initial strontium for a chondrite and the determination of a metamorphism or formation interval. *Earth Planet. Sci. Lett.* 7:33–43.

5.4. COMPACTION AGES

M. W. CAFFEE
Lawrence Livermore National Laboratory

and

J. D. MACDOUGALL
Scripps Institute of Oceanography

The time when meteoritic breccias were assembled in their final form is an important parameter for interpreting many of their properties. Several techniques have been devised for measuring such compaction ages but none are currently wholly satisfactory. Ages based on retention of fission tracks from ^{244}Pu suggest that carbonaceous-chondrite breccias were compacted within a few times 10^8 yr of solar-system formation.

5.4.1. INTRODUCTION

Chronometers based on isotopic records in meteorites allow us to "date" a variety of events. These dates may represent, in some meteorites, the actual time of formation of solid matter, or in others, the time of differentiation. The timing of these events is critical to understanding the formation of meteorites and the conditions of the early solar system. However, there are events that cannot be dated so easily by traditional methods. Chief among these is the time of formation of meteorite breccias (see Chapter 3.5). A breccia is essentially a mixture of rock fragments cemented together by fine-grained material. On the lunar surface, impacts are responsible for the compaction of breccias. Likewise, meteorite breccias are formed (or compacted) by impacts.

Another property shared by lunar and meteoritic breccias is exposure of a fraction of the fragments composing the breccia to the Sun, as evidenced by the presence of solar-wind-implanted ions and solar-flare tracks (Chapter 4.1). Beyond these two common properties, the analogy between lunar and meteoritic breccias breaks down. On the Moon, for example, the regolith is only ~ 10 m thick, thus regolith breccias constitute by volume only a small fraction of lunar material. Based on the fraction of chondrites exhibiting a brecciated structure and containing solar-wind-implanted gases, we conclude that this is not the case for meteorites. In some instances all of the members of a particular class of chondrites are breccias (Chapter 3.5; Goswami et al. 1984). Since some parent bodies consist largely of breccias, it follows that understanding the details of the formation of meteorite breccias is to some extent equivalent to understanding the formation of the parent body itself. Finally, an important aspect of meteorite breccias is the exposure of their constituents to the Sun prior to compaction. These irradiation records are a "snapshot" of the Sun at an earlier time. However, to make the most of these records we need to know when this exposure occurred.

When did meteoritic regolith breccias form, i.e., what is their compaction age? On the lunar surface the consolidation of breccias is an ongoing process; it is possible to find both very ancient and very young breccias. The situation is not so clear for meteorite breccias. Although there are probably both young and old meteorite breccias, the circumstances under which they formed differ from those of the lunar surface. Certainly breccia formation in the early solar system occurred under different circumstances. The cratering rate in the asteroid belt 4.55 Gyr ago was probably higher than that of the contemporary asteroid belt (Housen et al. 1979, and references therein). Also, the relative velocity of the colliding bodies 4.55 Gyr ago was considerably less than the relative velocity of asteroids in the contemporary asteroid belt. These and other factors may have resulted in increased regolith growth on asteroids in the early solar system. While not all meteorite regolith breccias need to be this ancient, even those meteorite breccias formed more recently encounter different circumstances from those on the Moon. For example, the size and chemical composition of the parent body directly influence the physics of cratering. Consequently, it is not possible to determine the exact circumstances under which a regolith existed simply by analogy with the Moon. What is needed is a chronometer from which the compaction age of a particular breccia can be determined. Unfortunately, the events responsible for brecciation (impacts and the shocks accompanying them) do not always reset the standard isotopic clocks. Thus, it is necessary to resort to novel and sometimes indirect techniques to date these events.

5.4.2. TECHNIQUES AND RESULTS

Fission-track Methods

One technique used to determine the compaction age of a breccia is the fission-track method. When a heavy element, such as ^{238}U or ^{244}Pu undergoes spontaneous fission, the fission fragments produce lattice damage (see Chapter 4.1 for a discussion of tracks). These damage trails (tracks) can be seen directly by transmission electron microscopy or chemically etched to produce a hole visible by optical microscopy. If, for example, a U-poor phase is in physical contact with a U-rich phase, the number of fission tracks in the U-poor phase is, *in principle,* proportional to the amount of time they have been in contact (Macdougall and Kothari 1976). If the U-poor material did not record any tracks before coming in contact with the U-rich material and if the event responsible for producing the breccia is also responsible for bringing the U-rich and U-poor phases into contact, then the compaction age is proportional to the track density in the U-poor mineral. There were two known heavy elements present in the early solar system that produced fission tracks: ^{238}U and now extinct ^{244}Pu ($t_{1/2}$ = 82 Myr). Since Pu-fission tracks are indistinguishable from U-fission tracks, the compaction age can only be determined if the ^{244}Pu/^{238}U ratio of the early solar system is known *and* if this ratio has not been disturbed or fractionated by some unspecified process. There are a number of instances where Pu has been fractionated from U. Lugmair and Marti (1977), for example, observed that in mineral separates from Angra Dos Reis, the Pu/U ratio varied but the Pu/Nd ratio was constant. The best current estimate of the PU/U ratio of the early solar system is that of Hudson et al. (1988), who determine a Pu/U ratio, based on the relative abundance of Pu- and U-derived fission Xe in St. Séverin, of 0.007 ± 0.0001. Galactic cosmic rays also produce tracks that can be mistaken for fission tracks, but studying meteorites with low cosmic-ray exposure ages largely eliminates this problem.

Macdougall and Kothari (1976) examined actinide-poor olivines in five CM2 chondrites using the fission-track technique. Assuming a Pu/U ratio of 0.007, the compaction ages range from 4.3 to 4.55 Gyr (Fig. 5.4.1). It should be noted that a factor of two change in the Pu/U ratio produces ~ 100 Myr change in the compaction age. The simplest, and perhaps most optimistic, interpretation of these results is that the events responsible for the formation of breccias on the CM2 parent body occurred ≥ 4.3 Gyr ago. Because some olivines in the CM2 chondrites contain solar-flare tracks, a corollary is that the exposure of portions of the CM2 parent body regolith to the Sun occurred before 4.3 Gyr ago. However, there are several problems that warrant examination. For example, it is possible that the differences in average track density between the meteorites represent Pu/U fractionation rather than an age spread of 200 Myr. Since CM2 chondrites are chemically primitive and have undergone only closed-system alteration, it is likely that there is not substan-

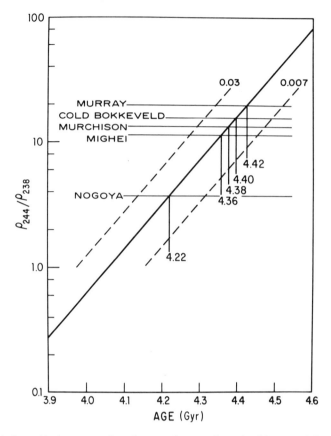

Fig. 5.4.1 A graphical representation of compaction ages determined for several CM meteorites using the fission-track method (Macdougall and Kothari 1976). The ordinate is the ratio of the track density due to ^{244}Pu to that due to ^{238}U. The U contribution was calculated based on the average measured U concentration in the matrix; the "excess" tracks were then assumed to be due to ^{244}Pu fission. Diagonal lines correspond to different initial ^{244}Pu/^{238}U values (0.0154 for the solid line, and 0.03 and 0.007 for the dashed lines, as marked). The values given are averages for many grains from each meteorite, and the calculated ages are for an initial ratio of 0.0154. As explained in the text, a more appropriate value is probably 0.007.

tial Pu/U fractionation, however, no rigorous demonstration of this exists. Another potential problem is the observation that within a single meteorite, the track density varies among the grains studied. One possible explanation for this variation is a difference in the track recording efficiency or retention in the actinide-depleted minerals. Yet another possible explanation is that the matrix material has some phases that have different actinide contents; for example the Pu concentration may vary throughout the matrix. For the above cases, averaging the track density, as was done by Macdougall and Kothari, gives a meaningful compaction age.

A more serious problem, however, is the possibility that one of the underlying assumptions of the technique is flawed, namely the assumption that the tracks recorded in the olivine grains were obtained in a regolith. It is possible, although Macdougall and Kothari considered it unlikely, that the grains existed prior to compaction in a gas of roughly solar composition and that during that time they recorded some fission tracks. In such an instance, the onset of track retention would not correspond to the formation of the breccia. This problem has not been well addressed but it may be possible to place some qualitative constraints on its plausibility. There are two issues: the track-recording efficiency of a grain in a nebular gas and the very existence of the grain in the nebular gas to begin with. Macdougall and Kothari estimated that the track-recording efficiency in a gas would be $\sim 10^3$ less than for a grain in matrix material. If we suppose that a non-negligible portion of the tracks observed in the olivines were acquired in the solar nebula, the required residence time would be at least tens of Myr and perhaps > 100 Myr. Based on the track records alone it is not possible to rule this out; however, this length of time is long compared to accretion time scales that predict the growth of small grains to m-sized bodies in $\sim 10^5$ to 10^6 yr (Chapter 6.4). Looking at this from a different point of view, if a grain existed in the nebula for 10^5 yr and its track-recording efficiency is 10^{-3} that for a grain in matrix, the number of Pu-fission tracks accumulated during this time would not substantially affect the calculated ages. The second issue is whether the track-bearing grains ever coexisted with the nebular gas. If these silicate grains condensed directly from the solar nebula, they could have resided in the gas for some time, thus acquiring Pu-fission tracks. However, they may have formed as a result of some other process, for example, fragmentation of larger entities, perhaps chondrules or aggregates (Chapter 10.4), and never resided in the nebular gas at all. While this issue is currently unresolved, there is hope that further petrographic and chemical analyses of carbonaceous chondrites will shed some like on these processes.

An alternative to studying fission tracks in actinide-depleted material adjacent to actinide-rich material is to find actinide-rich grains, such as phosphates. The fission-track density in the phosphates represents the length of time the mineral has retained tracks. If the event responsible for the compaction of the breccia annealed existing tracks, then the compaction age would be proportional to the present track density. Relative to other minerals, such as pyroxenes and feldspars, the track-retention temperatures in phosphates are low, so it is possible, *in principle,* that brecciation could reset the fission-track record. Kothari and Rajan (1982) studied fission tracks in phosphates from four chondrites. The samples were from meteorites that contain xenolithic clasts (Chapter 3.5). Three of the four meteorites (Weston, Fayetteville and St. Mesmin) are classified as gas rich and some of the clasts from these three were dated using ^{40}Ar-^{39}Ar and K-Ar techniques. One clast from St. Mesmin had a K-Ar age of 1.4 Gyr (Schultz and Signer 1977). If the event responsible for resetting these radiometric clocks also annealed any ancient tracks, then the track retention age should be ~ 1.4 Gyr. Somewhat surprisingly, in all the samples Kothari and Rajan studied, they obtained model fission-track ages of ≥ 4.3 Gyr. These are old ages, so ^{244}Pu decay contributed to the track inventory, hence the ages are dependent on the Pu/U ratio at the onset of track retention. Based on the study of De Chazal et al. (1985) it is likely that this is a serious problem for phosphates and may render the quantitative determination of a track-based compaction age almost impossible. However, there are too many tracks to be accounted for by U-decay alone, so it is probably safe to conclude that ^{244}Pu was still extant when the phosphate grains started retaining tracks. Since Kothari and Rajan studied clasts having young radiometric ages, they concluded that the final brecciation event did not reset the track clock and that the event they dated was the time at which the phosphates cooled to the point where tracks were quantitatively retained.

A slightly different method of setting some constraints on the time of formation of breccias involves the track records of U- and Pu-rich phosphates adjacent to actinide-depleted detectors, such as olivines, pyroxenes and feldspars (Pellas and Storzer 1981). Unlike the two previous methods, from which model fission-track ages are determined, this technique determines a cooling rate. In cooling minerals, the difference in the ^{244}Pu track density between different sets of track detectors (e.g., phosphate and olivine) defines a time differential between the different minerals. Pellas and Storzer applied this technique to H-, L- and LL-chondrites. Their results are most dramatic for the H chondrites. The cooling rate decreases with increases petrographic type, implying that type 3s cooled most rapidly. Pellas and Storzer took this as evidence that the H-chondrite parent body had an "onion-shell" structure where the lower petrographic types resided in the near-surface regions. Pellas and Storzer further concluded that the rapid cooling rate of H 4 material precludes the existence of an insulating regolith cover. Such a regolith must therefore have formed after the type-4 material cooled. Given the slower cooling rate of the type-6 material it is possible that this material was not excavated by impacts for 100 Myr after the formation of the parent body. If this interpretation is correct, then this technique establishes an upper limit to the time of breccia formation.

In principle, it is possible to obtain model fission-track ages for these adjacent grains, in similar fashion to the method of Macdougall and Kothari (1976). Again, the biggest problem is Pu/U fractionation in the phosphates: this problem may be surmountable but a technique for determining the degree of Pu/U fractionation, such as one based on the apparent constancy of the Pu/Nd ratio, would require considerable development.

Radiometric Ages

Considering the number of brecciated meteorites, it is somewhat amazing that, with only a few exceptions, most radiometric techniques yield ages of ~ 4.55 Gyr. Among those few exceptions are dates for several clasts that yield ages of ~ 3.5 Gyr from Kapoeta (Rajan et al. 1979; Papanastassiou et al. 1974), 1.4 Gyr for St. Mesmin (Schultz and Signer 1977) and 3.6 Gyr for Plainview (Keil et al. 1980). These ages have been taken as upper limits to the time of brecciation for these meteorites. If it is assumed that the brecciation process is gentle enough to reset chronometers in only a few meteorites, it follows that breccia formation in other chondrites may be relatively recent. This necessarily implies that the gas-rich meteorites could have been exposed to the Sun very late in their evolution. In light of the observation of excess ^{21}Ne in solar-flare-rich grains from Kapoeta that Caffee et al. (1987) have proposed may require an enhanced particle flux, these radiometric ages are important. One plausible cause of this increased production rate is an active early Sun (cf. Chapter 4.1). Since such an exposure must have occurred very early in the solar system, the existence of clasts in Kapoeta ~ 3.6 Gyr old presents an opportunity to test the active-Sun hypothesis. Specifically, any event capable of resetting isotopic clocks should anneal solar-flare-produced tracks. Therefore, if the grains containing solar-flare tracks and excess spallogenic Ne are ~ 3.6 Gyr old, irradiation by an active early Sun could be ruled out. Caffee et al. (1985) obtained ^{40}Ar-^{39}Ar ages on individual feldspar grains. Of the six grains analyzed (3 containing solar-flare tracks) all had ^{40}Ar-^{39}Ar ages ≥ 4.3 Gyr. This result does not prove an early irradiation, since the exposure could have occurred anytime after 4.3 Gyr; however, it is a necessary condition for the plausibility of this hypothesis.

Authigenic Phases

Carbonaceous chondrites are our best example of chemically unfractionated material; however, even in these meteorites there is ample evidence of non-nebular secondary processing (Chapter 3.4). The brecciated structure of CI chondrites indicates extensive mechanical reworking, both on a microscale and from clast to clast. Richardson (1978) proposed that the veins in several CI meteorites were deposited during impact brecciation and aqueous activity. Among the various phases filling these veins are carbonate and sulfate minerals. Macdougall et al. (1984) analyzed the Sr-isotopic composition and the Rb/Sr ratio in carbonates and sulphates from the CI meteorite Orgueil, a gas-rich breccia. The chronometer in this study is the ^{87}Sr/^{86}Sr ratio (Chapter 5.1). The lowest ^{87}Sr/^{86}Sr value measured for a carbonate is similar to the initial Sr-isotopic ratio in refractory inclusions from Allende (Gray et al. 1973; Minster et al. 1982). This suggests that the carbonates formed very early, most likely within a few Myr of the formation of the CI parent body (Fig. 5.4.2), implying that the regolith of the CI parent body is al-

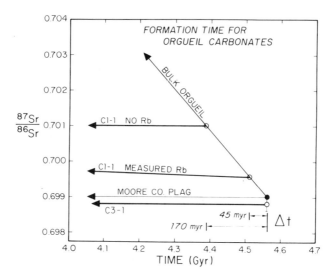

5.4.2. An illustration of the Sr-isotope method for determining the time of formation of low-Rb phases such as carbonates in the CI meteorites, in this case Orgueil (Macdougall et al. 1984). Evolution lines are shown for the bulk meteorite (^{87}Rb/^{86}Sr = 0.8), plagioclase from the Moore County achondrite, and sample C1-1, one of the carbonates measured by Macdougall et al. (1984). For the latter sample, two possible scenarios are shown, corresponding to the case in which all the measured Rb actually was a measurement blank (curve labeled "no Rb"), or assuming that all Rb measured was indigenous to the carbonate. One carbonate separate, C3-1, contained essentially no measurable Rb and had extremely low ^{87}Sr/^{86}Sr values; its evolution is shown for comparison. The parameter Δt is the time between 4.56 Gyr and the isolation of the carbonate from bulk Orgueil Sr.

most as ancient as the material comprising the CI parent body itself. Solar-flare and solar-wind records in CI chondrites would necessarily be close to 4.5 Gyr old.

5.4.3. SUMMARY

Determining the time of breccia formation establishes constraints on models of meteorite parent-body formation and may provide limits on the duration of the exposure of portions of the regolith to the Sun. Unfortunately, such compaction ages are elusive. Evidently, the standard radiometric clocks are rarely reset by the events that consolidate breccias. For this reason, much attention has focused on track methods. The technique of Macdougall and Kothari (1976) has considerable promise; however, more work is needed before its use as a reliable chronometer is accepted. Likewise, the Sr-isotopic composition of authigenic phases may be very informative, but this technique also needs further testing. These reservations notwithstanding, the initial results of these two methods as applied to carbonaceous chondrites provide some measure of evidence that these particular regoliths and the breccias formed from them were compacted early in the solar system, i.e. ≥ 4.3 Gyr ago.

On the other hand, there are at least a few clasts from ordinary chondrites and achondrites whose isotopic clocks have been reset, implying some breccia formation perhaps as late as 1.4 Gyr ago. Also, Pellas and Storzer (1981) cite track records in ordinary chondrites as evidence that their parent bodies, most notably those of the H chondrites, had an onion-shell structure. Furthermore, they deduce that the regolith could not have formed until after the parent body had largely cooled. These two lines of reasoning indicate that regolith formation for ordinary chondrites and achondrites occurred later than for carbonaceous chondrites.

Thus we are faced with a dilemma. If we accept all of these results as correct, then we must conclude that there were at least two epochs of breccia formation: one very early ($\gtrsim 4.3$ Gyr) for carbonaceous chondrites and a later (perhaps continuous) one for ordinary chondrites and achondrites. Alternatively, one or more of these chronometers may be leading us astray. Given the early stage of development of most of these techniques, it is difficult to speculate which, if any, of them are being improperly applied. In most cases, only a few samples have been analyzed. The obvious next step is therefore to apply these methods to more meteorites.

REFERENCES

Caffee, M. W., Hohenberg, C. M., Swindle, T. D., and Goswami, J. N. 1985. Pre-compaction irradiation of Kapoeta: ^{40}Ar/^{39}Ar ages of meteorite grains. *Meteoritics* 20:620–622 (abstract).

Caffee, M. W., Hohenberg, C. M., Swindle, T. D., and Goswami, J. N. 1987. Evidence in meteorites for an active early Sun. *Astrophys. J.* 313:L31–L35.

De Chazal, S., Crozaz, G., Bourot-Denise, M., and Pellas, P. 1985. Plutonium and uranium in individual crystals of merrilite and apatite of St. Severin. *Meteoritics* 19:216–217 (abstract).

Goswami, J. N., Lal, D., and Wilkening, L. L. 1984. Gas-rich meteorites: Probes for particle environment and dynamical processes in the inner solar system. *Space Sci. Rev.* 37:111–159.

Gray, C. M., Papanastassiou, D. A., and Wasserburg, G. J. 1973. The identification of early condensates from the solar nebula. *Icarus* 20:213–239.

Housen, K. R., Wilkening, L. L., Chapman, C. R., and Greenberg, R. 1979. Asteroidal regoliths. *Icarus* 39:317–351.

Hudson, G. B., Kennedy, B. M., Podosek, F. A., and Hohenberg, C. M. 1988. The early solar system abundance of ^{244}Pu as inferred from the St. Séverin chondrite. *Proc. Lunar Planet. Sci. Conf.* 19, submitted.

Keil, K., Fodor, R. V., Starzyk, P. M., Schmitt, R. A., Bogard, D. D., and Husain, L. 1980. A 3.6-b.y.-old impact-melt rock fragment in the Plainview chondrite: Implications for the age of the H-group chondrite parent body regolith formation. *Earth Planet. Sci. Lett.* 51:235–247.

Kothari, B. K., and Rajan, R. S. 1982. Fission track studies of xenolithic chondrites: Implications regarding brecciation and formation. *Geochim. Cosmochim. Acta* 46:1747–1754.

Lugmair, G., and Marti, K. 1977. Sm-Nd-Pu timepieces in the Angra Dos Reis meteorite. *Earth Planet. Sci. Lett.* 35:273–284.

Macdougall, J. D., and Kothari, B. K. 1976. Formation chronology for C2 meteorites. *Earth Planet. Sci. Lett.* 33:36–44.

Macdougall, J. D., Lugmair, G. W., and Kerridge, J. F. 1984. Early solar system aqueous activity: Sr isotope evidence from the Orgueil CI meteorite. *Nature* 307:249–251.

Minster, J.-F., Birck, J.-L., and Allègre, C. J. 1982. Absolute age of formation of chondrites studied by the ^{87}Rb ^{87}Sr method. *Nature* 300:414–419.

Papanastassiou, D. A., Rajan, R. S., Huneke, J. C., and Wasserburg, G. J. 1974. Rb-Sr ages and lunar analogs in a basaltic achondrite; Implications for early solar system chronologies. *Lunar Sci.* V:583–585 (abstract).

Pellas, P., and Storzer, D. 1981. ^{244}Pu fission track thermometry and its application to stony meteorites. *Proc. Roy. Soc. London* A374:253–270.

Rajan, R. S., Huneke, J. C., Smith, S. P., and Wasserburg, G. J. 1979. Argon 40-argon 39 chronology of lithic clasts from the Kapoeta howardite. *Geochim. Cosmochim. Acta* 43: 957–971.

Richardson, S. M. 1978. Vein formation in the CI carbonaceous chondrites. *Meteoritics* 13: 141–159.

Schultz, L., and Signer, P. 1977. Noble gases in the St. Mesmin chondrite: Implications to the irradiation history of a brecciated meteorite. *Earth Planet. Sci. Lett.* 36:363–371.

PART 6
The Early Solar System

6.1. CHONDRITES AND THE EARLY SOLAR SYSTEM: INTRODUCTION

JOHN F. KERRIDGE
University of California at Los Angeles

The preceding chapters have described what meteorites look like, what they are made of, where they come from, how old they are and what has happened to them since their formation. Before delving more deeply into their properties, we must place those properties in the appropriate astrophysical context.

6.1.1. ESTABLISHING A THEORETICAL CONTEXT

As its title implies, this book is about the insights that meteorites can provide about conditions and processes in the early solar system. Their value in this regard first became apparent when it was realized that the chondritic meteorites, in particular, were very old and closer to the Sun in chemical composition than any other known natural material (see, e.g., Patterson 1956; Suess and Urey 1956). However, it was not until the mid-1960s that it became common for chondritic properties to be discussed in a solar-nebular context (see, e.g., Wood 1963; Larimer and Anders 1967). Since that time, investigations into the nature of chondrites and theoretical modeling of the nebular and prenebular phases of solar-system history have both proceeded apace. The objective of this book is to review progress in the former area. A comparable review of progress in the theoretical area is beyond the scope of this book (see, instead, Black and Matthews 1985), but a relatively brief overview of that area is necessary in order to put the meteoritic record into perspective. Such an overview occupies the following three chapters.

Chapter 6.2 explores the collapse of a fragment of a dense molecular

Fig. 6.1.1. The Orion nebula, a region of contemporary star formation. North is at the top, east is to the left. At bottom left is the interaction front between the ionized nebula, a few hundred $M_\odot$ in size, and a giant molecular cloud, much larger in size. The contours near the top define a region of infrared emission from molecular H_2 at about 2000 K, buried within the molecular cloud behind the nebula. In about 10^5 yr, this region will probably look like the present-day Orion nebula. Photograph courtesy of B. Zuckerman.

cloud, such as that illustrated in Fig. 6.1.1, to form a flattened nebula rotating around the proto-Sun. Chapter 6.3 discusses the physical and chemical evolution of that nebula to the point where most of the potentially condensible material was in solid form. Chapter 6.4 then considers the agglomeration of that solid material into planetesimals, asteroids and planets. This stage is depicted in Fig. 6.1.2.

The subsequent chapters are devoted to the meteoritic, predominantly chondritic, record. The nature of that record is such that it is not practicable

Fig. 6.1.2. A cocoon nebula, perhaps the primordial solar nebula. Dust particles have condensed in the nebular disk and begun accreting into small planetesimals. Inhomogeneities in dust distribution block starlight from the inner nebula in some regions, but allow light to escape in certain directions. (A black-and-white reproduction of a painting by William K. Hartmann.)

to organize the material in a chronological fashion, as in Chapter 6.2 through 6.4. Instead, the underlying scheme is to proceed from the more general to the more specific, but the many exceptions to this trend will be apparent from the numerous cross-references.

REFERENCES

Black, D. C., and Matthews, M. S., eds. 1985. *Protostars and Planets II* (Tucson: Univ. of Arizona Press).
Larimer, J. W., and Anders, E. 1967. Chemical fractionations in meteorites—II. Abundance patterns and their interpretation. *Geochim. Cosmochim. Acta* 31:1239–1270.
Patterson, C. C. 1956. Age of meteorites and the earth. *Geochim. Cosmochim. Acta* 10:230–237.
Suess, H. E., and Urey, H. C. 1956. Abundances of the elements. *Rev. Mod. Phys.* 28:53–74.
Wood, J. A. 1963. Chondrites and chondrules. *Sci. Amer.* 209(4):65–82.

6.2. PROTOSTELLAR COLLAPSE, DUST GRAINS AND SOLAR-SYSTEM FORMATION

PATRICK CASSEN
NASA Ames Research Center

and

ALAN P. BOSS
Carnegie Institution of Washington

Attempts to understand the origin of the solar system come from two perspectives. The first is that of the insider looking out; the system components beyond the Earth are examined in detail, their properties categorized, and explanations are sought that make sense of the patterns and relations among them. The primitive meteorites are prime objects of study from this perspective, and most of this book is devoted to the discovery of what they have to reveal about the beginnings of our planetary system. The other perspective is that of the outsider looking in; we adopt this view by turning our attention to other stars that are now in their formative stages and by trying to understand the physical processes that govern all such events, under the assumption that our solar system is not unique. A compelling theory of the origin of the solar system must, of course, merge the fruits of both perspectives.

6.2.1. INTRODUCTION

The orbital characteristics of the planets strongly imply that the solar system formed out of a rotationally flattened nebula surrounding the protosun. Because this nebula was formed as a part of the same process that formed the Sun, its origin must be studied within the framework of the general theory of star formation. The purpose of this chapter is to describe the results of recent work on the theory of star formation and to relate it to ideas stimulated by the study of meteorites.

Conceptually speaking, it is a long path from the vast and tenuous reaches of the interstellar clouds in which stars are born to the chunks of rock that attract our attention by virtue of their recent arrival on Earth. However, we are guided and motivated by the evidence that the meteorites have retained some physical memory of the presolar environment (see Chapters 12.1, 13.1, 13.2 and 14.3). This chapter explores the relationship between meteoritical data and protostellar collapse: the formation of the protosolar nebula, and the evolution of the dust component of interstellar clouds prior to the formation of the nebula. It is generally thought that the parent bodies of the meteorites formed after the solar nebula was established; thus this chapter will not focus on hypotheses which attempt to form the parent bodies prior to solar nebula formation. Accumulation of dust grains within the solar nebula and the formation of parent bodies are discussed in Chapters 6.3 and 6.4.

We first describe observations of new stars that have masses similar to that of the Sun (and are therefore expected to become similar to the Sun in other respects), and the theories that attempt to cope with these observations. Then we focus on aspects of the theories that deal with collapse and nebular dynamics, with a view towards understanding the nature of the meteorite-forming environment, and understanding the fate of interstellar grains. It will be apparent that many aspects of these problems require further development.

6.2.2. OBSERVATIONS OF SOLAR-TYPE STAR FORMATION

Observations reveal that stars are born in dense molecular clouds. It is generally agreed that stars are the result of collapse, due to gravitational instability, of cloud material: molecular hydrogen gas, helium, and cosmic proportions of heavy elements mostly in the form of grains. The physical properties (mass, temperature, radiative environment, dynamical state, etc.) of molecular clouds in star-forming regions vary, and it is reasonable to suppose that this variation is in some way responsible for the spectrum of stellar masses (ranging from less than $0.1M_\odot$ to perhaps as high as $100M_\odot$). It is one of the primary goals of star-formation theory to explain the relationships between the properties of the clouds and the properties of the stars that form from them.

In discussing the origin of the solar system, it is important to restrict attention to facts and ideas relevant to the formation of low-mass stars, i.e., stars of about one or two solar masses or less. Conditions associated with the formation of higher-mass stars are different in several ways. First, because of their much greater luminosities, high-mass stars rapidly disrupt their surroundings, making it difficult to relate their observed environments to the actual birth conditions. (Stellar luminosity is a steep function of mass, for both main sequence and pre-main sequence evolution; see Sec. 6.2.3.) In fact, the radiation field of a high-mass star is intense enough to affect the dynamics of collapse and accretion, whereas the radiation field of a low-mass star is insufficient to affect the dynamics appreciably. Second, the conditions which result in the birth of high-mass stars may be essentially different from those that lead to low-mass star formation. The evidence for this comes from observations that there exist regions of exclusively low-mass star formation (e.g., the Taurus molecular cloud complex), and others that appear to be exclusively producers of high-mass stars (e.g., NGC 2362; G. H. Herbig 1987, personal communication). Of course, there are also regions in which a wide spectrum of masses have been or are being produced, but the idea that there is a tendency toward spatial segregation has gained support from recent attempts to characterize the molecular cloud and stellar populations of the Galaxy (see Shu et al. 1987, and references therein).

One of the best-studied regions of solar-type star formation is the Taurus-Auriga cloud complex, in which about a hundred stars with masses in the range 0.2 to 3.0 $M_\odot$ have been born in the last five million years or so (Jones and Herbig 1979; Cohen and Kuhi 1979). Most of these stars have masses between 0.5 and 1.0 $M_\odot$. Relatively few main sequence stars are seen, and there are no very massive stars. Although there can be no certainty that the Sun was born in a similar association, the properties of this group might offer a good representation of the early solar condition and environment. A possible objection to this proposition is the lack of any evolved stars to provide sources of the isotopic anomalies described in Chapter 14.3 (see also Chapter 15.1).

The Taurus-Auriga cloud itself is an irregular structure with a typical overall dimension of 10 pc ($= 3 \times 10^{19}$ cm), and is estimated to contain 8×10^3 $M_\odot$ of gas. Less than 10% of the mass of the cloud has been converted to stars, which have a mean number density of about 0.5/pc^3 (Jones and Herbig 1979; cf. Cohen and Kuhi [1979] who estimate a higher stellar density). As in most other dark clouds, the temperature is uniformly low, about 10 K. The complex as a whole is probably not gravitationally bound, although substructures within it undoubtedly are. Against the cloud mean density of 10^2 to 10^3 particles/cm^3, clumps or cores are observed (Myers and Benson 1983; Fig. 6.2.1) with mean densities an order of magnitude or two larger than the background, and size/density relationships consistent with inverse power law density profiles (Myers 1985). These cores typically contain 1 to 10 $M_\odot$ of gas. Most importantly, many of them (at least half) have embedded infrared sources, which have been convincingly identified as protostars. An interesting feature of the cores is that their rotation rates indicate that angular momentum was effectively removed from them during whatever process caused their contraction. It has been suggested that the cores are formed due to the slow growth of density perturbations which remain magnetically

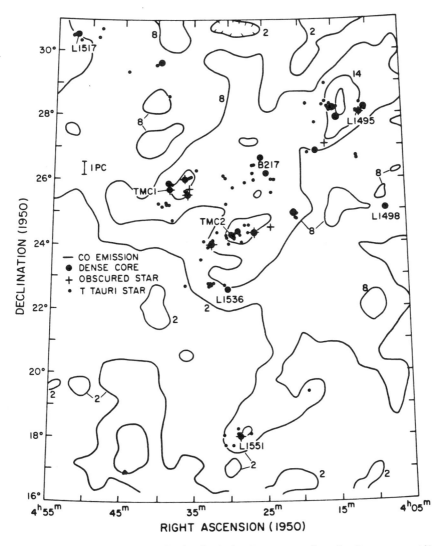

Fig. 6.2.1. Map of the Taurus molecular cloud, showing contours of gas density as represented by emission from CO and the locations of dense cores (observed in NH_3), embedded infrared sources, and optically visible T Tauri stars. L1551 is an embedded object surrounded by a bipolar flow (figure from Myers 1987).

coupled to the surrounding cloud, thereby allowing angular momentum to be transferred from the growing condensation.

One of the most basic astronomical observations relevant to the origin of the solar system is the fact that the Sun is a single star. Single stars appear to be in the minority, with most normal stars being members of both a close binary system and a more widely separated multiple system (Abt 1983). The absolute frequency of spectroscopic binaries is very hard to determine (Morbey and Griffin 1987). The fact that detection of spectroscopic binaries is limited by finite measuring accuracies, as well as the independent evidence for frequent occurrence of visual binaries, implies that we must assume that single stars are the exception rather than the rule.

This presents the problem that the formation of single stars with planetary systems may be a relatively rare event, so that observations of typical star-forming regions such as Taurus-Auriga (where primarily binary and multiple stars are probably being formed) may not necessarily be relevant for the formation of single stars. No binary protostars have been found as yet in the dense molecular cloud cores, which might be taken as reassuring evidence that these cores form single stars. However, the angular resolution of the infrared survey satellite used to detect the embedded protostellar objects is such that it could only detect binaries with separations greater than about 10,000 AU (1 AU = 1.5×10^{13} cm) at the distances of nearby star-forming regions, whereas essentially all known binaries have separations much less than this (typically 0.1 to 1000 AU). Hence it is quite uncertain whether or not an individual dense core has formed a single or binary protostar, and caution must be used in applying these observations to solar-system formation.

Considering that close binary systems have on the order of 1000 times the orbital angular momentum of the solar system, it is clear that the amount of angular momentum is a primary determinant for whether a dense cloud will collapse to form a single or binary protostar. Unfortunately, this quantity is poorly known for dense clouds, because rotational effects are hard to distinguish from other motions such as turbulence, collapse and outflow, and because even the most rapidly rotating clouds have Doppler shifts which are barely measurable with the most sensitive astronomical spectrometers. Goldsmith and Arquilla (1985) have reviewed the evidence for rotation in a sample of 16 clouds, and find the angular momentum per unit mass (J/M) to range between $\sim 10^{21}$ and 10^{24} cm^2/s. The present Sun has $J/M \approx 10^{16}$ cm^2/s. The J/M of the reconstituted solar system might have been as little as $\sim 3 \times 10^{18}$ cm^2/s (see Sec. 6.2.4), which is vastly smaller than the observed values for clouds. The Goldsmith and Arquilla data apply only to the most rapidly rotating clouds, however. Because of this, their data do not imply that a process which converts rapidly rotating clouds into single stars must be found; there appear to be dense cloud cores rotating so slowly as to have J/M initially much closer to that of the solar system. The more rapidly rotating clouds presumably form the more prevalent binary systems. An unbiased survey of cloud rotation rates is needed in order to assess whether or not the angular momentum of interstellar clouds is consistent with theoretical models of their collapse and astronomical constraints such as the apparent binary star frequency.

It has long been recognized that T Tauri stars possess strong winds, with mass-loss rates on the order of 10^{-8} $M_\odot$/yr (Kuhi 1964; DeCampli 1981). In addition, T Tauri stars appear to experience an FU Orionis phase at least once every 10^4 yr, during which the luminosity increases by a factor of 10 to 100, and the mass-loss rate increases by a factor of 100 to 1000 (Herbig 1977; Croswell et al. 1987). What has recently become apparent is that these winds have a bipolar form when they first become observable. Bipolar molecular outflows are seen at large distances ($\sim 10^{18}$ cm) from the star, even when the star itself is still obscured. Highly collimated optical jets are seen close to some stars. Emission from shocked gas (Herbig-Haro objects) with velocity vectors closely aligned with exciting T Tauri stars are also observed (Cohen and Schwartz 1980).

T Tauri stars exhibit excess infrared radiation, which, in most cases, can be unequivocally attributed to emission from heated circumstellar material. Adams and Shu (1986) have shown that typical T Tauri spectra are well represented by models that include radiation from protostellar disks. Other evidence for disks has often been cited (from radio, infrared, polarization and absorption-line studies; see review by Harvey 1985), but it is usually difficult to distinguish between true protostellar disks in the sense of the primitive solar nebula (i.e., rotationally supported) and other axisymmetric circumstellar configurations. In any event, there is ample evidence for some kind of rotational control of near-star structure. It is notable that in most cases where evidence of rotation exists, the alignments of bipolar flows are parallel to the implied rotational axis.

The abbreviated discussion given above hardly does justice to the vast amount of information gained in the last several years about young solar-type stars and the regions in which they are now forming. It is meant to provide the main observational rationale for the following picture of star formation in the Taurus-Auriga cloud and others like it (Shu 1984; Cassen et al. 1985; Shu et al. 1987). The densest regions of the clouds are the sites of star formation; these cores have condensed gradually, perhaps on a time scale of 10 Myr, transferring angular momentum by magnetic coupling to the rest of the cloud as they do. Because magnetic support is lost relative to that enjoyed by the rest of the cloud, the core eventually reaches a degree of concentration that is gravitationally unstable. A protostar begins to form, its accretion luminosity emerging from the dense core as an infrared source. At some stage of the star's growth, a protostellar wind is generated that is strong enough to push its way through the accreting material, sweeping up molecular gas to form the observed extended bipolar outflows. It is not clear whether winds powerful enough to reverse a full-blown accretion flow occur, or whether the accretion rate must diminish before the wind emerges. Nor is it known whether the wind is collimated very close to the star or starts out isotropic and becomes channeled along rotational poles by the surrounding large-scale structure (see Boss 1987b, and references therein). Nevertheless, observations of protostars and fully revealed T Tauri stars strongly suggest that the winds begin while accretion is still taking place and eventually lose their bipolar form as the obscuring material is removed. The fact that T Tauris are observed to have rotational velocities significantly lower than breakup speed (Vogel and Kuhi 1981), suggests that stellar angular momentum has been lost at an early stage, perhaps by magnetic coupling to the stellar wind. It is also likely that disks begin to form in the obscured stage, and that they remain for some time after the star is revealed.

Finally, we point out that there is no obvious observational clue as to why star-forming clouds like Taurus-Auriga began to form stars in such a prolific way several million years ago; that is, there is no apparent triggering mechanism, nor is there clear evidence that star formation is propagating in some way (although some regions of steller influence surely overlap). It has been argued that other star-forming regions, usually where more massive stars are being produced, do exhibit characteristics of triggered star formation (see, e.g., Lada et al. 1978; Klein et al. 1985). But at present both observations and theory suggest that stars in the Taurus-Auriga complex are being formed spontaneously, i.e., by the gradual evolution of a molecular cloud towards regional gravitational instability.

6.2.3. PROTOSTELLAR COLLAPSE

The progress made in developing the theory of protostellar collapse is outlined in this section. More details can be found in reviews by Shu et al. (1987) and Boss (1987a).

Spherically Symmetric Models

The pioneering calculations of protostellar collapse (see, e.g., Bodenheimer 1968; Larson 1969) involved the collapse of nonrotating, nonmagnetic clouds, constrained to spherical symmetry. The mathematical simplicity of a one-dimensional problem has allowed the evolution of protostars to be followed all the way from interstellar clouds to pre-main sequence stars (Winkler and Newman 1980; Stahler et al. 1981). The following description is specialized to solar-type protostars, i.e., protostars with the same mass and chemical composition as the Sun.

Before collapse, protostellar clouds are very cold (~ 10 K) and are so diffuse that they are transparent to infrared radiation. Because of the presence of cool dust grains, which radiate energy primarily in the infrared, contraction initially occurs isothermally. The dust grains absorb the thermal energy produced by compression of the contracting gas cloud and radiate this energy out of the cloud preserving the temperature at ~10 K. Isothermal contraction rapidly becomes dynamic collapse, where gravitational forces overwhelm thermal pressure, leading to rapid infall and supersonic speeds.

Collapse from a gravitationally unstable equilibrium may proceed from the inside out (Shu 1977). Material near the center contracts first, leaving the outer envelope stationary until a pressure wave front, traveling at the sound speed, arrives from the inner collapsing regions. Thus the increasing volume of gas within the wave front moves toward the core produced by the first material to collapse. The initial core is formed when the center of the cloud becomes dense enough to be opaque to infrared radiation, so that the compressional energy no longer freely exits the protostar (nonisothermal regime). The central temperature rises nearly adiabatically, and increased thermal pressure stops the collapse in the inner regions, while the outer envelope continues to infall. The first core gains mass and increases in temperature until $T \approx 2000$ to 3000 K, when the dissociation of molecular H_2 begins. The dissociation process removes energy from the thermal support, prompting a second contraction of the core to even higher densities and temperatures, until all the H_2 is dissociated. Initially small in mass (perhaps 0.01 the mass of the Sun for a solar-mass protostar), the final core grows by the collapse of the cloud envelope through an accretion shock, where the supersonically infalling gas and dust is compressed and nearly brought to rest.

The luminosity of the protostar initially derives mainly from the energy liberated at the accretion shock. When such a protostar is still embedded in a substantial envelope of gas and dust, absorption and reradiation of the energy from the accretion shock by the dust grains in the envelope produce a strong infrared source. As the envelope is accreted or otherwise cleared, the accretion shock evolves into the protostellar photosphere, visible in optical light. The accretion phase lasts for 10^5 to 10^6 yr. There follows a pre-main sequence contraction phase, lasting about 10^6 to 10^7 yr, until the star's central temperature and density become high enough to initiate sustained thermonuclear fusion, which defines the beginning of the long-lived ($\sim 10^{10}$ yr) main sequence phase.

This picture of protostellar evolution is basically consistent with astronomical observations of the luminosities and effective temperatures of embedded infrared sources found in molecular cloud cores (Beichman et al. 1986) and of T Tauri stars, i.e., young, pre-main sequence stars of solar type (Stahler 1983). The fundamentals of protostellar thermodynamics are thus well understood. However, in order to account for the detailed spectra of embedded protostars, the existence of binary stars, and solar-nebula formation, the effects of rotation must be included.

Axisymmetric Models

The simplest way to study rotation in protostellar formation is to assume symmetry about the rotation axis, which requires a 2-dimensional calculation.

Unfortunately, the mathematical complications associated with going from a spherically symmetric to an axisymmetric protostar (much less a fully 3-dimensional protostar) have not been completely overcome, with the result that at present there is only a partial understanding of the evolution of real protostars. Nevertheless, a preliminary picture can be drawn.

Different outcomes appear to be possible for the collapse of rotating clouds (Boss and Haber 1982), depending on their mass and angular momentum distributions, and how far they are from equilibrium when they become unstable. Roughly speaking, high J/M clouds collapse to form rings (Larson 1972; Black and Bodenheimer 1976; Tohline 1980), while low J/M clouds form disks (Norman et al. 1980). If rotational effects become dominant in an axisymmetric calculation, a centrifugal rebound away from the rotation axis produces an outward moving density wave which plows into inwardly moving matter, forming a self-gravitating, off-axis ring. The formation of rings in high J/M clouds is very reassuring from the point of view of binary star formation: a rotating ring is likely to fragment into a binary or multiple system (Larson 1972). However, such clouds appear to have far too much angular momentum to form the solar system.

If rotational effects do not produce a significant rebound during the collapse, the cloud flattens into a disk, because collapse parallel to the rotation axis is unhindered by centrifugal effects. The cores of low J/M ($< 10^{20}$ cm^2/s) clouds collapse into the nonisothermal regime to form pressure-supported, rotating, quasi-equilibrium disks (Tscharnuter 1978; Safronov and Ruzmaikina 1978; Boss 1984a). Terbey et al. (1984) studied the isothermal collapse of a slowly rotating cloud, starting from a strongly centrally condensed configuration ($\rho \propto r^{-2}$), and found that it produced a central protostar surrounded by a flattened disk. Presumably, the Sun formed from such a cloud.

It is clear that the angular momentum distribution of the presolar cloud must have been substantially altered to produce that which is seen in the present solar system, where virtually all of the angular momentum ($\approx 98\%$) resides in a tiny fraction ($\sim 0.1\%$) of the mass, remote from the Sun. This redistribution is unlikely to have occurred during the dynamic collapse of a slowly rotating cloud; rather, it probably awaited the formation of a central, rapidly rotating configuration like the protosolar disk. Angular momentum transport mechanisms are discussed in Sec. 6.2.4.

Three-Dimensional Models

Three-dimensional models of protostellar collapse are able to assess directly whether or not a cloud will fragment into a binary or multiple system or form a single protostar. These calculations have shown that it is very unlikely that the solar system formed from a cloud with high initial J/M. High J/M clouds and clouds initially far from thermal equilibrium fragment into binary or higher order systems (Bodenheimer et al. 1980; Boss 1986). Fragmentation may be rotationally driven in the first instance (e.g., following formation of a ring), or may occur directly from the rapid growth of initial density inhomogeneity in the second instance. Formation of the presolar nebula by formation in a 3-body protostellar system, with subsequent ejection of the presolar nebula following the orbital decay of the 3-body system, does not appear feasible, because the nebula ejected in such a scenario is likely to fragment into subsolar-mass protostars during its collapse and hence not form a single star like the Sun (Boss 1983). Hence the progenitor of the solar system appears to have been a slowly rotating cloud with $J/M < 10^{20}$ cm^2/s (Tscharnuter 1978; Boss 1985). Collapse from strongly centrally condensed dense clouds also appears to lead to the formation of single stars, independent of the initial angular momentum, though the total angular momentum is always relatively small for such clouds because of their reduced moments of inertia. These results must be considered tentative, however, because they are limited to fragmentation during the first dynamic collapse phase and during the formation of the first quasi-equilibrium core. Although further fragmentation or fission is unlikely, the subsequent evolution has yet to be calculated.

6.2.4. PRIMITIVE SOLAR NEBULA

The large-scale properties of the nebular disk that forms from the collapse of a rotating cloud depend on the mechanisms of angular momentum transport that operate within the disk. In the case of the solar system, it seems that some effective process caused the angular momentum of the original cloud to be concentrated in a relatively small fraction of the mass. Virtually all detailed models of the nebula that have been constructed so far suppose that its essential nature was that of a viscous accretion disk, in which angular momentum is transported outwards by turbulent viscosity, thereby allowing mass to flow inwards towards the Sun (Cameron 1978,1985; Lin and Papaloizou 1980; Lin and Bodenheimer 1982; Morfill 1985; Hayashi 1981; Ruden and Lin 1986; Cabot et al. 1987a,b). In most models, turbulent viscosity is attributed to convective instability. Indeed, Lin and Papaloizou (1980) showed that convection is expected if the disk is allowed to cool to space, and if grains (providing the requisite opacity) are present and distributed throughout the disk. The turbulent viscosities that have been calculated in the most recent studies (Ruden and Lin 1986; Cabot et al. 1987b) yield similar dissipation times for the nebula, 10^6 to 10^7 yr, although the two studies differ substantially in their details. In particular, Cabot et al. (1987b) found that the effective strength of the viscous stresses was a factor of 10^2 to 10^4 less than has been commonly assumed, and concluded that convectively driven turbulence was unimportant in the solar nebula, in which case some other means must be sought for generating turbulence or otherwise transporting angular momentum. The resolution of these differences will probably require numerical hydrodynamic calculations that resolve turbulent structure, a formidable but not impossible task.

It should be noted that an important property of viscous accretion disk models is their homogeneity. In these models, material is mixed over length scales comparable to the size of the disk itself, because the mixing process is the very same that spreads the disk; that is, even though the orbital radii of individual fluid elements suffer only small fluctuations on a short time scale, over the longer time scale that characterizes disk evolution, fluid elements may stochastically traverse much of the disk.

However, inhomogeneities in the elemental and isotopic properties of meteorites and spectroscopic properties of asteroids suggest that the solar nebula was not a well-mixed disk. Also, there are theoretical reasons to suspect that a viscous accretion disk model is not appropriate for the formation stage of the nebula. First, there is the complicated problem of grain settling; if grains do not remain suspended in the nebula, but accumulate rapidly at the midplane, or in particles larger than about 100 μm, the nebula opacity would be insufficient to sustain convection (Chapter 6.3). Second, even if small grains were abundant, convection in the nebula would be inhibited during the formation stage by the existence of the accretion shock. Shock heating of the gas from above tends to establish a stable thermal gradient unlike that in a disk which is allowed to cool to space. Finally, even if the nebula is turbulent in spite of the previous two considerations (if, for instance, some other hydrodynamic instability occurs), turbulent transport might not exceed the rate of growth of the nebula due simply to the addition of material with ever increasing angular momentum (Cassen and Moosman 1981; Cassen and Summers 1983). Under any of these circumstances, the initial growth of the nebula is not compensated by a loss of material to the protosun, as occurs in a viscous accretion disk, with the result that the disk can grow to be comparable in mass to the star. The expected result would be large-scale gravitational instability, leading to nonaxisymmetric structure: barred or spiral waveforms, or fragmentation. Gravitational torques between nonaxisymmetric distributions of matter in a differentially rotating solar nebula will result in the outward transport of angular momentum (Fig. 6.2.2). The rate of transport can then be comparable to that assumed in the early models of convectively turbulent accretion disks (Larson 1984; Boss 1984b).

Unlike turbulence, nonaxisymmetric structure need not lead to mixing. Thus, inhomogeneities that existed in the original cloud, or that were produced in the gas phase of the nebula, would be preserved to a much greater extent than in a turbulence-dominated disk.

A rigorous calculation of the 3-dimensional structure of the solar nebula as it forms by accretion has not yet been achieved, so an accurate evaluation of the efficiency of gravitational torques at transporting angular momentum is not possible. Because of this deficiency, the details of the final accumulation of the protosun from the solar nebula remain uncertain. Without a detailed

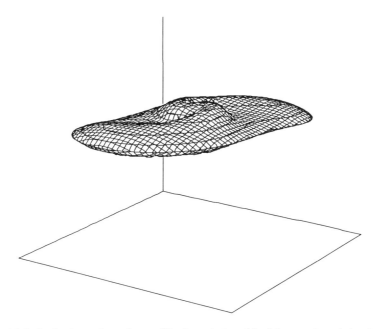

Fig. 6.2.2. Isodensity surface of a possible theoretical model of the presolar nebula, showing rotational flattening, a central condensation which becomes the protosun, and spiral structure capable of transporting angular momentum. The surface shown has a density of 10^{-12} g/cm^3 (figure adapted from Boss 1986).

knowledge of the evolution of angular momentum in the solar nebula, it is hard to make definitive statements of use to meteoritical models. For example, estimates of a relatively cool ($T \approx 1000$ K for Mercury; $T \approx 300$ K for Earth) nebula in the terrestrial planet region by Morfill et al. (1985) are based upon the temperatures experienced by matter with the same specific angular momentum as the terrestrial planets, at a moment in Tscharnuter's (1978) axially symmetric models when the protosun has accumulated about one half its mass. Nebula particles might experience significant changes in temperature because of radial motion inward or outward in the nebula associated with the loss or gain (respectively) of angular momentum caused by gravitational torques. Furthermore, in a nonaxisymmetric solar nebula model, the temperature in restricted regions might be considerably hotter than in an axisymmetric model, such as in the densest regions of large-scale bars or spiral density waves, with variations on the order of 1000 K at ~ 2 AU. Thus a conservative assessment is that current astrophysical estimates for the temperatures experienced by particles in the solar nebula should only be accepted provisionally.

Suppose that one tries to reconstruct the primitive solar nebula by redistributing the masses of the planets, augmented to solar abundances by the addition of the appropriate amounts of H and He, between the present planetary orbits (see, e.g., Weidenschilling 1977). The result is the so-called minimum-mass nebula. It has a mass of ~0.01 to $0.02 M_\odot$, and an angular momentum of about 5×10^{51} g cm^2/s. If this angular momentum was originally distributed smoothly in the protosolar cloud, the entire cloud could have collapsed to a rather compact configuration, < 1 AU in radius, before the collapse was halted by centrifugal force. The nebula would then have had to spread to the present size of the solar system, drawing on the angular momentum of the innermost material. If the spreading was the result of turbulent viscosity, the nebula would be very homogeneous.

On the other hand, a very different kind of evolution results if the mass and angular momentum greatly exceeded that of the minimum-mass nebula. With this in mind, recent developments regarding the population of comets in the solar system become important. It is generally believed that most comets reside in the Oort cloud, orbiting the Sun at distances $> 2 \times 10^4$ AU, and occasionally appearing in the terrestrial planet region when perturbed by passing stars, galactic tidal forces or clouds of interstellar gas (Oort 1950). The present mass of the Oort cloud is estimated to be < 10 $M_\oplus$ (Weissman 1985). This small amount of mass does not appreciably influence estimates of the minimum-mass nebula. However, there are recent suggestions that an inner Oort cloud exists, with comets occupying the region between Neptune and the classical Oort cloud. The evidence for the existence for this cloud, which is presently circumstantial, is summarized by Weissman (1985). Particularly notable are the observations of dust disks with sizes of hundreds of AU found around other stars (Smith and Terrile 1984; Aumann et al. 1984). Harper et al. (1984) and Weissman (1984) have suggested that this dust could come from cometary clouds similar to the proposed inner Oort cloud. Shoemaker and Wolfe (1984) have conducted a study of the dynamical evolution of planetesimals that might have occupied the Uranus-Neptune region of the early solar system. They concluded that ~90% of such planetesimals would have been perturbed out of the solar system, but ~10% could have ended up in an inner Oort cloud. Furthermore, they suggested that the present mass of the cloud could be as high as 200 $M_\oplus$; the original mass of planetesimals would have been thousands of $M_\oplus$, far exceeding the mass of the ice/rock components of the giant planets.

If the existence of such an inner Oort cloud is confirmed, estimates of the minimum mass and angular momentum of the primitive solar nebula must be revised upward by a large amount. The mass might have been on the order of $0.5 M_\odot$, with the angular momentum increased by a factor between 10 and 100 times the earlier estimates. Such a nebula resembles, at least in its gross properties, the models long advocated by Cameron (1973) and Cameron and Pine (1973), who pointed out their susceptibility to gravitational instability. Gravitationally unstable disks can propagate angular momentum via spiral density waves (or other nonaxisymmetric structure), with the consequent evolution then determined by the nature of the waves and their damping. Although the net result must be transfer of angular momentum outward and mass inward, the details could differ substantially from the evolution of a viscous accretion disk. A rigorous calculation of this evolution has yet to be done.

6.2.5. GRAIN GROWTH IN THE INTERSTELLAR MEDIUM AND DURING PROTOSTELLAR COLLAPSE

The elements that form interstellar grains were produced by nucleosynthesis in stellar interiors, then ejected into the interstellar medium by novae, supernovae, and stellar winds (see Chapters 14.1, 14.2 and 13.1). Astronomical observations of the obscuration of starlight imply that interstellar dust grains in diffuse clouds (~ 10 H/cm^3) primarily range in size between 0.001 and 0.1 μm; the average size is about a factor of two larger in dense clouds ($\sim 10^4$ H/cm^3). Elements heavier than H and He make up about 1 to 2% by mass of the interstellar medium. If all of these elements are locked into dust grains of mean radius 0.1 μm, then the ratio of the number density of dust grains to hydrogen atoms is about 10^{-11}. Given the evidence that some of these grains entered the solar nebula and found their way into meteorites (Chapters 13.1 and 13.2), it is useful to review the processes that are important for their evolution; see also Chapter 13.3.

Grain-destruction Mechanisms

Dust grains can be destroyed by several different processes. Collisions between grains and H or He ions can sputter atoms and molecules off the grains if the ion-grain relative velocity v_{ig} is on the order of 100 km/s; icy mantles (CH$_4$, H$_2$O) sputter at $v_{ig} <$ 100 km/s, while refractory material (Si, Ca) requires $v_{ig} >$ 100 km/s (Shull 1977). Collisions between grains can result in the evaporation of both grains when the grain-grain relative velocity $v_{gg} \approx$ 10 km/s, while collisions at $v_{gg} \approx$ 1 km/s can shatter the grains (Scalo 1977). Shattering is most destructive of large grains. Other destructive processes include sublimation, photodesorption by ultraviolet light, and cosmic ray sputtering. Estimated lifetimes for individual grains in quiescent molecular clouds are about 10^6 yr for pure CH$_4$ grains, and about 10^8 yr for pure H$_2$O grains (Draine and Salpeter 1979). These mean lifetimes are reduced by factors of about 10 for grains subjected to supernova shocks, where v_{ig} is large enough for efficient sputtering. Silicate grains in molecular clouds are largely indestructible, but silicate grains in diffuse clouds, subjected to supernova shocks, have lifetimes of about 10^8 yr (Draine and Salpeter 1979). These destruction rates must be compared with grain growth rates, in order to determine the steady-state grain population.

Grain-growth Mechanisms

Two processes are important for grain growth: gas accretion and sticking following collisions (i.e., coagulation). Grain accretion of gas-phase species appears to be highly efficient, because observations of the heavy-element content in the gaseous phase of the interstellar medium indicate abundances lower than those found in the Sun, implying that these elements are largely locked into dust grains. This observation is consistent with estimates of the time scale t_{ac} (e-folding time) for grain accretion of gas. Burke and Silk (1976) and Scalo (1977) found

$$t_{ac}/t_{ff} \approx \left(\frac{\rho}{\rho_s}\right)^{-1/2} \quad (1)$$

for dense interstellar clouds. Here $t_{ff} = [3\pi/(32G\rho)]^{1/2}$ is the free-fall time of an interstellar cloud, G is the gravitational constant, ρ is the cloud density, and $\rho_s = 10^{-20}$ g/cm^3. The free-fall time t_{ff} is the time needed for gravitational collapse to proceed to a point mass, in the absence of any retarding forces. Hence t_{ff} is a lower bound on the lifetime of any cloud in the absence of external disruptive processes. Thus grain accretion occurs on a time scale short compared to the average age of dense interstellar clouds, but is limited by the size of the gaseous phase reservoir.

The second mechanism for grain growth, sticking following grain-grain collisions, requires that collisions occur at small v_{gg} in order to avoid shattering or vaporization. A second requirement is that collisions occur frequently. A number of different mechanisms for producing differential grain motion (and thus collisions) have been investigated. For instance, turbulence can produce relative grain motion. The velocity dispersion of grains in a turbulent gas is determined by gas velocity fluctuations with characteristic times comparable to the slowing time (or coupling time) of the grains (Draine 1985). Coagulation by the resulting collisions appears to be effective in the interstellar medium on time scales of

$$t_{tc}/t_{ff} \approx 10 \left(\frac{\rho}{\rho_s}\right)^{1/4} \left(\frac{a}{a_s}\right)^{1/2}. \quad (2)$$

(Völk et al. 1980) where a is the grain radius and $a_s = 0.1$ μm. Note that t_{tc}/t_{ff} *increases* as the density increases, so that turbulent coagulation becomes less efficient (compared to collapse times) as contraction proceeds.

Coagulation as a result of random Brownian motion of grains can occur in the interstellar medium and during dynamic collapse. Burke and Silk (1976) and Scalo (1977) estimate the time scale for Brownian coagulation to be

$$t_{Bc}/t_{ff} \approx 10^4 \left(\frac{\rho}{\rho_s}\right)^{-1/2} \left(\frac{a}{a_s}\right)^{5/2}. \quad (3)$$

Evidently Brownian coagulation proceeds rather slowly even in dense clouds. However, the strong dependence on grain size a implies that very small grains will rapidly coagulate, ensuring that the mean grain size remains around 0.1μm.

Cloud Lifetimes

Next we compare the previous estimates for grain growth and destruction processes with estimates for the lifetimes of interstellar clouds and for the duration of protostellar collapse. The latter is well represented by the free-fall time used in the previous analysis. Elmegreen (1985) has summarized various types of evidence for the lifetimes of clouds, such as the time needed to assemble the clouds, and the inferred ages of their star-forming regions, and finds that estimated lifetimes vary between $\sim 10^5$ and $\sim 10^8$ yr, with smaller estimates generally applying to smaller-mass clouds. An upper bound for the age of interstellar clouds t_{isc} thus appears to be $\sim 10^8$ yr. In terms of the free-fall time, this upper bound is $t_{isc}/t_{ff} \approx 10^2 \, (\rho/\rho_s)^{1/2}$.

The estimates for grain destruction times imply that weakly bound components of grains (such as CH_4) may well be removed in diffuse clouds, and if the clouds are subjected to supernovae, stronger grain components will be removed as well. Because the estimated destruction times are comparable to the range of estimated cloud lifetimes (except in the case of silicate grains in molecular clouds), the degree of grain destruction in any given cloud will depend on the cloud's detailed history. Massive clouds with massive star formation may experience substantial grain destruction, while small-mass, quiescent clouds may experience little grain attrition. Even if grains are destroyed through evaporation or fragmentation, the gaseous phases relatively quickly reaccrete onto grains, and very small grain fragments quickly coagulate through Brownian motion, so that the grain population may be substantially reworked, but never completely destroyed.

For diffuse interstellar clouds ($\rho \approx 10^{-22}$ g/cm^3), the time scale for Brownian coagulation is about 10^4 times the estimated upper bound on cloud lifetimes, while the time scale for turbulent coagulation is comparable to the cloud lifetime. Thus in diffuse clouds, some coagulation as a result of any turbulent motions may occur, but because $t_{tc} \propto (a/a_s)^{1/2}$, growth slows considerably as the grains enlarge. This occurs because for a fixed cloud density and finite reservoir of grain mass, the grain number density must decrease as the mean size increases, thereby reducing the chances for further collisions and growth. For dense interstellar clouds ($\rho \approx 10^{-20}$ g/cm^3), the time scale for Brownian coagulation is about 100 times the estimated cloud lifetime, but the time scale for turbulent coagulation is only about one tenth the lifetime. Hence some grain growth in turbulent, dense interstellar clouds should occur, provided a long-lived source of turbulence is available. Again, this growth is naturally limited by the dependence on grain size. Hence, even in the most optimistic situation, a long-lived, turbulent, dense cloud, only modest grain growth (perhaps to sizes of 1 or 10 μm) can occur.

Cloud Collapse

Coagulation beyond sizes of 1 to 10 μm seems to require much higher densities than are found in interstellar clouds. While protostellar collapse provides such increased densities, the time available for growth is short; dynamic collapse occurs on the free-fall time scale. Turbulent coagulation may become relatively *less* rapid as the density increases, but Brownian coagulation becomes *more* important. Very little theoretical work has been done on grain growth during protostellar collapse. A numerical calculation of grain growth during protostellar collapse by Morfill et al. (1978) found the mean grain size to increase by a factor of 100 in the innermost regions of an axisymmetric model of protostar collapse.

Gravitational coagulation invokes the gravitational collapse of an interstellar cloud to provide relative grain motion due to the differential acceleration of gas drag (drag force $\propto a^2$, mass $\propto a^3$, $\rightarrow$ acceleration $\propto 1/a$. However, because gas-grain collisions are much more frequent than grain-grain collisions, even in dynamic collapse the grains are closely coupled to the gas. The relative gas-grain velocity is likely to be a small fraction of the sound speed c_s (~0.2 km/s) in dense clouds. These low velocities lead to negligible differential gas drag and hence negligible gravitational coagulation (Scalo 1977). This point can be simply illustrated. Due to their low abundance, dust grains sweeping through a cloud will hit 50 times as much gas mass as dust mass. Because the gas-grain collisions damp the relative gas-grain motion by the time a grain has hit its own mass in gas molecules, grain growth by coagulation occurs about 50 times too slowly to be effective.

It seems that only modest grain growth can occur during protostellar collapse, unless very large grains are present to begin with, which seems unlikely in view of our estimate of grain growth in quiescent interstellar clouds.

6.2.6. ENHANCED DUST-TO-GAS RATIOS

In this section we consider the possibility of achieving local enhancements of the dust-to-gas ratio prior to solar-nebula formation. While inhomogeneity is certain to be present at some level (i.e., by factors of order unity), the question of importance for meteorites is whether *very* large enhancements can be achieved. For example, Wood (1984) has suggested dust-to-gas enhancement factors of 1000 may be necessary to account for the oxygen fugacity inferred to be present during the formation of certain meteoritic inclusions. A possible means for achieving enhanced dust-to-gas ratios is gravitational sedimentation of dust grains. Sedimentation generally is enhanced for larger grains, and hence some sedimentation may be inferred from the observations showing that mean grain sizes increase slightly in dense clouds, though grain accretion of the gas phase and grain coagulation could also be involved.

Sedimentation is opposed by gas drag if the gas component is stable. Flannery and Krook (1978) and Harrison (1978) found that in typical interstellar clouds the dust grains could settle into a compact layer, with the dust-to-gas ratio increasing exponentially with a time constant of order 10^8 yr. In diffuse clouds which are transparent to ultraviolet light, radiative acceleration can reduce this time scale by a factor of ten (Flannery and Krook 1978). Achieving an increase of a factor of 1000 thus requires at least 10^9 yr for dense clouds, or at least 10^8 yr for ultraviolet-transparent clouds. Neglected physical effects, such as rotation, turbulence, and cloud-cloud collisions imply that these estimates are lower bounds. Harrison (1978) pointed out that supersonic turbulence (e.g., produced by strong stellar winds) in particular can increase the time scale for sedimentation by large factors. As the estimated upper bound on cloud lifetimes is 10^8 yr, and enhancements in diffuse clouds are most likely to be disrupted by turbulence, it appears that very large enhancements (factors of 1000) of the dust-to-gas ratio do not occur in the interstellar medium. More modest enhancements, perhaps by factors of 10, may be reasonable.

Napier and Humphries (1986) have suggested that radiative forces on dust grains in quiescent, dense clouds could result in the accumulation of dust assemblies of considerable (cometary) size. They derive a criterion for the radiative instability which is independent of gas drag, because initially there is no relative motion between gas and grains and hence no drag. However, gas drag must become important in the nonlinear phases of growth, and because gas and grains are expected to become tightly coupled, the radiative instability will be resisted by the stability of the gas.

There remains the possibility of achieving enhanced dust-to-gas ratios during protostellar collapse. Unfortunately, this problem has not yet been addressed by rigorous calculation. In essence, sedimentation of dust grains is a potentially runaway process, because while the infall of the gas is hindered by pressure gradients, the dust grains are impeded only by collisions with the gas. If sufficient time is available, it would seem that the dust-to-gas ratio could become very large; because the time available is in fact limited and quite short, the actual degree of enhancement during protostellar collapse is uncertain. However, application of Harrison's (1978) analysis implies that a significant enhancement of dust relative to gas can only be attained on a time scale long compared to the time available during protostellar collapse. Hence it seems unlikely that substantial dust-to-gas enhancements can be produced in interstellar clouds or during protostellar collapse.

6.2.7. EFFECTS OF THE PROTOSUN

Radiation and Shocks

In order to enter the solar nebula, an interstellar grain must survive two ordeals: the radiation field of the accreting star and disk, and passage through the accretion shock itself. The luminosity of an accreting (low-mass) protostar is approximately the rate of release of the gravitational energy of collapse:

$$L = \left(\frac{dM}{dt}\right)\left(\frac{GM}{R}\right) \qquad (4)$$

where R is the radius to which gas falls before encountering the accretion shock. A solar-mass star accreting at a rate of $10^{-5} M_\odot$/yr radiates on the order of 10 solar luminosities, at an effective temperature of several thousand degrees. The radiation heats gas and dust close to the protostar, with the result that infalling grains are destroyed within a certain radius. Both sublimation and chemical erosion contribute to the destruction. Stahler et al. (1980,1981) calculate that the radius of the grain-destruction front evolves from several times 10^{12} cm (~ 0.1 AU) to about 10^{13} cm (~ 1 AU) during the accumulation of $1 M_\odot$ at $10^{-5} M_\odot$/yr. They considered graphite grains to be the most resistant to destruction; silicates are destroyed at somewhat larger distances.

Figure 6.2.3 represents the dust-grain temperature as a function of distance for a specific case ($M = 1 M_\odot$, $dM/dt = 10^{-5} M_\odot$/yr, no rotation) calculated by Adams and Shu (1985). Grains are destroyed within 10^{13} cm and heated to 100 K or more within the entire planetary region. All grains that were likely to enter the solar system would be heated above the background interstellar temperature of 10 K. The effects of rotation and a slower accretion rate are to reduce this heating and allow grain survival to somewhat closer distances, but rigorous radiative transfer calculations have yet to be done (see Adams and Shu 1986).

If a grain survives the protostellar luminosity field, it must also avoid destruction in the accretion shock if it is to enter the solar nebula intact. Grain-grain collisions, chemical reactions, sputtering and evaporation can destroy grains in shocks. So far, detailed analyses of these processes have only been carried out for the low densities that characterize the interstellar medium (see, e.g., Hollenbach and McKee 1979). The application of these results to a nebular accretion shock suggests that grains that survive the luminosity field will not evaporate in the shock. However, certain destruction processes that are not necessarily effective at interstellar densities (such as chemical reactions) may become very important at the high densities of the accretion shock. Also, an extensive range of shock conditions over time and radial distance must be considered. Therefore detailed evolutionary models are needed to determine the full extent of grain processing by the accretion shock.

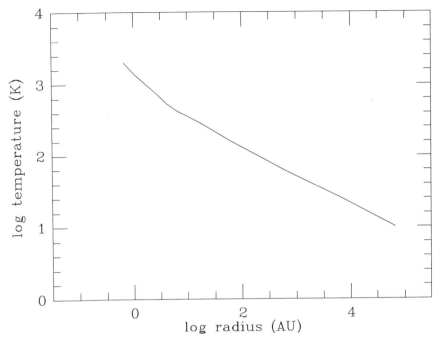

Fig. 6.2.3. Calculated temperature profile of the gas being accreted onto a $1 M_\odot$ protostar at a rate of $10^{-5} M_\odot$/yr. The heating is solely due to release of gravitational energy of the infalling matter. The infalling gas was assumed not to be rotating, so no disk is present. Within about 1 AU all dust grains are vaporized (figure adapted from Adams and Shu 1985).

Stellar Winds

A T Tauri phase (Sec. 6.2.2; see also Chapter 4.1) has long been invoked to blow away residual nebular gases and dust at an appropriate phase of planetary formation (Cameron 1973). Protostellar winds may process and return dust grains to the interstellar medium (see, e.g., Burke and Silk 1976). Unfortunately, little is known about the effects of such a wind on the solar nebula. Horedt's (1978) model implies that a $0.1 M_\odot$ nebula could be blown away within about 10^7 yr. However, Elmegreen (1978) suggests that the nebula is not blown away, but is dissipated into the Sun by the interaction of the wind on the faces of the disk. Given the uncertainties in the removal mechanism, the structure of the nebula, and the timing of the T Tauri flow, the problem of nebula removal cannot be considered to be solved.

Recent evidence for the prevalence of FU Orionis and bipolar stellar wind phases prior to the T Tauri phase suggests that these earlier phases could be even more energetic than the T Tauri phase (Lada 1985). While it is currently uncertain how the bipolar flows are generated, an energetic, isotropic stellar wind channeled by a rotationally or magnetically flattened cloud appears to be consistent with the observations. An isotropic wind strong enough to power the bipolar flows would be expected to erode the solar nebula at a rate even faster than the classical T Tauri wind. Clearly planetary accretion in the solar nebula must have begun prior to the generation of these stellar winds in the protosun.

6.2.8. SUMMARY AND FUTURE DIRECTIONS

Observations of low-mass protostars, though presently limited in spatial resolution, have provided the basis for understanding the formation of stars like the Sun. We can expect future observations to provide an even more detailed picture. Rigorous radiative transfer calculations will be necessary in order to use these observations for the precise determination of dust-destruction front locations, accretion-shock properties, mass-accretion rates, and other protostellar characteristics relevant to nebular formation and the thermal processing of solid material.

The theory of protostellar collapse has given us a good understanding of the physics of nebula formation, but one of the most outstanding problems of nebula evolution, the relative roles of gravitational and turbulent torques, has yet to be solved. The possible existence of a massive inner Oort cloud bears heavily on this question because it could force a major revision of estimates of the total mass and angular momentum of the presolar nebula. Future models should also incorporate boundary conditions provided by the theory of protostellar collapse, such as the radiation field of the protosun.

Evidence for the survival of interstellar grains in meteorites (see Chapters 13.1, 13.2 and 13.3) demands that a good theoretical understanding of the evolution of the interstellar grain population during star formation be developed. The application of orthodox grain-growth theory implies that the interstellar distribution of grain sizes is not modified appreciably, either during the interstellar cloud phase or during protostellar collapse. The same conclusion holds for the cosmic dust-to-gas ratio. Although grains should enter the nebula relatively unscathed in the vast regions of the outer nebula, those that enter in the terrestrial planet region can be significantly altered due to the stellar radiation field and/or the accretion shock. At present it is clearly difficult to relate directly the detailed physics and chemistry of meteorites to global models of the solar nebula, and exploring this relationship further constitutes a major scientific challenge.

Acknowledgments. We thank G. H. Herbig for discussions of star-forming regions, and W. K. Hartmann, D. J. Hollenbach, and M. Wolfire for their reviews of the manuscript. This work was partially supported under the auspices of a special NASA Astrophysical Theory Program, which supports a joint Center for Star Formation Studies at NASA-Ames Research Center,

U. C. Berkeley, and U. C. Santa Cruz. APB was partially supported by a National Science Foundation grant and a grant from the National Aeronautics and Space Administration.

REFERENCES

Abt, H. A. 1983. Normal and abnormal binary frequencies. *Ann Rev. Astron. Astrophys.* 21:343–372.
Adams, F. C., and Shu, F. H. 1985. Infrared emission from protostars. *Astrophys. J.* 296:655–669.
Adams, F. C., and Shu, F. H. 1986. Infrared spectra of rotating protostars. *Astrophys. J.* 308:836–853.
Aumann, H. H., Gillett, F. C., Beichman, C. A., de Jong, T., Houck, J. R., Low, F., Neugebauer, G., Walker, R. G., and Wasselius, P. R. 1984. Discovery of a shell around Alpha Lyrae. *Astrophys. J.* 278:L23–L27.
Beichman, C. A., Myers, P. C., Emerson, J. P., Harris, S., Mathieu, R., Benson, P. J., and Jennings, R. E. 1986. Candidate solar-type protostars in nearby molecular cloud cores. *Astrophys. J.* 307:337–349.
Black, D. C., and Bodenheimer, P. 1976. Evolution of rotating protostellar clouds. II. The collapse of protostars of 1, 2, and 5 $M_\odot$. *Astrophys. J.* 206:138–149.
Bodenheimer, P. 1968. The evolution of protostars of 1 and 12 solar masses. *Astrophys. J.* 153:483–494.
Bodenheimer, P., Tohline, J. E., and Black, D. C. 1980. Criteria for fragmentation in a collapsing rotating cloud. *Astrophys. J.* 242:209–218.
Boss, A. P. 1983. Fragmentation of a nonisothermal protostellar cloud. *Icarus* 55:181–184.
Boss, A. P. 1984a. Protostellar formation in rotating interstellar clouds. IV. Nonisothermal collapse. *Astrophys. J.* 277:768–782.
Boss, A. P. 1984b. Angular momentum transfer by gravitational torques and the evolution of binary protostars. *Mon. Not. Roy. Astron. Soc.* 209:543–567.
Boss, A. P. 1985. Three dimensional calculations of the formation of the presolar nebula from a slowly rotating cloud. *Icarus* 61:3–9.
Boss, A. P. 1986. Protostellar formation in rotating interstellar clouds. V. Nonisothermal collapse and fragmentation. *Astrophys. J. Suppl.* 62:519–552.
Boss, A. P. 1987a. Theory of collapse and protostar formation. In *Summer School on Interstellar Processes*, eds. D. Hollenbach and H. Thronson (Dordrecht: D. Reidel), pp. 321–348.
Boss, A. P. 1987b. Bipolar flows, molecular gas disks, and the collapse and accretion of rotating interstellar clouds. *Astrophys. J.*, 316:721–732.
Boss, A. P., and Haber, J. G. 1982. Axisymmetric collapse of rotating, interstellar clouds. *Astrophys. J.* 255:240–244.
Burke, J. R., and Silk, J. 1976. The dynamical interaction of a newly formed protostar with infalling matter: The origin of interstellar grains. *Astrophys. J.* 210:341–364.
Cabot, W., Canuto, V. M., Hubickyj, O., and Pollack, J. B. 1987a. The role of turbulent convection in the primitive solar nebula. I. Theory. *Icarus* 69:387–422.
Cabot, W., Canuto, V. M., Hubickyj, O., and Pollack, J. B. 1987b. The role of turbulent convection in the primitive solar nebula. II. Results. *Icarus* 69:423–457.
Cameron, A. G. W. 1973. Accumulation processes in the primitive solar nebula. *Icarus* 18:407–450.
Cameron, A. G. W. 1978. Physics of the primitive solar accretion disk. *Moon and Planets* 18:5–40.
Cameron, A. G. W. 1985. Formation and evolution of the primitive solar nebula. In *Protostars & Planets II*, eds. D. C. Black and M. S. Matthews (Tucson: Univ. of Arizona Press), pp. 1073–1099.
Cameron, A. G. W., and Pine, M. R. 1973. Numerical models of the primitive solar nebula. *Icarus* 18:377–406.
Cassen, P. M., and Moosman, A. 1981. On the formation of protostellar disks. *Icarus* 48:353–376.

Cassen, P. M., and Summers, A. 1983. Models of the formation of the solar nebula. *Icarus* 53:26–40.
Cassen, P., Shu, F. H., and Terebey, S. 1985. Protostellar disks and star formation: An overview. In *Protostars & Planets II*, eds. D. C. Black and M. S. Matthews (Tucson: Univ. of Arizona Press), pp. 448–483.
Croswell, K., Hartmann, L., and Avrett, E. H. 1987. Mass loss from FU Orionis objects. *Astrophys. J.* 312:227–242.
Cohen, M., and Kuhi, L. V. 1979. Observational studies of pre-main-sequence evolution. *Astrophys. J. Suppl.* 41:743–843.
Cohen, M., and Schwartz, R. D. 1980. A search for the exciting stars of Herbig-Haro objects. *Mon. Not. Roy. Astron. Soc.* 191:165–168.
DeCampli, W. M. 1981. T Tauri winds. *Astrophys. J.* 244:124–146.
Draine, B. T. 1985. Grain evolution in dark clouds. In *Protostars & Planets II*, eds. D. C. Black and M. S. Matthews (Tucson: Univ. of Arizona Press), pp. 621–640.
Draine, B. T., and Salpeter, E. E. 1979. Destruction mechanisms for interstellar dust. *Astrophys. J.* 231:438–455.
Elmegreen, B. G. 1978. On the interaction between a strong stellar wind and a surrounding disk nebula. *Moon and Planets* 19:261–277.
Elmegreen, B. G. 1985. Molecular clouds and star formation: An overview. In *Protostars & Planets II*, eds. D. C. Black and M. S. Matthews (Tucson: Univ. of Arizona Press), pp. 33–58.
Flannery, B. P., and Krook, M. 1978. The sedimentation of grains in interstellar clouds. *Astrophys. J.* 223:447–457.
Goldsmith, P. F., and Arquilla, R. 1985. Rotation in dark clouds. In *Protostars & Planets II*, eds. D. C. Black and M. S. Matthews (Tucson: Univ. of Arizona Press), pp. 137–149.
Harper, D. A., Lowenstein, R. F., and Davidson, J. A. 1984. On the nature of the material surrounding Vega. *Astrophys. J.* 285:808–812.
Harrison, E. R. 1978. Diffusion and sedimentation of dust in molecular clouds. *Astrophys. J.* 226:L95–L98.
Harvey, P. M. 1985. Observational evidence for disks around young stars. In *Protostars & Planets II*, eds. D. C. Black and M. S. Matthews (Tucson: Univ. of Arizona Press), pp. 484–492.
Hayashi, C. 1981. Structure of the solar nebula, growth and decay of magnetic fields and effects of magnetic and turbulent viscosities on the nebula. *Prog. Theor. Phys. Suppl.* 70:35–53.
Herbig, G. H. 1977. Eruptive phenomena in early stellar evolution. *Astrophys. J.* 217:693–715.
Hollenbach, D., and McKee, C. F. 1979. Molecule formation and infrared emission in fast interstellar shocks. I. Physical processes. *Astrophys. J. Suppl.* 41:555–592.
Horedt, G. P. 1978. Blow-off of the protoplanetary cloud by a T Tauri like solar wind. *Astron. Astrophys.* 64:173–178.
Jones, B. F., and Herbig, G. H. 1979. Proper motions of T Tauri variables and other stars associated with the Taurus-Auriga dark clouds. *Astrophys. J.* 84:1872–1889.
Klein, R. I., Whitaker, R. W., and Sandford, M. T., II. 1985. Processes and problems in secondary star formation. In *Protostars & Planets II*, eds. D. C. Black and M. S. Matthews (Tucson: Univ. of Arizona Press), pp. 340–367.
Kuhi, L. V. 1964. Mass loss from T Tauri stars. *Astrophys. J.* 140:1409–1433.
Lada, C. J. 1985. Cold outflows, energetic winds, and enigmatic jets around young stellar objects. *Ann. Rev. Astron. Astrophys.* 23:267–317.
Lada, C. J., Blitz, L., and Elmegreen, B. G. 1978. Star formation in OB associations. In *Protostars and Planets*, ed. T. Gehrels (Tucson: Univ. of Arizona Press), pp. 341–367.
Larson, R. B. 1969. Numerical calculations of the dynamics of a collapsing proto-star. *Mon. Not. Roy. Astron. Soc.* 145:271–295.
Larson, R. B. 1972. The collapse of a rotating cloud. *Mon. Not. Roy. Astron. Soc.* 156:437–458.
Larson, R. B. 1984. Gravitational torques and star formation. *Mon. Not. Roy. Astron. Soc.* 206:197–207.
Lin, D. N. C., and Bodenheimer, P. 1982. On the evolution of convective accretion disk models of the primordial solar nebula. *Astrophys. J.* 262:768–779.

Lin, D. N. C., and Papaloizou, J. 1980. On the structure and evolution of the primordial solar nebula. *Mon. Not. Roy. Astron. Soc.* 191:37–48.
Morbey, C. L., and Griffin, R. F. 1987. On the reality of certain spectroscopic orbits. *Astrophys. J.* 317:343–352.
Morfill, G. E. 1985. Physics and chemistry in the primitive solar nebula. In *The Birth and Infancy of Stars,* eds. R. Lucas, A. Omont, and L. R. Stora (Amsterdam: North Holland), pp. 693–792.
Morfill, G. E., Tscharnuter, W., and Völk, H. 1978. The dynamics of dust in a collapsing protostellar cloud and its possible role in planet formation. *Moon and Planets* 19:211–220.
Morfill, G. E., Tscharnuter, W., and Völk, H. J. 1985. Dynamical and chemical evolution of the protoplanetary nebula. In *Protostars & Planets II,* eds. D. C. Black and M. S. Matthews (Tucson: Univ. of Arizona Press), pp. 493–533.
Myers, P. C. 1985. Molecular cloud cores. In *Protostars & Planets II,* eds. D. C. Black and M. S. Matthews (Tucson: Univ. of Arizona Press), pp. 81–103.
Myers, P. C. 1987. Dense cores and young stars in dark clouds. In *Star Forming Regions, IAU Symp. 115,* eds. M. Peimbert and J. Jugaku (Dordrecht: D. Reidel), pp. 33–43.
Myers, P. C., and Benson, P. J. 1983. Dense cores in dark clouds. II. NH_3 observations and star formation. *Astrophys. J.* 266:309–320.
Napier, W. M., and Humphries, C. M. 1986. Interstellar planetesimals—II. Radiative instability in dense molecular clouds. *Mon. Not. Roy. Astron. Soc.* 221:105–117.
Norman, M. L., Wilson, J. R., Barton, R. T. 1980. A new calculation on rotating protostar collapse. *Astrophys. J.* 239:968–981.
Oort, J. H. 1950. The structure of the cometary cloud surrounding the solar system and a hypothesis concerning its origin. *Bull. Astron. Inst. Netherlands* 11:91–110.
Ruden, S., and Lin, D. N. C. 1986. The global evolution of the primordial solar nebula. *Astrophys. J.* 308:883–901.
Safronov, V. S., and Ruzmaikina, T. V. 1978. On the angular momentum transfer and the accumulation of solid bodies in the solar nebula. In *Protostars & Planets,* ed. T. Gehrels (Tucson: Univ. of Arizona Press), pp. 545–564.
Scalo, J. M. 1977. Grain size control in dense interstellar clouds. *Astron. Astrophys.* 55:253–260.
Shoemaker, E. M., and Wolfe, R. F. 1984. Evolution of the Uranus-Neptune planetesimal swarm. *Lunar Planet. Sci.* XV:780–787 (abstract).
Shu, F. H. 1977. Self-similar collapse of isothermal spheres and star formation. *Astrophys. J.* 214:488–497.
Shu, F. H. 1984. Star formation in molecular clouds. In *The Milky Way Galaxy,* IAU Symp. No. 106, eds. H. van Woerden, R. J. Allen, and W. B. Burton (Dordrecht: D. Reidel), pp. 561–566.
Shu, F. H., Adams, F. C., and Lizzano, S. 1987. Star formation in molecular clouds: Observation and theory. *Ann. Rev. Astron. Astrophys.,* in press.
Shull, J. M. 1977. Grain disruption in interstellar hydromagnetic shocks. *Astrophys. J.* 215:805–811.
Smith, B. A., and Terrile, R. J. 1984. A circumstellar disk around β Pictoris. *Science* 226:1421–1424.
Stahler, S. W. 1983. The birthline for low-mass stars. *Astrophys. J.* 274:822–829.
Stahler, S. W., Shu, F. H., and Taam, R. E. 1980. The evolution of protostars I. Global formulation and results. *Astrophys. J.* 241:637–654.
Stahler, S. W., Shu, F. H., and Taam, R. E. 1981. The evolution of protostars III. The accretion envelope. *Astrophys. J.* 298:727–737.
Terebey, S., Shu, F. H., and Cassen, P. 1984. The collapse of the cores of slowly rotating isothermal clouds. *Astrophys. J.* 286:529–551.
Tohline, J. E. 1980. Ring formation in rotating protostellar clouds. *Astrophys. J.* 236:160–171.
Tscharnuter, W. M. 1978. Collapse of the presolar nebula. *Moon and Planets* 19:229–236.
Vogel, S. N., and Kuhi, L. V. 1981. Rotational velocities of pre-main-sequence stars. *Astrophys. J.* 245:960–976.
Völk, H. J., Jones, F. C., Morfill, G. E., and Söser, S. 1980. Collisions between grains in a turbulent gas. *Astron. Astrophys.* 85:316–325.

Weidenschilling, S. J. 1977. The distribution of mass in the planetary system and solar nebula. *Astrophys. Space Sci.* 51:153–158.
Weissman, P. R. 1984. The Vega particulate shell: Comets or asteroids? *Science* 224:987–989.
Weissman, P. R. 1985. Cometary dynamics. *Space Sci. Rev.* 41:299–349.
Winkler, K.-H., and Newman, M. J. 1980. Formation of solar-type stars in spherical symmetry. II. Effects of detailed constitutive relations. *Astrophys. J.* 238:311–325.
Wood, J. A. 1984. On the formation of meteoritic chondrules by aerodynamic drag heating in the solar nebula. *Earth Planet. Sci. Lett.* 70:11–26.

6.3. A REVIEW OF SOLAR NEBULA MODELS

JOHN A. WOOD
Harvard-Smithsonian Center for Astrophysics

and

GREGOR E. MORFILL
Max-Planck-Institut für Physik und Astrophysik

Since 1962, when A. G. W. Cameron introduced his model of the protosolar nebula to meteoriticists as the environment in which chondrite-forming processes occurred, astrophysicists have constructed increasingly sophisticated disk models. By now, protosolar nebula models (including Cameron's) are substantially different from the original Cameron (1962) nebula. This chapter briefly discusses the major changes in thinking that have occurred, and elaborates on the concept favored by astrophysicists for the last decade, that of a viscous accretion-disk nebula. The properties of recent accretion-disk models that are most relevant to chondrite-forming processes are noted.

6.3.1. INTRODUCTION

Most of the properties of primitive chondritic meteorites were established in the solar nebula: one of the main reasons for studying meteorites is to learn about the nebula. This endeavor brings together two communities of scientists, meteoriticists and astrophysicists, that speak different languages. Because of the difficulty of communication many meteoriticists have only an imperfect understanding of even the qualitative properties of the nebula models currently being developed by astrophysicists. This limits their ability to interpret the observations they make in meteorites. The present chapter reviews astrophysical nebula models in basic terms, and will discuss those aspects of the models that are particularly important to meteorite studies.

We begin at the familiar starting point, our understanding that stars and protostellar nebulae form from cold gas (~ 10 K, mostly H_2 and He) and dust (~ 0.1 μm grains of silicates mantled by ices of HCNO compounds) in interstellar clouds. The self-gravitational collapse of a particular volume of gas and dust gave rise to the Sun and solar nebula. In its initially extended state this material possessed a large amount of gravitational potential energy; as it collapsed most of this was converted into kinetic energy and then into heat. Also, in its initially extended state, the protosolar material would have had some net angular momentum because, among other reasons, it was part of a galaxy that rotates. If only a fraction of this angular momentum was retained as the interstellar material collapsed to solar system proportions, it guarantees that the condensed structure that was formed would have been rotating and would have had a more or less spun-out disk-like form.

The first semi-quantitative attempt to understand the origin and physical nature of this protosolar disk or nebula in modern astrophysical terms was made by Cameron (1962). His model, which strongly influenced the thinking of workers in cosmochemistry when this field was born in the 1960s, is reviewed in the following section.

6.3.2. ABRUPTLY COMPRESSED NEBULA MODELS

Cameron's 1962 Model

In this model (Cameron 1962), collapsing interstellar material successively fragmented into smaller and denser volumes until the latter became dense and opaque enough to retain the heat being generated in them by compression of the gas. When internal temperatures reached ~ 1000 K, fragmentation ceased; free-fall collapse also ceased, as internal pressures in the warming fragments became great enough to counterbalance gravitational forces. Quasi-stable structures in hydrostatic equilibrium were formed, which could continue contracting only as fast as they could lose heat by radiation from their surfaces. This stage was reached $\sim 10^6$ yr after the onset of collapse. The solar system was formed from one such structure (protostar), of mass 2 to 4 $M_\odot$ ($M_\odot$ = one solar mass or 1.992×10^{33} g).

The protosun continued to radiate energy and contract until, after a relatively short time, its dimension approached that of the present planetary system. By this time a critical fraction of its mass had reached a temperature (~ 1800 K) at which H_2 dissociates to a significant degree. Beyond this point, contraction no longer had to wait on the loss of energy by radiation, because an instant energy sink was at hand in the form of H_2 dissociation. The hydrostatic balance between gravity and pressure was lost; the protosun resumed free-fall collapse. (Approximate calculations based on the properties of gaseous polytropes led to conclusions like these; no computer calculations contributed to the 1962 model.)

As the protosun collapsed it rotated faster, because of the conservation of its angular momentum. Twisted magnetic field lines damped differential rotation in the protosun and made it rotate like a rigid sphere. As the protosun collapsed to less than the radius of the planetary system it became rotationally unstable at its equator, i.e., its equator was rotating at the Keplerian velocity; equatorial material was left behind in orbital motion while the rest of the protosun continued to collapse. Material shed in this fashion created an orbiting disk (nebula) as the protosun collapsed to roughly solar dimensions. (Collapse continued after the H_2 was completely dissociated: H and He ionization then came into play as energy sinks.)

Thus the nebula was formed in a very short time, the free-fall time from the periphery of the planetary system to the Sun (~ 10 yr). Most of the 2 to 4 $M_\odot$ of collapsing material was in this nebula at first; less than 1 $M_\odot$ collapsed directly to the vicinity of the Sun. As noted, the nebular material in this model was hot, >1800 K. Again, twisted magnetic field lines damped differential motions in the nebula, making it rotate as a rigid disk. This led to a redistribution of angular momentum and mass in the disk; gas at small radii was decelerated to less than the Keplerian velocity, so it migrated inward toward the Sun; gas at large radii was accelerated and spun outward.

The initial surface density of nebular material Σ within 3 AU from the rotation axis was $>\sim 10^5$ g/cm^2. The midplane pressure P was $\sim 1.3 \times 10^{-13} \Sigma^2 \sim 10^{-3}$ atm. (This is the source of the canonical 10^{-3} atm nebular pressure that is still often used by meteoriticists when they try to estimate conditions where the chondrites formed.) Cameron did not attempt to specify the thickness of the nebula, or the variation of pressure, temperature or density with radial distance in the nebula or height from the midplane.

The nebula cooled by radiation as $T = 24.5(\Sigma/t)^{1/3}$ (temperature T is in Kelvin; time t is in years), reaching a temperature of 1000 K in ~ 10 yr and 100 K in $\sim 10^4$ yr. Observational evidence indicated that new stars enter their T Tauri phase of evolution less than 4×10^6 yr after they are formed, so this placed an upper limit on the lifetime of the nebula.

This model had important consequences for cosmochemistry; such a hot nebula would have completely vaporized the dust component of the interstellar material that went into it. Subsequently, as the disk cooled, the metal vapors would have recondensed, possibly into minerals that could still be recognized in the chondritic meteorites. Lord (1965) and Grossman (1972) used the principles of equilibrium thermodynamics to predict what the minerals would be and in what sequence they would appear during condensation. When the Allende chondrite was found to contain Ca,Al-rich inclusions (CAIs) made up of exactly the minerals predicted at the most refractory end of the condensation sequence, these were taken to be initial condensates from the cooling nebula (see, e.g., Marvin et al. 1970). This seemed a brilliant vindication of the abruptly compressed nebula model as well as the thermodynamic calculations.

The Cameron and Pine 1973 Model

This work extended and quantified the concepts of Cameron (1962). The assumed prenebular setting was the same: a rigidly rotating 2 $M_\odot$ sphere of interstellar gas and dust that had contracted to solar-system dimensions. This was taken to have flattened into a disk in a very short time (less than the thermal and angular momentum transport time scales). A numerical (computer) analysis was made of the properties of the initially formed disk, taking into account (1) radial and vertical force balances in the gas; (2) conservation of the angular momentum each element of gas had in the prenebular contracting sphere, after it joined the nebula; and (3) radiative and convective heat transport in the disk.

It was found that the nebula had a cross section shaped like a squashed X: it was thin (~ 0.2 AU) near the rotation axis, thicker (~ 2 AU) near its periphery. At the nebula midplane the initial temperature of the gas was ~ 2500 K near the rotation axis; ~ 1000 K at 3 AU; and ~ 500 K at 8 AU. (8 AU is the radius most pertinent to meteoritic observations, because material there had the same specific angular momentum that present-day asteroids have. Material would have moved from there inward to 3 AU, the present asteroid belt, after the 2 $M_\odot$ nebula somehow lost half its mass.) The midplane gas pressure also decreased with radial distance: at 3 AU it was $\sim 10^{-4}$ atm (a value some meteoriticists have adopted) and $\sim 10^{-5}$ atm at 8 AU.

This disk was not stiffened by magnetic field lines. Its material was in differential rotation, everywhere revolving at approximately the Keplerian angular velocity. Cameron and Pine pointed out that such a dynamic system is not stable, but tends to evolve by redistributing its mass and angular momentum either by the agency of turbulent gas viscosity (discussed below) or via patterns of large-scale gas circulation. They estimated the time scale on which angular momentum distribution would change the character of the nebula importantly (~ 300 yr at 6 AU), but did not attempt to follow the evolution of the nebula from its initial state. They also estimated the time scale for substantial cooling of the nebula; this ranged from 1 to 30 yr at 1.5 AU, depending upon the opacity of the dust grains in the nebula and other model details.

6.3.3. VISCOUS ACCRETION DISKS

The above models make the assumption that interstellar material fell together to form the protosolar nebula very rapidly, effectively instantaneously. A crucial paper published by Larson (1969), in the year Allende fell, showed this is not what would have happened, and started solar nebula modeling down a new avenue. Larson numerically followed the collapse of a nonrotating gas sphere of uniform density and showed that it does not occur homologously (i.e., with uniform but increasing density). Instead, at first the center of the sphere drops out to form a small mass concentration; thereafter, material from farther and farther out in the original volume rains down on it (the collapse is nonhomologous). For a nominal 1 $M_\odot$ collapse model, material continues to be added to the central mass for $\sim 2 \times 10^5$ yr.

The same would be true when a rotating volume collapsed, except (as noted earlier) the central mass that formed would have a spun-out, disk-like form. Cameron (1976) observed that the time scale for disk growth ($\sim 10^5$ yr) is much longer than the time scale for disk evolution via mass and angular momentum transfer ($< 10^3$ yr in the inner solar system), so it is unrealistic to model a nebula by bringing it together abruptly and then letting it relax and evolve. Instead, the addition of new material and disk evolution should be treated as coupled processes from the beginning. The solar nebula was an example of the astrophysical concept of a *viscous accretion disk*. All modeling studies of the nebula carried out since 1973 have been in the framework of the viscous accretion disk concept.

The word "accretion" in "viscous accretion disk" refers to the continuing accretion of infalling material that has too much angular momentum to fall directly onto the prestellar central object, into a disk that forms around it in a plane perpendicular to the rotation axis of the system. The infalling material is decelerated abruptly, and compressed and heated, when it crosses accretion shock fronts that bound the top and bottom surfaces of the disk (Fig. 6.3.1).

"Viscous" is part of the term because the effective viscosity of the substance of the disk is crucial to its evolution. As material joins the disk, the latter evolves in a way that tends to offset its gains. Since adjacent annular elements of gas rotate at different Keplerian angular velocities, there is shear in the gas, and viscous friction tends to heat the gas and slow the rotation of the faster (inner) element while it speeds the rotation of the slower (outer) element. This brings about a continuing redistribution of mass and angular momentum in the disk; gas that is slowed to less than the orbital angular velocity at its radial distance from the mass center must move to a smaller radius; conversely, gas whose rotation is accelerated moves to a greater radius. Near the center of the system this process continually feeds disk gas into the infant star (which grows, becomes hotter and eventually initiates nuclear fusion). Thus the accretion disk can be thought of as a gateway, or as a holding pattern where infallen interstellar material resides for a limited time while most of its angular momentum is taken from it, after which it is fed into the central star. The angular momentum removed is transferred outward to the peripheral disk material, which is spun out to greater and greater radii.

The rate of disk evolution (radial transfer of mass and angular momentum) depends upon the effective viscosity of the substance of the nebula. If the solar nebula was axially symmetric, and if its gas was quiescent, so that

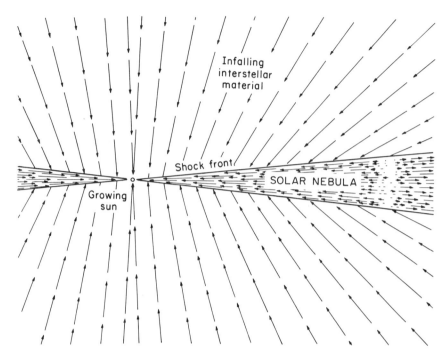

Fig. 6.3.1. Formation of an accretion disk by infall of interstellar material (schematic cross section). Turbulent viscosity redistributes mass and angular momentum in the disk. There is a "watershed" radius, inside of which disk material is transported inward to the Sun; outside this radius, material is spun out to greater radii.

only the molecular viscosity of the extremely thin disk gases was available to transmit momentum, the rate of disk evolution would have been extremely small, essentially zero. Since the disk did evolve and feed the Sun, some more effective form of viscosity must have been at work. The nature and strength of this viscosity are not known; this is a major element of uncertainty in the viscous accretion disk concept.

One possible form of "viscosity" is gravitational torques (Larson 1984; Cassen et al. 1985). Nonaxisymmetric density concentrations in the disk would apply torques to materials at greater and lesser radial distances, redistributing angular momentum among these materials. The density concentrations could take the form of a galaxy-like spiral or bar, or rings, or irregular patches; they could be generated by gravitational instability of the nebular gas, or by the gravitationally disturbing action of a rapidly rotating, triaxially asymmetric early protosun. This mechanism has begun to appear increasingly attractive to astrophysicists trying to understand the nebula.

It is also possible that magnetic forces exerted torques which redistributed important amounts of angular momentum in the nebula. However, only

Hayashi (1981) among current nebula modelists attaches importance to this mechanism.

A third possibility is that the viscosity of the disk was enhanced because the gas was turbulent, which greatly increases the mean velocity at which elements of gas interact and therefore the effective viscosity of the gas. The authors of most accretion-disk nebula models assume that this was the principal agency of viscosity. Turbulence can be induced by thermal gradients between the disk midplane and its surfaces, which drive vertical convective motions (Lin and Papaloizou 1980). Thermal gradients are established when the disk is thick enough and contains enough dispersed dust to be optically opaque, so that heat generated deep in the nebula by viscous friction is not radiated away but accumulates and warms the gas. The actual value of viscosity attributable to convective turbulence is not known; this constitutes a very formidable problem in fluid mechanics. Disk modelists tend to approximate the viscosity ν by use of the expression

$$\nu = K \alpha cL \qquad (1)$$

where K is ~ 1, c is the local sound speed, L the nebula half-thickness or scale height, and α is a dimensionless number that defines the equilibrium strength of subsonic turbulence in the gas. Various authors defend values for α ranging from 10^{-4} to $1/3$.

A Simple Accretion Disk Nebula Model

It is convenient to define three stages of evolution for a turbulent viscous disk. In Stage 1, the disk was still small and the rate at which it could feed material into the Sun was less than the rate at which new material was falling onto it; so it was growing, and presumably the rate at which mass and angular momentum were being transported within it was increasing. The stage defined as 2 was reached if and when the disk reached a quasi-steady state in which the rate of inflow of collapsing interstellar material was approximately balanced by the rate at which the disk fed material into the Sun. This would have been a self-regulating condition: if the infall rate exceeded the Sun-feeding rate, the thickness and opacity of the disk would increase; this would make the disk hotter at the midplane, increasing the convection rate and degree of turbulence and hence the effective viscosity of the nebula gases. The rate of disk evolution, including the rate at which material was fed into the central star, would then increase. Obviously the reverse of all this would occur if the star-feeding rate exceeded the infall rate. Stage 3 was the period after infall ceased but before the nebula was dissipated.

Of most importance to meteoriticists are the properties of the nebula in late Stage 2 (or late Stage 1, if there was no Stage 2) and Stage 3. Maximum temperatures and convective activity would have been reached, for materials that would end up at 3 AU, sometime in late Stage 2 (or 1); quiescence, grain settling and planetesimal accretion may have occurred in Stage 3.

To show how the properties of a nebula model are dependent upon various physical parameters, we have set up a simple Stage 2 accretion disk nebula model, having more or less average properties, in which the viscosity is caused by convectively driven gas-turbulence. Space constraints prohibit the derivation of this model, though it follows straightforwardly from a few simple physical principles (e.g., assuming hydrostatic pressure equilibrium perpendicular to the midplane; equating the local viscous heating rate, the vertical radiative heat flux, and radiative losses from the surface of the nebula). The derivation is a generalized version of that given by Morfill (1985).

Derivation of the model yields approximate expressions for the midplane temperature, photospheric temperature (i.e., the apparent radiative temperature of the nebula as viewed from the outside), surface density of the nebula (local value of the mass in it, per cm^2), density and pressure at the midplane, and half-thickness and optical depth of the nebula, as a function of radial distance r, in units of AU. (Optical depth is the number of times the intensity of radiation is diminished by a factor of e, by absorption and scattering, as it passes out of the nebula perpendicular to the midplane.) The expression for the midplane temperature T_m in K is

$$T_m \approx 54600 \; \alpha^{-1/3} M^{1/2} \dot{M}^{2/3} k_0^{1/3} r^{-3/2}. \tag{2}$$

Expressions for other properties of the nebula are shown in Fig. 6.3.2. In addition, expressions for the net disk mass (M_d, in $M_\odot$) and disk lifetime (t_d in yr) are

$$M_d \approx 5.2 \times 10^{-5} \alpha^{-2/3} \dot{M}^{1/3} R^2 \tag{3}$$

$$t_d \approx 10^6 \dot{M}^{-1}. \tag{4}$$

(The disk lifetime refers to the period when the disk is receiving interstellar material and passing it in to the Sun. It says nothing about the time when the final residual disk is dissipated.) Parameters appearing in these expressions are:

- α = a dimensionless number that defines the equilibrium strength of subsonic turbulence in the nebular gases (see above)
- M = mass of the sun, in present solar masses ($M_\odot$)
- $\dot{M}$ = rate at which mass is being fed from the collapsing cloud into the disk, and (in this quasi-steady state formulation) from the disk into the Sun, in units of $M_\odot/10^6$ yr

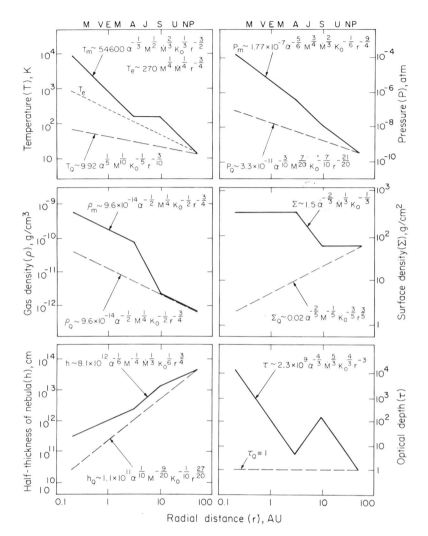

Fig. 6.3.2. Temperature, pressure, gas density, surface density, half-thickness and optical depth in the Simple Stage 2 Nebula Model. Expressions for these parameters are given, and are plotted as a function of radial distance r in AU. Subscripts m, e, and Q stand for midplane, photospheric and quiescent (Stage 3). The middle segments of solid curves represent conditions where both silicates and ices contribute importantly to opacity.

k_O = coefficient in expression for opacity (k) of the nebular material, $k = k_O T^2$
R = overall disk radius in AU.

To describe a mainstream accretion disk nebula, we adopt the following nominal values for these parameters:

α = 0.3 (at the high end of the range used by modelists)
M = 1 $M_\odot$ (i.e., the terminal stages of accretion)
$\dot{M}$ = 1 $M_\odot/10^6$ yr
k_O = 10^{-6} cm/g/K² for 160 K < T < 1600 K (opacity dominated by silicate and Fe metal grains)
k_O = 2×10^{-4} cm/g/K² for T < 160 K (opacity dominated by ices)
R = 50 AU.

Properties of this Simple Stage 2 Nebula Model are shown, as functions of radial distance, in Fig. 6.3.2.

When Infall Ceases: Stage 3

Being a quasi-steady-state system, the Simple Stage 2 Nebula Model assumes a constant inflow of interstellar material into the disk. After ~10^6 yr, inflow of the material that formed the solar system ended, after which the nebula would have evolved into a different state. Accretion of planetesimals, including meteorite parent bodies, was probably going on at this time, and other processes that left their mark in meteorites also may have occurred; therefore it is important to consider the terminal stages of disk evolution.

As the infall rate to the nebula diminished, the self-regulating qualities of the accretion disk discussed earlier would have acted to decrease the mass of the nebula and the inflow rate to the Sun. This evolution reached a crucial stage when, at any given radial distance, the amount of material in the nebula (the surface density) became small enough to render the nebula transparent, which allowed material near the midplane to radiate its heat directly to space. At this point, thermally driven convection and associated gas turbulence stopped, causing the effective viscosity of the gas to fall by a large factor. The transfer of mass and angular momentum and the generation of heat by viscous friction all but ceased.

Properties of the nebula at the time when (at a particular radius) it became transparent and ceased to convect and evolve can be estimated by specifying that $\dot{M}$ (the mass-accretion rate) was sufficiently small to lower the surface density to the point where τ (the optical depth to the midplane) = $\Sigma k/2$ = 1. Anticipating that the nebula was cold at this time, the higher value of opacity ($k_{ice} = 2 \times 10^{-4} T^2$) appropriate for ices is used. The expressions for surface density and midplane temperature then become Σ_Q and T_Q, shown in Fig. 6.3.2. These temperatures are low and will quickly fall lower, now that the nebula is transparent to radiation, finally reaching values in equilibrium with the ambient radiation field. The overall disk mass and the disk mass in the terrestrial zone (< 5 AU) at this time would be approximately

$$M_\mathrm{d} \sim 0.05\ \alpha^{-2/5} k_\mathrm{ice}^{-7/15} R^{13/5} \approx 0.26\ \mathrm{M}_\odot. \tag{5}$$

$$M_\mathrm{5AU} \sim 1.25 \times 10^{-4} \alpha^{-2/5} k_\mathrm{ice}^{-7/15} \approx 7 \times 10^{-4}\ \mathrm{M}_\odot. \tag{6}$$

M_5AU is only about 1/20 the present mass of the terrestrial planets (if their complement of volatile elements and compounds is also counted); it therefore seems clear that accretion of the planetesimals had begun, and a large amount of solid material had become incorporated in bodies too large to be moved by gas drag into the Sun, before the infall of interstellar material ended.

In view of our ignorance of the mechanism of final dispersal of the nebula, it cannot be said with certainty that Stage 3 was ever reached. The abrupt onset of a powerful solar outflow might have begun to strip the nebula at the same time that it turned back infalling interstellar material that had not yet reached the nebula.

6.3.4. REVIEW OF RECENT SOLAR NEBULA MODELS

This Section briefly reviews several recent models of the solar nebula. These are accretion disk models, and in general they are based on the concepts of viscous evolution outlined in Sec. 6.3.3. In the interest of brevity, we omit reviews of a number of historically important models that appeared in the decade after Cameron and Pine (1973), and concentrate on the models published by the same authors in *Protostars and Planets II* (Black and Matthews 1985). Also we do not attempt to describe each model completely, but instead stress its special features and major differences from the Simple Stage 2 Accretion Disk Nebula Model.

Cameron's 1985 Model

Distinctive qualities of this model are that Cameron (1985) makes a special effort to mesh it with the astronomical properties of young stellar objects; to do this he assumes a nonconstant accretion rate (which means the accretion rate always differed from the Sun-feeding rate, so a quasi-steady state was never really reached). During Stage 1, when the central object was still small, the growing disk was gravitationally unstable, and gaseous protoplanets formed which were precursors of the planets; but by the time peak temperatures were reached in the nebula, the atmospheres had been stripped from protoplanets in the inner solar system and the nebula could be treated as a continuum.

Two important parameters in any accretion disk nebula model are α, which defines the viscosity of the disk gas, and $\dot{M}$, the mass accretion rate;

both appear in Eq. (2), the expression for the Stage 2 midplane temperature at a given radial distance. Cameron assumes viscosity due to gas turbulence and justifies use of $\alpha = 0.24$, which is close to the value used in the Simple Stage 2 Nebula Model.

However, he argues that the period during which material accreted to the disk ($\dot{M}$ is inversely proportional to this) was not equivalent to the collapse time of Larson (1969) ($\sim 2 \times 10^5$ yr) or the value assumed for the Simple Stage 2 Nebula Model ($\sim 10^6$ yr), but only a fraction of those times, 1 to 2 $\times 10^4$ yr. This is because most of the collapse time would elapse before a 1 to 2 $M_\odot$ protosolar volume within a larger collapsing cloud became dense enough to fragment from its host and begin autonomous collapse (centered on a high-density fluctuation), and only then would collapse in the protosolar volume become nonhomologous so that an embryonic disk could form at its center. In the units defined above, the mean value for the rate of accretion onto the disk $\dot{M}$ was not 1 but 50 to 100 $M_\odot/10^6$ yr. Further, Cameron postulates that $\dot{M}$ was not constant but increased with time, reaching a peak value of ~ 200 $M_\odot/10^6$ yr just before infall ceased and the Sun approached a mass of 1 $M_\odot$. The evolutionary models calculated by Cameron indicate that such an accretion history is required for the model to reproduce the high luminosities displayed by the youngest T Tauri stars in the HR diagram.

From Eq. (2) it is clear that this enhanced value of $\dot{M}$ would produce very high midplane temperatures in the nebula during its last stage of accretion; the model predicts ~ 1500 K at 3 AU, and a pressure of $\sim 10^{-5}$ atm.

Model of Lin and Papaloizou

Lin and Papaloizou (1985) employ relatively low values for α of ~ 0.02. They explored the viscous evolution of disks by starting from an arbitrary initial condition consisting of a 0.33 $M_\odot$ disk, uniformly distributed out to a radius of 10^{14} cm (~ 7 AU) about a 0.67 $M_\odot$ Sun, then letting the disk evolve without additional accretion. They observed the redistribution of mass and angular momentum and the expansion of the disk perimeter that are characteristic of viscous disks, and noted that for the viscosity used, the disk evolution time scale is $\sim 10^6$ yr. This is not shorter than the accretion time scale assumed but approximately equal to it, which means the nebula would not have spent long as a quasi-steady-state Stage 2 system. The authors define an infall stage, during which the nebula increased in mass; then a viscous stage, after infall ceased, when the nebula evolved along the lines of their model calculations; and finally a clearing stage.

Lin and Papaloizou also modeled a quasi-steady-state disk, which makes comparison possible with the Simple Stage 2 Nebula Model. This condition would have been approached in their evolutionary scenario during the late phase of the infall stage. For $\alpha \approx 0.02$ and $\dot{M} = 1.6$ $M_\odot/10^6$ yr, they obtain a midplane temperature and pressure of ~ 280 K and $\sim 8 \times 10^{-7}$ atm at 3 AU, and relationships between T, P and radial distance generally similar to those of Fig. 6.3.2. Their evolutionary experiments, described above, showed that a disk with arbitrary initial properties tends to evolve toward this state.

Model of Morfill, Tscharnuter and Völk

The paper by Morfill et al. (1985) reflects the emphasis of the past work of these authors, which has not been on the properties or evolution of the solar nebula *per se* but on the collapse of interstellar material that formed the nebula; and on the transport and processing of solid grains during collapse and within the nebula, the effects of which may be preserved in chondritic meteorites. Numerical studies of the 2-dimensional hydrodynamic collapse of interstellar matter show that it was not fed evenly to the surfaces of the nebula; material from near the mean rotation axis of the collapsing volume fell in first, after which accreting material came from lower and lower latitude angles in the source volume. This tended to feed the nebula surface at progressively larger radial distances. The last stages of accretion consisted mostly of the addition of material from near the equatorial plane of the source volume, directly onto the periphery of the nebula.

The authors assume that viscosity in the nebula was caused by gas turbulence. Turbulence also would have had the effect of transporting solid grains vertically and radially in the disk. This is mostly a random process, wherein particles initially together become separated and can diffuse great distances in the gas; but to a lesser extent turbulence can transport ensembles of particles without separating them. Morfill et al. suggest that the evidence for temperature cycling in CAIs, and other of their properties (see Chapter 10.3), can be understood as resulting from turbulent transport in and out of the hot nebula zone at small radial distances, and ultimately out to $\sim$3 AU.

Morfill et al. (1985) use a straightforward nebula model as a tool in discussing the processes outlined above. The nebular properties it predicts are not spelled out, but are similar to those of the Simple Stage 2 Nebula Model of this chapter. The latter assumes $\alpha = 0.3$ and $\dot{M} = 1$ $M_\odot/10^6$ yr, and yields a midplane temperature and pressure of 160 K and $\sim 2 \times 10^{-7}$ atm at 3 AU.

Model of Hayashi, Nakazawa and Nakagawa

These authors (Hayashi et al. 1985) assume that a thin nebula has formed, and concern themselves mostly with the formation of planets in the nebula. The mass of the nebula is only 0.01 to 0.04 $M_\odot$, essentially the minimal amount of nebular material needed to make the planets. The special point of view of this group is and has been that the collisional accretion of particles to form the terrestrial planets occurred within the nebula, before it was dissipated; other workers have assumed most of the growth of the planets occurred in the absence of gas, after the nebula was removed.

Hayashi et al. do not treat growth of the nebula from collapsing interstellar material, or viscous evolution of the disk. The nebula is taken to have formed, and solids in it to have settled to the midplane, so the disk is transparent and nonturbulent (this corresponds to Stage 3 defined earlier). It is heated only by radiation from the Sun. $\dot{M} = 0$ and α is insignificant. In this nebula, the midplane temperature and pressure are ~160 K and ~ 5×10^{-7} atm at 3 AU.

Safronov and Ruzmaikina (1985) discuss nebula models only in general terms, and do not predict conditions in the presolar disk.

6.3.5. CHONDRITIC METEORITES IN THE CONTEXT OF ACCRETION DISK NEBULA MODELS

Thermal Processing in the Nebula

The evidence is irrefutable that chondritic material was processed at high temperatures before it accreted. Chondrules are clearly igneous objects, and Taylor et al. (1983) have marshalled the evidence that the melting occurred in the nebula (see also Chapter 9.3). Peak temperatures required for the melting of most chondrules are in the range 1700 to 1900 K (Chapter 9.2). CAIs also are generally understood to have condensed and/or otherwise been thermally processed in the nebula. Like chondrules, many CAIs were melted as dispersed objects (MacPherson et al. 1984; Chapter 10.3); peak temperatures of ~1700 K were required (Stolper and Paque 1986).

In addition, most chemical classes of chondrites are more or less depleted in the relatively volatile elements. This is broadly understood to be a temperature effect; chondrite accretion occurred after a high-temperature epoch, under circumstances where the last-to-condense minerals were underrepresented. Their lack has been attributed to physical fractionations that enhanced the local concentration of volatile-poor chondrules and CAIs either relative to gas with the potential of condensing the missing volatiles (see, e.g., Wai and Wasson 1977), or to fine-grained late condensates (silicates, organics, ices) that contained these volatiles (see e.g., Larimer and Anders 1967). This topic is discussed in more detail in Chapters 7.5 and 7.6.

Were temperatures in the nebula high enough to achieve these effects? The preceding review shows that only one recent nebula model, that of Cameron (1985), reaches temperatures at 3 AU, possibly high enough to melt and vaporize silicates. (However, Cameron [personal communication] points out that at the stage when these high temperatures are attained in his 1985 model, the rate of transport of disk gas inward toward the Sun is so great that it would not allow thermally processed chondrules and CAIs to accrete locally, but would quickly sweep them into the growing Sun.) Even the Cameron (1962) model, which embedded the idea of a totally vaporized nebula in the minds of so many meteoriticists, when quantified by Cameron and Pine (1973) was found to reach only ~500 K at 8 AU.

By now it is widely understood that hot nebula gases could not have been responsible for melting the chondrules anyway. This is because limits have been placed on the rates at which the chondrules cooled; dynamic crystallization experiments have established the range of cooling rates that reproduce the textures and mineral zonations observed in natural chondrules and CAIs. Cooling rates of 5 to 5000 K/hr have been found for chondrules (Chapter 9.2), and 0.5 to 50 K/hr for a subset of CAIs (Stolper and Paque 1986). In the case of chondrules, this is too fast to represent cooling of the nebula, or movement of material from a hot to a cooler region in it. (The constraint is less clear-cut for CAIs.)

The alternative is that some transient heat source, completely outside the workings of the accretion disk models discussed, operated pervasively in the nebula and (at least at ~3 AU) processed precursor silicate material into chondrules with a high degree of efficiency. Several transient high-energy processes have been proposed (e.g., lightning discharges in the nebula, aerodynamic heating of presolar dust aggregates as they fell into the surface of the nebula [see Chapter 9.3 and 9.4]), but so far none has survived close scrutiny. There are several advantages to a transient heat source, as opposed to a globally hot nebula, when it comes to interpreting the properties of chondrites:

1. Chondrules and CAIs can be cooled on an appropriately fast time scale;
2. There are opportunities to fractionate dust from gas in the nebula prior to the heating event: this seems required in order to account for high, noncosmic oxygen fugacities during chondrule/CAI formation (Chapter 7.7): volatilization in dust-enhanced zones would elevate O/H locally;
3. It is easier to understand the evidence in chondrules and CAIs of repeated thermal cycling (relict grains, discrete concentric igneous layerings on chondrules: multiple stages of mineralization in CAIs) if the heat source was transient;
4. If heating occurred in optically thin settings, the interesting situation would have been created where solids and/or liquids were cooler than the gas surrounding them (Arrhenius and De 1973).

It is not clear whether transient heating events suffice to explain the volatile depletions observed among chondrite chemical classes.

Pressure of the Nebular Gas

The recent nebula models reviewed predict midplane pressures of 10^{-7} to 10^{-5} atm at 3 AU. If the transient heating events that processed chondritic material, whatever their nature, operated elsewhere than at the midplane, the pressure would have been even less. However, meteoriticists, who were strongly influenced by the early Cameron models, still tend to assume a system pressure of 10^{-3} or 10^{-4} atm.

It is important to use a realistic value for pressure in attempting to understand meteoritic phenomena; pressure affects many aspects of chondrite formation, such as rates of condensation, evaporation, nucleation and iso-

TABLE 6.3.1
Approximate Condensation Temperature (K) of $MgAl_2O_4$ as a Function of Total Pressure P and Abundance of Condensable Elements

P (atm)		Condensable Elements/H, Relative to Cosmic Ratio		
	1x	10x	100x	1000x
10^{-3}	1493	1644	1873	2190
10^{-4}	1414	1547	1747	2022
10^{-5}	1343	1461	1637	1875
10^{-6}	1278	1384	1541	1748
10^{-7}	1220	1316	1456	1639

topic exchange. Details of the equilibrium condensation sequence also change with pressure; e.g., at high pressures forsterite and enstatite have higher condensation temperatures than Fe metal, but at low pressures the reverse is the case. The crossover occurs at $\sim 3 \times 10^{-5}$ atm (Grossman 1972).

Most importantly, the temperature range of the condensation sequence shifts with pressure. Even at 10^{-3} atm the temperature range for condensation of the most refractory minerals (1760–1360 K) in a system of cosmic composition is lower than the range of peak temperatures reached by chondrules (1900–1700 K), but at lower pressures the discrepency becomes even greater. This is shown in the first column of Table 6.3.1, which shows the effect of pressure on the condensation temperature of one of the principal refractory minerals, spinel ($MgAl_2O_4$). (If condensation/evaporation occurs at such low temperatures, why did not the chondrules vaporize when they were melted? They must have been saved by the relatively brief duration of their thermal experience. Evaporation experiments of Hashimoto [1983] showed that a chondrule would lose about half its mass in two hours, if held at 2000 K in a vacuum. This confirms the evidence from dynamic crystallization experiments that chondrules were heated and melted by transient events.)

On the other hand, the evidence from oxidation state of minerals in chondrules and CAIs is that the composition of nebular gas often was not cosmic (Chapter 7.7); presumably solid/gas fractionation prior to thermal processing sometimes enhanced the local abundances of condensed matter, and subsequent vaporization by a transient heating event elevated the partial pressures of condensable vapors, relative to H, to greater than cosmic values. This would have affected condensation temperatures in much the same way as an increase in total pressure; most of the pressure effects noted above actually result from variations in the partial pressures of condensable elements, not from the overall pressure ($\cong H_2$ pressure). Effects of such compositional enhancements are shown in Table 6.3.1.

Other Aspects of Accretion Disk Nebula Models

Besides temperature and pressure, accretion disks have other qualities that may have affected chondrite formation. These are no secret, but little explicit reference has been made to them in the interpretive discussions of meteoriticists, so we note them briefly here.

Accretion disks evolve; out to approximately the radial distance of Neptune there would have been a net inward flow of gas, sweeping small particles along with it. Aggregations of particles (≥ 10 cm) would not be swept with the gas, but would migrate inward at a different rate because of gas-drag effects (Weidenschilling 1977). Chondrite components may have formed at a greater radial distance than 3 AU, and may have been size fractionated as they migrated inward.

An accretion disk is probably turbulent, perhaps intermittently. Turbulence has the potentiality for transporting some solid material outward, against the current. It could also prevent small particles from settling to the midplane and concentrating; if intermittent, the turbulence could give rise to discrete episodes of settling, fractionation, concentration and accretion.

The interstellar material accreting to the disk does not necessarily have to be uniform in density, as the infall models always assume. There could be density lumps, and when one of these joins the nebula it can unbalance the mechanical and thermal workings of the disk in a way that translates into a discrete episode of chondrite formation.

Summary

1. Stars like the Sun form from gravitationally collapsing interstellar material. Most of the latter has too much angular momentum to fall directly into the growing star; instead it accumulates in an *accretion disk* that rotates about the star. The solar nebula of cosmochemists was an accretion disk.

2. The solar nebula accreted over a period of 10^4 to 10^6 yr; it did not collapse abruptly, so it was not uniformly heated to high temperatures.

3. Such protostellar disks are not dynamically stable. Interactions between the central star and the disk, and among mass elements of the disk, have the effect of transferring angular momentum outward and causing most of the material of the disk to migrate inward, ultimately to join the star (Sun). The mechanism of interaction (turbulent viscosity? gravitational torques? magnetic forces?) is not known. Turbulent viscosity is the mechanism that has been most extensively studied.

4. These interactions that drive disk evolution waste some energy, which is converted to heat and acts to raise the overall temperature of the disk. However, the energy dissipation associated with turbulent viscosity is not great enough to have heated the solar nebula at the radial distances where meteorites formed (now $\sim$ 3 AU) more than a few hundred K.

5. Therefore some other heating mechanism, as yet undiscovered, must have been responsible for the pervasive high-temperature (1500–2000 K) effects displayed by chondrules and CAIs in meteorites. Evidence in these chondrite components indicates that the heating mechanism was transient and local, rather than nebula-wide in scale.

Acknowledgments. This work was supported in part by a grant from the National Aeronautics and Space Administration. We are grateful to W. K. Hartmann and A. G. W. Cameron for reviewing the manuscript, and to A. P. Boss, P. Cassen, S. J. Weidenschilling and G. W. Wetherill for helpful discussions.

REFERENCES

Arrhenius, G., and De, B. R. 1973. Equilibrium condensation in a solar nebula. *Meteoritics* 8:297–313.

Black, D. C., and Matthews, M. S., eds. 1985. *Protostars & Planets II* (Tucson: Univ. of Arizona Press).

Cameron, A. G. W. 1962. The formation of the sun and planets. *Icarus* 1:13–69.

Cameron, A. G. W. 1976. The primitive solar accretion disk and the formation of the planets. In *The Origin of the Solar System*, ed. S. F. Dermott (New York: Wiley & Sons), pp. 49–74.

Cameron, A. G. W. 1985. Formation and evolution of the primitive solar nebula. In *Protostars & Planets II*, eds. D. C. Black and M. S. Matthews (Tucson: Univ. of Arizona Press), pp. 1073–1099.

Cameron, A. G. W., and Pine, M. R. 1973. Numerical models of the primitive solar nebula. *Icarus* 18:377–406.

Cassen, P., Shu, F. H., and Terebey, S. 1985. Protostellar disks and star formation. In *Protostars & Planets II*, eds. D. C. Black and M. S. Matthews (Tucson: Univ. of Arizona Press), pp. 448–483.

Grossman, L. 1972. Condensation in the primitive solar nebula. *Geochim. Cosmochim. Acta* 36:597–619.

Hashimoto, A. 1983. Evaporation metamorphism in the early solar nebula—evaporation experiments on the melt $FeO\text{-}MgO\text{-}SiO_2\text{-}CaO\text{-}Al_2O_3$ and chemical fractionations of primitive materials. *Geochem. J.* 17:111–145.

Hayashi, C. 1981. Structure of the solar nebula, growth and decay of magnetic fields and effects of magnetic and turbulent viscosities on the nebula. *Prog. Theor. Phys. Suppl.* 70:35–53.

Hayashi, C., Nakazawa, K., and Nakagawa, Y. 1985. Formation of the solar system. In *Protostars & Planets II*, eds. D. C. Black and M. S. Matthews (Tucson: Univ. of Arizona Press), pp. 1100–1153.

Hewins, R. H. 1983. Dynamic crystallization experiments as constraints on chondrule genesis. In *Chondrules and Their Origins*, ed. E. A. King (Houston: Lunar and Planet. Inst.), pp. 122–133.

Larimer, J. W., and Anders, E. 1967. Chemical fractionations in meteorites—II. Abundance patterns and their interpretation. *Geochim. Cosmochim. Acta* 31:1239–1270.

Larson, R. B. 1969. Numerical calculations of the dynamics of a collapsing proto-star. *Mon. Not. Roy. Astron. Soc.* 145:271–295.

Larson, R. B. 1984. Gravitational torques and star formation. *Mon. Not. Roy. Astron. Soc.* 206:197–207.

Lin, D. N. C., and Papaloizou, J. 1980. On the structure and evolution of the primordial solar nebula. *Mon. Not. Roy. Astron. Soc.* 191:37–48.

Lin, D. N. C., and Papaloizou, J. 1985. On the dynamical origin of the solar system. In *Protostars & Planets II*, eds. D. C. Black and M. S. Matthews (Tucson: Univ. of Arizona Press), pp. 981–1072.

Lord, H. C., III. 1965. Molecular equilibria and condensation in a solar nebula and cool stellar atmospheres. *Icarus* 4:279–288.

MacPherson, G. J., Paque, J. M., Stolper, E., and Grossman, L. 1984. The origin and significance of reverse zoning in melilite from Allende Type B inclusions. *J. Geol.* 92:289–305.

Marvin, U. B., Wood, J. A., and Dickey, J. S., Jr. 1970. Ca-Al rich phases in the Allende meteorite. *Earth Planet. Sci. Lett.* 7:346–350.

Morfill, G. E. 1985. Physics and chemistry in the primitive solar nebula. In *Birth and Infancy of Stars*, eds. R. A. Lucas, A. Omont, and R. Stora (Amsterdam: North-Holland), pp. 693–792.

Morfill, G. E., Tscharnuter, W., and Völk, H. J. 1985. Dynamical and chemical evolution of the protoplanetary nebula. In *Protostars & Planets II*, eds. D. C. Black and M. S. Matthews (Tucson: Univ. of Arizona Press), pp. 493–533.

Safronov, V. S., and Ruzmaikina, T. V. 1985. Formation of the solar nebula and planets. In *Protostars & Planets II*, eds. D. C. Black and M. S. Matthews (Tucson: Univ. of Arizona Press), pp. 959–980.

Stolper, E., and Paque, J. M. 1986. Crystallization sequences of Ca-Al-rich inclusions from Allende: The effects of cooling rate and maximum temperature. *Geochim. Cosmochim. Acta* 50:1785–1806.

Taylor, G. J., Scott, E. R. D., and Keil, K. 1983. Cosmic setting for chondrule formation. In *Chondrules and Their Origins*, ed. E. A. King (Houston: Lunar and Planet. Inst.), pp. 262–278.

Wai, C. M., and Wasson, J. T. 1977. Nebular condensation of moderately volatile elements and their abundances in ordinary chondrites. *Earth Planet. Sci. Lett.* 36:1–13.

Weidenschilling, S. J. 1977. Aerodynamics of solid bodies in the solar nebula. *Mon. Not. Roy. Astron. Soc.* 180:57–70.

6.4. FORMATION PROCESSES AND TIME SCALES FOR METEORITE PARENT BODIES

S. J. WEIDENSCHILLING
Planetary Science Institute

This chapter examines the transition from small particles suspended in the solar nebula to the planetesimals (asteroids) that became the parent bodies of meteorites. The motions of solid bodies in the nebula were dominated by the drag of the surrounding gas over a range of sizes ~ µm to km. The "classical" scenario for planetesimal formation involves settling to the central plane of the disk, followed by gravitational instability of the dust layer, but even a very low degree of turbulence in the gas would have prevented this process. Planetesimals probably grew by coagulation of grain aggregates that collided due to different rates of settling and drag-induced orbital decay. Their growth was accompanied by radial transport of solids, possibly sufficient to deplete the primordial mass in the asteroid zone, but with relatively little mixing. The formation of asteroid-sized planetesimals was probably rapid, on a time scale < 1 Myr.

6.4.1. INTRODUCTION

The undifferentiated meteorites contain evidence of complex histories of aggregation of diverse small components (grains, chondrules, refractory inclusions) into larger parent bodies (planetesimals). The formation of planetesimals was probably ubiquitous in the solar nebula, but only in the asteroid belt was their accretion halted before a planet-sized body could form. Thus, the asteroids are a remnant of the original planetesimal population of that region of the solar system, albeit to some degree thermally altered and collisionally evolved. In this chapter we examine some processes that may have been involved in planetesimal formation. Emphasis is on time scales in the context of solar-nebula models (cf. Chapter 6.3) and the formation of the asteroids.

At some early stage, the solid particles that were destined to compose meteorites coexisted with the H-He gas that dominated the nebula's mass. Their motions were controlled or strongly influenced by drag until large ($\gtrsim$ km-sized) bodies formed or the gas was removed. The lifetime of the nebula is poorly constrained by either theoretical models or astronomical observations. There is a general belief that the solar nebula was removed by some activity of the early Sun, e.g., the "T Tauri solar wind" (Elmegreen 1978; Horedt 1978) or a strong ultraviolet flux (Sekiya et al. 1980). Such activity appears to be common among pre-main-sequence stars (Rydgren and Cohen 1985), but the mechanisms involved have not been identified, and its duration cannot be estimated with confidence. It is plausible to identify the active stage with the convective track of pre-main-sequence evolution, which lasted $\sim$ 10 Myr after the Sun attained its present mass (Iben 1965). Hayashi et al. (1985) assume that the nebula was driven outward, and estimate a minimum lifetime of a few Myr based on the energy required for its escape. A lifetime $\sim$ 10 Myr is consistent with the spread of model formation ages for meteorites inferred from isotopic measurements (Chapters 5.2 and 15.3), which are interpreted as representing an extended interval of condensation and accretion. However, it will be shown that dynamical time scales for grain aggregation and planetesimal formation are much shorter, probably $\lesssim$ 1 Myr. The shorter time scale is supported by observations of T Tauri stars that indicate relatively low concentrations of μm-sized circumstellar dust, suggesting that larger bodies have formed (Imhoff 1978; Kuhi 1978; Rydgren and Cohen 1985).

Planetesimal formation presumably involved some degree of particle coagulation and settling toward the central plane of the nebular disk. In order to compare the various physical processes that may have been involved, we assume a suite of properties for the nebula. The values in Table 6.4.1 are assumed for $a = 2.77$ AU, the heliocentric distance of Ceres. These values generally fall within the range of properties for low-mass ($\lesssim 0.1$ M$_\odot$) disk models (Chapter 6.3). They are not meant to be definitive, but only to provide rough estimates and comparisons of time scales for processes affecting solid particles. For this purpose, we also assume a uniform density of 1 g/cm^3 for solid particles, although actual densities probably varied with particle size due to varying degrees of compaction during accretion. Initially, grains and gas are assumed to be well mixed through the thickness of the disk, as a plausible outcome of the initial collapse of the presolar cloud (Chapter 6.2). The thickness of the disk is determined by the balance between thermal pressure in the gas and the component of the Sun's gravity perpendicular to the plane of the disk. The half-thickness H is approximately c/Ω, where c is the mean thermal velocity of the gas molecules, and Ω the local orbital frequency.

TABLE 6.4.1
Nominal Nebular Parameters*

a	heliocentric distance	2.77 AU
T	temperature	300 K
ρ	gas density	10^{-10} g/cm^3
c	mean molecular speed	1.7×10^5 cm/s
λ	gas mean free path	19 cm
ν	kinematic viscosity	1.6×10^6 cm^2/s
V_k	Kepler velocity	1.8×10^6 cm/s
$\Delta V/V_k$	fractional deviation from V_k	5×10^{-3}
ρ_s	particle density	1 g/cm^3
f	solids/gas mass ratio	0.0034

*Illustrative parameters only.

For solar abundances with silicates and metal fully condensed, the dust/gas mass ratio is $f = 0.0034$ (Podolak and Cameron 1974), although this ratio may vary locally with time due to settling.

It will be shown below that the relative velocities of solid bodies in the solar nebula are primarily controlled by aerodynamic drag forces exerted by the nebular gas. The range of sizes for which drag is dominant extends roughly from μm to km, or more than eight orders of magnitude in diameter (for smaller and larger sizes, velocities are dominated by thermal motion and gravitational scattering, respectively). An important parameter that controls the behavior of a particle is t_e, its response time to the drag force. Formally, t_e is defined as mv/F_D, where m is the particle mass, v is the particle's velocity relative to the gas, and F_D is the drag force. The appropriate expression for F_D, and hence for t_e, depends on the size and density of the particle, the density and viscosity of the gas, and on v (Weidenschilling 1977a).

An important property of the nebula is a radial pressure gradient that partially supports the gas and causes it to rotate at slightly less than the Kepler frequency (Weidenschilling 1977a). The existence of such a pressure gradient is not due solely to an assumed gradient in the surface density of the gaseous disk Σ_g. For a low-mass ($\lesssim 0.1$ M$_\odot$) disk, the pressure in the central plane is $\sim \Sigma_g \Omega c/4$. Because Ω and c decrease with increasing heliocentric distance, even disk models with uniform surface density have radial pressure gradients and non-Keplerian rotation. Typical nebular models have a fractional deviation from the Keplerian velocity $\Delta V/V_k$ of a few times 10^{-3}. The effect of this deviation on a solid body depends on its size, or more precisely, on the ratio of t_e to its orbital period. Small particles, with $t_e \ll 1/\Omega$, are strongly coupled to the gas and move with its angular frequency, with a radial drift ve-

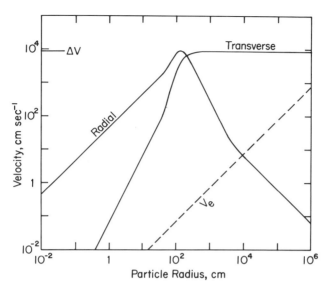

Fig.6.4.1. Radial and transverse velocity components, relative to the surrounding gas, of a solid body in the nominal model nebula. Small particles with radii $\ll 10^2$ cm have predominantly radial motion. Large bodies have a transverse "headwind" of magnitude ΔV and a smaller radial velocity due to orbital decay. In this example, $\Delta V = 9 \times 10^3$ cm/s, and a particle density of 1 g/cm³ is assumed. The peak radial velocity, equal to ΔV, occurs for a particle of radius $\simeq 10^2$ cm; the changes in slope are due to transitions between drag regimes. Also shown is the escape velocity from the body's surface (dashed line).

locity that increases with particle size. Large bodies, with $t_e \gg 1/\Omega$, move in Keplerian orbits that decay gradually due to drag from their motion with respect to the gas; the decay rate decreases with size. There is a complex transition regime near $t_e \simeq 1/\Omega$, where radial velocities reach a maximum value equal to ΔV for roughly meter-sized bodies (Fig. 6.4.1). For our purposes, "small" and "large" particles are those significantly less than or greater than meter-sized.

6.4.2. PARTICLE SETTLING

If the gas is without turbulence, small particles settle toward the central plane of the disk due to the vertical component of the Sun's gravity, $g_z = GM_\odot z/a^3 = \Omega^2 z$. For particles smaller than the mean free path of the gas molecules (a few tens of cm at the nominal density of 10^{-10} g/cm³, with pressure $\simeq 10^{-6}$ bar), the Epstein drag law is appropriate, with $t_e = s\rho_s/\rho c$, where s is the particle radius, ρ_s its density, and ρ the gas density. The settling rate is simply the terminal velocity at which the drag force per unit mass equals g_z:

$$\frac{dz}{dt} = g_z t_e = \frac{s\rho_s \Omega^2 z}{\rho c}. \tag{1}$$

Note that dz/dt is proportional to z; for small particles there is no overshoot of the central plane. The characteristic settling time is

$$\tau_z = \frac{z}{(dz/dt)} = \frac{\rho c}{s\rho_s \Omega^2}. \tag{2}$$

For the nominal parameters, $\tau_z \simeq 280/s$ yr or $\tau_z \simeq$ a few Myr for μm-sized grains. Because τ_z is independent of z, settling is homologous for particles of a given size; i.e., an initially uniform layer would remain uniform while becoming thinner and denser. τ_z is formally the e-folding time for settling; an interval $\simeq 7\tau_e$ is required to reach a dust/gas ratio of unity. During vertical settling, particles also have a radial drift velocity inward toward the Sun (Weidenschilling 1977a), with

$$da/dt = 2a\Omega^2 t_e (\Delta V/V_k). \tag{3}$$

In a time τ_z, a particle moves inward $\tau_z da/dt = 2(\Delta V/V_k)a$, or 0.01 a for the nominal value of $\Delta V/V_k$. The total distance traveled radially will approach 0.1 a by the time the dust/gas ratio reaches unity (Fig. 6.4.2).

In principle, the dust layer may become dense enough to fragment into gravitationally bound condensations (Safronov 1969; Goldreich and Ward 1973). The critical density is $\sim 3M_\odot/2a^3$, corresponding to a half-thickness $z_c \simeq f\rho GH/3\Omega^2$ and a dust/gas ratio $\simeq 10^3$. For the nominal parameters, $z_c/H \simeq 4 \times 10^{-6}$, assuming all of the condensed solids are in the dust layer (homologous settling). For a surface density $\Sigma_s = 3$ g/cm^2 for the dust layer, the critical wavelength for the instability yields planetesimals ~ 10 km in size. A time scale of $\sim 12\ \tau_z$ is required to reach the critical density. Formation of such "Goldreich-Ward planetesimals" is a popular scenario in cosmogony, and is generally assumed to require no nongravitational sticking of particles. However, even a very slight amount of turbulence in the gas would preclude this process.

6.4.3. EFFECT OF TURBULENCE ON SETTLING

A particle embedded in turbulent gas has an rms random velocity given by

$$V_s \simeq \frac{V_t}{[1 + (t_e/t_{ko})^{1/2}]} \tag{4}$$

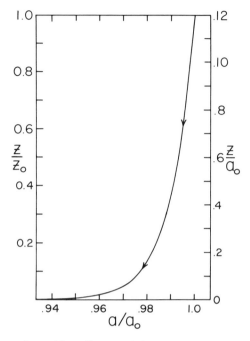

Fig. 6.4.2. Trajectory of a particle settling toward the central plane of the disk while undergoing simultaneous radial drift. This path is unaffected by coagulation and independent of size for small ($\lesssim$ a few cm) particles. All trajectories are identical when z is normalized to the initial value (left-hand scale); the right-hand scale gives actual z values for $z_o = c/\Omega$.

where V_t is the turbulent velocity in the largest eddies, and t_{ko} is their turnover time scale (Völk et al. 1980). The largest eddies are assumed to have a size $\simeq$ the disk thickness. In the rotating nebula, $t_{ko} \simeq 1/\Omega$. For $s \lesssim 10$ cm, $t_e \ll t_{ko}$, so $V_s \simeq V_t$; i.e., "small" particles follow the turbulent motion of the gas. Equating the mean z-component of V_s, $V_t/\sqrt{3}$, to the settling velocity from Eq. (1) gives the characteristic thickness of a turbulent dust layer

$$\frac{z'}{H} \simeq \frac{\rho V_t}{\sqrt{3} s \rho_s \Omega}. \tag{5}$$

In order to have any enhancement of solids over solar abundances, it is required that $z'/H < 1$, or $V_t \lesssim 3 \times 10^2$ s cm/s, e.g., cm-sized bodies with turbulent velocities less than a few meters per second. In order to reach a dust/gas ratio of unity, i.e., an enhancement of a few times 10^2 over solar abundances, $z'/H \simeq f$, giving $V_t \lesssim s$ cm/s for the nominal parameters. For μm-sized grains, this implies turbulent velocities of only tens of meters *per year*, which is implausibly low.

Conditions are even more stringent for reaching the dust density of gravitational instability. Near the critical density, the gravity of the dust layer exceeds the vertical component of solar gravity. At the top of the layer, $g_z \simeq 2\pi G \Sigma_s$, where Σ_s is the surface density of the dust layer, and $dz/dt \simeq 2\pi G \Sigma_s s \rho_s / \rho c$. Equating this to $V_t/\sqrt{3}$ gives the condition $V_t \lesssim 0.1\,s$ cm/s for $\Sigma_s = 3$ g/cm^2. This result differs from the random velocities of ~ 10 cm/s (independent of particle size) in a marginally unstable layer according to Goldreich and Ward (1973) and Weidenschilling (1980). That value is based on the implicit assumption that the distance from the central plane attained by a particle is controlled only by gravity rather than drag. That assumption implies that $t_e \gg 1/\Omega$, which is the case only for roughly meter-sized or larger bodies.

It seems implausible that the solar nebula was so lacking in turbulence that small grains could ever form planetesimals directly by gravitational instability. Even if turbulence is initially absent, formation of a layer with dust/gas $\gtrsim 1$ would induce localized turbulence that could prevent further settling. When the local density of dust exceeds that of the gas, the gas within the layer is dragged by the dust, and the entire layer tends to acquire essentially Keplerian rotation. Nakagawa et al. (1986) assumed that this would halt radial drift, and yield a final settling stage with vertical motion (Fig. 6.4.3). However, shear between the dust layer and the surrounding gas produces a turbulent boundary layer. By analogy with an Ekman boundary layer produced by a rotating disk, we expect eddy velocities $\sim \Delta V/30$, or a few m/s in the boundary layer. Such velocities are more than adequate to stir the dust layer; Eq. (5) implies particle sizes $\gtrsim 10^2$ cm would be required for further settling. One can use the methods of Weidenschilling (1980) to show that neither density stratification nor viscous damping is sufficient to damp turbulence within the dust layer, regardless of the assumed particle size (larger particles are more effective in damping eddies, but there will be too few of them). Thus, a possible result of settling in the disk is the formation of a turbulent particle layer with dust/gas ratio of order unity, and turbulent velocities of a few m/s. The energy source to maintain this turbulence is the inward motion of the dust layer in the Sun's gravity field (Weidenschilling 1980). This condition should persist until collisions result in accretion of bodies large enough to become decoupled from the turbulence ($t_e \gtrsim 1/\Omega$, i.e., sizes $\gtrsim 10^2$ cm), with depletion of the smaller particles so that the dust/gas ratio falls below unity. Therefore, some mechanism of coagulation is a necessary prelude to formation of planetesimals.

6.4.4. COAGULATION MECHANISMS

In general, it is a straightforward process to derive formal expressions for the rate at which particles collide, albeit with some simplifying assumptions. However, collisions do not necessarily result in coagulation. The latter

FORMATION OF METEORITE PARENT BODIES

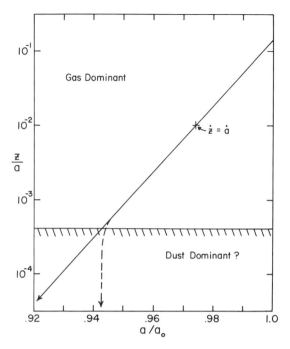

Fig. 6.4.3. A settling trajectory similar to that in Fig. 6.4.2, with the z-coordinate on a logarithmic scale. The + marks the transition from predominantly vertical to radial motion at $z/a = 2\Delta V/V_k$. At $z/a \leq f$, the local dust density may exceed that of the gas, causing that layer to attain Keplerian rotation and ending radial drift (dashed line). Nonhomologous settling due to coagulation will delay this transition, and turbulence may prevent further settling, as described in the text.

depends on the sticking mechanism, e.g., van der Waals bonding, electrostatic attraction, ferromagnetism, chemical reactions and sticky coatings, and upon such factors as relative velocity, particle sizes, densities, structure (fluffy vs smooth) and mechanical strength. Weidenschilling (1980) has discussed the possible role of van der Waals bonding in particle aggregation in the solar nebula. This process is generally insensitive to particle composition, and should yield unfractionated assemblages of grains. The strength of van der Waals bonding appears adequate to produce cm-sized aggregates, but some other mechanism may be needed to form larger bodies. The varied iron/silicate ratios of ordinary chondrites suggest a possible role of ferromagnetism, but textural evidence (e.g., clumps of Fe grains) is lacking. Actually, these chondrites are depleted in Fe compared with CI abundances, indicating selective loss of metal (Chapter 7.4). The high density of Fe, as well as its strength and magnetic properties, would seem to favor its rapid settling and accretion relative to silicates, if there was any dependence on composition. The missing Fe might have been locked up in an earlier generation of planetesimals before the ordinary chondrites accreted, but no complementary Fe-enriched undifferentiated material has been identified. Possibly all such material was heated and melted shortly after accretion, or some may still exist in a region of the asteroid belt not sampled by meteorites. It is also possible that it was removed from the belt by orbital decay due to gas drag (see below). In any case, it is not clear how metal-silicate fractionation could have been effective at some locations and not at others, so that we also have specimens with "cosmic" abundances. Further laboratory work is needed to examine grain coagulation mechanisms and evidence for their operation on components of meteorites.

For the purpose of calculating time scales for various aggregation processes, we simply assume that particle collisions result in sticking, yielding a lower limit to coagulation time scales. If one assumes a "sticking efficiency" less than unity, the time scales are inversely proportional to its value. In general, sticking efficiency depends on the relative velocity, and may become zero if an impacting body dislodges more than its own mass (Hartmann 1985), or the energy density exceeds the impact strength (Weidenschilling 1984). In the following sections we examine various processes that produce collisions of particles, starting with those relevant for small grains and proceeding toward larger objects.

6.4.5. THERMAL COAGULATION

A particle of mass m will have a mean thermal velocity $\bar{v} = (3kT/m)^{1/2}$, where T is the temperature and k is Boltzmann's constant. The number of particles per unit volume is $N = 3f\rho/4\pi\rho_s s^3$. The mean time between collisions of identical particles is

$$\tau_{th} = \frac{\pi \rho_s^{3/2} s^{5/2}}{6f\rho(3kT)^{1/2}}. \tag{6}$$

Assuming that collisions result in sticking, τ_{th} is also the time scale to double the mass of a typical particle. For the nominal parameters, $\tau_{th} \simeq 10^{11} s^{5/2}$ yr. This is only ~ 10 yr for μm-sized grains, but the thermal coagulation time scale increases rapidly with particle size, due to the decrease of both $\bar{v}$ and N. Here we have assumed that metal and silicates are fully condensed; if only refractory elements of lower abundance were condensed at some early stage (Boynton 1985), their coagulation time scale would be longer. For thermal motion, the relative velocities of grains and aggregates decrease with increasing size. Thermal coagulation quickly depletes the smallest particles, yielding a narrow distribution of sizes. Under these conditions, growth is self-similar, i.e., the larger aggregates are clusters of clusters (etc.) of smaller aggregates, having characteristics of a fractal (Donn 1987). However, it will be shown below that nonthermal processes, such as settling and turbulence, dominate relative velocities of particles at sizes $\gtrsim 10^{-3}$ cm. Those processes yield velocities and growth rates that increase with size, broadening the size distribution. Under those conditions growth is not self-similar; the larger aggregates preferentially accrete much smaller ones. Also, the impact velocities eventually become great enough to rearrange the fragile bonds between individual grains and cause compaction. Thus it appears unlikely that macroscopic fractal-like structures would form, although a definite conclusion will require quantitative modeling and numerical simulations.

6.4.6. DIFFERENTIAL SETTLING AND COAGULATION

The possibility of coagulation changes the nature of particle settling in the nebula from that discussed above, and shortens its time scale. We consider first the idealized case without turbulence. If a grain that is somewhat larger than its neighbors (but still small in the sense that $t_e \ll 1/\Omega$) settles through a field of smaller ones, its velocity relative to them is approximately the settling velocity given by Eq. (1). As it encounters the smaller grains and sweeps them up, its size and settling velocity increase. The rate of mass gain is $dm/dt \simeq \pi s^2 \delta(dz/dt)$, where δ is the space density of matter in the form of small grains. Taking $\delta = f\rho$, the characteristic growth time is

$$\tau_g = m/(dm/dt) \simeq 4c/3f\Omega^2 z. \qquad (7)$$

Because τ_g is smaller for larger values of z, aggregate growth begins well away from the central plane and proceeds as a "raining out" from the top layers of the disk (Weidenschilling 1980; Nakagawa et al. 1981,1986). Taking $z(\max) \simeq c/\Omega$, $\tau_g(\min) \simeq 4/3f\Omega$, independent of ρ, c or the initial grain size s_o (note that τ_g is also independent of the assumed density of the aggregates). For the nominal parameters, $\tau_g(\min) \simeq 300$ yr. During descent to the central plane, a grain aggregate can grow to a size $s_f = f\rho z(\max)/4\rho_s$, $\simeq 0.3$ cm due to purely vertical motion (radial drift will allow the particle to grow to larger sizes). The time to reach this size is $\sim 3\,\tau_g \ln(s_f/s_o)$. For s_o, one can choose the size at which $\tau_g(\min)$ equals the thermal coagulation time scale of Eq. (6), giving $s_o \simeq 2(6\rho^2 kT/\pi^2\Omega^2\rho_s^3)^{1/5}$. For the nominal parameters, $s_o \simeq 2 \times 10^{-4}$ cm, giving a total time $\simeq 20\tau_g \simeq 6 \times 10^3$ yr for aggregate growth with descent to the central plane.

Near the central plane ($z/a < 2\Delta V/V_k \simeq 10^{-2}$), radial drift due to non-Keplerian motion of the gas is more important than vertical settling. Using da/dt from Eq. (3) in place of dz/dt gives a growth time (for $\delta = f\rho$) of

$$\tau_g(z=0) \simeq \frac{2c}{3fa\Omega^2(\Delta V/V_k)}. \qquad (8)$$

For the nominal parameters, $\tau_g(z = 0) \simeq 3 \times 10^3$ yr, but the actual time scale for growth will be shorter, because the local density of solids near the central plane increases due to settling. Numerical simulations (Weidenschilling 1980) that include both vertical and radial settling by populations of competing aggregates typically show significant concentration of solids (dust/gas ≥ 1) near the central plane after $\sim 10\tau_g(\min)$ or a few times 10^3 yr. In contrast to the case of settling without coagulation, this concentration is nonhomologous, with the dense layer at that time containing only a small fraction (~ 1 to 10%) of the total surface density of solids. The grain aggregates are also larger than predicted by the expression for s_f, and may reach sizes of tens of cm. Such bodies are too widely spaced to force the gas into local Keplerian rotation, and thus a turbulent shear layer would not form until further settling of smaller particles occurs. Their further growth was not calculated, but would probably occur by mutual collisions rather than gravitational instability. Radial drift during this stage of settling with growth (cf. Eq. 3) amounts to about 0.1 times the original distance from the Sun.

6.4.7. TURBULENT COAGULATION

Various solar-nebula models (Chapter 6.3) include turbulence in the gaseous disk. The source of the turbulence may be infall of prestellar matter, convective instability, or other mechanisms. Turbulence may be global or local in scale, and either steady or intermittent. While turbulence can inhibit settling of solid particles to the central plane of the disk, it can enhance the rate of collisions and coagulation. Völk et al. (1980) derived expressions for the rms relative velocity of particles embedded in a turbulent medium. There are several expressions that differ from the diffusion velocity of Eq. (4), because particle motions are correlated to a degree that depends on their individual values of t_e/t_{ko} (cf. Weidenschilling 1984). For two equal-sized particles with $t_e/t_{ko} \ll 1$, their relative velocity is

$$V_{\rm rel} \simeq (t_e/t_{ko})^{1/2} V_t \qquad (9)$$

The collision (coagulation) time scale is then

$$\tau_{t1} \simeq \frac{4}{9fM} \left[\frac{s\rho_s}{\rho c \Omega} \right]^{1/2} \qquad (10)$$

where $M \simeq 4 V_t/3c$ is the Mach number of the turbulence, and we have assumed $\delta = f\rho$ and $t_{ko} \simeq 1/\Omega$. For the nominal parameters, $\tau_{t1} \approx 5 s^{1/2}/M$ yr.

Equation (10) formally yields very rapid coagulation for small particles in the presence of vigorous turbulence (some nebular models have M as large as 1/3, although Cabot et al. (1987) maintain that convection can only produce turbulent velocities of a few percent of the sound speed). However, its derivation assumes a spectrum of eddies extending to arbitrarily small length scales, which formally implies an infinite rate of energy dissipation. Actually, there is a lower size cutoff, or "inner scale," for turbulent eddies. If the largest eddies have a length scale L, the dissipation rate per unit mass is $\varepsilon \simeq V_t^3/L$, or $\sim 0.4 M^3 c^2 \Omega$ for $L \simeq c/\Omega$. The smallest eddies have a size scale $l \simeq (\nu^3/\varepsilon)^{1/4}$ and a turnover time scale $t_{ks} \simeq (\nu/\varepsilon)^{1/2}$, where ν is the kinematic viscosity. For the nominal parameters, $l \simeq 10^4/M^{3/4}$ cm, or typically ~ 1 km. Taking $\nu = c\lambda/2$, where λ is the mean free path,

$$t_{ks} \simeq \left[\frac{\lambda}{M^3 c \Omega}\right]^{1/2}. \qquad (11)$$

A particle responds to the smallest eddies if $t_e \lesssim t_{ks}$, or if $s \lesssim s' \simeq (\rho/\rho_s)(5\lambda c/M^3\Omega)^{1/2}$. Equations (9) and (10) are applicable only to particles larger than s'. For nominal parameters, $s' \simeq 10^{-2}$ cm for $M = 1/3$, and $\simeq 1$ cm for $M = 0.01$. The corresponding values for $\tau_{t1}(s')$ are ~ 1 yr and 400 yr, respectively.

Particles smaller than s' have $t_e < t_{ks}$, and their motions will be highly correlated. Bodies of equal size within the same eddy would have negligible relative velocity induced by turbulence. However, the gas within the eddy is accelerated, and different-sized particles will drift at different rates, causing collisions. The velocity scale in the smallest eddies is $u \simeq (\nu\varepsilon)^{1/4}$, causing drift velocities $V_d \simeq ut_e/t_{ks}$. Assuming that a larger particle sweeps up smaller ones, a derivation similar to that of Eq. (7) gives a growth time scale

$$\tau_{t2} \simeq \frac{2}{f}\left[\frac{\lambda}{M^9 c \Omega^3}\right]^{1/4} \qquad (12)$$

or $\tau_{t2} \simeq 1/M^{9/4}$ yr, independent of particle size or density, for the nominal parameters. The turbulent coagulation time scale for particles smaller than s' is shorter than that due to vertical settling and radial drift (Eqs. 7 and 8) only for turbulence Mach numbers greater than a few times 10^{-2}. Smaller turbulent velocities would have little effect on particle aggregation.

Vigorous turbulence can cause rapid coagulation of μm-sized grains, but prevents the formation of large (cm-sized) aggregates by causing their disruption in collisions. Suppose that aggregates have an effective impact strength of E erg/cm^3. Then the collision of 2 equal-sized bodies results in disruption when $V_{rel}^2 = 8E/\rho_s$. From Eq. (9), $V_{rel} \propto s^{1/2}$, and the nominal parameters give disruption at $s \simeq 10^{-7} E/\rho_s M^2$. A plausible value of E for loose aggregates is $\sim 10^5$ erg/cm^3, which yields maximum values of $s \simeq 0.1$ cm for $M = 1/3$, and $s \simeq 4$ cm for $M = 0.05$. More elaborate numerical simulations (Weidenschilling 1984) confirm that strong turbulence leads to a dead-end steady state in which aggregates are destroyed as fast as they form. Thus, turbulence must decay to $M \lesssim$ a few times 10^{-2} in order for planetesimals to form.

6.4.8. COLLISIONAL GROWTH AND RADIAL MIGRATION

All of the processes discussed thus far have not produced any bodies larger than a few cm in size. We have argued against the formation of planetesimals by gravitational instability in a dust layer; the alternative appears to be collisional coagulation. In the size range from cm to hundreds of meters, the dominant cause of collisions is differential motions induced by gas drag on particles of different sizes. Their relative velocities can be as large as ΔV, approaching 10^4 cm/s. However, such values are reached only for small particles ($s \ll 10^2$ cm) encountering large ones ($s \gg 10^2$ cm). Collisions between such unequal bodies will tend to result in the smaller one being embedded in the larger. Bodies of similar size will always have low relative velocities. Thus, collisional disruption is less likely than for random motions induced by turbulence, which allows high-speed collisions between equal-sized bodies. We have noted that small particles in the Epstein drag regime have predominantly radial motion (Eq. 3) with drift velocities proportional to their sizes, leading to collisional growth (Eq. 8). Radial velocities reach a peak value equal to ΔV for $t_e = 1/\Omega$ at $s \simeq 10^2$ cm, and then decrease with increasing size (Fig. 6.4.1). Larger bodies in Keplerian orbits have velocities of magnitude ΔV relative to the local gas, in the direction tangential to their orbits, allowing them to sweep up small particles that are coupled to the gas. In moving through a field of small particles with space density δ, the large body gains mass at a rate $dm/dt = \pi s^2 \delta \Delta V$. Its size increases at a rate $ds/dt = \delta \Delta V/4\rho_s$. The space density δ, which here refers to particles less than a few cm in size, is uncertain. A lower limit is $f\rho$, but by the time settling and coagulation have produced any meter-sized or larger bodies, the dust/gas ratio is probably of order unity near the central plane. Moreover, we have suggested that once $\delta \simeq \rho$, shear between the dust layer and surrounding gas would tend to induce enough turbulence to prevent further settling. Taking $\delta \simeq \rho$ yields $ds/dt \simeq 7/\rho_s$ cm/yr, allowing formation of km-sized planetesimals in $\sim 10^4$ yr for the nominal parameters.

During this growth, a planetesimal spirals inward on a decaying orbit. How far will it travel, i.e., how much radial mixing will occur? For large bodies with $t_e \gg 1/\Omega$, the rate of inward motion is

$$da/dt = 2a(\Delta V/V_k)/t_e. \qquad (13)$$

For bodies of size $s \simeq 10^2$ to 10^3 cm, the Stokes drag law is appropriate, and can be applied without serious error up to $s \simeq 10^5$ cm. For Stokes drag, $t_e = 2s^2\rho_s/9\rho\nu$. Combining the expressions for ds/dt and da/dt, we have $ds/da = s^2 V_k \delta/36 a\rho\nu$, which integrates to

$$\frac{1}{s_o} - \frac{1}{s_f} = \frac{V_k \delta}{36\rho\nu} \ln(a_o/a_f) \tag{14}$$

where s_o, s_f are initial and final sizes, and a_o, a_f are initial and final heliocentric distances. For growth to arbitrarily large size, a_f approaches a limiting value,

$$\left[\frac{a_f}{a_o}\right] = \exp\left[\frac{-36\rho\nu}{V_k \delta s_o}\right]. \tag{15}$$

If we take $s_o = 10^2$ cm (smaller bodies will not be in Keplerian orbits) and $\delta = \rho = 10^{-10}$ g/cm^3, then $a_f/a_o \simeq 2/3$, i.e., a meter-sized "seed" body would decrease its heliocentric distance by a few tens of percent before it grew so large that orbital decay due to drag became negligible. Most of the mass of the final body would be acquired *en route*, predominantly at distances near a_f, so a mass-weighted distance for radial transport due to drag would be significantly less. The limiting value of a_f/a_o does depend strongly on δ, which may have varied with time or location in the nebula. In principle, orbital decay during this stage of accretion could have significantly depleted the total mass in the asteroid belt. If we take $\delta = f\rho$ (no enhancement of dust/gas due to settling), all bodies would spiral into the Sun before reaching a_f. Thus, a quiescent nebula that allows small particles to settle until $\delta \simeq \rho$ appears to be needed for any planetesimals to survive. Since a_f/a_o is independent of the final size, inward spiraling during accretion would not be accompanied by significant *differential* mass transport; i.e., it appears possible to preserve radial zoning of composition in the solar nebula. Radial zoning of asteroid compositional types, perhaps reflecting nebular conditions, is observed in the present-day asteroid belt (Gradie and Tedesco 1982).

We have assumed unit collisional efficiency, i.e., all grains in a column with a cross section equal to that of the accreting body actually collide with it. Whipple (1971) pointed out that the flow of gas around the larger body could deflect small grains, causing them to be swept past it instead of being accreted. This effect would increase the time scale for growth. Whipple also suggested that it could cause preferential accretion of chondrules compared with the smaller grains of matrix material. In order for grains to be deflected, their response time to the drag force must be comparable to, or smaller than, the encounter time with the large body. The encounter time is $\sim S\Delta V$, where S is the radius of the large body. We infer that the small particles are deflected if $s/S \lesssim \rho c/\rho_s \Delta V \simeq 10^{-9}$. Thus, bodies in the size range ~ 1 km to $\gtrsim 10^2$ km may accrete chondrule-sized objects ($s \simeq 0.1$ cm) more efficiently than μm-sized dust. However, quantitative segregation would require the "matrix" grains to be encountered individually, rather than as larger aggregates. From the arguments given above, coagulation of the grains should have preceded the formation of asteroid-sized bodies. There was probably some variation in accretion efficiency of grain aggregates and chondrules or inclusions due to differences in density and mean size, but any such effect was probably subtle. Still, it should not be automatically assumed that any meteorite, however primitive, represents an unbiased sample of the solid matter from some localized region of the solar nebula.

Wood (1985) argues against any aerodynamic fractionation of chondrules and grains, based on the composition of the Murchison CM2 chondrite. This meteorite has different Fe/Si ratios in the chondrules and matrix, but an overall ratio close to the solar value. Wood interprets this as due to quantitatively comprehensive accretion of chondrules and condensates formed by the same heating event. (However, the relationship between CM matrix and "condensates" is unclear; see Chapter 3.4.) According to Wood, the lack of separation between mm-sized chondrules and 10-μm dust particles implies prompt accretion, perhaps "within hours or minutes" of the chondrule-forming event. The time scale depends on the assumed spatial scale of the heating event; from Eq. (3) we infer a relative drift rate between the two components of a few cm/s, or $\lesssim 10^3$ km/yr. If the solar nebula was sufficiently homogeneous so that local events produced similar chondrule and condensate compositions over a radial range of $\sim 10^{-2}$ AU, the interval between chondrule formation and accretion could have been $\sim 10^3$ yr. Note that a large body moves relative to the gas and small particles with a transverse velocity $\Delta V \simeq 10^{-2}$ AU/yr. Thus, a layer of meter thickness formed by sweeping up small components represents a *transverse* length scale $\gtrsim 0.1$ AU for a local dust/gas ratio of unity. Comprehensive accretion of dust (and chondrules) from smaller spatial scales requires correspondingly higher dust/gas concentrations.

6.4.9. GRAVITATIONAL ACCRETION

Equation (15) refers only to growth of large bodies in Keplerian orbits by sweeping up small particles that are strongly coupled to the gas. Eventually the small particles will be depleted, with most of the mass in large ($\gg 1$ m) bodies. Their further growth is governed by their mutual gravitational perturbations rather than by gas drag. In a system of gravitating bodies, it is customary to express their relative velocity in terms of the escape velocity from the surface of an individual body, $V_e = (2Gm/s)^{1/2}$. Safronov (1969) showed that an "equilibrium" velocity distribution exists with the mean random velocity V_r proportional to V_e. The Safronov number Θ is defined such that $V_r^2 = V_e^2/2\Theta$, where Θ typically has a value of a few units, and V_e refers to a body of the size that dominates the mass distribution.

We can compare V_r with the velocity V_s induced by turbulence (Eq. 4) to determine the size at which gravitational perturbations become more important than turbulence. We assume that the appropriate drag law is that for large Reynolds numbers, with $t_e = 6s\rho_s/\rho V$, where the particle-gas relative velocity is the larger of V_t or ΔV (for nominal parameters, they are equal at $M \simeq 0.07$). If $V_t > \Delta V$, the transition size s' is

$$s' \simeq \frac{3}{8}(\rho\Theta/\pi G\rho_s^2\Omega)^{1/3}Mc \qquad (16)$$

and

$$s' \simeq \frac{3}{16}(c^2\rho\Theta\Delta V/6\pi\Omega G\rho_s^2)^{1/3}M^{2/3} \qquad (17)$$

for $V_t < \Delta V$. Taking $\Theta = 5$, Eq. (16) gives $s' \simeq 10^6$ cm at $M = 1/3$, and Eq. (17) gives $s' \simeq 2 \times 10^5$ cm at $M = 0.05$. Thus, planetesimals larger than a few km have relative velocities that are not significantly affected by turbulence. In a laminar nebula, large bodies of different sizes have relative velocities due to different rates of orbital decay. Equating the decay rate from Eq. (13) to V_r gives

$$s' \simeq a(\Delta V/V_k)\left[\frac{\Theta\Omega^2\rho^2}{12\pi G\rho_s^3}\right]^{1/4} \qquad (18)$$

or $s' \simeq 10^4$ cm for the nominal parameters. Thus, even in a laminar nebula, bodies must approach kilometer sizes before their growth rates can be modeled as gravitational accretion. Once the mean size exceeds this value, gas drag has little direct effect on planetesimals. It is sometimes assumed that relative velocities (or orbital eccentricities) must have remained low until the gas had dissipated. However, the gas densities expected in a low-mass nebula ($\sim 10^{-10}$ g/cm^3) are not a serious obstacle. At velocities of a few km/s, the dynamic pressure, $0.5\,\rho V^2$, is too low to disrupt even weak bodies, and damping rates are low ($t_e \simeq 10^4$ yr for $s = 10$ km). If some mechanism, e.g., gravitational perturbations by a massive body, can pump up eccentricities, then planetesimals can attain high velocities in the presence of gas. Damping due to drag may actually increase the effects of resonances in velocity stirring (Weidenschilling and Davis 1985).

A planetesimal (or planetary embryo) accreting other bodies by gravitational encounters gains mass at the rate

$$\frac{dm}{dt} = \delta V_r \pi s^2(1 + V_e^2/V_r^2). \qquad (19)$$

The factor $(1 + V_e^2/V_r^2) = (1 + 2\Theta)$ represents the enhancement of the geometric cross section by gravitational focusing. The random velocity V_r is proportional to the eccentricity and inclination of the orbits of bodies in the planetesimal swarm. The thickness of the swarm is proportional to the mean inclination, or to V_r. Safronov (1969) showed that the space density can be expressed as $\delta \simeq 4\Sigma_s/V_r P$, where Σ_s is the surface density of the planetesimal swarm and P is the orbital period. Both Σ_s and δ refer only to those bodies large enough to be in Keplerian orbits. Using this expression for δ, the growth rate of a planetesimal is

$$\frac{ds}{dt} = \frac{\Sigma_s(1 + 2\Theta)}{\rho_s P}. \tag{20}$$

From Eqs. (13) and (20), using $t_e \simeq 6\rho_s s/\rho \Delta V$ for large Reynolds numbers (applicable for $s \gtrsim 1$ km), we find the change in orbital radius during growth is

$$\frac{a_f}{a_o} = \left[\frac{s_f}{s_o}\right]^{-P\rho V_k(\Delta V/V_k)^2/3\,\Sigma_s(1+2\Theta)} \tag{21}$$

For nominal parameters, with $\Sigma_s = 3$ g/cm² and $s_f/s_o = 100$, $a_f/a_o \simeq 0.97$; i.e., orbital decay during this stage of growth to asteroidal size amounts to only a few percent of the initial distance. For lower values of Σ_s, there is more orbital decay; e.g., $\Sigma_s = 0.1$ g/cm² gives $a_f/a_o \simeq 0.4$ for $s_f/s_o = 100$. However, the effective value of Θ was probably large during accretion of the asteroids (see below).

The major uncertainties in evaluating Eq. (20) are the appropriate values of Σ_s and Θ. The present asteroid belt has $\Sigma_s \simeq 10^{-3}$ g/cm², but with that value and $\Theta = 5$ it would require several times 10 Gyr to form a Ceres-sized body ($s \simeq 500$ km). There must have been more mass present in the asteroidal zone originally. Assuming that there once was enough matter to make an Earth-sized planet, or interpolating between the inferred surface densities for Earth's and Jupiter's zones (Weidenschilling 1977b) gives $\Sigma_s \simeq 3$ g/cm². That value gives $ds/dt \simeq 3$ cm/yr, or $\gtrsim 10$ Myr to make Ceres-sized objects. This is probably an overestimate of the actual growth time, because Safronov's evaluation of Θ assumes that the size distribution of the growing bodies is a power law with a shallow slope; i.e., most of the mass is in the largest bodies. In such a scenario, an interval of ~ 10 Myr should produce not one, but $\sim 10^3$ Ceres-sized objects in the asteroid belt. It seems unlikely that all but one of these bodies could have been removed. Mutual collisions among bodies of this size would have resulted in their accretion rather than disruption. Numerical simulations of collisional evolution (Chapman and Davis 1975; Davis et al. 1979,1985) indicate that the present asteroid belt's size distribution could not be the product of an original population that contained many more bodies of sizes a few hundred km or larger than there are today.

A few bodies may have reached the sizes of the largest asteroids in a much shorter time than the ~ 10 Myr implied by Safronov's scenario. Numerical simulations of accretion (Greenberg et al. 1978) do not yield a size distribution with a power-law slope. Instead, a small number of bodies undergo rapid growth, while most of the mass remains in bodies near the original size (Wetherill and Stewart 1987). Relative velocities remain of the order of the escape velocity of the smaller bodies, which dominate the mass distribution. The number of bodies per logarithmic increment of size falls off steeply at somewhat larger sizes, but the slope of this distribution may eventually become shallow for the largest bodies due to their runaway growth. The effective value of Θ for encounters between the smaller bodies remains of the order of a few units. However, a body much larger than the mean size has an escape velocity much greater than the mean random velocity. The growth rate of such a body can be expressed as

$$\frac{ds}{dt} \simeq \frac{\Sigma_s(1+2\Theta s^2/s_0^2)}{\rho_s P} \tag{22}$$

where s_o refers to the initial mean size that dominates the mass distribution. For the larger bodies, the effective value of Θ is increased by the factor s^2/s_o^2. For $2\Theta s^2/s_0^2 \gg 1$, Eq. (22) can be integrated, formally yielding $s \to \infty$ after a time $\tau_\infty \simeq \rho_s P s_o/2\Sigma_s \Theta$. For $s_o = 1$ km and $\Sigma_s = 3$ g/cm^2, $\tau_\infty \simeq 3 \times 10^4$ yr. The assumptions break down well before that time, but this argument does indicate that runaway growth could produce a few large objects on a time scale of $\sim 10^4$ yr. More elaborate numerical simulations by Greenberg et al. (1978) confirm this behavior and time scales of order 10^4 to 10^5 yr for growing a Ceres-sized object. While their particle-in-a-box model cannot be extended to late-stage accretion of a planet-sized body, it appears to be valid for the early stage discussed here.

Regardless of the actual time scale for accretion in the asteroid zone, the process must have been interrupted before bodies larger than Ceres formed. It was also necessary to increase eccentricities and inclinations, and to remove a large fraction (perhaps more than 99%) of the original mass of solids, in order to produce the present belt. These phenomena are probably related, and may have a single cause. It is generally suspected that they were due to the formation of Jupiter, but the responsible mechanisms have not been clearly identified. Any such scenario has a problem with time scale, as it must be assumed that Jupiter itself formed more rapidly than Ceres (cf. Chapter 2.1).

Safronov (1969) and Weidenschilling (1975) suggested that the asteroidal zone was depleted in mass by an external bombardment of Jupiter-scattered planetesimals in eccentric orbits. Davis et al. (1979) pointed out that high-speed impacts would be more effective at destroying asteroid-sized targets than they would be for stirring up their velocities; the target bodies could not survive to reach the observed random velocities of a few km/s. They suggested that the bombarding population included one or more Earth-sized bodies. Such an object could stir up asteroidal orbits by gravitational scattering before being removed by collision with Jupiter or ejection from the solar system. The pumping up of velocities would cause mutual collisions between asteroids to result in disruption, thereby halting accretion.

This scenario was examined in more detail by Davis et al. (1985). They estimated that stirring of the asteroids to present velocities by encounters with Jupiter-scattered bodies would require a time scale of ~ 1 Myr, much longer than the runaway growth time to form Ceres. If the stirring process itself took longer than the accretion time, it would have had to start well before the onset of accretion in the asteroid zone. Such a scenario would imply higher relative velocities during accretion, preventing runaway growth from occurring. While the longer growth time scale allows more gravitational scattering by Jupiter-scattered bodies, the resulting size distribution would have too many Ceres-sized asteroids (or none, if stirring was too effective). Davis et al. (1985) modeled the collisional evolution of several hypothetical asteroid populations after they achieved random relative velocities $\simeq 5$ km/s. Only runaway-growth type initial distributions, with most of the mass in the smaller ($\lesssim 10$ km) bodies, but also containing a few bodies up to the size of Ceres, were found to evolve to a reasonable match to the present asteroid belt. A wide range of initial masses, up to at least a few hundred times the present mass, appears to be compatible with this outcome. However, the extensive collisional evolution of an initially massive belt would imply that asteroids of all sizes had experienced impacts large enough to shatter them. The largest survivors would be "rubble piles" held together by gravity; the smaller asteroids would be mostly fragments of disrupted parent bodies. Spectrophotometry indicates that Vesta has an essentially intact basaltic crust, and therefore has not experienced a major impact by a body large enough to shatter it (diameter $\gtrsim 100$ km) since its differentiation. Davis et al. found that this condition is met only by low-mass "initial" belts, i.e., having no more than a few times the present mass at the time random velocities were pumped up or Vesta's crust formed, whichever occurred later. Only the runaway-growth size distribution with a shallow slope at sizes > 100 km, containing relatively few bodies large enough to shatter Vesta, yields a reasonable probability for Vesta to remain intact until the present.

There are unsolved problems with this scenario. It is conceivable that Vesta differentiated late, erasing evidence of earlier major impacts. However, collisional depletion of a massive belt would require more than the ~ 10 Myr that the most plausible heat sources, short-lived radionuclides or intense solar activity, were effective (if Vesta is the eucrite parent body, it must have differentiated early; cf. Chapter 5.2). If a runaway-growth size distribution was established while the asteroid zone contained a planetary mass of condensed matter, then the subsequent removal of the excess had to preserve the size distribution at sizes $\gtrsim 10$ km. Davis et al. (1985) showed that "Gaussian" distributions depleted at diameters $\lesssim 100$ km would not evolve to the present belt; there would be too few small asteroids produced by collisions between the larger ones. Possibly our understanding of collisional outcomes and scaling laws is flawed, and future work may modify these conclusions, but it seems likely that the "missing mass" was never incorporated into bodies more than a few km in size.

The short time scale for runaway growth implies that the mass was removed, or accretion halted by other means, on a time scale $< 10^5$ yr. Wetherill and Stewart (1987) suggest that perturbations by Jupiter, perhaps through secular resonances, could stir up relative velocities by a small amount (eccentricities $\simeq 10^{-3}$) and retard runaway growth. Such a process would then allow the longer ~ 10 Myr time scale for mass removal and additional stirring to the present relative velocities. A quantitative model for this scenario has not been developed. It may be possible to show that the formation of Jupiter or some other event could produce the requisite amount of stirring in a short time. Still, if the initial size distribution of the asteroids represented a "frozen" stage of incipient runaway growth, then the stirring event also had to be precisely timed. Had it begun $\sim 10^4$ yr earlier or later, accretion would have produced no body as large as Ceres, or else one or more that were much larger. It is much more probable that rapid runaway growth would go to completion, or be prevented entirely, than that it would be halted while in progress.

An alternative possibility is that the missing mass was removed before the onset of gravitational accretion. The earlier depletion of solid matter might occur during the m-to-km growth stage. Equation (15) shows that the amount of orbital decay in this stage is sensitive to δ, the space density of solids. In principle a major change in δ, perhaps associated with condensation of water ice, could produce a discontinuity in the amount of radial motion and open a gap in the distribution of solids in the disk. Because the growth rate in Eq. (22) is proportional to Σ_s, prior removal of solids could yield slow runaway growth on a time scale of 1 to 10 Myr, provided that velocities remained low. It may not be realistic to assume that low velocities could be maintained while Jupiter was forming (cf. Chapter 2.1), but this scenario would allow Jupiter to form later, on a longer time scale, and make a "frozen" runaway somewhat more plausible. Development of a quantitative model of this type is beyond the scope of this chapter. There is also need for much more work to understand the role that external influences, particularly by Jupiter, played in the formation of the asteroids. Until this is accomplished, the chronology of formation of the parent bodies of meteorites cannot be tied to that of the planets.

6.4.10. CONCLUSIONS

Some tentative conclusions can be drawn from consideration of the "micro-astrophysical" environment of small solid bodies in the solar nebula. Highly turbulent disk models (Mach numbers $\gtrsim 0.1$) do not allow accretion

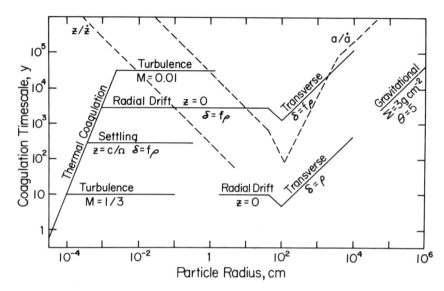

Fig. 6.4.4. Time scales for coagulation ($m/\dot{m}$) in the nominal model nebula. Approximate particle size ranges for validity of the various mechanisms are shown. Particle bulk density is assumed to be 1 g/cm³. Except where otherwise noted, the space density of accretable matter is taken to be $\delta = f\rho = 3.4 \times 10^{-13}$ g/cm³. Also shown (dashed lines) are time scales for vertical settling ($z/\dot{z}$) and radial motion ($a/\dot{a}$) due to drift or orbital decay. Thermal coagulation dominates for sizes $\lesssim 10^{-3}$ cm. Vigorous turbulence ($M = 1/3$) yields rapid coagulation of small particles, but the process is terminated by collisional breakup at ~ 0.1 cm. Turbulent coagulation is slower than that due to differential settling and radial drift for M a few times 10^{-2}. In a laminar nebula, vertical settling gives the shortest growth time scales at large values of $z \simeq c/\Omega$, but particles reach the vicinity of the central plane by the time sizes approach 1 cm. Time scales for further growth near $z = 0$ depend strongly on the space density of dust; values are shown for $\delta = f\rho$ and $\delta = \rho$. Gravitational accretion dominates for sizes $\gtrsim 1$ km; the case shown is for $\Sigma = 3$ g/cm² and $\Theta = 5$.

of solid bodies larger than ~ 1 cm, or any significant concentration of condensates toward the central plane. Very small turbulent velocities are enough to prevent gravitational instability in a dust layer. Even in a globally laminar nebula, shear between a dense dust layer and the surrounding gas may cause local turbulence in the layer and prevent the layer from settling beyond the state where dust/gas $\simeq 1$. Thus, planetesimals probably formed by collisional coagulation, rather than gravitational instability.

Figure 6.4.4 summarizes the time scales for various coagulation processes in the solar nebula and the ranges of particle size for which they are applicable. Gas drag dominates relative velocities in the range from a few μm to a few km. If the turbulence Mach number is less than a few times 10^{-2}, the dominant cause of particle collisions is differential settling and radial drift velocities caused by non-Keplerian rotation of the gas. Particle coagulation yields significant concentration of solids toward the central plane with formation of bodies up to a meter in size on a time scale of a few times 10^3 yr. There is no plausible model for formation of meter-sized or larger bodies far from the central plane. Further growth from meter- to km-sized bodies requires $\sim 10^3$ to 10^4 yr. Formation of objects of the size of the larger asteroids is controlled by gravitational forces rather than gas drag, and requires $\sim 10^4$ to 10^5 yr, or longer if the mass in the asteroid zone was depleted before this stage. There is no current astrophysical model of the solar nebula that yields extended condensation of solids from a hot gas over ~ 10 Myr or a similar interval between condensation and accretion, as has been proposed from the spread of isotopic model ages (cf. Chapter 15.3). On the other hand, there is no dynamically plausible scenario for rapid accretion of planetesimals on the time scale of a few orbital periods.

Gas drag causes radial drift and orbital decay of solid matter during planetesimal formation. Descent to the central plane and growth to meter-sized bodies probably involves a change in heliocentric distance of $\sim 10\%$. The greatest change in distance probably occurs during growth from meter to kilometer sizes. The amount of orbital decay is highly uncertain, but may exceed several tens of percent, and in principle could be responsible for the low mass of the asteroid belt. The amount of differential motion is much smaller, i.e., gas drag produces little radial mixing during accretion. Note that bodies of sizes ~ 0.1 and 10^4 cm have essentially identical radial velocities (cf. Fig 6.4.1), although their transverse velocities relative to the gas are very different. There is no need to postulate very rapid accretion in order to argue against radial mixing. The amount of orbital decay experienced by planetesimals of sizes $\gtrsim 10$ km was probably only a few percent before dissipation of the solar nebula.

Conditions for accretion of the asteroids are poorly constrained. Collisional evolution models suggest that much of the original mass of condensed matter in the asteroid zone was removed before the onset of gravitational accretion, i.e., before the formation of bodies $\gtrsim 10$ km in size. However, such an inference is presently quite speculative, as there are no quantitative models for such a scenario. A slightly more robust conclusion is that the mass was removed before asteroid random velocities were pumped up to a few km/s. The mechanism for this velocity pumping, and the extent to which Jupiter influenced the accretion of the asteroids, are unclear. There is still very little evidence to constrain the chronology of formation of the asteroids relative to that of the planets.

Acknowledgments. This work was supported by the NASA Planetary Geophysics and Geochemistry Program. I wish to thank R. P. Binzel, H. Campins, C. R. Chapman, D. R. Davis, W. K. Hartmann, C. Patterson, D. Spaute, and G. Wetherill for helpful discussions and comments on the manuscript.

REFERENCES

Boynton, W. V. 1985. Meteoritic evidence concerning conditions in the solar nebula. In *Protostars & Planets II*, eds. D. C. Black and M. S. Matthews (Tucson: Univ. of Arizona Press), pp. 772–787.

Cabot, W., Canuto, V. M., Hubickyj, O., and Pollack, J. 1987. The role of turbulent convection in the primitive solar nebula. II. Results. *Icarus* 69:423–457.

Chapman, C. R., and Davis, D. R. 1975. Asteroid collisional evolution: Evidence for a much larger early population. *Science* 190:553–556.

Davis, D. R., Chapman, C. R., Greenberg, R., and Harris, A. 1979. Collisional evolution of asteroids: Populations, rotations, and velocities. In *Asteroids*, ed. T. Gehrels (Tucson: Univ. of Arizona Press), pp. 528–557.

Davis, D. R., Chapman, C. R., Weidenschilling, S. J., and Greenberg, R. 1985. Collisional history of asteroids: Evidence from Vesta and the Hirayama families. *Icarus* 62:30–53.

Donn, B. (abstract) 1987. Grain formation and accretion: Initial stages. *Lunar Planet. Sci.* XVIII 243.

Elmegreen, B. 1978. On the interaction between a strong stellar wind and a surrounding disk nebula. *Moon and Planets* 19:261–277.

Goldreich, P., and Ward, W. R. 1973. The formation of planetesimals. *Astrophys. J.* 183:1051–1061.

Gradie, J., and Tedesco, E. 1982. Compositional structure of the asteroid belt. *Science* 216:1405–1407.

Greenberg, R., Wacker, J. F., Hartmann, W. K., and Chapman, C. R. 1978. Planetesimals to planets: Numerical simulation of collisional evolution. *Icarus* 35:1–26.

Hartmann, W. K. 1985. Impact experiments. I. Ejecta velocity distributions and related results from regolith targets. *Icarus* 63:69–98.

Hayashi, C., Nakazawa, K., and Nakagawa, Y. 1985. Formation of the solar system. In *Protostars & Planets II*, eds. D. C. Black and M. S. Matthews (Tucson: Univ. of Arizona Press), pp. 1100–1153.

Horedt, G. P. 1978. Blow-off of the protoplanetary cloud by a T Tauri-like solar wind. *Astron. Astrophys.* 64:173–178.

Iben, I. 1965. Stellar evolution. I. The approach to the main sequence. *Astrophys. J.* 141:993–1018.

Imhoff, C. L. 1978. T Tauri star evolution and evidence for planetary formation. In *Protostars and Planets*, ed. T. Gehrels (Tucson: Univ. of Arizona Press), pp. 699–707.

Kuhi, L. 1978. Spectral characteristics of T Tauri stars. In *Protostars and Planets*, ed. T. Gehrels (Tucson: Univ. of Arizona Press), pp. 708–717.

Nakagawa, Y., Nakazawa, K., and Hayashi, C. 1981. Growth and sedimentation of dust grains in the primordial solar nebula. *Icarus* 45:517–528.

Nakagawa, Y., Sekiya, M., and Hayashi, C. 1986. Settling and growth of dust particles in a laminar phase of a low-mass solar nebula. *Icarus* 67:375–390.

Podolak, M., and Cameron, A. G. W. 1974. Models of the giant planets. *Icarus* 22:123–148.

Rydgren, A. E., and Cohen, M. 1985. Young stellar objects and their circumstellar dust: An overview. In *Protostars & Planets II*, eds. D. C. Black and M. S. Matthews (Tucson: Univ. of Arizona Press), pp. 371–385.

Safronov, V. S. 1969. *Evolution of the Protoplanetary Cloud and Formation of the Earth and the Planets* (Moscow: Nauka), NASA TTF-677.

Sekiya, M., Nakazawa, K., and Hayashi, C. 1980. Dissipation of the primordial terrestrial atmosphere due to irradiation of the solar far-UV during T Tauri stage. *Prog. Theor. Phys.* 66:1301–1316.

Völk, H., Jones, F., Morfill, G., and Röser, S. 1980. Collisions between grains in a turbulent gas. *Astron. Astrophys.* 87:316–325.

Weidenschilling, S. J. 1975. Mass loss from the region of Mars and the asteroid belt. *Icarus* 26:361–366.

Weidenschilling, S. J. 1977a. Aerodynamics of solid bodies in the solar nebula. *Mon. Not. Roy. Astron. Soc.* 180:57–70.

Weidenschilling, S. J. 1977b. The distribution of mass in the planetary system and solar nebula. *Astrophys. Space Sci.* 51:153–158.

Weidenschilling, S. J. 1980. Dust to planetesimals: Settling and coagulation in the solar nebula. *Icarus* 44:172–189.
Weidenschilling, S. J. 1984. Evolution of grains in a turbulent solar nebula. *Icarus* 60:553–567.
Weidenschilling, S. J., and Davis, D. R. 1985. Orbital resonances in the solar nebula: Implications for planetary accretion. *Icarus* 62:16–29.
Wetherill, G. W., and Stewart, G. 1987. Factors controlling early runaway growth of planetesimals. *Lunar Planet. Sci.* XVIII:1077 (abstract).
Whipple, F. 1971. Accumulation of chondrules on asteroids. In *Physical Studies of Minor Planets*, ed. T. Gehrels, NASA SP-267, pp. 251–256.
Wood, J. 1985. Meteoritic constraints on processes in the solar nebula: An overview. In *Protostars & Planets II,* eds. D. C. Black and M. S. Matthews (Tucson: Univ. of Arizona Press), pp. 687–702.

PART 7
Chemistry of Chondrites and the Early Solar System

7.1. THE COSMOCHEMICAL CLASSIFICATION OF THE ELEMENTS

JOHN W. LARIMER
Arizona State University

In cosmochemistry the elements are usually classified in terms of volatility, the most important chemical property governing their abundance in planetary material. At the same time, Goldschmidt's geochemical classification of the elements is commonly retained to describe the mineralogical, or major planetary, siting of the elements. For present purposes, the elements have been divided into four groups: refractory, moderately volatile, highly volatile and siderophile. Each element is grouped according to its predicted volatility in a solar gas, and its abundance and distribution in chondrites. Refractory elements are predicted to condense at higher temperatures than Fe, Mg and Si (1300–1400 K, at $P_T = 10^{-4}$ atm); moderately volatile elements are predicted to condense at lower temperatures than Fe, Mg and Si but at $T > 670$ K, the temperature at which S begins to condense as FeS; and highly volatile elements condense or become trapped at $T < 670$ K. The siderophile group includes all elements that are more easily reduced to metal than Fe; these elements tend to behave as a coherent group and occur primarily as alloys in the FeNi-metal phase though their abundance and distribution also appear to depend in part on their volatility.

7.1.1. HISTORICAL PERSPECTIVE

In cosmochemistry the elements are classified according to their behavior in systems with solar composition. This contrasts with the classification used in geochemistry where the elements are classified according to their behavior in the terrestrial environment. Even so, the cosmochemical classification closely parallels the geochemical classification scheme proposed by

Goldschmidt in 1922 (Goldschmidt 1954). The similarity is due to the fact that he classified many elements on the basis of their behavior in meteorites, reasoning with great foresight that an element's behavior in the Earth as a whole should resemble its behavior in meteorites bearing metal, sulfides and silicates.

Goldschmidt divided the elements into four groups: lithophile, siderophile, chalcophile and atmophile. The terms have Greek roots which loosely translated mean, respectively: rock- (or silicate-) loving, metal-loving, sulfide-loving and gas- (or atmosphere-) loving. These descriptive terms have proven extremely useful, once committed to memory, and the widespread, long-lived use of this scheme attests to its elegant simplicity. As will become evident, these terms are widely used in cosmochemistry because, with few exceptions, the elements behave similarly in both the terrestrial and cosmic environment.

There is, however, one important difference. From the time of Urey's (1952) pioneering work, it has been generally accepted that the most important chemical fractionation process in the history of planetary matter is the separation of condensed nebular dust from the associated gas during accretion. Moreover, it now appears that fractionation processes involving partially condensed dust occurred several times prior to accretion. The distribution of each element between dust and gas is related to its cosmic volatility. To understand this important stage in the history of planetary matter, each element's cosmic volatility, as well as its "geochemical" tendencies must be taken into account.

In most current literature in cosmochemistry and meteoritics, including this book, the cosmic volatility of the elements is expressed in relative terms: refractory, moderately volatile and highly volatile. The division is based on a combination of empirical and theoretical evidence drawn from each element's observed behavior in chondritic meteorites and its calculated condensation (or evaporation) temperature in a solar gas. Admittedly, this is not an ideal basis for classifying the elements but it does serve a useful purpose. One difficulty is that a few deviant elements behave as volatiles in one type of chondrite and display more refractory behavior in another type of chondrite. More often than not, these are also elements whose condensation behavior is difficult to predict.

A distressing aspect of the literature for the newcomer to the field is the seemingly random mix of geochemical and volatility terms. It is not unusual to see terms combined: e.g., refractory-lithophile, moderately volatile-siderophile, etc. Although it is not reassuring, this simply reflects the current state of affairs which even among those who work in the field is often a confusing source of contention. We hope the classification adopted here will at least help to clarify the following discussions in this book.

7.1.2. A COSMOCHEMICAL CLASSIFICATION OF THE ELEMENTS

We begin by dividing the elements into four groups: refractory, siderophile, moderately volatile and highly volatile (Fig. 7.1.1). In three cases, the grouping is based on relative volatility and, in the fourth case, the geochemical group of siderophile elements is retained. To the extent that a rationale exists, the distinctive chemical and physical characteristics of the metals set them apart as a group in a more definitive manner than does their volatility. The chemical behavior of the remaining elements is less important than their volatility in controlling their abundance. The perceptive reader will note, however, that the refractory group is dominated by lithophile elements, the moderately volatile group includes many chalcophile elements and the highly volatile group, as might be expected, consists largely of atmophile elements.

This subdivision based on volatility is best discussed in terms of the condensation sequence of the elements (Fig. 7.1.2). (A brief discussion of how this sequence is calculated is given in section 7.1.3.) The point to be emphasized here is the natural divisions which separate the elements according to their volatility. There are two benchmarks: the condensation of Mg-silicates along with FeNi-metal at 1300 to 1400 K and the condensation of S as FeS at 670 K. Those elements that condense at higher temperatures than Mg-silicates and FeNi-metal are classified refractory, those expected to con-

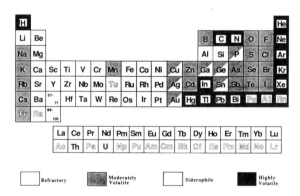

Fig. 7.1.1. Cosmochemical classification of the elements grouped according to their volatility and geochemical behavior. Refractory elements display similar fractionation patterns in chondritic material and condense at higher temperatures than Mg-silicates or Fe($>$ 1300 K). Moderately volatile elements display a more varied fractionation pattern: they do not vary in abundance within chondrite groups; they are not fractionated relative to CI abundances by more than a factor of 5; and they condense between Fe or the Mg-silicates and S (1300 $<$ T $<$ 670 K). The highly volatile elements do vary in abundance within chondrite groups, being more strongly depleted in the higher petrographic types, and several vary in abundance by factors of 1000 or more. They condense below the P_T-independent temperature where FeS forms ($<$ 670 K).

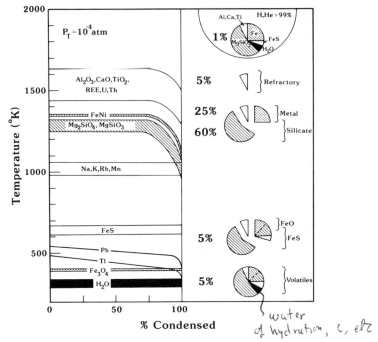

Fig. 7.1.2. The condensable fraction of solar matter comprising about 1% of the total mass of the Sun. It is expected to condense over the temperature range 1800 to 300 K at $P_T = 10^{-4}$ atm. Elements belonging to the various cosmochemical groups appear to be concentrated in components which were blended together in different proportions just prior to or during accretion of the chondrite parent bodies.

dense at temperatures between the Mg-silicates and S are classified moderately volatile and those that condense at even lower temperatures are classified highly volatile. We can now consider the elements group by group.

Refractory Elements

This group includes most of the elements on the left side of the periodic table plus Al, Lanthanides and Actinides (Fig. 7.1.1). These elements are predicted to be the first to condense from a cooling solar gas and the last to vaporize as interstellar dust is heated. They behave as a coherent group in most chondrites, that is, they all increase or decrease in abundance by the same factor. When added together with their complement of O they represent about 5% by mass of the total condensable matter in CI chondrites.

Siderophile Elements

These elements are grouped together because they all occur as metals, usually alloyed in FeNi-metal. A simple test for whether an element should be classified siderophile is to ascertain whether it is more easily reduced to metal than Fe or, in simple terms, whether it is more noble than Fe. Elements that meet this criterion are classified siderophile. In addition, several elements with mixed chemical tendencies are usually classified siderophile. These include: P, which occurs as both phosphides and phosphates; Ga, Ge and W, which also have lithophile tendencies; and Ag, As, Cu, Mo, Sb and Sn, which have chalcophile tendencies. Some of these elements, as well as more volatile Bi, Hg, Pb and Tl, usually are classified lithophile or chalcophile in terrestrial geology but display much stronger siderophile tendencies in extraterrestrial environments where the O and S abundances are low.

Iron is the key element owing to its great abundance and the extent to which it forms oxides and sulfides in cosmic systems. Iron is difficult to classify geochemically because a significant fraction occurs as FeO and FeS as well as Fe° making it an important lithophile, chalcophile and siderophile element.

Since siderophile elements are grouped on the basis of similar chemical behavior, rather than volatility, it becomes necessary at times to subdivide them on the basis of their position in the condensation sequence. The division follows the scheme outlined above: "refractory siderophile" implies condensation at higher temperatures than Fe, "moderately volatile siderophile" implies condensation at temperatures between that of Fe and FeS formation and "highly volatile siderophile" implies condensation at temperatures below FeS formation.

Moderately Volatile Elements

These elements fall on the right side of the periodic table except for the alkali metals (Li, Na, K and Rb) and Mn. The alkali metals and Mn, along with the halogens (F and Cl), are classified lithophile while the remainder are classified chalcophile or siderophile in geochemistry.

Moderately volatile elements obviously display diverse chemical behavior and, except for S, they occur as minor or trace components in one of the abundant minerals. In some cases the host phase has yet to be identified and even when it is known, the key data on the thermodynamics of solubility are usually lacking. For these reasons condensation temperatures are difficult to predict. Classification of moderately volatile elements therefore is based largely on their abundance patterns in chondritic meteorites. A convenient empirical division between the moderately and highly volatile elements is that moderately volatile elements are *never depleted by more than a factor of 5* in any chondrite relative to CI abundances and moderately volatile elements display *no detectable variations within a chondrite group* while the highly volatile elements vary by factors of 3 to > 1000 within groups of chondrites.

Highly Volatile Elements

These elements display the greatest variation in abundance, both between and within chondrite groups. Some chondrites contain the full solar complement of elements such as Bi, In, Pb and Tl while in others these same elements are depleted by over a factor of 1000. The extremely volatile noble gases are strongly depleted in all chondrites, but are much less depleted in carbonaceous chondrites than in any other type and they too display variations by factors of 1000 or more (see Chapter 7.9). Highly volatile elements are of special interest because they may provide information on the temperature at the time of accretion. They are predicted to condense from a solar gas at temperatures less than 670 K or, in the case of the noble gases, to become trapped somehow in appreciable quantities at these low temperatures.

Carbon, N and O are included but the cosmochemistry of these elements is complex. Like the noble gases, these elements are depleted relative to their solar abundances in all types of chondrites. One of the most important reactions in cosmochemistry is:

$$CO + 3H_2 \rightarrow CH_4 + H_2O. \qquad (1)$$

This reaction is predicted to proceed to the right at low temperatures (< 500 K) but will do so only in the presence of a suitable catalyst. Metallic FeNi is a suitable catalyst in the nebular setting. However, the product of the catalytically induced reaction is not just CH_4 but a variety of compounds dominated by straight-chained hydrocarbons, a pattern that significantly is also observed in C chondrites (Hayatsu and Anders 1981; see also Chapter 10.5). To the extent that the reaction does not proceed to the right, the gas becomes supersaturated in C leading to the possibility of graphite precipitation (Lewis and Prinn 1980).

A key variable in cosmic systems is the C/O ratio of the gas. In a gas with the composition of the Sun with C/O (~ 0.6) less than 1, some C may condense as complex organic compounds at $T = 200-400$ K via catalytically induced reactions (see Chapter 10.5). A larger fraction condenses at even lower temperatures (< 120 K) as CO-clathrates or solid C (Lewis and Prinn 1980). Much of the N in carbonaceous chondrites must also be in the organic compounds. However, in a more reduced gas with C/O $\simeq 1$, a significant fraction of C and some N are much more refractory and condense at $T > 1300$ K as graphite, SiC, AlN and TiN (Larimer and Bartholomay 1979).

Oxygen is regarded as a highly volatile element even though a small fraction condenses as refractory silicate material and a much larger fraction in the form of MgFe-silicates. Nonetheless, O is highly volatile because the largest fraction condenses as hydrated silicates or water ice at $T < 350$ K.

COSMOCHEMICAL CLASSIFICATION OF ELEMENTS 381

7.1.3. NEBULAR CHEMISTRY

Since this classification of the elements is based on both their predicted behavior in a solar gas and their observed behavior in chondritic meteorites, a brief discussion of each source of information is required.

Thermochemical Calculations

Let us first consider the problem of gas-dust equilibrium in a gas with solar composition. The condensation temperature of an element reflects its cosmic volatility, which is determined by several factors: the vapor pressure of the element, its tendency to react with other elements to form gaseous or solid compounds, its tendency to form solid solutions or alloys, the elemental abundances which determine the partial pressures in the gas, and the total pressure on the system (Grossman and Larimer 1974).

A typical calculation is the condensation of metallic Fe:

$$Fe(g) \rightarrow Fe(s). \tag{2}$$

The partial pressure of Fe, $p(Fe)$, in the gas is simply the mole fraction of Fe in the gas multiplied by an assumed total pressure:

$$p(Fe) = \frac{E(Fe)}{1/2 \; E(H)} \times P_T \tag{3}$$

where the mole fraction of Fe is approximated by the solar Fe abundance, E(Fe), divided by 1/2 the solar H abundance. The mole fraction of Fe actually equals the solar Fe abundance divided by the total number of moles of *all* other gaseous species. However, since H_2 is so much more abundant than all other species combined, the approximation introduces a negligible error. The factor 1/2 is required since most H occurs as H_2, except at high temperatures (~ 2000 K) where a fraction of H_2 dissociates to monatomic $H(g)$. Condensation begins when $p(Fe)$ equals the vapor pressure of Fe. Vapor pressure equations have the form:

$$\ln p^\circ(Fe) = A/T - B \tag{4}$$

where A and B are the heat and entropy of vaporization divided by the gas constant. Once an element begins to condense, its partial pressure drops as some fraction α of the element condenses leaving a fraction $(1 - \alpha)$ in the gas:

$$p(Fe) = \frac{(1 - \alpha) \; E(Fe)}{1/2 \; E(H)} \times P_T. \tag{5}$$

TABLE 7.1.1
The 50% Condensation Temperatures of Some Selected Elements in a Gas of Solar Composition at $P_T = 10^{-4}$ atm

Element	Temperature (K)	Element	Temperature (K)
REFRACTORY		MODERATELY VOLATILE	
Os	1814		
Al	1650	Au	1225
Sc	1644	Mn	1190
Ir	1610	As	1157
Ti	1549	Rb	1080
U	1540	K	1000
Ca	1518	Na	970
La	1500	Ga	920
Pt	1411	Sb	912
Eu	1290	Ge	825
COMMON ELEMENTS		F	736
		IRON SULFIDE FORMATION	
Ni	1354		
Mg	1340	S (as FeS)	648
Fe	1336	HIGHLY VOLATILE	
Si	1311	Pb	520
		Bi	472
		In	470
		Tl	448
		IRON OXIDE FORMATION	
		Fe_3O_4	400

The equation describing condensation can then be written:

$$\ln(1 - \alpha) = A/T - B - \ln \frac{E(\text{Fe})}{1/2\ E(\text{H})} - \ln P_T. \tag{6}$$

The temperatures at which 50% of an element is condensed have proven useful for comparative purposes; these temperatures are obtained by setting $\alpha = 0.5$. The 50% condensation temperatures for some key elements are listed in Table 7.1.1. Note also that Eq. (6) can be used to calculate partial evaporation during heating.

For elements which form compounds, in the gas or condensate, the reactions are more complicated. For example, consider the condensation of olivine:

$$2\text{Mg}(g) + \text{SiO}(g) + 3\text{H}_2\text{O} \rightarrow \text{Mg}_2\text{SiO}_4(s) + 3\text{H}_2. \tag{7}$$

The algebra is not much more complicated than it is for the case of Fe; an equilibrium constant for the reaction can be written:

$$\ln K = \frac{a(Mg_2SiO_4)\, p^3(H_2)}{p^2(Mg)\, p(SiO)\, p^3(H_2O)\, P_T^3} \tag{8}$$

and ln K as a function of T can be obtained from thermodynamic data:

$$\ln K = A/T - B. \tag{9}$$

However, in contrast to the simple case of Fe, here each partial pressure must be computed taking into account the distribution of each constituent element in the gas phase (the fraction of O as H_2O, CO, SiO, etc.) as well as the fraction present in other solid species (Mg as $MgAl_2O_4$, etc.). Thus, while the algebra involved is no more complicated, it does become more tedious making the job of computing nebular chemistry ideally suited for a computer. The results of such a study are schematically summarized in Fig. 7.1.2.

An additional complication arises in the case of elements that condense as solid solutions or alloys, frequently as dilute solutes in a major host phase. To compute these gas-solid equilibria, it is necessary to reformulate Eqs. (4) and (8). The tacit assumption in those equations is that only pure phases condense, where the mole fraction (N) = activity (a) = 1. However, in solutions, N is always < 1. Moreover, $N = a$ only in the special case of ideal solution; for nonideal solutions, an activity coefficient (γ) must be introduced:

$$a = \gamma N \tag{10}$$

where for ideal solutions $\gamma = 1$ and for non-ideal solutions:

$$\gamma \rightarrow 1 \text{ as } N \rightarrow 1. \tag{11}$$

Values for the activity coefficients can be obtained if the heats and entropies of solution are known; however, such data are rather sparse. This is an area where a considerable amount of experimental work is needed. Nonetheless, the assumption of ideal solution is often made and, though recognized to be inadequate, seems to provide useful estimates particularly for the refractory elements (Fegley and Palme 1985; Kornacki and Fegley 1986).

Besides condensation, several other important reactions can be studied. The chemistry of Fe provides illustrative examples. Gaseous Fe first condenses to FeNi grains, whose composition and condensation temperature vary slightly with pressure. As these grains cool, they acquire increasing amounts of moderately volatile metals such as Au and Cu. At 670 K, Fe reacts with $H_2S(g)$ to form FeS:

$$Fe + H_2S \rightarrow FeS + H_2. \qquad (12)$$

This reaction is pressure independent; the reaction temperature depends only on the H_2S/H_2 ratio in the gas. From the onset of condensation, a small but continually increasing amount of Fe reacts with H_2O in the gas to form FeO which, if equilibrium is maintained, becomes incorporated into the Mg-silicates:

$$Fe + MgSiO_3 + H_2O \rightarrow FeSi_{0.5}O_2 + MgSi_{0.5}O_2 + H_2. \qquad (13)$$

The Fe^{+2} content of the silicates is predicted to increase markedly between about 500 and 400 K. The question of whether equilibrium can be maintained at these temperatures is problematical, and deserves further study, because not only must a gas-solid reaction occur but the solids must continue to exchange material. Moreover, several other factors could affect the extent to which Fe is oxidized: the rate at which CO and H_2 react to form H_2O and CH_4 (Lewis and Prinn 1980); the C/O ratio (Larimer and Bartholomay 1979); and the gas/dust ratio (Wood 1984). Finally, any metallic Fe left unreacted when the temperature falls to 400 K combines with gaseous H_2O to form Fe_3O_4:

$$3Fe + 4H_2O \rightarrow Fe_3O_4 + 4H_2. \qquad (14)$$

This is another pressure-independent reaction but here the temperature is fixed by the H_2O/H_2 ratio.

Variables and Uncertainties in Thermochemical Calculations

The calculations in Table 7.1.1 and Fig. 7.1.2 are based on two assumptions: chemical equilibrium and a total pressure ($P_T = 10^{-4}$ atm). The assumption of chemical equilibrium has generated considerable controversy. On one extreme, the argument is made that the calculations are misleading, or serve little purpose, because equilibrium is unlikely to be achieved in the nebula. But this argument is specious; the question of whether or not equilibrium was achieved cannot even be debated intelligently if the equilibrium configuration is unknown. The reason for assuming complete equilibration is to provide a base for comparison: a compositional and mineralogical configuration against which observations drawn from chondrites can be compared in order to understand where and why equilibrium was or was not achieved. At the other extreme, arguments sometimes are made based, for example, on a specific temperature like those listed in Table 7.1.1. Such arguments invite skepticism; in general, this list of temperatures provides a guide to the order of condensation but the exact numerical values have little significance.

A potentially useful approach to the question of equilibration is to take into account reaction and diffusion-rate data. The available data are meager, however, making this an area which deserves considerable attention. Larimer and Anders (1967) emphasized the importance of the cooling rate vs reaction and diffusion rates in determining the final configuration by considering two scenerios: a slowly cooling, completely equilibrated system and a rapidly cooling, incompletely equilibrated system. A different approach has been developed by Prinn and Fegley (1987) who considered reaction-rate data available on several key gas-gas, gas-solid and solid-solid reactions. They note, for example, that reactions between solid silicates and H_2O in the gas to form hydrated silicates are complex and sluggish. These are important reactions because they determine the amount of H_2O acquired by a body and whether it is acquired in the form of hydrated silicates or via the direct condensation of H_2O.

Note also that this presentation of the thermodynamic calculations is not meant to imply that all material in the nebula passed through a single, all-inclusive vaporization event followed by condensation during monotonic cooling. The calculations could equally well be used to consider the partial vaporization of dust in the nebula. Figure 7.1.2 is simply a means of displaying the temperature ranges over which the various solids would be in equilibrium with a gas of solar composition; in effect, it is a phase diagram for the nebula.

The choice of what pressure to use in the calculations is somewhat arbitrary, the only constraints coming from current astrophysical models of the nebula. A wide range of plausible pressures have been studied and the relationships are well understood. Increasing P_T increases the partial pressures, as indicated in Eq. (3), and results in shifting most of the gas-solid reactions to higher temperatures; decreasing the pressure has just the opposite effect. In a few cases P_T variations result in reversals in the sequence of reactions. One important switch involves FeNi-metal and the Mg-silicates; at pressures higher than about 5×10^{-5} atm, the FeNi-metal condenses at higher temperatures than the silicates while at lower pressures the silicates condense before metal.

Another source of uncertainty in the calculations is the quality of the available thermodynamic data. Typically these uncertainties fall in the range of 1 to 5% which translates into an uncertainty in condensation temperature of about 10 to 20 K. Obviously, the general pattern in Fig. 7.1.2 is unlikely to change but some of the details might alter as better data become available. It is also possible that an important compound has been overlooked; this appears to be a remote possibility at present but should not be forgotten.

Elemental Fractionations in Chondrites

A cursory examination of the compositional variations among chondritic meteorites leaves the impression that they are minor and hence of little importance. However, these seemingly inconsequential chemical variations constitute one of the best-documented primitive features of chondrites. Except for the most volatile elements (H, He, etc.), CI chondrites are remarkably similar to the Sun in composition (Anders and Ebihara 1982), but in all other chondrites at least a few elements occur in different proportions. These differences fall into patterns that clearly are not the result of planetary processes, but instead suggest more primitive, nebular processes.

A characteristic feature of chondritic fractionation is that the groups of elements move in unison. If one element is enriched or depleted, then all other elements in that group are enriched or depleted as well, and by nearly the same factor. Each class of chondrites appears to contain its own blend of these groups of elements. A good example is the siderophile elements which are usually alloyed in the FeNi phase. Their coherent behavior suggests that the different chondrite groups each acquired its own characteristic amount of metal bearing all the siderophile elements. The one property shared by the elements in each of the other three groups is volatility: (1) refractory ($T > 1300$ K); (2) moderately volatile ($1300 > T > 670$ K); and (3) highly volatile ($T < 670$ K). The temperature ranges are those in which elements in each group are predicted to condense from a solar gas at $P = 10^{-4}$ atm and reflect the cosmic volatilities within each group of elements.

These chondritic fractionations are usually discussed in the framework of a model in which chondrites consist of a mixture of components. Each component carries its own cosmochemical group of elements. Such a model has considerable observational support, the key observation being that chondrites consist of several discrete components that can be studied microscopically, separated and analyzed. Each component displays a characteristic mineralogy and contains a characteristic group of elements. For example, the CaAl-inclusions carry the refractory elements and contain a suite of minerals similar to those predicted to be the stable solid phases in a gas/dust mixture with solar composition at high temperatures, $T > 1300$ K (see Chapter 10.3). The mineralogy and elemental abundances of the various components are consistent, to a first approximation, with the formation temperatures inferred from thermochemical calculations.

The importance of characterizing the chemistry and mineralogy in the various components lies in the information they contain regarding conditions in the nebula and the nature of the accretion process. Each chondrite parent body apparently acquired its own characteristic mix of components, with the proportions varying from body to body. The proportions of each component, as they occur in an unfractionated nebula, are shown in Fig. 7.1.2. CI chondrites are the only samples of planetary material in which the components occur in their solar proportions (see Fig. 1.1.1). All other chondrites acquired the components in distinctly nonsolar proportions, as schematically illustrated in Fig. 7.1.3.

Compared to refractory and siderophile elements, moderately and highly

COSMOCHEMICAL CLASSIFICATION OF ELEMENTS

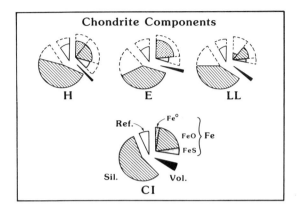

Fig. 7.1.3. The various groups of chondrites each containing their own special mix of the nebular components. The CI chondrites evidently contain the components in their solar proportions but all other groups contain a characteristic blend that varies markedly from the solar mix. Shading corresponds to that in Fig. 7.1.2.

volatile elements are more strongly fractionated and in a different manner. In C chondrites, all moderately and highly volatile elements move in unison. Anders (1964) noted that this pattern is consistent with the observation that these meteorites consist of two materials: volatile-poor chondrules and volatile-rich matrix-mixed together in different proportions. CI chondrites contain ~ 0% chondrules and are not depleted in volatiles; CM chondrites contain ~ 50% chondrules and are depleted in volatiles by ~ 50%, etc. In ordinary and enstatite chondrites, the pattern is more complex. Moderately volatile elements still move in unison but the least volatile of these are less depleted than the most volatile. However, since ordinary and enstatite chondrites also are a mix of chondrules and matrix, it seems reasonable to explain the basic pattern in terms of the same model, allowing for variable depletion factors in chondrules.

Thus, in addition to the components illustrated (Fig. 7.1.3), it is necessary to superimpose two more: chondrules and matrix. This is best modeled by simply dividing the silicate and metal components into two fractions; one fraction (chondrules) that carries all of the elements which condense between 1400 and 1000 K but little or none of the elements that condense at lower temperatures, and another fraction (matrix) that carries the full complement of elements that condense between 1300 and 600 K. This grouping is consistent with the observation that a portion of the silicate and metal resides in chondrules and another portion resides in the finer-grained matrix.

Wasson (Wasson and Chou 1974; Wai and Wasson 1977) has argued that the volatile-element patterns are better interpreted in terms of fractionation during condensation while the nebular gas was dissipating and the nebular dust was settling. In this model, the more volatile elements are expected to condense on progressively finer-grained dust which would settle more slowly toward the mid plane of the nebula and hence be less efficiently accreted in the parent bodies. These contrasting views along with the data on which they are based are discussed in more detail in the following chapters.

The highly volatile elements behave in a unique manner in ordinary and enstatite chondrites. Instead of being depleted by the same factor within a chondrite group, these elements are fractionated relative to each other and tend to be more depleted in meteorites of higher petrographic type. The extent to which each of these elements is depleted appears to be a function of its volatility: elements that are highly volatile tend to be more strongly depleted than those that are less volatile. This may reflect accretion during condensation (Larimer and Anders 1967), loss during thermal metamorphism (Dodd 1981) or some combination of the two processes. If the volatile content was established at the time of accretion, then it may be possible to obtain quantitative estimates of the accretion temperature providing that the condensation temperatures are known (see Chapter 7.6).

Acknowledgments. This work was supported by a grant from the National Aeronautics and Space Administration. I am indebted to B. Fegley and H. Palme for constructive reviews.

REFERENCES

Anders, E. 1964. Origin, age and composition of meteorites. *Space Sci. Rev.* 3:583–704.
Anders, E., and Ebihara, M. 1982. Solar-system abundances of the elements. *Geochim. Cosmochim. Acta* 46:2363–2380.
Dodd, R. T. 1981. *Meteorites* (Cambridge: Cambridge Univ. Press).
Fegley, B., and Palme, H. 1985. Evidence for oxidizing conditions in the solar nebula from Mo and W depletions in refractory inclusions in carbonaceous chondrites. *Earth Planet. Sci. Lett.* 72:311–326.
Goldschmidt, V. M. 1954. *Geochemistry* (London: Oxford Univ. Press).
Grossman, L., and Larimer, J. W. 1974. Early chemical history of the solar system. *Rev. Geophys. Space Phys.* 12:71–101.
Hayatsu, R., and Anders, E. 1981. Organic compounds in meteorites and their origin. In *Cosmo- and Geochemistry*, vol. 99, *Topics in Current Chemistry*, pp. 1–37.
Kornacki, A. S., and Fegley, B., Jr. 1986. The abundance and relative volatility of refractory trace elements in Allende Ca, Al-rich inclusions: Implications for chemical and physical processes in the solar nebula. *Earth Planet. Sci. Lett.* 79:217–234.
Larimer, J. W., and Anders, E. 1967. Chemical fractionations in meteorites—II. Abundance patterns and their interpretation. *Geochim. Cosmochim. Acta* 31:1239–1270.
Larimer, J. W., and Bartholomay, M. 1979. The role of carbon and oxygen in cosmic gases: Some applications to the chemistry and mineralogy of enstatite chondrites. *Geochim. Cosmochim. Acta* 43:1455–1466.
Lewis, J. S., and Prinn, R. G. 1980. Kinetic inhibition of CO and N_2 reduction in the solar nebula. *Astrophys. J.* 238:357–364.
Prinn, R. G., and Fegley, B., Jr. 1987. The atmospheres of Venus, Earth and Mars: A critical comparison. *Ann. Rev. Planet. Sci.* 15:171–212.
Urey, H. C. 1952. *The Planets: Their Origin and Development* (New Haven: Yale Univ. Press).

Wai, C. M., and Wasson, J. T. 1977. Nebular condensation of moderately volatile elements and their abundances in ordinary chondrites. *Earth Planet. Sci. Lett.* 36:1–13.

Wasson, J. T., and Chou, C. L. 1974. Fractionation of moderately volatile elements in ordinary chondrites. *Meteoritics* 9:69–84.

Wood, J. A. 1984. On the formation of meteoritic chondrules by aerodynamic drag heating in the solar nebula. *Earth Planet. Sci. Lett.* 70:11–26.

7.2. ELEMENTAL VARIATIONS IN BULK CHONDRITES: A BRIEF REVIEW

GREGORY W. KALLEMEYN
University of California at Los Angeles

The three classes of chondrites, carbonaceous, ordinary and enstatite, are clearly delineated into nine separate groups: CI, CM, CO, CV, H, L, LL, EH and EL, by their elemental abundances. Several analytical techniques are available for determining these elemental abundances, often to a high degree of precision. They include wet-chemical methods, X-ray fluorescence analysis, fused-bead analysis, instrumental and radiochemical neutron-activation analysis and isotope-dilution mass spectroscopy. Refractory lithophile abundances show distinct trends in the carbonaceous chondrites where $CV > CM \approx CO > CI$, and in the ordinary chondrites where $H > L,LL$. Greater siderophile-element abundances clearly delineate EH enstatite chondrites from EL, and also distinguish L ordinary chondrites from LL. The four carbonaceous-chondrite groups are also readily distinguished by their volatile-element abundances.

Chondritic meteorites are basically rocks with near-solar abundances of non-volatile elements. Three classes of chondrites can be distinguished by differences in major lithophile/Si ratios: carbonaceous, enstatite and ordinary chondrites. The classes can be subdivided into nine clearly distinguished chondrite groups using more complete compositional data: carbonaceous into CI, CM, CO and CV; enstatite into EH and EL; and ordinary into H, L and LL (Chapter 1.1 and Appendix 3).

Several different analytical techniques are used for determining elemental concentrations in chondrites. The traditional method for determining bulk composition of major elements, e.g., Si, Al, Mg and Fe, is the gravimetric or *wet-chemistry method* (Jarosewich 1966). This technique can also provide useful information about the oxidation state of Fe, i.e., the amount of metallic Fe present. The drawbacks of this technique are the relatively large sample sizes required (several g) and the very high level of wet-chemistry skills needed to provide consistently accurate and precise data.

An alternative method for determining major elements is *X-ray fluorescence analysis* (Von Michaelis et al. 1969). The XRF method can provide very precise major-element data, but also usually requires large sample sizes. A large amount of a meteorite is usually ground to a powder, then fused with a fluxing agent, reground and formed into a pellet for analysis. The sample is excited with an intense beam of X-rays and fluoresces X-rays characteristic of its composition. Standards are run and compared for determining elemental concentrations. The final sample is usually unusable for other types of analysis. Furthermore, samples with significant amounts of metal may not fuse properly, producing poor results for Fe.

A similar technique, known as the *fused-bead method* (Jezek et al. 1979) uses only a small (a few mg) amount of powder, often a split from a larger amount of homogenized powder used for another type of analysis. This sample is melted to a glass, then mounted, polished and analyzed for characteristic X-rays induced by the electron beam of an electron microprobe. The high temperatures involved in forming the glass preclude the measurement of elements such as Na which are partially volatilized. Furthermore, it is sometimes difficult or impossible to produce fused beads of uniform composition for samples with hydrated silicates or significant amounts of metal.

An excellent technique for determining major, minor and trace elements in chondrites is *neutron activation analysis* (NAA). A meteorite sample is irradiated with neutrons, converting some stable isotopes to radioactive isotopes. The radiations emitted by the sample are measured and compared to irradiated standards for determining elemental concentrations. The NAA technique follows one of two methods: *radiochemical neutron activation* (RNAA), where individual elements or groups of elements are chemically separated for analysis after the irradiation (Sundberg and Boynton 1977), or *instrumental neutron activation analysis* (INAA), where the sample is analyzed whole (Kallemeyn and Wasson 1981). Both RNAA and INAA methods can be used on very small samples; e.g., chondrules or inclusions < 1 mg, and can routinely determine concentrations to the ng/g level. The RNAA method can determine concentrations to the pg/g level since the separated elements are usually free from interferences by other radionuclides. It has the disadvantage that the samples are destroyed by the chemical separations performed on them, and the time and effort spent performing these chemical separations usually limit the number of elements determined to < 20. The INAA method has the advantage that it requires little sample handling since the samples are not chemically processed, but counted directly on a γ-ray detector after irradiation. This method is essentially nondestructive (albeit the sample is radioactive) and can routinely determine concentrations of ≥ 25 elements in a sample. It lacks the extreme sensitivity of the RNAA method. The principle disadvantage of both NAA methods is that some important elements, e.g., Si, S, C and P, are very difficult or impossible to determine.

Another method for analyzing trace elements in chondrites is the *isotope-dilution mass-spectroscopy method* (Evensen et al. 1978). A sample is spiked with known amounts of the elements to be determined and analyzed on a mass spectrometer. Normal sample sizes range from 50 to 300 mg, although samples as small as chondrules can be analyzed. This technique is generally used for analyzing rare-earth elements (REE), which can be measured to a very high degree of precision. Some of the REEs cannot be determined, though, because they are monoisotopic.

The carbonaceous chondrite class of meteorites has four established groups: CI, CM, CO and CV, that can be distinguished both petrographically and compositionally. Abundances of refractory lithophile elements show the trend: CV > CM ≈ CO > CI, which does not fully distinguish the groups (Kallemeyn and Wasson 1981). It is necessary to use volatile elements to classify them fully. Abundances of moderately volatile elements show an initial trend: CI > CM > CO > CV which changes to CI > CM > CV > CO for the more highly volatile elements, those below the condensation-temperature range of S (Anders et al. 1976; Kallemeyn and Wasson 1981; Krähenbühl et al. 1973). The highly volatile elements may also provide a compositional basis for separating the oxidized and reduced subgroups of the CV chondrites (Kallemeyn and Wasson 1982; McSween 1977). The oxidized subgroup appears to have a higher content of highly volatile elements.

The two enstatite chondrite groups can easily be distinguished by their differences in lithophile- and siderophile-element abundances. Except for the most refractory lithophile elements, all lithophile- and siderophile-element abundances show a general trend: EH > EL (Hertogen et al. 1983; Kallemeyn and Wasson 1986). Although individual members from any chondrite group can sometimes show fractionated REE patterns (Nakamura and Masuda 1973), the EL chondrites seem to be the only group to show a fractionated REE pattern as a whole (Kallemeyn and Wasson 1986).

Among the three ordinary chondrite groups, the H chondrites can be distinguished clearly from the L and LL chondrites using compositional data, but the differences between L and LL are rather subtle. Abundances of refractory lithophile elements show a trend: L ≈ LL > H, while the H chondrites have higher siderophile-element abundances. The main compositional difference distinguishing between the L and LL chondrites is the somewhat higher abundance of siderophile elements in L. The groups are generally indistinguishable for the more volatile elements (Wai and Wasson 1977). The highly volatile elements for all three groups tend to correlate with petrographic type.

Acknowledgments. The author wishes to thank J. Wasson for providing the time and facilities for producing this chapter, and W. Boynton for a speedy and helpful review.

REFERENCES

Anders, E., Higuchi, H., Ganapathy, R., and Morgan, J. W. 1976. Chemical fractionations in meteorites—IX. C3 chondrites. *Geochim. Cosmochim. Acta* 40:1131–1139.

Evensen, N. M., Hamilton, P. J., and O'Nions, R. K. 1978. Rare-earth abundances in chondritic meteorites. *Geochim. Cosmochim. Acta* 42:1199–1212.

Hertogen, J., Janssens, M.-J., Takahashi, H., Morgan, J. W., and Anders, E. 1983. Enstatite chondrites: Trace element clues to their origin. *Geochim. Cosmochim. Acta* 47:2241–2255.

Jarosewich, E. 1966. Chemical analyses of ten stony meteorites. *Geochim. Cosmochim. Acta* 30:1261–1265.

Jezek, P. A., Sinton, J. M., Jarosewich, E., and Obermeyer, C. R. 1979. Fusion of rock and mineral powders for electron microprobe analysis. *Smithsonian Contrib. Earth Sci.* 22:46–52.

Kallemeyn, G. W., and Wasson, J. T. 1981. The compositional classification of chondrites: I. The carbonaceous chondrite groups. *Geochim. Cosmochim. Acta* 45:1217–1230.

Kallemeyn, G. W., and Wasson, J. T. 1982. The compositional classification of chondrites: III. Ungrouped carbonaceous chondrites. *Geochim. Cosmochim. Acta* 46:2217–2228.

Kallemeyn, G. W., and Wasson, J. T. 1986. Compositions of enstatite EH3, EH4,5 and EL6 chondrites: Implications regarding their formation. *Geochim. Cosmochim. Acta* 50:2153–2164.

Krähenbühl, U., Morgan, J. W., Ganapathy, R., and Anders, E. 1973. Abundance of 17 trace elements in carbonaceous chondrites. *Geochim. Cosmochim. Acta* 37:1353–1370.

McSween, H. Y. 1977. Petrographic variations among carbonaceous chondrites of the Vigarano type. *Geochim. Cosmochim. Acta* 41:1777–1790.

Nakamura, N., and Masuda, A. 1973. Chondrites with peculiar rare-earth patterns. *Earth Planet. Sci. Lett.* 19:429–437.

Sundberg, L. L., and Boynton, W. V. 1977. Determination of ten trace elements in meteorites and lunar materials by radiochemical neutron activation analysis. *Anal. Chim. Acta* 89:127–140.

Von Michaelis, H., Erlank, A. J., Willis, J. P., and Ahrens, L. H. 1969. The composition of stony meteorites: I. Analytical techniques. *Earth Planet. Sci. Lett.* 5:383–394.

Wai, C. M., and Wasson, J. T. 1977. Nebular condensation of moderately volatile elements and their abundances in ordinary chondrites. *Earth Planet. Sci. Lett.* 36:1–13.

7.3. REFRACTORY LITHOPHILE ELEMENTS

JOHN W. LARIMER
Arizona State University

and

JOHN T. WASSON
University of California at Los Angeles

Refractory lithophile elements are the last to evaporate and the first to condense from a mix of elements in solar proportions. In bulk samples of chondrites, if one refractory lithophile element is enriched or depleted then all the others are too, and by nearly the same factor. They are enriched in Ca-Al inclusions by factors of 15 to 25 times the CI-chondrite levels; however, these enrichments do not always parallel CI abundances. In CV, CM and CO chondrites the uniform refractory-element enrichments relative to CI chondrites suggest that these meteorites contain an excess of material resembling "average" CAIs. While by definition, "refractory" includes only elements more refractory than Fe, Mg and Si, in H, L, LL and E chondrites, Mg and Si are depleted in a parallel manner but not to the same extent as the more refractory-lithophile elements. This depletion in refractory elements along with Mg and Si indicates that these meteorites accreted with less than their full complement of a refractory component containing CAI-like material plus olivine. The compositions of amoeboid olivine aggregates suggest that they could represent such material. The complex elemental and isotopic composition, mineralogy and texture displayed by the CAIs suggest extensive high-temperature processing of incompletely evaporated presolar material. The widespread fractionation of Mg and Si indicates temperatures in a nebular setting high enough to evaporate a large fraction, possibly all, of the olivine.

7.3.1. INTRODUCTION

The refractory lithophile elements are contained in a suite of oxide and silicate minerals that are stable at high temperatures in cosmic systems. They are the last elements to vaporize as interstellar dust is heated and the first to condense from a cooling solar gas. For this reason, these elements play a central role in understanding the extent to which presolar matter was heated, thermally processed and partially or completely vaporized in the nebula. All refractory lithophile elements are enriched in Ca,Al-rich inclusions (CAIs; see Chapter 10.3) and, with one exception, are uniformly enriched or depleted in chondrite groups. The uniform fractionation pattern implies the existence of nebular, refractory lithophile components, with characteristics resembling CAIs, that existed at all chondrite formation locations.

None of the refractory lithophile elements are particularly abundant; Al, Ca and Ti are the most abundant but even added together with their complement of O they make up only ~ 4 wt.% of an anhydrous CI chondrite. The bulk of planetary matter in the inner solar system consists of metallic Fe, FeO, MgO and SiO_2. Owing to their large abundance and relatively refractory nature, Fe, Mg and Si have traditionally been regarded as key elements in understanding high-temperature processes in the early solar system. However, now that this large group of elements is recognized to be more refractory and, of equal significance, to display elemental and isotopic fractionations, they have become a focus of attention in cosmochemistry. The term refractory lithophile as used here refers exclusively to elements which exist as stable condensates at temperatures above those where most Fe, Mg and Si would be vaporized. Refractory siderophile elements (W, Re, Os, Ir, etc.) are discussed in Chapter 7.4.

These elements behave as a coherent group in a variety of natural settings. Refractory elements are depleted in interstellar gas, presumably because they are highly concentrated in the associated interstellar dust (Field 1974). They are also enriched by nearly uniform factors of 15 to 25 relative to CI abundances in CAIs (Chapter 10.3). In bulk chondrite analyses, refractory lithophile elements behave as a coherent group, displaying abundances of 0.6 to 1.4 times CI abundances. Several of these elements are of interest to students of igneous petrology, the melting and differentiation processes by which volcanic rocks form in planetary settings. In igneous processes, when ionic charge and size effects control an element's behavior, these elements become enriched or depleted relative to one another resulting in abundance patterns distinctly different from those in chondritic meteorites. Since the characteristic feature of the chondritic pattern is coherent behavior as a group, the processes involved could not have been igneous; instead the evidence indicates that the processes were more primitive and took place prior to or during accretion of the chondritic parent bodies.

This coherent behavior made these elements useful for classification purposes long before their significance as a group of cosmochemically similar elements was recognized. As discussed in Chapters 1.1 and 7.2, the Al/Si, Ca/Si and Mg/Si ratios cluster into groups in a manner which is consistent with the other diagnostic indicators used for classification. The observation that refractory elements move in unison, i.e., increase or decrease by equivalent amounts, was first noted by Larimer and Anders (1970). Such behavior suggested a common cause, presumably related to their refractory nature. The interpretation proposed was that each group of meteorites acquired its own unique amount of a component carrying all the refractory elements in solar proportions.

In 1969, the Allende meteorite with its large, abundant CAIs fell in northern Mexico. Petrographic studies of these inclusions revealed that their dominant minerals are those predicted to be first to condense from a cooling solar gas (see, e.g., Grossman 1980). Moreover, chemical analyses showed that refractory lithophile elements are enriched in CAIs by nearly constant factors of 15 to 25 relative to CI chondrites (Wänke et al. 1974; Grossman 1980). An obvious inference is that CAIs represent relict samples of a nebular component, one which formed at very high temperatures, and that similar components were blended into each chondritic parent body in specific and characteristic amounts. Indeed, small amounts of this material are observed in other chondrite groups (Bischoff and Keil 1984; Bischoff et al. 1985).

The mineralogical and chemical evidence suggesting that CAIs formed at high temperatures piqued interest in their isotopic composition. Curiously, the data obtained raised questions about the peak temperature and led to the idea of a cool (< 1500 K) as opposed to a hot (> 2000 K) nebula. The key element discovered to display anomalous isotopic composition was O (Clayton et al. 1973). The patterns observed suggest that CAIs, minerals within CAIs, chondrules and even the various groups of chondrites, achondrites, stony irons, etc. contain variable amounts of ^{16}O (see Chapter 12.1). This suggests the presence of a nebular component highly enriched in ^{16}O. A widely accepted interpretation is that some O retained its nucleogenetic signature implying the existence of a presolar component that survived thermal processing. The material most likely to survive thermal processing in the nebula are the phases containing the most refractory elements.

While recent experiments raise the possibility that nebular processes can produce anomalous O, other isotopic data impose additional constraints. These issues are discussed in detail in Chapter 12.1 but bear on the question of interest here: the elemental-fractionation patterns. In some cases, the isotopic patterns in CAIs reflect mass-dependency (e.g., Mg, Si, Ca) suggesting partial volatilization (see Chapter 10.3). In other cases, however, the pattern is not mass dependent (e.g., Ti), implying preserved nucleogenetic signatures, while still others display a mix of mass- and nonmass-dependent patterns (elements in "FUN" inclusions; see Chapter 14.3). These are just the isotopic effects found among the refractory elements; it is also necessary to keep in mind, particularly when considering large-scale, nebula-wide processes, that highly volatile elements (H, C, Ne and Xe) also display isotopic signatures that may reflect nucleogenetic or interstellar processes implying preserved, highly volatile presolar material (see Chapters 13.1 and 13.2).

7.3.2. FRACTIONATION PATTERNS

Data Base

Owing to their importance as diagnostic indicators in classifying meteorites, as well as their cosmochemical significance, the refractory lithophile elements have been extensively studied over the past 10 to 15 years. A variety of analytical techniques have been used, many of which have been specifically developed or modified to study meteorites (Chapter 7.2). The choice of which technique to use is governed by sample size, the elements of interest and, ultimately, the question posed. When the question posed involves classification or trends between meteorite families, or when small-scale inhomogeneities are best avoided by analyzing relatively large (~ 1 g) samples, standard wet-chemical or x-ray fluorescence techniques are commonly used. However, when questions arise involving trace-element distribution between individual minerals (~ 1 μg) extracted from a small inclusion, instrumental neutron-activation (INAA) or isotope dilution are more commonly employed.

The Chondrite Patterns

The abundance and distribution of refractory-lithophile elements in meteoritic material are best discussed as a two-part problem. On the one hand, there is the question of how each chondrite group acquired its own characteristic amount of these elements and, on the other, there is the problem of how refractory elements became selectively concentrated in CAIs. Moreover, an interesting possibility is that the CAIs, however they formed, together comprise the refractory nebular component that was incorporated into chondritic planetesimals in different and characteristic amounts.

Let us begin by examining the data on the abundances of refractory lithophile elements in the different groups of chondrites followed by a brief discussion of the more pertinent data from the CAIs. In Sec. 7.3.3, we evaluate the various interpretations and, in Sec. 7.3.4, consider what might be productive directions for future research. The simplest way to illustrate the abundance patterns in the different groups of chondritic meteorites is by means of histograms; in Fig. 7.3.1, element/Si atomic ratios are plotted against the number of analyses. As is evident, the major elements (Mg/Si, Al/Si and Ca/Si) are readily resolved into three main clans: carbonaceous, ordinary and enstatite. This clear resolution forms the basis of the chemical classification of chondrites as discussed in Chapter 1.1 (Fig. 1.1.6).

Here, however, we are primarily interested in patterns. Note that in all cases, element/Si ratios are highest in carbonaceous, intermediate in ordinary

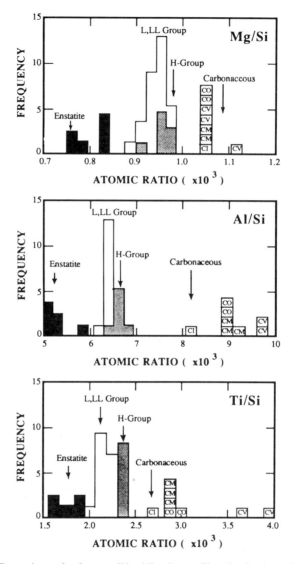

Fig. 7.3.1. Comparison of refractory lithophile element/Si ratios in the various chondrite groups. Note a consistent pattern in which the ratio in C chondrites > H, L, LL chondrites > E chondrites. There is no resolvable fractionation in Mg/Si among C chondrites but the C groups can be resolved on Al/Si, Ca/Si, Ti/Si, etc. plots (figure from Ahrens et al. 1969).

and lowest in enstatite chondrites. On Al/Si and Ti/Si histograms, it is also possible to resolve the CI, CM-CO and CV peaks among carbonaceous chondrites, but these are not resolved on the Mg/Si plot.

Before considering other data, the variable Mg/Si ratio should be clarified. While by definition "refractory lithophile elements" includes only elements that are more refractory than Fe, Mg or Si, appreciable amounts of Mg and Si, but significantly not Fe, apparently were associated with the more refractory elements in some nebular components. The association is evident in the correlation of Mg/Si with Al/Si and Ti/Si through the C—(H, L, LL)—E sequence in Fig. 7.3.1. Some of the most refractory minerals in cosmic systems contain Mg and Si (e.g., $MgAl_2O_4$, $Ca_2Al_2SiO_7$) but little Fe, though stoichiometry dictates that the amount of Mg and Si in these minerals is always less than Al and Ca. More than 95% of the Mg and Si will be present in olivine and pyroxene.

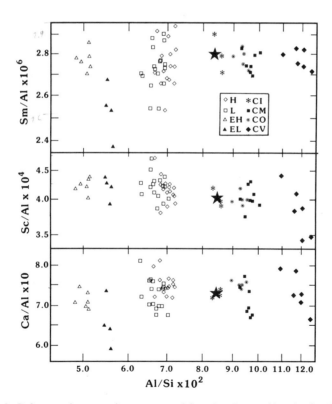

Fig. 7.3.2. Refractory-element ratios, represented here by element/Al ratios in chondrites. When data from a single analyst (Kallemeyn, see text) are used these ratios are nearly constant (± 15%). The comparatively inhomogeneous E chondrites show the greatest scatter. The Al/Si ratio does vary as also indicated in Fig. 7.3.1, presumably because the abundance of more volatile Si varies.

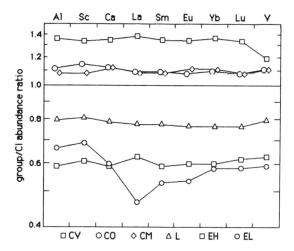

Fig. 7.3.3. Enrichment or depletion of all refractory elements to the same extent within each chondrite group. A notable exception is the EL group where the variation might be related to the observation that some lithophile elements are highly concentrated in CaS (Kallemeyn and Wasson 1981,1986, and unpublished data).

Additional data are presented in Figs. 7.3.2 and 7.3.3. The analyses plotted in Fig. 7.3.2 were collected by one analyst, G. W. Kallemeyn, using INAA and indicate the scatter expected in bulk analyses obtained by this method (Kallemeyn and Wasson 1981,1982,1986; Sears et al. 1982). The scatter increases slightly among minor and trace elements but, with few exceptions, the ratios plotted are constant within each group of chondrites to within ±15%. The greatest scatter is in the notoriously inhomogeneous enstatite chondrites. In Fig. 7.3.3, element/Si ratios in the various chondrite groups are normalized to those in CI chondrites. Except for the EL chondrites, all refractory lithophile elements appear to be enriched or depleted in the various chondrite groups by remarkably constant factors. The anomalous fractionation in the EL chondrites may or may not be related to the refractory nature of the elements involved, as discussed below.

The CAI Patterns

Although the well-studied Allende CAIs are generally enriched in all refractory lithophile elements, detailed analytical and petrographic studies discussed in other chapters have revealed a rich variety of elemental and isotopic compositions, textures and mineralogies. It is not our intention to review the details here, but rather to emphasize those chemical aspects that relate to the question of elemental fractionation and the processes involved.

Kornacki and Fegley (1986) summarized the available data on 25 elements in 97 CAIs, representing the more abundant and presumably more

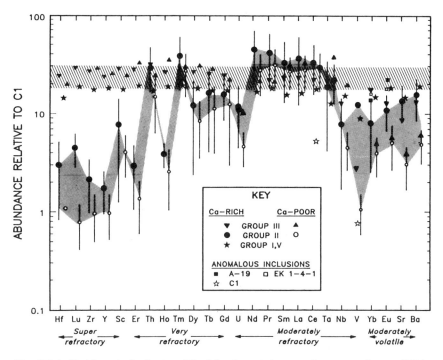

Fig. 7.3.4. Enrichment of refractory lithophile elements by a nearly constant factor of 20 in most CAIs. The exceptions are the Group II inclusions which are depleted in both the most refractory and most volatile of these elements (figure modified after Kornacki and Fegley 1986).

important types. In Fig. 7.3.4, a modified version of their Fig. 7.3.1, the elements are arranged according to best estimates of their condensation temperatures. Two patterns are evident: the CAIs in groups I, III and V are enriched in all refractory elements by factors of 20 to 30 relative to CI abundances while those in group II are less enriched in both the most and least refractory elements but are enriched by the common factor of 20 to 30 in elements of intermediate volatility. (CAIs are assigned to groups on the basis of their REE contents; see Chapter 10.3.) Kornacki and Wood (1983) found that group II CAIs are more abundant volumetrically than any other variety. They were first recognized to be chemically unique by their fractionated REE pattern (Tanaka and Masuda 1973; Mason and Martin 1977) which parallels the overall pattern in Fig. 7.3.4 and has become an important factor in interpreting CAIs (Boynton 1975). Since group II inclusions are so abundant, whole-rock analyses of Allende might be expected to display a subdued group II pattern. Fegley and Kornacki (1984) argue that the whole-rock data do in fact display a slight fractionation implying that the component carrying the most and least refractory elements, those "missing" in group II CAIs, are under-represented in Allende. In our opinion, the data are still equivocal on this point.

Another volumetrically abundant group of inclusions that may bear on the question of refractory-lithophile fractionation are the amoedoid olivine inclusions (Kornacki and Wood 1984; see also Chapter 10.3). In general, these inclusions tend to be uniformly enriched in refractory elements though the enrichments tend to be smaller than those in CAI (Grossman et al. 1979); presumably this reflects dilution of CAI material by the olivine.

7.3.3. INTERPRETATION OF THE FRACTIONATION PATTERNS

Chondritic Fractionation

The chondritic fractionation pattern is most simply explained by a model in which each chondrite group acquired its own proportion of a refractory component with a composition approximated by mixing CAI-like material plus the olivine required to account for the variations in Mg/Si (Larimer and Anders 1970). The ratio of CAI to olivine in this refractory component apparently varied at different nebular locations. Assume that each parent body accreted from a volume of nebular gas in which the condensable elements were originally present in CI proportions. Those chondrites which are enriched in refractory elements relative to CI, such as CV chondrites, must contain an excess amount of refractory component while those which are depleted, i.e., H, L, LL and E chondrites, must contain less than their share of refractory component. Although this explanation is almost certainly oversimplified, the model serves a useful purpose in that it can readily be tested by further observations and analyses.

One argument in favor of blending into the various chondritic parent bodies slightly different amounts of a refractory-rich component is that all refractory lithophile elements vary by constant factors (Figs. 7.3.1, 7.3.2 and 7.3.3). Furthermore, refractory elements are enriched in many CAIs by a nearly constant factor (Fig. 7.3.4). A useful device for testing this model and for obtaining some insights on the direction and extent of fractionation is a mixing diagram, as suggested independently by Kerridge (1979) and Larimer (1979). Such diagrams have long been used to interpret isotopic ratios but are adopted here to interpret elemental ratios. Two element pairs are plotted on an x-y diagram with one of the three elements selected to serve as a common denominator.

In constructing such diagrams, the first question to consider is which element to use for normalization, or as the common denominator. We adopt Al for normalization because it is generally well determined and because Al is the most refractory of the more abundant elements; if any element is condensed, then Al can reasonably be expected to be condensed. Also in contrast to the other possible choice Ca, Al appears more retentive during terrestrial weathering.

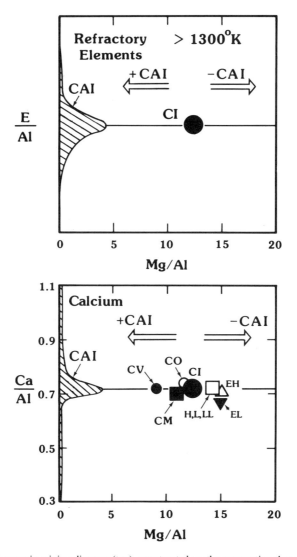

Fig. 7.3.5. A generic mixing diagram (top), constructed on the assumption that chondrites are mixtures of two materials. One is mixed with CI and the other with CAI-like composition. This may be compared to the observed patterns in chondritic meteorites (bottom). The composition of the various C group chondrites evidently reflects an addition of CAI-like material to CI material whereas the H, L, LL and E chondrites do not contain their full solar complement of CAI-like material.

A generic mixing diagram, applicable to any refractory lithophile element, is presented in Fig. 7.3.5 along with a diagram that illustrates the behavior of Ca and Mg in chondritic material. This diagram should also be compared to Fig. 7.3.2 where the data for a number of minor and trace elements are presented. First, consider a simple model in which all chondritic material is derived from a nebula with CI composition. In principle, the refractory-element/Al ratio in all other chondrites can be obtained by adding or subtracting a refractory component *RC* which contains its full complement of all refractory lithophile elements in solar proportions but relatively little Mg, Si and even less of the more volatile elements. Chondrites formed from a mixture of CI plus excess *RC* have compositions falling between CI and the *RC* field (e.g., the CV group). Conversely, chondrites formed from CI but lacking its full complement of *RC* material have bulk compositions shifted away from the *RC* field (e.g., the EH group). This mixing model thus explains the varying concentrations of all those refractory lithophile elements that occur in constant, solar proportions in chondrite groups, i.e., all except a few elements in the EL group.

The most uncertain aspect of this diagram is the composition of the refractory component. Here, we have schematically plotted McSween's (1977) data on 270 inclusions from 19 C chondrites obtained by the electron microprobe, defocused beam technique. The data indicate that in all but 40 cases, the Ca/Al ratio falls within ± 15% of the solar ratio with an almost equal distribution of high and low ratios, though there is a tendency for the most refractory-rich inclusions to have systematically low Ca/Al ratios. However, some (13) inclusions with nonsolar ratios are coarse grained, which increases the possibility of sampling problems that are inherent in this technique.

Let us now turn to Mg and Si which tend to follow the refractory elements (Fig. 7.3.1) but with important differences. Both Mg and Si are deficient in CAIs relative to Al and the Mg/Al and Si/Al ratios vary markedly among chondrite groups. A mixing diagram for these two elements (Fig. 7.3.6) provides some additional insights into the fractionation process. When the averaged data from chondrite groups are plotted, two, rather than one, mixing lines are indicated. C-group chondrites define one line which passes through the CAI field, near the origin, while the ordinary and enstatite chondrites define a second line which does not pass through the CAI field, intercepting the Mg/Al axis at ~ 9. It is interesting, and almost assuredly significant, that the two lines intersect at the CI point, which was noted independently by Kerridge and Larimer. These relations are often discussed in terms of all C chondrites having the same Mg/Si ratio (1.07) while the ratio is lower in ordinary (0.95) and enstatite (0.85 to 0.75) chondrites. The mixing diagram is another way of illustrating these relations and relating them to Al.

The simplest model to explain these fractionations requires two different refractory components: one with a composition resembling CAI and the other

Larimer

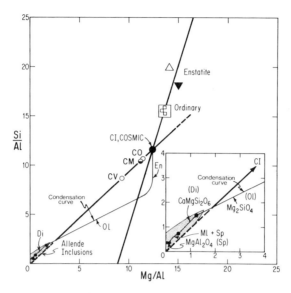

Fig. 7.3.6. The mixing diagram for Al, Mg and Si. This indicates a complexity that was not evident in Fig. 7.3.5. C chondrites can still be explained as consisting of a mixture of CAI-like and CI material. Since CAIs contain little Mg and Si, the addition does not appreciably change the Mg/Si ratio. However, to generate the composition of H, L, LL and E chondrites from CI composition requires loss of a component that contained all the refractory lithophile elements in solar proportions plus a significant fraction of Mg and Si. Condensed dust with the required composition would be present in the nebula when all the refractory elements plus a fraction of olivine were condensed. Significantly, the two fractionation lines, one for C and one for H, L, LL and E chondrites cross at the CI composition point (figure modified after Larimer 1979).

a mixture of CAI-like material plus olivine (Mg_2SiO_4). C-chondrite compositions are then explained as reflecting different admixtures of a *RC* having ~CAI composition to material with CI composition. In Fig. 7.3.6 the compositions of CM, CO and CV chondrites fall on a line with a slope of 1 because the major change in abundance among the three elements considered is the addition of Al.

To derive ordinary and enstatite chondrites from CI composition, a refractory component enriched in Mg and Si relative to CAI must be added or subtracted. For example, a material which falls on the extension of the mixing line at higher Si/Al and Mg/Al ratios than E chondrites could be addded to CI composition. But material with such an unusual composition is not known to exist nor is there an obvious way to form such material. Alternatively, ordinary and enstatite chondrites could be derived from CI composition by subtracting, or losing, material which falls on the opposite extension of the mixing line, material with Mg/Al and Si/Al ratios lower than the CI values.

Material with such a composition is found in CV and LL chondrites. In CV chondrites, it occurs as amoeboid olivine inclusions (AOIs) or aggregates (AOAs) which, for our purpose, can be regarded as a mixture of olivine and fine-grained CAI-like material. A chemical signature of this component is also found in chondrules from Semarkona (Grossman and Wasson 1983), an LL 3.0 chondrite, indicating that it was present in the region of ordinary-chondrite formation.

Superimposed on the mixing diagram (Fig. 7.3.6) is a track of the composition of condensed material in equilibrium with a solar gas as a function of temperature. The key feature is that Mg_2SiO_4 becomes stable just below the temperature where all refractory minerals are condensed. Since the proportion of Mg/Si in olivine is 2, the composition of the condensed material follows a line with a slope of 1/2. If aggregates of olivine and precursor refractory grains, resembling amoeboid olivine aggregates, formed in a reservoir with CI composition and were only partially accreted, then the remaining dust would have the composition of ordinary and enstatite chondrites.

Several important points can be gleaned from Fig. 7.3.6: (1) there are two fractionation trends which intersect at the CI point strongly suggesting that all chondrites are derived from this composition, even though they may have originated at widely separated locations; and (2) at least two refractory components are required to account for the abundances of the refractory elements in the different groups of chondrites. One refractory component with a composition similar to that of average CAIs evidently was accreted in excess amounts to produce the abundances observed in carbonaceous chondrites. Another refractory component with the composition of average CAI-like material plus olivine apparently was not fully accreted to produce the ordinary and enstatite chondrites.

The fraction of material gained or lost can also be estimated by regarding the trends on Fig. 7.3.6 as mixing lines (Larimer 1979). For example, consider the ordinary and enstatite chondrites where material evidently was lost. Let α = fraction of material lost. A series of equations may then be written:

$$Al_{CI} - \alpha Al_{RC} = Al_{O(E)} \qquad (1)$$

$$Mg_{CI} - \alpha Mg_{RC} = Mg_{O(E)} \qquad (2)$$

etc. where the number of atoms of each element in the refractory component, CI and ordinary (or enstatite) chondrites are indicated by the subscripts *RC*, *CI* and *O(E)*. Consider a volume of the nebula that originally contained the elements in CI proportions (Anders and Ebihara 1982). Assuming that Al is completely condensed and entirely resides in the refractory component at the time of fractionation ($Al_{CI} = Al_{RC}$), Eq. (2) may be divided by Eq. (1) and solved for α:

TABLE 7.3.1
Fractions of Refractory Component (α) and Lithophile Elements [$f(X)$] that Did Not Accrete in the Ordinary- and Enstatite-Chondrite Regions

Element/Element Ratio (atoms)	CI	Ordinary	Enstatite	Refractory Component
Al	8.49×10^4	5.20×10^4	3.53×10^4	8.49×10^4
Mg	1.075×10^6	7.10×10^5	5.27×10^5	9.51×10^5
Si	1.00×10^6	7.80×10^5	6.75×10^5	5.61×10^5
Mg/Al	12.66	13.65	14.85	11.10
Si/Al	11.78	15.06	19.04	6.60
α	0	0.39	0.58	
f (Al, Ca, Ti, etc.)		0.39	0.58	
f (Mg)		0.34	0.51	
f (Si)		0.22	0.33	

$$\alpha = \frac{(Mg/Al)_{O(E)} - (Mg/Al)_{CI}}{(Mg/Al)_{O(E)} - (Mg/Al)_{RC}}. \tag{3}$$

The Mg/Al and Si/Al ratios in the refractory component are obtained from the intersection of the fractionated and the condensation lines; $(Mg/Al)_{RC} \sim 11.10$ and $(Si/Al)_{RC} \sim 6.60$ and the ratios in ordinary and enstatite chondrites are obtained from the data discussed above. The data used and the results of the calculations are presented in Table 7.3.1. The fraction α of refractory component that did not accrete in the regions of ordinary and enstatite chondrites is approximately 0.39 and 0.58. From the inferred composition of the refractory component, the fraction f of each element that did not accrete can be readily computed: f(Al, Ca, Ti, etc.) = 0.39, f(Mg) = 0.34 and f(Si) = 0.22 for ordinary chondrites and f(Al, etc.) = 0.58, f(Mg) = 0.51 and f(Si) = 0.33 for the average E chondrite, where the fractionation is more variable. A significant fraction of nebular material evidently was not accreted by the ordinary and E-chondrite parent bodies, 1/3 to 1/2 Mg, 1/5 to 1/3 Si plus even larger fractions of Al and the other refractory elements. These fractionations have important implications regarding the geochemistry and evolution of chondritic material.

Before turning to the implications, however, the non-CI refractory-lithophile pattern in EL chondrites (Fig. 7.3.3) requires additional discussion. The pattern is reproducible, thus cannot be explained by sampling variations. Many of the refractory lithophiles reside in CaS, which is especially prone to weathering, thus some caution must be exercised in drawing conclusions from these data. However, the EL data plotted in Fig. 7.3.3 include only falls—even slightly weathered finds were excluded—and the mineralogically similar EH chondrites do not show fractionated patterns. These arguments

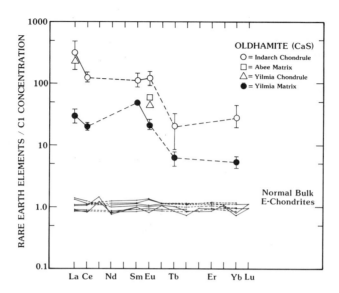

Fig. 7.3.7. Enrichment of REE in CaS extracted from enstatite chondrites. Note that light REE are more enriched than the heavy REE. Evidently these and several other refractory lithophile elements occur in the sulfides of the highly reduced E chondrites (figure modified after Larimer and Ganapathy 1987).

and the reproducibility of the EL patterns indicate that the fractionations originated in the nebula. Larimer and Ganapathy (1987) report high concentrations of REE in CaS, with a significant enrichment of light over heavy REE (Fig. 7.3.7) and an overall decrease in concentration with increasing degree of metamorphism. The observed EL fractionation thus is consistent with loss of unmetamorphosed (i.e., nebular) CaS prior to or during accretion.

The data suggest that the carriers of these nominally lithophile elements during the primitive fractionation event were phases, such as CaS, that are unique to enstatite chondrites. Several of these minerals are only stable at high temperatures in a gas somewhat more reducing than a gas with solar composition (Larimer and Bartholomay 1979). If CaS was in fact the carrier at the time of fractionation then the inference would be that the reducing environment necessary to form CaS existed before the fractionation of refractory lithophile elements at the EL formation location. Kallemeyn and Wasson (1986) suggest that EL chondrites formed closer to the Sun than EH chondrites and that, since the radial velocity of solids relative to the gas increased with decreasing distance from the Sun, this led to enhanced sorting of solid grains on the basis of size and density.

CAI Fractionations

Some CAIs are thought to be incompletely evaporated residues of presolar solids, formed during infall into the solar nebula (Chou et al. 1976; Wood 1981; cf. Chapter 10.3). The widespread isotopic anomalies in CAIs support this view. Many large, coarse-grained and evidently once-molten, CAIs are difficult to explain as equilibrium condensates but are more readily understood in terms of heating and partial volatilization of large, volatile-rich presolar lumps. On the other hand, fine-grained, porous CAIs with fractionated refractory-lithophile element patterns evidently reacted more extensively with nebular gases, yet managed to exchange incompletely those elements which retain isotopic anomalies.

The elemental fractionation patterns among the different types of CAIs are similar in some ways—and different in others—to that found in the different chondrite groups. The key difference is that refractory lithophile elements behave coherently among different chondrite groups but display fractionations among the different types of CAIs. The chondrite pattern implies that all the refractory lithophile elements were condensed and gained or lost as a group, in one or more unfractionated nebular components. But in some CAIs, it appears that only a "super-refractory" subset was condensed at the time of an even more primitive fractionation event. The common interpretation is that the nebula was once hot enough to volatilize all but these most refractory elements.

The first hint of elemental fractionation in CAIs was the discovery of a fractionated REE pattern (Tanaka and Masuda 1973; Mason and Martin 1977). The common REE pattern in geochemistry is one where concentration varies as a smooth function of mass, or ionic radius, with an occasional Eu anomaly owing to the ease with which it is reduced to the $+2$ state. But in some CAIs a unique, irregular pattern is found. Boynton (1975) calculated the relative volatilities of REE and Y in a solar gas and showed that this observed CAI pattern could be reproduced under a special set of circumstances. The interesting requirement is that these CAIs formed from a solar gas from which the most refractory REE had been removed, presumably by condensation, and before the most volatile (Eu, Yb) had condensed (Fig. 7.3.8). The implication is that nebular gas was once hot enough to vaporize all but the most refractory REE and their host phases. Since the amount of "super-refractory" material is very small, this poses an interesting problem in accounting for the preservation of the isotopic anomalies found throughout the various types of CAIs (see Chapter 10.3).

If such super-refractory material formed, perhaps some relict might still exist either as a separate inclusion or as a fragment of a larger inclusion. In fact, such inclusions have been found, bearing the complementary trace-element pattern, in the Essebi, Murchison and Ornans chondrites (Boynton et al. 1980; Palme et al. 1982; Ekambaram et al. 1984; Davis 1984; El Goresy

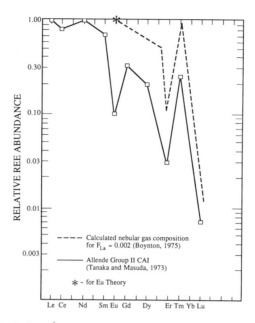

Fig. 7.3.8. The highly fractionated REE pattern in Group II CAIs. This pattern matches that calculated for a solar gas after a fraction of the most refractory REE have condensed. The Group II CAIs are also depleted in the most volatile REE, Eu and Yb, indicating some intermediate temperature of formation. The CAI pattern thus is similar to that in Fig. 7.3.4 (figure modified after Boynton 1975).

et al. 1984). Evidently, super-refractory inclusions are more common in some chondrites than in others, indicating that formation conditions varied in time or space.

Kornacki and Fegley (1986) divide the refractory lithophile elements into four subgroups, based on predicted condensation temperatures and fractionation behavior in CAIs (Fig. 7.3.4). This arrangement is subject to revision because the calculated condensation temperatures of many trace elements are uncertain. With this caveat, the pattern is of interest since Group II inclusions with their fractionated REE pattern are also depleted in other super-refractory elements. These authors interpret all CAIs, including these of Group II, as evaporative residues rather than condensates, and attribute the fractionation patterns not just to volatility but to crystal-chemical and kinetic effects as well. One appealing feature of this interpretation is that it explains the survival of isotopic anomalies in inclusions with Group II REE patterns.

7.3.4. IMPLICATIONS AND FUTURE WORK

The abundance and distribution of refractory lithophile elements in chondritic meteorites and CAIs indicate that, at least locally, protoplanetary material was heated to very high temperatures. Three elemental fractionations can be resolved:

1. The fractionations found within the CAIs indicate temperatures at least high enough to evaporate a large fraction of the refractory elements.
2. The fractionation of refractory lithophile elements as a group requires a setting in which only these elements were condensed (at temperatures high enough to evaporate most Mg-silicates) in order to aggregate into objects resembling CAIs which eventually were nonuniformly incorporated into parent bodies.
3. The fractionation of Mg and Si, two of the most abundant elements in chondritic material, indicates that high temperatures were prevalent throughout the region of chondrite formation.

While this chapter has focused on elemental fractionations, any discussion of peak temperatures in the nebula must also take into account the widespread isotopic anomalies observed in chondrites. Since the discovery of ^{16}O anomalies (Clayton et al. 1973), the conventional thought has been to attribute the effects to presolar relicts of nucleogenetic processes. Though this view may change (Chapter 12.1), the exotic chemical mechanisms suggested to date do not explain all the anomalies. It appears that at least a vague, isotopic memory of presolar material exists in the meteorites. A key question is whether this presolar material accreted onto the nebular disk after the peak temperature was reached. As discussed in Chapter 13.1, the best way to resolve this question is to isolate and characterize the carrier phases. Our current assessment is that the existence of the O, Ti and FUN anomalies together with the mass fractionation of Mg and Si in CAIs (Chapters 10.3 and 14.3) is evidence of incomplete evaporation.

A variety of models have been proposed to explain the elemental and isotopic abundances of refractory lithophile elements in chondritic material. Recent models emphasize CAI and isotopic data. These models tend to be characterized by a nebula cool enough to preserve the presolar memory but hot enough to vaporize most of the dust and melt the remainder (Chapter 6.3). Experimental data suggest that some CAIs cooled rapidly (from above the solidus to 200° below it) in at most a matter of days (Stolper and Paque 1986; see also Chapter 10.3). Since the entire nebula could not possibly cool so rapidly, local heating events are required in all models. The fractionation of refractory elements, including Mg and Si, between the chondrite groups usually is attributed to differential settling of condensed dust (Larimer and Anders 1970; Wai and Wasson 1977), though transport of CAIs through the nebula has also been suggested (Chapter 6.3).

In Table 7.3.2 several viable models are evaluated according to how well they account for six categories of pertinent data. The model long favored by most meteoriticists of a hot, totally vaporizing nebula that cools monotonically suffers in its inability to account for any presolar memory and the rapid

TABLE 7.3.2
Evaluation of Models Suggested to Explain Observations

Observations	Optional Models				
	Statically Hot Nebula	Episodically Hot Nebula	Warm Nebula + Flash Heating	Drag Heating	Mass Transport
1. Mg/Si fractionation	+	+	−	−	−
2. Refrac. Elem. fractionation	+	+	−	−	−
3. CAI formation	+	+	?	+	+
4. CAI fractionation	+	+	−	−	−
5. Presolar isotope anomalies	−	+	+	+	+
6. Rapid cooling of CAI	−	+	+	+	+

cooling rates of CAIs. On the other hand, a cool nebula except for local heating events cannot easily account for differences in refractory-element content among chondrite parent bodies. In mass transport models, high-temperature processing occurs in the inner solar system (<1 AU) and the refractory material then moves to the asteroid belt. Even if such a model proves to be dynamically plausible, it cannot readily explain fractionation among different CAIs nor how the various components are reassembled to produce unfractionated whole-rock refractory-element patterns. Besides, while material additions are inherent in this model, and explain the admixture of a refractory component to C chondrites, the origin of ordinary and E chondrites which involved loss of material with a specific composition is not an inherent feature of such a model.

The most satisfying model is a compromise, it adopts the positive aspects of several other models. We call it the episodically hot-nebula model. To account for Group II CAI composition requires temperatures high enough to vaporize most, and occasionally all, of the refractory elements. Differential settling of dusty residual aggregates, which still carried some presolar memory, plus fresh condensate is required to account for the CAI fractionations as well as the differences between chondrite parent bodies. A flash-heating mechanism, perhaps arising during deceleration of infalling parcels of interstellar debris, continually processed the solids. The most refractory minerals either failed to evaporate or quickly recondensed. Conditions varied locally as indicated by differences in the grain size and composition of olivine (coarse, low Fe^{+2} in CV; fine, high Fe^{+2} in LL3 chondrites) and differences in the minerals associated with the refractory lithophile elements (olivine in EH, LL, CV; low-Ca pyroxene in CO, CM: Rubin and Wasson 1986). Further cooling led to condensation of the remaining olivine, pyroxene and, eventually, all the more volatile materials.

An important, though infrequently mentioned, constraint on nebular processes is the extensive fractionation of Mg and Si. If these elements were fully condensed at the time of fractionation, it would be necessary to find a mechanism that discriminates between high temperature ($T > 1300$ K) and somewhat lower temperature olivine ($T < 1290$ K). One proposal is that Mg and Si were only partially condensed at the time of this widespread fractionation event; another is that the grain size of high-temperature olivine varied locally.

A curious corollary of Mg-Si fractionation is that there is no correlation with Fe/Si fractionation even though FeNi-metal and olivine condense together over a wide range of P-T conditions. Kerridge (1979) suggested as possible explanations that either Fe condensation was inhibited by kinetic effects, or FeNi and olivine were present as discrete grains and one of the many physical differences between metal and silicate led to different fractionation processes, perhaps accentuated by differences in grain size. Yet another possibility is local enrichment in the dust-to-gas ratio. Since the dust carries a substantial amount of condensed O, the increase in condensation (or evaporation) temperatures of oxides and silicates is greater (at constant P_T) with increasing dust enrichments than for Fe which condenses in elemental form. However, this requires evaporation of dust without significant admixture of the vapor with the pre-existing nebular gas.

Many unanswered questions remain. No samples of the refractory component inferred to exist, but evidently not accreted, in the ordinary- and E-chondrite source regions have been found; it may have resembled the amoeboid olivine inclusions. The distributions of REE among different chondrite groups needs further study as it is not clear whether existing whole-rock data for ordinary and E chondrites are consistent with fractionation of a single refractory component (Evensen et al. 1978). A more adequate inventory of CAIs with fractionated elemental abundances is needed, which requires analyses of petrographically studied samples complemented with measurements of isotopic composition. The fractionation of refractory lithophile elements in E chondrites is poorly understood; additional studies on unequilibrated E chondrites should provide new clues. There is no shortage of interesting problems for future research.

Acknowledgments. This work was supported by a grant from the National Aeronautics and Space Administration (JWL) and a National Science Foundation grant (JTW). We are indebted to B. Fegley and H. Palme for constructive reviews.

REFERENCES

Ahrens, L. H., von Michaelis, H., Erlank, A. J., and Willis, J. P. 1969. Fractionation of some abundant lithophile elements in chondrites. In *Meteorite Research,* ed. P. M. Millman (Dordrecht: D. Reidel), pp. 166–173.
Anders, E., and Ebihara, M. 1982. Solar-system abundances of the elements. *Geochim. Cosmochim. Acta* 46:2363–2380.
Bischoff, A., and Keil, K. 1984. Al-rich objects in ordinary chondrites: Related origin of carbonaceous and ordinary chondrites and their constituents. *Geochim. Cosmochim. Acta* 48:693–709.
Bischoff, A., Keil, K., and Stöffler, D. 1985. Perovskite-hibonite-spinel-bearing inclusions and Al-rich chondrules and fragments in enstatite chondrites. *Chem. Erde.* 44:97–106.
Boynton, W. V. 1975. Fractionation in the solar nebula: Condensation of yttrium and the rare earth elements. *Geochim. Cosmochim. Acta* 39:569–584.
Boynton, W. V., Frazier, R. M., and Macdougall, J. D. 1980. Identification of an ultra-refractory component in the Murchison meteorite. *Lunar Planet. Sci.* XI:103–105 (abstract).
Chou, C.-L., Baedecker, P. A., and Wasson, J. T. 1976. Allende inclusions: Volatile-element distribution and evidence for incomplete volatilization of presolar solids. *Geochim. Cosmochim. Acta* 44:85–94.
Clayton, R. N., Grossman, L., and Mayeda, T. 1973. A component of primitive nuclear composition in carbonaceous meteorites. *Science* 182:485–488.
Davis, A. M. 1984. A scandalously refractory inclusion in Ornans. *Meteoritics* 19:214 (abstract).
Ekambarum, V., Kawabe, K., Tanaka, T., Davis, A. M., and Grossman, L. 1984. Chemical compositions of refractory inclusions in the Murchison C2 chondrite. *Geochim. Cosmochim. Acta* 48:2089–2105.
El Goresy, A., Palme, H., Yabuki, H., Nagel, K., Herwerth, I., and Ramdohr, P. 1984. A calcium-aluminum-rich inclusion from the Essebi (CM2) chondrite: Evidence for captured spinel-hibonite spherules and for an ultra-refractory rimming sequence. *Geochim. Cosmochim. Acta* 48:2283–2298.
Evensen, N. M., Hamilton, P. J., and O'Nions, R. K. 1978. Rare-earth abundances in chondritic meteorites. *Geochim. Cosmochim. Acta* 42:1199–1212.
Fegley, B., Jr., and Kornacki, A. S. 1984. The origin and mineral chemistry of Group II inclusions in carbonaceous chondrites. *Lunar Planet. Sci.* XV:262–263 (abstract).
Field, G. B. 1974. Interstellar abundances: Gas and dust. *Astrophys. J.* 187:453–459.
Grossman, J. N., and Wasson, J. T. 1983. Refractory precursor components of Semarkona chondrules and the fractionation of refractory elements among chondrites. *Geochim. Cosmochim. Acta* 47:759–771.
Grossman, L. 1980. Refractory inclusions in the Allende meteorite. *Ann. Rev. Earth Planet. Sci.* 8:559–608.
Grossman, L., Ganapathy, R., Methot, R. L., and Davis, A. M. 1979. Trace elements in the Allende meteorite—IV. Amoeboid olivine aggregates. *Geochim. Cosmochim. Acta* 43:817–829.
Kallemeyn, G. W., and Wasson, J. T. 1981. The compositional classification of chondrites— I. The carbonaceous chondrite groups. *Geochim. Cosmochim. Acta* 45:1217–1230.
Kallemeyn, G. W., and Wasson, J. T. 1982. The compositional classification of chondrites— III. Ungrouped carbonaceous chondrites. *Geochim. Cosmochim. Acta* 46:2217–2228.
Kallemeyn, G. W., and Wasson, J. T. 1986. Compositions of enstatite (EH4, EH4,5 and EL6) chondrites: Implications regarding their formation. *Geochim. Cosmochim. Acta* 50:2153–2164.
Kerridge, J. F. 1979. Fractionation of refractory lithophile elements among chondritic meteorites. *Proc. Lunar Planet. Sci. Conf.* 10:989–996.
Kornacki, A. S., and Fegley, B., Jr. 1986. The abundance and relative volatility of refractory trace elements in Allende Ca,Al-rich inclusions: Implications for chemical and physical processes in the solar nebula. *Earth Planet. Sci. Lett.* 79:217–234.
Kornacki, A. S., and Wood, J. A. 1984. Petrography and classification of Ca,Al-rich and olivine-rich inclusions in the Allende CV3 chondrite. *Proc. Lunar Planet. Sci. Conf.* 14, *J. Geophys. Res. Suppl.* 89:B573–B587.

Larimer, J. W. 1979. The condensation and fractionation of refractory lithophile elements. *Icarus* 40:446–454.

Larimer, J. W., and Anders, E. 1970. Chemical fractionations in meteorites—III. Major element fractionations in chondrites. *Geochim. Cosmochim. Acta* 34:367–387.

Larimer, J. W., and Bartholomay, M. 1979. The role of carbon and oxygen in cosmic gases: Some applications to the chemistry and mineralogy of enstatite chondrites. *Geochim. Cosmochim. Acta* 43:1455–1466.

Larimer, J. W., and Ganapathy, R. 1987. The trace element chemistry of CaS in enstatite chondrites and some implications regarding its origin. *Earth Planet. Sci. Lett.* 84:123–134.

Mason, B., and Martin, P. M. 1977. Geochemical differences among the components of the Allende meteorite. *Smithsonian Contrib. Earth Sci.* 19:84–95.

McSween, H. Y. 1977. Chemical and petrographic constraints on the origin of chondrules and inclusions in carbonaceous chondrites. *Geochim. Cosmochim. Acta* 41:1777–1790.

Palme, H., Wlotzka, F., Nagel, K., and El Goresy, A. 1982. An ultra-refractory inclusion from the Ornans meteorite. *Earth Planet. Sci. Lett.* 61:1–12.

Rubin, A. E., and Wasson, J. T. 1986. Chondrules in the Murray CM2 meteorite and compositional differences between CM-CO and ordinary chondrite chondrules. *Geochim. Cosmochim. Acta* 50:307–315.

Sears, D. W., Kallemeyn, G. W., and Wasson, J. T. 1982. The compositional classification of chondrites: II. The enstatite chondrite groups. *Geochim. Cosmochim. Acta* 46:597–608.

Stolper, E., and Paque, J. M. 1986. Crystallization sequences of Ca-Al inclusions from Allende: The effects of cooling rates and maximum temperatures. *Geochim. Cosmochim. Acta* 50:1785–1806.

Tanaka, T., and Masuda, A. 1973. Rare-earth elements in matrix, inclusions and chondrules of the Allende meteorite. *Icarus* 19:523–530.

Wai, C. M., and Wasson, J. T. 1977. Nebular condensation of moderately volatile elements and their abundances in ordinary chondrites. *Earth Planet. Sci. Lett.* 36:1–13.

Wänke, H., Baddenhouse, H., Palme, H., and Spettel, B. 1974. On the chemistry of the Allende inclusions and their origin as high-temperature condensates. *Earth Planet. Sci. Lett.* 23:1–7.

Wood, J. A. 1981. The interstellar dust as a precursor of Ca,Al-rich inclusions in carbonaceous chondrites. *Earth Planet. Sci. Lett.* 56:32–44.

7.4. SIDEROPHILE ELEMENT FRACTIONATION

JOHN W. LARIMER
Arizona State University

and

JOHN T. WASSON
University of California at Los Angeles

The variation in the metal-to-silicate ratio in chondrites can be attributed in part to variable oxidation levels (Fe^{+2}/Fe^o ratios) and in part to variable total Fe (Ni, Co, etc.) abundances. The variation in bulk Fe abundance implies nebular fractionation of Fe relative to Si in the source regions of the various chondrite groups. CI-normalized chondrite-group abundances of minor and trace siderophile elements are similar to those of Fe, but show important intergroup and interelement differences. Their fractionation appears to be related to their cosmic siderophile character, i.e., condensation into metal or oxides, and to their cosmic volatility. In the CM, CO and CV groups, refractory-siderophile/Ni ratios are enhanced to about the same degree as the refractory-lithophile/Si ratios, but variations in the Ir/Ni ratio among ordinary chondrites suggest the existence of a refractory siderophile nebular component distinct from the refractory lithophile component. Elements more volatile than Fe also display abundances that tend to track Fe/Si variations implying that they were already condensed and thus that nebular temperatures were <800 K at the time of the metal/silicate fractionation. With the possible exception of H3 chondrites, there is no correlation between siderophile-element content and petrographic type among the ordinary-chondrite groups, and thus no evidence for systematic change in siderophile abundance with parent-body burial depth.

7.4.1. INTRODUCTION

It is generally accepted that the density differences between Mercury, Earth and Moon reflect different metal/silicate ratios. The cause of these differences is a key question in modeling the formation of the planets and satellites, but despite considerable effort, no consensus on the cause has been reached. These differences in planetary density and the implied metal/silicate fractionation were recognized long before the much smaller differences in Fe/Si ratios in chondrites were discovered. It was not until Urey and Craig (1953), in a survey of superior bulk-chemical analyses, discovered the hiatus in Fe/Si ratios between the H- and L-group chondrites that meteorite parent bodies joined the list of objects displaying metal/silicate fractionations.

Attention then focused on estimating the Fe/Si ratio in mean solar-system matter; at that time the Fe/Si ratios in chondrites appeared to be greater than that of the Sun. This discrepancy was eventually resolved with the discovery of a ten-fold error in the oscillator strength of the Fe I lines used to determine solar abundance (Garz and Kock 1969). The current solar Fe/Si ratio coincides with the CI value to within the uncertainty in the solar value. With this change, as well as revisions and new determinations of many other elements, the once-questioned similarity between the Sun and CI chondrites can no longer be doubted; they contain the elements in virtually the same proportions (Anders and Ebihara 1982).

Today, a key question is how and when the differences in Fe/Si ratio arose. In the Earth, Moon and planets, the compositional record of planetary formation processes has been altered by post-accretional events. Iron and Si have been redistributed by igneous differentiation and core formation leaving mantle and crustal rocks almost devoid of the easily reduced siderophile elements. Evidently, during core formation such elements were efficiently (though not totally) extracted from the silicate mantles. The only accessible clues regarding the abundances of siderophile elements in the Earth, Moon and planets are contained in the small amount of residual siderophiles that remain in crust and mantle rocks.

For this reason the proposition that the variation in Fe/Si between the planets and satellites is the product of the same process that produced the differences in chondrites is difficult to prove or disprove. What is clear is that the record of nebular events is better preserved in chondrites because they have not been subjected to the igneous and core-forming processes characteristic of more evolved bodies. In chondrites, the abundances of many siderophile elements increase and decrease along with Fe, indicating that they were alloyed in the metal at the time of nebular fractionation. These elements are similar chemically yet differ markedly in volatility: relative to the common or reference elements Fe, Co and Ni, they range from refractory (Ir, Os, Re, etc.) to moderately volatile (Au, As, etc.). A proper understanding of their abundance and distribution in chondrites holds promise of revealing the timing, the *P-T* conditions and the causal mechanisms of their fractionations.

Before continuing, we need to describe our picture of the solar nebula and clarify some terms. We use the following simple cosmochemical picture of the nebula:

1. The mean composition at all locations was originally solar;
2. The condensable elements were distributed in a variety of solid phases prior to accretion;
3. Portions of one or more phases that are similarly fractionated, presumably by physical mechanisms, are designated components;
4. Chondrites formed by the agglomeration of these components;
5. The components were not usually agglomerated in their solar proportions, thus the products are generally fractionated relative to solar (i.e., CI) composition.

One goal in cosmochemistry is to characterize the components and to understand the fractionation mechanisms at the locations where the various chondrite parent bodies formed.

The term metal-silicate fractionation, refers to the abundance of siderophile elements (Au, Ir, Ni, etc.) relative to that of lithophile elements (Mg, Si, etc.). Although this may seem to imply that certain elements were associated with the "metal" during fractionation, it is not always clear whether the metal still existed in its original form. Some siderophile elements may have been present in nonmetallic phases such as oxides or sulfides which formed by "corrosion" of the metal prior to fractionation. Also, while the term metal-silicate fractionation would seem to imply two discrete components, in reality there may have been several metallic components which differed in composition (e.g., Ga/Ir ratio). In addition, fine-grained silicates may have been dispersed in the metal or fine-grained metal in the silicates which would lead to less clear-cut patterns than the term metal-silicate fractionation implies.

Iron is an especially interesting element because in most chondrite groups it is distributed among three phases: metal, FeS and silicates. This distribution reflects differences in nebular processes from location to location, and continued careful investigation may eventually tell us much about the temperature history and the mechanics of agglomeration at these different locations.

7.4.2. THE ABUNDANCE AND DISTRIBUTION OF IRON IN CHONDRITES

Prior (1916) noted two relationships that came to be known as Prior's Rules: the lower the metal content of a chondrite, (1) the higher the Ni content of the metal and (2) the higher the FeO content of the associated FeMg-silicates. These rules are still valid in a broad sense but were formulated when analytical techniques were not sufficiently precise to resolve the differences in bulk Fe/Si ratios among chondrites. They apply only to variations in Fe°/FeO which are now recognized to be superimposed on variations in total Fe abundance. Although prior's rules offer the student the unifying concept that a decrease in the fraction of Fe present as Fe° in the metal is more or less compensated for by an increase in the fraction of Fe as FeO in the silicates, an unfortunate byproduct was the persistence of the view that the different groups of chondrites were genetically related by simple redox processes.

The concept of a simple redox process should have been abandoned when Urey and Craig (1953) demonstrated that, in addition to the differences in oxidation state, the more reduced H-group chondrites also have a much higher Fe/Si ratio than the more oxidized L-group chondrites. These relations are revealed in a Urey-Craig diagram (Fig. 7.4.1). Here the total Fe content

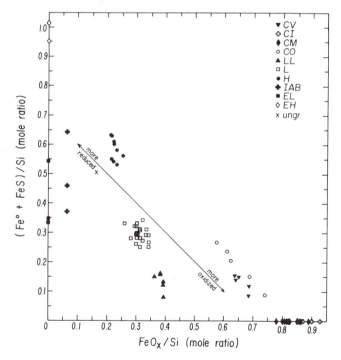

Fig. 7.4.1. Reduced Fe (Fe° and FeS) plotted against oxidized Fe (FeO$_x$) on a Urey-Craig plot. Chondrites with the same Fe/Si ratio form arrays with slopes of -1. On older versions of this diagram (see, e.g., Fig. 1.1.7), the H, L and LL chondrites formed such arrays, but this largely resulted from experimental errors. We use only Jarosewich data for H, L and LL falls of petrographic types 5 and 6. In the H and L groups there is evidence of slopes steeper than -1, indicating that bulk Fe/Si is decreasing with increasing FeO$_x$.

is divided into the fractions present as Fe°, FeS and FeO and normalized to Si on an atomic basis. The FeS and Fe° contents are combined on the vertical axis because virtually all S is present as FeS in metal-bearing chondrites (all groups except CM and CI). The Fe present as FeO in silicates and Fe_3O_4 is shown as FeO_x. A small portion of the Fe in CM and CI chondrites is present as FeS but a truer picture of their oxidation state is obtained by assigning all Fe to FeO_x. The complementary relation between FeO and Fe° plus FeS contents is illustrated by the tendency of some groups (e.g., CV and CO) to scatter along lines of constant Fe/Si ratio and thus with a slope of -1. The H and L groups are clearly resolved in terms of Fe/Si ratio. The CV, CO and CM chondrites have Fe/Si ratios similar to those in the H-group and the reduced enstatite chondrites display a wide range of Fe/Si ratios.

On other versions of this diagram (see, e.g., Fig. 1.1.7) the H, L and LL groups spread along lines with a slope of -1 but the ranges in FeO, at least in part, are an artifact resulting from analytical errors; FeO is usually calculated from the difference between total Fe and (Fe° + FeS) contents. The scatter decreases markedly when only data from type 5 and type 6 falls acquired by one analyst (E. Jarosewich) are plotted (Fig. 7.4.1). The slopes within each group are somewhat steeper than -1.

The change in the proportions of Fe° and FeO is also reflected in compositional changes among the constituent minerals. An example is illustrated

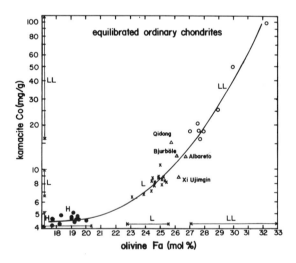

Fig. 7.4.2. Compositional changes among Fe-bearing minerals of chondrites reflecting redox-dependent variations in the distribution of Fe between metal and silicate. The increase in the Co content of kamacite with increasing oxidation of Fe is amplified by small decreases in the bulk Fe/Si ratio through the H-L-LL sequence but mainly by large changes in the kamacite/taenite ratios through this sequence. The meteorites shown as triangles are intermediate between L and LL chondrites in composition.

in Fig. 7.4.2. In going from H to LL chondrites, or left to right on the diagram, the proportion of total Fe as FeO increases while the proportion of Fe° decreases. The increase in the FeO/(FeO + MgO) ratio in olivine reflects the increasing proportion of FeO. The increase in the Co content of the kamacite in part reflects the decreasing proportion of Fe°, but the effect is amplified because Co preferentially partitions into the low-Ni kamacite phase, and the fraction of the metal present as kamacite decreases with increasing degree of Fe oxidation. Only Co and Fe concentrate in kamacite; all other siderophiles prefer taenite and, like Ni, their concentrations in the two phases are less affected by changes in oxidation state.

7.4.3. ABUNDANCE AND DISTRIBUTION OF OTHER SIDEROPHILE ELEMENTS

Abundances of the remaining siderophile elements roughly parallel, but do not duplicate, that of Fe. This reflects competing factors: the elements are easily reduced to metal, yet they differ in their reactivities with O and S and in their cosmic volatilities. Several siderophile elements form stable oxides, sulfides or silicates under certain nebular or parent-body conditions and some evidently condense in one mineral site but are redistributed in response to changing nebular conditions or during thermal metamorphism and concentrate in some other site. Let us begin by examining the overall abundance patterns in chondrites.

Abundances of Siderophile Elements

In Fig. 7.4.3, the abundances of nine siderophile elements in eight chondrite groups are compared to their abundances in CI chondrites. The elements are ordered according to their nebular condensation temperatures from refractory Os and Ir to common elements Ni, Co and Fe to moderately volatile Au, As, Ga and Sb. The data are normalized to Ni, the most abundant siderophile element after Fe, which, because it is more easily reduced, is always more highly concentrated in the metal than Fe. Note that normalizing to Ni, rather than the traditional Si, masks the effects of the metal/silicate fractionation; i.e., H, L and LL chondrites all appear to have the same Fe, Ni and Co contents whereas in fact the Fe, Ni and Co contents of these meteorites decrease in the order H > L > LL (Fig. 7.4.1). Normalizing to Ni allows us to assess more readily the extent of interelement siderophile fractionations.

These patterns reveal important clues. A key feature is that the closely related elements Co, Fe and Ni display little fractionation relative to each other, whereas the more refractory and more volatile elements are generally enriched or depleted. This indicates that Co, Ni and Fe were originally highly concentrated in metal and therefore coherent at the time of the metal/silicate fractionation. The remaining siderophile elements were not as highly concen-

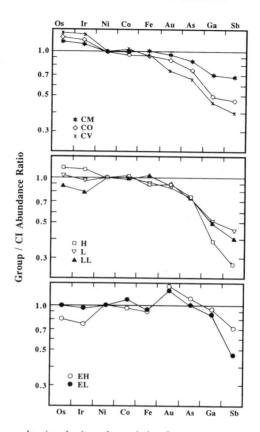

Fig. 7.4.3. Diagrams showing the irregular variation from group to group of Ni-normalized siderophile-element contents of chondrites. Fe, Ni and Co appear little fractionated relative to one another suggesting concentration in a single component that, presumably originated as metal. The other elements are usually fractionated relative to these three suggesting incomplete condensation in common metal. Appreciable fractions condensed in other components that were fractionated during agglomeration and accretion (data from Wasson and Kallemeyn 1987).

trated in this common metal; evidently they condensed or became incorporated into other phases or components.

Some portion of the refractory siderophile elements may have been present in refractory components (Chou et al. 1973) that were incorporated into the chondrite planetesimals in different proportions (see Chapter 7.3). Some CAIs are as enriched in refractory siderophile elements as they are in refractory lithophile elements, by the common factor of ~ 20 relative to CI chondrites (Fig. 7.4.4; see also Chapter 10.3). However, siderophile elements display larger variations in concentration suggesting a more heterogeneous distribution. They are often concentrated in μm-sized metal nuggets that con-

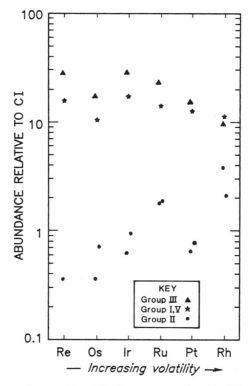

Fig. 7.4.4. Different refractory-siderophile-element patterns found in different types of Allende CAIs. In CAI groups I, III and V, they are enriched by factors of 10 to 20, though variation in enrichment factor indicate more heterogeneous distribution than is the case for refractory lithophile elements. In group II inclusions, concentrations of the more refractory elements are 2.4 times lower than those of the more volatile elements, similar to the pattern displayed by refractory lithophiles (diagram modified after Kornacki and Fegley 1986).

tain Pt, Ir, etc. in the percent range (Palme and Wlotzka 1976; Wark and Lovering 1976). These nuggets occur as dispersed metal grains in the major oxide phases and sometimes are enclosed in, and associated with, complex aggregates of FeNi grains, sulfides, phosphates, oxides or silicates. These complex aggregates commonly contain high concentrations of volatile elements and display peculiar textures suggesting that they formed independently of the host inclusions, which prompted El Goresy et al. (1977) to call them Fremdlinge (see also Chapter 10.3).

In some CAIs, the refractory metals display fractionated patterns; Mo, W and Pt are depleted relative to the others (Fig. 7.4.5). Platinum is one of the most volatile refractory elements and Mo and W become increasingly volatile with increasing oxidation state. Fegley and Palme (1985) therefore suggested that CAIs displaying this fractionation pattern formed under conditions much more oxidizing than expected in the nebular gas. These large

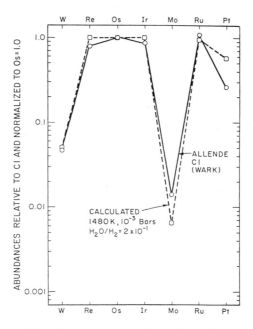

Fig. 7.4.5. Refractory-siderophile-element pattern for an Allende CAI compared with that calculated for an environment that was much more oxidizing than a gas of solar composition. The most volatile (Pt) is depleted along with Mo and W which become increasingly volatile with increasing oxidation state (diagram modified after Fegley and Palme 1985).

interelement variations make it difficult to predict the extent to which the refractory siderophile elements should covary with refractory lithophile elements.

In ordinary chondrites, refractory siderophile elements appear to be both enriched and depleted relative to CI abundances (Fig. 7.4.3). Some fraction of the refractory elements appear to have been associated with the metal since they all display the same fractionation pattern: H > L > LL. However, they are also fractionated relative to Fe, Ni and Co; for example, the Ir/Ni ratio in H group is 15% greater than the CI ratio, in the L group it is about the same as CI and in the LL group, it is 20% lower than CI. This indicates some additional complexity, e.g., one involving variable siderophile-element concentrations in the metal condensing at the various formation locations (Wasson 1972). Similarly complex explanations seem required for EH and EL fractionations.

Elements more volatile than Co, Ni and Fe pose other unresolved problems (Fig. 7.4.3). In ordinary chondrites, As and Au covary with Ni. The obvious inference is that during metal/Si fractionation, both were concentrated in the metal, though not in their full cosmic proportions. In contrast, Ge (not shown), Ga and Sb are fractionated in just the opposite sense to the metal/Si fractionation, decreasing in abundance in the order LL > L > H. This suggests that a significant fraction of these elements was in the silicates during the metal/silicate fractionation, consistent with the observation that a large fraction of Ga (and Ge) is found in the silicates of unequilibrated ordinary chondrites while an increasingly greater amount is found in the metal in types 4, 5 and 6 (Chou and Cohen 1973). Their overall depletion relative to CI abundances requires that a sizeable fraction was in a component (gas or fine-grained metal) that was not agglomerated into planetesimals.

In the E chondrites, moderately volatile elements covary with Ni indicating that they were concentrated in the metal during metal/Si fractionation. Three of these four elements, Au, As and Ga are slightly more depleted (~ 6%) in the low-Fe (EL) than in the high-Fe (EH) chondrites while the Sb/Ni ratio is much lower in the EL group. Evidently, Ga condensed as metal rather than silicates in the more reducing environment under which these meteorites originated. The high element/Ni ratios of Au and (to a smaller extent) As in E chondrites are the only known cases where moderately volatile elements are enriched over CI abundances.

The CM, CO and CV chondrites are enriched in refractory and depleted in moderately volatile siderophile elements relative to CI abundances. The patterns are straightforward: the refractory elements are enriched in the order CV > CO ~ CM > CI and the moderately volatile abundances decrease in the same order, but with CM much less depleted than CO. The refractory-siderophile pattern is similar to that of the refractory lithophile elements implying the same mechanism: a progressively greater admixture of a refractory component (see Chapter 7.3). An interesting complication is that the refractory-siderophile/Ni enhancements are a few percent smaller than the refractory-lithophile/Si enhancements in the CM and CO chondrites.

7.4.4. SIDEROPHILE ABUNDANCES IN IRON METEORITES

Iron meteorites are thought to have originated by melting and gravitational segregation of dense molten metal from less dense silicates. They may be derived from the cores of asteroids or from large pools of metal that did not completely segregate to form a core. Since iron meteorites have passed through a melting and freezing cycle, they cannot be regarded as pristine samples of nebular matter. Nonetheless, a large fraction of siderophile elements are expected to be highly concentrated in the molten metal and swept into the core along with the more abundant Fe and Ni. The mean composition of iron meteorites from an entire core can then be regarded as a representative sample of the siderophile elements of that parent body. Iron-meteorite compositions are therefore of interest because they enlarge the spectrum of potential parent-body compositions beyond the limited number of bodies that are represented by the chondrites.

However, a complication arises in attempting to relate the composition

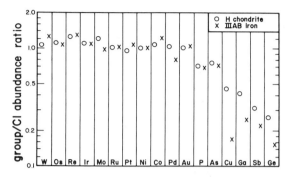

Fig. 7.4.6. The estimated composition of the melt from which IIIAB iron meteorites evolved. Note the similarity to that of H-group chondrites. Other iron-meteorite groups show similar patterns, with refractory-siderophile abundances within about a factor of 2 of CI levels, moderately volatile siderophiles more variable and always below CI levels and different degrees of oxidation (i.e., different Fe/Ni ratios). (Diagram from Willis 1980.)

of an iron meteorite to that of its parent body. During melting and freezing, unequal partitioning between the molten and solid metal has led to extensive elemental fractionation in iron meteorites. Fractional crystallization during solidification of the core is thought to be responsible for most elemental correlations within the different iron-meteorite groups (Scott 1972; Kelly and Larimer 1977; Wasson et al. 1980). If this is correct, it is possible to reconstruct the bulk compositions of the parent melts using simple models.

The result of one such reconstruction is presented in Fig. 7.4.6, where the composition of the parent melt of group IIIAB iron meteorites is compared to the siderophile-element abundances in H chondrites (Willis 1980). The similarity is striking; the only minor difference is that Cu, Ga, Sb and Ge are slightly deficient in the iron meteorites relative to the chondrites. But these elements may not have fully partitioned into the metal, some portion may have partitioned into the silicates or sulfides. Other groups of iron meteorites display similar patterns; some are depleted, others enriched in the moderately volatile elements compared to this particular group.

7.4.5. INTERPRETATION OF ABUNDANCE PATTERNS

The fractionation patterns of the siderophile elements in bulk chondrites are more complex than those of the refractory lithophile elements discussed in Chapter 7.3. The complications undoubtedly are due in part to the widely different cosmic volatilities of the elements and in part to partitioning of some of these elements into components other than the FeNiCo metal component. Several different nebular fractionation mechanisms seem required to account for the various patterns.

Let us begin untangling the problem by citing those points on which there is general agreement. There is evidence for several types of fractionation: some portion of the refractory siderophiles Ir, Os, etc. was in a separate component; to a good approximation, the common siderophiles Fe, Ni and Co are fractionated as a group relative to lithophiles Al, Mg, Si, etc.; and the moderately volatile elements (Au, As) tend to follow Fe, Ni and Co but generally are more depleted and display patterns that differ from group to group. In addition to the fractionations of refractory and common siderophile components, a mechanism for depleting the moderately volatile elements is required.

Fractionation of Fe, Ni and Co

We develop alternative approaches to discuss metal/silicate fractionation mechanisms: the first makes use of Al normalization, and illustrates possible links with refractory elements; the second employs a Si normalization and emphasizes links with the common elements. In Fig. 7.4.7, we show the fractionation on an Al-normalized mixing diagram similar to Figs. 7.3.5 and 7.3.6. Figure 7.4.7a is generic in that any of the three common elements Fe, Co or Ni could be plotted; it illustrates the combined effects of refractory-lithophile- and siderophile-element fractionations. As discussed in Chapter 7.3, the abundances of refractory lithophile elements suggest that chondrites can be derived from bulk nebular materials having CI composition: elemental abundances in CM, CO and CV chondrites indicate an addition of material resembling CAIs to CI composition while those in H, L, LL, EH and EL

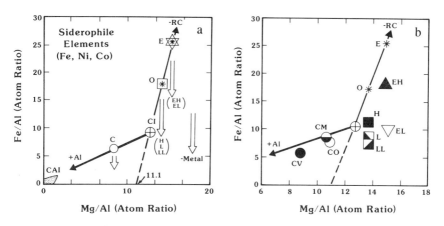

Fig. 7.4.7. The extent and direction of metal/silicate fractionation revealed on refractory-element mixing diagrams. In (a), variable abundances of a CAI-like component are indicated by the arrow marked + Al. The effects of removing a Mg-rich refractory lithophile component (Mg/Al = 1.0) from CI material is indicated by the arrow marked − RC. The Fe data from chondrites are plotted in (b). Data that fall below the lines could be explained by loss of a metal component (or gain of silicates).

chondrites indicate removal from CI materials of a refractory component (RC) having a composition similar to CAIs augmented with Mg_2SiO_4. The result of this fractionation is to shift the composition of the parental material away from CI composition to points such as E and O along the $-$ RC line or C along the $+$ Al line. The loss of metal (or gain of common silicate) component shifts the composition vertically downward.

The Fe data for the various chondrite groups are shown in Fig. 7.4.7b; the Co and Ni data yield similar patterns. The CV, CM and CO groups fall slightly below the $+$ Al line while the H, L, LL, EH and EL groups fall considerably below the $-$ RC line. The fractionation is always downward from the two reference lines. In principle, the pattern could be explained in terms of either the addition of silicate material, the loss of metal or some combination of the two. Which explanation is correct remains an unresolved problem, though to simplify discussion the fractionation is usually couched in terms of metal loss.

In Fig. 7.4.8, the compositions of the nine chondrite groups are dis-

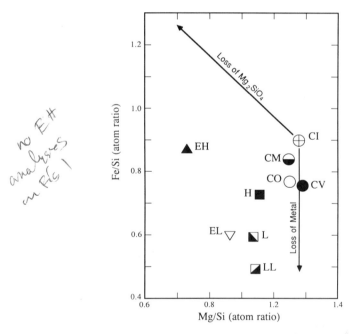

Fig. 7.4.8. An Fe/Si vs Mg/Si diagram showing the chondrite groups falling into siderophile-rich groups with Fe/Si atom ratios between 0.76 and 0.87, and siderophile-poor groups with Fe/Si atom ratios between 0.49 and 0.60. If the low Mg/Si ratios result from incomplete accretion of olivine, the fact that the EH and H chondrites plot well below a Mg_2SiO_4-loss trajectory drawn through the CI point implies that they also incompletely accreted the common-siderophile carrier, probably Fe-Ni metal.

played on an Fe/Si vs Mg/Si diagram. Again, taking the approach that the different groups formed in a nebula having CI bulk composition, we show two trajectories leading away from the CI point: a line with -1 slope shows the effect of incomplete accretion of Mg_2SiO_4, a vertical line the effect of incomplete accretion of metal (or its oxidized/sulfurized equivalent). On this plot, it appears that the similarity in Fe/Si ratios between EH, H and CI chondrites is fortuitous. If the low Mg/Si ratios were produced by Mg_2SiO_4 loss, then metal loss is also required to explain the position of these points below the Mg_2SiO_4-loss trajectory. However, perhaps there is some significance to this ratio that we do not yet understand.

Another approach is to attempt to evaluate the amount of Fe associated with the silicate fraction. Anders (1964) developed an approach that was modified by Larimer and Anders (1970); the implications were reviewed by Kerridge (1977). The Fe fraction in the nonmetallic components can be inferred from the extent of Fe/Ni fractionation. In CM, CO, CV, EH and EL chondrites, the Fe/Ni ratio is close to the CI value suggesting that both elements were highly concentrated in the common siderophile component during fractionation. But in H, L and LL chondrites, the Fe/Ni ratio changes systematically with H < L < LL. On Fe/Mg vs Ni/Mg diagrams the data form a linear array intercepting the Fe/Mg axis at about 0.14. If the nonmetallic components are assumed to be the same at the H, L and LL locations, and Ni was totally concentrated in the common metal component, the intercept yields the Fe present as FeO and/or FeS during fractionation.

Refractory Siderophile Elements

The fractionation of refractory siderophile elements Ir, Os, etc. is complex (Figs. 7.4.3 and 7.4.4). In CM and CO chondrites, Ni-normalized refractory-siderophile abundance ratios are slightly higher than those of Mg-normalized refractory lithophile elements. This suggests that refractory siderophiles were added along with the refractory-lithophile materials but not to the same extent. Perhaps the missing refractory siderophiles were present as metal grains that served as condensation nuclei for the common siderophiles. This provides a qualitative explanation but quantitatively the uncertainties are too large to test this simple model, not to mention more complex models.

The refractory-siderophile-element pattern in ordinary chondrites is deceptively simple (Fig. 7.4.3). Relative to Ni, these elements are progressively depleted in the order H > L > LL. The pattern thus appears to follow that of FeNiCo/Si, suggesting a link to the common metal/silicate fractionation. However, since the extent to which Ir, Os, etc. are fractionated exceeds that of Ni, the refractory-siderophile/Ni ratio in the common siderophile component would have to exceed the CI value. It seems more plausible that much of the Ir, Os, etc. was in a refractory siderophile component; a study of chondrules from the highly unequilibrated LL3.0 chondrite Semarkona yielded evidence for such a component (Grossman and Wasson 1983). Because $(Ir/Ni)_H > (Ir/Ni)_{CI}$, the refractory lithophile and siderophile components were distinct at the H-chondrite location. The refractory siderophile component was accreted more efficiently than the common siderophile component, whereas the refractory lithophiles were accreted less efficiently than the common lithophiles.

There is also evidence for a refractory siderophile component that is associated with the silicates. Dispersed, fine-grained metal particles associated with silicates in unequilibrated ordinary chondrites have high Ir/Ni ratios (Chou et al. 1973; Rambaldi 1977a,b). Uniform contents of refractory-siderophile grains in silicates cannot explain the bulk H-L-LL trends, since the LL meteorites have the highest lithophile/siderophile ratio, but the lowest Ir/Ni ratio. Either the fraction of trapped refractory metal was much higher in H silicates or, more likely, a refractory siderophile component only indirectly associated with the silicates (i.e., mainly present as metal) dominated the Ir/Ni fractionation.

In enstatite chondrites, the refractory-siderophile/Ni ratios are lower in the EH group than in the EL group. In a qualitative sense the depletion in the EH group is consistent with an appreciable fraction of the Ir, Os, etc. being associated with refractory lithophile elements, but refractory-lithophile/Si ratios are much lower than the refractory-siderophile/Ni ratios.

Moderately Volatile Siderophile Elements

At least two processes are required to account for the abundances of moderately volatile siderophile elements: one to explain the general depletion relative to CI abundances and another to explain the correlations with Fe, Ni and Co. The correlations with common siderophiles are generally agreed to be related to the metal/silicate fractionation. In both the ordinary and enstatite chondrites, elements such as As and Au tend to increase or decrease in concert with Fe, Ni and Co suggesting that they were in phases associated with the common metal at the time of metal/silicate fractionation. It is significant and worth emphasizing that these correlations imply that moderately volatile elements had already condensed at the time of the metal/silicate fractionation in the source regions of both ordinary and enstatite chondrites.

The depletion of moderately volatile elements in all chondrite groups except CI has been interpreted in two ways. In one model, the depletion is attributed to the correlated variation in chondrule/matrix ratio (Anders 1964; Larimer and Anders 1967,1970). Each chondrite group is considered to contain its own blend of volatile-depleted chondrules and undepleted matrix. The extent to which depletion is correlated with cosmic volatility is regarded as a byproduct of the chondrule-forming process; more volatile elements were more efficiently outgassed (or less efficiently recondensed) than more refractory elements. In the other model, greater significance is attached to the inverse correlation between elemental abundance and cosmic volatility. Moderately volatile elements are considered to have condensed on fine-grained dust that settled more slowly and was depleted in the midplane materials at the time of chondrule formation in the midplane. The greater the volatility of an element, the larger the fraction condensed onto fine grains that did not settle to the midplane (Wasson and Chou 1974; Wai and Wasson 1977; Wasson 1985).

In both models, the depletion of moderately volatile elements is attributed to their volatility. The contentious issue is whether chondrites are deficient in these elements as a result of volatile loss during chondrule formation, or whether the chondrites (and chondrules) largely preserve the moderately-volatile-element contents of the precursors. Neither model easily explains the one exception to the generality that the moderately volatile elements are always depleted relative to CI abundances: the high abundance of Au in E chondrites (Fig. 7.4.3).

Metal/Silicate Fractionation During Accretion?

Several attempts have been made to determine whether the siderophile-element content varies within individual groups of chondrites (Müller et al. 1971; Dodd 1976, 1981; Jarosewich and Dodd 1985; Heyse 1978; Morgan et al. 1985), most recently by Sears and Weeks (1986). The purpose of these studies is to determine whether correlations exist between petrographic type, oxidation state, siderophile-element abundance and metal composition. To the extent that correlations can be demonstrated to exist, and to be primary, some information regarding the nature of the accretion process might be obtained. The key is to identify and demonstrate that a particular feature is indeed primary.

Müller et al. (1971) reported weak negative correlations between siderophile abundances and the FeO/(FeO + MgO) ratios in the ferromagnesian silicates within each group of ordinary chondrites. They inferred that each group represented a small portion of a grand "continuous-fractionation-sequence" through the H, L and LL groups. The H-L-LL trend through the Jarosewich data plotted in Fig. 7.4.1 tend to support this view.

Another hypothesis offered by Dodd (1976) is that, within the ordinary-chondrite groups, siderophile content increases with increasing petrographic type. Recently, Sears and Weeks (1986) summarized data bearing on this question; data from the appendix of their paper are plotted in Fig. 7.4.9. The error bars correspond to 70% confidence limits on the mean based on Student's t formulation. The only evidence for a relationship between siderophile content and type is observed for the H group, where type-3 siderophile abundances appear to be significantly lower than those in the higher types. However, even this trend is suspect. One of their H3 chondrites (Tieschitz) seems as closely related to L as to H chondrites, and four of the seven H3-chondrite samples are more or less weathered finds from which siderophiles may have been lost by leaching.

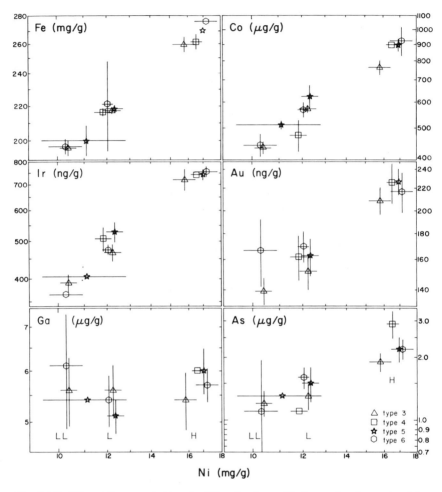

Fig. 7.4.9. Siderophile concentrations in different petrographic types of ordinary chondrites tabulated by Sears and Weeks (1986). Bars show 70% uncertainty limits. With the possible exception of low siderophile abundances in H3 chondrites, there is no evidence indicating a significant variation in siderophile-element abundance as a function of petrographic type.

Nonetheless, the lack of correlations among LL3-6, L3-6 and H4-6 is of some interest; the metal/Si ratios seem to have been established independently of the degree of metamorphism now recorded. There is no support for models calling for metal/Si fractionation during sequential accretion of layers onto the parent bodies. In fact, there are no strong arguments for models linking the agglomeration of chondrites to their petrographic grade.

More comprehensive studies are needed to determine the degree to which siderophile abundances fit the continuous-fractionation-sequence model, and whether abundances are in any way related to petrographic type.

7.4.6. IMPLICATIONS AND FUTURE WORK

The siderophile elements form a chemically coherent set but differ widely in their cosmic volatility. Cosmic volatility has apparently played an important role in determining their abundances in chondritic material and has led to minor but significant fractionations that were not anticipated in early, simple models. At least three fractionations can be resolved:

1. Formation and fractionation of a component bearing elements that are more refractory than Fe, Ni or Co.
2. Formation of a common FeNiCo component that separated from the common silicates to variable degrees.
3. Depletion (except Au in E chondrites) of moderately volatile elements by factors up to 5, the abundance falling with increasing volatility.

It is difficult to assess whether the fractionations occurred simultaneously during settling of components to the nebular midplane, sequentially during cooling and condensation of the nebula followed by alteration events including chondrule formation and parent-body metamorphism or by more complex scenarios, e.g., onset of condensation earlier at one location than another with interzonal mixing of gas yielding enhanced condensation at the first location and depletions at the other.

A refractory siderophile component seems required to explain the trends in bulk siderophiles among closely related chondrite groups. But unlike the refractory lithophile elements, the full cosmic complement of refractory siderophile elements was not present in the refractory lithophile component, and at most locations the component accounted for only a minor fraction of the accreted refractory siderophiles.

Siderophile-lithophile fractionations appear to have occurred in all source regions. The volatile siderophile elements Ga, Ge and Sb, appear to have been involved in the fractionations, thus the fractionation processes occurred after these elements were condensed. They presumably condensed in part as alloys in the common metal, in part as trace components in one or more silicate phases, and possibly also in minor components such as sulfides. Inadequate thermodynamic data on their behavior as solutes hinders the accurate assessment of 50%-condensation temperatures. Recent estimates (compiled in Wasson 1985) suggests a range of 800 to 950 K at $pH_2 = 10^{-4}$ atm, which is not unreasonable. However, additional work needs to be done on identifying their host phases and determining the necessary thermodynamic data. Similar data on other moderately volatile siderophile elements such as Ag, Cu and Sn are also required. Such data are needed not only to constrain the conditions under which the metal/Si fractionation occurred but also to understand the cause of the general depletion of the moderately volatile elements. This topic is considered in more detail in Chapter 7.5.

Obviously, much remains to be learned about, and from, the siderophile elements. Many details of their abundance and distribution in chondrites need to be studied in order to understand better how they condensed, the nature of nebular fractionation processes and whether the same processes were instrumental in the formation of the terrestrial planets. Attempts should be made to isolate and characterize pure nebular components from the most unequilibrated chondrites. Additional analytical and thermodynamic data on the moderately volatile elements are required to constrain further the conditions that led to the formation of the various components. Careful measurements should be carried out to confirm whether the fractionations commonly observed among the groups of a single clan (e.g., the ordinary chondrites) can also be resolved as well as trends among the members of the individual groups.

Acknowledgments. We would like to thank B. Fegley for a helpful review, and G. Kallemeyn for diagrams and the use of unpublished data. This work was supported in part by grants from the National Aeronautics and Space Administration.

REFERENCES

Anders, E. 1964. Origin, age and composition of meteorites. *Space Sci. Rev.* 3:583–714.
Anders, E., and Ebihara, M. 1982. Solar system abundances of the elements. *Geochim. Cosmochim. Acta* 46:2363–2380.
Chou, C. L., and Cohen, A. J. 1973. Gallium and germanium in the metal and silicates of L- and LL-chondrites. *Geochim. Cosmochim. Acta* 37:315–327.
Chou, C. L., Baedecker, P. A., and Wasson, J. T. 1973. Distribution of Ni, Ga, Ge and Ir between metal and silicate portions of H-group chondrites. *Geochim. Cosmochim. Acta* 37:2159–2171.
Dodd, R. T. 1976. Iron-silicate fractionation within ordinary chondrite groups. *Earth Planet. Sci. Lett.* 28:479–484.
Dodd, R. T. 1981. *Meteorites: A Petrologic-Chemical Synthesis.* (Cambridge: Cambridge Univ. Press).
El Goresy, A., Nagel, K., Dominik, B., and Ramdohr, P. 1977. Fremdlinge: Potential presolar material in Ca-Al-rich inclusions of Allende. *Meteoritics* 12:215–216 (abstract).
Fegley, B., and Palme, H. 1985. Evidence for oxidizing conditions in the solar nebula from Mo and W depletions in refractory inclusions in carbonaceous chondrites. *Earth Planet. Sci. Lett.* 72:311–326.
Garz, T., and Kock, M. 1969. Experimental oscillator strengths for Fe I lines. *Astron. Astrophys.* 2:274–279.
Grossman, J. N., and Wasson, J. T. 1983. The compositions of chondrules in unequilibrated chondrites: An evaluation of models for the formation of chondrules and their precursor materials. In *Chondrules and Their Origins,* ed. E. A. King (Houston: Lunar and Planetary Inst.), pp. 88–121.
Heyse, J. V. 1978. The metamorphic history of LL-group ordinary chondrites. *Earth Planet. Sci. Lett.* 40:365–381.
Jarosewich, E., and Dodd, R. T. 1985. Chemical variations among L-chondrites—IV. Analyses with petrographic notes of 13 L-group and 3 LL-group chondrites. *Meteoritics* 20:23–36.
Kelly, W. R., and Larimer, J. W. 1977. Chemical fractionations in meteorites—VIII. Iron meteorites and the cosmochemical history of the metal phase. *Geochim. Cosmochim. Acta* 41:93–111.
Kerridge, J. F. 1977. Iron: Whence it came, where it went. *Space Sci. Rev.* 20:3–68.
Kornacki, A. S., and Fegley, B., Jr. 1986. The abundance and relative volatility of refractory

trace elements in Allende Ca, Al-rich inclusions: Implications for chemical and physical processes in the solar nebula. *Earth Planet. Sci. Lett.* 79:217–234.

Larimer, J. W., and Anders, E. 1967. Chemical fractionation in meteorites—II. Abundance patterns and their interpretation. *Geochim. Cosmochim. Acta* 31:1215–1238.

Larimer, J. W., and Anders, E. 1970. Chemical fractionations in meteorites—III. Major element fractionations in chondrites. *Geochim. Cosmochim. Acta* 34:367–387.

Morgan, J. W., Janssens, M.-J., Takahashi, H., Hertogen, J., and Anders, E. 1985. H-chondrites: Trace element clues to their origin. *Geochim. Cosmochim. Acta* 49:249–270.

Müller, O., Baedecker, P. A., and Wasson, J. T. 1971. Relationship between siderophile-element content and oxidation state of ordinary chondrites. *Geochim. Cosmochim. Acta* 35:1121–1137.

Palme, H., and Wlotzka, F. 1976. A metal particle from a Ca, Al-rich inclusion from the meteorite Allende, and the condensation of refractory siderophile elements. *Earth Planet. Sci. Lett.* 33:45–60.

Prior, G. T. 1916. On the genetic relationship and classification of meteorites. *Mineral. Mag.* 18:26–44.

Rambaldi, E. R. 1977a. The content of Sb, Ge and refractory siderophile elements in metals of L-group chondrites. *Earth Planet. Sci. Lett.* 33:407–419.

Rambaldi, E. R. 1977b. Trace element content of metal from H- and L-group chondrites. *Earth Planet. Sci. Lett.* 36:347–358.

Scott, E. R. D. 1972. Chemical fractionation in iron meteorites and its interpretation. *Geochim. Cosmochim. Acta* 36:1205–1236.

Sears, D. W. G., and Weeks, K. S. 1986. Chemical and physical studies of type 3 chondrites—VI. Siderophile elements in ordinary chondrites. *Geochim. Cosmochim. Acta* 50:2815–2832.

Urey, H. C., and Craig, H. 1953. The composition of stone meteorites and the origin of the meteorites. *Geochim. Cosmochim. Acta* 4:36–82.

Wai, C. M., and Wasson, J. T. 1977. Nebular condensation of moderately volatile elements and their abundances in ordinary chondrites. *Earth Planet. Sci. Lett.* 36:1–13.

Wark, D. A., and Lovering, J. F. 1976. Refractory/platinum metal grains and Allende calcium-aluminum-rich clasts (CARC's): Possible exotic presolar material? *Lunar Planet. Sci.* VIII:912–914 (abstract).

Wasson, J. T. 1972. Formation of ordinary chondrites. *Rev. Geophys. Space Phys.* 10:711–759.

Wasson, J. T. 1985. *Meteorites: Their Record of Early Solar-System History* (New York: W. H. Freeman).

Wasson, J. T., and Chou, C. L. 1974. Fractionation of moderately volatile elements in ordinary chondrites. *Meteoritics* 9:69–84.

Wasson, J. T., and Kallemeyn, G. W. 1987. Compositions of the chondritic meteorite groups. *Phil. Trans. Roy. Soc. London*, in press.

Wasson, J. T., Willis, J., Wai, C. M., and Kracher, A. 1980. Origin of iron meteorite groups IAB and IIICD. *Z. Naturforschung* 35a:781–795.

Willis, J. 1980. The Mean Compositions of Iron Meteorite Parent Bodies. Ph.D. Thesis, Univ. of California, Los Angeles.

7.5. MODERATELY VOLATILE ELEMENTS

H. PALME
Max-Planck-Institut für Chemie

J. W. LARIMER
Arizona State University

and

M. E. LIPSCHUTZ
Purdue University

Moderately volatile elements are elements with condensation temperatures between those of Mg-silicates and FeS. CI chondrites have solar abundances of these elements while all other types of chondritic meteorites have lower contents. The extent of this depletion increases with calculated nebular volatility. Parent bodies of differentiated meteorites, including iron meteorites, as well as the inner planets are also depleted in moderately volatile elements. It appears therefore that most of the solid material in the inner solar system has significantly lower contents of moderately volatile elements than CI chondrites. Fractionation of these elements occurred either by volatilization or incomplete condensation shortly after formation of the solar system. Both processes require temperatures in excess of 1000 K. The lack of meteorites and planetary materials with enrichments in moderately volatile elements seems to exclude local heat sources. The depletion patterns of moderately volatile elements in chondritic meteorites are roughly comparable, although absolute depletions are quite variable. There is no obvious correlation of the abundances of moderately volatile elements with any other property of chondritic meteorites, such as, oxidation state or content of refractory elements.

7.5.1. INTRODUCTION

The term "moderately volatile" describes those elements that are predicted to condense in a gas of solar composition between 1250 and 650 K at a pressure of 10^{-4} bar. Their cosmic volatility places them between FeNi-metal and Mg-silicates at the high-temperature extreme and FeS at the low-temperature extreme (Chapter 7.1). Since the extent to which a particular element varies in abundance in chondritic material appears to increase with increasing volatility, these elements display greater variations in abundance than those discussed in Chapters 7.3 and 7.4. In chondrites they vary in abundance by up to a factor of 5 and there is evidence of even larger variations in other planetary objects. On the other hand, they display little or no variation with petrographic type (metamorphic grade), in contrast with the highly volatile elements which do vary with petrographic type, as discussed in Chapter 7.6.

This greater variability in chondritic material makes the more abundant, moderately volatile elements useful for classification purposes. An illustrative example is Na (Fig. 7.5.1). The Na/Mg ratios in different types of chondrites vary in a manner that not only permits resolution of the different chondrite types but clearly indicates that the CI chondrites are the best match to solar photospheric values. This diagram also demonstrates that the variation in abundance of refractory Ca is distinct from the variations in moderately volatile Na.

The elements defined as moderately volatile display a variety of geo-

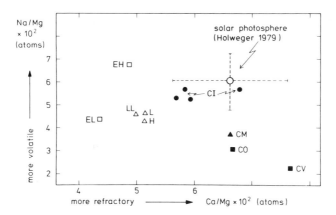

Fig. 7.5.1. Elemental abundances in CI chondrites are in good agreement with abundances in the solar photosphere. Moderately volatile elements, such as Na or S, are diagnostic for CI chondrites. The spread in the refractory element Ca is sufficient to separate the different types of chondritic meteorites. CI data are as follows: Anders and Ebihara (1982); Kallemeyn and Wasson (1981); Palme et al. (1981b); Mason (1979). Other data: see Figs. 7.5.2 and 7.5.6. Solar data: Holweger (1979).

chemical properties: chalcophile, siderophile and lithophile. They tend to vary as a group, however, suggesting that volatility is ultimately responsible for the variations and not geochemical character. Nonetheless, these elements are fractionated relative to one another to a much greater extent than the groups of refractory or siderophile elements. It is useful to compare the extent to which the various elements are fractionated relative to each other in terms of their cosmic volatility. A convenient measure of cosmic volatility is the condensation temperature, or, by convention, the 50% condensation temperature defined as that temperature where 50% of the atoms of an element are expected to be condensed.

In order to facilitate comparisons between different groups of meteorites, all elements are normalized to Si. The element/Si ratios will then be divided by the same ratios in CI chondrites (i.e., mainly Orgueil). If the resulting ratio is <1, the element is considered to be depleted relative to Orgueil, i.e., relative to average solar-system abundances. Elements with similar Si-normalized ratios as Orgueil are designated as undepleted. After normalization to Si and CI abundances, elemental abundances can be plotted vs condensation temperatures, e.g., Fig. 7.5.2. below. The more or less continuous decrease in abundance with decreasing condensation temperature observed for most meteorite groups is called a depletion sequence.

7.5.2. GENERAL PROPERTIES OF MODERATELY VOLATILE ELEMENTS

Condensation Temperatures

Condensation temperatures for moderately volatile elements are listed in Table 7.5.1. Included are some undepleted elements, from Ni to Cr. The division between these undepleted and the moderately volatile elements is somewhat arbitrary since ratios of undepleted elements vary within chondritic meteorites. However, these variations cannot be easily fitted into the scheme of the moderately volatile elements. The Mg/Si ratio in all groups of carbonaceous chondrites is, for example, nearly constant (Fig. 7.3.1), while abundances of moderately volatile elements decrease in the order CI > CM > CO > CV.

The condensation temperatures in Table 7.5.1 are from a compilation by Wasson (1985). All elements with condensation temperatures above that of S are listed, with the exception of Cl and Br. The condensation temperature of Br derived from the calculations of Fegley and Lewis (1980) is, however, only 357 K, as compared to 690 K given in Wasson's compilation. Furthermore Keays et al. (1971) and Morgan et al. (1985) have shown that Br concentrations in ordinary chondrites vary with petrographic type in a fashion similar to the variations of the highly volatile elements. Bromine should therefore be considered a highly volatile element (Chapter 7.6). The situation with respect to Cl is unclear. The condensation temperature of Cl is 863 K,

TABLE 7.5.1
Condensation Temperatures (10^{-4} bar) and Geochemical Classification of Moderately Volatile Elements (from Li to S)

Condensation Temperature[a] (K)	Siderophile	Chalcophile	Lithophile[b]	
			comp.	incomp.
1354	Ni			
1351	Co			
1340			Mg	
1336	Fe			
1334	Pd			
1311			Si	
1277	Cr		Cr	
1225				Li
1225	Au			
1190			Mn	
1157	As			
1151	P			P
1080				Rb
1037	Cu	Cu		
1000				K
970			Na	
952	Ag			
918	Ga			Ga
912	Sb			
825	Ge			
736			F	
720	Sn			
684		Se		
680	Te	?		?
660		Zn		Zn
648		S		

[a] 50%-condensation, from a compilation by Wasson (1985).
[b] The lithophile elements are divided into those that are compatible and incompatible during igneous differentiation.

according to Fegley and Lewis (1980); this temperature is also given in Wasson's compilation. Whether the Cl content is variable in the different petrographic types of chondrites is not clear. There are not enough high-quality data in the literature to decide this question (see the compilation of Mason 1979).

All elements with condensation temperatures below S, i.e., highly volatile elements, including noble gases and C, show large concentration variations in ordinary chondrites (see Fig. 1 in Keays et al. 1971) and concentrations generally decrease with increasing petrographic type; possible reasons are discussed in Chapter 7.6. It is important to realize that the abundance trends of moderately volatile elements may reflect completely different processes from those affecting the highly volatile elements.

The condensation temperatures in Table 7.5.1 are less well known than one might infer from the apparently precise values. They depend on the thermodynamic properties of the elements and their relevant compounds. In particular, since all moderately volatile elements (except P and S) condense in solid solution with major phases such as forsterite or metal, the activities of these elements in their host phases are of great importance. The activity coefficients for Ge and Ga in γ-Fe are, for example, more than two orders of magnitude lower than unity, thus considerably facilitating condensation (see references given in Wasson 1985). An example for this type of condensation calculation is given in Table 7.5.2. Variations in the activity coefficients γ, of Pd in γ-Fe of 0.1 to 10 produce a spread in condensation temperatures of 60 K. Thermodynamic data suggest a value for γ of around 0.4, resulting in a condensation temperature of 1342 K, slightly higher than the temperature listed by Wasson (1985). At the 50%-condensation temperature of Pd, 3.7% of Fe and 71.5% of Ni would be condensed (assuming ideal solid solution), thus reversing the condensation sequence of Fe and Pd (Table 7.5.1). Variations in condensation temperatures of this magnitude should not be taken too seriously, however. Calculated condensation temperatures may depend on various other parameters such as O fugacity and the inclusion of all relevant gas compounds and solid phases (see, e.g., Wai and Wasson 1977). Condensation temperatures generally increase with increasing total pressure; but S is an exception. The condensation temperatures of FeS and chalcophile trace elements, such as Se and Zn, are independent of pressure (see Chapter 7.1). The volatility sequence given in Table 7.5.1 is therefore somewhat pressure

TABLE 7.5.2
Condensation Temperature of Pd
($p = 10^{-4}$ atm)

	Activity Coefficient (γ)	50% Condensation Temperature (K)
$H_2O/H_2 = 5 \times 10^{-4}$ [a]		
multielement	1	1322
alloy	0.1	1359
	10	1300
Fe-Ni-alloy	1	1321
Most likely γ[b]	0.4	1342
$H_2O/H_2 = 0.1$		
FeNi-alloy	1	1320

[a] Solar-nebula O fugacity.
[b] Hultgren et al. 1973.

dependent. Uncertainties in cosmic abundances do not affect condensation temperatures of elements dissolved in a major phase.

Geochemical Classification and Significance in Planets

In Table 7.5.1, we have classified the elements according to their geochemical properties. Some elements may display either lithophile, siderophile or chalcophile character, depending on O and S fugacity (e.g., Cr). It is obvious from this table that the sequence in condensation temperatures is independent of the geochemical properties of the elements.

The moderately volatile lithophile elements Li, P, Rb and K should be considered as incompatible elements, with a strong tendency to be enriched in partial melts or in late crystallizing liquids during igneous processes. These elements are of great geochemical importance (see Chapter 7.8). Constant ratios of K/U or K/La in samples from a single planet can be used to calculate the K content if the U concentration is known. Uranium, as a refractory element is always found in constant proportions with the major elements Ca and Al in bulk planets. Independent estimates of Ca and Al thus allow the calculation of the K content. Differentiated planets (Earth, Moon, Mars and the eucrite parent body) appear to have their own characteristic ratios of K/La (Wänke 1981). This ratio is in all cases significantly below the solar ratio, suggesting loss of K as the primary cause for this variation.

The variable K and Rb contents of differentiated planetary bodies must be seen in the context of a general depletion of moderately volatile and highly volatile elements. As we shall see, most undifferentiated (i.e., chondritic) meteorites show some depletion of K and other volatiles. One should therefore not be surprised to find similar depletions in planets. There is also the possibility that moderately volatile and highly volatile elements were lost during planetary accretion processes. One way to distinguish between these two possibilities is to study the pattern of volatile-element concentrations in undifferentiated and in differentiated materials. Loss during or immediately after accretion presumably occurred at higher O fugacity compared with solar-nebula processes, responsible for variations in volatile elements in undifferentiated meteorites.

Another process that affected the concentrations of moderately volatile elements observed in the accessible parts of planets involves formation of an Fe or FeS core. This would completely scavenge strongly siderophile elements like Au and Pd and to a lesser extent other siderophile and/or chalcophile elements such as Ge, Co, Cu or P, depending on their metal-silicate partition coefficients and the O fugacity prevailing during core formation. The fractions of these elements remaining in the planetary mantle are sensitive indicators of the type and extent of core formation. Alternatively, inhomogeneous accretion could lead to a chondritic pattern of moderately volatile elements in a planetary mantle. Depletion of moderately volatile elements would then provide a characteristic signature for the accreting material, having been established by processes in the solar nebula.

In summary, moderately volatile elements, in particular those that are lithophile or partly lithophile, may provide important constraints for the mechanism of accretion and for the origin of protoplanetary material (see also Chapter 7.8.). This alone is incentive enough to establish variations of these elements in undifferentiated meteorites. If we can identify the processes responsible for the variations in their concentrations, we may be able to distinguish processes that occurred in the early solar nebula from initial differentiation processes within planets.

7.5.3. ABUNDANCE PATTERNS IN CHONDRITIC METEORITES

Carbonaceous Chondrites

Abundances of moderately volatile elements in CM chondrites, normalized to Si and CI, are plotted vs condensation temperatures in Fig. 7.5.2.

There is a clear trend of decreasing abundances with decreasing condensation temperatures. This has been pointed out by Wasson and Chou (1974) and Wai and Wasson (1977). Abundances of highly volatile elements (e.g.,

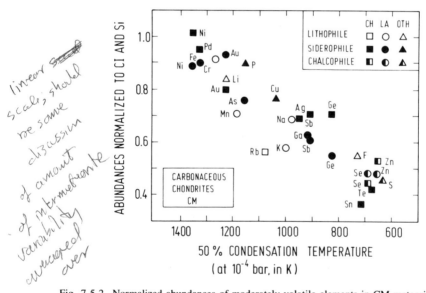

Fig. 7.5.2. Normalized abundances of moderately volatile elements in CM meteorites vs condensation temperatures giving an approximately linear relationship. The large difference in Ge between LA and CH reflects a difference in the measured CM content *and* in the CI value, used for normalization by the two different groups. Data are as follows: CH: Wolf et al. (1980); LA: Kallemeyn and Wasson (1981); OTH: Mason (1979), Dreibus et al. (1979) and unpublished data from MPI-Mainz.

In, Bi and Tl) with still lower condensation temperatures are similar to those of the most volatile elements plotted here (Zn, Te, Se and S). Including these elements in Fig. 7.5.2 would result in a horizontal continuation of the correlation to lower temperatures (see Chapter 7.6). This flat portion of the curve has been emphasized by Anders (1971, 1977a).

The quality of the analytical data plotted in Fig. 7.5.2 varies somewhat depending on the analytical procedures. Accuracies should be better than 10% for most elements. Sample heterogeneity should not play a major role for CM chondrites because the chondrules and inclusions in them are generally very small. Differences between CM chondrites are also small except for the alkali elements Na, K and Rb, as discussed below. For comparison we have plotted, in a few cases, data from two laboratories. Differences in Ni, Au, Ge, Se and Zn determined at University of Chicago and UCLA give some impression of the overall accuracies involved in analytical procedures, sampling and occasionally in different normalization values (e.g., Ge).

From Fig. 7.5.2 it appears that there is a single trend for siderophile and lithophile elements, although Mn and the alkali elements Rb and K are somewhat low. Figure 7.5.3 shows the depletion sequence for CO and CV chondrites compared to that of CM. All patterns are very similar, despite larger absolute depletions in CO and CV. Even small details in the CM pattern are reproduced in the CO patterns: the dip in Mn, the high P, and the low Rb and K contents. Abundances of moderately volatile elements in CM in all cases are higher than those of the CV and CO chondrites. Ornansites (CO chondrites), on the other hand, have higher abundances than the Vigarano-type chondrites (CV chondrites) for all moderately volatile elements less volatile than F. Elements with higher volatility than F have lower abundances in COs

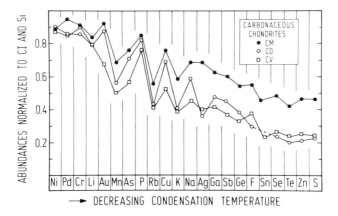

Fig. 7.5.3. Comparison of moderately volatile elements in different types of carbonaceous chondrites. The three patterns are very similar, suggesting that they are the result of the same type of process. Data are as follows: same as Fig. 7.5.2, Takåhashi et al. (1978) and Anders et al. (1976).

than in CVs. This is also true for the highly volatile elements (Takahashi et al. 1978; Kallemeyn and Wasson 1981; see also Chapter 7.6). The only exception among the moderately volatile elements is Ag. However, this element is difficult to analyze and the results of radiochemical neutron-activation analyses show considerable scatter. Silver contents for CVs range, for example, from 91 $\mu g/g$ (Allende) to 158 $\mu g/g$ for Mokoia (Anders et al. 1976; Takahashi et al. 1978). Hence, deviation of Ag may be due to analytical or sampling uncertainties.

The similarities in these abundance patterns strongly suggest that the same process was responsible for the depletion of moderately volatile elements in all types of carbonaceous chondrites.

Figure 7.5.4 is an Fe vs Mn plot for carbonaceous chondrites. Each group has its own characteristic ratio. The CVs have slightly lower Fe and Mn contents than the COs. The narrow compositional clustering of the Ornansites is remarkable. This and similar plots for other elements show that the CO group is compositionally very uniform. In part, this may be due to the smaller and more uniform size of chondrules and inclusions in COs compared with the CVs (Dodd 1981). The CMs are similar in this respect to the COs, but they form a compositionally less well-defined group than the COs.

Alkali Elements in Carbonaceous Chondrites. The most variable elements in carbonaceous chondrites are the alkali elements Na, K, Rb and the halogens Br and Cl. Any carbonaceous chondrite can be immediately grouped

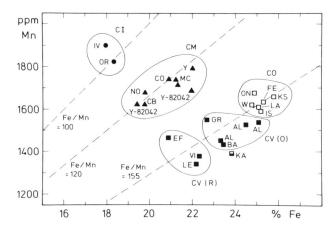

Fig. 7.5.4. Fe-Mn plot for carbonaceous chondrites. Each group has its own characteristic ratio. Meteorites: CI: Ivuna, Orgueil; CM: Cold Bokkeveld, Cochabamba, Murchison, Nogoya, Yamato 74662, Yamato 82042; CO: Felix, Isna, Kainsaz, Lancé, Ornans, Warrenton; CV: Allende, Bali, Efremovka, Grosnaja, Leoville, Vigarano, Karoonda. Data are from MPI-Mainz.

according to its Au, Mn, Se or Zn content. This is not so with Na, since its bulk Na content and those of associated elements vary in CM and CV chondrites. These variations cannot be seen in Figs. 7.5.2 and 7.5.3, since meteorites with low Na and K contents are excluded from averaging. In CV chondrites, low Na contents are restricted to the reduced subgroup of the CVs, as defined by McSween (1977). These meteorites are called reduced because they contain more metal than magnetite, in contrast to the oxidized subgroup, whose members have more magnetite than metal. The reduced CV subgroup consists of four meteorites (Vigarano, Arch, Efremovka and Leoville), which except for Vigarano, are all finds. Since Kallemeyn and Wasson (1982) have provided some evidence for Na loss during terrestrial weathering, there is the possibility that low Na contents were produced on the surface of the Earth. However, Vigarano and the two CM meteorites, Murray and Murchison, are falls with low Na contents (see also Goles et al. 1967). The rather uniform

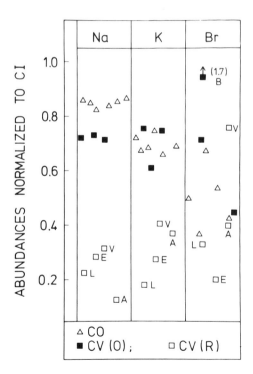

Fig. 7.5.5. Abundances of Na, K and Br in CO and CV carbonaceous chondrites. COs and members of the oxidized subgroup of CV have similar Na and K contents and somewhat variable Br. All meteorites of the reduced subgroup of CV have uniformly low Na and K concentrations. Br is also lower, except in Vigarano. Meteorities: Arch, Leoville, Efremovka, Vigarano, Bali. Data are from MPI-Mainz.

depletion in Na, K and Br of all members of the reduced CV subgroup also speaks against terrestrial weathering (Fig. 7.5.5). Furthermore, the depletion sequence is the same in all cases: Na, K, Br and Cl. Fuchs et al. (1973) have shown that the Na concentration in interior pieces of the Murchison meteorite is variable, ranging from 0.19 to 0.71 wt.% Na_2O. Recent loss of K, which behaves very similar to Na, should result in excessively high K-Ar and ^{39}Ar-^{40}Ar ages. This is not observed (Dominik and Jessberger 1979). The variability of Na and K contents in carbonaceous chondrites was originally noted by Edwards and Urey (1955) and later confirmed by Nichiporuk and Moore (1974). Variable alkali and halogen contents may be related to incomplete sampling of the carrier phases of these elements in carbonaceous chondrites (Spettel et al. 1978). These carrier phases are, in part, fine-grained spinel-rich aggregates with high concentrations of Na, K, Rb, Br and Cl.

Arsenic and Gold in Carbonaceous Chondrites. Besides the alkali elements, there is some evidence for inhomogeneous distribution of As and Au in Allende. Chemical analyses of a dark inclusion in Allende and a large related fragment demonstrate huge variabilities in the contents of As, Au, Na and K between the different lithologies. These variations obviously do not correlate with condensation temperatures, since elements with very low condensation temperatures, such as Se and Zn, show little scatter among the samples listed in Table 7.5.3. Interestingly, the same elements display somewhat unusual behaviour in ordinary and enstatite chondrites (Figs. 7.5.6 and 7.5.7).

TABLE 7.5.3

Large Variations in the Contents of Moderately Volatile Elements Au, As, K, and Na in Different Lithologies of Allende[a]

µg/g	Fragment		Dark Inclusion	Bulk (average 4 samples)
	102.4 mg	311.7 mg	264.6 mg	
Au	2.45	1.43	0.144	0.110
Mn	1804	1660	2050	1510
As	6.8	4.3	1.8	1.33
P	1140		1510	1112
K	781	566	85	290
Na	7207	4940	770	3430
Ga	6.3	7.0	8.3	6.1
Se	5.7	6.9	9.1	8.6
Zn	124	139	160	132

[a] Note the uniform contents of the more volatile elements Se and Zn (Palme et al. 1985).

MODERATELY VOLATILE ELEMENTS 447

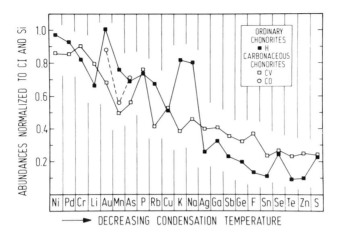

Fig. 7.5.6. Moderately volatile elements in H and CV chondrites. The high Au, Na and K contents of H chondrites do not fit into the pattern of the other moderately volatile elements. Data are the same as in previous figures; compilation by Palme et al. (1981a).

Ordinary Chondrites

The depletion sequence for moderately volatile elements in ordinary chondrites is roughly similar to that for carbonaceous chondrites, as shown in Fig. 7.5.6, where average concentrations of moderately volatile elements in H chondrites are compared to those in CV chondrites. Since the contents of moderately volatile elements do not depend on the petrographic type, average concentrations from type 3 to 6 are plotted in Fig. 7.5.6. Basically similar depletion sequences are obtained for L and LL chondrites with some modifications arising from the lower-metal contents of these meteorites, as discussed in Chapter 7.4 (see also compilation by Wasson and Chou 1974).

In detail, the abundance pattern of H chondrites may be divided into two parts. Elements more volatile than Na (i.e., with lower condensation temperatures) have lower abundances in H chondrites than in CV chondrites. Elements less volatile than Ag have in many cases higher abundances in H than in CV and CO chondrites. This is particularly true for the alkali metals (Na, K and Rb) and for Mn and Au. Arsenic and P have similar abundances in H and CO chondrites. The chalcophile elements Se and S have somewhat higher concentrations in H chondrites than do other elements of similar volatility, but are similar to the CO abundances. This could reflect a lower total pressure during condensation, since the condensation temperatures of chalcophiles (S, Se etc.) are not pressure dependent while all other elements would have lower condensation temperatures at lower pressures.

However, in drawing such comparisons it should be noted that the normalizing element Si may also be fractionated. Larimer (1979) has attempted to relate the composition of CI chondrites and those of other chondritic groups by fractionation of olivine and a refractory-enriched component (see Chapter 7.3). To account for the lower Mg/Si ratio in ordinary chondrites, olivine might be removed from material with CI composition. Reconstituting the original composition of H chondrites, by adding back the 25% or so of Si that seems to be missing, would result in a 25% lowering of all the H chondrite points in Fig. 7.5.6. This would leave only Na, K, Rb and Mn slightly enriched in H chondrites relative to CO chondrites.

Enstatite Chondrites

Enstatite chondrites present a somewhat different picture. The high-Fe group of the enstatite chondrites (EH), comprising the petrographic types from 3 to 5 (Sears et al. 1982), often has concentrations of moderately volatile elements between those of CI and CM chondrites (Fig. 7.5.7). Notable exceptions are Au, P and As which are the least volatile among the moderately volatile siderophile elements and are enriched above the CI level in EH chondrites. Several laboratories have confirmed the high level of Au in the EH chondrites (Binz et al. 1974; Hertogen et al. 1983; Kallemeyn and Wasson

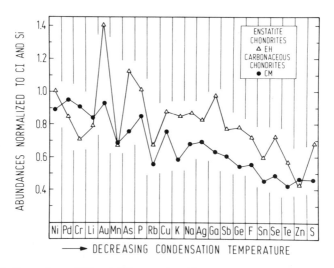

Fig. 7.5.7. Comparison of EH and CM meteorites. The contents of Au, As and P in EH chondrites are above the CI level. The pattern from P to S is, with few exceptions, remarkably similar for the reduced enstatite chondrites (EH) and the oxidized carbonaceous chondrites (CM). Either the oxidation state was not relevant, when these patterns were established or the different oxidation states of CM and EH were fixed after moderately volatile elements were lost or incompletely condensed. Data are the same as in previous Figs.; Kallemeyn and Wasson (1986) and Hertogen et al. (1983).

1986). The last authors normalized their data to Mg and since the Mg/Si ratio is 50% lower in EH chondrites than in CI chondrites, the Au enhancement is even more pronounced. Nevertheless, the overall pattern of siderophiles is similar to the other groups, in that abundances decrease with decreasing condensation temperatures. There is no clear trend for lithophile elements; Cr, Mn and Li are depleted by 30%, while Na and K are only slightly below CI levels. Fluorine is depleted similar to Cr, Mn or Rb. Only Zn has a significantly lower depletion factor. Some caution, however, is appropriate in interpreting these data, because compositional variations among enstatite chondrites are larger than those in other types of chondrites.

The pattern for moderately volatile elements matches the CI-CM pattern for most elements with condensation temperatures below Mn. This similarity in the patterns of highly oxidized and highly reduced meteorites may indicate either that the process responsible for the volatile depletion is independent of O fugacity or that the present oxidation state of the two groups of meteorites was established after the volatiles were lost or incompletely condensed. The EL6 (and sometimes also EH5) chondrites (not shown here) have significantly lower contents of moderately volatile elements than EH3 and EH4 chondrites and may reflect metamorphic loss (Binz et al. 1974; Biswas et al. 1980; Sears et al. 1982).

The major-element composition of E chondrites is even more fractionated than that of H chondrites. Correcting for this fact in the same way as for H chondrites would reduce all abundances by about a factor of 2. This would indicate a higher degree of volatile depletion for enstatite chondrites than conventionally assumed.

Ungrouped Meteorites

There are meteorites with chondritic major-element composition but different textures, mineralogies, degrees of oxidation and O-isotopic compositions from those characteristic of the chondrite groups. The meteorites Acapulco, Pontlyfni, Mount Morris, Winona, Enon, Tierra Blanca and ALHA 77081, and the chondritic inclusions of the aubrites Cumberland Falls and ALHA 78113 are intermediate between H and E chondrites in their degree of oxidation. They generally have significantly higher contents of moderately volatile elements than do H chondrites as shown in Fig. 7.5.8 for Acapulco. Normalized Mn, Au, As, Ga, Sb and Ge abundances are very similar to those in CI and significantly above the H chondrite abundances. The low Cu and Se contents of Acapulco indicate separation of a sulfide phase (Palme et al. 1981a). It must also be mentioned that silicate inclusions in IAB-iron meteorites are very similar to these meteorites in several respects (Palme et al. 1981a; Kallemeyn and Wasson 1985). The Zn contents of IAB-silicate inclusions are, for example, around 200 μg/g compared with 243 μg/g in Acapulco-type 30 to 50 μg/g in H chondrites (Bild et al. 1977). The Acapulco-type meteorites and the IAB inclusions are coarse-grained rocks with equilibrated

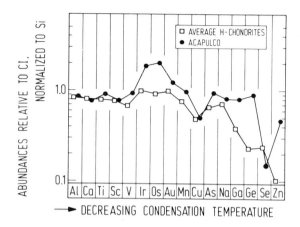

Fig. 7.5.8. Moderately volatile elements in Acapulco are near the C1 level (Au, Mn, As, Na, Ga, Ge). The low Cu and Se contents indicate sulfide removal. Zn is also much higher than in H chondrites. This pattern is typical of meteorites more reduced than H and more oxidized than E chondrites. Silicate inclusions in IAB meteorites are related to these meteorites. Data are from Palme et al. (1981a).

texture and uniform mineral compositions. If they ever had chondrules, the record of them is now completely erased.

The chondritic inclusions in the more reduced Cumberland Falls and ALHA 78113 aubrites contain less Ag, As, Au, Co and Sb than Acapulco but much more Rb and Ga, indicating a more lithophile-rich/siderophile-poor sample of parental material (Verkouteren and Lipschutz 1983). This material is primitive since chondrules in the inclusions are quite distinct, like those of type 3 chondrites. More volatile elements indicate higher formation temperatures for this chondritic matter than for other primitive chondrites (see Chapter 7.6).

In summary, ungrouped chondritic meteorites often have higher contents of moderately volatile elements than ordinary chondrites, but none of these meteorites matches CI levels of moderately volatile elements, as perhaps best exemplified by the Zn abundances of these meteorites.

The Planetary Aspect of Moderately Volatile Elements

Some fundamental questions in planetology are concerned with the relationship of planets to meteorites. Are all planets ultimately composed of average solar-system material, i.e., CI chondrites? Were the fractionation processes that produced the different meteorite classes, also operative in generating protoplanetary material? The bulk composition of planets estimated by Anders (1977b), for example, is based on the assumption that planets consist of the same components as meteorites, only in different proportions,

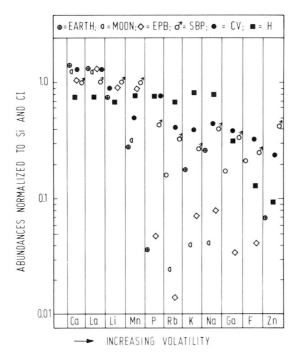

Fig. 7.5.9. The extent of variation of moderately volatile elements in Earth, Moon, eucrite parent body (EPB) and Shergotty parent body (SPB) compared to H and CV chondrites. Planets are depleted in moderately volatile elements. Alkali-element depletion in EPB and Moon is much more pronounced than in CV chondrites. Low P and Ga may indicate removal into the core. Low Mn in the Earth's mantle and in the Moon is interpreted in a similar way by Wänke et al. (1984). Figure is adapted from Dreibus and Wänke (1982).

i.e., protoplanetary material was subject to the same nebular fractionation processes as meteorites (see also Lewis 1974; Morgan and Anders 1980).

Refractory elements are the best example of such a component. Ratios among refractory lithophile elements are the same in undifferentiated meteorites and in bulk planets (see, e.g., Jochum et al. 1986). The amounts of the refractory component vary in different types of chondritic meteorites and also in planets. The range of these variations appears to be of a similar magnitude in chondritic meteorites and in planets.

Although there is no uniquely defined component of moderately volatile elements, one may ask if the degree of fractionation of these elements is comparable in undifferentiated meteorites and planets. In Fig. 7.5.9 the abundances of moderately volatile elements in planets are compared with those in undifferentiated meteorites. It is obvious that all moderately volatile elements for which reasonable estimates exist are depleted in the planets. The extent of the depletion is comparable to that in chondritic meteorites. Depletions of the alkali elements Na, K and Rb in the eucrite parent body (EPB) and in the Moon, however, are much more pronounced than in chondritic meteorites. One could therefore ascribe the low alkali contents to processes occurring, or conditions prevailing, during or shortly after accretion of these bodies (see, e.g., Mittlefehldt 1987).

The relatively low Mn content of the mantle of the Earth may reflect special conditions during formation of the Earth, rather than an initial depletion, because it appears to be correlated with the depletions of the much less volatile elements, Cr and V (Wänke et al. 1984). The low Ga and P contents, estimated for the EPB are in part due to removal of these elements into its core.

Most iron meteorites are probably fragments of FeNi cores or pods in larger parent bodies. During initial differentiation of these parent planetesimals when the FeNi cores formed, siderophile elements would have more or less quantitatively partitioned into the metal phase. Additional partial melting or fractional crystallization of the metal phase would have fractionated siderophile elements (see, e.g., Kelly and Larimer 1977). From abundance patterns in iron meteorites, it is possible to estimate the original abundances of siderophiles in the FeNi cores. Figure 7.5.10 demonstrates the same depletion sequence for siderophile elements in the parent body of IIAB irons and in H chondrites. The IVB irons show an even stronger depletion with decreas-

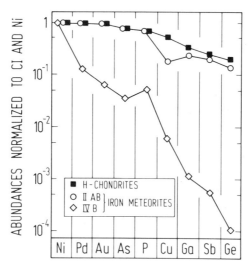

Fig. 7.5.10. Abundances of moderately volatile siderophile elements in two groups of iron meteorites compared to H chondrites. Depletions of siderophiles imply depletions of lithophiles in the silicate fractions of the corresponding parent bodies. Depletions in the IVB meteorites are orders of magnitude larger than those in chondritic meteorites. Data are from Willis (1980).

ing condensation temperature. Germanium, for example, is depleted by a factor of 10^{-4} compared to Ni. Similar depletions should be expected for lithophile elements in the silicate part of the corresponding parent planets.

It appears that most iron meteorites derive from planetesimals with some depletion of moderately volatile elements (Scott 1979). Some groups, e.g. IAB, show little or no depletion. However, there is no group of iron meteorites with an apparent overabundance of volatile or moderately volatile elements.

In conclusion, undifferentiated meteorites (chondrites), differentiated meteorites and bulk planets are similar in that they are depleted in moderately volatile and highly volatile elements, relative to the CI chondrites.

7.5.4. SUMMARY OF OBSERVATIONS OF MODERATELY VOLATILE ELEMENTS IN METEORITES

In assessing the significance of moderately volatile elements, it is appropriate to start by summarizing the observational constraints.

1. All groups of undifferentiated meteorites (i.e., chondritic meteorites) show some depletion of moderately volatile elements relative to the abundances in CI chondrites. The only exceptions are Au, As and P in EH chondrites.
2. Samples from differentiated planetary bodies (Earth, Moon, Mars and the eucrite parent body) are depleted in moderately volatile elements. This is best documented in the alkali elements Li, Na, K, Rb and Cs.
3. Most iron meteorites are depleted in moderately volatile siderophile elements. The extent of the depletions is variable.
4. Depletions in CI-normalized abundances of moderately volatile elements show a dependence upon their 50% condensation temperatures.
5. Depletion of moderately volatile elements occurred very early in the evolution of the solar system. This can be inferred from initial Sr-isotope ratios and from Ag-isotope ratios (Chapters 5.2 and 15.1).
6. Quite specific statements can be made for chondritic meteorites, since they were not affected by core formation and igneous processing, as follows:
6a. The extent of depletion and the ratios among depleted elements are variable and different for each group of chondritic meteorites. There is no single, well-defined component of moderately volatile elements, comparable to the component of refractory elements. Moderately volatile siderophile elements show a closer relationship with condensation temperatures than do lithophiles;
6b. In most cases, the extent of the depletion increases from CI (undepleted) to EH, CM, CO, CV and H chondrites. The only moderately volatile elements with higher abundances in H chondrites than in CO or CV

chondrites, are Mn and Au and the alkali elements Na, K and Rb. Elements more volatile than Sn have higher abundances in CV than CO;

6c. There is no correlation of the degree of depletion of moderately volatile elements with any other property of chondritic meteorites such as degree of oxidation, O-isotopic composition or content of refractory elements.

Some of these points are illustrated in Fig. 7.5.11, where Si- and CI-normalized abundances (depletion factors) for representative elements are plotted for various chondrite groups. The meteorite groups are arranged in the order of decreasing equilibrium O fugacities, with CI meteorites being the most oxidized, and enstatite chondrites the most reduced chondritic meteorites. The contents of refractory elements increases from CI through CM and CO to CV then drops to H and E chondrites. Evidently, there is no correlation of the abundance of moderately volatile elements with the contents of refractories.

Figure 7.5.11 also demonstrates quite clearly the variety of patterns. The two lithophile elements Mn and Na have a similar pattern. Another lithophile

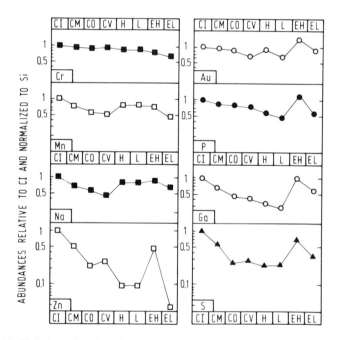

Fig. 7.5.11. Variations of moderately volatile elements in different types of chondritic meteorites. Meteorite groups are arranged in the order of increasing degrees of oxidation. Five different patterns for seven moderately volatile elements are found, indicating that there is not a single well-defined component of moderately volatile elements. There is no correlation of the depletion of moderately volatile elements with any other property of chondritic meteorites.

element Li, behaves differently: its abundance decreases from CV to H chondrites (not shown in Fig. 7.5.11).

The pattern of Au is different from those of the two other siderophiles P and Ga. The S pattern is similar to P and Ga, except for the same abundance of S in H and L chondrites; S evidently was not affected by the metal-silicate fractionation (see Chapter 7.4).

The patterns of S and Zn are very similar, although the extent of fractionation is significantly greater for Zn. The similarity in the pattern of S and Zn may be a reflection of their similar condensation temperatures (Table 7.5.1). In the solar nebula, Zn condenses predominantly as sulfide (Wai and Wasson 1977). Sulfides are, however, not the host phase for Zn in carbonaceous and ordinary chondrites. In these meteorites Zn resides in the silicate phase (Nishimura and Sandell 1964).

7.5.5. INTERPRETATIONS

Constraints on Solar-System Formation

Most samples of planetary material, including nearly all groups of chondritic meteorites, are depleted in moderately volatile elements. The depletion sequence of these elements is roughly similar in the different types of chondritic meteorites. This is particularly true for siderophile elements. This may indicate that the same process is ultimately responsible for the depletion of moderately volatile elements in the various types of meteorites. We have mentioned before that chondritic meteorites have acquired their share of volatile elements very early in their history. It is also known from extensive investigations of CAIs and similar materials in CM, CO and CV chondrites that at least some solid matter was processed at high temperatures in the early solar system. There is ample evidence for evaporation and condensation processes recorded in the chemistry and mineralogy of these materials (Chapter 10.3). One may accordingly explain the depletion of moderately volatile elements either by incomplete condensation or by evaporation of already condensed matter or by a combination of both. This requires that the temperature, on a broad scale, was at least 1200 K throughout the early solar system. An upper limit to the temperature of fractionation of solid material in the solar nebula is given by the unfractionated refractory elements in planets and bulk meteorites. These temperatures are in the range of 1500 K (Jochum et al. 1986). The CAIs in carbonaceous chondrites provide evidence for locally much higher temperatures (see Chapter 10.3).

Whatever caused these depletions, it is remarkable that there are only depletions, never enrichments of moderately volatile elements. This is true for undifferentiated and differentiated meteorites as well as for planets. One might think that if we find abundant solid matter which has lost or not acquired its full complement of volatile elements, we should also find some material high in volatile or moderately volatile elements, with complementary element patterns. However, with few exceptions—an H6 meteorite (Higuchi et al. 1977), an enstatite chondrite (Ganapathy and Larimer 1980) and H chondrite regolith breccias (Lipschutz et al. 1983)—this is not the case. Since only a small fraction of the total mass of the solar system is contained in planets and meteorites, one may confidently conclude that the missing elements were either blown out of the solar system or were swept up by the Sun. An appreciable enrichment of these elements in the Sun is not observed.

Important constraints for the formation of the solar system can be deduced from the abundances of moderately volatile and volatile elements:

1. Evaporation or incomplete condensation of the moderately volatile elements requires temperatures in the inner solar system on the order of 1200 K. Because there is no apparent correlation of the extent of depletion with any other property of meteorites, it is difficult to specify conditions for the loss of volatiles more precisely. In particular, there is no supporting evidence that decreasing depletion of moderately volatile elements simply reflects formation at increasing heliocentric distances.
2. Moderately volatile and volatile elements that were either vaporized or incompletely condensed were not sampled at other locations; in particular, there is no net enrichment or depletion of these elements in the Sun.
3. The lack of meteorites enriched in moderately volatile elements makes it unlikely that the missing volatiles were transported outwards, away from the Sun. Recondensation on cooler material should produce volatile-rich meteorites. The lack of volatile enrichment in the Sun (relative to CI) indicates that the amount of additional volatiles in the Sun is so small that it cannot be detected.
4. The deficiency in volatile-rich meteorites argues against a local heat source for vaporization of volatile elements (e.g., impacts).

Mechanism of Fractionation of Moderately Volatile Elements

As pointed out before, there are two possibilities to account for the depletion of moderately volatile and highly volatile elements in meteoritic matter: fractionation during evaporation and fractionation during condensation (post-accretionary processes are not considered here; they are discussed in Chapter 7.6).

Let us begin with the first possibility. The rather uniform concentrations of highly volatile elements in carbonaceous and in many type 3 ordinary chondrites place severe constraints on the evaporation model. It is very unlikely that evaporation can produce such a pattern for the following reasons:

a. Fractionation of moderately volatile elements during evaporation implies temperatures that are so high that highly volatile elements should be completely lost.
b. Elements residing in very different phases in meteorites such as Zn, Te and Se are unlikely to be lost at exactly the same level by evaporation.

c. Heating experiments by Lipschutz and coworkers on many primitive chondrites (see Chapter 7.6) demonstrate huge differences in volatilization of elements with similar cosmochemical volatilities (i.e., condensation temperatures). A good example are the different retentivities for Zn and Se.

The only way that evaporation could be a realistic possibility to fractionate moderately volatile elements is to lose these elements only from a fraction of the meteorite and leave the rest unaffected. This is the basis for the 2-component model originally proposed by Wood (1963) and later modified by Anders (1964) and Larimer and Anders (1967). One component would contain volatile elements, the other one, having been heated to high temperatures would have lost these elements more or less completely. The gradual depletion of moderately volatile elements in chondritic meteorites thus could be explained by incomplete loss. In this type of model, it is assumed that the two components are present in meteorites as matrix and chondrules. Chondrules having formed during melting of matrix, completely lost their volatiles and some fraction of their moderately volatiles during heating. The attractiveness of this model stems from the fact that there is indeed an approximate correlation between the degree of depletion of volatiles and the content of chondrules in carbonaceous chondrites. CI chondrites have no chondrules and the proportion of chondrules increases from CM through CO and CV to H3 chondrites parallel to the decrease of highly volatile and some moderately volatile elements (e.g., Ga and Zn; Fig. 7.5.11). In detail, there are problems with this correlation (see Takahashi et al. 1978 for a detailed discussion). An extreme example is the CM meteorite Yamato 82042. Texture and petrography of this Antarctic meteorite are similar to those of CI meteorites. Despite the complete absence of chondrules, Y82042 exactly matches CM meteorites with respect to chemical composition (see Fig. 7.5.4 for Fe and Mn). This is also true for the two moderately volatile elements Se and Zn (Grady et al. 1987).

Intuitively one might think that it should be easy to verify experimentally the 2-component model by determining the content of volatile elements in chondrules. This is, however, not such a simple task, mainly for the following two reasons (Grossman and Wasson 1983; Chapter 9.1):

1. Chondrules are by no means free of volatiles; instead, they span a wide range of compositions, also with respect to volatile elements;
2. Chondrules occasionally have enrichments of volatiles on their surface, possibly arising from recondensation.

The siderophile elements among the moderately volatiles show in all meteorite classes a more regular pattern of decreasing abundances with decreasing condensation temperatures than do the lithophiles. In the 2-component model, the abundances of volatile metals would be governed by incomplete loss during formation of coarse metal simultaneously with the melting of chondrules from a fine-grained matrix (Anders 1964).

The second possibility to fractionate moderately volatile elements, is by incomplete condensation. Wasson and Chou (1974) have suggested that dissipation of gas during condensation led to progressively lower contents of volatiles as condensation proceeded. Other possibilities include settling of dust to the nebular midplane during condensation (Anders et al. 1976) or incomplete collection of volatile-rich, fine-grained dust (Wasson 1985).

Lack of equilibrium between gas and solid matter at lower temperatures, when solid material has grown to sizeable chunks, has been proposed as a possible mechanism to fractionate volatile elements (Blander and Abdel-Gawad 1969; Wasson 1985).

7.5.6. OUTLOOK

Fractionation of moderately volatile and highly volatile elements was a major process in the early solar system. This is, for example, reflected in the variable concentrations of Rb and associated variations in initial Sr-isotopic ratios in chondritic meteorites and in parent bodies of differentiated meteorites, i.e., planetesimals and larger planets (Chapter 5.2).

A better understanding of the processes that led to the depletion of volatiles would place stronger constraints on the conditions of formation of solid material in the solar system. In particular, it should be possible to clarify the question whether evaporation or incomplete condensation was the major processes in establishing the elemental abundance patterns observed in primitive meteorites and planets. It has been mentioned before that O partial pressures would have been much higher during evaporation than during condensation. Elements, whose volatility is dependent on O fugacity are therefore of special interest. The abundances of Cr, P, Ga and Zn may turn out to be diagnostic. In addition, refractory elements that become volatile under sufficiently oxidizing conditions (i.e., Os, Ru, Re, W and Mo) are potentially valuable indicators for maximum O partial pressures during fractionation processes in the solar nebula. The gross similarity in the depletion patterns of carbonaceous and enstatite chondrites clearly indicates that a fundamental process is ultimately responsible for the fractionation of these elements.

REFERENCES

Anders, E. 1964. Origin, age and composition of meteorites. *Space Sci. Rev.* 3:583–714.
Anders, E. 1971. Meteorites and the early solar system. *Ann. Rev. Astron. Astrophys.* 9:1–34.
Anders, E. 1977a. Critique of "Nebular condensation of moderately volatile elements and their abundances in ordinary chondrites" by Chien M. Wai and John T. Wasson. *Earth Planet. Sci. Lett.* 36:14–20.
Anders, E. 1977b. Chemical composition of the Moon, Earth and eucrite parent body. *Phil. Trans. Roy. Soc. London* A285:23–40.

Anders, E., and Ebihara, M. 1982. Solar-system abundances of the elements. *Geochim. Cosmochim. Acta* 46:2363–2380.
Anders, E., Higuchi, H., Ganapathy, R., and Morgan, J. W. 1976. Chemical fractionations in meteorites—IX C3 chondrites. *Geochim. Cosmochim. Acta* 40:1131–1139.
Bild, R. W. 1977. Silicate inclusions in group IAB irons and a relation to the anomalous stones Winona and Mt. Morris (Wis.). *Geochim. Cosmochim. Acta* 41:1439–1456.
Binz, C. M, Kurimoto, R. K., and Lipschutz, M. E. 1974. Trace elements in primitive meteorites—V. Abundance patterns of thirteen trace elements and interelement relationships in enstatite chondrites. *Geochim. Cosmochim. Acta* 38:1579–1606.
Biswas, S., Walsh, T., Bart, G., and Lipschutz, M. E. 1980. Thermal metamorphism of primitive meteorites—XI. The enstatite meteorites: Origin and evolution of a parent body. *Geochim. Cosmochim. Acta* 44:2097–2110.
Blander, M., and Abdel-Gawad, M. 1969. The origin of meteorites and the constrained equilibrium condensation theory. *Geochim. Cosmochim. Acta* 33:701–716.
Dodd, R. T. 1981. *Meteorites* (Cambridge: Cambridge Univ. Press), p. 49.
Dominik, B., and Jessberger, E. K. 1979. ^{40}Ar-^{39}Ar dating of Murchison, Allende and Leoville whole rock. *Lunar Planet. Sci.* X:306–308 (abstract).
Dreibus, G., and Wänke, H. 1982. Parent body of the SNC-meteorites: Chemistry, size and formation. *Meteoritics* 17:207–208 (abstract).
Dreibus, G., Spettel, B., and Wänke, H. 1979. Halogens in meteorites and their primordial abundances. In *Origin and Distribution of the Elements*, ed. L. H. Ahrens (New York: Pergamon), pp. 33–38.
Edwards, G., and Urey, H. C. 1955. Determination of alkali metals in meteorites by a distillation process. *Geochim. Cosmochim. Acta* 7:154–168.
Fegley, B., Jr., and Lewis, J. S. 1980. Volatile element chemistry in the solar nebula: Na, K, F, Cl, Br, and P. *Icarus* 41:439–455.
Fuchs, L. H., Olsen, E., and Jensen, K. J. 1973. Mineralogy, mineral-chemistry, and composition of the Murchison (C2) meteorite. *Smithsonian Contrib. Earth Sci.* 10:1–39.
Ganapathy, R., and Larimer, J. W. 1980. A meteoritic component rich in volatile elements: Its characterization and implications. *Science* 207:57–59.
Goles, G. G., Greenland, L. P., and Jérome, D. Y. 1967. Abundances of chlorine, bromine and iodine in meteorites. *Geochim. Cosmochim. Acta* 31:1771–1787.
Grady, M. M., Graham, A. L., Barber, D., Ayhner, R., Kurat, G., Ntaflos, T., Ott, U., Palme, H., and Spettel, B. 1987. Yamato-carbonaceous chondrite with CM affinities. *Proc. of the Eleventh Symposium on Antarctic Meteorites, 1986* (Tokyo: Natl. Inst. of Polar Research), pp. 162–178.
Grossman, J. N., and Wasson, J. T. 1983. The compositions of chondrules in unequilibrated chondrites: An evaluation of models for the formation of chondrules and their precursor materials. In *Chondrules and Their Origins*, ed. A. King (Houston: Lunar and Planetary Inst.), pp. 88–121.
Hertogen, J., Janssens, M.-J., Takahashi, H., Morgan, J. W., and Anders, E. 1983. Enstatite chondrites: Trace element clues to their origin. *Geochim. Cosmochim. Acta* 47:2241–2255.
Higuchi, H., Ganapathy, R., Morgan, J. W., and Anders, E. 1977. "Mysterite": A late condensate from the solar nebula. *Geochim. Cosmochim. Acta* 41:843–852.
Holweger, H. 1979. Abundances of the elements in the Sun: Introductory report. In *Les Éléments et leur Isotopes dans l'Univers, XXIInd Liége International Astrophysical Symp.* (Liége: Univ. de Liége, Inst. d'Astrophysique), pp. 117–138.
Hultgren, R., Desai, P. D., Hawkins, D. T., Geiser, M., and Kelley, K. K. 1973. *Selected Values of the Thermodynamic Properties of Binary Alloys* (Metals Park, Ohio: American Society for Metals), p. 858.
Jochum, K. P., Seufert, H. M., Spettel, B., and Palme, H. 1986. The solar-system abundances of Nb, Ta, and Y, and the relative abundances of refractory lithophile elements in differentiated planetary bodies. *Geochim. Cosmochim. Acta* 50:1173–1183.
Kallemeyn, G. W., and Wasson, J. T. 1981. The compositional classification of chondrites: I. The carbonaceous chondrite groups. *Geochim. Cosmochim. Acta* 45:1217–1230.
Kallemeyn, G. W., and Wasson, J. T. 1982. The compositional classification of chondrites: III. Ungrouped carbonaceous chondrites. *Geochim. Cosmochim. Acta* 46:2217–2228.

Kallemeyn, G. W., and Wasson, J. T. 1985. The compositional classification of chondrites. IV. Ungrouped chondritic meteorites and clasts. *Geochim. Cosmochim. Acta* 49:261–270.

Kallemeyn, G. W., and Wasson, J. T. 1986. Composition of enstatite (EH3, EH4, 5 and EL6) chondrites: Implications regarding their formation. *Geochim. Cosmochim. Acta* 50:2153–2164.

Keays, R. R., Ganapathy, R., and Anders, E. 1971. Chemical fractionations in meteorites—IV. Abundances of fourteen trace elements in L-chondrites; implications for cosmothermometry. *Geochim. Cosmochim. Acta* 35:337–363.

Kelly, W. R., and Larimer, J. W. 1977. Chemical fractionation in meteorites—VIII. Iron meteorites and the geochemical history of the metal phase. *Geochim. Cosmochim. Acta* 41:93–111.

Larimer, J. W. 1979. The condensation and fractionation of refractory lithophile elements. *Icarus* 40:446–454.

Larimer, J. W., and Anders, E. 1967. Chemical fractionations in meteorites—II. Abundance patterns and their interpretations. *Geochim. Cosmochim. Acta* 31:1239–1270.

Lewis, J. S. 1974. The temperature gradient in the solar nebula. *Science* 186:440–443.

Lipschutz, M. E., Biswas, S., and McSween, H. Y., Jr. 1983. Chemical characteristics and origin of H chondrite regolith breccias. *Geochim. Cosmochim. Acta* 47:169–179.

Mason, B. 1979. Cosmochemistry. Part 1. Meteorites. In *Data of Geochemistry, Sixth Edition*, ed. M. Fleischer, Geol. Survey Prof. Paper 440-B-1 (Washington, DC: U.S. Govt. Print. Office).

McSween, H. Y., Jr. 1977. Petrographic variations among carbonaceous chondrites of the Vigarano type. *Geochim. Cosmochim. Acta* 41:1777–1790.

Mittlefehldt, D. W. 1987. Volatile degassing of basaltic achondrite parent bodies: Evidence from alkali elements and phosphorus. *Geochim. Cosmochim. Acta* 51:267–278.

Morgan, J. W., and Anders, E. 1980. Chemical composition of Earth, Venus and Mercury. *Proc. Natl. Acad. Sci. U.S.A.* 77:6973–6977.

Morgan, J. W., Janssens, M.-J., Takahashi, H., Hertogen, J., and Anders, E. 1985. H-chondrites: Trace element clues to their origin. *Geochim. Cosmochim. Acta* 49:247–259.

Nichiporuk, W., and Moore, C. B. 1974. Lithium, sodium and potassium abundances in carbonaceous chondrites. *Geochim. Cosmochim. Acta* 38:1691–1701.

Nishimura, M., and Sandell, E. B. 1964. Zinc in meteorites. *Geochim. Cosmochim. Acta* 28:1055–1079.

Palme, H., Schultz, L., Spettel, B., Weber, H. W., Wänke, H., Christophe Michel-Lévy, M., and Lorin, C. 1981a. The Acapulco meteorite. Chemistry, mineralogy and irradiation effects. *Geochim. Cosmochim. Acta* 45:727–752.

Palme, H., Suess, H. E., and Zeh, H. D. 1981b. Abundances of the elements in the solar system. In *Landolt-Börnstein, Group IV: Astronomy Astrophysics and Space Research, Volume 2*, Astronomy and Astrophysics, Extension and Supplement to Volume I, Subvolume a, ed. K.-H. Hellwege (Berlin: Springer-Verlag), pp. 257–272.

Palme, H., Kurat, G., Brandstätter, F., Burghele, A., Huth, J., Spettel,B., and Wlotzka, F. 1985. An unusual chondritic fragment from the Allende meteorite. *Lunar Planet. Sci.* XVI:645–646 (abstract).

Scott, E. R. D. 1979. Origin of iron meteorites. In *Asteroids*, ed. T. Gehrels (Tucson: Univ. of Arizona Press), pp. 892–925.

Sears, D. W., Kallemeyn, G. W., and Wasson, J. T. 1982. The compositional classification of chondrites: II. The enstatite groups. *Geochim. Cosmochim. Acta* 46:597–608.

Spettel, B., Palme, H., and Wänke, H. 1978. The anomalous behaviour of Na and K in carbonaceous chondrites. *Meteoritics* 13:636–639 (abstract).

Takahashi, H., Janssens, M.-J., Morgan, J. W., and Anders, E. 1978. Further studies of trace elements in C3 chondrites. *Geochim. Cosmochim. Acta* 42:97–106.

Verkouteren, R. M., and Lipschutz, M. E. 1983. Cumberland Falls chondritic inclusions—II. Trace element contents of forsterite chondrites and meteorites of similar redox state. *Geochim. Cosmochim. Acta* 47:1625–1633.

Wänke, H. 1981. Constitution of terrestrial planets. *Phil. Trans. Roy. Soc. London* A303:287–302.

Wänke, H., Dreibus, G., and Jagoutz, E. 1984. Mantle chemistry and accretion history of the Earth. In *Archean Geochemistry*, eds. A. Kröner et al. (Berlin: Springer-Verlag), pp. 1–24.

Wai, C. M., and Wasson, J. T. 1977. Nebular condensation of moderately volatile elements and their abundances in ordinary chondrites. *Earth Planet. Sci. Lett.* 36:1–13.

Wasson, J. T. 1985. *Meteorites* (New York: W. H. Freeman and Company).

Wasson, J. T., and Chou, C.-L. 1974. Fractionation of moderately volatile elements in ordinary chondrites. *Meteoritics* 9:69–84.

Willis, J. 1980. The mean composition of iron meteorite parent bodies. Ph.D. thesis, Univ. of California, Los Angeles.

Wolf, R., Richter, G. R., Woodrow, A. B., and Anders, E. 1980. Chemical fractionations in meteorites—XI. C2 chondrites. *Geochim. Cosmochim. Acta* 44:711–717.

Wood, J. A. 1963. On the origin of chondrules and chondrites. *Icarus* 2:152–180.

7.6. HIGHLY LABILE ELEMENTS

MICHAEL E. LIPSCHUTZ
Purdue University

and

DOROTHY S. WOOLUM
California State University at Fullerton

Certain elements of high lability are very responsive to thermal processes, being either highly volatile during primary nebular condensation or highly mobile (easily volatilized) by post-accretionary metamorphic or shock heating. Because of this high lability, contents of such elements are generally at trace or ultratrace ($\mu g/g$ to pg/g) levels and they provide sensitive markers of meteorites' thermal histories even at relatively low temperatures. Data for highly labile elements indicate that different thermal processes were important in the genesis of each of the chondritic groups and a discussion of each is given. Contents of highly labile elements in a given group of contemporary falls differ from those of the same group that fell in Antarctica > 0.1 Myr ago. This difference is due either to a time-dependent change in meteorite sources or, less likely, orbital variation of the meteorite flux to Earth.

7.6.1. INTRODUCTION

An element can be fractionated only if some physical or chemical process occurs that alters its physical state *and* if the system if open to transfer of the element in its new state. Almost by definition, easily fractionated elements are present in nearly all meteorites only at trace and ultratrace levels, i.e., at μg/g to pg/g concentrations. This occurs because such elements are so labile that any genetic process easily causes their loss. Either very little of the element enters from the nebula during primary condensation or, if it does enter, a considerable portion can be lost during post-accretionary parent-body heating. In any event, only very small amounts of such elements are left in the meteorite.

In the condensed state, physical properties of trace constituents can differ considerably from those of macroscopic quantities of these same elements. For weighable quantities of matter, it is possible to identify an atom's nearest neighbors, specify bond lengths and strengths and establish thermodynamic properties, hence volatility, of the system. In contrast, highly volatile trace elements are not generally found in a single host site but rather are dispersed throughout the meteorite. Atomic nearest neighbors cannot be specified and, therefore, neither can bond lengths nor strengths. Thermodynamic properties of such trace elements in the condensed state are unknown and possibly unpredictable so that the response of a trace element to a thermal episode needs to be established experimentally.

It is important to distinguish between a trace element's volatility in a primary, nebular context from its mobility, i.e., the ease of its vaporization and loss during open-system post-accretionary heating of meteoritic parent material. Operationally, the ratio of an element's mean atomic abundance in equilibrated ordinary chondrites—which are demonstrably metamorphosed (perhaps in a closed system)—relative to that in CI chondrites (Chapter 1.1), coupled with information from abundance ratios in carbonaceous chondrites, provide a qualitative measure of its volatility and we adopt that practice here. Calculations of elemental volatilities are model-dependent and the volatility order, even for the same model, may differ slightly from one publication to another of a particular research group. Condensation sequences calculated by various authors are listed in Chapter 7.1. It should be noted that in making condensation calculations, a variety of assumptions are made including the identity of the condensed phase (see also Chapter 7.5). Since each element is assumed to condense as a pure phase (metal, oxide or sulfide)—which, in reality, cannot be expected—the calculations must necessarily constitute imperfect models. The reader should, therefore, keep this in mind in comparing experimental data with theoretical curves.

Opportunities for post-accretionary heating are depicted in Fig. 3.1.1 and listed to the right of it. Meteorites studied today could have experienced one or more of these fractionations. For trace element vapor-phase mobilization, it is essentially immaterial whether a given high temperature is attained during thermal metamorphism or, after a high-intensity shock event on decompression in a well-insulated region. Loss can occur if high temperatures are sustained for days and if the system is an open one. To disentangle thermal histories, we interpret elemental abundance patterns, interrelationships between groups of elements, and relationships between trace-element abundances and other chemical and physical records of high temperatures in the meteorites.

Lipschutz and coworkers demonstrated that trace elements can be vaporized from chondritic matter heated under temperature and ambient atmospheric conditions reasonable for early solar-system objects (Chapter 3.3 and 3.6). They found that loss does not occur at a sharply defined temperature but, rather, takes place over a broad range, indicating a variety of trace-element host sites. The loss pattern is only slightly dependent on the nature of the material being heated (i.e., its particulate size or oxidation state). Treating the loss as reflecting diffusion, apparent activation energies can be calculated (Bart et al. 1980). The low apparent activation energies for Bi and Tl loss, for example, are inconsistent with diffusion, according to Alaerts and Anders (1979), and indicate desorption. In fact, however, diffusion of Tl and other ions is known to occur with low apparent activation energies in a variety of systems (Bart et al. 1980). The apparent activation energies are temperature dependent and provide a measure of the ease of vaporization and loss of highly mobile species. For example, data for the Allende CV3 chondrite (Table 7.6.1) indicate some surprises. These especially include the high lability of elements like Bi and Tl that are capable of forming chemical bonds, relative to Ar isotopes and, indeed, all noble gases but He (Bart et al. 1980). These experiments yield important information on trace-element vaporization from solid or liquid but not on solid/liquid fractionation in parent bodies. Many meteorites apparently were heated under conditions such that trace-element vaporization occurred; some show evidence for trace-element transport in a liquid phase, perhaps at temperatures above 988° C, the FeS-Fe eutectic temperature.

While highly volatile trace elements are generally highly mobile, and *vice versa,* no calculated condensation sequence (Chapter 7.1) corresponds in a 1:1 fashion with the mobility order determined in laboratory experiments (Table 7.6.1). A major impetus for analyzing these elements is to determine the genetic process(es) responsible for establishing their contents in the meteoritic group under study. So as not to bias the reader unduly toward some particular fractionation cause, we generally will refer to high elemental lability unless primary volatility or secondary mobilization is specifically intended.

The highly labile elements are Bi, Cd, Cs, Hg, In, Pb, Se, Te, Tl and Zn. Depending on circumstances, other elements (e.g., Ag, Ga and Rb) may be included with these (cf. Chapter 7.5). The highly labile elements give an enormous amount of information on thermal histories at relatively low temperatures. Most of these have been studied extensively for the past 2 decades, mainly by radiochemical neutron activation analysis. Exceptions are Pb (which must be determined separately since it cannot be measured by neutron activation) and Hg (because its extreme lability causes it to contaminate meteorites easily). However, Jovanovic and Reed (1985, and references cited

TABLE 7.6.1
Apparent Activation Energies for Loss of Trace Elements and Noble Gas Nuclides from Heated Allende

Element or Nuclide	Temperature Range (°C)	Q (kcal/mol)	Ref[b]
Ag	700–1000	26	d
	1000–1200	2.1	d
	1200–1315	28	d
^{36}Ar	900–1000	45	b
^{38}Ar	900–1000	45	b
^{40}Ar	500–1000	8.1	b
As	1000–1100	98	d
	1100–1393	18	d
Bi	400– 600	25	a,c
	600–1000	3.0	a,c
	1000–1315	12	a,c
C	700–1000	19	b
Cd	400– 900	8.0	d
Co	1200–1393	140	a,c,d
Cs	700– 900	38	d
	1000–1393	14	d
Cu	700–1000	18	d
	1000–1100	200	d
	1100–1393	16	d
Ga	900–1200	63	a,c
^{3}He	500– 900	13	b
^{4}He	500–1000	13	b
In	600–1268	21	a,c
^{84}Kr	900–1000	45	b
^{20}Ne	500–1000	13	b
^{21}Ne	700–1000	20	b
^{22}Ne	700–1000	18	b
Rb	700–1000	37	d
	1000–1393	14	d
Sb	700–1393	29	d
Se	600– 900	11	a,c
	1000–1100	120	a,c
	1100–1315	15	a,c
Te	600– 900	39	d
	900–1393	12	d
Tl	400– 600	25	a,c,d
	600– 900	2.4	a,c,d
	900–1315	8.6	a,c,d
^{132}Xe	900–1000	44	b
Zn	700– 800	59	d
	900–1393	16	d

[a] Table taken from Ngo and Lipschutz (1980).
[b] References: (a) Ikramuddin and Lipschutz (1975); (b) Herzog et al. (1979); (c) Bart et al. (1980); (d) Ngo and Lipschutz (1980).

therein) used stepwise heating on neutron-activated Hg released by chondrites, to address the contamination question and to discuss chondrite thermal histories.

Little is known directly about labile trace-element siting, since determinations of these elements normally involve bulk analyses. The geochemical (siderophile/lithophile/chalcophile) affinity of these elements is normally inferred indirectly, from factor analyses or by analogy with terrestrial examples. Both approaches have obvious pitfalls. Some studies determined trace-element sitings directly, by analyses of physically or chemically separated minerals (see, e.g., Curtis and Schmitt 1979). Such studies are preferable to indirect geochemical-affinity assignments. However, it is difficult, in direct studies, to avoid external contamination in separations, and one must assess the inherent purity of the separates: contaminants include grains of other phases from the sample and veins or inclusions carried within the mineral grains themselves. Further, in analyzing bulk mineral separates, information on possible zoning and intergrain chemical variability of the host phase is lost. Most of these problems are largely obviated where techniques for in-situ microdistribution determinations by, for example, ion probe, proton probe, nuclear microprobe or particle-track radiography of trace elements are feasible (see, e.g., Burnett and Woolum 1983). Such techniques have the added advantage that trace element data may be interpreted more confidently in a petrographic context. Except for Bi and Pb, no such in-situ microdistribution studies of highly labile trace elements in meteorites have been made.

Woolum et al. (1976) devised nuclear particle-track techniques for determining Bi and ^{208}Pb microdistributions at μg/g levels for 100 μm-sized grains. In general, studies of polished sections of chondritic meteorites showed no strong Bi or Pb localizations relative to bulk. In particular, ^{208}Pb in chondrites showed little tendency to concentrate in troilite (FeS). This contrasts with terrestrial experience, where Pb is chalcophile, and with iron meteorites, where Pb is found in troilite-graphite nodules.

Larimer's (1973) equilibrium condensation calculations predicted that about 95% of the Bi and Pb would condense as pure phases. The microdistributions of these elements in E, CV3 and unequilibrated ordinary chondrites (UOC) provided no evidence for this. However, while Bi and Pb microdistributions qualitatively appeared uniform, detailed track-by-track mapping of Bi for E chondrites and for Allende indicated that it is not randomly distributed (Woolum et al. 1979): more 2- to 5-track associations appeared on the microscale than would be expected for a totally random Bi distribution. These data are consistent with a model in which 90% of the Bi resides in 10^{-16} g Bi "point" sources and the rest in 10^{-14} g "point" sources. If the point sources are pure Bi, as suggested by Larimer's predictions, they are 10^2 Å and 10^3 Å in size, respectively. To the extent that equilibrium thermodynamic calculations are valid for condensation of 10^2 to 10^3 Å grains, the data agree with predictions for nebular condensation. However, existence of such grains can be argued only if one is found, and no Bi grains could be located in the meteorite sections analyzed.

The one case where strong Bi and Pb localizations were observed is in the L3 chondrite, Khohar. These localizations were associated with some Fe, Ni metal grains that are particularly poor (only a few percent) in Ni and that appeared finely polycrystalline (Woolum and Burnett 1981). Siderophile behavior was previously unrecorded, although it had been anticipated by Larimer's (1973) nebular condensation calculations. The fact that Bi- and Pb-rich Fe,Ni metal appeared finely polycrystalline raises the possibility that Bi and Pb were adsorbed on metal surfaces rather than dissolved in metal. Woolum and Burnett (1981) argue that the Bi and Pb localizations were most reasonably obtained in a reheating event (probably shock-induced) on the Khohar parent body, and not in the nebula. Hence, in the one case where labile element siting can be established, the localizations appear to be secondary and unrelated to primary cosmochemical behavior.

If it is assumed that the solar nebula was wholly vaporized, every sample of solar-system matter experienced at least one chemical fractionation episode during primary condensation (Fig. 3.1.1). In principle, then, a sample's composition can give information on nebular condensation conditions if post-accretionary heating was absent or occurred under closed conditions. If such heating took place under open conditions, the resulting fractionation must be compensated for if a sample's composition is to provide information on nebular condensation conditions. Of course, data for highly mobile elements can yield direct information on post-accretionary heating processes. There is no reason to believe that primary nebular condensation of all solar-system objects or, indeed, even that all chondrites occurred under the same conditions. Only one nebular condensation model, that of Larimer (1973), predicts detailed trends for even a few highly labile elements. Hence, we will compare our experimental results to its predictions even though the model explicitly assumes a nebula of fixed (cosmic) composition. Differences in nebular physical (pressure, temperature) and chemical (oxygen fugacity, composition) conditions almost certainly were instrumental in forming primitive parent material for each sort of meteorite chemical group. Variations may even have occurred during formation of primitive parental material for one group. Furthermore, all chondrites need not have experienced similar post-accretionary thermal processing.

That post-accretionary thermal processing has occurred, at least for some meteorites, is not really debated. The principal question is whether metamorphism was of the open-system type, where labile elements were lost, or of the closed-system type, where labile elements could be redistributed but not lost. As we shall see, labile element depletion correlates strongly with petrographic type in some meteorite groups; the interpretation of this is disputed, sometimes vigorously. Some (see, e.g., Larimer and Anders 1967) hold that there is no causal relation between labile-element content and metamorphic grade and that the relation is due to incomplete condensation of the equilibrium, gas-to-dust variety in the solar nebula. Meteorites of higher petrographic type were accreted first, at high temperature, and were isolated, perhaps by deep burial in an accreting parent body. Material surviving in the nebula to lower temperatures incorporated the lower-temperature condensates and provided the volatile-rich source material for the lower petrographic types. It is assumed that where metamorphism was involved, it was of the closed-system type. The main opposing view to this condensation model assumes a causal relationship between labile element content and metamorphic type (see, e.g., Wood 1967; Dodd 1969). In this, labile-element contents were established primarily during thermal metamorphism of the open-system type. Higher petrographic types are derived from lower ones and, relative to the latter, are volatile poor because they suffered greater loss of volatiles during more severe or prolonged metamorphism. Shock heating can also cause mobile-element loss (Walsh and Lipschutz 1982; Huston and Lipschutz 1984). There are variations on these general views as well (see, e.g., Blander and Abdel Gawad 1969), but these illustrate the main differences. That these different views still exist is due in part to the lack of decisive tests by which to decide unambiguously between them and probably also because neither view is correct in every detail.

Three lines of evidence argue that post-accretionary heating of the open-system sort can be and has been important for parent objects of at least some chondritic groups. Chondritic gas permeabilities are sufficiently high that trace elements would have been mobilized under any reasonable parent-body model (Sugiura et al. 1983). Gas-retention ages of non-Antarctic L chondrites demonstrate collisional heating 500 Myr ago, sufficient to lose radiogenic ^{4}He and ^{40}Ar (Chapters 3.6 and 5.3). Since ^{40}Ar is less mobile than are several trace elements (Table 7.6.1; Bart et al. 1980), their shock-induced mobilization is expected and, as will be seen, observed. Finally, chalcogenide (S/Se/Te) trends are unique in E4–6 chondrites and are essentially duplicated by heating experiments on Abee (E4) chondrite (Fig. 7.6.1); Te, and other elements, seemingly were mobilized by post-accretionary processes (Biswas et al. 1980). Trends in enstatite achondrites, and other evolved meteorites, indicate redistribution of chalcogenides and other elements in their parent bodies (Biswas et al. 1980). Since enstatite achondrites may be related by simple chemical processing to primitive E chondrites, we summarize some data for them later.

With this introduction, we may now consider the current status of data for highly labile elements in various chondritic groups and their implications for the genesis of each group. For simplicity, we will generally discuss elemental atomic abundances or their ranges, normalized to cosmic (CI) values and, at times, 2-element correlations. In the literature, 2-element correlations mainly involve Bi, In or Tl. To simplify, we choose here to compare only 1 pair, In vs Tl, involving elements with quite different apparent activation

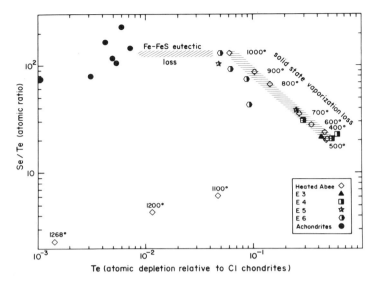

Fig. 7.6.1. Se/Te atomic ratios vs Te depletion (relative to Cl) in E3–6 chondrites, enstatite achondrites (aubrites) and Abee (E4) chondrite samples heated at 400–1268° C. The near superposition of heated Abee data (reflecting preferential solid-state Te vaporization) and E3–6 chondrite results suggests volatilization loss during open-system metamorphism of parent-body material, initially containing Abee-like quantities of highly labile elements. Like Se and Te, all volatile/mobile elements in E3,4 chondrites are near Cl levels and are systematically depleted in E5,6 chondrites. Above 1000° C, Se vaporizes from a condensed phase much more readily than does Te. Data for non-Antarctic aubrites no longer parallel heated Abee trends but rather are displaced leftward from the chondrite trend, consistent with removal of chalcogenides having compositions fixed at about 1000° C. Removal of Se, Te and other labile calcogenides apparently occurred in Fe-FeS eutectic, formed and separated from an enstatite chondrite region heated in an open-system manner at 988° C.

energies for loss during post-accretionary heating (Table 7.6.1). The apparent activation energies for Bi and Tl are quite similar so that during late heating, a point on a Bi vs Tl plot will move diagonally left and down (at 45°) while one on an In vs Tl (or In vs Bi) plot will move horizontally leftward. It is this response to high temperatures that has prompted so much study of these elements. Other ways in which data for highly labile elements can be interpreted are by statistical techniques; either in the patterns of significant 2-element relationships, by factor or cluster analysis or by testing population means. For simplicity, we will not generally discuss those here.

Recently, Dennison et al. (1986) suggested that the Antarctic meteorites of a given type generally derive from parent bodies or regions different from those yielding contemporary non-Antarctic falls. If this suggestion continues to be verified by additional data, non-Antarctic and Antarctic chondrites of a given group (congeners) might well reflect different genetic processes compositionally. Accordingly, we will consider each non-Antarctic chondritic group first, then special sorts of chondrites including the question of the Antarctic / non-Antarctic difference. Non-Antarctic finds often are contaminated and we will not consider trace element data for such meteorites unless they constitute the only available samples of a given group. In that event, this will be pointed out.

7.6.2. NON-ANTARCTIC CARBONACEOUS CHONDRITES

The major genetic process reflected by bulk trace element contents in carbonaceous chondrites seems to be primary nebular condensation. While CI and CM2 chondrites show evidence for post-accretionary aqueous alteration, this apparently took place under conditions closed to trace-element

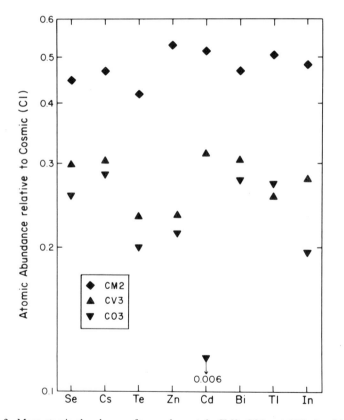

Fig. 7.6.2. Mean atomic abundances of trace elements in CM2, CO3 and CV3 chondrites relative to those in Cl chondrites. Highly volatile elements (to the right) show a rather flat pattern for CM2 and CV3 chondrites, suggesting that they represent mixtures of volatile-rich and -poor components in different proportions (essentially 1:1 for CM2 and 1:2 for CV3). The pattern for CO3 also suggests a similar, but possibly modified, origin.

transport. If such alteration occurred under open conditions, highly soluble trace elements would either be highly enriched or depleted, corresponding to aqueous transport in, or leaching out. In fact, only Br, I and perhaps B show evidence for hydrothermal redistribution in any Cl chondrite (Anders and Ebihara 1982). On average, irrespective of geochemical behavior (Fig. 7.6.2), mean Cl-normalized abundances of highly (and moderately) volatile elements in CM2, CV3 and CO3 chondrites are 0.48, 0.29 and 0.24, respectively (Anders et al. 1976; Wolf et al. 1980). These authors interpret their data in terms of a "two-component model" in which, for example, CM2 chondrites contain 48% Cl-like material condensed at low temperature ("matrix"), admixed with 52% similar material partly to totally devolatilized by a high-temperature nebular episode. CV3 and CO3 chondrites would then represent 29%/71% and 24%/76% mixtures, respectively, in this model. Some moderately mobile elements (As, Au, Rb and Sb) would not have been vaporized completely from the high-temperature portions of these chondrites. Such elements would therefore have abundances intermediate between the undepleted (Cl) values of refractory elements and values typical of highly volatile ones (Anders et al. 1976). Abundances for highly volatile elements in CR2 chondrites are lower than those in CM2 (Wolf et al. 1980).

This model for CM2, CO3 and CV3 chondrites explicitly assumes that all highly (and moderately) volatile elements are equally depleted in a given carbonaceous chondrite subgroup. A controversial exchange resulted (Anders 1977; Wasson 1977) when Wai and Wasson (1977) argued that CM2/Cl ratios for 19 elements denoted by them as moderately volatile were not constant but decreased from 0.9 to 0.4, more or less according to putative condensation temperatures. Wai and Wasson (1977) attributed this decrease to a "volatile-loss" model in which volatiles were lost as gas prior to nebular condensation or during incomplete agglomeration of condensed dust. The elements considered by Wai and Wasson (1977) ranged from undepleted ones to strongly depleted Te and Zn. Effects of such a "volatile-loss" process should be most evident for highly volatile elements but measurements for additional CM2 chondrites and additional more volatile elements revealed only the horizontal, 2-component trend line (Wolf et al. 1980). Since every highly volatile element (except Cd) is depleted to roughly the same extent in each of the petrographic types of carbonaceous chondrite, each such element obviously correlates in a 1:1 direct, statistically significant manner with every other one.

Few carbonaceous chondrites of petrographic types 4–6 exist and nearly all are finds. The few data for highly labile elements in these chondrites (Takahashi et al. 1978; Kallemeyn and Wasson 1982) may be compromised by terrestrial weathering (cf. Bart et al. 1980). In any event, these data are insufficient to determine the process(es) important in establishing the compositional characteristics.

7.6.3. NON-ANTARCTIC ENSTATITE CHONDRITES

There is less consensus on the genetic process(es) primarily responsible for labile element trends in E chondrites than in any other chondritic group. In this unique chondritic group, contents of siderophiles (and other elements) differ markedly in E3–6 samples—just as they do in ordinary chondrites, where Fe (and other elements) compositionally distinguish, for example, H from L chondrites (Chapter 1.1). For this reason, E3–5 chondrites are called EH by some and E6 chondrites, EL. Some authors (see, e.g., Kallemeyn and Wasson 1986) argue that abundances of highly labile elements in E3–6 chondrites must then reflect the same primary processes that fractionated siderophiles (and other elements) during nebular condensation. However, attributions to primary processes do not account in any way for the specific, unique trends exhibited by highly labile elements in E chondrites.

Contents of highly labile elements do not depend upon the degree of oxidation or siderophile makeup of ordinary chondrite groups but can vary with petrographic type (see below). Since ordinary chondrites present no evidence that highly mobile elements necessarily condensed and accreted with nebular siderophiles, a valid working hypothesis is that siderophile-rich and -poor parent materials condensed with essentially Cl-like proportions of mobile elements. According to this view (Biswas et al. 1980), highly mobile elements were then lost during solid-state, open-system metamorphism of the E-chondrite parent object(s).

Abundances of every highly mobile element in E3–6 chondrites show trends like those of Se and Te in Fig. 7.6.1 (Binz et al. 1974). The most mobile elements are the most variable and are distinctly more abundant in E3,4 than in E5,6 chondrites. The E4 chondrites Indarch and Abee have systematically Cl-like contents of labile elements. (Abee, a monomict breccia, contains at least one very volatile-rich component [Ganapathy and Larimer 1980].) Since mobility order and volatility order are not identical, the fact that elemental variability in E chondrites is highest for the most mobile elements argues that the data predominantly reflect parent-body metamorphism (Biswas et al. 1980). In reporting their data, Hertogen et al. (1983) concluded that trends for refractory siderophiles indicated metamorphic fractionation into metal but could not decide whether labile-element patterns in E chondrites indicated fractionation during nebular condensation/accretion or metamorphic loss in the parent body as proposed by Binz et al. (1974) and Biswas et al. (1980). The only known E3 chondrite fall, Qingzhen, has volatile/mobile trace element contents within the E4 range, often near the bottom of it. This contrasts sharply with the situation in ordinary chondrites where contents in type 3 chondrites are very much higher than those in types 4–6 (see below).

Compositional ranges of mobile trace elements in E chondrites are essentially reproduced by week-long heating of Abee at temperatures of 400 to

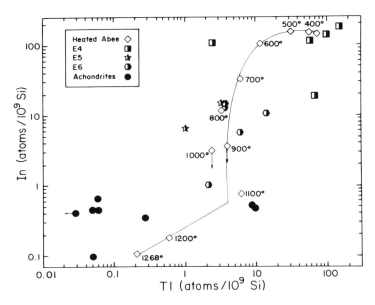

Fig. 7.6.3. Comparison of In vs Tl data for non-Antarctic enstatite chondrites and aubrites with results for Abee samples heated for 1 week at 400–1268° C in a low-pressure environment. All chondrite and achondrite results agree acceptably with In vs Tl data for heated Abee samples. The experimental data here are unnormalized and this figure should be compared with Fig. 7.6.4 below.

1000° C under ambient conditions initially 10^{-5} atm H_2 (Biswas et al. 1980). Specific 2-element correlation trends, such as those involving chalcogenides (Fig. 7.6.1) or In vs Tl (Fig. 7.6.3) in E chondrites, or patterns of statistically significant interelement relationships (Biswas et al. 1980) are also essentially duplicated by results from heated Abee samples. Figs. 7.6.3 and 7.6.4 depict In vs Tl data for enstatite meteorites and heated Abee samples. In the former, the symbols essentially reflect the analytical results directly, converted to abundances only to compensate for Si differences in the various types of enstatite meteorites. Experimental points in the latter (and in all subsequent In vs Tl plots) are Ga-normalized as described in the caption to Fig. 7.6.4, for model-dependent comparison with a family of calculated condensation curves. To be sure, Hertogen et al. (1983) concluded that because of cosmochemical differences, condensation curves used for ordinary chondrites were not strictly applicable to E chondrites. Hence, the disagreement between theory and experiment in Fig. 7.6.4 is not unexpected; it is included here as a reference, against which trends for other chondritic groups can be compared.

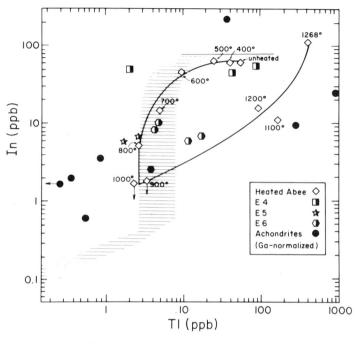

Fig. 7.6.4. Comparison of In vs Tl data for non-Antarctic enstatite chondrite and achondrite falls and heated Abee samples with trends predicted by a condensation model of a gaseous nebula of cosmic composition at pressures of 5×10^{-5} to 5×10^{-4} atm (shaded area). All data are normalized using Ga results as required by the condensation model which explicitly assumes that highly volatile elements are introduced into chondritic matter in the matrix fraction. According to the model, the amount of matrix can be quantified by normalizing highly volatile-element concentrations to that of a moderately volatile one like Ga, and, since matrix volatiles are assumed to be at cosmic levels, multiplying by the Cl Ga content. Agreement (cf. Fig. 7.6.3) with the condensation curve might be improved if a different nebular composition were chosen. Proponents of the condensation model consider this necessary. The existing model treats all nebular regions as uniform and initially chemically equivalent to cosmic. If post-accretion alteration(s) took place under open-system conditions, the condensation model is inapplicable.

7.6.4. NON-ANTARCTIC UNEQUILIBRATED ORDINARY CHONDRITES

Contents of highly labile elements are very variable both in unequilibrated and equilibrated ordinary chondrites, respectively (Figs. 7.6.5 and 7.6.6), and their mean abundances show no systematic variation with chemical group: H, L or LL (Binz et al. 1976; Lipschutz et al. 1983). Ranges of elemental contents overlap but mean contents of several elements (Bi, Cd, Cs, In and Tl) in unequilibrated ordinary chondrites are 1 to 2 orders of

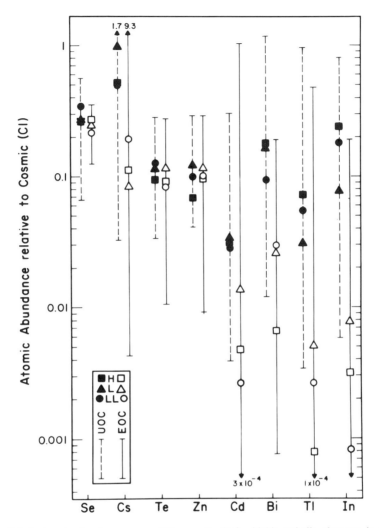

Fig. 7.6.5. Atomic-abundance ranges (Cl-normalized) for highly volatile elements in non-Antarctic unequilibrated and equilibrated ordinary chondrites (UOC and EOC, respectively). Elements are ordered by presumed condensation temperature assumed here to be represented by decreasing Cl-normalized abundance in EOC, the Bi-Tl-In triad being a unit. Mean values (arithmetic for Se, Te and Zn—because of their smaller variation—and geometric for all others) are shown for H, L and LL chondrites of each sort. Mean volatile element contents in UOC or EOC show no systematic variation with chemical group. While ranges for UOC and EOC overlap markedly, mean contents of Cs, Cd, Bi, Tl and In differ markedly, UOC containing more of each.

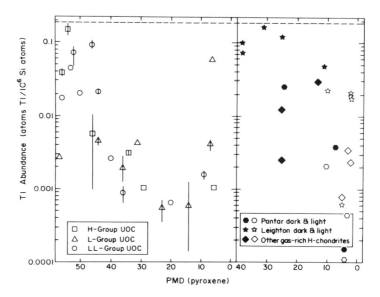

Fig. 7.6.6. Thallium relationship with pyroxene disequilibrium parameter (percent mean deviation) in unequilibrated ordinary chondrites (UOC) at left and coexisting dark and light portions of gas-rich H chondrites at right. Vertical lines for UOC encompass ranges for replicate analyses. Regardless of chemical type, Tl data for UOC define a single trend (cf. Fig. 7.6.5). Most data for gas-rich H chondrites, particularly the dark, gas-containing regions, are 10 to 100 times higher than those for most UOC, some being at the Cl levels indicated by the horizontal dashed line at the top of the figure. The UOC and gas-rich H chondrites must reflect different genetic processes.

magnitude higher (cf. Fig. 7.6.5) than those of their equilibrated congeners, i.e., petrographic type 5 or 6 representatives of the respective chemical group (Lipschutz et al. 1983). Contents of Bi, In and Tl correlate directly with the disequilibrium parameter (expressed as percent mean deviation; see Chapter 1.1) for ferromagnesian silicates (Fig. 7.6.6) and vary with TL sensitivity (Chapter 1.1) and contents of C and trapped Ar and Xe (Binz et al. 1976; Anders and Zadnik 1985). These trends, involving as they do such disparate parameters, must primarily reflect physical, not chemical, fractionation during primary nebular processes. However, as noted earlier, Woolum and Burnett (1981) found evidence for Bi and Pb redistribution by reheating in the L3 chondrite Khohar.

7.6.5. NON-ANTARCTIC EQUILIBRATED L CHONDRITES

We believe that the predominant process responsible for mobile trace-element trends in these chondrites is shock heating. Most L chondrites experienced uniquely severe shock heating 500 to 650 Myr ago, manifested by

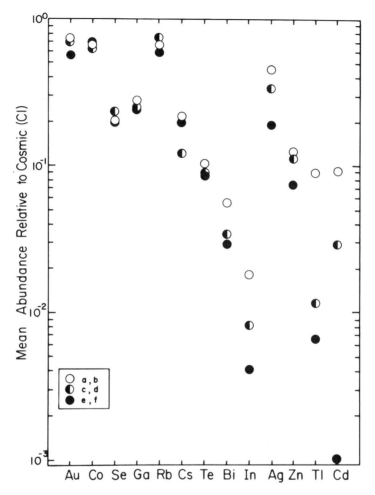

Fig. 7.6.7. Mean trace-element atomic abundances (Cl-normalized) in non-Antarctic, equilibrated L4–6 chondrites (Walsh and Lipschutz 1982) with different amounts of shock loading: ≤ 20 GPa (open symbols); 20–35 GPa (half-filled symbols); ≥ 35 GPa (filled symbols). Elements are ordered from the left by increasing mobility in representative chondritic matter heated at 1000°C in the laboratory. Absolute Cl-normalized means reflect both shock and preshock processes: the latter fix relative positions of lightly shocked samples along the ordinate. Means for the 6 least mobile elements do not vary systematically with degree of shock loading. Mean contents of the 7 most mobile elements (Te-Cd) decrease in the order, lightly > moderately > heavily shocked. There is some tendency for the difference between means for lightly (≤ 20 GPa) and heavily (≥ 35 GPa) shocked samples to become greater as mobility of these 7 elements increases. The inverse relationship between mean mobile-element content and shock history suggests that preshock concentration levels were lowered during slow cooling of collisional debris.

mineralogic and textural changes and loss of radiogenic ^{40}Ar and, of course, ^{4}He (Chapter 3.6). From laboratory heating experiments, we would expect that chondrites heated strongly enough to lose ^{40}Ar, would lose more labile elements as well (Table 7.6.1). Hence, when elements are ordered by mobility, as in Fig. 7.6.7, trace-element contents of only the more mobile elements should decrease progressively with degree of shock loading. L chondrites show this: mean contents of Te and elements more mobile than it (Fig. 7.6.7) decrease with increasing shock loading estimated from petrologic indicators (Walsh and Lipschutz 1982). L chondrites, whose petrographic characteristics place them in shock facies d-f, exhibit substantial ^{40}Ar (and ^{4}He) loss and *vice versa*. Siderophile and mobile trace elements in such strongly shocked L chondrites are significantly depleted relative to mildly shocked (< 22 GPa) ones because of transport in shock-formed FeS-Fe eutectic and/ or by vaporization (Huston and Lipschutz 1984). The compositional distinction between mildly and strongly shocked L chondrites is also evident in other ways. For example, in 2-element plots involving Bi, In or Tl, data for strongly shocked samples are systematically displaced toward lower Bi and Tl values relative to mildly shocked samples. This trend is consistent with those found in primitive chondrites subjected to simulated, post-shock heating in the laboratory (Walsh and Lipschutz 1982; Huston and Lipschutz 1984).

Labile-element contents in the relatively few unshocked or mildly shocked L chondrites do not vary extensively with petrographic type (Walsh and Lipschutz 1982; Huston and Lipschutz 1984; Lingner et al. 1987) suggesting that metamorphic processes were not important in establishing pre-shock genetic trends, but this must be examined further. On the other hand, current predictions do not appear to agree well with the admittedly sparse data for mildly shocked L chondrites. Figure 7.6.8 illustrates data for meteorites that escaped substantial post-accretionary shock heating with predictions of the condensation model for In vs Tl (Keays et al. 1971).

Proponents of condensation models always assume overall similarity in thermal histories of all ordinary chondrites and the closed-system nature of post-accretionary processes (see, e.g., Keays et al. 1971; Laul et al. 1973; Wai and Wasson 1977; Wasson 1977; Takahashi et al. 1978; Morgan et al. 1985). If shock-heating largely determined contents of highly labile elements in L chondrites, these assumptions are clearly contradicted.

7.6.6. NON-ANTARCTIC EQUILIBRATED H CHONDRITES

The thermal history of equilibrated H chondrites, as reflected by contents of labile trace elements is very different from that of L chondrites and mainly reflects primary processes (Morgan et al.1985; Lingner et al. 1987). Heavily shock-loaded H chondrites are uncommon and samples exhibiting petrographic effects indicative of substantial shock do not necessarily show ^{40}Ar loss and *vice versa* (Lingner et al. 1987). A large number of moderately and

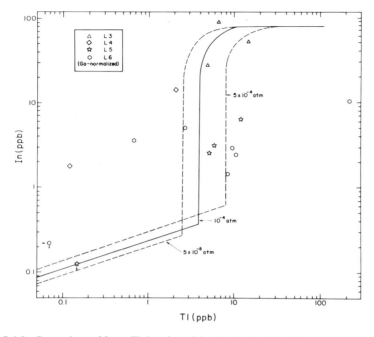

Fig. 7.6.8. Comparison of In vs Tl data for mildly shocked (≤ 22 GPa) non-Antarctic L3–6 chondrites with trends predicted for condensation from a nebula of cosmic composition at pressures of 5×10^{-5}, 10^{-5} and 5×10^{-4} atm. As in Fig. 7.6.4, all experimental data are Ga-normalized as required by the condensation model, which assumes that post-accretionary processes are closed-system. More heavily shocked samples (not illustrated here but cf. Huston and Lipschutz [1984]) fall in regions predicted from the apparent activation energies Q for mobilization of these elements: In vs Tl points distribute to the left of the curve since Q for Tl $< Q$ for In, i.e., shock-heated samples lose Tl more easily than they lose In. The reader should judge the agreement of experiment with condensation-curve prediction; the predicted curves could be altered by a change in nebular composition and/or condensation conditions.

highly labile elements are significantly depleted in H5 compared with H4 chondrites and/or H6 relative to H5 (Lingner et al. 1987). Relatively few elements vary significantly with degree of shock loading in H chondrites but 1 highly and 6 moderately labile elements have significantly *higher* mean contents in H chondrites exhibiting substantial ^{40}Ar loss than in those evidencing no such loss (Lingner et al. 1987). This can be interpreted as indicating that such elements were never present in H chondrite parent material to be lost in post-accretionary processes and points to higher-temperature formation conditions for H than for L chondrites.

This could accord with data obtained by proponents of condensation models but, as mentioned earlier, these proponents assume overall similarities in the thermal histories of H, L and LL chondrites (see, e.g., Wasson 1977; Morgan et al. 1985). The seemingly random dispersal of H chondrite points

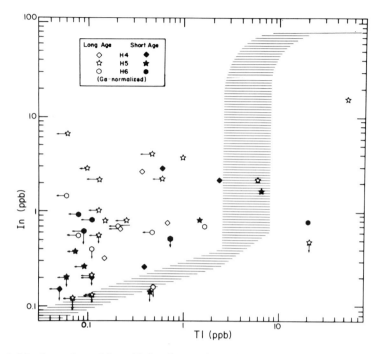

Fig. 7.6.9. Comparison of In vs Tl data for non-Antarctic, equilibrated H4–6 chondrite falls with trends predicted for condensation from a nebula of cosmic composition at pressures between 5×10^{-5} and 5×10^{-4} atm (shaded area). As in Fig. 7.6.4, for example, all experimental data are Ga-normalized. H chondrites of short (≤ 3.8 Gyr) and long (> 3.8 Gyr) K/Ar radiometric ages are intermixed randomly on In vs Tl plots as are strongly and moderately shocked samples. This argues that volatiles were lost from H chondrites very early in their history so that such elements could not be affected by later events. This contrasts sharply with the situation in L chondrites (cf. Huston and Lipschutz 1984).

with long and short gas-retention ages is illustrated in a plot of In vs Tl (Fig. 7.6.9) and contrasts sharply with data for L chondrites (Huston and Lipschutz 1984). The reader can judge the extent to which Ga-normalized data for H chondrites accord with the trend predicted (Morgan et al. 1985) to arise during nebular condensation (Fig. 7.6.9; cf. Figs. 7.6.4 and 7.6.8). Clearly, views about the detailed nature of genetic process(es) responsible for establishing compositional trends of highly labile elements in H and L chondrites are controversial.

Some H chondrites—especially regolith breccias (see below and Chapter 3.5)—have large contents of highly labile trace elements for reasons that are not well understood. Another unique H chondrite is Supuhee, supposed to contain thus far unidentified "mysterite", postulated as a late nebular condensate having relatively enormous proportions of highly volatile elements (see Davis et al. 1977; Morgan et al. 1985).

7.6.7. NON-ANTARCTIC EQUILIBRATED LL CHONDRITES

Few data for highly labile trace elements in LL4–7 chondrites have been published and these are interpreted as according with predictions of theoretical condensation curves (Laul et al. 1973).

7.6.8. ANTARCTIC METEORITES

Since 1969, large numbers of meteorites have been discovered in various parts of Antarctica, the largest numbers having been recovered in Queen Maud Land and Victoria Land (Lipschutz and Cassidy 1986). These populations differ in mean terrestrial age (0.1 and 0.3 Myr, respectively) and distribution. In terms of meteorite type (by number or mass), Antarctic populations differ from those of contemporary falls (Dennison et al. 1986). For example, the H/L ratio for Antarctic samples is 3, and 1 for falls (and non-Antarctic finds). Contents of many moderately and highly labile elements in Antarctic H4–6 chondrites and contemporary falls differ significantly, as do, for example, 2-element trends (Dennison and Lipschutz 1987). These and other physical differences (in, e.g., shock history, cosmogenic ^{53}Mn contents and TL properties) indicate that sample populations from the Victoria Land and Queen Maud Land collection areas and contemporary falls of H4–6 chondrites may derive from different extraterrestrial parent populations with different genetic histories. The In vs Tl relationships for Antarctic H chondrites (Fig. 7.6.10) illustrate this: the data (Dennison and Lipschutz 1987) are distributed very differently from those (Lingner et al. 1987) of contemporary H4–6 chondrite falls (Fig. 7.6.9). The reader should compare each data set against the other as well as against the theoretical condensation curves of Morgan et al. (1985), here acting also as fiducial marks. The differences are not confined to H chondrites: other meteorite groups exhibit Antarctic/non-Antarctic compositional differences involving moderately and highly labile elements.

Needless to say, the suggestion that Antarctic and non-Antarctic meteorites of a given type differ compositionally, hence genetically in some way, is very controversial. The difference implies that the flux of extraterrestrial material to Earth has varied in space or time over the 10^5 to 10^6 yr separating the two populations. According to conventional ideas (Wetherill 1986), this span is too short to expect variations in asteroid-belt sources: however, other possibilites exist (Dennison and Lipschutz 1987). If the differences continue to be verified, the implication is that a genetic process important in the history of some group of contemporary meteorite falls may well have been of greater or lesser importance during formation of the same type of Antarctic meteorites. If so, we will have a comparative look at the same nebular region where two (or more) sorts of the given meteorite group formed.

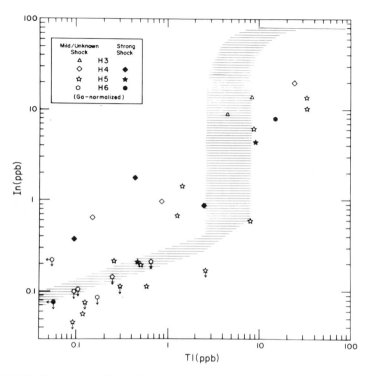

Fig. 7.6.10. Comparison of In vs Tl data for Antarctic, equilibrated H4–6 chondrites with trends predicted for condensation of a nebula of cosmic composition at pressures between 5×10^{-5} and 5×10^{-4} atm (shaded area). As in Fig. 7.6.4, for example, all experimental data are Ga-normalized as required by the condensation model. As in the case of non-Antarctic H4–6 chondrite falls (Fig. 7.6.9), but unlike the situation in L4–6 chondrites (Huston and Lipschutz 1984), Antarctic H chondrites with mild or strong shock loading are intermixed. The pronounced difference in the data distribution for non-Antarctic (Fig. 7.6.9) and Antarctic populations point to different thermal and/or genetic histories.

7.6.9. CUMBERLAND FALLS INCLUSION

The numerous large dark inclusions in this enstatite achondrite exhibit textural and petrographic characteristics of a type 3 chondrite and possess some mineralogic characteristics of enstatite chondrites and some of ordinary chondrites. As with many chondritic groups, contents of highly labile elements in these inclusions are quite variable, although typically less so than in ordinary or E5,6 chondrites (Verkouteren and Lipschutz 1983). From Fig. 7.6.11, it may be seen that mean contents of the 5 most volatile elements are even lower than those of such highly evolved samples as equilibrated ordinary and E5,6 chondrites. Other moderately to strongly labile elements, like Rb, Cs or Te are at CM2 levels. From this unique, chondritic, trace-element pattern and other information, Verkouteren and Lipschutz (1983) concluded that

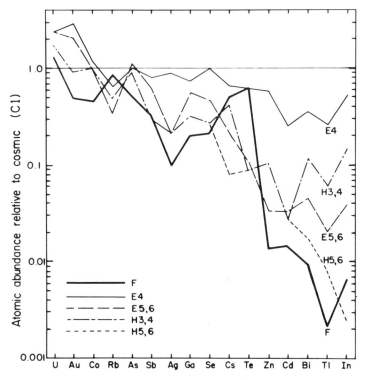

Fig. 7.6.11. Mean atomic abundances (Cl-normalized) for chondritic inclusions in the Cumberland Falls aubrite (here denoted as F) compared with those for unequilibrated and equilibrated E and H chondrites. (Volatile-element means for L and LL chondrites are quite similar to those for H chondrites. Means for these elements in CM2 and CV3 samples would parallel the horizontal Cl line but would lie somewhat lower, at 0.5 and 0.2–0.3, respectively, as in Fig. 7.6.2). Volatile-element contents for each of these groups (including the chondritic inclusions) are quite variable and concentration ranges for each would overlap markedly. To the extent that the means are representative of "average" material of that chondritic group, the pattern for these inclusions is quite distinct, especially in their low volatile-element contents and high Rb, Cs and Te. The most volatile elements are depleted in inclusions well below values for E and H (and C, L and LL) chondrites of similar low (3) petrographic type. In most cases, depletions are even more severe than those of equilibrated E and ordinary chondrites, pointing to unusually high formation temperatures for the inclusions' parent material.

trends for highly labile elements are primary and that these inclusions represent a newly recognized suite of primitive chondrite: the forsterite chondrites. They proposed that these inclusions formed under different redox conditions and, from the low contents of highly labile elements, at condensation temperatures higher than those typical of other chondrites. This conclusion has been criticized by Kallemeyn and Wasson (1985) who hypothesize that the inclusions were originally shock-heated LL-chondrite fragments, equilibrated to a greater or lesser extent with the enstatite-achondrite host.

7.6.10. REGOLITH BRECCIAS

Most regolith breccias are H chondrites and contents of highly labile trace elements in them are very variable and can be very high in dark, gas-containing portions (Bart and Lipschutz 1979; Lipschutz et al. 1983). In some samples, contents of a few elements exceed those of Cl chondrites and unequilibrated H3 chondrites. Light, gas-free portions of H-chondrite regolith breccias contain highly labile elements in amounts typical of those of other H5–6 chondrites. Highly labile elements whose abundances correlate with the percent mean deviation disequilibrium parameter for ferromagnesian silicates in unequilibrated ordinary chondrites (UOC) (Chapter 1.1), also correlate—but less well—with this parameter in at least one regolith breccia, Leighton (Fig. 7.6.6). Trend lines for UOC and regolith breccias are different, however (Lipschutz et al. 1983) and most data for gas-rich H chondrites, particularly the dark, gas-containing regions, are 10 to 100 times higher than in UOC with similar percent mean deviation values (Fig. 7.6.6). Tentatively, trends for highly labile elements in the dark portions of regolith breccias are suggested as primary condensation features, although these trends might reflect mobile element deposition on the surface of an H-chondrite parent (Lipschutz et al. 1983). Dark portions of the only non-H-group chondritic regolith breccia studied thus far, the LL chondrite St. Mesmin, show no enrichment of highly labile trace elements (Bart and Lipschutz 1979; Lipschutz et al. 1983). Earlier suggestions that much of each highly labile trace element was introduced into regolith breccias in a carbonaceous-chondrite carrier (Wilkening 1976) seems disproven at least for H-chondrite regolith breccias since Cl-normalized data for labile elements do not form a horizontal array. Furthermore, some elements are present at >Cl levels, an impossibility if carbonaceous chondrites are mixed with material more depleted in volatiles (Bart and Lipschutz 1979; Lipschutz et al. 1983).

7.6.11. SUMMARY

Data for highly labile elements in meteorites can give valuable clues to the condensation history of parental material and/or subsequent thermal episodes in the parent bodies. These elements are uniquely sensitive to temperatures that could have been reached in meteorites or their parent bodies by a variety of causes. Hence, interpretation of data for such elements is often controversial because of debates about which thermal process these elements are recording.

Proponents of condensation models assume that post-accretionary processes caused no net loss of highly labile trace elements despite undoubted evidence for heating. But experiments have shown that such elements can be easily lost during such heating and, in the case of enstatite chondrites, unique trace-element trends are duplicated by data from artificially heated primitive material. Contents of highly labile elements are significantly lower in L chondrites showing evidence for strong shock heating than in those that escaped such heating. Some data for UOC and H chondrites indicate labile-element redistribution during metamorphism. Hence, the assumption that post-accretionary heating of chondritic parent material was always closed-system is highly questionable.

Contents of highly labile elements in other chondritic groups seem unaffected by post-accretionary heating so that they may well reflect conditions during condensation under various sets of conditions. However, except for carbonaceous chondrites, existing condensation theories do not account for detailed compositional trends involving highly labile elements. Considering developments in this area in the last decade or so, it may well be that this situation will be improved in the near future.

Acknowledgments. Grants from the National Aeronautics and Space Administration are gratefully acknowledged.

REFERENCES

Alaerts, L., and Anders, E. 1979. On the kinetics of volatile loss from chondrites. *Geochim. Cosmochim. Acta* 43:547–553.

Anders, E. 1977. Critique of "Nebular condensation of moderately volatile elements and their abundances in ordinary chondrites" by Chien M. Wai and John T. Wasson. *Earth Planet. Sci. Lett.* 36:14–20.

Anders, E., and Ebihara, M. 1982. Solar system abundances of the elements. *Geochim. Cosmochim. Acta* 46:2363–2380.

Anders, E., and Zadnik, M. G. 1985. Unequilibrated ordinary chondrites: A tentative subclassification based on volatile-element content. *Geochim. Cosmochim. Acta* 49:1281–1291.

Anders, E., Higuchi, H., Ganapathy, R., and Morgan, J. W. 1976. Chemical fractionations in meteorites—IX. C3 Chondrites. *Geochim. Cosmochim. Acta* 40:1131–1139.

Bart, G., and Lipschutz, M. E. 1979. On volatile element trends in gas-rich meteorites. *Geochim. Cosmochim. Acta* 43:1499–1504.

Bart, G., Ikramuddin, M., and Lipschutz, M. E. 1980. Thermal metamorphism of primitive meteorites—IX. On the mechanism of trace element loss from Allende heated up to 1400° C. *Geochim. Cosmochim. Acta* 44:719–730.

Binz, C. M., Kurimoto, R. K., and Lipschutz, M. E. 1974. Trace elements in primitive meteorites—V. Abundance patterns of thirteen trace elements and interelement relationships in enstatite chondrites. *Geochim. Cosmochim. Acta* 38:1579–1606.

Binz, C. M., Ikramuddin, M., Rey, P., and Lipschutz, M. E. 1976. Trace elements in primitive meteorites—VI. Abundance patterns of thirteen trace elements and interelement relationships in unequilibrated ordinary chondrites. *Geochim. Cosmochim. Acta* 40:59–71.

Biswas, S., Walsh, T., Bart, G., and Lipschutz, M. E. 1980. Thermal metamorphism of primitive meteorites—XI. The enstatite meteorites: Origin and evolution of a parent body. *Geochim. Cosmochim. Acta* 44:2097–2110.

Blander, M., and Abdel-Gawad, M. 1969. The origin of meteorites and the constrained equilibrium condensation theory. *Geochim. Cosmochim. Acta* 33:701–716.

Burnett, D. S., and Woolum, D. S. 1983. In-situ trace element microanalysis. *Ann. Rev. Earth Planet. Sci.* 11:329–358.

Curtis, D. B., and Schmitt, R. A. 1979. The petrogenesis of L6 chondrites: Insights from the chemistry of minerals. *Geochim. Cosmochim. Acta* 43:1091–1103.

Davis, A. M., Grossman, L., and Ganapathy, R. 1977. Chemical characterization of a "mysterite"-bearing clast from the Supuhee chondrite. *Geochim. Cosmochim. Acta* 41:853–856.

Dennison, J. E., and Lipschutz, M. E. 1987. Chemical studies of H chondrites—II. Weathering effects in the Victoria Land, Antarctic population and comparison of two Antarctic populations with non-Antarctic falls. *Geochim. Cosmochim. Acta* 51:741–754.

Dennison, J. E., Lingner, D. W., and Lipschutz, M. E. 1986. Antarctic and non-Antarctic meteorites: Different populations. *Nature* 319:390–393.

Dodd, R. T., Jr. 1969. Metamorphism of the ordinary chondrites: A review. *Geochim. Cosmochim. Acta* 33:161–203.

Ganapathy, R., and Larimer, J. W. 1980. A meteorite component rich in volatile elements: Its characterization and implications. *Science* 207:57–59.

Hertogen, J., Janssens, M.-J., Takahashi, H., Morgan, J. W., and Anders, E. 1983. Enstatite chondrites: Trace element clues to their origin. *Geochim. Cosmochim. Acta* 47:2241–2255.

Huston, T. J., and Lipschutz, M. E. 1984. Chemical studies of L chondrites, III. Mobile trace elements and $^{40}Ar/^{39}Ar$ ages. *Geochim. Cosmochim. Acta* 48:1319–1329.

Jovanovic, S., and Reed, G. W., Jr. 1985. The thermal release of Hg from chondrites and their thermal histories. *Geochim. Cosmochim. Acta* 49:1743–1751.

Kallemeyn, G. W., and Wasson, J. T. 1982. The compositional classification of chondrites: III. Ungrouped carbonaceous chondrites. *Geochim. Cosmochim. Acta* 46:2217–2228.

Kallemeyn, G. W., and Wasson, J. T. 1985. The compositional classification of chondrites: IV. Ungrouped chondritic meteorites and clasts. *Geochim. Cosmochim. Acta* 49:261–270.

Kallemeyn, G. W., and Wasson, J. T. 1986. Compositions of enstatite (EH3, EH4,5 and EL6) chondrites: Implications regarding their formation. *Geochim. Cosmochim. Acta* 50:2153–2164.

Keays, R. R., Ganapathy, R., and Anders, E. 1971. Chemical fractionations in meteorites—IV. Abundances of fourteen trace elements in L-chondrites; implications for cosmothermometry. *Geochim. Cosmochim. Acta* 35:337–363.

Larimer, J. W. 1973. Chemical fractionations in meteorites—VII. Cosmothermometry and cosmobarometry. *Geochim. Cosmochim. Acta* 37:1603–1623.

Larimer, J. W., and Anders, E. 1967. Chemical fractionations in meteorites—II. Abundance patterns and their interpretation. *Geochim. Cosmochim. Acta* 31:1239–1270.

Laul, J. C., Ganapathy, R., Anders, E., and Morgan, J. W. 1973. Chemical fractionations in meteorites—VI. Accretion temperatures of H-, LL- and E-chondrites, from abundance of volatile trace elements. *Geochim. Cosmochim. Acta* 37:329–345.

Lingner, D., Huston, T. J., Hutson, M., and Lipschutz, M. E. 1987. Chemical studies of H chondrites—I. Mobile trace elements and gas retention ages. *Geochim. Cosmochim. Acta* 51:727–739.

Lipschutz, M. E., and Cassidy, W. A. 1986. Antarctic meteorites: A progress report. *Eos, Trans. AGU* 67:1339–1341.

Lipschutz, M. E., Biswas, S., and McSween, H. Y., Jr. 1983. Chemical characteristics and origin of H chondrite regolith breccias. *Geochim. Cosmochim. Acta* 47:169–179.

Morgan, J. W., Janssens, M.-J., Takahashi, H., Hertogen, J., and Anders, E. 1985. H chondrites: Trace element clues to their origin. *Geochim. Cosmochim. Acta* 49:247–259.

Ngo, H. T., and Lipschutz, M. E. 1980. Thermal metamorphism of primitive meteorites—X. Additional trace elements in Allende (C3V) heated to 1400° C. *Geochim. Cosmochim. Acta* 44:731–739.

Sugiura, N., Brar, N. S., and Strangway, D. W. 1983. Degassing of meteorite parent bodies. *Proc. Lunar Planet. Sci. Conf.* 14, *J. Geophys. Res. Suppl.* 89:B641–B644.

Takahashi, H., Gros, J., Higuchi, H., Morgan, J. W., and Anders, E. 1978. Volatile elements in chondrites: Metamorphism or nebular fractionation? *Geochim. Cosmochim. Acta* 42:1859–1869.

Verkouteren, R. M., and Lipschutz, M. E. 1983. Cumberland Falls chondritic inclusions—II. Trace element contents of forsterite chondrites and meteorites of similar redox state. *Geochim. Cosmochim. Acta* 47:1625–1633.

Wai, C. M., and Wasson, J. T. 1977. Nebular condensation of moderately volatile elements and their abundances in ordinary chondrites. *Earth Planet. Sci. Lett.* 36:1–13.

Walsh, T. M., and Lipschutz, M. E. 1982. Chemical studies of L chondrites—II. Shock-induced trace element mobilization. *Geochim. Cosmochim. Acta* 46:2491–2500.

Wasson, J. T. 1977. Reply to Edward Anders: A discussion of alternative models for explaining the distribution of moderately volatile elements in ordinary chondrites. *Earth Planet. Sci. Lett.* 36:21–28.

Wetherill, G. W. 1986. Unexpected Antarctic chemistry. *Nature* 319:357–358.

Wilkening, L. L. 1976. Carbonaceous chondrite xenoliths and planetary-type noble gases in gas-rich meteorites. *Proc. Lunar Sci. Conf.* 7:3549–3559.

Wolf, R., Richter, G., Woodrow, A. B., and Anders, E. 1980. Chemical fractionations in meteorites. XI C2 chondrites. *Geochim. Cosmochim. Acta* 49:711–717.

Wood, J. A. 1967. Criticism of a paper by H. E. Suess and H. Wänke, "Metamorphosis and equilibration in chondrites." *J. Geophys. Res.* 72:6379–6383.

Woolum, D. S., and Burnett, D. S. 1981. Metal and Pb/Bi microdistribution studies of an L3 chondrite: Their implications for a meteorite parent body. *Geochim. Cosmochim. Acta* 45:1619–1632.

Woolum, D. S., Burnett, D. S. and August, L. S. 1976. Lead-bismuth radiography. *Nucl. Instr. Meth.* 138:655–662.

Woolum, D. S., Bies-Horn, L., Burnett, D. S., and August, L. S. 1979. Bismuth and ^{208}Pb microdistributions in enstatite chondrites. *Geochim. Cosmochim. Acta* 43:1819–1828.

7.7. OXIDATION STATE IN CHONDRITES

ALAN E. RUBIN
University of California at Los Angeles

BRUCE FEGLEY
Massachusetts Institute of Technology

and

ROBIN BRETT
United States Geological Survey

The chemistry, mineralogy and oxidation state of planet-forming materials in the solar nebula were influenced strongly by the O fugacity (fO_2) of nebular gas. Information on fO_2 in the solar nebula may be preserved in relatively primitive samples that originally equilibrated with nebular gas and may not have been significantly altered by subsequent processes. Petrologic and chemical data can be used to arrange the major chondritic groups in order of decreasing degree of oxidation: CI > CM > CV > CO > LL > L > H > EL > EH. Many ungrouped chondrites as well as the silicate inclusions in IAB irons lie between H and E chondrites in this sequence. Refractory Ca,Al-rich inclusions (CAIs) in carbonaceous chondrites are depleted in Mo and W relative to other refractory siderophiles of similar volatility, consistent with oxidation of the refractory metals at high temperatures. Some Fremdlinge in CAIs contain V-rich magnetite and scheelite ($CaWO_4$) that apparently formed at higher fO_2 than other phases within the same inclusion. The enrichment and subsequent vaporization of silicate dust in certain regions of the solar nebula may have led to the locally high-fO_2 conditions required to have produced these Mo and W depletions and oxidized mineral assemblages. Oxygen-fugacity measurements can also be used to estimate the pressure at which meteorites cooled, thus *providing constraints on parent-body size. Such data indicate that chondrites were derived from parent bodies at least 30 to 70 km in diameter.*

7.7.1. INTRODUCTION

The chondritic meteorites exhibit a large range of oxidation states, ranging from the FeO-poor enstatite chondrites to the magnetite- and water-bearing CI chondrites. In general, oxidation-state variations among chondrite groups are reflected in the oxidation state of chondrites' constituent Fe-Mg-Si-O-bearing minerals, principally olivine, pyroxene and metal. The oxidation state of this mineral assemblage reflects the O fugacity (fO_2) that prevailed during the final equilibration of the major ferromagnesian minerals with each other and with a gas phase. (The O fugacity is identical to the O partial pressure of an ideal gas.) Reactions exemplified by:

$$3Fe + 2O_2 (g) = Fe_3O_4 \qquad (1)$$

$$2MgSiO_3 + 2Fe + O_2 (g) = Mg_2SiO_4 + Fe_2SiO_4 \qquad (2)$$

and

$$Mg_2SiO_4 + 2FeSiO_3 = Fe_2SiO_4 + 2MgSiO_3 \qquad (3)$$

which distribute Fe between metal, oxide and silicate phases, were presumably involved in this process. It is evident from Le Chatelier's Principle and from thermochemical data (see, e.g., Larimer 1968*a*) that, for objects of similar Fe content, the more-oxidized samples are characterized by higher concentrations of FeO in silicates and by the presence of magnetite (Fe_3O_4), whereas the less-oxidized samples are characterized by lower concentrations of FeO in silicates, the absence of magnetite and the presence of metallic Fe.

However, these considerations do not tell us where or how the observed oxidation states of the chondrites were established. One possibility is that the observed distribution of Fe between metal, oxide and silicate phases was originally established in the solar nebula by reactions between nebular gas and solid grains or liquid droplets and was unaltered by subsequent events. Another possibility is that the original oxidation states were altered and reset by gas-solid reactions associated with metamorphism and outgassing reactions on chondrite parent bodies. A third, more plausible, possibility is that a combination of nebular and parent-body processes have influenced to varying extents the oxidation states displayed by chondrites. In fact, the observed heterogeneity of the chondrites and the internal variation in oxidation states argue strongly for a mixture of nebular and planetary processes.

In this chapter we review the different oxidation-state indicators in chondrites and the methods used to estimate and infer O fugacities. Implications for the relative influences of nebular and parent-body processes are explored.

7.7.2. METHODS FOR MEASURING O FUGACITY

The intrinsic O fugacity of an equilibrium mineral assemblage is a function of both temperature and total pressure. Intrinsic fO_2 is related to the partial pressure of O in equilibrium with the mineral assemblage in a manner analogous to the way the activity of a component is related to concentration. Methods for measuring intrinsic fO_2 include:

1. The solid electrolyte-electrochemical method (Sato 1972). When properly done, this method is the most precise and accurate because the intrinsic fO_2 of the meteorite is measured directly as a function of temperature. Determinations of fO_2 in meteorites are listed (or referenced) in Brett and Sato (1984), Hewins and Ulmer (1984), Kozul et al. (1986) and the Basaltic Volcanism Study Project (1981).
2. Calculations based on mineralogical compositions determined by electron microprobe. As Eqs. (2) and (3) show, the proportion of oxidized Fe in mafic silicates reflects a meteorite's oxidation state. Such data, when combined with experimental and thermodynamic data, allow fO_2 to be calculated as a function of temperature. Calculations of fO_2 in ordinary chondrites based on olivine-metal-pyroxene equilibria (Williams 1971) are in good agreement with the determinations using the electrochemical method. Examples of such calculations are listed in Table 7.7.1.
3. Measurement of Ti^{3+}/Ti^{4+} in minerals using electron spin resonance (Live et al. 1986b). This method is useful for Ca,Al-rich inclusions (CAIs) that contain Ti-bearing minerals such as fassaitic pyroxene and

TABLE 7.7.1
Types of Mineral Equilibria Used to Calculate fO_2 in Meteorites as a Function of Temperature.

Equilibria	Reference
Fe-Ti oxide minerals	Harlow et al. (1982); McSween and Stolper (1978); Reid and Bunch (1975); Smith and Hervig (1979).
Phosphate-phosphide minerals	Olsen and Fuchs (1967); Olsen and Fredriksson (1966).
Pyroxene-metal-silica	Gooley and Moore (1976); Holmes and Arculus (1982).
Diopside-troilite-enstatite-metal-silica	Olsen and Fuchs (1967).
System Fe-Ca-Si-O-S	Larimer and Buseck (1974); Mueller and Saxena (1977).
System Fe-Si-O	Larimer (1968a).

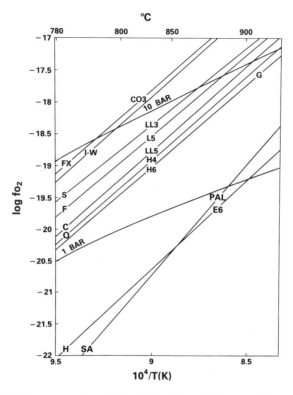

Fig. 7.7.1. Plot of linear regression relations for log fO_2 vs $1/T$ for several meteorites, with the iron-wüstite buffer shown for reference. The graphite surface is also shown at 1 bar and 10 bar pressure. Fx = Felix, S = Semarkona, F = Farmington, C = Cherokee Springs, O = Ochansk, G = Guarena, H = Hvittis, SA = Salta. Type of meteorite is also shown; PAL = pallasite.

hibonite. As in method (2), experimental and thermodynamic data are used to calculate fO_2.

4. Phase equilibria methods in which the fO_2 of the synthetic assemblage is varied until the assemblage has the exact mineral composition of the meteorite at a given temperature (see, e.g., Stolper 1977). This method can be precise but is very time-consuming.

Figure 7.7.1 shows intrinsic fO_2 determinations of several meteorites (Brett and Sato 1984). Measurements of intrinsic fO_2 as a function of temperature of 2 little-metamorphosed chondrites, Semarkona (LL3) and Felix (CO3), show considerable scatter relative to five other chondrites of higher metamorphic grade. The bulk fO_2 of Semarkona and Felix decreased with time during the experiment. Brett and Sato interpreted these results to mean that the minerals constituting these meteorites formed over a range of fO_2

conditions and tended to equilibrate with each other during fO_2 measurement: the most oxidized minerals dominated the early fO_2 determinations and tended to equilibrate with other phases, causing a lowering of apparent fO_2 with time. The fact that meteorites of higher metamorphic grade do not exhibit this phenomenon suggests that equilibration with respect to fO_2 takes place between grades 3 and 4 of metamorphism. However, more experimental data are required to confirm this.

7.7.3. OXIDATION-STATE INDICATORS IN CHONDRITES

Intergroup Redox Variations

The clearest mineralogic indicator of the oxidation state of meteorite whole rocks is the occurrence of Fe in different oxidation states. Iron exists as a metal (Fe^0) alloyed with Ni and Co, as ferrous Fe (Fe^{2+}) in silicates, oxides and sulfides, and as ferric Fe (Fe^{3+}) primarily in oxides. As the oxidation state of a meteorite group increases, the proportion of oxidized Fe (e.g., FeO in silicates) increases at the expense of metallic Fe; similarly, because Fe oxidizes more readily than Ni or Co, metallic Fe-Ni becomes increasingly depleted in Fe and enriched in Ni and Co. These relationships are sometimes referred to as "Prior's rules" after G. T. Prior who studied chemical variations in meteorites in the early part of this century.

The variation in oxidation state among the ordinary chondrites is reflected by several mineralogical and chemical properties such as the compositions of olivine, pyroxene and kamacite, the bulk FeO/(FeO + MgO) ratio and the bulk FeO/(total Fe) ratio. These properties indicate that, among ordinary chondrites, H chondrites are the most reduced, L chondrites are intermediate and LL chondrites are the most oxidized. In going from H to L to LL chondrites, the FeO contents of olivine and pyroxene increase, kamacite becomes richer in Co, and FeO/(FeO + MgO) and FeO/(total Fe) both increase (Table 7.7.2). Figure 1.1.8 shows the variation in FeO content of olivine and pyroxene in the ordinary-chondrite groups; Fig. 7.4.2 illustrates the variation in kamacite and olivine compositions in ordinary chondrites. Magnetite is extremely rare to absent in H and L chondrites, but is present in several unequilibrated LL chondrites (e.g., Semarkona, Ngawi, Chainpur and Allan Hills A77278).

Carbonaceous chondrites are more oxidized than ordinary chondrites. CI chondrites are the most oxidized of all meteorites. They contain abundant H_2O, ~ 10 wt.% magnetite and no metal. CM chondrites have a somewhat lower oxidation state; they contain less H_2O and roughly 0.1 the amount of magnetite of CI chondrites. Some CM magnetite occurs in association with metal. CV chondrites comprise two subgroups; the oxidized subgroup, which includes Allende, averages ≤ 0.1 vol.% metal and ~ 2 vol.% magnetite; the reduced subgroup, which includes Leoville and Vigarano, averages 2 vol.% metal and considerably less magnetite (~ 0.3 vol.%) (McSween 1977a). CO

TABLE 7.7.2
Mean Compositional Properties of Ordinary-Chondrite Groups

	H	L	LL	ref[a]
Olivine (mol% Fa)	18.8	24.6	28.5	1
Low-Ca pyx (mol% Fs)	17.2	21.3	24.1	1
FeO/(FeO+MgO) (mol%)	17	22	27	2
Metallic Fe-Ni (mg/g)	168	77	~40	3
FeO/(total Fe)	0.38	0.66	0.88	4
Kamacite (mg/g Co)	4.4	8.9	~22	3,5
Bulk Fe (mg/g)	265	215	183	5
Bulk Ni (mg/g)	15.7	12.0	10.4	5
Kamacite/taenite	~5	~1	~0.4	3

[a] References: (1) Gomes and Keil (1980); (2) Wasson (1985); (3) Afiattalab and Wasson (1980); (4) E. Jarosewich, unpublished data; (5) G. W. Kallemeyn, unpublished data (1987).

chondrites average 2.5 wt.% metal (McSween 1977b) and contain only minor magnetite (primarily inside chondrules).

At the opposite extreme are the highly reduced enstatite chondrites. The EH (high-Fe) group averages 22 wt.% metallic Fe-Ni. Very little FeO occurs in the silicates: pyroxene contains < 1 mol% $FeSiO_3$ (i.e., Fs < 1). EL chondrites contain about the same amount of metal, but even less FeO in the pyroxene. Both groups are so reduced that metallic Si is dissolved in the kamacite. EH kamacite averages ~ 3 wt.% Si, whereas EL kamacite averages only one-third as much. Magnetite is absent from both groups.

The presence of highly reduced refractory minerals (e.g., oldhamite, CaS; osbornite, TiN; sinoite, Si_2N_2O) in enstatite chondrites is in accord with the above observations. Experimental data on CaS stability (Larimer 1968b) and theoretical calculations of mineral stabilities under low O fugacities (Larimer and Bartholomay 1979) show that these accessory minerals are only stable under highly reducing conditions comparable to those required for formation of essentially FeO-free silicates and Si-bearing metal.

From petrologic data, all of the major chondrite groups can be arranged in order of oxidation state. From highest to lowest, they are: CI > CM > CV > CO > LL > L > H > EL > EH. The relative positions of EL and EH are uncertain. Bulk compositional data are consistent with this ordering: the atomic ratio of oxidized Fe to Si decreases from CI to E chondrites. This is illustrated in Fig. 7.4.1; for the present discussion, we need only consider the projection of the points onto the abscissa of that diagram. The mean FeO_x/Si ratio of each chondrite group falls on the abscissa in the same order of oxidation state as that derived above from petrological considerations.

The silicate inclusions in IAB irons are approximately chondritic in bulk composition and can be added to the list. IAB pyroxene (Fs 4–8) is low in FeO, indicating that these inclusions are intermediate in oxidation state between H and E chondrites. Three closely related meteorites, Winona, Mount Morris (Wisconsin) and Pontlyfni are similar in oxidation state and bulk and O-isotopic compositions to IAB silicates; they probably belong to group IAB (Kallemeyn and Wasson 1985). Certain other ungrouped meteorites are also intermediate in oxidation state between ordinary and enstatite chondrites; they include Kakangari, Suwahib Buwah, Acapulco, Allan Hills A77081, Willaroy and the chondrule-bearing silicates in the Netschaëvo IIE iron.

The ordering of chondrite groups by oxidation state is based principally on the nature of interchondrule material, some of which, e.g., magnetite in CI and CM chondrites, probably formed by secondary alteration. If we restrict our attention to chondrules (which are nebular products; see Chapter 9.3), a more muddled sequence emerges. Chondrules with reduced (i.e., FeO-poor) olivine and pyroxene phenocrysts constitute 100% of the chondrules in EH chondrites, $\geq 90\%$ of those in CV, CM and CO chondrites, and only 50 to 90% of those in ordinary chondrites (Table 9.1.1). The reason for these differences is unclear.

Although chondrite groups vary in overall oxidation state, individual chondrites of petrographic types 2 and 3 are highly disequilibrated assemblages that can preserve materials formed under different fO_2 conditions. For example, Si- and Cr-bearing metallic Fe-Ni grains occur in CM2 (e.g., Murchison), CO3 (e.g., Allan Hills A77307), LL3 (e.g., Bishunpur) and EH3 (e.g., Qingzhen) chondrites. These grains are probably relics of high-temperature, low-fO_2 conditions in the early solar nebula. They must have been prevented from oxidizing (via reaction with H_2O), possibly because they were encapsulated in larger silicate grains (see, e.g., Rambaldi et al. 1980).

Intragroup Redox Variations in Ordinary Chondrites

Individual equilibrated chondrites within a single chondrite group can also have different oxidation states. Because diffusion in olivine is relatively rapid, olivine compositions are a useful indicator of oxidation state. Figure 7.7.2 shows the distributions of Fa in olivines from 10 H-group chondrites (8 of which are from China). The distribution in each chondrite is narrow; σ_{Fa} ranges from 0.3 mol% in Dhajala, Changxing and Lunan to 0.9 mol% in Changde. Although Dhajala (type 3.8) has sharply delineated chondrules with glassy-to-cryptocrystalline mesostases, its olivine is rather uniform. In contrast, low-Ca pyroxene in Dhajala is heterogeneous (see Fig. 1 of Noonan et al. 1976). A few olivine grains in Anlong, Changde, Enshi and Xingyang are obviously not in equilibrium with the majority. Scott et al. (1985) found that at least 25% (and probably most) type 4 to 6 ordinary chondrites contain rare olivine and/or pyroxene grains with Fe/(Fe + Mg) ratios that differ significantly from the majority. Such chondrites are probably breccias that were assembled after individual components were equilibrated during metamorphism.

Several of the 10 chondrites (e.g., Allegan, Lunan and Xingyang) have

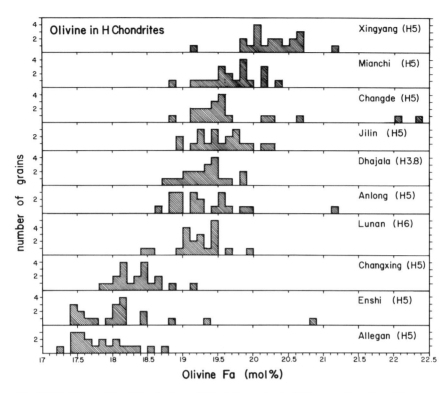

Fig. 7.7.2. Compositional distributions of 20 olivine grains in 10 H-group chondrites. The narrow distributions result from equilibrium at specific oxidation states; i.e., the oxidation state of Xingyang is considerably higher than that of Allegan. The occurrence of a few olivine grains with aberrant compositions suggests that several of these chondrites may be breccias that formed after metamorphic equilibration.

mean olivine compositions that differ significantly from one another. These chondrites are not in equilibrium with each other and must have been derived from different metamorphic terranes on the H parent body or bodies. It is unclear whether the material in each terrane preserved oxidation states acquired in the nebula during agglomeration or whether local parent-body processes affected terranes differently. In the latter case, reducing agents such as graphite could have been responsible. However, Scott et al. (1986) found that mean FeO contents of olivine and low-Ca pyroxene in ordinary chondrites increase by 3 to 5% from type 4 to 6 and attributed these properties to nebular and accretionary processes.

7.7.4. INFERENCES ABOUT SOLAR-NEBULAR PROCESSES

The available data on chondrite oxidation states and intrinsic O fugacities can be interpreted in terms of nebular processes, parent-body processes or a combination of both. It is reasonable to expect that relatively unaltered primitive samples that originally equilibrated with nebular gas and still retain the imprint of nebular processes preserve information about nebular O fugacity variations and fO_2 values. Similarly, it is reasonable to expect that metamorphosed meteorites may help constrain parent-body processes. These two propositions are useful operating assumptions which we employ in our discussion in this section and the following one.

Before trying to interpret the oxidation state of relatively unaltered primitive chondrites and their components in terms of a solar-nebular origin, we need to know how the fO_2 of nebular gas was regulated and what the expected oxidation states are of solid materials that equilibrated with the gas.

The fO_2 of an H_2-rich gas with relative solar atomic proportions of $H(1350):O(1):C(0.6)$ is controlled by the reaction:

$$H_2 + 0.5\ O_2 = H_2O. \tag{4}$$

The equilibrium fO_2 for Reaction (4) is given by the expression:

$$\log_{10} fO_2 = 2\log_{10} (H_2O/H_2) + 5.59 - 25{,}598/T \tag{5}$$

which utilizes the latest JANAF (1986) data and is valid from 300 to 2500 K. For reference, the (H_2O/H_2) number ratio for solar-composition gas is $\sim 5 \times 10^{-4}$ (because all C is present as CO) and the fO_2 at 1600 K is $\sim 10^{-17}$ bar (independent of total pressure in the P,T region where CO is the dominant C-bearing gas).

It is clear from the relative abundances of C and O that the (H_2O/H_2) ratio in gas of solar composition is also affected by equilibria between oxidized and reduced C-bearing gases (but not by graphite formation which cannot occur at temperatures > 470 K or pressures $> 10^{-7.6}$ bar at equilibrium in a gas of solar composition [Lewis et al. 1979]). These equilibria are exemplified by the reaction:

$$CO + 3H_2 = CH_4 + H_2O \tag{6}$$

which shifts to the right with decreasing temperature at constant pressure. Thermodynamic calculations (see, e.g., Prinn and Fegley 1988) show that the destruction of CO (which is the major C-bearing gas at high temperatures) by Reaction (6) is 50% complete at ~ 710 K (10^{-3} bar total pressure) or at ~ 520 K (10^{-6} bar total pressure). If thermochemical equilibrium is maintained, the complete conversion of CO to CH_4 ultimately increases the (H_2O/H_2) ratio to $\sim 1 \times 10^{-3}$.

However, the kinetics of the CO to CH_4 conversion are sufficiently slow (relative to the rates of nebular radial mixing or thermal evolution) so that Reaction (6) probably did not go to completion at low temperatures (Lewis and Prinn 1980). In fact, the CO to CH_4 conversion is quenched at such a high temperature that essentially all C is present as CO; CH_4 does not form (Prinn and Fegley 1988).

The oxidation state of FeO-bearing silicates in equilibrium with solar-nebular gas and metallic Fe can be calculated using experimental and thermodynamic data for Reactions (2) and (4). If we assume ideal solution for Fe and Mg in olivine, the data of Larimer (1968a) yield the expression:

$$\log_{10} X_{Fa} = \log_{10} (H_2O/H_2) - 0.74 + 1690/T \qquad (7)$$

which relates the mole fraction of fayalite in olivine coexisting with metallic Fe to the (H_2O/H_2) ratio and temperature of the coexisting gas phase.

An important implication of Eq. (7) is that olivines that equilibrated with solar-composition gas $(H_2O/H_2 \simeq 5 \times 10^{-4})$ at high temperatures will be FeO-poor while olivines that continued to equilibrate with the gas phase down to low temperatures will be FeO-rich. For example, the calculated fayalite content of olivine in equilibrium with solar-composition gas at 1000 K is only ~ 0.4 mol%, but is much larger (~ 22 mol%) at 500 K. It is evident from experimental and thermodynamic data on the exchange reaction (Eq. 3) that similar FeO enrichments are also expected in low-Ca pyroxenes at low temperatures.

We can use this basic background to evaluate different models for the origin of the olivines found in primitive, little-metamorphosed chondrites. These olivines have heterogeneous compositions that vary from Fa 0.3 to Fa > 90 mol% in type-3 ordinary chondrites (e.g., Semarkona), CO3 chondrites (e.g., Colony) and CM2 chondrites (e.g., Mighei). The high-fayalite olivines occur primarily in fine-grained matrix material, and, in some cases, in matrix-like rims around metal grains (see, e.g., Scott et al. 1984; Chapter 10.2). For present purposes, these fayalitic olivines are potentially the most interesting, and we will concentrate on them. (The origin of coarser-grained olivine in the same meteorites is discussed in Chapter 10.4.)

The simplest model for explaining the origin of such fayalite-rich olivines (Fa $\simeq 60-90$) is equilibration with solar-composition gas at low temperatures. Taken at face value, Eq. (7) indicates that temperatures of ~ 420 to 440 K are required to produce olivines with 60 to 90 mol% fayalite. Although this model is qualitatively appealing, it implicitly assumes intimate contact and complete equilibration of three different solids (metallic Fe, low-Ca pyroxene and olivine) down to temperatures of ≤ 500 K. This is questionable, especially in light of the slow rate of solid-state diffusion at these low temperatures.

An alternative model involves the melting and rapid cooling of highly oxidized precursor materials, as in, for example, the formation of chondrules. In this scenario, which has been recently revived by Wood (1985), different degrees of partial equilibration of rapidly cooling olivines with nebular gas, assumed to have a solar (H_2O/H_2) ratio, are reflected in the observed heterogeneity of olivine compositions. However, in some cases surviving relict grains in chondrules are more reduced than adjacent chondrule phenocrysts (Wood 1988). Furthermore, if chondrules were formed from oxidized (FeO-rich) precursors, they must have lost considerable amounts of metal during melting and reduction because, in general, chondrules are metal poor. For example, the average Fe/Mg ratio of chondrules in Semarkona (LL3) is only 40% of that of the whole rock. The paucity of metallic spherules relative to chondrules in chondrites suggests that most chondrules probably did not form from highly oxidized precursors.

Grossman and Wasson (1983a,b) proposed a third possible model in which the observed olivine heterogeneity is ascribed, at least in part, to variations in the grain sizes of nebular condensates. They proposed that, with decreasing temperature, finer-grained materials (with higher-surface/volume ratios) were preferentially enriched in FeO relative to coarser-grained materials (with smaller surface/volume ratios). Thus, this model attempts to remedy the kinetic inhibition of low-temperature gas-solid and solid-solid reactions by postulating grain sizes that were small enough to permit equilibration. One potential pitfall to this model is that agglomeration of fine grains may have produced coarser, chemically unreactive grains faster than FeO incorporation could proceed.

However, another alternative model may not suffer from such kinetic constraints. Examination of Eq. (7) shows that high-fayalite olivines could be formed at high temperatures (where reactions between gas and solids proceed much faster) under more oxidizing conditions. For example, an increase in the (H_2O/H_2) ratio from the canonical solar value of $\sim 5 \times 10^{-4}$ to a much higher value of $\sim 10^{-1}$ increases the fayalite content of olivine from ~ 0.1 mol% to ~ 29 mol% at 1400 K; more detailed calculations by Palme and Fegley (1987) confirm this general result. The formation of fayalitic olivines by high-temperature oxidation is both theoretically plausible (Palme and Fegley 1987) and consistent with observational data (Nagahara 1984; Hua et al. 1987); it is also consistent with the high-temperature oxidation of refractory mineral assemblages such as the CAIs found in carbonaceous chondrites. (See Chapter 10.3 for detailed discussion of CAIs.)

Several research groups have proposed that CAIs actually contain evidence for oxidizing conditions in the solar nebula. These proposed indicators are summarized in Table 7.7.3, which also shows whether or not these indicators of high-temperature oxidation are potentially reversible (i.e., if they can be reset by either more or less oxidizing conditions). The concept of reversibility is important because the different indicators may be recording fO_2 values that are characteristic of separate events.

The most common evidence in CAIs for oxidizing conditions in the solar nebula involves depletions of Mo and W (both of which are readily oxidized) relative to other refractory siderophiles of similar volatility that are far less

TABLE 7.7.3
Proposed Oxygen-Fugacity Indicators in Ca,Al-Rich Inclusions in Carbonaceous Chondrites

Proposed Indicator	Reversible?	Ref.[a]
Barium depletions	No	1
Cerium depletions	No	1–3
Color of hibonite	Yes	4
Fremdlinge minerals (e.g., scheelite)	No	5,6
Refractory metal (Mo, W) depletions	No	7
Siderophile behavior of Hf, Nb, Ta, Zr	No	8
(Ti^{3+}/Ti^{4+}) ratios in hibonite	Yes	9,10
(Ti^{3+}/Ti^{4+}) ratios in Ti-bearing pyroxenes	No	11
Uranium depletions	No	1
Vanadium depletions	No	1

[a] References: (1) Davis et al. (1982); (2) Boynton (1984); (3) Fegley (1986); (4) Ihinger and Stolper (1986); (5) Armstrong et al. (1985a); (6) Bischoff and Palme (1987); (7) Fegley and Palme (1985); (8) Fegley and Kornacki (1984); (9) Live et al. (1986b); (10) Live et al. (1986a); (11) Beckett (1986).

readily oxidized (Re, Os, Ir, Ru and Pt). In fact, 77% of all CAIs (20/26) for which INAA data of the 7 refractory metals are available (H. Palme and B. Spettel, personal communication, 1986) show Mo depletions at the 95% confidence level; 15% show W depletions at the 95% confidence level. These statistics are consistent with the production of Mo depletions at lower fO_2 than W depletions (Fig. 7.7.3).

The refractory-metal abundance patterns displayed in Fig. 7.7.3 are the result of an interplay between the vapor pressures of Mo and W and the tendency of these two metals to be oxidized in the gas phase by reactions such as:

$$Mo + H_2O = MoO + H_2 \qquad (8)$$

and

$$W + H_2O = WO + H_2. \qquad (9)$$

Fegley and Palme (1985) discussed these factors in detail and showed that the characteristic nature of the Mo- and W-depletion patterns can be produced only by oxidation at high temperatures. In particular, the observed depletions do not correlate with the degree of CAI alteration and are not unique to Allende but are observed in 6 of 7 carbonaceous chondrites (Allende, Arch, Efremovka, Grosnaja, Leoville and Ornans but not Essebi) from which CAIs have been analyzed. Fegley and Palme (1985) showed that, at an assumed total pressure of 10^{-3} bar, the (H_2O/H_2) ratios indicated by the Mo and W depletions in CAIs range from slightly larger than the solar value ($\sim 5 \times 10^{-4}$) to about 10^{-1}. Figure 7.7.4 illustrates a highly fractionated metal pattern observed

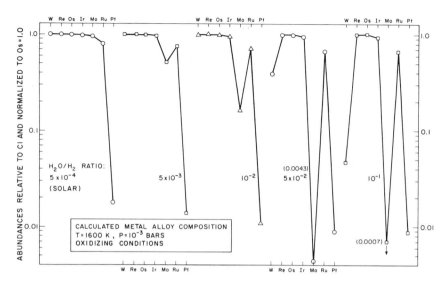

Fig. 7.7.3. Mo and W depletions in refractory metal alloys as a function of O fugacity at constant temperature and pressure. The O fugacity is expressed as the (H_2O/H_2) ratio, and can be calculated from $\log_{10} K$ (H_2O, g) using Eq. (5).

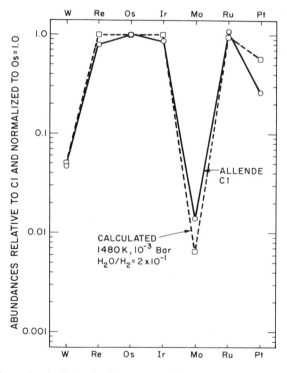

Fig. 7.7.4. Observed and calculated refractory-metal abundances in Allende FUN inclusion Cl. The observed composition is from EDS/SEM analyses of metal grains by D. A. Wark.

in the Allende FUN inclusion C1 and a calculated match to it. (FUN inclusions are discussed in detail in Chapter 14.3.) Such highly fractionated patterns were found to be typical of equilibration at (H_2O/H_2) ratios of $\sim 10^{-1}$ and not limited to FUN inclusions.

Similar (H_2O/H_2) ratios in the range of 10^{-2} to 10^{-1} for equilibration of hibonite from the Allende Blue Angel inclusion with nebular gas were suggested by Ihinger and Stolper (1986). They had proposed earlier that the color of hibonite (ideally $CaAl_{12}O_{19}$ but, in many cases, containing Ti, V, Fe, Mg, Si and Cr in coupled substitutions) could be used as an O-fugacity indicator in CAIs. Related work by Live et al. (1986a) suggested that the Ti^{3+}/Ti^{4+} ratio in meteoritic hibonites could be used as an fO_2 indicator. Results for the spinel-hibonite inclusion SH-7 give an upper limit on fO_2 that is 10^3 to 10^4 times more oxidizing than solar gas. This upper limit is comparable to the fO_2 values estimated from Blue Angel hibonite and from the Mo and W depletions in CAIs.

Thus, high-temperature oxidation in the solar nebula is indicated by the fayalitic olivines found in primitive, little-metamorphosed chondrites and by several features of CAI trace-element chemistry and mineralogy. However, no completely satisfactory mechanisms have been proposed to account for such large implied fO_2 variations in the solar nebula. Several of the most popular (although not necessarily correct) suggestions are reviewed below.

One frequently proposed mechanism for attaining high fO_2 in the nebula is by the enhancement of the ratio of silicate-dust to gas. Simple mass-balance arguments illustrate why this mechanism is popular. Approximately 20% of all O resides in anhydrous silicate dust (taken as MgO + SiO_2 + FeO + CaO + Al_2O_3), $\sim 20\%$ is in $H_2O(g)$, and the remaining $\sim 60\%$ is in $CO(g)$. This latter O is essentially locked up in the gaseous CO, which is thermodynamically stable at high temperatures and kinetically stable against conversion to CH_4 at low temperatures (Lewis and Prinn 1980).

Table 7.7.4 shows the effects of silicate-dust enrichments on O fugacity in the solar nebula where the O fugacity at any temperature can be calculated from the tabulated (H_2O/H_2) ratio using Eq. (7). In order for the dust to affect fO_2, it must first be vaporized. Dust enrichments of 50 to 500 times greater than solar are required to produce (H_2O/H_2) ratios corresponding to the observed fayalite compositions in olivines in primitive chondrites (Nagahara 1984; Palme and Fegley 1987; Hua et al. 1987), the observed Mo and W depletions in CAIs (Fegley and Palme 1985) and the observed colors of meteoritic hibonites (Ihinger and Stolper 1986). Because approximately equal amounts of O are present in silicate dust and in water, the required enrichment factors for ice-plus-dust mixtures are about a factor of 2 smaller than the dust-enrichment factors unless ice and dust are present in grossly nonsolar proportions. Dust enrichment could conceivably occur at the nebular midplane due to gravitational settling and collisions or at the boundary of the nebular surface with interstellar space due to infalling parcels of interstellar grains.

TABLE 7.7.4
Effects of Dust Enrichments on (H_2O/H_2) Ratios in the Solar Nebula[a]

Dust Enrichment	X^b Dust	Mass % Dust	H_2O/H_2^c
1 (Solar)	2.0×10^{-4}	0.5	5×10^{-4}
50	0.01	20	1.6×10^{-2}
250	0.05	55	7.6×10^{-2}
500	0.09	71	0.15
750	0.13	78	0.23
1000	0.16	83	0.30
10^4	0.67	98	3.0^d
10^5	0.95	99.8	30
10^6	0.993	99.98	300

[a] Dust = $MgO + SiO_2 + FeO + CaO + Al_2O_3$.
[b] X = mol fraction = (mol dust)/(mol dust + gas).
[c] Excludes the ~60% of O present in $CO(g)$ which is thermodynamically stable at high T and kinetically stable against conversion to CH_4 at low T.
[d] At dust enrichments of ≳3200, H_2O becomes the dominant H reservoir. Calculations presented for higher enrichments assume an infinite H supply.

Dust enrichment has several other consequences in addition to increasing the O fugacity. These include increases in the opacity, viscosity and coagulation rate in the dust-enriched region of the nebula. This mechanism also requires that (1) a powerful but, as yet, unidentified heat source capable of vaporizing at least the MgO- and SiO_2-bearing phases in the dust be operative after dust enrichment has occurred, and (2) the total amount of dust is only partially vaporized so as to maintain high opacity and slow cooling rates. Furthermore, extreme dust-to-gas fractionations are required to generate the large fO_2 values that may be required at high temperatures to produce some of the oxidized phases such as scheelite ($CaWO_4$) and magnetite which are observed in Fremdlinge (Armstrong et al. 1985a; El Goresy et al. 1978; Bischoff and Palme 1987). In addition, the presence of the low-temperature phases observed in Fremdlinge such as Ge-bearing alloys and sulfides (see, e.g., Armstrong et al. 1985b; El Goresy et al. 1978) is a potential problem unless they were incorporated into the Fremdlinge after the high-temperature oxidizing event, possibly during a second oxidizing event at low temperatures.

7.7.5. INFERENCES ABOUT PARENT-BODY PROCESSES

Pressure and Depth of Origin of Meteorites

The irons, mesosiderites, achondrites and ordinary chondrites all probably had fO_2 values of the graphite surface (i.e., the fO_2 value that is in equilibrium with an equilibrium assemblage of C, CO and CO_2) at the pressures at which these meteorites cooled (Brett and Sato 1984). Oxygen-fugacity measurements of these meteorites can provide information on the pressures at which they cooled and hence provide constraints on parent-body size. The pressures obtained seem reasonable; those for chondrites are consistent with the results of Pellas and Storzer (1981) who used the independent method of fission-track thermometry. Nevertheless, the fO_2 method of pressure determination must receive further confirmation before it can be fully accepted.

The fO_2 of mixtures of CO, CO_2 and graphite is remarkably sensitive to pressure (especially at geologically low pressures), whereas the fO_2 of other condensed species (such as Fe-bearing silicates) and gas is relatively insensitive to pressure. Sato (1978) refined the log fO_2 vs $1/T$ plots at various pressures for equilibria involving C-CO-CO_2, taking into account the nonideal behavior of the gases involved (Fig. 7.7.5). If a meteorite contains no graphite, CO-CO_2-C equilibria have no relevance. If graphite is not the C-bearing solid phase, then the activity of C will be less than unity, and the fO_2 of this phase with CO + CO_2 will be somewhat higher than that for equilibria in-

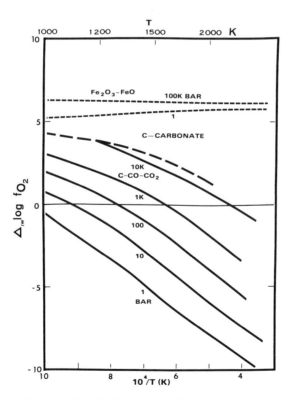

Fig. 7.7.5. Isobars in the graphite-CO-CO_2 system as a function of T and Δ_{IW} log fO_2. The latter function is log fO_2 of the graphite surface minus log fO_2 of IW. Equilibria involving carbonate dominate at higher pressures, as shown (figure after Brett and Sato 1984). IW = iron-wüstite.

volving graphite at a given T and P. If a meteorite contains graphite and shows no evidence of reduction, then it lay at or below the fO_2 at which graphite is stable with $CO + CO_2$ (Fig. 7.7.5; see also Fig. 4 of Sato 1979). By measuring the fO_2 in equilibrium with the meteorite at a given temperature at low pressure, it is possible to arrive at a minimum pressure for equilibration of the meteorite at that temperature (the pressure for C-CO-CO$_2$ equilibrium—i.e., the graphite surface—corresponding to the fO_2 and T). However, if the meteorite was undergoing reduction at the given temperature, then it lay on the graphite surface, and the precise pressure can be determined if the fO_2 at the temperature is known.

Enstatite Chondrites

Because enstatite chondrites were probably not reduced by graphite during metamorphism, they most likely lay beneath the graphite surface. Inspection of Fig. 7.7.1 indicates that if enstatite chondrites were reduced, the pressure must have been ≤ 1 bar. This is in accord with their proposed reduction in the solar nebula (Larimer and Bartholomay 1979).

Iron Meteorites

Calculations by Olsen and Fuchs (1967), based on phosphate-phosphide equilibria, suggest that the fO_2 values of iron meteorites lie within the range of ordinary chondrites. Prinz et al. (1983) reported that some iron meteorites appear to have been reduced by C; fO_2 measurements or calculations should be made on these meteorites, because applying the calculations of Olsen and Fuchs (1967) to Fig. 7.7.5 suggests a pressure of only a few bars. However, such meteorites may not have been reduced.

Mesosiderites

Thermodynamic calculations by Harlow et al. (1982) based on Fe-Ti-O equilibria indicate that the fO_2 of mesosiderites lies in the upper range of that of chondrites. Agosto et al. (1980) suggested that some mesosiderites were reduced by C or P. If it was C, then inspection of Fig. 7.7.5 indicates that mesosiderites formed at a pressure of about 20 bar. If reaction by C did not occur, then this pressure is a minimum. Snellenburg et al. (1979) deduced log fO_2 of -20 at 840° C for 4 mesosiderites of low metamorphic grade. This corresponds to pressures of a few bars.

Achondrites

Oxygen-fugacity data on achondrites provide little information on pressure during cooling. Brett et al. (1977) deduced a minimum pressure during cooling of 50 to 80 bar for the unique achondrite, Angra dos Reis. This corresponds to a depth of 6 to 10 km in a parent body 250 km in diameter having a specific gravity of 3.5 and to a depth of 1000 to 1600 m in a lunar-sized body.

Ordinary Chondrites

Larimer (1968a) suggested that metamorphism of chondrites essentially took place in a closed system. He pointed out, however, that metallic Fe decreases slightly with percent-mean-deviation of Fa in olivine within individual chondrite groups; this suggests that some reduction occurred during metamorphism. Dodd (1974) showed that metal in H3 chondrites tends to contain more Ni than that in H4–6 chondrites. When coupled with the fact that C in ordinary chondrites decreases with increasing metamorphic grade (see, e.g., Mason 1979), this observation suggests that a small amount of reduction by C took place in ordinary chondrites during metamorphism. The reaction probably did not go to completion during metamorphism because C dissolved in the metal, lowering its activity.

The fact that some reduction by C appears to have taken place, suggests that the ordinary chondrites lay at or close to the graphite surface during their metamorphic histories. Inspection of Fig. 7.7.5 indicates that the graphite surface within the T-fO_2 range of ordinary chondrites is at a pressure of about 3 to 5 bar at 800° C, and about 8 to 20 bar at 900° C. The valid temperature to use for the pressure estimate is the one at which graphite ceased to equilibrate with the meteorite (probably closer to 900° C). Thus, the chondrites were probably at pressures of 3 to 20 bar. The *minimum* parent body (having a specific gravity of 3.5) required is 30 to 70 km in diameter.

Range of Parent-Body Redox Conditions

Carbon is almost ubiquitous in meteorites. A range of redox conditions as a function of temperature and depth is a natural consequence of the buffering effect of the graphite surface in a parent body (Fig. 7.7.5). Increasing depth (pressure) within the parent body favors higher fO_2, as does increasing temperature. However, the effect of pressure on iron-wüstite (IW) and Fe^{2+}/Fe^{3+} reactions is negligible; the fO_2 of these boundaries increases more steeply than that of the graphite surface as a function of temperature (Fig. 7.7.5). (The IW buffer is the fO_2 value that is in equilibrium with metallic Fe and wüstite ($Fe_{1-x}O$) as a function of temperature.) A parent body that was heated sufficiently to cause equilibration with respect to fO_2 would have a redox state in its interior that reflected the T/P gradient. A steep T/P gradient would produce a thin reduced surface layer, an oxidized near-surface layer and an interior that was increasingly more reduced with depth (Fig. 7.7.6).

The final oxidation state of the body depends on the T/P gradient at the temperature at which equilibration ceased during cooling. If equilibration was not achieved due to insufficient temperature, or insufficient time of metamorphism, unequilibrated meteorites would have resulted. The situation would have become complicated if all the graphite was consumed, in which case, one of the various buffers involving Fe species would have taken over. If metallic Fe-Ni was present, C would have dissolved in it. When all graphite

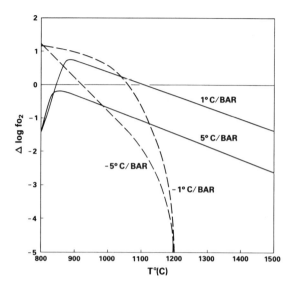

Fig. 7.7.6. The T-Δlog fO_2 regime within four hypothetical meteorite parent bodies lying on the graphite surface. Δlog fO_2 is the log fO_2 of the graphite surface minus log fO_2 of IW. Two parent bodies have positive thermal gradients (1°C/bar, 5°C/bar) with a surface temperature of 800° C. The other two bodies have negative thermal gradients ($-1°$ C/bar, $-5°$ C/bar) and a surface T of 1200°C (figure after Brett and Sato 1984).

was dissolved, the activity of C would have been lowered. The system would lie below this C (metal) surface at equivalent P and T, and would be buffered by the Fe species.

In principle, meteorites lying on the graphite surface could have experienced quite different fO_2 regimes and have resided relatively close to one another in the same parent body, separated by a temperature and pressure gradient (Brett and Sato 1984).

7.7.6. SUMMARY

The existing data on chondrite oxidation states and intrinsic O fugacities lend themselves to a variety of interpretations. Such ambiguities notwithstanding, it is worthwhile to present a summary of tentative conclusions based on our current understanding of the data.

1. A variety of oxidation states are exhibited by the chondritic meteorites. Petrologic and chemical data can be used to arrange the major chondrite groups in order of oxidation state; from highest to lowest, this order is: CI > CM > CV > CO > LL > L > H > EL > EH. Many ungrouped chondrites as well as the silicate inclusions in IAB irons lie between H and E chondrites in this sequence.

2. Intrinsic O fugacity measurements on chondrite whole-rock samples display a corresponding ordering of oxidation states. Again, from highest to lowest, this order is generally CO > LL ≃ L ≃ H > E. However, this sequence is based on a much smaller number of samples than that deduced from petrologic and chemical data, and may be subject to experimental errors.
3. Relatively unaltered primitive materials that originally equilibrated with the solar-nebular gas and still retain the imprint of nebular processes may preserve information on the O fugacity of nebular gas.
4. Fayalitic olivines coexisting with metallic Fe found primarily in the matrix of primitive, little-metamorphosed chondrites may preserve information on the O fugacity of local nebular gas.
5. Formation of these fayalitic olivines may have occurred at low temperatures (≤ 500 K) in a gas of solar composition or at higher temperatures in a more oxidizing gas. Neither alternative is unambiguously favored by the available experimental, observational and theoretical data. The former alternative may not be kinetically feasible, whereas the latter requires an unspecified mechanism for producing a highly oxidizing region.
6. Several different oxidation-state indicators are present in Ca,Al-rich inclusions in carbonaceous chondrites: e.g., Mo and W depletions relative to other refractory siderophiles of similar volatility; Ti^{3+}/Ti^{4+} ratios in hibonite; depletions of Ba, Ce, U and V. These indicators are consistent with oxidation at high temperatures in the solar nebula.
7. Metamorphosed chondrites and igneous meteorites that were substantially altered by metamorphic reactions, outgassing and igneous processes may preserve information on the oxidation state and size of their parent bodies. If ordinary chondrites underwent some reduction by graphite, we can deduce that they were derived from parent bodies at least 30 to 70 km in diameter.

It is likely that a combination of experimental, observational and theoretical work in some key areas will aid our understanding of nebular and parent-body processes that may have influenced the oxidation state of chondrites. The following studies are particularly important in this regard:

1. Detailed observational studies of the zoning patterns and trace-element contents of individual fayalite-rich olivine grains from primitive, little-metamorphosed chondrites.
2. Intrinsic O-fugacity measurements of individual components (e.g., chondrules and CAIs) and separated phases (e.g., CAI pyroxene and hibonite; matrix olivine grains) from primitive little-metamorphosed chondrites. Such measurements will provide much needed data for calibrating indirect fO_2 estimates and for testing theoretical models.

3. Laboratory studies of the kinetics of FeO incorporation into olivine and low-Ca pyroxene at low temperatures and at the solar (H_2O/H_2) ratio.
4. Theoretical models of the relative rates of gas-solid and solid-solid reactions that are important for the incorporation of FeO into silicates.
5. More detailed theoretical models of nebular radial mixing and thermal evolution.

These and related studies hold great potential to help unravel the physical and chemical conditions prevailing in the solar nebula and early solar system. The most challenging aspect of these studies is in distinguishing nebular effects from parent-body processes.

Acknowledgments. We thank D. W. G. Sears, J. T. Wasson, J. N. Grossman, M. Sato, J. W. Larimer, H. Palme and F. Wlotzka for helpful comments, and G. W. Kallemeyn and E. Jarosewich for use of unpublished data.

REFERENCES

Afiattalab, F., and Wasson, J. T. 1980. Composition of the metal phases in ordinary chondrites: Implications regarding classification and metamorphism. *Geochim. Cosmochim. Acta* 44: 431–446.

Agosto, W. N., Hewins, R. H., and Clarke, R. S., Jr. 1980. Allan Hills A77219, the first Antarctic mesosiderite. *Proc. Lunar Planet. Sci. Conf.* 11:1027–1045.

Armstrong, J. T., El Goresy, A., and Wasserburg, G. J. 1985a. Willy: A prize noble Ur-Fremdling—Its history and implications for the formation of Fremdlinge and CAI. *Geochim. Cosmochim. Acta* 49:1001–1022.

Armstrong, J. T., Hutcheon, I. D., and Wasserburg, G. J. 1985b. Ni-Pt-Ge-rich Fremdlinge: Indicators of a turbulent early solar nebula. *Meteoritics* 20:603–604 (abstract).

Basaltic Volcanism Study Project. 1981. *Basaltic Volcanism on the Terrestrial Planets* (New York: Pergamon Press).

Beckett, J. R. 1986. The origin of Ca, Al-rich inclusions from carbonaceous chondrites: An experimental study. Ph.D. thesis, Univ. of Chicago.

Bischoff, A., and Palme, H. 1987. Composition and mineralogy of refractory-metal-rich assemblages from a Ca,Al-rich inclusion in the Allende meteorite. *Geochim. Cosmochim. Acta* 51:2733–2748.

Boynton, W. V. 1984. Cosmochemistry of the rare earth elements: Meteorite studies. In *Rare Earth Element Geochemistry,* ed. P. Henderson (Amsterdam: Elsevier), pp. 63–114.

Brett, R., and Sato, M. 1984. Intrinsic oxygen fugacity measurements on seven chondrites, a pallasite, and a tektite and the redox state of meteorite parent bodies. *Geochim. Cosmochim. Acta* 48:111–120.

Brett, R., Huebner, J. S., and Sato, M. 1977. Measured oxygen fugacities of the Angra dos Reis achondrite as a function of temperature. *Earth Planet. Sci. Lett.* 35:363–368.

Davis, A. M., Tanaka, T., Grossman, L., Lee, T., and Wasserburg, G. J. 1982. Chemical composition of HAL, an isotopically-unusual Allende inclusion. *Geochim. Cosmochim. Acta* 46:1627–1651.

Dodd, R. T. 1974. The metal phase in unequilibrated ordinary chondrites and its implications for calculated accretion temperatures. *Geochim. Cosmochim. Acta* 38:485–494.

El Goresy, A., Nagel, K., and Ramdohr, P. 1978. Fremdlinge and their noble relatives. *Proc. Lunar Sci. Conf.* 9:1279–1303.

Fegley, B., Jr. 1986. A comparison of REE and refractory metal oxidation state indicators for the solar nebula. *Lunar Planet. Sci.* XVII:220–221 (abstract).

Fegley, B., Jr., and Kornacki, A. S. 1984. The geochemical behavior of refractory noble metals and lithophile trace elements in refractory inclusions in carbonaceous chondrites. *Earth Planet. Sci. Lett.* 68:181–197.

Fegley, B., Jr., and Palme, H. 1985. Evidence for oxidizing conditions in the solar nebula from Mo and W depletions in refractory inclusions in carbonaceous chondrites. *Earth Planet. Sci. Lett.* 72:311–326.

Gomes, C. B., and Keil, K. 1980. *Brazilian Stone Meteorites* (Albuquerque: Univ. of New Mexico Press).

Gooley, R. C., and Moore, C. B. 1976. Native metal in diogenite meteorites. *Amer. Mineral.* 61:373–378.

Grossman, J. N., and Wasson, J. T. 1983a. Refractory precursor components of Semarkona chondrules and the fractionation of refractory elements among chondrites. *Geochim. Cosmochim. Acta* 47:759–771.

Grossman, J. N., and Wasson, J. T. 1983b. The compositions of chondrules in unequilibrated chondrites: An evaluation of models for the formation of chondrules and their precursor materials. In *Chondrules and Their Origins,* ed. E. A. King (Houston: Lunar and Planetary Inst.), pp. 88–121.

Harlow, G. E., Delaney, J. S., Nehru, C. E., and Prinz, M. 1982. Metamorphic reactions in mesosiderites: Origin of abundant phosphate and silica. *Geochim. Cosmochim. Acta* 46:339–348.

Hewins, R. H., and Ulmer, G. C. 1984. Intrinsic oxygen fugacities of diogenites and mesosiderite clasts. *Geochim. Cosmochim. Acta* 48:1555–1560.

Holmes, R. D., and Arculus, R. J. 1982. Metal-silicate redox reactions: Implications for core mantle equilibrium and the oxidation state of the upper mantle. *Lunar Planet. Sci.* XIII:45–46 (abstract).

Hua, X., Adam, J., Palme, H., and El Goresy, A. 1987. Fayalite-rich rims around forsteritic olivines in CAIs and chondrules in carbonaceous chondrites: Types, compositional profiles and constraints of their formation. *Lunar Planet. Sci.* XVIII:443–444 (abstract).

Ihinger, P. D., and Stolper, E. 1986. The color of meteoritic hibonite: An indicator of oxygen fugacity. *Earth Planet. Sci. Lett.* 78:67–79.

JANAF Thermochemical Tables, third ed. 1986. Eds. M. W. Chase, C. A. Davies, J. R. Downey, D. J. Frurip, R. A. McDonald, and A. N. Syverud, *Suppl. 1 to Jour. Phys. Chem. Ref. Data* 14.

Kallemeyn, G. W., and Wasson, J. T. 1985. The compositional classification of chondrites: IV. Ungrouped chondritic meteorites and clasts. *Geochim. Cosmochim. Acta* 49:261–270.

Kozul, J., Ulmer, G. C., and Hewins, R. 1986. Allende inclusions are oxidized! *Eos, Trans. AGU* 67:300 (abstract).

Larimer, J. W. 1968a. Experimental studies on the system Fe-MgO-SiO$_2$-O$_2$ and their bearing on the petrology of chondritic meteorites. *Geochim. Cosmochim. Acta* 32:1187–1207.

Larimer, J. W. 1968b. An experimental investigation of oldhamite CaS; and the petrologic significance of oldhamite in meteorites. *Geochim. Cosmochim. Acta* 32:965–982.

Larimer, J. W., and Bartholomay, M. 1979. The role of carbon and oxygen in cosmic gases: Some applications to the chemistry and mineralogy of enstatite chondrites. *Geochim. Cosmochim. Acta* 43:1455–1466.

Larimer, J. W., and Buseck, P. R. 1974. Equilibration temperatures in enstatite chondrites. *Geochim. Cosmochim. Acta* 38:471–477.

Lewis, J. S., and Prinn, R. G. 1980. Kinetic inhibition of CO and N$_2$ reduction in the solar nebula. *Astrophys. J.* 238:357–364.

Lewis, J. S., Barshay, S. S., and Noyes, B. 1979. Primordial retention of carbon by the terrestrial planets. *Icarus* 37:190–206.

Live, D., Beckett, J. R., Tsay, F.-D., Grossman, L., and Stolper, E. 1986a. Ti^{3+} in meteoritic and synthetic hibonite: A new oxygen barometer. *Lunar Planet. Sci.* XVII:488–489 (abstract).

Live, D., Tsay, F. D., Beckett, J. R., and Stolper, E. 1986b. Determination of oxygen fugacities in the early solar system. *Eos, Trans. AGU* 67:1071 (abstract).

Mason, B. 1979. Chapter B, Cosmochemistry, Part 1. Meteorites. In *Data of Geochemistry,* sixth ed., ed. M. Fleischer, U.S. Geol. Survey Prof. Paper 440-B-1.

McSween, H. Y. 1977a. Petrographic variations among carbonaceous chondrites of the Vigarano type. *Geochim. Cosmochim. Acta* 41:1777–1790.

McSween, H. Y. 1977b. Carbonaceous chondrites of the Ornans type: A metamorphic sequence. *Geochim. Cosmochim. Acta* 41:477–491.

McSween, H. Y., and Stolper, E. M. 1978. Shergottite meteorites I: Mineralogy and petrography. *Lunar Planet. Sci.* IX:732–734 (abstract).

Mueller, R. F., and Saxena, S. A. 1977. *Chemical Petrology* (New York: Springer-Verlag).

Nagahara, H. 1984. Matrices of type 3 ordinary chondrites—primitive nebular records. *Geochim. Cosmochim. Acta* 48:2581–2595.

Noonan, A. F., Fredriksson, K., Jarosewich, E., and Brenner, P. 1976. Mineralogy and bulk, chondrule, size-fraction chemistry of the Dhajala, India, chondrite. *Meteoritics* 11:340–343 (abstract).

Olsen, E., and Fredriksson, K. 1966. Phosphates in iron and pallasite meteorites. *Geochim. Cosmochim. Acta* 30:459–470.

Olsen, E., and Fuchs, L. H. 1967. The state of oxidation of some iron meteorites. *Icarus* 6:242–253.

Palme, H., and Fegley, B., Jr. 1987. Formation of FeO-bearing olivines in carbonaceous chondrites by high temperature oxidation in the solar nebula. *Lunar Planet Sci.* XVIII:754–755 (abstract).

Pellas, P., and Storzer, D. 1981. ^{244}Pu fission track thermometry and its application to stony meteorites. *Proc. Roy. Soc. London* A374:253–270.

Prinn, R. G., and Fegley, B., Jr. 1988. Solar nebula chemistry: Origin of planetary, satellite, and cometary volatiles. In *Planetary and Satellite Atmospheres: Origin and Evolution*, eds. S. K. Atreya, J. B. Pollack and M. S. Matthews (Tucson: Univ. of Arizona Press), in press.

Prinz, M., Nehru, C. E., Delaney, J. S., and Weisberg, M. 1983. Silicates in IAB and IIICD irons, winonaites, lodranites, and Brachina: A primitive and modified primitive group. *Lunar Planet. Sci.* XIV:616–617 (abstract).

Rambaldi, E. R., Sears, D. W., and Wasson, J. T. 1980. Si-rich Fe-Ni grains in highly unequilibrated chondrites. *Nature* 287:817–820.

Reid, A. M., and Bunch, T. E. 1975. The nakhlites, part II: Where, when, and how. *Meteoritics* 10:317–324.

Sato, M. 1972. Intrinsic oxygen fugacities of iron-bearing oxide and silicate minerals under low total pressure. *Geol. Soc. Amer. Mem.* 135:289–307.

Sato, M. 1978. A possible role of carbon in characterizing the oxidation state of a planetary interior and originating a metallic core. *Lunar Planet. Sci.* IX:990–992 (abstract).

Sato, M. 1979. The driving mechanism of lunar pyroclastic eruptions inferred from the oxygen fugacity behavior of Apollo 17 orange glass. *Proc. Lunar Planet. Sci. Conf.* 10:311–325.

Scott, E. R. D., Rubin, A. E., Taylor, G. J., and Keil, K. 1984. Matrix in type 3 chondrites—Occurrence, heterogeneity and relationship with chondrules. *Geochim. Cosmochim. Acta* 48:1741–1757.

Scott, E. R. D., Lusby, D., and Keil, K. 1985. Ubiquitous brecciation after metamorphism in equilibrated ordinary chondrites. *Proc. Lunar Planet. Sci. Conf.* 16, *J. Geophys. Res. Suppl.* 90:D137–D148.

Scott, E. R. D., Taylor, G. J., and Keil, K. 1986. Accretion, metamorphism, and brecciation of ordinary chondrites: Evidence from petrologic studies of meteorites from Roosevelt County, New Mexico. *Proc. Lunar Planet. Sci. Conf.* 17, *J. Geophys. Res. Suppl.* 91:E115–E123.

Smith, J. V., and Hervig, R. L. 1979. Shergotty meteorite: Mineralogy, petrography, and minor elements. *Meteoritics* 14:121–142.

Snellenburg, J. W., Nehru, C. E., Caulfield, J. B. D., Zucker, S., and Prinz, M. 1979. Petrology of temperature and oxygen fugacity indicating mineral assemblages in four low grade mesosiderites. *Lunar Planet. Sci.* X:1137–1138 (abstract).

Stolper, E. 1977. Experimental petrology of eucritic meteorites. *Geochim. Cosmochim. Acta* 41:587–611.

Wasson, J. T. 1985. *Meteorites: Their Record of Early Solar-System History* (New York: W. H. Freeman).

Williams, R. J. 1971. Equilibrium temperatures, pressures, and oxygen fugacities of the equilibrated chondrites. *Geochim. Cosmochim. Acta* 35:407–411.

Wood, J. A. 1985. Meteoritic constraints on processes in the solar nebula. In *Protostars & Planets II*, eds. D. C. Black and M. S. Matthews (Tucson: Univ. of Arizona Press), pp. 687–702.

Wood, J. A. 1988. Chondritic meteorites and the solar nebula. *Ann. Rev. Earth Planet. Sci.* 16, in press.

7.8. PLANETARY COMPOSITIONS

S. R. TAYLOR
Australian National University

The components of meteorites (metal, sulfide, silicate) have long been considered as representative of the building blocks of the terrestrial planets. An examination of the density, major- and trace-element composition (including refractory/volatile-element ratios), oxygen isotopes and noble gases reveals, however, that for the presently available population of meteorites, none of these properties match those of the terrestrial planets. Differences both between planets and with meteorites for many of these parameters argue against much lateral mixing in the nebula during planetary accretion. This implies that the planets accumulated from rather narrow concentric zones, and that the present zonal structure of the asteroid belt may be an analogue for the structure of the inner portions of the solar nebula during the accretion of the terrestrial planets. The meteoritic evidence for high temperatures in the asteroid belt contrasts with the astrophysical evidence for a low-temperature nebula. This paradox may be resolved by localized heating during infall of material to the median plane of the nebula.

7.8.1. THE METEORITIC ANALOGY

Meteorites have long provided us with analogues for planetary compositions. The internal structures of the terrestrial planets are explained most rationally by metallic cores overlain by silicate mantles and the presence of metallic, sulfide and silicate phases in meteorites has lent credence to these models.

In this section, the possible relevance of the present population of meteorites as building blocks of the solid planets is discussed. The objective is to assess whether any classes of meteorites, either singly, or in combination, might provide suitable material from which the terrestrial planets could be constructed. Although somewhat peripheral to the overall thrust of this book, the topic raises many fascinating if dimly understood problems associated with the formation of the planets. For obvious reasons, this chapter concentrates on the rocky inner planets, with some comments on the satellites of the outer planets.

The question of the composition of the cores of the large outer planets, which might have been put together from planetesimals, cannot yet be addressed, except in the most general terms. One might expect that they are comprised of CI-type material plus ices on account of their distance from the Sun. However, the existence of fractionated meteorites in the asteroid belt with their evidence of at least local high temperatures, must give pause to such conjectures. If chondrules formed by heating of interstellar material falling into the median plane of the solar nebula (Wood 1986,1987), then volatile-element depletion might be only weakly correlated with heliocentric distance and could extend beyond the "snow line" at about 4 AU, which is usually ascribed to removal of the gaseous components of the inner solar nebula due to early intense solar activity. (In the present context, the term *gaseous* is used to refer to H, C, N, O and the noble gases. The other elements are classified as *very volatile:* e.g., Bi, Tl; *volatile:* e.g., Rb, Cs; *moderately volatile:* e.g., K, Mn; *moderately refractory:* e.g., V, Eu; *refractory:* e.g., Ca, Al, U, La; and *super-refractory:* e.g., Zr, Sc, as well as being categorized as lithophile, chalcophile and siderophile from their entry into silicate, sulfide or metal phases in meteorites.)

7.8.2. MODELS FOR PLANETARY ORIGIN

Our ideas about the primordial solar nebula are strongly controlled by the evidence from meteorites, which influence much of our thinking about the origin of the inner planets. A major uncertainty which cannot be addressed here is the role of cometary material in planetary formation. In the present context, it is taken as equivalent in nongaseous element content to that of CI meteorites, and so is not readily distinguishable. The question of cometary contributions to planetary atmospheres is addressed in Chapter 7.10.

A primary assumption is that the composition, for the nongaseous elements (i.e., excluding H, C, N, O and the noble gases), of the primordial solar nebula is given by that of CI carbonaceous chondrites. The rationale for this is the similarity between the abundances in the solar photosphere and in the CI chondrites (Fig. 7.8.1; see also Fig. 1.1.1). This implies a broad similarity in the composition of the nebula, on which many second-order heterogeneities (e.g., O isotopes, K/U ratios, asteroidal belt zoning) are superimposed. Although this has been a useful hypothesis, it has been clear for a long time that the compositions of the inner planets differ in detail from that of the CI meteorites, and that substantial chemical fractionation from CI composi-

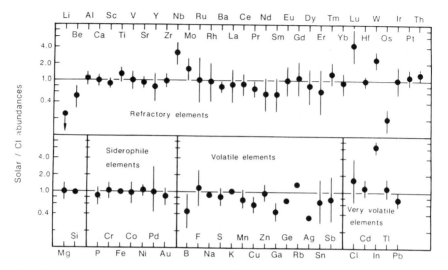

Fig. 7.8.1. Comparison between the elemental abundances in CI carbonaceous chondrites with those in the solar photosphere. Apart from the depletion of Li in the Sun, where it is consumed in nuclear reactions, there is little significant difference between the two data sets. The discrepancies for Nb, Lu, W, Os, Ga, Ag and In are probably due to lack of precision in the solar data (figure adapted from Anders and Ebihara 1982, and Taylor and McLennan 1985, Fig. 11.1).

tions occurred before the final accretion of the terrestrial planets. Models which have used CI compositions to construct the inner planets (see, e.g., Ringwood 1966; Ringwood and Anderson 1977) have encountered numerous problems (e.g., see discussion in Lewis and Prinn, 1984, p.193).

One basic question about the accretion of the inner planets concerns the size of the accreting material. Did the planets accrete directly from dust and the dispersed material of the nebula, or are they the end product of the accretion of a hierarchical succession of bodies? Five observations seem to point to the latter process.

1. Individual meteorites do not consist of single lithic components (e.g., mineral grains, Ca,Al-rich inclusions [CAI], chondrules or metallic particles) but are always mixtures of several components, or are fragments of larger bodies (Wasson 1985).
2. There is ample observational evidence from the battered surfaces of planets and satellites for the impact of innumerable large (>100 km) bodies throughout the solar system.
3. The obliquities of the planets are consistent with the collision of very large objects (>1000 km).
4. The most reasonable hypothesis for lunar origin demands an impacting object of 0.1 to 0.2 Earth mass at a rather late stage in accretional his-

tory, when both it and the Earth had already formed a core (Benz et al. 1987).
5. Accretion from a dusty nebula might be expected to lead to rather similar planetary compositions rather than the observed chemical and isotopic diversity, both of meteorites and planets.

These comments point to planetary accretion from a hierarchy of massive objects rather than from the infall of dust. A corollary is that internal fractionation may occur in such bodies before their final sweepup into the planets. The presently available meteorites represent some sample of this planetesimal population from which the planets were built; it is the purpose of this chapter to enquire what this relationship is.

Hypotheses for the formation of the terrestrial planets fall into several classes. The giant-gaseous-protoplanet theory calls for the condensation of gaseous protoplanets directly from the nebula (Cameron 1978). In this scenario, meteorites play little identifiable role as primary building blocks. It seems inherent in the hypothesis that fractionation of volatile from refractory elements would be minimal, at least in the accretional stages so that the inner planets might be essentially of CI composition, analogous perhaps to that of comets. This does not seem to be the case, although processes connected with the dissipation of the gaseous envelopes might be invoked to accomplish some separation of the more volatile elements. The apparent absence of any trace of the original gaseous components of the nebula in the Earth (Ozima and Podosek 1983, p.208; see Sec. 7.8.10 and Chapter 7.10) is difficult to account for in this theory. A deduction from this evidence is that the terrestrial planets accumulated in a gas-free environment and this seems to be a fairly firm constraint (Ozima and Podosek 1983; Lewis and Prinn 1984).

The other hypotheses for planetary origin divide themselves into two classes: those which form the planets as the result of condensation from an initially hot solar nebula and those in which the planets represent the final stages of the accretion of a hierarchy of smaller bodies. The latter scenario is most readily addressed with evidence from the meteorites, which are generally thought to be left-over fragments from the latter process.

Condensation from an initially hot nebula is usually referred to as the equilibrium condensation theory (see, e.g., Lewis 1973; Goettel and Barshay 1978). It postulates that each planet forms within rather narrow concentric zones. This theory encounters various difficulties. Many meteorites are rather complex mixtures of high- and low-temperature minerals, rather than being comprised of a simple sequence of minerals which had condensed from a hot nebula. There is ample evidence for isotopic heterogeneities and other irregularities that might be homogenized in a high-temperature nebula (see, e.g., Clayton et al., 1985, p.765; see also Chapter 12.1). Astrophysical theory likewise predicts cool rather than hot stellar nebulae (Gehrels 1978; Black and Matthews 1985; see also Chapter 6.3). Heating of an inner zone by early intense solar activity (flares, strong solar winds, enhanced ultraviolet flux), which seems necessary to account both for the volatile element depletion in the inner planets, and for depleting that zone in H, He and the noble gases, is unlikely to have extended much beyond 2 to 3 AU. Most accretion-disk models, for example, predict gas temperatures at 3 AU of less than a few hundred K (Chapter 6.3). Among supporting evidence from meteorites for these low temperatures is the presence of apparently interstellar organic matter in CI and CM2 chondrites, which would decompose at temperatures in excess of ~700 K (see Chapter 13.2). Accordingly, the concept of an initial hot (~2000 K) nebula which cooled monotonically, producing en route the classical condensation sequences of minerals (see, e.g., Grossman and Larimer 1974) does not seem consistent with several lines of evidence. Since the meteorites contain abundant evidence of high-temperature processes (see, e.g., Wasson 1985; Boynton 1985), this raises the well-known paradox which will be discussed later (Sec. 7.8.13).

A related hypothesis of planetary origin is the heterogeneous accretion model (see, e.g., Turekian and Clark 1969). This proposes that as each phase condenses in the nebula, it is accreted, so that the planets grow like a layer-cake. Many of the difficulties which beset the homogeneous accretion model apply equally to this scenario (see especially comments in *Basaltic Volcanism Study Project* 1981, pp.646–647).

In contrast to these somewhat theoretical models, the remaining hypotheses rely more heavily on meteoritic analogies. Many scenarios have been based on the two components (fine-grained matrix and coarser metal particles, chondrules etc.) which are intimately mixed in many meteorites (Wood 1962). Sometimes these were characterized as volatile-rich and volatile-poor, or refractory (Anders 1964,1971), and similar scenarios continue to appear (see, e.g., Wänke et al. 1984). A more complex version, involving seven components based on meteorite mineralogy has been advanced by Anders and coworkers in several publications (see, e.g., Ganapathy and Anders 1974; Morgan and Anders 1980). The philosophy behind this approach is that the variations in the compositions of the inner planets are similar to those observed in chondrites (both show volatile element depletion, for example). Inherent in this approach is the concept that varying mixtures of the seven components could account for the complexity of both meteorites and planets, without seeking specific matches between them.

The more recent models (Wänke 1981; Wänke et al. 1984; Ringwood 1979) employing this general scenario have tended to simplify the components to two: one, a volatile-rich, and the other, a high-temperature volatile-poor, refractory and metal-rich (and so reduced) component. Such models attempt in this way to account for the obvious presence of reduced metal, oxidized phases and some volatiles in the inner planets, but do not shed much light on the source or origins of the postulated two components in the nebula. Several properties (e.g., O isotopes, noble-gas contents, etc.) indicate that neither the volatile nor the refractory component can be matched with a specific meteorite class. Although the volatile component is commonly identified with CI, this is ruled out on many grounds (e.g., O isotopes). The E chondrites are the most obvious candidates for the refractory component but although they are reduced, they are not volatile-depleted. One might expect that if the planets were built out of these two components, both planetary and meteoritic compositions should show considerably more homogeneity than is the case.

In the following sections, the properties of meteorite groups are examined to investigate their possible relationship both to planetary compositions and to the various scenarios which attempt to account for their existence.

7.8.3. DENSITY

The densities of the inner planets are given in Table 7.8.1. This lists both actual and uncompressed densities. The latter allow for comparison with chondritic densities. The uncompressed densities are uncertain within ±10 to 20% because of assumptions about mineral phases, and the extrapolations to Mbar pressures and high temperatures for Venus and the Earth. The chondrite data are for those groups (EH, EL, IAB, H, L, LL, CV and CO) which do not contain hydrated silicates. These values are shown in Fig. 7.8.2. The uncompressed density for Mars falls within the range for chondrites, but the values for the Earth and Venus barely overlap the meteorite field, while that for Mercury lies far outside the meteorite data. The fact that the meteoritic and planetary data for this fundamental property do not match is a significant constraint on the nature of the material which accreted to form the inner planets.

TABLE 7.8.1
Densities of the Terrestrial Planets[a]

	Density (g/cm^3)	Uncompressed Density (g/cm^3)
Earth	5.514 ±0.005	4.0[b]
Venus	5.245 ±0.005	4.0
Mars	3.934 ±0.004	3.7
Mercury	5.435 ±0.008	5.2
Moon	3.344 ±0.003	3.3

[a] Table adapted from *Basaltic Volcanism Study Project* 1981, Chapter 4.
[b] Fegley and Cameron (1987) give a value of 4.45 g/cm^3 for the uncompressed density of the Earth, citing Lewis (1972) as the source.

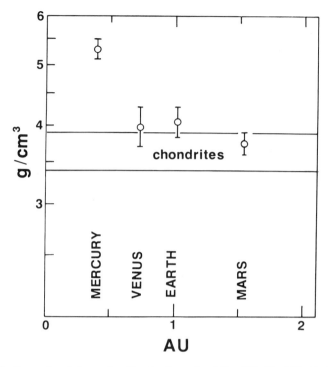

Fig. 7.8.2. Comparison between densities for those chondrites (EH, EL, IAB, H, L, LL, CV and CO groups) which do not contain hydrated mineral phases, and for the inner planets (recalculated to 298 K and 1 bar pressure). The planetary data are plotted at their relative distance from the Sun (figure adapted from Wasson 1985, Fig. IX-1).

7.8.4. MAJOR-ELEMENT CHEMISTRY

The FeO/(FeO + MgO) ratio is not known for the Earth with great precision (Table 7.8.2) since there is a major uncertainty both on account of the metallic core and the unknown Fe and Mg contents of the lower mantle. Was the metal in the core produced by reduction following accretion (see, e.g., Ringwood 1966) or does it represent metal which was present in the accreting planetesimals, so that the metal-silicate equilibrium was established in a low-pressure environment external to the Earth?

With these uncertainties in mind, one can compare the FeO/(FeO + MgO) ratios for various chondrite classes with that of the Earth (Fig. 7.8.3). None of the meteorite classes match that of the Earth, which falls between the H group and the silicate inclusions in the IAB irons. Comparisons with other major elements suffer from the lack of knowledge about the composition of the lower mantle. The bulk Earth ratios (e.g., Mg/Si) are commonly assumed to be chondritic, but this begs the question.

TABLE 7.8.2
Element Ratios for the Earth and Meteorites[a]

Meteorite Class	Ref/Si[b]	FeO/(FeO/MgO)
CI	1.0	45
CV	1.35	35
CM	1.13	43
CO	1.10	35
H	0.79	17
L	0.77	27
LL	0.76	27
IAB[c]	0.7	6
EL	0.60	0.05
EH	0.59	0.05
Earth	0.8	9

[a] Table adapted from Wasson (1985).
[b] Refractory element/Si ratio.
[c] Inclusion.

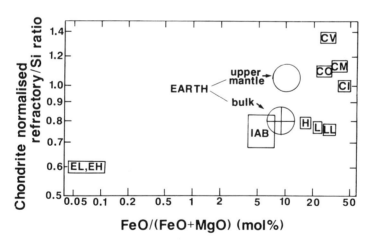

Fig. 7.8.3. Variation in refractory-element/Si ratios with FeO/(FeO + MgO) ratios for chondrite groups and the Earth (figure adapted from Wasson 1985, Fig. IX-8).

7.8.5. REFRACTORY-ELEMENT/Si RATIOS

Wasson (1985) has used the CI-normalized refractory-element/Si ratio as a useful parameter to classify the chondrites, and this forms the vertical axis in Fig. 7.8.3 (see also Figs. 7.8.4 and 7.8.5). Data are given in Table 7.8.2 and show the wide variation in this parameter among chondrites. The best match for the Earth is with the ordinary chondrites (H, L, LL and IAB). Both the fully oxidized (CI) and reduced (EL, EH) are distinct from terrestrial values, although a mix of the two groups might provide an appropriate mantle ratio. The range in refractory element abundances, relative to somewhat more volatile Si, indicates high-temperature processing of these meteorite classes (Wasson 1985; Boynton 1985) and constitutes primary evidence for at least local high temperatures in the nebula. If they have not moved far from their original position in the nebula, as is consistent with the zoned structure of the asteroid belt, then this observation is at variance with the other evidence for low temperatures and for lack of homogenization in the nebula. (However, some resolution of this paradox may be possible: see Sec. 7.8.13.)

7.8.6. REFRACTORY/VOLATILE-ELEMENT RATIOS

From a geochemical point of view, one of the most significant observations is the depletion in volatile elements which appears to be common to all four terrestrial planets. This depletion is well shown on a plot of K (a mod-

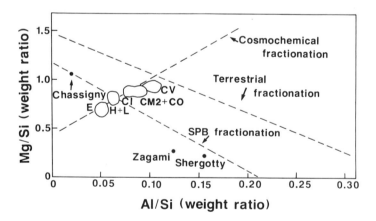

Fig. 7.8.4. Relationships between Al/Si and Mg/Si ratios for chondrites, the Earth and the SNC meteorites (= Mars?) (figure adapted from Wänke et al. 1986, Fig. 5). According to these authors the intersection of the cosmochemical and planetary trends gives the bulk planetary values. A straightforward interpretation of this diagram is that the Earth has higher Mg/Si ratios than any common chondrite group, and that Mars has the same Mg/Si and Al/Si ratios as the H and L chondrites. However, it should be noted that in the case of the Earth, the terrestrial fractionation line is derived from only upper mantle data.

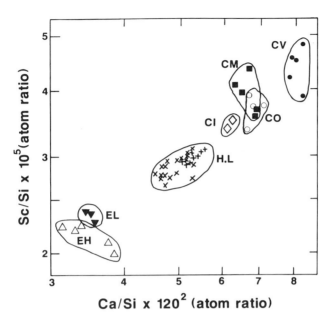

Fig. 7.8.5. The correlation between Ca/Si and Sc/Si ratios in chondrite groups (figure adapted from Wasson 1985, Fig. II-3). Scandium is a super-refractory and Ca is a refractory element. Both have been fractionated relative to less refractory silicon during chondrite formation.

erately volatile element) vs U (a refractory element). This is one of the few geochemical measurements available for Earth, Venus and Mars, as well as for the meteorites, and it provides some crucial information (Fig. 7.8.6).

CI ratios are approximately 60,000, while terrestrial ratios are 10,000. Is the depletion of K observed in terrestrial samples a bulk planetary effect, or has the element been extracted into the core, as suggested by Lewis (1971) and Murthy and Hall (1972)? This question can be addressed only indirectly from the information available on the Earth. However, K appears to be depleted, relative to U, in the surface rocks of Mars and Venus, as well as on the Earth. Five measurements of K/U ratios for Venus (Surkov 1981,1984) indicate an overlap with those for the terrestrial surface. The K/U ratios for the SNC meteorites appear to be somewhat higher (1.5×10^4), consistent with a higher volatile content for that planet, but still a factor of four lower than CI values. This conclusion is supported by the Nd and Sr isotopic systematics for both planets (Fig. 7.8.7) assuming the SNC connection with Mars (see Chapter 7.10).

Accordingly, it is a firm conclusion that K, and presumably the other volatile elements, are depleted in the surficial rocks of these planets. The central pressure in Venus ($r = 6070$ km) is 2.8 to 2.9 Mbar, only slightly less than that in the Earth's core (3.6 Mbar), so that K might be hidden in the

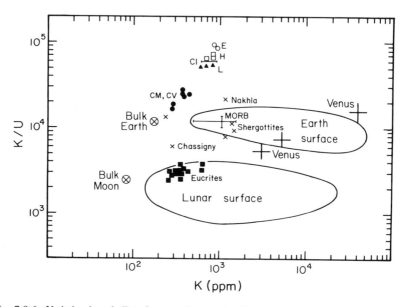

Fig. 7.8.6. Variation in volatile/refractory-element abundances in the inner solar nebula, illustrated by the ratio of K (moderately volatile) to U (refractory) for chondrites, Earth, Venus, SNC meteorites (= Mars), Moon and the eucrites (figure adapted from Taylor 1987). Both elements are concentrated in residual melts during intraplanetary fractionation processes, and thus tend to preserve their bulk planetary ratios. The depletion of moderately volatile relative to refractory elements appears to have been widespread in the inner solar system, relative to the source regions of the common chondrites.

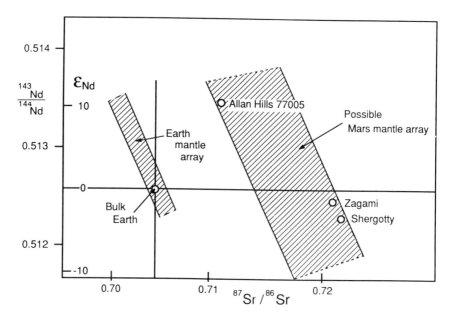

Fig. 7.8.7. The relationship between Nd and Sr-isotopic systematics for the Earth and the SNC meteorites. If the SNC meteorites come from Mars, then that planet has a higher content of the volatile element Rb relative to the refractory element Sr, by a factor of about two (figure adapted from Taylor and McLennan 1987, Fig. 15).

core of that planet. However, the central pressure in Mars ($r = 3390$ km) is only 400 kbar. Recent studies by Liu (1986) indicate that this pressure is insufficient to allow K to enter a Martian core. Thus the depletion of K in Mars must be due to some other process.

Elements of the atomic weight of K cannot be lost from the terrestrial planets, even at elevated temperatures once they have reached their present size. If they were boiled off in some manner during accretion, the K/U ratio should vary with planetary size. This does not appear to be the case. In addition to K, many other volatile elements are depleted in the Earth relative to CI abundances. It is unlikely that they are all hidden in a metallic core. The most reasonable explanation is that these volatile elements were depleted in the precursor planetesimals from which the inner planets accumulated.

The common classes of chondrites (E, H, L, LL and CI) all have high K/U ratios and so are unsuitable building blocks for the inner planets, although the CM, CO and CV classes have lower ratios, approaching the values for Mars.

The Rb/Sr-isotopic systematics also indicate that the Earth is depleted in volatile Rb relative to refractory Sr. No samples derived from the mantle bear any evidence of having been in an Rb/Sr environment as high as that of CI since T_O. Since Rb has closely similar properties to K, it is unlikely that either K or Rb are present in the mantle in CI concentrations. The presence of a CI component in the Earth, required in the two-component models of Wänke et al. (1984) and Ringwood (1979) should be apparent from Sr-isotopic systematics but is not obvious.

It is argued above that the terrestrial depletion in K, and other volatile elements, was an inherent feature of the incoming planetesimals. However, according to Liu (1986), if core-mantle equilibrium were attained in an initially molten Earth, then K would enter the core. This apparently did not happen. The implications for planetary accretion and core-mantle relationships in the terrestrial planets are considerable since Liu's data imply that there was little or no high-pressure core-mantle equilibrium during accretion and core-mantle separation. This is consistent with other lines of evidence, such as the high siderophile abundances in the upper mantle, although a popular hypothesis suggests that these may have been added as a later veneer following core separation (see, e.g., Morgan 1986).

A further consequence may be noted. The metallic core of the Earth contains about 10% of a light element. The two current contenders are O and S. If high-pressure core-mantle equilibrium was not attained in the early Earth, then it seems unlikely that O entered the core, since this requires Mbar pressures, as is the case for K. Sulfur then becomes the most viable candidate for the light element in the Earth's core. The often-repeated argument that the Earth must be depleted in S because that element is more volatile than K, fails to recognize that most of the S accreting to the Earth in planetesimals is combined with Fe as troilite (FeS).

The only meteorites which have very low K/U ratios are the eucrites, which have ratios less than those of the Earth, and which overlap with those of the Moon. The K/U ratios for lunar samples are very low (2500), consistent with a second stage of volatile loss during lunar formation. The Moon appears to have a unique chemical composition among the satellites and to be so far removed in composition from that of the primordial nebula, and of most meteorites, except for the eucrites, that it does not seem profitable to discuss the possibility that any of the present population of meteorites were its building blocks. It would be a very difficult task to assemble a circumterrestrial ring composed exclusively of eucrite parent-body planetesimals from which the bone-dry refractory-rich volatile-poor Moon might be formed (see Sec. 7.8.11). This, however, raises the question of the process by which the eucrites lost their volatile elements. This apparently happened close to T_0, since the eucrite ages are about 4.55 Gyr. Thus, at least one asteroid went through a similar sequence of volatile- and siderophile-element depletion to that experienced by the Moon. Possibly a large collisional event is required not only to account for the Moon but also for the eucrite parent body. The canonical view is that asteroids with surfaces of achondritic composition are rare (Chapman and Gaffey [1979] list 4 Vesta, 44 Nysa, 64 Angelina, possibly 434 Hungaria and a few others) so that the processes which produced the eucrite parent body appear to have been uncommon in the early solar nebula, at least in the region of the asteroid belt.

7.8.7. VANADIUM, CHROMIUM AND MANGANESE

Relative to CI abundances, moderately volatile Cr and Mn are depleted with respect to more refractory V in CAIs, in refractory chondrules from ordinary and enstatite chondrites (Kurat et al. 1985), in the CO, CM and CV classes of chondrites (Kallemeyn and Wasson 1981) and in the Earth and Moon (Wänke et al. 1984). In contrast, the eucrites have effectively chondritic ratios, while the SNC meteorites (Mars?) have CI abundances of Mn, but are depleted in both V and Cr to the same extent. The SNC values are consistent with the separation of a sulfide-rich core (Newsom and Drake 1987). The separation of an Fe-rich core in the eucrite parent body has not fractionated these elements from CI patterns. The depletions in the order V-Cr-Mn observed in the Earth, Moon and various meteorites listed above are in the order of increasing volatility of the elements, and are the reverse of those expected on the basis of their siderophile characteristics (Newsom and Drake 1987). These observations indicate that the Earth-Moon abundances are not unique, but probably reflect the compositions of the accreting planetesimals both to the Earth and to the Mars-sized impactor, whose mantle was most likely the source of the protolunar material. Vanadium, Cr and Mn in the CO, CM or CV meteorites match the terrestrial abundances. However, O-isotope data, K/U ratios and other parameters rule out these meteorites as major contributors to the Earth.

7.8.8. SIDEROPHILE TRACE ELEMENTS

There are high upper-mantle abundances in the Earth of Re, Au, Ni, Co and the platinum group elements (Ru, Rh, Pd, Os, Ir, Pt), indicating that the present upper mantle was never in equilibrium with the core. The distribution of the platinum group elements appears to be rather uniform, and they are present in approximately CI proportions (Morgan 1986). Three hypotheses have been advanced to explain this pattern: (a) inefficient core formation (Jones and Drake 1986); (b) Fe-S-O rather than Fe-Ni equilibrium (Brett 1984); and (c) late meteoritic bombardment, following core formation (Morgan 1986). Luck and Allègre (1983) and Morgan (1985) have studied the terrestrial Os/Re ratio, finding that it has a value of 11.8. As shown in Fig. 7.8.8, this ratio matches only those of CO3, CV3 and E group chondrites. The CI group ratios do not match.

From these data, Morgan (1985) concluded that a late infall of CV3 chondritic material was probably responsible for the upper-mantle siderophile-element abundance pattern, because of supporting evidence from the noble-gas data (Wacker and Anders 1984). More recent estimates, based on S and Se abundances point to the CM2 meteorites as a more likely source (Morgan 1986). Clearly, however, the infall of such material does not con-

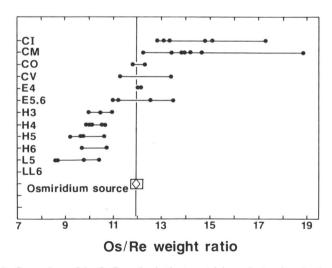

Fig. 7.8.8. Comparison of the Os/Re ratios in the terrestrial mantle (as given by the Os source ratio of 11.8) with those of various chondrite groups (figure adapted from Morgan 1985, Fig. 1). Overlap of Os/Re ratios with the terrestrial value occurs only for CO3, CV3 and E group chondrites.

tribute much toward answering the basic question of what were the building blocks of the planets, since it is not possible to construct the Earth either from CV3 or CM2 material, because of the O-isotope evidence. If this scenario is correct, it provides us with the insight that the bodies which were among the last to be accreted were not related to those responsible for providing the main bulk of the planets. The variety of explanations offered for the terrestrial upper-mantle siderophile-element abundance relationships, and their resemblance to CI patterns must give pause to claims that there is a unique terrestrial siderophile-element signature.

A similar scenario of late accretion of volatile-rich planetesimals is often invoked to account for the volatile (e.g., H_2O) inventory of the Earth. Morgan (1986, p.12,385) concludes that an addition of 10^{25} g of CM2 material between 4.5 and 3.8 Gyr would "provide all of the carbon found in the terrestrial crust, oceans and atmosphere—of approximately the right isotopic composition—and a substantial amount of the Earth's water." However, a late cometary influx could be an equally viable source. Late infall of planetesimals to planets which are essentially complete is like adding icing to a cake: the decoration may give little insight about the composition of the interior.

7.8.9. OXYGEN ISOTOPES

The well-known O-isotope plot is shown in Fig. 7.8.9 and needs little further comment (see, e.g., Clayton 1977; Clayton et al. 1985). This figure clearly separates various O-isotope reservoirs. In the present context, two observations may be made. Only the EH and EL groups overlap with the terrestrial values. H, L, LL (all long-standing favorite candidates as planetary builders) and IAB groups show no overlap with terrestrial values. Assuming that the SNC meteorites come from Mars, they form a separate group, which again does not overlap with any of the other meteorite groups or with that of the Earth-Moon system. This is additional evidence for lack of homogeneity, and hence a cool nebula at distances of 1 to 2 AU. CO, CM and CV groups are clearly very distinct from the terrestrial value. Wasson (1985) suggests that the separation of these latter groups from those which are closer to the terrestrial values indicates that they formed in the outer solar system. If so, the V-Cr-Mn depletions in these meteorites (Kallemeyn and Wasson 1981) indicate that some volatile depletion occurred far out in the nebula, beyond the region now occupied by the terrestrial planets and asteroids. This could be consistent with the Wood-Morfill scenario (Chapter 6.3) of heating of material as it falls onto the midplane of the nebula; information about the composition of the rocky portion (assumed generally to be equivalent to that of CI) of the satellites of the outer planets could answer this question.

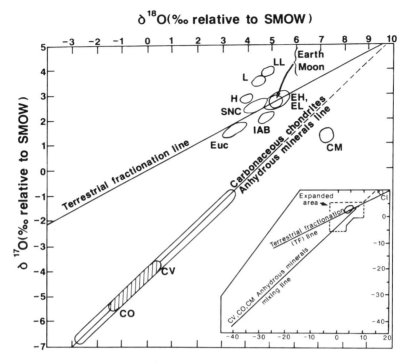

Fig. 7.8.9. The relationship between $\delta^{17}O$ and $\delta^{18}O$ for the Earth and the various meteorite groups (figure adapted from Clayton et al. 1976; Clayton 1977; Anders 1987).

7.8.10. THE NOBLE GASES

Two basic patterns are commonly identified. The so-called *solar pattern* is that of the solar wind, and is taken as representative both of the Sun and of the primordial solar nebula (refer to Figs. 7.8.10 and 7.10.1). Little of this material seems to have been incorporated in the Earth. Thus, "no obvious evidence for a major component of primordial or captured solar-composition gas can be found on (the terrestrial) planets" (Lewis and Prinn 1984, p.383). Likewise, "if the Earth's noble gases were ever, in any fundamental sense, more primitive than they are now, the traces of these primitive gases have vanished or remain hidden" (Ozima and Podosek 1983, p.208). This evidence is generally interpreted to indicate that the primordial solar-nebula gases had dissipated by the time that the Earth, and presumably the other terrestrial planets accreted from planetesimals.

The so-called *planetary pattern* is that found in the chondrites (see Chapter 7.9), and is distinguished by showing a depletion of the light gases (e.g., Ne) relative to the solar wind. There is a general resemblance between the

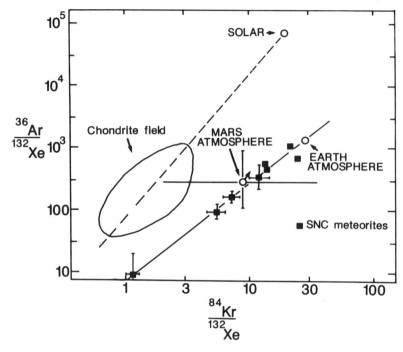

Fig. 7.8.10. Plot of trapped $^{36}Ar/^{132}Xe$ vs $^{84}Kr/^{132}Xe$, showing that the meteorite data and planetary (terrestrial, Martian and SNC) data form separate arrays (figure adapted from Ott and Begemann 1985).

relative abundances of the noble gases in chondrites and in the atmospheres of both the Earth and Mars for Ne, Ar and Kr (refer to Fig. 7.10.1). However, there is a significant difference in the abundances of the heavier noble gases. Xenon on the Earth is relatively depleted by a factor of about 20. A number of explanations have been advanced to account for this discrepancy. Thus, it has been proposed that the "missing" Xe is trapped in Antarctic ice, or in shales or has not been outgassed from the mantle to the same extent as Ne, Ar or Kr. None of these suggestions appear adequate (see, e.g., Wacker and Anders 1984). The most probable explanation is that the noble gases in the Earth and Mars were not derived from typical chondrites. Of the present classes, the only chondrites with high Ar/Xe and Kr/Xe ratios are the CO3 and E4 types. These are, however, ruled out as major contributors to the compositions of the inner planets on other grounds.

A plot of $^{36}Ar/^{132}Xe$ vs $^{84}Kr/^{132}Xe$ (Fig. 7.8.10) shows that most meteorite data form a distinct population, which may pass through the solar data. The terrestrial, Martian and SNC (= Mars?) data form a separate linear array (Ott and Begemann 1985), consistent with the existence of distinct reservoirs and thus with an inhomogeneous nebula.

The noble gas abundances in Venus and Mars present another unresolved problem. The concentration ratios are 75/1/0.1 for Venus, Earth and Mars, which are not readily accommodated by current models (see discussions by Lewis and Prinn 1984).

7.8.11. THE MOON AND MERCURY

These bodies represent special cases in the solar system: Mercury on account of its high density which suggests an iron/silicate ratio about twice that of the other inner planets, and the Moon because of its bone-dry refractory composition and low density, indicative of a low Fe content. Explanations for the peculiar nature of both bodies have a long history and much effort has been expended in attempts to fit one or both into overall schemes of planetary formation, but without conspicuous success. Recent progress in our understanding of the importance of massive impacts during the closing stages of planetary accretion enables us to understand the origins of both bodies, as well as many other features of the solar system (e.g., planetary obliquities and retrograde motion of Venus).

The Moon is seen in this context as being derived from the mantle of an impactor of about 0.1 to 0.2 Earth mass, which collided with the Earth (Benz et al. 1987). The preference for low-velocity collisions in this model, coupled with the V-Cr-Mn and O-isotope similarities point to an origin for the impactor in the same general region as the Earth. Such a scenario accounts not only for the unusual chemistry, but also for the high angular momentum of the Earth-Moon system, the lunar orbit, and other properties (Taylor 1987).

The two current hypotheses which explain the high density of Mercury are: (a) high-temperature evaporation of the silicate mantle (Fegley and Cameron 1987); or (b) collisional removal of the mantle (Cameron and Benz 1987). In the high-temperature vaporization scenario, loss of about 80% of the silicate is required, with nearly complete depletion of the alkali elements. The recent discovery of a Na cloud around Mercury (Hunten et al. 1988), coupled with the evidence from reflectance spectra of a lunar-highlands-like plagioclase(?) crust (Johnson et al. 1986) seems at variance with the requirements for nearly total loss of alkalies in evaporation models. Removal of much of the silicate mantle during a collision with a body about 1/6 Mercury's mass (Cameron and Benz 1987) can account at present for the exceedingly sparse geochemical observations on Mercury.

Such models are in keeping with current thinking which emphasizes the stochastic nature of planetary and satellite formation.

7.8.12. SATELLITES OF THE OUTER PLANETS

The icy satellites of Jupiter, Saturn and Uranus, about which we have little compositional information, may be composed of material approaching

CI in composition. The rationale for this conclusion is that volatile-refractory element fractionation appears to have been less likely in the colder outer regions of the solar nebula beyond a snow line at about 4 or 5 AU (water ice condenses at 160 K at nebula pressures; Fegley, personal communication). This assumes that the volatile depletion in the inner regions of the primordial nebula was connected with early intense solar activity. This scenario is also consistent with the composition of comets, assuming Halley to be typical. This view receives some credence from the small-scale analogue of the solar system in the Jovian system, in which the decrease in density outwards from Jupiter of the Galilean satellites is commonly attributed to mild warming of the proto-Jovian nebula, so that water ice became more abundant in the sequence Europa, Ganymede and Callisto (see, e.g., Malin and Pieri 1986; McKinnon and Parmentier 1986). However, if heating and volatile loss occurred during early infall of material into the median plane of the nebula (see, e.g., Wood 1986,1987; see also Chapter 6.3), then some volatile-element depletion may extend out beyond the snow line.

7.8.13. DISCUSSION

This summary of the interrelationships between the compositions of the chondritic meteorites and the terrestrial planets reveals few points of similarity. There is little overlap in density. The major element chemistry is not particularly diagnostic, because the silicate chemistry of the inner planets and that of the chondritic meteorites are broadly similar. Thus, as one of the authors of *Basaltic Volcanism Study Project* (1981, p.648) remarked, "One can call the Earth 'chondritic' in a general way, but it cannot be matched up with a specific chondrite class."

Comparison of refractory-element/Si ratios is hampered by uncertainties in the terrestrial data, because the upper-lower mantle compositional problem is unresolved. The refractory/volatile-element ratios are more useful, because K/U ratios are known to some degree of confidence both for the inner planets (except Mercury) and for the meteorites. They display no overlap between the chondrites and the inner planets. Vanadium-Cr-Mn relationships are mostly related to volatile depletion and are not unique to the Earth, but indicate probable differences in core formation between the Earth and Mars. Siderophile trace-element data are complicated by the possibility of a late addition of material to the terrestrial upper mantle. The O-isotope data show that, except for the EH and EL groups, there is no overlap with terrestrial values. Likewise the SNC data, assumed to represent Martian values, occupy a distinct field remote from the common chondrites. This important observation rules against nebula-wide mixing or homogenization during accretion of the planets. The noble-gas data indicate that no known meteorite class has appropriate elemental or isotopic compositions close to those of the inner planets.

Although the inner planets are chondritic in a broad sense, it does not appear possible to build the Earth and the other terrestrial planets out of the building blocks as supplied by the currently sampled population of meteorites. If, as appears reasonable, these provide us with an adequate sample of the inner asteroid belt, then there were substantial differences between that region and the zone sunwards of about 2 AU, in which the terrestrial planets accumulated. The most significant difference between these regions appears to have been a greater depletion of the volatile elements in the latter. However, the presence of the volatile-depleted eucrites, derived either from Vesta at 2.36 AU (Consolmagno and Drake 1977) or more likely from the innermost asteroid belt between 2.17 and 2.25 AU (Wetherill 1987; see also Chapter 2.1) serves to remind us that volatile-element depletion was not simply related to heliocentric distance.

A major paradox is that the astrophysical and isotopic evidence suggests a cool nebula (Chapter 6.3), whereas the petrological evidence from the meteorites (see, e.g., Boynton 1985; Wasson 1985; see also Chapters 7.3 and 10.3) indicates high temperatures, at least locally. The meteorites are derived from a region between 2 and 4 AU. There is really no indication that they were closer to the Sun at an early stage, or that they are typical samples of the planetesimals from which the inner planets accumulated, or that the zonal structure in the asteroid belt is not a primary feature (Chapter 2.1). They are mostly less strongly depleted in volatile elements, as shown by K/U ratios, than the inner planets, which seems at odds with the variation in refractory/Si ratios. E chondrites, placed closest to the Sun in many scenarios (Smith 1979) retained their complement of volatiles, but have the lowest refractory/Si ratios.

Some resolution of this paradox may be possible through the work of Wood (1985,1986,1987), Anders (1987) and Wood and Morfill (see Chapter 6.3). They conclude that thermal processing of chondritic material occurred during discrete local events, on short time scales (a few years rather than 10^5 yr for hot-nebula models), mostly connected with the infall of material into the median plane of the nebula.

In conclusion, it seems clear that we have not identified candidate building blocks for the terrestrial planets from among the present population of meteorites. Instead, these provide us with valuable information, both on the state of the early solar nebula, and on the nature of the asteroid belt. If there is little identifiable input into the inner planets from the meteorites from the asteroid belt, then this has important implications for planetary accretion models. One corollary is that the terrestrial planets accumulated from rather narrow (perhaps < 0.5 AU) concentric zones in the solar nebula. Thus, the present zonal structure in the asteroid belt (Gradie and Tedesco 1982) may be an analogue for the structure of the nebula, at least within 4 to 5 AU of the Sun. This view receives some support from the differences in composition between the inner planets, particularly Earth and Mars, which differ in K/U ratios, Mg/(Mg + Fe) ratios, V-Cr-Mn abundances and O isotopes, all of which indicate accretion from differing populations of planetesimals. The Earth-Moon relationship also is explicable if little material is derived from outside the terrestrial neighborhood. The most likely conditions for lunar origin by the large-impactor hypothesis involve low-velocity collisions of objects of about 0.12 to 0.17 Earth mass (Benz et al. 1987). Although the Moon and the terrestrial mantle differ significantly in composition, the similarity in O isotopes between Earth and Moon and various other coincidences (e.g., Cr, V and Mn abundances) are consistent with an origin for both bodies in the same portion of the nebula. Models which call upon nebula-wide mixing of planetesimals during planetary accretion receive little support from this study, which seems to require a nebula with some compositional zoning, analogous to that of the present asteroid belt.

Acknowledgments. I wish to thank E. Anders, J. Morgan and R. Rudnick for reviews, and J. Cowley for drafting the figures.

REFERENCES

Anders, E. 1964. Origin, age and composition of meteorites. *Space Sci. Rev.* 3:583–714.
Anders, E. 1971. Meteorites and the early solar system. *Ann. Rev. Astron. Astrophys.* 9:1–34.
Anders, E. 1987. Local and exotic components of primitive meteorites, and their origin. *Phil. Trans. Roy. Soc. London,* in press.
Anders, E., and Ebihara, M. 1982. Solar system abundances of the elements. *Geochim. Cosmochim. Acta* 46:2363–2380.
Basaltic Volcanism Study Project. 1981. *Basaltic Volcanism on the Terrestrial Planets* (New York: Pergamon Press), 1286 pp.
Benz, W., Slattery, W. L., and Cameron, A. G. W. 1987. The origin of the moon and the single impact hypothesis II. *Icarus* 71:30–45.
Black, D. C., and Matthews, M. S., eds. 1985. *Protostars & Planets II* (Tucson: Univ. of Arizona Press), 1293 pp.
Boynton, W. V. 1985. Meteoritic evidence concerning conditions in the solar nebula. In *Protostars & Planets II,* eds. D. C. Black and M. S. Matthews (Tucson: Univ. of Arizona Press), pp. 772–787.
Brett, R. 1984. Chemical equilibration of the Earth's core and upper mantle. *Geochim. Cosmochim. Acta* 48:1183–1188.
Cameron, A. G. W. 1978. The primitive solar accretion disk and the formation of the planets. In *The Origin of the Solar System,* ed. S. F. Dermott (New York: Wiley), pp. 49–73.
Cameron, A. G. W., and Benz, W. 1987. Planetary collision calculations: Origin of Mercury. *Lunar Planet. Sci.* XVIII:151–152 (abstract).
Chapman, C. R., and Gaffey, M. J. 1979. Reflectance spectra for 277 asteroids. In *Asteroids,* ed. T. Gehrels (Tucson: Univ. of Arizona Press), pp. 655–687.
Clayton, R. N. 1977. Genetic relations among meteorites and planets. In *Comet, Asteroids, Meteorites: Interrelations, Evolution and Origins,* ed. A. H. Delsemme (Toledo: Univ. of Toledo Press), pp. 545–550.
Clayton, R. N., Onuma, N., and Mayeda, T. K. 1976. A classification of meteorites based on oxygen isotopes. *Earth Planet. Sci. Lett.* 30:10–18.
Clayton, R. N., Mayeda, T. K., and Molini-Velsko, C. A. 1985. Isotopic variations in solar system material: Evaporation and condensation of silicates. In *Protostars & Planets II,* eds. D. C. Black and M. S. Matthews (Tucson: Univ. of Arizona Press), pp. 755–771.
Consolmagno, G. J., and Drake, M. J. 1977. Composition and evolution of the eucrite parent body: Evidence from rare earth elements. *Geochim. Cosmochim. Acta* 41:1271–1282.
Fegley, B., and Cameron, A. G. W. 1987. A vaporization model for iron/silicate fractionation in the Mercury protoplanet. *Earth Planet. Sci. Lett.* 82:207–222.

Ganapathy, R., and Anders, E. 1974. Bulk compositions of the Moon and Earth estimated from meteorites. *Proc. Lunar Sci. Conf.* 5:1181–1206.

Gehrels, T., ed. 1978. *Protostars and Planets* (Tucson: Univ. of Arizona Press), 756 pp.

Goettel, K. A., and Barshay, S. S. 1978. The chemical equilibrium model for condensation in the solar nebula: Assumptions, implications and limitations. In *The Origin of the Solar System*, ed. S. F. Dermott (New York: Wiley), pp. 611–627.

Gradie, J., and Tedesco, E. 1982. Compositional structure of the asteroid belt. *Science* 216:1405–1407.

Grossman, L., and Larimer, J. W. 1974. Early chemical history of the solar system. *Rev. Geophys. Space Phys.* 12:71–101.

Hunten, D. M., Shemansky, D. E., and Morgan, T. M. 1988. The Mercury atmosphere. In *Mercury*, eds. C. R. Chapman, F. Vilas, and M. S. Matthews (Tucson: Univ. of Arizona Press), in press.

Johnson, R. E., Nelson, M. L., Hawke, B. R., Bell, J. F., and Cintala, M. J. 1986. Mineralogical implications of Mercury's sodium cloud. Mercury Conf. Abstracts, Tucson, Ariz.

Jones, J. H., and Drake, M. J. 1984. The chemical signature of core formation in the Earth's mantle. *Lunar Planet. Sci.* XV:413–414 (abstract).

Kallemeyn, G. W., and Wasson, J. T. 1981. The compositional classification of chondrites—I. The carbonaceous chondrite groups. *Geochim. Cosmochim. Acta* 45:1217–1230.

Kurat, G., Palme, H., Brandstatter, F., Spettel, B., and Perelygin, V. P. 1985. Allende chondrules: Distillations, condensations and metasomatisms. *Lunar. Planet. Sci.* XVI:471–472 (abstract).

Lewis, J. S. 1971. Consequences of the presence of sulfur in the core of the Earth. *Earth Planet. Sci. Lett.* 11:130–134.

Lewis, J. S. 1972. Metal/silicate fractionation in the solar system. *Earth Planet. Sci. Lett.* 15:286–290.

Lewis, J. S. 1973. Chemistry of the planets. *Ann. Rev. Phys. Chem.* 24:339–351.

Lewis, J. S., and Prinn, R. G. 1984. *Planets and Their Atmospheres: Origins and Evolution* (New York: Academic Press), 470 pp.

Liu, L.-G. 1986. Potassium and the Earth's core. *Geophys. Res. Lett.* 13:1145–1148.

Luck, J.-M., and Allègre, C. L. 1983. $^{187}Re/^{187}Os$ systematics in meteorites and cosmochemical consequences. *Nature* 302:130–132.

Malin, M. C., and Pieri, D. C. 1986. Europa. In *Satellites*, eds. J. A. Burns and M. S. Matthews (Tucson: Univ. of Arizona Press), pp. 689–717.

McKinnon, W. B., and Parmentier, E. M. 1986. Ganymede and Callisto. In *Satellites*, eds. J. A. Burns and M. S. Matthews (Tucson: Univ. of Arizona Press), pp. 718–763.

Morgan, J. W. 1985. Osmium isotope constraints on Earth's late accretionary history. *Nature* 317:703:705.

Morgan, J. W. 1986. Ultramafic xenoliths: Clues to Earth's late accretionary history. *J. Geophys. Res.* 91:12,375–12,387.

Morgan, J. W., and Anders, E. 1980. Chemical composition of the Earth, Venus and Mercury. *Proc. Natl. Acad. Sci. U.S.* 77:6973–6977.

Murthy, V. R., and Hall, H. T. 1972. The origin and chemical composition of the Earth's core. *Phys. Earth Planet. Int.* 6:123–130.

Newsom, H. E., and Drake, M. J. 1987. Formation of the moon and terrestrial planets: Constraints from V, Cr and Mn abundances in planetary mantles and from new partitioning experiments. *Lunar. Planet. Sci.* XVIII:716–717.

Ott, U., and Begemann, F. 1985. Are all the "martian" meteorites from Mars? *Nature* 317:509–513.

Ozima, M., and Podosek, F. A. 1983. *Noble Gas Geochemistry* (Cambridge: Cambridge Univ. Press), 367 pp.

Ringwood, A. E. 1966. Chemical evolution of the terrestrial planets. *Geochim. Cosmochim. Acta* 30:41–104.

Ringwood, A. E. 1979. *Origin of the Earth and Moon* (Berlin: Springer-Verlag).

Ringwood, A. E., and Anderson, D. L. 1977. Earth and Venus-comparative study. *Icarus* 30:243–253.

Smith, J. V. 1979. Mineralogy of the planets: A voyage in space and time. *Mineral Mag.* 43:1–89.

Surkov, Y. A. 1981. Natural radioactivity of the moon and planets. *Proc. Lunar Planet. Sci.* 12:1377–1386.

Surkov, Y. A. 1984. New data on the composition, structure and properties of Venus rock obtained by Venera 13 and 14. *Proc. Lunar Planet. Sci. Conf.* 14, *J. Geophys. Res.* 89:B393–B402.

Taylor, S. R. 1987. The unique lunar composition and its bearing on the origin of the moon. *Geochim. Cosmochim. Acta* 51, in press.

Taylor, S. R., and McLennan, S. M. 1985. *The Continental Crust: Its Composition and Evolution* (Oxford: Blackwell), 312 pp.

Taylor, S. R., and McLennan, S. M. 1987. The significance of the rare earths in geochemistry and cosmochemistry. In *Handbook on the Physics and Chemistry of Rare Earths*, eds. K. A. Gschneidner, Jr., and L. Eyring (Amsterdam: North-Holland) pp. 1297–1309.

Turekian, K. K., and Clark, S. 1969. Inhomogeneous accretion of the Earth from the primitive solar nebula. *Earth Planet. Sci. Lett.* 6:346–348.

Wacker, J. F., and Anders, E. 1984. Trapping of xenon in ice: Implications for the origin of the Earth's noble gases. *Geochim. Cosmochim. Acta* 48:2373–2380.

Wänke, H. 1981. Constitution of the terrestrial planets. *Phil. Trans. Roy. Soc. London* A303:287–302.

Wänke, H., Dreibus, G., and Jagoutz, E. 1984. Mantle chemistry and accretion history of the Earth. In *Archean Geochemistry*, eds. A. Kroner et al. (Berlin: Springer-Verlag), pp. 1–24.

Wänke, H., Dreibus, G., Jagoutz, E., Palme, H., Spettel, B., and Weckwerth, G. 1986. ALHA 77005 and on the chemistry of the Shergotty parent body (Mars). *Lunar Planet. Sci.* XVII:919–920.

Wasson, J. T. 1985. *Meteorites: Their Record of Early Solar-System History* (New York: Freeman), 267 pp.

Wetherill, G. W. 1987. Dynamical relationships between asteroids, meteorites and Apollo-Amor objects. *Phil. Trans. Roy. Soc. London*, in press.

Wood, J. A. 1962. Chondrules and the origin of the terrestrial planets. *Nature* 194:127–130.

Wood, J. A. 1985. Meteoritic constraints on processes in the solar nebula. In *Protostars & Planets II*, eds. D. C. Black and M. S. Matthews (Tucson: Univ. of Arizona Press), pp. 687–702.

Wood, J. A. 1986. High temperatures and chondrule formation in a turbulent shear zone beneath the nebula surface. *Lunar Planet. Sci.* XVII:956–957 (abstract).

Wood, J. A. 1987. Was chondritic material formed during large-scale, protracted nebular evolution or by transient local events in the nebula? *Lunar Planet. Sci.* XVIII:1100–1101 (abstract).

7.9. TRAPPED NOBLE GASES IN METEORITES

TIMOTHY D. SWINDLE
University of Arizona

The most volatile elements in meteorites are the noble (or inert) gases. Although they are far less abundant than condensible elements, they are present in measurable quantities in virtually all meteorites. The trapped noble gases in meteorites come in two main varieties, usually referred to as "solar" and "planetary." The solar noble gases are implanted solar-wind or solar-flare materials, and thus their relative elemental abundances provide a good estimate of those of the Sun. The planetary noble gases have relative elemental abundances similar to those in the terrestrial atmosphere, but there are also important distinctions. At least one other elemental pattern ("subsolar") and several isotopic patterns have also been identified. Understanding the relationship among all these components continues to be a challenging problem.

7.9.1. INTRODUCTION

Early interest in the measurement of noble gases in meteorites came about because they were so rare. This meant that even a relatively inefficient process that produced a noble-gas atom might be detectable. The first applications focused on ^{4}He (from U decay) and ^{40}Ar (from ^{40}K decay), which provided dating schemes. These and other noble-gas components generated *in situ* continue to provide valuable information on the history of the early solar system.

Gerling and Levskii (1956) opened a related field of study when they discovered that the Staroe Pesyanoe aubrite contained noble gases that could not possibly be produced *in situ* (or ascribed to terrestrial contamination). This was followed by Reynolds' (1960) discovery that carbonaceous chondrites contain Xe that is isotopically distinct from that in the Earth's atmosphere. Noble gases such as these are generally referred to as trapped or primordial noble gases, implying that they were incorporated in the meteorite when it formed. Although this is not always true (see Sec. 7.9.3), it is still useful to make the distinction between trapped noble gases and those produced *in situ*.

The study of trapped noble gases has contributed to our knowledge of the early solar system in several ways. First, when studying noble-gas components produced *in situ*, it is necessary to make accurate corrections for the trapped noble gas (see, e.g., Chapters 4.1 and 15.3). Second, the search for explanations of variations in the isotopic composition of trapped components has led to evidence that some presolar grains were probably incorporated into solar-system material without melting (see Chapter 13.1). Third, meteoritic trapped noble gases provide a framework for interpreting variations in the elemental and isotopic compositions of the atmospheres of the terrestrial planets (Chapter 7.10). Finally, of course, we can learn something about what processes resulted in the incorporation of noble gases in solids. This chapter will focus on the last topic, presenting the elemental and isotopic structures present in trapped noble gases and discussing theories and experiments relating to their origin. This chapter is also intended to provide some of the framework used in the other chapters cited above.

The most comprehensive recent review of solar-system noble gases is in Ozima and Podosek (1983), while the state of knowledge about the incorporation of noble gases in solids has been discussed most recently by Zadnik et al. (1985).

7.9.2. RELATIVE ELEMENTAL-ABUNDANCE PATTERNS

Trapped noble gases in meteorites come in two main varieties, usually referred to as "solar" and "planetary" (Signer and Suess 1963; Pepin and Signer 1965). The solar noble gases are so named because their relative elemental abundances are similar to the best estimates of those of the Sun. The planetary noble gases have relative elemental abundances similar to the terrestrial atmosphere; but there are also important distinctions, and it is not at all clear how planetary noble gases, defined in terms of meteorites, relate to noble gases in planets (Chapter 7.10). At least one other elemental pattern ("subsolar") and several isotopic patterns, all of which may or may not be related to the solar or planetary components, have also been identified. Data from samples representative of several elemental abundance patterns are given in Table 7.9.1.

Solar Noble Gases

The similarity between solar noble gases and the Sun can be seen on a plot such as Fig. 7.9.1, where the elemental ratios (normalized to ^{36}Ar) are

TABLE 7.9.1
Elemental Compositional Data

Component	^{36}Ar cm^{3}STP/g	$\dfrac{^4\text{He}}{^{20}\text{Ne}}$	$\dfrac{^{20}\text{Ne}}{^{36}\text{Ar}}$	$\dfrac{^{36}\text{Ar}}{^{84}\text{Kr}}$	$\dfrac{^{84}\text{Kr}}{^{130}\text{Xe}}$	Ref.[a]
Solar						
Bulk solar system	11.9	630	40	3390	145	1
Solar wind (A1 foil)	—	570	28	—	—	2
Lunar soil (12001)	2.25×10^{-4}	260	34	1430	46	3
Pesyanoe	1.36×10^{-6}	390	16	2270	46	4
Planetary						
Allende (CV3)	1.7×10^{-7}	—	0.25	106	6.7	5
Q	4.5×10^{-4}	9.8	0.08	98	8.2	5
X	1.26×10^{-6}	203	4.3	126	4.7	5
Dimmitt (H3)	4.7×10^{-8}	—	4.25	78	4.7	6
Bruderheim (L5)	5.5×10^{-9}	—	—	67	5.1	7
Other						
Earth atmosphere	2.1×10^{-8}	0.27	0.52	48	183	7
Haverö (ureilite)	7.1×10^{-5}	40	0.034	247	17.2	8
S. Oman (subsolar)	7.6×10^{-6}	—	0.003	447	36	9

[a] References: 1: Anders and Ebihara 1982 (amount normalized to 17% Si); 2: Geiss 1973; 3: Eberhardt et al. 1972 (ilmenite, 10.9μm); 4: Marti 1969; 5: Lewis et al. 1975 (3CS4 used for X component); 6: Moniot 1980; 7: Ozima and Podosek 1983, Table 5.2; 8: Weber et al. 1976 ("primordial" component); 9: Crabb and Anders 1981.

compared with the bulk solar-system ratios (Anders and Ebihara 1982). Several things are worth noting. First, the pattern in Pesyanoe is almost identical to that of lunar soils. Both these patterns are much flatter (i.e., less fractionated) than any of the other patterns plotted, which is what led to the name "solar" for this component. Even so, these patterns are not flat; the heavy noble gases are systematically more abundant (less depleted) than the lighter gases. Also, there is a difference in absolute abundances between lunar soils and Pesyanoe (Table 7.9.1), and both have considerably less noble gas than bulk solar-system material.

Solar noble gases are acquired by direct implantation of solar-wind gases (see Sec. 7.9.3 below), which accounts for the similarity with the Sun's elemental pattern. The slight fractionation probably occurs either in the implantation process or by diffusive loss later in the material's history, because Al foils directly exposed to the solar wind during the Apollo missions have a flatter pattern for He, Ne and Ar (Fig. 7.9.1; Geiss 1973). The difference in absolute abundances between Pesyanoe (and other meteorites with solar noble gases) and lunar soils is the result of longer exposures and higher fluxes experienced by lunar soils.

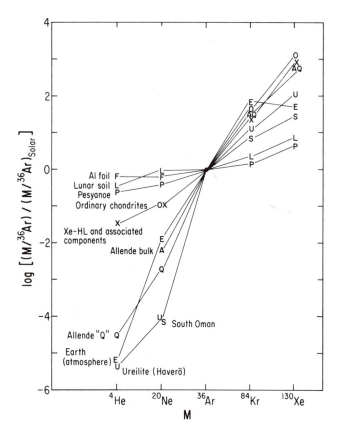

Fig. 7.9.1. Relative elemental abundances of various solar-system noble-gas components, normalized to bulk solar-system abundances (Anders and Ebihara 1982; all data from Table 7.9.1). Points for the ordinary chondrites Bruderheim and Dimmitt (0) are indistinguishable on this scale.

Planetary Noble Gases

The planetary noble-gas pattern, illustrated by data for Allende (CV3), Dimmitt (H3) and Bruderheim (L5) in Fig. 7.9.1, is characterized by a systematic fractionation, with the heavy gases enhanced (less depleted) relative to the lighter gases. There is more variation in the shape of the planetary noble-gas pattern than there is for the solar pattern, which means that planetary noble gases are not a single component, but should probably be considered a family of components. In particular, there are variations in the Ne/Ar ratio (see, e.g., Alaerts et al. 1979a), illustrated in Fig. 7.9.1 by the variation between bulk Allende and the two ordinary chondrites plotted. There is also a definite kink at Ne—the He/Ne ratio (for example, in component X) shows much less fractionation than do any of the ratios involving Ar, Kr or Xe.

Furthermore, the bulk of the planetary Ar, Kr and Xe resides in different mineral sites than the major portion of the planetary Ne, which is associated with isotopically anomalous Ar, Kr and Xe (Sec. 7.9.3). This led to the suggestion (Sabu and Manuel 1980b, and references therein) that the light and heavy noble gases are decoupled, or, more precisely, that there are two planetary noble-gas components, one (called "Y" or "Q") containing most of the Ar, Kr and Xe, and the other ("X" or "Anomalous") containing most of the He and Ne, and some Ar, Kr and Xe with anomalous isotopic compositions. However, the two components must be mixed in remarkably uniform proportion in most meteorites and must have similar relative abundances of Ar, Kr and Xe (Frick 1977). The idea that planetary noble gases are mixtures has gained acceptance, although the specific model that Sabu and Manuel (1980b) advocate is considered unlikely, in part because they argue that no He or Ne accompanies the Q component (but see Sec. 7.9.4 below).

The planetary pattern is similar to that in the terrestrial (or Martian) atmosphere for Ne, Ar and Kr, but He and Xe are much less abundant in the atmosphere of the Earth (Fig. 7.9.1). He and Xe are also depleted in the Martian atmosphere, if the gases in the shergottite EETA 79001 provide a reliable sample of the atmosphere on Mars (see Chapter 7.10). The He depletion is no problem because He escapes from planetary atmospheres. The Xe depletion is more difficult to explain. There are also isotopic differences in Xe, with the terrestrial atmosphere (and the gas in EETA 79001) depleted in light isotopes in a pattern suggesting strong mass-dependent fractionation. It is logical to suspect that the Xe depletion and fractionation are coupled. But any mass-dependent process affecting all the noble gases, which resulted in a depletion of the heaviest gas (Xe) would produce isotopic patterns enhanced in the light isotopes. Few models have attempted to address both these questions. One possibility, discussed in Chapter 7.10, is that a mass-dependent atmospheric process fractionated the Xe and removed virtually all of the other noble gases, then was followed by outgassing from the interior of all the noble gases except Xe. Even if it can be shown that retention of Xe in the core or deep mantle of a planet is possible, the similarity between Earth and Mars (or whatever the SNC parent body is) means that the fractionation and outgassing processes would have had to have operated to the same extent on both bodies. An alternative is that the Xe depletion and isotopic fractionation both occurred before planetary accretion (Ozima and Nakazawa 1980), but no single mechanism that can quantitatively account for both has been suggested (see Sec. 7.9.5).

Unlike the case for solar noble gases, where there is little doubt as to their origin, the mechanism(s) that produced the planetary noble-gas pattern is still a matter of debate. Several theories will be discussed in Sec. 7.9.5.

540   T. D. SWINDLE

Subsolar Noble Gases

Some enstatite chondrites seem to contain both planetary noble gases and noble gases with an elemental pattern that does not match either the solar or planetary pattern. This pattern, which is most pronounced in South Oman (Fig. 7.9.1), is flatter than the planetary noble-gas pattern for the heavy noble gases, which led Crabb and Anders (1981) to call it a "subsolar" pattern. However, the Ne/Ar ratio is lower than either the solar or planetary pattern, so Wacker and Marti (1983) suggested that the component be referred to only as "argon-rich." Since there may be other Ar-rich components that are not related to the component in enstatite chondrites, the term subsolar will be used in this chapter.

7.9.3. LOCATION OF METEORITIC NOBLE GASES

In this section, I will discuss the location of meteoritic noble gases, both on a macroscopic (which meteorites contain what components) and microscopic (where within individual meteorites the noble gases are sited) scale.

Solar Noble Gases

Although it had been noted that the solar noble-gas pattern was similar to the relative elemental abundances in the Sun, the method of incorporation was a matter of debate until Eberhardt et al. (1965a,b) and Hintenberger et al. (1965) demonstrated that solar noble gases are sited on or near the surfaces of grains. Experiments on grain-size separates showed that the amount of solar gas was inversely correlated with grain size and etching experiments on mineral separates showed that most of the gas was removed with the first few percent of the mass, both indicating that the gas was close to the surface. These experiments were taken as confirmation that the solar noble gases were implanted directly by the solar wind, as suggested by Suess et al. (1964). The surface-correlated component in lunar soils subsequently was found to be nearly identical to the meteoritic solar gases in terms of relative elemental abundances, isotopic structures and surface location (see, e.g., Eberhardt et al. 1972). In meteorites, solar noble gases are found in the "gas-rich" meteorites, which also show other signs of direct exposure to the Sun, including radiation-damage tracks from the passage of solar flare particles (see Chapter 4.1).

Planetary Noble Gases

Planetary noble gases are found in greatest abundance in carbonaceous chondrites. Ordinary chondrites also contain planetary noble gases, but the abundance decreases with increasing petrographic type (Table 7.9.1), suggesting that some gas was lost during metamorphism (Zähringer 1968). Metamorphic loss cannot explain the overall fractionation pattern, because the pattern is the same for meteorites that vary in noble-gas content by several orders of magnitude.

Little was known about the siting of planetary noble gases until 1975, and what was known seemed paradoxical. Planetary noble-gas abundances generally correlated with the abundance of volatiles (in particular, C [Otting and Zähringer 1967]), but meteoritic samples generally had to be heated to rather high temperatures to extract the planetary noble gases.

The first real breakthrough came when Lewis et al. (1975) treated a sample of the Allende meteorite with acids (HF/HCl) to remove the major silicate minerals. They found that the C-rich material remaining, representing only about 0.5% of the original mass, contained more than half the original noble gases. A further treatment with HNO_3 (later experiments used other oxidizing reagents, such as H_2O_2, $HClO_4$ or atomic O) removed only 8% of the remaining mass, but more than 90% of the Ar, Kr and Xe. The remaining fraction was isotopically distinct from the original sample. Further experiments demonstrated that this treatment produced similar results for other carbonaceous chondrites and for unequilibrated ordinary chondrites (Matsuda et al. 1980, and references therein). The results of a typical experiment are summarized in Fig. 7.9.2.

Lewis et al. (1975) proposed that most of the planetary noble gases resided in a phase "Q" that was insoluble in HF/HCl but soluble in HNO_3, perhaps an Fe-, Cr-sulfide. On the other hand, Reynolds et al. (1978) argued

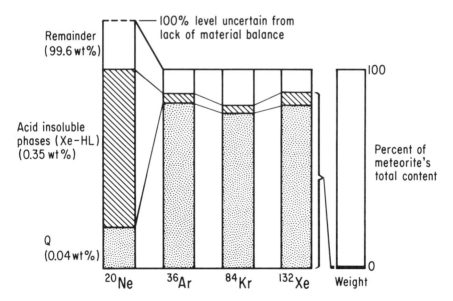

Fig. 7.9.2. Distribution of noble gases among components in a typical acid-dissolution experiment (data from Lewis et al. 1977; figure adapted from Reynolds et al. 1978).

that there was no need to invoke any carrier other than C, and later experiments showed that under various physical and chemical treatments of HF/HCl residues, noble-gas contents do not correlate with Fe, Cr or S content (Ott et al. 1981,1984). There now seems to be general agreement that the planetary noble gases reside in C, in sites that are readily accessible under oxidizing conditions (presumably near the surface) but that are quite retentive during stepwise heating experiments. Although Q turned out not to be a separate phase, the term is still often used in discussing the HF/HCl insoluble, oxidizable sites in C. The noble-gas component residing in those sites will be referred to in this chapter as "Q-type" noble gases. The isotopically anomalous component, in the sites that are not accessible to oxidizing reagents, will be referred to as "X." (Note that the Xe in this component has been given several different names. It is referred to in Chapter 13.1, for example, as "Xe-HL.")

Subsolar Noble Gases

The component referred to as subsolar noble gases has only been studied extensively in enstatite chondrites, but Crabb and Anders (1981) pointed out that there are other meteorites, including CO3s, the more reduced CV3s, and some ureilites (discussed below), aubrites and unequilibrated ordinary chondrites, with Ar/Xe ratios higher than is typical for planetary gases. It has also been suggested that addition of a subsolar component might explain the higher-than-planetary Ar/Xe ratio in the Venusian atmosphere (Crabb and Anders 1981; Wacker and Marti 1983). However, the lower limit to Venus' Ar/Xe ratio is currently lower than the ratio in the Martian or terrestrial atmosphere, so the high Ar/Xe ratio may be related to the more general problem of the depletion of Xe in terrestrial planets (see Sec. 7.9.2 above). If Venus does contain subsolar noble gases, that would mean that this component is found in the most reducing of the three planets so far observed, in the most reduced chondrites (the enstatite chondrites) and some of the most reduced achondrites (aubrites and ureilites). Thus, its origin might be related to the presence of reducing conditions (Crabb and Anders 1981), although it is not clear how this would affect noble-gas acquisition.

Some hints about the siting of the subsolar component in enstatite chondrites have been obtained (Crabb and Anders 1981,1982; Wacker and Marti 1983). In EL chondrites, the subsolar component is associated either with enstatite or with some phase so closely associated with enstatite that acid dissolution and density separations are not sufficient to distinguish it. In EH chondrites, about half the subsolar component seems to be in enstatite. In the other chondrite classes where a component with a high Ar/Xe ratio has been suggested, it is found in the main (silicate) fraction, the HF/HCl solubles. Although an origin involving precompaction solar-wind irradiation was suggested by Wacker and Marti (1983), the subsolar component is not surface correlated and, of course, the Ne/Ar ratio is much lower than would be expected for solar-wind irradiation. Crabb and Anders (1982) have suggested that the subsolar component arises from adsorption of solar noble gases on enstatite, and that the depletion in Ne (and He) is the result of diffusive loss.

Ureilites

It is appropriate at this point to discuss the fascinating ureilite meteorites. They are C-rich igneous achondrites, containing concentrations of heavy noble gases comparable to the carbonaceous chondrites (Göbel et al. 1978). The noble gases are apparently concentrated in the C, in both diamonds (Göbel et al. 1978) and amorphous C grains. The noble gases in the latter sites bear some notable similarities to Q-type planetary gases (Ott et al. 1985b), including their low Ne/Ar ratio (Fig. 7.9.1), Ne, Ar and Xe isotopic signatures (Sec. 7.9.4) and carbonaceous carriers. However, the ureilite gases are not sited near the surfaces of carrier grains (Wacker 1986) and they have a higher Ar/Xe ratio (Fig. 7.9.1). Ureilite noble gases also are similar to the subsolar (enstatite chondrite) gases in their high Ar/Xe ratios, Xe isotopic signature and occurrence in reduced material, but have different Ne and Ar isotopic compositions and a different carrier phase. It is not clear whether the ureilites contain modified planetary or subsolar noble gases, or merely record processes similar to those that affected one or both of the other components (see Ott et al. 1985b; Wacker 1986). Since the ureilites have apparently experienced a complex igneous history (Goodrich et al. 1987), it would not be surprising if the noble-gas elemental and isotopic compositions had been affected. Progress in understanding ureilite noble gases could come on two fronts: increased understanding of the processes that created ureilites, and isolation of the carrier of the Ar-rich component(s) in other meteorite classes.

7.9.4. ISOTOPIC PATTERNS

Isotopic variations often (but not always) go along with the elemental variations. In this section, the isotopic structure of various trapped noble-gas components will be summarized. The order of presentation (Ne, He, Xe, Kr, Ar) is selected because He and Kr isotopic variations are usually interpreted in terms of better-studied Ne and Xe components, respectively, while it is not clear whether the rather subtle Ar isotopic variations are more closely related to the He and Ne or to the Kr and Xe variations.

Neon

Neon has exactly three stable isotopes, so any Ne composition can be conveniently represented on a three-isotope diagram such as Fig. 7.9.3. To first order, meteoritic Ne consists of mixtures of three components: spallation, a component associated with planetary noble-gas elemental patterns, and a component associated with solar gas. The latter two components were given the "temporary" designations of Ne-A and Ne-B by Pepin (1967), who noted

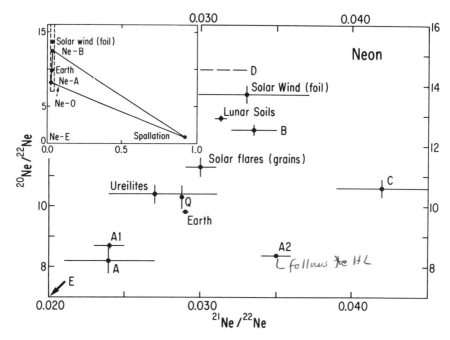

Fig. 7.9.3. Neon-isotopic compositions (from Table 7.9.2). Dashed line on inset shows region plotted in main figure. Note that no ^{21}Ne/^{22}Ne ratio is given for Ne-D (main figure), and that Ne-O (inset) is only required to lie along a line in the direction of the arrow.

that analyses of bulk meteorite samples all fell in a triangle on a plot such as the inset in Fig. 7.9.3 (solid lines).

However, a more detailed study led Black and Pepin (1969) to propose the existence of three more components, and Ne components continue to proliferate (Table 7.9.2), although some of these have been convincingly shown to be equivalent to one another.

Spallation Ne. Spallation Ne is plotted as a single point on Fig. 7.9.3, which is roughly true for bulk meteorite samples. However, the isotopic composition of spallation Ne actually can vary considerably, depending on the elemental composition of target material, the energy spectrum of the incident particles and the shielding depth (Chapter 4.1). In discussing trapped Ne components, the important point to remember is that spallation Ne contains roughly equal amounts of all three isotopes, whereas in most trapped components, ^{21}Ne is much less abundant than the other two isotopes. Thus, the relative amount of ^{21}Ne is much more difficult to determine in trapped components than the ^{20}Ne/^{22}Ne ratio.

Ne-E. Black and Pepin (1969) found that when Ne was extracted from carbonaceous chondrites by stepwise heating (analyzing the gas released at progressively higher temperatures), some extractions fell below the "normal" triangle on a three-isotope plot. They suggested that this was caused by a ^{22}Ne-rich component they designated Ne-E. Later experiments showed that there are actually two components extremely rich in ^{22}Ne, neither of which shows any evidence of ^{20}Ne or ^{21}Ne. This component is generally attributed to decay of ^{22}Ne in presolar grains that were not volatilized during meteorite formation (for a more complete discussion, see Chapter 13.1).

Solar Ne. Pepin (1967) identified Ne-B as solar Ne, but further experiments showed that there are apparently several solar Ne components.

A major source of solar Ne is the solar wind. This has been measured directly by analyzing the foils that were exposed to the solar wind during the Apollo missions. Surprisingly, it does not quite match two other components that are expected to be dominated by solar wind: meteoritic Ne-B and Ne in lunar soils. Possible reasons for the discrepancy include fractionation during implantation in mineral grains and subsequent regolith processing (Eberhardt et al. 1972), secular variations in the Ne isotopic composition of the solar wind (Pepin 1980), or mixing with another component (Black 1972a).

Black (1972a) suggested that one of the other Ne components identified in meteorites, Ne-C, represented solar-flare Ne, and that compositions between Ne-C and solar-wind Ne could be generated by mixing. Wieler et al. (1986, 1987) confirmed the existence of a distinct solar-flare component by analyzing the noble gas released by progressively etching away the surface layers of grains from lunar soils and gas-rich meteorites. Just below the surface, they found a component that they argued was probably what had been identified as Ne-C. In addition, in some etching experiments on lunar grains, the first (surface) Ne liberated matches the isotopic composition of the solar wind (Wieler et al. 1986; Frick and Pepin 1981), although the bulk isotopic compositions are more like Ne-B. Both these observations support Black's mixing hypothesis. However, the source of the difference between the Ne isotopic compositions of the solar wind and solar energetic particles is not understood (cf. Wieler et al. 1986; Mewaldt et al. 1984).

Black and Pepin (1969) also identified a Ne-D, which Black (1972a) attributed to solar wind from the premain-sequence Sun. The existence of a Ne component with a ^{20}Ne/^{22}Ne ratio higher than the modern solar wind is controversial (see, e.g., Hohenberg et al. 1970).

Planetary Ne. Pepin (1967) defined the composition of Ne-A, planetary Ne, as the intersection of a line passing through Ne-B and terrestrial Ne with the lower edge of the normal triangle in Fig. 7.9.3. Dramatic confirmation of the reality of Ne-A as a distinct component came with the discovery that many acid residues contained nearly pure Ne-A (see, e.g., Lewis et al.

1975; Reynolds et al. 1978). However, some acid residues from Murchison, Cold Bokkeveld and Allende are characterized by an isotopic composition richer in ^{21}Ne than Ne-A. This was originally attributed to spallation from a precompaction irradiation (Srinivasan et al. 1977; Reynolds et al. 1978), but more recent evidence has pointed to the existence of another trapped component, Ne-A2, differing from the original Ne-A (Ne-A1) mainly at ^{21}Ne (Alaerts et al. 1980; Ott et al. 1981).

The origin of Ne-A is uncertain. It could be related to solar Ne by severe mass-dependent fractionation (Sabu and Manuel 1980*a*) or mixing with Ne-E (Black 1972*b*), but the near-solar He/Ne ratio is hard to explain in the former scenario, and the concentration of Ne-A in phases distinct from those that carry Ne-E is difficult to reconcile with the latter. Ne-A follows Xe-HL in acid dissolution experiments, so it, like the anomalous heavy noble gases, is most commonly interpreted as a distinct nucleosynthetic component. The relationship of Ne-A1 to Ne-A2 has not been addressed.

The normal (Q) heavy noble gases that are lost in the HNO_3 treatment in acid-dissolution experiments are accompanied by very little Ne. However, etching of HF/HCl residues with HNO_3 usually results in a decrease in the ^{20}Ne/^{22}Ne ratio, suggesting that a component with a higher ^{20}Ne/^{22}Ne ratio than Ne-A is lost. The similarity of the calculated composition of this component (Q-Ne) to Ne-C led Alaerts et al. (1979*b*) to suggest that the carbonaceous chondrites that originally defined Ne-C contained Q-Ne instead of solar-flare Ne. Along the same lines, the near-surface siting of Q-Ne suggests that it might actually be solar-flare Ne (Ne-C). However, there are no other signs of the precompaction irradiation of these grains that would be required (Ott et al. 1981).

At least two other Ne components have been suggested (and named). Ureilite Ne, Ne-U (Ott et al. 1985*a*), is very similar to Ne-C except for a lack of ^{21}Ne, and may be related to Q-Ne (Ott et al. 1985*a,b*). Ne-O, postulated to explain a straight (mixing?) line on a three-isotope plot of some Orgueil samples, is probably unique to Orgueil (Eberhardt 1978), if it can even be considered a distinct component. Neon is very underabundant in the subsolar component in enstatite chondrites (Crabb and Anders 1981), but the isotopic composition is consistent with Ne-A.

Helium

Far fewer distinct isotopic compositions have been defined for He. This is at least partly because there are only two He isotopes, one of which (^{3}He) is often dominated by spallation, and the other of which (^{4}He, the α particle) is often radiogenic. Trapped He components therefore are frequently identified by their correlation with Ne components. There is definitely a difference between solar He (solar wind, lunar soil or He-B) and planetary He (He-A). Black (1972*a*) also identified He-C and He-D, associated with Ne-C and Ne-D, but they are poorly defined and even their existence is controver-

TABLE 7.9.2
Isotopic Compositional Data for He, Ne and Ar

Component	^{3}He/^{4}He ($\times 10^{-4}$)	^{20}Ne/^{22}Ne	^{21}Ne/^{22}Ne	^{36}Ar/^{38}Ar	Ref.[a]
Solar					
B	3.9(3)	12.52(18)	0.0335(15)	5.37(12)	1
Lunar soil (12001)	3.7(1)	12.9(1)	0.0313(4)	5.33(3)	2
Solar wind					
Foils	4.3(3)	13.7(3)	0.033(4)	—	3
71501 surface	—	13.26–13.56	0.032	5.4–5.8	4
D	1.5(1.0)	14.5(1.0)	—	6(1)	1
Solar energetic particles (solar flare)					
Etched grains	2.4	11.3(3)	0.030(1)	5.20–5.35	4,5
C	4.1(8)	10.6(3)	0.042(3)	4.1(8)	1
Planetary					
A	1.43(20)	8.2(4)	0.024(3)	5.31(5)	6
A1	—	8.70(11)	0.024(1)	—	7
A2	—	8.37(3)	0.035(1)	—	8
Q	—	10.3(4)	≡0.029	5.29(4)	9
Other					
Earth atmosphere	0.0140(1)	9.80(8)	0.0290(2)	5.320(13)	10
E(H)	—	<0.2	<0.003	—	11
E(L)	—	<0.01	<0.001	—	11
Orgueil (O)	—	>6.4	—[b]	—	12
Ureilites (U)	<5	10.4(3)	0.027(3)	5.26(6)	13
Subsolar	—	8–9	—	5.46(4)	14

[a] References: 1: Black 1972a; 2: Eberhardt et al. 1972; 3: Geiss 1973; 4: Wieler et al. 1986; 5: Wieler et al. 1987 (He only); 6: Black 1972b (Ne), Reynolds et al. 1978 (He, Ar); 7: Alaerts et al. 1980; 8: Ott et al. 1981; 9: Alaerts et al. 1979b (Ne), Lewis et al. 1975 (Ar); 10: Compilation of Ozima and Podosek 1983 (Table 2.2); 11: Jungck and Eberhardt 1979; 12: Eberhardt 1978; 13: Ott et al. 1985a (He, Ne), Göbel et al. 1978 (Ar); 14: Crabb and Anders 1981.
[b] ^{20}Ne/^{21}Ne = 52 ± 2.

sial. For example, Yaniv and Marti (1981) identified a solar-flare He component (in a lunar sample) with a ^{3}He/^{4}He ratio a factor of 100 higher than Black's He-C, while Wieler et al. (1987) used their etching technique on the meteorite Fayetteville and determined a solar-flare He with a ^{3}He/^{4}He ratio a factor of two lower than He-C. There are also variations in the He-isotopic composition of the solar wind on time scales ranging from hours to billions of years.

The differences among various He-isotopic compositions could be the result of mass-dependent fractionation or incorporation of interstellar grains (like Ne-E), but most can also be explained by processes within the Sun (see review by Geiss and Bochsler 1981). Satellite experiments and the Apollo foil experiments both suggest that the ^{3}He/^{4}He ratio can change on a time scale of hours. Although the satellite data indicate that variations can be as large as an order of magnitude (Ogilvie et al. 1980), they also give an average composition that agrees with the average value (weighted by exposure time) of the foils given in Table 7.9.2. On a longer time scale, lunar breccias (rocks produced by compaction of ancient soils) give ^{3}He/^{4}He ratios about 10 percent lower than those of soils (Eberhardt et al. 1972), suggesting a possible secular change in the ^{3}He/^{4}He ratio of the Sun over the last few Gyr, perhaps indicating a buildup of ^{3}He as result of incomplete H burning (Schatzman 1970; Eberhardt et al. 1970,1972). Finally, Geiss and Reeves (1972) suggested that modern solar He should be enhanced considerably in ^{3}He over primordial solar He, because the Sun would have converted virtually all of its deuterium to ^{3}He during its premain-sequence evolution. If planetary He (He-A) is assumed to be primordial solar He (Black 1972b), the difference between it and modern solar He leads to a calculated primordial D/H ratio of 2×10^{-5} (Geiss and Bochsler 1981), which is astrophysically reasonable.

Xenon

In Situ Components. Xenon has the most stable isotopes, nine, of any of the noble gases, which aids in identifying distinct components, but it also has the most *in situ* components. The most prominent is radiogenic ^{129}Xe from the decay of now-extinct ^{129}I (Chapter 15.3). Fission (e.g., of ^{244}Pu or U) can produce the heavy isotopes (^{131}Xe to ^{136}Xe) and spallation-produced Xe can be significant, particularly at the rare light isotopes. In addition, neutron-capture effects are sometimes important for ^{128}Xe (from ^{127}I) and ^{131}Xe (from ^{130}Ba). Some of these components produce serious complications for certain samples, but trapped components can be accurately determined at most isotopes (^{129}Xe being the most common exception).

Solar Xe. Xenon in the solar wind is not abundant enough to be measured in the foil experiments, but measurements of the surface-correlated component in lunar soils and gas-rich meteorites are generally consistent with one another (Table 7.9.3). Various calculated compositions do have differences that could be related to the compositions given in Table 7.9.3 by mass-dependent fractionation of less than 0.5% per amu (Bogard et al. 1974). Furthermore, lunar breccias frequently contain surface-correlated fission Xe and ^{129}Xe, apparently released from the lunar interior and implanted in lunar soils (see review by Swindle et al. 1986), so ^{129}Xe and the heavy isotopes may be overestimated (Pepin and Phinney 1978). Geiss and Bochsler (1981) argue that as much as 20% of the lunar surface-correlated Xe might not be from the solar wind, but they also conclude that isotopic ratios are not likely to be affected much (^{129}Xe/^{132}Xe might be reduced by 2%, for example).

Planetary Xe. The isotopic composition of planetary Xe was originally thought to be equal to the "average value for carbonaceous chondrites"

TABLE 7.9.3
Isotopic Compositional Data for Xe

Component	^{124}Xe	^{126}Xe	^{128}Xe	^{129}Xe	^{130}Xe	^{131}Xe	^{132}Xe	^{134}Xe	^{136}Xe	Ref.[a]
Solar										
Lunar soil (12001)	2.90(7)	2.59(9)	50.4(3)	635.4(1.7)	≡100	498.8(1.1)	606.2(1.6)	223.9(8)	181.8(6)	1
Pesyanoe (1000°C)	2.97(13)	3.21(13)	50.3(1.1)	630.9(9.4)	≡100	495.2(6.7)	606.1(7.3)	221.2(3.2)	178.8(2.8)	2
Planetary										
AVCC	2.85(5)	2.51(4)	50.7(4)	628.7(2.9)	≡100	504.3(2.8)	615.0(2.7)	235.9(1.3)	198.8(1.2)	3
U-Xe	2.95	2.54	50.9	628.7	≡100	499.6	604.8	212.9	166.3	3
Xe-HL	0.84(8)	0.53(11)	1.7(6)	?	≡0	19.4(3.3)	24.9(1.8)	71.2(8)	≡100	3
Other										
Earth atmosphere	2.337(7)	2.180(11)	47.15(5)	649.6(6)	≡100	521.3(6)	660.7(5)	256.3(4)	217.6(2)	4
Kenna	2.89(3)	2.54(2)	50.8(2)	635.8(1.5)	≡100	502.7(1.0)	613.6(1.1)	231.3(6)	191.6(6)	5
Subsolar (S. Oman)	2.97(10)	2.62(9)	51.1(6)	—	≡100	503.4(3.7)	606.4(3.7)	228.3(2.1)	187.7(1.7)	6

[a] References: 1: Eberhardt et al. 1972; 2: Marti 1969; 3: Pepin and Phinney 1978; 4: Ozima and Podosek 1983, Table 2.2; 5: Wilkening and Marti 1976; 6: Crabb and Anders 1981 (^{129}Xe has a substantial contribution from decay of ^{129}I).

(AVCC) (see Eugster et al. 1967; Pepin and Phinney 1978). Dramatic variations from this composition are observed in acid-treated samples (see, e.g., Lewis et al. 1975), and it is now clear that AVCC is not a distinct component, but rather a mixture of other components. However, whole-rock samples of carbonaceous and ordinary chondrites show only subtle variations from AVCC composition, which means the mixture that goes into planetary Xe must be nearly constant, and so it is still useful to consider AVCC.

There is little, if any, difference between solar and AVCC Xe at the light isotopes, but AVCC is enriched in the heavy isotopes (Table 7.9.3, Fig. 7.9.4). This difference, and the most prominent variations from the AVCC composition, can be represented as additions of a single heavy-isotope component. In addition, there is a unique pattern of light-isotope enrichments that accompanies the heavy-isotope component (Fig. 7.9.4). Variations in the ratio of the light- to the heavy-isotope component are rare, if they are present at all (Schelhaas et al. 1985). The heavy- and light-isotope enriched component(s) has variously been referred to as (1) X; (2) CCF (carbonaceous chondrite fission, since the heavy-isotope pattern resembles a fission-yield spectrum); (3) H (heavy) and L (light) or HL; or (4) DME_H (demineralized, etched heavy) and DME_L. It was the search for a pure sample of this component that led to the acid-dissolution experiments that have proven to be so useful. This search, and the experiments that led to the conclusion that Xe-HL is a distinct nucleosynthetic component, are discussed in more detail in Chapter 13.1.

Xenon in the subsolar component in enstatite chondrites is very similar to solar or planetary Xe at the light isotopes, but seems to contain less of the heavy-isotope component than does AVCC (Fig. 7.9.4). Ureilite Xe matches AVCC except at the heavy isotopes (i.e., it is similar to the Q-type part of planetary Xe) and is also quite similar to the subsolar Xe in South Oman.

The most detailed model of the relationship of Xe-isotopic compositions is that of Pepin and Phinney (1978), based on multidimensional correlations. They concluded that there was a single Xe component, U-Xe, from which solar and AVCC (and other carbonaceous chondrites') Xe were derived by addition of various amounts of H- and L-Xe. Pure U-Xe is hard to find, but Niemeyer and Zaikowski (1980) reported a composition similar to U-Xe in one extraction from an H_2O_2-treated sample of Murray. This model has proven to be a useful framework for discussion of isotopic variations, but it does have some unattractive aspects (Ozima and Podosek 1983). For example, L-Xe must be subtracted from U-Xe to explain some meteorites' compositions, and H-Xe must be present in significant quantities in solar Xe, but not in terrestrial Xe. Furthermore, this model does not try to explain why H-Xe and L-Xe are so closely correlated with each other and with U-Xe in most samples.

Terrestrial Xe, with its apparent strong mass-dependent fractionation (Fig. 7.9.4), is discussed in Chapter 7.10. Two other Xe components that contribute small fractions of the total Xe are discussed in Chapter 13.1. One

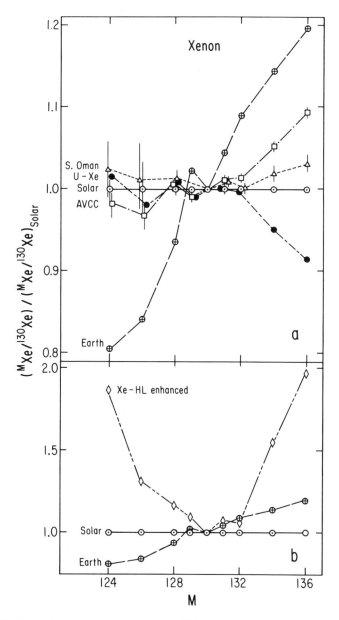

Fig. 7.9.4. Xenon isotopic compositions (from Table 7.9.3), plotted relative to solar values (surface-correlated component in lunar soil 12001). Uncertainties in measurements of solar ratio are plotted as error bars on solar ratio, rather than compounding them with the uncertainties on each of the other individual measurements. "Xe-HL enhanced" sample in (b) is a carbonaceous residue, sample 111, from Frick (1977), which is probably not a pure sample of Xe-HL. Note difference in scale for (b). (See also Fig. 13.1.5.)

is s-Xe, apparently produced by s-process nucleosynthesis, the other is rich in ^{124}Xe, possibly monisotopic.

Krypton

Krypton has nearly as many stable isotopes (seven) as Xe does, and has no *in situ* components as important as radiogenic ^{129}Xe or fission Xe, but observed variations in trapped composition have been far less dramatic for Kr than for Xe (note the vertical scales on Figs. 7.9.4 and 7.9.5). For this reason, and because the isotopic variations seen in Kr usually correlate with Xe variations, Kr is seldom considered, except to constrain models based on Xe. There are some Kr components that are produced *in situ*, the most prominent being spallation (which again has the largest relative effect at the rare light isotopes), fission (which produces primarily ^{86}Kr) and neutron capture (which produces ^{80}Kr and ^{82}Kr from Br).

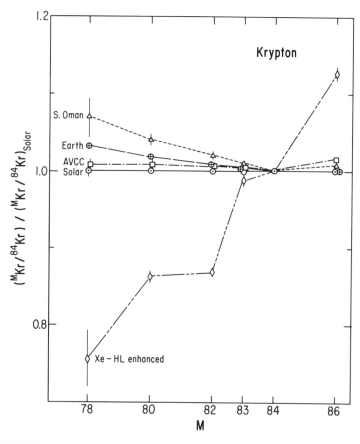

Fig. 7.9.5. Krypton isotopic compositions (from Table 7.9.4), plotted as in Fig. 7.9.4.

TABLE 7.9.4
Isotopic Compositional Data for Kr

Component	^{78}Kr	^{80}Kr	^{82}Kr	^{83}Kr	^{84}Kr	^{86}Kr	Ref.[a]
Solar							
Lunar soil (12001)	0.593(5)	3.89(2)	20.05(8)	20.09(7)	≡100	30.50(7)	1
Planetary							
AVCC	0.597(5)	3.92(3)	20.15(8)	20.17(8)	≡100	30.98(8)	2
Anomalous	0.448(22)	3.35(3)	17.40(9)	19.86(17)	≡100	34.36(31)	3
Other							
Earth atmosphere	0.609(2)	3.960(2)	20.22(1)	20.14(2)	≡100	30.52(3)	4
Ureilites (Kenna)	0.601(8)	3.959(14)	20.27(5)	20.22(5)	≡100	30.94(3)	5
Subsolar (S. Oman)	0.634(16)	4.05(3)	20.45(9)	20.27(7)	≡100	30.73(17)	6

[a] References: 1: Eberhardt et al. 1972; 2: Eugster et al. 1967; 3: Frick 1977 (Allende 111, enhanced in anomalous Kr and Xe-HL, but not a pure sample of these components); 4: Compilation in Ozima and Podosek 1983, Table 2.2; 5: Wilkening and Marti 1976; 6: Crabb and Anders 1981.

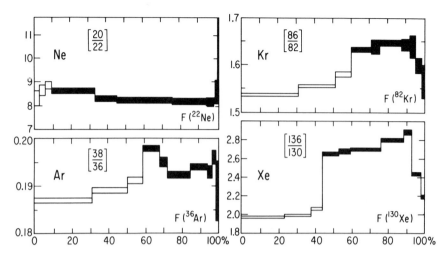

Fig. 7.9.6. Selected isotopic ratios plotted vs cumulative release fraction of denominator for stepwise heating experiment on a carbonaceous residue from Orgueil (adapted from Frick and Moniot 1977). Vertical width of bars denotes uncertainty. Solid bars represent extractions at temperatures of 900° C or higher, which clearly contain a component with isotopically distinct Ar, Kr and Xe.

Solar and planetary Kr (again represented by an AVCC composition) differ little, if at all, except at the heaviest isotope, ^{86}Kr, which is enriched by 1 to 2% in AVCC Kr (Table 7.9.4, Fig. 7.9.5). The same isotope is enriched in Kr accompanying the anomalous Xe in acid-dissolution experiments (Fig. 7.9.6). However, the heavy-isotope enrichment is accompanied by a depletion in light isotopes (Fig. 7.9.5), instead of the enrichment in light isotopes seen in Xe. Subsolar (enstatite chondrite) and terrestrial Kr may be enriched in the light isotopes, fractionated by about 1.0% and 0.5% per amu, respectively (Fig. 7.9.5). As with Xe (see above), calculated solar Kr compositions vary by about 0.5% per amu from the composition given in Table 7.9.4.

Argon

Argon has three stable isotopes, but one of them, ^{40}Ar, is almost always dominated by decay of ^{40}K (see Chapter 5.3). Furthermore, the ^{40}Ar/^{36}Ar ratio in the terrestrial atmosphere (295.5) is higher than in most suggested meteoritic trapped components by a factor of 1000 or more, so that contamination is a potentially severe problem. In the ureilite Dyalpur, the ^{40}Ar/^{36}Ar ratio is $(2.9 \pm 1.7) \times 10^{-4}$ (Göbel et al. 1978), about six orders of magnitude smaller than on Earth. The solar ^{40}Ar/^{36}Ar ratio might be even lower (Anders and Ebihara 1982), but measurement is difficult. Many lunar samples have ^{40}Ar/^{36}Ar ratios of one or more, reflecting the presence of "orphan" ^{40}Ar, similar to the excess surface-correlated Xe. Even the lunar samples with the lowest $^{40}Ar/^{36}Ar$ ratios probably contain substantial orphan ^{40}Ar (Eberhardt et al. 1972).

The $^{36}Ar/^{38}Ar$ ratio seems like a promising place to look for variations among trapped components, particularly since cosmic-ray-induced effects are generally smaller than for Ne (or He). Such variations are surprisingly hard to find (Table 7.9.2). Meteorites rich in either planetary or solar gas generally have $^{36}Ar/^{38}Ar$ ratios of 5.3 to 5.4, indistinguishable from each other or the terrestrial atmosphere. This may be partly coincidental, since two etching experiments have resulted in Ar with a higher $^{36}Ar/^{38}Ar$ ratio, 5.4 to 5.8, being released with Ne that is isotopically consistent with unfractionated solar wind (Frick and Pepin 1981; Wieler et al. 1986). Black (1972a) proposed Ar components related to Ne components C and D, but Ar-C, attributed to solar flares, does not match the lunar-soil measurements of Wieler et al. (1986), which give $^{36}Ar/^{38}Ar$ ratios of 5.20 to 5.35, very similar to normal Ar.

There is also evidence for variations associated with the heavy noble gases. In stepwise heating on etched acid residues, Ar with a low $^{36}Ar/^{38}Ar$ ratio, 4.8 to 5.1, accompanies Xe-HL and the isotopically anomalous Kr (Fig. 7.9.6; see also Lewis et al. 1975). The subsolar component in South Oman seems to be isotopically lighter than solar or planetary Ar ($^{36}Ar/^{38}Ar$ = 5.46 ± 0.04), and is associated with Kr and Xe that may be isotopically light. Ureilites have indistinguishable $^{36}Ar/^{38}Ar$ ratios of 5.26 ± 0.06 (Göbel et al. 1978) despite variable Ar/Xe ratios, some of which would seem to indicate that a substantial fraction of the Ar might be from the subsolar component.

Summary

If planetary noble gases were derived from solar noble gases, first-order comparison of Ar, Kr and Xe compositions suggests that the process was accompanied by little, if any, isotopic fractionation. There is, however, evidence for addition of one or more components (e.g., heavy Xe, ^{86}Kr), plausibly from unmelted interstellar grains (Chapter 13.1). Therefore, it is logical to attribute Ne-isotopic variations as well to mixing, rather than mass-dependent fractionation (although the larger mass difference for Ne isotopes makes it more likely that some components are related to others by mass fractionation).

The conclusion that there is little fractionation in Ar, Kr and Xe may have to be altered if the solar noble gas in meteorites and lunar soils is a mixture of isotopically distinct solar-wind and solar-flare gas. This possibility is strongly suggested by the Ne data (Black 1972a; Wieler et al. 1986), and the Ar data of Wieler et al. (1986) and Frick and Pepin (1981) could be interpreted in the same way. If this interpretation is correct, which composition is most representative of the bulk Sun: the solar wind, the solar flares, or the more easily observed time-averaged sum? If it is not the sum, then there may have been some isotopic fractionation associated with the formation of the planetary component (Pepin 1986).

The subsolar component in enstatite chondrites is poorly defined isotopically, but may be related to the solar component by mass-dependent fractionation that enhanced the lighter isotopes. The Ne- and Ar-isotopic differences between ureilites and the subsolar component suggests that there may be a family of Ar-rich components, not just one.

7.9.5. THE ORIGIN OF PLANETARY NOBLE GASES

It took less than 10 yr from the time of the discovery of solar noble gases to explain convincingly their origin and method of incorporation (Sec. 7.9.3). On the other hand, 25 yr after the discovery of planetary noble gases, there is much that is still unknown or controversial. Some workers argue that all of the planetary noble gases were derived from solar noble gases (Sabu and Manuel 1980a,b), others that the planetary gases were trapped from a molecular cloud by interstellar grains (Huss and Alexander 1987; see also Yang and Anders 1982; Ott et al. 1981). However, components such as Ne-E are impossible to account for in the first scenario, and the second requires the assumption that the isotopic composition of the surface of the Sun (solar wind, solar flares) is not representative of the bulk solar system. Therefore, it is generally assumed that the bulk of the planetary noble gases (the Q-type gases) were derived from solar noble gases, but that some of gases (including Ne-E and the gases associated with Xe-HL) were incorporated from unmelted interstellar material. The origin of these "foreign" components is discussed in Chapter 13.1. This section will discuss the portion that might be derived from solar noble gases.

Any model that derives Q-type planetary noble gases from solar noble gases must explain (1) the absolute amounts of each noble gas, and the fractionated relative elemental-abundance pattern, (2) the retentive near-surface siting of planetary noble gases, and (3) the largely unfractionated isotopic patterns. Several models are discussed below.

Adsorption

The most promising model, or at least the model that has received the most attention recently (Zadnik et al. 1985; Pepin 1986; Chapter 7.10), involves adsorption. Fanale and Cannon (1972) demonstrated that adsorption at very low temperatures (about 100 K) on materials with large surface areas might be able to account for the abundance of planetary gases. Furthermore, adsorption also seems capable of producing the elemental fractionation necessary to explain the planetary noble-gas pattern, with a variety of different substrates doing the adsorbing (Wacker 1987; Yang and Anders 1982, and references therein; Fig. 7.10.1). However, adsorption as proposed by Fanale and Cannon (1972) seems unlikely to result in the observed retentive siting of planetary gases (without postulating some effective fixing mechanism) and the required temperature seems unreasonably low.

Later experiments have shown that adsorption of noble gases is more complicated than had been realized, leading to unexpected properties of the adsorbed gas. For example, Niemeyer and Leich (1976) crushed a lunar sample in room atmosphere and found that most of the atmospheric noble-gas contamination was released at temperatures above 1000° C. Also, since adsorption of noble gases is most efficient for Xe, and is progressively less efficient for the lighter gases, the contaminant gas did not have atmospheric relative elemental abundances. Evidence of terrestrial atmospheric contamination was also found in 700 to 1000° C extractions of Allende samples by Srinivasan et al. (1978), again suggesting that adsorption at low (room) temperatures can lead to gas that is retentively sited. This was demonstrated more directly by Wacker et al. (1985), who found that carbon black that had been exposed to Xe at temperatures of about 100° C acquired some gas that was retained during months of exposure to vacuum at the same temperature but was released when a few percent of the mass of the sample was removed by an oxidizing acid (see also Yang et al. 1982). Exposure at higher temperatures led to gas that was retained until comparably high extraction temperatures. These properties provide a remarkably good match to the properties of Q-type planetary noble gas in meteorites (Zadnik et al. 1985).

What is the nature of the adsorption mechanism that can yield noble gases in sites that are thermally retentive, but susceptible to etching of surface material? Since amorphous C is known to contain micropores of atomic dimensions, Wacker et al. (1985) suggest that those properties are the result either of diffusion into internal labyrinths of micropores, from which a random walk to the exterior is extremely slow (Podosek et al. 1981), or "activated entry," where there are constricted pores connecting internal and external sites, which a given atom has only a low probability of traversing in any single encounter. If the solar-system noble gases diffused into C grains that already contained the Xe-HL and the associated Ar and Kr, the Q-type gases would be nearly inseparable from Xe-HL and associated gases, explaining the relative constancy of the AVCC composition (Zadnik et al. 1985). Furthermore, (normal) Ne might be able to diffuse throughout the entire grain, while the normal heavier gases remained concentrated in the near-surface regions, which would account for the correlation of the bulk of the planetary Ne with Xe-HL and associated Ar and Kr components.

The major remaining problem for this adsorption model is the absolute abundance of noble gases: the best samples of Wacker et al. (1985) fall short by about six orders of magnitude. Zadnik et al. (1985) proposed several reasons why the laboratory samples might have adsorbed less gas than their meteoritic counterparts, including failure to reach equilibrium in the laboratory experiments and the posibility that meteoritic material might have more active adsorption sites because of structural disorder or radiation damage. None of these proposals has yet been tested.

Other Models

Although adsorption into (more properly, onto internal surfaces of) C is a promising candidate explanation for the properties of Q-type planetary noble gases, it still should be considered no more than the leading candidate, particularly because of the abundance problem. A number of other models have been proposed to account for the origin of planetary noble gases, and one (or more) of these might turn out to be correct in some situations. To first order, each of these models can explain at least some aspects of planetary noble gases. However, to higher order each of these models has serious drawbacks. The same, of course, can be said of the adsorption model described above. Theories for the origin of planetary noble gases have been critically reviewed by Ozima and Podosek (1983), so only the main features (and major drawbacks) of these models will be presented.

The first proposal (before the isotopic composition of solar noble gas was known) was that planetary gas was the residue of gravitational escape from large bodies (Suess 1949). However, the predicted isotopic fractionations are not observed, and gravitational escape would usually produce much larger elemental fractionations than are observed. Thus this model has largely been abandoned.

Another possibility is that, as minerals formed in the solar nebula, noble gases were incorporated either as a result of solubility or adsorption followed by occlusion during grain growth. The results from synthesizing enstatite (Kirsten 1968) did not even give the right direction of elemental fractionation. Serpentine (Zaikowski and Schaeffer 1979) did better, but the distribution coefficients (the fraction of the ambient gas that ended up in the sample) were seven or eight orders of magnitude too low to account for the abundances of noble gases in carbonaceous chondrites. The absolute abundance was also a problem for enstatite, magnetite and other minerals (see Yang and Anders 1982, and references therein). Carbonaceous materials showed more promise (Frick et al. 1979; Niemeyer and Marti 1981; Dziczkaniec et al. 1981) but still generally fell six or more orders of magnitude short of accounting for the planetary gases. Furthermore, the trapping in C is probably dominated by adsorption anyway (Niemeyer and Marti 1981).

Yet another model suggests that the elemental fractionation is the result of loss of neutral species from a partially ionized gas (Jokipii 1964; Arrhenius and Alfvén 1971; Hostetler 1981). In this "ambipolar diffusion" scenario, the fractionation is the result of differences in first ionization potential. As first noted by Weber et al. (1971), elemental depletions for ureilites seem to correlate with first ionization potential (Fig. 7.9.7). However, Fig. 7.9.7 also shows that there is not such a correlation for the planetary noble gases in Allende or ordinary chondrites, the subsolar noble gases in South Oman or

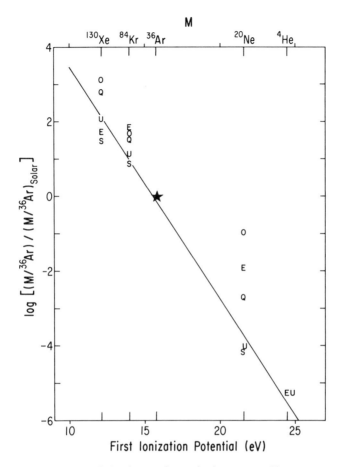

Fig. 7.9.7. Relative elemental abundances of several solar-system noble-gas components (symbols as in Fig. 7.9.1) plotted vs first ionization potential (from tabulation by Ozima and Podosek 1983). The line is a least-squares fit to ureilite data.

the terrestrial atmosphere (when Jokipii proposed this mechanism, the then-accepted values for atmospheric and solar-system noble-gas abundances did give a correlation). Also, this process leads to lower partial pressures of the noble gases, so an incorporation efficiency even greater than that required for adsorption or solubility models is demanded. Implantation of ionized gas might be a suitable mechanism, although low-energy ion implantation leads to isotopic fractionation (see Bernatowicz and Hagee 1987).

At least two other models that attempt to account for the noble-gas elemental patterns in planets probably do not apply to meteorites, but the difference between the terrestrial atmosphere and "planetary" elemental and isotopic structures for the heavy noble gases suggest that this may not be a bad feature. One of these models, by Sill and Wilkening (1978), has Ar, Kr and Xe dissolved in a clathrate of methane in water. This would require equilibrium in the solar nebula down to temperatures of about 80 K, far lower than meteorite parent bodies are believed to have experienced, but plausible for comets, whose impacts could then have transferred the dissolved gas to a planetary atmosphere. Another source would be required for Ne. Another model, by Ozima and Nakazawa (1980) invokes diffusional equilibrium of each noble gas in rather porous planetesimals several 100 km in radius, followed by trapping of the interior gas as the planetesimal grew. This model requires larger bodies than the meteorite parent bodies are believed to have been. Even as an explanation for planetary atmospheres, it requires a sharply peaked distribution in size. Both models have trouble reproducing the low Xe/Kr ratio of the terrestrial atmosphere without invoking a hiding place for much of the Xe.

7.9.6. FUTURE WORK

From the foregoing discussion, it should be apparent that there are models for most aspects of trapped noble gases that are at least plausible, if not proven. However, there are still some intriguing questions, even about the solar component that has been fairly well understood for more than 20 yr. Some of the major questions that should be amenable to testing in the near future are the following:

Solar Gases

The major area of interest in solar noble gases is in the isotopic relationship of surface-correlated gases, solar wind, solar flares and the bulk solar system. Solar wind and solar flares seem to have different Ne compositions (Table 7.9.2). The data for Ar are less clear, with Wieler et al. (1986) arguing that the variation is no more than 3%, and Frick and Pepin (1981) arguing for at least a 5 to 10% variation. No comparable measurements of the less abundant heavier gases (Kr and Xe) have yet been made. If the bulk solar system's isotopic composition for Ar, Kr and Xe differs substantially from the surface-correlated gases in lunar and meteoritic samples, then models of the origin of planetary and atmospheric noble-gas structures will require mechanisms that do produce isotopic fractionation (see, e.g., Pepin 1986).

Planetary Noble Gases

Adsorption models (particularly the model of Zadnik et al. 1985) seem to come close to explaining the properties of Q-type planetary noble gases, but there are questions that need to be answered before some of these long-standing problems can be considered solved. First, it must be verified that the retentive adsorption observed by Zadnik et al. (1985) results in the desired isotopic behavior. However, the crucial question is whether this form of adsorption, at the temperatures required to produce the desired elemental-fractionation behavior (Wacker 1987), can be demonstrated to be efficient enough to account for the observed abundance of planetary noble gases. The origin and siting of Xe-HL continue to be an area of active research (Chapter 13.1).

Subsolar Noble Gases

The major question about the proposed subsolar component is whether the Ar-rich component(s) seen in other meteorites is the same as the component seen in enstatite chondrites. Many ureilites have Ar-rich elemental abundances, but the isotopic structures and siting seem more closely related to Q-type planetary noble gases. Thus there may be a variety of Ar-rich components, each unique to a particular class of meteorite, or a family of related Ar-rich components. Separation of the carrier phase of the Ar-rich component in other meteorites (e.g., CO, reduced CV) would be a major step toward answering this question, particularly if detailed isotopic compositions could be obtained.

Acknowledgments. I gratefully acknowledge a thorough and thoroughly beneficial review by E. Anders. This work was supported by a grant from the National Aeronautics and Space Administration.

REFERENCES

Alaerts, L., Lewis, R. S., and Anders, E. 1979a. Isotopic anomalies of noble gases in meteorites and their origins—III. LL-chondrites. *Geochim. Cosmochim. Acta* 43:1399–1415.

Alaerts, L., Lewis, R. S., and Anders, E. 1979b. Isotopic anomalies of noble gases in meteorites and their origins—IV. C3 (Ornans) carbonaceous chondrites. *Geochim. Cosmochim. Acta* 43:1421–1432.

Alaerts, L., Lewis, R. S., Matsuda, J.-I., and Anders, E. 1980. Isotopic anomalies of noble gases in meteorites and their origins—VI. Presolar components in the Murchison C2 chondrite. *Geochim. Cosmochim. Acta* 44:189–209.

Anders, E., and Ebihara, M. 1982. Solar system abundances of the elements. *Geochim. Cosmochim. Acta* 46:2363–2380.

Arrhenius, G., and Alfvén, H. 1971. Fractionation and condensation in space. *Earth Planet. Sci. Lett.* 10:253–267.

Bernatowicz, T. J., and Hagee, B. E. 1987. Isotopic fractionation of Kr and Xe implanted in solids at very low energies. *Geochim. Cosmochim. Acta* 51:1599–1611.

Black, D. C. 1972a. On the origins of trapped helium, neon and argon isotopic variations in meteorites—I. Gas-rich meteorites, lunar soil and breccias. *Geochim. Cosmochim. Acta* 36:347–375.

Black, D. C. 1972b. On the origins of trapped helium, neon and argon isotopic variations in meteorites—II. Carbonaceous meteorites. *Geochim. Cosmochim. Acta* 36:377–394.

Black, D. C., and Pepin, R. O. 1969. Trapped neon in meteorites—II. *Earth Planet. Sci. Lett.* 6:395–405.

Bogard, D. D., Hirsch, W. C., and Nyquist, L. E. 1974. Noble gases in Apollo 17 fines: Mass fractionation effects in trapped Xe and Kr. *Proc. Lunar Sci. Conf.* 5:1975–2003.

Crabb, J., and Anders, E. 1981. Noble gases in E-chondrites. *Geochim. Cosmochim. Acta* 45:2443–2464.

Crabb, J., and Anders, E. 1982. On the siting of noble gases in E-chondrites. *Geochim. Cosmochim. Acta* 46:2351–2361.

Dziczkaniec, M., Lumpkin, G. R., Donohoe, K., and Chang, S. 1981. Plasma synthesis of carbonaceous material with noble gas tracers. *Lunar Planet. Sci.* XII:246–248 (abstract).

Eberhardt, P. 1978. A neon-E rich phase in Orgueil: Results of stepwise heating experiments. *Proc. Lunar Planet. Sci. Conf.* 9:1027–1051.

Eberhardt, P., Geiss, J., and Grögler, N. 1965a. Ueber die Verteilung der Uredelgase in Meteoriten Khor Temiki. *Mineral Petrog. Mitt.* 10:535–551.

Eberhardt, P., Geiss, J., and Grögler, N. 1965b. Further evidence on the origin of trapped gases in the meteorite Khor Temiki. *J. Geophys. Res.* 70:4375–4378.

Eberhardt, P., Geiss, J., Graf, H., Grögler, N., Krähenbühl, U., Schwaller, H., Schwarzmüller, J., and Stettler, A. 1970. Trapped solar wind noble gases, exposure age and K/Ar-age in Apollo 11 lunar fine material. *Proc. Apollo 11 Lunar Sci. Conf.*, pp. 1037–1070.

Eberhardt, P., Geiss, J., Graf, H., Grögler, N., Mendia, M. D., Mörgeli, M., Schwaller, H., Stettler, A., Krähenbühl, U. and von Gunten, H. R. 1972. Trapped solar wind noble gases in Apollo 12 lunar fines 12001 and Apollo 11 breccia 10046. *Proc. Lunar Sci. Conf.* 3:1821–1856.

Eugster, O., Eberhardt, P., and Geiss, J. 1967. Krypton and xenon isotopic composition in three carbonaceous chondrites. *Earth Planet. Sci. Lett.* 3:249–257.

Fanale, F. P., and Cannon, W. A. 1972. Origin of planetary primordial rare gas: The possible role of adsorption. *Geochim. Cosmochim. Acta* 36:319–328.

Frick, U. 1977. Anomalous krypton in the Allende meteorite. *Proc. Lunar Sci. Conf.* 8:273–292.

Frick, U., and Moniot, R. K. 1977. Planetary noble gas components in Orgueil. *Proc. Lunar Sci. Conf.* 8:229–261.

Frick, U., and Pepin, R. O. 1981. Study of solar wind gases in a young lunar soil. *Lunar Planet. Sci.* XII:303–305 (abstract).

Frick, U., Mack, R., and Chang, S. 1979. Noble gas trapping and fractionation during synthesis of carbonaceous matter. *Proc. Lunar Planet. Sci. Conf.* 10:1961–1973.

Geiss, J. 1973. Solar wind composition and implications about the history of the solar system. *13th Intl. Cosmic Ray Conf.* 5:3375–3398.

Geiss, J., and Bochsler, P. 1981. On the abundances of rare ions in the solar wind. In *Solar Wind Four*, ed. H. Rosenbauer, (Garching: Max-Planck Institut), pp. 403–414.

Geiss, J., and Reeves, H. 1972. Cosmic and solar system abundances of deuterium and helium-3. *Astron. Astrophys.* 18:126–132.

Gerling, E. K., and Levskii, L. K. 1956. On the origin of the rare gases in stony meteorites. *Doklady Akad. Nauk SSSR (Geochemistry)* 110:750.

Göbel, R., Ott, U., and Begemann, F. 1978. On trapped noble gases in ureilites. *J. Geophys. Res.* 83:855–867.

Goodrich, C. A., Jones, J. H., and Berkley, J. L. 1987. Origin and evolution of the ureilite parent magmas: Multi-stage igneous activity on a large parent body. *Geochim. Cosmochim. Acta* 51:2255–2273.

Hintenberger, H., Vilcsek, E., and Wänke, H. 1965. Über die Isotopenzusammensetzung und über den Sitz der leichten Uredelgase in Steinmeteoriten. *Z. Naturforsch.* 20a:939–945.

Hohenberg, C. M., Davis, P. K., Kaiser, W. A., Lewis, R. S., and Reynolds, J. H. 1970. Trapped and cosmogenic rare gases from stepwise heating of Apollo 11 samples. *Proc. Apollo 11 Lunar Sci. Conf.*, pp. 1283–1309.

Hostetler, C. J. 1981. A possible common origin for the rare gases on Venus, Earth, and Mars. *Proc. Lunar Planet. Sci. Conf.* 12:1387–1393.

Huss, G. R., and Alexander, E. C. 1987. On the presolar origin of the "normal planetary" noble gas component in meteorites. *Proc. Lunar Planet. Sci. Conf.* 17, *J. Geophys. Res. Suppl.* 92:E710–E716.

Jokipii, J. R. 1964. The distribution of gases in the protoplanetary nebula. *Icarus* 3:248–252.

Jungck, M., and Eberhardt, P. 1979. Neon-E in Orgueil density separates. *Meteoritics* 14:439–441 (abstract).

Kirsten, T. 1968. Incorporation of rare gases in solidifying enstatite melts. *J. Geophys. Res.* 73:2807–2810.

Lewis, R. S., Srinivasan, B., and Anders, E. 1975. Host phase of a strange xenon component in Allende. *Science* 190:1251–1262.

Lewis, R. S., Gros, J., and Anders, E. 1977. Isotopic anomalies of noble gases in meteorites and their origins—2. Separated minerals from Allende. *J. Geophys. Res.* 82:779–792.

Marti, K. 1969. Solar-type xenon: A new isotopic composition of xenon in the Pesyanoe meteorite. *Science* 166:1263–1265.

Matsuda, J.-I., Lewis, R. S., Takahashi, H., and Anders, E. 1980. Isotopic anomalies of noble gases in meteorites and their origins—VII. C3V carbonaceous chondrites. *Geochim. Cosmochim. Acta* 44:1861–1874.

Mewaldt, R. A., Spalding, J. D., and Stone, E. C. 1984. A high-resolution study of the isotopes of solar flare nuclei. *Astrophys. J.* 280:892–901.

Moniot, R. K. 1980. Noble-gas-rich separates from ordinary chondrites. *Geochim. Cosmochim. Acta* 44:253–271.

Niemeyer, S., and Leich, D. A. 1976. Atmospheric rare gases in lunar rock 60015. *Proc. Lunar Sci. Conf.* 7:587–597.

Niemeyer, S., and Marti, K. 1981. Noble gas trapping by laboratory carbon condensates. *Proc. Lunar Planet. Sci. Conf.* 12:1177–1188.

Niemeyer, S., and Zaikowski, A. 1980. I-Xe age and trapped Xe components of the Murray (C-2) chondrite. *Earth Planet. Sci. Lett.* 48:335–347.

Ogilvie, K. W., Coplan, M. A., Bochsler, P., and Geiss, J. 1980. Abundance ratios of $^4He^{++}/^3He^{++}$ in the solar wind. *J. Geophys. Res.* 85:6021–6024.

Ott, U., Mack, R., and Chang, S. 1981. Noble-gas-rich separates from the Allende meteorite. *Geochim. Cosmochim. Acta* 45:1751–1788.

Ott, U., Kronenbitter, J., Flores, J., and Chang, S. 1984. Colloidally separated samples from Allende residues: Noble gases, carbon and an ESCA study. *Geochim. Cosmochim. Acta* 48:267–280.

Ott, U., Löhr, H. P., and Begemann, F. 1985*a*. Trapped neon in ureilites—A new component. In *Isotopic ratios in the Solar System* (Toulouse: Centre National d'Etudes Spatiales, Cepadues-Editions), pp. 129–136.

Ott, U., Löhr, H. P., and Begemann, F. 1985*b*. Trapped noble gases in 5 more ureilites and the possible role of Q. *Lunar Planet. Sci.* XVI:639–640 (abstract).

Otting, W., and Zähringer, J. 1967. Total carbon content and primordial rare gases in chondrites. *Geochim. Cosmochim. Acta* 31:1949–1960.

Ozima, M., and Nakazawa, K. 1980. Origin of rare gases in the earth. *Nature* 284:313–316.

Ozima, M., and Podosek, F. A. 1983. *Noble Gas Geochemistry.* (Cambridge: Cambridge Univ. Press).

Pepin, R. O. 1967. Trapped neon in meteorites. *Earth Planet. Sci. Lett.* 2:13–18.

Pepin, R. O. 1980. Rare gases in the past and present solar wind. In *Proc. Conf. Ancient Sun,* eds. R. O. Pepin, J. A. Eddy, and R. B. Merrill (New York: Pergamon Press), pp. 411–421.

Pepin, R. O. 1986. A model for the origin of noble gas distributions in meteorites and planetary atmospheres. *Meteoritics* 21:483–484 (abstract).

Pepin, R. O. and Phinney, D. 1978. Components of xenon in the solar system. Preprint.

Pepin, R. O., and Signer, P. 1965. Primordial rare gases in meteorites. *Science* 149:253–265.

Podosek, F. A., Bernatowicz, T. J., and Kramer, F. E. 1981. Adsorption of xenon and krypton on shales. *Geochim. Cosmochim. Acta* 45:2401–2415.

Reynolds, J. H. 1960. Isotopic composition of primordial xenon. *Phys. Rev. Lett.* 4:351–354.

Reynolds, J. H., Frick, U., Neil, J. M., and Phinney, D. L. 1978. Rare-gas-rich separates from carbonaceous chondrites. *Geochim. Cosmochim. Acta* 42:1775–1797.

Sabu, D. D., and Manuel, O. K. 1980*a*. The neon alphabet game. *Proc. Lunar Planet. Sci. Conf.* 11:879–899.

Sabu, D. D., and Manuel, O. K. 1980*b*. Noble gas anomalies and synthesis of the chemical elements. *Meteoritics* 15:117–138.

Schatzman, E. 1970. CERN lecture notes.

Schelhaas, N., Ott, U., and Begemann, F. 1985. Trapped noble gases in some type 3 chondrites. *Meteoritics* 20:753 (abstract).

Signer, P., and Suess, H. E. 1963. Rare gases in the sun, in the atmosphere, and in meteorites. In *Earth Science and Meteorites,* eds. J. Geiss and E. D. Goldberg (Amsterdam: North-Holland), pp. 241–272.

Sill, G. T., and Wilkening, L. L. 1978. Ice clathrate as a possible source of the atmospheres of the terrestrial planets. *Icarus* 33:13–22.

Srinivasan, B., Gros, J., and Anders, E. 1977. Noble gases in separated meteoritic minerals: Murchison (C2), Ornans (C3), Karoonda (C5), and Abee (E4). *J. Geophys. Res.* 82:762–778.

Srinivasan, B., Lewis, R. S., and Anders, E. 1978. Noble gases in the Allende and Abee meteorites and a gas-rich mineral fraction: Investigation by stepwise heating. *Geochim. Cosmochim. Acta* 42:183–198.

Suess, H. E. 1949. Die Häufigkeit der Edelgase auf der Erde und im Kosmos. *J. Geol.* 57:600–607.

Suess, H. E., Wänke, H., and Wlotzka, F. 1964. On the origin of gas-rich meteorites. *Geochim. Cosmochim. Acta* 28:595–607.

Swindle, T. D., Caffee, M. W., Hohenberg, C. M., and Taylor, S. R. 1986. I-Pu-Xe dating and the relative ages of the Earth and Moon. In *Origin of the Moon*, eds. W. K. Hartmann, R. J. Phillips, and G. J. Taylor (Houston: Lunar and Planetary Inst.) pp. 331–358.

Wacker, J. F. 1986. Noble gases in the diamond-free ureilite, ALHA 78019: The roles of shock and nebular processes. *Geochim. Cosmochim. Acta* 50:633–642.

Wacker, J. F. 1987. Laboratory simulation of meteoritic noble gases: Sorption of Ne, Ar, Kr, and Xe on carbon. In *Papers Presented to the 50th Meeting of the Meteoritical Society*, Newcastle-upon-Tyne, England, p. 180 (abstract).

Wacker, J. F., and Marti, K. 1983. Noble gas components in clasts and separates of the Abee meteorite. *Earth Planet. Sci. Lett.* 62:147–158.

Wacker, J. F., Zadnik, M. G., and Anders, E. 1985. Laboratory simulation of meteoritic noble gases. I. Sorption of xenon on carbon: Trapping experiments. *Geochim. Cosmochim. Acta* 49:1035–1048.

Weber, H. W., Hintenberger, H., and Begemann, F. 1971. Noble gases in the Haverö ureilite. *Earth Planet. Sci. Lett.* 13:205–209.

Weber, H. W., Begemann, F., and Hintenberger, H. 1976. Primordial gases in graphite-diamond-kamacite inclusions from the Haverö ureilite. *Earth Planet. Sci. Lett.* 29:81–90.

Wieler, R., Baur, H., and Signer, P. 1986. Noble gases from solar energetic particles revealed by closed system stepwise etching of lunar soil materials. *Geochim. Cosmochim. Acta* 50:1997–2019.

Wieler, R., Baur, H., Benkert, J. P., Pedroni, A., and Signer, P. 1987. Noble gases in the meteorite Fayetteville and in lunar ilmenite originating from solar energetic particles. *Lunar Planet. Sci.* XVIII:1080–1081 (abstract).

Wilkening, L. L., and Marti, K. 1976. Rare gases and fossil particle tracks in the Kenna ureilite. *Geochim. Cosmochim. Acta* 40:1465–1473.

Yang, J., and Anders, E. 1982. Sorption of noble gases by solids, with reference to meteorites. III. Sulfides, spinels, and other substances; on the origin of planetary gases. *Geochim. Cosmochim. Acta* 46:877–892.

Yang, J., Lewis, R. S., and Anders, E. 1982. Sorption of noble gases by solids, with reference to meteorites. I. Magnetite and carbon. *Geochim. Cosmochim. Acta* 46:841–860.

Yaniv, A., and Marti, K. 1981. Detection of stopped solar flare helium in lunar rock 68815. *Astrophys. J.* 247:L143–L146.

Zadnik, M. G., Wacker, J. F., and Lewis, R. S. 1985. Laboratory simulation of meteoritic noble gases. II. Sorption of xenon on carbon: Etching and heating experiments. *Geochim. Cosmochim. Acta* 49:1049–1059.

Zähringer, J. 1968. Rare gases in stony meteorites. *Geochim. Cosmochim. Acta* 32:209–237.

Zaikowski, A., and Schaeffer, O. A. 1979. Solubility of noble gases in serpentine: Implications for meteoritic noble gas abundances. *Earth Planet. Sci. Lett.* 45:141–154.

7.10. PLANETARY ATMOSPHERES

D. M. HUNTEN
University of Arizona

R. O. PEPIN
University of Minnesota

and

T. C. OWEN
State University of New York at Stony Brook

Loss of gases from the terrestrial planets is discussed in the context of meteorites and their parent bodies. The present inventories of C, N and noble gases are discussed and tabulated for these bodies and for icy ones as well. Gases trapped in SNC meteorites are adopted as representing Mars. The principal loss processes (large impact, thermal escape and nonthermal escape) are outlined, with emphasis on recent developments such as diffusion limits and fractionation in hydrodynamic escape. A series of case studies consider the effects of hydrodynamic escape, and D and noble gases on Venus.

7.10.1. INTRODUCTION

Although planetary atmospheres are not the topic of this book, they have some important relationships to it. One component of the noble gases found in meteorites is named "planetary" because of its (perhaps misleading) resemblance to the Earth's atmosphere (Sec. 7.10.2 and Chapter 7.9). For the same reason, it has often been suggested that most of the volatiles found on planets were accreted late in the formation process from volatile-rich meteorites, with some contribution from comets. Regardless of the validity of this particular idea, the planets must have accreted out of smaller bodies, and the volatiles offer particularly useful clues to this process because their inventories can often be established with some accuracy. Finally, fractionation processes such as escape that could determine proportions and isotope ratios in meteoritic gases have received most of their study in the context of planets.

The next section discusses the data base for the inner planets and the small objects. Section 7.10.3 is devoted to evolutionary processes such as impacts and escape. Applications to the three major terrestrial planets are given in Sec. 7.10.4.

Hydrogen loss, both thermal and nonthermal, from the present terrestrial planets has been reviewed by Hunten and Donahue (1976). The entire gamut of known nonthermal processes is discussed by Hunten (1982). Mars and Earth are compared by Anders and Owen (1977), and Venus, Earth and Mars by Pollack and Yung (1980) and Donahue and Pollack (1983). Thermal escape from planetesimals (Donahue 1986) and hydrodynamic escape (Hunten et al. 1987) are reviewed here for the first time. More general books, containing relevant material, are Walker (1977) and Lewis and Prinn (1984).

7.10.2. VOLATILE INVENTORIES: INNER PLANETS AND SMALL BODIES

Atmospheric Composition

In this chapter, we are concerned primarily with the extent to which we can relate the gases that planetary atmospheres now contain (or used to contain) to the gases found in meteorites and/or comets. Thus we confine ourselves to C, N and the noble gases in our compilation (Table 7.10.1). Isotope ratios are discussed in Sec. 7.10.4. In considering the data in Table 7.10.1, it is important to remember that we can give reasonably accurate estimates only for the quantities of volatiles in the atmosphere and crust of the Earth, and for just the atmospheres of Mars and Venus. Mars in particular may harbor a large amount of gas in its regolith (Fanale 1976), while all three planets could have significant amounts of C and N sequestered in their interiors.

The primary sources of information about the composition of the atmosphere of Venus are the Pioneer Venus Mission (Hoffman et al. 1980) and the Venera 13 and 14 Missions (Istomin et al. 1982; Mukhin et al. 1982). Subsequent treatments of the former data set have led to the discovery that D is enriched on Venus by a factor 100 compared with Earth (McElroy et al. 1982; Donahue et al. 1982).

In the case of Mars, measurements of noble gases in the SNC meteorites may provide a better data base than the original Viking measurements. Otherwise, the basic references for direct analyses of atmospheric composition are Nier and McElroy (1977), McElroy et al. (1977) and Owen et al. (1977).

TABLE 7.10.1
Abundances of C, N and Noble Gases in the Solar System (g/g-Solar Composition), Two Classes of Volatile-Rich Meteorites (g/g-Meteorite), and Terrestrial Planet Atmospheres (g/g-Planet)

	^{12}C	^{14}N	^{20}Ne	^{36}Ar	^{84}Kr	^{130}Xe
[a]SOLAR SYSTEM	$3.90 \pm 0.03(-03)$[g]	$9.42 \pm 0.03(-04)$	$2.24 \pm 0.25(-03)$	$8.97 \pm 0.36(-05)$	$5.84 \pm 0.11(-08)$	$8.07 \pm 1.36(-10)$
METEORITES						
[b]CI Carbonaceous chondrites	$3.70 \pm 0.70(-02)$	$1.51 \pm 0.31(-03)$	$2.89 \pm 0.77(-10)$	$1.25 \pm 0.10(-09)$	$3.57 \pm 0.15(-11)$	$7.0 \pm 1.9(-12)$
[c]Enstatite chondrites	$1.50 - 7.00(-03)$	$1.33 - 9.46(-04)$	$\sim 0 - 6.80(-11)$	$1.47 - 1220(-11)$	$5.35 - 639(-13)$	$1.05 - 27.8(-13)$
PLANETARY ATMOSPHERES						
[d]Venus	$2.60 \pm 0.04(-05)$	$2.20 \pm 0.50(-06)$	$2.9 \pm 1.3 (-10)$	$2.51 \pm 0.97(-09)$	$4.72^{+0.57}_{-3.40}(-12)$	$8.9^{+2.5}_{-6.8}(-14)$
[e]Earth	$1.50 \pm 0.40(-05)$	$1.45 \pm 0.55(-06)$	$1.00 \pm 0.01(-11)$	$3.45 \pm 0.01(-11)$	$1.66 \pm 0.02(-12)$	$1.40 \pm 0.02(-14)$
[f]Mars	$1.11 \pm 0.20(-08)$	$7.30 \pm 1.90(-10)$	$4.38 \pm 0.74(-14)$	$2.16 \pm 0.55(-13)$	$1.76 \pm 0.28(-14)$	$2.08 \pm 0.41(-16)$

[a] Average of Anders and Ebihara (1982) and Cameron (1982), except for Ne; Ne from ^{36}Ar and $^{20}Ne/^{36}Ar \simeq 45$ in the solar wind (Bochsler and Geiss 1977).
[b] Mazor et al. (1970) for noble gases; Kerridge (1985) for C and N.
[c] Crabb and Anders (1981) for noble gases; Grady et al. (1986) and Kung and Clayton (1978) for C and N.
[d] von Zahn et al. (1983) for C, N, Ne and Ar; Donahue and Pollack (1983) for Kr; Donahue (1986) for Xe.
[e] Donahue and Pollack (1983) for C and N (minimum outgassed inventories, their Table IV).
[f] Owen et al. (1977) for C and N; Hunten et al. 1987 and references therein) for SNC noble-gas data.
[g] Power of ten multipliers in parentheses.

The first two refer to measurements from the Viking entry mass spectrometers, the third to results from the gas chromatograph mass spectrometer on each of the landers. The only new direct measurement is the recent discovery of HDO in the Martian atmosphere by Owen et al. (1987) from ground-based spectroscopy. The analysis of this observation is still in progress at this writing but an enrichment of D/H on Mars compared with SMOW on Earth seems evident.

Meteorites

Atmophilic elements and compounds trapped in meteorites are of specific interest in studies of planetary-atmospheric evolution because they may have contributed directly to atmospheric inventories by accretion of bodies similar to contemporary meteorite populations during planetary growth. Although usually assumed to derive from a late-accreting veneer (e.g., see, Anders and Owen 1977, for Mars and Earth), meteoritic volatiles liberated by impact degassing at earlier stages could have been released to form a co-accreting atmosphere once the planet had grown large enough (Tyburczy et al. 1986). There may also be more indirect relationships between meteoritic and atmospheric volatiles. Major components of both reservoirs were quite possibly taken up from the early solar nebula by adsorption on nebular dust grains that later formed planetary bodies, and by gravitational accretion of nebular gas on large protoplanets. If some of the fractionating mechanisms that operated in the processing of meteoritic and atmospheric volatiles from a common initial nebular source to their contemporary compositions were similar, comparative study of these offers clues to the nature of the mechanisms and the environments in which they occurred. Fractionation in hydrodynamic escape, discussed below, may be an important example.

In general, volatiles in meteorites may derive from a variety of different sources:

1. Nuclear spallation or fission reactions, induced by galactic cosmic-ray or solar-particle irradiation of meteoritic bodies in space (Chapter 4.1);
2. *In-situ* natural radioactive decay of certain extant (long-lived) or extinct (short-lived) radionuclides (Chapter 15.1);
3. Implantation of solar-wind and solar-flare ions into meteoritic grain surfaces exposed to solar-particle radiation on parent-body regoliths or in space (Chapter 4.1);
4. The primordial solar nebula (Chapter 7.9);
5. Meteoritic grains of extrasolar origin, carrying species sampled from presolar reservoirs (Chapter 13.1).

The processes responsible for the first three of these components are for the most part straightforward and firmly established. However, the bulk of the gases in the most volatile-rich meteorites, the carbonaceous and enstatite chondrites, are not due to spallation, radioactive decay or direct trapping of solar wind or flares. They must in some way have been incorporated into carrier grains from gases in the nebula or in prenebular environments. The physical and chemical processes that generated the mass distributions of volatiles observed in these meteorites from ambient gases in the nebula or elsewhere are poorly understood. It appears likely that, for nebular gases, the processing included mass fractionation of elements and isotopes, and gas adsorption on the surfaces of nebular dust (see Chapter 7.9).

Most of the heavy noble gases—Ar, Kr and Xe—in the carbonaceous chondrites are probably of "local" origin, incorporated from the gas phase of the early solar nebula. But major fractions of He and Ne, and particular components of the heavier gases, appear rather clearly to derive from prenebular environments enriched in the products of stellar nuclear processing. Since these gases are to a certain extent separable and thus unmixed with other components, they must have been carried into the solar system, and eventually into meteorite assemblages, in presolar grains that escaped volatilization and gas-phase homogenization in the early nebula. The detection, characteristics and possible origins of these "exotic" species, including C and N as well as noble gases, are discussed in Chapter 13.1. Here, for the reasons given above, we are mainly interested in bulk compositions in volatile-rich meteorites, and in the "local" component for evidence regarding the composition of the nebular gases that may have contributed to atmospheric inventories directly.

Abundances of noble gases, N and C in CI and enstatite chondrites, and in the Sun and terrestrial planets, are set out in Table 7.10.1. Helium is omitted since it is only weakly bound gravitationally and escapes thermally from the atmospheres of these planets. Relative-abundance patterns for the noble gases, normalized to the solar composition, are compared in Fig. 7.10.1, where we have included data from other classes of carbonaceous meteorites. The rough similarity of the meteoritic and planetary patterns has fostered the view that planets acquired their volatiles primarily from meteorite infall, and was historically responsible for an unfortunate choice of terminology—the generic designation of meteoritic gases that display the CI-type of pattern as the "planetary" component. This nomenclature has tended to obscure the fact that profound compositional differences exist between the meteoritic and planetary reservoirs. One example, evident in Fig. 7.10.1, is the relative abundance of Xe to Kr. The thesis that the Xe deficit for Earth is due to removal of terrestrial atmospheric Xe by adsorption on sediments, and subsequent sequestering in shales and slates in the geologic column, is becoming difficult to defend. Xenon abundances in sedimentary rocks are far too low to make up the deficiency [Bernatowicz et al. 1985], and in any case such a process is unlikely to have depleted atmospheric Xe in three very different environments to the similar extents implied by the relatively tight grouping of

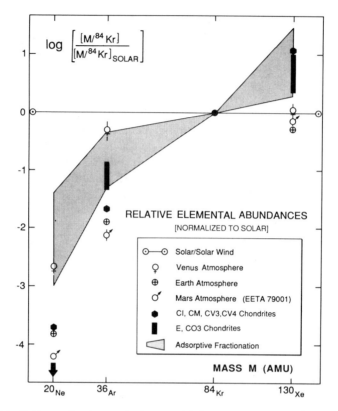

Fig. 7.10.1. Noble-gas abundance ratios in terrestrial-planet atmospheres and volatile-rich meteorites, plotted with respect to solar relative abundances. Data from Table 7.10.1, and from Mazor et al. (1970), Srinivasan et al. (1977), Alaerts et al. (1979) and Matsuda et al. (1980) for CM, CV3, CV4 and CO3 chondrites. Elemental ratios of noble gases adsorbed on various substrates in laboratory experiments on meteorite powder (Fanale and Cannon 1972), on carbonaceous materials (Frick et al. 1979; Niemeyer and Marti 1981; Wacker and Anders 1986), and on other solids (Yang et al. 1982; Yang and Anders 1982a,b) tend to fall within the shaded area of the figure when plotted relative to ambient gas-phase compositions. Several exceptions to this general adsorptive-fractionation trend have been reported, e.g., Ne/Kr up to ∼1 by Wacker and Anders (1986).

the planetary Xe/Kr ratios in Figure 7.10.1. The isotopic evidence is even more compelling, particularly for the heaviest gases: relevant data for Venus are lacking, but Kr and Xe compositions for Earth, Mars and the carbonaceous chondrites are all distinctly different in ways that point clearly to mass fractionation (Owen et al. 1976; Pepin and Phinney 1978; Pepin 1988).

The most likely message of Fig. 7.10.1, and the associated isotopic patterns, would seem to be that the noble gases now in meteorites and planets were derived from the solar composition, in part separately, through different degrees of processing by the same kinds of fractionating mechanisms. End states might then be crudely similar but would not be expected to be identical. A general overprint of elemental fractionation is clear in Fig. 7.10.1, although some anomalies resist this interpretation. A process that could have generated part of the fractionation, in carriers of both meteoritic and planetary volatiles, is adsorption of nebular gas on dust. Laboratory demonstrations of adsorptive fractionation, on carbonaceous and other substrates, are shown in the shaded region of Fig. 7.10.1. This origin is consistent with what is known about the local component in carbonaceous chondrites (named Q by the Chicago group): it is located on surfaces of carbonaceous grains—or more precisely in sites from which it is easily removed by oxidation (see Chapter 7.9)—and its relative abundance pattern, as noted earlier, is essentially identical to the bulk CI composition in Fig. 7.10.1 except for Ne/Kr (which is only about 15% of the bulk ratio [Alaerts et al. 1980]). The trends in Fig. 7.10.1 are clearly appropriate for adsorptive fractionation, but there remains one major unsolved problem: laboratory gas-solid partition coefficients for single-stage adsorption are many orders of magnitude too small, assuming

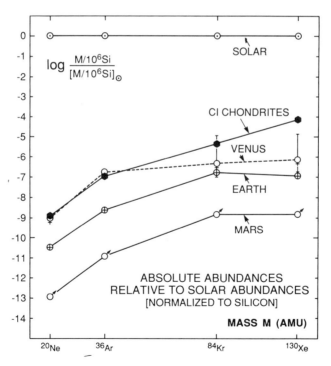

Fig. 7.10.2. Depletions of absolute noble-gas abundances in terrestrial-planet atmospheres (atoms per 10^6 Si atoms-planet) and CI carbonaceous chondrites with respect to solar noble-gas/Si ratios. Noble-gas data from Table 7.10.1.

reasonable nebular gas pressures of 10^{-4} to 10^{-6} bar, to account for the amounts of the Q-component in meteoritic carrier phases (see also Chapter 7.9). Multistage adsorption on shocked or growing nebular dust grains may provide a way out of this difficulty (Mazor et al. 1970; Fanale and Cannon 1972).

Although gas loadings of meteoritic carriers appear very high relative to these laboratory measures of adsorption, noble gases in meteorites and planets are actually grossly depleted with respect to solar abundances. Absolute abundances, normalized to Si, are plotted in Fig. 7.10.2. Depletions in the planetary-atmospheric inventories range from 6 to 9 orders of magnitude for Kr and Xe up to 9 to 13 orders of magnitude for Ne. Planet-to-planet comparisons are especially interesting. Argon, for example, is lower on the Earth than on Venus by a factor of about 70, and is further depleted on Mars relative to Earth by a factor of 200—just opposite to the trend in volatile content that one might expect with increasing distance from the Sun (Lewis 1973).

A particularly vexing problem that must be faced in trying to account for such apparent trends is the uncertainty in deciding which set of facts really tells us something meaningful and which is simply a coincidence. For example, the relatively small mass of Mars and its proximity to the asteroid belt may be more significant than its distance from the Sun in determining the evolution of its atmosphere, as we shall see in Sec. 7.10.3. The impact process that delivers volatiles to the planets can also remove them (Lange and Ahrens 1982), and in the case of Mars removal may have been particularly important.

The SNC Meteorites

The eight stones of the SNC group, named after Shergotty, Nakhla and Chassigny, the type specimens for the three subclasses, are oddities among meteorites. They seem, on petrologic, chemical and isotopic grounds, to differ distinctly from other classes of stony meteorites; they are radically young for meteorites (their chronology is complicated, but they probably crystallized from melts within the past 1.5 Gyr; see also Chapter 5.3); and they are most readily understood as products of differentiation within a large, thermally active planet. Indirect arguments for their origin on Mars had been advanced earlier (Wasson and Wetherill 1979; Wood and Ashwal 1981), but it was a discovery by Bogard and Johnson (Bogard 1982; Bogard and Johnson 1983) that provided the first direct and quantitative support for the hypothesis. They found that glassy nodules embedded in the basaltic matrix of the EETA 79001 shergottite contain a noble-gas component unique to meteorites, that resembled in striking detail the compositional pattern measured by the Viking spacecraft for noble gases in the Martian atmosphere. Work in several laboratories has since demonstrated that relative abundances and isotopic compositions of the noble gases, N, and CO_2 trapped in the EETA 79001 glass are completely consistent with Viking data. The EETA 79001 and Viking data sets are tabulated and compared in Hunten et al. (1987). These comparisons constitute a compelling geochemical case for the origin of EETA 79001 and its SNC relatives on Mars, although a contrary view has been expressed by Pellas (1987) and it is clear that important dynamical and statistical questions remain to be resolved (Pepin 1985; Pellas 1987; Vickery and Melosh 1987).

Accepting that they are in fact Martian derivatives, the importance of the SNC meteorites in the context of this chapter lies in the laboratory data they have yielded on the elemental and isotopic abundances of noble gases in the atmosphere of Mars that were measured with much lower precision, or not at all, by Viking. The data for Mars in Table 7.10.1 are hybrid: CO_2 and N_2 abundances from Viking, noble-gas abundances from EETA 79001. It is important to note that the correspondence of the EETA 79001 noble-gas data with nominal values from Viking (Hunten et al. 1987) is such that the abundances in Figs. 7.10.1 and 7.10.2 represent both data sets. If Viking data alone were used, the plotted points would simply grow large error bars. This is not the case with noble-gas isotopic compositions, where Viking data, except for $^{40}Ar/^{36}Ar$ and $^{129}Xe/^{132}Xe$, are nondiagnostic or absent.

The Moon

The surface of the Moon, unshielded by an atmosphere of significant stopping power and thus exposed directly to solar corpuscular radiation, is a natural laboratory for study of solar-wind and flare atoms implanted into regolith dust, and for investigation of the dynamics of a ballistic atmosphere. Fractionating processes abound in the regolith, the regolith-solar radiation interface, and the tenuous lunar atmosphere. Solar-wind ions, initially implanted to depths of a few 10^2 Å in exposed grains, are mobilized and fractionated by thermal pulses from meteorite impacts that crater the surface at scales ranging from less than 1 μm to a few 10^2 km. Diffusive loss from impact-heated materials preferentially depletes the lighter isotopes in retained gases and enriches them in the transient atmosphere. Ion sputtering has a similar effect, and can preferentially accelerate lighter species to lunar-escape velocities. Atmospheric atoms suffer interesting fates in a model proposed by Manka and Michel (1971). Eventually photoionized by solar ultraviolet radiation, about 60% of them are then accelerated by the solar-wind $\mathbf{v} \times \mathbf{B}$ electric field along trajectories that miss the surface; these ions are entrained in the solar-wind flow past the Moon and are lost. To the extent that the atmosphere is enriched in lighter species thermally diffusing from surface materials, this leads to a proportionally higher loss of these species from the lunar system. The remaining $\sim$ 40% of the accelerated atmospheric ions impact the Moon and are shallowly implanted and thus retrapped in grain surfaces. Ionization times are comparable to the rotation period of the Moon, and so atoms supplied to the atmosphere, say, from a local impact event, are pumped back into the surface on a global scale.

These and other fractionating processes operating through lunar history in, on and above the regolith have made it extraordinarily difficult to extract, with confidence, unaltered solar-wind compositions from lunar-soil data (Pepin and Phinney 1978). Signer et al. (1977) have pointed out the advantages of using pure mineral grains separated from the bulk soil for this purpose. Produced mostly by micrometeorite spalling of local rocks, these grains have been exposed to the solar wind in the upper regolith for times short enough to escape extensive thermal processing by impact. Research groups in Zürich and Minneapolis are actively pursuing this approach, for both solar-flare (Wieler et al. 1986) and solar-wind (Frick et al. 1987) compositions. In the latter work, separated grains of lunar ilmenite have yielded an elemental composition for solar-wind noble gases that agrees very well with the Anders and Ebihara (1982) solar composition and even better with Cameron's (1982) estimates. In considering implantation of gas from the early solar wind in preplanetary material as a possible source of noble gases for planetary atmospheres, the abundances in these grains should be a useful guide.

Comet Nuclei and Icy Planetesimals

Ever since the idea gained currency that the volatiles that ultimately formed planetary atmospheres might have been brought to planetary surfaces by late-accreting matter, comets have frequently been suggested as important contributors. A continuing problem with this hypothesis has been the absence of detailed information about the composition of comet nuclei. In particular, the important clues provided by the relative abundances of the noble gases and their isotopes have been totally lacking, since no noble gases have been detected in cometary spectra, nor have they been found by the mass-spectrometric analyses of gases from Giacobini-Zinner (Ogilvie et al. 1986) or Halley (Krankowsky et al. 1986).

Similarly, attempts to calculate the number of comets and the mass of cometary material that impacted the inner planets during the final stages of their formation have not led to definitive estimates (see, e.g., Fernández and Ip 1983). Current theories for formation of the outer planets and the scattering of residual planetesimals (cometesimals) into the Oort Cloud suggest that significant bombardment of the forming inner planets by icy material must have occurred (Safronov 1969; Shoemaker and Wolfe 1982, 1984). Thus it is appropriate to continue to try to evaluate this potential source of volatiles.

Past efforts to approach this problem with theoretical models have been hindered by the absence of experimental data on the trapping of gases by ice at low temperature. Sill and Wilkening (1978) specifically invoked clathrate hydrates as the gas-carrying form of ice, following an earlier suggestion by Miller (1961). Although the physical chemistry of clathrate hydrates lends itself to predictive calculations (see, e.g., Lunine and Stevenson 1985), it is not yet obvious that ice would actually assume this form in the conditions existing in the outer solar nebula.

However, it now seems that this question is moot, in that laboratory experiments have demonstrated that amorphous ice formed at appropriately low temperatures and pressures behaves as a microporous solid that can trap very large quantities of gas (Bar-Nun et al. 1987; Mayer and Pletzel 1986). Some highlights of this work are the difficulty in trapping H_2, Ne and He (temperatures below 25 K are required) and the apparent absence of fractionation of those gases that are trapped (this point is still being studied with increasing varieties of gas mixtures). Thus in principle there seems to be a way here to discriminate against three cosmically common volatiles in a carrier that could be expected to contribute to the atmospheres of the inner planets.

In addition to the need for more laboratory work, however, the potential role of ices in bringing in volatiles requires testing by direct measurement of the volatile inventories of comets. At this writing, there is still no general consensus among the Halley investigators about the elemental composition of the gas and dust examined by their instruments. Preliminary analysis of the measurements of D/H in the coma of Comet Halley was reported as $5 \times 10^{-5} < D/H < 5 \times 10^{-4}$ (Eberhardt et al. 1987). This range certainly brackets the terrestrial value of 1.6×10^{-4} for the oceans, and the range of uncertainty may be reduced. However, meteoritic values of D/H in this range are also prevalent; hence even an exact coincidence between Halley and terrestrial ocean water would not constitute proof that the oceans were produced by cometary impact. Perhaps the most important result for our purposes is the confirmation of previous groundbased investigations of comets that indicated a value of C/N closer to solar than planetary in both gas and dust (Delsemme 1982,1985; Jessberger et al. 1986). If this result holds up to continued analysis of the Halley data, it essentially rules out comets as the *major* source of volatiles in the inner-planet atmospheres, unless a plausible mechanism for depleting N by a similar factor on all three planets can be found. Nevertheless, comets must have made some contribution; we still have the problem of determining how significant this contribution was. This problem is made more difficult by the fact that the primitive meteorites *by themselves* could supply all the volatiles (see, e.g., Anders and Owen 1977).

7.10.3. EVOLUTION OF ATMOSPHERES

Although a considerable number of evolutionary processes have been studied, this discussion is limited to those that can noticeably affect the composition of an entire planet (or preplanetary object). Processes like photosynthesis and burial of reduced C are therefore excluded. Moreover, not all escape processes are of equal interest, because some of them are inherently unable to handle large quantities of material. A substantial change of composition can only occur if most of the original reservoir is lost, even for a fully efficient fractionation. For the more usual case of low efficiency, even more must be lost. As an example, to double a D/H ratio, no more than half the original H can remain; if some of the D is also removed, as is the case in most physical processes, the fraction of H remaining must be very much less than half. For significant evolution, a process of limited power must act on a small reservoir or act for a long time, or both.

Of the mechanisms discussed below, two stand out in their ability to handle large quantities of gas. The rate of hydrodynamic escape is limited only by energy input. Escape (by any mechanism) from large numbers of small bodies benefits from the large total area.

Small and Large Impacts

For the present purpose, a *small* impact is one whose immediate effect on the atmosphere is essentially local and does not extend far beyond the horizon; a *large* impact disrupts, heats and partially vaporizes the entire planet. No attempt will be made to define a boundary; there may even be an entire intermediate region. There is little doubt that a projectile of diameter less than the atmospheric scale height is in the small category. Confining themselves to the direct effect on the atmosphere, Cameron (1983) argues that the quantity of atmosphere ejected by such impacts is small, and Walker (1986) finds that this quantity is approximately equal to the content of the volume swept out by the projectile, as long as the escape velocity of the planet is 10 km/s or less. Although they stress a different part of the process, Ahrens and O'Keefe (1987) find a similar result. With such a low efficiency, it is necessary to balance the lost material against the volatiles brought in by the projectile. Lange and Ahrens (1982) have studied this for water vapor; other evolutionary models based on these results remain to be worked out.

The last few years have seen a surge of interest in formation of the Moon as the outcome of a large impact (Hartmann and Davis 1975; Cameron and Ward 1976; Cameron 1985; Benz et al. 1986, 1987; Melosh and Sonett 1986; Stevenson 1987). In principle, one of the attractive features is the likelihood that large quantities of volatiles are lost from the vaporized and molten material that forms the Moon. However, this is no more than a matter of plausibility; there are no quantitative studies, and an investigation of the effect on the atmosphere should probably await more work on the evolution of the event itself. The effect on the proto-Earth is even more difficult to evaluate, although a considerable loss of volatiles could occur.

An attractive feature of this mechanism for atmospheric loss is its ability to account for the thinness of the Martian atmosphere, as Cameron (1983) has pointed out. Even when one allows for the escape of N (McElroy 1972) and appropriate sequestration of carbon dioxide, CO_2, the reconstructed volatile inventory of Mars is much smaller (per g of planet) than the inventories of

Earth or Venus (McElroy et al. 1977; Anders and Owen 1977). Cameron (1983) suggests that the small final mass of Mars and its proximity to the asteroid belt are factors that will increase the vulnerability of the planet to massive atmospheric loss from impact and the duration of time (during the first Gyr of the solar system) when such impacts could occur.

Some support for this idea can be garnered from the present ratios of $^{40}Ar/^{36}Ar$ on Mars (~ 3000) and Earth (~ 300). The radiogenic Ar on Mars is about a factor 10 less abundant per g of planet than on the Earth, which can be understood from the weaker tectonic activity and hence lower degassing experienced by that smaller planet. But to account for the hundredfold smaller amount of ^{36}Ar on Mars, Anders and Owen (1977) had to invoke a lower initial volatile content, by simple analogy with the Moon (as another small body depleted in volatiles). In fact, owing to its greater distance from the Sun in the solar nebula, one could reasonably expect Mars to form with a *larger* proportion of volatiles than either Venus or Earth (Lewis 1972). A late-occurring massive impact on Mars that stripped off the contemporary atmosphere neatly solves this problem. In this view, the present atmosphere arose from the veneer that accreted subsequent to that event, smaller in mass relative to the planet itself than the veneers accumulated by the Earth and Venus over longer periods of time. Volatiles that had not yet outgassed from the Martian interior could also add to this tertiary atmosphere. But such arguments, while plausible, still lack quantitative verification by a detailed model for the impact event(s). The thorniest issue is the duration of the high-temperature phase of the event, which could be rather short. There is little doubt that gases and some vapors can escape thermally once they reach the exobase; the question is how rapidly they can be transported there. The system that is best understood is the present terrestrial atmosphere, where the diffusion limit clearly controls the escape rate of H (see the discussion following Eq. 5 below). If the hot phase of the Moon-forming event lasted only a few 10^2 or a few 10^3 yr, as suggested by Stevenson (1987), the inhibiting effect of diffusion is probably very large.

Escape, Hydrodynamic and Other

After several decades of belief in the supremacy of thermal escape as formulated by Jeans (1925), we have witnessed the arrival of several new ideas. Loss of H from the Earth is only partly thermal, being dominated by the nonthermal process of charge exchange with the plasmaspheric ions (Cole 1966; Tinsley 1973; Liu and Donahue 1974a,b). The actual loss rate is limited by diffusion through the stratosphere and mesosphere and the cold trap at the tropopause (Hunten 1973; Hunten and Donahue 1976). Although thermal escape from Mars is important for H, nonthermal processes dominate the loss of heavier elements and appear to have caused a major evolution of the N-isotopic ratio. Yet other nonthermal processes are important at Venus (McElroy et al. 1977; Kumar et al. 1983) and Mars (McElroy 1972). General reviews of nonthermal mechanisms appear in Hunten (1982) and Chamberlain and Hunten (1987). A process that may have been important just after the departure of the solar nebula is hydrodynamic escape of H at a rate fast enough to drag away heavier elements (Hunten et al. 1987; Zahnle and Kasting 1986). The same concept has been applied to Io (Hunten 1985), where the loss of S may be assisted by a rapid flow of O.

Diffusion-limited loss of H and blowoff-assisted (hydrodynamic) escape of heavy gases are different aspects of the same process, the mutual drag between two gases that are flowing with respect to each other. If the heavy major gas is stationary, it holds back the light minor gas; if a large flux of a light major gas is escaping at supersonic speed, it drags along heavy minor gases. The drag is conveniently described by the diffusion parameter b, the product of the diffusion coefficient and the total number density. Although it varies somewhat with the gas pair being considered, and as the 0.7 power (typically) of temperature T, $b \simeq 2 \times 10^{-19}$/cm s when H is one of the pair. If two gases have vertical fluxes F_1 and F_2, the coupled diffusion equations may be written:

$$\frac{dn_1}{dz} = -\frac{m_1 g}{kT} n_1 + \frac{1}{b}(n_1 F_2 - n_2 F_1) \qquad (1)$$

$$\frac{dn_2}{dz} = -\frac{m_2 g}{kT} n_2 + \frac{1}{b}(n_2 F_1 - n_1 F_2) \qquad (2)$$

where the subscripts 1 and 2 refer to the light and heavy gas, g is gravity, m is the molecular weight, k is Boltzmann's constant, and n is number density. For simplicity, temperature gradients, eddy diffusion and variation of gravity with height have been omitted. If both fluxes are zero, we recover the hydrostatic equations for diffusive equilibrium, where each species has its own scale height $\mathcal{H}_i = kT/m_i g$. The sum of Eqs. (1) and (2) represents total hydrostatic equilibrium, even when the fluxes are nonzero; they can also be combined to give an expression for their mutual diffusion:

$$\frac{1}{X_2} \frac{dX_2}{dz} = \frac{1}{b}\left(F_1 - \frac{X_1}{X_2} F_2\right) - (m_2 - m_1)\frac{X_1 g}{kT}. \qquad (3)$$

Here X represents mole fraction.

To discuss diffusion-limited escape, we interchange all subscripts, set F_2 to zero, and take X_1 very small, so that X_2 can be set equal to 1. The result can be written

$$F_1 = b[X_1(m_2 - m_1)g/kT - dX_1/dz]. \qquad (4)$$

The maximum, or "limiting" flux is obtained by setting the last term to zero. Clearly, a positive value of the gradient dX_1/dz gives a smaller flux. A negative value means that X_1 in the previous term decreases upwards, and thus also reduces the flux. The limiting flux is

$$F_1 = bX_1(1 - m_1/m_2)/\mathcal{H}_2 \simeq bX_1/\mathcal{H}_2 \tag{5}$$

where $\mathcal{H}_2$ is the pressure scale height of the heavy gas and the approximation is good when m_1 is much less than m_2. The escape flux does not exceed the value given by Eq. (5), which is controlled by the mole fraction at the level (homopause) where molecular diffusion starts to dominate. In practice, the actual H fluxes for Earth and Mars do approximately obey this relation. The densities at the top of the atmospheres adjust themselves so that the actual escape processes can do their job.

In hydrodynamic escape, it is X_1 that is essentially unity and X_2 that is small. Otherwise the argument is similar: a consistent flow field with nonzero F_2 requires the gradient on the left side of Eq. (3) to be zero, and a relation follows between F_2 and F_1:

$$F_2 = X_2 F_1 (m_c - m_2)(m_c - m_1) \tag{6}$$

where the crossover mass is defined by

$$m_c = m_1 + kTF_1/(bgX_1). \tag{7}$$

Expression (6) gives a negative flux for $m_2 > m_c$, which requires a boundary condition unlikely to be found in nature. What actually happens is a zero flux and a negative value of dX_2/dz in Eq. (3). Thus, constituents with mass greater than m_c are not lost, but instead have their scale heights raised by the drag force.

The novel and interesting thing about Eq. (6), in addition to the cutoff at m_c, is that the flux F_2 has a strong dependence on m_2. The value of this cutoff is dependent primarily on the flux F_1, normally of H_2 or H, and inversely on the gravity of the object. There is also some temperature dependence, close to $T^{0.3}$, since b normally varies approximately as $T^{0.7}$. The flux required is very large, and requires a very large heat input to the top of the atmosphere, which would not be available from the contemporary Sun except for a body around 1000 km radius. However, it is likely that the early Sun was much more luminous at the relevant ultraviolet wavelengths.

All these matters are discussed in much more detail by Hunten et al. (1987) and Zahnle and Kasting (1986). Some applications are given in Sec. 7.10.4.

A different approach has been explored by Donahue (1986), who has examined loss from planetesimals small enough (10^{24} g or $\sim$ 500 km diameter) so that thermal loss is substantial. The atmosphere is assumed to be essentially pure noble gases with solar abundance ratios. Under these conditions, escape is dominated by the gas, or isotope, that is most abundant at the exobase and therefore controls its position. The effect is to equalize the isotopic ratios for the lightest remaining element. Two classes of object are needed, in different proportions, to account for the noble-gas endowments of all three terrestrial planets, with one glaring exception, on Venus: the model fails to account for the observed ratios of $^{36}Ar/^{38}Ar$ and planet-to-planet differences in abundances and isotopic compositions of Kr and Xe, and does not explain why the very small meteorites exhibit similar fractionation patterns to those found in planetary atmospheres. Nevertheless, Donahue's judgment that his inability to account for the Ar isotope ratio is "probably fatal to the hypothesis" seems unusually harsh. Most scenarios do not come nearly as close to matching all the data.

We now return to nonthermal mechanisms, most of which produce high-velocity atoms that have no trouble escaping if they are directed upward and do not suffer a collision. However, the latter requirement confines the effects to the exosphere, and an important mass fractionation arises indirectly because there is a strong diffusive separation of the layers below and including the exosphere. Acting over the age of the planets on a small total inventory such as that of N on Mars, a nonthermal process can still be very effective, as McElroy (1972) first pointed out. Nitrogen molecules are dissociated by impact of electrons with energies up to a few 10 eV, the result of photoionization. The resulting atoms usually have enough kinetic energy for one of them to escape. Scenarios that could lead to the observed 40% enrichment of the heavier isotope are discussed by McElroy et al. (1977).

Charge exchange between a hot proton and a thermal H atom interchanges their identities but not their kinetic energies (to a good approximation). The resulting fast atom may then be able to escape. This process, though an important contributor to the total loss from Earth, has not had much effect on the huge reservoir of the oceans. Venus, with its tiny reservoir, is a different matter (see Sec. 7.10.4).

A third example is Io, whose SO_2 atmosphere is evidently being swept away by sputtering, the impact by ions of O and S in a plasma torus (McGrath and Johnson 1987). It seems unlikely that the processes operating in this strange environment would have been important in the early solar system, but that environment too was probably stranger than we now imagine. Solar-wind sputtering at Mars and Venus is discussed by Watson et al. (1980); it would be important only in the unlikely event that most of the incident flux actually reaches the atmosphere.

A comprehensive review appears in Hunten (1982), and much of it is reproduced in Chamberlain and Hunten (1987). Here we have focused on the subset of processes that are known or suspected to be causing substantial fractionation.

7.10.4. CASE STUDIES: THE INNER PLANETS

Introduction

We introduce the discussion of the individual terrestrial planets by some general remarks; many of the new ideas of the past decade have come from comparative studies. This approach is a feature of the discussions of Mars by Anders and Owen (1977) and of Venus by Donahue and Pollack (1983). There are two striking facts: compared with solar values, all three planets are very deficient in all the noble gases; and the lighter elements are more depleted than the heavier ones (Fig. 7.10.2).

Most workers have approached the problem with the view that the elemental and isotopic ratios for the noble gases have not changed since accretion, except for He. They have thus placed the onus on the source or a mixture of sources. These include solar wind or nebula, either accreted directly or as a component of incoming planetesimals, and trapped gases fractionated by such processes as diffusive loss. The idea of fractionation during hydrodynamic escape is just beginning to be studied, and no self-consistent scenario has been published. We shall begin with a review of this new idea.

Applications of Hydrodynamic Escape

It is already clear from the work of Zahnle and Kasting (1986) and Hunten et al. (1987) that if adequate H and an appropriate energy source were present early enough, hydrodynamic escape could readily have generated the fractionated noble-gas patterns observed in terrestrial planets and meteorites. For example, Hunten et al. (1987) showed that hydrodynamic loss of an early H-rich terrestrial atmosphere driven by the strongly enhanced output of extreme ultraviolet radiation from the young Sun proposed by Zahnle and Walker (1982) would have been capable of severely fractionating the isotopes of Xe, from initially meteoritic abundances to their present terrestrial composition. Sasaki and Nakazawa (1986) independently reached an identical conclusion. It seems likely, twenty-five years after the mass-fractionation relationship between meteoritic and terrestrial Xe was first noted (Krummenacher et al. 1962), that an astrophysically realistic process for the fractionation has finally been identified. These results from Xe raise a significant question: in what environments, and to what extents, might hydrodynamic escape have acted to establish more general features of present-day volatile distributions in planetary atmospheres?

Isotopic signatures of nonradiogenic Xe on Earth and Mars (using SNC data for Mars) appear to be those of fractionated residuals of primordial atmospheres in which volatiles were present initially in solar and carbonaceous-chondrite compositions, respectively (Hunten et al. 1987; Swindle et al. 1986). If episodes of hydrodynamic escape were responsible for producing these current Xe inventories and compositions, they would have been severe enough to sweep away almost completely the initial atmospheric inventories of lighter gases. These two observations suggest that current inventories of Kr, Ar and Ne on the Earth and Mars were supplied to their atmospheres by planetary outgassing at a later time in the hydrodynamic escape episode, when the solar extreme ultraviolet flux had decreased according to a power-law dependence on time (Zahnle and Walker 1982). Once in the atmosphere, these outgassed noble gases, initially of solar isotopic composition, were partially lost and fractionated to their present isotopic abundances. The relative elemental abundances in the outgassed components required to match the observed atmospheric patterns are similar to those generated in the laboratory by adsorption of solar-composition gases on grain surfaces (Fig. 7.10.1; see also Sec. 7.10.2 and Chapter 7.9). In this approach, the noble gases ultimately outgassed from the planets are assumed to originate in adsorption of solar gases on nebular dust grains that later accreted into the planetesimals subsequently captured by the growing protoplanets.

This scenario, involving two basic processes—adsorption of nebular gas on dust and hydrodynamic escape—accounts reasonably well for the mass distributions of noble gases in the contemporary terrestrial and Martian atmospheres. Venus appears to be a special and simpler case in that its atmospheric noble gases, to the extent they are known, are elementally and isotopically consistent with fractionation during hydrodynamic loss of a dense, solar-composition primordial atmosphere, except for Ne. Outgassed components could certainly be present, but, except perhaps for Ne, they do not seem to be dominant constituents of the very large Venusian noble-gas inventories.

The most straightforward sources for C and N on the three terrestrial planets are extant meteorite classes: CI carbonaceous chondrites for Mars (already invoked to account for the Martian Xe-isotopic composition) and a few 10 km of late-accreting enstatite-chondrite veneer on Venus and Earth (Pepin 1988). Added at the tail ends of the hydrodynamic-loss episodes, when the crossover mass is near that of CO_2, E-chondrite C could be slightly fractionated and N more severely fractionated, both to compositions that are well within error of contemporary elemental and isotopic abundances in the two planets.

There are several possible sources for the large initial abundances of H and noble gases that are required to be present in the primordial atmospheres of the terrestrial planets for hydrodynamic escape to work. For Mars, roughly adequate amounts of H (as H_2O) and Xe are supplied by accretion of carbonaceous-chondrite material. For Earth and Venus, however, the initial noble-gas inventories must have been solar, not meteoritic, in composition. One possibility is direct capture of a primordial atmosphere from surrounding nebular gas, a process that would not have contributed significantly to the smaller Mars unless temperatures were very low (Hunten 1979). Capture from the nebula is attractive because the amounts of H thus supplied are comparable to the amounts that must eventually be lost during hydrodynamic escape. However, it is unclear, and perhaps doubtful, that the nebula still existed at the times (on the order of 1 to 10 Myr) when the Earth-Venus protoplanets had grown large enough to accrete significant amounts of ambient gas gravitationally. A more specific problem involves Ne on Venus. A simple model of fractionation of initially solar noble gases by hydrodynamic escape can reproduce, within present uncertainties, the observed noble-gas compositions in the Venusian atmosphere *except* for the abundance of Ne, where the model value is 2 orders of magnitude too high. A terminal hydrodynamic escape episode affecting only Ne, followed by outgassing or capture of solar Ne, can technically account for this discrepancy but is somewhat contrived and unconvincing. An alternative involving cometary infall, and perhaps early solar-wind irradiation as well, is described below.

Venus Deuterium Enrichment

The most striking thing about the atmospheric composition of Venus is the large D/H ratio, 0.016 (McElroy et al. 1982; Donahue et al. 1982). This is a factor ~ 100 greater than on Earth, which is already a factor 10 greater than the cosmic ratio 1.5×10^{-5}. Perhaps even more significant is the large abundance of Ne and Ar, with an elemental pattern substantially different from that of the Earth. A critical discussion of the Pioneer Venus results is given by von Zahn et al. (1983), and attempts at explanation are summarized by Mukhin (1983) and Donahue and Pollack (1983).

The large enhancement of D was not discovered in time to be reviewed by Donahue and Pollack, but was discussed briefly in the discovery papers (McElroy et al. 1982; Donahue et al. 1982). Greatest interest has centered on the possibility that Venus might have started with an Earth-equivalent quantity of water (colloquially called "an ocean" even though much of it may have been in the form of vapor). McElroy et al. pointed out a remarkable nonthermal process that can eject H but not D with escape speed. The dissociative recombination of O_2^+ produces fast O atoms that cannot escape but that can collide with an H atom, giving it nearly twice its own speed. If a process so efficient at fractionating always dominated the loss, the original amount of water would be about 100 times the present amount, or a liquid equivalent of ~ 10 m. Anything less efficient would imply a larger endowment. A broader examination by Kumar et al. (1983) concluded that the dominant process even now is probably charge exchange of H atoms with hot protons, the same one that dominates on the Earth. There is still a large bias in favor of H over D; however, it is necessary to process somewhat more material to obtain the same end result. Extrapolation into the past gives a somewhat larger minimum initial endowment, still much less than an ocean. Kumar et al. pointed out that the actual amount could have been much greater if there is a continuing supply of unfractionated H (such as juvenile water) to dilute the enriched material.

In order for such large quantities of H to be lost, the amount in the upper atmosphere must follow the amount near the surface, and such an assumption was made by Kumar et al. It has recently been questioned by Krasnopolsky (1985) and Yatteau (1983), who present a model in which the upper atmosphere is desiccated by the sulfuric acid clouds. Their conclusion is that Venus probably started out "dry," i.e., with only a small fraction of an ocean. However, it remains true that an observed D/H ratio can only lead to a lower bound on the initial amount.

Grinspoon and Lewis (1987) have presented a model based on continual cometary impact, giving an inventory of H that, on average, is in steady state with fractionating escape. With reasonable parameters the Venusian H abundance and D/H ratio are reproduced; there is no need to regard them as "fossils" of the initial state of the planet. A similar conclusion is implicit in the finding by Kumar et al. (1983) that "juvenile water" had to be added to the system to keep the D/H ratio as small as observed.

Venus: Noble Gases

The abundances of the noble gases in the atmosphere of Venus pose a twofold problem when compared with Earth, Mars and the chondritic meteorites. Neon and the primordial isotopes of Ar are highly overabundant per g of rock. In addition, the pattern of relative abundances of Ne, Ar, Kr and Xe is distinctly different from the so-called planetary pattern (Fig. 7.10.2). Whereas the ratio $^{36}Ar/^{20}Ne$ is about the same on Venus and Earth, the ratio $^{36}Ar/^{84}Kr$ on Venus resembles the solar proportion rather than the planetary one.

Did Venus begin with a solar abundance distribution from which sufficient Ne was lost to create a terrestrial value of $^{20}Ne/^{36}Ar$? Or was this dual pattern set by contributions from two different sources, one for the Ne and one for the other noble gases? In either case, we seem to be dealing with a remarkable coincidence. Wetherill (1981) has explored the first possibility in detail. Owen (1986) has proposed a dual-source model that builds on Wetherill's basic approach. (An alternative scenario proposed by Pollack and Black [1982] was made prior to the discovery that the heavier noble gases exhibit a solar-like abundance pattern on Venus; this subsequent result requires some modification of their hypothesis.)

Inspection of Figs. 7.10.1 and 7.10.2 indicates that, while the amount of evolved gas per g of rock is anomalously high for Ar and Ne on Venus, the Kr and Xe are not so different from the terrestrial values. Wetherill (1981) took the position that the grains of material that ultimately formed Venus were irradiated by an intense early solar wind. The solar-wind pattern of noble-gas abundances was thus implanted in the material that later formed the planet, with the exception of Ne, which was able to escape. The fact that the escape of Ne occurred in such a way as to end up with $^{20}Ne/^{36}Ar$ so close to the terrestrial value must simply be regarded as a coincidence. This coincidence will haunt any attempt to account for the simultaneous presence of a nearly terrestrial value for this abundance ratio and a solar value for the ratio of heavier noble-gas abundances.

The dual source model assumes that this coincidence occurred as a result of the *selective build up* of Ne by solar-wind implantation rather than the *selective loss* of Ne through diffusive escape. The idea is to assume that Wetherill's (1981) argument is basically correct, but that the Ne we find on Venus today represents most of the Ne that was originally implanted in the material that formed the planet. Hence the other noble gases that were contributed by the solar wind are negligible fractions of the abundances we now find in the atmosphere. Some other source is required to provide the Ar, Kr and Xe. This source should account for the unusually high abundance of Ar.

This interpretation is compatible with the measured ratio of ^{20}Ne/^{22}Ne = 11.9 ± 0.7 (Istomin et al. 1982). Wieler et al. (1986) give data for ^{20}Ne/^{22}Ne in 11 bulk examples of lunar soil. These provide a mean value of 12.3 for this ratio. This lunar Ne is attributed to a mixture of solar-wind and solar-flare Ne. The latter has a lower Ne-isotopic ratio than the former, recent measurements suggesting 11.3 ± 0.2 (Wieler et al. 1986) and 13.7 ± 0.3 (Bochsler and Geiss 1977), respectively. The model for the early solar nebula and its evolution as given by Wetherill (1981) would still apply. The requirements for implantation are made easier by the fact that we now need a factor ~200 less implanted gas from the solar wind since Ne is 20 times more abundant than Ar in the Sun, but 10 times less abundant than Ar on Venus. In other words, by picking Ne as the index element and assuming that the abundance we now find on Venus is close to the original value, we require much less trapping of solar-wind gas. The argument still works even if the Ne/Ar ratio in the grains is less than it is in the solar wind. There is still a problem in accounting for the way in which the material forming Venus was irradiated by this early wind while that forming the Earth was not, as Wetherill (1981) emphasized in his original paper. One solution is to accept that both planets in fact experienced this irradiation, but that the record on Earth was lost by the Moon-forming impact.

These various arguments provide a plausible explanation for the total abundance and the isotopic composition of the Ne on Venus. But what about the Ar, Kr and Xe? A possible solution can be found by invoking an impact with an icy planetesimal. As pointed out in Sec. 7.10.2, ice forming at low temperature in a solar mixture of gases appears to trap the gases without changing their proportions. However, the ice will not be able to retain Ne unless it is formed and maintained at $T < 25$ K.

Hence we can supply a solar mixture of the heavier noble gases to the atmosphere of Venus "simply" by having an icy planetesimal of a suitable size (or a swarm of smaller planetesimals) crash into the planet. How big does this planetesimal have to be? That is, how much ice do we need? We can use Titan to scope the problem. Let us assume that the N on Titan was originally trapped as N_2 in the ices that formed the satellite, from a cosmic mixture of gases. This cannot be strictly true, because Titan formed in the proto-Saturnian nebula, where the mixture of gases is expected to be different from that in the solar nebula itself (Prinn and Fegley 1981). Nevertheless, elemental abundances should have been preserved. Titan is certainly deficient in Ne (Broadfoot et al. 1981; Owen 1982) as the laboratory data discussed in Sec. 7.10.2 above would predict, providing added support for using Titan as an icy planetesimal analogue in this discussion. We further assume that all of the trapped N is now in Titan's atmosphere, except for the 20% that has escaped or has been deposited as organic material on Titan's surface (Strobel 1982; Yung et al. 1984), and the 5 to 10% that may be dissolved in a global ocean of ethane (Lunine et al. 1983). This is a conservative approach, in that there may well be much more N still trapped inside the satellite. We then assume that Ar will scale with N, according to cosmic abundances. This translates into an icy planetesimal with a diameter of about 150 km to supply the Ar on Venus.

Although this may seem large compared with the 8×12 km dimensions of the nucleus of Halley's comet (see, e.g., Keller et al. 1986), it is actually less than the diameter of Phoebe (200 km), commonly regarded to be a captured icy object, and within the range of values being considered for the distant "asteroid" Chiron. We should also remember that our ideas about the dimensions of comet nuclei depend on those comets we have seen. Large comets, like large asteroids, may be rare but not nonexistent. The great comet of 1729, for example, was visible to the naked eye although it came no closer to the Sun than 4 AU (Marsden 1986). Scaling from Comet Schwassmann-Wachmann I in full flower, the 1729 comet would have had a nucleus again on the order of 100 to 200 km diameter.

This approach to a solution of the Venus noble-gas enigma has several consequences, at least one of which can be tested. It requires, first of all, a unique event, or a unique set of circumstances, always an unattractive characteristic. Why did only Venus suffer a large icy impact? One can respond that Earth and Mars were probably also bombarded by icy planetesimals, but that on these planets early history was wiped out by subsequent, larger impacts of nonicy bodies. According to Wetherill (1985), all of the inner planets were vulnerable to such large-scale impacts. Thus, uniqueness is transformed into a question of timing: a large impact on Venus was *followed* by an icy planetesimal impact; on the Earth it was not. Mars was pelted repeatedly by smaller asteroidal bodies that could blow off the atmosphere (or large fractions of it) because of the planet's smaller mass (Cameron 1983). Hence a record of early icy impacts would be lost.

But do the icy planetesimals really carry Ar, Kr and Xe in solar relative abundances, and no Ne? Or is it possible that Ne with the proper isotope ratio and abundance to account for Venus is actually found in comet nuclei as a result of their formation at an appropriately low temperature? For example, temperatures could have been such that Ne was partially retained in comets, say with an efficiency 1% of that for heavier noble gases. Infall of such comets could have provided the primordial Venusian atmosphere for the scenario of hydrodynamic escape discussed earlier; the Ne problem would vanish. This is the requirement of the model that can be tested. Future studies of relatively new comets could certainly search for cometary Ar and Ne, and the atmosphere of prototypical Titan could be studied by an orbiting ultraviolet telescope to see if the predicted amount of Ar is actually present. Those experiments will tell us whether the scenario suggested here is viable or not.

REFERENCES

Ahrens, T. J., and O'Keefe, J. D. 1987. Impact on the Earth, ocean, and atmosphere. *J. Impact Eng.*, in press.

Alaerts, L., Lewis, R. S., and Anders, E. 1979. Isotopic anomalies of noble gases in meteorites and their origins. IV. C3 (Ornans) carbonaceous chondrites. *Geochim. Cosmochim. Acta* 43:1421–1432.

Alaerts, L., Lewis, R. S., Matsuda, J., and Anders, E. 1980. Isotopic anomalies of noble gases in meteorites and their origins. VI. Presolar components in the Murchison C2 chondrite. *Geochim. Cosmochim. Acta* 44:189–209.

Anders, E., and Ebihara, M. 1982. Solar-system abundances of the elements. *Geochim. Cosmochim. Acta* 46:2363–2380.

Anders, E., and Owen, T. 1977. Mars and Earth: Origin and abundance of volatiles. *Science* 198:453–465.

Bar-Nun, A., Dror, J., Kochavi, E., and Laufer, D. 1987. Amorphous water ice and its ability to trap gases. *Phys. Rev.* B35:2427–2435.

Benz, W., Slattery, W. L., and Cameron, A. G. W. 1986. The origin of the Moon and the single-impact hypothesis. *Icarus* 66:515–535.

Benz, W., Slattery, W. L., and Cameron, A. G. W. 1987. The origin of the Moon and the single-impact hypothesis II. *Icarus* 71:30–45.

Bernatowicz, T. J., Kennedy, B. M., and Podosek, F. A. 1985. Xe in glacial ice and the atmospheric inventory of noble gases. *Geochim. Cosmochim. Acta* 49:2561–2564.

Bochsler, P., and Geiss, J. 1977. Elemental abundances in the solar wind. *Trans. IAU* XVIB:120–123.

Bogard, D. D. 1982. Trapped noble gases in the EETA 79001 shergottite. *Meteoritics* 17:185–186 (abstract).

Bogard, D. D., and Johnson, P. 1983. Martian gases in an Antarctic meteorite? *Science* 221:651–654.

Broadfoot, A. L., Sandel, B., Shemansky, D. E., Holberg, J. B., Smith, G. R., Strobel, D. F., McConnell, J. C., Kumar, S., Hunten, D. M., Atreya, S. K., Donahue, T. M., Moos, H. W., Bertaux, J. L., Blamont, J. E., and Pomphrey, R. B. 1981. Extreme ultraviolet observations from Voyager I encounter with Saturn. *Science* 211:206–211.

Cameron, A. G. W. 1982. Elemental and nuclidic abundances in the solar system. In *Essays in Nuclear Astrophysics*, eds. C. A. Barnes, D. D. Clayton, and D. N. Schramm (Cambridge: Cambridge Univ. Press), pp. 23–43.

Cameron, A. G. W. 1983. Origin of the atmospheres of the terrestrial planets. *Icarus* 56:195–201.

Cameron, A. G. W. 1985. Formation of the prelunar accretion disk. *Icarus* 62:319–327.

Cameron, A. G. W., and Ward, W. R. 1976. The origin of the Moon. *Lunar Sci.* 7:120–122 (abstract).

Chamberlain, J. W., and Hunten, D. M. 1987. *Theory of Planetary Atmospheres*, Second Ed. (New York: Academic Press).

Cole, K. D. 1966. Theory of some quiet magnetospheric phenomena related to the geomagnetic tail. *Nature* 211:1385–1387.

Crabb, J., and Anders, E. 1981. Noble gases in E-chondrites. *Geochim. Cosmochim. Acta* 45:2443–2464.

Delsemme, A. H. 1982. Chemical composition of cometary nuclei. In *Comets*, ed. L. L. Wilkening (Tucson: Univ. of Arizona Press), pp. 85–130.

Delsemme, A. H. 1985. What we do not know about cometary ices: A review of the incomplete evidence. In *Ices in the Solar System*, eds. J. Klinger, D. Benest, A. Dollfus and R. Smoluchowski (Dordrecht: D. Reidel), pp. 505–517.

Donahue, T. M. 1986. Fractionation of noble gases by thermal escape from accreting planetesimals. *Icarus* 66:195–210.

Donahue, T. M., and Pollack, J. B. 1983. Origin and evolution of the atmosphere of Venus. In *Venus*, eds. D. M. Hunten, L. Colin, T. M. Donahue, and V. I. Moroz (Tucson: Univ. of Arizona Press), pp. 1003-1036.

Donahue, T. M., Hoffman, J. H., Hodges, R. R., and Watson, A. J. 1982. Venus was wet: A measurement of the ratio of D to H. *Science* 216:630–633.

Eberhardt, P., Hodges, R. R., Krankowsky, D., Berthelier, J. J., Schulte, W., Dolder, U., Lammerzahl, P., Hoffman, J. H., and Illiano, J. M. 1987. The D/H and $^{18}O/^{16}O$ isotopic ratios in Comet Halley. *Lunar Planet. Sci.* 18:252–253 (abstract).

Fanale, F. P. 1976. Martian volatiles: Their degassing history and geochemical fate. *Icarus* 28:179–202.

Fanale, F. P., and Cannon, W. A. 1972. Origin of planetary primordial rare gas: The possible role of adsorption. *Geochim. Cosmochim. Acta* 36:319–328.

Fernandez, J. A., and Ip, W.-H. 1983. On the time evolution of the cometary influx in the region of the terrestrial planets. *Icarus* 54:377–387.

Frick, U., Becker, R. H., and Pepin, R. O. 1987. Solar wind record in the lunar regolith: Nitrogen and noble gases. *Proc. Lunar Planet. Sci. Conf.* 18, in press.

Frick, U., Mack, R., and Chang, S. 1979. Noble gas trapping and fractionation during synthesis of carbonaceous matter. *Proc. Lunar Planet. Sci. Conf.* 10:1961–1973.

Grady, M. M., Wright, I. P., Carr, L. P., and Pillinger, C. T. 1986. Compositional differences in enstatite chondrites based on carbon and nitrogen stable isotope measurements. *Geochim. Cosmochim. Acta* 50:2799–2813.

Grinspoon, D. H., and Lewis, J. S. 1988. Comet impacts and the evolution of the water and deuterium abundances on Venus. *Icarus*, in press.

Hartmann, W. K., and Davis, D. R. 1975. Satellite-sized planetesimals. *Icarus* 24:504–515.

Hoffman, J. H., Oyama, V. I., and Von Zahn, U. 1980. Measurements of the Venus lower atmosphere composition: A comparison of results. *J. Geophys. Res.* 85:7871–7881.

Hunten, D. M. 1973. The escape of light gases from planetary atmospheres. *J. Atmos. Sci.* 30:1481–1494.

Hunten, D. M. 1979. Capture of Phobos and Deimos by protoatmospheric drag. *Icarus* 37:113–123.

Hunten, D. M. 1982. Thermal and nonthermal escape mechanisms for terrestrial bodies. *Planet. Space Sci.* 30:773–783.

Hunten, D. M. 1985. Blowoff of an atmosphere and possible application to Io. *Geophys. Res. Lett.* 12:271–273.

Hunten, D. M., Pepin, R. O., and Walker, J. C. G. 1987. Mass fractionation in hydrodynamic escape. *Icarus* 69:532–549.

Hunten, D. M., and Donahue, T. M. 1976. Hydrogen loss from the terrestrial planets. *Ann. Rev. Earth Planet. Sci.* 4:265–292.

Istomin, V. G., Grechnev, K. V., and Kochnev, V. A. 1982. Preliminary results of mass-spectrometric measurements on board the Venera 13 and Venera 14 probes. *Pisma Astron. Zh.* 8:391–398.

Jeans, J. H. 1925. *The Dynamical Theory of Gases* (Cambridge: Cambridge Univ. Press), Chapter 15.

Jessberger, E. K., Kissel, J., Fechtig, H., and Krueger, F. R. 1986. On the average chemical composition of cometary dust. In *Comet Nucleus Sample Return*, Proc. ESA Workshop, Canterbury, United Kingdom.

Keller, H. U., Arpigny, C., Barbieri, C., Bonnet, R. M., Cazes, S., Coradini, M., Cosmovici, C. B., Delamere, W. A., Huebner, W. F., Hughes, D. W., Jamar, C., Malaise, D., Reitsema, H. J., Schmidt, H. U., Schmidt, W. K. H., Seige, P., Whipple, F. L., and Wilhelm, K. 1986. First Halley multicolour camera imaging results from Giotto. *Nature* 321:320–329.

Kerridge, J. F. 1985. Carbon, hydrogen and nitrogen in carbonaceous chondrites: Abundances and isotopic compositions in bulk samples. *Geochim. Cosmochim. Acta* 49:1707–1714.

Krankowsky, D., Lämmerzahl, P., Herrwerth, I., Woweries, J., Eberhardt, P., Dolder, U., Herrmann, U., Schulte, W., Berthelier, J. J., Illiano, J. M., Hodges, R. R., and Hoffman, J. H. 1986. In situ gas and ion measurements at Comet Halley. *Nature* 321:326–330.

Krasnopolsky, V. A. 1985. Total injection of water vapor into the Venus atmosphere. *Icarus* 62:221–229.

Krummenacher, D., Merrihue, C. M., Pepin, R. O., and Reynolds, J. H. 1962. Meteoritic krypton and barium versus the general isotopic anomalies in xenon. *Geochim. Cosmochim. Acta* 26:231–249.

Kumar, S., Hunten, D. M., and Pollack, J. B. 1983. Nonthermal escape of hydrogen and deuterium from Venus and implications for loss of water. *Icarus* 55:369–389.

Kung, C.-C., and Clayton, R. N. 1978. Nitrogen abundances and isotopic compositions in stony meteorites. *Earth Planet. Sci. Lett.* 38:421–435.

Lange, M. A., and Ahrens, T. M. 1982. The evolution of an impact-generated atmosphere. *Icarus* 51:96–120.

Lewis, J. S. 1972. Low temperature condensation from the solar nebula. *Icarus* 16:241–252.

Lewis, J. S. 1973. Chemistry of the planets. *Ann. Rev. Phys. Chem.* 24:339–351.

Lewis, J. S., and Prinn, R. G. 1984. *Planets and Their Atmospheres: Origin and Evolution* (New York: Academic Press).

Liu, S. C., and Donahue, T. M. 1974. Realistic model of hydrogen constituents in the lower atmosphere and escape flux from the upper atmosphere. *J. Atmos. Sci.* 31:2238–2242.

Liu, S. C., and Donahue, T. M. 1974. The aeronomy of hydrogen in the atmosphere of the Earth. *J. Atmos. Sci.* 31:1118–1136.

Liu, S. C., and Donahue, T. M. 1974. Mesospheric hydrogen related to exospheric escape mechanisms. *J. Atmos. Sci.* 31:1466–1470.

Lunine, J. I., and Stevenson, D. J. 1985. Thermodynamics of clathrate hydrate at low and high pressures with applications to the outer solar system. *Astrophys. J. Suppl.* 58:493–531.

Lunine, J. I., Stevenson, D. J., and Yung, Y. L. 1983. Ethane ocean on Titan. *Science* 222:1229–1230.

Manka, R. H., and Michel, F. C. 1971. Lunar atmosphere as a source of lunar surface elements. *Proc. Lunar Sci. Conf.* 2:1717–1728.

Marsden, B. 1986. *Catalogue of Cometary Orbits*, Fifth Ed. (Cambridge: Smithsonian Astrophysical Observatory).

Matsuda, J., Lewis, R. S., Takahashi, H., and Anders, E. 1980. Isotopic anomalies of noble gases in meteorites and their origins. VII. C3V carbonaceous chondrites. *Geochim. Cosmochim. Acta* 44:1861–1874.

Mayer, E., and Pletzer, R. 1986. Astrophysical implications of amorphous ice—a microporous solid. *Nature* 319:298–300.

Mazor, E., Heymann, D., and Anders, E. 1970. Noble gases in carbonaceous chondrites. *Geochim. Cosmochim. Acta* 34:781–820.

McElroy, M. B. 1972. Mars: An evolving atmosphere. *Science* 175:443–445.

McElroy, M. B., Kong, T. Y., and Yung, Y. L. 1977. Photochemistry and evolution of Mars' atmosphere: A Viking perspective. *J. Geophys. Res.* 82:4379–4388.

McElroy, M. B., Prather, M. J., and Rodriguez, J. M. 1982. Escape of hydrogen from Venus. *Science* 215:1614–1615.

McGrath, M. A., and Johnson, R. E. 1987. Magnetospheric plasma sputtering of Io's atmosphere. *Icarus* 69:519–531.

Melosh, H. J., and Sonett, C. P. 1986. When worlds collide: Jetted vapor plumes and the moon's origin. In *Origin of the Moon*, eds. W. K. Hartmann, R. J. Phillips, and G. J. Taylor (Houston: Lunar and Planetary Inst.) pp. 621–642.

Miller, S. L. 1961. The occurrence of gas hydrates in the solar system. *Proc. Natl. Acad. Sci. U.S.* 47:1798–1808.

Mukhin, L. M. 1983. The problem of rare gases in the Venus atmosphere. In *Venus*, eds. D. M. Hunten, L. Colin, T. M. Donahue, and V. I. Moroz (Tucson: Univ. of Arizona Press), pp. 1037–1042.

Mukhin, L. M., Gel'man, B. G., Lamonov, N. I., Mel'nikov, V. V., Nenarokov, D. F., Okhotnikov, B. P., Rotin, V. A., and Khokhlov, V. N. 1982. Gas-chromatographical analysis of chemical composition of the atmosphere of Venus done by Venera 13 and Venera 14 probes. *Pisma Astron. Zh.* 8:399–403.

Niemeyer, S., and Marti, K. 1981. Noble gas trapping by laboratory carbon condensates. *Proc. Lunar Planet. Sci. Conf.* 12:1177–1188.

Nier, A. O., and McElroy, M. B. 1977. Composition and structure of Mars' upper atmosphere: Results from the neutral mass spectrometers on Viking 1 and 2. *J. Geophys. Res.* 82:4341–4349.

Ogilvie, K. W., Coplan, M. A., Bochsler, P., and Geiss, J. 1986. Ion composition results during the International Cometary Explorer encounter with Giacobini-Zinner. *Science* 232: 374–377.

Owen, T. 1982. The composition and origin of Titan's atmosphere. *Planet. Space Sci.* 30:833–838.

Owen, T. 1986. Update of the Anders-Owen model for martian volatiles. In *The Evolution of the Martian Atmosphere*, LPI Tech. Rept. No. 87–01, p. 91.

Owen, T., Biemann, K., Rushneck, D. R., Biller, J. E., Howarth, D. W., and Lafleur, A. 1976. The atmosphere of Mars: Detection of krypton and xenon. *Science* 194:1293–1295.

Owen, T. C., Biemann, K., Rushneck, D. R., Biller, J. E., Howarth, D. W., and Lafleur, A. L. 1977. The composition of the atmosphere at the surface of Mars. *J. Geophys. Res.* 82:4635–4639.

Owen, T. C., Lutz, B. L., de Bergh, C., and Maillard, J.-P. 1987. Detection of HDO in the atmosphere of Mars: Its abundance and the value of D/H. *Bull. Amer. Astron. Soc.* 19:817–818 (abstract), and *Science*, submitted.

Pellas, P. 1987. Les météorites "SNC" sont-elles des échantillons de la planète Mars? *Bull. Soc. Geol. France* 3(1):21–29.

Pepin, R. O. 1985. Meteorites: Evidence of martian origins. *Nature* 317:473–475.

Pepin, R. O. 1988. Atmospheric compositions: Key similarities and differences. In *Planetary and Satellite Atmospheres: Origin and Evolution*, eds. S. K. Atreya, J. B. Pollack, and M. S. Matthews (Tucson: Univ. of Arizona Press), in press.

Pepin, R. O., and Phinney, D. 1978. Components of xenon in the solar system. Preprint.

Pollack, J. B., and Black, D. C. 1982. Noble gases in planetary atmospheres: Implications for the origin and evolution of atmospheres. *Icarus* 51:169–198.

Pollack, J. B., and Yung, Y. L. 1980. Origin and evolution of planetary atmospheres. *Ann. Rev. Earth Planet. Sci.* 8:425–487.

Prinn, R. G., and Fegley, B. J. 1981. Kinetic inhibition of CO and N_2 reduction in circumplanetary nebulae: Implications for satellite composition. *Astrophys. J.* 249:308–317.

Safronov, V. S. 1969. *Evolution of the Protoplanetary Cloud and Formation of the Earth and the Planets* (Moscow: Navka). In Russian. Trans. NASA TT-F-677, 1972.

Sasaki, S., and Nakazawa, K. 1986. Terrestrial Xe fractionation due to escape of primordial H_2-He atmosphere. In *Abstracts for Japan-U.S. Seminar on Terrestrial Rare Gases*, pp. 68–71.

Shoemaker, E. M., and Wolfe, R. F. 1982. Cratering time scales for the Galilean satellites. In *Satellites of Jupiter*, ed. D. Morrison (Tucson: Univ. of Arizona Press), pp. 277–339.

Shoemaker, E. M., and Wolfe, R. F. 1984. Evolution of the Uranus-Neptune planetesimal swarm. *Lunar Planet. Sci.* 15:780–781 (abstract).

Signer, P., Baur, H., Derksen, U., Etique, P., Funk, H., Horn, P., and Wieler, R. 1977. Helium, neon, and argon records of lunar soil evolution. *Proc. Lunar Sci. Conf.* 8:3657–3683.

Sill, G. T., and Wilkening, L. L. 1978. Ice clathrate as a possible source of the atmospheres of the terrestrial planets. *Icarus* 33:13–22.

Srinivasan, B., Gros, J., and Anders, E. 1977. Noble gases in separated meteoritic minerals: Murchison(C2), Ornans(C3), Karoonda(C5), and Abee(E4). *J. Geophys. Res.* 82:762–778.

Stevenson, D. J. 1987. Origin of the Moon—the collisional hypothesis. *Ann. Rev. Earth Planet. Sci.* 15:271–315.

Strobel, D. F. 1982. Chemistry and evolution of Titan's atmosphere. *Planet. Space Sci.* 30:838–848.

Swindle, T. D., Caffee, M. W., and Hohenberg, C. M. 1986. Xenon and other noble gases in shergottites. *Geochim. Cosmochim. Acta* 50:1001–1015.

Tinsley, B. A. 1973. The diurnal variation of atomic hydrogen. *Planet. Space Sci.* 21:686–691.

Tyburczy, J. A., Frisch, B., and Ahrens, T. J. 1986. Shock-induced volatile loss from a carbonaceous chondrite: Implications for planetary accretion. *Earth Planet Sci. Lett.* 80:201–207.

Vickery, A. M., and Melosh, H. J. 1987. The large-crater origin of SNC meteorites. *Science,* 237:738–743.
Von Zahn, U., Kumar, S., Niemann, H., and Prinn, R. 1983. Composition of the Venus atmosphere. In *Venus,* eds. D. M. Hunten, L. Colin, T. M. Donahue, and V. I. Moroz (Tucson: Univ. of Arizona Press), pp. 299–430.
Wacker, J. F., and Anders, E. 1986. Trapping of noble gases by carbon: Applications to meteorites and planets. In *Abstracts for Japan-U.S. Seminar on Terrestrial Rare Gases.*
Walker, J. C. G. 1977. *Evolution of the Atmosphere* (New York: Macmillan).
Walker, J. C. G. 1986. Impact erosion of planetary atmospheres. *Icarus* 68:87–98.
Wasson, J. T., and Wetherill, G. W. 1979. Dynamical, chemical and isotopic evidence regarding the formation locations of asteroids and meteorites. In *Asteroids,* ed. T. Gehrels, (Tucson: Univ. of Arizona Press), pp. 926–974.
Watson, C. C., Haff, P. K., and Tombrello, T. A. 1980. Solar wind sputtering effects in the atmospheres of Mars and Venus. *Proc. Lunar Planet. Sci. Conf.* 11:2479–2502.
Wetherill, G. W. 1981. Solar wind origin of ^{36}Ar on Venus. *Icarus* 46:70–80.
Wetherill, G. W. 1985. Occurrence of giant impacts during the growth of the terrestrial planets. *Science* 228:877–879.
Wieler, R., Baur, H., and Signer, P. 1986. Noble gases from solar energetic particles revealed by closed system stepwise etching of lunar soil minerals. *Geochim. Cosmochim. Acta* 50:1997–2017.
Wood, C. A., and Ashwal, L. D. 1981. SNC meteorites: Igneous rocks from Mars? *Proc. Lunar Planet. Sci. Conf.* 12:1359–1375.
Yang, J., and Anders, E. 1982*a*. Sorption of noble gases by solids, with reference to meteorites. II. Chromite and carbon. *Geochim. Cosmochim. Acta* 46:861–875.
Yang, J., and Anders, E. 1982*b*. Sorption of noble gases by solids, with reference to meteorites. III. Sulfides, spinels, and other substances; on the origin of planetary gases. *Geochim. Cosmochim. Acta* 46:877–892.
Yang, J., Lewis, R. S., and Anders, E. 1982. Sorption of noble gases by solids, with reference to meteorites. I. Magnetite and carbon. *Geochim. Cosmochim. Acta* 46:841–860.
Yatteau, J. H. 1983. Some issues related to evolution of planetary atmospheres. Ph.D. thesis, Harvard.
Yung, Y. L., Allen, M., and Pinto, J. P. 1984. Photochemistry of the atmosphere of Titan: Comparison between model and observations. *Astrophys. J. Suppl.* 55:465–506.
Zahnle, K. J., and Kasting, J. F. 1986. Mass fractionation during transonic hydrodynamic escape and implications for loss of water from Venus and Mars. *Icarus* 68:462–480.
Zahnle, K. J., and Walker, J. C. G. 1982. The evolution of solar ultraviolet luminosity. *Rev. Geophys. Space Phys.* 20:280–292.

PART 8
Magnetic Fields in the Early Solar System

8.1. MAGNETIC STUDIES OF METEORITES

NAOJI SUGIURA
University of Tokyo

and

DAVID W. STRANGWAY
University of British Columbia

The intensity of natural remanent magnetization (NRM) is nearly proportional to the intensity of saturation remanent magnetization in achondrites and carbonaceous chondrites. The NRM in these meteorites is stable against alternating field demagnetization. We consider this NRM to reflect magnetic fields that were present in the early solar system. The correlation between the NRM intensity and the intensity of saturation remanent magnetization in ordinary chondrites is weak. Since the coercive force of kamacite in ordinary chondrites is very small, it is possible that the extraterrestrial remanence in many ordinary chondrites is dominated by a soft, spurious remanence. Theoretically, interplanetary magnetic fields of appropriate strength could have existed for 10 Myr in the early solar system, causing meteorites to be magnetized during the accretion and cooling stages.

8.1.1. INTRODUCTION

Magnetic properties of meteorites have been studied to establish physical conditions in the early solar system. In this chapter, we review recent studies of magnetic minerals, natural remanent magnetization (NRM), paleomagnetic field intensity, and implications for the history of parent bodies and for the nature of magnetic fields in the early solar system. Earlier reviews on meteorite magnetism have been published by Gus'kova (1972), by Levy and Sonett (1978) and by Nagata (1979). For more details on rock magnetism in general, refer to O'Reilly (1984).

8.1.2. MAGNETIC MINERALS IN METEORITES

The main magnetic minerals found in meteorites are listed in Table 8.1.1. They are usually identified by thermomagnetic analysis which measures saturation magnetization as a function of temperature (Fig. 8.1.1). In cases where solid solutions are formed, magnetic properties such as Curie temperature (Néel temperature, or crystallographic transition temperature), and saturation magnetization (Js) are a function of composition, while the coercive force (Hc) is, in addition, dependent on crystal defects, grain size and the nature of any intergrowth texture. The coercive force shown in the table is a typical value found in meteorites in which the listed mineral is the main magnetic phase present.

Kamacite is, by far, the most abundant and common magnetic mineral in meteorites. It is found in all types of meteorites except for CI chondrites and SNC meteorites. It is important as a carrier of remanence when present in very small grains. At grain size > 100 μm, its coercive force is very small and it acquires spurious remanence easily.

TABLE 8.1.1
Magnetic minerals

Mineral	Tc,Tn,Tt[a] (°C)	Js (AM^2/KG)	Hc (10^{-4}T)	Occurrence[b]
Kamacite	$Tt = 500 - 770$[c] ($0-770$)[d]	<220	5	O,E,CM,CO,CV, HED,U,I,M,P
Taenite	0–600	<150	—	O,E,CO,I,M,P
Tetrataenite	$Tt = 550$	160	1000	O,I,M,P
Troilite	$Tn = 320$	0	—	O,E,CM,CO,CV, HED,U,I,M,P
Pyrrhotite	320	<19.5	—	(CI,SNC)
Cohenite	215	128	—	(E,U,I)
Schreibersite	<443	<155.4	—	(O,E,I,M,P)
Magnetite	580	92	100	CI,CM,CO,CV
Titanomagnetite	<580	<92	500	SNC
PCP	—	weak	—	CM

[a] Tc, Tn and Tt are Curie, Néel and other transition temperatures.
[b] U,I,M and P denote ureilites, iron, mesosiderites and pallasites, respectively.
[c] $\alpha-\gamma$ transition temperature which depends on nickel concentration.
[d] $\gamma-\alpha$ transition temperature which depends on nickel concentration.

MAGNETIC STUDIES OF METEORITES

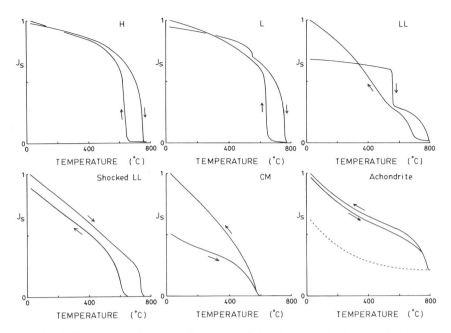

Fig. 8.1.1. Thermomagnetic curves of meteorites. Saturation magnetization is plotted against temperature. The curve for LL chondrites shows a sudden drop in Js at 550° C which is due to tetrataenite. The curves for L and H chondrites are essentially the same as that of LL chondrites but with a smaller amount of tetrataenite. Some ordinary chondrites which do not show the typical transition of tetrataenite at 550° C are considered to have cooled rapidly after a shock event. Saturation magnetization of CM chondrites generally increases after the heating due to production of magnetite from less magnetic material. In the case of achondrites, a large fraction of Js is due to paramagnetism (shown by the broken curve).

Taenite is not very common in chondrites, because tetrataenite is the stable phase below 320° C (Reuter et al. 1986). Its presence is detected by thermomagnetic analysis, when the thermomagnetic curve shows a reversible Curie point at $> 550°$ C (see Fig. 8.1.1; shocked LL). The presence of taenite in a meteorite indicates that the meteorite cooled rapidly, probably after a shock event. Taenite is formed in the laboratory when tetrataenite is heated above 550° C (Wasilewski 1982).

Tetrataenite is a fairly common mineral and abundant in LL chondrites. Its magnetic properties are not yet fully understood. The saturation magnetization of tetrataenite disappears abruptly at 550° C (see Fig. 8.1.1: LL chondrite). This was interpreted as a Curie temperature in the earlier studies (see, e.g., Wasilewski 1982). It is probable that this is due to the order-disorder transition from tetrataenite to taenite. The Curie temperature of tetrataenite is probably much higher than the order-disorder transition temperature (550° C), judging from the shape of the saturation magnetization vs temperature curve.

The coercive force and the remanent coercive force of nearly pure tetrataenite single crystals have been reported to be 4 to 76.5 mT and 25.5 to 148 mT, respectively (Nagata et al. 1986a). These values are quite high when compared with those of the other common magnetic minerals found in meteorites. Synthetic tetrataenite has a coercive force of 490 mT (Néel et al. 1964). The difference may be due to partial ordering in the natural samples. The remanent coercive force (and occasionally the coercive force) of the ordinary chondrites exceed the values found in single crystals of tetrataenite (Nagata et al. 1986a). Ordinary chondrites often contain cloudy taenite which is a microscopic mixture of tetrataenite and a Ni-poor phase (Reuter et al. 1986). This microstructure is probably responsible for the very high remanent coercive force (0.2 T) of ordinary chondrites. Since cloudy taenite is abundant, it is expected to be an important carrier of stable remanence in ordinary chondrites.

Troilite is ideally antiferromagnetic, which means that it cannot carry any remanence. There are a few studies (see, e.g., Pearce et al. 1976) which indicate that when it is in contact with metallic Fe, there is a significant magnetic interaction between these minerals, resulting in anomalous partial TRM (thermoremanent magnetization) acquisition behavior. More detailed studies are needed on the magnetic properties of this mineral as it occurs in meteorites.

Pyrrhotite is found in meteorites which are deficient in metallic iron, i.e., CI and SNC meteorites. Unstable S-containing minerals such as PCP (poorly characterized phase, see Chapter 3.4) could break down and produce pyrrhotite during heating experiments. Since it is strongly anisotropic, it is capable of carrying stable remanence.

Cohenite is fairly abundant in E chondrites. In Abee, it is the major carrier of the remanence (Sugiura and Strangway 1983). Cohenite is a common accessory mineral in iron meteorites and ureilites, but we do not know if it is an important carrier of NRM in these meteorites.

Schreibersite, $(FeNi)_3P$, is a fairly common mineral in iron meteorites. Its magnetic properties such as coercive force are not well known and we do not know if it is an important carrier of NRM.

Magnetite is the major magnetic mineral in carbonaceous chondrites. It has a fairly high coercive force because of the fine grain size and could be the major carrier of NRM in C chondrites. It is an accessory mineral in Type 3 ordinary chondrites. Since its Curie temperature overlaps with the transition temperature of tetrataenite, it is difficult to discover if magnetite carries some of the NRM in Type 3 ordinary chondrites. Since most C and Type 3 ordinary chondrites have not been reheated to the Curie temperature of magnetite (Dodd 1981), it is unlikely that the NRM carried by magnetite is of total TRM origin.

Titanomagnetite is found in SNC meteorites (Stolper and McSween 1979). It is usually subdivided by ilmenite lamellae and carries stable NRM, but sometimes acquires large viscous remanence magnetization in the geomagnetic field.

PCP (poorly characterized phase) in CM chondrites is known to be ferromagnetic (Fuchs et al. 1973). Little is known about its magnetic properties. According to thermomagnetic analysis by Watson et al. (1975), the saturation magnetization is probably much weaker than that of magnetite. PCP could be the carrier of the very stable NRM found in CM meteorites.

In summary, magnetic minerals in meteorites are quite often different from those in terrestrial rocks. We are only beginning to investigate them adequately. In particular, tetrataenite, troilite, cohenite, schreibersite and PCP must be studied systematically.

8.1.3. HYSTERESIS PARAMETERS OF METEORITES

Hysteresis parameters (J_s:saturation magnetization; J_{rs}:saturation remanence; H_c:coercive force; H_{cr}:remanence coercive force) have been measured on many meteorites. In Fig. 8.1.2, H_c is plotted against J_{rs}/J_s for various meteorites. High values of H_c and J_{rs}/J_s indicate a high degree of magnetic stability. The position in this diagram is mainly determined by the kind of magnetic minerals in the meteorite, and by the grain size of the magnetic minerals. Theoretically, J_{rs}/J_s should be proportional to H_c (for the same magnetic material) at low values of H_c, and asymptotically approach 0.5 to 0.8 at higher values of H_c. Minor deviation (a factor of 2) from the strict proportionality can be attributed to the shape anisotropy of magnetic grains. The main trends observed in Fig. 8.1.2 are explained by mixing of kamacite and tetrataenite, or by differences in the grain size of kamacite. Data points for several achondrites are distinctively displaced upwards. They probably contain a magnetic mineral of small J_s, but it has not yet been identified (Nagata and Funaki 1984).

Among chondrites, carbonaceous chondrites are the most reliable as ancient magnetic recorders. So far, no detectable differences in these parameters have been found among the subtypes of carbonaceous chondrites. Thermomagnetic characteristics are, however, distinct within each subgroup (Rowe 1976). The H_c values are tightly clustered around 0.015 T, which suggests that if magnetite is the main magnetic material, the mean grain size is submicroscopic (Heider et al. 1987). Electron microscope studies (Kerridge 1970; Hyman and Rowe 1986) showed that the grain size of magnetite ranges from < 1 μm to > 15 μm.

LL, L and H chondrites form a single trend in Fig. 8.1.2a. LL chondrites tend to be located on the upper right, while H chondrites tend to be on the lower left. This trend is explained as a result of mixing of kamacite and tetrataenite. Kamacite is the unstable (low H_c, low J_{rs}/J_s) component and tetrataenite is the stable component (high H_c, high J_{rs}/J_s). (The trend within the LL and L groups may require different degrees of ordering of tetrataenite.)

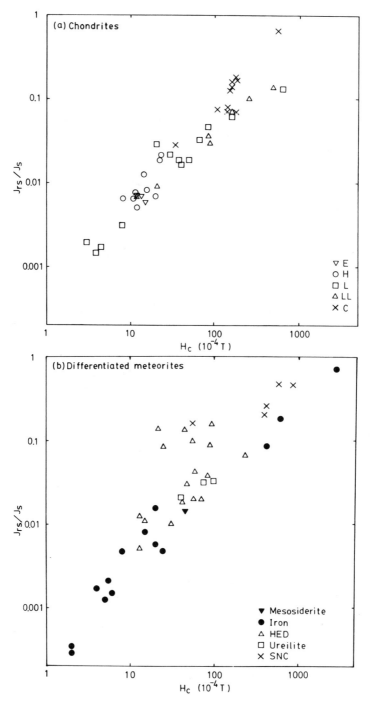

Fig. 8.1.2. Jrs/Js vs Hc diagram for (a) chondrites and (b) differentiated meteorites. Generally, the larger these parameters, the more reliable the paleomagnetic record (figure from Sugiura and Strangway 1987).

Four L chondrites are located near the lower-left corner of the diagram. Thermomagnetic analysis showed that either they do not contain tetrataenite (Nagata 1979c) or their tetrataenite has a distinctively lower transition temperature ($< 550°$ C: Nagata and Sugiura 1976; Nagata and Funaki 1981). In either case, it is an indication of complicated thermal history. E chondrites form a tight cluster on the trend shown in Fig. 8.1.2a. Although E chondrites do not contain tetrataenite, their Hc is similar to that of H chondrites, and significantly higher than that of shock-reheated L chondrites. It seems that microscopic texture in kamacite (e.g., cohenite exsolution) is responsible for the relatively high value of Hc.

Among differentiated meteorites, SNC meteorites have the highest Jrs/Js and Hc values. Titanomagnetite grains which are finely divided by ilmenite lamellae are probably responsible for the high coercive force of SNC meteorites (Cisowski 1986). The trend formed by iron meteorites (and a mesosiderite) is essentially the same as that of LL, L and H chondrites, and is explained as a mixing of kamacite and tetrataenite. Three iron meteorites located on the upper right, are Ni-rich ataxites and contain a large amount of tetrataenite (Nagata et al. 1986b). The main magnetic mineral in HED (howardite, eucrite and diogenite) achondrites (excluding those which are off the trend) is kamacite (Duke 1965). The difference in Hc within this group is probably explained by the difference in the grain size of kamacite (the smaller the grains, the higher the Hc). At present, no systematic difference in hysteresis properties is observed among the HED subgroups. Ureilites, whose main magnetic mineral is kamacite, are also located within the field of HED achondrites. Fine-grained kamacite possibly produced by reduction of silicates, may be responsible for the relatively high Hc values of these achondrites.

8.1.4. NATURAL REMANENT MAGNETIZATION

In Fig. 8.1.3, the NRM intensity is plotted against the intensity of saturation remanent magnetization (Jrs) for various meteorites. This diagram tells more about the strength of the NRM than the plot of NRM vs susceptibility diagram used by Sonett (1978). Susceptibility is more sensitive to the amount of large grains while saturation remanence is more sensitive to the amount of small grains which carry the NRM. One of the difficult problems in meteorite magnetism is that researchers cannot be sure what may have happened to the meteorites between the time of the fall to the Earth and the time that the samples were measured. In some museums, magnets may have been used to identify meteorites. Direct contact with magnetized iron (or steel) products could happen more often. Exposure of a meteorite to such an artificial magnetic field can produce a large isothermal remanence, resulting in an anomalous NRM. The diagrams in Fig. 8.1.3 are constructed by choosing the smaller value of NRM intensity, if there are duplicate NRM measurements on the same meteorite (about 10% of the data). The NRM and Jrs are positively correlated for all differentiated meteorites. The correlation is very good within the group of ureilites and within the group of diogenites (Fig. 8.1.3b). This good correlation suggests that the problem of exposure to artificial field is not a serious one for most differentiated meteorites. Two eucrites which are displaced to the upper left from the general trend may have been exposed to a strong artificial magnetic field. The NRM of many achondrites are fairly stable (Fig. 8.1.4) against alternating field (AF) demagnetization (ureilites: Brecher and Furhman 1979b; eucrites: Nagata 1980; SNC meteorites: Cisowski 1986), because of the small grain size of their magnetic minerals. Thus, the NRM in these meteorites is probably of extraterrestrial origin, unless they are affected by viscous remanent magnetization (VRM) acquired in the geomagnetic field. (VRM is a time-dependent magnetization whose intensity is proportional to $\log(t)$, where t is the time for which a sample is ex-

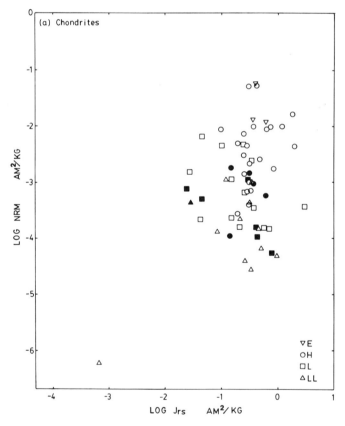

Fig. 8.1.3. NRM vs Jrs diagram for (a) ordinary and enstatite chondrites (Antarctic chondrites are shown by solid symbols), (b) differentiated meteorites, and (c) carbonaceous chondrites (figure from Sugiura and Strangway 1987).

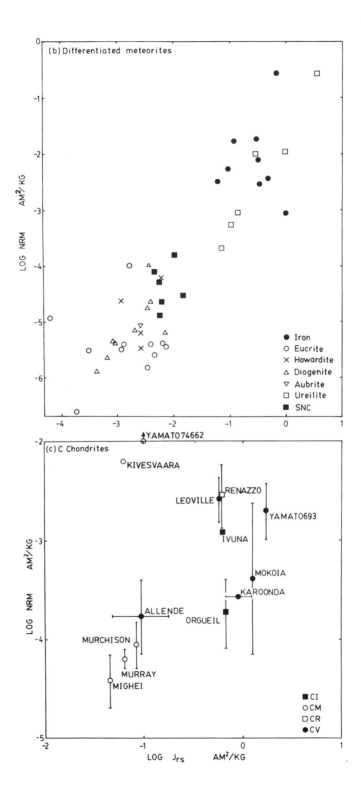

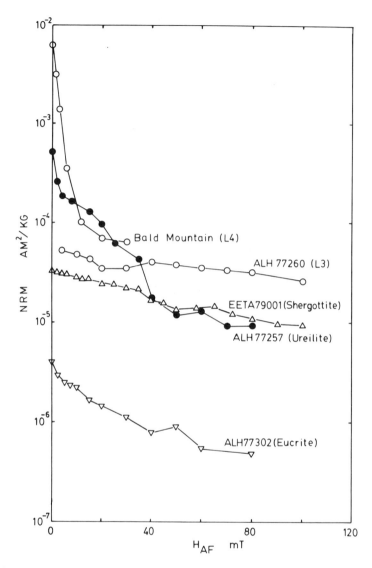

Fig. 8.1.4. Alternating field demagnetization curves of NRM for various achondrites and for two L chondrites. NRM in achondrites is generally quite stable against AF demagnetization. The AF demagnetization curve for Bald Mountain shows rapid decrease in intensity at < 10 mT, which is common to many ordinary chondrites with a high NRM/Jrs ratio. The NRM in ALH77260, whose NRM/Jrs ratio is small, is quite stable against AF demagnetization (data from Brecher and Ranganayaki 1975; Nagata 1980; Nagata and Funaki 1982; Collinson 1986).

posed to a magnetic field.) VRM could be fairly resistant against alternating field demagnetization. The best way to erase VRM in NRM is to thermally demagnetize the NRM. Because of experimental difficulties, thermal demagnetization has not been done in most meteorite studies. Experiments of viscous decay and acquisition of VRM (Nagata 1980; Collinson 1986) indicate that some of the HED achondrites and SNC meteorites are magnetically quite viscous (acquire large VRM in the geomagnetic field). The magnetic viscosity of ureilites is not well known, but the presence of abundant fine Fe grains suggests viscous magnetization may be important in ureilites. Thermal demagnetization of NRM or long-term viscous magnetization experiments are required to assure that the NRM in these meteorites is of extraterrestrial origin.

The correlation between NRM and Jrs for iron meteorites is rather weak. We think this lack of correlation is caused by exposure to a weak artificial magnetic field, because the coercive force of kamacite in iron meteorites is very small. For more details, see the following discussion on ordinary chondrites.

The NRM vs Jrs diagram for ordinary chondrites and E chondrites (Fig. 8.1.3a) is quite different from that for differentiated stony meteorites. The strong positive correlation between NRM and Jrs is not observed among these chondrites. A possible explanation of the lack of correlation would be that many ordinary chondrites have been exposed to an artificial magnetic field and acquired large isothermal remanences. This explanation may seem rather contradictory to the observation that few differentiated stony meteorites seem to have acquired large isothermal remanence. The key to this apparent contradiction is the difference in coercive force Hc between these meteorites. Differentiated stony meteorites contain fine-grained kamacite (or titanomagnetite in the case of SNC meteorites) whose Hc is larger than 3 mT. In contrast, the Hc of kamacite in ordinary chondrites is only 0.5 mT. (The Hc shown in Fig. 8.1.2a is larger because of the presence of tetrataenite.) Therefore, exposure to an artificial magnetic field of, say 1 mT, does not produce much remanence in achondrites, but produces large remanence in ordinary chondrites. There are two lines of circumstantial evidence in support of this explanation: (1) Antarctic chondrites which were collected recently and have less chance of exposure to a large magnetic field, have relatively small NRM (less than 2×10^{-3} Am^2/Kg for ordinary chondrites; Fig. 8.1.3a); and (2) alternating field demagnetization experiments reveal that the NRM in chondrites with small NRM/Jrs is more stable than that in chondrites with large NRM/Jrs. As an example, the AF demagnetization curve of ALH77260 which has the lowest NRM/Jrs ratio among L chondrites is compared with the AF demagnetization curve of the Bald Mountain L chondrite which has the highest NRM/Jrs ratio (Fig. 8.1.4).

A detailed study of the Olivenza (LL5) chondrite on the possibility of magnetic contamination (i.e., exposure to an artificial magnetic field) has been reported by Collinson (1985). If the above explanation is correct, then the ratios of the probable extraterrestrial NRM to Jrs are about 10^{-3}, 10^{-4} and 10^{-4} for H, L and LL chondrites, respectively. In comparison, the ratio is 2×10^{-3} for differentiated stony meteorites. There are other possible explanations of the scattered NRM vs Jrs diagram for ordinary chondrites. NRM in some chondrites may have been partly erased by extraterrestrial shock demagnetization. NRM intensity in some chondrites may have increased in the geomagnetic field due to viscous remanence, shock remanence or chemical remanence during weathering. Magnetic properties of individual meteorite samples have to be carefully examined to discover if any of the above mechanisms could explain the NRM properties of the meteorite. Unfortunately, such detailed studies are scarce.

NRM in carbonaceous chondrites is the most stable NRM among the various groups of meteorites. Viscous remanence is not very important in most carbonaceous chondrites (Brecher and Arrhenius 1974), and because of the high coercivity, exposure to a small artificial magnetic field is not a serious problem in most cases. Therefore, the NRM in carbonaceous chondrites is considered to provide the best evidence for the presence of magnetic fields in the early solar system. Fig. 8.1.3c shows that the NRM and the Jrs correlate fairly well and the ratio NRM/Jrs is about 10^{-3}.

On the whole, the NRM data suggest that a preterrestrial remanence is preserved in meteorites, although the origin of NRM in many meteorites is not fully understood. In the following, some of the contentious issues are summarized.

Heating during the fall through the Earth's atmosphere

Stacey et al. (1961) argued that heating during ablation in the Earth's atmosphere has a negligible effect on remanent magnetization in the interior of stony meteorites. The difference in the NRM directions of the fusion crust and the interior (see, e.g., Nagata 1979b) supported this argument. But a study of the Wellman chondrite (Hamano and Matsui 1983) showed that the remanence directions of the fusion crust and the interior are related. It was argued that the remanence was acquired during the fall through the atmosphere. It may be that ordinary chondrites, which contain low-coercivity kamacite, are susceptible to such a secondary effect. In any case, it seems unlikely that very stable NRM (e.g., NRM in carbonaceous chondrites which cannot be erased by a field of 0.1 T) was produced during ablation. The directional heterogeneity of NRM within an Allende specimen (Sugiura and Strangway 1985) also provides evidence that the NRM is of extraterrestrial origin.

Shock effects

Many meteorites have been strongly shocked (Chapter 3.6). Remanent magnetization is partially erased if a shock event occurs in the absence of a magnetic field. If a shock is very strong, then post-shock temperature can exceed the Curie temperature of magnetic materials, and the subsequent cooling (in the absence of magnetic field) should completely erase the NRM. (In the presence of a field, a new TRM would be produced.) Although not many ^{40}Ar-^{39}Ar dates are available, it seems that strong shock events causing partial Ar loss occurred fairly recently (Bogard et al. 1976). Therefore, there must have been no magnetic field at the time of the shock events. Nevertheless, many strongly shocked meteorites (Brecher et al. 1977; Brecher and Furhman 1979a) seem to have fairly strong NRM. This is puzzling. Generation (or amplification) of magnetic fields associated with the shock event is not a satisfactory explanation because post-shock cooling is slow. It is possible that a parent body was strongly magnetized. The local magnetic field generated by a remanent magnetization of 2.5×10^{-4} AM^2/Kg is about 0.1 μT, which seems to be smaller than the magnetic field required to produce the NRM in shocked chondrites (see Sec. 8.1.5 below). More detailed studies of shocked chondrites are needed to solve this problem. In the case of ureilites and SNC meteorites, which are strongly shocked, there are other possible explanations for the presence of strong NRM. Since ^{40}Ar-^{39}Ar ages are not known for ureilites, the shock may have occurred in the early solar system, where interplanetary magnetic fields may have existed. SNC meteorites may have been shocked on Mars which might have had a magnetic field at the time of shock events.

Heterogeneous magnetization

It is known that NRM directions of some meteorites are inhomogeneous among subsamples. In some cases, there are reasonable explanations. In the case of iron meteorites (Brecher and Albright 1977), the remanence seems to be controlled by the crystallographic orientation of the Widmanstätten structure. In an Allende specimen, the NRM direction appears to have been disturbed by mechanical deformation (Sugiura and Strangway 1985). But many ordinary chondrites also show scattered NRM directions among subsamples (Brecher et al. 1977; Funaki et al. 1981; Sugiura and Strangway 1982). Inhomogeneity in NRM direction is pronounced when tetrataenite is abundant in the sample. Therefore, a possible explanation of the inhomogeneity is that highly anisotropic tetrataenite acquired nearly random NRM whose direction is controlled by the crystallographic orientation (Sugiura and Strangway 1982). In the case of very primitive type 3 chondrites (e.g., Chainpur), the directional inhomogeneity of NRM may be due to preaccretional remanence (Sugiura and Strangway 1982). Many ordinary metamorphosed ordinary chondrites may, although not easily recognized, actually be breccias formed after metamorphism (Scott et al. 1985). If such breccias were assembled without much reheating, random NRM directions will result. Among the heterogeneously magnetized meteorites, Bjurböle is such a breccia (Scott et al. 1985); ALH76009 may also be such a breccia (Funaki et al. 1981). It is important to know which of the above explanations is correct, because it is directly related to the origin of the NRM in these chondrites.

Tetrataenite

All the magnetic studies of ordinary chondrites (and iron meteorites) before Wasilewski (1982) suffer from the lack of recognition of the importance of tetrataenite as a carrier of NRM. Thermal demagnetization of NRM in ordinary chondrites (Sugiura and Strangway 1982; Nagata et al. 1986a) clearly shows that tetrataenite is the main carrier of the NRM in L and LL chondrites. Apart from the problem that NRM directions tend to be scattered in tetrataenite-containing chondrites, it is not yet clear how tetrataenite acquires remanent magnetization. According to the Fe-Ni phase diagram (Reuter et al. 1986), Fe-Ni alloy with 10% nickel is expected to experience the following events during cooling from the peak metamorphic temperature. At about 700° C, kamacite starts to nucleate and the Ni concentration in taenite increases. Kamacite at this temperature is already ferromagnetic and acquires remanent magnetization if there is an ambient magnetic field, but taenite is still paramagnetic. If equilibrium is maintained, taenite becomes ferromagnetic at about 360° C when the Ni concentration is about 40%. If there is the usual Ni concentration gradient in taenite, it will be the Ni-rich rim that first becomes ferromagnetic. Since the adjacent kamacite is already ferromagnetic, the remanence in taenite may be affected by ferromagnetic interactions between kamacite and taenite. As the temperature drops below 320° C, taenite transforms to tetrataenite. Depending on the relative strength of the external magnetic field and the remanent magnetization in the precursor taenite, tetrataenite could acquire (on the average) a remanence either in the direction of the precursor remanence or in the direction of the external magnetic field. (In each grain the NRM direction may be deflected to the direction of the c-axis.) The interior of taenite (cloudy taenite) decomposes to tetrataenite and martensite (body-centered cubic phase with 11% Ni) which are both ferromagnetic. The NRM in this region may be very complicated because of the magnetic interactions with the tetrataenite rim and within the cloudy taenite. Detailed experimental work is needed to find out if tetrataenite (and cloudy taenite) are faithful recorders of the paleomagnetic field.

Carrier of NRM in Carbonaceous chondrites

In spite of the fact that carbonaceous chondrites provide the best paleomagnetic evidence (Brecher and Arrhenius 1974) for the presence of a magnetic field in the early solar system, the carriers of the NRM in carbonaceous chondrites have not been identified.

Allende has been studied in detail (see, e.g., Banerjee and Hargraves 1972; Sugiura et al. 1979; Wasilewski and Saralker 1981; Nagata and Funaki 1983; Sugiura and Strangway 1985). Magnetite and Fe-Ni alloy are the main magnetic phases according to the thermomagnetic analyses, but the NRM of a bulk sample is completely demagnetized at 320° C (Sugiura et al. 1979). Therefore, the carrier of NRM is not likely to be either magnetite or Fe-Ni alloy. As 320° C corresponds to the Curie temperature of pyrrhotite, and some of the sulfide minerals seem to be deficient in iron compared with troilite, Nagata and Funaki suggested that pyrrhotite is the NRM-carrying mineral. However, the NRM-carrying mineral is unstable when heated to 75° C– 150° C (Wasilewski and Saralker 1981). It seems that the NRM-carrying mineral is a sulfide which decomposes between 75 and 150° C to pyrrhotite. The NRM in Allende is probably a chemical remanent magnetization (CRM) associated with a sulfidation event (Wasilewski and Saralker 1981).

Carriers of NRM in other carbonaceous chondrites are less well known. A large fraction of NRM in Orgueil and Murchison is demagnetized when heated to 150–200° C (Banerjee and Hargraves 1972). Such demagnetization behavior suggests that the carrier is not the main magnetic mineral, magnetite. Poorly characterized phase (PCP) is a prime candidate as the carrier of NRM in CM chondrites (Bunch and Chang 1980; see also Chapter 3.4).

8.1.5. PALEOINTENSITY

Paleomagnetic field intensities are best known for achondrites because their NRM is probably TRM whose properties are well understood. The paleointensity estimates for C and O chondrites are made on a less certain basis.

Achondrites

Because magnetic properties of achondrites easily change by laboratory heating (Brecher et al. 1979), the conventional Thellier's method of paleomagnetic field determination has not been used for achondrites. (In Thellier's method, the NRM loss during zero-field heating and the acquisition of partial TRM during heating in a magnetic field are compared.) It has been observed that alternating field demagnetization characteristics of anhysteretic remanent magnetization (ARM, which is analogous to thermoremanence, but does not require heating) and NRM are similar for many achondrites (Sugiura 1977; Nagata 1980; Nagata and Dunn 1981). In such cases, the ARM method of paleointensity determination can be employed (Stephenson and Collinson 1974). The method requires a calibration factor which is the ratio of TRM to ARM. Usually the ratio is assumed to be about 1.3 but the ratio for a eucrite was found experimentally to be 6.80 (Nagata and Dunn 1981). Assuming this calibration factor applies to all achondrites, the paleointensities for three achondrites range from 1.1 μT to 1.9 μT with the average value of 1.6 μT.

For the ureilite ALH 77257, a paleointensity of 1.7 μT (recalculated with the calibration factor of 6.8) was obtained with the ARM method (Nagata 1980). By contrast, a much higher paleointensity (140 μT) was obtained for Goalpara by Brecher and Furhman (1979b) by comparing the AF demagnetization of NRM and of TRM (van Zijl method). The difference is not attributable to the difference in the method, but to the difference in the NRM/Jrs of these ureilites. More studies are needed to find the cause of this difference.

The same ARM method was applied to an SNC meteorite ALH 77005, and a paleointensity of 1 μT was obtained with a calibration factor of 1.3 (Nagata 1980). Since another sample from the same meteorite did not have a stable remanence (Collinson 1986), further studies are needed to confirm this paleointensity; confirming this is important because it could record a paleofield on Mars. Thellier's method of paleointensity determination was applied to Shergotty (Cisowski 1986), but the result was inconclusive.

Ordinary chondrites

It is not known what method of paleointensity determination can be best applied to ordinary chondrites. The ARM method is not used mainly because the NRM and the ARM have different AF demagnetization behavior (Sugiura 1977). The van Zijl method was applied by Brecher and coworkers (see, e.g., Brecher and Ranganayaki 1975; Brecher and Leung 1979), but this is not a suitable method to apply to ordinary chondrites, because tetrataenite (the main carrier of the stable remanence) is destroyed by heating. Thellier's method could be applied in the temperature range < 550° C where tetrataenite is not destroyed. Paleointensities estimated with this method by Nagata and Sugiura (1977) range from 10 to 68 μT, while those reported by Westphal and Whitechurch (1983) range from 48 to 188 μT. There are two problems with this method: (1) the NRM in the temperature range < 550° C is carried by kamacite which can easily acquire a spurious remanence, and it is not easy to distinguish this artificial NRM from the extraterrestrial NRM; and (2) because of the multidomain behavior of kamacite, the estimated paleointensities tend to be larger than those expected theoretically (Westphal 1986). At present, there is no ideal method of paleointensity determination for ordinary chondrites.

Carbonaceous chondrites

Paleointensity determinations on carbonaceous chondrites have been made mainly by Thellier's method based on the assumption that the NRM is of TRM origin. As discussed earlier, the NRM may be of CRM origin. Therefore, the paleointensities obtained by Thellier's method may be considered slightly smaller than the actual paleomagnetic field (Banerjee and Hargraves 1972). The paleointensities range from 100 μT for Allende to 18 μT for Murchison. Wasilewski (1981) questions the validity of these paleointensities, because irreversible changes occur at temperatures as low as 50° C. It should perhaps be emphasized that the temperature used in the paleointensity determinations is mostly < 150° C. The ARM method with an assumed calibration factor 1.3, yielded a paleointensity of 73 μT (Nagata 1979a) for Allende. Since the calibration factor cannot be precisely known, and since the main magnetic materials, magnetite and Fe-Ni alloy, contribute to the ARM but not to the NRM, the paleointensity must be considered as a rough estimate. Nevertheless, it agrees well with the value estimated from Thellier's method.

Among other meteorites, most enstatite chondrites are strongly magnetized and the estimated paleointensities range from 700 to 1600 μT (Sugiura and Strangway 1982,1983). These values are unusually high and duplicate measurements are needed to confirm these results. Most iron meteorites are poor recorders of paleomagnetic field, and no reliable paleointensity has been obtained (Brecher and Albright 1977).

8.1.6. IMPLICATIONS FOR THERMAL, CHEMICAL AND MECHANICAL HISTORY

Paleomagnetism has been applied to many geological problems on the Earth. For instance, the temperature dependence of NRM directions can be used to infer the thermal history of a geological body, or if the age of the body is known, the mechanical history (such as continental drift) can be revealed. In principle, the paleomagnetism of meteorites can also be used to study thermal, chemical and mechanical histories of meteorites.

In the case of the Allende meteorite, a detailed study revealed the following history (Sugiura and Strangway 1985). Chondrules seem to have acquired remanence before accretion. Some time after (or during) accretion, a sulfidation event occurred and homogeneously remagnetized the whole meteorite, but a fraction of the preaccretion remanence survived. The temperature at the time of the sulfidation event is not well known but certainly $< 320°$C. After the sulfidation event, a mechanical deformation, probably by shock, took place which deformed the matrix and rotated the chondrules slightly.

As discussed earlier, scattered NRM directions in ordinary chondrites may be a result of brecciation which occurred after metamorphism. The scattered NRM directions in very primitive chondrites are probably due to preaccretional remanence. This means that magnetization acquired by chondrules that predate formation of the meteorite is random and that they were not remagnetized during or after assembly of the meteorite. This is strong evidence for early solar-system fields, through better understanding of the magnetic properties of tetrataenite are desirable to verify this conclusion.

Presence of tetrataenite can be easily detected by magnetic measurements. Since the presence of tetrataenite means slow cooling below 320°C (and the absence probably means quick cooling below 320°C after a shock event), it is possible to draw an inference on the thermal and shock history of meteorites from the measurement of magnetic properties.

8.1.7. SUMMARY OF THE PALEOMAGNETISM OF METEORITES

In our view, carbonaceous chondrites are the best recorders of paleomagnetic fields. The NRM is most probably of extraterrestrial origin. Achondrites (including ureilites and SNC meteorites) come second. Because they seem to have a fairly high magnetic viscosity, care must be taken to remove the viscous remanence acquired in the geomagnetic field. LL chondrites come third. Their NRM is very stable, but they are directionally scattered. Until we understand the magnetic properties of tetrataenite well, the interpretation of the paleomagnetic information in LL chondrites remains ambiguous. In the case of H and L chondrites, possible extraterrestrial remanence may be swamped with a soft, spurious remanence. There are not many detailed studies on enstatite chondrites. Most iron meteorites seem to be very poor recorders of paleomagnetic field.

8.1.8. ORIGIN OF THE MAGNETIC FIELDS

The origin of the magnetic fields which magnetized meteorites is not well known. Achondrites may have been magnetized by a magnetic field generated in the core of the parent body. But, chondrites and the chondrules within them were probably magnetized by interplanetary magnetic fields.

Theoretical studies on the effects of a magnetic field in the solar nebula (see, e.g., Umebayashi and Nakano 1984) show that fairly strong magnetic fields (30 μT) are expected to have existed in the solar nebula, and the Sun itself could have a field of 0.1 T on the surface. The solar magnetic field is, as discussed by Levy and Sonett (1978), not strong enough to produce the remanence found in carbonaceous chondrites. The intensity of magnetic field in the solar nebula (30 μT), which was inherited from the interstellar cloud, is estimated to be about the right strength needed to magnetize the carbonaceous chondrites. Ip (1984) considered the possibility of a large magnetic field generated by interaction of the T-Tauri wind and the solar nebula. It was estimated that the intensity of the field could be 20 to 60 μT. Such interplanetary magnetic fields would be expected to decay with the dispersal of the nebular gas, which probably occurred on a time scale of 10^7 Myr.

Acknowledgment. Financial support for the research was provided by Natural Science and Engineering Research Council.

REFERENCES

Banerjee, S. K., and Hargraves, R. B. 1972. Natural remanent magnetizations of carbonaceous chondrites and the magnetic field in the early solar system. *Earth Planet. Sci. Lett.* 17:110–119.

Bogard, D. D., Husain, L., and Wright, R. J. 1976. ^{40}Ar-^{39}Ar dating of collisional events in chondrite parent bodies. *J. Geophys. Res.* 81:5664–5678.

Brecher, A., and Albright, L. 1977. The thermoremanence hypothesis and the origin of magnetization in iron meteorites. *J. Geomag. Geoelectr.* 29:379–400.

Brecher, A., and Arrhenius, G. 1974. The paleomagnetic record in carbonaceous chondrites: Natural remanence and magnetic properties. *J. Geophys. Res.* 79:2081–2106.
Brecher, A., and Furhman, M. 1979a. Magnetism, shock and metamorphism in chondritic meteorites. *Phys. Earth Planet. Int.* 20:350–360.
Brecher, A., and Furhman, M. 1979b. The magnetic effects of brecciation and shock in meteorites. II. The ureilites and evidence for strong nebula magnetic fields. *Moon and Planets* 20:251–263.
Brecher, A., and Leung, L. 1979. Ancient magnetic field determination on selected chondritic meteorites. *Phys. Earth Planet. Int.* 20:361–378.
Brecher, A., and Ranganayaki, R. P. 1975. Paleomagnetic systematics of ordinary chondrites. *Earth Planet. Sci. Lett.* 25:57–67.
Brecher, A., Stein, J., and Furhman, M. 1977. The magnetic effects of brecciation and shock in meteorites: I. The LL-chondrites. *The Moon* 17:205–216.
Brecher, A., Furhman, M., and Stein, J. 1979. The magnetic effects of brecciation and shock in meteorites: III. The achondrites. *Moon and Planets* 20:265–279.
Bunch, T. E., and Chang, S. 1980. Carbonaceous chondrites, II. Carbonaceous chondrite phyllosilicates and light element geochemistry as indications of parent body processes and surface conditions. *Geochim. Cosmochim. Acta* 44:1543–1578.
Cisowski, S. M. 1986. Magnetic studies on Shergotty and other SNC meteorites. *Geochim. Cosmochim. Acta* 50:1043–1048.
Collinson, D. W. 1985. Magnetic properties of the Olivenza chondrite—a cautionary tale. *Meteoritics* 20:629.
Collinson, D. W. 1986. Magnetic properties of Antarctic shergottite meteorite EETA 79001 and ALHA 77005: Possible relevance to a martian magnetic field. *Earth Planet. Sci. Lett.* 77:159–164.
Dodd, R. T. 1981. *Meteorites: A Petrologic-Chemical Synthesis* (Cambridge: Cambridge Univ. Press).
Duke, M. B. 1965. Metallic irons in basaltic achondrites. *J. Geophys. Res.* 70:1523–1527.
Fuchs, L. H., Olsen, E., and Jensen, K. J. 1973. Mineralogy, mineral chemistry, and composition of the Murchison (C2) meteorite. *Smithson. Contrib. Earth Sci.* 10:1–39.
Funaki, M., Nagata, T., and Momose, K. 1981. Natural remanent magnetization of chondrules, metallic grains and matrix of antarctic chondrite, ALH-769. *Proc. 6th Symp. Antarctic Meteorites*, pp. 300–315.
Gus'kova, E. G. 1972. *Magnetic Properties of Meteorites* (Leningrad: Nauka). In Russian. English trans., NASA TTF-792, 1976.
Hamano, Y., and Matsui, T. 1983. Natural remanent magnetization of Wellman meteorite. *Geophys. Res. Lett.* 10:861–864.
Heider, F., Dunlop, D. J., and Sugiura, N. 1987. Magnetic properties of hydrothermally recrystallized magnetite crystals. *Science* 236:1287–1290.
Hyman, M., and Rowe, M. W. 1986. Saturation magnetization measurements of carbonaceous chondrites. *Meteoritics* 21:1–22.
Ip, W.-H. 1984. Magnetic field amplification in the solar nebula through interaction with the T-Tauri wind. *Nature* 312:625–626.
Kerridge, J. F. 1970. Some observations on the nature of magnetite in the Orgueil meteorite. *Earth Planet. Sci. Lett.* 9:299–306.
Levy, E. H., and Sonett, C. P. 1978. Meteorite magnetism and early solar system magnetic fields. In *Protostars and Planets,* ed. T. Gehrels (Tucson: Univ. of Arizona Press), pp. 516–532.
Nagata, T. 1979a. Meteorite magnetism and the early solar system magnetic field. *Phys. Earth Planet. Int.* 20:324–341.
Nagata, T. 1979b. Natural remanent magnetization of fusion crust of meteorites. *Proc. 4th Symp. Antarctic Meteorites*, pp. 253–272.
Nagata, T. 1979c. Magnetic properties of Yamato -7301(j), -7305(k), and -7304(m) chondrites in comparison with their mineralogical and chemical compositions. *Proc. 3rd. Symp. Antarctic Meteorites*, pp. 250–269.
Nagata, T. 1980. Paleomagnetism of Antarctic achondrites. *Proc. 5th Symp. Antarctic Meteorites*, pp. 233–242.
Nagata, T., and Dunn, J. R. 1981. Paleomagnetism of Antarctic achondrites (II). *Proc. 6th Symp. Antarctic Meteorites*, pp. 333–344.

Nagata, T., and Funaki, M. 1981. Magnetic properties of Antarctic stony meteorites Yamato-74115 (H5), -74190 (L6), -74354 (L6), and -74646 (L6). *Proc. 6th Symp. Antarctic Meteorites*, pp. 316–332.

Nagata, T., and Funaki, M. 1982. Magnetic properties of tetrataenite-rich stony meteorites. *Proc. 7th Symp. Antarctic Meteorites*, pp. 222–250.

Nagata, T., and Funaki, M. 1983. Paleointensity of the Allende carbonaceous chondrite. *Proc. 8th Symp. Antarctic Meteorites*, pp. 403–434.

Nagata, T., and Funaki, M. 1984. Notes on magnetic properties of Antarctic polymict eucrites. *Proc. 9th Symp. Antarctic Meteorites*, pp. 319–326.

Nagata, T., and Sugiura, N. 1976. Magnetic characteristics of some Yamato meteorites—magnetic classification of stone meteorites. *Memoirs Nat. Inst. of Polar Res.*, Ser. C., No. 10, pp. 30–58.

Nagata, T., and Sugiura, N. 1977. Paleomagnetic field intensity derived from meteorite magnetization. *Phys. Earth Planet. Int.* 13:373–379.

Nagata, T., Funaki, M., and Danon, J. A. 1986a. Magnetic properties of tetrataenite-rich meteorites II. *Proc. 10th Symp. Antarctic Meteorites*, pp. 364–381.

Nagata, T., Funaki, M., and Danon, J. A. 1986b. Magnetic properties of tetrataenite in Ni rich ataxites. Abstract for "11th Symp. Antarctic Meteorites," p. 143.

Nêel, L., Pauleve, J., Pauthenet, R., Laugier, J., and Dautreppe, D. 1964. Magnetic properties of an iron-nickel single crystal ordered by neutron bombardment. *J. Appl. Phys.* 35:873–876.

O'Reilly, W. 1984. *Rock and Mineral Magnetism* (Glasgow: Blackie & Son Ltd.).

Pearce, G. W., Hoye, G. S., Strangway, D. W., Walker, B. M., and Taylor, L. A. 1976. Some complexities in the determination of lunar paleointensities. *Proc. Lunar Sci. Conf.* 7:3271–3297.

Reuter, K. B., Williams, D. B., and Goldstein, J. I. 1986. The Fe-Ni phase diagram below 350° C. Abstract for 49th Meteoritical Soc. Meeting, D-9.

Rowe, M. W. 1976. Correlation of thermomagnetic behavior of carbonaceous chondrites with chemical petrographic classification. *Geochem. J.* 10:215–218.

Scott, E. R. D., Lusby, D., and Keil, K. 1985. Ubiquitous brecciation after metamorphism in equilibrated ordinary chondrites. *Proc. Lunar Planet. Sci. Conf.*, 16, *J. Geophys. Res.* Suppl. 90:D137–D148.

Sonett, C. P. 1978. Evidence for a primordial magnetic field during the meteorite parent body era. *Geophys. Res. Lett.* 5:151–154.

Stacey, F. D., Lovering, J. F., and Parry, L. G. 1961. Thermomagnetic properties, natural magnetic moment, and magnetic anisotropies of some chondritic meteorites. *J. Geophys. Res.* 66:1523–1534.

Stephenson, A., and Collinson, D. W. 1974. Lunar magnetic field paleointensities determined by an anhysteretic remanent magnetization method. *Earth Planet. Sci. Lett.* 23:220–228.

Stolper, E., and McSween, H. Y., Jr. 1979. Petrology and origin of the shergottite meteorites. *Geochim. Cosmochim. Acta.* 43:1475–1498.

Sugiura, N. 1977. Magnetic properties and remanent magnetization of stony meteorites. *J. Geomag. Geoelectr.* 29:519–539.

Sugiura, N., and Strangway, D. W. 1982. Magnetic properties of low-petrologic grade non-carbonaceous chondrites. *Proc. 7th Symp. Antarctic Meteorites*, pp. 260–280.

Sugiura, N., and Strangway, D. W. 1983. A paleomagnetic conglomerate test using the Abee E4 meteorite. *Earth Planet Sci. Lett.* 62:169–179.

Sugiura, N., and Strangway, D. W. 1985. NRM directions around a centimeter-sized dark inclusion in Allende. *Proc. Lunar Planet. Sci. Conf.*, 15, *J. Geophys. Res. Suppl.* 90:C729–C738.

Sugiura, N., and Strangway, D. W. 1987. Hysteresis and NRM properties of meteorites. Abstract for 12th Symp. on Antarctic Meteorites.

Sugiura, N., Lanoix, M., and Strangway, D. W. 1979. Magnetic fields of the solar nebula as recorded in chondrules from the Allende meteorite. *Phys. Earth Planet. Int.* 20:342–349.

Umebayashi, T., and Nakano, T. 1984. Origin and dissipation of magnetic fields in protostars and in the primitive solar nebula. *Proc. 16th ISAS Lunar Planet. Symp.*, pp. 118–121.

Wasilewski, P. 1981. New magnetic results from Allende C3(V). *Phys. Earth Planet. Int.* 26:134–148.

Wasilewski, P. J. 1982. Magnetic characterization of tetrataenite and its role in the magnetization of meteorites. *Lunar Planet. Sci.* XIII: 843–844 (abstract).

Wasilewski, P., and Saralker, C. 1981. Stable NRM and mineralogy in Allende: Chondrules. *Proc. Lunar Planet Sci.* 12B: 1217–1227.

Watson, D. E., Larson, E. E., Herndon, J. M., and Rowe, M. W. 1975. Thermomagnetic analysis of meteorites, 2. C2 chondrites. *Earth Planet. Sci. Lett.* 27: 101–107.

Westphal, M. 1986. Natural remanent magnetization, thermoremanent magnetization and reliability of paleointensity determination on H chondrites. *Phys. Earth Planet. Int.* 43: 300–306.

Westphal, M., and Whitechurch, H. 1983. Magnetic properties and paleointensity determination of seven H-group chondrites. *Phys. Earth Planet. Int.* 31: 1–9.

PART 9
Chondrules

9.1. PROPERTIES OF CHONDRULES

JEFFREY N. GROSSMAN
United States Geological Survey

ALAN E. RUBIN
University of California at Los Angeles

HIROKO NAGAHARA
University of Tokyo

and

ELBERT A. KING
University of Houston

Chondrules are the most abundant constituent of most groups of chondritic meteorites. The properties of chondrules are consistent with formation in the solar nebula from primitive material, and thus provide constraints on the nature of the nebula. Chondrules show a wide variety of textures, mineral abundances, mineral chemistries, bulk and isotopic compositions and sizes. Each major chondrite group (carbonaceous, ordinary and enstatite) contains chondrules with a different distribution of these properties, generally reflecting variations among the formation regions of the groups; there was little mixing of material between the groups. Chondrules contain a record of many important features of the solar nebula. The high-temperature events that produced the chondrules may have spanned <10 Myr. The nebular environment allowed for a short cooling period. As chondrules cooled, they became magnetized by nebular fields. Chondrule rims show that dust was present in the vicinitty of chondrule formation, and that multiple melting events were possible. Diverse mineral assemblages were present prior to chondrule formation. This material (chondrule *precursors) included many large grains, showed variable degrees of Fe-oxidation, and encompassed many isotopic reservoirs. The properties of these nebular materials changed systematically from the enstatite to ordinary to carbonaceous chondrite formation regions.*

9.1.1. INTRODUCTION

Most of the meteorites that fall on the Earth are chondrites (see Chapter 1.1). The preceding chapters of this volume have shown that the chondrites are primitive clastic rocks that formed early in the history of the solar system (Chapter 5.2). Most of them have compositions that are not greatly different from that of the total condensible matter in the Sun (Chapters 1.1 and 7.1). Most of the relatively small differences in composition between the nine major chondrite groups (CI, CM, CO, CV, H, L, LL, EH and EL) apparently were caused by the fractionation (separation) of various solid components and gas that existed in the nebula prior to the final accretion of the parent bodies. Thus, chondrites are the logical places to search for surviving examples of the solids that were present in these early stages of nebular history.

Many chondrites have experienced secondary processes such as thermal and shock metamorphism (Chapters 3.3 and 3.6), and aqueous activity (Chapter 3.4). All but one of the chondrite groups contain members that have escaped extensive thermal metamorphism. These are the petrographic type-3 H, L, LL, EH, CO and CV chondrites, the CM chondrites (type 2) and the CI chondrites (type 1). In turn, these groups all contain members that are not heavily altered by shock. Only the EL group contains no known unmetamorphosed members. Aqueous activity has extensively affected all of the CI and CM chondrites (Chapter 3.4), and may have affected (to a much smaller degree) some type-3 ordinary (Chapter 3.4), CV (Bunch and Chang 1980) and EH (El Goresy 1985) chondrites. Chondrites that have undergone extensive changes by these processes are not the ideal objects in which to search for primitive nebular material. Therefore, most of the discussion in this chapter is based on those chondrites that are thought to have escaped most secondary processing. Unmetamorphosed members of each chondrite group are shown in Fig. 9.1.1.

The chondrites contain several readily identifiable petrographic components. In ordinary chondrites, one component comprises a major fraction of the rock: these are the *chondrules,* mm-sized, igneous-textured spheroids, mostly composed of silicate minerals. In carbonaceous and enstatite chondrites, chondrules occupy smaller volumes; they are a major component in CM, CO, CV and EH groups, and are absent from the CI group. The most important characteristic of chondrules, as the term is used herein, is that they were once either fully or partially molten droplets. In carbonaceous and ordinary chondrites, another major component is a fine-grained porous mixture of minerals known as *matrix* (Chapter 10.2). Enstatite chondrites contain little matrix. The other components of chondrites include: large aggregates of opaque minerals (mostly metal and troilite), abundant in ordinary and enstatite chondrites, but rare in the more oxidized carbonaceous groups; small aggregates or isolated grains of opaque phases (including metal, sulfides, chromite, magnetite and other oxides); non-igneous-textured objects or *inclusions*, dominated by silicate and oxide minerals, most abundant in the CM, CO and CV carbonaceous chondrites (Chapter 10.3); and, isolated coarse silicate grains (Chapter 10.4). Chondrites that have experienced processing in a regolith also may contain clasts of other chondritic material (Fig. 1.1.3a; see also Chapter 3.5). Table 9.1.1 summarizes the relative abundances of these components in unbrecciated and unequilibrated members of each chondrite group.

Radiometric dating of individual chondrules shows that they formed at about the same time as the chondrites themselves accreted and were metamorphosed (see below and Chapter 15.3). Objects with this characteristic cannot be presolar materials. In this chapter, we present the evidence leading to the conclusion that most chondrules are the melting products of preexisting solids *in the solar nebula*. Chondrules will be described in terms of their textures, mineralogy, physical properties and chemical and isotopic compositions in order to place constraints on how solid material may have formed and evolved in the early solar system. Chapter 9.2 discusses laboratory attempts to simulate the conditions of chondrule formation. Chapter 9.3 combines the physical and experimental evidence to evaluate various models of chondrule formation. Finally, Chapter 9.4 discusses the energetics of chondrule formation in the solar nebula. Chapters in Part 10 of the book continue with a discussion of the other important components of chondrites. It is important to bear in mind that little is known about relationships between the constituents of chondrites, including whether they formed in the same regions or at the same time.

9.1.2. GOALS OF RESEARCH ON CHONDRULES

There are two fundamentally different settings in which chondrules may have formed: they may be products formed in the early solar nebula; or, they may have formed on asteroidal or planetary bodies during or shortly after accretion. Depending on which of these settings is correct, chondrules may constrain different aspects of solar-system evolution. If chondrules formed on parent bodies then they are indicators of the conditions (pressure, temperature, composition, etc.) that existed at the time of their formation on those bodies. If a mechanism for forming chondrules on parent bodies could be deduced, it would constrain the early evolution of these bodies. However, if chondrules are nebular products, they provide information about the conditions in the early solar nebula.

The initial goal in the study of chondrules is to use their properties to ascertain whether most chondrules formed in the nebula or on parent bodies. If this can be done, then many questions concerning that setting can also be studied. *Energy source:* What kinds of energy sources are consistent with the properties of chondrules? *Environment:* When chondrules were molten, in which kind of environment were they situated (P, T, gas/dust ratio, gas composition, magnetic field)? Did chondrules from different chondrite groups experience the same conditions? *Precursors:* If chondrules are secondary objects, then what kind of material existed prior to their formation? Were chondrules formed by the melting of low-temperature or high-temperature solids or both? Were the solids oxidized or reduced? Were they coarse-grained or fine-grained? How could the precursors have formed?

9.1.3. DEFINITIONS AND DIFFICULTIES

Chondrule is defined here in a *general* sense to be an object formed as an isolated droplet of molten or partially molten material. The strongest evidence for melting is the presence of a glassy or cryptocrystalline mesostasis and an igneous texture. The best evidence for formation as an isolated droplet is a spheroidal shape indicative of the original tensional surface of the liquid. In addition, any preferred orientation of crystals with respect to the present surface of the object is strong evidence for formation as an isolated object.

There are objects in chondrites that do not satisfy these criteria, but nonetheless can be called chondrules by inference. Broken chondrules are quite common even in the least metamorphosed and unshocked chondrites. Objects with textures identical to those found in known chondrules, but which lack evidence for a droplet surface, are probably fragments of chondrules. Objects that formed with a low degree of partial melting may never have developed a smooth tensional surface, nor a good igneous texture. These objects also should be classified as chondrules, but may be difficult to differentiate from objects that never reached solidus temperatures.

An additional difficulty is that not all chondrules may be related. Tektites fit this general definition of chondrule, as do glass spherules in the lunar regolith, atmospheric ablation spherules and sleet. Likewise, there are many processes that can plausibly have created chondrules during different stages of the solar nebula, on early-accreted planetesimals, on asteroids and on planetary bodies. However, different kinds of material and different conditions may have existed in each setting. Thus, there may be many populations of unrelated chondrules mixed together in chondrites.

As a working hypothesis, one must assume that the majority of chondrules were formed in just one of the potential settings by one dominant process. *Chondrule,* in a *restricted* sense, refers only to those objects formed by that dominant process. Restrictive criteria are used in most studies of chondrules; unfortunately, there are no well-established criteria by which "true" chondrules can be differentiated from "false" ones. Previous workers have used wildly different and often unstated criteria for deciding which objects to call chondrules. The fact that any such criteria are likely to be model dependent further complicates the dilemma, but seems unavoidable.

In the following discussion, one restrictive criterion has been used to eliminate objects that might not be chondrules. This is based on the hypothesis that related chondrules should show a continuum of textures, mineralogies, chemical and isotopic compositions, sizes and other properties. Populations of objects with distinctly different properties from those of most chondrules are not used as constraints on chondrule formation, although any complete model must account for all of the objects found in a single place.

All nonchondritic droplets can be excluded from true chondrules on the basis of this restriction. For example, lunar chondrules and chondrule-like objects have been identified in Apollo 14 and 15 samples by King et al. (1972), Kurat et al. (1972) and others. Some achondrites contain glassy spherules similar to those in the lunar rocks (Brownlee and Rajan 1973). Such spherules have also been found in breccias from terrestrial impact craters (Fredriksson et al. 1973; Graup 1981). These are all believed to be formed by impact processes in regoliths. Tiny hypervelocity impact craters are observed on the surfaces of the lunar and achondritic spherules, but not on spherules in chondrites. This (and other) differences between the two populations indicates that the nonchondritic spherules are not true chondrules.

The coarse-grained, type-B Ca-Al inclusions (CAI) (see Chapter 10.3), most abundant in CO and CV chondrites, are also not true chondrules. Although they formed as isolated molten droplets, their mineralogical, isotopic and physical properties (including rims) are quite distinct from those of most chondrules (Grossman and Wasson 1983a). Data from those inclusions are not included in this chapter.

9.1.4. MINERALOGY AND PETROGRAPHY OF CHONDRULES

Texture and Classification

Every chondrite contains chondrules displaying a wide variety of textures (Fig. 9.1.1). As with igneous rocks, the texture of a chondrule depends on composition, extent of melting, peak temperature, cooling rate, O fugacity, secondary (thermal or aqueous) alteration and other parameters. Because chondrules formed as isolated droplets, probably in a low gravitational field, spin, surface physics, collisions and accretion may also be important, whereas crystal settling and convection may not be important.

As a first step toward understanding the effects of these parameters, it is desirable to classify chondrules. Several classification schemes are in common use for olivine-pyroxene-rich chondrules. Perhaps the most useful scheme is that proposed by Gooding and Keil (1981). It allows most chondrules to be categorized unambiguously on the basis of texture and modal

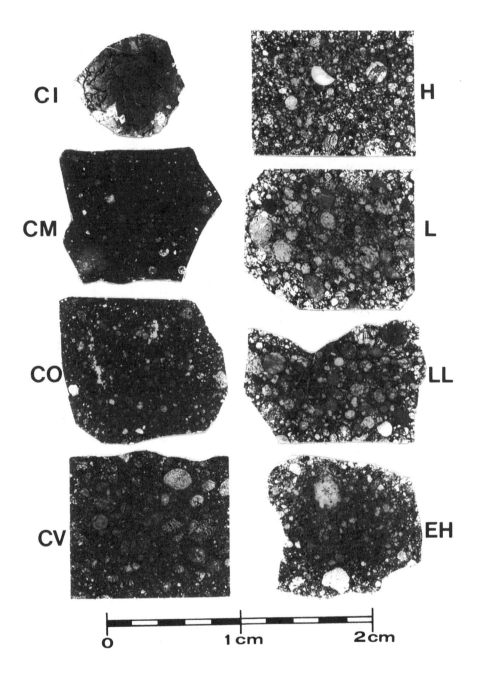

mineralogy. A second scheme by McSween (1977a), allows classification on the basis of texture and mineral chemistry.

Gooding and Keil (1981) defined two broad categories of chondrules: *porphyritic* and *nonporphyritic*. Porphyritic chondrules have phenocrysts of olivine and/or low-Ca pyroxene surrounded by glassy or microcrystalline mesostasis. Phenocrysts may be as large as the radius of the chondrule. Porphyritic chondrules may be subdivided according to the proportion of olivine and pyroxene phenocrysts, giving the categories porphyritic olivine (PO), porphyritic pyroxene (PP) and porphyritic olivine-pyroxene (POP). Other workers have found it useful to add the modifiers metal-rich and metal-poor to these groups to describe the mineralogy further. Phenocryst morphology is extremely variable in porphyritic chondrules. Olivine crystals range from euhedral to anhedral or rounded, and may be sharply faceted, deeply embayed, or skeletal. Poikilitic textures, in which olivine crystals are enclosed in larger low-Ca pyroxene crystals commonly are observed. In some chondrules, the olivine has a morphology known as "barred," where parallel plates of olivine are in optical continuity, in some cases surrounded by a spherical olivine shell; these are called barred-olivine (BO) chondrules. BO textures are so distinctive that they usually are considered separately from porphyritic chondrules.

Nonporphyritic textures described by Gooding and Keil include radial pyroxene (RP), cryptocrystalline (C) and granular olivine-pyroxene (GOP). The RP and C chondrules commonly are referred to as "droplet" or "molten-drop" chondrules because they most clearly preserve the original tensional surface of the molten chondrule. These chondrules are dominated by fan-like arrays of pyroxene crystals radiating from one or more points on the surface. In the C chondrules, the pyroxene is too fine grained to be seen in an optical

---

Fig. 9.1.1. Photographs in transmitted light of thin sections of unequilibrated members of the major (chemical) chondrite groups. Each meteorite is an observed fall, and is texturally representative of the least metamorphosed, least altered members of its group. Nearly all of light-colored objects in the H, L, LL and EH photographs are chondrules and chondrule-fragments; in CM, CO and CV, about one-third of the light-colored objects are inclusions. CI: Ivuna, USNM section (<10 μm thick) prepared by R. Beauchamp, Battelle Laboratories: no chondrules are present; most visible features result from aqueous activity (veins, dark matrix, light mineral grains). CM: Murray, USNM 1769-7: small chondrules and inclusions are set in abundant dark matrix; effects of aqueous alteration are extensive, but less than for many other CM chondrites. CO: Ornans, USNM 1105-2: chondrules are similar in size to those in CM chondrites, but are more abundant. CV: Vigarano (reduced subgroup), USNM 6295-4: CV chondrites contain the largest chondrules of any group. H: Sharps (type 3.4), USNM 640-4: chondrules are generally small, although a few mm-sized examples are visible. L: Mezö-Madaras (type 3.7), USNM 4838-1: L-group chondrules are intermediate in size between H and LL. LL: St. Mary's County (type 3.3), USNM 5623-2: chondrules are abundant and large. EH: Kota-Kota, USNM 6001: this chondrite has been terrestrially weathered, but retains a typical EH3 texture; chondrules are nearly all small, with a few obvious exceptions.

microscope at low magnification (≤ 2 μm), and may or may not show a radiating structure; many have multiple extinction domains. The granular chondrules contain closely packed, fine-grained anhedral crystals of olivine and/or pyroxene in a glassy mesostasis, and are texturally similar to some porphyritic chondrules. A population of objects called "dark-zoned" chondrules by Dodd and Van Schmus (1971) resembles GOP chondrules, but does not contain evidence for melting (Nagahara 1983). Rubin (1984) renamed these objects "coarse-grained lumps" because they are not chondrules by the current definition.

Transitional textural types exist between all of these categories, including between the porphyritic and nonporphyritic groups. The diversity of chondrule textures is documented in photomicrographs by Tschermak (1885), Wasson (1974,1985), Gooding and Keil (1981), Nagahara (1983) and Grossman (1983).

McSween (1977a) devised a classification scheme based on a combination of textural and mineral-chemical properties. Modifications to the scheme were made by McSween et al. (1983) and Scott and Taylor (1983). There is a relationship between the textures of certain chondrules and the compositions of their minerals. Porphyritic chondrules may be divided into groups with low-FeO and high-FeO in their silicates. The type-I (FeO-poor) chondrules tend to have glassy mesostases, pyroxene crystals aligned with the chondrule surface; many are metal rich. Most chondrules with poikilitic textures are type I. The type-II (FeO-rich) chondrules tend to have larger phenocrysts set in a darker or microcrystalline mesostasis. A third group, type III, is equivalent to the combined RP and C groups of Gooding and Keil (1981). Barred olivine and porphyritic chondrules are found in both type-I and type-II categories.

A problem with the McSween system is that many chondrules, especially in ordinary chondrites, have intermediate FeO contents or ambiguous textures. Thus, not all chondrules can be classified in this scheme.

Distribution of Chondrule Types. The three ordinary-chondrite groups are indistinguishable from one another in their distributions of chondrule textural types. However, other chondrite groups contain different distributions of chondrule types (Table 9.1.1). The categories in Table 9.1.1 are essentially those of Gooding and Keil (1981), but the porphyritic types are divided both into FeO-rich and FeO-poor groups (in light of the petrographic differences described by McSween [1977a]), and into PO-POP-PP groups. FeO-poor porphyritic chondrules are particularly abundant in CM, CO, CV and EH chondrites. FeO-rich porphyritic chondrules are most abundant in ordinary chondrites. PO chondrules decrease and PP chondrules increase from carbonaceous to ordinary to enstatite chondrites. Nonporphyritic chondrules (RP + C) are nearly absent from CV chondrites, and increase in abundance from CM-CO carbonaceous to ordinary to enstatite chondrites, paralleling the

TABLE 9.1.1
Abundances of Petrographic Constituents and Characteristics of Chondrules in CM, CO, CV, Ordinary (Combined H, L, LL) and EH Condrites.[a]

	EH[1]	OC[2]	CV[3]	CO[4]	CM[5]
Petrographic Constituents:					
Vol.% Matrix	< 5	10–15	40–50	30–40	~60
Vol.% Inclusions	≪ 1	< 1	6–12	10–15	~5
Vol.% Chondrules[b]	15–20	65–75	35–45	35–40	≤ 15
Chondrule Characteristics:					
Diameters (mm)	0.2	*0.3–0.9*[c]	*1.0*	*0.2–0.3*	0.3
% Porphyritic (P)	81	81	94	96	≥ 90
% P w/low FeO	100	50–90	94	97	90
% P w/high FeO	0	10–50	6	3	10
% P w/PO texture	0.2	28	90	—	—
% P w/POP texture	5	60	8	—	—
% P w/PP texture	95	12	2	—	—
% Nonporphyritic (NP)	19	15	0.3	2	*3–8*
% NP w/RP,C texture	92	80	*100*	—	—
% NP w/GOP texture	8	20	*0*	—	—
% Barred (BO)	≤ 0.1	4	6	2	present
Chondrules w/relict grains:					
% with isolated dusty olivines	present	*5–15*	*< 1*	*1–2*	*0–1*
% with poikilitic olivine	*5–10*	*20–30*	*20–50*	*~30*	*10–30*

[a] Italic designate numbers estimated by the authors. Superscripts to the five chondrites indicate the sources of data given in the table footnotes.
[b] Including fragments of chondrules; abbreviations for chondrule types: RP = radial pyroxene, C = cryptocrystalline, BO = barred olivine, PO = porphyritic olivine, POP = porphyritic olivine-pyroxene, PP = porphyritic pyroxene, GOP = granular olivine-pyroxene.
[c] Sizes differ among the three ordinary chondrite groups. Individual group means (estimates) are H: 0.3 mm, L: 0.6–0.8 mm, LL: 0.9 mm.
Sources of data: (1) EH constituents: Okada (1975), Rubin (1983); diameters and chondrule types: Rubin and Grossman (1987, unpublished data). (2) OC constituents: Huss et al. (1981), chondrule vol.% estimated as 100% − (% matrix + % opaques); diameters: Rubin (Rubin and Keil 1984 and unpublished data), King and King (1979); chondrule types: Gooding and Keil (1981), Gooding (1979). (3) CV constituents: McSween (1977b); chondrule types: McSween (1977b), Simon and Haggerty (1980), Rubin and Wasson (1987b). (4) CO constituents and chondrule types: McSween (1977c); diameters: King and King (1978), King (unpublished data). (5) CM constituents: McSween (1979); chondrule diameters: Rubin and Wasson (1986); chondrule types (porph.): roughly estimated from data of Wood (1967).

change in PP abundances. Thus, the modal mineralogy and mineral chemistry change systematically with chondrite group.

Modal and Normative Mineralogy

Most chondrules are dominated by the phases olivine, low-Ca-pyroxene and feldspathic glass, with olivine and pyroxene accounting for more than half the volume, and feldspathic material plus opaque phases comprising the

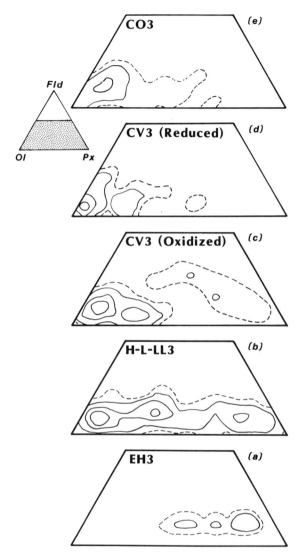

Fig. 9.1.2. Normative compositions (CIPW) of porphyritic chondrules from major chondrite groups. Ol = olivine, Px = total pyroxenes, Fld = total feldspar, feldspathoids. Contour lines were drawn by the method outlined by Dodd and Teleky (1967), and Dodd (1974). The number of samples in areas equal to 4% of the diagram were calculated. Thus, the lines labeled *1* bound those areas that contain >1% of the samples. The solid lines are at evenly spaced intervals; the dashed lines show the contour corresponding to half of the interval between solid lines. Major-element data sources: (a) Grossman et al. (1985); (b) McSween (1977*a*), Dodd (1978*a,b*), Lux et al. (1981), Nagahara (1981*b*); (c,d) McSween (1977*b*); (e) McSween (1977*c*). Porphyritic chondrules become progressively richer in normative olivine in the sequence EH3–UOC–CV3 (oxidized)–CV3 (reduced)–CO3.

remainder. In unequilibrated chondrites, most pyroxene grains are clinoenstatite; crystalline feldspar is rare; the most abundant opaque minerals are troilite and kamacite.

In general, the mineralogy of chondrules reflects that of the chondrite as a whole. There are relatively few compilations in the literature of modal mineral data in chondrules. However, broad-beam electron microprobe studies of major-element compositions have been done, and allow normative mineral compositions to be calculated. Figure 9.1.2 compares the normative mineralogies of porphyritic chondrules from carbonaceous, ordinary and enstatite chondrites. Olivine-rich chondrules dominate in the carbonaceous chondrites with pyroxene-rich chondrules being somewhat rare. The opposite is true in enstatite chondrites. Ordinary chondrites show a wide spectrum of chondrule mineralogies.

BO and (RP, C) chondrules, as expected, are dominated by normative olivine and pyroxene, respectively (Fig. 9.1.3). Both of these groups overlap with fields occupied by porphyritic chondrules, but extend into regions essentially devoid of porphyritic chondrules. Chondrules with >20% normative feldspar are almost exclusively barred, and chondrules with <2% normative olivine are almost exclusively nonporphyritic.

The chondrules from each group span wide ranges of normative mineral abundances. However, the chondrules do not define trends on Fig. 9.1.2 that can be attributed to simple igneous differentiation processes. For example, as pointed out by Dodd (1971), crystal-liquid separation for highly olivine-normative melts should produce trends toward increased pyroxene and feldspathic components with decreasing olivine component. Instead, the chondrules define a wide band of olivine/pyroxene ratios with no relation to the feldspathic component. If any igneous differentiation signatures are present

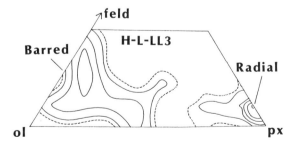

Fig. 9.1.3. Normative mineral compositions (CIPW) in barred olivine (BO) and radial pyroxene (RP) chondrules in unequilibrated ordinary chondrites. See Fig. 9.1.2 for explanation of methods and sources of data. RP chondrules are generally richer in normative pyroxene than porphyritic pyroxene chondrules. BO chondrules span the same range as porphyritic olivine chondrules, but extend to much more feldspathic positions along the Ol-Fld join.

in chondrule sets they must be overshadowed by other sources of variation. Displacements on these diagrams could also be caused by primary (precursor) variations between chondrules or by fractionation processes such as fractional evaporation.

Minor phases that commonly are found in the olivine-pyroxene-rich chondrules include Ca-rich pyroxene (pigeonite, diopside), Fe-Ni metal (kamacite, taenite and tetrataenite), troilite, chromite and Mg-Al spinel. Many chondrites that contain unusual phases in their matrices and in isolated metal-sulfide grains also contain these phases in their chondrules. For example, EH3 chondrules contain phases such as niningerite, oldhamite, perryite, silica, caswellsilverite and Cr-sulfides; CV3 chondrules contain magnetite, pentlandite, awaruite and, in a few cases, nepheline and sodalite; type-3 ordinary chondrites such as Semarkona contain magnetite, carbides and pentlandite. The converse of this is not always true: chondrules may contain phases that are not found in the matrix. For example, low-Ni metal in oxidized CV chondrites is mostly limited to chondrule interiors. Olivine in enstatite chondrites also is limited to chondrule interiors.

Chondrules that are dominated by phases other than olivine and pyroxene are rare. Glass-rich chondrules occur in many chondrites. Chondrules rich in metal-sulfide (Gooding and Keil 1981), silica (Brigham et al. 1986), chromite (Christophe Michel-Lévy 1981), anorthite, spinel, (Bischoff and Keil 1983) and other phases have been described. These show clear evidence for formation as molten droplets. Due to the scarcity of such objects, it is unclear how they are related to "true" chondrules. They do not occur in sufficient abundance to have much effect on mass- or number-weighted means of chondrule properties. Their low abundances also make it difficult to decide whether the objects fit the restrictive definition of chondrule, or whether they have discrete properties. Thus, we exclude these objects from our discussion of general properties of chondrules.

Mineral Chemistry

Ranges of Mineral Compositions. Chondrules from each group of unequilibrated chondrites show a wide range in the composition of all minerals. The Fe/(Fe + Mg) ratio of olivine and pyroxene in chondrules from individual ordinary and carbonaceous chondrites may range from near 0 to 60 mol %. Minor elements in olivine and pyroxene also show wide concentration ranges.

The mean compositions of phases in chondrules vary with chondrite group. As stated above, FeO-rich chondrules are rare in carbonaceous and enstatite chondrites. Thus, the major-element contents of olivine and pyroxene are different in these groups from those in ordinary chondrites. Rubin (1986) showed that differences in mean mineral compositions extend to many minor elements. For some elements, differences between groups can be attributed simply to variations in bulk composition, with partition coefficients staying constant. However, elements such as Cr, Mn and Fe occur in different oxidation states in some of the chondrite groups, and thus partition differently among phases. Calcium, which occurs in only one oxidation state, partitions between olivine, pyroxene and glass differently in each chondrite group; these variations are more difficult to explain, and may be due to the presence of unmelted (relict) phases.

Gooding (1983) reported that the Co contents of taenite in chondrules from H chondrites were systematically different from those in L plus LL chondrites. This weak trend is the only mineralogical difference thus far reported among chondrules from ordinary chondrite groups.

Chondrules of a single textural-mineralogical class show similar mineralogical trends in the different chondrite groups. Scott and Taylor (1983) showed that type-I (FeO-poor) PO chondrules from H, L, LL, CM, CO and CV chondrites occupy coincident fields on a CaO vs FeO plot of olivine compositions. The same is true for type-II (FeO-rich) porphyritic chondrules. This does not imply, however, that all chondrules of one type formed in one region and were then distributed into each chondrite-formation region. It will be shown below that other properties, notably the O-isotopic composition, of these same chondrules differ from one chondrite group to another. This observation implies that only the conditions that control the mineralogical and textural properties of chondrules were similar in each region of chondrite formation.

Evidence for Rapid Cooling and Disequilibrium. The textures of the olivine and pyroxene in RP, C and BO chondrules resulted from rapid crystal growth caused either by rapid cooling or by undercooling. In these chondrule types, needle-like or skeletal crystals are present. Although porphyritic chondrules generally have equant phenocrysts, many have dendrites in their mesostases. Such dendrites may result from rapid crystal growth. In addition, isotropic glass attests to rapid cooling in chondrules.

Most mineral assemblages in chondrules are not products of equilibrium crystallization (see, e.g., Keil and Fredriksson 1964; Kurat 1967,1969). Olivine and pyroxene grains generally are zoned. The zoning usually follows normal igneous trends, i.e., FeO, MnO and CaO increase from core to rim. Some chondrules contain SiO_2-rich glass together with olivine. These properties, like chondrule textures, are consistent with rapid crystallization. Both the zoning and the silica-rich glass may result from the failure of growing solids to maintain equilibrium with residual liquid. Zoning profiles are related to the textural-mineralogical classes of chondrules. Olivines in type-II PO chondrules tend to be strongly zoned, whereas olivines in BO and type-I PO chondrules tend to have less pronounced zoning.

Some chondrules show evidence for complicated conditions of crystal growth. These chondrules may contain grains that have either reverse or oscillatory zoning. In others, coexisting olivine and pyroxene do not have an equilibrium distribution of FeO. An extreme case of disequilibrium exists in chondrules that contain olivine having a bimodal distribution of FeO contents, as discussed in the next section.

Relict Grains: Surviving Precursor Material

Types of Relict Grains. Relict grains are chondrule precursor solids that survived the high-temperature period without melting (see also Chapter 10.4). These have been recognized in chondrules in type-3 members of all chondrite groups. Some olivine and pyroxene grains appear "dusty" in transmitted light due to the presence of numerous blebs of low-Ni metallic Fe or chromite grains (Fredriksson et al. 1969; Nagahara 1981*a;* Rambaldi 1981). Other olivine and pyroxene grains are anomalously large compared with normal grains in the same chondrule, and may be compositionally different (Kracher et al. 1984). Some olivine grains have compositions that are far from that expected to crystallize from coexisting liquids. Grains of forsteritic olivine that are rich in refractory elements and that show blue cathodoluminescence also may be relict grains (Steele 1986).

Dusty olivine grains are found in porphyritic chondrules from most of the chondrite groups. Dusty olivines contain numerous sub-μm to 10-μm blebs of Fe-metal concentrated in the centers of the grains surrounded by inclusion-free mantles. Olivine coexisting with metal ranges in composition from Fa_{5-20} and typically has <0.1 wt. % CaO. Because it is clear that the Fe blebs formed by reduction of an FeO component of the olivine, one can calculate that many of the olivines had more than 30 mol % fayalite prior to reduction. Mantle olivine has much less FeO, mostly in the range Fa_{1-10}, and much more CaO, 0.1 to 0.5 wt. %. Nondusty olivine grains in the outer portions of the chondrules resemble mantles in composition. The mantles and the clear grains probably crystallized from the chondrule melt, with the mantles growing epitaxially on relict cores. Their CaO and FeO contents are typical of type-I porphyritic chondrules (Scott and Taylor 1983). The dusty olivine grains are probably relict; Fe-metal grains may have formed when the chondrule melted in the presence of an unknown reducing agent, possibly H_2 (in the surrounding gas) or C (Rambaldi 1981). In some chondrules, the relict (dusty) olivine and the clear grains have the same FeO and different CaO contents, indicating that Fe reduction reached equilibrium, but there was insufficient time to equilibrate Ca between the relict crystal and surrounding melt.

Olivine crystals that are poikilitically enclosed in pyroxene grains in porphyritic chondrules may resemble dusty olivine grains. Many of these grains contain small metal blebs, and are high in FeO and low in CaO. This similarity suggests that these also may be relict grains. Other poikilitic olivine grains that do not contain metal also may be relict grains by analogy. However, Lofgren and Russell (1986) showed that similar textures can be produced by the growth of olivine from tiny grains during the heating of chondrule precursors. It is not clear whether olivine that formed in this way could become depleted in CaO and enriched in FeO, as is observed.

Several other types of grains may be relict, but are not FeO-rich. Anomalously large olivines sometimes are much poorer in FeO than surrounding small grains. Kracher et al. (1984) described a large magnesian olivine (Fa_2) and small ferroan olivines (Fa_{33}) in the same chondrule. Other chondrules contain olivine with a much higher Mg/Fe ratio than could ever crystallize from a liquid with the bulk composition of the chondrule (i.e., from a completely molten chondrule). Undercooling can produce olivines that have lower Mg/Fe ratios than equilibrium olivine, but not olivine with higher Mg/Fe.

Nearly pure forsterite that cathodoluminesces blue is found as isolated grains in the matrices and less commonly inside chondrules of CV, CO and CM chondrites (Steele 1986; see also Chapter 10.4). These tend to be rich in the refractory elements Al, Ca, Ti and V; many are mantled by olivine that is poor in refractories, but rich in Fe, Mn and Cr. The unusual compositions of these grains suggest that they too are relict. It is not known whether these grains correspond to the other types of relict magnesian olivine discussed above.

Abundance of Relict Grains. Different chondrite groups contain different amounts of relict-grain-bearing chondrules (Table 9.1.1). In type-3 ordinary chondrites, 5 to 15% of the chondrules bear dusty olivine, ~10% have poikilitic olvine with metal inclusions, and 20 to 30% have clear poikilitic olivine grains. As many as 20% may have olivine with unusually high Mg/Fe ratios, but without further evidence of the grains being relict. As many as 70% of the porphyritic chondrules in ordinary chondrites may contain some form of relict grains. Experimental studies reviewed in Chapter 9.2 suggest that *all* porphyritic chondrules were incompletely melted. Given that 85% of all chondrules are porphyritic, it is apparent that most chondrules were never completely molten.

In carbonaceous chondrites, chondrules showing evidence for relict grains are less common. The fractions containing dusty olivine and poikilitic olivine are shown in Table 9.1.1; the fraction with blue luminescing oliving is not known. However, it is important to note that many CV chondrites may have experienced late-stage oxidation (the oxidized group of McSween [1977*b*]), and dusty metal may have been destroyed. Also, chondrules that contain amounts of normative pyroxene sufficient to allow the development of a poikilitic texture are less common in carbonaceous groups than in ordinary chondrites. Thus, chondrules with relict grains may be much more difficult to identify.

Olivine is present in small amounts in EH3 chondrites; most such occurrences are relict grains. Much of the olivine is poikilitically enclosed by pyroxene in chondrules, although some is present as discrete grains. Metal blebs are quite common in both types of olivine. Some of these olivines are inferred to have contained 20 to 30 mol% Fa prior to reduction. More than 5% of the chondrules in EH3 chondrites contain relict olivine. Some FeO-rich pyroxene grains are also present in EH3 chondrules. These relict grains have been reduced to form enstatite, Fe-metal and SiO_2. Initial compositions ranged from Fs_5 to Fs_{30}.

Origin of Relict Grains. Many of the relict grains do *not* appear to be relicts from earlier generations of chondrules. FeO-rich olivines in normal, type-II porphyritic chondrules are strongly zoned, with magnesian cores and Fe-rich rims. If FeO-rich relict grains started out with this zoning, and then were immersed in the chondrule melt, they should still preserve much of their original zoning profile. Assuming a peak temperature of 1770 K, a grain size of 100 μm, a cooling rate of 100 K/hr (conservatively slow; see Chapter 9.2) and experimentally determined diffusion coefficients for Mg and Fe in olivine, there is insufficient time to homogenize the olivine composition; thus, a doubly zoned grain should form (the zoned core surrounded by the normally zoned overgrowth of mantle olivine).

The CaO content of the relict FeO-rich olivines is also inconsistent with formation in an earlier generation of chondrules. Most porphyritic chondrules have olivine with >0.1 wt. % CaO. Relict-grain compositions are therefore outside the range of that in chondrule olivine in ordinary and carbonaceous chondrites. No obvious process would lower the CaO content of relict olivines during heating.

Similarly, the CaO, Al_2O_3 and TiO_2 contents of the relict olivines described by Steele (1986) appear to be too high to have originated inside typical porphyritic chondrules.

Equilibrated chondrites contain chondrules with low-CaO, unzoned olivine grains. Thus equilibrated chondrites are a potential source for this type of relict grain. Scott et al. (1983) reported that many unequilibrated chondrites actually contain entire equilibrated chondrules, derived by brecciation in a parent body. Some type-3 chondrites (Semarkona and Tieschitz) do not contain any equilibrated chondrules, yet they have many chondrules with relict low-CaO olivine grains. From this, one must conclude that the equilibrated chondrites were not the source of the relict grains.

If earlier generations of "normal" chondrules or chondrites were not the source for relict grains, then where did they originate? An unlikely possibility is that they came from generations of chondrules that were chemically different and were totally destroyed by subsequent recycling. Other possibilities are: the grains formed in the nebula by condensation; the grains formed from nebular condensates by an unknown process prior to chondrule formation; or they are interstellar grains (see also Chapter 10.4).

9.1.5. PHYSICAL PROPERTIES OF CHONDRULES

Chondrule Sizes

Ordinary Chondrites. Chondrules range in size from <1 μm to ≥1.2 cm. Dodd (1976) made the first systematic measurements of the size-frequency distributions of chondrules. He measured the apparent sizes of all silicate particles >100 μm in diameter in petrographic thin sections of 16 type-3 ordinary chondrites. The grain size-frequency distributions were nearly log-normal, with H chondrites being finer-grained than L and LL chondrites.

King and King (1979) reported the grain size-frequency characteristics of spherical, whole chondrules (termed "fluid-drop chondrules") in thin sections of 11 unequilibrated ordinary chondrites. The results for these selected chondrules were similar to those of Dodd (1976). The range of median values for the size distributions was about 300 μm. Rubin and Keil (1984) found a mean apparent diameter of ~800 μm for BO, RP and C chondrules in thin sections of two L3 chondrites.

The most accurate measurements of the dimensions of separated chondrules are those of Hughes (1978), who analyzed 955 "near-spherical" chondrules from Bjurböle (L4). He found the chondrule size-frequency distribution to have a mean of 750 μm. Hughes obtained similar results using thin-section techniques for Bjurböle and Chainpur (LL3.4) chondrules.

Carbonaceous Chondrites. The apparent diameters of chondrules and chondrule fragments in thin sections of 11 CM and CO chondrites (King and King 1978) show nearly identical size-frequency distributions. These authors found that CM chondrules and fragments range from 170 to 220 μm in mean diameter, and CO chondrules and fragments range from 180 to 200 μm. Rubin and Wasson (1986) found that the mean apparent diameter of 100 whole Murray (CM) chondrules is 270 μm. Measurements by King and King (1978) showed that chondrules and chondrule fragments are much larger in CV chondrites than in CM and CO, ranging from 270 to 670 μm in mean diameter. Unbroken CV chondrules may have a mean diameter as large as 1000 μm. Several chondrites have extraordinarily large chondrules compared with other related meteorites, e.g., Renazzo (ungrouped carbonaceous chondrite) and Inman (L3).

Evidence for Size Sorting. Chondrule sizes vary among the different chondrite groups (Table 9.1.1). However, there is no evidence that differences were produced by size-sorting processes operating on a single population. The distinct populations of objects in carbonaceous, ordinary and enstatite chondrites, and the chemical and isotopic compositions of the bulk meteorites preclude the possibility that the chondrite groups are derived from different size fractions of one initial distribution. In addition, there are no coarse-grained, chemical equivalents of CM, CO or EH chondrite groups, nor are there fine-grained equivalents of H, L, LL or CV chondrite groups. It is much more likely that chondrules formed with different size distributions in each chondrite formation location.

However, there is some variation present within each chondrite group suggesting that size sorting occurred locally. Rubin and Keil (1984) showed that each ordinary chondrite has the same size distribution of BO and RP + C chondrules, but that the mean sizes and the proportions of the two groups of chondrules vary from one chondrite to another. Rubin et al. (1982) found extreme examples of this phenomenon in clasts containing very small (0.2 to 74 μm) "microchondrules" in Piancaldoli (LL3) and Rio Negro (L regolith breccia). Christophe Michel-Lévy (1987) described microchondrule-bearing clasts in Mezö-Madaras (L3.7). Fruland et al. (1978) also described clasts in carbonaceous chondrites that contain populations of chondrules smaller than those in the host meteorite.

The sorting of chondrules and other particles in the nebula by gas drag was suggested by Dodd (1976), Rubin et al. (1982) and Rubin and Keil (1984). Spattering of droplets from a parent melt spherule was shown by King (1982,1983) to produce chondrules in restricted size ranges. Other authors have suggested that sorting might be accomplished by settling in temporary gas plumes on planetary surfaces during impacts.

Chondrule Shapes

Many authors have remarked that whole chondrules tend to be nearly spherical (see, e.g., Hughes 1978) or circular in thin section (see, e.g., King and King 1978). Gooding (1983) reported that chondrules in LL3 chondrites tend to be less spherical than those in H3 and L3 chondrites: the ratio of the maximum to minimum dimension averages is ~ 0.8 for LL3 chondrites and ~ 0.85 for H3/L3, but the differences may not be statistically significant. The general sphericity of chondrules indicates that most were solid at the time they agglomerated to form chondrites.

Ellipsoidal chondrules also occur in undeformed chondrites. King and King (1978) reported elongations of up to $2.36\times$ in chondrules from carbonaceous chondrites; Rubin (unpublished data) observed an ellipsoidal RP chondrule in Semarkona (LL3) with an elongation of $2.3\times$. These chondrules were probably deformed by spinning while molten.

Several authors have suggested that some chondrules may have been plastic or partially molten *in situ*. Strong evidence for this is found in Tieschitz (H3.6) (Hutchison et al. 1979; Hutchison and Bevan 1983). In this meteorite, chondrules and their dark rims have been deformed by the impingement of adjacent objects, presumably while the rock was hot. The implications of these observations are not clear. Tieschitz has a unique texture in which feldspathic material, "white matrix," is present between chondrules. If Tieschitz accreted within hours (or minutes) of chondrule formation, then serious constraints are placed on mechanisms for chondrule formation. Tieschitz may also have had an unusual accretion history that resulted in reheating of chondrules.

Remanent Magnetism of Chondrules

Chondrules preserve a record of variable-strength magnetic fields that were present during chondrule formation (see also Chapter 8.1). Lanoix et al. (1977,1978), Sugiura et al. (1979) and Sugiura and Strangway (1985) measured the natural remanent magnetism in separated chondrules from Allende (CV3). Similar studies in ordinary chondrites are reported by Funaki et al. (1981) and Nagata and Funaki (1983). The chondrules show a high-temperature, randomly oriented magnetic component, apparently acquired when chondrules cooled through the Curie temperatures of their magnetic minerals. The magnetic fields recorded in chondrules were probably features of the solar nebula. The fields associated with lightning or impact events, both of which are potential chondrule-forming processes, are too transient to have magnetized the chondrules even if they cooled extremely quickly by blackbody radiation into space. This subject is clearly an area requiring further research.

9.1.6. CHEMICAL PROPERTIES OF CHONDRULES

The chemical compositions of chondrules provide some of the most important constraints on problems involving their formation and thermal histories. The key questions addressed by chemical data are:

1. What were the preexisting solids that melted to form chondrules?
2. Were the same kinds of precursors present at each nebular location?
3. How long were chondrules hot (a constraint on the environment and energy source of chondrule formation)?
4. How are chondrules, a major constituent of chondrites, related to the components hypothesized to have caused the chemical fractionations between chondrite groups (see Chapters 7.1 through 7.6)?

Grossman and Wasson (1983*a*) have reviewed literature data on chondrule compositions. They emphasized the importance of combining chemical and petrographic data in order to be certain that the objects under study are chondrules and that there are no sampling biases. As of 1983, chemical-petrographic studies of chondrules in unequilibrated ordinary chondrites had been done by Gooding (Gooding et al. 1980; Gooding 1983) and Grossman (Grossman et al. 1979; Grossman and Wasson 1982,1983*b*), both of whom used neutron-activation analysis of whole separated chondrules. Other studies of chondrules in unequilibrated ordinary and carbonaceous chondrites had been done by McSween (1977*a,b,c;* McSween et al. 1983), Dodd (1978*a,b*), Snellenburg (1978), Evensen et al. (1979), Lux et al. (1980,1981), Simon and Haggerty (1980), Fujimaki et al. (1981) and Nagahara (1981b); these studies all involved electron-microprobe measurements of chondrules in polished thin sections, and are somewhat limited by sample inhomogeneity and problems inherent in broad-beam analyses of complex materials. No studies had been made of chondrules from type-3 enstatite chondrites, partly because the existence of these meteorites was not widely recognized at that time.

Since 1983, combined chemical-petrographic studies using neutron-activation analysis have been extended for ordinary chondrites (Kurat et al. 1983; Wilkening et al. 1984), and completed for other chondrite groups: EH3 chondrules (Grossman et al. 1985); CO3 chondrules (Rubin and Wasson 1987a); and, CV3 chondrules (Kurat et al. 1985; Rubin and Wasson 1987b). In addition, Rubin and Wasson (1986) completed a microprobe study of rare petrographically unaltered CM chondrules. In this section, the data reviewed by Grossman and Wasson (1983a) will be summarized together with the newer data on other chondrite groups.

Compositional Diversity

Major-element (concentrations >10 mg/g) compositions of most chondrules are dominated by the elements Si, Mg, Al, Ca, Fe and O. Minor elements (concentrations 0.5 to 10 mg/g) generally include Na, S, K, Ti, Cr, Mn and Ni.

Major element variability is represented in Fig. 9.1.4. The ranges of porphyritic chondrules from major unequilibrated chondrite groups are shown on a SiO_2-MgO-FeO ternary diagram. Also shown in Fig. 9.1.4 are the bulk compositions of the chondrite groups. All Fe present in metal and sulfide is excluded from chondrule and chondrite analyses. The most obvious difference between chondrules and their host chondrite group is that the chondrules are depleted in FeO (except in EH chondrites which have very little FeO). In ordinary and enstatite chondrites, chondrules are depleted in metal and sulfide compared with the whole rock; thus, the depletion in total Fe is greater than that of FeO.

The Mg/Si ratio in average chondrules varies systematically with chondrite bulk composition. In ordinary and enstatite chondrites, chondrules and whole rocks both are depleted in Mg relative to CI chondrites. In CV chondrites, the chondrules are greatly enriched in Mg, and the whole rock is marginally enriched in Mg. CO chondrules are somewhat enriched in Mg, but the whole rock has a solar Mg/Si ratio.

Minor and trace elements also range widely in chondrules. Lithophile elements cover ranges of about 1 order of magnitude in all chondrite groups; siderophiles and chalcophiles vary by more than 2 orders of magnitude. In Fig. 9.1.5, the compositional ranges (2 sigma) and mean compositions of chondrules separated from unequilibrated members of EH, LL, CV and CO chondrite groups are plotted together with the bulk composition of the host chondrite group (data are normalized to Mg). All data are from the same

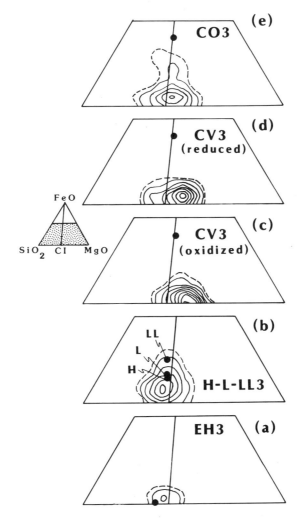

Fig. 9.1.4. Major-element compositions of porphyritic chondrules from major-chondrite groups. See Fig. 9.1.2 for explanation of methods and sources of data. The near-vertical line represents the solar (CI) MgO/SiO$_2$ ratio. The solid dots show the host-chondrite compositions. FeO is measured Fe in oxides, and excludes metal and sulfides in all cases except for the EH3 chondrules. In the EH3 group, only total Fe was available, and this is shown converted to FeO (these chondrules may contain subequal amounts of Fe and FeO). The chondrules show large changes in MgO/SiO$_2$ in the same directions as smaller deviations from the solar ratio in the host chondrites. FeO contents of chondrules are lower than the host rock in all groups except, possibly, EH3.

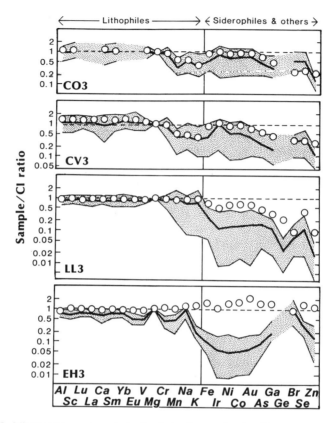

Fig. 9.1.5. Minor and trace elements in chondrules from major-chondrite groups. Data are normalized to Mg and to CI chondrites. Lithophiles, on the left, and siderophiles and chalcophiles, on the right, are arranged in order of volatility. Iron is placed out of sequence because it partitions between metal and silicates in all groups of chondrules. The stippled patterns bound the ranges of > 90% of the chondrules, the bold lines show the arithmetic means, and the circles show the host-chondrite composition. Chondrules from every group are on average depleted in siderophiles and chalcophiles compared to whole rock. Smaller depletions are present in volatile lithophiles except in CO3 chondrules. EH3 and CV3 chondrules are also depleted slightly in refractory lithophiles. All data are from UCLA. (a) Qingzhen, EH3 (Grossman et al. 1985); (b) Semarkona, LL3 (Grossman and Wasson 1983b, 1985); (c) Allende, CV3 (Rubin and Wasson 1987b); (d) Ornans, CO3 (Rubin and Wasson 1987a).

laboratory (UCLA). Again, the most obvious difference between chondrules and the host chondrite group is the depletion in siderophile and chalcophile elements, although different groups show different depletions. The greatest depletions are seen in the most reduced chondrite groups. Also evident are small depletions in chondrules of moderately volatile lithophile elements relative to nonvolatile elements compared to the host chondrites.

Grossman and Wasson (1982,1983a,b) showed that the large variations in the compositions of individual chondrules are highly systematic. Elements of similar volatility and mineral affinity vary together. From the exact nature of these variations and the relations between composition and mineralogy, it is possible to determine how elements behaved during chondrule melting, and what kind of material was melted. This is shown in the next three subsections.

Open- or Closed-system Melting?

Relative to the whole chondrite, chondrules show depletions in siderophile, volatile lithophile and chalcophile elements. It is important to determine whether chondrule precursors were originally depleted in these elements, or whether they might have been partially lost during chondrule formation. A possible loss mechanism for siderophiles and chalcophiles is separation of immiscible metal/sulfide phases from silicates during melting (Grossman and Wasson 1983a). Elements may also be susceptible to partial evaporative loss under the conditions of chondrule melting (King 1983).

Chondrules containing metal and sulfide grains concentrated near their surfaces commonly are reported. Grossman and Wasson (1985) found a chondrule apparently frozen in the process of losing metal and sulfide. Beads of these phases were located at the ends of the major axis of the elliptical chondrule. This chondrule must have been spinning, but it solidified before the metal/sulfide could be thrown out. However, chondrites do not contain enough metal and sulfide droplets in their matrices to account for all of these phases that might have been lost from chondrules. Because most chondrules are depleted in siderophiles, there should be at least one isolated metal droplet for every chondrule. But, most of the extra-chondrule Fe and Ni in many type-3 chondrites is in the form of fine-grained matrix (which contains tiny metal and sulfide grains), irregular masses of metal and sulfide, and, in type 3.0 to 3.4 ordinary chondrites, metallic chondrules and large aggregates (see, e.g., Gooding and Keil 1981; Rambaldi and Wasson 1981,1984; Grossman and Wasson 1985). Therefore, metal/silicate liquid separation may not be the most important mechanism that caused chondrules to be poor in siderophiles.

Partial evaporative loss of elements from hot chondrules must have occurred to some extent. In many laboratory experiments, volatile lithophiles (Na, K and, in extreme cases, Si) are lost from silicate liquids (see, e.g., Gooding and Muenow 1976; Donaldson 1979; Tsuchiyama et al. 1981). In other experiments, siderophile elements such as Fe, Os and Ir are observed to be volatile. The amount of evaporative loss of an element should be a function of droplet size, thermal history, O fugacity and bulk composition. There have been several attempts to link these properties to the volatile contents of chondrules. Dodd and Walter (1972) and Grossman and Wasson (1983a) failed to find any correlation between Na depletion and chondrule size in ordinary chondrites. Likewise, correlations have not been found between chondrule composition (i.e., Mg/Si, Fe/Si major element ratios) and volatile contents, or between texture and volatile contents.

Dodd and Walter (1972) observed that some chondrules containing abundant metallic Fe and low-Fa olivine tend to be low in Na. These chondrules probably are type-I, metal-rich types in the McSween (1977a) classification system; the metal is likely to have a high Fe/Ni ratio in this type of chondrule. At silicate liquidus temperatures in gases of near-solar composition, FeO will be reduced by H_2 to form Fe-metal and H_2O. Chondrules that remained at these temperatures long enough might experience both reduction and volatile loss, although the relative kinetics of the two processes in chondrules are not well understood. This type of reduced, volatile-poor chondrule is rare in ordinary chondrites, but may be more abundant in CV chondrites.

Most chondrules contain substantial amounts of FeO. Sodium content is not related to the oxidation state of Fe. Figure 9.1.6 shows this on Na_2O-Al_2O_3-FeO ternary diagrams for major chondrite groups. The chondrules span wide ranges of Na/Al ratios with the average ratio near to or just below that in the bulk rock; however, there is no relation between Na/Al and FeO content. Furthermore, there is no relationship between depletions of volatile lithophiles and depletions of volatile siderophiles (Grossman and Wasson 1983a). All of this evidence leads to the conclusion that most chondrules did not suffer extensive fractionation by volatilization.

Lack of Igneous Fractionation Trends

As was shown in the mineralogy section (Sec. 9.1.1), chondrules as a group do not show trends in their normative mineral abundances that are consistent with fractional crystallization of olivine. However, if many igneous systems of different bulk composition contributed to the population of chondrules, then fractionation trends might be masked by these bulk variations.

The trace-element chemistry of chondrules, however, indicates that the precursors of chondrules did not experience fractional crystallization processes. These processes could be expected to produce fractionated rare-earth-element abundance patterns. In fact, they could fractionate any elements that reside in different igneous phases. Instead, chondrules display elemental fractionations that are related to the cosmochemical affinities and not to the mineralogical affinities of the elements. Because metamorphism can fractionate elements in ways similar to igneous processes, these volatility-related fractionation effects can only be seen in chondrules from the least metamorphosed chondrites. In Fig. 9.1.7, the rare-earth abundance patterns from a suite of Semarkona (LL3.0) chondrules are shown. The rare earths show only very small fractionations or Eu-anomalies. In Fig. 9.1.8, scatter diagrams between five refractory and semi-refractory elements that might be fractionated by fractional crystallization are shown. Of the elements in Fig. 9.1.8, all of those plotted on the abscissa partition between pyroxene and glass; the ordinate, Al, is almost entirely in glass in chondrules (Rubin 1986). If low Al contents

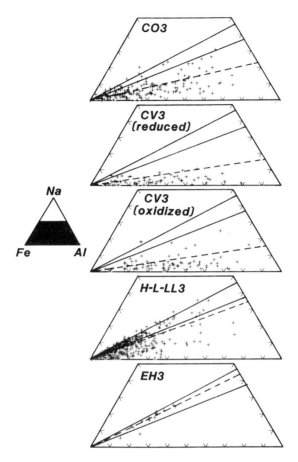

Fig. 9.1.6. Bulk concentrations of chondrules from major chondrite groups shown in the system Fe-Al-Na. Iron data are for oxides only except in EH3. Sources of data are those listed for Fig. 9.1.2. The upper solid line shows the Na/Al ratio of albite (1:1 molar), the lower solid line shows the solar Na/Al ratio, and the dashed line shows the bulk chondrite group Na/Al ratio. Prolonged heating in a gas of solar composition should cause rapid depletion of Na and reduction of FeO to Fe, possibly accompanied by metal loss. Trends toward increasing Fe depletion with decreasing Na/Al ratio are not present except for a weak relationship in the CV3 chondrules. Most groups of chondrules have a Na/Al ratio very close to that of the bulk rock.

are indicative of the concentration of crystals relative to residual liquid, then all of these trends should turn upwards on the left side where the pyroxene component is dominant. Instead, the good correlations span the range of Al contents. The slopes in each case are nearly those of the unfractionated CI chondrites.

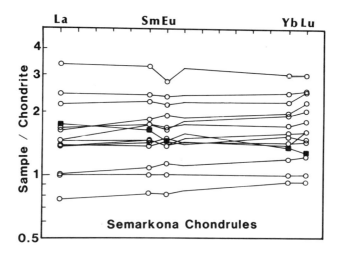

Fig. 9.1.7. Rare-earth-element abundance patterns in 13 Semarkona chondrules. Data are from Grossman and Wasson (1983b), normalized to CI chondrites, but not to Mg as in Fig. 9.1.5. Most chondrules have nearly flat rare-earth patterns, indicating that they were not derived from coarse-grained igneous precursors that experienced partitioning of rare earths into multiple phases.

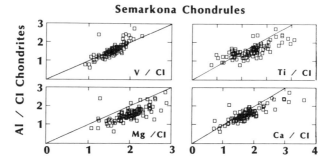

Fig. 9.1.8. Refractory-element correlations in 130 Semarkona chondrules. Data are from Grossman and Wasson (1983b) and unpublished data of J. Grossman. Calcium, Ti, V and Mg are plotted vs Al, normalized to CI chondrites. These elements, and all other measured refractory-lithophile elements, intercorrelate strongly with slopes near 1 (i.e., most chondrules have CI ratios of the elements). These trends indicate that few chondrules form by the fragmentation (with or without remelting) of coarse-grained rocks with chondrule-like mineralogy. If they were formed in that way, the trends should be more scattered, especially at the low-Al end.

In summary, the precursors of chondrules do not appear to have been igneous rocks that underwent fractional crystallization. They could also not have been coarse-grained igneous rocks that were fractionated by nonrepresentative sampling of phases; this process would produce the same kinds of fractionation effects as fractional crystallization.

Chemical Precursors

The covariations of elements in chondrules can be used to define chemical components of chondrules. For those elements for which chondrules behaved as closed systems, theoretical components can be equated with actual precursor material, and the mineralogy of the precursors can be inferred. For elements that were subject to loss by partial evaporation, a component may either be related to the composition of the evaporated fraction or to that of the precursor material. The same is true for elements subject to loss by separation of immiscible liquids.

The important assumptions behind all such interpretations of chemical components are:

1. The chondrules have not been changed by secondary processes such as thermal metamorphism (causing chemical exchange with surrounding material), chemical alteration (used here to mean subsolidus reactions of chondrule minerals with gas), or aqueous activity;
2. The chondrules are all true chondrules in the restrictive sense, all formed by the same process in the same setting.

If these assumptions are incorrect, then inferred components may result from secondary processing or may be meaningless products of correlations between similar but unrelated populations of objects.

Table 9.1.2 summarizes the properties of precursor components that have been inferred for chondrules in each chondrite group. Grossman and Wasson (1982,1983*b*,1985) studied chondrules from Chainpur (LL3.4) and Semarkona (LL3.0). They found similar trends in both chondrites. Rubin and Wasson (1987*b*) studied a suite of Allende (CV3) chondrules and their surrounding coarse-grained rims. They also inferred several chemical components, all of which seemed to be shared by chondrules and their rims. Chondrule-precursor components have not yet been inferred for the reduced subgroup of CV chondrites. Rubin and Wasson (1986) also attempted to infer the chemical components of chondrules from a CM chondrite, Murray. Data were gathered by microprobe; analysis is complicated by extensive aqueous alteration and a small sample population. Rubin and Wasson (1987*a*) obtained INAA data for chondrules from Ornans (CO3). They found similar components to those inferred for Murray chondrules. The CM and CO results are summarized together in Table 9.1.2. Grossman et al. (1985) studied a small suite of Qingzhen (EH3) chondrules. They reported several chemical

TABLE 9.1.2
Characteristics of Chondrule Precursor Material in Major Chondrite Groups.

Precursor material		EH[b]	LL[c]	CV[d]	CO[e] CM[f]
Olivine:	Composition	FeO poor		wide range	FeO-rich
	Assoc. elements	refractory		not linked	nonrefractory
	Abundance	minor	half		major
Pyroxene:	Composition	FeO poor	FeO rich	unknown FeO	FeO poor
	Assoc. elements	nonrefractory		Mn, Cr	refractory, Mn, Cr
	Abundance	major	half	very minor	minor
Metal:	Composition	complex	normal	Au As indep.	Cr bearing
	Abundance		variable, increasing to right		
Ir-Os-Ru:	In metal	yes	yes	yes	yes
	In silicates	no?	yes	yes	yes
Sulfides:	Composition	many phases		mostly troilite	
	Assoc. phases	metal, silic.		metal	
Alkalis:	Correlations	Al	Ga?		Al, Ga
Zinc:	Correlations	none	none	none	none

[a] Order of chondrite groups is based on similarity of precursor components.
[b] Qingzhen: Grossman et al. (1985).
[c] Semarkona: Grossman and Wasson (1983b, 1985).
[d] Allende: Rubin and Wasson (1987a).
[e] Ornans: Rubin and Wasson (1987b).
[f] Murray: Rubin and Wasson (1986).

components, although interpretations are again complicated by the possibility of secondary processing in this EH3 meteorite.

It is now clear that chondrules in different chondrite groups formed from different precursor materials. However, in each group the compositions of the precursors are generally related to volatility and to the affinity of specific elements for metal, silicate or sulfide phases.

Several trends may be present among inferred precursors going in the direction EH-LL-CV-(CO-CM) as shown in Table 9.1.2. Independent olivine-rich and pyroxene-rich assemblages were found to be precursors of chondrules in every group (the studies all used the $MgO + FeO/SiO_2$ ratio to determine mineralogy). Across the four chondrite groupings, olivine-rich assemblages change from FeO-poor to FeO-rich, from a minor component to a major component, and are associated with refractory minerals (a refractory component) in EH and LL chondrites, but not in carbonaceous chondrites. Pyroxene-rich assemblages show more complex variations: these change from a major component to a minor component, from FeO-poor (EH) to FeO-rich (LL) back to FeO-poor (CO, CM), and show strong links to Cr and Mn only in carbonaceous chondrites.

All chondrule sets have two different types of metal-bearing precursors, represented by a common (chondritic) siderophile component and a refractory siderophile component. In most groups, this precursor mainly contains metal phases; in EH chondrules, other reduced phases (silicides, phosphides) may also be present. In addition to the siderophile elements, this component contains some Cr in the CM-CO grouping. Common metal components increase in abundance across the groups as listed in Table 9.1.2. The refractory-metal precursors (shown in Table 9.1.2 as Ir-Os-Ru) are minor in abundance, and may be associated with refractory silicates in most groups. The independent behavior of two elements, Au and As, in CV chondrules may indicate a third type of metal precursor in that group.

Sulfides (mostly FeS) are part of the common metal-precursor assemblages in all groups. Like metal, the sulfides in EH chondrule precursors are mineralogically complex.

The alkalis form a chemical component in all groups, and are associated with refractory lithophiles (Al) in EH and CO-CM, but not in LL or CV. The most likely explanation for the behavior of the alkalis is that they were present in feldspathic minerals in refractory precursors; in some chondrite groups the abundance of alkalis was limited by the availability of Al for the formation of feldspathic minerals.

Other elements have not been linked to precursor components in any chondrite group. These are represented by Zn in Table 9.1.2. The possible causes of this behavior include: nearly equal concentrations of the element in several components; open-system chondrule formation; or, simply, nonuniform distribution of the element among precursor grains.

In summary, there are regular changes in the precursor material of chondrules in each chondrite group. These changes may be linked to changes in the composition of mineral assemblages with position in the nebula.

9.1.7. ISOTOPIC PROPERTIES OF CHONDRULES

Radiometric Ages and Formation Intervals

Chondrules from unequilibrated chondrites have ages that are indistinguishable from those of bulk unequilibrated chondrites. A review of this subject is found in Swindle et al. (1983) (see also Chapter 15.3). Hamilton et al. (1979) measured a whole-chondrule Rb-Sr isochron for Parnallee (LL3.4); chondrules define an age of 4.53 ± 0.02 Gyr with a primitive $^{87}Sr/^{86}Sr$ ratio. A similar study for Allende (CV3) chondrules by Tatsumoto et al. (1976) failed to define an isochron; the oldest model ages of chondrules in this study approach 4.5 Gyr, the same value as the model age of the whole meteorite (Chapter 5.2). Chondrules from Chainpur (LL3.4) have been studied by Podosek (1970), Herrwerth et al. (1983) and Swindle et al. (1986) by $^{40}Ar - ^{39}Ar$ techniques. Similar to the Rb-Sr data for Allende chondrules, these data do not show well-defined Ar plateau ages: the K-Ar system seems to have been disturbed (Chapter 5.3). However, in many chondrules the high-temperature release of Ar approaches an apparent age of about 4.5 Gyr.

Chondrules also contain iodine-correlated ^{129}Xe, which is evidence that they contained primordial ^{129}I when they formed. This has been shown for chondrules in Chainpur, Allende and Semarkona (Podosek 1970; Swindle et al. 1983; Swindle and Grossman 1987; see also Chapter 15.3). In Chainpur, chondrules range over a factor of 4 in their initial $^{129}I/^{127}I$ ratios, indicating a spread of about 38 Myr in formation intervals (-4 to $+32$ Myr, relative to the formation interval of the Bjurböle standard). Several chondrules in Chainpur and most chondrules in Allende do not produce well-defined isochrons. These chondrules, as well as some of the relatively young chondrules in Chainpur may have suffered a later disturbance of the I-Xe system. Semarkona, the least metamorphosed of these three chondrites, shows a narrower range of I-Xe ages in its chondrules (-4 to $+6$ Myr relative to Bjurböle for 6 chondrules). Some of the spread in ages for Semarkona may be due to late-stage alteration processes. Thus, the chondrules may have all formed (melted) within a few Myr.

When taken as a whole, sets of chondrules from Chainpur, Allende and three measured equilibrated chondrites, Bjurböle (L/LL4), Allegan (H5) and Bruderheim (L6) all seem to have formation ages that are several Myr younger than coexisting matrix fractions. It is possible that this is in part due to late-stage processing of the chondrules because many chondrules show evidence for I-Xe disturbance. Future studies of the most unequilibrated chondrites may provide an answer to this question.

No convincing correlations between chondrule texture and age have been documented. Although Herrwerth et al. (1983) reported that some chondrule types are up to 200 Myr older than other types by $^{40}Ar - ^{39}Ar$ dating techniques, this is contradicted by the I-Xe and Rb-Sr isotopic data.

Oxygen Isotopes

Oxygen isotopic ratios have been measured in individual chondrules from all of the groups of unequilibrated chondrites (Clayton et al. 1981, 1983a,b; Gooding et al. 1983; Clayton and Mayeda 1985; Clayton, unpublished data 1987—for Ornans [CO3] chondrules). These data are summarized in Fig. 9.1.9 on a standard 3-isotope plot (see also Fig. 12.1.6). Chondrules from each group are heterogeneous, and do not lie on fractionation trends with a slope of 0.5. Thus, it is clear that the chondrules are derived from the

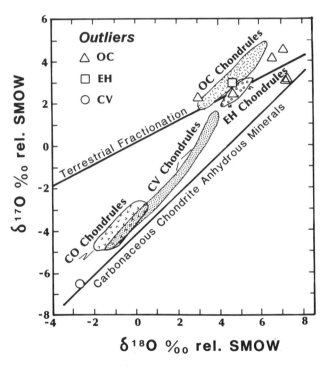

Fig. 9.1.9. Oxygen-isotopic trends in separated chondrules and chondrule composite samples shown on a $\delta^{17}O$ vs $\delta^{18}O$ diagram. Sources of data: Qingzhen (EH3), Clayton and Mayeda (1985); Allende (CV3), Clayton et al. (1983a); ordinary chondrites, Clayton et al. (1981,1983b), Gooding et al. (1983); Ornans (CO3), Clayton (1987, personal communication). The balloons show the fields occupied by most chondrules, and were drawn by eye to exaggerate the evident trends and the clustering of data. Outlying points are all shown. Three fields (carbonaceous, ordinary and enstatite) are clearly distinct, although the ordinary chondrite field overlaps with the EH3 field. The chondrule populations are in similar positions to the host-chondrite groups, and all define trends with slopes different from 0.52 (mass-dependent fractionation). See also Fig. 12.1.6.

mixing of at least two distinct O-bearing components in each chondrite group. Other components may be required if the effects of mass fractionation were small (see also Chapter 12.1).

The fields occupied by chondrules from ordinary, carbonaceous and enstatite chondrites show little overlap, and are near to the positions of the host chondrite groups. H, L and LL chondrules are indistinguishable. There is a small amount of overlap between EH and ordinary chondrite chondrules, and substantial overlap between CO and CV chondrules (only four CO chondrules and composites have been measured, and two lie inside the CV field). Therefore, chemical and mineralogical similarities between chondrules from ordinary, carbonaceous and enstatite chondrites are not due to the mixing together of chondrule populations in variable proportions in each group. Instead, the chondrules in each group must have formed from different components, either in different locations or at different times.

Therefore, the primary question posed by these isotopic data is, what components mixed together in each chondrite group? The three mixing trends for the different chondrule groups all seem to converge near one segment of the terrestrial-fractionation line. This suggests that one component may be common to all chondrule groups. Clayton et al. (1983a) suggested that this component was nebular gas, leading to the conclusion that chondrules from each group formed from a unique solid reservoir (three reservoirs, all different) that partially equilibrated with the same gas (Chapter 12.1). This equilibration could have occurred when chondrules were hot. If so, then there should be correlations between chemical, physical and petrological properties with the degree of isotopic exchange.

Several attempts have been made to correlate other properties of chondrules with O-isotope data. Clayton et al. (1983a) and McSween (1985) found that barred olivine chondrules in Allende (CV3) lie closer to the terrestrial fractionation line than porphyritic chondrules. The textures of barred chondrules are commonly interpreted to indicate complete melting, whereas the other chondrule types may not have been fully molten. These particular Allende chondrules also are rich in FeO compared to the other chondrules, and therefore have lower melting temperatures. These facts are consistent with the hypothesis that some chondrules exchanged O with nebular gas at high temperature. Other data, however, raise questions about this hypothesis. Chondrules that thoroughly exchanged with nebular gas at high temperature should have experienced the greatest reduction of FeO to metallic Fe. Yet, the chondrules that underwent the greatest exchange also have the most FeO (an oxidizing gas cannot account for this because most chondrules contain too little metal to produce the observed FeO). Clayton et al. (1983a) showed that Si isotopes lie along a mass fractionation line for the same chondrules; they suggested that this was caused by high-temperature volatilization of Si, but the degree of Si-isotopic fractionation is not correlated with position on the O-isotope trend.

Gooding et al. (1983) and Clayton and Mayeda (1985) showed that there is not a strong relationship in the ordinary and enstatite chondrites between chondrule texture and isotopic composition, apparently contradicting the high-temperature gas-liquid exchange hypothesis. However, in size-sorted composites of chondrules from Dhajala (H3.8), the smallest chondrules, which might have exchanged most completely with a gas while they were hot, are indeed nearest the terrestrial fractionation line. Unfortunately, silicates in Dhajala chondrules are mostly equilibrated due to thermal metamorphism. Small chondrules might also be the most susceptible to alteration by secondary processes. Gooding et al. (1983) showed that in chondrules from ordinary chondrites, $(CaO + Al_2O_3)/MgO$ is correlated with distance from the terrestrial fractionation line. This chemical parameter should not be sensitive to partial evaporative changes because none of the elements are volatile. They concluded that both O reservoirs in ordinary-chondrite chondrules were in the form of preexisting solids. In fact, the Allende data of Clayton et al. (1983a) could also be interpreted in terms of two solid precursors if an FeO-rich precursor was also poor in ^{16}O.

In summary, chondrules from different chondrite groups (EH, H-L-LL, CO and CV) do not share all of the same O-bearing precursor components. There is some evidence that chondrules from every chondrite group exchanged with a common gas during melting. Alternatively, they may have sampled some isotopically similar solid precursors with compositions near the terrestrial fractionation line.

9.1.8. CHONDRULE SURFACE FEATURES

Craters and Compound Chondrules

Many chondrules have spherical indentations on their surfaces, possibly formed when a plastic (cooling) chondrule collided with a solid chondrule at low velocity (Gooding and Keil 1981). Some of these "craters" may also be places where immiscible metal and sulfide liquid collected due to the molten chondrule's spin (Grossman and Wasson 1985). There is no simple way to distinguish between these possibilities if the object that once occupied the crater is gone. In some cases, a blob of metal and sulfide remains in the indentation, while for other chondrules, a second chondrule is still attached. The latter are known as "compound" chondrules. Oddly, the smaller member of most documented compound chondrule pairs is the one that was plastic at the time of the collision; one might expect that the larger chondrule would have the higher probability of being plastic at the time of the collision because it might cool more slowly. In less common cases, the smaller chondrule was solid at the time of collision, and is completely enclosed in the larger chondrule.

From the number of chondrules in ordinary chondrites that encountered a second chondrule shortly after melting, Gooding and Keil (1981) estimated the number density of chondrules in the formation area. They assumed that chondrules would have the correct plasticity between 900 K and several hundred K below the liquidus. They further assumed velocities limited to 100 m/s, and the very rapid cooling rates (short durations of plasticity) of Nelson et al. (1972). They found that porphyritic chondrules formed at >100 times lower number densities than nonporphyritic chondrules. Reasonable values of N (number density) for porphyritic chondrules were found to be several/m^3, although this number is strongly dependent on the choice of parameters. Slower cooling rates, as more recently determined, would lower N. Elimination of indentations formed by metal separation would further lower N, especially for nonporphyritic chondrules, which are more depleted in metal than are porphyritic chondrules (Gooding et al. 1980; Grossman and Wasson 1982).

A remarkably high fraction of compound chondrules are composed of chemically and petrographically similar chondrules (Lux et al. 1981), although some compound pairs are quite dissimilar (e.g., one shown in Scott and Taylor 1983). Of the dissimilar pairs, most are barred chondrules inside porphyritic chondrules. Porphyritic-nonporphyritic pairs are almost unknown. One reason for this may be that chondrule precursors were homogeneous within local formation regions. Alternatively, the similar pairs may represent a recoalescence of different droplets formed from the melting of one precursor assemblage. If the latter possibility is true, then the number densities of nonporphyritic chondrules (which contain the most compound chondrules) may be reduced further.

Rims

Several different types of rims occur around chondrules in unequilibrated chondrites. Some rims may have developed *in situ* in a parent body, but most seem to predate accretion (see also Chapter 10.4). Detailed study of these rims can help place important constraints on chondrule-formation models and parent-body processes.

Fine-grained Rims. These rims with mineralogy similar to that of chondrite matrix, occur around many ordinary, CO and CV chondrules (Cristophe Michel-Lévy 1976; Ashworth 1977; Allen et al. 1980; Scott et al. 1984). Those in ordinary chondrites generally are 10 to 30 μm wide and consist of sub-μm sized grains of MgO-rich low-Ca pyroxene, FeO-rich olivine, amorphous-feldspathic material and variable (in some cases, large) amounts of sulfide and/or metal. The bulk compositions of these rims (determined by broad-beam electron-microprobe analysis) resembles that of matrix material. Individual rims are fairly homogeneous in composition; typical compositions are ~50 mg/g SiO$_2$, ~50 mg/g MgO and ~100 mg/g FeO (Scott et al. 1984). Different rims from the same chondrite can have significantly different compositions; for example, two different rims in Piancaldoli (LL3) have mean compositions of 160 mg/g and 420 mg/g FeO, respectively.

Evidence that fine-grained rims formed prior to accretion and not in a planetary regolith includes the relatively narrow compositional ranges of individual rims compared to the large inter-rim variations in the same meteorite (Scott et al. 1984). If chondrules had acquired their rims as dusty coatings in a regolith, impact-gardening processes would have homogenized their compositions and there should now be much less interchondrule variation in rim composition. In addition, clasts and mineral grains known to have formed in regoliths are not typically rimmed.

Grossman and Wasson (1987) studied chondrules and their rims in Semarkona (LL3.0) by INAA and found similarities in abundances of siderophile and chalcophile elements. They suggested that much of the metal and sulfide in these rims was derived from the interior chondrules.

The occurrence of fine-grained rims around chondrules, isolated mineral grains, chondritic clasts (in accretionary breccias such as Leoville, CV3), and refractory and mafic inclusions indicates that the rims are composed of nebular dust that was available at the disparate locations where chondrites agglomerated (Chapter 10.4). Such dust may also have been available in the zones where chondrules and refractory inclusions formed.

Dark halos surround chondrules and inclusions in CM2 chondrites (Bunch and Chang 1980). They consist principally of phyllosilicates, carbonaceous matter and fine-grained sulfides; they appear dark when viewed in thin section in transmitted light. Aqueous alteration may have formed dark halos from fine-grained chondrule rims *in situ*.

Coarse-grained Rims. These rims surround $\sim 50\%$, $\sim 10\%$ and $<1\%$ of the chondrules in CV3, ordinary and CO3 chondrites, respectively (Rubin 1984). Rim thicknesses average 150 μm in ordinary and 400 μm in CV3 chondrites; mean grain sizes are ~ 4 μm in ordinary and ~ 10 μm in CV3 chondrites. Rims consist primarily of FeO-rich olivine and pyroxene (low-Ca pyroxene in ordinary chondrites, low-Ca pyroxene and diopside in CV3 chondrites) amidst blebs and discontinuous shells of sulfide and accessory metallic Fe-Ni. Many rims in CV3 chondrites contain nepheline and sodalite; a few (in ordinary and CV3 chondrites) contain microcrystalline plagioclase. The occurrence of metallic Fe-Ni droplets in a few rims implies heating to 1250 K.

The occurrence of fine-grained rims surrounding many coarse-grained rims indicates that coarse-grained rims also are preaccretionary.

Rubin and Wasson (1987*b*) analyzed by INAA 13 coarse-grained rims in Allende (CV3) including 9 chondrule-rim pairs. They found that, in general, rims are more similar to the mean chondrule composition than to the particular chondrule they enclose. Rims as a group are also more homogeneous than chondrules, suggesting that rims formed from greater numbers of well-mixed fine-grained particles. Because Allende matrix is similar in composition to some coarse-grained rims, it seems likely that matrix was derived from precursor materials similar to those that formed these rims.

Clayton et al. (1987) found that chondrules and coarse-grained rims lie along an ^{16}O-mixing line, and that each rim is depleted in ^{16}O relative to its enclosed chondrule. Coarse-grained rims and matrix material appear to have been formed from dust that was finer grained and poorer in ^{16}O than the chondrule precursor material.

Chondrules with coarse-grained rims appear to have been reheated following acquisition of dusty coatings. This observation restricts plausible chondrule-forming mechanisms to those that allow multiple episodes of heating.

Fe-, Ca-rich Rims. Rims around 3 to 5% of the chondrules in Kainsaz (CO3) consist of layers of FeO-rich olivine and high-Ca pyroxene (Kring 1987; Kring and Wood 1987). Some of these rims are surrounded by a layer of pigeonite, metal and sulfide. The sharp boundaries between chondrules and rims, and the igneous textures of some of these rims suggest that molten silicate accreted to solid chondrules. If this is true, then such rimmed objects are a special type of compound chondrule in which the molten member of the pair wet the surface of the solid member. Kring and Wood (1987) interpret monominerallic rims as nebular condensates added to the chondrules during fluctuations in nebular conditions.

Rims of similar mineralogy to those just described also occur around some CAIs. Kring (1987) suggested that solid chondrules and CAIs coexisted in a high-temperature environment where they acquired rims.

9.1.9. CONCLUSIONS

The chemical, isotopic, mineralogical and physical properties of chondrules vary among the major chondrite groups. Most of these differences probably result from the different conditions in the solar nebula at each chondrite-formation region. Thus, it is now becoming possible to infer structural properties of the nebula based on data other than the bulk compositions of chondrites.

The bulk- and mineral-chemical variations among chondrules from individual meteorites seem to mirror the chemical fractionations of the host chondrite groups. This has been interpreted to mean that chondrules formed by sampling the same nebular components that were separated to cause the intergroup fractionations (Grossman and Wasson 1983*b*, 1985). However, isotopic data indicate that there was little mixing of the chondrules from each chondrite formation region. This is a very significant conclusion, for it indicates that chondrules themselves are *not* one of the components involved in the intergroup fractionations. Moreover, because O is a major constituent of all of the major chondrule precursors, it becomes clear that even the chondrule precursors were not shared among formation regions. Thus, there is no complementarity among the chondrite groups. Chemically and isotopically variable material was formed at each location, and was then fractionated to produce the chondrite groups.

Acknowledgments. Technical assistance by S. S. Sorensen, and helpful discussions with C. P. Sonett are greatly appreciated. Thanks go to P. A. Baedecker, R. Brett and an anonymous referee for their reviews, and to G. J. MacPherson and K. Fredriksson of the Smithsonian Institution for providing thin sections.

REFERENCES

Allen, J. S., Nozette, S., and Wilkening, L. L. 1980. A study of chondrule rims and chondrule irradiation records in unequilibrated ordinary chondrites. *Geochim. Cosmochim. Acta* 44:1161–1175.
Ashworth, J. R. 1977. Matrix textures in unequilibrated ordinary chondrites. *Earth Planet. Sci. Lett.* 35:25–34.
Bischoff, A., and Keil, K. 1983. Ca-Al-rich chondrules and inclusions in ordinary chondrites. *Nature* 303:588–592.
Brigham, C. A., Yabuki, H., Ouyang, Z., Murrell, M. T., El Goresy, A., and Burnett, D. S. 1986. Silica-bearing chondrules and clasts in ordinary chondrites. *Geochim. Cosmochim. Acta* 50:1655–1666.
Brownlee, D. E., and Rajan, R. S. 1973. Micrometeorite craters discovered on chondrule-like objects from Kapoeta meteorite. *Science* 182:1341–1344.
Bunch, T. E., and Chang, S. 1980. Carbonaceous chondrites—II. Carbonaceous chondrite phyllosilicates and light element geochemistry as indicators of parent body processes and surface conditions. *Geochim. Cosmochim. Acta* 44:1543–1577.
Christophe Michel-Lévy, M. 1976. La matrice noire et blanche de la chondrite de Tieschitz (H3). *Earth Planet. Sci. Lett.* 30:143–150.
Christophe Michel-Lévy, M. 1981. Some clues to the history of H-group chondrites. *Earth Planet. Sci. Lett.* 54:67–80.
Christophe Michel-Lévy, M. 1987. Microchondrules in the Mezö-Madaras and Krymka unequilibrated chondrites. *Meteoritics* 22, in press (abstract).
Clayton, R. N., and Mayeda, T. K. 1985. Oxygen isotopes in chondrules from enstatite chondrites: Possible identification of a major nebular reservoir. *Lunar Planet. Sci.* XVI:142–143 (abstract).
Clayton, R. N., Mayeda, T. K., Gooding, J. L., Keil, K., and Olsen, E. J. 1981. Redox processes in chondrules and chondrites. *Lunar Planet. Sci.* XII:154–156 (abstract).
Clayton, R. N., Onuma, N., Ikeda, Y., Mayeda, T. K., Hutcheon, I. D., Olsen, E. J., and Molini-Velsko, C. 1983a. Oxygen isotopic compositions of chondrules in Allende and ordinary chondrites. In *Chondrules and Their Origins,* ed. E. A. King (Houston: Lunar and Planetary Inst.), pp. 37–43.
Clayton, R. R., Mayeda, T. K., Molini-Velsko, C. A., and Goswami, J. N. 1983b. Oxygen and silicon isotopic composition of Dhajala chondrules. *Meteoritics* 18:282–283 (abstract).
Clayton, R. N., Mayeda, T. K., Rubin, A. E., and Wasson, J. T. 1987. Oxygen isotopes in Allende chondrules and coarse-grained rims. *Lunar Planet. Sci.* XVIII:187–188 (abstract).
Dodd, R. T. 1971. The petrology of chondrules in the Sharps meteorite. *Contrib. Mineral. Petrol.* 31:201–227.
Dodd, R. T. 1974. The petrology of chondrules in the Hallingeberg meteorite. *Contrib. Mineral. Petrol.* 47:97–112.
Dodd, R. T. 1976. Accretion of the ordinary chondrites. *Earth Planet. Sci. Lett.* 30:281–291.

Dodd, R. T. 1978a. The composition and origin of large microporphyritic chondrules in the Manych (L-3) chondrite. *Earth Planet. Sci. Lett.* 39:52–66.
Dodd, R. T. 1978b. Compositions of droplet chondrules in the Manych (L-3) chondrite and the origin of chondrules. *Earth Planet. Sci. Lett.* 40:71–82.
Dodd, R. T., and Teleky, L. S. 1967. Preferred orientation of olivine crystals in porphyritic chondrules. *Icarus* 6:407–416.
Dodd, R. T., and Van Schmus, W. R. 1971. Dark-zoned chondrules. *Chem. Erde* 30:59–69.
Dodd, R. T., and Walter, L. S. 1972. Chemical constraints on the origin of chondrules in ordinary chondrites. In *On the Origin of the Solar System,* ed. H. Reeves (Paris: CNRS), pp. 293–300.
Donaldson, C.H. 1979. Composition changes in a basalt melt contained in a wire loop of $Pt_{80}Rh_{20}$: Effects of temperature, time, and oxygen fugacity. *Mineral. Mag.* 43:115–119.
El Goresy, A. 1985. The Qingzhen reaction: Fingerprints of the EH planet? *Meteoritics* 20:639 (abstract).
Evensen, N. M., Carter, S. R., Hamilton, P. J., O'Nions, R. K., and Ridley, W. I. 1979. A combined chemical-petrological study of separated chondrules from the Richardton meteorite. *Earth Planet. Sci. Lett.* 42:223–236.
Fredriksson, K., Jarosewich, E., and Nelen, J. 1969. The Sharps chondrite—New evidence on the origin of chondrules and chondrites. In *Meteorite Research,* ed. P. M. Millman (Dordrecht: D. Reidel), pp. 155–165.
Fredriksson, K., Noonan, A., and Nelen, J. 1973. Meteoritic, lunar and lonar impact chondrules. *The Moon* 7:475–482.
Fruland, R. M., King, E. A., and McKay, D. S. 1978. Allende dark inclusions. *Proc. Lunar Planet. Sci. Conf.* 9:1305–1329.
Fujimaki, H., Matsu-ura, M., Sunagawa, I., and Aoki, K. 1981. Chemical compositions of chondrules and matrices in the ALH-77015 chondrite (L3). In *Proc. of the Sixth Symposium on Antarctic Meteorites* (Tokyo: National Inst. of Polar Research), pp. 161–174.
Funaki, M., Nagata, R., and Momose, K. 1981. Natural remanent magnetizations of chondrules, metallic grains and matrix of an Antarctic chondrite, ALH-769. *Mem. Natl. Inst. Polar Res. Special Issue* 20:300–315.
Gooding, J. L. 1979. Petrogenetic Properties of Chondrules in Unequilibrated H-, L-, and LL-Group Chondritic Meteorites. Ph.D. Thesis, Univ. of New Mexico, Albuquerque.
Gooding, J. L. 1983. Survey of chondrule average properties in H-, L-, and LL-group chondrites: Are chondrules the same in all unequilibrated ordinary chondrites? In *Chondrules and Their Origins,* ed. E. A. King (Houston: Lunar and Planetary Inst.), pp. 61–87.
Gooding, J. L., and Keil, K. 1981. Relative abundances of chondrule primary textural types in ordinary chondrites and their bearing on conditions of chondrule formation. *Meteoritics* 16:17–43.
Gooding, J. L., and Muenow, D. W. 1976. Activated release of alkalis during the vesiculation of molten basalts under high vacuum: Implications for lunar volcanism. *Geochim. Cosmochim. Acta* 40:675–686.
Gooding, J. L., Keil, K., Fukuoka, T., and Schmitt, R. A. 1980. Elemental abundances in chondrules from unequilibrated chondrites: Evidence for chondrule origin by melting of pre-existing materials. *Earth Planet. Sci. Lett.* 50:171–180.
Gooding, J. L., Mayeda, T. K., Clayton, R. N., and Fukuoka, T. 1983. Oxygen isotopic heterogeneities, their petrological correlations, and implications for melt origins of chondrules in unequilibrated ordinary chondrites. *Earth Planet. Sci. Lett.* 65:209–224.
Graup, G. 1981. Terrestrial chondrules, glass spherules and accretionary lapilli from the suevite, Ries Crater, Germany. *Earth Planet. Sci. Lett.* 55:407–418.
Grossman, J. N. 1983. A Chemical and Petrographic Study of Chondrules from the Chainpur (LL3.4) and Semarkona (LL3.0) chondrites. Ph.D. Thesis, Univ. of California, Los Angeles.
Grossman, J. N., and Wasson, J. T. 1982. Evidence for primitive nebular components in chondrules from the Chainpur chondrite. *Geochim. Cosmochim. Acta* 46:1081–1099.
Grossman, J. N., and Wasson, J. T. 1983a. The compositions of chondrules in unequilibrated chondrites: An evaluation of models for the formation of chondrules and their precursor materials. In *Chondrules and Their Origins,* ed. E. A. King (Houston: Lunar and Planetary Inst.), pp. 88–121.
Grossman, J. N., and Wasson, J. T. 1983b. Refractory precursor components of Semarkona chon-

drules and the fractionation of refractory elements among chondrites. *Geochim. Cosmochim. Acta* 47:759–771.
Grossman, J. N., and Wasson, J. T. 1985. The origin and history of the metal and sulfide components of chondrules. *Geochim. Cosmochim. Acta* 49:925–939.
Grossman, J. N., and Wasson, J. T. 1987. Compositional evidence regarding the origins of rims on Semarkona chondrules. *Geochim. Cosmochim. Acta* 51, in press.
Grossman, J. N., Kracher, A., and Wasson, J. T. 1979. Volatiles in Chainpur chondrules. *Geophys. Res. Lett.* 6:597–600.
Grossman, J. N., Rubin, A. E., Rambaldi, E. R., Rajan, R. S., and Wasson, J. T. 1985. Chondrules in the Qingzhen type-3 enstatite chondrite: Possible precursor components and comparison to ordinary chondrite chondrules. *Geochim. Cosmochim. Acta* 49:1781–1795.
Hamilton, P. J., Evensen, N. M., and O'Nions, R. K. 1979. Chronology and chemistry of Parnallee (LL-3) chondrules. *Lunar Planet. Sci.* X:494–495 (abstract).
Herrwerth, I., Müller, N., Jessberger, E. K., and Kirsten, T. 1983. ^{40}Ar-^{39}Ar dating of individual chondrules and the siting of trapped argon. *Meteoritics* 18:311 (abstract).
Hughes, D. W. 1978. A disaggregation and thin section analysis of size and mass distributions of the chondrules in the Bjurböle and Chainpur meteorites. *Earth Planet. Sci. Lett.* 38:391–400.
Huss, G. R., Keil, K., and Taylor, G. J. 1981. The matrices of unequilibrated ordinary chondrites: Implications for the origin and history of chondrites. *Geochim. Cosmochim. Acta* 45:33–51.
Hutchison, R., and Bevan, A. W. R. 1983. Conditions and time of chondrule accretion. In *Chondrules and Their Origins*, ed. E. A. King (Houston: Lunar and Planetary Inst.), pp. 162–179.
Hutchison, R., Bevan, A. W. R., Agrell, S. O., and Ashworth, J. R. 1979. Accretion temperature of the Tieschitz, H3, chondritic meteorite. *Nature* 280:116–119.
Keil, K., and Fredriksson K. 1964. The iron, magnesium, and calcium distribution in coexisting olivines and rhombic pyroxenes of chondrites. *J. Geophys. Res.* 69:3487–3515.
King, E. A. 1982. Refractory residues, condensates and chondrules from solar furnace experiments. *Proc. Lunar Planet. Sci. Conf.* 13, J. Geophys. Res. 87:A429–A434.
King, E. A. 1983. Reduction, partial evaporation, and spattering: Possible chemical and physical processes in fluid drop chondrule formation. In *Chondrules and Their Origins*, ed. E. A. King (Houston: Lunar and Planetary Inst.), pp. 180–187.
King, E. A., Butler, J. C., and Carman, M. F. 1972. Chondrules in Apollo 14 samples and size analyses of Apollo 14 and 15 fines. *Proc. Lunar Sci. Conf.* 3:673–686.
King, T. V. V., and King, E. A. 1978. Grain size and petrography of C2 and C3 carbonaceous chondrites. *Meteoritics* 13:47–72.
King, T. V. V., and King, E. A. 1979. Size frequency distributions of fluid drop chondrules in ordinary chondrites. *Meteoritics* 14:91–96.
Kracher, A., Scott, E. R. D., and Keil, K. 1984. Relict and other anomalous grains in chondrules: Implications for chondrule formation. *Proc. Lunar and Planet. Sci. Conf.* 14, J. Geophys. Res. Suppl. 89:B559–B566.
Kring, D. A. 1987. Fe,Ca-rich rims around magnesian chondrules in the Kainsaz (CO3) chondrite. *Lunar Planet. Sci.* XVIII:517–518 (abstract).
Kring, D. A., and Wood, J. A. 1987. Fe,Ca-rich and Mg-rich chondrule rims in the Kainsaz (CO3) chondrite: Evidence of fluctuating nebular conditions. *Meteoritics* 22, in press (abstract).
Kurat, G. 1967. Formation of chondrules. *Geochim. Cosmochim. Acta* 31:491–502.
Kurat, G. 1969. The formation of chondrules and chondrites and some observations on chondrules from the Tieschitz meteorite. In *Meteorite Research* ed. P. M. Millman (Dordrecht: D. Reidel), pp. 185–190.
Kurat, G., Keil, K., Prinz, M., and Nehru, C. E. 1972. Chondrules of lunar origin. *Proc. Lunar Sci. Conf.* 3:707–722.
Kurat, G., Pernicka, E., and Herrwerth, I. 1983. Chondrules from Chainpur (LL-3): Reduced parent rocks and vapor fractionation. *Earth Planet. Sci. Lett.* 68:43–56.
Kurat, G., Palme, H., Brandstätter, F., Spettel, B., and Perelygin, V. P. 1985. Allende chondrules: Distillations, condensations, and metasomatisms. *Lunar Planet. Sci.* XVI:471–472 (abstract).
Lanoix, M., Strangway, D. W., and Pearce, G. W. 1977. Anomalous acquisition of thermo-

remanence at 130° C in iron and paleointensity of the Allende meteorite. *Proc. Lunar Sci. Conf.* 8:869–701.

Lanoix, M., Strangway, D. W., and Pearce, G. W. 1978. The primordial magnetic field preserved in chondrules of the Allende meteorite. *Geophys. Res. Lett.* 5:73–76.

Lofgren, G., and Russell, W. J. 1986. Dynamic crystallization of chondrule melts of porphyritic and radial pyroxene composition. *Geochim. Cosmochim. Acta* 50:1715–1726.

Lux, G., Keil, K., and Taylor, G. J. 1980. Metamorphism of the H-group chondrites: Implications from compositional and textural trends in chondrules. *Geochim. Cosmochim. Acta* 44:841–855.

Lux, G., Keil, K., and Taylor, G. J. 1981. Chondrules in H3 chondrites: Textures, compositions and origins. *Geochim. Cosmochim. Acta* 45:675–685.

McSween, H. Y. 1977a. Chemical and petrographic constraints on the origin of chondrules and inclusions in carbonaceous chondrites. *Geochim. Cosmochim. Acta* 41:1843–1860.

McSween, H. Y. 1977b. Petrographic variations among carbonaceous chondrites of the Vigarano type. *Geochim. Cosmochim. Acta* 41:1777–1790.

McSween, H. Y. 1977c. Carbonaceous chondrites of the Ornans type: A metamorphic sequence. *Geochim. Cosmochim. Acta* 41:477–491.

McSween, H. Y. 1979. Alteration in CM carbonaceous chondrites inferred from modal and chemical variations in matrix. *Geochim. Cosmochim. Acta* 43:1761–1770.

McSween, H. Y. 1985. Constraints on chondrule origin from petrology of isotopically characterized chondrules in the Allende meteorite. *Meteoritics* 20:523–540.

McSween, H. Y., Fronabarger, A. K., and Driese, S. G. 1983. Ferromagnesian chondrules in carbonaceous chondrites. In *Chondrules and Their Origins* ed. E. A. King (Houston: Lunar and Planetary Inst.), pp. 195–210.

Nagahara, H. 1981a. Evidence for secondary origin of chondrules. *Nature* 292:135–136.

Nagahara, H. 1981b. Petrology of chondrules in ALH-77015 (L3) chondrite. *Mem. Natl. Inst. Polar Res. Special Issue* 20:145–160.

Nagahara, H. 1983. Texture of chondrules. *Mem. Natl. Inst. Polar Res. Special Issue* 30:61–83.

Nagata, T., and Funaki, M. 1983. Paleointensity of the Allende carbonaceous chondrite. *Mem. Natl. Inst. Polar Res. Special Issue* 30:403–434.

Nelson, L. S., Blander, M., Skaggs, S. R., and Keil, K. 1972. Use of a CO_2 laser to prepare chondrule-like spherules from supercooled molten oxide and silicate droplets. *Earth Planet. Sci. Lett.* 14:338–344.

Okada, A. 1975. Petrological studies of the Yamato meteorites Part 1. Mineralogy of the Yamato meteorites. *Mem. Natl. Inst. Polar Res. Special Issue* 5:14–66.

Podosek, F. A. 1970. Dating of meteorites by the high-temperature release of iodine-correlated ^{129}Xe. *Geochim. Cosmochim. Acta* 34:341–365.

Rambaldi, E. R. 1981. Relict grains in chondrules. *Nature* 293:558–561.

Rambaldi, E. R., and Wasson, J. T. 1981. Metal and associated phases in Bishunpur, a highly unequilibrated ordinary chondrite. *Geochim. Cosmochim. Acta* 45:1001–1015.

Rambaldi, E. R., and Wasson, J. T. 1984. Metal and associated phases in the highly unequilibrated ordinary chondrites Krymka and Chainpur. *Geochim. Cosmochim. Acta* 48:1885–1897.

Rubin, A. E. 1983. The Adhi Kot breccia and implications for the origin of chondrules and silica-rich clasts in enstatite chondrites. *Earth Planet. Sci. Lett.* 64:201–212.

Rubin, A. E. 1984. Coarse-grained chondrule rims in type 3 chondrites. *Geochim. Cosmochim. Acta* 48:1779–1789.

Rubin, A. E. 1986. Elemental compositions of major silicic phases in chondrules of unequilibrated chondritic meteorites. *Meteoritics* 21:283–293.

Rubin, A. E., and Grossman, J. N. 1987. Size-frequency distributions of EH3 chondrules. *Meteoritics* 22, in press.

Rubin, A. E., and Keil, K. 1984. Size-distributions of chondrule types in the Inman and Allan Hills A77011 L3 chondrites. *Meteoritics* 19:135–143.

Rubin, A. E., and Wasson, J. T. 1986. Chondrules in the Murray CM2 meteorite and compositional differences between CM-CO and ordinary chondrite chondrules. *Geochim. Cosmochim. Acta* 50:307–315.

Rubin, A. E., and Wasson, J. T. 1987a. Chondrules in the Ornans (CO3) meteorite and the timing

of chondrule formation relative to nebular fractionation events. *Geochim. Cosmochim. Acta* 51, submitted.
Rubin, A. E., and Wasson, J. T. 1987b. Chondrules, matrix and coarse-grained rims in the Allende meteorite: Origin, interrelationships and possible precursor components. *Geochim. Cosmochim. Acta* 51:1923–1937.
Rubin, A. E., Scott, E. R. D., and Keil, K. 1982. Microchondrule-bearing clast in the Piancaldoli LL3 meteorite: A new kind of type 3 chondrite and its relevance to the history of chondrules. *Geochim. Cosmochim. Acta* 46:1763–1776.
Scott, E. R. D., and Taylor, G. J. 1983. Chondrules and other components in C, O, and E chondrites: Similarities in their properties and origins. *J. Geophys. Res.* 88:B275–B286.
Scott, E. R. D., Taylor, G. J., and Keil, K. 1983. Type 3 ordinary chondrites—Metamorphism, brecciation and parent bodies. *Meteoritics* 18:393–394 (abstract).
Scott, E. R. D., Rubin, A. E., Taylor, G. J., and Keil, K. 1984. Matrix material in type 3 chondrites—Occurrence, heterogeneity and relationship with chondrules. *Geochim. Cosmochim. Acta* 48:1741–1757.
Simon, S. B., and Haggerty, S. E. 1980. Bulk compositions of chondrules in the Allende meteorite. *Proc. Lunar Planet. Sci. Conf.* 11:901–927.
Snellenburg, J. W. 1978. *A Chemical and Petrographic Study of the Chondrules in the Unequilibrated Ordinary Chondrites, Semarkona and Krymka*. Ph.D. Thesis, State University of New York at Stony Brook.
Steele, I. M. 1986. Compositions and textures of relic forsterite in carbonaceous and unequilibrated ordinary chondrites. *Geochim. Cosmochim. Acta* 50:1379–1395.
Sugiura, N., and Strangway, D. W. 1985. NRM directions around a centimeter sized dark inclusion in Allende. *Proc. Lunar Planet. Sci. Conf.* 15, *J. Geophys. Res.* 90:C729–C738.
Sugiura, N., Lanoix, M., and Strangway, D. W. 1979. Magnetic fields of the solar nebula as recorded in chondrules from the Allende meteorite. *Phys. Earth Planet. Int.* 20:342–349.
Swindle, T. D., and Grossman, J. N. 1987. I-Xe studies of Semarkona chondrules: Dating alteration. *Lunar Planet. Sci.* XVIII, pp. 982–983 (abstract).
Swindle, T. D., Caffee, M. W., Hohenberg, C. M., and Lindstrom, M. M. 1983. I-Xe studies of individual Allende chondrules. *Geochim. Cosmochim. Acta* 47:2157–2177.
Swindle, T. D., Caffee, M. W., and Hohenberg, C. M. 1986. I-Xe and ^{40}Ar-^{39}Ar ages of Chainpur chondrules. *Lunar Planet. Sci.* XVII:857–858 (abstract).
Tatsumoto, M., Unruh, D. M., and Desborough, G. A. 1976. U-Th-Pb and Rb-Sr systematics of Allende and U-Th-Pb systematics of Orgueil. *Geochim. Cosmochim. Acta* 40:617–634.
Tschermak, G. 1885. *Die mikroskopische Beschaffenheit der Meteoriten erlautert durch photographische Abbildungen*. (Stuttgart: Schweizerbart'sche Verlag).
Tsuchiyama, A., Nagahara, H., and Kushiro, I. 1981. Volatilization of sodium from silicate melt spheres and its application to the formation of chondrules. *Geochim. Cosmochim. Acta* 45:1357–1367.
Wasson, J. T. 1974. *Meteorites* (Berlin: Springer-Verlag).
Wasson, J. T. 1985. *Meteorites: Their Record of Early Solar-System History* (New York: W. H. Freeman).
Wilkening, L. L., Boynton, W. V., and Hill, D. H. 1984. Trace elements in rims and interiors of Chainpur chondrules. *Geochim. Cosmochim. Acta* 48:1071–1080.
Wood, J. A. 1967. Olivine and pyroxene compositions in type II carbonaceous chondrites. *Geochim. Cosmochim. Acta* 31:2095–2108.

9.2. EXPERIMENTAL STUDIES OF CHONDRULES

R. H. HEWINS
Rutgers University

Dynamic crystallization experiments on several chondrule compositions have shown that the abundance of heterogeneous nuclei controls chondrule textures. Barred and radiating textures form from melts at or above the liquidus containing very few, if any, nuclei. Melts not initially superheated develop microporphyritic texture (just below liquidus, several nuclei) or granular texture (well below liquidus, abundant nuclei). Spherules cooled faster than ~2000° C/hr tend to develop more spherulitic or acicular crystals than those in the most typical chondrules. Cooling rates of 100 to 2000° C/hr (but not lower) provide olivine zoning and apparent Fe/Mg partitioning coefficients similar to what are observed in chondrules. Matching chondrule textures and phase compositions in the laboratory is hampered by gravitational settling of crystals. The absence in nature of perfectly glassy chondrules, which can be simulated very easily, shows that chondrule melts were not much superheated. The dominance of granular textures in Mg-rich olivine chondrules, porphyritic textures in Fe-rich olivine chondrules and radiating textures in Si-rich chondrules indicates a common limit to the initial temperature. Very few chondrules can have experienced a temperature above about 1600° C.

9.2.1. INTRODUCTION

There have been several attempts to make synthetic chondrules in the laboratory because, if the textures and mineral compositions of natural chondrules can be duplicated exactly, the conditions of their formation may be defined. Dynamic crystallization experiments have the potential to define not only the cooling history of chondrules, but also the nature of the heating mechanism, the state of the precursor materials and ambient conditions. Early attempts to simulate chondrules, reviewed by Hewins (1983), led to considerable disagreements about the cooling history. More recent work (see, e.g., Lofgren and Russell 1986) has shown that the disagreements between various investigators are partly due to different starting materials and assumptions (melting procedures) as well as different cooling histories. Critical variables include initial temperature and heating time, because these factors determine the number of nuclei.

Since chondrules are broadly defined as solidified melt droplets (Chapter 9.1), every object produced in a melting experiment has been regarded as a chondrule analogue. The range of cooling rates claimed for chondrules has therefore tended to be almost the entire range used in all laboratory experiments. Any olivine-rich spherule has tended to be classified automatically as either porphyritic or barred, although a much more complex terminology was used in studies of olivine morphology not specifically applied to chondrules (Donaldson 1976). This work documented the increasingly skeletal nature of olivine crystals grown from superheated liquids with increased cooling rate or initial supercooling. Although rare chondrules may resemble spherules produced under the most extreme laboratory conditions, it is clear that chondrule origins can be better understood only if agreement is reached on what is required to duplicate the most common types of chondrule. It is necessary to know exactly what the properties of the typical chondrules are and to check synthetic droplets very strictly for match to all these properties. In this chapter, olivine-rich, pyroxene-rich and Ca-Al-rich chondrules and chondrule analogues are compared, with emphasis on where different investigations have led to disagreements. Textures typical of porphyritic olivine (PO) and barred olivine (BO) chondrules (Figs. 9.2.2 and 9.2.3 below) are taken as those illustrated by Scott and Taylor (1983), Nagahara (1983a) and Weisberg (1987). Data on mineral compositions for chondrules are rarely reported in complete form, giving details of zoning and grain-to-grain variation. However, information on Ca contents of olivine (Scott and Taylor 1983), Fe-Mg partitioning between olivine and glass (Taylor and Cirlin 1986) and zoning in olivine (Miyamoto et al. 1986) is available.

9.2.2. EXPERIMENTAL TECHNIQUES

The majority of the experiments were performed on 9- or 10-component analogues to natural chondrule compositions at O fugacities 1/2 log unit below the iron-wüstite buffer curve, consistent with the small amounts of metal found in chondrules. The exception is the pioneering study of Planner (1979) who used SiO_2-MgO-"FeO" mixtures and more oxidizing conditions. The compositions used are shown in projection in Fig. 9.2.1. The technique involved mounting pellets on platinum wire loops, in some cases Fe-plated to prevent loss of Fe from the charges, which allow the formation of droplets after melting. The pellets were heated above or below the liquidus tempera-

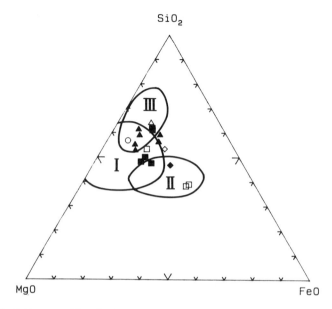

Fig. 9.2.1. Bulk compositions in wt.% used by Planner (1979) (open squares); Tsuchiyama et al. (1980a) and Tsuchiyama and Nagahara (1981) (filled squares); Lofgren and Russell (1985,1986) (filled triangles); Hewins et al. (1981) (open triangle); Bell (1986) (filled diamond); Radomsky et al. (1986) (open diamond); Radomsky (1988) (open circle). Fields are for types I (granular olivine), II (porphyritic or barred olivine) and III (excentroradial pyroxene) analyzed chondrules from McSween (1977). (Figure after Radomsky 1988.)

ture for an initial time period (minutes to hours) and then cooled through a range of controlled cooling rates before being quenched. The final temperatures were for the most part near the solidus but, because of relatively rapid cooling, abundant glass is preserved in most runs. Some glass is preserved even with "slow" cooling down to 600°C in some experiments. Most synthetic spherules are 2 to 4 mm in diameter, a little larger than average chondrules.

Early investigators assumed certain restricted conditions in attempting to duplicate chondrules. Blander et al. (1976), for example, used a laser to melt the samples and therefore their investigation was confined to very high initial temperatures, rapid heating and very high cooling rates. The resulting fine radial olivine dendrites in pyroxene-rich initial compositions showed that the majority of natural chondrules must have experienced less extreme conditions. Planner (1979) therefore tried lower cooling rates for synthesizing porphyritic spherules. Planner and Keil (1982) and Hewins et al. (1981) were influenced by techniques used in phase-equilibrium studies to ensure reproducible results. Their initial conditions were restricted to temperatures above the liquidus with reasonably long heating times so as to ensure total melting.

At about the same time, Tsuchiyama et al. (1980*a*) used a fixed initial temperature, above the liquidus for two bulk compositions but below it for their most olivine-rich one, and extremely short heating times, which are expected in some models for chondrule formation. In general, these studies did not state explicitly the assumptions made about the heating of chondrule precursors. None of these investigations examined the influence *on a single composition* of initial temperatures above and below the liquidus, or of variations in heating times. All of these investigations were felt to throw light on the cooling history of natural chondrules, and yet a porphyritic-olivine spherule typical of one investigation might consist of a single large skeletal olivine crystal or hopper (Donaldson 1976) and of another many, many tiny olivine granules.

Tsuchiyama and Nagahara (1981), Lofgren and Russell (1985) and Radomsky et al. (1986) used a range of initial temperatures above and below the liquidus, because of clear evidence that this factor influenced nucleation in silicate melts, as summarized by Lofgren (1983). It should be remembered that liquidus temperatures quoted in chondrule studies are generally apparent values, based on disappearance of crystals for a given time at temperature, and could be up to $10°C$ above true liquidus temperature. The apparent liquidus temperature is the appropriate reference for the given experiments, which tend to show that textures alone are not highly sensitive to cooling rate. Current studies, for example those of Radomsky et al. (1986), Bell (1986) and Radomsky (1988), are therefore emphasizing the combination of crystal morphology and compositional factors, especially zoning, which appears to constrain the cooling rates well. Both the heating history and the cooling history must be defined to allow the origin of chondrules to be understood. Individual charges do not permit the identification of specific initial temperatures or heating times, because both factors reduce the numbers of nuclei. However, by comparing all populations of synthetic spherules to natural chondrules, the possible range of heating conditions can be limited.

9.2.3. PORPHYRITIC OLIVINE CHONDRULES

The most abundant chondrules are the porphyritic olivine (PO) or porphyritic olivine + pyroxene (POP) variety with granular and microporphyritic textures. The appearance of pyroxene in experimental analogues was sometimes prevented by quenching from relatively high final temperatures. Textures assumed to be typical of PO (Fig. 9.2.2a,b) are those shown by Scott and Taylor 1983, Figs. 1–3) and Nagahara (1983*a*, Fig. 5). The olivine crystals in these chondrules are polyhedral or, less commonly, hoppers or granular (Donaldson 1976). In unequilibrated ordinary chondrites, such chondrules show normally zoned olivine grains, with Ca and Fe increasing towards the rims (Tsuchiyama et al. 1980*b*; Scott, personal communication, 1985; Miyamoto et al. 1986; Radomsky 1988).

a) b) ··· natural
c) ·· f) experimental

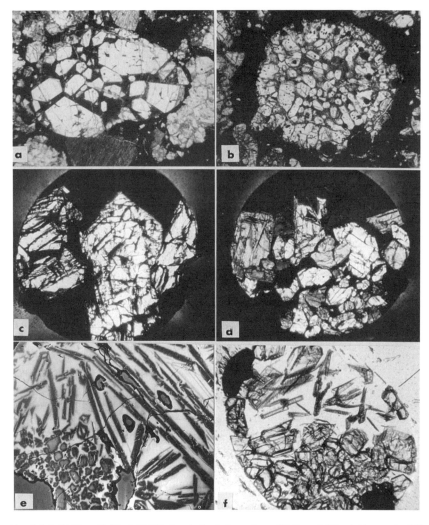

Fig. 9.2.2 (a) Type II (Fe-rich) PO chondrule, 1.3 mm long; (b) Type I (Mg-rich) granular PO chondrule, 0.8 mm; (c) Porphyritic-olivine spherule, 2 mm, initial T = liquidus, rate = 10° C/hr; (d) PO spherule, 2 mm, initial T subliquidus, rate is 10° C/hr, with granular cumulate layer overlain by microporphyritic zone with euhedral-to-hopper olivine; (e) Transitional PO/BO spherule, 4 mm (BSE). Gravitational settling produces PO zone and later nucleation in undercooled residual liquid yields parallel olivine, with a thin transition containing hoppers, initial T subliquidus, rate is 1000° C/hr; (f) Nucleation sequence with settling, 4mm, subliquidus, 100° C/hr: granular, euhedral, hopper, branching and lattice olivine. Photographs (a) and (b) are from Scott and Taylor (1983); (c) through (f) are from Radomsky (1987) and Radomsky and Shanahan (personal communication).

Planner and Keil (1982) produced, from a very simple liquid (Si-Fe-Mg-O only), porphyritic olivine spherules generally containing a very small number of large hopper olivine crystals (equant, amoeboid or elongated). These resemble a few rare (truly) porphyritic chondrules but not the microporphyritic/granular majority with many small polyhedral olivine grains. However, Planner (1979) did reproduce a microporphyritic texture in three runs (with similar thermal history to runs producing large hopper grains) apparently because chance tiny gas bubbles acted as nuclei. The textures more commonly produced by Planner result from the presence of many fewer nuclei than in typical PO chondrules. This indicates that the initial heating (30 minutes at about 20° above the liquidus) was more thorough than that experienced by most PO chondrules. These liquids, which had very few nuclei, did not develop barred-olivine (BO) texture, probably because the cooling rates were relatively low, about 300° C/hr. Higher rates, however, produced several elongated hoppers per spherule. The absence of a feldspathic component in this liquid probably made viscosities, growth rates and textures somewhat different from natural and other synthetic cases. Planner and Keil (1982) argued that an isothermal step is required in the cooling cycle because PO chondrules contain olivine more ferroan than calculated. Both the olivine composition and the number of crystals could be better matched, alternatively, by a lower initial temperature. Zoning in their olivine is extreme, compared with other experiments and chondrules, possibly because growth rates would be higher in liquids without a feldspathic component.

Tsuchiyama et al. (1980a) adopted a very short, and conceivably very appropriate, heating time of about two minutes at 1600° C. This was sufficient to melt some bulk compositions totally but not their "sample 1" which produced charges they described as porphyritic. The latter runs contain both fine granular olivine which survived the heating and skeletal blades grown during cooling, a combination not common in PO chondrules. The run which most approaches the typical natural chondrule texture with a single population of crystals is the one made at the lowest rate, 1200° C/hr. Tsuchiyama and Nagahara (1981) extended this work in experiments on a rather similar bulk composition, except for higher Fe/Mg ratio, with a liquidus temperature of about 1580° C. Porphyritic textures are confined to runs with subliquidus initial temperatures and the best match to natural PO is for the run with the highest initial subliquidus temperature. With slightly lower initial temperatures, the tenth-of-a-millimeter polyhedral olivine crystals are replaced by very fine granules. Cooling rates, in the range between 1200 and 4800° C/hr, were not specifically identified. The additional presence of bladed olivine in equivalent runs which were quenched, instead of cooled rapidly, suggests that the lowest of these rates is most appropriate to PO chondrules.

Lofgren and Russell (1985) produced textural analogues to PO chondrules with near-liquidus and subliquidus initial temperatures. The crystals are smaller, more numerous and more elongated for higher cooling rates, with

100° C/hr producing a better match to PO than 5 or 1900° C/hr. Crystals are also more numerous for lower initial temperatures, and Lofgren and Russell emphasized that the need for nuclei in reproducing PO points to remelting of preexisting crystals during chondrule formation. Taylor and Cirlin (1986) noted that the apparent Fe/Mg partition coefficient for olivine vs melt is less than the equilibrium value for PO in unequilibrated chondrites and also for the most rapidly cooled charges of Lofgren and Russell (1985). Although textural criteria favor 100° C/hr, Fe/Mg partitioning would be more consistent with cooling rates near 1900° C/hr. A significant complication in duplicating chondrules first became apparent in this study (Lofgren and Russell 1985). Olivine grown during the initial heating (analogous to relict grains in chondrules) tended to settle under gravity to the bottom of the charge and become isolated from the liquid. This settling clearly must influence the ability of the crystals to react with the liquid and hence the final shape, size, abundance and composition of the crystals. The requirements of morphology and partitioning put the cooling rate of PO in the range between 100 and 1900°/hr, with failure to obtain a more precise estimate possibly due to the influence of gravitational settling.

Bell (1986) used a type II (ferroan) PO chondrule composition (McSween 1977; see also Chapter 9.1) and produced PO analogues for initial near-liquidus and subliquidus temperatures. Glassy spherules were formed for initially superheated liquids for all cooling rates (down to 10° C/hr), demonstrating the need for survival of nuclei during chondrule heating because chondrules always contain some crystalline material. Gravitational settling is very obvious in these runs, with olivine growing up into the liquid from the top of the cumulate. Olivine in the bottom half of the charges is zoned for 100° C/hr and 1000° C/hr cooling rates, with similar Ca but higher Fe (Fo_{78-60}) than in PO chondrule olivine for which zoning data are available. Zoning may be enhanced by development of a thick boundary layer depleted in Mg in the liquid adjacent to settled olivine and by the difficulty of reequilibrating the somewhat larger crystals developed in the synthetic spherules. Liquidus temperatures fall sharply with increasing Fe/Mg ratios, so that it is much more difficult to erase zoning in ferroan olivine than in Mg olivine. Available zoning data are for Fo_{92-80} cores (Scott, personal communication, 1985; Miyamoto et al. 1986) and, since Bell (1986) reported zoning for 100° C/hr for more ferroan olivine, these natural chondrules probably cooled at rates higher than 100° C/hr.

Radomsky (1988) and Radomsky and Shanahan (personal communication) performed an analogous set of experiments to those of Bell, except that a type I (magnesian) PO chondrule composition was used with 2 mm and 4 mm spherules. They produced PO textural analogues from liquidus and subliquidus temperatures at cooling rates of 10 to 100° C/hr (Fig. 9.2.2c,d). Subliquidus 1000° C/hr runs are also PO but transitional to BO because of late nucleation near the top of the charge (Fig. 9.2.2e). Natural chondrules

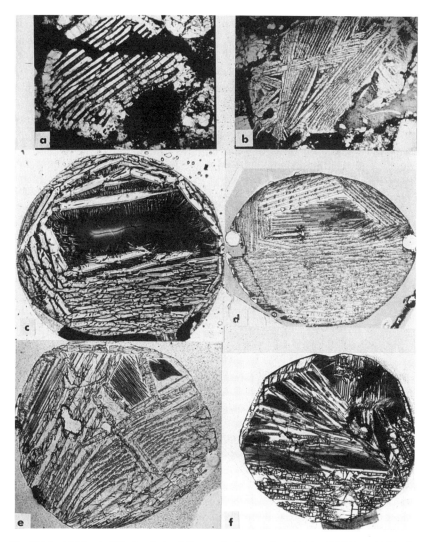

Fig. 9.2.3. (a) Classic BO chondrule, 0.9 mm, single grouplet with rim (Weisburg 1987); (b) BO chondrule with multiple grouplets, 1.8 mm (Weisberg 1987); (c) BO spherule, 2 mm, Si-rich composition, initial $T = 30°$ C above liquidus, cooled at 328° C/hr to 1459° and held for 1 hour (Planner 1979); (d) BO spherule, 3 mm, initial $T = 20°$ C above liquidus, 100° C/hr (Lofgren and Russell 1985); (e) BO spherule, 4 mm, cooled from liquidus at 1000° C/hr (Radomsky 1988); (f) BO spherule, 4 mm, cooled at 500° C/hr from superheated melt containing one large olivine relict. Coarse barred olivine overgrew the seed (adjacent to electron raster burns); fine parallel and lattice olivine later grew in liquid portion of charge (Radomsky 1988).

cooled at this rate would be PO with no settling and consequently no late nucleation. In general, gravitational settling is very obvious in these runs which contain a hierarchy of olivine textural types (Radomsky and Hewins 1987). Above the cumulate zone, there are later nucleated hoppers and near the top of the charges some open hoppers (belt-buckle-shaped), branching, parallel or lattice-work olivine (Fig. 9.2.2f). The hierarchy results from removal of nuclei from the liquid as grains crystallized and settled, followed by more rapid growth of new crystals when embryos graduated to nuclei after a period of supercooling (relative to the liquidus temperature of the fractionated liquid). The very finest lattice olivine (e.g., Fig. 9.2.3e,f) may have grown during quenching. The 100° C/hr runs show moderate zoning (Radomsky 1988). Runs at 10° C/hr have essentially unzoned olivine grains (Fo_{96-95}) which show equilibrium Fe/Mg partition coefficients. Apparent partition coefficients of less than the equilibrium value, as found in PO chondrules (Taylor and Cirlin 1986), are observed at higher cooling rates. Calcium contents of olivine tend to increase with increasing cooling rate, although the effect is not systematic. Compositional data would permit 100 to 1000° C/hr cooling rates for magnesian PO chondrules. Radomsky et al. (1986) also produced PO textures from a porphyritic olivine-pyroxene (POP) composition by quenching from above the onset of pyroxene crystallization, a process which seems unlikely in nature.

The thermal history of PO chondrules is fairly well constrained (Table 9.2.1). The heating cycle must allow the survival of abundant nuclei after melting. Temperatures reached can be between a few degrees above the liquidus to about thirty degrees below within the range 1600 to 1550° C for different Fe/Mg ratios (Figs 9.2.4 and 9.2.5). With a two-minute heating time (Tsuchiyama et al. 1980a; Tsuchiyama and Nagahara 1981), the initial temperature is critical and temperatures not much below the liquidus give very fine-grained granular olivine. Conditions producing good textural analogues are so restricted that it seems unlikely that natural heating times can be shorter than 2 minutes. With longer heating times, fewer nuclei survive and charges finer grained than typical chondrules are not as readily produced. Further-

TABLE 9.2.1
Conditions for Barred Olivine and Porphyritic Chondrule Textures and Phase Compositions, Suggested by Experiments

Expt.[a]	SiO_2	MgO	T_L°C	BO(ΔT)	PO(ΔT)	Rate °C/hr
T81	43.6	22.1	1575	0 to +15	0 to −25	1200
L85	48.7	32.7	1580	0 to +20	0 to −30	100–1900
Bell	43.0	23.8	1548	0 to ?	0 to −30	100–1000
R88	47.9	30.5	1597	0 to +5	0 to −30	100–1000

[a]Experiment identification: T81 = Tsuchiyama & Nagahara 1981; L85 = Lofgren & Russell 1985; Bell = Bell 1986; R88 = Radomsky 1988.

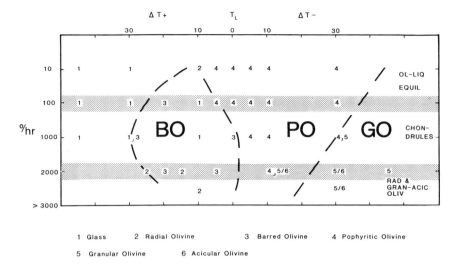

Fig. 9.2.4. Textures of highly olivine-normative experimental charges relative to initial temperature (compared to apparent liquidus) and cooling rate for the experiments of Tsuchiyama et al. (1980a), Tsuchiyama and Nagahara (1981), Lofgren and Russell (1985), Radomsky et al. (1986), Bell (1986) and Radomsky (1988). Chondrules are restricted to moderate rates by phase compositions.

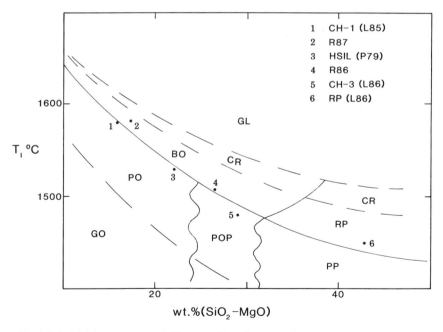

Fig. 9.2.5. Initial temperature vs bulk-composition diagram to show textures in spherules cooled at about 1000° C/hr. The apparent liquidus curve is based on experiments by Planner (1979), Lofgren and Russell (1985, 1986), Radomsky et al. (1986), and Radomsky and Hewins (1987).

more, liquidus runs tend to be barred and superheated runs tend to be glassy. The heating for PO chondrules must therefore always be subliquidus, unless totally molten droplets can later pick up crystals and associated nuclei. For the cooling cycle, rates in excess of 100° C/hr up to 2000° C/hr are permitted from compositional data but 1000° C/hr seems to be the limit from a morphological point of view for magnesian chondrules. A lower initial temperature coupled with 1000° C/hr or even faster cooling might produce a good analogue even for these PO chondrules. If gravitational settling could be inhibited, some runs at 1900° C/hr (Lofgren and Russell 1985) might be more similar to PO.

9.2.4. BARRED OLIVINE CHONDRULES

The exact morphology of the classic BO chondrule comprising one olivine crystal, with smooth wide bars optically continuous with a solid rim (Nagahara 1983a, Fig. 1a; Weisberg 1987, Fig. 1a), has yet to be reproduced (Fig. 9.2.3). The texture corresponds to a single linked parallel growth crystal in the terminology of Donaldson (1976). However, many synthetic spherules resemble those BO chondrules which contain multiple olivine grouplets (Fig. 9.2.3b) much more than any other textural type and are sufficiently similar to tell us the difference in conditions required to produce BO rather than PO chondrules. The experimental analogues have been produced with both Si-poor (olivine-rich) and Si-rich (pyroxene-rich) bulk compositions. Since natural BO chondrules in UOC average about 46% SiO_2 and Si-rich ones are rare (Weisberg 1987), the more important experiments on Si-poor compositions are described first.

Tsuchiyama et al. (1980a) produced subparallel olivine dendrites akin to barred olivine by cooling superheated droplets at very high rates. The charges that experienced the highest cooling rates contain very large numbers of thin bars with two directions interspersed (rather than in discrete grouplets) perhaps more like crystallographic branching olivine than LPG (linked parallel growth) or parallel plate (Donaldson 1976). A run at 3000° C/hr produced two grouplets with very broad bars, but the numerous projections (secondary dendrite arms) seen on the bars are uncommon in coarse BO chondrules. Totally glassy runs were frequently formed in duplicate runs at cooling rates down to 3000° C/hr and highly olivine-normative glass chondrules are unknown. The synthetic olivine bars are strongly zoned, whereas natural BO is very rarely zoned (Miyamoto et al. 1986). All of the evidence indicates that BO chondrules must have formed at lower cooling rates and possibly lower initial temperatures than those of the radial olivine spherules of Tsuchiyama et al. (1980a). Tsuchiyama and Nagahara (1981) produced very similar textures to those described above, when the melt which otherwise (subliquidus runs) produced PO analogues was superheated and rapidly cooled.

Lofgren and Russell (1985) produced analogues to BO chondrules with little to no zoning in olivine from melt (which otherwise yielded PO spherules) when it was slightly superheated and cooled at rates of 100 to 1900° C/hr (Fig. 9.2.3). At 1900° C/hr, the spherules contain multiple grouplets of parallel plate or LPG olivine (Fig. 9.2.3d); at 100° C/hr, the texture consists of hopper and parallel olivine with abundant glass. In his set of runs producing PO analogues, Bell (1986) produced a texture transitional to BO, with several small LPG crystals, in a run from 5° below the liquidus at 1000° C/hr. Radomsky (1988) produced a BO spherule from the liquidus at 1000° C/hr, which contained half-a-dozen grouplets with minor irregularities on the arms, plus some late-nucleated LPG olivine (Fig. 9.2.3e). This run had slightly more nuclei than a single-crystal BO chondrule, i.e., it was not quite melted enough, and may have been cooled at a slightly greater rate than the classic BO chondrule (which is more ferroan than the synthetic). However, this spherule is very similar to many multiple-grouplet BO chondrules (Weisberg 1987). As in the subliquidus runs of Bell (1986), the lower cooling rates (10 and 100° C/hr) produced PO analogues, but Radomsky's 1000° C/hr run from below the liquidus produced a transitional BO/PO texture (Fig. 9.2.2e), because of late nucleation in supercooled liquid after the settling of polyhedra grown during heating. Superheated melts in both studies (Bell 1986; Radomsky and Hewins 1987) produced totally glassy charges. The condition of the melt before cooling cannot be directly measured and a discussion of the theory of graduation of embryos to nuclei is beyond the scope of this review (see Lofgren 1983; Lofgren and Russell 1986). However, it is clear that the main difference between BO and PO is the very small number of nuclei available in BO. Radomsky (1987) repeated his normal runs with charges containing large olivine seeds. When such charges were totally melted except for the large relict grain and cooled at 500° C/hr, coarse barred olivine overgrew this seed (Fig. 9.2.3f) and finer parallel and lattice olivine nucleated later in the liquid part of the charge. This shows that a single grouplet of bars (a single crystal) is obtained from a single nucleus.

BO textures from Si-poor melts require an initial heating very close to the liquidus temperature (Fig. 9.2.4) so as to obtain very few nuclei, ideally only one. BO formed at higher temperature than PO, or at higher rates from the same (liquidus) temperature, or both, based on textures (Fig. 9.2.4). As compositional constraints rule out the lowest rates, most natural BO formed from higher initial temperatures than PO, at comparable cooling rates (Fig. 9.2.5 and Table 9.2.1). Cooling rates must be moderate, close to 1000° C/hr. Slightly lower rates (down towards 100° C/hr) may be effective in slightly more superheated runs.

BO analogues formed from Si-rich melts include those of Planner (1979). He crystallized only olivine from a highly pyroxene-normative melt (HSil) superheated by up to 50° C for cooling rates of 300 to 400° C/hr with or without an isothermal stage. The dominant morphology is LPG, more akin to BO than PO. One run contains 4 or 5 olivine grouplets very similar to

TABLE 9.2.2
Conditions for Radial Olivine/Barred Olivine Chondrules from Pyroxene-Rich Compositions

Expt.[a]	SiO$_2$	MgO	T_L°C	ΔT	Rate °C/hr
P79	53.2	31.0	1530	+20 to +50	300
T81	53.5	22.1	1445	0 to +100	230–4800
L86	55.2	28.9	1505	5 to +20	100
R86	49.6	23.1	1504	0 to +25	1000

[a] P79 = Planner 1979; T81 = Tsuchiyama and Nagahara 1981; L86 = Lofgren & Russell 1986; R86 = Radomsky et al. 1986.

olivine in a BO chondrule (Fig. 9.2.3c). High cooling rates with this composition produced glass.

Tsuchiyama et al. (1980a) produced exceedingly fine radial olivine from very rapidly cooled superheated Si-rich melt. With cooling rates of 1200 to 4800°C/hr, Tsuchiyama and Nagahara (1981) reported comparable radial-barred olivine (RO) with a superheated melt of similar composition. A run at 500°C/hr was closer in appearance to BO chondrules. Lower rates produced barred-radial pyroxene.

Lofgren and Russell (1986) studied pyroxene-rich compositions which crystallized small amounts of olivine near the liquidus. Superheated runs cooled at 100°C/hr produced long dendritic olivine crystals akin to those in BO chondrules. Radomsky et al. (1986) used a pyroxene-olivine-porphyry composition and produced textures akin to BO, but indicative of more rapid growth, in runs cooled at 1000°C/hr from near-liquidus and superheated melts. Highly superheated melts produced glass, even at 10°C/hr, unless seed olivine grains were added, in which cases radial and barred olivine could be produced. Parallel olivine blades were found intergrown with pyroxene in such runs formed at 20 to 100°C/hr. These runs, however, were more like pyroxene porphyry overall than the BO chondrules with pyroxene of Nagahara (1983a) and Weisberg (1987). The most Si-rich BO chondrules probably cooled at rates of 100 to 1000°C/hr after being moderately superheated (Table 9.2.2).

9.2.5. PYROXENE-RICH CHONDRULES

The earliest experiments on pyroxene-rich compositions failed to crystallize pyroxene because melts were too superheated and/or cooling rates were too high (Planner 1979; Tsuchiyama et al. 1980). Tsuchiyama and Nagahara (1981) showed that a superheated melt which produced barred-radial olivine at higher cooling rates grew increasingly coarse radial pyroxene with rates from 46 to 5°C/hr. Hewins et al. (1981) also documented increase of pyroxene dendrite width grown from slightly superheated melts with cooling rates from 3000 to 10° C/hr. There is a considerable mismatch between the results of Tsuchiyama and Nagahara (1981) and Hewins et al. (1981), because neither group examined the effect of initial temperature. The higher the initial temperature, the more nuclei and embryos are destroyed, the faster crystals will grow after some undercooling and the narrower such crystals will tend to be. The lowest cooling rates from the higher initial temperature and the highest cooling rates from the lower initial temperature suggested for chondrules in these studies are therefore suspect.

Experiments by Hewins have produced residual glass interstitial to pyroxene with cooling rates as low as 50° C/hr and final temperatures of 600° C. This shows that a quenching episode (Planner and Keil 1982) is not essential to produce chondrule glass.

Lofgren and Russell (1986) performed experiments on four Si-rich compositions. At the highest cooling rates from near-liquidus melts and at 5 to 3100° C/hr for superheated melts, they produced charges with fine radial pyroxene similar to some chondrules loosely called "glassy" or cryptocrystalline. Such pyroxene-bearing chondrules do not *require* cooling rates of 3000° C/hr or higher (Hewins et al. 1981) but could instead have been highly superheated. Barred/dendritic pyroxene spherules were formed at 100° C/hr with near-liquidus initial temperatures. Excentroradial pyroxene (RP) chondrules with olivine phenocrysts could be formed from subliquidus initial temperatures at very high cooling rates (Lofgren and Russell 1986). Porphyritic pyroxene (PP) spherules were formed from near-liquidus initial temperatures and low cooling rates and from subliquidus temperatures and a wide range of cooling rates. PP chondrules could have cooled as slowly as 5° C/hr if the temperature was near liquidus or faster than 100° C/hr for lower initial temperatures. Texturally, the only requirement is existence of more nuclei than for excentroradial pyroxene chondrules and as yet no compositional parameters have been identified that would permit a precise estimate of the cooling rate.

Radomsky et al. (1986) found that porphyritic pyroxene-olivine spherules could be produced at any cooling rate (10–1000° C/hr) below the liquidus and at the lowest cooling rates from near-liquidus temperatures. The requirement for the texture is an abundance of nuclei, or time for development of nuclei and growth of equant crystals. This suite of experiments like others showed very clearly the importance of nuclei and nucleation in generating textures. Runs terminated before the appearance of pyroxene show a granular olivine cumulate below a porphyry consisting of late-nucleated olivine hoppers. Porphyritic textures did not form in superheated runs or runs rapidly heated from the liquidus, and glasses were produced from highly superheated melts. With identical conditions and a seed crystal present, textures were distinctly modified. Glasses did not form with the highest initial temperatures: LPG olivine overgrew the seed to produce a barred texture. In subliquidus runs the olivine-seed crystal was accompanied by many olivine granules (formed not directly on the seed but on abundant associated nuclei) which settled to the bottom of the charge.

Olivine in POP spherules is generally enclosed poikilitically by pyroxene. Nagahara (1983b) synthesized such textures at 5° C/hr, Radomsky et al. (1986) at 10 to 1000° C/hr and Lofgren and Russell (1986) at 5 to 3100° C/hr. In short, all textures in POP compositions are more sensitive to the presence of nuclei, i.e., initial temperature and heating time, than they are to cooling time. Olivine was zoned in POP spherules only in those formed at the highest rates, however, and by analogy with PO chondrules this indicates cooling rates close to 1000° C/hr. The comparison is difficult, however, in the case of tiny olivine grains in PP chondrules, where little information on zoning is available.

9.2.6. CALCIUM-ALUMINUM-CHONDRULES

Type B Ca-Al-rich inclusions (CAI) (Clarke et al. 1970; see also Chapter 10.3) fit the general definition of chondrules (Chapter 9.1) and there exist small numbers of chondrules intermediate in composition between CAI and PO chondrules (Bischoff and Keil 1984). CAI compositions are refractory in terms of solid-vapor equilibria, but they melt at similar temperatures to most chondrules. Many CAI cooled from subliquidus temperatures at very low rates, at most about 10° C/hr (Stolper and Paque 1986). Their thermal history is clearly different from that of Fe-Mg-rich chondrules. Derivation from normal-chondrule precursor material is unlikely because it would have required extreme superheating, followed by an extended isothermal event at about 1400° C or collisions which allowed them to pick up nuclei. This hypothetical event or the heating of refractory primordial material occurred in an environment (or at a time) where slower cooling than for normal chondrules was possible. It is most likely that the low cooling rates for CAI indicate a sufficiently different environment that exclusion of CAI from the restricted chondrule definition given in Chapter 9.1 is justified.

9.2.7. DISCUSSION

Olivine-rich chondrules of the classic types represent melts heated to temperatures very close to their liquidi, with BO textures formed for initial temperatures up to about 20° C above the liquidus, PO textures formed for initial temperatures down to about 30° C below the liquidus and granular olivine for lower initial temperatures (Table 9.2.1, Figs. 9.2.4 and 9.2.5). The difference in texture is entirely due to the number of nuclei surviving the heating event. These liquidus temperatures range from about 1600 to 1550° C, influenced mainly by Fe/Mg ratio. A maximum effective initial temperature of about 1600° C (Fig. 9.2.5) and an average initial temperature closer to 1500° C would explain why type I (magnesian) PO chondrules are predominantly granular, why type II (ferroan) PO are predominantly microporphyritic, why BO chondrules are mainly ferroan (type II composition) and why cryptocrystalline and glassy chondrules are rare except for pyroxene-rich compositions (McSween 1977; Scott and Taylor 1983; Chapter 9.1).

Very finely granular GO chondrules, sometimes called lithic chondrules, clearly were not heated to as high initial temperatures as classic PO chondrules. The very rare chondrules with spherulitic olivine may well have been strongly superheated and not necessarily rapidly cooled. It is exceptionally easy to produce glass with olivine-rich compositions at cooling rates down to $10°$ C/hr, provided that the melt is sufficiently superheated by more than $10°$ C (Radomsky 1987). The total absence of pure-glass magnesian chondrules can be explained in either of two ways. By chance the peak temperature was about $1600°$ C (Fig. 9.2.5), never much above the liquidus so that embryos were not destroyed and at least one nucleus was available. Alternatively, many spherules were superheated and totally melted, but collided with solids such that sufficient associated nuclei survived the thermal equilibration. The temperature of $1600°$ C would then be an effective initial temperature before cooling rather than necessarily a true peak temperature. Many chondrules contain relict grains (Chapter 9.1). Compound chondrules formed by accretion of partially liquid droplets are well known. Chondrules totally enclosing other chondrules, i.e., largely liquid droplet enclosing largely solid chondrule, are rare but known. It also appears likely that chondrule droplet coalescence could have occurred, as in palisade-bearing CAI (Wark and Lovering 1982). Therefore, large unrecognized relict grains with associated nuclei might have been inserted inside chondrule droplets in many cases. (If only very fine material was added, it would have been held by surface tension to initiate crystallization from the rim, which is rarely observed.)

The seeding model to explain the near coincidence of peak heating temperatures with chondrule liquidus temperatures fails when composition-texture relations are considered. Type I (Mg-rich) PO are mostly granular rather than truly porphyritic and BO are not highly magnesian (Fig. 9.2.6; McSween 1977; Weisberg 1987). Magnesium-poor chondrules are mainly radial and spherulitic not porphyritic pyroxene (Fig. 9.2.6). There is no reason that Si-rich droplets should fail to pick up nuclei and Mg-rich chondrules always should. The conclusion that the heating process was limited to near-liquidus temperatures is therefore discussed in more detail below. Heating time does not appear to be critical but it becomes more difficult to match natural textures with very short (2 minute) initial heating.

Radomsky (1988) has calculated liquidus temperatures from bulk compositions for a suite of texturally characterized chondrules, using an algorithm from MAGMAE.FOR by J. Longhi. The temperatures are essentially colinear when plotted against wt.% MgO for porphyritic, barred and other nonporphyritic chondrules (Fig. 9.2.6). The suite of chondrules was divided by Radomsky into four MgO ranges and histogram bars showing textural variation were attached to the liquidus curve, so as to fix porphyritic textures requiring

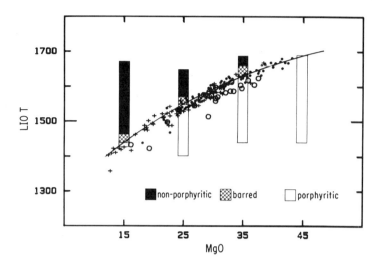

Fig. 9.2.6. Liquidus temperatures for porphyritic (dots), barred-olivine (open circles) and non-porphyritic, including cryptocrystalline and radial-pyroxene (plus symbols) chondrules calculated by Radomsky (1988). The bars giving the percentages of textural types within a 10% MgO range are attached to the liquidus curve to show the proportions of chondrules formed from lower initial temperatures (i.e., porphyritic). A maximum initial temperature of about 1600° C is indicated.

subliquidus initial temperatures below the line (Fig. 9.2.6). The similar alignment of these bars indicates a common thermal history for all composition ranges, with a maximum initial temperature a little over 1600° C. Because of the common initial temperatures, the bulk composition (MgO content) becomes an important control of chondrule texture, as originally suggested by Tsuchiyama et al. (1980a).

PO and BO chondrule cooling rates can be restricted to the range between 100 and 2000° C/hr. Morphological constraints favor the lower rates and composition constraints the higher rates. As pointed out by Miyamoto et al. (1986), olivine zoning requires fast cooling from below the liquidus rather than slow cooling from above the liquidus, an alternative for PO suggested by Hewins (1983). The morphological match at 1000° C/hr might be improved in the absence of gravitational settling of early olivine grains, which permits nucleation of a later generation of more dendritic olivine after some undercooling. 1000° C/hr is a good estimate of the average cooling rate of classic olivine-rich chondrules.

The 1000° C/hr cooling rate is not high enough to allow Na to be retained by chondrules based on the experiments of Tsuchiyama et al. (1981) assuming hot reducing nebular gases. However, chondrules are not significantly depleted in Na (Chapter 9.1) and Na loss is much reduced at high O_2 fugacities (Tsuchiyama et al. 1981). Chondrules must have been generated by transient heating of solids in a cool gas, and O_2 fugacity may have been locally high because of evaporation in a dust-ice cluster or generally high because of reactions between nebular-gas species at low temperature (see Chapter 7.7). There is evidence of some Na being added to some chondrules from the gas (Ikeda 1982) and, as with CAI alteration, this requires high O_2 fugacity.

Somewhat less definitive thermal histories can be deciphered for the more Si-rich chondrules. Pyroxene-porphyry textures can form from subliquidus temperatures (1500–1425° C) at essentially any cooling rate but the zoned olivine in POP chondrules requires approximately 1000° C/hr. Excentroradial pyroxene (RP) requires near-liquidus initial temperatures (1550–1450° C) and the cooling rates are not well constrained. The abundance of porphyritic textures in olivine-normative chondrules and of radiating to cryptocrystalline textures in more Si-rich chondrules shows that initial temperatures above 1500° C were common. The rare Si-rich BO chondrules tend to occur instead of RP for higher initial temperatures (1580–1500° C) and higher cooling rates.

The moderate (1000° C/hr) cooling rates required for chondrules rules out nebular condensation and lightning strike for chondrule genesis (Hewins 1983). The substantiation of a limit to chondrule heating presents a major additional constraint on possible chondrule heating mechanisms, such as friction between droplets and nebula (Y. Ikeda quoted in Tsuchiyama et al. 1980*a;* Wood 1984). With such a slow-heating model, the maximum temperature might be governed by the abundance of chondrules and dust, which would control the ability of droplets to cool by radiation.

9.2.8. CONCLUSIONS

The most informative chondrule simulations are those that have used different initial temperatures as well as different cooling rates on the same bulk composition. Such studies have been able to match the textures and mineral zoning of natural chondrules reasonably well, although gravitational settling in experimental charges makes detailed comparisons difficult. The textures are controlled by the number of nuclei surviving melting, and zoning preservation depends on cooling rate. The dynamic crystallization experiments indicate an upper limit to the initial temperature and moderate cooling rates. With very short heating times, it is extremely difficult to match chondrule textures.

Most chondrules cooled from a little below 1600° C at about 1000° C/hr. This requires a cold environment and a sufficient number density of chondrules to retard radiative cooling. Nebular-cooling and flash-heating models are very hard to reconcile with the experimental data.

Acknowledgments. This chapter has benefited from discussion with J. N. Grossman, sharing ideas and figures with P. M. Radomsky, access to runs made by T. Shanahan, and a detailed review by G. E. Lofgren. The study was partially supported by NASA. Photographs were kindly supplied by G. E. Lofgren, H. N. Planner, P. M. Radomsky, E. R. D. Scott and M. K. Weisberg.

REFERENCES

Bell, D. J. 1986. Experimental crystallization of an olivine-normative chondrule composition. B.S. thesis, Rutgers Univ.

Bischoff, A., and Keil, K. 1984. Al-rich objects in ordinary chondrites: Related origin of carbonaceous and ordinary chondrites and their constituents. *Geochim. Cosmochim. Acta* 48:693–709.

Blander, M., Planner, H. N., Keil, K., Nelson, L. S., and Richardson, N. L. 1976. The origin of chondrites: Experimental investigation of metastable liquids in the system. Mg_2SiO_4-SiO_2. *Geochim. Cosmochim. Acta* 40:889–896.

Clarke, R. S., Jr., Jarosewich, E., Mason, B., Nelen, J., Gómez, M., and Hyde, J. R. 1970. The Allende, Mexico, meteorite shower. *Smithsonian Contrib. Earth Sci.* 5.

Donaldson, C. H. 1976. An experimental investigation of olivine morphology. *Contrib. Mineral. Petrol.* 57:187–213.

Hewins, R. H. 1983. Dynamic crystallization experiments as constraints on chondrule genesis. In *Chondrules and Their Origins*, ed. E. A. King (Houston: Lunar and Planetary Inst.), pp. 122–133.

Hewins, R. H., Klein, L. C., and Fasano, B. V. 1981. Conditions of formation of pyroxene excentroradial chondrules. *Proc. Lunar Planet. Sci. Conf.* 12B:1123–1133.

Ikeda, Y. 1982. Petrology of the ALH 77003 chondrite (C3). *Mem. Natl. Inst. of Polar Res. Special Issue* 25:34–65.

Lofgren, G. E. 1983. Effect of heterogeneous nucleation on basaltic textures: A dynamic crystallization study. *J. Petrol.* 24:229–255.

Lofgren, G. E., and Russell, W. J. 1985. Dynamic crystallization experiments on chondrule melts of porphyritic olivine composition. *Lunar Planet. Sci.* XVI:499–500 (abstract).

Lofgren, G. E., and Russell, W. J. 1986. Dynamic crystallization of chondrule melts of porphyritic and radial pyroxene composition. *Geochim. Cosmochim. Acta* 50:1715–1726.

McSween, H. Y., Jr. 1977. Chemical and petrographic constraints on the origin of chondrules and inclusions in carbonaceous chondrites. *Geochim. Cosmochim. Acta* 41:1843–1860.

Miyamoto, M., McKay, D. S., McKay, G. A., and Duke, M. B. 1986. Chemical zoning and homogenization of olivines in ordinary chondrites and implications for thermal histories of chondrules. *J. Geophys. Res.* 91:12,804–12,816.

Nagahara, H. 1983a. Texture of chondrules. *Mem. Natl. Inst. Polar Res. Special Issue* 30:61–83.

Nagahara, H. 1983b. Chondrules formed through incomplete melting of the pre-existing mineral clusters and the origin of chondrules. In *Chondrules and Their Origins*, ed. E. A. King (Houston: Lunar and Planetary Inst.), pp. 211–222.

Planner, H. N. 1979. Chondrule thermal history implied from olivine compositional data. Ph.D. thesis, Univ. of New Mexico.

Planner, H. N., and Keil, K. 1982. Evidence for the three-stage cooling history of olivine-porphyritic fluid droplet chondrules. *Geochim. Cosmochim. Acta* 46:317–330.

Radomsky, P. M. 1988. Dynamic crystallization experiments on magnesian olivine-rich and pyroxene-olivine chondrule composition. M.S. thesis, Rutgers Univ.

Radomsky, P. M., and Hewins, R. H. 1987. Dynamic crystallization experiments on an average type I (MgO-rich) chondrule composition. *Lunar Planet. Sci.* XVIII:808–809 (abstract).

Radomsky, P. M., Turrin, R. P., and Hewins, R. H. 1986. Dynamic crystallization experiments on a pyroxene-olivine chondrule composition. *Lunar Planet. Sci.* XVII:687–688 (abstract).

Scott, E. R. D., and Taylor, G. J. 1983. Chondrules and other components in C, O and E chondrites: Similarities in their properties and origins. *Proc. Lunar Planet. Sci. Conf.* 14, *J. Geophys. Res. Suppl.* 88:B275–B286.

Stolper, E., and Paque, J. M. 1986. Crystallization sequences of Ca-Al-rich inclusions from Allende: The effects of cooling rate and maximum temperature. *Geochim. Cosmochim. Acta* 50:1785–1806.

Taylor, L. A., and Cirlin, E.-H. 1986. Olivine/melt Fe/Mg K_D's <0.3: Rapid cooling of olivine-rich chondrules. *Lunar Planet. Sci.* XVII:879–880 (abstract).

Tsuchiyama, A., and Nagahara, H. 1981. Effects of precooling thermal history and cooling rate on the texture of chondrules: A preliminary report. *Mem. Natl. Inst. Polar Res. Special Issue* 20:175–192.

Tsuchiyama, A., Nagahara, H., and Kushiro, I. 1980a. Experimental reproduction of textures of chondrules. *Earth Planet. Sci. Lett.* 48:155–165.

Tsuchiyama, A., Nagahara, H., and Kushiro, I. 1980b. Investigations on the experimentally produced chondrules: Chemical compositions of olivine and glass and formation of radial pyroxene chondrules. *Mem. Natl. Inst. of Polar Res. Special Issue* 17:83–94.

Tsuchiyama, A., Nagahara, H., and Kushiro, I. 1981. Volatilization of sodium from silicate melt spheres and its application to the formation of chondrules. *Geochim. Cosmochim. Acta* 45:1357–1367.

Wark, D. A., and Lovering, J. F. 1982. Evolution of Ca-Al-rich bodies in the earliest solar system: Growth by incorporation. *Geochim. Cosmochim. Acta* 46:2595–2607.

Weisberg, M. K. 1987. Barred olivine chondrules in ordinary chondrites: Petrologic constraints and implications. *Proc. Lunar Planet. Sci. Conf.* 17, *J. Geophys. Res. Suppl.* 91:E663–E678.

Wood, J. A. 1984. On the formation of meteoritic chondrules by aerodynamic drag heating in the solar nebula. *Earth Planet. Sci. Lett.* 70:11–26.

9.3. FORMATION OF CHONDRULES

JEFFREY N. GROSSMAN
United States Geological Survey

The mineralogical, chemical, isotopic and physical properties of chondrules constrain the energy source, the environment and the precursor material involved in chondrule formation. Brief energetic events efficiently heated clumps of solid material from low temperatures to near liquidus temperatures, but rarely hotter. Cooling was also rapid, but not nearly as fast as by blackbody radiation into space. The O fugacity and dust/gas ratio during chondrule formation were probably not much greater than expected in an unfractionated gas of solar composition, although existing data are somewhat ambiguous. The solids that clumped together prior to chondrule formation were chemically and isotopically heterogeneous. These materials contained volatile elements, and were mixtures of coarse and fine grains. Chondrule precursors included solids that may have condensed from nebular gas as well as large grains of unknown origin, but they were not composed of material formed by crystal-liquid fractionation processes. Planetary models for chondrule formation that rely upon impact heating or asteroidal volcanism for the generation of chondrule melts fail to satisfy many of these constraints. Similarly, nebular models that form chondrule liquids by direct condensation from a gas cannot be reconciled with existing data. The process of chondrule formation must have taken place in a dusty solar nebula prior to or during the accretion of planetesimals. The details of chondrule formation, e.g., the amount of dust, the efficiency of clumping and melting, the average intensity of heating, and the proportions of various precursor assemblages, were somewhat different in the formation location of each chondrite group.

9.3.1. INTRODUCTION

Virtually every environment that is or ever was thought to be appropriate for the generation of heat in the early solar system has been proposed as a setting for chondrule formation. These environments are of two types: "planetary," i.e., on meter-sized or larger bodies in the solar system; and, "nebular," i.e., in the pre- or syn-accretionary solar system.

A variety of possible planetary settings, in rough chronological order of publication, are listed in Table 9.3.1. The proposed nebular settings, listed in Table 9.3.2, are divided into two types: the first produces melts by the direct condensation of liquids from hot gases in the nebula; the other involves the remelting of solids in transient events.

The energy for melting in most of the planetary models comes from processes known to occur on meteorite parent bodies. The exceptions are Sorby's model, based on incorrect speculations about the nature of the Sun, and Podolak and Cameron's condensation model, involving the formation of hypothetical giant gaseous protoplanets. From meteoritic data, we can make reasonable assumptions about what kinds of precursor or target material would be available on these bodies to form chondrules. Thus, the strongest test of these models is to see if the chemical/isotopic data for real chondrules

TABLE 9.3.1
Planetary Settings and Mechanisms Proposed for Chondrule Formation

Body	Mechanism	Reference
Sun	ejection from surface in prominences	Sorby 1877.
Unknown	magmatism/volcanism	Tschermak 1883; Borgström 1904; Merrill 1920, 1921.
Planetary, large asteroids	magmatism/volcanism	Ringwood 1959; Fredriksson and Ringwood 1963.
Planetary	metamorphism	Fermor 1938; Mason 1960.
Asteroids, large bodies	impact melting	Urey and Craig 1953; Fredriksson 1963; Urey 1967; Wlotzka 1969; Dodd 1971.
Planetesimals (meter-sized)	melting during collisions	Wasson 1972; Kieffer 1975.
Giant protoplanets	condensation of droplets in atmospheres	Podolak and Cameron 1974.
Planetesimals (molten)	splashing in collisions	Zook 1980, 1981; Leitch and Smith 1982.

TABLE 9.3.2
Representative Nebular Models Proposed for Chondrule Formation

Precursor Material	Energy Source	Assumptions	Reference
		Condensation Models	
Gas	hot nebula		Suess 1949.
Gas	hot nebula		Wood 1962.
Gas	hot nebula	"constrained equilibrium"	Blander and Katz 1967.
Gas	hot nebula	high dust/gas ratio	Wood and McSween 1976.
Gas	hot nebula	hydrogen depletion	Herndon and Suess 1977.
		Secondary Melting Models	
Nebular dust	lightning	turbulent nebula	Whipple 1966; Cameron 1966.
Nebular dust	radiational heat in solar events	T-Tauri phase of Sun	Herbig 1978.
Nebular dust	relativistic electrons	reconnecting magnetic fields	Sonett 1979.
Interstellar grains	chemical energy	large clumps of amorphous interstellar solids	D. Clayton 1980.
Nebular dust	lightning	differential rotation of stratified nebula	Rasmussen and Wasson 1982.
Interstellar grains	frictional heat	large clumps falling into nebula	Wood 1983, 1984.

fit with the assumptions that can be made for target materials. Because melting processes of this type are fairly well understood, the models can also be tested by the physical properties of chondrules.

In contrast, all of the nebular models rely on hypothetical energy sources. Furthermore, the solar nebula no longer exists, and observations of other nebulae are difficult. In this category, only the condensation models make testable predictions about the compositions of chondrules. The strongest test for the majority of these models is their consistency with current astrophysical and cosmochemical models for the early solar system.

9.3.2. CONSTRAINTS ON THE PROCESS OF CHONDRULE FORMATION

Chapter 9.1 describes the mineralogical, chemical, isotopic and physical properties of chondrules. In this section, the properties of chondrules are used to derive generalized constraints on the energy source that led to the formation of melts, on the environment in which melting occurred, and on the nature of the material from which chondrules formed.

Constraints on the Energy Source for Chondrule Formation

Heating Intensity is Restricted. All of the models for chondrule formation except those that involve condensation require an energy source to heat silicates at least up to their solidus temperatures. Condensation models require a heat source that vaporizes solids in the nebula over some large region, and are discussed separately. The initial temperature of silicates (before chondrule formation) has an upper limit of ~ 650 K, the temperature at which the FeS present in chondrules (and presumed to be present among the precursors) would have evaporated. Wood (personal communication, 1978) has argued that much of the sulfide and metal present in chondrules, even in those from highly unequilibrated chondrites, may be secondary (metamorphic) in origin. I strongly disagree with this opinion based on petrographic evidence, e.g., sulfides are in many cases enclosed by glass or clear crystals (Grossman et al. 1979; Grossman and Wasson 1982,1983a); but, if Wood is correct, then the upper limit temperature would be ~ 900 K, the temperature at which the Na and K in the precursors would have evaporated. The energy required to heat and melt chondrules with compositions in the range of those listed in Chapter 9.1 from 300 K up to their liquidus temperatures near 1800 K is ~ 1500 J/g (1.5×100^{10} erg/g) (Wasson 1985). About half as much energy will bring the silicates up to the point of incipient melting.

From the experimental studies of chondrules (Chapter 9.2), and from the textures of real chondrules, most chondrules appear to have been at least 80% molten, but few were totally molten. Thus, the energy source seems to have deposited a rather narrow range of energy per unit mass in chondrules, somewhat less than 1500 J/g. The amount of energy deposited in chondrules does not seem to have been a function of the mass of the precursor assemblage. If there was such a relationship, then composition and texture should be related to mass; yet this is not observed (Chapter 9.1).

Duration of Heating is Probably Brief. If the time scale for the heating of chondrule precursors to the melting point could be determined, this would be an excellent constraint on the nature of the energy source. The experimental data in Chapter 9.2 suggest that many chondrules were heated up rapidly enough that a large number of crystal nuclei were present at the peak temperature, even when this temperature was above or near the liquidus. Relict grains also, by definition, survived chondrule heating. However, it is difficult to use this information to *quantify* the heating time scale because the initial grain size and the mineralogy of the precursors are not known. It is also possible that some solids were accreted by the chondrule during the high-temperature period.

Processes such as volatilization and reactions with surrounding gas integrate the entire high-temperature period from initial heating to final cooling. Thus, the time scales for these processes are constraints on the energy source as well as the environment in which cooling occurred. These time scales provide upper limits for the heating duration, and are discussed below.

Efficiency of Chondrule Formation is High. In some regions the energy source for chondrule formation must have been very efficient at converting precursor material into liquid droplets. Although it is probable that sorting processes separated chondrules from finer-grained constituents in some chondrite formation locations, the expected complementary material, namely matrix-rich chondrites, is never observed (O-isotopic evidence rules out the possibility that matrix-rich CI or CM chondrites are complementary to ordinary chondrites). It is conceivable that collisions preferentially destroy matrix-rich chondrites, or that they do not survive passage through the Earth's atmosphere. However, at least five CI chondrites, all of which are very fragile, have survived and been collected. It is also conceivable that a large amount of fine-grained material never accreted, and was swept out of the solar system with the gaseous portions of the nebula. This too may not have had a great effect on some chondrite groups. Wood (1985) pointed out that chondrules and matrix in Murchison (CM) have very different Fe/Si ratios, yet the whole rock has a near-solar Fe/Si ratio. This suggests that the two components were (nearly) representatively sampled during accretion. Similar arguments can be made for H chondrites, although these meteorites have experienced more lithophile-element fractionation.

Constraints on the Environment During Chondrule Formation

Thermal Environment. Chondrule textures are consistent with experimentally determined cooling rates in the range of 10^2 to a few $\times 10^3$ K/hr (Chapter 9.2). Given that chondrules were heated instantaneously, this implies that they did not cool by freely radiating into space. In that environment, mm-sized spheres at liquidus temperatures would initially cool at rates on the order of 10^6 K/hr. Rather, cooling must have taken place in an environment that would prevent rapid cooling. This could be achieved if chondrules were immersed in a "thermal bath" such as that envisioned by Wood (1984), where many chondrules formed simultaneously and irradiated each other. Rapid cooling could also be prevented if chondrules entered or formed in a zone of hot gas. However, Wood (1985) also pointed out that it is doubtful that chondrules could have traversed such a zone rapidly enough to allow them to cool as fast as experimentally determined cooling rates. Note that this argument rules out *any* condensation model for the generation of chondrule melts.

Oxygen Fugacity and its Relation to Volatile Loss. Tsuchiyama et al. (1981) did a series of experiments showing that the rate at which Na is lost from molten droplets is a function of temperature, bulk composition, droplet size and O fugacity f_{O_2}. From these experiments, it is possible to calculate the fractional loss of Na from a chondrule cooling at a constant rate at fixed f_{O_2}. Unfortunately, the experimental data only cover temperatures above the liquidus and relatively high O fugacities. Therefore, a large amount of extrapolation is required to use the experiments to constrain chondrule-formation processes.

The graph shown in Fig. 9.3.1 illustrates the amount of Na lost from a melt with composition similar to radial pyroxene (*RP*) chondrules (Tsuchiyama et al. 1981, composition 1) at different O fugacities as a function of peak temperature. Heating is assumed to be instantaneous. Calculations are shown for 1-mm diameter droplets cooled near the maximum (2000 K/hr) rate believed to be appropriate for chondrules (Chapter 9.2). Oxygen fugacity is assumed to remain constant during the cooling interval (a crude assumption; f_{O_2} should actually decrease by ~1 log-unit in the cooling interval). Curves with f_{O_2} lower than 10^{-10} atm (10 μPa), and parts of curves at temperatures lower than 1750 K are extrapolated from laboratory data. Calculations using the SILMIN program of Ghiorso (1985) indicate that the liquidus for this composition is around 1770 K (within the expected range for *RP* chondrules).

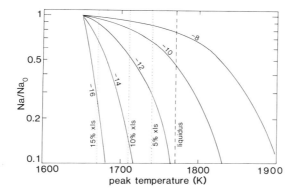

Fig. 9.3.1. Plot of fraction of the initial Na remaining in a chondrule after cooling vs peak-heating temperature. Heating is instantaneous. Curves for different O fugacities in the surrounding gas are based on the experimental data of Tsuchiyama et al. (1981), and are calculated for 1-mm spherules cooling at 2000 K/hr. Vertical dashed lines show the temperatures at which 0, 5, 10 and 15% of the melt would crystallize as olivine under equilibrium fractional-crystallization conditions (calculated using the method of Ghiorso [1985]). It is assumed in the calculation that Na loss would be prevented by the presence of 20% crystals, but this effect was ignored for the cooling droplets when there were < 20% crystals. In this case, a chondrule heated to near the liquidus temperature of ~1770 K will lose half of its Na by the time it cools to 1650 K unless f_{O2} is higher than ~10^{-10} atm. If only 5% cyrstallization is sufficient to prevent Na loss, the f_{O2} would need to be above ~10^{-12} atm. These values of f_{O2} are far higher than for an unfractionated solar gas at the same temperature, and conflict with f_{O2} estimates based on other methods (see text).

At the lowest temperature (1650 K) shown in Fig. 9.3.1, about 20% of the melt would have crystallized (as olivine) during equilibrium fractional crystallization. At this point, it is assumed in the calculation that Na loss instantaneously stops. In reality, the curves below 1770 K may be shifted upwards due to the presence of crystals in the melt (i.e., volatile loss would be suppressed). The magnitude of this effect is uncertain. The formation of a solid crust at the surface of the droplet would also hinder Na loss, but there is very little textural evidence in chondrules for the occurrence of this phenomenon.

From the compositional data in Chapter 9.1, it is clear that most natural chondrules have not lost much Na: they have Na/Mg ratios near to the wholerock ratio (that is, near to the solar ratio). For the sake of this discussion, a worst-case assumption is that the typical chondrule lost 50% of its Na during melting, although the real average may be closer to 5%. For radial chondrules heated to temperatures near their liquidi and cooled at 2000 K/hr, the Tsuchiyama et al. (1981) data in Fig. 9.3.1 seem to indicate that f_{O_2} in the region had to be in the range of 10^{-10} to 10^{-11} atm depending on the effect of 5 to 10% crystallization on volatile loss. If the cooling rate was 200 K/hr (the slowest rate suggested in Chapter 9.2), all of the curves in Fig. 9.3.1 shift to lower temperatures, and even higher f_{O_2} would be required to prevent Na loss.

The f_{O_2} of a system of solar composition in the range of *RP* liquidus temperatures would be 10^{-16} atm assuming that Mg, Si, CA and Al oxides were fully condensed (Wai and Wasson 1977) or 1/2 log-unit higher assuming the magnesian silicates were fully vaporized. Thus, we have an apparent paradox. Either the chondrules formed at distinctly nonnebular f_{O_2} (a minimum of 5 orders-of-magnitude more oxidizing), or the experimentally determined cooling rates are too low by 1 to 2 orders-of-magnitude, or the Tsuchiyama et al. (1981) data are not applicable. Higher-than-nebular f_{O_2} could be achieved by vaporizing a large amount of silicates during chondrule formation: if dust was concentrated to 250 times the amount expected to condense from a volume of solar gas, and then was vaporized during high-temperature events, H_2O/H_2 in the gas would rise to ~300 times its low-temperature ratio, and the f_{O_2} at 1750 K could be increased to ~10^{-10} (Chapter 7.7). This seems implausible, however, because unless the surrounding gas stayed very hot the refractory and magnesian oxides would rapidly recondense.

Other petrochemical data from chondrules suggest that the surrounding gas was fairly reducing, not oxidizing. Johnson (1986), following the arguments of earlier work by Wood (see, e.g., 1967, 1985), assumed that the chondrules with the lowest FeO contents most closely approached equilibrium between mafic silicates and nebular gas at high temperatures. She concluded that the gas was somewhat enriched in O (corresponding to the evaporation of silicates in a system < 10 times enriched in dust). However, the f_{O_2} of the gas was not nearly as high as that which seems to be required by the Na-volatility data.

Thus, chondrules provide *ambiguous* data on the nature of the surrounding gas during the melting events. This is clearly an area requiring further research. The most probable resolution of the contradictory data, in my opinion, will be that chondrules formed in an environment without a greatly elevated H_2O/H_2 ratio, that the inferred cooling rates are not greatly wrong, but that the rate at which volatiles are lost is imperfectly understood.

Presence of Dust and Other Solid Material. The presence of matrix-like rims around many chondrules means that after solidification, chondrules encountered and accreted dust. Taylor et al. (1984) argue that these rims were not formed by processes in accreting planetesimals, and must, therefore, be nebular. Rubin (1984) interpreted coarse-grained rims around chondrules as the product of reheating such dusty rims. This implies that high-temperature events affected individual objects multiple times. The greater abundance of objects that underwent substantial rim accretion and reheating in CV3 chondrites compared with other groups indicates that there were different chondrule-forming environments in different places. Either the CV region was dustier, or the frequency or intensity of high-temperature events was greater in the ordinary-chondrite region (and completely melted newly accreted material, incorporating it into the chondrules), or both.

The frequency of compound and cratered chondrules gives an indication of the number density of chondrules during typical melting events. As discussed in Chapter 9.1, these data can at best provide rough (model-dependent) estimates of the number of chondrules/m³. In a system of solar composition at 600 K, the solid/gas ratio will be $\sim 7 \times 10^{-3}$ g/g. At a total pressure of 10^{-3} atm (100 Pa), there should be ~ 0.4 mg/m³ of solids, corresponding to ~ 1 average-sized ordinary-chondrite chondrule/m³, assuming no residual dust. Gooding and Keil (1981) concluded that porphyritic chondrules in ordinary chondrites formed in regions where there were 1 to 100 chondrules/m³. Given the uncertainties in the Gooding and Keil (1981) model, the statistical data seem reasonably consistent with chondrule formation in a system with solid/gas ratio 1 to 10 times the unfractionated solar value; this conclusion is based on the assumption that the chondrite/matrix ratio in ordinary chondrites is representative of the clumped-solids/dust ratio of the nebula.

Constraints on the Precursors of Chondrules

Physical evidence. The most practical method of determining what kinds of materials were melted to form chondrules is to search for unmelted remnants of the precursors. Relict (unmelted) grains inside chondrules are, by definition, examples of chondrule precursors. These grains represent residual precursor material that was most resistant to incorporation in the melt by virtue of large grain size or high melting temperature. The identification of pristine samples of chondrule precursors is more difficult. It is logical to assume that they are present in at least minor amounts among the unmelted components of chondrites. However, no good candidates for unaltered chondrule precursor material have yet been identified.

The relict grains, discussed in Chapter 9.1, show that the precursors of chondrules contained some very large crystals of olivine. Some of these could not be the products of the destruction of earlier generations of chondrules because their compositions are unlike those of igneous olivines in most chondrules. Thus, there was a wide spectrum of grain sizes among chondrule precursors. Once again, as was concluded above based on the rate of Na loss, the frequent presence of relict crystals (of any size) precludes an origin of chondrule melts by direct condensation from the vapor phase unless large volumes of coarse solids were raining into a hot chondrule-formation region from other parts of the nebula or from interstellar space.

The relict grains also indicate that there was a wide range of mineral chemistry among chondrule precursors. Both magnesian and once-ferroan (reduced during melting) olivines are present as relict grains. Some relict olivines are also enriched in refractory elements.

Finally, relict grains show that certain types of material were common to the formation locations of nearly all chondrites. In fact, none of the varieties of relict grains seem to be limited to chondrules from just one chondrite group. The statistics on the frequency with which certain relicts occur in different groups are poor, but seem to show that the precursors represented by the relict grains were not uniformly distributed.

Chemical Evidence. The chemical components of chondrules, tabulated in Chapter 9.1, were shown to be mainly due to variations in the proportions of different precursor assemblages in protochondrules. Only the alkali-element components may be artifacts of open-system melting (that is, open to loss of the most volatile elements). These chemical components are largely consistent with physical evidence about the nature of chondrule precursors presented above, and provide further information about precursors that have been largely destroyed by incorporation into chondrule melts.

Characteristics of one of the chemical components shared by chondrules from several chondrite groups are generally similar to those of forsteritic, refractory-rich relict olivines (Steele 1986). Grossman and Wasson (1983a) identified an FeO-poor, refractory lithophile component, the presence of which is independent of observable relict grains. They postulated that this must represent coarse-grained olivine precursors. The large grain size of these hypothetical precursors is a model-dependent conclusion based on the assumption that FeO entered chondritic silicates at low temperatures by the oxidation of Fe-Ni metal (for chondrules, this is prior to melting); large grains should be the most resistant to diffusive entry of FeO. Thus, the hypothetical chemical component is very similar to the later-discovered relict material. MacPherson (Chapter 10.3) suggests that this precursor material may be identical with or related to the (nonigneous) amoeboid olivine aggregates most abundant in carbonaceous chondrites (see also Chapter 7.3).

The chemical data for refractory-lithophile-rich components of chondrules show that this precursor material had near-solar ratios of lithophiles less volatile than Mg and V in most chondrite groups. Rare-earth-element patterns are generally flat.

Other chemical components of chondrules are difficult to match with relict grain populations. The FeO-, SiO_2-rich, refractory-poor precursors of chondrules in ordinary chondrites are a poor match for observed FeO-rich relict grains. The chemical component is postulated to be fine grained and pyroxene rich, whereas the relict grains tend to be olivine rich and coarse grained. However, fine-grained, FeO-rich, MgO-poor material is present in ordinary chondrites in another form: opaque matrix. While the compositional match between the chemical component and matrix is imperfect, it seems reasonable that the two are in some way connected. FeO-rich, coarse-grained olivine may in fact have been a precursor component of ordinary-chondrite chondrules, but its effect on chondrule compositions seems to have been statistically unresolvable.

The compositions of chondrules from all chondrite groups clearly show that the precursor material contained volatile elements. Not clear from the compositional data is the physical form many of these elements had in precursor grains. Sodium and K, moderately volatile-lithophile elements, were essentially undepleted in chondrule precursors compared with other lithophile elements (relative to whole-rock abundances). Data are lacking for lithophile elements more volatile than K (e.g., Rb and Cs). In several chondrite groups, there seems to be a link between alkalis and Al, suggesting that feldspathic material was a precursor component. In chondrules from ordinary chondrites, there is no relationship between alkalis and Al, but the Na/Al atom ratio is *in all cases* ≤ 1, suggesting there may have been a feldspathic component perhaps modified by a small amount of Na loss.

Many volatile siderophile and chalcophile elements were also present among chondrule precursors. The abundances of volatile siderophiles such as Au, As and Ga are similar to the abundances of nonvolatile elements of similar geochemical affinity, e.g., Ni and Co. In most cases, volatile chalcophile elements such as Se and S are present in chondrules at levels comparable with siderophile elements.

9.3.3. EVALUATION OF MODELS PROPOSED FOR CHONDRULE FORMATION

Planetary Models: Inconsistency with Constraints

Impact Melting in Regoliths. Impact on planetary bodies at first appears to be an excellent mechanism for generating chondrules. The process certainly happened, and provides enough energy to melt target material. Such melting events would involve rapid heating. Cooling could be sufficiently fast, but could be prevented from reaching extremely high rates by imbedding chondrules in an opaque cloud or an ejecta blanket. The target material may be heterogeneous and would be at the appropriate (low) temperature.

However, this type of planetary model fails to meet many of the other constraints presented above. Taylor et al. (1983) showed how similar environments, the lunar regolith and achondritic regolith breccias, contain a completely different population of objects from that observed in chondrites. The low production of melt spherules coupled with the high production of comminuted material and agglutinates is inconsistent with the apparent high efficiency of the chondrule-formation process (Kerridge and Kieffer 1977). Taylor et al. (1983) also point out that the uniformly old ages of chondrules are inconsistent with impact melting on large bodies because this process should have continued at least as long as bombardment in the lunar highlands (until 3.9 Gyr ago). The same authors go on to question whether formation in a regolith could produce rims around chondrules or the mostly spherical shapes of chondrules.

In addition, the paleomagnetic fields recorded by chondrules were high, and are randomly oriented with respect to the host chondrites (see Chapter 8.1). If chondrules cooled in a regolith, these fields should be oriented in similar directions, and should be lower in strength.

Impacts Between Small Bodies. Collisions between meter-sized bodies in the nebula could produce melts if the impact velocity is high enough (> 3 km/s; Kieffer 1975). However, this is not a viable model for chondrule formation. The efficiency of this process in producing molten droplets is too low, like the case for impacts into regoliths. Also, the high velocities required between objects in adjacent orbits could not be maintained in the nebula (e.g., see Wood and McSween 1976). Sonett (1979) points out that this type of chondrule-forming process also fails because it leads to the breakup of the colliding planetesimals, thus aborting chondrule formation.

Planetary Igneous Processes. The volcanic models for chondrule formation, as well as Dodd's (1971,1978,1981) variant on the impact-melting model, require the presence of magma bodies on asteroids or planetesimals where chondrules formed. In volcanic models, chondrules form by the ejection of droplets during eruption of magma. In Dodd's model, many chondrules (metal-poor porphyritic chondrules) are the products of the breakup of igneous rocks formed in earlier impacts, while others are remelted igneous-rock fragments. The appeal of these models is that they use processes that are known to happen elsewhere in the solar system. However, these models also cannot satisfy many constraints.

Igneous rocks fail to satisfy the constraints on the precursor material of chondrules presented above. The breakup of coarse-grained igneous rocks should produce a population of chondrules with interelement correlations consistent with equilibrium partitioning of elements among igneous phases. In particular, this process should fractionate elements that are cosmochemically refractory from each other, resulting in scattered Ca/Al or Ti/Al ratios, non-flat rare-earth patterns, anticorrelated Al and Mg (in felsic and mafic components), O-isotopic compositions lying along a slope-1/2 mass-fractionation line, and other predictable effects *none of which are observed.* One possible way to avoid these inconsistencies is to produce the chondrules in a single chondrite by mixing together material from many planetesimals that were themselves chemically and isotopically heterogeneous. This may allow large chemical and isotopic variations between the planestesimals to overshadow smaller, predictable fractionation effects in each local igneous system. However, this too seems unlikely, as there should still be many chondrules with fractionated rare-earth patterns.

Another serious flaw of models involving any type of planetary igneous activity is the lack of preservation of the expected igneous rocks. In the ordinary chondrites, rare clasts of coarse-grained "igneous rocks" have been reported (see, e.g., Hutchison 1982). However, these do not have the appropriate composition to be chondrule-precursor rock fragments, generally being Na poor. Grossman and Rubin (1986) speculated that these clasts may, in fact, be large chondrules themselves. Cm-sized impact-melt clasts occur in many chondrite breccias. Dodd (1981) suggests that these types of microporphyritic clasts may be suitable precursor igneous rocks for forming chondrules. However, Rubin (1985) showed that clasts like these are texturally different from typical porphyritic chondrules. Finally, as pointed out by Taylor et al. (1983), there are no known achondrites with a bulk composition like that of chondrules; all achondrite groups have either much higher or much lower Ca/Si, Al/Si ratios and very much lower Na/Si and K/Si ratios than chondrules and chondrites. The O-isotopic characteristics of achondrite groups are also inconsistent with those of chondrules.

Molten Planetesimals. This scenario for chondrule formation is a special type of igneous model that begins with a population of planetesimals that are partially molten due to heating from the decay of ^{26}Al (Zook 1980). Low-velocity collisions between these objects result in the release of liquid droplets that quickly solidify into chondrules. This makes more sense astrophysically than impact melting due to collisions between small bodies, and satisfies the high-efficiency constraint.

Taylor et al. (1983) made several arguments as to why this model must be incorrect. They pointed out that neither of the two O-isotope trends defined by chondrules in carbonaceous chondrites and chondrules in ordinary chondrites can be produced by the collision of two well-mixed molten planetesimals. This process should result in either homogeneous chondrules or in a bimodal distribution of isotopic compositions depending on whether the two planetesimals were isotopically identical or not. However, if *many* molten planetesimals contributed to the objects in individual chondrites then the observed isotopic trends could be produced if the planetesimals possessed the correct distribution of isotopic compositions. Other arguments by Taylor et al. (1983) are similar to those used against igneous models in the preceding section, and perhaps are stronger evidence against this model: there are no surviving examples of large volumes of the parental melt, and chondrules do not show the expected igneous fractionation trends. Furthermore, metal and sulfide would quickly segregate from the silicate liquid in the molten planetesimal making it impossible to produce metal-bearing chondrules.

Condensation Models

All of the condensation models listed in Table 9.3.2 require a hot gaseous nebula which, upon cooling, produces liquid silicate droplets. In most of these papers, it was recognized that under "nebular" pressure/composition conditions liquids are not stable equilibrium phases at any temperature. Thus, the models all have built-in speculations about how other conditions could be achieved, e.g., periods of transient high pressures (Wood 1962), nonsolar gas composition (Herndon and Suess 1977), nonequilibrium ("constrained equilibrium") condensation (Blander and Katz 1967; Blander 1983).

These models were originally considered plausible because it was more generally accepted that the solar nebula underwent a high-temperature period that vaporized most of the dust. While contemporary models for the evolution of the nebula indicate that there was no widespread hot period (Chapter 6.3), it remains a possibility that such conditions existed locally within the nebula. Thus, condensation models may still be considered in this revised context.

Condensation models for chondrule formation fail to satisfy many, if not most, of the constraints outlined above. Several of the more important failings are listed below.

1. It does not seem possible to rationalize high cooling rates (thousands of K/hr) with the production of the melt in a large region of hot gas.
2. Chondrules have volatile elements as *primary* constituents. The only way to incorporate elements (such as Na, K, Ga, S and Zn) in condensation processes is by *secondary* processing below the solidus temperature.
3. Chondrules are chemically and isotopically diverse, but condensation would tend to make rather uniform chondrules.
4. The coarse-grained precursors indicated by the presence of relict grains would evaporate in a high-temperature nebula.

Remelting of Dust in the Nebula: Summary

The failures of the planetary models and the high-temperature condensation models for chondrule formation leave a relatively cool, dusty solar nebula as the only conceivable setting for chondrule formation. By fitting together the constraints on chondrule formation, we can derive the following image of this nebula. The term "region," as used below in the context of chondrule formation, includes both the position in the nebula and the time at which a particular group of chondrules or chondrites formed.

Clumps of solid matter (chondrule precursors) formed in each region where chondrites were to form except, perhaps, in the CI-formation region. The temperature was below 650 K. The sizes of the clumps varied in a seemingly nonsystematic way in the regions where the chondrite groups formed, decreasing in the order $CV > LL > L > H \cong CO \cong CM \geqslant EH$. Clumping, as indicated by the variable amounts of matrix in chondrites, may have been more efficient in some formation regions than in others: in order of apparent decreasing efficiency, $EH > LL \cong L \cong H > CV > CO \cong CM$.

Coarse grains (> 100 μm in size) that did not originate in earlier generations of chondrules were present in many precursor clumps; the origin of these grains remains a mystery. Also present in the clumps were heterogeneous assemblages of grains that had chemical and isotopic compositions characteristic of the formation region. Chemical evidence suggests that many of these solids may themselves have formed by the earlier condensation of dust from gas in high-temperature processing.

Transient events heated these clumps, imparting a limited amount of energy to the solids. In many cases, the temperature of the clumps rose to near the liquidus temperature. The average intensity of the heating may have varied between formation regions. Heating and subsequent cooling were rapid enough to prevent volatilization of Na and K. Experiments suggest that the peak temperatures could not have been maintained for more than a few seconds, and cooling took place at rates up to thousands of K/hr. However, cooling was apparently not able to proceed nearly as fast as perfect radiative cooling into cold space.

The surrounding gas had an H_2O/H_2 ratio close to that calculated for a gas of solar composition. The total amount of solids was probably close to that which would coexist with an unfractionated gas below 1000 K. Another characteristic of the environment was the presence of fairly strong magnetic fields.

After solidification, dust often accreted to the surfaces of the newborn chondrules. Some regions appear to have been dustier than others; rims are much thicker and more abundant around CV chondrules than around chondrules from other groups. Sometimes the dust-coated chondrules experienced reheating or remelting, most likely by the same process that caused the original chondrule to form.

It is not known how other chondritic components fit into this picture of chondrule formation in the nebula. Chondrite matrix could be related to the dust from which chondrules formed, and may have evolved chemically and isotopically in the nebula or after accretion. The inclusions with an igneous origin, especially those found in CV chondrites, are not constrained by many of the arguments used here for chondrules: they are low in volatile elements, they can be much more massive, they do not have the same kinds of rims, they may have formed from different precursor material, and they cooled at different rates. Some of the nonigneous inclusions such as amoeboid olivine aggregates may be related to chondrule precursors, but other nonigneous inclusions are not clearly related to chondrules.

The models for chondrule formation by nebular remelting (listed in Table 9.3.2) differ from each other in two ways: the solids that get melted are different, and the energy sources are different. Most of the models listing nebular dust as the precursor material for chondrules are nonspecific about what this material might be. Other authors have suggested that it was nebular condensates, chondrite matrix or interstellar grains. Two of the models (D. Clayton 1980; Wood 1983,1984) require that the precursors are interstellar dust. As discussed earlier in this chapter and by Grossman and Wasson (1983a,b), matrix and interstellar dust would probably not have the correct composition or variability to be chondrule precursors. If this is true, then the infall model of Wood, and the Clayton model are incorrect. However, it is not clear why *any* of the three postulated types of precursors would contain coarse grains.

A discussion of the plausibility of the various nebular heat sources continues in the next chapter.

"At the same time so little is positively known respecting the original constitution of the solar system, that all these conclusions must to some extent be looked upon as only provisional"—*H. C. Sorby* (1877).

Acknowledgments. I thank A. Rubin and an anonymous referee for their reviews, and R. Hewins for numerous helpful discussions.

REFERENCES

Blander, M. 1983. Condensation of chondrules. In *Chondrules and Their Origins,* ed. E. A. King (Houston: Lunar and Planetary Inst.), pp. 1–9.

Blander, M., and Katz, J. L. 1967. Condensation of primordial dust. *Geochim. Cosmochim. Acta* 31:1025–1034.

Borgström, L. H. 1904. The Shelburne meteorite. *Trans. Roy. Astron. Soc. Canada*, pp. 69–94.

Cameron, A. G. W. 1966. The accumulation of chondritic material. *Earth Planet. Sci. Lett.* 1:93–96.

Clayton, D. D. 1980. Chemical energy in cold-cloud aggregates: The origin of meteoritic chondrules. *Astrophys. J.* 239:L37–L41.

Dodd, R. T. 1971. The petrology of chondrules in the Sharps meteorite. *Contrib. Mineral. Petrol.* 31:201–227.

Dodd, R. T. 1978. The composition and origin of large microporphyritic chondrules in the Manych (L-3) chondrite. *Earth Planet. Sci. Lett.* 39:52–66.

Dodd, R. T. 1981. *Meteorites: A Petrologic-Chemical Synthesis* (New York: Cambridge Univ. Press).

Fermor, L. L. 1938. Garnets and their role in nature. *Indian Assoc. Adv. Sci. Spec. Publ.* 6:87–91.

Fredriksson, K. 1963. Chondrules and the meteorite parent bodies. *Trans. N.Y. Acad. Sci.* 25:756–769.

Fredriksson, K., and Ringwood, A. E. 1963. Origin of meteoritic chondrules. *Geochim. Cosmochim. Acta* 27:639–641.
Ghiorso, M. S. 1985. Chemical mass transfer in magmatic processes. I. Thermodynamic relations and numerical algorithms. *Contrib. Mineral. Petrol.* 90:107–120.
Gooding, J. L., and Keil, K. 1981. Relative abundances of chondrule primary textural types in ordinary chondrites and their bearing on conditions of chondrule formation. *Meteoritics* 16:17–43.
Grossman, J. N., and Rubin, A. E. 1986. The origin of chondrules and clasts bearing calcic plagioclase in ordinary chondrites. *Lunar Planet. Sci.* 17:293–294 (abstract).
Grossman, J. N., and Wasson, J. T. 1982. Evidence for primitive nebular components in chondrules from the Chainpur chondrite. *Geochim. Cosmochim. Acta* 46:1081–1099.
Grossman, J. N., and Wasson, J. T. 1983a. Refractory precursor components of Semarkona chondrules and the fractionation of refractory elements among chondrites. *Geochim. Cosmochim. Acta* 47:759–771.
Grossman, J. N., and Wasson, J. T. 1983b. The compositions of chondrules in unequilibrated chondrites: An evaluation of models for the formation of chondrules and their precursor materials. In *Chondrules and Their Origins*, ed. E. A. King (Houston: Lunar and Planetary Inst.), pp. 88–121.
Grossman, J. N., Kracher, A., and Wasson, J. T. 1979. Volatiles in Chainpur chondrules. *Geophys. Res. Lett.* 6:597–600.
Herbig, G. H. 1978. Some aspects of early stellar evolution that may be relevant to the origin of the solar system. In *The Origin of the Solar System*, ed. S. F. Dermott (New York: Wiley and Sons), pp. 219–235.
Herndon, J. M., and Suess, H. E. 1977. Can the ordinary chondrites have condensed from a gas phase? *Geochim. Cosmochim. Acta* 41:233–236.
Hutchison, R. 1982. Meteorites—Evidence for the interrelationships of materials in the solar system 4.55 Ga ago. *Earth Planet. Sci. Lett.* 29:199–208.
Johnson, M. C. 1986. The solar nebula redox state as recorded by the most reduced chondrules of five primitive chondrites. *Geochim. Cosmochim. Acta* 50:1497–1502.
Kerridge, J. F., and Kieffer, S. W. 1977. A constraint on impact theories of chondrule formation. *Earth Planet. Sci. Lett.* 35:35–42.
Kieffer, S. W. 1975. Droplet chondrules. *Science* 189:333–340.
Leitch, C. A., and Smith, J. V. 1982. Petrography, mineral chemistry and origin of Type I enstatite chondrites. *Geochim. Cosmochim. Acta* 46:2083–2098.
Mason, B. 1960. Origin of chondrules and chondritic meteorites. *Nature* 186:230–231.
Merrill, G. P. 1920. On chondrules and chondritic structure in meteorites. *Proc. Natl. Acad. Sci. USA* 8:449–472.
Merrill, G. P. 1921. On metamorphism in meteorites. *Geol. Soc. Amer. Bull.* 32:395–414.
Podolak, M., and Cameron, A. G. W. 1974. Possible formation of meteoritic chondrules and inclusions in the precollapse Jovian protoplanetary atmosphere. *Icarus* 23:326–333.
Rasmussen, K. L., and Wasson, J. T. 1982. A new lightning model for chondrule formation. In *Papers Presented to the Conference on Chondrules and Their Origins* (Houston: Lunar and Planetary Inst.), p. 53 (abstract).
Ringwood, A. E. 1959. On the evolution and densities of the planets. *Geochim. Cosmochim. Acta* 15:257–283.
Rubin, A. E. 1984. Coarse-grained chondrule rims in type 3 chondrites. *Geochim. Cosmochim. Acta* 48:1779–1789.
Rubin, A. E. 1985. Impact melt products of chondritic material. *Rev. Geophys.* 23:277–300.
Sonett, C. P. 1979. On the origin of chondrules. *Geophys. Res. Lett.* 6:677–680.
Sorby, H. C. 1877. On the structure and origin of meteorites. *Nature* 15:495–498.
Steele, I. M. 1986. Compositions and textures of relic forsterite in carbonaceous and unequilibrated ordinary chondrites. *Geochim. Cosmochim. Acta* 50:1379–1395.
Suess, H. E. 1949. Zur Chemie der Planeten und Meteoritenbindung. *Z. Electrochem.* 53:237–241.
Taylor, G. J., Scott, E. R. D., and Keil, K. 1983. Cosmic setting for chondrule formation. In *Chondrules and Their Origins*, ed. E. A. King (Houston: Lunar and Planetary Inst.), pp. 262–268.
Taylor, G. J., Scott, E. R. D., Keil, K., Boynton, W. V., Hill, D. H., Mayeda, T. K., and

Clayton, R. N. 1984. Primitive nature of ordinary chondrite matrix materials. *Lunar Planet. Sci.* 15:848–849.
Tschermak, G. 1883. Beitrag zur Classification der Meteoriten. *Sitzber. Akad. Wiss. Wien, Math.-Naturw. Cl.* 85(1):347–371.
Tsuchiyama, A., Nagahara, H., and Kushiro, I. 1981. Volatilization of sodium from silicate melt spheres and its application to the formation of chondrules. *Geochim. Cosmochim. Acta* 45:1357–1367.
Urey, H. C. 1967. Parent bodies of the meteorites. *Icarus* 7:350–359.
Urey, H. C., and Craig, H. 1953. The composition of the stone meteorites and the origin of the meteorites. *Geochim. Cosmochim. Acta* 4:36–82.
Wai, C. M., and Wasson, J. T. 1977. Nebular condensation of moderately volatile elements and their abundances in ordinary chondrites. *Earth Planet. Sci. Lett.* 36:1–13.
Wasson, J. T. 1972. Formation of ordinary chondrites. *Rev. Geophys. Space Phys.* 10:711–759.
Wasson, J. T. 1985. *Meteorites: Their Record of Early Solar-System History* (New York: W. H. Freeman).
Whipple, F. L. 1966. Chondrules: Suggestions concerning their origin. *Science* 153:54–56.
Wlotzka, F. 1969. On the formation of chondrules and metal particles by shock melting. In *Meteorite Research*, ed. P. M. Millman (Dordrecht: D. Reidel), pp. 174–183.
Wood, J. A. 1962. Chondrules and the origin of the terrestrial planets. *Nature* 197:127–130.
Wood, J. A. 1967. Chondrites: Their metallic minerals, thermal histories, and parent planets. *Icarus* 6:1–49.
Wood, J. A. 1983. Formation of chondrules and CAI's from interstellar grains accreting to the solar nebula. *Mem. Natl. Inst. Polar Res. Special Issue* 30:84–92.
Wood, J. A. 1984. On the formation of meteoritic chondrules by aerodynamic drag heating in the solar nebula. *Earth Planet. Sci. Lett.* 70:11–26.
Wood, J. A. 1985. Meteoritic constraints on processes in the solar nebula. In *Protostars & Planets II*, eds. D. C. Black and M. S. Matthews (Tucson: Univ. of Arizona Press), pp. 687–702.
Wood, J. A., and McSween, H. Y. 1976. Chondrules as condensation products. In *Comets, Asteroids, Meteorites: Interrelations, Evolution, and Origins*, ed. A. H. Delsemme (Toledo: Univ. of Toledo), pp. 365–373.
Zook, H. A. 1980. A new impact model for the generation of ordinary chondrites. *Meteoritics* 15:390–391 (abstract).
Zook, H. A. 1981. On a new model for the generation of chondrules. *Lunar Planet. Sci.* 12:1242–1244 (abstract).

9.4. ENERGETICS OF CHONDRULE FORMATION

E. H. LEVY
University of Arizona

Meteorite chondrules apparently were formed as a result of localized, transient heating events in the protoplanetary nebula. Such transient events, which seem to have heated the chondrules to temperatures in excess of 1500° C for not more than a few minutes, are not easily explicable in terms of the canonically accepted evolutionary processes of the nebular disk. Thus the occurrence of extraordinary dynamical processes may be indicated by the presence of chondrules and, consequently, the existence of chondrules poses questions fundamental to our understanding of protoplanetary and protostellar systems. This chapter briefly considers the gross energetics, as well as some related questions, of chondrule formation and the implications for several previously proposed sources of chondrule formation energy, including gravitational infall of the nebula, energy derived from solid-body impacts within the nebula, and energy liberated by dissipative evolution of the nebula itself.

9.4.1. INTRODUCTION

The Chondrule Problem

The primary challenge posed by the existence of meteorite chondrules is to understand what mechanism could have transiently heated and melted small beads of rock in an otherwise cool region of the solar nebula. The heating involved in chondrule formation entailed a transient temperature change in excess of 1000° C, to a melting point exceeding 1500° C (Chapter 9.2). In this chapter we examine basic aspects of some possible heating mechanisms. For this purpose, we take as given that chondrules are solar-nebula products and that they are not direct nebular condensates. We also take as given that chondrule formation occurred as the result of transient heating of preexisting assemblages of solid matter. The weight of the evidence on these particulars, while not conclusive, seems to point in those directions (Chapter 9.3). Radioisotope ages of chondrules indicate that their formation is associated, within a few Myr, with the formation and accumulation of the chondrites themselves (Herzog et al. 1973; Caffee et al. 1982; Swindle et al. 1983; Chapter 15.3). Alternative ideas, including the possibility that chondrules were brought to the solar system intact from elsewhere, or that they formed as a result of planetary processes on larger bodies, while probably not logically excludable at this time, are not considered in this chapter, wherein we concentrate on the implications, for gross nebular energetics, of the assumption that chondrules formed in the nebula.

Several aspects of chondrule formation, as it is presently understood, deserve special mention. The fact that chondrules retain large amounts of relatively volatile elements such as Na, the textures of chondrules, and their mineralogical structures seem to indicate that the chondrule heating events were short, of the order of minutes in duration (Chapters 9.2 and 9.3), although even much shorter times have been proposed by some investigators. The most important point here is that chondrule-heating events were very short in comparison with any gross nebular dynamical time scales. For example, one might imagine that chondrules could have been melted close to the Sun and subsequently transported to larger heliocentric distances by nebular convection. However, the time scale for such a trip, depending on random motion at speeds of about 1 km/s, would have involved heating over an interval of many years rather than one of minutes. Similarly, heating in high-temperature interiors of conceivable protoplanetary subnebulae also suffers from severe time scale (as well as dynamical) problems. Thus, it seems that chondrule heating must have involved unusual localized and transient phenomena in the nebula.

Evidence suggests that the heating source was only marginally adequate for melting. From the prevalence of partial melts, unmelted residual structures and grains, and other indications, it appears that the temperature rise of chondrules frequently stalled at about the temperature of incipient melting (Rambaldi 1981; Chapter 9.2). In some cases, only the lower-melting-point components of chondrules may have liquified, while, in other instances, the temperature may not have held high long enough to melt the entire chondrule. This stalling of the temperature rise, if real, places an important constraint on any chondrule-heating mechanism (see Chapter 9.3). It probably cannot completely be ruled out, however, that the apparent temperature stall could be an artifact resulting from the total destruction of those chondrules which were heated to higher temperatures. But, in that case, many objects that had been heated well above the point of incipient melting, but below the temperatures at which they would be destroyed, would also be expected to have survived. Indeed, in the absence of some real temperature-stalling mechanism, these latter objects might be expected to be the most prevalent. Altogether then, it seems that the apparent stall in temperature during chondrule heating resulted from some specific physical aspect of the chondrule-forming process.

Any theory of chondrule formation must deal with the fact that a large amount of matter must have been processed to chondrules. Because such a large fraction of primitive meteorite mass is composed of chondrules, it seems reasonable to take 10^{24} g (roughly the mass of the asteroid belt) as a measure of the total mass of chondrules. Considering the likelihood that many chondrules could have been lost from the solar system or incorporated into planets or the Sun, this number is more likely to err on the low side than on the high side.

It is also remarkable that chondrules are so abundant in meteorites; many chondritic meteorites are dominantly made of chondrules—essentially as closely packed spheres with an intervening fine-grained matrix. This suggests either that more than half of the nebular dust was processed to chondrules, at least in the region of the nebula where ordinary chondrites formed, or that some process concentrated the chondrules and led to their preferential incorporation into certain meteorites. It is essential to understand which of these occurred in order ultimately to understand the energetics of the chondrule-formation process and also to understand solar-system accretion processes.

Implications of the Chondrule Problem

Assuming that chondrules formed in the solar nebula, then their existence may imply large departures of the nebula from the simple and quiescent object of our theoretical fantasies and fairy tales. A large fraction of preplanetary matter (at least in some parts of the nebula) was processed to states far from the average thermal equilibrium of its surroundings.

The fact that the protoplanetary nebula might have been such a dynamically interesting object should not surprise us. In this respect, the nebula probably adhered to the cosmical norm. Numerous similar examples can be cited; one is the Sun's corona. The corona displays many unpredicted, energetic dynamical phenomena. The very existence of the solar corona (as well as other stellar coronae) is surprising on the basis of simple thermodynamics and requires the intermediary of mechanical and collective effects to heat the corona to its unlikely high temperature. The explosive dynamical behaviors of planetary magnetospheres and the stormy behaviors of planetary atmospheres are other important examples. Unexpected explosive outbursts characterize many cosmical systems. Apparently, nature finds many creative ways to expend energy in those systems that have it to spare. The existence of chondrules may be telling us that the protoplanetary nebula followed this cosmical norm.

From another point of view, it is instructive to remember that the formation of the solar system was a close-to-home example of star formation. Astronomical observations of star-forming regions show protostars to be far from simple, quiescent objects. For example, bipolar outflows around objects thought to be protostars indicate that protostellar disks are likely to be very dynamically active systems. The origin of such surprising dynamical behavior is not yet understood, but I suspect that chondrules (and other disequilibrium manifestations in primitive solar-system matter) may be connected with the larger-scale manifestations of dynamical disequilibrium that seem generally to be associated with star formation.

It is worth remembering, too, that other properties of meteorite matter also point toward the occurrence of transient high-energy and high-temperature events in the protoplanetary nebula. Oxygen-isotopic anomalies have been shown to result from energetic discharges in gas (Thiemens and Heidenreich 1983; Chapter 12.1), although it is still problematical whether the major O anomalies observed in primitive meteorites actually arose this way. In addition, high-temperature Ca- and Al-rich inclusions and their apparently quickly cooked rims seem also to point to transient heating events in an otherwise cool nebula (Boynton and Wark 1987; Wark and Boynton 1987).

9.4.2. CHONDRULE ENERGETICS

The formation of a chondrule requires the supply of sufficient energy to raise the material through $\sim 1000°$ C, and to take it through a melting phase transition, in the face of some heat loss during the heating event. Crudely, this requires about 5×10^{10} erg/g of chondrule matter, assuming that the temperature rise is sufficiently rapid that a very much larger amount of energy is not lost during the heating event. Thus, I will define *the first energy problem of chondrules* to be 5×10^{10} erg/g; this number is conservative and depends on the apparent fact that the chondrule-making process is rapid so that a far larger amount of energy is not radiated away during the heating event. Taking, as above, the nominal total chondrule production to be in excess of 10^{24} g, we get *the second energy problem of chondrules* which is at least 5×10^{34} erg/nebula. Recall that this latter number is highly uncertain, and likely to be low; the reader can multiply it, along with any of the following consequences, by whatever factor he or she chooses.

In this chapter we are concerned with the grossest aspects of the energy sources that might have driven chondrule formation. We consider three possible sources of energy: energy derived from the original gravitational infall of the nebula; energy derived from solid-body impacts within the nebula; and energy liberated by the dissipative evolution of the disk nebula itself. While not an exhaustive list, our consideration of these three candidate energy sources will outline the general scope of the energy problems posed by chondrule formation.

Infall

The major portion of the free energy liberated during star formation is the gravitational potential energy of collapse. In order to see how much of such gravitational energy might be available for melting chondrules, consider that a dust clump, falling from infinity to the nebula at a distance R from a one-solar-mass Sun, liberates gravitational energy in the amount

$$\Delta E = \frac{GM}{R}. \quad (1)$$

With $R = 3$ AU (to guess at the heliocentric distance typical of chondrule formation), $\Delta E \simeq 3 \times 10^{12}$ erg/g. If the material ends up in circular orbit at 3 AU, then half of this energy had to have been dissipated during the accretion. In principle, this is plenty of energy in comparison with the approximately 5×10^{10} erg/g needed to melt chondrules.

However, the maximum temperature to which an incoming dust glob is heated depends on the details of the inflow and, especially, the time over which the energy is dissipated. The chondrule dust balls cannot have been melted in a hot phase of the nebula itself, even had one existed at several astronomical units from the center, inasmuch as that would have violated the time scale constraints. In a scenario suggested by Wood (1984), the incoming dust and gas move together until the gas is decelerated in an accretion shock. Thereafter, the dust finds itself moving at high velocity through the decelerated gas and is heated by friction with the gas. An attribute of this chondrule-formation mechanism is that a lot of material can be processed. The heat for melting is derived directly from the gravitational energy of the incoming chondrule precursor material; 10^{24} g are as easily melted to chondrules as is one gram. The adequacy of this accretional mechanism for making chondrules then hinges entirely on the details, and is not restricted by energy supply.

Wood finds that, if the accreting matter has an ordinary cosmic ratio of dust to gas, then it is optically thin. In that case, the frictionally heated dust cools rapidly by radiating to free space and the temperature rises to only a few hundred °C, except in the case of relatively massive nebulae, exceeding 2 $M_\odot$. In order for accretional heating to work with a lower-mass nebula, Wood finds that the dust-to-gas ratio must be amplified locally, by a factor of several hundred to a thousand over its normal cosmic value, in an accreting blob from which chondrules are made. The linear dimensions of each such blob must exceed 3×10^{12} cm in order to create sufficient optical depth to blanket the loss of heat and allow the chondrules to melt. The need for such effective thermal blanketing raises again the question of the time scale for chondrule heating events. It is not clear whether all of the constraints can be satisfied. If the incoming blob is sufficiently well thermally blanketed to generate high temperatures, then the cooling time will be long. The problem is particularly complicated by the fact that the opacity, dominated as it is by the grains which become the chondrules in this scenario, may vary as the grains melt and collapse from loose aggregates into tightly bound chondrules. This would work in the right direction, but a more quantitative treatment than has thus far been given is needed before real conclusions can be drawn.

Because chondrule masses are in the mg range, chondrule formation through melting during nebular accretion requires that interstellar dust must have already accumulated to that mass range in interstellar clouds before accretion into the nebula. There is insufficient time for this accumulation to happen during the heating event. It is problematical whether interstellar dust could already have accumulated to such high-mass aggregations already before its fall into the nebula, although some studies suggest this possibility (Cameron 1975; Morfill et al. 1978), if the surface sticking coefficients of the interstellar dust grains are high.

Perhaps more problematical than the prior existence of mg-mass chondrule precursors in the accretion flow, is the maintenance of their integrity during the heating event. Interstellar dust grains are likely to be coated with ice and organic-rich volatile mantles. In the course of a heating event, these mantles would vaporize well before the temperature had risen high enough to melt the silicates. It is not clear that a mg agglomeration of loosely held subμm interstellar grains would maintain its structural integrity in the face of rapid sublimation of the ice and organic mantle that surrounds each grain. If mantle vaporization disrupted the grains at the beginning of the heating event then chondrules could not form through this process.

Altogether, the gravitational energy of nebular accretion is sufficient to melt chondrules. However, this energy is not delivered rapidly enough by frictional interaction with the nebular gas to permit the dust to be heated to silicate-melting temperatures, unless a large region of dust-laden space is thermally blanketed to inhibit the speedy escape of heat. However, the many details are at least arguable. The several hundred to a thousand-fold amplification of the local dust abundance over its normal cosmic value seems to be a severe constraint, as does the need to keep the grain aggregates together during the rapid sublimation of their ice and organic mantles at the beginning of the melting event. It is not clear that a large thermally blanketed chondrule-forming region will cool rapidly enough to agree with the inferred short intervals of sustained high temperature and rapid chondrule cooling rates.

Impact

Melting of rock droplets in collisions between small protoplanetary bodies has been put forward as a mechanism to account for chondrule formation. In order for this process to work, the specific kinetic energy of a collision must exceed the few $\times$ 10^{10} erg/g needed to melt chondrular matter; very roughly $\rho(\delta v)^2 \geq 10^{11}$, or $\delta v \geq 3 \times 10^5$ cm/s, assuming that the dissipated collisional energy is distributed generally throughout the colliding objects and not concentrated into a very small fraction of the smaller mass. In fact, the energy in such a collision is not necessarily uniformly distributed throughout the colliding mass. But 3 km/s still seems to represent an approximate threshold impact speed for the production of significant quantities of melt (Kieffer 1976). The 3 km/s velocity dispersion is to be compared with an orbital speed of about 18 km/s. Therefore, the orbits of colliding bodies would have to have been rather well scrambled if collisions were to account for chondrule melting. If chondrule formation occurred before the substantial development of planet formation, then the orbit scrambling would have to have been produced by gravitational interactions within the asteroid belt itself. Because gravitational interactions tend to scramble orbits to relative velocities comparable to the escape velocity of the largest scatterer, it is worth noting that 3 km/s corresponds to the escape speed from a body larger than the Earth's Moon. Such a body would contain as much or more mass than the entire present asteroid belt. The fate of such an object is problematical, although it probably could have been lost from the system or destroyed. If chondrule formation occurred after major-planet formation, then the needed high collisional velocities could have been produced by the gravitational disturbance of Jupiter. Indeed, the present asteroid belt contains relative velocities as high as the required value or larger.

Several major problems with a collisional origin of chondrules assert themselves. It seems unlikely, at least to me, that such a mechanism could make chondrules of such a large fraction of meteorite matter and then mix it with apparently primitive grains, without an extremely effective sorting mechanism. It is difficult to see, without resorting to highly contrived mechanisms, how such effective sorting could take place in such a scrambled, collision-dominated environment. Finally, although some chondrule-like objects are found in other material that has clearly been heavily processed by collisions, like the lunar regolith, such material is dominated by other impact products, notably agglutinates, that are absent from chondrites (Kerridge and Kieffer 1977). In the absence of further compelling information, this last fact alone counts as a heavy burden against collisional chondrule-production theories (see also Chapter 9.3).

Kieffer (1976) suggests that 3 km/s collisions between small objects (ranging from somewhat smaller than 1 cm to about 20 cm in diameter) could have produced chondrules without suffering from the objections based on observed differences between meteorite chondrules and the other material known to have been processed by impacts. As we have already mentioned, 3 km/s dispersion velocities are within the range of what is observed today in the asteroid belt; the basic energy supply is not, on its face, a problem. However, an important challenge to the tenability of such an impact mechanism is the maintenance of a 3 km/s velocity dispersion in a population of such small bodies. At 3 km/s, particles move supersonically in the cool nebular gas and therefore dissipate kinetic energy very rapidly. Moreover, even a 20-cm-diameter particle has a projected mass density of only some 60 g/cm². In an ambient gas density of, say, 10^{-9} g/cm³, the stopping distance would have been shorter than 10^{11} cm, even for the largest particles considered. In the presence of nebular gas, it would seem to be difficult to maintain the high velocity dispersion. Jupiter apparently captured a massive gas envelope *after* accretion had proceeded to very high masses; thus, gas was abundant in the nebula (and in the region near the asteroid belt) well past the time of major planetary accretion. If the proposed collisional mechanism is to have worked, then it seems likely that the collisions would have to have occurred after the dissipation of the nebular gas. It remains to be shown that a workable scenario for the subsequent accumulation of chondritic meteorites can emerge from the confines of the various constraints.

Tapping the Nebula

The evolution of the protosolar nebula involved the dispersal of orbital energy through viscous or other dissipative forces. Some of this energy could have gone into making chondrules. For the most part, however, in the usual picture of nebular evolution, this energy was dissipated directly as low-grade heat. Indeed, this energy of dissipation dominates the thermal state of a nebula, but apparently produced in the protosolar nebula, bulk temperatures in the putative chondrule-formation region that were too low on the scale we are considering here. In order for the dissipated energy to have made chondrules, some of it must have been "beneficiated": converted to an ordered, low-entropy state from which it could have been available, through the intermediary of collective effects, to drive transient high-temperature events. Experience with a variety of cosmical systems suggests that the mechanisms most commonly responsible for the transient and rapid release of energy involve the generation and explosive dissipation of magnetic field structures.

Consider the amount of energy released during the dissipative evolution of an accretion disk such as the protosolar nebula. In magnitude, the energy of an orbiting gram of matter is

$$E \approx \frac{GM}{r} \qquad (2)$$

and the energy liberated as the gram of matter evolves dissipatively through a change in heliocentric radius Δr is given by

$$\Delta E \approx \frac{GM}{r^2} \Delta r. \qquad (3)$$

Thus, a gram of matter at about 3 AU from the Sun liberates approximately 10^{12} erg of energy as it evolves through a change in heliocentric radius equal to about 1 AU. This is, as it must be, comparable to the energy of accretion at 3 AU.

Energy Conversion Efficiency. Altogether, 1 $M_\oplus$ processed through 1 AU at 3 AU from the Sun during the time of chondrule formation would evolve some 10^{40} erg; this compares with the 5×10^{34} that is needed to make our nominal 10^{24} g of chondrules—requiring an efficiency of about 10^{-5}. Of course, much more than an Earth's equivalent of total mass was likely to have been processed through the chondrule-manufacturing zone; but, at the same time, we are similarly uncertain, probably by several orders of magnitude, in our estimates of the total mass of chondrules that was made in the nebula. These numbers could easily be imagined to take on other values, demanding a chondrule-manufacturing efficiency somewhat different from the above number.

The question of energy efficiency for chondrule production is an important one. Assuming that chondrules form as a result of local, transient events which depart far from thermodynamic equilibrium, then one question that immediately pushes itself forward is: With what efficiency do other known cosmical systems convert energy to high-temperature states far from the prevailing thermal equilibria? Only a few systems have been studied in sufficient detail to permit such an estimate, and all of these are within the solar system.

Consider the Sun. The Sun emits some 4×10^{33} erg/s of thermal energy at a temperature of about 5000 K. In addition, however, the Sun emits energy in several high-energy forms at effective "temperatures" far above this thermal base. The Sun's corona is heated to temperatures in excess of 10^6 K, most likely by a variety of dissipative magnetic processes. The energy required to maintain the high coronal temperature and to support other coronal energetic phenomena, including flares and the solar wind, can be estimated to be about 10^{29} erg/s (Withbroe and Noyes 1977). Thus, in the Sun, nearly one part in 10^4 of the total energy goes into the "nonthermal" corona and into very energetic, high-temperature, transient phenomena. This energy-conversion efficiency seems to be in the range of what is needed if nebular dissipation energy is to account for the formation of chondrules. However, it is also important to remember that only a fraction of the energy that is converted to some high-temperature phase will ultimately make it into chondrular matter—putting yet another lien against the ultimate source of chondrule-manufacturing energy.

As another example, consider the Earth's magnetosphere. The magnetosphere is distorted far from its nominal equilibrium by the varying stress of the solar wind. The solar-wind energy impinging on the magnetosphere during times of disturbed interplanetary conditions is of the order of 10^{22} erg/s. During a disturbed time, some 10^{24} erg may be dissipated in a geomagnetic storm in half a day, (Chapman 1964; Axford 1964). This nonthermal energy appears, at the rate of about 10^{19} erg/s, in the form of energetic particles and hot plasma, suggesting conversion efficiencies from the solar-wind energy input, though somewhat ill defined, of about 10^{-3}.

There is considerable uncertainty about the collective processes that could have produced transient, high-temperature, nonthermal events as a result of the nebula's general dissipative evolution. However, the foregoing comparison with a few of the more familiar systems, such as the Earth's magnetosphere and the Sun, suggests that the efficiencies of energy conversion that would be needed if the chondrule-formation mechanism were energized by the evolution of the nebula are within the range of similar efficiencies observed in cosmical objects.

Lightning. Whipple (1966) suggested that lightning discharges in the turbulent, differentially rotating nebula could produce localized transient heating that would melt dust accumulations to produce chondrules. This is an attractive idea, inasmuch as lightning is observed to accompany turbulent conditions in planetary atmospheres.

At least two basic questions are important in assessing the likelihood that lightning discharges could produce chondrules in the protoplanetary nebula. The first is whether macroscopic separations of electric charge could have occurred in the nebula that were sufficiently large to produce lightning discharges; the second is whether a sufficient amount of energy could be deposited by the lightning to account for chondrule formation. These are complicated questions; indeed, lightning is by no means fully understood even in the Earth's atmosphere where it has been intensively studied for many years.

Lightning occurs as a result of macroscopic electric-charge separations when the induced electric field is intense enough to produce electrical breakdown in the intervening medium. In order to understand the possible role of lightning in producing meteorite chondrules, it is necessary first to understand the mechanism by which large-scale electric-charge separation might occur in a protoplanetary nebula. In the Earth's atmosphere, lightning is produced in a variety of circumstances ranging from active, precipitating thunderstorms to dry dust devils and volcanic eruptions. In addition, static electrical potentials build up to high levels, resulting in electrical discharges, in a variety of industrial processes that involve the rapid motion of dielectric media. The common perception seems to be that any rapid and turbulent motions of particulate-laden gas will produce frictional charge separation, thus setting the stage for lightning. This is not obviously the case.

One of the major uncertainties in understanding the origin of terrestrial lightning remains the mechanism of charge separation. Various kinds of theories have been proposed, ranging from precipitation-induced charge separation to the convective separation of preexisting atmospheric charge distributions. Vertical charge transport by precipitation seems to provide the basic mechanism for generating large-scale atmospheric electricity (Iribarne and Cho 1980). In addition to causing spectacular localized and transient lightning discharges, this charge-transport mechanism is thought to provide the driving power that maintains a more or less steady background atmospheric electric field, corresponding to a persistent vertical space-charge distribution in the Earth's atmosphere. Altogether, while atmospheric convection may interact with the resulting background electric charge to produce secondary charge transport (and even lightning), it seems likely to me, as well as to others (see, e.g., Malan 1963), that the basic mechanism that sustains terrestrial atmospheric electricity and lightning is differential vertical charge transport on precipitating particles of condensed matter.

This question of the physics of electric-charge separation is of fundamental importance in understanding what conditions in the protoplanetary nebula could have produced lightning and how much energy might have been available to make chondrules from lightning. Although other possibilities could be imagined, assume that the charge separation in the nebula was produced as electrically charged dust grains or larger dust aggregations settled to the nebula's midplane. The dust settled, under the influence of gravity. The first question is whether gravity acting on the dust was strong enough to overcome the restoring electrical force, which acts against the charge separation, until the electric field exceeded that needed to cause a breakdown discharge. Assuming further that the settling dust was charged to an electrical potential of about 10 volts (as a result of photo-ejection of electrons, say, presuming the presence of sufficiently many energetic photons in the nebula), then it can be shown that the vertical component of nebular gravity was large enough to induce a breakdown discharge if the settling dust had already accumulated into at least mg-mass globs. Thus, it is reasonable to believe that lightning could have occurred in the nebula. It is also interesting that, to make lightning, the settling dust must already have gathered into accumulations with chondrule-like masses.

The total amount of energy available for lightning generation through this process is limited by the gravitational energy liberated as dust settled to the nebula midplane in the local vertical component of gravitational acceleration. If the local vertical gravity is dominated by the attraction of a central, 1 $M_\odot$ object, then by settling to the midplane from about 1 AU above it, the dust can, in principle, yield energy in the amount of about 5×10^{10} erg/g. This, in fact, is the same amount of energy as that needed to melt the dust to make chondrules, but only at a very high efficiency.

Realistically, we must expect that many orders of magnitude more energy would have to have been delivered in the form of lightning than would actually have been delivered to the chondrules. Therefore, extracting enough energy in the form of lightning from the nebula seems to require either repeated cycling of evaporation, condensation and subsequent precipitation, or the gravitational settling of orders of magnitude more solid matter than ultimately ended up in chondrules. It is the repetitive cycle of evaporation, condensation and precipitation that produces the repeated lightning discharges in the Earth's atmosphere, and that is thought to generate the background terrestrial atmospheric electricity. However, in the nebula, the energy of dust settling may have been available only once. There are possible ways to get around this problem. One might envision successive episodes of dust stirring and settling. Alternatively, in the vicinity of an evaporation-condensation boundary (the ice-water vapor boundary probably occurred near the asteroid formation region in the solar nebula), it is possible to imagine that water may have cycled through many repeated episodes of evaporation, condensation and precipitation.

Altogether, getting enough energy from lightning to produce chondrules is problematical but at least possible if many orders of magnitude more dust settled than was processed to chondrules or if dust settling occurred in repetitive cycles. However, the problem of electrification to produce lightning in the nebula remains substantially unsolved. Without more detailed discussion here, it is probably safe to say that the reasonable possibility of making chondrules from lightning has not been demonstrated thus far.

Magnetic Flares. The most prevalent cosmical processes of rapid, transient energy release seem to be associated with the phenomenon of magnetic flaring. This phenomenon is thought to occur in a variety of systems, ranging from planetary magnetospheres to stellar coronae, accretion disks and extragalactic radio sources, although even moderately detailed observations are available only for the Earth's magnetosphere and the Sun (Kennel et al. 1979; Sturrock et al. 1986).

Briefly, explosive magnetic flares occur in magnetic-field-dominated plasma systems in which fluid motion distorts the magnetic field, producing magnetic structures or topologies that are far from the equilibrium that the field would adopt in a vacuum if left to its own devices. Generally, the magnetic field is prevented from evolving toward the relaxed, vacuum configuration by the constraints imposed on its topology by the high electrical conductivity of its embedding fluid, which, to use a pictorial metaphor, prevents the magnetic lines of force from passing through one another.

For example, the solar magnetic field passes from the convection zone through the photosphere and into the corona. In the convection zone, the forces associated with the fluid's motion overwhelm the magnetic field stresses, and thus the evolution of the magnetic field is largely determined by the fluid motion pushing the field lines around. In much of the corona, on the other hand, the magnetic field stresses are far greater than the other forces acting on the fluid. Crudely speaking, the coronal magnetic field tries to evolve toward its lowest energy state, vacuum configuration that still matches the magnetic boundary condition at the photosphere, while the boundary condition itself evolves in response to subphotospheric fluid motions which drag the magnetic field lines. The free evolution of the coronal magnetic field is inhibited by the high electrical conductivity of the coronal gas, which does not allow the coronal magnetic field to diffuse through the fluid rapidly enough to keep up with the evolving conditions at the photosphere below. Consequently, the coronal field structure develops locally "tangled" regions, with the magnetic field lines hung up on one another and with large localized magnetic field gradients and intense electrical currents. At these singular places, the magnetohydrodynamic idealization of the fluid breaks down and the magnetic field structures can collapse explosively, leading to a rearrangement of the field-line topology and the rapid release of energy in the form of hot plasma and energetic particles. This processes seems to be responsible for solar activity, involving magnetic field intensities of several hundred gauss. An essentially similar process seems to be responsible for magnetospheric activity, involving magnetic fields having intensities of a few times 10^{-4} gauss.

Meteorite evidence indicates that the protoplanetary nebula carried a relatively strong magnetic field in the range of 1 to 10 gauss (see overview in Levy and Sonett 1978, Ch. 8.1), which might have been generated in the disk itself by a magnetohydrodynamic dynamo (Levy 1978), if the electrical conductivity had been raised to a high enough value by nonthermal processes (see, e.g., Consolmagno and Jokipii 1978). It is possible that such a magnetic field could have produced flares in a corona above the faces of the protoplanetary disk. A nebular magnetic field, with the intensity inferred from the meteorite paleomagnetic analyses, could produce flares in a tenuous disk corona with energy releases that could melt chondrules in the disk below.

A magnetic flare system delivers energy to the flare site at the Alfvén speed. The effectiveness of the flare process for liberating energy rapidly depends on a high magnetic field strength and a low ambient mass density, so that the local Alfvén speed is high enough to deliver magnetic energy rapidly to the flare. The need for low ambient mass density requires that the disk flares occur moderately high in the disk's corona, for much the same reason as solar flares occur in the Sun's corona. It is arguable, even given the meteorite evidence suggesting adequately strong magnetic fields in the disk, whether the disk's coronal field would be adequately strong at the required altitudes above the disk. The same process would not work with flares in the disk itself, inasmuch as the high mass density of the gas there precludes the release of energy at a *rate* high enough to melt silicates (cf. Sonett 1979).

The underlying source of energy for such disk-corona magnetic flares would be the gravitational energy released during the evolution of the disk itself, in the case that the magnetic field provided a major part of the stress and dissipation involved in the nebula's evolution. That basic energy source is easily sufficient to the problem.

9.4.3. DISCUSSION AND CONCLUSIONS

The existence of chondrules still poses a fundamental challenge to our understanding of the conditions under which primitive matter was processed and accumulated in the protoplanetary nebula. The fact that chondrules were processed through very high temperatures in large, short-lived excursions away from the thermodynamic conditions prevailing in the nebula, probably indicates that the nebula was "dynamically interesting" in that it exhibited unexpected transient dynamical phenomena. In this respect then, the nebula was not unlike most other complicated physical systems that we can scrutinize in detail. Physical systems with much energy to expend on their way toward thermal equilibrium frequently give rise to dynamically interesting dissipative structures with surprisingly high levels of organization (Prigogine 1980). Without observational provocation, we would not have predicted the host of similar phenomena that we confront in well-observed cosmical systems: solar and stellar coronae, solar flares, geomagnetic storms, terrestrial lightning storms, tornadoes and hurricanes, are more homely examples of such surprises. It is interesting—but, *a posteriori* not surprising—that we may confront similarly unexpected behavior in protoplanetary and protostellar nebulae.

The formation and evolution of the nebula liberated sufficient energy to account for chondrule formation many times over. However, the fact that the immediate chondrule-forming energy was liberated in transient, very-high-temperature events, forces a much closer look at the energetics and other details of any potential chondrule-forming process. On the basis of the cursory and critical examination that was given here to several major prevailing ideas about chondrule formation, it is our view that there is not yet a wholly satisfactory basis, one which falls easily within our prevailing understanding of the protoplanetary nebula and its environment, for understanding the formation of chondrules.

Other astronomical manifestations of star formation point to unexpected, energetic behavior in protostars, high-velocity outflows and Herbig-Haro objects being among the more obvious of these. Other manifestations in meteorites (high-temperature Ca- and Al-rich inclusions and possibly some isotopic anomalies, for example) also point to the occurrence of episodic and large deviations of some nebular environments from their local equilibrium states. Thus, the chondrules are not an isolated manifestation in need of an *ad hoc* explanation. Rather, the chondrules seem to provide another piece of evidence pointing toward the need for a deeper and more comprehensive picture of the complex evolutionary dynamics of cosmical disk systems, probably including episodic and dynamical processes that are not currently part of the standard picture. The existence of chondrules in meteorites and the other manifestations of transient energetic events, are among those provocative mysteries and conceptual gaps that will continue to provoke our understanding in this area and move it forward.

Acknowledgments. I am grateful to W. V. Boynton and T. D. Swindle for helpful comments. This work was supported in part by a grant from the National Aeronautics and Space Administration.

REFERENCES

Axford, W. I. 1964. Viscous interaction between the solar wind and the Earth's magnetosphere. *Planet. Space Sci.* 12:45–53.

Boynton, W. V., and Wark, D. A. 1987. Origin of CAI rims—I: The evidence from rare earth elements. *Lunar Planet. Sci.* XVIII:117–118 (abstract).

Caffee, M. W., Hohenberg, C. M., and Swindle, T. D. 1982. I-Xe ages of individual Bjurbole chondrules. *Proc. Lunar Planet. Sci. Conf.* 13, *J. Geophys. Res. Suppl.* 87:A303–A317.

Cameron, A. G. W. 1975. Clumping of interstellar grains during formation of the primitive solar nebula. *Icarus* 24:128–133.

Chapman, S. 1964. The energy of magnetic storms. *Geophys. J. Roy. Astron. Soc.* 8:514–536.

Consolmagno, G. J., and Jokipii, J. R. 1978. ^{26}Al and the partial ionization of the solar nebula. *Moon and Planets* 19:253–259.

Herzog, G. F., Anders, E., Alexander, E. C., Jr., Davis, P. K., and Lewis, R. S. 1973. Iodine-129/Xenon-129 age of magnetite from the Orgueil meteorite. *Science* 180:489–491.

Iribarne, J. V., and Cho, H.-R. 1980. *Atmospheric Physics* (Dordrecht: D. Reidel).

Kennel, C. F., Lanzerotti, L. J., and Parker, E. N. 1979. *Solar System Plasma Physics* (Amsterdam: North-Holland).

Kerridge, J. F., and Kieffer, S. W. 1977. A constraint on impact theories of chondrule origin. *Earth Planet. Sci. Lett.* 35:35–42.

Kieffer, S. W. 1976. Droplet chondrules. *Science* 189:333–339.

Levy, E. H. 1978. Magnetic field in the primitive solar nebula. *Nature* 276:481.

Levy, E. H., and Sonett, C. P. 1978. Meteorite magnetism and early solar-system magnetic fields. In *Protostars & Planets,* ed. T. Gehrels (Tucson: Univ. of Arizona Press), pp. 516–532.

Malan, D. J. 1963. *Physics of Lightning* (London: English Universities Press).

Morfill, G., Roser, W., Tscharnuter, W., and Völk, H. 1978. The dynamics of dust in a collapsing protostellar cloud and its possible role in planet formation. *Moon and Planets* 19:211–220.

Prigogine, I. 1980. *From Being to Becoming: Time and Complexity in the Physical Sciences* (San Francisco: W. H. Freeman).

Rambaldi, E. R. 1981. Relict grains in chondrules. *Nature* 293:558–561.

Sonett, C. P. 1979. On the origin of chondrules. *Geophys. Res. Lett.* 6:677–680.

Sturrock, P. A., Holzer, T. E., Mihalas, D. M., and Ulrich, R. K. 1986. *Physics of the Sun* (Dordrecht: D. Reidel).

Swindle, T. D., Caffee, M. W., Hohenberg, C. M., and Lindstrom, M. M. 1983. I-Xe studies of individual Allende chondrules. *Geochim. Cosmochim. Acta* 47:2157–2177.

Thiemens, M. H., and Heidenreich, J. E., III. 1983. The mass independent fractionation of oxygen: A novel isotope effect and its possible cosmochemical implications. *Science* 219:1073–1075.

Wark, D. A., and Boynton, W. V. 1987. Origin of CAI rims—II. The evidence from refractory metals, major elements and mineralogy. *Lunar Planet. Sci.* XVIII:1054–1055 (abstract).

Whipple, F. L. 1966. Chondrules: Suggestion concerning the origin. *Nature* 153:54–56.

Withbroe, G. L., and Noyes, R. W. 1977. Mass and energy flow in the solar chromosphere and corona. *Ann. Rev. Astron. Astrophys.* 15:363–387.

Wood, J. A. 1984. On the formation of meteoritic chondrules by aerodynamic drag heating in the solar nebula. *Earth Planet. Sci. Lett.* 70:11–26.

PART 10
Primitive Material Surviving in Chondrites

10.1. POTENTIAL SIGNIFICANCE OF PRISTINE MATERIAL

JOHN F. KERRIDGE
University of California at Los Angeles

In this chapter, we consider what sort of information can be derived about conditions and processes in the early solar system from the study of meteoritic components that appear to have survived as discrete entities from that epoch. We discuss the criteria that might be used to identify such entities and suggest that it is probably desirable for those criteria to be independent of the properties being studied. The following four chapters describe examples of different types of possibly primitive chondritic material that have been investigated, the term "primitive" here implying nebular origin, excluding presolar material.

10.1.1. POTENTIAL SIGNIFICANCE OF PRIMITIVE MATERIAL IN METEORITES

Two lines of evidence have pointed towards chondrites as being the most primitive meteorites, hence the best candidates to serve as probes of early solar-system processes. We have seen in Part 7 how whole-rock samples of chondrites exhibit elemental abundance patterns that seem to have been produced by nebular, as distinct from planetary, fractionation processes. Similarly, it is clear from Chapter 9.3 that chondrules were formed in the nebula and not on a planetary surface. We have also seen in both cases how detailed study of the meteoritic record can yield information about conditions in the nebula. However, it will have been apparent that the complexities in the different records are still far from being fully decyphered; we do not know, for example, what processes led to fractionation of metal from silicate, nor do we know either the nature or the pervasiveness of the energy source that melted the chondrules. Furthermore, chondrules, though obviously very old, nonetheless clearly represent material that was reprocessed in the nebula; just what the precursor material consisted of is not yet clear, though it must have been very primitive. Pending resolution of such issues, the meteoritic record based on bulk elemental composition and chondrule petrology affords us a perception of the nebula 4.56 Gyr ago that is blurred and incomplete. What more informative lines of evidence into the nebular environment might be potentially available?

One obvious means of probing the early solar nebula is by identifying within a meteorite a mineral grain, or aggregate of grains, that was actually formed in the nebula. Because, for many minerals, their stability fields in terms of temperature, pressure, O fugacity, etc., are well known, knowledge that a specific mineral had a nebular origin enables us, in principle, to define those parameters for a particular time and place in the nebula, provided the appropriate age and formation location are known. The caveat is important: we seldom, if ever, know the time and place of origin of a meteoritic mineral grain or aggregate in any other than broad terms. In addition, because of the prevalence of secondary processes in the history even of apparently primitive chondrites, a nebular origin cannot simply be assumed for individual constituents of such meteorites. It follows that a nebular origin needs to be demonstrated in each case, and there is little agreement in the community as to what constitutes an adequate demonstration. In cases where the objective is to test a model, such as a condensation theory, there is a natural tendency towards circularity of argument: a single observation is employed both as the proof of nebular origin and as the confirmation of theory. It is this editor's opinion that the criterion for primitiveness of material should be independent of the ultimate use to which that material is put, but contributions to this book reflect a variety of approaches to this issue.

Primitive material surviving in meteorites may be used to study several different aspects of nebular history. (In this context, "primitive" means that the material acquired its present form, i.e. chemical composition and/or crystal structure, in the nebula. It does not include presolar material that survived entry into the protosolar system: such material is discussed in Chapters 13.1 and 13.2.) One of the most commonly studied topics concerns the question of whether the nebula was ever wholly vaporized and, if so, how subsequent cooling and condensation proceeded. This topic has already been addressed in Part 7 where whole-rock elemental fractionation patterns were compared with the predictions of equilibrium condensation theory. In a similar fashion, a fairly well-defined sequence of minerals, formed either by direct condensation or by back reaction between earlier-formed condensate and nebular gas, can be predicted by such theory. Those predictions can then be tested using observations such as the crystallization sequence in a mineral aggregate, the ubiquitous rimming of one mineral by another or the minor- and trace-element content of an individual mineral grain.

Many of the individual constituents of chondrites have been postulated to have had an independent existence in the nebula prior to accumulation of the chondrite parent asteroids. In the following four chapters we assess the support for such postulated origins and the information about the nebula that may thereby be obtained. It must be emphasized, however, that the individual entities described in these chapters are best regarded as candidates for identification as primitive material. Which of these, if any, actually qualify for such a title is a matter of judgment.

10.2. PRIMITIVE MATERIAL SURVIVING IN CHONDRITES: MATRIX

E. R. D. SCOTT
University of New Mexico

D. J. BARBER
University of Essex

C. M. ALEXANDER
Open University

R. HUTCHISON
British Museum (Natural History)

and

J. A. PECK
Harvard University

A logical place to search for surviving pristine nebular material is in the fine-grained matrices of ordinary and carbonaceous chondrites of petrographic type 3. Unfortunately, many of these chondrites have experienced brecciation, thermal metamorphism and aqueous alteration so that interpreting individual features in terms of specific nebular conditions and/or processes is difficult. It follows that the origin and evolutionary history of such matrix phases are controversial and a consensus is difficult to define. In this chapter, therefore, after summarizing the salient mineralogical, petrographic, chemical and isotopic features of matrix in apparently primitive chondrites, we shall attempt to pro-

vide an overview both of areas of agreement and of topics that are currently in dispute.

10.2.1. INTRODUCTION

The matrix in chondrites of petrographic type 3 is an opaque fine-grained mixture of mainly anhydrous ferromagnesian silicates with lesser amounts of hydrous silicates, oxides, sulfides and carbonaceous and other phases. Matrix is among the most abundant components of chondrites (Table 10.2.1; cf. Table 9.1.1). Mean grain sizes in different chondrite groups vary from 0.1 to 10 μm being larger in CV3 matrix than in that of ordinary chondrites. Unfortunately, there is no precise, accepted definition of what constitutes chondrite matrix. There is a traditional view (see, e.g., Van Schmus 1969; Dodd 1981) that "matrix" embraces everything external to optically defined chondrules and inclusions, and this continues to be used in some recent papers (see, e.g., Allen et al. 1980; King and King 1981). However, it is important to note that many type-3 chondrites have silicate chondrules and clasts and metal/sulfide objects rimmed with fine-grained opaque material that is included in this broad definition of matrix. Some workers (see, e.g., Huss et al. 1981; Nagahara 1984; Scott et al. 1984) take the view that, within the nonrim materials between chondrules that are covered by this all-embracing definition, there exists in most ordinary chondrite (OC) matrices an opaque, very fine-grained component (< 5 μm in size) that is distinct in nature from the coarser-grained fragments; these researchers apply the term matrix only to this ultrafine component. However, there is evidence that this size distinction is not applicable to all OCs; for example, in Bishunpur (LL3) there is a continuous distribution of grain sizes from 200 μm down to 1 μm in diameter (Alexander 1987). Yet another definition of matrix is that of McSween (1979*b*), who excludes optically identifiable Fe-Ni metal and sulfides and magnetite.

The differences between these definitions are not trivial but, because our purpose is to try to use matrix as a probe of the early solar system, we can afford to adopt a relatively loose definition: matrix in type-3 ordinary and carbonaceous chondrites is taken as the fine-grained, predominantly silicate, material interstitial to macroscopic, whole or fragmented, entities such as chondrules, inclusions and large isolated mineral (i.e., silicate, metal, sulfide and oxide) grains (see Chapter 10.4). (Note that the term matrix is also used for the fragmental material between coarser rock fragments in chondritic or achondritic regolith breccias; see Chapter 3.5.)

Chondrite matrices can contain materials derived from a wide variety of sources including presolar grains, solar-nebular condensates and finely ground lithic material such as chondrules and inclusions. However, unravelling the origins of specific matrix components is complicated because chondrites are not unaltered aggregates of solar-nebular products that were simply lithified during accretion: they are rocks that were lithified after accretion in asteroids that experienced significant heating, alteration and impact processing. Because of their fine-grained nature, matrix materials were easily modified in their parent asteroids, and are difficult to characterize in the laboratory. Recent analyses by electron microscopy and electron-microprobe techniques (see, e.g., Ashworth 1977,1980,1981; Huss 1979; Huss et al. 1981; Housley and Cirlin 1983; Nagahara 1984; Scott et al. 1984; Alexander 1987) have revealed an extraordinary diversity of matrix materials; no two chondrites appear to have identical matrix material, and the material in one kind of chondrite cannot have formed directly from that in another. A feature common to many type-3 carbonaceous and ordinary chondrites is the presence of abundant FeO-rich olivine or its alteration products. The least-metamorphosed and the most-metamorphosed type-3 ordinary chondrites tend to have more magnesian olivine. The highly reduced EH chondrites, and the unique type-3 chondrite Kakangari, also lack FeO-rich olivine. Nevertheless, the overall similarities in matrix mineralogy suggest that the same processes produced the matrix material of each type of chondrite. Likewise, the similarity in mineralogy and petrology of chondrules in CM, CO, CV, EH, H, L and LL chondrites suggest that the chondrules in each group have similar origins (Scott and Taylor 1983).

10.2.2. MATRIX MINERALOGY AND PETROLOGY

Matrices of type-3 carbonaceous and ordinary chondrites are porous aggregates of fine-grained minerals. In most chondrite groups, those minerals are commonly olivine, high- and low-Ca pyroxene, sulfides, metallic Fe-Ni, feldspars and/or feldspathoids, and carbonaceous material (see, e.g., Wark 1979; McSween and Richardson 1977; Huss et al. 1981; Housley and Cirlin 1983). Phyllosilicates, magnetite and carbonates are present in CI and CM2 matrices and have been identified in several type-3 CV and ordinary chondrites. CI and CM2 matrix also contains tochilinite (see Chapter 3.4) and macromolecular organic material (see Chapter 10.5). Table 10.2.1 summarizes the main mineralogical features of matrix in various kinds of chondrites, including results of some limited studies of Kakangari and types-4 and -5 carbonaceous chondrites; these matrices are unique in that matrix silicates have compositions close to those of chondrule silicates. Not listed in Table 10.2.1 are minor occurrences of other minerals, such as phosphates, spinel, chromite and C-bearing minerals.

Chondrite matrices are not homogeneous mixtures. Texturally discrete lumps of matrix are present in many chondrites. Matrix material may form compositionally and texturally distinct rims on chondrules, aggregates, inclusions and metallic Fe-Ni grains (see, e.g., Bunch and Chang 1980; Allen et al. 1980; Scott et al. 1984; MacPherson et al. 1985). In ordinary chondrites, the rims are typically 10 to 30 μm wide and are finer grained than other matrix occurrences; e.g., chondrule rims in Chainpur (LL3) have mean grain

TABLE 10.2.1
Mineralogy and Abundance of Matrix in Chondrites

Chondrite	Matrix (vol%)	References[a]	Minerals	References[a]
CI1	>95	1	phyllosilicates (serpentine), magnetite, dolomite, pyrrhotite, sulfates	7, 1
CM2	55–85	2	phyllosilicates (serpentine), tochilinite, calcite, aragonite, magnetite, epsomite, pentlandite, pyrrhotite	7
CO3	30–40	2	olivine (Fa 30–60), phyllosilicates	8, 1
CV3	35–50	2, 3	olivine (Fa 40–60), high-Ca pyroxene (Fs 10–50, Wo 45–50), nepheline, sodalite, pentlandite, troilite, magnetite, phyllosilicates	9
C4–5	50–80	4	olivine (Fa 30–40), plagioclase (An 20–90) high-Ca and low-Ca pyroxene, magnetite, pentlandite, pyrrhotite	4, 10
EH3	< 2		not characterized	
Kakangari[b]	~ 50	5	enstatite (Fs 1–10), troilite, Fe,Ni	11
H3,L3,LL3	5–15	6	olivine (Fa 20–70), low-Ca pyroxene (Fs 1–20), glass, troilite, Fe,Ni, magnetite	6, 12

[a] References: 1: Dodd 1981; 2: McSween 1979a; 3: McSween 1977; 4: Scott and Taylor 1985; 5: Mason and Wiik 1966; 6: Huss et al. 1981; 7: Barber 1985; 8: Christophe Michel-Lévy 1969; 9: this Chapter; 10: Brearley et al. 1987b; 11: Graham and Hutchison 1974; 12: Nagahara 1984.
[b] Unique type-3 chondrite.

sizes of 0.06 μm (Ashworth 1977; see also Chapter 9.1). Analogous features in CM2 and CV3 chondrites have been described as "halos" by Bunch and Chang (1980). CV3 chondrites also contain some dark inclusions that consist largely of matrix material with a few chondrules and inclusions, and chondritic clasts rimmed by matrix material (see, e.g., Fig. 1 of Fruland et al. 1978; Fig. 25 of Clarke et al. 1970; Bunch and Chang 1983; Kracher et al. 1985). Type-3 OCs tend to have smaller matrix lumps and matrix-rich chondritic clasts (Scott et al. 1984).

Detailed petrographic studies have been made of matrix material in type-3 OC and CV chondrites. The matrices of CI and CM2 chondrites are so heavily affected by aqueous alteration (Chapter 3.4) that studies of them will not be reviewed in this chapter. The matrix of EH chondrites is either very low in abundance or nonexistent; no data are published.

Ordinary Chondrites

Ashworth (1977) distinguished two forms of opaque, fine-grained material in type-3 ordinary chondrites: the *dark rim material* around chondrules and the more porous interchondrule matrix that he termed *clastic*. Figure 10.2.1 shows some examples of matrix textures found in H-group chondrites by Ashworth (1981). Scott et al. (1982) described *translucent glassy-looking matrix* in several chondrites as being lower in opaque minerals and having lower FeO/MgO ratios than opaque matrix. Huss et al. (1981) defined matrix that had been thermally metamorphosed as another type called *recrystallized matrix*.

Alexander (1987) examined the clastic texture of the submicrometer-sized fraction of OC matrix by means of analytical electron microscopy. In Bishunpur (LL3), the clastic material is cemented within a partially amorphous groundmass that has a composition similar to that of chondrule mesostases. However, in Chainpur, for example, the groundmass is very fine grained (<0.1 µm), densely packed olivine (Fa_{50}). This is essentially the same nonclastic material described by Ashworth (1977). Alexander also

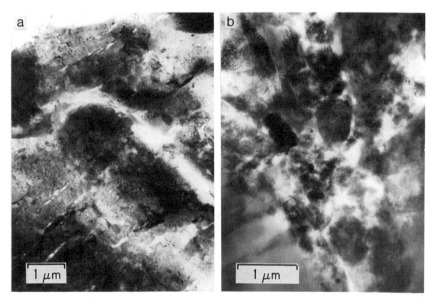

Fig. 10.2.1. Transmission electron micrographs, taken at 1 MeV, of matrix material in H-group chondrites. (a) *White* matrix in Tieschitz, consisting of coarse nepheline with inclusions and cracks; (b) deformationally compacted matrix between two relatively large grains, bottom left and bottom right, in Bremervorde (H3) (figure from Ashworth 1981).

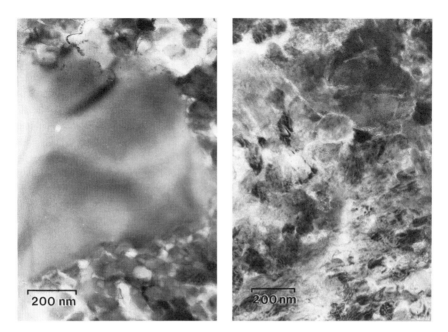

Fig. 10.2.2. Transmission electron micrographs, taken at 200 keV, of fine-grained materials in type-3 ordinary chondrites. (a) Olivine grain with overgrowths of olivine groundmass in the matrix of Tieschitz; (b) olivine groundmass comprising rim material in Krymka (figure from Alexander 1987).

showed that both clastic and nonclastic materials are present in rims. In those cases where the nonclastic component is crystalline, it is dominated by olivines with grain sizes < 0.2 μm, which are similar in composition and microstructure to the matrix groundmass in the given meteorite. The clastic component, which generally has grain sizes > 0.2 μm, comprises olivine and twinned monoclinic low-Ca pyroxene. In many instances the clastic olivines have overgrowths with textures similar to groundmass (Fig. 10.2.2).

The most abundant silicates in the matrix of type-3 OCs are Fe-rich olivine (Fa_{30-60}), magnesian low-Ca pyroxene and very fine-grained amorphous or uncharacterized FeO-rich assemblages (the "fluffy" particles of Nagahara [1984]). Variable amounts of albitic feldspar, calcic pyroxene and FeO-poor olivine are also present. Hutchison et al. (1987) showed that some of the fine-grained FeO-rich silicates in Semarkona (LL3.0) are composed of smectite, a phyllosilicate. Huss et al. (1981) found that the feldspar grows during recrystallization of matrix, but Nagahara (1984) reported that some albite grains have textures inconsistent with such an origin and that some albite is present in the least-metamorphosed chondrites.

Olivine compositions in matrix changed during metamorphism within

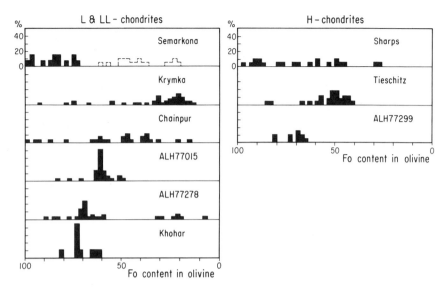

Fig. 10.2.3. Chemical compositions of olivine in fine-grained matrices. The size of the grains is mostly < 5 μm. Dotted portions of the Semarkona diagram show the Mg/(Mg + Fe) ratios of Fe- and Mg-rich fluffy particles (figure from Nagahara 1984).

the type-3 ordinary chondrites (see Sec. 10.2.5). This is shown in Fig. 10.2.3, taken from Nagahara (1984); similar results were found by Huss et al. (1981). In L/LL3 chondrites, average olivine is quite fayalitic (Fa_{70-85}) in Krymka (LL3.1), and decreases smoothly towards equilibrated L- and LL-chondrite values as metamorphic effects increase up to Khohar (L3.6). The exception to this trend is Semarkona, which has low-FeO olivine and two unusually abundant oxidized-Fe-bearing phases, smectite and magnetite. A similar pattern may be present in the H3 chondrites, as Sharps (H3.4) has considerable amounts of low-FeO olivine, Tieschitz (H3.6) averages near Fa_{50}, and Allan Hills A77299 (H3.7) has less fayalitic olivine; however, there are no known H3.0–H3.3 chondrites, which would be needed to make a thorough comparison.

Wood (1962) showed that the matrix of Tieschitz contains the nonsilicate minerals troilite, Fe-Ni metal and magnetite. Later studies have shown that metal and troilite are ubiquitous in OC matrix. These occur as intergrown grains with sizes similar to those of the silicates (Nagahara 1984). Huss et al. (1981) found magnetite in the opaque, i.e. fine-grained, matrices of several chondrites (e.g., Semarkona and Ngawi [LL3]). Nagahara (1984) found that isolated euhedral-to-irregular magnetite grains are present in the matrices of most L/LL3 chondrites, especially Semarkona, but did not find magnetite in H3 matrices. She also described magnetite occurring as an alteration product of metal grains in matrix. Chromite and apatite are mainly present in recrystallized matrix (Huss et al. 1981). Semarkona also contains a variety of additional nonsilicate minerals, including spinel, pentlandite, pyrrhotite, carbides, Ni-poor maghemite and calcite (Huss et al. 1981; Nagahara 1984; Hutchison et al. 1987).

Studies on Sharps (H3) (Fredriksson et al. 1969) and Tieschitz (Wood 1962; Kurat 1969; Christophe Michel-Lévy 1976) showed that their matrices contain more Fe^{2+} and C than other silicate portions of the meteorites. Tieschitz also contains a unique kind of chondritic matrix material known as *white matrix*. This material, which is rich in Al_2O_3, SiO_2 and alkalies, is located between the normal opaque matrix rims on chondrules (Christophe Michel-Lévy 1976; Hutchison et al. 1979; Wlotzka 1983). Another rare matrix component is the C-rich matrix of some type-3 chondritic clasts that is also found in OC regolith breccias. These clasts have no silicate matrix material; instead chondrules are separated by poorly graphitized C containing significant amounts of Fe both in solid solution and as minute metal grains (Scott et al. 1981; Brearley et al. 1987a).

Clasts of pyroxene and olivine were identified as major matrix components in OCs by Ashworth (1977, 1981). He showed that the low-Ca pyroxene almost invariably had a twinned monoclinic structure, also characteristic of chondrule pyroxene, that probably results from inversion of protopyroxene (Binns 1970; Iijima and Buseck 1975; Ashworth 1980). Matrix low-Ca pyroxenes also fall within the same compositional range (Fujimaki et al. 1981) as chondrule pyroxenes (Rubin 1986).

Lumps of matrix are sometimes observed included within chondrules (see, e.g., Kurat 1970; Scott et al. 1984); some examples are illustrated in Fig. 10.2.4. Their mineralogy is like that of "normal" matrix, although mineral proportions vary widely, leading to large variations in chemical composition (see Sec. 10.2.3).

CV3 Chondrites

Matrix in CV3 chondrites has an average grain size of about 5 μm, an average porosity of about 10% and, in general, a random fabric. It is a complex assemblage that varies from chondrite to chondrite in the identity and relative proportions of its components but in each case includes, in order of decreasing abundance, olivine, pyroxene, nepheline and sodalite. Some CV3 matrices also contain andradite, awaruite (or high-Ni taenite), magnetite, pentlandite, troilite and phyllosilicates. Backscattered scanning-electron images of typical matrix areas in five chondrites are shown in Fig. 10.2.5. (These chondrites were chosen for study because they appeared petrographically to be minimally altered; see Sec. 10.2.5 and Chapter 3.4.) In each micrograph, the two dominant phases, olivine (ol) and clinopyroxene (cpx), have been identified.

Olivine is the dominant matrix phase in each of the five chondrites considered here. It has an average composition of around Fa_{50} in each case, but

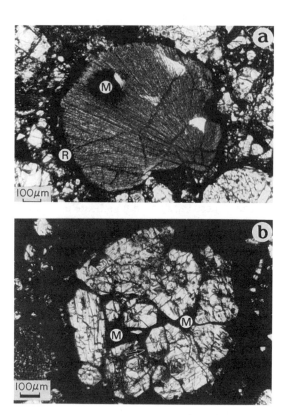

Fig. 10.2.4. Transmitted-light micrographs of chondrules containing opaque matrix material. (a) Radial pyroxene chondrule in Ngawi. The matrix material forms a rounded lump (M) 200 μm in diameter and an irregular rim (R); (b) porphyritic pyroxene chondrule in Tieschitz. The matrix material inside the chondrule (M) occupies most of the space between low-Ca pyroxene phenocrysts; there is also a small amount of translucent mesostasis, some of which is marked with an arrow. Compositions of the matrix material are shown in Fig. 10.2.8; (a) and (b) are labeled chondrules 2 and 6, respectively (figure from Scott et al. 1984).

its compositional range, i.e., its apparent degree of equilibration, varies significantly. Mean olivine compositions are given in Table 10.2.2. Homogeneity based on Fa content decreases in the order Allende, Grosnaja, Vigarano, Mokoia and Kaba (Fig. 10.2.6). Olivine has two dominant forms: subhedral plates with dimensions of about 15×2 μm; and anhedral chips < 3 μm in size. Acicular or spindle-shaped olivine has also been reported as a major constituent (Green et al. 1971; MacPherson et al. 1985; Bunch et al. 1986). The relative proportions of the different forms vary among the chondrites, but there is no detectable systematic difference in the compositions of the different forms. Plates are not found intergrown, and neither the plates nor the

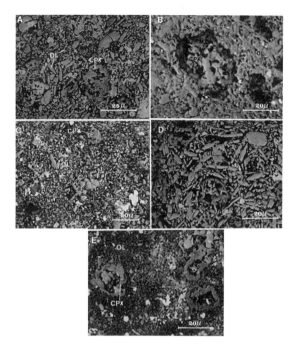

Fig. 10.2.5. Backscattered scanning electron micrographs of representative areas of matrix in (A) Allende, (B) Grosnaja (note blurring of grain boundaries resulting from shock metamorphism), (C) Vigarano, (D) Mokoia and (E) Kaba.

TABLE 10.2.2
Mean Compositions of CV3 Matrix Olivine, in wt.% Oxides[a]

	SiO_2	TiO_2	Al_2O_3	Cr_2O_3	FeO	MnO	NiO	MgO	CaO	Total
1	34.08	0.08	0.29	0.14	39.68	0.27	0.21	24.49	0.20	99.95
2	34.67	0.03	0.30	0.15	40.02	0.26	N/M	23.97	0.16	99.56
3	35.15	0.02	0.80	0.14	40.65	0.30	0.13	22.56	0.33	100.08
4	35.83	0.06	0.27	0.16	35.89	0.33	0.08	27.90	0.10	100.77
5	36.35	0.02	0.67	0.11	35.27	0.38	0.22	27.58	0.21	100.81

[a] 1: Allende; 2: Grosnaja; 3: Vigarano; 4: Mokoia; 5: Kaba. N/M = not measured.

chips exhibit chemical zoning. Some olivine grains contain sub-μm inclusions of opaque minerals, such as magnetite and pentlandite, and of Al-bearing glass, that appear as dark spots in backscattered-electron images.

Clinopyroxene is the second most abundant phase in CV3 matrix material. Its composition in each chondrite covers the range from hedenbergite to diopside; mean compositions are given in Table 10.2.3. Its most common occurrence is as 25 μm-sized porous grains that are ring shaped in cross sec-

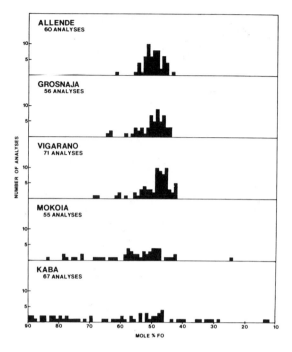

Fig. 10.2.6. Results of electron-microprobe analyses of CV3 matrix olivines. The histograms display the relative degree of olivine equilibration in each matrix. The range of matrix olivine compositions increases, and the degree of chemical equilibration decreases, steadily from Allende to Kaba.

TABLE 10.2.3
Mean Composition of CV3 Matrix Clinopyroxene in wt.% of Oxides[a]

	SiO_2	TiO_2	Al_2O_3	Cr_2O_3	FeO	MnO	MgO	CaO	NaO	Total
1	50.81	0.03	1.05	0.07	13.78	0.23	10.22	23.37	0.12	99.69
2	49.53	0.02	0.77	0.16	19.47	0.25	6.56	22.65	N/M	99.41
3	49.76	0.08	1.43	0.12	16.44	0.16	8.82	22.49	0.09	99.39
4	49.92	0.07	0.61	0.08	17.11	0.31	7.80	23.21	0.10	99.21
5	49.06	0.03	0.64	0.09	20.78	0.27	5.27	22.61	0.08	99.28

[a] 1: Allende; 2: Grosnaja; 3: Vigarano; 4: Mokoia; 5: Kaba. N/M = not measured.

tion and concentrically zoned around a large central hole. There are two types of zoning profiles, measured from the grain interiors to the surfaces: type A, diopside (Fe-poor) to hedenbergite (Fe-rich) and type B, hedenbergite to diopside. In many cases subhedral blocky grains of andradite occur embedded in type A hedenbergite. Also subhedral grains of magnetite and troilite in some cases occur nested in the inner pores or embedded in the walls of both

TABLE 10.2.4
Representative Analyses of Opaque Phases in CV3 Matrix in Wt. % of Elements[a]

	S	Cu	Co	Ni	Fe	Total
1	0.14	0.16	1.92	64.78	30.06	97.06
2	0.01	0.34	1.98	44.75	50.58	97.66
4	0.01	0.00	0.02	0.04	65.64	65.70
7	33.24	0.00	1.04	21.90	41.84	98.04
13	35.53	0.03	0.06	0.27	60.91	96.79

[a] 1: Allende Awaruite; 2: Vigarano Taenite; 4: Mokoia Magnetite; 7: Grosnaja Pentlandite; 13: Kaba Troilite.

types of clinopyroxene grains. The walls themselves are highly porous, giving the grains an overall web-like appearance.

Representative compositions of the five opaque minerals commonly found in CV3 matrix are given in Table 10.2.4. Two or three of these minerals (but in no observed case all five) are found in each matrix. The relative abundances of the different minerals vary widely from chondrite to chondrite. Each mineral occurs in < 7 μm-sized grains that exhibit a variety of forms. The surfaces of some grains are pitted, suggesting that they have been subjected to corrosion. Opaque minerals occur both as isolated grains and as intergrown clumps containing up to several hundred grains.

Sodalite and nepheline occur in all five meteorites as anhedral grains, 2 to 5 μm in size. The two minerals do not appear to be intergrown and one does not seem to be the alteration product of the other as in many cases on Earth. The feldspathoids that occur in rims around chondrules and inclusions poikilitically enclose a number of small, anhedral olivine chips. Such rims or halos are enriched in C and S with respect to the bulk matrix, with C apparently associated with olivine grains (Bunch and Chang 1980). Similar halos surround chondrules and inclusions in CM chondrites.

Phyllosilicates occur in low abundance in CV3 chondrites, generally as anhedral, porous grains about 5 μm in size. They occur as discrete grains, sometimes including sub-μm blebs of olivine, but not as rims on large olivine grains. Phyllosilicates in CV3 chondrites are not only petrographically distinct from those in CI and CM chondrites (Chapter 3.4) but also chemically distinct, being enriched in Ca, Al, Cr and the alkalies (Bunch and Chang 1980).

Along with chondrules, CAIs and other mineralogically and texturally distinct inclusions, CV 3 matrix contains a huge variety of fine-grained lithic inclusions which themselves consist primarily of some or all of the components of the matrix proper. The proportions of these components vary considerably among different matrix inclusions. *Dark inclusions* exhibit considerable variety in size and texture, ranging from aphanitic lumps devoid of CAIs and refractory and mafic inclusions, to miniature xenolithic chondrites. However, most or all of these inclusions are genetically related to the CV3 host material (Fruland et al. 1978). At least in Allende, some dark inclusions are substantially enriched in C relative to the bulk meteorite (Heymann et al. 1987).

CO3 Chondrites

Matrix constitutes about 35 vol.% of a CO3 chondrite. It is highly olivine normative (Rubin and Wasson 1988), but detailed petrographic data are lacking. Rubin et al. (1985) observed FeO-rich olivine, metallic Fe-Ni and troilite in the matrix of Colony; pyroxene is probably also present.

10.2.3. CHEMICAL COMPOSITION OF MATRIX

Chondrite matrices have been analyzed *in situ* by broad-beam electron microprobe, and a limited number of samples have been analyzed by neutron-activation and wet-chemical analysis (Table 10.2.5).

Ordinary Chondrites

Examples of electron-microprobe analyses of OC matrix are those by Ikeda (1980), Huss et al. (1981), Matsunami (1984), Scott and Taylor (1983), Nagahara (1984), Scott et al. (1984) and Hutchison et al. (1987). Note that the definition of matrix has not remained constant among these different studies.

Matrices in type-3 OCs have FeO/(FeO + MgO) ratios that decrease with increasing degree of metamorphism, i.e., petrographic grade (see Chapter 3.3), as matrix mineral compositions approach those of chondrules (see,

TABLE 10.2.5
Chemical Analyses of Chondrite Matrices

Chondrites	Technique[a]	Reference
CV3 (Allende)	chemical	Clarke et al. 1970
CM2 (Murchison)	probe	Fuchs et al. 1973
CI, CM, CR, CO, CV (32)	probe	McSween and Richardson 1977
H-L-LL3 (14)	probe	Huss et al. 1981
H-L-LL3, CO3 (5)	probe	Ikeda et al. 1981
H3 (ALHA77299)[b]	INAA	Taylor et al. 1984
H-L-LL3, CV3 (9)	probe	Scott et al. 1984
LL3 (Semarkona)	INAA	Grossman 1985
CV3 (Allende)	INAA	Rubin and Wasson 1987
CO3 (Ornans)	INAA	Rubin and Wasson 1988
CV3 (5)	probe	this Chapter

[a] Probe, broad-beam electron microprobe analysis; chemical, wet chemical analyses; INAA, instrumental neutron activation analysis.
[b] One matrix lump.

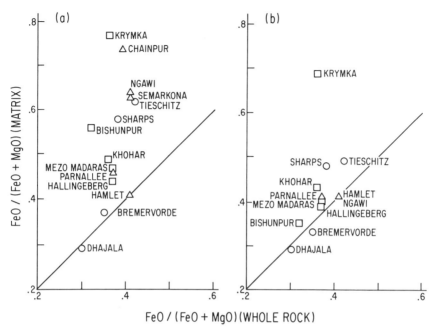

Fig. 10.2.7. FeO/(FeO + MgO) ratios plotted for (a) opaque matrix, and (b) recrystallized matrix in type-3 ordinary chondrites vs the same ratios for the respective whole-rock data. Data and definitions of matrix are from Huss et al. (1981); whole-rock data are from the literature.

e.g., Huss et al. 1981). One manifestation of this effect is a systematic difference in FeO/(FeO + MgO) between opaque matrix and recrystallized matrix (Fig. 10.2.7) (Huss et al. 1981). Scott et al. (1984) showed that different occurrences of matrix (e.g., lumps and rims) were chemically inhomogeneous and varied widely in composition within a single chondrite as well as between different type-3 OCs (Fig. 10.2.8). They attributed these variations to variable proportions of FeO-rich olivine and a mixture of Ca-, K- and Na-bearing feldspathic minerals, low-Ca pyroxene and less-fayalitic olivine. The existence of material enriched in Na, K and other volatiles, as well as rare-earth elements, has been demonstrated by Rambaldi et al. (1981) using ultrafine-grained fractions obtained by sieving OCs of types 3, 4 and 5. It is unclear, however, how this material is related to matrix material.

Scott et al. (1984) showed that mean concentrations of Mg, Na, Al and Ca in individual matrix occurrences showed fivefold variations within a single chondrite. There was no evidence of a correlation between the compositions of fine-grained rims and their enclosed chondrules, nor any systematic compositional differences between rims and other matrix occurrences (see also Huss 1979; Allen et al. 1980; Ikeda et al. 1981). Wilkening et al. (1984) and Wilkening and Hill (1985) found that rims are moderately enriched in sider-

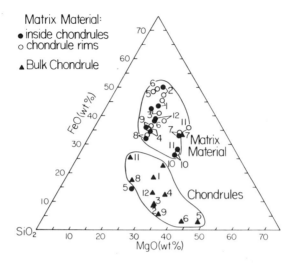

Fig. 10.2.8. Compositions of lumps of matrix material inside 12 chondrules plotted with compositions of the same chondrules and their rims. Lumps inside chondrules generally have similar compositions to rims; chondrules 5, 6 and 11 are exceptions. Both rims and lumps are more FeO-rich than bulk chondrules, except for chondrule 7, in which all three values are similar. Chondrule assignments: 1, Inman (LL3); 2 and 12, Ngawi; 3–5, Piancaldoli; 6 and 7, Tieschitz; 8 and 9, Semarkona; 10 and 11, Hallingeberg (L3) (figure from Scott et al. 1984).

ophile and several moderately volatile elements with respect to chondrule interiors and that rim compositions are therefore basically similar to those of interchondrule matrix. Similarly, Alexander et al. (1985) showed that the mean chemical composition of rims in Tieschitz corresponds closely with that found by Huss et al. (1981) for "opaque matrix."

Neutron-activation analyses of matrix samples are too few to permit generalizations about the trace-element chemistry of matrix. A single, mm-sized matrix lump from an H3 chondrite showed elemental abundances of 19 lithophile and siderophile elements within 20% of CI values (Taylor et al. 1984). The composition of this lump, which contains 20 vol.% of magnesian olivine fragments $>10\,\mu$m in size set in typical matrix material, is shown in Fig. 10.2.9, together with the average composition of 70 chondrules from OCs. Except for siderophiles, which are grossly depleted in chondrules, element/Mg ratios for the matrix lump and chondrules appear similar, though element/Si ratios would probably not be the same for matrix and chondrules because matrix tends to have lower Mg/Si values (see, e.g., Huss et al. 1981; Grossman and Wasson 1983).

In Semarkona, Grossman (1985) found that concentrations of nonvolatile lithophiles (i.e., Al, Ca, Sc, REE, Mg, Cr and Mn) were lower in matrix than in chondrules by a factor of about 2. Semarkona matrix was enriched in Na and K relative to chondrules by a factor of about 1.5.

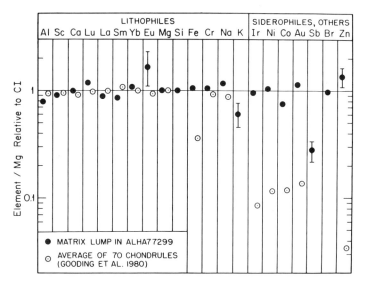

Fig. 10.2.9. Elemental abundances, normalized to Mg and CI chondrites, in a lump of matrix material from an H3 chondrite (Allan Hills A77299) determined by neutron-activation analysis. With rare exceptions, the matrix sample is unfractionated relative to CI chondrites and resembles the mean composition of chondrules in type-3 OCs, except that siderophiles are depleted in chondrules (figure from Taylor et al. 1984).

CV3 Chondrites

Mean matrix compositions for five representative meteorites are given in Table 10.2.6. Rubin and Wasson (1987) analyzed 3 mg of Allende matrix by neutron-activation analysis and found refractory-lithophile contents, relative to Mg, in the range 0.85 to 1.3 times the CI values (0.54 to 0.86 × bulk CV values). Contents of Na and K were 0.3 to 0.4 times the CI values (0.5 to 0.8 × bulk CV values).

CO3 Chondrites

Using neutron-activation analysis, Rubin and Wasson (1988) analyzed 7.5 mg of CO3 matrix obtained by sieving a disaggregated sample of Ornans from which the chondrules had been removed. They found it to have moderately high abundances, normalized to Mg, of siderophiles and chalcophiles (1.2 to 1.5 × bulk CO values) and lithophiles within 25% of bulk values.

Thus, available neutron-activation analytical data for different chondrite groups suggest that matrix and whole-chondrite compositions are surprisingly similar for a wide variety of elements.

TABLE 10.2.6
Mean Compositions of Bulk CV3 Matrix, in Wt. % Oxides[a]

	Na_2O	MgO	Al_2O_3	SiO_2	Cr_2O_3	MnO	FeO	NiO	SO_3	Cl	K_2O	CaO	TiO_2	Total
1	0.30	17.83	2.35	29.98	0.37	0.20	33.61	1.46	2.70	0.08	0.04	2.35	0.06	91.52
2	0.41	16.51	2.85	30.76	0.38	0.25	36.39	1.46	2.39	0.08	0.03	3.28	0.08	94.87
3	0.36	16.17	3.47	27.93	0.39	0.22	36.96	1.46	0.80	0.09	0.06	1.69	0.09	88.69
4	0.53	18.11	2.09	28.82	0.40	0.16	32.30	1.36	2.79	0.05	0.08	1.87	0.03	89.22
5	0.78	17.99	3.32	27.74	0.37	0.23	28.99	1.02	2.18	0.18	0.19	2.17	0.07	85.23

[a] 1: Allende; 2: Grosnaja; 3: Vigarano; 4: Mokoia; 5: Kaba.

10.2.4. OXYGEN-ISOTOPIC COMPOSITION OF MATRIX

The available O-isotopic data for matrices, chondrules and objects that might be related to matrix in type-3 chondrites are shown in Fig. 10.2.10 (cf. Figs. 1.1.11 and 9.1.9). In CV3 chondrites, all the analyzed components, which are from Allende, plot close to the line defined by minerals in Allende CAIs (see Fig. 12.1.1). Two matrix samples and five dark inclusions, which are probably rich in matrix material, plot on the ^{16}O-poor end of this line, with ^{18}O/^{16}O ratios that are higher than those of most Allende porphyritic chondrules. The range of ^{18}O/^{16}O ratios for matrix-like, coarse-grained rims (Rubin 1984) overlaps that of the porphyritic chondrules that they enclose, but each rim is typically 1 to 2‰ enriched in ^{18}O relative to its chondrule. The spread of analyses along the refractory-inclusion line implies that matrix, chondrules and coarse-grained rims participated in an exchange between isotopically distinct reservoirs or that they formed from materials that had exchanged O isotopes. The distinct differences in O-isotopic composition be-

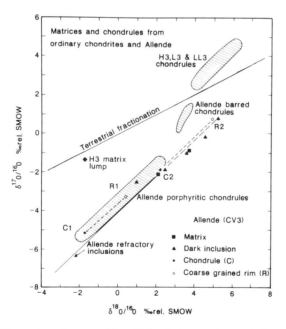

Fig. 10.2.10. Oxygen-isotopic composition of chondrules, matrices and other components in type-3 ordinary chondrites and Allende. The sole analyzed matrix sample from an ordinary chondrite plots below the terrestrial fractionation line and could not have been derived by comminution of chondrules in OCs, which plot above the terrestrial line. In Allende, matrix samples and dark inclusions, which are probably rich in matrix, partially overlap the range of porphyritic chondrules, but their mean compositions are quite distinct. Coarse-grained rims on Allende chondrules, which are probably derived from matrix material, also plot close to the line defined by Allende refractory inclusions. (Data from Clayton et al. 1977, 1983, 1987.)

tween chondrules and coarse-grained rims (or matrix) imply that they were not derived directly from one another.

Also shown in Fig. 10.2.10 is the O-isotopic composition of the matrix lump from A77299, whose bulk chemical composition is plotted in Fig. 10.2.9, compared with those of chondrules in type-3 OCs. The composition of the lump might have been shifted almost parallel to the terrestrial fractionation line by contamination with O from ^{18}O-depleted Antarctic ice, but its preterrestrial composition must have been below the terrestrial fractionation line. It could, therefore, only be related to OC chondrule material by invoking exchange with some other isotopic reservoir. Grossman et al. (1987) analyzed matrix material from Semarkona and found that its matrix is much closer in O-isotopic composition to its chondrules.

10.2.5. METAMORPHISM AND ALTERATION OF CHONDRITE MATRIX

It is now well established that even the most primitive chondrites have been subjected to either thermal or aqueous processing, or both, during the asteroidal stage of their history (see Chapter 3.3 and 3.4). In this section, we explore the chemical and petrographic effects that such processing has had on matrix material.

Ordinary Chondrites

Huss et al. (1981) and Scott et al. (1984) pointed out that even in type-3 OCs, evidence for mild metamorphic equilibration between matrix and chondrules can be found (Sears et al. 1980; see also Chapter 3.3). Huss et al. reported that changes in matrix texture occurred without major changes in the textures of chondrules, but found that there were significant correlated changes in the compositions of chondrule olivines and those in the matrix. (Metamorphic equilibration leads to a progressive decrease in the FeO/(FeO + MgO) ratios of both bulk matrix and its olivine; see Chapter 3.3.) We note that Fig. 10.2.3 shows that the change in matrix olivine composition is continuous all the way down through the least-metamorphosed chondrites, although Sharps and Semarkona break the trends. Thus, it is possible that no chondrite preserves its initial mineralogy unchanged by metamorphic processes.

Alexander (1987) noted that the effects of metamorphism are most pronounced for the finest-grained olivines in the matrix of type-3 OCs, and concluded that the groundmass of both rims and matrix crystallized during limited metamorphism. Earlier, Ashworth (1977) had concluded that rims, which exhibit less porosity than matrix, had textures indicative of grain growth in the solid state, perhaps assisted by infilling through gas-solid condensation. However, the interpretation of metamorphic effects in type-3 OCs is greatly

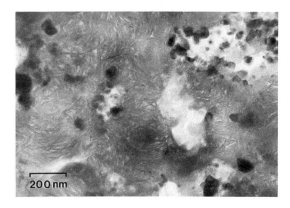

Fig. 10.2.11. Transmission electron micrograph, taken at 200 keV, showing the effect of aqueous alteration in the matrix of Semarkona. The fibrous material is smectite and the darker grains and clusters are maghemite.

complicated by the fact that many, if not most, of them are breccias of components with diverse metamorphic and other histories (see, e.g., Scott 1984).

Unlike the case for thermal metamorphism, the effects of aqueous alteration have only recently been recognized in OCs, although known in carbonaceous chondrites for many years (see Chapter 3.4). In Semarkona matrix, the smectite, calcite and maghemite all probably formed during preterrestrial aqueous alteration, probably of fine-grained interstitial matrix material (Hutchison et al. 1987). The magnetite, pyrrhotite and pentlandite in Semarkona may also have been produced by secondary alteration, although perhaps prior to the aqueous alteration. Alteration products in Semarkona are illustrated in Fig. 10.2.11. What is not clear is whether Semarkona once had abundant fayalitic olivine, like Krymka, which was destroyed by alteration leaving only magnesian olivine, or whether Semarkona preserves the more primitive material. Ikeda et al. (1981) reported a matrix component in several Antarctic OCs with a chemical signature similar to that of CM matrix. A similar finding had been made earlier in Tieschitz by Kurat (1969).

CV3 Chondrites

The monotonic increase in homogeneity of matrix olivine from Kaba to Allende in Fig. 10.2.6 is suggestive of progressive thermal metamorphism acting on an initially heterogeneous olivine population, as discussed for ordinary chondrites above and in Chapter 3.3. Evidence bearing on this issue is ambiguous. The observation that matrix clinopyroxenes occur in highly disequilibrated assemblages whose compositions do not vary in any systematic way with matrix olivine homogeneity argues against metamorphism, but not definitively because olivines equilibrate more rapidly than pyroxenes (see, e.g., Freer 1981). The common presence of relatively heterogeneous matrix inclusions embedded within more homogeneous host matrix rules out metamorphic equilibration of the latter after lithification, but leaves open the possibility of post-accretional metamorphism prior to final lithification. Similar arguments apply to the occurrence, in relatively homogeneous matrix such as that of Allende, of Mg-rich chondrule olivine in direct contact with Fe-rich matrix olivine, as such contacts could have postdated any metamorphism. Such mixing of material with putatively different thermal histories is consistent with the brecciated nature of at least some of the CV3 chondrites (see Chapter 3.5). The presence of solar-wind-implanted noble gases in Allende, Mokoia and Vigarano indicates that those chondrites, at least, contain ingredients that existed separately in the regolith of the CV3 parent asteroid.

Whether or not the trends in Fig. 10.2.6 are due to progressive thermal metamorphism, the possibility must be considered that matrix olivine is itself a secondary mineral, formed by alteration of some earlier-formed phase, possibly nebular in origin. Housley and Cirlin (1983) suggested that Fe-rich olivine in Allende chondrules, chemically identical to that in the matrix, shows a reaction relationship with clinoenstatite. They argued that the olivine, including that in the matrix, was formed by the reaction

$$Fe + 1/2O_2 + MgSiO_3 = FeMgSiO_4 \qquad (1)$$

with the enstatite and metal originating in chondrules. They proposed that this reaction occurred in the CV parent body and that strains resulting from the volume change involved in the reaction cleaved the enstatite into the forms now exhibited by the olivine. However, the occurrence of fine- and coarse-grained rims of likely nebular origin (see Chapter 9.1) around $>50\%$ of Allende chondrules (Rubin 1984) suggests that little matrix could have been formed from chondrules through chemical reactions on the parent body.

The debate over the cause of the olivine mineral chemistry has a significance that transcends simply the origin of the olivine itself. Olivine is the most easily altered of the matrix minerals, either by aqueous alteration or by thermal metamorphism. Consequently, if matrix olivines have remained unaltered since accretion of the CV asteroid, the rest of the matrix constituents are probably also pristine.

10.2.6. ORIGIN OF MATRIX MATERIAL

Between 10 and 20 years ago, meteoriticists tended to believe that chondrites are made of primitive material that formed in the solar nebula. Some cosmochemical models required that chondrites formed by the mixing of two components: devolatilized chondrules and primitive dust consisting of nebular condensates (see, e.g., Larimer and Anders 1967; Grossman and Larimer 1974). Other models also invoked production of chondrites by mixing chondrules and fine-grained dust from the nebula, but with both chondrules and dust containing volatiles (see, e.g., Wasson and Chou 1974; Wai and Wasson 1977). Later models for the solar nebula would incorporate surviving interstellar material into the collection of nebular dust. In addition, solid objects in the nebula certainly would have collided and fragmented, thereby forming more dust. Matrix, in such models, would be composed of compacted nebular dust.

Ideally, then, pristine matrix could be searched for condensates, interstellar grains and clastic material, its volatiles could be analyzed, and other properties determined in order to learn about what really happened in the solar nebula. However, it is clear from the data and observations reviewed in this chapter that most chondrites do not preserve pristine matrix material. Heating, alteration and impacts have modified the textures, compositions and physical properties of matrix phases. One extreme view of matrix is that no nebular dust survives, and that material currently observed as matrix is solely the product of comminution and/or alteration of chondrules, inclusions, metal grains and whatever dust might have been present (see, e.g., Housley and Cirlin 1983). However, several crucial questions have not been thoroughly addressed:

1. Which chondrites have matrix showing the least secondary processing?
2. In those chondrites, are there any pristine grains left?
3. What properties of altered grains might still be evidence for nebular processes (e.g., a grain might have the morphology of a condensate but a composition resulting from secondary exchange or alteration?

Only with these questions answered can one proceed to learn about the solar nebula by the study of matrix. The discussion in the remainder of this chapter should be read in the knowledge of how poorly these questions have been answered, and should therefore be regarded as a guide to further research rather than as a comprehensive review of a well-understood field.

Fayalitic olivine is a major component of matrix in some of the least-altered, least-metamorphosed chondrites. Is this pristine material or a secondary product? Table 10.2.7. lists the various origins that have been proposed over the past five years for this material.

Huss et al. (1981) thought that matrix olivine represents a low-temperature nebular product. Nagahara (1984) agreed with that view because the texture and mineral chemistry argued against production of matrix phases by comminution or alteration of chondrules. She therefore tried to link the parageneses of OC matrix minerals to those predicted by condensation theory. A few matrix relationships could thereby be documented as seemingly consistent with an equilibrium condensation sequence: magnesian pyroxene was formed after calcic pyroxene, and FeO-rich olivine was formed as mantles around the magnesian pyroxene. Some important predicted relationships such as pyroxene formed from olivine were not observed. In the context of a nebular condensation model, many of the observed relationships required some-

TABLE 10.2.7
Proposed Origins for FeO-rich Olivine in Matrices of Chondrites and Interplanetary Dust

Occurrence	Origin	Reference
H3, L3, LL3	nebular condensate	Nagahara 1984
	metamorphic crystallization	Ashworth 1977
	comminuted chondrule fragments	Alexander 1987
	crystallization from chondrule melts	Hutchison and Bevan 1983
CV3	pristine nebular condensate	MacPherson et al. 1985
	processed nebular condensate	Kornacki and Wood 1984
	regolith comminution of altered aggregates	Bunch et al. 1986
	planetary alteration of chondrule enstatite	Housley and Cirlin 1983
Interplanetary dust	annealing of presolar amorphous condensates	Rietmeijer 1986

what contrived situations. For example, FeO-rich olivine often surrounds FeO-poor olivine. In order for the FeO-rich olivine to have formed from pyroxene in the nebula, Fe metal and pyroxene must have been in direct contact. Formation of the FeO-rich olivine around FeO-poor olivine therefore requires that three phases, olivine, pyroxene and metal, were in contact in the nebula. Furthermore, formation of olivine with FeO greater than Fa_{50} would have required disequilibrium condensation of silica followed by reaction of that phase with metal.

The similarity between the composition of FeO-rich olivines in the matrix, in the outer rims of FeO-poor chondrules and in certain aggregates in Allende suggests that these ingredients experienced similar processing. Peck (1984; see also Peck and Wood 1987) therefore proposed that the ranges in matrix-olivine composition reflect variations present in the different nebular regions, sampled by each chondrite, within which the olivine itself condensed.

Nebular condensation is also suggested in some cases by the form of matrix olivine plates, which can resemble that of vapor-condensed olivine grains found within vugs in terrestrial volcanic rocks. The shape of olivine crystals in CV3 matrix, elongated along the [001] direction, is taken by some (see, e.g., MacPherson et al. 1985) as evidence that they are nebular condensates. Bunch et al. (1986), on the other hand, argue that such crystals were formed by cleavage of olivine in altered aggregates, but several arguments have been advanced against such a comminution origin. Matrix olivine, besides being much finer grained than that in chondrules, also has a very different composition. Chondrule grains would have had to fracture many times, in a remarkably regular fashion, to produce matrix olivine plates and chips, which are consistently the same size to within a few μm. This seems unlikely, because Mg-rich olivine, such as that in chondrules, has poor cleavage. Furthermore, the chondrule olivine must then have been enriched in Fe, Cr, Al and Ti and depleted in Mg in order to resemble matrix olivine. Because intact olivine chondrules do not have equilibration rims with the composition of matrix olivine (Peck and Wood 1987), the chondrule fragments would have had to have been very efficiently separated from intact chondrules, chemically altered to matrix-olivine compositions, and then recombined with chondrules. Such a scenario seems improbable, suggesting that the observations favor a nebular origin, but the remarkably high abundances of minor elements in matrix olivines, at least in CV3 chondrites, are inconsistent with equilibrium condensation. A supersaturated vapor or an amorphous condensate therefore seems to be a more plausible precursor to matrix olivine.

Although the evidence for comminution during formation of CV3 matrix appears equivocal, Ashworth (1977), Alexander et al. (1985) and Alexander (1987) have shown that matrix grains in type-3 OCs, such as Chainpur and Bishunpur, obey cumulative power-law distributions appropriate to fragmentation processes, suggesting that, at least in those cases, comminution of chondrules contributed a significant fraction of the matrix components. Further support for this view comes from the finding that the compositions of the low-Ca pyroxenes in the matrix of the type-3 OCs studied by Alexander et al. are essentially identical to those of the low-Ca pyroxenes in chondrules (see also Fujimaki et al. 1981). The coarser olivine grains also tend to fall within the compositional range of OC chondrule olivine (see, e.g., Alexander 1987), but this does not offer such a strong constraint.

Oxygen-isotopic data do not argue for a chondrule origin for matrix material unless the chondrules were fragmented in the nebula and subsequently reacted with O in the nebular gas. Whether chondrules exhibit sufficient evidence for fragmentation prior to accretion to satisfy this condition is not clear but seems unlikely. There is good evidence that coarse-grained rims on chondrules were derived from matrix material and some evidence that they were heated in the nebula by the chondrule-forming process (Rubin 1984; Rubin and Wasson 1987; Clayton et al. 1987). The intimate association of matrix material with igneous-textured material of similar composition in certain chondrule-like inclusions in ordinary chondrites (Recca et al. 1984; Scott et al. 1984) also suggests that some matrix material was melted by the chondrule-forming process, and that chondrule precursor material partly resembled material observed now as matrix. However, fine-grained rims can also be observed around noritic and anorthositic clasts in some chondrites (Hutchison 1982).

To summarize, current evidence is ambiguous as to the relative contributions of primary nebular material and secondary products to the present make-up of chondritic matrix; it is clear that secondary processing has been important, but also that there are primary nebular features preserved. It is still not obvious which, if any, chondrites contain pristine nebular material in their matrices. The best evidence for matrix formation in the nebula, rather than on parent bodies, is that coarse-grained rims represent matrix material that was heated with chondrules. This, of course, presupposes that chondrules had a nebular origin (Chapter 9.3). The O-isotopic data for those rims are also more consistent with nebular processes than with those on parent bodies. Enrichment of matrix in some volatile elements, and fractionated major-element ratios, e.g., Mg/Si, also seem to require a nebular origin. Mineral chemistry and textural relationships in type-3 ordinary and CV chondrites are difficult to explain in terms of parent-body processes, but neither do they fit in with simple nebular models. The importance of secondary processes in matrix formation is exemplified by the occurrence of clastic materials in all chondrite matrices, the presence of alteration products in many matrices, and the metamorphic effects shown even by many chondrites of low petrographic grade. Many of the data on matrix materials are controversial or ambiguous. The isotopic compositions of O in OC matrix materials are poorly known and conflicting. The origin of the most abundant matrix component, the fayalitic olivine, is not well understood. The morphologies exhibited by olivine in many chondrites are difficult to interpret. Clearly, many more detailed studies are necessary before robust conclusions can be drawn about conditions and processes in the early solar nebula.

Acknowledgments. We thank G. J. Taylor, A. J. Brearley, I. D. R. Mackinnon and F. J. M. Rietmeijer for valuable discussions, and A. E. Rubin for permission to quote unpublished work. Helpful reviews were provided by J. N. Grossman and A. E. Rubin. This work was partly supported by a grant to K. Keil from the National Aeronautics and Space Administration.

REFERENCES

Alexander, C. M. 1987. The matrices and rims of U.O.C.s: Origins, metamorphism and alteration. Ph.D. thesis, Univ. of Essex, Colchester, UK.

Alexander, C., Barber, D. J., Davis, M., and Hutchison, R. 1985. Relationships between dark rims, interchondrule matrix, and chondrules in U.O.C.'s. *Meteoritics* 20:600 (abstract).

Allen, J. S., Nozette, S., and Wilkening, L. L. 1980. A study of chondrule rims and chondrule irradiation records in unequilibrated ordinary chondrites. *Geochim. Cosmochim. Acta* 44:1161–1175.

Ashworth, J. R. 1977. Matrix textures in unequilibrated ordinary chondrites. *Earth Planet. Sci. Lett.* 35:25–34.

Ashworth, J. R. 1980. Chondritic thermal histories: Clues from electron microscopy of orthopyroxene. *Earth Planet. Sci. Lett.* 46:167–177.

Ashworth, J. R. 1981. Fine structure in H-group chondrites. *Proc. Roy. Soc. London* A374:179–194.

Barber, D. J. 1985. Phyllosilicates and other layer-structured materials in stony meteorites. *Clay Min.* 20:415–454.

Binns, R. A. 1970. Pyroxenes from non-carbonaceous chondritic meteorites. *Mineral Mag.* 37:649–669.

Brearley, A. J., Scott, E. R. D., and Keil, K. 1987a. Carbon-rich aggregates in ordinary chon-

drites: Transmission electron microscope observations of Sharps (H3) and Plainview (H regolith breccia). *Meteoritics* 22:338–339.
Brearley, A. J., Mackinnon, I. D. R., and Scott, E. R. D. 1987b. Electron petrography of fine-grained matrix in the Karoonda C4 carbonaceous chondrite. *Meteoritics* 22:339–340.
Bunch, T. E., and Chang, S. 1980. Carbonaceous chondrites—II: Carbonaceous chondrite phyllosilicates and light element geochemistry as indicators of parent body processes and surface conditions. *Geochim. Cosmochim. Acta* 44:1543–1577.
Bunch, T. E., and Chang, S. 1983. Allende dark inclusions: Samples of primitive regoliths. *Lunar Planet. Sci.* XIV:75–76 (abstract).
Bunch, T. E., Chang, S., Cassen, P., and Reynolds, R. 1986. Allende: Profile of parent body growth. *Lunar Planet. Sci.* XVII:89–90 (abstract).
Christophe Michel-Lévy, M. 1969. Etude minéralogique de la chondrite CIII de Lancé. In *Meteorite Research*, ed. P. M. Millman (Dordrecht: D. Reidel), pp. 492–499.
Christophe Michel-Lévy, M. 1976. La matrice noir et blanche de la chondrite de Tieschitz (H3). *Earth Planet. Sci. Lett.* 30:143–150.
Clarke, R. S., Jr., Jarosewich, E., Mason, B., Nelen, J., Gomez, M., and Hyde, J. R. 1970. The Allende, Mexico, meteorite shower. *Smithsonian Contrib. Earth Sci.* 5:1–53.
Clayton, R. N., and Mayeda, T. K. 1977. Anomalous anomalies in carbonaceous chondrites. *Lunar Sci.* VIII:193–195 (abstract).
Clayton, R. N., Onuma, N., Ikeda, Y., Mayeda, T. K., Hutcheon, I., Olsen, E. J., and Molini-Velsko, C. 1983. Oxygen isotopic compositions of chondrules in Allende and ordinary chondrites. In *Chondrules and Their Origins*, ed. E. A. King (Houston: Lunar and Planetary Inst.), pp. 37–43.
Clayton, R. N., Mayeda, T. K., Rubin, A. E., and Wasson, J. T. 1987. Oxygen isotopes in Allende chondrules and coarse-grained rims. *Lunar Planet. Sci.* XVIII:187–188 (abstract).
Dodd, R. T. 1981. *Meteorites: A Petrologic-Chemical Synthesis* (Cambridge: Cambridge Univ. Press).
Fredriksson, K., Jarosewich, E., and Nelen, J. 1969. The Sharps chondrite—New evidence on the origin of chondrules and chondrites. In *Meteorite Research*, ed. P. M. Millman (Dordrecht: D. Reidel), pp. 155–165.
Freer, R. 1981. Diffusion in silicate minerals and glasses: A data digest and guide to the literature. *Contrib. Mineral. Petrol.* 76:440–454.
Fruland, R. M., King, E. A., and McKay, D. S. 1978. Allende dark inclusions. *Proc. Lunar Planet. Sci. Conf.* 9:1305–1329.
Fuchs, L. H., Olsen, E., and Jensen, K. J. 1973. Mineralogy, mineral-chemistry, and composition of the Murchison (C2) meteorite. *Smithsonian Contrib. Earth Sci.* 10:1–39.
Fujimaki, H., Matsu-ura, M., Sunagawa, I., and Aoki, K. 1981. Chemical compositions of chondrules and matrices in the ALH-77015 chondrite (L3). In *Proc. 6th Symp. Antarctic Meteorites, Mem. Natl. Inst. Polar Res.* Spec. Issue 20:161–174.
Graham, A. L., and Hutchison, R. 1974. Is Kakangari a unique chondrite? *Nature* 251:128–129.
Green, H. W., Radcliffe, S. V., and Heuer, A. H. 1971. Allende meteorite: A high voltage electron petrographic study. *Science* 172:936–939.
Grossman, J. 1985. Chemical evolution of the matrix of Semarkona. *Lunar Planet. Sci.* XVI:302–303 (abstract).
Grossman, J. N., and Wasson, J. T. 1983. Refractory precursor components of Semarkona chondrules and fractionation of refractory elements among chondrites. *Geochim. Cosmochim. Acta* 47:759–771.
Grossman, J. N., Clayton, R. N., and Mayeda, T. K. 1987. Oxygen isotopes in the matrix of the Semarkona (LL3.0) chondrite. *Meteoritics* 22, in press.
Grossman, L., and Larimer, J. W. 1974. Early chemical history of the solar system. *Rev. Geophys. Space Phys.* 12:71–101.
Heymann, D., Van Der Stap, C. C. A. H., Vis, R. N., and Verheul, H. 1987. Carbon in dark inclusions of the Allende meteorite. *Meteoritics* 22:3–15.
Housley, R. M., and Cirlin, E. H. 1983. On the alteration of Allende chondrules and the formation of matrix. In *Chondrules and Their Origins*, ed. E. A. King (Houston: Lunar and Planetary Inst.), pp. 145–161.

Huss, G. R. 1979. The matrix of unequilibrated ordinary chondrites: Implications for the origin and history of chondrites. M.S. thesis, Univ. of New Mexico.
Huss, G. R., Keil, K., and Taylor, G. J. 1981. The matrices of unequilibrated ordinary chondrites: Implications for the origin and history of chondrites. *Geochim. Cosmochim. Acta* 45:33–51.
Hutchison, R. 1982. Meteorites—Evidence for the interrelationships of materials in the solar system of 4.55 Ga ago. *Phys. Earth Planet. Int.* 29:199–208.
Hutchison, R., and Bevan, A. W. R. 1983. Conditions and time of chondrule accretion. In *Chondrules and Their Origins*, ed. E. A. King (Houston: Lunar and Planetary Inst.), pp. 162–179.
Hutchison, R., Bevan, A. W. R., Agrell, S. O., and Ashworth, J. R. 1979. Accretion temperature of the Tieschitz, H3, chondritic meteorite. *Nature* 280:116–119.
Hutchison, R., Alexander, C. M. O., and Barber, D. J. 1987. The Semarkona meteorite: First recorded occurrence of smectite in an ordinary chondrite, and its implications. *Geochim. Cosmochim. Acta* 51:1875–1882.
Iijima, S., and Buseck, P. R. 1975. High resolution electron microscopy of enstatite. 1: Twinning, polymorphism, and polytypism. *Amer. Mineral.* 60:758–770.
Ikeda, Y. 1980. Petrology of Allan Hills-764 chondrite (LL3). In *Proc. 5th Symp. Antarctic Meteorites, Mem. Natl. Inst. Polar Res.*, Special Issue 17:50–82.
Ikeda, Y., Kimura, M., Mori, H., and Takeda, H. 1981. Chemical compositions of matrices of unequilibrated ordinary chondrites. In *Proc. 6th Symp. Antarctic Meteorites, Mem. Natl. Inst. Polar Res.*, Special Issue 20:124–144.
King, T. V. V., and King, E. A. 1981. Accretionary dark rims in unequilibrated ordinary chondrites. *Icarus* 48:460–472.
Kornacki, A. S., and Wood, J. A. 1984. The mineral chemistry and origin of inclusion matrix and meteorite matrix in the Allende CV3 chondrite. *Geochim. Cosmochim. Acta* 48:1663–1676.
Kracher, A., Keil, K., Kallemeyn, G. W., Wasson, J. T., Clayton, R. N., and Mayeda, T. K. 1985. The Leoville (CV3) accretionary breccia. *Proc. Lunar Planet. Sci. Conf.* 16, *J. Geophys. Res. Suppl.* 90:D123–D135.
Kurat, G. 1969. The formation of chondrules and chondrites and some observations on chondrules from the Tieschitz meteorite. In *Meteorite Research*, ed. P. M. Millman (Dordrecht: D. Reidel), pp. 185–190.
Kurat, G. 1970. Zur Genese der Ca-Al reichen Einschlusse im Chondriten von Lancé. *Earth Planet. Sci. Lett.* 9:225–231.
Larimer, J. W., and Anders, E. 1967. Chemical fractionations in meteorites—II. Abundance patterns and their interpretation. *Geochim. Cosmochim. Acta* 31:1239–1270.
MacPherson, G. J., Hashimoto, A., and Grossman, L. 1985. Accretionary rims on inclusions in the Allende meteorite. *Geochim. Cosmochim. Acta* 49:2267–2279.
Mason, B., and Wiik, H. B. 1966. The composition of the Bath, Frankfort, Kakangari, Rose City, and Tadjera meteorites. *Amer. Mus. Novitates* 2272:1–24.
Matsunami, S. 1984. The chemical compositions and textures of matrices and chondrule rims of eight unequilibrated ordinary chondrites. In *Proc. 9th Symp. Antarctic Meteorites. Natl. Inst. Polar Res.* 35:126–148.
McSween, H. Y., Jr. 1977. Petrographic variations among carbonaceous chondrites of the Vigarano type. *Geochim. Cosmochim. Acta* 41:1777–1790.
McSween, H. Y., Jr. 1979a. Are carbonaceous chondrites primitive or processed? A review. *Rev. Geophys. Space Phys.* 17:1059–1078.
McSween, H. Y., Jr. 1979b. Alteration in CM carbonaceous chondrites inferred from modal and chemical variations in matrix. *Geochim. Cosmochim. Acta* 43:1761–1770.
McSween, H. Y., Jr., and Richardson, S. M. 1977. The composition of carbonaceous chondrite matrix. *Geochim. Cosmochim. Acta* 41:1145–1161.
Nagahara, H. 1984. Matrices of type 3 ordinary chondrites—Primitive nebular records. *Geochim. Cosmochim. Acta* 48:2581–2595.
Peck, J. A. 1984. Origin of the variation in properties of CV3 meteorite matrix and matrix clasts. *Lunar Planet. Sci.* XV:635–636 (abstract).
Peck, J. A., and Wood, J. A. 1987. The origin of ferrous zoning in Allende chondrule olivines. *Geochim. Cosmochim. Acta* 51:1503–1510.

Rambaldi, E. R., Fredriksson, B. J., and Fredriksson, K. 1981. Primitive ultrafine matrix in ordinary chondrites. *Earth Planet. Sci. Lett.* 56:107–126.

Recca, S. I., Scott, E. R. D., Taylor, G. J., and Keil, K. 1984. Fine-grained millimeter-sized objects in type 3 ordinary chondrites and their relation to chondrules and matrix. *Meteoritics* 19:296–297 (abstract).

Rietmeijer, F. J. M. 1986. Olivines and iron-sulfides in chondritic porous aggregate U2015*B formed at low-temperature during annealing of amorphous precursor materials. *Meteoritics* 21:492–493 (abstract).

Rubin, A. E. 1984. Coarse-grained chondrule rims in type 3 chondrites. *Geochim. Cosmochim. Acta* 48:1779–1789.

Rubin, A. E. 1986. Elemental compositions of major silicic phases of unequilibrated chondritic meteorites. *Meteoritics* 21:283–294 (abstract).

Rubin, A. E., and Wasson, J. T. 1987. Chondrules, matrix and coarse-grained chondrule rims in the Allende meteorite. *Geochim. Cosmochim. Acta* 51:1923–1937.

Rubin, A. E., and Wasson, J. T. 1988. Chondrules in the Ornans CO3 meteorite and the timing of chondrule formation relative to nebular fractionation events. *Geochim. Cosmochim. Acta,* in press.

Rubin, A. E., James, J. A., Keck, B. D., Weeks, K. S., Sears, D. W. G., and Jarosewich, E. 1985. The Colony meteorite and variations in CO3 chondrite properties. *Meteoritics* 20:175–196.

Scott, E. R. D. 1984. Classification, metamorphism and brecciation of type 3 chondrites from Antarctica. *Smithsonian Contrib. Earth Sci.* 26:73–94.

Scott, E. R. D., and Taylor, G. J. 1983. Chondrules and other components in C, O, and E chondrites: Similarities in their properties and origins. *Proc. Lunar Planet. Sci. Conf.* 14, *J. Geophys. Res. Suppl.* 88:B275–B286.

Scott, E. R. D., and Taylor, G. J. 1985. Petrology of types 4–6 carbonaceous chondrites. *Proc. Lunar Planet. Sci. Conf.* 15, *J. Geophys. Res. Suppl.* 90:C699–C709.

Scott, E. R. D., Rubin, A. E., Taylor, G. J., and Keil, K. 1981. New kind of type 3 chondrite with a graphite-magnetite matrix. *Earth Planet. Sci. Lett.* 56:19–31.

Scott, E. R. D., Taylor, G. J., and Maggiore, P. 1982. A new LL3 chondrite, Allan Hills A79003, and observations on matrices in ordinary chondrites. *Meteoritics* 17:65–75.

Scott, E. R. D., Rubin, A. E., Taylor, G. J., and Keil, K. 1984. Matrix material in type 3 chondrites—Occurrence, heterogeneity and relationship with chondrules. *Geochim. Cosmochim. Acta* 48:1741–1757.

Sears, D. M., Grossman, J. M., Melcher, C. L., Ross, L. M., and Mills, A. A. 1980. Measuring the metamorphic history of unequiliberated ordinary chondrites. *Nature* 287:791–795.

Taylor, G. J., Scott, E. R. D., Keil, K., Boynton, W. V., Hill, D. H., Mayeda, T. K., and Clayton, R. N. 1984. Primitive nature of ordinary chondrite matrix materials. *Lunar Planet. Sci.* XV:848–849 (abstract).

Van Schmus, W. R. 1969. The mineralogy and petrology of chondritic meteorites. *Earth Sci. Rev.* 5:145–184.

Wai, C. M., and Wasson, J. T. 1977. Nebular condensation of moderately volatile elements and their abundances in ordinary chondrites. *Earth Planet. Sci. Lett.* 36:1–13.

Wark, D. A. 1979. Birth of the presolar nebula: The sequence of condensation revealed in the Allende meteorite. *Astrophys. Space Sci.* 65:275–295.

Wasson, J. T., and Chou, C.-L. 1974. Fractionation of moderately volatile elements in ordinary chondrites. Meteoritics 9:69–84.

Wilkening, L. L., and Hill, D. H. 1985. Fine-grained chondrule rims. *Meteoritics* 20:785–786 (abstract).

Wilkening, L. L., Boynton, W. V., and Hill, D. H. 1984. Trace elements in rims and interiors of Chainpur chondrules. *Geochim. Cosmochim. Acta* 48:1071–1080.

Wlotzka, F. 1983. Composition of chondrules, fragments and matrix in the unequilibrated ordinary chondrites Tieschitz and Sharps. In *Chondrules and Their Origins,* ed. E. A. King (Houston: Lunar and Planetary Inst.), pp. 296–318.

Wood, J. A. 1962. Metamorphism in chondrites. *Geochim. Cosmochim. Acta* 26:739–749.

10.3. PRIMITIVE MATERIAL SURVIVING IN CHONDRITES: REFRACTORY INCLUSIONS

GLENN J. MacPHERSON
Smithsonian Institution

D. A. WARK
University of Arizona

and

JOHN T. ARMSTRONG
California Institute of Technology

The previous chapter addressed the question of whether meteorite matrix consists of physically identifiable primitive material preserved from the earliest stages of solar-system formation. Embedded within the matrix of many carbonaceous and some ordinary and enstatite chondrites is a class of unusual objects that are also strongly suspected of being very primitive—refractory inclusions. Refractory inclusions are characterized by mineralogy and bulk compositions that have been correlated with the first, i.e. highest-temperature, condensation products predicted to form from a hot and cooling solar nebula. They may thus preserve some of the most primitive material in the solar system. Their isotopic compositions reveal evidence of (presolar) nucleosynthetic processes that contributed to production of the elements from which the solar system was formed. After their formation, many refractory inclusions experienced complex histories involving a variety of processes, including melting, recrystallization and alteration. These have obscured much of the textural and other information concerning their formational process(es). The best clues to their origin are probably preserved in their trace-element and isotopic signatures.

10.3.1. INTRODUCTION

Refractory inclusions (= Ca-Al-rich Inclusions = CAI) in chondritic meteorites are < 1 mm- to > 1 cm-sized objects whose mineralogy is dominated by compounds having very high vaporization temperatures ($\gtrsim 1300$ K; see, e.g., Grossman and Larimer 1974), primarily oxides and silicates of Al and Ca with minor Ti and Mg. Relative to ferromagnesian chondrules and to bulk chondrites, the bulk compositions of refractory inclusions are enriched in Al, Ca and Ti and also in refractory trace elements such as rare-earth elements (REE), Sc, Y and Pt-group elements; they are depleted in volatile elements such as Na and Fe. CAI range in shape from highly irregular to nearly spheroidal, and in color from white to pink or even pale sky-blue. They are most abundant in carbonaceous chondrites (though absent from CI chondrites) and are rarer in ordinary and enstatite chondrites.

Refractory inclusions were known to exist in carbonaceous chondrites prior to 1969 (see, e.g., Christophe Michel-Lévy 1968), but it was only after the fall, in that year, of the very large Allende CV3 meteorite with its abundant cm-sized inclusions that the attention of the meteoritic community was drawn to these unusual objects. What immediately attracted such attention is that the mineralogy of these objects bears a remarkable resemblance (see, e.g., Marvin et al. 1970) to the first phases predicted by thermodynamic calculations to condense out of a cooling gas of solar composition (see, e.g., Lord 1965; Grossman 1972). With the additional recognition of the extreme ages of the inclusions (see, e.g., Gray et al. 1973; Tatsumoto et al. 1976; Chen and Tilton 1976), and of trace-element abundance patterns that resemble those predicted by gas-solid partitioning (Grossman 1973; Boynton 1975), refractory inclusions were initially interpreted by many workers to be *bona-fide* samples of primary, high-temperature condensate material formed during the birth of the solar nebula. Moreover, it is in refractory inclusions that evidence was first found for exotic presolar isotopic components (see, e.g., Clayton et al. 1973; Lee et al. 1977).

In the ensuing years since the fall of Allende, we have come to recognize that the CAI are a very diverse group of complex objects. In fact, many have had such complex histories that only trace-element and isotopic signatures of their original state have been sufficiently preserved to testify to their ultimate mode of formation. Nonetheless, the study of and debate over CAI properties and genesis continues to be intense—and for good reason. What is at stake is far more important than simply how this obscure and volumetrically minor class of mineralogic oddities formed. At a minimum, CAI preserve many clues to conditions and processes that prevailed during the formation and earliest evolution of the solar nebula. A resolution of the debate over how CAI originally formed—whether by condensation from a "hot nebula" or by some other process(es) in a largely cold nebula—may replace real constraints on the physics of protostellar nebulae in general and our own presolar nebula in particular. Finally, there is compelling isotopic evidence that these objects preserve information about the sources of matter that went into formation of the solar system and even about the nucleosynthetic processes that ultimately formed some of that matter (see Chapter 14.3).

Our purpose here is to provide an overview of the chemical and petrologic properties of refractory inclusions, and also of the evidence bearing on the processes that may have formed them. An exhaustive review of the isotopic properties of CAI is, however, beyond the scope of this Chapter. Chapters 12.1, 14.3 and 15.1 treat that subject in more detail, as do the following papers (and references therein): Wasserburg and Papanastassiou (1982), D. Clayton (1982), Niederer and Papanastassiou (1984), Niederer et al. (1985), Wasserburg (1985) and R. Clayton et al. (1985); some older but useful reviews are those by Podosek (1978), Lee (1979), Begemann (1980) and several sections in a paper by Grossman (1980). We will concern ourselves with certain isotopic properties of inclusions that bear most directly on processes involved in inclusion formation and evolution, such as isotopic mass fractionation and late disturbance of isotope systems. A separate topic of interest here is isotopic evidence for relict grains in inclusions.

The first half of this review (Secs. 10.3.2–10.3.6) outlines the general properties of refractory inclusions in general and of those from each of the chondrite classes specifically. The second half (Sec. 10.3.7 onward) is dedicated to an appraisal of the current thinking regarding various primary and secondary processes which have been proposed to explain CAI genesis and evolution.

10.3.2. OVERVIEW

Components, Mineralogy and Structure

It is convenient to describe refractory inclusions in terms of four texturally defined structural "components": (1) primary refractory silicate and oxide phases; (2) complex metal + sulfide + oxide grains enclosed within inclusions; (3) thin multilayered rim sequences that mantle inclusions; and (4) fine-grained secondary phases that replace the primary phases.

Primary Silicates and Oxides. We define as "primary" those phases that have historically been correlated with the highest-temperature major condensate materials that equilibrium thermodynamic calculations predict to form out of a cooling gas of solar composition (see, e.g., Grossman 1972). An example of a primary melilite crystal that is partially replaced by "secondary" phases (see below) is shown in Fig. 10.3.1. Primary phases include melilite (mostly gehlenite-åkermanite solid solution), spinel, perovskite, hibonite and fassaite (a pyroxene solid solution consisting of diopside, Ca-Tschermak's molecule $CaAl_2SiO_6$, Ti-Tschermak's molecule $CaTi^{4+}Al_2O_6$, and a trivalent Ti-bearing component that might be expressed as $CaTi^{3+}$-

PRIMITIVE MATERIAL: REFRACTORY INCLUSIONS   749

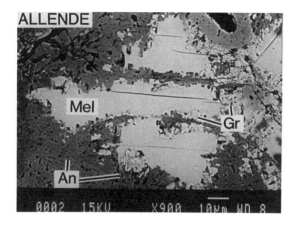

Fig. 10.3.1. Back-scattered electron photomicrograph of an altered primary melilite (Mel) crystal in an Allende (CV3) Type A inclusion. Blocky grossular (Gr) and fibrous anorthite (An) replace the embayed and corroded relict of a single melilite crystal. Alteration has preferentially penetrated the crystal along the direction of cleavage.

$AlSiO_6$) (see Appendix 1 for chemical formulae not given here). At least some anorthite is also primary (see below). These major primary phases are known to be carriers (to varying degrees) of the ^{16}O-rich component (Clayton et al. 1977) in inclusions, and Al-rich ones such as anorthite are also known to contain excess ^{26}Mg correlated with $^{27}Al/^{24}Mg$—suggestive of *in situ* decay of the now-extinct short-lived nuclide ^{26}Al (Lee et al. 1977; see also Chapter 15.1). Rarer primary phases in some inclusions are corundum, the unnamed Ca aluminate $CaAl_4O_7$, thorianite, cordierite, beckelite, celsian and rhönite (see, e.g., Fuchs 1969, 1971; Lovering et al 1979; Christophe Michel-Lévy et al. 1982; Bar-Matthews et al. 1982). Refractory metal nuggets and the complex metal-sulfide-phosphate-oxide bodies known as "Fremdlinge" may also be primary, at least in part.

Analyses of the major silicate and oxide phases by electron microprobe generally reveal them to be almost free of other components: only Fe, Cr and V occasionally reach the weight-percent level in their oxide form, primarily in spinel and hibonite. Scandium-rich phases have been reported in inclusions from Efremovka (Ulyanov et al. 1982), Essebi (El Goresy et al. 1984) and Ornans (Davis 1984).

Fremdlinge and Refractory Metal Nuggets. These are particles within CAI that are highly enriched in refractory siderophile elements (e.g., Re, W, Mo and the Pt-group metals, Pt, Pd, Ru, Os, Ir and Rh) and, in some cases, in refractory lithophile elements (e.g., Zr and Sc) as well. Such grains were first described by Wark and Lovering (1976) and Palme and Wlotzka (1976). *Fremdlinge* (Fig. 10.3.2) are complex bodies with multiple metal, sulfide,

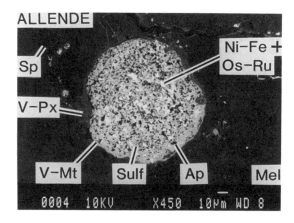

Fig. 10.3.2. Back-scattered electron photomicrograph of a large Fremdling in an Allende (CV3) Type B inclusion. The spheroidal body is a highly porous mixture of Ni-Fe metal (bright), tiny Os-Ru metal nuggets associated with the Ni-Fe but not discernable in this low magnification photo, V-rich magnetite (V-Mt), sulfide (Sulf), Apatite (Ap; dark), and is enclosed in a thin shell of V-rich pyroxene (V-pyx). (Sp. = Spinel.) The entire Fremdling is enclosed in a large melilite (Mel) crystal. This Fremdling is described in detail by Armstrong et al. (1985a).

phosphate, silicate and oxide phases: Ni-Fe metal, pentlandite, troilite, whitlockite, pyrrhotite, scheelite, molybdenite, various alloys of the Pt-group metals, V-rich pyroxene and magnetite, apatite, sodalite, nepheline, baddeleyite, zirconolite and many other bizarre phases have been reported. El Goresy et al. (1978) give many excellent photos and descriptions. The largest known Fremdling is only ~ 1 mm in diameter (Armstrong et al. 1985b), but most are < 50 μm. *Refractory metal nuggets* are generally μm sized or smaller and consist of single-phase pure noble metals or their alloys. Most of the Fremdlinge and metal nuggets that have been studied are from Allende inclusions (see references above, and: Wark and Lovering 1978; Fuchs and Blander 1980; Blander et al. 1980; Wark 1987; Bischoff and Palme 1987), with a few from Leoville (El Goresy et al. 1978; Armstrong et al. 1984b) and Bali (Armstrong et al. 1985c). In CV3 meteorites, Fremdlinge and refractory metal nuggets are known mostly from the pyroxene-rich Type B inclusions (defined below), but are present also in the Type A inclusions as well.

The origins of refractory metal nuggets and, especially, Fremdlinge, remain controversial and a thorough statement of current thinking would be a long chapter in itself. Indeed, the three authors of this review are not in agreement on the subject. Most workers would agree that high-temperature condensation processes were probably involved at some point in the formation of these objects. However, there are two essential problems: (1) how grains within single inclusions can have diverse and extremely fractionated refractory siderophile elemental ratios, particularly when the bulk inclusions themselves have roughly chondritic ratios; and (2) how Fremdlinge assemblages that consist of high- and low-temperature phases could have survived (without equilibrating) the high-temperature processing that has affected their host inclusions. Grossman (1980) gives a fairly objective statement of the problems and arguments involved. Other, more partisan statements include those of Wark and Lovering (1978), Fuchs and Blander (1980), Fegley and Kornacki (1984) and Armstrong et al. (1985a, 1987). The recent experimental work of Blum et al. (1987) confirms earlier suggestions (see, e.g., Palme and Wlotzka 1976) that complex oxidation and sulfidization reactions in, and exsolution from, originally homogeneous metal droplets can account for at least some of the puzzling features of Fremdlinge.

Rim Sequences. These are thin multilayered bands (typically 20 to 50 μm in total thickness with each layer 5 to 10 μm thick) that form the outer boundaries of most inclusions; an example is shown in Fig. 10.3.3. Individual layers are composed of 1 to 3 phases, and are surprisingly constant in thickness around the perimeters of even the most irregularly shaped inclusions. Rims were noted on inclusions in Vigarano as early as 1968 (Christophe Michel-Lévy 1968), but the first systematic descriptions were of rims on Allende inclusions by Wark and Lovering (1977). In consequence, rims are commonly referred to as Wark-Lovering rims.

An idealized rim sequence on CV3 inclusions is more or less as follows,

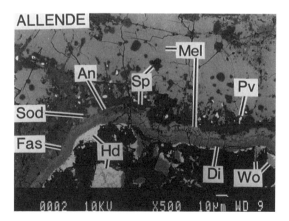

Fig. 10.3.3. Back-scattered electron photomicrograph of a segment of a Wark-Lovering rim sequence on an Allende (CV3) Type A inclusion. Sequentially overlying the interior melilite are layers of: (1) spinel (Sp) + perovskite (Pv); (2) melilite (Mel) that is partially replaced by anorthite (An) and sodalite (Sod); (3) pyroxene that grades outward from fassaite (Fas; lighter grey) to diopside (Di; darker grey); (4) hedenbergite (Hd; bright); and (5) wollastonite (Wo). The dark region at bottom is a cavity.

from innermost layer (directly overlying an inclusion) to outermost (in contact with the meteorite matrix):

1. Spinel + perovskite ± hibonite ± fassaite;
2. Melilite or its alteration products anorthite ± nepheline ± sodalite;
3. Pyroxene, grading outward in composition from fassaite to Al-diopside;
4. Hedenbergite ± andradite or olivine ± diopside.

Actual rims rarely contain all of these layers. Rims on CM inclusions tend to be simpler: Fe-rich phyllosilicate innermost, followed by a pyroxene layer similar to layer (3) above. The phyllosilicate layer is probably a secondary alteration feature, but the precursor has not been identified (see, e.g., MacPherson et al. 1983). Rims on rare melilite-rich CM inclusions are more complex (*ibid.*):

1. Spinel + perovskite + hibonite;
2. Melilite;
3. Anorthite;
4. Pyroxene.

The rim on a melilite-rich Essebi inclusion (El Goresy et al. 1984) has similar mineralogy, plus feldspathoids and a Sc-pyroxene, but is even more complex. In general, melilite and anorthite rim layers occur only on melilite-rich inclusions, suggesting that the rim-forming process involved the inclusion interiors. Inclusions in ordinary chondrites are also rimmed (Bischoff and Keil 1984) but systematic patterns have not been established. Inclusions in enstatite chondrites are apparently not rimmed (Bischoff et al. 1985*a*), although statistics are limited.

Some systematic differences among rims from different meteorites of the same type are observed. Recent comparisons (Korina et al. 1982; Fahey et al. 1985*b*; Paque 1985; Davis et al. 1986) of rim sequences on Allende inclusions with those on inclusions from Vigarano and Efremovka suggest that the rims from the three CV3s were probably similar initially, but those from Allende have experienced intense alteration of the same kind as has affected the Allende inclusion interiors. Alteration is only slight in the Vigarano and Efremovka rims, just as the inclusion interiors themselves are relatively unaltered. The formation of rims thus predates alteration of the inclusion interiors (cf., MacPherson et al. 1981).

Rims have been the subject of considerable attention, for two reasons. First, because they consist of one or more monomineralic (or nearly so) layers in a consistent sequence, Wark and Lovering (1977) suggested that rims might be depositional sequences that preserve time-stratigraphic records of condensation and aggregation. Second, some of the constituent layers are composed of the same refractory phases as found in the interiors of the inclusions they enclose, e.g., spinel and hibonite. Such features might be interpreted as recording, for example, a second episode of condensation or an episode of volatilization. Whatever their origin, the presence of rims clearly indicates that CAI have experienced a more complex history than simply condensation and aggregation of grains.

The result has been a lengthy discussion in the literature of rims and their origins, and as of this writing rims remain something of an enigma. Most workers would probably agree that multiple events and processes were involved. In a sequence of abstracts, Boynton and Wark (e.g., 1987) have suggested that rims initially formed as volatilization residues followed by later secondary modification (see Sec. 10.3.11 below). Murrell and Burnett (1987) have reviewed models for rim formation and the relevant evidence.

Secondary (Alteration) Phases. ["feldspathoids"] These occur in a very fine-grained mixture of material that surrounds, fills veins in and replaces the corroded relicts of the primary phases noted above (e.g., Fig. 10.3.1). Secondary minerals include nepheline, sodalite, calcite, grossular, monticellite, zeolites, titanomagnetite, kirschsteinite, fine-grained anorthite (distinct from coarse-grained primary anorthite), wollastonite and (mostly in CM meteorites) Fe-bearing phyllosilicates. Much secondary alteration involved the introduction into CAI of volatile components such as alkalies, Fe, halogens, CO_2 and H_2O.

Melilite seems to be the phase most susceptible to alteration. Direct evidence indicates extensive corrosion and replacement of melilite by secondary phases (Fig. 10.3.1; Allen et al. 1978; MacPherson et al. 1981). Moreover, the inclusions with the greatest volume percent of secondary phases are melilite rich.

Alteration has not affected inclusions in different meteorites to equal degrees or in the same way. Melilite-bearing inclusions in Allende and Grosnaja, for example, are highly altered whereas those in Vigarano and Efremovka are much less so. CV3 alteration assemblages include alkali- and halogen-bearing feldspathoids; such phases are rare in CM inclusions which contain, instead, abundant calcite and Fe-rich phyllosilicates (Macdougall 1979; MacPherson et al. 1983). Magnesium isotopic studies by ion microprobe indicate that the formation of at least some secondary phases post-dated that of the primary phases by 2 Myr or more (Hutcheon and Newton 1981; however, see also Brigham et al. 1986); O-isotopic studies of mineral separates from altered inclusions (Clayton et al. 1977) suggest that the reservoir from which the secondary phases formed was much closer to being "solar" than that from which the primary phases formed.

Intergroup Comparisons

No widely accepted classification scheme has yet been developed for CAI that transcends the boundaries of the meteorite groups containing them. This will change, we hope, as systematic studies are made of inclusions from CO3, CM, ordinary and enstatite chondrites. Most work to date has been on

TABLE 10.3.1
Refractory Inclusions in Chondrites

	CV	CO	CM	OC	EC[a]
Inclusion Types					
Type B	common	none	none	none	none
Type A	common	common	rare	rare	?
Amoeboid Ol. Aggregates	common	common	common	none	?
Spinel-Pyx Aggregates	common	common	common	rare?	rare
Spinel-Hib-Pv (all kinds)	common	common	common	rare?	?
Inclusion Sizes					
>2mm	common	none	none	none	none
1mm–2mm	common	rare?	rare	rare	none
<1mm	common	common	common	common	rare

[a] So few inclusions have been described from enstatite chondrites that all types should be considered rare until it is demonstrated to the contrary.

the large CV3 inclusions, especially those from Allende, and even there the proposed classifications remain contested.

Nomenclature aside, it seems useful here to attempt some generalizations about CAI before subjecting the reader to descriptions, in the following sections, of the myriad inclusion varieties and the complexities of each. Table 10.3.1, which compares the CAI populations of the different chondrite classes, serves as the basis for the following discussion.

CV3s have the largest inclusions, commonly > 1 cm in Allende and Leoville, and also the greatest diversity of inclusion types. Whereas CV3s contain inclusion types similar to most of those in the other chondrite classes, the reverse does not hold. Thus, for example, primary anorthite and fassaite are major constituents of many CV3 inclusions but are rare in other meteorite types. Melilite is abundant in many CV3 inclusions.

CMs contain abundant inclusions, but virtually all of them are smaller than 1 mm. The primary mineralogy of these inclusions is dominated by spinel; perovskite and hibonite are common accessories, but melilite is rare. Conversely, the important phase corundum has been well documented only in inclusions from CMs.

Inclusions in CO3s apparently are intermediate between those in the CMs and CV3s: they are small (< 2 mm) like those in the CMs but differ in having melilite as a common constituent.

CAI in ordinary chondrites are not abundant. Most are very small objects, < 1.5 mm in maximum size, that resemble inclusions in CMs. Melilite is rare. There are also chondrules that contain accessory aluminous spinel and fassaite ("Al-rich chondrules").

Refractory objects in enstatite chondrites are even smaller on average than those in ordinary chondrites and of similar type with the exception that, in the case of Al-rich chondrules, the coexisting ferromagnesian phase is enstatite. Inclusions in enstatite chondrites are not rimmed.

One way of summarizing the above observations is that there seem to be a limited number of basic inclusion types, that the CV3s contain representatives of virtually all of them, and that CO3s, CMs, ordinary and enstatite chondrites have successively more restrictive populations of inclusions. Only spinel-rich inclusions ($\pm$ accessory hibonite and perovskite) and small inclusions occur throughout. Alteration and rimming (except in E chondrites) are also general, although the style of each varies as noted earlier. However, very limited (except from Allende) data indicate systematic differences in the O-isotopic compositions of inclusions from different chondrites, implying that intergroup differences do not simply reflect differences in sampling of a single inclusion population from region to region in the solar nebula. Similar inclusion-forming processes occurring in isotopically distinct gaseous reservoirs (nebula regions?) are indicated.

10.3.3. INCLUSIONS IN CV3 METEORITES

A. Classification

Refractory inclusions in the CV3 meteorites have been classified according to chemical and petrographic characteristics. Martin and Mason (1974) classified Allende inclusions chemically, based on the relative abundances of rare-earth elements (REE); Mason and Martin (1977) later expanded on that earlier work (see also Taylor and Mason 1978; Mason and Taylor 1982). They recognized five REE patterns in refractory inclusions (a sixth, Group IV, is composed of ferromagnesian objects and not considered here), but in essence there are two main types: those (Groups I, III, V, VI) in which the more refractory REE (Gd, Tb, Dy, Ho, Er and Lu) are *not* fractionated with respect to the less refractory REE (La, Ce, Pr, Nd, Sm and Tm) and those (Group II) in which the more and less refractory REE *are* strongly fractionated from one another. The unfractionated groups are distinguished from one another by the presence or absence of positive or negative anomalies in Yb and/or Eu, the two most volatile REE; all of these patterns are shown in Fig. 10.3.4. These groupings distinguish inclusions with various formation closure temperatures or fractionation histories.

Various petrographic classifications have been proposed. Grossman (1975) first distinguished between those inclusions whose minerals and textures are coarse enough to be easily studied with a petrographic microscope ("coarse-grained") and those that are too fine-grained to do so (the "fine-grained inclusions" of Grossman and Ganapathy [1975]). Grossman then fur-

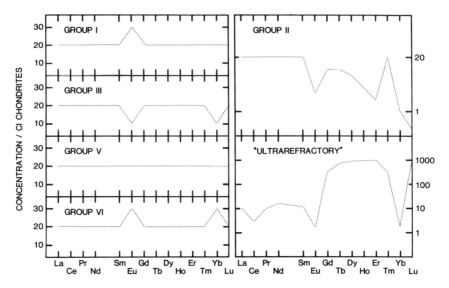

Fig. 10.3.4. The basic rare-earth-element (REE) patterns observed in refractory inclusions. All of the patterns shown are highly schematic and do not represent actual data. The vertical scales are meant to serve only as approximate references to enrichment factors; actual data for each type can show considerable variation in enrichment factors. The unfractionated patterns at left differ only in the presence or absence of Eu and Yb anomalies. The fractionations evident in the patterns at right are the result of volatility differences among the various REE.

ther divided the "coarse-grained inclusions" into two types based on the abundance of fassaite: Type A, composed mostly of melilite with very little pyroxene, and Type B which have abundant pyroxene. An anorthite-rich variety, which he labeled as Type I, were thought to be intermediate between A and B (but see below). Another variety, the forsterite-bearing inclusions were first described by Blander and Fuchs (1975) and Dominik et al. (1978); see also Clayton et al. (1984). Forsterite-bearing inclusions so closely resemble those of Type B in mineralogy, texture and chemistry that Wark et al. (1987) argue that these types form a continuum. Wark and Lovering (1977) subdivided Type B into B1 and B2 (see Fig. 10.3.5a,b), the former having a melilite-rich outer mantle that is lacking in B2. Fine-grained inclusions have been subdivided into spinel-rich fine-grained inclusions (Grossman and Ganapathy 1975,1976) and amoeboid olivine aggregates (Grossman and Steele 1976).

Kornacki and Wood (1984) proposed a different petrographic classification, based on the relative abundances of three textural components that they recognize in CV3 inclusions: *rimmed* or *"concentric" objects; chaotic material* that occurs between individual rimmed objects within aggregates of rimmed objects; and *matrix-like material*.

None of these three classifications is entirely satisfactory. Mason and

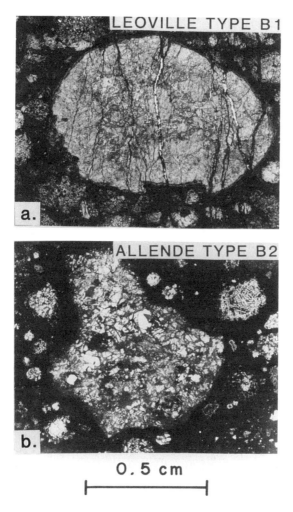

Fig. 10.3.5. Type B inclusions in CV3 meteorites. Both photographs taken with transmitted, plane-polarized light, and both are at the same scale. (a) A B1 inclusion from Leoville (CV3). The lighter outer mantle consists mostly of melilite, and the darker interior is melilite, fassaite, anorthite and spinel. (b) A B2 inclusion from Allende (CV3). The inclusion has a uniform distribution of melilite, fassaite, anorthite and spinel; there is no mantle. The difference in shape between these two objects is not diagnostic.

Taylor (1982) concluded on the basis of their bulk chemical data that there is a continuum between the Type A and B varieties, rather than a dichotomy as Grossman implied. More important, however, is that there is little correlation between the groupings discriminated by chemical vs petrographic classifications. Group II REE patterns, for example, are known in Type A, Type B, fine-grained and coarse-grained inclusions (although *many* Group II inclusions are in fact fine grained and spinel rich). For this reason, Wark (1985) proposed a hybrid chemical/petrographic classification.

The classification of Kornacki and Wood (1984) does not succeed in discriminating the fundamental mineralogical, trace-element or major-element chemical differences of the other classifications. Primarily, it emphasizes that Allende inclusions are all altered ("chaotic material"), that fine-grained inclusions of all types are very complex and share some common features, and that rims are common on inclusions (or some of their constituent parts: "rimmed concentric objects").

For convenience, we will herein refer to inclusions mostly by their petrographic types (à la Grossman, Wark and Lovering, etc.), and to a lesser extent by their REE groups (Martin and Mason, etc.). Because Grossman's mineralogic criteria for classifying the types may be too restrictive, as Mason and Taylor have noted, we suggest that petrographic classification might better be based on the phase fields in which CAI bulk compositions plot on Stolper's (1982) normative anorthite-gehlenite-forsterite-spinel projection diagram (Fig. 10.3.6). Since most refractory inclusions contain abundant spinel, projecting from spinel is a convenient and valid way of representing CAI compositions on a two-dimensional diagram. In effect a liquidus phase diagram, it is nonetheless useful for displaying the compositional variations of

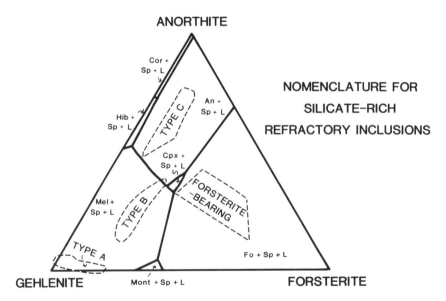

Fig. 10.3.6. Schematic classification scheme for silicate-rich refractory inclusions. Dashed lines enclose the observed bulk-composition fields of the principal CV3 coarse-grained inclusion types. The diagram shows the liquidus phase relationships as projected from spinel onto the plane anorthite-gehlenite-forsterite (Stolper 1982). Bulk-composition data from Beckett (1986), Wark (1981,1987) and Wark et al. (1987).

most silicate-rich refractory inclusions regardless of whether they have ever been molten or not. Note that the bulk compositions of primary silicate-deficient inclusions, such as spinel-rich fine-grained inclusions and most CM inclusions, cannot be accurately represented on this diagram.

The major CV3 inclusion types are distributed as shown on Fig. 10.3.6. As can be seen, this diagram effectively discriminates the major petrographic types and shows their bulk compositional relationships (note that we have used Type C to refer to Grossman's Type I, as explained below). The reader should be aware of difficulties in determining CAI bulk compositions, specifically: (1) the inherent inaccuracy of some analytical techniques, e.g., defocused electron-beam analysis; (2) many CAI are extremely heterogeneous, making it hard to obtain a representative sample; and (3) many CAI are greatly altered, which must be corrected for to obtain their primary compositions (Wark 1981). These problems are discussed in Grossman (1980), Stolper (1982), and Wark and Lovering (1982a).

"FUN" Inclusions. One other category of CAI exists whose members are distinguished solely by their isotopic properties. These inclusions contain fractionated and unidentified nuclear isotopic effects, i.e., FUN (Wasserburg et al. 1977); for discussions of the isotopic characteristics of FUN inclusions, see Clayton (1978), Lee (1979; and Chapter 14.3), Grossman (1980), and Clayton et al. (1984,1985).

No physical, petrographic, bulk chemical or mineral chemical property has been found by which to recognize FUN inclusions (but cf. Brigham et al. 1988). To date, only six examples are known to have mass fractionated O, all of them from CV3 chondrites; five are from Allende and one is from Vigarano, and each has acquired its own name in the literature. Two of these are Type B (Cl, EK 1-4-1), one is unique (HAL) and three are forsterite-bearing inclusions (TE, CG-14, USNM 1623-5); for descriptions see Gray et al. (1973), Nagasawa et al. (1982), Allen et al. (1980), Clayton et al. (1984), MacPherson et al. (1987) and Wark and Wasserburg (1980).

Textural Characteristics

The most thoroughly studied of all CAI are the cm-sized Type B in CV3 chondrites (Fig. 10.3.5a,b). With individual crystals commonly 1 mm or more in length, these objects are readily accessible to study by multiple analytical techniques including bulk as well as microbeam methods. Detailed descriptions and interpretations of Type B inclusions from Bali, Allende, Vigarano and Leoville are given by Kurat et al. (1975), MacPherson and Grossman (1981), Wark and Lovering (1982a,b), El Goresy et al. (1985) and Christophe Michel-Lévy (1986) among others. The major primary phases are melilite, fassaite, anorthite and spinel. Approximately one half of the Ti in the fassaite is Ti^{3+}, indicating that the fassaite formed under oxidation conditions at least as reducing as $\log fO_2 = -18$ to -20 (Beckett and Grossman 1986). Large complex Fremdlinge are common in Allende Type B (see, e.g., Armstrong et al. 1985a). Allende, Leoville and Bali Type Bs show secondary alteration (nepheline, grossular, etc.), but to a lesser extent than other inclusion types in the same meteorites. Those in Vigarano are largely unaltered.

Most Type Bs show strong evidence of having solidified from partially molten droplets (see Sec. 10.3.13), although there is also evidence that other processes played a supplemental role in their evolution, e.g., the entrapment of "xenoliths" by the molten droplets (Wark and Lovering 1982b; El Goresy et al. 1985). Experimental work on liquids of Type B composition has shown that natural Type Bs cooled at rates of 1 to 50° C/hr, from maximum temperatures on the order of 1400 to 1450° C; these limits in turn constrain the nature of the cooling environment (see review by Stolper and Paque 1986).

Type A inclusions have received less attention than those of Type B, in part because most are smaller, finer grained and more highly altered (especially in Allende) than Type B. A sampling of detailed studies of Allende Type A can be found in Grossman (1975), Allen et al. (1978), MacPherson and Grossman (1984) and Wark (1986). Described examples of Type A inclusions from other CV3s include several from Leoville (see, e.g., Christophe Michel-Lévy et al. 1982; Christophe Michel-Lévy 1986), one from Vigarano (MacPherson 1985; Davis et al. 1986), and one from Kaba (Fegley and Post 1985).

Type A embraces a more diverse group of objects than Type B. The most common variety of Type A is referred to as "Fluffy" Type A (FTA; see MacPherson and Grossman [1984] for numerous illustrations) for reasons explained below. They are irregular in shape (many consist of multiple independently rimmed objects partly or completely separated by meteorite matrix), range in size from < 1 mm to > 2 cm, and consist predominantly of Al-rich melilite (Åk 0 to 33), with minor and highly variable proportions (even within a single inclusion) of spinel, hibonite, perovskite and fassaite. The rare phase $CaAl_4O_7$ has been reported in FTAs from Allende (Paque 1987), Vigarano (Davis et al. 1987a) and Leoville (Christophe Michel-Lévy et al. 1982). Phases in FTAs show considerable grain-to-grain variations in composition (MacPherson and Grossman 1984), a feature not seen in Type B.

FTAs in Allende contain up to 70 to 80% by volume of secondary alteration products, prompting MacPherson and Grossman (1984) to speculate that the prealteration inclusions must have been highly porous (cf., Wark and Lovering 1982b) with respect to the altering medium. It was for this reason, and because of the aggregate-like structures and highly irregular shapes of most of them, that MacPherson and Grossman coined the term "Fluffy Type A" in allusion to a resemblance to composite fluffy snowflakes. However, the intense alteration in the Allende FTAs obscures original grain shapes, grain-to-grain contacts and textures. Recent studies of Vigarano FTAs (MacPherson 1985) have shown that those inclusions are identical to their Allende counterparts excepting only that the alteration products are largely missing. The textures are consequently much clearer and are revealed to be largely polygonal granular, suggestive of solid-state recrystallization; in fact, there is some evidence to suggest multiple melilite growth stages (Davis et al. 1987b). The high porosity that MacPherson and Grossman predicted for the Allende FTA protoliths does in fact exist locally in the Vigarano inclusions, but is clearly due to intense post-crystallization brecciation—the inclusions do not resemble loose clumps of condensate grains. Nevertheless, in spite of this recrystallization, FTAs retain certain features that may be remnants of primary condensation; these features and their possible significance will be discussed later (see Sec. 10.3.8).

Only a few rare Type As have features suggestive of melt solidification—rounded shapes, axiolitic textures (suggestive of crystallization from the rim inwards; MacPherson and Grossman 1981), and relatively Mg-rich melilite that shows a wide compositional range (MacPherson and Grossman 1984). Grossman (1975) described a "compact" spheroidal Type A, TS2-Fl, whose texture and mineralogy are similar to those of Type B except in being pyroxene poor and anorthite free.

The Type I inclusions of Grossman (1975) were so-named because of their "intermediate" nature relative to Types A and B inclusions. Recent detailed work has shown, to the contrary, that the bulk compositions of Type I are melilite deficient and more anorthite normative than both A and B (see review by Wark 1987). Wark therefore proposed renaming them Type C to avoid any erroneous implication of "intermediate." The textures of Type C are dominated by anorthite, which forms euhedral laths in the coarse-grained poikilitic to ophitic varieties and a fine-grained groundmass in the finer-grained examples. Alteration, even in Allende examples, is minimal because of the paucity of melilite. Because of the ophitic textures in some, and the fact that their crystallization sequences are broadly consistent with those predicted by melt/solid phase equilibria for such bulk compositions (Paque and Stolper 1984), Type C are thought to have solidified from melts (see Sec. 10.3.13 below).

Forsterite-bearing inclusions (see references cited above) are characterized by poikilitic textures: equant grains of Mg-rich olivine and smaller grains of spinel are inhomogeneously distributed within fassaite crystals. The olivine contains > 1.0 wt. % CaO. Mg-rich melilite of a composition rarely seen in Type B occurs in some inclusions. There is little secondary alteration. Overall, the mineralogy and mineral chemistry of these objects indicates their bulk compositions to be Mg-, Si-rich and Ca-, Al- and Ti-poor relative to Types B and A.

Considerable evidence indicates that forsterite-bearing inclusions solidified from melt droplets (see Sec. 10.3.13). Also, some examples contain evidence for partial volatilization of Mg and Si while the inclusions were partially molten (see Sec. 10.3.11).

The fine-grained, spinel-rich inclusions in CV3s have generated considerable interest ever since Tanaka and Masuda (1973) first found in one such Allende inclusion a fractionated REE pattern (now known as Group II). Wark and Lovering (1977) first described their essential petrographic characteristics. Other Allende examples are described by Armstrong and Wasserburg (1981), MacPherson and Grossman (1982), Kornacki and Wood (1984) and Hashimoto and Grossman (1985). Similar inclusions have been described in Mokoia by Cohen et al. (1983). A series of chemical and petrographic studies of an Efremovka fine-grained inclusion are given by Ulyanov (1984), Boynton et al. (1986) and Wark et al. (1986).

Figure 10.3.7a,b shows a typical example of a spinel-rich fine-grained inclusion. The dominant feature is numerous small (≥ 1 μm-sized) grains of spinel, sometimes intergrown with perovskite and hibonite, each of which is rimmed onion-skin-like by successive concentric layers of feldspathoids $\pm$

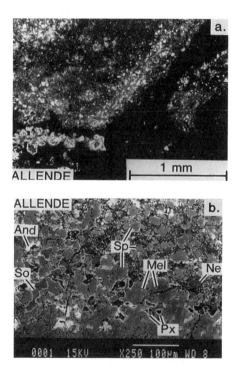

Fig. 10.3.7. A spinel-rich, fine-grained, Group II inclusion from Allende (CV3). (a) Transmitted-light photomicrograph of a portion of the inclusion; note the vague concentric structure, with coarser material near the outer margin and finer-grained dark material in the interior. (b) Back-scattered electron photomicrograph. Irregularly shaped spinel (Sp) grains, many of which have rims of pyroxene (Px), are dispersed in a porous matrix of sodalite (So), nepheline (Ne) and andradite (And). Some of the spinel grains have tiny inclusions of melilite (Mel) within them.

anorthite and pyroxene. Melilite is very rare. The inclusions commonly have a macroscopic concentric color zonation, clearly visible on cut meteorite surfaces, that reflects modal abundance variations of the constituent phases. In the Allende and Mokoia inclusions, the interstices between the rimmed spinel bodies are filled with abundant fine-grained material consisting of nepheline, grossular, sodalite and Fe-bearing garnet and pyroxene. In the Efremovka inclusion El4, the texture is more compact, feldspathoids are absent, and the spinel grains are mantled only by anorthite and pyroxene.

Most studies of fine-grained spinel-rich inclusions have interpreted them as aggregates of independently formed grains (Wark and Lovering 1977; MacPherson and Grossman 1982; Brigham et al. 1985; but, cf., Kornacki and Wood 1985).

Amoeboid olivine aggregates (AOAs; Fig. 10.3.8a,b) are themselves not, strictly speaking, *refractory* inclusions. Their major minerals are olivine, feldspathoids and pyroxene. Enclosed within them, however, are refractory nodules consisting of spinel ± perovskite ± melilite (Fig. 10.3.8b; see also Hashimoto and Grossman 1987). These generally spheroidal nodules, mostly less than 0.1 mm in diameter, are compact in texture and have rims of feldspathoids and aluminous diopside. They are similar in appearance and mineral chemistry to the spinel nodules in many spinel-rich fine-grained inclusions (above) and to spinel-rich spherules in CM meteorites. Descriptions of Allende AOAs are given in Grossman and Steele (1976), Bar-Matthews et al. (1979), Kornacki et al. (1983), Kornacki and Wood (1984) and Hashimoto

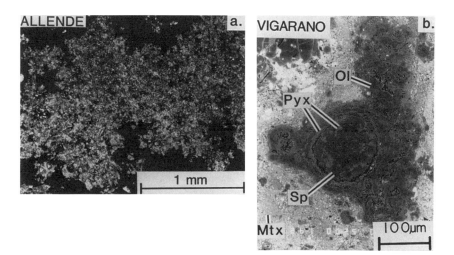

Fig. 10.3.8. Amoeboid olivine aggregates. (a) Transmitted light photomicrograph of a large example from Allende (CV3). (b) Back-scattered electron photomicrograph of a refractory nodule inside a small fragment of an amoeboid olivine (Ol) aggregate in Vigarano (CV3). Mtx = meteorite matrix; Sp = Spinel; Pyx = Pyroxene.)

Fig. 10.3.9. Back-scattered electron photomicrograph of a porous aggregate of hibonite (Hib) crystals in the Vigarano (CV3) meteorite. (Mtx = meteorite matrix.)

and Grossman (1987); Allende and Mokoia AOAs are described in Cohen et al. (1983) and Kornacki et al. (1983).

Another variety of fine-grained inclusion consists solely of a porous aggregate of hibonite crystals (Fig. 10.3.9). MacPherson and Grossman (1982) described a similar object from Allende. Little is known about these rare objects.

C. Bulk Chemistry

Grossman (1980) has summarized the essential chemical characteristics of "coarse-grained" refractory inclusions in Allende, and compiled an extensive reference list for sources of published major- and trace-element compositions. To this list can be added the papers by Mason and Taylor (1982), Wark and Lovering (1982a; Type B, including data for one Vigarano and one Leoville inclusion), Wark et al. (1987) and Clayton et al. (1984) for forsterite-bearing inclusions, and Wark (1987) for Type C inclusions. Trace-element and some major-element data for AOAs are given in Grossman et al. (1979), and for spinel-rich fine-grained inclusions in Grossman and Ganapathy (1976) and Mason and Taylor (1982). Kornacki and Fegley (1986) have compiled and reviewed trace-element data for many varieties of Allende inclusions. Representative and average major element bulk compositions of some CV3 inclusion types are given in Table 10.3.2.

Many inclusions are rather uniformly enriched in all refractory elements to 15 to 30 times CI chondritic abundances. A few "less refractory" elements such as Rh, Eu, Yb and V, and some elements whose volatility is increased under oxidizing conditions such as Mo and W, are occasionally lower in abundance. The majority of these unfractionated inclusions are of the coarse-grained variety; a few are fine grained. The absence of fractionation between refractory lithophiles and refractory siderophiles, and between compatible and incompatible refractory lithophiles, indicates that the bulk compositions of the inclusions were not produced by igneous or other planetary fractionation processes (Wänke et al. 1974). Fractionations that do exist between elements of similar chemical behavior (e.g., Au and Ir, or Yb and Lu) can be

TABLE 10.3.2
Major Element Compositions of CV3 and CM Refractory Inclusions[a]

	CV3				CM		
	1	2	3	4	5	6	7
SiO_2	29.1	32.9	25.1[b]	33.7	—	—	~6.7[b]
Al_2O_3	29.6	26.7	37.6	26.6	91.5	75.1	55.7
TiO_2	1.3	1.5	1.0	1.3	<1.9	2.5	1.8
FeO	0.6	0.9	1.7	2.3	0.6	0.4	0.3
MgO	10.2	11.5	4.3	13.1	~0.5	17.1	12.3
CaO	28.8	25.9	29.4	21.6	6.2	4.8	23.0
Na_2O	0.18	0.50	0.8	1.1	0.03	<0.03	0.02
K_2O	0.01	0.02	—	0.05	—	—	—
Cr_2O_3	0.04	0.05	0.02	0.1	<0.01	0.08	0.05
MnO	—	—	0.01	—	<0.01	<0.01	<0.01
V_2O_3	—	—	0.09	—	<0.03	0.29	0.12
NiO	0.06	0.04	0.03	0.08	<0.01	<0.01	0.02
SUM	99.89	100.01	100.05	99.93	<100.73	100.27	100.01

[a] (1) Type B1; average of 12 microprobe analyses, from Wark and Lovering (1982a); (2) Type B2; average of 5 microprobe analyses, from Wark and Lovering (1982a); (3) Fluffy Type A; bulk analysis of CG-11, from Davis et al. (1978); (4) Spinel-rich fine-grained inclusion; bulk analysis from Clarke et al. (1970); (5) Corundum-hibonite inclusion; bulk analysis of BB-5 from Ekambaram et al. (1984b); (6) Spinel-hibonite spherule; bulk analysis of BB-3 from Ekambaram et al. (1984b); (7) Melilite-rich inclusion; bulk analysis of MUM-3 from Ekambaram et al. (1984b).
[b] SiO_2 by difference.

readily explained in terms of volatility differences. Grossman (1973) showed that average Allende coarse-grained CAI compositions correspond to the first ~5% of condensable material predicted to solidify out of a solar gas at a pressure of 10^{-3} atm, at temperatures in excess of 1400 K. Alternatively, these compositions could also represent the last 5% of material residual from a fractional volatilization process. Therefore, although there is evidence (e.g., deviation of observed compositions or crystallization sequences from those predicted) to suggest that many unfractionated CAI in their present forms are not direct condensates, nonetheless their bulk compositions indicate that they or their parental material must at some point have passed through a high-temperature condensation or volatilization process.

A number of inclusions, particularly many fine-grained ones, have strongly fractionated (Group II) refractory-element patterns, depleted in both the most refractory and least refractory lithophiles, and also in all siderophiles. A few of these Group II inclusions are coarse grained, but most are fine grained and spinel rich or amoeboid olivine aggregates. As discussed later in Sec. 10.3.8, Group II inclusions are very important because they must have formed by a condensation process.

The fact that Group II compositions occur in a number of different textural inclusion types suggests that the Group II-forming process was independent of the processes that produced the different inclusion types.

The trace-element chemistry of CV3 inclusions illustrates a paradox concerning the conditions under which CAI formed. Fegley and Palme (1985) showed that the depletions (relative to other refractory elements) of Mo and W in many refractory inclusions cannot be understood in terms of condensation of a "canonical" solar nebular gas. Rather, the observed abundances require condensation under 3 to 4 orders of magnitude higher O fugacity than the canonical solar nebula. The problem is that many inclusions showing Mo and W depletions also contain Ti^{3+} as a major component of the primary pyroxene. Stolper et al. (1982) and Beckett and Grossman (1986) showed experimentally that pyroxenes having the Ti^{3+}/Ti^{4+} ratios found in natural inclusions must have formed under oxidation conditions at least as reducing as inferred for the canonical solar nebula (log $fO_2 \approx -18$ to -20). The resolution to this discrepancy may lie in recognizing that the chemistry and mineralogy of a given inclusion, or of different inclusions, were established by different processes under different conditions (Wark and Wlotzka 1982; Fegley and Kornacki 1984; Beckett et al. 1987). Specifically, the W and Mo depletions may have been established in an early high-temperature event under oxidizing conditions, whereas melting was a subsequent event that occurred under conditions that were sufficiently reducing to stabilize Ti^{3+}. To make matters even more complicated, there is also evidence that the late-stage alteration took place under very oxidizing conditions. Ihinger and Stolper (1986) showed that the orange-pleochroic hibonite characteristic of inclusions in Allende formed by secondary oxidation of originally blue hibonite that formed under reducing conditions. Thus, a variety of evidence seems to indicate that inclusions formed and evolved by a sequence of processes under a variety of conditions.

D. Isotopic Characteristics

The literature on isotopes in CV3 inclusions is long and complex. Wasserburg (1985) has summarized the state of knowledge concerning short-lived nuclides in CAI and other solar system material (see also Chapter 15.1). Isotope anomalies are discussed in Chapters 12.1 and 14.3. See also other references noted in the introduction to this chapter. We limit ourselves here to brief discussions of the O and Mg-Al systems, and also we are in general concerned with systems showing evidence for mass-dependent fractionation that may be the result of condensation or volatilization processes.

With the exception of the FUN inclusions, Allende inclusions of all petrographic varieties define a single mixing line with unit slope (e.g., Fig. 10.3.10) on a three-isotope diagram for O (Clayton et al. 1973). Analyses of

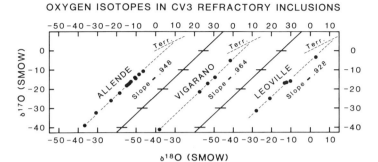

Fig. 10.3.10. Oxygen-isotopic data for bulk refractory inclusions from the CV3 meteorites Allende, Vigarano and Leoville. Representative Allende data are from Clayton et al. (1977). Vigarano and Leoville data from Clayton et al. (1986,1987). The line marked "Terr." is a fractionation line that includes terrestrial materials. Data are plotted relative to Standard Mean Ocean Water (SMOW). (See also Chapter 12.1.)

mineral separates from inclusions have shown that the phases within each inclusion define the same line as do the bulk inclusions. Spinel and pyroxene generally show greater enrichment in ^{16}O than do melilite, anorthite or the secondary phases (Clayton et al. 1977; see also Fig. 12.1.1). This mixing line has been interpreted in terms of diffusion-controlled solid-state exchange of O with a different reservoir from that in which the inclusions initially formed (see review by Grossman 1980). Leoville inclusions (Fig. 10.3.10) define a mixing line that is slightly displaced from the Allende line at ^{16}O-poor compositions and convergent with it at the ^{16}O-rich end (Clayton et al. 1986). Interpreted in the context of the Allende mixing model outlined above, the Leoville data indicate that these inclusions exchanged O with a somewhat different reservoir than did the Allende CAI. Most Vigarano inclusions (Fig. 10.3.10) lie on a line displaced slightly to the ^{18}O-poor side of the Allende mixing line (Clayton et al. 1987). The significant observation here is that the Vigarano inclusions are virtually free of the alteration phases that are so abundant in Allende. Therefore, the observation that they disperse along a similar mixing line to that of the Allende CAI shows that the isotopic exchange that produced the mixing line was not connected with alteration (see Sec. 10.3.14).

In Type B1 inclusions, there is a well-defined correlation between $^{26}Mg/^{24}Mg$ and $^{27}Al/^{24}Mg$, attributable to the *in situ* decay of the short-lived radionuclide ^{26}Al. Data from a number of B1s define a line with a slope corresponding to $(^{26}Al/^{27}Al)_o = 5 \times 10^{-5}$. In Types B2, C and A, however, analyses of individual phases do not define a unique correlation line. In the case of B2s, ion-microprobe studies have shown that some interior grains do fall on the B1 "isochron" while others nearer the edge of the inclusion and some grain boundaries indicate much lower initial $^{26}Al/^{27}Al$ ratios. Hutcheon

(1982) and Armstrong et al. (1984a) interpret the data in terms of late-stage disturbance of the isotopic systems in which there was exchange of bulk Mg or ^{26}Mg. Those authors also note that the type of exchange process may not have been the same for all inclusions. Magnesium-isotopic data for several spinel-rich fine-grained inclusions are given in Brigham et al. (1984, 1985,1986). The data obtained so far hint at the *in situ* decay of ^{26}Al, but again the trends are not well defined and cannot be interpreted in terms of simple single-stage decay. Ion-microprobe studies of Mg isotopic mass fractionation in phases in Wark-Lovering rims (Fahey et al. 1985b; Davis et al. 1986) have shown rims to be isotopically lighter and more nearly normal (relative to terrestrial) than the inclusions they mantle. Since other features of rims are suggestive of volatilization (see Sec. 10.3.11), this observation seems to indicate that rimming involved isotopic exchange with a different reservoir from that in which the inclusions initially formed. Collectively, all the data suggest that considerable isotopic heterogeneity existed in the early solar nebula, and that a number of processes acted on the inclusions after they were formed to disturb the isotopic systems.

There are small but consistent, correlated, mass-dependent isotopic fractionation effects in CV3 inclusions that are related to inclusion type but are not correlated with nuclear anomalies other than in O (i.e., non-FUN inclusions); the elements involved are Ca, Mg and Si (see Niederer and Papanastassiou 1984; review by Clayton et al. 1985). Fine-grained inclusions tend to be enriched in the light isotopes of Si and Mg and either light or heavy Ca; coarse-grained inclusions are enriched in the heavy isotopes of Si and Mg and the light isotopes of Ca. Cumulatively, the data suggest (see Sec. 10.3.11) that the inclusions evolved by a complex series of processes that probably included both condensation and volatilization.

10.3.4. INCLUSIONS IN CO3 METEORITES

A. Classification

No classification scheme has yet been proposed for CO3 inclusions. Most authors have referred the properties of CO3 inclusions to the existing classifications for CM and CV3 inclusions. We will do likewise here.

B. Textural Characteristics

There have been numerous brief descriptions of CO3 refractory inclusions, but few systematic or detailed studies. Among the meteorites from which inclusions have been described are Lancé (Kurat 1970,1975; Frost and Symes 1970; Kurat and Kracher 1977), Isna (Methot et al. 1975), Ornans (Noonan et al. 1977; Davis 1984,1985; Davis and Hinton 1985,1986), Colony (Rubin et al. 1985) and ALH 77003 (Ikeda 1982). Examples of several types of CO3 inclusions are shown in Fig. 10.3.11, all from Ornans.

Melilite-rich Type A CAI (Fig. 10.3.11a) are common in CO3s. They

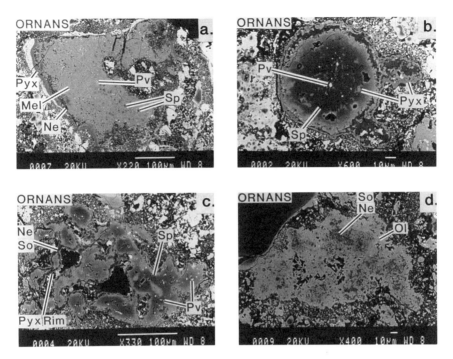

Fig. 10.3.11. Back-scattered electron photomicrographs of inclusions from the Ornans (CO3) meteorite. (a) A melilite-rich Type A inclusion. Enclosed within melilite are tiny spinel and perovskite grains. A rim of pyroxene and nepheline mantles the inclusion. (b) A spinel-perovskite spherule. Note the pyroxene in the rim as well as in the interior. Dark band in rim consists of nepheline and sodalite. (c) Spinel-pyroxene aggregate. Note the highly irregular, chain-like nature of this object. The spinel encloses perovskites, and in turn is rimmed by nepheline, sodalite and pyroxene. (d) Amoeboid olivine aggregate. Note the irregular shape, porous texture and dark alteration zones of nepheline and sodalite. Abbreviations are as follows: Pyx = Pyroxene; Mel = Melilite; Ne = Nepheline; Pv = Perovskite; Sp = Spinel; So = Sodalite; Ol = Olivine.

are similar in their properties to CV3 Type A but smaller (most are < 0.5 mm; rarely, as large as ~ 1 mm); their shapes can be sinuous, irregular, compact or, rarely, spheroidal.

The sinuous variety consists of gehlenitic melilite, Åk 0 to 5, enclosing bands or chains of Fe-free spinel and accessory perovskite (see, e.g., Kurat 1975, his Fig. 5). The more compact variety of CO3 Type A (e.g., Fig. 10.3.11a) is characterized by dense polygonal-granular mosaics of gehlenite-rich melilite (Åk 1 to 30) enclosing accessory spinel, perovskite and, more rarely, hibonite. Davis (1985) reported the occurrence of Pt metal nuggets in the interior of a melilite-rich Ornans inclusion. Wark-Lovering-type rims consist mostly of diopsidic pyroxene and lesser feldspathoids.

The descriptions by Davis indicate that the more compact melilite inclusions are less altered than their irregular and sinuous counterparts. In either case, alteration is generally most intense in the outer zones, and consists mostly of feldspathoids; the same is true in Lancé (Kurat 1975).

Other types of melilite-bearing inclusions include a spinel-free variety (in Ornans; Davis 1985) and a melilite "chondrule" in Colony consisting of gehlenite blades in a fine-grained mesostasis (Rubin et al. 1985).

An example of a refractory spherule from Ornans is shown in Fig. 10.3.11b. It has a core of spinel + perovskite + pyroxene mantled by a rim of feldspathoids and diopsidic pyroxene. Frost and Symes (1970) described a similar spherule from Lancé. Several hibonite-rich blue spherules were reported in Colony by Rubin et al. (1985). A unique chondrule in Lancé (Kurat 1975) consists entirely of two hibonite crystals enclosed in a sphere of extremely Ca-Al-rich glass.

Another common type of inclusion consists of irregular chains and aggregates of spinel nodules ($\pm$ accessory perovskite and hibonite), each rimmed by aluminous diopside (e.g., Fig. 10.3.11c). Spinel in Isna inclusions is uniformly Fe rich ($\sim 22\%$ FeO; Methot et al. 1975), in contrast to the inhomogeneous spinels reported from some other CO3s (e.g., Ornans; Davis 1985). In all respects, the spinel-rich inclusions in CO3 meteorites resemble the "nodular" and "banded" spinel-pyroxene inclusions described from the Murchison CM meteorite (MacPherson et al. 1983; see below).

Amoeboid olivine aggregates (Fig. 10.3.11d) in CO3s are similiar to those in CV3s but much smaller (generally <0.6 mm). They consist mostly of a granular matrix of olivine that encloses occasional nodules of spinel plus diopside (not shown in the figure).

Of particular interest and significance are several so-called ultrarefractory inclusions reported from Ornans. One of these, the inclusion OSCAR (Davis 1984), is a fassaite-hibonite-spinel-perovskite inclusion in which many of the phases (and especially the fassaite) are exceedingly enriched in Sc. An ultrarefractory Fremdling in Ornans was reported by Palme et al. (1982).

C. Bulk Chemistry

Little is known about the bulk compositions of CO3 inclusions, other than what has been inferred from mineral chemistry and modal abundances. The only two inclusions for which bulk chemical data exist are nonrepresentative: the ultrarefractory inclusions RNZ (Palme et al. 1982) and OSCAR (Davis 1984). The OSCAR composition is reconstructed using modal mineral proportions, ion- and electron-microprobe analyses of the individual phases. Davis (1985) inferred, from a paucity of fassaite and the Al-rich nature of the melilite, that Ornans inclusions are more enriched in refractory elements than most CV3 inclusions.

The term "ultrarefractory" refers to the very highest-temperature fraction (the first $\sim 1\%$) of matter predicted to condense out of a cooling solar nebular gas. The hallmark of such a hypothetical component is an enrichment of the most refractory over the less refractory REE, and fractionation of the most refractory REE and refractory metals relative to one another, the exact complement to a Group II REE pattern (see Fig. 10.3.4, and Sec. 10.3.8). The Ornans inclusions RNZ and OSCAR both show such REE patterns. RNZ is also highly enriched in refractory siderophile elements.

D. Isotopic Characteristics

The only isotopic data for CO3 inclusions are ion-microprobe analyses of individual phases in inclusions from Ornans (Davis and Hinton 1986) and Lancé (Fahey et al. 1986; Fahey and Zinner 1987). Some Ornans inclusions show evidence for the *in situ* decay of ^{26}Al while others do not. The inclusion OSCAR has normal Ti isotopes within the analytical uncertainty of the ion probe, suggesting that, in spite of its bizarre Sc-rich chemistry, OSCAR probably did form in the solar system. The Lancé inclusion described by Fahey and co-workers shows large Ti and Ca anomalies in all of the phases analyzed. Measured excesses of ^{26}Mg do not correlate with Al/Mg and are inhomogeneously distributed, suggesting possible late (post-^{26}Al decay) disturbance of the Mg-Al system in CO3 inclusions just as in the CV3 inclusions. Fahey and Zinner (1987) searched for, but did not find, anomalous isotopic effects in Fe that might be expected to correlate with the Ca and Ti anomalies.

10.3.5. INCLUSIONS IN CM METEORITES

A. Classification

Fuchs et al. (1973) first noted many of the inclusion types present in Murchison, the most thoroughly studied CM. Later detailed surveys of CM refractory inclusions by Macdougall (1979,1981) and MacPherson et al. (1983,1984*a*) established the essential petrographic characteristics of most of the inclusion types. The nomenclature used here follows that developed by the latter authors, which incorporates and expands on Macdougall's earlier work.

B. Textural Characteristics

CM refractory inclusions are small: the largest known is ~2 mm in maximum dimension (Hashimoto et al. 1986) and all others are ≤ 1 mm. The analytical difficulties posed by such small sizes are partly responsible for the paucity (until recently) of bulk chemical and isotopic studies on these objects.

The most easily recognized CM inclusions contain hibonite with a very characteristic and spectacular sky-blue color, visible on cut meteorite surfaces and in thin section. There are two principal varieties of hibonite inclusions: irregularly shaped spinel-hibonite inclusions, and spinel-hibonite spherules. Within the category of irregularly shaped inclusions are a wide variety of objects. Most have spinel and hibonite in subequal abundances, with accessory perovskite (e.g., Fig. 10.3.12). Others consist entirely of loose aggre-

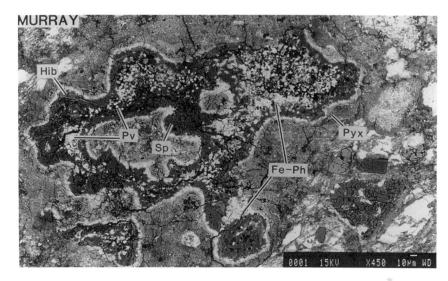

Fig. 10.3.12. Back-scattered electron photomicrograph of an irregularly shaped spinel-hibonite-perovskite inclusion from Murray (CM). Perovskite (Pv) is concentrated towards the interior of the loop-like inclusion. A rim of Fe-rich phyllosilicate (Fe-Ph) and pyroxene (Pyx) mantles the inclusion. (Hib = Hibonite; Sp = Spinel.)

Fig. 10.3.13. Back-scattered electron photomicrograph of a loose aggregate of hibonite (Hib) crystals in Murray (CM). The inclusion is partly mantled by Fe-rich phyllosilicate (Fe-Ph), which also occurs within cavities in hibonite.

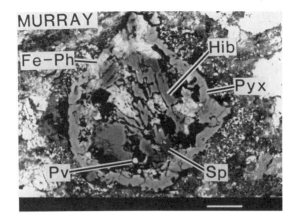

Fig. 10.3.14. Back-scattered electron photomicrograph of a small spinel-hibonite spherule in Murray (CM). Note Fe-rich-phyllosilicate (Fe-Ph) in both the interior and rim. (Pv = Perovskite; Sp = Spinel; Hib = Hibonite; Pyx = Pyroxene.)

gates of hibonite crystals (e.g., Fig. 10.3.13; cf. Fig. 10.3.9). In both cases, these inclusions can take the form of a "train" of clumps of crystals that are separated from one another by intervening meteorite matrix. Each clump may be individually rimmed. A particularly important subgroup of hibonite-rich inclusions include the Blue Angel (Armstrong et al. 1982), and SH-5 and SH-6 (MacPherson et al. 1984a). These very porous aggregates of tiny, perfectly euhedral hibonite crystals are among the best candidates for texturally preserved solar-nebula condensates, as discussed in Sec. 10.3.8.

Spinel ($\pm$ hibonite) spherules are small (rarely as large as 300 μm) and consist of spinel + hibonite $\pm$ perovskite or spinel + perovskite (e.g., Fig. 10.3.14). Many of these spherules contain radial sprays of hibonite crystals that project inward from the inclusion margins. This feature, together with the spheroidal shapes and compact textures of the inclusions, suggests that the inclusions solidified from melt droplets (Macdougall 1981; MacPherson et al. 1983). Two spherules are known so far that contain corundum in addition to hibonite and perovskite: BB-5 (Bar-Matthews et al. 1982) and GR-1 (MacPherson et al. 1984a). GR-1 also contains unusual and tiny Ca-Ti-Sc-Zr-Y-rich grains.

Spinel-pyroxene inclusions are probably the most numerous of all CM refractory inclusion types. These range in structure from highly sinuous to banded to nodular (Fig. 10.3.15; see also MacPherson et al. 1983,1984a), but the sequence of mineralogy is always the same: a core of Fe-poor spinel, mantled by a rim of pyroxene that grades outward in composition from fassaite to nearly pure diopside. A layer of forsterite mantles the pyroxene in some instances (Fig. 10.3.16), in which case the inclusions resemble the refractory nodules within amoeboid olivine aggregates in CV3 meteorites (cf.,

Fig. 10.3.15. Back-scattered electron photomicrograph of a nodular spinel-pyroxene inclusion in Mighei (CM). Note the thick pyroxene (Pyx) rim and patchy Fe-rich phyllosilicate (Fe-Ph) between the pyroxene and the spinel (Sp).

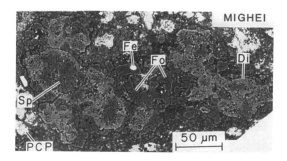

Fig. 10.3.16. Back-scattered electron photomicrograph of a nodular spinel-pyroxene-olivine inclusion in Mighei (CM). Rare grains of Fe metal are enclosed in the olivine. Abbreviations as used previously.

Fig. 10.3.8b). Separating the pyroxene from the spinel is the ubiquitous layer of Fe-rich phyllosilicate. Some examples contain accessory hibonite; with increasing proportions of that phase they grade into the irregularly shaped spinel-hibonite inclusions.

Among the rarest CM inclusions are melilite-rich ones (MacPherson et al. 1983; El Goresy et al. 1984). The former authors described one (MUM-1) that consists largely of gehlenite-rich melilite that encloses euhedral crystals of hibonite, spinel and perovskite. The rim sequence consists of a thick layer of hibonite + perovskite + spinel + calcite, followed sequentially outwards by layers of melilite, anorthite and pyroxene. Despite its rarity as a major phase in CM inclusions, melilite does occur sporadically as a minor phase in some spinel-hibonite spherules (MacPherson et al. 1984a); it also occurs in the hibonite-rich Blue Angel inclusion (Armstrong et al. 1982).

The freeze-thaw disaggregation procedure that has been used to recover large numbers of Murchison inclusions (see, e.g., MacPherson et al. 1980) has the unfortunate side effects of removing objects from their context and of breaking up fragile inclusions. These drawbacks are particularly evident in the case of some interesting single hibonite crystals and crystal fragments recovered by this technique. These hibonites have low Ti and Mg contents and show no evidence for the *in situ* decay of ^{26}Al. Detailed ion-microprobe analyses of a large collection of such hibonite crystals (Fahey et al. 1987*a*) have revealed a complex assortment of trace-element and isotopic signatures that, taken together, point to a very heterogeneous early solar nebula. Thus these crystals are very interesting, and it is desirable to know what their textural context is within the meteorite. Because of the recovery procedure, unfortunately, it is not clear whether the hibonites were originally sited in some kind of inclusion prior to disaggregation (see, e.g., Hutcheon et al. 1980; MacPherson et al. 1983) or whether they actually occurred as crystals and crystal fragments isolated in the matrices of their host CM meteorites. Ion-microprobe analyses of an isolated crystal found in place within the Murchison matrix (MH88; Macdougall and Phinney 1979) showed no excess ^{26}Mg and low bulk Mg, indicating that single isolated crystals are one viable source for the extracted hibonites.

C. Chemical Characteristics

The general absence of primary silicate phases, and the abundance of aluminous phases such as spinel and hibonite, demonstrate qualitatively that CM inclusions are enriched in Al and depleted in Si relative to CV3 inclusions (Macdougall 1979). Several examples of CM inclusion bulk compositions are given in Table 10.3.2. Bulk analyses of Murchison inclusions by instrumental neutron-activation analysis (Ekambaram et al. 1984*a,b*,1985) show that most CM inclusions have REE abundance patterns broadly similar to either Group II or III CV3 inclusions. Similar results were obtained by ion-microprobe analyses of numerous individual phases (Fahey et al. 1987*a*). Five CM inclusions are known to have ultrarefractory REE patterns: MH-115 (Boynton et al. 1980*a,b*), SH-2 (Ekambaram et al. 1984*b*), BBT-10 and SH-5 (Ekambaram et al. 1984*a*) and the corundum-bearing inclusion GR-1 (Hinton and Davis 1986).

Another corundum-bearing inclusion, BB-5, does *not* have an ultrarefractory REE pattern (Ekambaram et al. 1984*b*) even though, like GR-1, it contains the highest-temperature phase assemblage known in refractory inclusions. This unexpected feature illustrates how petrographic characteristics commonly are not correlated with chemical (or isotopic) ones.

Ekambaram et al. (1984*b*) showed that, relative to CV3 inclusions, CM inclusions commonly are: (1) more enriched in refractory lithophiles; (2) more enriched in the most refractory siderophile elements (Os, Re) relative to Ir and Ru; and (3) more fractionated in their REE patterns. Macdougall

(1979) and Ekambaram et al. (1984b) concluded that CM inclusions stopped equilibrating with the nebular gas at higher temperatures than did the CV3 inclusions.

D. Isotopic Characteristics

Only two CM refractory samples have been analyzed in bulk for O isotopes: SH-7, a hibonite-rich spinel-free inclusion (Hashimoto et al. 1986), and "2C10c", an acid residue of bulk meteorite composed mostly of spinel (Clayton and Mayeda 1984). Fahey et al. (1987b) reported ion-microprobe analyses of O isotopes in four separated hibonite grains from Murray and Murchison. Oxygen isotopes in nonrefractory components from Murchison, consisting mostly of olivine and pyroxene separated from the bulk meteorite by heavy-liquid density techniques, were reported by Clayton and Mayeda (1984).

The Murchison spinel and olivine-pyroxene density separates define a mixing line on a three-isotope O diagram that is similar to but slightly displaced from the well-known CV3 mixing line (Clayton and Mayeda 1984), converging with it at the ^{16}O-rich (spinel) end and diverging at the ^{16}O-poor end. The CM hibonites analyzed by ion microprobe all cluster around the CV3 mixing line, but the analytical uncertainties by this technique are as yet too great to indicate whether these grains lie on the spinel-olivine-pyroxene CM line or the CV3 mixing line. The hibonite-rich inclusion SH-7, however, plots well off and to the left of both the CM and CV3 mixing lines. This deviation from the mixing lines is in the opposite sense from FUN inclusions (which plot to the right of the CV3 mixing line). As Hashimoto et al. (1986) pointed out, there are as yet insufficient data for O in CM inclusions to interpret adequately the composition of SH-7. Collectively, the Murchison data reinforce the idea that several nebular reservoirs are required to explain the isotopic variations of all CAI.

The Ti- and Mg- isotopic characteristics of hibonite-bearing CM inclusions have been reviewed by Fahey et al. (1987a; see also Fahey et al. 1985a; Hinton et al. 1987). In brief, those authors made the following observations. First, only a relatively small fraction of the hibonites that they analyzed contain excess ^{26}Mg attributable to the *in situ* decay of ^{26}Al. These data support earlier similar observations (see, e.g., Hutcheon et al. 1980). There is no consistent correlation of ^{26}Mg excesses with textural or other isotopic properties: Hutcheon et al. (1986) suggested that hibonites that coexist with other phases (spinel and melilite) commonly show ^{26}Mg excesses, but Fahey et al. (1987a) noted that this correlation is not perfect. Second, many hibonites show anomalous Ti isotopes, especially excesses of ^{50}Ti; similar to suggestions made for CV3 inclusions (Niederer et al. 1981,1985; Niemeyer and Lugmair 1984), Fahey et al. concluded that at least four isotopically distinct components are required to account for all of the observed Ti data. They also concluded, however, that there is no evidence that the hibonite grains or subgrains within them are presolar in origin (see Sec. 10.3.15 below).

10.3.6. INCLUSIONS IN ORDINARY AND ENSTATITE CHONDRITES

Although studies of refractory inclusions have concentrated almost exclusively on those in carbonaceous chondrites, such objects and "Al-rich chondrules" are well-known (albeit rare) constituents of ordinary and enstatite chondrites as well. Recent studies by Bischoff and colleagues (see, e.g., Bischoff and Keil 1984; Hinton and Bischoff 1984; Bischoff et al. 1985a) provide the most comprehensive sources of descriptions and analytical information, and there have been scattered other published reports as well (see, e.g., Noonan 1975; Noonan et al. 1978; Nagahara and Kushiro 1982).

A. Classification

Examination of the published descriptions of inclusions in ordinary and enstatite chondrites suggests that the existing classifications used for inclusions in CM meteorites can be applied to these as well.

B. Textural Characteristics

Ordinary Chondrites. Refractory spherules have been reported in several ordinary chondrites (Nelen et al. 1978; Noonan et al. 1978; Bischoff and Keil 1984). Two spherules from Dhajala (H3.8) consist largely of spinel and ilmenite, rims of fassaitic pyroxene, and abundant secondary nepheline and sodalite. The one illustrated in Noonan et al. is encased in a mantle of olivine. Judging from the shapes of the ilmenite crystals in the spherule illustrated by Bischoff and Keil (their Fig. 3d), it seems likely that this phase is replacing perovskite. These two objects closely resemble spherules found in CO3s (cf., Fig. 10.3.11b) and within many amoeboid olivine inclusions in CV3s (cf., Fig. 10.3.8b).

Inclusions that resemble spinel-hibonite inclusions in CM chondrites are rare, but one spectacular example is the blue hibonite-spinel-gehlenite-perovskite inclusion found in Semarkona (LL3.0) by Bischoff and Keil (1984) and which is shown here in Fig. 10.3.17. Its textures resemble those of CM spinel-hibonite spherules while its melilite-rich mineralogy is similar to the inclusion MUM-1 (see above).

Bischoff and Keil also describe an irregularly shaped spinel-rich inclusion that is identical to the spinel-pyroxene aggregates in CM meteorites. It consists of nodules of spinel, with minor perovskite, all encased in a continuous rim of fassaitic pyroxene.

The Al-rich chondrules described by Nagahara and Kushiro (1982) and Bischoff and Keil (1984) are a distinct group of objects. Although in size, shape and texture they resemble normal ferromagnesian chondrules, they differ in major-element bulk chemistry and are significantly more enriched in refractory elements. Their bulk compositions do not overlap those of coarse-grained inclusions in CV3s or refractory spherules in CMs, being less Ca-Al-

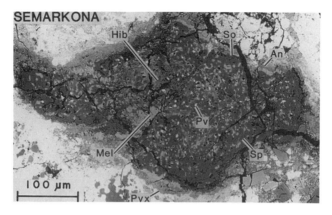

Fig. 10.3.17. Back-scattered electron photomicrograph of a hibonite-perovskite-spinel-melilite inclusion from the Semarkona (LL3) chondrite. This inclusion is described by Bischoff and Keil (1984). At upper right (near "An") and lower center (near "Pyx") can be seen the vestiges of a melilite-anorthite-pyroxene rim sequence. Abbreviations as used previously.

rich than either of those groups of inclusions. However, in mineralogy and major-element chemistry they are indistinguishable from Al-rich chondrules in CV3 chondrites (Wark 1987).

Among a diversity of other kinds of refractory objects that have been found in ordinary chondrites (see references given above), one that is especially worth noting is the isolated hibonite crystal found enclosed in the Dhajala matrix and analyzed by Hinton and Bischoff (1984). The significance of this crystal lies in its isotopic composition showing evidence of excess ^{26}Mg, as discussed below.

Enstatite Chondrites. The only descriptions of refractory inclusions in enstatite chondrites are of a single Al-rich chondrule in Qingzhen by Rambaldi et al. (1984) and a survey of 30 Al-rich objects in 4 enstatite chondrites by Bischoff et al. (1985a).

The Qingzhen object described by Rambaldi et al., and 13 of those described by Bischoff et al., can be described as Al-rich chondrules whose primary mineralogy is dominated by enstatite and which have accessory fassaite or calcic plagioclase. Eleven Al-rich "fragments" described by Bischoff et al. are probably broken pieces of Al-rich chondrules.

The remaining six objects are five irregular inclusions and one isolated spinel grain. The members of this limited population of inclusions are smaller in maximum size than their counterparts in ordinary chondrites, being generally less than 140 μm in size. Most are spinel rich with accessory perovskite; one contains hibonite as a major phase in addition to spinel. All have porous textures, leading Bischoff et al. to conclude that they had never been melted.

The inclusions are all extensively altered to alkali- and halogen-rich secondary phases. In two respects, these objects differ from their OC and CM counterparts: they lack rims, and the spinels do not contain Fe.

C. Bulk Chemistry

All major-element bulk chemical data that presently exist for Al-rich objects in ordinary and enstatite chondrites have been obtained by electron-microprobe analyses of thin sections. Al-rich chondrules from both enstatite and ordinary chondrites are similar, having positively correlated Ca and Al with Ca/Al wt. ratios lower than the CI value of 1.1. Those from enstatite chondrites are generally lower in bulk Fe than those from ordinary chondrites, reflecting the absence of Fe from the silicate and oxide phases in the former.

While the Al-rich chondrules from both chondrite groups are more Ca-Al-rich than ordinary ferromagnesian chondrules, they are distinctly Ca-Al-poor and Si-Mg-rich relative to coarse-grained inclusions in CV3s (Nagahara and Kushiro 1982; Bischoff and Keil 1984; Bischoff et al. 1985a). Al-rich inclusions in enstatite and ordinary chondrites are similar to one another in bulk composition but, in contrast to the chondrules, show very low and decreasing Ca with increasing Al. The Ca/Al ratios of the most Al-rich inclusions are much lower than 1.10.

Some trace-element data for Al-rich objects in ordinary chondrites are available, mostly for Al-rich chondrules. Three chondrules from Dhajala (Boynton et al. 1983) are somewhat enriched in refractory lithophile elements (3 to 8 times CI chondrites), with unfractionated REE patterns and depletions in refractory siderophiles. Three chondrules from Ybbsitz (H4) (Bischoff et al. 1985b) show similar levels of enrichments of the refractory lithophile elements, but with REE patterns that are fractionated in favor of the light REE and positive Eu anomalies that together suggest an important plagioclase component in the precursor material for the chondrules. The spinel-ilmenite inclusion from Dhajala noted above (Bischoff and Keil 1984) has a strongly fractionated Group II REE pattern but with much lower Eu than is typical for most Group II inclusions (Boynton et al. 1983). Hinton et al. (1985) reported ion-microprobe analyses of REE in the single crystal of hibonite from Dhajala for which isotopic analyses were previously made (see below). That grain shows a smooth fractionation of REE, in favor of the light REE, with a very large negative Ce anomaly and a slight Pr anomaly. Such a pattern was previously observed in the Allende FUN inclusion HAL, and is interpreted to indicate that both the HAL and Dhajala hibonites formed under much more oxidizing conditions than the canonical solar-nebular values.

D. Isotopic Characteristics

The only isotopic data that currently exist for Al-rich objects in ordinary chondrites (there are no data for those in enstatite chondrites) are ion-microprobe analyses of Mg and Ca in plagioclase and hibonite in several Type

3 chondrites (Hinton and Bischoff 1984; Hinton et al. 1984). Plagioclase grains in two Al-rich chondrules (in Sharps (H3.4) and ALHA 77299 (H3.7)) and one normal ferromagnesian chondrule (in Krymka (L3)) show no excesses of ^{26}Mg. However, a hibonite clast from Dhajala shows distinct excess ^{26}Mg that is correlated with ^{27}Al/^{24}Mg, providing clear evidence for the *in situ* decay of ^{26}Al and corresponding to an initial ^{26}Al/^{27}Al ratio of 8.4 × 10^{-6}. These results imply that ^{26}Al was present during the formation of at least this ordinary-chondrite fragment. However, because of the much lower initial ^{26}Al/^{27}Al ratio than that inferred for most carbonaceous chondrite inclusions (~5 × 10^{-5}), it is possible that this hibonite formed as much as 2 Myr later than those in the carbonaceous chondrites (see also Chapter 15.1). Alternatively, scenarios involving nonuniform abundances of ^{26}Al or later mobilization of Mg cannot be ruled out. Calcium isotopes in the Dhajala hibonite are mass fractionated in favor of the heavy isotopes (Hinton et al. 1984).

10.3.7. PROCESSES THAT HAVE AFFECTED CAI: BACKGROUND

The main framework for considering early solar-nebular evolution has been the equilibrium condensation model, in which thermodynamics is used to calculate what condensed species would have formed out of the hot and cooling protosolar nebula. Assuming thermodynamic equilibrium (nucleation and condensation occurring at the equilibrium temperature) and plausible physical conditions (mainly pressure) in the nebula, the sequence of condensed phases that form as temperature falls can be directly calculated from the thermodynamic properties of all possible condensing and gaseous species (*if known*).

Equilibrium condensation is not without its problems in explaining the formation of refractory inclusions. A number of workers have pointed out that, in the tenuous environment of the early solar nebula, equilibrium condensation of crystalline solids would be kinetically unfavorable relative to the subcooled condensation of either liquids (Blander and Fuchs 1975) or amorphous solids (see, e.g., Nuth and Donn 1983*a*). This problem would be particularly acute at the highest temperatures (i.e., during CAI formation) where few opportunities for heterogeneous nucleation would be present.

A second difficulty is the "hot nebula" problem. Assuming that the condensable species in the gas were brought into the solar nebula largely as interstellar dust grains which became vaporized by ambient high nebular temperatures, then condensation must have occurred during subsequent cooling of the gas. Many recent physical models for the presolar nebula do not predict temperatures in excess of 2000 K except within 1 AU or so of the center. As a consequence, refractory condensates could form only in that inner zone, requiring that they then be transported radially outward at least as far as the asteroid belt—the presumed accretionary region for chondrites. To avoid these problems (and others), an alternative model for CAI is that they formed by volatilization of less refractory matter in response to *local* transient heating events.

Equilibrium condensation makes specific predictions about the phases that condense and the order in which they do so. Deviations of observed mineralogy and crystallization sequences from those predicted pose difficulties for a condensation interpretation. Examples include the absence from most CAI (Kornacki and Wood 1985) of the predicted phase $CaAl_4O_7$ (Fegley 1982), and the observation that spinel is generally enclosed within melilite when the reverse crystallization order is predicted (see, e.g., MacPherson and Grossman 1984).

In fact, it has in general proven difficult to attribute many features of CAI to formation through simple equilibrium condensation. The data presented in the preceding sections reveal the great diversity and complexity of CAI. It seems obvious now that multiple processes and protracted histories are required to explain not only the diversity of CAI but also the great complexity of individual inclusions, yet this recognition was slow in coming. Much disagreement over CAI genesis can be traced to the vagaries of textural interpretation. First, there is the problem of recognizing what a condensate should look like. Compounding this difficulty is the fact that many CAI *do* look like solidified melts. Then, there is the problem that different scientists studying different examples of a diverse assortment of objects arrive at different conclusions. Finally, most CAI have experienced secondary events whose signatures obscure primary features; such secondary processes include rimming, alteration and metamorphism.

The most frustrating problem has been that textural properties of refractory inclusions seldom correlate with isotopic and bulk chemical properties. This dichotomy of petrographic vs isotopic/bulk-chemical evidence led to conflicting interpretations for CAI origins. It now seems likely, however, that secondary reprocessing eradicated primary textures more thoroughly than it did the chemical and isotopic signatures. From this point of view, the different forms of analysis are complementary: evidence for the ultimate genesis of the inclusions is best preserved in bulk chemical and isotopic systematics, whereas textural properties are roadmaps of subsequent CAI history. With this introduction to the pitfalls of interpreting CAI genesis, we turn in the following sections to a discussion of various primary and secondary processes that may have participated in that genesis.

10.3.8. PROCESSES: VAPOR-SOLID CONDENSATION

Equilibrium condensation in the context of the solar nebula has come to mean vapor-solid condensation, that is, no intermediate liquid is formed (see below) because of the low pressures involved.

In full equilibrium calculations (see, e.g., Grossman 1972) the changing composition of the gas as condensation proceeds is taken into account, and it is assumed that all solids remain in contact (and equilibrium) with the gas as the temperature falls. At the opposite extreme is fractional condensation, in which solid species are instantly removed and prevented from equilibrating with the system. Neither end-member is realistic, since there is ample evidence for preservation of very high-temperature phases and partial back reaction of those phases with the gas at lower temperatures.

Equilibrium condensation has scored some notable successes and failures in predicting the chemical and textural properties of CAI. Without question the greatest success has been in explaining trace-element patterns in inclusions. The model has not succeeded well in explaining CAI textures, but this failing may be irrelevant to the extent that original textures were obliterated by later melting and other reprocessing.

We noted earlier (Sec. 10.3.3) that the major-element compositions of CV3 inclusions are broadly compatible with the predictions of equilibrium condensation models for the first $\sim 5\%$ of matter to condense out of a solar gas. We also noted that the elemental fractionation patterns indicate gas-solid fractionation rather than planetary processes. Grossman (1975), Beckett et al. (1980), Kerridge (1981) and Stolper (1982) have discussed the correlation between major-element bulk compositions of CV3 inclusions and the predicted bulk compositions of total condensed solids. The agreement of Type A bulk compositions with the predicted solar-nebula condensation trajectory is fair, but Type B show significant discrepancies that are progressively worse going from B1 to B2. These discrepancies can be partially accounted for by condensation of a complex pyroxene solid solution for which thermodynamic data do not exist. Stolper concluded that the observed and calculated trends are reconcilable "to first order", if the pyroxene that condenses is fassaite rather than diopside. The bulk compositions of Type C inclusions are completely at variance with the predicted solar-nebular trajectory (see Sec. 10.3.9).

The bulk compositions of spinel-hibonite-rich, melilite-free CM inclusions are a major problem for equilibrium condensation models. Spinel is predicted to condense after, not before, melilite. Since hibonite should react with the gas to form melilite, the formation of a spinel-hibonite assemblage without melilite is difficult to understand. This is the same problem posed by melilite-rich CV3 inclusions with spinel enclosed by melilite, since the reverse enclosing relationship is expected on the basis of condensation calculations. This unresolved problem has been the subject of considerable debate (see, e.g., Kornacki and Wood 1985; Ekambaram et al. 1984*b*).

The unfractionated abundances of trace elements in many CV3 Group I inclusions have been explained by Grossman (1973) in terms of equilibrium condensation calculations, as being due to the fact that all of these elements are completely condensed above the accretion temperature of the inclusions, and that the inclusions very efficiently "scavenged" them (Grossman et al.

1977). Boynton (1975) and Davis and Grossman (1979) specifically considered the REE, and showed that the abundance patterns of REE in refractory inclusions can be successfully explained in terms of a condensation model in which the condensation temperatures of the elements are controlled by their relative volatilities. The REE abundances in unfractionated (e.g., Group I) inclusions reflect essentially complete condensation of all REE except Eu and Yb into those inclusions. However, as these unfractionated REE patterns would also be characteristic of the residue from a partial volatilization process, they do not allow discrimination between a condensation or volatilization origin for the inclusions.

The case for the fractionated Group II inclusions is far more restrictive. Boynton (1975) calculated that the very highest-temperature fraction of material to condense from a cooling solar-nebular gas (the "ultrarefractory component") would show large enrichments of the most refractory over less refractory REE and that the most refractory REE would themselves be fractionated from one another. The early removal of this component from contact with the gas leaves a residual gas whose subsequent condensation would produce Group II REE abundances. The critical feature of Boynton's model is that the Group II material *cannot be solely a volatilization residue,* as it would then necessarily have an ultrarefractory pattern, i.e., the reverse of what is actually observed. If volatilization was involved in the genesis of Group II inclusions, then these inclusions were formed by the recondensation of the early-evaporated portion. Either way, the trace-element fractionation in Group II inclusions is powerful evidence that condensation played a role in their genesis. The hypothetical ultrarefractory component had not yet been observed in inclusions at the time Boynton wrote his 1975 paper, but was eagerly sought afterwards because it would provide strong support for the condensation model. Since then, several such ultrarefractory inclusions have been found in CM and CO3 chondrites as we have already noted. Ironically, it has yet to be found in CV3 inclusions, the very population of objects whose Group II subset caused the existence of this component to be proposed in the first place.

The equilibrium condensation model has not been notably successful in explaining textural features of CAI. Beyond the fact that there is little agreement on what condensation textures should look like, it is becoming increasingly clear that most CAI have experienced multiple kinds of post-formation reworking that obscured many textural traces of the formational process.

Nonetheless, there are a few cases for which good textural arguments can be made in favor of condensation of crystalline solids. Grossman et al. (1975) documented euhedral crystals of nepheline, sodalite and other silicates lining cavities in an Allende fine-grained inclusion. Similar and even more spectacular examples are illustrated in Allen et al. (1978). Especially in the case of the wollastonite needles illustrated by Allen and co-workers, the physical attributes of the crystals are those of "whiskers," a morphology that is common during laboratory growth of vapor-condensed phases onto a substrate. The critical observation that all of the above phases project into open cavities indicates that they formed by heterogeneous nucleation on the walls of those cavities, and grew in a medium that permitted unobstructed crystal growth. Since the phases in these examples all formed as part of the secondary alteration of primary silicates at temperatures estimated to be 670° to 800° C (Hutcheon and Newton 1981), the medium was probably the hot gas of the solar nebula. The existence of such crystals, therefore, is powerful evidence that vapor-solid condensation in the solar nebula resulted at least locally in the formation of highly crystalline solids, similar in morphology to laboratory-produced whisker crystals grown during vapor condensation.

Similar evidence among the highest-temperature, *primary* phases in inclusions is rather more difficult to come by. The best examples are the Murchison inclusions Blue Angel (Armstrong et al. 1982) and SH-6 (MacPherson et al. 1984a, see their Fig. 2). The core of SH-6 consists entirely of highly euhedral hibonite crystals, all smaller than ~ 10 μm in maximum size, that rest loosely next to one another and show no mutual intergrowths. As in the case of the Allende secondary minerals described above, these hibonites fit the conception most people have of what a vapor-condensed crystal should look like: tiny, and highly perfect in morphology.

Allen et al. (1978) and MacPherson and Grossman (1984) proposed that certain features in Fluffy Type A inclusions in CV3 meteorites are best explained if the inclusions are aggregates of vapor-solid condensate grains. These features include: (1) highly irregular, often convoluted shapes; (2) reversely-zoned melilite crystals (monotonic increase of Al/Mg from core to rim) that have not been successfully explained by any model other than vapor-solid condensation (cf., the different type of reverse zoning in Type B melilite; MacPherson et al. 1984b); (3) large grain-to-grain chemical variations in the spinels; (4) concentrically zoned distribution of mineral phases, with the most refractory phase, hibonite, occurring in clusters of grains enclosed within a mantle of melilite (see also Wark [1986], who has made a detailed case of this sort for a complex Allende inclusion); (5) intense alteration, up to 70 volume percent or more, in the Allende examples on which the studies were done. MacPherson and Grossman (1984) argued that the convoluted shapes and intense alteration indicated the inclusions to be aggregates whose original extreme porosity (fluffiness) allowed the intense alteration relative to other Allende inclusions.

Recent work on unaltered inclusions from Vigarano supports this concept to a point. However, MacPherson (1985) showed that in some Vigarano FTAs at least, the high degree of porosity that is indeed present is due to intense brecciation; where not brecciated, the melilite shows dense polygonal-granular textures suggestive of solid-state recrystallization. Nonetheless, in spite of what may well prove to be a general metamorphic overprint on Type A inclusions, the evidence cited above suggests that FTAs still preserve vestiges of their condensation origins. Recent ion-microprobe studies (Davis et al. 1986) on the same inclusion described by MacPherson (1985) showed that some spinel grains are not in Mg-isotopic equilibrium with the melilite enclosing them and thus probably predate melilite and the inclusion.

Our general conclusion is that good chemical and sparse textural evidence indicate that vapor-solid condensation played a major role in the genesis of many refractory inclusions in the early solar nebula. The condensation probably reflected neither perfect equilibrium nor perfect fractionation. The ambiguity of textural evidence, both pro and con, regarding the role of vapor-solid condensation is at least partially a reflection of the extensive reprocessing that most inclusions have undergone since their formation. Interpretations of inclusion genesis based only on textural evidence can be misleading.

10.3.9. PROCESSES: LIQUID CONDENSATION

Because of the complexities of predicting the nature and composition of a condensed multicomponent liquid, most condensation calculations have not directly considered liquids at all. Those models that do (see, e.g., Wagner and Larimer 1978) are deliberately simplified owing to the complexities of the potential melts. The majority of condensation calculations have used indirect arguments to show that melts played little if any role during condensation in the solar nebula (see, e.g., Grossman and Clark 1973).

Because of the kinetic barriers to crystalline nucleation at very low pressures, as noted above, Blander and Fuchs (1975) proposed that condensation proceeded by formation of supercooled liquids at temperatures far below those at which equilibrium condensation would be expected. Their model was proposed in part because the Allende CAI they examined have textures interpreted to be indicative of liquid solidification. Although those authors argued in favor of metastable liquid condensation, they nonetheless recognized that the melting could have been a later event unrelated to condensation.

There is certainly evidence (see Sec. 10.3.13) that many inclusions did solidify from molten droplets, but it is not clear how those melts formed. In the case of Type B inclusions in CV3s, the strongest evidence against the Blander and Fuchs metastable liquid-condensation model comes from dynamic crystallization experiments (Paque and Stolper 1983; Stolper and Paque 1986) on liquids of Type B bulk composition. Those authors showed that melilite displays dendritic textures when it crystallizes rapidly from substantially subcooled liquids such as those predicted by the metastable liquid-condensation model. Such textures are not observed in natural Type Bs. Stolper and Paque were able to simulate natural Type B textures only in runs in which the cooling rate was slow and in which solid spinel grains were present in the melt when melilite started to crystallize. We therefore conclude, as did Stolper and Paque, that the metastable liquid-condensation model cannot explain the textures of Type B inclusions.

Wark (1987) has recently proposed that the plagioclase-rich Type C inclusions may be liquid condensates. He pointed out that whereas the bulk compositions of Type C do not fall anywhere near the calculated trajectory for vapor-condensed solids, or for evaporative residues, they are more closely approximated by an equilibrium melt-condensation model developed by Wagner and Larimer (1978). Few would argue that Type Cs were never molten (see Sec. 10.3.13), and the deviation of their bulk compositions from the predictions of the equilibrium vapor-solid condensation model cannot be denied. Evidence that Type C might have a separate origin from Type B is the observation that Type C apparently do not show the depletions in Mo and W that were noted by Fegley and Palme (1985) in most Bs, suggesting at least that the bulk compositions of the two groups of inclusions were established under different oxidation conditions. Finally, it is possible in principle for liquid condensation to account for the Group II REE signature of some Type C inclusions just as well as vapor-solid condensation, providing the gas has already been depleted in the most refractory REE.

However, even the model of Wagner and Larimer (1978) requires nebular pressures 2 to 3 orders of magnitude higher than most modern astrophysical models for protosolar nebulae allow. To get around this problem, Wark (1987) suggested that local vaporization of large amounts of dust increased the partial pressures of condensable components in the gas. It is clear that more work is needed to demonstrate the viability of equilibrium liquid condensation under nebular pressures.

Stolper and Paque (1986) showed that the slow cooling rates ($\leq 50°$ C/hr) inferred for Type B pose a problem for equilibrium liquid-condensation models: inclusions would necessarily cool at the rate of the bulk nebula, a concept totally at variance with estimated nebular cooling times on the order of 10^4 yr (see, e.g., Cameron 1978). If inferred CAI cooling rates at all reflect those of the bulk nebula as opposed to those of local subnebular cooling environments (see, e.g., Stolper and Paque 1986; MacPherson et al. 1984b), this problem may exist for Type C as well. Cooling rates of Type Cs need to be experimentally evaluated.

Finally, Beckett and Grossman (1988) have postulated a mechanism whereby vapor-solid condensation can explain the bulk compositions of Type C, involving a reaction of melilite and spinel in spinel-rich Type A inclusions with a silica-rich gas to form anorthite and diopside instead of anorthite and åkermanitic melilite as would normally be expected. The problem here is why Type C evolved differently from Type B, unless Type B was not derived from Type A. In any event, the viability of this model would remove Wark's principal objection to the vapor-solid condensation model.

For now, we can only conclude that Type C poses a problem for all models. Whether liquid-condensation models will prove viable in this or any other case will be seen when more sophisticated versions of that model are developed.

10.3.10. PROCESSES: CONDENSATION OF AMORPHOUS SOLIDS

Nuth and Donn (1983a) argued that there is no *a priori* way to predict the chemical or physical properties of solar-nebular condensates. Their arguments are based on laboratory experiments in which condensation in simple systems led to the formation of amorphous solids rather than any crystalline species. However, several caveats should be kept in mind in considering the applicability of these results to refractory inclusions. First, the experiments to date have been conducted on Al-free silicate materials; the oxide phases (e.g., spinel) encountered in Al-rich refractory liquids are known to nucleate much more easily than silicates and presumably this would apply to condensation as well. Second, it is not yet known what the compositions of the amorphous condensates would be in a complex chemical system. It will have to be shown that modeling the condensation of amorphous solids can equal or surpass the equilibrium-condensation theory in predicting some of the chemical properties of solar-system materials.

There is a problem with extrapolating the laboratory results of Nuth and Donn, achieved over very short time scales (days or less), to processes in the early solar nebula that may have taken many orders of magnitude longer. This scaling problem is beautifully illustrated in their own work (Nuth and Donn 1983b, their Fig. 7). X-ray diffraction spectra of their run products show that amorphous materials anneal very rapidly into crystalline solids over very short time periods, on the order of hours to days at temperatures above 1000 K. Given the putative very slow cooling time of the solar nebula, on the order of 10^4 yr as noted earlier, it seems likely that annealing would have kept pace with nucleation and resulted in the formation of highly crystalline solids. Crystals formed in this way at the highest temperatures would serve as nucleation sites for subsequent condensation. Thus, once any condensation begins, there should be essentially no difference between the predictions of the Nuth and Donn model and those of equilibrium-condensation theory for the solar-nebula case. Moreover, the existence of very long time scales in the solar nebula would tend to counteract the nucleation problems during condensation anyway. Finally, there is the possibility that solid noble-metal nuggets or refractory oxide grains were already present and would have served as heterogeneous nucleation sites. Thus, while we agree in principal that astrophysical environments are hard to model by theoretical calculations, they are also hard to model in the laboratory. Our conclusion regarding condensation of amorphous solids is that more experimental and theoretical work is needed to establish whether this model is truly relevant to solar-nebular condensation; at this time it presents no demonstrated advantages over the equilibrium-condensation model in terms of explaining CAI bulk compositions or mineralogy.

10.3.11. PROCESSES: VOLATILIZATION

A variety of terms—volatilization, distillation, evaporation, evaporation metamorphism—has been used to refer to a fractionation process whereby materials are progressively enriched in refractory components by the loss, due to intense heating, of volatile constituents to the vapor. The process has been documented experimentally (Notsu et al. 1978; Hashimoto et al. 1979). Although sublimation (transformation of a solid directly to vapor) is a variety of volatilization, most discussions of volatilization in a solar-nebular context refer to the intense fractional evaporation of a melt. We follow that usage here.

Volatilization is frequently invoked as an alternative to condensation to produce refractory inclusions, for a variety of reasons: (1) to overcome difficulties posed by the equilibrium-condensation model, such as the nucleation and "hot nebula" problems; (2) because so many CAI do seem to have once been molten; (3) because incomplete volatilization of interstellar dust allows preservation of isotopic heterogeneities in grains rather than in the gas; (4) because of a correlation between Type B textures and composition (Wark and Lovering 1982a); and others.

Grossman and Clark (1973) argued that pressures in the region of the solar nebula where many CAI formed were less than $\sim 2 \times 10^{-3}$ atm, and that molten CAI were produced by reheating and melting of solid condensates. Rapid heating of solid particles will produce melts even at pressures lower than 10^{-3} atm, and which in the latter case are then unstable against evaporation (I. Kushiro, personal communication, 1986). Given the apparent abundance of once-molten CAI (see Sec. 10.3.13), therefore, volatilization must be seriously considered as a CAI-forming or -modifying process.

It has proven no easier to make a strong case for volatilization than it has for condensation. Most textural evidence is ambiguous or in conflict with isotopic evidence. Isotopic studies have suggested that volatilization might have occurred in some refractory inclusions, but the evidence is not straightforward and is rarely correlated with any petrographic features. To date, only two cases exist for which volatilization is virtually proven; the rest remain open to question.

Clayton et al. (1984) described an inclusion (A16S3; a photograph is shown in Clayton et al. 1977) in which two distinct populations of melilite are present: very Al-rich grains directly adjacent to the margin of the inclusion, accompanied by anorthite, and very Mg-rich melilite in the core of the inclusion (anorthite absent). These two melilite populations lie on opposite sides of the liquidus minimum in the binary join åkermanite-gehlenite, which extends relatively unchanged into more complex systems. Molini-Velsko (1983) showed that the Si isotopes in A16S3 are mass fractionated, with the mantle being enriched in heavy isotopes relative to the inclusion core. No simple igneous fractionation process within a homogeneous liquid droplet can account for the two melilite compositions, nor can any igneous process account for the isotopic fractionation of the magnitude observed. Clayton et al. (1984) proposed that the liquid droplet from which A16S3 solidified suffered volatilization loss of Mg and Si from its exterior, resulting in a radial composition gradient within the droplet. The volatilization must have happened rapidly in order to preserve this compositional gradient against melt diffusion prior to solidification.

An even stronger case has recently been made (MacPherson et al. 1987) for a Vigarano inclusion that also (coincidentally?) happens to be a forsterite-bearing variety. This object, USNM 1623-5, also has an Al-rich margin that contains gehlenitic melilite + hibonite + spinel, while the core contains nearly pure åkermanite + forsterite + fassaite + spinel. Petrologically, therefore, 1623-5 poses the same problem for simple igneous processes as Al6S3. However, 1623-5 is also a FUN inclusion, with O, Si and Mg showing large isotopic mass-fractionation effects in both the margin and the core (Clayton et al. 1987; MacPherson et al. 1987). Therefore, the margin is cogenetic with the core and not simply a late addition. Ion-probe studies by MacPherson et al. (1987) show that the Al-rich phases in the margin are even more mass fractionated in favor of heavy Mg and Si isotopes than are the core phases. Only a short-lived, intense volatilization episode can account for the properties of the mantle relative to the core; however, volatilization cannot account for the very large isotopic fractionation inherent in the core itself (i.e., the F in FUN), so volatilization has acted here to modify an already existing object. The same is probably true for the inclusion Al6S3 noted above. Therefore, although the question of how these two objects came into being in the first place is not directly answered, the combined petrologic and isotopic evidence clearly shows that volatilization did take place and did *modify* (cf. below) their properties. Moreover, we now have, for the first time, type examples against which to evaluate the role of volatilization in other CAI.

Another forsterite-bearing inclusion contains a different kind of hint that volatilization may have been active. MacPherson et al. (1981) illustrated an inclusion nicknamed ALVIN (their Fig. 5a), that is riddled by small spherical cavities interpreted to be vesicles. Recent experimental work (I. Kushiro, personal communication, 1986) suggests that one fingerprint by which to identify liquids that are (or were) unstable against volatilization is the presence of vesicles due to escaping volatile species, presumably Si- and Mg-rich components in the case of a refractory liquid (Notsu et al. 1978; Hashimoto et al. 1979). At the temperatures required for melting ALVIN in the solar nebula, > 1200 K, the most likely source of volatiles is vaporization of the melt itself. This conclusion is supported by the presence of wollastonite needles lining the vesicles in ALVIN, suggesting the reaction of the trapped vapor with the residual melt upon cooling and condensation onto the surfaces of the vesicle walls. However, any volatilization that took place in ALVIN seemingly did not progress far, since there is no evidence of Al enrichment near its margin as in the inclusions described above. No systematic isotopic study of ALVIN has been made.

In an important paper, Niederer and Papanastassiou (1984) carefully examined the systematics of Ca- and Mg-isotopic mass fractionation in CV3 inclusions. They showed that there are correlated effects for the two elements, and that the nature of the correlation depends on inclusion type. Specifically, coarse-grained inclusions tend to be enriched in heavy Mg and light Ca, while fine-grained inclusions tend to be enriched in light Mg. The observation that the fractionation effects for Mg and Ca tend to be opposite in sign led Niederer and Papanastassiou to conclude that CAI formed by a complex series of processes that included both condensation and volatilization. Similar conclusions were reached by Clayton et al. (1985) who extended the study of isotopic systematics to include Si. However, two important points must be kept in mind with respect to the implications of these two studies for volatilization. First, the enrichment of isotopically light Ca in coarse-grained CAI is not well understood and cannot be explained by volatilization. Second, O-isotopes are very rarely mass fractionated (except in FUN inclusions) and thus do not correlate with fractionation effects in Ca, Mg and Si; yet, being lighter, O should show effects of comparable magnitude. This may be the effect of later exchange of O, as Clayton et al. (1985) suggest, but it provides no support for a volatilization model. Finally, as Niederer and Papanastassiou (1984) point out, there is as yet no general evidence for significant correlation between isotopic mass-fractionation effects and CAI bulk composition; this is contrary to what would be expected were volatilization the only process involved, given the magnitudes of the fractionation effects. The correlated effects seen in the two forsterite-bearing inclusions noted above are exceptions, and even in those cases the effects are localized in the inclusion margins.

Two cases where good petrologic arguments have led to the proposal of volatilization are the spinel-perovskite portions of Wark-Lovering rims (see, e.g., Murrell and Burnett 1987) and the unusual corundum-rich inclusion GR-1 (MacPherson et al. 1984*a*). In both cases, however, subsequent Mg-isotopic analyses using the ion microprobe have shown that the proposed volatilization residues are isotopically lighter than the starting material from which they formed (Hinton et al. 1984; Davis et al. 1986; Hinton and Davis 1986; Fahey et al. 1985*b*). Volatilization is expected to enrich the residue in heavy, not light, isotopes. Thus, although re-equilibration of the putative residues with an isotopically light (near normal) reservoir cannot be ruled out, the isotopic data would seem to argue against volatilization as the cause of refractory rims or the inclusion GR-1. In the case of rims, however, Boynton and Wark (1987, and previous work cited therein) showed that REE in the rims they separated and analyzed are enriched relative to the respective host inclusions, but parallel (except for Eu) the REE patterns of those hosts. These results are strong evidence that the rims were somehow derived from the host inclusions, thus ruling out an origin by later condensation onto the CAI surfaces. Boynton and Wark concluded that volatilization was the process that formed the rims, and that the Mg-isotopic evidence cited above does indeed imply re-equilibration of the isotopes during or after volatilization. An ion-microprobe study by Davis et al. (1986) reached a similar conclusion. Therefore, a volatilization origin for rims remains favored in spite of the isotopic evidence. More isotopic and trace-element work is needed to settle the question of rim origin once and for all.

Kurat et al. (1975) proposed volatilization of a liquid droplet to explain the enrichment in refractory components of the outer relative to inner portions of a large Type B1 inclusion in Bali. However, MacPherson and Grossman (1981) and Stolper (1982) have argued that the mineral-chemical patterns seen in this and other B1 inclusions are simply a predictable consequence of the inward-proceeding fractional crystallization of a molten droplet of appropriate composition.

For different reasons, Wark and Lovering (1982a) and Wark et al. (1987) proposed that all Type B and all forsterite-bearing inclusions, respectively, formed as volatilization residues. That these proposals are controversial is best illustrated by the fact that two of the authors of this chapter are in disagreement on the subject. The absence (noted above) of a general correlation between isotopic mass fractionation and CAI bulk composition is taken as evidence by one of us (GJM) that volatilization was not the fundamental CAI-forming mechanism. A correlation of petrologic properties with composition (Wark and Lovering 1982a) is taken by DAW to indicate the contrary. Forsterite-bearing inclusions are obviously a special case here, since we have gone to some length to show that volatilization did occur in three of them. However, the reader is cautioned that even in the case of those three, the evidence only clearly points to *local* volatilization: it is not clear whether volatilization created the overall refractory character of the inclusions in the first place. Certainly, volatilization cannot explain the extreme isotopic mass fractionation inherent in the Vigarano FUN inclusion 1623-5: gas-phase kinetic isotope effects are required.

J. A. Wood (1981) and collaborators (see, e.g., Kornacki and Fegley 1984; Kornacki and Wood 1985) have proposed that fine-grained spinel-rich CAI originated by ablation and volatilization of partially melted interstellar dust aggregates. This model breaks down in attempting to explain the isotopic mass fractionation of Mg and Si in such objects, which tends to favor the light isotopes (Niederer and Papanastassiou 1984; Clayton et al. 1985), contrary to what would be expected for a volatilization residue. Moreover, there is the problem noted earlier for Group II material (i.e., many spinel-rich fine-grained inclusions) that, to be explained as volatilization residues, the starting material must itself have been a condensate from an ultrarefractory-REE-depleted gas.

On balance, we conclude that although many tantalizing pieces of petrologic and isotopic evidence indicate volatilization has played an important role in the evolution of some or even many CAI, in only a few rare cases do combined isotopic and petrologic evidence prove it. The question of whether volatilization was the *primary* process by which the fundamental refractory character of any inclusions was established must remain open here for reasons noted above.

10.3.12. PROCESSES: METAMORPHISM

We define metamorphism here as solid-state recrystallization, with no implication for the cause of that recrystallization. Planetary burial is one possibility, as are shock recrystallization and simple subsolidus heating in a nebular environment. Even though alteration is also metamorphism in the sense that it involves recrystallization, we treat it separately in the next section.

Meeker et al. (1983) proposed that some features observed in Type B1 inclusions are products of planetary metamorphism. This specific proposal has not received wide acceptance primarily because Type Bs are the inclusions for which an igneous origin seems most unambiguously established on other grounds (see Sec. 10.3.13).

Yet, apart from the B1s, there is good evidence that solid-state recrystallization (not necessarily planetary in origin) has occurred in some or even many inclusions. The best examples are the FTAs from CV3 and CO3 meteorites. As noted above, many melilite crystals in Allende FTAs show intense kink-banding (suggestive of deformation), and melilite crystals commonly occur in polygonal-granular clumps showing 120° triple-grain junctions. Numerous observations of this feature (MacPherson and Grossman 1984; Teshima and Wasserburg 1985; Wark 1986) indicate that it characterizes many Type As; it has generally been interpreted as indicating metamorphic recrystallization. The cause of that recrystallization is open to question.

Armstrong et al. (1982) proposed that calcite in the Blue Angel (Murchison) inclusion represents the effect of planetary metamorphism superimposed on earlier nebula condensation that produced the primary refractory phases. Their argument is based primarily on the conditions necessary to stabilize calcite, and they conclude that only planetary burial could explain the presence of that phase. MacPherson et al. (1983) disagreed with this idea, but the origin of calcite remains unsettled.

We conclude that there is compelling evidence to indicate that some kind of recrystallization has occurred in many inclusions. However, this idea needs far more investigation regarding conditions, timing and mechanisms.

10.3.13. PROCESSES: MELTING

One of the earliest disputes regarding refractory inclusions was whether they were ever molten. Some early studies concluded that the answer was yes, based on the crystallization sequences, textures and rounded shapes of many large inclusions in the CV3 meteorites Bali (Kurat et al. 1975) and Allende (Clarke et al. 1970; Blander and Fuchs 1975). The opposing view was that all inclusions are pristine aggregates of vapor-to-solid condensate grains. The view now is that some inclusions were indeed melted, from precursor material that may have been solid condensate grains.

The strongest case by far has been made for the Type B inclusions from CV3 meteorites. MacPherson and Grossman (1981) showed that the texturally inferred sequence of crystallization for one Type B1 is in agreement with that predicted for its bulk composition on the basis of experimentally determined phase equilibria. Stolper (1982) conducted experimental phase equilibrium studies on liquids of Type B bulk composition and showed that the mineralogy and crystallization sequences of Type B in general are consistent with a molten origin. Dynamic crystallization experiments have successfully duplicated some of the textures and mineral-chemical features observed in natural Type B inclusions (Paque and Stolper 1983,1984; Stolper and Paque 1986). Indeed, those studies not only reaffirmed that Type Bs were probably molten, but also showed that melt solidification must have started from temperatures below the liquidus such that solid spinel grains were present in the melt. Later experiments showed that the textures and certain mineral-chemical features such as melilite zoning in some Type Bs can only be understood in terms of dynamic crystallization of a molten droplet cooling at relatively slow cooling rates, on the order of $< \sim 1$ to $50°$ C/hr (MacPherson et al. 1984b; Stolper and Paque 1986). Finally, measurements of the trace-element distributions in individual phases in Type B inclusions (Mason and Martin 1974; Nagasawa et al. 1977) show that the phases are in internal equilibrium and formed from a common reservoir. Moreover, those distributions are in agreement with experimentally determined crystal/liquid partition coefficients for the same phases in synthetic liquid systems (Nagasawa et al. 1980). Thus textural, experimental and trace-element evidence are all consistent with and indicative of a once-molten origin for Type B.

Superimposed on this seemingly simple model for Type B are some complexities that have yet to be adequately explained. Meeker et al. (1983) noted that melilite crystals in many Type Bs enclose irregularly shaped fassaite grains, which those authors interpreted as indicating a *metamorphic* replacement of original pyroxene by melilite rather than melt solidification. This interpretation has been disputed (MacPherson et al. 1984b; Paque and Stolper 1984).

Wark and Lovering (1982b) and El Goresy et al. (1985) described Allende Type B inclusions with "xenoliths" enclosed within them. In both studies, these features were interpreted as foreign solid particles entrapped by the inclusions during their molten stage. There may be alternative interpretations for how these "xenoliths" originated, but at the very least their presence seems to indicate a more complex history than simple melt solidification for their host inclusions.

Even more than Type B, Type C inclusions have generally been interpreted as the result of melt solidification. Such ready acceptance of an igneous model may be attributable to the subophitic to ophitic textures of many of them, which are directly analogous to those of terrestrial rocks of known igneous origin. As Stolper (1982, p. 2169) observed ". . . the association of ophitic texture with crystallization from a melt is an involuntary reflex for petrologists. . . ." Once again, the texturally deduced sequence of crystallization in ophitic-textured Type C (spinel and anorthite early, followed by pyroxene) is in accord with the predictions of melt-solid phase equilibria for a liquid of such a bulk composition (Paque and Stolper 1984; Wark 1987).

Forsterite-bearing inclusions also contain evidence for having been melted. As summarized in Clayton et al. (1984), the essential points are: (1) high Ca contents in the olivines indicate temperatures in excess of the solidus temperatures for the bulk compositions of these inclusions; and (2) some forsterite-bearing inclusions are vesicular. The inferred order of crystallization, spinel followed sequentially by olivine and pyroxene, is consistent with the order predicted for liquids of these bulk compositions. Some of the inclusions have spheroidal shapes. Collectively, these features all indicate that forsterite-bearing inclusions were once at least partially molten (Clayton et al. 1984; Wark et al. 1987).

Macdougall (1981) argued on the basis of textures and spheroidal shapes that the spinel-hibonite spherules in CM meteorites were once molten. Although no experimental work has been done on compositions such as these, Macdougall's interpretation is very reasonable and has been advocated for similar inclusions by others. Experimental confirmation of this interpretation is desirable, because the implied temperatures of melting are in excess of 1550°C for many spinel-hibonite spherules (MacPherson et al. 1983) and in excess of 1800°C for inclusions such as the corundum-bearing GR-1 (MacPherson et al. 1984a). These temperatures are far in excess of those allowable in most astrophysical models for the solar nebula and may require very energetic local heat sources to produce melting.

Having concluded that many CAI passed through a molten or partially molten stage, we now face the problem of how those melts originated—by liquid condensation or melting (possibly followed by volatilization) of solid precursors or as evaporative residues. Stolper and Paque (1986) effectively showed that Type B cannot have originated as subcooled liquid condensates, since experimental evidence indicates that such a process would give rise to textures not observed in natural inclusions. Grossman and Clark (1973) showed that some refractory "chondrules" in CV3 meteorites cannot have formed as equilibrium liquid condensates, as the pressures required to produce the observed crystallization sequences (spinel first, at temperatures of 1773 K) would necessarily be ~ 1 atm, far in excess of any reasonable pressure estimates for the solar nebula. The work of Stolper (1982) showed that spinel is the liquidus phase for most Type Bs as well, again at temperatures in excess of 1773 K. Therefore, the arguments of Grossman and Clark (1973) may apply to Type B as well. From the perspective of bulk composition, evaluating *equilibrium* liquid condensation as a viable model for once-molten inclusions in general is difficult because detailed model predictions are not yet available. We can only restate that, in spite of some discrepancies (see, e.g., Kerridge 1981), the compositions of Types A and B CAI are broadly in agreement with the compositions predicted for equilibrium vapor-solid condensates. The bulk compositions of Type C inclusions are at variance with such predictions, as noted earlier; they may be liquid condensates (Wark 1987), although Beckett and Grossman (1988) have proposed an alternative model that may reconcile Type C bulk compositions with vapor-solid condensation.

We can confidently conclude here that many CAI solidified from melt droplets, and we can also conclude that many of those melts probably originated by melting of solid precursors. Those precursors may have been primary condensates, or they might themselves have been derived from condensates. The melts may have experienced volatilization. However, the heat source for such melting remains conjecture. A different sort of problem concerns the cooling rates inferred for Type B: the rates are far too slow for a droplet radiating into a near vacuum, and far too fast to be indicative of the cooling rate of the solar nebula as a whole (MacPherson et al. 1984*b;* Stolper and Paque 1986). Therefore, the setting for melting is not clearly understood.

10.3.14. PROCESSES: SECONDARY ALTERATION

Throughout the course of this chapter we have referred to secondary alteration in CAI largely in the negative context of its obscuring effects on primary chemistry and textures. (Note that in Part 3 of this book, the term secondary is used to denote alteration that took place after accretion of planetesimals; in this chapter we adopt a broader definition that includes nebular alteration.) This reflects our natural bias in seeking to decipher the ultimate genesis of the inclusions. In fact, however, secondary phases apparently record the first condensation of many important elements in the solar system, e.g., the alkalies and the halogens. Indeed, we noted in Sec. 10.3.8 how some of the best textural evidence for that process is to be found in the beautiful euhedral crystals of wollastonite, nepheline and grossular that line cavities in CV3 inclusions. Presumably because these phases were condensing at a time when nebular melting and recrystallization of CAI had largely ceased, the evidence for condensation below ~ 1000 K is much better preserved than that for condensation of the "primary phases" at higher temperatures.

Many interesting questions are posed by the alteration assemblages. For example, a number of secondary phases such as wollastonite and grossular are not predicted to condense from a solar gas; conversely, some phases that are predicted to have condensed, such as albite, are not found. We have already noted how Armstrong et al. (1982) invoked planetary burial to explain the presence of calcite in the Blue Angel CM inclusion. Part of the answer, in the case of phases that should not be present but are, may be that secondary alteration took place largely inside inclusions where local concentration gradients induced reactions that would not otherwise have occurred (see, e.g., MacPherson et al. 1981).

Another problem is posed by the different patterns of alteration in CAI from different chondrite groups and, for that matter, in CAI from different members of a single chondrite group (e.g., Vigarano vs Allende). The differences suggest different settings for alteration; were all of the settings nebular, or were some planetary? If nebular, what do the differences reflect—time, distance from the Sun or local chemical environments?

A few definitive statements that we can make about the alteration process are as follows:

1. Theoretical (see, e.g., Fegley and Lewis 1980) as well as petrologic (see, e.g., Nord et al. 1982; Hutcheon and Newton 1981) evidence suggests that the formation of feldspathoids, grossular and Fe-bearing pyroxene occurred below ~ 1100 K;
2. The presence of phases such as andradite and hedenbergite indicates oxidizing conditions for CV3 alteration;
3. Alteration was not the process that caused differential depletion of ^{16}O in different phases within CV3 inclusions (Clayton et al. 1987).

The timing of alteration is poorly defined. Some ion-microprobe evidence (Hutcheon and Newton 1981) suggests that alteration might have occurred as much as 2 to 4 Myr after formation of the primary phases, while other ion-probe studies (Brigham et al. 1986) have indicated a much shorter time interval (in both cases making the questionable assumption that differences in initial $^{26}Al/^{27}Al$ can be attributed to age differences).

The systematic study of alteration in CAI is an open field needing much more attention from theoretical, experimental and observational approaches.

10.3.15. THE SEARCH FOR RELICT GRAINS

The "exotic" isotopic components that have been found in meteorites, such as short-lived radionuclides and ^{16}O, may have condensed into solid grains within the expanding shells of novae or supernovae and been carried through the interstellar medium into our solar system by those grains. Consequently, ever since the discovery of these isotopic components, considerable optimism has existed that we might be able to locate and recognize discrete relict interstellar grains by virtue of the isotopic signatures they carry. Much effort has been devoted to this search.

Bulk isotopic studies have shown that no significant mass fraction of presolar *grains* can be preserved in inclusions even though presolar isotopic *components* are clearly indicated. Wasserburg and Papanastassiou (1982) and Wasserburg (1985) have summarized evidence that the inferred decay of short-lived isotopes in the early solar system occurred *in situ* within materials that were processed and crystallized in the solar system, not before (but see Clayton 1982). Podosek (1978) has also reviewed this topic, including a succinct summary of the case for O, and specifically addressed at some length the question of preserved interstellar grains (see also Chapter 15.1). Collectively, the isotopic evidence implies (cf., Wood 1981) that interstellar grains were largely evaporated and recondensed by processes in the solar system, but that regions of the solar-nebular gas nonetheless locally preserved some isotopic heterogeneities; recondensation of these different gas regions gave rise to the solid objects we now see in meteorites.

Even so, there is a good possibility that volumetrically minor amounts of presolar grains may be preserved. The best case for physically identifiable presolar grains associated with refractory inclusions has been made by Zinner and Epstein (1987). They examined portions of acid-insoluble Murchison residues with an ion microprobe and showed that highly anomalous C ($^{13}C/^{12}C$ > 8 times terrestrial values) is concentrated along with Si in μm-sized grains within spinel, chromite and Fe-oxide crystals—the chief components of the acid residues. Because the C and Si are present in roughly equal proportions, Zinner and Epstein identified the grains as SiC. They speculated that the grains are of true presolar origin, formed in the tenuous outer atmospheres of red-giant stars (see also Chapter 13.1).

Hutcheon et al. (1983) also used an ion microprobe to measure the isotopic composition of a Murchison hibonite inclusion, MH-8, which had previously been shown (Macdougall and Phinney 1979) to contain large mass-dependent fractionation effects in Mg isotopes. Hutcheon et al.'s measurements revealed extreme variations on a scale of tens of nm in Mg isotopic fractionation within individual crystals of hibonite and spinel. Such fluctuations could be taken to indicate the presence of tiny, isotopically exotic "seed" crystals of hibonite and spinel within the presently observed larger ones. Unfortunately, the Ti isotopes showed no such variations and were "disappointingly normal." These inferred subgrains possess some of the expected properties of interstellar grains, but they are not physically recognizable as distinct from their host crystals.

Fahey et al. (1987*a*) considered the case of hibonite grains from CM meteorites. Numerous ion-microprobe measurements revealed an amazing variability of isotopic and trace-element compositions among those grains, particularly in the Ti isotopes. Nonetheless, Fahey et al. concluded that the hibonites are *not* themselves interstellar grains, based on: (1) variable and highly fractionated Pu/Th ratios that are far from expected values for condensates from either supernovae or red-giant envelopes; (2) inferred ($^{26}Al/^{27}Al)_o$ ratios within the range observed for all other refractory inclusions; (3) the Ti-isotope scatter, as great as it is, still clusters around average solar-system values and is far from any pure nucleosynthetic component.

There is no evidence that entire inclusions can themselves be presolar. Grossman (1980) gives a brief review of age data for Allende inclusions, which virtually always yield ages close to 4.5 Gyr or else show evidence for more recent disturbance of the radiogenic systems. Several inclusions have apparent ^{40}Ar-^{39}Ar ages in excess of 4.5 Gyr (Jessberger and Dominik 1979; Jessberger et al. 1980; Herzog et al. 1980) the significance of which is disputed. Villa et al. (1983) attributed the results to ^{39}Ar recoil loss during irradiation of the samples, and concluded that there is no evidence for presolar ages of any inclusions.

Up to this point, our discussion has focused on the search for presolar grains as such. One approach to finding such grains may be to search for relict pre-*inclusion* grains. Such entities may or may not be presolar, a testable proposition, but they can be physically identified. Possible candidates for pre-inclusion grains might include: (1) perovskite grains in Type B inclusions (Stolper 1982); (2) spinel grains in some Type A inclusions that are not in isotopic equilibrium with the melilite that encloses them (Davis et al. 1986); and (3) clusters of spinel-hibonite grains in Type A inclusions that show textural evidence of being resorbed by the melilite enclosing them, such as the example shown in Fig. 10.3.18.

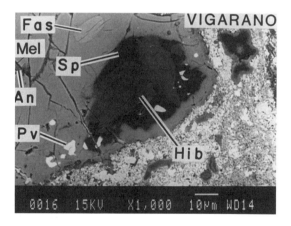

Fig. 10.3.18. Back-scattered electron photomicrograph of a portion of a Vigarano (CV3) Type A inclusion, showing a dense clump of hibonite (Hib), spinel (Sp) and perovskite (Pv) enclosed within melilite (Mel). The rounded border of the clump cuts across hibonite-spinel boundaries, suggesting that the clump is being resorbed. The clump is thus preinclusion; such objects are candidates for presolar grains, although there is as yet no evidence to indicate that they are anything but preinclusion. (Fas = Fassaite; An = Anorthite.)

10.3.16. CONCLUSIONS

As many others have noted, the fall of the Allende meteorite with its large refractory inclusions had a profound effect on our understanding of the origin and evolution of the solar system. Not least of Allende's important attributes is its large size: we have learned over the past nearly 20 yr better ways to study refractory inclusions, and yet we have made only a small dent in the amount of Allende material. Moreover, many other similar meteorites have yet to be studied at all. We are now able to make more efficient use of the other, scarcer chondrites with their generally smaller CAI.

The ideal approach that many workers now follow is to do as many kinds of analysis on each individual inclusion as possible. Consuming an entire inclusion for one kind of measurement is seldom productive, because multiple lines of evidence are necessary to constrain the origin of these objects. Obviously, with very small inclusions such as those in CMs, a multiple-analysis approach can be exceedingly difficult. Fortunately, even though the non-Allende inclusions are smaller, our analytical techniques have improved and it is now possible to work with much smaller samples than was feasible 20 yr ago. A major role in much future work will be played by the ion microprobe, with its ability for <10 μm resolution of trace-element and isotopic signatures.

Acknowledgments. This chapter has benefitted greatly from discussions with A. M. Davis, A. Fahey, J. Grossman, H. Palme and E. Zinner. J. Beckett kindly provided us with a preprint of his paper on Type C inclusions as well as data from his dissertation. The patient reviews by A. Davis, J. Paque, E. Stolper and J. Wood are greatly appreciated. We absolve all of the above individuals of any guilt by association with ideas expressed herein.

REFERENCES

Allen, J. M., Grossman, L., Davis, A. M., and Hutcheon, I. D. 1978. Mineralogy, textures, and mode of formation of a hibonite-bearing Allende inclusion. *Proc. Lunar Planet. Sci. Conf.* 9:1209–1233.

Allen, J. M., Grossman, L., Lee, T., and Wasserburg, G. J. 1980. Mineralogy and petrography of HAL, an isotopically-unusual Allende inclusion. *Geochim. Cosmochim. Acta*, 44: 685–699.

Armstrong, J. T., and Wasserburg, G. J. 1981. The Allende Pink Angel: Its mineralogy, petrology, and the constraints of its genesis. *Lunar Planet. Sci.* XII:25–27 (abstract).

Armstrong, J. T., Meeker, G. P., Huneke, J. C., and Wasserburg, G. J. 1982. The Blue Angel: I. The mineralogy and petrogenesis of a hibonite inclusion from the Murchison meteorite. *Geochim. Cosmochim. Acta*, 46:575–595.

Armstrong, J. T., Hutcheon, I. D., and Wasserburg, G. J. 1984a. Disturbed Mg isotopic systematics in Allende CAI. *Lunar Planet. Sci.* XV:15–16 (abstract).

Armstrong, J. T., Hutcheon, I. D., and Wasserburg, G. J. 1984b. Fremdlinge in Leoville and Allende CAI: Clues to post-formation cooling and alteration. *Meteoritics* 19:186–187 (abstract).

Armstrong, J. T., El Goresy, A., and Wasserburg, G. J. 1985a. Willy: A prize noble Ur-Fremdling—Its history and implications for the formation of Fremdlinge and CAI. *Geochim. Cosmochim. Acta*, 49:1001–1022.

Armstrong, J. T., Hutcheon, I. D., and Wasserburg, G. J. 1985b. Zelda revealed. *Lunar Planet. Sci.* XVI:15–16 (abstract).

Armstrong, J. T., Hutcheon, I. D., and Wasserburg, G. J. 1985c. Ni-Pt-Ge-rich Fremdlinge: Indicators of a turbulent early solar nebula. *Meteoritics* 20:603–604 (abstract).

Armstrong, J. T., Hutcheon, I. D., and Wasserburg, G. J. 1987. Zelda and Company: Petrogenesis of sulfide-rich Fremdlinge and constraints on solar nebula processes. *Geochim. Cosmochim. Acta* 51:3155–3173.

Bar-Matthews, M., MacPherson, G. J., and Grossman, L. 1979. An SEM-petrographic study of amoeboid olivine aggregates in Allende. *Meteoritics* 14:342 (abstract).

Bar-Matthews, M., Hutcheon, I. D., MacPherson, G. J., and Grossman, L. 1982. A corundum-rich inclusion in the Murchison carbonaceous chondrite. *Geochim. Cosmochim. Acta* 46:31–41.

Beckett, J. R. 1986. The Origin of Calcium-, Aluminum-Rich Inclusions from Carbonaceous Chondrites: An Experimental Study. Ph.D. Thesis, Univ. of Chicago.

Beckett, J. R., and Grossman, L. 1986. Oxygen fugacities in the solar nebula during crystallization of fassaite in Allende inclusions. *Lunar Planet. Sci.* XVII:36–37 (abstract).

Beckett, J. R., and Grossman, L. 1988. The origin of Type C inclusions from carbonaceous chondrites. *Earth Planet. Sci. Lett.*, submitted.

Beckett, J. R., MacPherson, G. J., and Grossman, L. 1980. Major element compositions of coarse-grained Allende inclusions. *Meteoritics* 15:263 (abstract).

Beckett, J. R., Live, D., Tsay, F., Grossman, L., and Stolper, E. 1987. Ti^{3+} in meteoritic and synthetic hibonite: A new oxygen barometer. *Geochim. Cosmochim. Acta*, in press.

Begemann, F. 1980. Isotopic anomalies in meteorites. *Rept. Prog. Phys.* 43:1309–1356.

Bischoff, A., and Keil, K. 1984. Al-rich objects in ordinary chondrites: Related origin of carbonaceous and ordinary chondrites and their constituents. *Geochim. Cosmochim. Acta* 48:693–709.

Bischoff, A., Spettel, B., and Palme, H. 1985b. Trace elements in Al-rich chondrules from Ybbsitz (H4). *Meteoritics* 20:609–610 (abstract). 51:2733–2748.

Bischoff, A., Keil, K., and Stöffler, D. 1985a. Perovskite-hibonite-spinel-bearing inclusions and Al-rich chondrules and fragments in enstatite chondrites. *Chem. Erde* 44:97–106.

Bischoff, A., Spettel, B., and Palme, H. 1985b. Trace elements in Al-rich chondrules from Ybbsitz (H4). *Meteoritics* 20:609–610 (abstract).

Blander, M., and Fuchs, L. H. 1975. Calcium-aluminum-rich inclusions in the Allende meteorite: Evidence for a liquid origin. *Geochim. Cosmochim. Acta* 39:1605–1619.

Blander, M., Fuchs, L. H., Horowitz, C., and Land, R. 1980. Primordial refractory metal particles in the Allende meteorite. *Geochim. Cosmochim. Acta* 44:217–223.

Blum, J. D., Armstrong, J. T., Hutcheon, I. D., and Wasserburg, G. J. 1987. Fremdlinge and the cooling of CAI: Observational and experimental constraints from the coexistence of NiFe and RuOs. *Lunar Planet. Sci.* XVIII:97–98 (abstract).

Boynton, W. V. 1975. Fractionation in the solar nebula: Condensation of yttrium and the rare earth elements. *Geochim. Cosmochim. Acta* 39:569–584.

Boynton, W. V., and Wark, D. A. 1987. Origin of CAI rims—I. The evidence from the rare earth elements. *Lunar Planet. Sci.* XVIII:117–118 (abstract).

Boynton, W. V., Frazier, R. M., and Macdougall, J. D. 1980a. Identification of an ultra-refractory component in the Murchison meteorite. *Lunar Planet. Sci.* XI:103–105 (abstract).

Boynton, W. V., Frazier, R. M., and Macdougall, J. D. 1980b. Trace element abundances in ultra-refractory condensates from the Murchison meteorite. *Meteoritics* 15:269 (abstract).

Boynton, W. V., Hill, D. H., Wark, D. A., and Bischoff, A. 1983. Trace elements in Ca, Al-rich chondrules in the Dhajala (H3) chondrite. *Meteoritics* 18:270–271 (abstract).

Boynton, W. V., Wark, D. A., and Ulyanov, A. A. 1986. Trace elements in Efremovka fine-grained inclusion E14: Evidence for high temperature, oxidizing fractionations in the solar nebula. *Lunar Planet. Sci.* XVII:78–79 (abstract).

Brigham, C. A., Papanastassiou, D. A., and Wasserburg, G. J. 1984. Mg isotopic measurements in fine-grained Ca-Al-rich inclusions. *Meteoritics* 19:198–199 (abstract).

Brigham, C. A., Papanastassiou, D. A., and Wasserburg, G. J. 1985. Mg isotopic heterogeneities in fine-grained Ca-Al-rich inclusions. *Lunar Planet. Sci.* XVI:93–94 (abstract).

Brigham, C. A., Hutcheon, I. D., Papanastassiou, D. A., and Wasserburg, G. J. 1986. Evidence

for ^{26}Al and Mg isotopic heterogeneity in a fine-grained CAI. *Lunar Planet. Sci.* XVII:85–86 (abstract).
Brigham, C. A., Hutcheon, I. D., Papanastassiou, D. A., and Wasserburg, G. J. 1988. Isotopic heterogeneity and correlated isotope fractionation in purple FUN inclusions. *Lunar Planet. Sci.* XIX:132–133 (abstract).
Cameron, A. G. W. 1978. Physics of the primitive solar accretion disk. *Moon and the Planets* 18:5–40.
Chen, J. H., and Tilton, G. R. 1976. Isotopic lead investigations on the Allende carbonaceous chondrite. *Geochim. Cosmochim. Acta* 40:635–643.
Christophe Michel-Lévy, M. 1968. Un chondre exceptionnel dans la météorite de Vigarano. *Bull. Soc. Fr. Minéral. Cristallogr.* 91:212–214.
Christophe Michel-Lévy, M. 1986. Étude comparative des chondrites carbonées d'Allende et de Léoville. *Bull. Muséum National Histoire Naturelle, Paris* 8:89–147.
Christophe Michel-Lévy, M., Kurat, G., and Brandstätter, F. 1982. A new calcium-aluminate from a refractory inclusion in the Leoville carbonaceous chondrite. *Earth Planet. Sci. Lett.* 61:13–22.
Clarke, R. S., Jr., Jarosewich, E., Mason, B., Nelen, J., Gómez, M., and Hyde, J. R. 1970. The Allende, Mexico, meteorite shower. *Smithsonian Contrib. Earth Sci.* 5.
Clayton, D. D. 1982. Cosmic chemical memory: A new astronomy. *Quar. J. Roy. Astron. Soc.* 23:174–212.
Clayton, R. N. 1978. Isotopic anomalies in the early solar system. *Ann. Rev. Nucl. Part. Sci.* 28:501–522.
Clayton, R. N., and Mayeda, T. K. 1984. The oxygen isotope record in Murchison and other carbonaceous chondrites. *Earth Planet. Sci. Lett.* 67:151–161.
Clayton, R. N., Grossman, L., and Mayeda, T. K. 1973. A component of primitive nuclear composition in carbonaceous chondrites. *Science* 182:485–488.
Clayton, R. N., Onuma, N., Grossman, L., and Mayeda, T. K. 1977. Distribution of the presolar component in Allende and other carbonaceous chondrites. *Earth Planet. Sci. Lett.* 34:209–224.
Clayton, R. N., MacPherson, G. J., Hutcheon, I. D., Davis, A. M., Grossman, L., Mayeda, T. K., Molini-Velsko, C., and Allen, J. M. 1984. Two forsterite-bearing FUN inclusions in the Allende meteorite. *Geochim. Cosmochim. Acta* 48:535–548.
Clayton, R. N., Mayeda, T. K., and Molini-Velsko, C. A. 1985. Isotopic variations in solar system material: Evaporation and condensation of silicates. In *Protostars & Planets II*, eds. D. C. Black and M. S. Matthews (Tucson: Univ. of Arizona Press), pp. 755–771.
Clayton, R. N., Mayeda, T. K., Palme, H., and Laughlin, J. 1986. Oxygen, silicon, and magnesium isotopes in Leoville refractory inclusions. *Lunar Planet. Sci.* XVII:139–140 (abstract).
Clayton, R. N., Mayeda, T. K., MacPherson, G. J., and Grossman, L. 1987. Oxygen and silicon isotopes in inclusions and chondrules from Vigarano. *Lunar Planet. Sci.* XVIII:185–186 (abstract).
Cohen, R. E., Kornacki, A. S., and Wood, J. A. 1983. Mineralogy and petrology of chondrules and inclusions in the Mokoia CV3 chondrite. *Geochim. Cosmochim. Acta* 47:1739–1757.
Davis, A. M. 1984. A scandalously refractory inclusion in Ornans. *Meteoritics* 19:214 (abstract).
Davis, A. M. 1985. Refractory inclusions in the Ornans C3O chondrite. *Lunar Planet. Sci.* XVI:165–166 (abstract).
Davis, A. M., and Grossman, L. 1979. Condensation and fractionation of rare earths in the solar nebula. *Geochim. Cosmochim. Acta* 43:1611–1632.
Davis, A. M., and Hinton, R. W. 1985. Trace element abundances in OSCAR, a scandium-rich refractory inclusion from the Ornans meteorite. *Meteoritics* 20:633–634 (abstract).
Davis, A. M., and Hinton, R. W. 1986. Magnesium and titanium isotopic compositions and trace element chemistry of refractory inclusions in the Ornans carbonaceous chondrite. *Lunar Planet. Sci.* XVII:154–155 (abstract).
Davis, A. M., Grossman, L., and Allen, J. M. 1978. Major and trace element chemistry of separated fragments from a hibonite-bearing Allende inclusion. *Proc. Lunar Planet. Sci. Conf.* 9:1235–1247.
Davis, A. M., MacPherson, G. J., and Hinton, R. W. 1986. Rims revealed: Ion microprobe

analysis of individual rim layers in a Vigarano Type A inclusion. *Meteoritics* 21:349-351 (abstract).
Davis, A. M., MacPherson, G. J., Hinton, R. W., and Laughlin, J. R. 1987a. An unaltered Group I fine-grained inclusion from the Vigarano carbonaceous chondrite. *Lunar Planet. Sci.* XVIII:223-224 (abstract).
Davis, A. M., Hinton, R. W., and MacPherson, G. J. 1987b. Relict grains in a Vigarano refractory inclusion. *Meteoritics* 22, in press.
Dominik, B., Jessberger, E. K., Staudacher, Th., Nagel, K., and El Goresy, A. 1978. A new type of white inclusion in Allende: Petrography, mineral chemistry, $^{40}Ar/^{39}Ar$ ages, and genetic implications. *Proc. Lunar Planet. Sci. Conf.* 9:1249-1266.
Ekambaram, V., Sluk, S. M., Grossman, L., and Davis, A. M. 1984a. Trace elements in high-temperature inclusions from Murchison. *Meteoritics* 19:222-223 (abstract).
Ekambaram, V., Kawabe, I., Tanaka, T., Davis, A. M., and Grossman, L. 1984b. Chemical compositions of refractory inclusions in the Murchison C2 chondrite. *Geochim. Cosmochim. Acta* 48:2089-2105.
Ekambaram, V., Kawabe, I., Tanaka, T., Davis, A. M., and Grossman, L. 1985. Erratum. *Geochim. Cosmochim. Acta* 49:1293.
El Goresy, A., Nagel, K., and Ramdohr, P. 1978. Fremdlinge and their noble relatives. *Proc. Lunar Planet. Sci. Conf.* 9:1279-1303.
El Goresy, A., Palme, H., Yabuki, H., Nagel, K., Herrwerth, I., and Ramdohr, P. 1984. A calcium-aluminum-rich inclusion from the Essebi (CM2) chondrite: Evidence for captured spinel-hibonite spherules and for an ultra-refractory rimming sequence. *Geochim. Cosmochim. Acta* 48:2283-2298.
El Goresy, A., Armstrong, J. T., and Wasserburg, G. J. 1985. Anatomy of an Allende coarse-grained inclusion. *Geochim. Cosmochim. Acta* 49:2433-2444.
Fahey, A., and Zinner, E. 1987. Determination of the Fe isotopic ratios in terrestrial minerals and a Lancé hibonite-hercynite inclusion. *Lunar Planet. Sci.* XVIII:277-278 (abstract).
Fahey, A., Goswami, J. N., McKeegan, K. D., and Zinner, E. 1985a. Evidence for extreme ^{50}Ti enrichments in primitive meteorites. *Astrophys. J.* 296:L17-L20.
Fahey, A., Zinner, E., Crozaz, G., Kornacki, A. S., and Ulyanov, A. A. 1985b. REE and isotopic studies of a coarse-grained CAI from Efremovka. *Meteoritics* 20:643-644 (abstract).
Fahey, A., Zinner, E., and Kurat, G. 1986. Anomalous Ca and Ti in a hercynite-hibonite inclusion from Lancé. *Meteoritics* 21:359-361 (abstract).
Fahey, A., Goswami, J. N., McKeegan, K. D., and Zinner, E. 1987a. ^{26}Al, ^{244}Pu, ^{50}Ti, REE, and trace element abundances in hibonite grains from CM and CV meteorites. *Geochim. Cosmochim. Acta* 51:329-350.
Fahey, A., Goswami, J. N., McKeegan, K. D., and Zinner, E. 1987b. More isotopic measurements in CM hibonites: Carbon, oxygen and silicon. *Lunar Planet. Sci.* XVIII:279-280 (abstract).
Fegley, M. B. 1982. Hibonite condensation in the solar nebula. *Lunar Planet. Sci.* XIII:211-212 (abstract).
Fegley, B., Jr., and Kornacki, A. S. 1984. The geochemical behavior of refractory noble metals and lithophile trace elements in refractory inclusions in carbonaceous chondrites. *Earth. Planet. Sci. Lett.* 68:181-197.
Fegley, B., Jr., and Lewis, J. S. 1980. Volatile element chemistry in the solar nebula: Na, K, F, Cl, Br, and P. *Icarus* 41:439-455.
Fegley, B., Jr., and Palme, H. 1985. Evidence for oxidizing conditions in the solar nebula from Mo and W depletions in refractory inclusions in carbonaceous chondrites. *Earth. Planet. Sci. Lett.* 72:311-326.
Fegley, B., Jr., and Post, J. E. 1985. A refractory inclusion in the Kaba CV3 chondrite: Some implications for the origin of spinel-rich objects in chondrites. *Earth Planet. Sci. Lett.* 75:297-310.
Frost, M. J., and Symes, R. F. 1970. A zoned perovskite-bearing chondrule from the Lancé meteorite. *Mineral Mag.* 37:724-726.
Fuchs, L. H. 1971. Occurrence of wollastonite, rhönite, and andradite in the Allende meteorite. *Amer. Mineral.* 56:2053-2068.
Fuchs, L. H. 1971. Occurrence of wollastonite, rhönite, and andradite in the Allende meteorite. *Amer. Mineral.* 56:2053-2068.

Fuchs, L. H., and Blander, M. 1980. Refractory metal particles in refractory inclusions in the Allende meteorite. *Proc. Lunar Planet. Sci. Conf.* 11:929–944.
Fuchs, L. H., Olsen, E., and Jensen, K. J. 1973. Mineralogy, mineral chemistry, and composition of the Murchison (C2) meteorite. *Smithsonian Contrib. Earth Sci.* 10.
Gray, C. M., Papanastassiou, D. A., and Wasserburg, G. J. 1973. The identification of early condensates from the solar nebula. *Icarus* 20:213–239.
Grossman, L. 1972. Condensation in the primitive solar nebula. *Geochim. Cosmochim. Acta* 36:597–619.
Grossman, L. 1973. Refractory trace elements in Ca-Al-rich inclusions in the Allende meteorite. *Geochim. Cosmochim. Acta* 37:1119–1140.
Grossman, L. 1975. Petrography and mineral chemistry of Ca-rich inclusions in the Allende meteorite. *Geochim. Cosmochim. Acta* 39:433–454.
Grossman, L. 1980. Refractory inclusions in the Allende meteorite. *Ann. Rev. Earth Planet. Sci.* 8:559–608.
Grossman, L., and Clark, S. P., Jr. 1973. High-temperature condensates in chondrites and the environment in which they formed. *Geochim. Cosmochim. Acta* 37:635–649.
Grossman, L., and Ganapathy, R. 1975. Volatile elements in Allende inclusions. *Proc. Lunar Sci. Conf.* 6:1729–1736.
Grossman, L., and Ganapathy, R. 1976. Trace elements in the Allende meteorite—II. Fine-grained, Ca-rich inclusions. *Geochim. Cosmochim. Acta* 40:967–977.
Grossman, L., and Larimer, J. W. 1974. Early chemical history of the solar system. *Rev. Geophys. Space Phys.* 12:71–101.
Grossman, L., and Steele, I. M. 1976. Amoeboid olivine aggregates in the Allende meteorite. *Geochim. Cosmochim. Acta* 40:149–155.
Grossman, L., Fruland, R. M., and McKay, D. S. 1975. Scanning electron microscopy of a pink inclusion from the Allende meteorite. *Geophys. Res. Lett.* 2:37–40.
Grossman, L., Ganapathy, R., and Davis, A. M. 1977. Trace elements in the Allende meteorite—III. Coarse-grained inclusions revisited. *Geochim. Cosmochim. Acta* 41:1647–1664.
Grossman, L., Ganapathy, R., Methot, R. L., and Davis, A. M. 1979. Trace elements in the Allende meteorite—IV. Amoeboid olivine aggregates. *Geochim. Cosmochim. Acta* 43:817–829.
Hashimoto, A., and Grossman, L. 1985. SEM-petrography of Allende fine-grained inclusions. *Lunar Planet. Sci.* XVI:323–324 (abstract).
Hashimoto, A., and Grossman, L. 1987. Alteration of Al-rich inclusions inside amoeboid olivine aggregates in the Allende meteorite. *Geochim. Cosmochim. Acta* 51:1685–1704.
Hashimoto, A., Kumazawa, M., and Onuma, N. 1979. Evaporation metamorphism of primitive dust material in the early solar system. *Earth Planet. Sci. Lett.* 43:13–21.
Hashimoto, A., Hinton, R. W., Davis, A. M., Grossman, L., Mayeda, T. K., and Clayton, R. N. 1986. A hibonite-rich Murchison inclusion with anomalous oxygen isotopic composition. *Lunar Planet. Sci.* XVII:317–318 (abstract).
Herzog, G. F., Bence, A. E., Bender, J., Eichhorn, G., Maluski, H., and Schaeffer, O. A. 1980. $^{39}Ar/^{40}Ar$ systematics of Allende inclusions. *Proc. Lunar Planet. Sci. Conf.* 11:959–976.
Hinton, R. W., and Bischoff, A. 1984. Ion microprobe magnesium isotope analysis of plagioclase and hibonite from ordinary chondrites. *Nature* 308:169–172.
Hinton, R. W., and Davis, A. M. 1986. Trace elements and calcium isotopes in the Murchison corundum-hibonite inclusion GR-1. *Lunar Planet. Sci.* XVII:344–345 (abstract).
Hinton, R. W., MacPherson, G. J., and Grossman, L. 1984. Magnesium and calcium isotopes in hibonite-bearing CAIs. *Meteoritics* 19:240–241 (abstract).
Hinton, R. W., Davis, A. M., and Scatena-Wachel, D. E. 1985. Ion microprobe determination of REE and other trace elements in meteoritic hibonite. *Lunar Planet. Sci.* XVI:352–353 (abstract).
Hinton, R. W., Davis, A. M., and Scatena-Wachel, D. E. 1987. Large negative ^{50}Ti anomalies in refractory inclusions from the Murchison carbonaceous chondrite—Evidence for incomplete mixing of neutron-rich supernova ejecta into the solar system. *Astrophys. J.* 313:400–428.
Hutcheon, I. D. 1982. Ion probe magnesium isotopic measurements of Allende inclusions. *Amer. Chem. Soc. Symp. Series* 176:95–128.
Hutcheon, I. D., and Newton, R. C. 1981. Mg isotopes, mineralogy, and mode of formation of secondary phases in C3 refractory inclusions. *Lunar Planet. Sci.* XII:491–493 (abstract).

Hutcheon, I. D., Bar-Matthews, M., Tanaka, T., MacPherson, G. J., Grossman, L., Kawabe, I., and Olsen, E. 1980. A Mg isotope study of hibonite-bearing Murchison inclusions. *Meteoritics* 15:306–307 (abstract).
Hutcheon, I. D., Steele, I. M., Wachel, D. E. S., Macdougall, J. D., and Phinney, D. 1983. Extreme Mg fractionation and evidence of Ti isotopic variations in Murchison refractory inclusions. *Lunar Planet. Sci.* XIV:339–340 (abstract).
Hutcheon, I. D., Armstrong, J. T., and Wasserburg, G. J. 1986. Mg isotopic studies of CAI in C3V chondrites. *Lunar Planet. Sci.* XVII:372–373 (abstract).
Ihinger, P. D., and Stolper, E. 1986. The color of meteoritic hibonite: an indicator of oxygen fugacity. *Earth Planet. Sci. Lett.* 78:67–79.
Ikeda, Y. 1982. Petrology of the ALH-77003 chondrite (C3). *Mem. Natl. Inst. Polar Res. Special Issue* 25:34–65.
Jessberger, E. K., and Dominick, B. 1979. Gerontology of the Allende meteorite. *Nature* 277:554–556.
Jessberger, E. K., Dominick, B., Staudacher, Th., and Herzog, G. F. 1980. ^{40}Ar-^{39}Ar ages of Allende. *Icarus* 42:380–405.
Kerridge, J. F. 1981. Compositions of calcium, aluminum-rich inclusions: Constraints on models of nebular condensation. *Lunar Planet. Sci.* XII:534–536 (abstract).
Korina, M. I., Nazarov, M. A., and Ulyanov, A. A. 1982. Efremovka CAI's: Composition and origin of rims. *Lunar Planet. Sci.* XIII:399–400 (abstract).
Kornacki, A. S., and Fegley, B., Jr. 1984. Origin of spinel-rich chondrules and inclusions in carbonaceous and ordinary chondrites. *Proc. Lunar Planet. Sci. Conf.* 14, *J. Geophys. Res. Suppl.* 89:B588–B596.
Kornacki, A. S., and Fegley, B., Jr. 1986. The abundance and relative volatility of refractory trace elements in Allende Ca,Al-rich inclusions: implications for chemical and physical processes in the solar nebula. *Earth Planet. Sci. Lett.* 79:217–234.
Kornacki, A. S., and Wood, J. A. 1984. Petrography and classification of Ca,Al-rich and olivine-rich inclusions in the Allende CV3 chondrite. *Proc. Lunar Planet. Sci. Conf.* 14, *J. Geophys. Res. Suppl.* 89:B573–B587.
Kornacki, A. S., and Wood, J. A. 1985. Mineral chemistry and origin of spinel-rich inclusions in the Allende CV3 chondrite. *Geochim. Cosmochim. Acta* 49:1219–1237.
Kornacki, A. S., Cohen, R. E., and Wood, J. A. 1983. Petrography and classification of refractory inclusions in the Allende and Mokoia CV3 chondrites. *Mem. Natl. Inst. Polar Res. Special Issue* 30:45–60.
Kurat, G. 1970. Zur Genese der Ca-Al-reichen Einschlüsse im Chondriten von Lancé. *Earth Planet. Sci. Lett.* 9:225–231.
Kurat, G. 1975. Der kohlige Chondrit Lancé: Eine petrologische Analyse der komplexen Genese eines Chondriten. *Tschermaks Min. Petr. Mitt.* 22:38–78.
Kurat, G., and Kracher, A. 1977. A new type of Ca-Al-Na-rich inclusions with an igneous texture in the Lancé carbonaceous chondrite. *Meteoritics* 12:283 (abstract).
Kurat, G., Hoinkes, G., and Fredriksson, K. 1975. Zoned Ca-Al-rich chondrule in Bali: New evidence against the primordial condensation model. *Earth Planet. Sci. Lett.* 26:140–144.
Lee, T. 1979. New isotopic clues to solar system formation. *Rev. Geophys. Space Phys.* 17:1591–1611.
Lee, T., Papanastassiou, D. A., and Wasserburg, G. J. 1977. Aluminum-26 in the early solar system: Fossil or fuel? *Astrophys. J.* 211:L107–L110.
Lord, H. C., III. 1965. Molecular equilibria and condensation in a solar nebula and cool stellar atmospheres. *Icarus* 4:279–288.
Lovering, J. F., Wark, D. A., and Sewell, D. K. B. 1979. Refractory oxide, titanate, niobate and silicate accessory mineralogy of some Type B Ca-Al-rich inclusions in the Allende meteorite. *Lunar Planet. Sci.* X:745–747 (abstract).
Macdougall, J. D. 1979. Refractory element-rich inclusions in CM meteorites. *Earth Planet. Sci. Lett.* 42:1–6.
Macdougall, J. D. 1981. Refractory spherules in the Murchison meteorite: Are they chondrules? *Geophys. Res. Lett.* 8.966–969.
Macdougall, J. D., and Phinney, D. 1979. Magnesium isotopic variations in hibonite from the Murchison meteorite: An ion microprobe study. *Geophys. Res. Lett.* 6:215–218.
MacPherson, G. J. 1985. Vigarano refractory inclusions: Allende unaltered, and a possible link with C2 inclusions. *Meteoritics* 20:703–704 (abstract).

MacPherson, G. J., and Grossman, L. 1981. A once-molten, coarse-grained, Ca-rich inclusion in Allende. *Earth Planet. Sci. Lett.* 52:16–24.
MacPherson, G. J., and Grossman, L. 1982. Fine-grained spinel-rich and hibonite-rich Allende inclusions. *Meteoritics* 17:245–246 (abstract).
MacPherson, G. J., and Grossman, L. 1984. "Fluffy" Type A Ca-, Al-rich inclusions in the Allende meteorite. *Geochim. Cosmochim. Acta* 48:29–46.
MacPherson, G. J., Bar-Matthews, M., Tanaka, T., Olsen, E., and Grossman, L. 1980. Refractory inclusions in Murchison: Recovery and mineralogical description. *Lunar Planet. Sci.* XI:660–662 (abstract).
MacPherson, G. J., Grossman, L., Allen, J. M., and Beckett, J. R. 1981. Origin of rims on coarse-grained inclusions in the Allende meteorite. *Proc. Lunar Planet. Sci. Conf.* 12B:1079–1091.
MacPherson, G. J., Bar-Matthews, M., Tanaka, T., Olsen, E., and Grossman, L. 1983. Refractory inclusions in the Murchison meteorite. *Geochim. Cosmochim. Acta* 47:823–839.
MacPherson, G. J., Grossman, L., Hashimoto, A., Bar-Matthews, M., and Tanaka, T. 1984*a*. Petrographic studies of refractory inclusions from the Murchison meteorite. *Proc. Lunar Planet. Sci. Conf.* 15, *J. Geophys. Res. Suppl.* 89:C299–C312.
MacPherson, G. J., Paque, J. M., Stolper, E., and Grossman, L. 1984*b*. The origin and significance of reverse zoning in melilite from Allende Type B inclusions. *J. Geol.* 92:289–305.
MacPherson, G. J., Davis, A. M., Laughlin, J. R., and Hinton, R. W. 1987. Isotopic heterogeneity in a forsterite-rich FUN inclusion. *Meteoritics* 22, in press.
Martin, P. M., and Mason, B. 1974. Major and trace elements in the Allende meteorite. *Nature* 249:333–334.
Marvin, U. B., Wood, J. A., and Dickey, J. S., Jr. 1970. Ca-Al-rich phases in the Allende meteorite. *Earth Planet. Sci. Lett.* 7:346–350.
Mason, B., and Martin, P. M. 1974. Minor and trace element distribution in melilite and pyroxene from the Allende meteorite. *Earth Planet. Sci. Lett.* 22:141–144.
Mason, B., and Martin, P. M. 1977. Geochemical differences among components of the Allende meteorite. *Smithsonian Contrib. Earth Sci.* 19:84–95.
Mason, B., and Taylor, S. R. 1982. Inclusions in the Allende meteorite. *Smithsonian Contrib. Earth Sci.* 25.
Meeker, G. P., Wasserburg, G. J., and Armstrong, J. T. 1983. Replacement textures in CAI and implications regarding planetary metamorphism. *Geochim. Cosmochim. Acta* 47:707–721.
Methot, R. L., Noonan, A. F., Jarosewich, E., deGasparis, A. A., and Al-Far, D. M. 1975. Mineralogy, petrology and chemistry of the Isna (C3) meteorite. *Meteoritics* 10:121–131.
Molini-Velsko, C. A. 1983. Isotopic Composition of Silicon in Meteorites. Ph.D. Thesis, Univ. of Chicago.
Murrell, M. T., and Burnett, D. S. 1987. Actinide chemistry in Allende Ca-Al-rich inclusions. *Geochim. Cosmochim. Acta* 51:985–999.
Nagahara, H., and Kushiro, I. 1982. Calcium-aluminum-rich chondrules in the unequilibrated ordinary chondrites. *Meteoritics* 17:55–63.
Nagasawa, H., Blanchard, D. P., Jacobs, J. W., Brannon, J. C., Philpotts, J. A., and Onuma, N. 1977. Trace element distribution in mineral separates of the Allende inclusions and their genetic implications. *Geochim. Cosmochim. Acta* 41:1587–1600.
Nagasawa, H., Schreiber, H. D., and Morris, R. V. 1980. Experimental mineral/liquid partition coefficients of the rare earth elements (REE), Sc and Sr for perovskite, spinel and melilite. *Earth Planet. Sci. Lett.* 46:431–437.
Nagasawa, H., Blanchard, D. P., Shimizu, H., and Masuda, A. 1982. Trace element concentrations in the isotopically unique Allende inclusion, EK 1-4-1. *Geochim. Cosmochim. Acta* 46:1669–1673.
Nelen, J. A., Noonan, A. F., and Fredriksson, K. 1978. A CAI in Clovis, an impact droplet. *Meteoritics* 13:573–577.
Niederer, F. R., and Papanastassiou, D. A. 1984. Ca isotopes in refractory inclusions. *Geochim. Cosmochim. Acta* 48:1279–1294.
Niederer, F. R., Papanastassiou, D. A., and Wasserburg, G. J. 1981. The isotopic composition of titanium in the Allende and Leoville meteorites. *Geochim. Cosmochim. Acta* 45:1017–1031.
Niederer, F. R., Papanastassiou, D. A., and Wasserburg, G. J. 1985. Absolute isotopic abundances of Ti in meteorites. *Geochim. Cosmochim. Acta* 49:835–851.

Niemeyer, S., and Lugmair, G. W. 1984. Titanium isotopic anomalies in meteorites. *Geochim. Cosmochim. Acta* 48:1401–1416.
Noonan, A. F. 1975. The Clovis (no. 1), New Mexico, meteorite and Ca,Al and Ti-rich inclusions in ordinary chondrites. *Meteoritics* 10:51–59.
Noonan, A. F., Nelen, J., Fredriksson, K., and Newbury, D. 1977. Zr-Y oxides and high-alkali glass in an amoeboid inclusion from Ornans. *Meteoritics* 12:332–335.
Noonan, A. F., Nelen, J., and Fredriksson, K. 1978. Ca-Al-Na rich inclusions and aggregates in H-group and carbonaceous chondrites. *Meteoritics* 13:583–587.
Nord, G. L., Jr., Huebner, J. S., and McGee, J. J. 1982. Thermal history of a Type A Allende inclusion. *Eos: Trans. AGU* 63:462 (abstract).
Notsu, K., Onuma, N., Nishida, N., and Nagasawa, H. 1978. High temperature heating of the Allende meteorite. *Geochim. Cosmochim. Acta* 42:903–907.
Nuth, J., and Donn, B. 1983a. Nucleation theory is *not applicable* to the condensation of refractory grains in the primitive solar nebula. *Lunar Planet. Sci.* XIV:570–571 (abstract).
Nuth, J., and Donn, B. 1983b. Laboratory studies of the condensation and properties of amorphous silicate smokes. *Proc. Lunar Planet. Sci. Conf.* 13, *J. Geophys. Res. Suppl.* 88: A847–A852.
Palme, H., and Wlotzka, F. 1976. A metal particle from a Ca,Al-rich inclusion from the meteorite Allende, and the condensation of refractory siderophile elements. *Earth Planet. Sci. Lett.* 33:45–60.
Palme, H., Wlotzka, F., Nagel, K., and El Goresy, A. 1982. An ultra-refractory inclusion from the Ornans carbonaceous chondrite. *Earth Planet. Sci. Lett.* 61:1–12.
Paque, J. M. 1985. Refractory inclusions in the Efremovka meteorite and a comparison with other carbonaceous chondrites. *Meteoritics* 20:725–726 (abstract).
Paque, J. M. 1987. $CaAl_4O_7$ from Allende Type A inclusion NMNH 4691. *Lunar Planet. Sci.* XVIII:762–763 (abstract).
Paque, J. M., and Stolper, E. 1983. Experimental evidence for slow cooling of Type B CAI's from a partially molten state. *Lunar Planet. Sci.* XIV:596–597 (abstract).
Paque, J. M., and Stolper, E. 1984. Crystallization experiments on a range of Ca-Al-rich inclusion compositions. *Lunar Planet. Sci.* XV:631–632 (abstract).
Podosek, F. A. 1978. Isotopic structures in solar system materials. *Ann. Rev. Astron. Astrophys.* 16:293–334.
Rambaldi, E. R., Rajan, R. S., Housley, R. M., and Wang, D. 1984. Oxidized, refractory and alkali-rich components in Qingzhen enstatite chondrite: Implications and their origin. *Lunar Planet. Sci.* XV:661–662 (abstract).
Rubin, A. E., James, J. A., Keck, B. D., Weeks, K. S., Sears, D. W. G., and Jarosewich, E. 1985. The Colony meteorite and variations in CO3 chondrite properties. *Meteoritics* 20:175–196.
Stolper, E. 1982. Crystallization sequences of Ca-Al-rich inclusions from Allende: An experimental study. *Geochim. Cosmochim. Acta* 46:2159–2180.
Stolper, E., and Paque, J. 1986. Crystallization sequences of Ca-Al-rich inclusions from Allende: The effects of cooling rate and maximum temperature. *Geochim. Cosmochim. Acta* 50:1785–1806.
Stolper, E., Paque, J., and Rossman, G. R. 1982. The influence of oxygen fugacity and cooling rate on the crystallization of Ca-Al-rich inclusions from Allende. *Lunar Planet. Sci.* XIII:773–774 (abstract).
Tanaka, T., and Masuda, A. 1973. Rare-earth elements in matrix, inclusions, and chondrules of the Allende meteorite. *Icarus* 19:523–530.
Tatsumoto, M., Unruh, D. M., and Desborough, G. A. 1976. U-Th-Pb and Rb-Sr systematics of Allende and U-Th-Pb systematics of Orgueil. *Geochim. Cosmochim. Acta* 40:617–634.
Taylor, S. R., and Mason, B. 1978. Chemical characteristics of Ca-Al-rich inclusions in the Allende meteorite. *Lunar Planet. Sci.* IX:1158–1160.
Teshima, J., and Wasserburg, G. J. 1985. Textures, metamorphism and origin of Type A CAI's. *Lunar Planet. Sci.* XVI:855–856 (abstract).
Ulyanov, A. A. 1984. On the origin of fine-grained Ca,Al rich inclusions in the Efremovka carbonaceous chondrite. *Lunar Planet. Sci.* XV:872–873 (abstract).
Ulyanov, A. A., Korina, M. I., Nazarov, M. A., and Sherbovsky E. Ya. 1982. Efremovka CAI's: Mineralogical and petrological data. *Lunar Planet. Sci.* XIII:813–814 (abstract).

Villa, I. M., Huneke, J. C., and Wasserburg, G. J. 1983. ^{39}Ar recoil losses and presolar ages in Allende inclusions. *Earth Planet. Sci. Lett.* 63:1–12.

Wagner, R. D., and Larimer, J. W. 1978. Condensation and stability of oxide/silicate melts. *Meteoritics* 13:651 (abstract).

Wänke, H., Baddenhausen, H., Palme, H., and Spettel, B. 1974. On the chemistry of the Allende inclusions and their origin as high temperature condensates. *Earth Planet. Sci. Lett.* 23:1–7.

Wark, D. A. 1981. The pre-alteration compositions of Allende Ca-Al-rich condensates. *Lunar Planet. Sci.* XII:1148–1150 (abstract).

Wark, D. A. 1985. Combined chemical/petrological classification of Ca-Al-rich inclusions. *Lunar Planet. Sci.* XVI:887–888 (abstract).

Wark, D. A. 1986. Evidence for successive episodes of condensation at high temperatures in a part of the solar nebula. *Earth Planet. Sci. Lett.* 77:129–148.

Wark, D. A. 1987. Plagioclase-rich inclusions in carbonaceous chondrite meteorites: Liquid condensates? *Geochim. Cosmochim. Acta* 51:221–242.

Wark, D. A., and Lovering, J. F. 1976. Refractory/platinum metal grains in Allende calcium-aluminum-rich clasts (CARC'S): Possible exotic presolar material? *Lunar Sci.* VII:912–914 (abstract).

Wark, D. A., and Lovering, J. F. 1977. Marker events in the early solar system: Evidence from rims on Ca-Al-rich inclusions in carbonaceous chondrites. *Proc. Lunar Sci. Conf.* 8:95–112.

Wark, D. A., and Lovering, J. F. 1978. Refractory/platinum metals and other opaque phases in Allende Ca-Al-rich inclusions (CAI's). *Lunar Planet. Sci.* IX:1214–1216 (abstract).

Wark, D. A., and Lovering, J. F. 1982a. The nature and origin of type B1 and B2 Ca-Al-rich inclusions in the Allende meteorite. *Geochim. Cosmochim. Acta* 46:2581–2594.

Wark, D. A., and Lovering, J. F. 1982b. Evolution of Ca-Al-rich bodies in the earliest solar system: Growth by incorporation. *Geochim. Cosmochim. Acta* 46:2595–2607.

Wark, D. A., and Wasserburg, G. J. 1980. Anomalous mineral chemistry of Allende FUN inclusions C1, EK-1-4-1 and Egg 3. *Lunar Planet. Sci.* XI:1214–1216 (abstract).

Wark, D. A., and Wlotzka, F. 1982. The paradoxical metal compositions in LEO-1, a Type B1 Ca-Al-rich inclusion from Leoville. *Lunar Planet. Sci.* XIII:833–834 (abstract).

Wark, D. A., Kornacki, A. S., Boynton, W. V., and Ulyanov, A. A. 1986. Efremovka fine-grained inclusion E14: Comparisons with Allende. *Lunar Planet. Sci.* XVII:921–922 (abstract).

Wark, D. A., Boynton, W. V., Keays, R. R., and Palme, H. 1987. Trace element and petrologic clues to the formation of forsterite-bearing Ca-Al-rich inclusions in the Allende meteorite. *Geochim. Cosmochim. Acta* 51:607–622.

Wasserburg, G. J. 1985. Short-lived nuclei in the early solar system. In *Protostars & Planets II*, eds. D. C. Black and M. S. Matthews (Tucson: Univ. of Arizona Press), pp. 703–737.

Wasserburg, G. J., and Papanastassiou, D. A. 1982. Some short-lived nuclides in the early solar system—A connection with the placental ISM. In *Essays in Nuclear Astrophysics*, eds. C. A. Barnes, D. D. Clayton, and D. N. Schramm (Cambridge: Cambridge Univ. Press), pp. 77–140.

Wasserburg, G. J., Lee, T., and Papanastassiou, D. A. 1977. Correlated oxygen and magnesium isotopic anomalies in Allende inclusions: II. Magnesium. *Geophys. Res. Lett.* 4:299–302.

Wlotzka, F., and Wark, D. A. 1982. The significance of zeolite and other hydrous alteration products in Leoville Ca-Al-rich inclusions. *Lunar Planet. Sci.* XIII:869–870 (abstract).

Wood, J. A. 1981. The interstellar dust as a precursor of Ca, Al-rich inclusions in carbonaceous chondrites. *Earth Planet. Sci. Lett.* 56:32–44.

Zinner, E., and Epstein, S. 1987. Heavy carbon in individual oxide grains from the Murchison meteorite. *Earth Planet. Sci. Lett.* 84:359–368.

10.4. PRIMITIVE MATERIAL SURVIVING IN CHONDRITES: MINERAL GRAINS

IAN M. STEELE
The University of Chicago

Besides chondrules and various kinds of polymineralic inclusion, carbonaceous chondrites commonly contain, embedded in their matrices, isolated grains of mafic silicates and metallic iron. Most of the silicate grains probably originated in chondrules, but some appear to predate chondrule formation and may have formed as individual grains in the solar nebula. If that was the case, their compositions suggest some departure from equilibrium condensation from a gas of solar composition. Metal-grain compositions are broadly suggestive of nebular formation but the exact nature of the conditions in which they were formed remains problematical.

10.4.1. INTRODUCTION

Mineral grains occur in many textural variations within the chondrites but the discussion to follow will emphasize unattached grains which do not occur in chondrules or in other texturally distinct features such as aggregates. These single grains grade with respect to size into matrix material, but grains finer than about 20 μm are not considered here since microprobe analysis is very uncertain and identification by standard microscope techniques is seldom possible.

The source of these mineral grains is particularly important with respect to the formation processes of meteorites. Possibilities include disaggregation of chondrules or aggregates and survival of single crystals of minerals of either presolar or nebular origin. Any of these may have been modified prior to or during aggregation of the meteorite. The specific intent of this discussion is to distinguish between the above processes based on chemical and mineralogic criteria in order to recognize those grains formed in the solar nebula.

10.4.2. CHONDRULES AS A SOURCE OF SINGLE MINERAL GRAINS

Casual observation of chondrules shows that although most are complete, many are partial indicating that at some stage chondrules were mechanically degraded. It is natural to expect that in some cases this process continued, to result in single, angular mineral grains. The resulting component would be represented among the isolated grains of the present meteorite. In this case, the chemical features of the mineral grains should match those of the chondrule grains. An alternative to mechanical breakdown of chondrules is simple alteration of the intergranular material to the point where single grains are released from a chondrule, preserving their original morphology. Examples of altered chondrules are present in meteorites (see Fig. 3.4.4) and many individual mineral grains are not angular, suggesting that this latter process does occur. However, if grains were to grow by condensation from a vapor, a similar morphology might result.

The relation between the individual mineral grains and the grains of chondrules has been discussed with reference to both textures and chemistry. Olivine ranging in composition from nearly pure forsterite to Fe-rich compositions ($\sim Fo_{50}$) is the most abundant phase in chondrules and among single mineral grains and thus has been most studied. Olivine grains which occur within the matrix are often euhedral (Fig. 10.4.1) which led Fuchs et al. (1973) to propose that those euhedral grains found in Murchison (CM2) formed by direct condensation of nebular gas. Olsen and Grossman (1974) described the surface morphology of some euhedral grains from Murchison and concluded that the surface features were similar to those found on terrestrial olivines formed by condensation from a vapor phase. While similar features may be present, this evidence cannot be considered diagnostic because only a few crystals were examined and textural growth features are in general not well documented. In contrast to those textural suggestions of vapor growth, McSween (1977) concluded that the single olivine grains in CO3 meteorites have major-element concentration ranges which match those of olivine grains of the chondrules in these meteorites. In addition, he described glass inclusions within those single olivine grains whose compositions match closely the intergranular glass of the chondrules. While the above chemical and inclusion features do show a good match, a better comparison using the minor elements of the olivine would have been even more convincing. In addition, the morphological studies of Olsen and Grossman (1974) apply to a CM2 meteorite whereas McSween's observations were made on CO3 meteorites. Richardson and McSween (1978) continued the comparison of single

Fig. 10.4.1. Scanning electron micrograph of individual olivine grain separated from the Murchison CM2 carbonaceous chondrite. Note euhedral nature of the grain. Width of field of view, 200 μm. Micrograph courtesy of J. D. Macdougall.

olivine grains with olivine grains in chondrules. From their textural observations of CM2 meteorites, they concluded that by various alteration processes, especially of the mesostasis glass, euhedral olivine grains in porphyritic chondrules could be released to be eventually incorporated into the matrix as single euhedral grains. Olsen and Grossman (1978) emphasized that there are a variety of features in Murchison, namely chondrules formed by melting and aggregates of grains which show no evidence of melting. They pointed out that aggregates are much more common than chondrules and that based on grain-size distribution, glass compositions and crystal morphology, the isolated olivine grains do not match with chondrules but rather with the grains of aggregates.

The above studies were based mainly on textural features and major-element (Mg/Fe) compositions of the olivine grains. Desnoyers (1980) made a similar study of olivine in Niger(I) (CM2) but emphasized the minor elements Ca, Mn and Cr as well as major elements in the olivines from chondrules, aggregates and single grains. All compositional variables showed the same trends in each textural type of olivine (Fig. 10.4.2), and he concluded that the aggregates are portions of chondrules which have partially or wholly lost their original shapes and that individual olivine grains originated in chondrules. The mainly textural evidence presented by McSween (1977) and Rich-

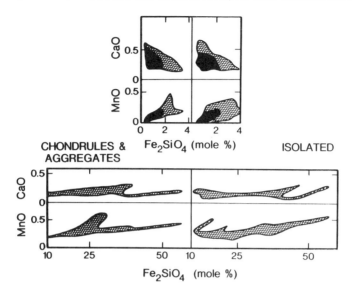

Fig. 10.4.2. Compositions of isolated olivine grains (right) compared with those in chondrules and aggregates (left) for the CM2 chondrite Niger I. The upper part of the figure shows data for forsterites, the lower part those for fayalitic olivines. The heavily shaded areas represent the greatest concentration of data points. Note the general similarity between each plot on the right with its counterpart on the left, indicating that isolated olivines were mainly derived from chondrules and aggregates. Data from Desnoyers (1980).

ardson and McSween (1978) was now complemented by more extensive chemical data for olivine in CM2 meteorites. Nagahara and Kushiro (1982) compared chondrule and single olivines from ALH-77307 (CO3) and recognized two populations of olivine (Mg- and Fe-rich) in both. They concluded that the single olivine grains originated from chondrules and were later incorporated into the meteorite. Roedder (1981) made extensive observations on the characteristics of melt inclusions in Murchison olivines in chondrules and single grains and concluded that the trapped melt inclusions resulted from crystallization of olivine from a gas-bearing melt and not vapor growth.

The evidence from the above studies clearly supports the hypothesis that ascribes the origin of single olivine grains to the breakdown of chondrules, and sometimes aggregates, producing both angular and subhedral to euhedral single grains now found embedded in CM2 and CO3 matrices. These single grains retain the chemical and textural signatures of their chondrule source. Because these studies all sample a relatively small population of grains, one cannot rule out the possibility that a minor fraction of the single mineral grains originate from a source other than chondrules and aggregates. It has been recognized that a small fraction of grains within chondrules do not have the same textural and chemical features of the host chondrules and are thought to represent grains which existed prior to chondrule formation (Nagahara 1981; Rambaldi 1981); these are termed relict grains (see also Chapter 9.1). An important consideration is whether these grains are represented among the single mineral grains where original chemical and textural features may be better preserved in contrast to having been modified or erased during their incorporation into chondrules at high temperatures.

10.4.3. FEATURES OF RELICT OLIVINE GRAINS IN CHONDRULES

The recognition of relict grains has generally been based on textural features which contrast with other grains in the chondrule. These features include a distinct difference in grain size, shape, boundary or internal appearance of the relict grain relative to coexisting grains. Figure 10.4.3 shows examples where a grain within a chondrule is distinguished by one or more of the above features. Similar examples are illustrated by Nagahara (1981,1983), Rambaldi (1981) and Kracher et al. (1984). The textural dichotomy is taken to indicate that one texture was not developed during formation of the chondrule but was inherited, although possibly modified, during formation of the chondrule. While this may be true, it is important to note that diverse textures can be produced in synthetic chondrules (see, e.g., Lofgren et al. 1979) from the same starting composition by varying the rate of cooling and presumably other unknown factors which would affect the melt structure (see also Chapter 9.2).

In conjunction with textural recognition, complementary compositional data are frequently used as supporting evidence of a relict origin. Chemical criteria which have been used include: (1) zoning trends in the relict which are in conflict with those expected for normal crystal-liquid fractionation from the bulk chondrule composition; (2) minor-element contents of presumed relict and coexisting grains which indicate different environments of formation; (3) a relict grain composition which is not in equilibrium with the bulk composition of the chondrule under assumed conditions of formation. The absence of chemical indicators does not necessarily prove that a particular grain is not a relict because a relict grain may in fact have a composition similar to that of the host grains. This might be especially true if an early suite of chondrules were recycled into a later suite.

Most descriptions of relict grains in chondrules are of olivine probably because this is the most refractory common mineral and hence is more likely to survive the melting event associated with chondrule formation. Two distinct types of olivine have been claimed to be relict grains in chondrules and these can be distinguished quite easily by optical examination; one type is usually described as dusty or cloudy due to a large number of micrometer-

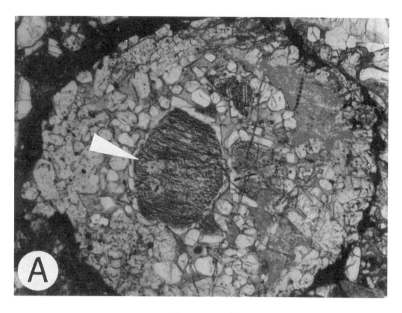

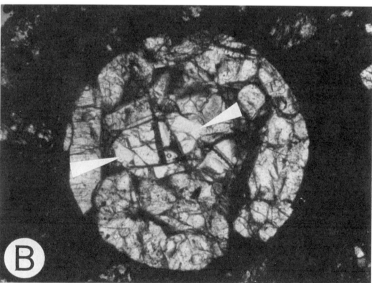

Fig. 10.4.3. Relict olivine grains in chondrules of chondrites: (A) Chainpur (UOC), a dusty olivine grain with unusually good morphology. The grain is surrounded by a rim of clear olivine; (B) Tieschitz (UOC), a single crystal of forsterite in Fe-rich chondrule. The indicated grain shows blue cathodoluminescence and is distinctly clear in transmitted light relative to other olivine grains. The width of each photograph is 0.9 mm, and both are photographed in plane polarized transmitted light.

sized opaque inclusions while the second type is unusually clear and rarely has inclusions. A third type has been described from enstatite chondrites.

Dusty olivine has been described and illustrated by Nagahara (1981,1983), Rambaldi (1981) and Kracher et al. (1984); see also Fig. 10.4.3a. These grains have been recognized in chondrules of type 3 chondrites and less commonly in carbonaceous chondrites; specific occurrences are in Krymka, Chainpur, Bishunpur, St. Mary's County, ALHA-77015 (all UOC) and Murchison, Ornans and Vigarano (carbonaceous). These dusty grains are usually larger than other coexisting olivine grains and have irregular boundaries with the dusty regions surrounded by a clear rim often with a sharp boundary.

The second type of olivine (Fig. 10.4.3b) has been documented by Steele (1986a,b), Palme et al. (1986) and briefly by Kracher et al. (1984), and has been found in all meteorites studied including Chainpur, Krymka, Tieschitz, Semarkona, Sharps, ALHA 76004, Prairie Dog Creek (all UOC), Orgueil, Murchison, Mighei, Boriskino, Belgica 7904, Vigarano, Allende, Ornans, Felix (all carbonaceous). In contrast to the dusty olivine, this second type is very clear and often can be recognized petrographically but is most easily recognized by its rather brilliant and characteristic blue cathodoluminescence. The composition of this blue olivine is unusual with FeO ranging from an absolute low of 0.25 wt.% to about 1.5 wt.% and with high concentrations (relative to nearly all olivine) of Al, Ca, Ti and V and low Mn and Ni. These grains form cores of grains with sharp transition to a normally zoned rim, and have a core as shown only by cathodoluminescence which is euhedral or subhedral with sometimes delicate and complex textures.

The third type of olivine has been described from enstatite chondrites (Rambaldi and Wang 1982; Leitch and Smith 1982). Within enstatite-rich chondrules of Qingzhen, Indarch and Kota-Kota, small round olivine grains occur within some enstatite laths; these olivines show Fe-rich cores relative to the rim (Rambaldi and Wang 1982). In cathodoluminescence, these round olivines appear either orange or blue with the blue having similar compositions to the clear type of forsterite described above (Leitch and Smith 1982).

10.4.4. SINGLE OLIVINE GRAINS AND THEIR RELATION TO RELICT GRAINS

As pointed out above, a critical question is whether the relict olivines found in chondrules are represented among the single mineral grains. If they are present among single mineral grains and derived from chondrules, these grains will show the same chemical and textural features as they show in chondrules, assuming that no selection factors operated during their preservation. If present, and not showing the same features, a strong case can be made that they represent a prechondrule suite which was directly incorporated into the meteorite and was not involved in chondrule formation.

In the several descriptions above of dusty olivines, no mention was given of an equivalent suite of grains occurring as single grains. If indeed this is the case, the dusty grains all were incorporated into chondrules and none avoided this fate. Alternatively, the chondrule-forming process caused the dusty texture as concluded by Rambaldi and Wasson (1981) and Kracher et al. (1984). In contrast, blue luminescing olivine grains are common among the single olivine grains (Steele 1986*a*). However, an important difference does occur; single blue luminescing grains show sharp textures in luminescence often with euhedral outlines while the blue grains in chondrules show embayments, poorly defined textural features, and numerous filled fractures (Steele 1986*a, b*). The simplest explanation is that the original blue olivine grains have partially reacted during chondrule formation. Because these reaction features are rarely present in single olivine grains, most blue luminescing single grains apparently did not originate by disaggregation of chondrules; that is, they are grains that were not incorporated into chondrules but simply were trapped in the meteorite and are good candidates for primitive (i.e., prechondrule) material.

10.4.5. SINGLE PYROXENE GRAINS

Pyroxene is the second major coarse-grained silicate occurring within the chondrites both in chondrules and as single grains. Reports of relict pyroxene are rare but resemble those for olivine. Thus, Rambaldi (1981) reported large twinned enstatite grains in Bishunpur (UOC) chondrules with relatively Fe-rich cores compared to the rim (reverse zoning); dusty pyroxenes were not described. Within the C and UOC meteorites, cathodoluminescence has revealed Mg-rich pyroxene (enstatite) frequently with internal textures as shown by contrasting luminescence colors. Although no systematic description has been made either of luminescence or of chemical zoning, the observed features are similar to those of olivine. The enstatite chondrites, based on chemistry and cathodoluminescence, show several types of pyroxene. The brilliant cathodoluminescence reveals a blue luminescing pyroxene included within red luminescing pyroxene. Within chondrules the two colors have been shown to represent chemically distinct pyroxenes (Leitch and Smith 1982) but there is controversy over whether or not a continuum of compositions exists and whether the luminescence color reflects minor element concentrations (McKinley et al. 1984). Because two distinct types of enstatite occur in the chondrules, the included blue could represent relict grains which are now incorporated into and surrounded by the red, minor-element-rich enstatite. The sharp boundaries between these two types of grain suggest that the conditions of formation were distinctly different. McKinley et al. (1984) present data showing that the luminescence color of single enstatite grains embedded in matrix does not accurately reflect the minor-element content, and thus simple observation of luminescence is not satisfactory for inferring chemical differences or similarities between grains in chondrules and single grains. Additional studies are required to compare compositions with quantitative cathodoluminescence measurements. As suggested for olivine in Sec. 10.4.4, a majority of the pyroxene single grains may be derived from chondrules and a minor fraction may have bypassed incorporation in chondrules.

10.4.6. METAL

Iron metal is a relatively common phase occurring both within chondrules and as single grains. Some metal in primitive unequilibrated ordinary and carbonaceous chondrites contains detectable levels of Cr and Si which are known to enter the metal under reducing conditions (see also Chapter 7.7). Metal is often contained completely within silicates, especially olivine, but larger grains do occur enclosed only by matrix. Grossman and Olsen (1974) analyzed metal beads included within forsterite single crystals in 9 CM2 chondrites. While Si data were not considered reliable due to secondary fluorescence, Ni, Co and Cr were all detected. The presence of Cr was interpreted as indicating that all samples crystallized under reducing conditions. These authors assumed that the enclosing silicates were formed by condensation and therefore that the composition of the enclosed metal possibly represented gas-metal equilibrium prior to condensation of forsterite. However, the arguments outlined in Sec. 10.4.2 above point to most of the single olivine grains as being derived from chondrules. If the olivine formed during the melting event, the included metal composition probably represents chondrule-forming conditions rather than nebular conditions. Although few details are given with regard to technique, the measurement of low levels of Co in Fe-metal must be done with care. Likewise, the olivine of CM2 meteorites usually contains about 0.2 to 0.3 wt.% Cr and secondary fluorescence may be a problem for small metal grains. Grossman et al. (1979) reported a single large metal grain from Murchison which because of its size allowed reliable Si and Cr measurements. Their measured Si and Cr contents were 0.12 and 0.73 wt.%, respectively. Errors were not quoted but the Cr value was obtained by energy-dispersive analysis with its associated larger error relative to wavelength dispersive measurements. On the basis of thermodynamic calculations, Grossman et al. concluded that a similar composition can be derived for metal condensing from a solar nebular gas if the metal does not equilibrate completely with the gas. An uncertainty in their conclusions was how the metal was in fact separated from the gas, because there are no associated phases and grains of this size are not observed within silicates.

Rambaldi et al. (1980) and Rambaldi and Wasson (1981,1984) described metal from many textural associations in Bishunpur, Krymka and Chainpur (all UOC) and in particular noted metal surrounded by troilite within chondrules whose silicates are highly reduced as indicated by their low-Fe compositions. This metal showed 0.1 to 0.7 wt.% Si with lower values at the edge relative to the core, suggesting Si loss at the boundaries; Cr was present in some grains. Their conclusion was that these grains represent clusters that formed in the nebula and incompletely reacted with silicates during chondrule formation due in part to shielding by a troilite rim. Another occurrence of metal is as clusters or, rarely, single crystals surrounded by matrix that is similar to that described by Olsen and Grossman (1978) but with undetected Si (< 0.03 wt.%).

10.4.7. SUMMARY AND IMPLICATIONS FOR PRECHONDRULE CONDITIONS

Several mineral phases appear to have survived the events up to and including chondrule formation. The same suite of single mineral grains also can be found within chondrules but modified to various extents as a result of the melting event. Because of the relatively unmodified features of the isolated grains, they possibly represent some of the best examples of primitive material found in meteorites. These grains of forsterite, enstatite and metal often are not easily recognized and frequently may have been overlooked due to similarities with more common grains derived from chondrules.

Based on these surviving grains, some inferences may be made regarding nebular conditions. The internal textures revealed in certain olivines and pyroxenes by cathodoluminescence are consistent with crystal growth by condensation (Steele 1986a). The uniform compositions of the blue luminescing interiors within single grains, among grains within one meteorite, and among grains from the primitive meteorite types, suggest similar conditions of formation. All grains show a distinct and sharp boundary both in luminescence and in composition which can be interpreted as a widespread event affecting these grains. The compositions of these grains after this boundary formed are distinctly different between the CM2 meteorites and the CV3-UOC meteorites suggesting that distinct physicochemical regions formed in the nebula producing different meteorite types. The FeO contents of blue luminescing interiors, though low, are nonetheless substantially higher than anticipated for equilibrium condensation from a gas of solar composition. Similarly, the high Si and Cr contents of metal grains are broadly consistent with the reducing conditions expected in the early nebula, but the correspondence with compositions predicted by condensation theory is not exact.

Acknowledgment. This work was supported by a grant from the National Aeronautics and Space Administration.

REFERENCES

Desnoyers, C. 1980. The Niger (I) carbonaceous chondrite and implications for the origin of aggregates and isolated olivine grains in C2 chondrites. *Earth Planet. Sci. Lett.* 47:223–234.

Fuchs, L. H., Olsen, E., and Jensen, K. J. 1973. Mineralogy, mineral-chemistry and composition of the Murchison (C2) meteorite. *Smithsonian Contrib. Earth Sci.* 10:1–39.

Grossman, L., and Olsen, E. 1974. Origin of the high-temperature fraction of C2 chondrites. *Geochim. Cosmochim. Acta* 38:173–187.

Grossman, L., Olsen, E., and Lattimer, J. M. 1979. Silicon in carbonaceous chondrite metal: Relic of high-temperature condensation. *Science* 206:449–451.

Kracher, A., Scott, E. R. D., and Keil, K. 1984. Relict and other anomalous grains in chondrules: Implications for chondrule formation. *J. Geophys. Res. Suppl.* 89:B559–B566.

Leitch, C. A., and Smith, J. V. 1982. Petrography, mineral chemistry and origin of Type I enstatite chondrites. *Geochim. Cosmochim. Acta.* 46:2083–2097.

Lofgren, G. E., Grove, T. L., Brown, R. W., and Smith, D. P. 1979. Comparison of dynamic crystallization techniques on Apollo 15 quartz normative basalt. *Proc. Lunar Planet. Sci. Conf.* 10:423–438.

McKinley, S. G., Scott, E. R. D., and Keil, K. 1984. Composition and origin of enstatite in E chondrites. *J. Geophys. Res.* 89:B567–B572.

McSween, H. Y. 1977. On the nature and origin of isolated olivine grains in carbonaceous chondrites. *Geochim. Cosmochim. Acta* 41:411–418.

Nagahara, H. 1981. Evidence for secondary origin of chondrules. *Nature* 292:135–136.

Nagahara, H. 1983. Textures of chondrules. *Mem. Natl. Inst. Polar Res.* 30:61–83.

Nagahara, H., and Kushiro, I. 1982. Petrology of chondrules, inclusions and isolated olivine grains in ALH-77307 (CO3) chondrite. *Mem. Natl. Inst. Polar Res.* 25:66–77.

Olsen, E., and Grossman, L. 1974. A scanning electron microscope study of olivine crystal surfaces. *Meteoritics* 9:243–254.

Olsen, E., and Grossman, L. 1978. On the origin of isolated olivine grains in Type 2 carbonaceous chondrites. *Earth Planet. Sci. Lett.* 41:111–127.

Palme, H., Spettel, B., and Steele, I. 1986. Trace elements in forsterite-rich inclusions in Allende. *Lunar Planet. Sci.* XVII:640–641 (abstract).

Rambaldi, E. R. 1981. Relict grains in chondrules. *Nature* 293:558–561.

Rambaldi, E. R., and Wang, D. 1982. Relict grains in enstatite chondrites. *Meteoritics* 17:272.

Rambaldi, E. R., and Wasson, J. T. 1981. Metal and associated phases in Bishunpur, a highly unequilibrated ordinary chondrite. *Geochim. Cosmochim. Acta* 45:1001–1016.

Rambaldi, E. R., and Wasson, J. T. 1984. Metal and associated phases in Krymka and Chainpur: Nebular formational processes. *Geochim. Cosmochim. Acta* 48:1885–1897.

Rambaldi, E. R., Sears, D. W., and Wasson, J. T. 1980. Si-rich Fe-Ni grains in highly unequilibrated chondrites. *Nature* 287:817–820.

Richardson, S. M., and McSween, H. Y. 1978. Textural evidence bearing on the origin of isolated olivine crystals in C2 carbonaceous chondrites. *Earth. Planet. Sci. Lett.* 37:485–491.

Roedder, E. 1981. Significance of Ca-Al-rich silicate melt inclusions in olivine crystals from the Murchison type II carbonaceous chondrite. *Bull. Mineral.* 104:339–353.

Steele, I. M. 1986a. Compositions and textures of relic forsterite in carbonaceous and unequilibrated ordinary chondrites. *Geochim. Cosmochim. Acta* 50:1379–1395.

Steele, I. M. 1986b. Cathodoluminescence and minor elements in forsterites from extraterrestrial samples. *Amer. Mineral.* 71:966–970.

10.5. ORGANIC MATTER IN CARBONACEOUS CHONDRITES, PLANETARY SATELLITES, ASTEROIDS AND COMETS

JOHN R. CRONIN and SANDRA PIZZARELLO
Arizona State University

and

DALE P. CRUIKSHANK
University of Hawaii

Although meteorites are depleted in C, H and N relative to their cosmic abundances, many apparently primitive chondrites contain significant quantities of those elements, mostly in the form of organic matter. Organic matter seems to be present also in comets, numerous asteroids and some planetary satellites. There is good reason to believe that the meteoritic organic matter, and by inference that in comets, asteroids and satellites, was synthesized abiotically early in solar-system history. It seems likely that more than one process contributed to the organic population and that several different environments were involved, possibly including dense interstellar clouds, the solar nebula and the surface regions of planetesimals. In common with the inorganic fraction of meteorites, the organic fraction may have been subject to alteration at various stages in its history. Nonetheless, potentially powerful constraints on conditions and processes active in the earliest solar system can be derived from many of the extant properties of the organic matter. Such studies also have considerable relevance to questions of prebiotic evolution and the origin of life. We review here the available information on (a) the organic constituents of carbonaceous chondrites with particular emphasis on the Murchison meteorite, and (b) possibly similar organic material on the surfaces of asteroids and satellites and in comets.

10.5.1. METEORITIC ORGANIC MATTER

The presence of organic matter in carbonaceous chondrites was noted with great interest by the early chemists who examined these meteorites (Berzelius 1834; Berthelot 1868). However, nineteenth century analytical methods were generally inadequate for characterizing organic materials in the amounts available from meteorite specimens. By the turn of the century, such efforts apparently had been discontinued and nothing appears in the literature between 1899 and 1953 when Mueller published the results of his work on the organic component of the Cold Bokkeveld meteorite. The early literature is accessible through a contemporary review by Cohen (1894) and an interesting historical account by Vdovykin (1967).

By the late 1950s, a rising interest in space science and the development of new analytical methods led to a renewal of efforts to characterize meteoritic organic matter. Amino acids were the object of several studies and between 1961 and 1965 analyses were made on several carbonaceous chondrites with generally positive results. However, these results were soon discredited when it was discovered that the amino acids and the amounts reported were essentially those to be expected from handling and from contamination during meteorite storage (see review by Hayes 1967 for details). This outcome, although discouraging at the time, served to alert later investigators to the pitfalls of terrestrial contamination. Subsequently, isoprenoid hydrocarbons and long-chain fatty acids were also found to be contaminants (Studier et al. 1972; Yuen and Kvenvolden 1973). Studies of meteorite organics carried out during the period 1953–1969 have been reviewed by Briggs and Mamikunian (1963), Hayes (1967), Vdovykin (1967), Baker (1971), Anders et al. (1973) and Nagy (1975).

The year 1969 brought a second renewal of interest in the organic chemistry of meteorites. Several laboratories were poised to begin the search for organic compounds in returned lunar samples, but with an appreciation of the contamination problems that had plagued the meteorite analyses. By this time, sensitive methods had been developed for quantification of the D- and L-enantiomers of the amino acids, a capability that made it possible to recognize terrestrial contamination on the basis of its contribution of an excess of L-amino acids. At this propitious time two meteorite falls occurred, a CM chondrite near Murchison, Australia, and a CV chondrite near Pueblito de Allende, Mexico. These meteorites were promptly analyzed and Murchison was found to contain both amino acids and aliphatic hydrocarbons (Kvenvolden et al. 1970). The discovery of both nonbiological and biological amino acids as racemic mixtures provided the first persuasive evidence that these compounds are constituents of a carbonaceous chondrite. The unusual $^{13}C/^{12}C$ ratios obtained for the extractable nonpolar organic matter carried the same implication. Over the last 17 years, the Murchison meteorite has been the subject of many more analyses and the results comprise the principal focus of this review. Analytical data obtained from other carbonaceous chondrites have been included where they complement or extend the Murchison data and seem not to have been compromised by terrestrial contamination. Other recent reviews are by Hayatsu and Anders (1981) and by Mullie and Reisse (1987).

Amino Acids

Free amino acids are easily extracted into water from a powdered meteorite and then analyzed directly by ion-exchange chromatography or, after desalting and conversion to volatile derivatives, by gas chromatography (GC) or combined gas chromatography-mass spectrometry (GC-MS) (Kvenvolden et al. 1970). Lawless and Peterson (1975) observed that when the Murchison meteorite is refluxed with D_2O, deuterium is found in some of the extracted amino acids. However, this D incorporation is apparently the result of H-isotope exchange rather than the formation of new C-D bonds (Pereira et al. 1975). Thus, free amino acids preexist in the meteorite as opposed to being produced by the extraction process.

The amino acid content of the Murchison extract is significantly increased by acid hydrolysis (Cronin and Moore 1971). The exact nature of these acid-labile precursors is unknown, although they have been partially characterized by chromatographic methods (Cronin 1976a). They appear to be mainly low-molecular-weight derivatives in which the amino group, but not the carboxyl group, of the amino acid is modified (Cronin 1976b). Although dipeptides have been tentatively identified in a hot-water extract of Murchison they can account for only ~1% of the acid-labile amino acid precursors (Walters 1983).

$$\underset{\text{Acyclic monoamino alkanoic acid}}{\overset{R}{\underset{|}{NH_2-CH-COOH}}} \qquad \underset{\text{N-Alkyl monoamino alkanoic acid}}{\overset{R}{\underset{|}{R-NH-CH-COOH}}} \qquad \underset{\text{Cyclic monoamino alkanoic acid}}{\overset{CH_2}{\underset{NH-CH-COOH}{CH_2 \diagdown CH_2}}} \qquad \underset{\text{Monoamino alkandioic acid}}{\overset{COOH}{\underset{|}{\overset{(CH_2)_n}{\underset{|}{NH_2-CH-COOH}}}}}$$

The first analyses of Murchison led to the positive identification of 22 amino acids [a] (Kvenvolden et al. 1970; Kvenvolden et al. 1971; Lawless 1973;

---

[a] Positive identification is based on a mass spectral match between the meteorite component and an authentic standard; tentative identification indicates that a suggestive mass spectrum was obtained without an authentic standard for confirmation.

Pereira et al. 1975; Buhl 1975). Since 1975, 40 more amino acids have been positively identified and 12 have been tentatively identified (Cronin et al. 1981; Cronin et al. 1985; Cronin and Pizzarello 1983,1986). There is now evidence to support the identification of a total of 74 amino acids in the Murchison extract. These amino acids comprise two general classes: (a) cyclic and acyclic monoamino alkanoic acids (including N-alkyl derivatives), and (b) monoamino alkandioic acids. Eight of the protein amino acids (glycine, alanine, valine, leucine, isoleucine, proline, aspartic acid and glutamic acid) have been identified along with 11 less common biological amino acids (e.g., β-alanine and γ-aminobutyric acid). The remaining 55 amino acids have no apparent terrestrial source. As might be expected, given the large number of amino acids identified within only two general classes, there appears to be little structural selectivity. For example, all of the 20 possible C_2 through C_5 acyclic primary monoamino alkanoic acids are present (Cronin et al. 1985) as are all of the 33 possible acyclic primary α-monoamino alkanoic acids through seven C atoms (Cronin and Pizzarello 1986).

Some interesting quantitative relationships among the acyclic monoamino alkanoic acids can be seen in the data of Fig. 10.5.1A. The meteoritic content of these amino acids drops with increasing C number from C_2 to C_3,

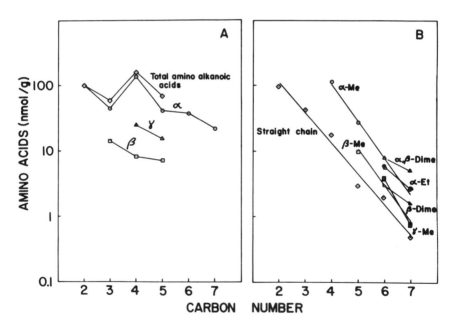

Fig. 10.5.1. Amino-acid concentration in the Murchison meteorite related to carbon number. (A) Total amino alkanoic acids and subtotals of various amino position isomers. (B) α-Amino alkanoic acids plotted as homologous series.

rises to a maximum at C_4 and drops again at C_5. α-Amino acids are the most abundant of the amino position isomers at each C number and their abundance parallels that of the total monoamino alkanoic acids, including the maximum at C_4. However, if the α-amino acids are plotted as homologous series on the basis of C chain structure, e.g., α-methyl, straight-chain, etc., plots of log concentration versus C number show a smooth decline with a slope of about -0.7. It can be seen from Fig. 10.5.1B that within the α-amino alkanoic acid series, the concentrations of branched-chain species greatly exceed those of the straight-chain isomers. Thus, the maximum in the abundance plots at C_4 can be understood in terms of this predominance of branched-chain forms and the fact that chain branching becomes possible only with C_4 and higher amino acids. Plots of the abundance of individual β- and γ-amino position isomers show similar exponential declines with increasing C number. In terms of total abundance, γ-amino acids are second to α-amino acids, and are more abundant than the β-amino isomers. The total concentration of amino acids in the Murchison meteorite can be estimated at about 0.6 μmol/g (60 ppm).

The chirality of the Murchison amino acids has been of considerable interest. Early GC analyses of volatile diastereomeric derivatives (e.g., N-TFA-(+)-2-butyl esters) showed them to be nearly racemic when extracted from uncontaminated interior samples (Kvenvolden et al. 1970; Oró et al. 1971; Pollock et al. 1975). These results provided a powerful argument for an extraterrestrial origin of the Murchison amino acids.[a] Conflicting results have been reported recently (Engel and Nagy 1982); however they are controversial and may reflect the effects of terrestrial contamination (Bada et al. 1983).

The Murchison amino acids include among those positively identified, seven pairs of diastereomers, i.e., amino acids with two chiral centers. In one instance, 2-amino-3,4-dimethyl pentanoic acid/allo-2-amino-3,4-dimethyl pentanoic acid, it has been possible to quantitate the individual diastereomers (Cronin and Pizzarello 1986). The diastereomer ratio is rather large, a finding that may have implications for the origin of the Murchison amino acids (see below).

The skepticism that pervaded the pre-1970 literature regarding meteoritic amino acids has been dispelled by the findings that the Murchison amino acids (a) are racemic, (b) include many that have no known terrestrial source, and (c) are isotopically unusual (Chang et al. 1978; Epstein et al. 1987). Now that the Murchison amino acids are well defined and established as indigenous compounds, it is interesting to inquire whether such a suite of amino acids is generally characteristic of carbonaceous chondrites. Ion-exchange chromatography has been used to screen 13 additional carbonaceous chon-

---

[a] Although in most instances finding a meteoritic amino acid to be racemic can be taken as evidence for an extraterrestrial origin, it is not necessarily indicative of an achiral synthesis. Radioracemization provides a possible mechanism for the conversion of a single enantiomer to a DL-mixture (Bonner et al. 1979).

drites for amino acids. The meteorites examined were distributed by class as follows: CI(1), CM(8), CV(3) and CR(1). Amino acids were found to be abundant only in meteorites having a phyllosilicate matrix. They occur in the CV chondrites, which have largely anhydrous silicate matrices, in amounts at least two orders of magnitude lower than in Murchison. Whether the amino acids found in CV chondrites are indigenous is not clear since those identified thus far have terrestrial sources and isotopic and/or chirality tests have not been applied. It is interesting that in a serial sampling experiment done with an Allende (CV) stone, the amino-acid content initially declined with depth below the surface but reached a constant interior concentration, a finding that suggests indigenous amino acids (Harada and Hare 1980). The single CI chondrite analyzed (Orgueil) shows traces of what are believed to be indigenous amino acids (α-aminoisobutyric acid, β-aminoisobutyric acid, β-amino-n-butyric acid and sarcosine) underlying what is probably extensive biological contamination (Lawless et al. 1972). This finding is not surprising given the long terrestrial residence (>120 yr) of this meteorite and the decrepit condition of the specimen analyzed.

A general qualitative similarity can be observed between the Murchison chromatogram and those of the other eight CM chondrites. The similarity is especially marked among Murchison, Murray, Santa Cruz and Crescent, the more recent falls. Crescent, which fell in 1936, is the oldest of this group. There is a significant difference in total amino-acid content between this group and an older subset (Cold Bokkeveld, Mighei and Nogoya) which have terrestrial ages between 98 and 149 yr. The older group appear to be depleted in amino acids by a factor of about 4. The older meteorites are particularly depleted in the nonbiological and biologically uncommon amino acids but retain relatively high levels of aspartic acid, glutamic acid, β-alanine and γ-aminobutyric acid. The replacement of indigenous amino acids by biologically common ones (or their decarboxylation products) might be expected as a consequence of microbial contamination of the meteorite. Furthermore, experiments in which Murchison stones were heated for various time periods have shown similar changes, i.e., alteration of the amino-acid content and composition to be more like that of an older CM chondrite. These results suggest that spontaneous decomposition may also be a factor. It seems possible that all of the CM chondrites had, at the time of their fall, an amino-acid composition similar to that of the Murchison meteorite and that either spontaneous decomposition, microbial alteration, or both have altered that composition over the subsequent years (Cronin and Pizzarello 1983).

During the last few years, six carbonaceous chondrites recovered from Antarctica have been analyzed for amino acids. Of the five CM meteorites in this group, one gave results very similar to those obtained with Murchison (Shimoyama et al. 1985), two appear to be qualitatively similar but depleted by 50 to 90% (Kotra et al. 1979; Cronin et al. 1979), and two are devoid of amino acids (Shimoyama and Harada 1984). All of the samples appear to be nearly free of terrestrial contaminants. However, the variation in amino acid content, particularly the absence of amino acids in two specimens, suggests the possibility of losses due to weathering, particularly periodic exposure to water (Holzer and Oró 1979).

Aliphatic Hydrocarbons

Higher (Nonvolatile) Series. Within two years of the Murchison fall, the results of three independent analyses of the nonvolatile aliphatic hydrocarbons were reported. In each case, a pulverized sample was extracted with benzene-methanol, stripped of solvent, fractionated by silica-gel chromatography and analyzed by GC-MS. Kvenvolden et al. (1970) used a relatively nonpolar benzene-MeOH (9:1) solvent at room temperature with sonication, and a packed-column GC system of comparatively low resolution. Oró et al. (1971) and Studier et al. (1972) used refluxing benzene-MeOH (3:1)for extraction and high-resolution capillary GC columns. In each case the primary extract was found to contain a complex mixture of aromatic and aliphatic compounds. The aliphatic fraction, which was separated by silica-gel chromatography, contained nonvolatile alkanes, alkenes and cycloalkanes of C number greater than about C_{10}. Any lower hydrocarbons that might have been extracted from the meteorite would have been lost during evaporation of the solvent.

An important question that arises in comparing the results of these three analyses centers on the extent to which the numerous chain isomers that are possible for the higher alkanes are actually represented in the meteorite. The chromatograms of Kvenvolden et al. (1970) have the appearance of symmetrical envelopes with peak intensities in the C_{16} to C_{18} n-alkane elution positions. Such a chromatogram implies the presence of a large number of overlapping isomeric alkanes at each C number. They found the order of abundance to be bicyclics > alkenes > tricyclics > alkanes (at the highest C numbers the tricyclics outweigh the alkenes). Oró et al. (1971), using a capillary column, were able to resolve numerous peaks, suggesting more isomeric specificity than did the earlier results. They found the alkanes to be dominant, with the minimally branched-chain species, i.e., methyl- and dimethyl-alkanes, most abundant. Studier et al. (1972) used an extraction procedure quite similar to that of Oró et al. and a GC system of very high resolution for analysis of the extract. Their chromatogram of the aliphatic fraction is relatively simple, with major peaks corresponding to the normal alkane isomers and only a few smaller intervening peaks which were identified as the slightly branched isomers, e.g., the 2- and 3-methyl substituted straight-chain alkanes.

Studier et al. (1972) have attributed the apparent isomeric complexity inferred from the analysis of Kvenvolden et al. (1970) to the inadequate resolution of their GC system. This position is supported by experiments carried out by Oró et al. (1970) in which meteorite extracts that ran as an unresolved "hump" on packed columns gave a series of well-resolved discrete peaks similar to those reported by Studier et al. (1972) when reanalyzed by capillary GC. Furthermore, the finding of a higher aliphatic hydrocarbon fraction in Murchison that is biased strongly toward the normal, methyl- and dimethyl-branched alkanes is consistent with earlier results with other carbonaceous chondrites (Gelpi and Oró 1970). The Murchison data add significance to these earlier results which, at the time, were thought to include a biological contribution.

Lower (Volatile) Series. The lower aliphatic hydrocarbons of Murchison and other carbonaceous chondrites have also been of considerable interest. Eck et al. (1966) calculated hydrocarbon compositions for various CHONS mixtures assuming a limited thermodynamic equilibrium from which graphite was excluded on the basis of the sluggish kinetics of its formation. Those calculations predicted a predominance of methane followed by higher homologues in sharply declining amounts together with formation of aromatic hydrocarbons. A number of investigators have analyzed the lower alkanes of carbonaceous chondrites motivated, at least in part, by an interest in the validity of this prediction (Studier et al. 1965; Studier et al. 1968; Belsky and Kaplan 1970; Studier et al. 1972; Yuen et al. 1984).

Analysis of the lower hydrocarbons is difficult because they are volatile and occur in the meteorite only as a result of being immobilized, e.g., within crystals or between crystal boundaries, adsorbed to grain surfaces, or dissolved in the organic coating of grains (Belsky and Kaplan 1970). The analyst is faced with the dilemma of choosing conditions for disruption of the meteorite that will release a representative, contamination-free sample of these compounds, but not lead to their production by degradation of the other organic materials present. As might be expected, there is a good deal of variation in the analytical results. Studier et al. (1972) estimate a methane/ethane ratio > 30 and perhaps as high as 700. On the other hand, Belsky and Kaplan (1970) found ratios between 2.5 and 30 for Murray, a meteorite very similar in all respects to Murchison, and Yuen et al. (1984) found essentially equal amounts of methane and ethane in the light hydrocarbons released from Murchison by freeze-thaw disaggregation. The more drastic extraction method of Studier et al. (1972) released 100 to 1000 times more methane than did the gentler method of Yuen et al. (1984). An obvious difficulty in interpretation arises from not knowing which procedure, if any of the three, liberates a representative sample of the original product mixture.

With respect to aliphatic hydrocarbons in the C_3 to C_8 range, Studier et al. (1972) found branched-chain alkanes but *no* straight-chain isomers, in contrast with the results for the nonvolatile, higher members of the series. Yuen et al. (1984) report the presence of propane, *n*-butane and *n*-pentane along with isobutane. Belsky and Kaplan (1970) also found the lower *n*-

alkanes, including heptane, as well as the branched-chain isomers and the lower alkenes.

The recent work of Yuen et al. (1984) is particularly significant in that they isolated individual members of the lower hydrocarbon series for $^{13}C/^{12}C$ measurements. They found methane and its homologues through the butanes to be isotopically heavier than their terrestrial counterparts, clearly indicating their extraterrestrial origin. More interestingly, they discovered a downward trend in $^{13}C/^{12}C$ with increasing C-chain length. These results have important implications with regard to the production mechanism (see below).

The lower aliphatic hydrocarbons trapped in carbonaceous chondrites may represent only a fraction of the meteorite's original content of such compounds. Given the amounts of nonvolatile hydrocarbons in Murchison, it would follow that the meteorite originally contained much larger quantities of the volatile lower hydrocarbons if the exponential decline in amount with increasing C number that characterizes several other classes of compounds is also applicable to the hydrocarbons. The loss of a significant amount of less firmly held volatile hydrocarbons is suggested by the common observation that carbonaceous chondrites have a specific "bitumen-like" odor immediately after falling (Vdovykin 1967). In the case of the Murchison meteorite, Lovering et al. (1971) refer in their initial description to an "obvious volatile hydrocarbon component." The opportunities for preterrestrial loss would probably have been even greater.

In summary, the Murchison meteorite contains a complex mixture of alkenes and cyclic and acyclic alkanes. Over 140 specific compounds in these classes have been identified ranging from methane to C_{20} hydrocarbons. The lower hydrocarbons include both straight- and branched-chain isomers. The nonvolatile fraction appears to be dominated by straight-chain and slightly branched-chain isomers, with the methyl and dimethyl forms predominant in the latter category. In contrast to the amino acids, structural diversity does not appear to be complete among the higher alkanes. The total concentration is probably in excess of 35 ppm (Kvenvolden et al. 1970).

Aromatic Hydrocarbons

The first analysis of aromatic hydrocarbons in Murchison was made by Oró et al. (1971), who examined the benzene eluate from the same silica-gel fractionation from which the aliphatic fraction was obtained. They identified naphthalene, phenanthrene/anthracene and alkyl derivatives of bicyclic and tricyclic aromatics. Shortly thereafter, Pering and Ponnamperuma (1971) reported the identification of 14 aromatic compounds, and the general characterization of nine others, in a similarly fractionated Murchison extract. They found a mixture dominated by unbranched, polynuclear aromatic hydrocarbons (naphthalene, acenaphthene, phenanthrene, fluoranthene and pyrene) with lesser amounts of their methyl- and dimethyl-substituted homologues.

Studier et al. (1972) analyzed a Murchison benzene-methanol extract both before and after fractionation of silica gel and identified a suite of aromatic hydrocarbons similar to that reported by Pering and Ponnamperuma. However, their results suggest a more complex mixture in terms of greater numbers of alkyl-substituted species. Finally, Basile et al. (1984) obtained polycyclic aromatics from Murchison employing an extraction and fractionation procedure very much like that used by Studier et al. (1972). They report the presence of 24 polycyclic aromatic hydrocarbons, some of which were specifically identified while others were only generally identified, e.g., as a methylphenanthrene. Their results are comparable to those of Studier et al. (1972), with the exception that some of the more volatile compounds are missing, perhaps as a result of evaporation.

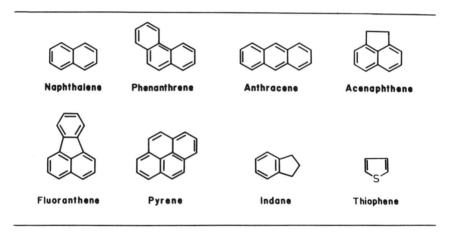

It has been shown that significant losses of hydrocarbons with boiling points less than 300° C occur during the evaporation of benzene-methanol extracts (Nooner and Oró 1967). Consequently, benzene and many alkyl-substituted benzenes cannot be analyzed using the extraction techniques employed in the studies just described. As a result, attempts have been made to determine these compounds by direct volatilization. Levy et al. (1973) subjected a crushed Murchison sample to sequential five-minute heating periods of 150°, 300° and 430° C. At the higher temperatures, i.e., at 300° C and above, where pyrolysis occurs, benzene, substituted benzenes, naphthalene and a few other aromatics were detected. However, at 150° C, where the primary process is vaporization, the main products were CO_2 and H_2O with only traces of aromatic compounds. They found less than 10 ppm (about 0.04% of the Murchison organic matter) to be released under these conditions. Studier et al. (1972) also analyzed the volatiles released from Murchison at 150° C in a progressive heating experiment. They identified benzene, several alkyl benzenes (*t*-butyl benzene prominent), indanes and naphthalene after prolonged heating of a pulverized sample (64 hr). These results, taken together, indicate that Murchison contains little or no free, low-boiling aromatics. However, disintegration of the matrix or degradation of the insoluble C effected by KOH treatment (Studier et al. 1972) liberates a suite of otherwise volatile trapped aromatic compounds.

In addition to the aromatic hydrocarbons, heteroaromatic compounds, such as thiophenes (sulfur) and N-containing aromatics have been detected (Studier et al. 1972; Basile et al. 1984) as well as aromatic ketones (Basile et al. 1984) and halogenated aromatics (Studier et al. 1972). Some questions regarding formation by secondary reactions (thiophenes) and contamination (halogenated aromatics) remain to be settled. However, the finding of thiophenes in a room-temperature extract of Murchison argues for their preexistence (Studier et al. 1972).

In summary, it is clear that the Murchison meteorite contains an extensive suite of aromatic and alkyl-substituted aromatic compounds. Pering and Ponnamperuma (1971) have estimated a total content of extractable aromatics in the 15 to 28 ppm range. Among the lower-molecular-weight compounds, all possible isomeric forms are accounted for both in terms of ring substitution position (e.g., all of the xylenes, methyl ethyl benzenes, trimethyl benzenes and dimethyl ethyl benzenes have been found) and substituent chain branching (e.g., all of the isomeric propyl and butyl benzenes have been found). Quantitative relationships within homologous series of these compounds are difficult to evaluate due to evaporation loss of the lower members when using the extraction technique, and contributions from pyrolysis of the insoluble organic material when samples are heated to vaporize the higher members using the direct-volatilization technique.

Carboxylic Acids

Nearly identical suites of C_2 through C_8 alkanoic acids have been obtained from powdered samples of the Murchison and Murray meteorites by extraction with methanolic KOH (Yuen and Kvenvolden 1973). Eleven acids have been positively identified and four others tentatively identified in the Murchison extract using GC-MS. All of the six possible C_4 and C_5 acids have been either positively or tentatively identified, as have six or possibly seven of the eight isomeric C_6 acids. Like the amino acids, when log concentration is plotted vs C number, a smooth decline is seen within homologous series. Long-chain fatty acids are found only in extracts of exterior samples and thus appear to be contaminants. Branched-chain species, especially the 2-methyl isomers, are more abundant than the straight-chain forms, but their dominance is not as pronounced as in the case of the amino acids (Lawless and Yuen 1979). The alkanoic acids are 10 to 20 times more abundant than the corresponding amino acids and 10 to 100 times more abundant than the trapped alkanes of the same C number (Yuen et al. 1984). The true ratio of alkanoic acids to alkanes may actually be somewhat lower. The comparison is based on amounts released by freeze-thaw disaggregation, a mild method that may not maximize alkane release. The total content of alkanoic acids in Murchison exceeds 4 μmol/g (> 300 ppm).

Shimoyama et al. (1986) have obtained qualitatively similar results for the carboxylic acids in an Antarctic CM meteorite, Yamato 791198. They observed the straight-chain acids through C_{12} and found them to be the most abundant isomers at each C number. The individual carboxylic acids of this meteorite are 10 to 100 times less abundant than in Murchison.

Yuen et al. (1984) have determined the C-isotope composition of six alkanoic acids from Murchison and found a smooth decrease in $^{13}C/^{12}C$ with increasing C number for the straight-chain isomers. The decline in $^{13}C/^{12}C$ parallels that found for the lower alkanes.

$CH_3-(CH_2)_n-COOH$	$HOOC-(CH_2)_n-COOH$	$R-\overset{\overset{\displaystyle OH}{\mid}}{CH}-COOH$
Carboxylic acids	Dicarboxylic acids	Hydroxycarboxylic acids

Dicarboxylic Acids

Lawless et al. (1974) isolated a structurally diverse suite of dicarboxylic acids from Murchison by hot-water extraction, silica-gel chromatography and extraction into ether. Seventeen dicarboxylic acids were positively identified by GC-MS of their methyl and butyl esters and several others were tentatively identified. They include both branched- and straight-chain isomers, as well as saturated and unsaturated forms, with chain lengths through nine C atoms. All structural isomers with two, three or four C atoms are present and six of thirteen possible isomers with five and six C atoms were identified. Three of the positively identified dicarboxylic acids have chiral centers and in one case, methylsuccinic acid, both enantiomers were identified and shown to be present in about equal amounts. Whether these compounds occur in the meteorite as salts or some type of nonvolatile precursor is unknown except for oxalic acid which occurs, at least in part, as whewellite, its Ca salt (Fuchs et al. 1973). Lawless et al. (1974) did not quantitatively analyze the dicarboxylic acids, although they estimated them to be present at concentrations 10 to 100 times greater than the amino acids. In contrast, Peltzer et al. (1984) reported concentrations for six of the Murchison dicarboxylic acids that are comparable to the concentrations of the more abundant amino acids; their data indicate a declining content of the straight-chain series with increasing C number and a prevalence of branched-chain isomers.

Hydroxycarboxylic Acids

Peltzer and Bada (1978) obtained a suite of α-hydroxycarboxylic acids from the Murchison meteorite by extraction with hot water followed by ion exchange and silica-gel chromatographic purification. Seven α-hydroxy acids were positively identified by GC-MS of the methyl esters. These correspond to the complete set of isomers with five or fewer C atoms. None of the longer-chain homologues were observed. With the exception of glycolic acid, which may have been depleted because of its volatility, the concentrations of these compounds are similar to the analogous amino acids. The total content is approximately 0.15 μmol/g (15 ppm). A declining content is seen as the C number increases within the straight-chain series. The straight-chain isomers are slightly more abundant than the branched-chain compounds. Volatile diastereomeric derivatives were prepared for four of the five chiral hydroxy acids and enantiomer ratios (D/L) of 0.82 to 0.93 were obtained by GC, results consistent with an abiotic origin.

Peltzer et al. (1984) have pointed out that the presence of both amino and hydroxy acids in Murchison suggests a Strecker-cyanohydrin synthesis in an aqueous, ammonia-containing medium. These authors have performed calculations of ammonium-ion and hydrogen cyanide concentration on a presumed Murchison parent body based on their analytical data for amino-, hydroxy- and dicarboxylic acids, plus rate and equilibrium constants for the appropriate reactions. If NH_4^+ and HCN provided the dominant buffer components for the parent-body aqueous phase, a pH of 9 ± 1 would be expected. Interestingly, DuFresne and Anders (1962) suggested a pH in the 8 to 10 range on the basis of the observed mineral assemblage and the Eh-pH relationship (see also Chapter 3.4).

Nitrogen Heterocycles

The well-known biological role of purines and pyrimidines as coding elements of ribonucleic acid (RNA) and deoxyribonucleic acid (DNA) provoked an early interest in the question of whether these compounds and/or related N-heterocycles are present in carbonaceous chondrites (Hayatsu 1964). The first such analyses carried out on the Murchison meteorite were those of Folsome et al. (1971,1973) who examined charcoal adsorbates of hot-water and hot-formic-acid extracts by GC-MS of trimethylsilyl (TMS) derivatives. They found mainly 4-hydroxypyrimidine and two isomeric methyl-4-hydroxy pyrimidines. Data were also obtained allowing tentative identifications of a few other nonbiological pyrimidines, hexahydropyrimidines and quinazoline. Interestingly, none of the biologically occurring purines or pyrimidines was detected.

Adenine, Guanine, Xanthine, Hypoxanthine, Uracil

Pyridine, Aniline, Quinoline, Isoquinoline

Urea, Guanylurea

Shortly thereafter, Hayatsu et al. (1975) reported the results of a similar study. However, they employed somewhat different procedures and obtained quite divergent results. Their methods included extractions with 3-6 M HCl or trifluoroacetic acid and different procedures for the preparation, wash and elution of the charcoal adsorbent. Their analytical methods were direct-probe MS without derivatization, paper chromatography and ultraviolet spectrophotometry. They found the principal solutes after water and formic-acid extraction, the solvents used by Folsome et al. (1973), to be aliphatic amines and C_2-C_6 alkyl pyridines. The 4-hydroxypyrimidines were not detected. Using the stronger acids (HCl and trifluoroacetic acid) as solvents, they obtained two of the biological purines, adenine and guanine, plus melamine, cyanuric acid, urea and guanylurea.

The conflicting results of Folsome et al. and Hayatsu et al. were difficult to rationalize. Similar extraction procedures seemed to yield pyrimidines but no purines in one laboratory and purines but no pyrimidines in the other. Happily, more recent work by van der Velden and Schwartz (1977) appears to provide an explanation. These investigators used a sensitive HPLC system to analyze N-heterocycles after their extraction into a series of solvents similar to those used in the prior studies, and adsorption and elution from charcoal columns. Their identifications were based on the coincidence of retention times with standards on two columns and appropriate extinction ratios at two wavelengths. By these criteria, xanthine, hypoxanthine and guanine were identified in the formic acid extract. No pyrimidines were detected. However, some results suggested that the 4-hydroxypyrimidines reported by Folsome et al. (1971) were artifacts of their silylation procedure.

Subsequently, Stoks and Schwartz (1979) used HPLC to analyze water and formic-acid extracts of Murchison, Murray and Orgueil that had been desalted and fractionated with charcoal and a cation exchanger, a procedure that allows the isolation of uracil and thymine from a mixture of several purines and pyrimidines. Uracil was identified in all three meteorites using the criteria previously employed and also by mass spectrometry.

Finally, Stoks and Schwartz (1981a) added reversed-phase chromatography to their evolving isolation and analytical methodology. The previously reported purines (Hayatsu et al. 1975; van der Velden and Schwartz 1977) were confirmed by MS, and adenine was identified by both chromatographic and MS criteria. Furthermore, neither the hydroxypyrimidines nor the s-triazines (melamine, ammeline, ammelide and cyanuric acid) reported by Hayatsu et al. (1975) were detected above their respective detection limits of 10 and 50 ppb. Stoks and Schwartz (1981b) presented additional experimental evidence suggesting that s-triazines could have been formed from urea and guanylurea during the analysis and thus might be procedural artifacts like the 4-hydroxypyrimidines. The inability of Folsome et al. (1971,1973) to detect the purines and uracil was attributed to insufficient sensitivity.

Stoks and Schwartz (1982) have also identified a suite of *basic* N-heterocyclic compounds in water and formic-acid extracts of the Murchison meteorite. These compounds were prepared for GC and GC-MS analysis by repeated extraction between ethyl acetate and both dilute aqueous acid and base. Positive identifications were made of 2,4,6-trimethylpyridine, quinoline, isoquinoline, 2-methyl- and 4-methylquinoline. In addition, N-methylaniline, a suite of 12 additional alkyl pyridines, and a suite of 14 methylquinolines and/or isoquinolines were tentatively identified. Hayatsu et al. (1975) also observed a series of C_2-C_6 alkyl-substituted pyridines in a formic-acid extract. However, they suspected contamination since they were observed in only one of the two samples studied.

In summary, Murchison appears to contain several classes of basic and neutral N-heterocycles, including purines, pyrimidines, quinolines/isoquinolines and pyridines. The latter two groups are structurally diverse and contain a large number of isomeric alkyl derivatives. Taken together they may total about 7 ppm. In contrast, only one pyrimidine and four purines have been identified at a total concentration in the meteorite of about 1.3 ppm (Stoks and Schwartz 1981a). Hayatsu et al. (1975), using more rigorous extraction conditions, obtained 20 to 60 times more guanine and adenine, respectively. These compounds are apparently released by hydrolysis of the macromolecular C. All of the purines and the pyrimidine found are common biological constituents and no biologically unknown or unusual analogues accompany them. Isotopic measurements have not been reported and stereoisomeric criteria are not applicable since the compounds are achiral. Thus, although blank runs testify to contamination-free procedures, the possibility that these compounds originated in terrestrial microorganisms in the sample should be kept in mind.

Amines and Amides

Hayatsu et al. (1975), using the extraction and analytical procedures described previously, identified guanylurea (30-45 ppm) and tentatively identified urea (25 ppm) and some substituted ureas, perhaps including phenylureas, in a 6 M HCl extract of Murchison. They suggested, on the basis of paper chromatography and mass spectra, that aliphatic amines were also present in the hydrolyzed water extract.

Jungclaus et al. (1979) later tentatively identified 10 aliphatic amines in aqueous extracts of Murchison using gas chromatography of both the free amines and their dinitrophenyl derivatives, as well as ion-exchange chromatography. This suite of amines showed several of the characteristics of the amino acids: their content is increased about two-fold by acid hydrolysis; their concentrations decline sharply with increasing C number within the straight-chain homologous series; the branched-chain isomers are more abundant than the straight-chain isomers; and all possible isomers are present, at least for the primary amines through C_4. The concentrations of individual amines are comparable to those of the corresponding amino acids. Their total content is in excess of 0.2 μmol/g (8 ppm).

Alcohols and Carbonyl Compounds

Jungclaus et al. (1976) have identified series of lower alcohols, aldehydes and ketones in aqueous extracts of the Murchison and Murray meteorites. Alcohols and aldehydes through C_4 and ketones through C_5 were found. Declining concentrations with increasing C number were observed in the straight-chain homologous series of each class. All possible isomers appear to be present through C_4.

General Characteristics

Comparison of the analytical data for the various classes of soluble organic compounds of the Murchison meteorite allows the recognition of several general characteristics (see Table 10.5.1).

1. *Complete structural diversity.* At a given C number, all stable isomeric forms are present for most classes of compounds, at least among the lower homologues. This suggests synthesis by a random combination of C atoms, free of directing influences, for example, by catalysts. The purines and pyrimidines and the higher aliphatic hydrocarbons are possible exceptions.
2. *No enantiomeric preference.* All chiral compounds analyzed have been

TABLE 10.5.1
Concentrations and Molecular Characteristics of Soluble Organic Compounds of Meteorites[a]

Class	1 Concentration (ppm)	2 Compounds Identified	3 Chain Length	4 Homologous Decline	5 Branched- or Straight-Chain Predominance	6 Structural Diversity	7 Chirality
Amino acids	60	74	C_2–C_7	yes	Br	yes	R
Aliphatic hydrocarbons	>35	140	C_1–$C_{\geq 23}$	?	$<C_{10}$ Br $>C_{10}$ St	yes no	?
Aromatic hydrocarbons	15–28	87	C_6–C_{20}	NA	(Br)	yes	?
Carboxylic acids	>300	20	C_2–C_{12}	yes	Br	yes	?
Dicarboxylic acids	>30	17	C_2–C_9	yes	Br	yes	R
Hydroxycarboxylic acids	15	7	C_2–C_6	yes	St	yes	R
Purines & Pyrimidines	1.3	5	NA	NA	NA	no	NA
Basic N-heterocycles	7	32	NA	NA	NA	yes	?
Amines	8	10	C_1–C_4	yes	Br	yes	?
Amides	55–70	>2	NA	NA	NA	yes	?
Alcohols	11	8	C_1–C_4	yes	?	yes	?
Aldehydes & Ketones	27	9	C_1–C_5	yes	?	yes	?
Total	≥560	411					

[a] NA: not applicable; Br: branched; St: straight; R: racemic; ?: unknown.

found to be nearly racemic, i.e., to occur as nearly equimolar mixtures of the enantiomers.
3. *Concentrations decline with increasing C number in homologous series.* This observation is quite general, although apparent loss of the lower aliphatic hydrocarbons makes it impossible to evaluate for the alkanes as a whole.
4. *Branched-chain isomers are predominant.* This appears to be generally true, although the higher aliphatic hydrocarbons and the hydroxy acids are exceptions.

Insoluble Carbon

Most of the organic matter of carbonaceous chondrites is found in an insoluble residue that persists after prolonged treatment of the crushed meteorite with a series of refluxing solvents that may include nonpolar organic solvents, water and aqueous acids such as the HF-HCl mixtures that dissolve or partially dissolve the inorganic constituents of the meteorite matrix.

This C-rich fraction is heterogeneous. Even after prolonged and vigorous HF-HCl treatment, it retains a few percent of inorganic matter comprised of chromite, pentlandite and spinel grains (Lewis et al. 1975). In CI and CM chondrites, the most abundant organic component is a macromolecular material containing H, N, O, S and perhaps halogens, in addition to C. Various terms have been used in the literature in reference to this component, although it is most commonly called either "organic polymer" or "kerogen-like" material. In fact, it has little in common structurally with organic polymers or, with respect to its formation, with kerogens. It will be referred to here as macromolecular carbon.

In general, the insoluble carbon of CO and CV chondrites includes a smaller fraction of macromolecular carbon than is found in CI and CM chondrites. The Allende meteorite appears to be a limiting case in which the insoluble carbon is fundamentally different from macromolecular carbon. The principal insoluble component in this meteorite appears to be an elemental C differing from graphite mainly in its poorly ordered structure (Green et al. 1971; Breger et al. 1972; Smith and Buseck 1981; Lumpkin 1981). The insoluble carbon of the CV chondrite Grosnaja appears to contain macromolecular carbon as well as a more graphitic component (Vdovykin 1967). Thus, the insoluble carbon of carbonaceous chondrites must be viewed as a mixture of components, differing in the extent of their aromatic character and structural order. The content of macromolecular carbon is maximal in CI chondrites, dominates the insoluble carbon in CM chondrites, and is variable among the CV and CO chondrites, declining to a very low level in some meteorites such as Allende and Kainsaz.

In addition to macromolecular carbon, the insoluble carbon of the carbonaceous chondrites also contains multiple minor carbonaceous phases that

ORGANIC MATTER IN CARBONACEOUS CHONDRITES 837

have been recognized on the basis of their content of isotopically anomalous noble gases and their own unusual light-element isotopic compositions (Lewis and Anders 1983). These components are discussed in detail in Chapter 13.1.

Macromolecular Carbon. The composition of dry preparations obtained from Murchison after exhaustive HF-HCl digestion (approximately 4%

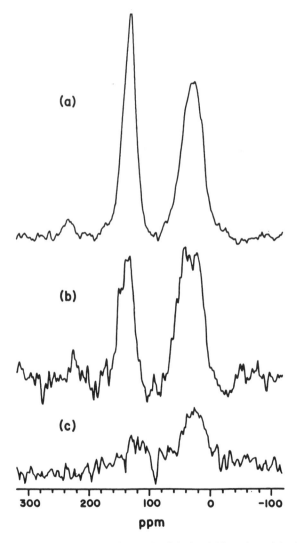

Fig. 10.5.2. Natural abundance ^{13}C NMR spectra of the insoluble carbon of the Orgueil, Murchison and Allende meteorites. (a) Orgueil (C:68.46%); (b) Murchison (C:6.65%); (c) Allende (C:15.26%). The cross polarization-magic angle spinning technique was used to obtain spectra of HF-HCl demineralized samples.

residual ash) is $C_{100}H_{71}N_3O_{12}S_2$ calculated on an ash-free basis (Hayatsu et al. 1977; Hayatsu et al. 1980). Zinner (Chapter 13.2) has revised the empirical formula to $C_{100}H_{48}N_{1.2}O_{12}S_2$ based on pyrolytic-release studies. The ^{13}C NMR spectrum (see Fig. 10.5.2b) of a partially demineralized (6.65 wt. % C) Murchison sample shows two broad features characteristic of sp^2-(olefinic/aromatic) and sp^3-hybridized (aliphatic) C atoms in a variety of structural environments (Cronin et al. 1987). Both the atomic H/C ratio and the ^{13}C NMR spectra are similar to those of the more aromatic (type III) terrestrial kerogens (Miknis et al. 1984).

Hayatsu et al. (1977,1980) have investigated the structure of the Murchison macromolecular carbon using a variety of degradative techniques. They conclude that the material is comprised of condensed aromatic, heteroaromatic and hydroaromatic ring systems in up to four-ring clusters, cross linked by short methylene chains, ethers, sulfides and biphenyl groups. Hayatsu et al. (1983) also note the similarity between their results with the Murchison macromolecular carbon and those obtained with vitrinite macerals of low-volatile bituminous coal or type III mature kerogen. Despite a general similarity to these terrestrial materials, significant differences in the detailed structure have been brought out by comparison of CuO-oxidation products (Hayatsu et al. 1980).

CI and CM chondrites contain μm-sized carbonaceous inclusions of various morphologies that are, in some cases, fluorescent. Rossignol-Strick and Barghoorn (1971) thoroughly surveyed the particles liberated by HF-digestion of the Orgueil meteorite and identified numerous hollow spheres as well as less abundant irregular objects having membranous and spiraled structures. They suggested that these objects are composed of macromolecular carbon and proposed abiotic schemes for their formation. The hollow spheres may have originated as coatings on mineral grains. Alpern and Benkheiri (1973) studied sections of the Orgueil meteorite by fluorescence microscopy and observed both small spherical bodies and larger, irregular objects formed around a central grain. More recently, similar fluorescent particles have been observed in Murchison sections (Deamer 1985). However, the fluorescent material can be extracted into chloroform suggesting that the fluorescence is not entirely attributable to macromolecular carbon. The chloroform-methanol extracts of Murchison were shown to form fluorescent droplets when dispersed in aqueous solution, a finding that has led Deamer (1985) to speculate on a possible role in prebiotic membrane formation.

10.5.2. ORIGIN OF METEORITIC ORGANIC MATTER

Twenty years ago, in reviewing the origin of meteorite organic compounds, Hayes (1967) stated, "This highly controversial subject is, at its simplest level, a question of whether the compounds observed are biogenic or abiogenic." It is encouraging to note that this question has been clearly resolved in favor of an abiogenic origin, at least in the case of uncontaminated specimens. Questions of origin, though still capable of generating controversy, are now posed in the context of the physicochemical processes that accompany the collapse of an interstellar cloud and the formation of a star and planetary system. They focus on possible sites of synthesis, such as the interstellar cloud, the solar nebula and planetesimals, and they inquire about synthetic pathways and the possibilities for secondary processing in these very different locales. It is important to note that, within the conventional context of a cooling nebula of solar composition, equilibrium conversion of C into organic compounds of greater complexity than methane would not have been straightforward (Fig. 10.5.3). At temperatures well above those at which organic compounds would have become thermodynamically stable, CH_4 and graphite would have become the stable phases. Furthermore, the reactions of C shown in Fig. 10.5.3 would have been kinetically inhibited (Lewis and Prinn 1980; Hayatsu and Anders 1981), ruling out production of organic compounds by purely equilibrium reactions. It follows that abiotic organic synthesis would have required the action of one or more nonequilibrium processes. Note also that the intrinsic properties of meteoritic organic matter are not obviously indicative of the environment, i.e., interstellar, nebular or planetary, in which it was formed.

Hypotheses dealing with the origin of organic compounds in meteorites are constrained by their ability to account for both the molecular and isotopic composition of the compounds in question and by their general consistency with meteorite petrology. In a few cases, laboratory models of proposed processes have been evaluated with respect to their ability to form an organic mixture with the molecular and isotopic characteristics of the meteorite organics. Two such models have been well studied, the Fischer-Tropsch type (FTT) process and the Miller-Urey (MU) synthesis, but several other possibilities have been considered in less detail. It may be an oversimplification to focus entirely on a single process as the source of meteorite organics. It is probably fair to say that most investigators believe that multiple processes and/or environments have contributed to the meteoritic organic matter.

The FTT process was first suggested in this context by Studier et al. (1968). It gives rise to a variety of organic compounds from reactions of CO, H_2 and NH_3 at the surface of catalytic particles, such as magnetite or phyllosilicates. (It should be noted that occurrences of these minerals in meteorites are now believed to have been formed by secondary aqueous alteration on the parent body; see Chapter 3.4.) The FTT process can be reasonably proposed whenever CO, H_2 and catalytic mineral surfaces are present at sufficiently high temperature. These requirements may have been commonly met, consequently the FTT process has been proposed as a significant mechanism in a variety of situations. Such proposals have received support from laboratory simulations that, in certain instances, form product mixtures that match the meteorite composition and carry a C-isotope fractionation pattern between

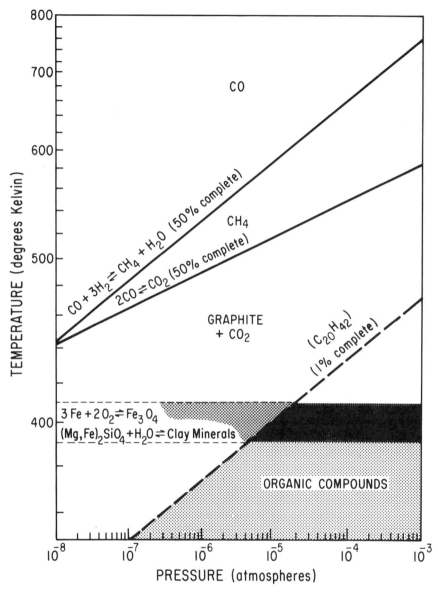

Fig. 10.5.3. Chemical state of C in the solar nebula assuming thermodynamic equilibrium. Conversion of CO to CH_4 and graphite during cooling of an initially hot nebula is believed to have been kinetically inhibited, permitting survival of CO to temperatures at which relatively complex hydrocarbons would have become stable. Production of such organic compounds may then have been catalysed by formation of magnetite and clay minerals from earlier condensed minerals, though clay formation in the nebula may also have been kinetically inhibited (Prinn and Fegley 1987). (Figure after Lewis and Anders 1983.)

organic C and CO_2, taken to be equivalent to carbonate, like that of carbonaceous chondrites. Anders and his colleagues have written extensively about the role of the FTT process in the synthesis of meteorite organics (see, e.g., Hayatsu and Anders 1981).

The MU synthesis involves the production and recombination of radical and ionic species in a reduced gas atmosphere under the influence of one or more of several possible energy sources, followed by secondary reactions in an aqueous phase (Miller 1955). With respect to meteorite organics, this process is usually assumed to have occurred at the surface of a parent body, perhaps the precursor of an asteroid (Peltzer et al. 1984). However, there are difficulties with this concept in relation to the maintenance of an appropriate atmosphere, cycling a large fraction of the mass of the body through the atmosphere, and distributing the organic products evenly throughout the body (E. Anders, personal communication). The MU synthesis has received support mainly from matches of the product mixture with the meteorite constituents (Wolman et al. 1972; Zeitman et al. 1974). It does not fractionate C isotopes between CO_2 and organic molecules to the extent that is seen in the Murchison meteorite (Hayatsu and Anders 1981). Miller et al. (1976) have discussed the MU synthesis with regard to its role in the formation of meteorite organic compounds.

Other photochemical processes have been suggested for the production of organic-rich grains that might ultimately contribute to the composition of carbonaceous chondrites. They involve photoionization of reduced gases either in the gas phase (Khare and Sagan 1973) or in interstellar grains (Greenberg 1984). In the latter case, synthesis of complex molecules is attributed to photoproduction of free radicals in the icy mantles surrounding the silicate cores of interstellar grains, followed by chain reactions among those radicals triggered by an event such as a grain-grain collision at moderate velocity. Such a sequence of events would be expected to occur many times during the lifetime of a grain, leading to accumulation of a substantial population of complex organics. The products obtained in laboratory simulations of these processes are not completely characterized, although the gas-phase experiments yield macromolecular materials called "tholins" (Sagan and Khare 1979) and the solid-state experiments have been shown to produce a complex organic residue that includes numerous simple acids, hydroxyacids, amides and glycols (Agarwal et al. 1985). In general, the photochemical production of simple radicals and ions is expected to produce a product mix of great structural variety.

A process that has recently become of interest in the context of prebiotic synthesis is the ion-molecule reaction. The occurrence of such reactions in dense interstellar clouds has been extensively investigated (see, e.g., Watson 1976; Herbst 1985), but how applicable this work is to conditions in the early solar system is not clear. In such a reaction scheme, an atom or molecular fragment is ionized by galactic cosmic radiation and can then react with neutral species at a temperature far below that at which two neutral species could react. It is this property that makes such reactions dominant in dense interstellar clouds, where temperatures are often close to 10 K. Note that it is these low temperatures, not the nature of the ion-molecule reaction itself, that lead to the pronounced isotopic fractionations that are commonly taken to be diagnostic of such reactions (see below and Chapter 13.2).

If secondary processes have significantly altered the original synthetic products, molecular analyses alone may be of limited value in elucidating the original process. In contrast, isotopic data may retain a characteristic imprint of the original process even after substantial alteration, though even in this case the possibility of secondary exchange must not be overlooked. Isotopic data for C, H and N have become available in increasing detail during the last 10 years and are now of considerable importance in understanding the origin of meteorite organic compounds. A review of meteorite light-element isotopic analyses has been provided by Pillinger (1984), and Mullie and Reisse (1987) have recently reviewed the origin question with emphasis on isotopic data. The determination of H-isotope ratios for the macromolecular carbon and, more recently, for a few soluble organic compounds has provided evidence for a genetic relationship between at least some meteorite organics and interstellar molecules (see Chapter 13.2).

In what follows we shall attempt to summarize the evidence bearing on the origin of several of the more extensively studied classes of organic compounds.

Carboxylic Acids. The suite of carboxylic acids found in the Murchison meteorite is rather typical of the soluble organic compounds in general. All structural isomers are present, at least among the lower homologues, and a decline in amount of 60 to 70% per unit increase in chain length is seen within the straight-chain series. Comparisons of the Murchison analytical data with the carboxylic acids formed by MU synthesis and by the FTT process failed to show close agreement in either case (Lawless and Yuen 1979). More recently, Epstein et al. (1987) have shown that total carboxylic acids in Murchison are moderately enriched in deuterium ($\delta D = 377\%o$) and have suggested a possible relationship with interstellar material. A hypothetical connection is readily made. Formic acid, the simplest of the carboxylic acids, has been identified in a dense cloud and methyl- and ethylcyanide, which can be hydrolytically converted to acetic acid and propionic acid, respectively, are also observed interstellar molecules (Pasachoff 1987).

Yuen et al. (1984) have determined the C-isotope composition of individual Murchison carboxylic acids and made the interesting observation that the ^{13}C content smoothly declines with increasing chain length (see Fig. 10.5.4). This finding is consistent with a synthetic scheme in which the higher acids are built up from lower homologues and in which C chains are formed by a kinetically controlled process, that is, one in which the C bond-making

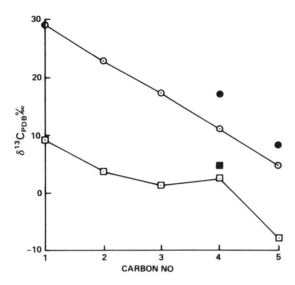

Fig. 10.5.4. Plots of $\delta^{13}C_{PDB}$ values of individual saturated hydrocarbons (□) and carboxylic acids (○) against carbon number. For example, 1 denotes methane and CO_2, 2 denotes ethane and acetic acid and so forth. The open symbols indicate values for the straight chain compounds and the filled symbols values for the branched isomers. (Figure from Yuen et al. 1984.)

reaction must have a significant activation energy. In contrast, D enrichment occurs in cold interstellar clouds where only reactions with negligible activation energies can occur. Thus the C- and H-isotope data taken together suggest C-chain formation in a warm environment, e.g., a dense cloud-diffuse cloud transition region (Herbst 1985), followed by hydrogenation by D-rich species or H exchange by ion-molecule reactions at very low temperature. A possible alternative is the synthesis of a reduced precursor(s) under cold cloud conditions, e.g., D-rich methane, followed by C-chain formation at a higher temperature without complete H loss. The latter step could be either a presolar or nebular process.

Lower Alkanes. As mentioned previously, the lower alkanes of carbonaceous chondrites present sampling difficulties that have given rise to considerable variation in the results of molecular analyses. As a result, it is difficult to draw inferences regarding their origin from the molecular analyses alone. The C-isotope analyses of Yuen et al. (1984) shown in Fig. 10.5.4 are more informative. The ^{13}C content ($\delta^{13}C$ = +2.4 to +9.2‰) of these alkanes confirms their extraterrestrial origin, and the decline in $\delta^{13}C$ from methane to butane suggests chain elongation of lower homologues in a kinetically controlled process as in the case of the carboxylic acids. G. Yuen and S. Chang (personal communication) have also determined $\delta^{13}C$ separately for the methyl and carboxyl groups of acetic acid isolated from the Murchison meteorite. The carboxyl group was found to have a $\delta^{13}C$ value similar to that of CO_2 in the meteorite whereas $\delta^{13}C$ of the methyl group was similar to that of methane. Assuming that the carboxyl group of all of the carboxylic acids is derived from CO_2, $\delta^{13}C$ for the alkyl group of each acid can be calculated. The alkyl groups of the carboxylic acids are thus found to be isotopically comparable to the corresponding alkanes. This similarity in ^{13}C content suggests a common origin for the C chains of the alkanes and the carboxylic acids. Thus one might speculate, even though H-isotope analyses have not yet been made on the lower alkanes, that they may also be of interstellar origin.

A straightforward synthesis of the lower alkanes by an FTT process seems to be inconsistent with the finding of Yuen et al. (1984) that both CO_2 ($\delta^{13}C = +29$) and these alkanes are isotopically substantially heavier than the coexisting CO ($\delta^{13}C = -32$), which might be assumed to represent the source gas for the putative FTT synthesis. This conclusion is based on the previously mentioned carbonate/organic ^{13}C fractionation that accompanies a simple FTT synthesis (Lancet and Anders 1970). If the trapped CO from Murchison that was isotopically analyzed by Yuen et al. (1984) is representative of the CO available for meteorite organic synthesis, which is by no means certain, its low $\delta^{13}C$ poses a general problem for the FTT process as an explanation for the formation of Murchison organic compounds that are isotopically heavier, i.e., with $\delta^{13}C$ values > -32. This would seem to be true for the carboxylic acid alkyl groups ($\delta^{13}C = -1$ to $+17$), the amino acids and possibly the higher alkanes (see below).

Higher Alkanes. Kvenvolden et al. (1970) proposed that the Murchison higher alkanes were produced by an MU synthesis because of the apparent multiplicity of chain isomers. However, analysis of the mixture at high resolution indicated, to the contrary, a synthesis that is impressive in its selectivity (Studier et al. 1972). Moreover, the straight-chain, methyl- and dimethyl-branched isomers that dominate the Murchison extract are the same ones that are the major products of the FTT process (Studier et al. 1968).

Carbon-isotope analyses have not been obtained on the higher alkanes, as such. However, Kvenvolden et al. (1970) reported $\delta^{13}C = +5$ for the Murchison benzene-MeOH extract, which includes nonvolatile aromatics and alkenes along with the higher alkanes. If this value is representative of the higher alkanes, the $\delta^{13}C$ difference of $+40$ between carbonate ($\delta^{13}C = +45.4$) and alkanes is consistent with the isotopic fractionation expected for the FTT synthesis (Lancet and Anders 1970). However, as mentioned previously, the $\delta^{13}C$ value obtained by Yuen et al. (1984) for the Murchison CO (-32) is difficult to reconcile with a simple FTT synthesis *if* this CO represents the source gas for the synthesis.

Amino Acids and Hydroxyacids. The results of the initial analyses of the Murchison amino acids were compared with the amino-acid products of both the FTT process (Studier et al. 1972) and the MU synthesis (Wolman et al. 1972). The comparison was viewed as favorable by the proponents of both processes. However, as more detailed analyses of the meteorite amino acids were done, some significant differences became apparent (Cronin and Pizzarello 1983).

The first isotopic analyses of amino acids were reported by Chang et al. (1978), who found that the Murchison amino acids were enriched in ^{13}C (δ^{13}C = +23 to +44) almost to the extent of the coexisting carbonate (δ^{13}C = 44.4). This finding is difficult to reconcile with a simple FTT synthesis in view of the substantial ^{13}C-fractionation between carbonate and organic matter discussed previously. However, Hayatsu and Anders (1981) have suggested that it could reflect recycling of ^{13}C-rich product CO_2 through the catalytic process.

Peltzer and Bada (1978) noted that their identification of α-hydroxy acids in Murchison suggests an MU synthesis. Moreover, the parallel between the abundances of these compounds and the Murchison α-amino acids suggests the formation of both groups from common precursors by way of a Strecker-cyanohydrin mechanism, the aqueous-phase component of amino-acid synthesis by the MU process (Peltzer et al. 1984).

$$RCHO + HCN \begin{matrix} \nearrow \\ +NH_3 \searrow \end{matrix} \begin{matrix} \underset{R-CH-CN}{\overset{OH}{|}} \\ \underset{R-CH-CN}{\overset{NH_2}{|}} \end{matrix} \xrightarrow{+H_2O} \begin{matrix} \underset{R-CH-C-NH_2}{\overset{OH\ \ O}{|\ \ \ ||}} \\ \underset{R-CH-C-NH_2}{\overset{NH_2\ O}{|\ \ \ ||}} \end{matrix} \xrightarrow[-NH_3]{+H_2O} \begin{matrix} \underset{R-CH-COOH}{\overset{OH}{|}} \\ \underset{R-CH-COOH}{\overset{NH_2}{|}} \end{matrix}$$

Strecker–Cyanohydrin Reaction

Several characteristics of the Murchison suite of amino acids led Cronin and Pizzarello (1986) to speculate as to a role for interstellar processes in their synthesis. Evidence for this hypothesis was recently provided by Epstein et al. (1987) who found the Murchison amino acids as a group to be quite D-rich (δD = 1370‰). As in the case of the carboxylic acids, this finding strongly suggests that the amino acids or their precursors were formed by ion-molecule reactions at low temperature. To date, amino acids have not been identified in interstellar space (Hollis et al. 1980; Snyder et al. 1983). However, their immediate precursors in a Strecker-cyanohydrin synthesis, aldehydes, hydrocyanic acid, ammonia and water, were among the first interstellar molecules detected by radioastronomy (Pasachoff 1987). Thus, the molecular analyses, the isotopic analyses and the observed relationship to the α-

hydroxy-acids can all be accommodated in a formation scheme in which D-rich interstellar precursors are converted to amino acids by a Strecker-cyanohydrin synthesis. The latter might be reasonably expected to have accompanied aqueous alteration on an asteroidal parent body (Bunch and Chang 1980; Peltzer et al. 1984; see also Chapter 3.4).

The recent finding of a substantial D enrichment in the Murchison amino acids and carboxylic acids raises the question whether other classes of soluble organic compounds are similarly enriched. The generally similar molecular characteristics of the soluble organic compounds suggests that common isotopic characteristics should not be unexpected. If this turns out to be the case, the synthetic processes leading to additional classes of soluble organic compounds may also have included interstellar ion-molecule reactions. This possibility should be kept in mind when considering the origin of the remaining classes of compounds.

Aromatic Hydrocarbons. The aromatic hydrocarbon data do not seem to support unambiguously any particular synthetic process and may in fact reflect a secondary aromatization independent of any primary synthetic process. The FTT process, for example, does not produce aromatic hydrocarbons as primary products. However, Studier et al. (1972) have pointed out that additional heating of the aliphatic primary products in the presence of the catalyst leads to their conversion to aromatic species. They rationalize the predominance of aromatics and the low levels of normal aliphatic hydrocarbons in the C_6–C_{10} range observed in Murchison on the basis of such a two-stage FTT synthesis. Thermodynamic calculations (Eck et al. 1966) suggest conditions under which aromatics (along with methane) will be the dominant organics formed. Studier et al. (1972) find aromatic/aliphatic ratios, e.g., benzene/hexane, ranging from 10 to 60. These values are lower by several orders of magnitude than the predicted ratios, suggesting that the aromatization process acted on only part of the aliphatic content or did not go to completion. It should be noted that an aromatic predominance has not been universally observed. Belsky and Kaplan (1970) failed to see abundant lower aromatics in Murray and several other CM chondrites, and Yuen et al. (1984), although reporting a significant quantity of benzene after freeze-thaw disaggregation of Murchison, find the result to be questionable on the basis of its irreproducibility and a terrestrial $\delta^{13}C$ value (G. U. Yuen, personal communication).

Pering and Ponnamperuma (1971) and Basile et al. (1984) found the higher, nonvolatile aromatic hydrocarbons to be similar to the products of a high-temperature thermal synthesis. Such a process, if it occurred, must have taken place early or have been localized in order to avoid degradation of other, more temperature-sensitive organic compounds (Hayatsu and Anders 1981).

Dicarboxylic Acids. The large number of isomers of the Murchison low-molecular-weight dicarboxylic acids suggested to Lawless et al. (1974) a random chemical synthesis. An electrical-discharge experiment produced a similar suite containing eight of the 17 dicarboxylic acids found in the meteorite (Zeitman et al. 1974). Peltzer et al. (1984) have pointed out that succinic acid, the most abundant dicarboxylic acid in both cases, is readily made from acrylonitrile, a major product of electrical discharge through a methane- and ammonia-containing atmosphere. It might be noted that acrylonitrile has also been identified in interstellar clouds (Pasachoff 1987). Dicarboxylic acids have not been sought among FTT products (Hayatsu and Anders 1981).

N-Containing Compounds. Again, the analytical data do not seem to discriminate between different synthetic processes. Hayatsu et al. (1975) found aliphatic amines and alkyl pyridines in water extracts of Murchison and various N-heterocycles, guanidines and ureas in hot-acid extracts. Earlier, several of these compounds had been shown to be products of an FTT synthesis carried out in the presence of ammonia (Hayatsu et al. 1972). These findings suggest a role for FTT reactions in the synthesis of the Murchison N-compounds. Stoks and Schwartz (1982) have noted that alkyl pyridines are produced from aldehydes and ammonia by a catalyzed reaction that occurs under FTT reaction conditions. Stoks and Schwartz (1981b) have also pointed out that many of the Murchison N-compounds can be produced by oligomerization of HCN followed by hydrolysis. HCN is an intermediate in the MU synthesis as well as an interstellar molecule. A clear choice among several plausible synthetic routes does not seem possible on the basis of the analytical data presently available. Kung et al. (1979) showed that neither FTT nor MU syntheses produced N-isotope fractionations comparable to the diversity observed in meteoritic organic matter.

Macromolecular Carbon. Speculation as to the origin of the macromolecular carbon has been based on comparisons of its structure with the insoluble carbonaceous matter produced by laboratory models of plausible synthetic schemes and, more recently, on the results of stable-isotope analyses. Anders et al. (1973) noted the similarity between mass spectra of macromolecular material obtained from Murchison and from an extended FTT synthesis. Later, Hayatsu et al. (1977) carried out a more detailed comparison of these materials that included the use of various degradation techniques. Both similarities and differences were observed. However, the authors pointed out a fundamental problem with the comparative structural approach in evaluating the role of an FTT process in the origin of the macromolecular carbon, i.e., that since the nature of the FTT product can vary with reaction conditions, observed discrepancies can probably be eliminated by cosmochemically plausible variations in the synthetic conditions or secondary processing, e.g., reheating. Since the actual conditions are impossible to establish in detail, a structural match or mismatch is not conclusive. Nevertheless, other observations suggest an FTT contribution to the formation of the macromolecular carbon, i.e., the presence of straight-chain alkyl fragments, the close association of the macromolecular carbon with potentially catalytic mineral grains (though see earlier comment on the secondary nature of the catalytically active lithology in CI and CM chondrites), and the $\delta^{13}C$ difference between carbonate and bulk macromolecular carbon.

Other laboratory models, e.g., electrical discharge (Miller 1955) or ultraviolet irradiation (Sagan and Khare 1979) of reduced gas atmospheres, also generate macromolecular products. However, these products are not structurally well characterized. Their elemental compositions suggest a more aliphatic or alicyclic structure than appears to be the case for the Murchison material (Hayatsu and Anders 1981).

Stable-isotope analyses have contributed significantly to current ideas about the origin of macromolecular carbon. The enrichment of a substantial fraction of this material in D is the principal evidence for a relationship with interstellar molecules (see Chapter 13.2). Kolodny et al. (1980) used an indirect plasma-combustion method to demonstrate that the D enrichment of carbonaceous chondrites first observed by Boato (1954) was concentrated in the organic matter. The work of Smith and Rigby (1981) showed that the insoluble carbon of Murchison was, in fact, D rich. In an attempt to localize further the D-rich moiety, Robert and Epstein (1982) carried out stepwise pyrolysis of the insoluble carbon of Murchison and other carbonaceous chondrites both before and after chemical oxidation. They found the H released from the unoxidized samples to be D rich and isotopically heterogeneous. The D-rich component was removed by prior oxidation indicating its organic character. These observations were incorporated by Robert and Epstein into a model in which D enrichment is a consequence of synthesis by ion-molecule reactions in an interstellar cloud as suggested by Kolodny et al. (1980).

It is of interest to know how many isotopically distinct components comprise the macromolecular carbon and to define their structural characteristics and individual isotopic ratios (Kerridge 1983; Yang and Epstein 1984). Recently, Kerridge et al. (1987) have used both stepwise pyrolysis and stepwise combustion techniques in an attempt to "dissect" the Murchison macromolecular carbon into isotopically unique components. After correcting for the contribution of the "exotic" carbons (Chapter 13.1), they were able to identify three or possibly four moieties that are isotopically distinctive and presumably structurally different. Two are believed to be predominantly aromatic and one or, possibly, two are aliphatic. All are D rich, but to varying degrees ($\delta D \approx 400$ to 1800), and unexceptional with respect to C ($\delta^{13}C = -22$ to > -12.4). It is interesting to recall that the Murchison CO_2, carbonate and polar organic compounds are significantly enriched in ^{13}C with respect to these macromolecular entities, a fact that may point to differences in origin.

The sequence of chemical events that has given rise to the macromolecular carbon of carbonaceous chondrites remains uncertain. However, the recognition of a possible contribution from surviving interstellar molecules has opened a new and promising avenue of investigation, namely, determining both the extent of their contribution and the number of isotopically distinguishable components that participated. The remainder of the macromolecular carbon, i.e., that lacking an interstellar isotopic signature, may be explicable in terms of the nebular and/or planetary processes that have stimulated so much discussion previously. Defining the interstellar contribution should provide a much clearer view of the molecular species that are to be properly assigned to these latter processes. Much work remains to be done in identifying, structurally characterizing and quantifying the isotopically distinct components of the macromolecular carbon, as well as the isotopic composition of the soluble organic compounds, in carbonaceous chondrites. Nevertheless, it is work that holds promise of providing an understanding of the origin and evolution of the organic matter of these meteorites that far exceeds the indistinct outlines that are now apparent.

10.5.3. SOLID ORGANIC MATTER ON PLANETARY SATELLITES, ASTEROIDS AND COMETS

Insofar as the meteorites come from some class or classes of solar-system body, a brief review of the evidence for organic matter elsewhere in the solar system is in order. The evidence for solid organic material on the surfaces of solar-system bodies is largely circumstantial. The clearest case is for the C-H stretch spectral feature at 3.4 μm in the spectrum of Comet Halley, both from groundbased observations and from spacecraft measurements, a subject to which we shall return below. Most of the evidence for organic matter on other bodies, such as asteroids, planetary satellites and comets other than Halley, comes from telescopic observations in the spectral range 0.3 to 4 μm. This relatively narrow wavelength window is imposed by the transmission properties of the Earth's atmosphere and by the energy distribution of the sunlight illuminating the bodies in the solar system.

Planetary Satellites

The most striking planetary satellite found to have large surface expanses of very low-albedo material is Saturn's Iapetus, long known to have one relatively high-albedo hemisphere and one that is low. The dark hemisphere is that centered on the direction of motion of the satellite in its orbit around Saturn, leading to speculation that the dark material is in some way related to the motion of Iapetus through space. Dark material might be deposited on the leading hemisphere by accretion from space, or some sputtering phenomenon might remove lighter-colored material from the surface, exposing a dark component in a deeper layer (Cruikshank et al. 1983; Bell et al. 1985; Squyers and Sagan 1983). While no distinct spectral signatures in the dark material have been observed, the leading hemisphere has an extremely red color (the spectral reflectance strongly increases toward longer wavelengths).

The materials considered by Bell et al. (1985) as candidates for the dark deposit (on the basis of laboratory comparisons) included the organic component of Murchison and an insoluble residue from terrestrial coal tar. Neither sample had the strong red color of Iapetus, but when mixed with clays of moderate Fe content (similar to meteorite clays) which do have strong colors, the resulting composite had both a color and absolute reflectance that matched the dark side of Iapetus reasonably well. Bell et al. suggested that the unusual reflectance of Iapetus could be matched by a mixture of about 10% organics and 90% hydrated silicates (clay), and that a small amount of elemental C may also be present.

Asteroids

The low-albedo asteroids have been classified according to their spectral reflectances or photometrically observed colors in the photovisual region of the spectrum (0.3 to 1.05 μm), with the information on their albedos usually coming from infrared radiometry at 10 and 20 μm. The three basic classes of low-albedo asteroids are C, P and D (see Chapter 2.1), representing the slope of the color curve in the spectral region noted above. C-type asteroids are usually neutral over most of the spectral region, with a downturn in the reflectance toward the violet end of the spectrum (see Fig. 2.1.3). In the near-infrared (1 to 2.5 μm), the C types show a range of slopes from flat (neutral) to slightly upward (reddish). No discrete spectral bands are seen in the C-type spectra in this region, but further into the infrared, Feierberg et al. (1985) have found that several C types show a distinct absorption band at 3 μm attributed to bound water in the mineral assemblages in the asteroid regolith. Some Cs have no bound water, and there is apparently considerable variability among them, with several degrees of hydration represented.

The P and D types are distinguished by their redder slopes in the near infrared; the Ds include the reddest known asteroids. There are presently no data on the water of hydration in the P and D types; their surfaces may contain too much opaque material to permit a detection of the bound-water band even if it exists.

Because the low-albedo asteroids have largely featureless spectra and since their albedos are similar to the C-bearing dark carbonaceous chondritic meteorites, the C, P and D types are usually thought to contain elemental C and/or complex C compounds in their regoliths.

The space distribution of the asteroids by spectral type bears on the question of their surface compositions. From studies by Gradie and Tedesco (1982), two clear trends have emerged. First, the low-albedo objects dominate the region in the asteroid main belt beyond about 3 AU, and second, there is a strong trend toward increased redness of the spectral reflectances with increasing heliocentric distance.

Various authors have suggested that the dominance of low-albedo types and the increased redness at large heliocentric distances are a result of a gradient in the temperature of the solar nebula that is preserved in the present distribution of the asteroids that formed at different distances from the Sun (see, e.g., Chapman et al. 1975; Anders 1978; Zellner 1979; Gradie and Tedesco 1982). The appearance of dark bodies at about 3 AU apparently marks the onset of a region of formation in the solar nebula of C-rich materials and complex organic compounds that color the materials of which the asteroids are composed.

Gradie and Veverka (1980) called attention to the possible presence of complex organics in the form of kerogens in the outermost asteroids. They simulated the photovisual reflectance of the red D-type asteroids by synthetic mixtures of clays and kerogens; the Trojan asteroid 624 Hektor, for example, was matched by a mixture of 85% montmorillonite clay and about 8 to 15% kerogen, plus some magnetite or carbon black. It was, in fact, this work that inspired the comparison of clay-kerogen mixtures with Iapetus, as discussed above.

Comets

Spectral observations of Comet Halley show the C-H stretch band at 3.4 μm in emission, indicating the presence of organic matter on the nucleus (Combes et al. 1986) and in the dust of the inner coma (Wickramasinghe and Allen 1986; Baas et al. 1986; Knacke et al. 1986). Specific organic compounds have not been identified, but it is widely presumed that complex aliphatic and aromatic hydrocarbons contribute to, or are responsible for, the low albedo of the nuclear surface and the dust, and that these same compounds give rise to the emission band at 3.4 μm. Furthermore, some of the interplanetary dust particles collected from the Earth's stratosphere, and which are widely presumed to originate chiefly from comets passing through the inner solar system, show the 3.4 μm C-H feature (Sandford and Walker 1985; see also Chapter 11.1).

Mass-spectral analysis by Vega 1 of 43 Halley dust particles, 10 pg or less in size, revealed a preponderance of chondritic cores coated with highly unsaturated organic material (Kissel and Krueger 1987). Those workers reported a number of unsaturated N-containing species (e.g., nitriles, aldimines, enamines) and inferred the presence of alkenes, alkynes and aromatic hydrocarbons, in addition to other nitrogenous species such as pyrrol, pyridine and pyrimidine. Biologically important amino acids, if present at all, were at least a factor of 30 less abundant than the pyrimidines and purines.

10.5.4. SUMMARY

1. CM chondrites contain C largely in molecular forms. About 70% is present as an insoluble macromolecular material and about 30% occurs as a

diverse suite of soluble organic compounds. CI chondrites may be similar to CM chondrites in this respect, whereas CV chondrites are relatively poor in organic compounds and contain most of their C in an elemental form differing from graphite mainly in its poorly ordered structure.

2. The soluble organic compounds of CM chondrites are found as aliphatic and aromatic hydrocarbons, alcohols, carbonyl compounds, monocarboxylic-, dicarboxylic-, hydroxy- and amino acids, aliphatic amines and amides and N-heterocycles.

3. Several of the classes of soluble organic compounds of CM chondrites show common characteristics. These are (i) complete structural diversity, (ii) enantiomeric equivalence, (iii) exponentially declining content within homologous series, and (iv) a predominance of branched-chain molecules.

4. The insoluble macromolecular carbon is comprised of condensed aromatic ring systems cross linked and substituted by short methylene chains and other functional groups.

5. Meteorite organic compounds have been commonly viewed as products of either a Fischer-Tropsch type process operating in the solar nebula or a Miller-Urey type process occurring on a meteorite parent body. Recently, D analyses of amino acids, carboxylic acids and macromolecular carbon suggest a relationship to interstellar organic compounds.

6. Dark material occurs throughout the planetary system beyond about 2.5 AU from the Sun. The color of this material ranges from neutral to red, represented by the slope of the reflectance curve from 0.3 to 2.5 μm wavelength.

7. The surfaces of comets, asteroids and some planetary satellites are probably covered in part by materials representative of the low-albedo substances found in primitive carbonaceous chondrites; these materials include the C-rich compounds, of which macromolecular carbon is the most important.

8. Carbonaceous material from the meteorites is itself the wrong color to match the solar-system bodies, but in combinations with silicates, some hydrated and some anhydrous, the overall colors can be matched.

9. Organic material must occur on the satellites and asteroids, but its direct identification depends upon further improvements in astronomical techniques, or upon *in situ* observations, such as those that have been accomplished for Comet Halley.

Acknowledgments. This work was supported by the National Aeronautics and Space Adminstration. We are grateful to E. Anders and J. Oró for thoughtful and constructive reviews and to G. Yuen for many helpful conversations.

REFERENCES

Agarwal, V. K., Schutte, W., Greenberg, J. M., Ferris, J. P., Briggs, R., Connor, S., Van de Bult, C. P. E. M., and Baas, F. 1985. Photochemical reactions in interstellar grains: Photolysis of CO, NH_3, and H_2O. *Origins of Life* 16:21–40.

Alpern, B., and Benkheiri, Y. 1973. Distribution de la matière organique dans la météorite d'Orgueil par microscopie en fluorescence. *Earth Planet. Sci. Lett.* 19:422–428.

Anders, E. 1978. Most stony meteorites come from the asteroid belt. In *Asteroids: An Exploration Assessment*, eds. D. Morrison and W. C. Wells, NASA CP-2053, pp. 57–75.

Anders, E., Hayatsu, R., and Studier, M. 1973. Organic compounds in meteorites. *Science* 182:781–789.

Baas, F., Geballe, T. R., and Walther, D. M. 1986. Spectroscopy of the 3.4 micron emission feature in Comet Halley. *Astrophys. J.* 311:L97–L101.

Bada, J. L., Cronin, J. R., Ho, M.-S., Kvenvolden, K. A., Lawless, J. G., Miller, S. L., Oró, J., and Steinberg, S. 1983. On the reported optical activity of amino acids in the Murchison meteorite. *Nature* 301:494–497.

Baker, B. L. 1971. Review of organic matter in the Orgueil meteorite. *Space Life Sci.* 2:472–497.

Basile, B. P., Middleditch, B. S., and Oró, J. 1984. Polycyclic aromatic hydrocarbons in the Murchison meteorite. *Org. Geochem.* 5:211–216.

Bell, J. F., Cruikshank, D. P., and Gaffey, M. J. 1985. The composition of the Iapetus dark material. *Icarus* 61:192–207.

Belsky, T., and Kaplan, I. R. 1970. Light hydrocarbon gases, C^{13}, and origin of organic matter in carbonaceous chondrites. *Geochim. Cosmochim. Acta* 34:257–278.

Berthelot, M. P. E. 1868. Sur la matière charbonneuse des météorites. *Comptes Rendus Hebd. Seances Acad. Sci.* 67:849.

Berzelius, J. J. 1834. Ueber Meteorsteine. *Ann. Phys. Chem.* 33:113–148.

Boato, G. 1954. The isotopic composition of hydrogen and carbon in the carbonaceous chondrites. *Geochim. Cosmochim. Acta* 6:209–220.

Bonner, W. A., Blair, N. E., and Lemon, R. M. 1979. The radioracemization of amino acids by ionizing radiation: Geochemical and cosmochemical implications. *Origins of Life* 9:279–290.

Breger, I. A., Zubovic, P., Chandler, J. C., and Clarke, R. S. 1972. Occurrence and significance of formaldehyde in the Allende carbonaceous chondrite. *Nature* 236:155–158.

Briggs, M. H., and Mamikunian, G. 1963. Organic constituents of the carbonaceous chondrites. *Space Sci. Rev.* 1:647–682.

Buhl, P. 1975. *An Investigation of Organic Compounds in the Mighei Meteorite*. Ph.D. Thesis, Univ. of Maryland, College Park.

Bunch, T. E., and Chang, S. 1980. Carbonaceous chondrites–II. Carbonaceous chondrite phyllosilicates and light element geochemistry as indicators of parent body processes and surface conditions. *Geochim. Cosmochim. Acta* 44:1543–1577.

Chang, S., Mack, R., and Lennon, K. 1978. Carbon chemistry of separated phases of Murchison and Allende meteorites. *Lunar Planet. Sci.* IX:157–158 (abstract).

Chapman, C. R., Morrison, D., and Zellner, B. 1975. Surface properties of asteroids: A synthesis of polarimetry, radiometry and spectrophotometry. *Icarus* 25:104–130.

Cohen, E. 1894. *Meteoritenkunde*, vol. 1 (Stuttgart: E. Schweizerbart'sche Verlagshandlung), pp. 159–169.

Combes, M., Moroz, V., Crifo, J. F., Bibring, J. P., Coron, N., Crovisier, J., Encrenaz, T., Sanko, N., Grigoriev, A., Bockelée-Morvan, D., Gispert, R., Emerich, C., Lamarre, J. M., Rocard, F., Krasnopolsky, V., and Owen, T. 1986. Detection of parent molecules in comet Halley from the VEGA experiment. In *Proc. 20th ESLAB Symp. on the Exploration of Halley's Comet*, vol. 1, ESA SP-250, pp. 353–358.

Cronin, J. R. 1976a. Acid-labile amino acid precursors in the Murchison meteorite. I. Chromatographic fractionation. *Origins of Life* 7:337–342.

Cronin, J. R. 1976b. Acid-labile amino acid precursors in the Murchison meteorite. II. A search for peptides and amino acyl amides. *Origins of Life* 7:343–348.

Cronin, J. R., and Moore, C. B. 1971. Amino acid analyses of the Murchison, Murray, and Allende carbonaceous chondrites. *Science* 172:1327–1329.

Cronin, J. R., and Pizzarello, S. 1983. Amino acids in meteorites. *Adv. Space Res.* 3:5–18.
Cronin, J. R., and Pizzarello, S. 1986. Amino acids of the Murchison meteorite. III. Seven carbon acyclic primary α-amino alkanoic acids. *Geochim. Cosmochim. Acta* 50:2419–2427.
Cronin, J. R., Pizzarello, S., and Moore, C. B. 1979. Amino acids in an Antarctic chondrite. *Science* 206:335–337.
Cronin, J. R., Gandy, W. E., and Pizzarello, S. 1981. Amino acids of the Murchison meteorite: I. Six carbon acyclic primary α-amino alkanoic acids. *J. Molec. Evol.* 17:265–272.
Cronin, J. R., Pizzarello, S., and Yuen, G. U. 1985. Amino acids of the Murchison meteorite: II. Five carbon acyclic primary β-, γ-, and δ-amino alkanoic acids. *Geochim. Cosmochim. Acta* 49:2259–2265.
Cronin, J. R., Pizzarello, S., and Frye, J. S. 1987. ^{13}C NMR spectroscopy of the insoluble carbon of carbonaceous chondrites. *Geochim. Cosmochim. Acta* 51:299–303.
Cruikshank, D. P., Bell, J. F., Gaffey, M. J., Brown, R. H., Howell, R., Beerman, C., and Rognstad, M. 1983. The dark side of Iapetus. *Icarus* 53:90–104.
Deamer, D. W. 1985. Boundary structures are formed by organic components of the Murchison carbonaceous chondrite. *Nature* 317:792–794.
DuFresne, E. R., and Anders, E. 1962. On the chemical evolution of the carbonaceous chondrites. *Geochim. Cosmochim. Acta* 26:1085–1114.
Eck, R. V., Lippincott, E. R., Dayhoff, M. O., and Pratt, Y. T. 1966. Thermodynamic equilibrium and the inorganic origin of organic compounds. *Science* 153:628–633.
Engel, M. H., and Nagy, B. 1982. Distribution and enantiomeric composition of amino acids in the Murchison meteorite. *Nature* 296:837–840.
Epstein, S., Krishnamurthy, R. V., Cronin, J. R., Pizzarello, S., and Yuen, G. U. 1987. Unusual stable isotope ratios in amino acid and carboxylic acid extracts from the Murchison meteorite. *Nature* 326:477–479.
Feierberg, M. A., Lebofsky, L. A., and Tholen, D. J. 1985. The nature of C-class asteroids from 3-μm spectrophotometry. *Icarus* 63:183–191.
Folsome, G. E., Lawless, J., Romiez, M., and Ponnamperuma, C. 1971. Heterocyclic compounds indigenous to the Murchison meteorite. *Nature* 232:108–109.
Folsome, G. E., Lawless, J. G., Romiez, M., and Ponnamperuma, C. 1973. Heterocyclic compounds recovered from carbonaceous chondrites. *Geochim. Cosmochim. Acta* 46:455–465.
Fuchs, L. H., Olsen, E., and Jensen, K. J. 1973. Mineralogy, crystal chemistry, and composition of the Murchison (C2) meteorite. *Smithsonian Contrib. Earth Sci.* 10:1–39.
Gelpi, E., and Oró, J. 1970. Organic compounds in meteorites—IV. Gas chromatographic-mass spectrometric studies on the isoprenoids and other isomeric alkanes in carbonaceous chondrites. *Geochim. Cosmochim. Acta* 34:981–994.
Gradie, J., and Tedesco, E. F. 1982. Compositional structure of the asteroid belt. *Science* 216:1405–1407.
Green, H. W., Radcliffe, S. V., and Heuer, A. H. 1971. Allende meteorite: A high-voltage electron petrographic study. *Science* 172:936–939.
Greenberg, J. M. 1984. Chemical evolution in space. *Origins of Life* 14:25–36.
Harada, K., and Hare, P. E. 1980. Analyses of amino acids from the Allende meteorite. In *Biogeochemistry of Amino Acids*, eds. P. E. Hare, T. C. Hoering, and K. King, Jr. (New York: Wiley), pp. 169–181.
Hayatsu, R. 1964. Orgueil meteorite: Organic nitrogen contents. *Science* 146:1291–1293.
Hayatsu, R., and Anders, E. 1981. Organic compounds in meteorites and their origins. In *Cosmo- and Geochemistry*, vol. 99, *Topics in Current Chemistry* (Berlin: Springer-Verlag), pp. 1–37.
Hayatsu, R., Studier, M. H., Matsuoka, S., and Anders, E. 1972. Origin of organic matter in early solar system—VI. Catalytic synthesis of nitriles, nitrogen bases and porphyrin-like pigments. *Geochim. Cosmochim. Acta* 36:555–571.
Hayatsu, R., Studier, M. H., Moore, L. P., and Anders, E. 1975. Purines and triazines in the Murchison meteorite. *Geochim. Cosmochim. Acta* 39:471–488.
Hayatsu, R., Matsuoka, S., Scott, R. G., Studier, M., and Anders, E. 1977. Origin of organic matter in the early solar system—VII. The organic polymer in carbonaceous chondrites. *Geochim. Cosmochim. Acta* 41:1325–1339.
Hayatsu, R., Winans, R. E., Scott, R. G., McBeth, R. L., Moore, L. P., and Studier, M. H.

1980. Phenolic ethers in the organic polymer of the Murchison meteorite. *Science* 207:1202–1204.
Hayatsu, R., Scott, R. G., and Winans, R. E. 1983. Comparative structural study of meteoritic polymer with terrestrial geopolymers coal and kerogen. *Meteoritics* 18:310 (abstract).
Hayes, J. M. 1967. Organic constituents of meteorites—A review. *Geochim. Cosmochim. Acta* 31:1395–1440.
Herbst, E. 1985. On the formation and observation of complex interstellar molecules. *Origins of Life* 16:3–19.
Hollis, J. M., Snyder, L. E., Suenram, R. D., and Lovas, F. J. 1980. Search for the lowest-energy conformer of interstellar glycine. *Astrophys. J.* 241:1001–1006.
Holzer, G., and Oró, J. 1979. The organic composition of the Allan Hills carbonaceous chondrite (77306) as determined by pyrolysis-gas chromatography-mass spectrometry and other methods. *J. Molec. Evol.* 13:265–270.
Jungclaus, G. A., Yuen, G. U., Moore, C. B., and Lawless, J. G. 1976. Evidence for the presence of low molecular weight alcohols and carbonyl compounds in the Murchison meteorite. *Meteoritics* 11:231–237.
Jungclaus, G., Cronin, J. R., Moore, C. B., and Yuen, G. U. 1979. Aliphatic amines in the Murchison meteorite. *Nature* 261:126–128.
Kerridge, J. F. 1983. Isotopic composition of carbonaceous-chondrite kerogen: Evidence for an interstellar origin of organic matter in meteorites. *Earth Planet. Sci. Lett.* 64:186–200.
Kerridge, J. F., Chang, S., and Shipp, R. 1987. Isotopic characterization of kerogen-like material in the Murchison carbonaceous chondrite. *Geochim. Cosmochim. Acta* 51:2527–2540.
Khare, B. N., and Sagan, C. 1973. Red clouds in reducing atmospheres. *Icarus* 20:311–321.
Kissel, J., and Krueger, F. R. 1987. The organic component in dust from comet Halley as measured by the PUMA mass spectrometer onboard Vega 1. *Nature* 326:755–760.
Knacke, R. F., Brooke, T. Y., and Joyce, R. R. 1986. Observations of 3.2–3.6 micron emission features in Comet Halley. *Astrophys. J.* 310:L49–L53.
Kolodny, Y., Kerridge, J. F., and Kaplan, I. R. 1980. Deuterium in carbonaceous chondrites. *Earth Planet. Sci. Lett.* 46:149–158.
Kotra, R. K., Shimoyama, A., Ponnamperuma, C., and Hare, P. E. 1979. Amino acids in a carbonaceous chondrite from Antarctica. *J. Molec. Evol.* 13:179–184.
Kung, C. C., Hayatsu, R., Studier, M. H., and Clayton, R. N. 1979. Nitrogen isotope fractionations in the Fischer-Tropsch synthesis and the Miller-Urey reaction. *Earth Planet. Sci. Lett.* 46:141–146.
Kvenvolden, K., Lawless, J., Pering, K., Peterson, E., Flores, J., Ponnamperuma, C., Kaplan, I. R., and Moore, C. 1970. Evidence for extraterrestrial amino acids and hydrocarbons in the Murchison meteorite. *Nature* 288:923–926.
Kvenvolden, K., Lawless, J. G., and Ponnamperuma, C. 1971. Nonprotein amino acids in the Murchison meteorite. *Proc. Natl. Acad. Sci. USA* 68:486–490.
Lancet, M. S., and Anders, E. 1970. Carbon isotope fractionation in the Fischer-Tropsch synthesis and in meteorites. *Science* 170:980–982.
Lawless, J. G. 1973. Amino acids in the Murchison meteorite. *Geochim. Cosmochim. Acta* 37:2207–2212.
Lawless, J. G., and Peterson, E. 1975. Amino acids in carbonaceous chondrites. *Origins of Life* 6:3–8.
Lawless, J. G., and Yuen, G. U. 1979. Quantification of monocarboxylic acids in Murchison carbonaceous meteorite. *Nature* 282:396–398.
Lawless, J. G., Kvenvolden, K. A., Peterson, E., Ponnamperuma, C., and Jarosewich, E. 1972. Evidence for amino acids of extraterrestrial origin in the Orgueil meteorite. *Nature* 236:66–67.
Lawless, J. G., Zeitman, B., Pereira, W. E., Summons, R. E., and Duffield, A. M. 1974. Dicarboxylic acids in the Murchison meteorite. *Nature* 251:40–41.
Levy, R. L., Grayson, M. A., and Wolf, C. J. 1973. The organic analysis of the Murchison meteorite. *Geochim. Cosmochim. Acta* 37:467–483.
Lewis, J. S., and Prinn, R. G. 1980. Kinetic inhibition of CO and N_2 reduction in the solar nebula. *Astrophys. J.* 238:357–364.
Lewis, R. S., and Anders, E. 1983. Interstellar matter in meteorites. *Sci. Amer.* 249:66–77.

Lewis, R. S., Srinivasan, B., and Anders, E. 1975. Host phase of a strange xenon component in Allende. *Science* 190:1251–1262.
Lovering, J. F., LeMaitre, R. W., and Chappell, B. W. 1971. Murchison C2 carbonaceous chondrite and its inorganic composition. *Nature* 230:18–20.
Lumpkin, G. R. 1981. Electron microscopy of carbonaceous matter in Allende acid residues. *Proc. Lunar Planet. Sci. Conf.* 12B:1153–1166.
Miknis, F. P., Lindner, A. W., Gannon, J., Davis, M. F., and Maciel, G. E. 1984. Solid State ^{13}C NMR studies of selected oil shales from Queensland, Australia. *Org. Geochem.* 7:239–248.
Miller, S. L. 1955. Production of some organic compounds under possible primitive Earth conditions. *J. Amer. Chem. Soc.* 77:2351–2361.
Miller, S. L., Urey, H. C., and Oró, J. 1976. Origin of organic compounds on the primitive Earth and in meteorites. *J. Molec. Evol.* 9:59–72.
Mueller, G. 1953. The properties and theory of genesis of the carbonaceous complex within the cold bokevelt meteorite. *Geochim. Cosmochim. Acta* 4:1–10.
Mullie, F., and Reisse, J. 1987. Organic matter in carbonaceous chondrites. In *Topics in Current Chemistry*, vol. 139 (Berlin: Springer-Verlag), pp. 85–117.
Nagy, B. 1975. *Carbonaceous Meteorites* (Amsterdam: Elsevier).
Nooner, D. W., and Oró, J. 1967. Organic compounds in meteorites—I. Aliphatic hydrocarbons. *Geochim. Cosmochim. Acta* 31:1359–1394.
Oró, J., Nooner, D. W., and Olson, R. J. 1970. Hydrocarbons in meteorites. In *Advances in Organic Geochemistry*, ed. G. D. Hobson (New York: Pergamon Press), pp. 507–521.
Oró, J., Gibert, J., Lichtenstein, H., Wikstrom, S., and Flory, D. A. 1971. Amino acids, aliphatic and aromatic hydrocarbons in the Murchison meteorite. *Nature* 230:105–106.
Pasachoff, J. M. 1987. *Astronomy: From the Earth to the Universe* (New York: CBS College Pub.).
Peltzer, E. T., and Bada, J. L. 1978. α-Hydroxycarboxylic acids in the Murchison meteorite. *Nature* 272:443–444.
Peltzer, E. T., Bada, J. L., Schlesinger, G., and Miller, S. L. 1984. The chemical conditions on the parent body of the Murchison meteorite: Some conclusions based on amino, hydroxy and dicarboxylic acids. *Adv. Space Res.* 4:69–74.
Pereira, W. E., Summons, R. E., Rindfleisch, T. C., Duffield, A. M., Zeitman, B., and Lawless, J. G. 1975. Stable isotope mass fragmentography: Quantitation and hydrogen-deuterium exchange studies of eight Murchison meteorite amino acids. *Geochim. Cosmochim. Acta* 39:163–172.
Pering, K. L., and Ponnamperuma, C. 1971. Aromatic hydrocarbons in the Murchison meteorite. *Science* 173:237–239.
Pillinger, C. T. 1984. Light element stable isotopes in meteorites—From grams to picograms. *Geochim. Cosmochim. Acta* 48:2739–2766.
Pollock, G. E., Chang, C.-N., Cronin, S. E., and Kvenvolden, K. E. 1975. Stereoisomers of isovaline in the Murchison meteorite. *Geochim. Cosmochim. Acta* 39:1571–1573.
Prinn, R. G., and Fegley, B. 1987. The atmospheres of Venus, Earth and Mars: A critical comparison. *Ann. Rev. Earth Planet. Sci.* 15:171–212.
Robert, F., and Epstein, S. 1982. The concentration and isotopic composition of hydrogen, carbon, and nitrogen in carbonaceous meteorites. *Geochim. Cosmochim. Acta* 46:81–95.
Rossignol-Strick, M., and Barghoorn, E. 1971. Extraterrestrial abiogenic organization of organic matter: The hollow spheres of the Orgueil meteorite. *Space Life Sci.* 3:89–107.
Sagan, C., and Khare, B. N. 1979. Tholins: Organic chemistry of interstellar grains and gas. *Nature* 277:102–107.
Sandford, S. A., and Walker, R. M. 1985. Laboratory infrared transmission spectra of individual interplanetary dust particles from 2.5 to 25 microns. *Astrophys. J.* 291:838–851.
Shimoyama, A., and Harada, K. 1984. Amino acid depleted carbonaceous chondrites (C2) from Antarctica. *Geochem. J.* 18:281–286.
Shimoyama, A., Harada, K., and Yanai, K. 1985. Amino acids from the Yamato-791198 carbonaceous chondrite from Antarctica. *Chem. Lett.* 1985:1183–1186.
Shimoyama, A., Naraoka, H., Yamamoto, H., and Harada, K. 1986. Carboxylic acids in the Yamato-791198 carbonaceous chondrite from Antarctica. *Chem. Lett.* 1986:1561–1564.

Smith, J. W., and Rigby, D. 1981. Comments on D/H ratios in chondritic organic matter. *Earth Planet. Sci. Lett.* 54:64–66.
Smith, P. P. K., and Buseck, P. R. 1981. Graphitic carbon in the Allende meteorite: A microstructural study. *Science* 212:322–324.
Snyder, L. E., Hollis, J. M., Suenram, R. D., Lovas, F. J., Brown, L. W., and Buhl, D. 1983. An extensive galactic search for conformer II glycine. *Astrophys. J.* 268:123–128.
Squyres, S. W., and Sagan, C. 1983. Albedo asymmetry of Iapetus. *Nature* 303:782–785.
Stoks, P. G., and Schwartz, A. W. 1979. Uracil in carbonaceous meteorites. *Nature* 282:709–710.
Stoks, P. G., and Schwartz, A. W. 1981a. Nitrogen-heterocyclic compounds in meteorites: Significance and mechanism of formation. *Geochim. Cosmochim. Acta* 45:563–569.
Stoks, P. G., and Schwartz, A. W. 1981b. Nitrogen compounds in carbonaceous meteorites: A reassessment. *Proc. 6th Internatl. Conf. on the Origin of Life,* ed. Y. Wolman (Dordrecht: D. Reidel), pp. 59–64.
Stoks, P. G., and Schwartz, A. W. 1982. Basic nitrogen-heterocyclic compounds in the Murchison meteorite. *Geochim. Cosmochim. Acta* 46:309–315.
Studier, M. H., Hayatsu, R., and Anders, E. 1965. Organic compounds in carbonaceous chondrites. *Science* 149:1455–1459.
Studier, M. H., Hayatsu, R., and Anders, E. 1968. Origin of organic matter in the early solar system—I. Hydrocarbons. *Geochim. Cosmochim. Acta* 32:151–173.
Studier, M. H., Hayatsu, R., and Anders, E. 1972. Origin of organic matter in the early solar system—V. Further studies of meteoritic hydrocarbons and a discussion of their origin. *Geochim. Cosmochim. Acta* 36:189–215.
Van der Velden, W., and Schwartz, A. W. 1977. Search for purines and pyrimidines in the Murchison meteorite. *Geochim. Cosmochim. Acta* 41:961–968.
Vdovykin, G. P. 1967. *Carbonaceous Matter in Meteorites (Organic Compounds, Diamonds, Graphite)* (Moscow: Nauka Press). In Russian. Trans. NASA-TT-F-582.
Walters, C., Kotra, R. K., and Ponnamperuma, C. 1983. Dipeptides in the Murchison and Yamato meteorites. *Abs. ACS Meeting* 186:GEOC 11.
Watson, W. D. 1976. Interstellar molecule reactions. *Rev. Mod. Phys.* 48:513–552.
Wickramasinghe, D. T., and Allen, D. A. 1986. Discovery of organic grains in comet Halley. *Nature* 323:44–46.
Wolman, Y., Haverland, W. J., and Miller, S. L. 1972. Nonprotein amino acids from spark discharges and their comparison with the Murchison meteorite amino acids. *Proc. Natl. Acad. Sci. USA* 69:809–811.
Yang, J., and Epstein, S. 1984. Relic interstellar grains in Murchison meteorite. *Nature* 311:544–547.
Yuen, G. U., and Kvenvolden, K. A. 1973. Monocarboxylic acids in Murray and Murchison carbonaceous meteorites. *Nature* 246:301–302.
Yuen, G., Blair, N., Des Marais, D. J., and Chang, S. 1984. Carbon isotope composition of low molecular weight hydrocarbons and monocarboxylic acids from Murchison meteorite. *Nature* 307:252–254.
Zeitman, B., Chang, S., and Lawless, J. G. 1974. Dicarboxylic acids from electrical discharge. *Nature* 251:42–43.
Zellner, B. 1979. Asteroid taxonomy and the distribution of the compositional types. In *Asteroids,* ed. T. Gehrels (Tucson: Univ. of Arizona Press), pp. 783–806.

PART 11
Micrometeorites

11.1 INTERPLANETARY DUST PARTICLES

JOHN P. BRADLEY
McCrone Associates, Inc.

SCOTT A. SANDFORD
NASA Ames Research Center

and

ROBERT M. WALKER
Washington University

Meteorites of a size that can be recovered from the surface of the Earth after atmospheric entry sample only a small fraction of the size range of particulate material in interplanetary space. For our purpose, another important part of that population consists of particles of interplanetary dust that are systematically collected in the stratosphere by high-altitude aircraft. This chapter first describes various ways that have been used to establish the extraterrestrial nature of different subsets of the collected stratospheric dust. Although the particles are small (~ 10 μm in diameter), recent advances in microanalytic techniques have made it possible to obtain detailed experimental information on the mineralogical and petrographic characteristics, the mid-infrared absorption spectra, the Raman spectra and the isotopic properties of individual particles. The ensemble of information and its implications for the origin of interplanetary dust are discussed. It is shown that, in some ways, the particles are less altered samples of solar-system material than so far found in meteorites. Many of the particles must come from comets but an unknown fraction also comes from asteroids. Small regions (≤ 1 μm) of isotopically distinct material are found, suggesting that at least part of the dust consists of interstellar-cloud material that predated the solar system. The relationship of these laboratory studies *of interplanetary dust to astronomical observations of dust in various astrophysical settings is also considered.*

11.1.1. INTRODUCTION

Many dust particles collected in the stratosphere originate in interplanetary space. Laboratory studies show that these interplanetary dust particles (IDPs) constitute a distinctive type of primitive solar-system material that exhibits both similarities and differences when compared with meteorites. The IDPs make up a diverse set of objects whose classification and properties do not fit naturally into the framework established for meteorites. Partly because of their small sizes ($\sim 10^{-3}$ cm in diameter), which have required the development of sophisticated experimental techniques, the laboratory study of IDPs is in its infancy compared to that of the objects described elsewhere in this book. For these reasons, and others mentioned below, a separate chapter is devoted to this field of research.

The dust particles described in this chapter are systematically collected by NASA and are available for study by qualified investigators. Plastic "flags" coated with silicone oil are deployed at altitudes ≥ 20 km from pylons mounted under the wings of a U-2 or similar high-altitude aircraft. The particles that impact the flags are subsequently removed and given a preliminary examination using a scanning electron microscope (SEM) equipped with an X-ray energy dispersive spectrometer (EDS). Catalogs describing specific cosmic dust particles are available from M. Zolensky, curator of cosmic dust samples, Johnson Space Center, Houston, Texas 77058.

We shall show that, in some ways, IDPs appear to be less altered samples of nebular and prenebular materials than those found in meteorites. Many, and possibly most, IDPs originate from comets, although a substantial contribution from asteroids may exist. The scientific questions that can be addressed by the study of IDPs also differ from those that can be posed for meteorites. For example, we know that interplanetary dust produces characteristic optical effects (e.g., zodiacal light, comet emission spectra, etc.), and we can compare laboratory measurements of IDPs with the remotely observed properties of interplanetary dust. We can also determine whether interplanetary dust and interstellar dust have common features.

The following questions guide the presentation of the material in this chapter:

a. How is it proven that certain dust particles are extraterrestrial?
b. Are all IDPs the same or can we delineate different classes?
c. How do IDPs compare with meteorites?
d. Do IDPs contain "primitive" materials that give information about nebular or prenebular processes?
e. What do we know about the conditions under which the constituents of the dust formed?

f. What can we say about the parent bodies of the dust?
g. How do the properties of IDPs compare with those of comet dust and interstellar dust?

We will return to each of these questions in the discussion section.

Recent reviews dealing with various aspects of dust studies include those by Fraundorf et al. (1982), Brownlee (1985), Mackinnon and Rietmeijer (1987) and Sandford (1987). The first summarized the situation on IDPs as it then existed; the second is a more general treatment of interplanetary dust collections, including extraterrestrial materials found in sea sediments; the third is a detailed account of the mineralogy and petrography of IDPs and the implications for their formation, and the fourth gives an extended treatment of possible astronomical sources of IDPs.

11.1.2. PROOF OF THE EXTRATERRESTRIAL ORIGIN OF CERTAIN STRATOSPHERIC DUST PARTICLES

At the start of the space age the flux of interplanetary dust particles was thought to be much higher than it is now known to be. It was considered to be so high, in fact, as to pose a major hazard for space travel. [Ironically, impacts from terrestrial contaminants introduced by man are becoming an increasing hazard in near-Earth orbit (Kessler and Cour-Palais 1978).] This perception stimulated attempts to collect interplanetary dust on Earth and in the stratosphere as well as to measure the flux of particles arriving at detectors in near-Earth orbit. This early period is aptly described by Hodge (1981) as follows: ". . . a patchwork of enthusiastic attempt, dreadfully bad luck, naive interpretations, and blissful amateurism." Almost all the purported interplanetary particles found during this period were terrestrial contaminants.

The modern era of laboratory studies of interplanetary dust was ushered in by Brownlee and his colleagues at the University of Washington, Seattle. This group collected particles from the stratosphere under extremely clean conditions, first using balloons and later, in the method used today, by high-altitude aircraft (Brownlee et al. 1976b).

The heritage of uncertainty left by the early work made it imperative to establish the extraterrestrial nature of this new class of materials. Attention first focused on the "chondritic" subset of particles[a] whose major- and minor-

---

[a]Chondritic particles are generally defined as those which have relative abundances of Mg, Al, S, Ca, Fe and Ni, measured with respect to Si, within a factor of three of those found in C-type meteorites. In practice, the definition is somewhat loose; particles with strong depletions or enrichments of specific elements have been classified as chondritic by different investigators. Other common particle types consist primarily of Fe, Ni and S (FSN), Al-only (probably aluminum oxide from solid fuel rocket exhaust [Brownlee et al. 1976a]), and Al' (Al with minor contributions of other elements). Terrestrial particles, particularly debris from volcanic eruptions, are a common and variable component of the stratospheric dust collection (Fraundorf et al. 1982a; Zolensky and Mackinnon 1985).

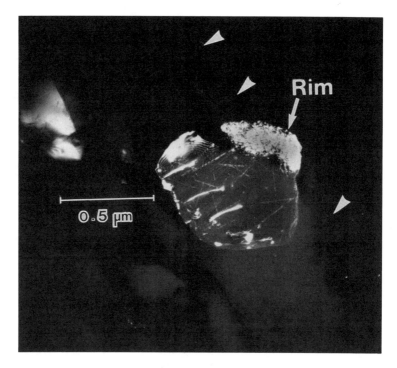

Fig. 11.1.1. Darkfield transmission electron micrograph of an olivine crystal within a thin-sectioned IDP. The linear features in this crystal are solar-flare tracks. Arrows show the position of the outer edge of the section. Note the presence of a rim on that part of the olivine that was exposed along the outer edge of the IDP. Similar rims seen on lunar soil grains have been attributed to irradiation by solar-wind ions.

elemental compositions strongly suggested an extraterrestrial origin (Ganapathy and Brownlee 1979). Noble-gas measurements (Rajan et al. 1977) established that chondritic particles contain large concentrations of He similar to those found in lunar soil samples that have been exposed to the solar wind. Subsequent measurements of Ne and Ar (Hudson et al. 1981) confirmed the presence of large concentrations of nonterrestrial noble gases in IDPs. Hydrogen-isotopic measurements demonstrated the existence of large D enrichments in some particles, providing further evidence for their interplanetary origin (Zinner et al. 1983; McKeegan et al. 1985). The final proof of the extraterrestrial origin of chondritic dust was the observation of high densities of fossil nuclear tracks in the silicate minerals within some particles (see Fig. 11.1.1) (Bradley et al. 1984a). These tracks are produced by the irradiation of the particles in space by heavy nuclei from solar flares (Fleischer et al. 1975).

Recently, O-isotope measurements of several refractory dust particles (McKeegan et al. 1986) have established that at least some nonchondritic particles are also extraterrestrial. Other types of nonchondritic particles in the stratospheric dust collection, for example, particles dominated by Fe, S and Ni (FSN) or those rich in low Z elements may also be samples of interplanetary dust.

11.1.3. PHYSICAL AND MINERALOGICAL PROPERTIES

External Morphologies and Densities

Typical collected IDPs have diameters from 1 to 50 μm and span a range of shapes and textures (see Figs. 11.1.2a,b). The particles are approximately equidimensional and the ratio of length to width rarely exceeds a factor of 1.5. Textures range from smooth, to highly reentrant or "fluffy." While IDPs consisting of single mineral grains are seen, the most common particles consist of aggregates of smaller grains (typically < 0.1 to 3 μm in diameter) in varying states of agglomeration.

The densities of the few chondritic IDPs that have been directly measured are between 0.7 and 2.2 g/cm^3 (Fraundorf et al. 1982b). The variation in densities is largely due to differences in particle porosities (Bradley and Brownlee 1986), rather than to differences in elemental compositions.

Electron-Beam Studies of IDPs: Old and New Techniques

Interplanetary dust particles are highly unequilibrated and mineralogically diverse. Instruments used to study them include the electron microprobe, scanning electron microscope (SEM), and the analytical scanning transmission electron microscope (AEM). The AEM has become the most widely used instrument because it offers optimum spatial resolution (usually 2 to 4 Å in conventional transmission mode) together with a variety of imaging and spectroscopic functions. The small amount of material in an IDP necessitates special experimental techniques for both sample preparation and analysis. In the past, IDPs were first hand picked from their collection "flags," cleaned of residual silicone oil with organic solvents, and then transferred to beryllium or Nuclepore substrates (see Figs. 11.1.2a,b), where properties like size, shape, morphology and chemical composition could be determined using SEM-EDS techniques.

Preparation of specimens for AEM studies is more difficult because the samples need to be thin (ideally ~1000 Å thick) to obtain crystallographic, structural and chemical information. The traditional means of achieving this was to take an IDP and crush it between polished surfaces to form a specimen composed of μm and ≤ μm sized grains, some of which were sufficiently thin for AEM analyses (Flynn et al. 1978; Fraundorf and Shirck 1979; Bradley and Brownlee 1983). The major drawback with all crush-and-dispersion methods is that much of the original structure and petrographic associations within an IDP are destroyed. Nonetheless, these procedures for a long time provided the only means of examining IDPs, and much of the analytical in-

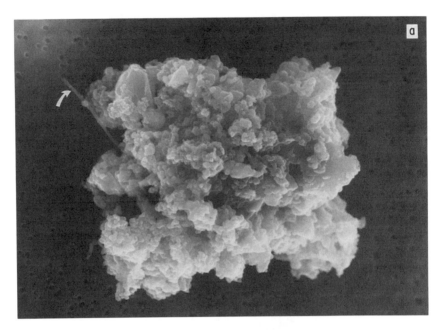

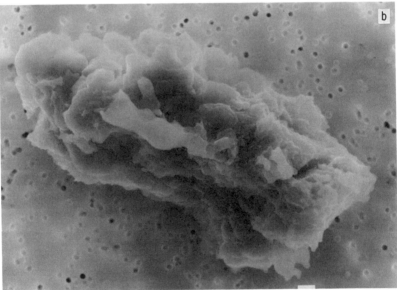

Fig. 11.1.2. Scanning electron micrographs of chondritic IDPs approximately 10 μm in size. (a) Highly reentrant, porous particle; the elongated fiber (arrowed) is an enstatite (MgSiO$_3$) single crystal that probably grew by direct condensation from a vapor. (b) Compact, less porous particle.

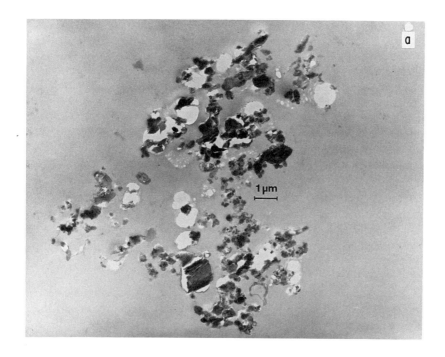

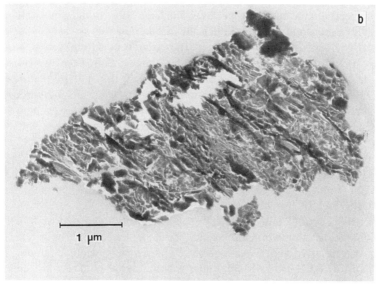

Fig. 11.1.3. Brightfield transmission electron micrographs of ultramicrotomed sections (~ 800 Å thickness) of chondritic IDPs (embedded in epoxy). (a) Anhydrous particle; note the high porosity throughout the section. (b) Layer-lattice silicate particle; note the low porosity (void spaces seen in the upper half of the IDP are an artifact of the sectioning procedure).

formation available today has been obtained from samples prepared in this manner.

A recent significant advance has been the development of a method for making thin-sections of IDPs (Bradley and Brownlee 1986). Sectioning is performed using an ultramicrotome equipped with a diamond knife. Using such an apparatus, sections from ~ 1 μm down to ~ 500 Å thickness can be cut (See Figs. 11.1.3a,b). The thickness of sections can be controlled to within ± 200 Å and, in principle, a 10 μm IDP can be sliced into 100 sections with complete preservation of its internal microstructure. With such sections, it is now possible to describe IDPs in terms of texture, porosity and petrographic associations. Quantitative analyses can be performed on ultramicrotomed sections with optimum precision and accuracy because the geometry and thickness of the specimen are well defined. For example, using a 50 Å diameter, 200 keV incident beam, a lateral resolution of better than 200 Å should be routine for 1000 Å sections (Goldstein et al. 1977).

Mineralogical Results from Electron-Beam Studies

Early studies of chondritic IDPs by electron microscopy showed that various minerals were present including pyroxene, olivine, layer-lattice silicates, iron-rich sulfides, magnetite and carbonaceous material (Brownlee 1978; Fraundorf 1981). It is now known that chondritic IDPs can be divided into subgroups using infrared measurements and/or electron microscopy. Infrared spectral transmission measurements have shown that almost all chondritic IDPs are dominated by either anhydrous minerals or layer-lattice silicates (Sandford and Walker 1985). The anhydrous group can be further subdivided into "pyroxene" and "olivine" classes, after the minerals that provide the best match for the observed infrared 10 μm spectral features.

TEM analysis confirms the division into layer-lattice and anhydrous-silicate classes and thin sections show that particles in these classes can be distinguished from one another on the basis of internal texture alone. Anhydrous IDPs consist of highly porous aggregates (Fig. 11.1.3a), whereas the layer-lattice silicate IDPs are low-porosity, compact objects (Fig. 11.1.3b). (It should be noted, however, that even low-porosity objects can appear fluffy in the SEM (Fig 11.1.2b). External morphology is, therefore, not sufficient to classify IDPs.)

In thin-section, anhydrous IDPs appear to be composed of four basic building blocks: single mineral grains, "tar balls," a carbonaceous phase and glass. The relative abundances of these species vary widely, ranging from abundant in one IDP to virtually absent within another. The most common single mineral species are olivine and pyroxene, and infrared measurements indicate that either one or the other tends to dominate in a given IDP. Olivine compositions vary from Fo_{40} to Fo_{100} and the grains occur as (~ 0.05 to 1 μm diameter) anhedral grains, euhedral crystals, platelets and almost spherical crystals (Bradley and Brownlee 1986; Christoffersen and Buseck 1986a).

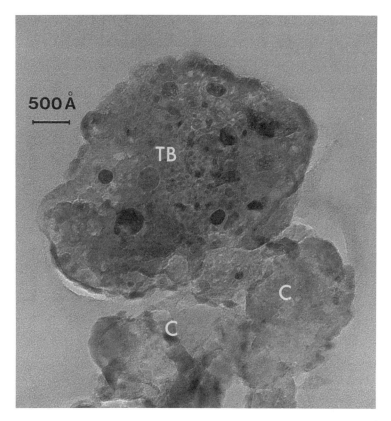

Fig. 11.1.4. Brightfield electron micrograph of a "tar ball" (labeled TB). Note the population of < 0.05 μm rounded crystals. Disordered carbonaceous material (labeled C) is also a common constituent of chondritic IDPs.

Reported pyroxenes include enstatite, hypersthene, fassaite, diopside and augite compositions (Bradley et al. 1983; Rietmeijer and McKay 1985,1986; Christoffersen and Buseck 1986b). Other less common single mineral constituents of anhydrous IDPs include magnetite, Fe-Ni alloy (kamacite), Fe-Ni carbides and chromite.

"Tar balls" consist of aggregates of extremely small (< 0.01 − 0.05 μm) rounded crystals embedded in a carbonaceous matrix (Fig. 11.1.4). It is remarkable that their overall compositions are themselves chondritic; possibly they are products of an accretional event prior to their incorporation into IDPs. Kamacite and olivine are the dominant minerals, together with minor amounts of Fe-rich sulfides. Carbonaceous material is widespread as a disordered material. At present, little is known about the molecular constitution of the carbon, although some results suggest that some IDPs contain hydrocarbons (McKeegan et al. 1985; Allamandola et al. 1987). Glass is also present in anhydrous IDPs and usually contains embedded olivine, pyroxene or Fe-rich sulfide grains (Bradley and Brownlee 1986).

IDPs dominated by layer-lattice silicates differ from their anhydrous ones in that they have low porosities (Fig. 11.1.3b). On the basis of electron-microscopic observations they may be subdivided into two classes, one with a layer-lattice silicate basal spacing of 10 to 12 Å, and the other with a spacing of ~ 7 Å. On the basis of these spacings, and quantitative elemental thin-film analyses, these layer silicates have been identified as smectite and serpentine, respectively. The smectite type is the most common and may actually account for over half of all chondritic IDPs (Sandford and Walker 1985). A typical smectite IDP is composed largely of two phases, one a noncrystalline glass-like phase whose composition varies widely, and the other a poorly crystallized Mg-rich smectite (Tomeoka and Buseck 1984, 1985a; Rietmeijer and Mackinnon 1985). The smectite may have formed by aqueous alteration of either glass or pyroxene. Other reported minerals in particles dominated by layer-lattice silicates include pyroxenes, olivine, and Mg- and Fe-rich carbonates (Tomeoka and Buseck 1986a; Sandford 1986a; Rietmeijer and McKay 1986; Bradley and Brownlee 1986). "Tar balls" and carbonaecous material are also found in smectite IDPs.

A variety of mineral species have so far only been observed as constituents of only one or two chondritic IDPs. Christoffersen and Buseck (1986b) identified a suite of refractory phases, including Mg-Al spinel, anorthite and perovskite within an IDP from the pyroxene infrared class. Sulfides containing Fe and Zn, and Zn only, have been observed (Christoffersen and Buseck 1986a), and a tetragonal Fe sulfide was reported by Bradley and Brownlee (1983). Fraundorf (1981) and Rietmeijer and Mackinnon (1985) have reported high-Sn grains within two different IDPs. Aluminum, Si-, Ti- and Bi-oxides, SiC, Fe phosphide, Fe-Cr sulfide, FeOOH, and $BaSO_4$ were identified within two smectite-containing particles (Mackinnon and Rietmeijer 1984; Rietmeijer and Mackinnon 1984,1985; Rietmeijer and McKay 1986; Rietmeijer 1985). Some of these phases are important constituents of carbonaceous chondrites (cf. Christoffersen and Buseck 1986b), while others have not previously been found in meteoritic materials. The status of these less common phases will be clarified as more IDPs are subjected to detailed electron-beam analyses.

Serpentine IDPs are the least abundant of the chondritic particles. Only three have been reported to date: two were collected in the stratosphere (Brownlee 1978; Bradley and Brownlee 1986), the third in space by impaction onto thermal blanket materials subsequently returned to Earth by the Solar Max Repair Mission (Bradley et al. 1986). The layer-lattice silicates in serpentine particles are better crystallized than those in the smectite class, and so far no coexisting glass-like phase has yet been identified. The dominant accessory phases are pentlandite and pyrrhotite together with minor magnetite. The magnetite is often embedded within carbonaceous material. No olivines, pyroxenes or "tar balls" have yet been seen in serpentine particles.

Chondritic IDPs and Primordial Grain-Forming Reactions

Several minerals within chondritic IDPs appear to have formed by direct crystal growth from a vapor. Enstatite crystals in IDPs exhibit unique morphologies and microstructures (Fraundorf 1981; Bradley et al. 1983), occurring as rods, ribbons and platelets (Fig. 11.1.5). Rods have round or rectangular cross sections and often have aspect ratios >20. Some of them contain axial screw dislocations. Ribbons are blade-shaped crystals with aspect ratios of 5 or more. Both the rods and ribbons are composed of clinoenstatite and are elongated along the crystallographic [100] direction. Terrestrial and meteoritic rock-forming pyroxenes, in contrast, when not equiaxial, are elongated along [001]. Platelets consist of flat, angular and disk-shaped crystals that are thin along [010] or [001]. They differ from rods and ribbons in that they consist of pervasively intergrown ortho-and clinopyroxene together with material possessing extreme stacking disorder. Such platelets have not previously been reported for any pyroxenes. In general, rods, ribbons and platelets, particularly rods with axial screw dislocations, are typical of crys-

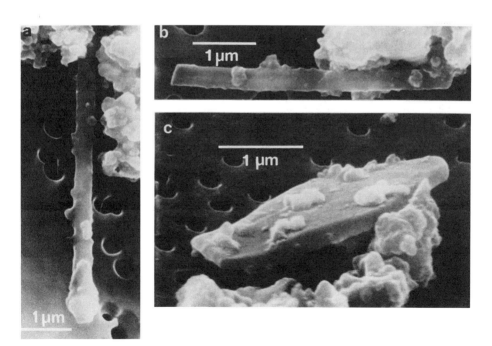

Fig. 11.1.5. Scanning electron micrographs of three unique pyroxene morphologies seen in IDPs: (a) an enstatite rod; (b) an enstatite ribbon; (c) an enstatite platelet (figure after Bradley et al. 1983).

tals grown directly from the vapor phase (Veblen and Post 1983). As a consequence, Bradley et al. (1983) concluded that these enstatite crystals in IDPs were almost certainly formed by primary condensation from a gas. Tomeoka and Buseck (1984) have reported low-Ni pentlandite grains within IDPs, which they suggest may also have formed by vapor-phase condensation.

Some C compounds in chondritic IDPs also appear to have been formed by gas-phase reactions (Bradley and Brownlee 1983; Christoffersen and Buseck 1983; Bradley et al. 1984b). Carbonaceous rims on grains, filamentous C and several Fe-Ni carbides have been found, and these are characteristic by-products of Fischer-Tropsch (FT) catalytic reactions between a C-containing gas (e.g., CO) and the active surfaces of grains (Anderson 1956). One particular carbide found in chondritic IDPs, hexagonal epsilon Fe-Ni carbide, has previously been seen only as a laboratory by-product of FT reactions. It is interesting to note that Hayatsu and Anders (1981) had previously proposed that organic C compounds in carbonaceous chondritics were formed as a result of FT catalytic reactions between CO and dust grains (see also Chapter 10.5).

Relationship to Primitive Meteorites

Most IDPs are similar to carbonaceous chondrites in that their bulk compositions are "chondritic," and most of the minerals found in IDPs have also been found in the meteorites. However, in spite of this superficial similarity, specific details of the crystallography, crystal chemistry, mineralogy and petrography within IDPs are different from those in meteorites. For example, the pyroxene whiskers and platelets, and certain C-bearing phases indicate that the products of primary grain-forming reactions are well preserved in IDPs, while these have not yet been seen in meteorites. The porosities of anhydrous IDPs are much higher than those of meteorites, which typically have porosities of only 20% or less by volume. Like meteorites, layer-lattice silicate IDPs are compact objects that contain phases that could have formed by aqueous processing. However, smectite is the dominant layer-lattice silicate in IDPs, whereas it is serpentine in CM carbonaceous chondrites (Bunch and Chang 1980; Barber 1985; Tomeoka and Buseck 1985b; see also Chapter 3.4). Only the serpentine class of IDPs is texturally and mineralogically similar to the minerals found in fine-grained matrices of carbonaceous chondrites, but these are the least abundant type of IDP. It should be noted that smectite-like, layer-lattice silicates have been observed as minor constituents within CI, CV3 and CM2 chondrites (Cohen et al. 1983; Tomeoka and Buseck 1986b), and more recently within unequilibrated ordinary chondrites (Alexander et al. 1986).

From the above discussion we conclude that IDPs differ from carbonaceous chondrites and other meteorites. Their delicate microstructures and lack of evidence for post-accretional processing suggest that many of them may represent the most pristine extraterrestrial materials yet studied.

11.1.4. OPTICAL PROPERTIES OF IDPs

The Mid-Infrared Spectra of IDPs

Almost all the chondritic IDPs that have been studied in the 2.5 to 25 μm (4000–400 cm^{-1}) spectral region have a dominant absorption feature at $\sim$ 10 μm (1000 cm^{-1}) (Sandford and Walker 1985). The position and shape of this band varies from particle to particle, but three broad groupings are evident. These spectral groups have been labeled olivines, pyroxenes and

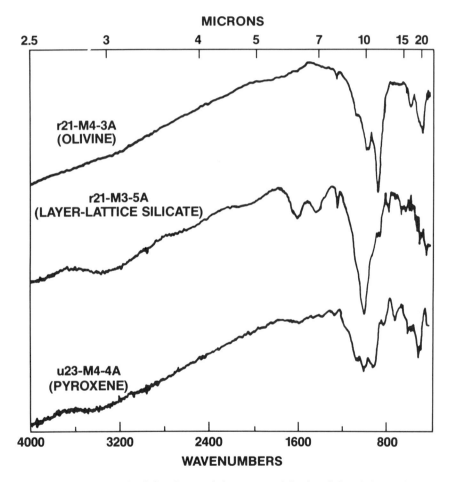

Fig. 11.1.6. Representative infrared transmission spectra of the three infrared classes that are seen in IDP spectra. From top to bottom the spectra are from representative members of the olivine class, the layer-lattice silicate class and the pyroxene class. The identifications of the various absorption bands are given in the text. The 7.9 and 12.5 μm bands seen in some of the spectra are probably due to incompletely removed silicone oil that is used in the process of collecting the IDPs.

layer-lattice silicates, by analogy with the 10 μm features seen in the infrared spectra of terrestrial mineral standards. The existence of these three major silicate types in IDPs was subsequently confirmed by transmission electron microscopy (Christoffersen and Buseck 1986a,b; Bradley and Brownlee 1986; Tomeoka and Buseck 1984). A typical spectrum from each of the three infrared types is shown in Fig. 11.1.6. The observed relative abundance of the olivine, pyroxene and layer-lattice silicate particles is roughly 1:1:2, respectively.

The spectra of IDPs in the layer-lattice silicate group have additional bands at 3.0, 3.4, 6.0, 6.8, 7.9 and 12.5 μm (3330, 2940, 1670, 1470, 1270 and 800 cm^{-1}, respectively). A weak band near 11.4 μm (880 cm^{-1}) is also seen in some spectra. The bands centered at 3.0 and 6.0 μm are characteristic of terrestrial layer-lattice silicates and are attributed to adsorbed and absorbed water. The 3.0 μm band is due to O-H stretching vibrations, and the 6.0 μm band to H-O-H bending vibrations. The depths of these bands relative to the 9.8 μm silicate band are variable and may reflect differing water loss during atmospheric entry, but could also be due to the presence of different amounts

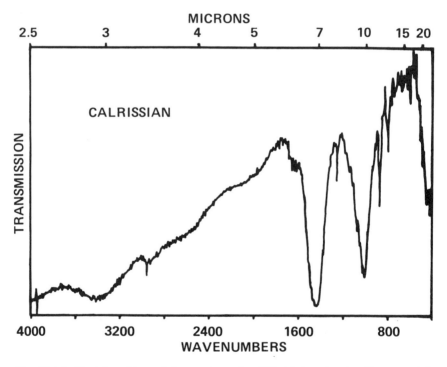

Fig. 11.1.7. The infrared transmission spectrum of an IDP rich in carbonates. The strong 6.8 μm absorption band is due to CO_3 asymmetric stretching vibrations. The carbonate identification is further demonstrated by the presence of the 11.4 μm CO_3 scissors vibration band. The narrow 3.4, 7.9 and 12.5 μm bands are probably due to silicone oil contamination.

of initial water or to differential adsorption of water since atmospheric entry. The lack of a narrow O-H structural band at 2.7 μm (3700 cm^{-1}) in most of the spectra implies that the layer-lattice silicates are relatively disordered.

The 6.8 and 11.4 μm band depths appear to be correlated and these bands are attributed to the asymmetric stretch and scissors vibrations, respectively, of CO_3 anions in carbonate minerals (see Fig. 11.1.7). The presence of calcite in these IDPs has been confirmed by electron diffraction studies (Tomeoka and Buseck 1986a). Acid-dissolution experiments have verified that the entire band is produced by carbonates and that only a minor component can be due to acid-insoluble hydrocarbons (Sandford 1986a).

The 7.9, 12.5 and possibly the 3.4 μm band depths also appear to be correlated. Such bands are characteristic of materials containing hydrocarbon functional groups. Unfortunately, the silicone oil used to collect the IDPs also exhibits features at these same locations, and the bands may be due to contamination. The 3.4 μm band is, however, substantially wider in many of the IDP spectra than the same band in the spectrum of pure silicone oil, implying that some part of the spectral feature may be due to hydrocarbons original to the particles.

In spite of the differences in the mineral structures, infrared spectra of IDPs in the layer-lattice silicate infrared class are similar to those of the CM carbonaceous chondrites (Sandford 1984), although on the average the IDP spectra have stronger 6.8 μm carbonate features.

Visible Properties of IDPs

The spectral properties of IDPs have not yet been quantitatively measured at ultraviolet and visible wavelengths. However, the chondritic IDPs appear black when observed under a microscope, indicating that their albedo is low. The black color is probably caused by carbonaceous materials in which the silicate grains in IDPs are generally imbedded.

Raman and Luminescence Spectra of IDPs

In the laser Raman microprobe technique, the sample is illuminated with visible light and the spectrum of the scattered light is measured. Two distinct spectral phenomena are observed in IDPs: Raman bands and, in many of the particles, a broad emission excess (luminescence) in the red portion of the spectrum (Wopenka 1987).

The Raman spectra of IDPs generally show no lines characteristic of the silicates present in the particles. Instead, the spectra are dominated by broad, strong bands at Raman shifts at about 1355 and 1600 Δcm^{-1} (corresponding to 7.38 and 6.25 μm; see Fig. 11.1.8) due to C-C vibrations of amorphous C. These are similar to the bands seen in the Raman spectra of certain meteorites (Christophe-Michel-Levy and Lautie 1981; Wopenka and Sandford 1984). In addition, some of the spectra (see, e.g., Essex in Fig. 11.1.8) show a broad band between 3330 and 2220 Δcm^{-1} (3.0 and 4.5 μm) due to second-

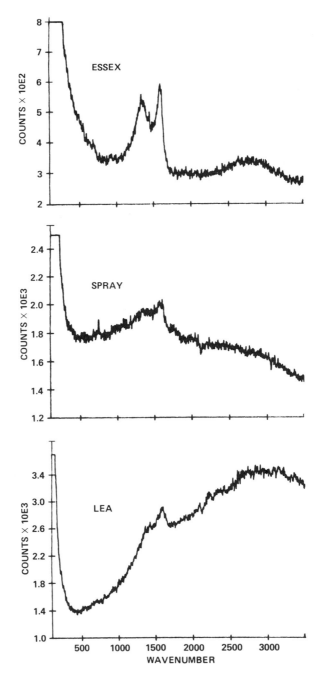

Fig. 11.1.8. Laser Raman microprobe measurements of three IDPs. The horizontal axis is the Raman shift from the incident 5145 Å laser line in relative wavenumbers (Δcm^{-1}). The first-order Raman features at $\sim$ 1355 and 1600 Δcm^{-1} are characteristic of amorphous C as is the second-order feature at $\sim$ 2800 Δcm^{-1} seen most clearly in the spectrum of Essex. In contrast, the spectrum of Lea is dominated by a broad luminescence centered at about 6200 Å. The Raman signals and luminescence are variable among IDPs and have no obvious relationship to their infrared absorption spectra.

order Raman lines of amorphous C, and a weak band at about 1460 Δcm^{-1} (6.85 μm) also associated with C-C vibrations. Both the major Raman features and luminescence can be attributed to carbonaceous material. A comparison of the Raman features of IDPs with those produced by laboratory synthesized materials (Nemanich and Solin 1979) suggests that the carbonaceous material contains microcrystalline aromatic domains $\leq$25 Å in size (Allamandola et al. 1987).

Comparison of the Spectral Properties of IDPs with those of Astrophysical Objects

Before the advent of interplanetary dust collection programs and space probes, information about cosmic dust was provided only by telescopic observations. This is still an important method for studying interplanetary dust and will probably always be the major technique used to study interstellar dust. For this reason, it is interesting to compare the spectral properties of IDPs with those of the dust seen in different astrophysical objects.

As discussed in a later section, comets are prime candidates for the major sources of IDPs. At present, only two comets (Kohoutek and Halley) have been examined in the infrared with sufficient resolution and signal-to-noise ratios to allow meaningful comparisons with the 10 μm features in the IDP spectra. As shown in Fig. 11.1.9, none of the three individual IDP infrared spectral types match the cometary data well. The failure to match with individual IDP spectra suggests that cometary features may be produced by the superposition of spectra from a variety of particle types. Figure 11.1.10 shows the comparison between the 10 μm spectra of comets Kohoutek and Halley, with mixtures of IDP spectra of the three infrared classes. In deriving the spectral features of comet dust, it is necessary to subtract a blackbody continuum (600 K for Kohoutek and 320 K for Halley); it should be kept in mind that the blackbody temperature is a somewhat adjustable parameter.

A reasonable fit to the inverted 10 μm emission feature from comet Kohoutek is obtained by adding approximately equal numbers of IDP absorption spectra from the pyroxene and layer-lattice silicate infrared groups (Sandford and Walker 1985). No composite spectrum involving a large contribution from the olivine infrared class matches the Kohoutek data. The wide variation in the 10 μm band shape in the spectra of pyroxene-type IDPs suggests that the Kohoutek band might also be matched if enough spectra from more IDPs of the pyroxene type were summed. The major discrepancy in the fit is due to the presence of minor detailed features in the IDP data that are not seen in the Kohoutek spectrum. The absence of these features in Kohoutek may be due to the smearing of bands when seen in emission (Rose 1979).

The astronomical data show differences between comets Halley (Bregman et al. 1987) and Kohoutek, and different combinations of IDP spectra are necessary to fit the different comet data. In contrast to Kohoutek, the

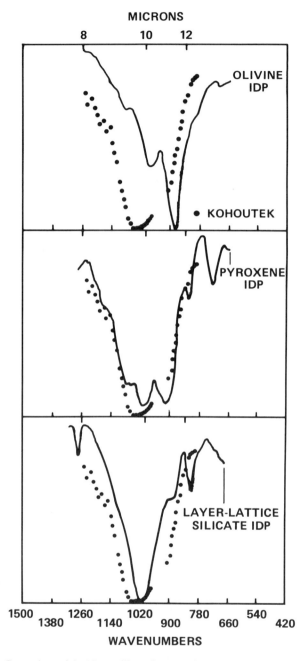

Fig. 11.1.9. Comparison of the 10 μm silicate features of the three IDP infrared spectral classes with inverted comet Kohoutek emission data. The solid points are cometary data taken from Merrill (1974) with a 600 K blackbody contribution removed. No individual IDP spectrum matches the cometary data in both position and band profile (figure from Sandford and Walker 1985).

INTERPLANETARY DUST PARTICLES

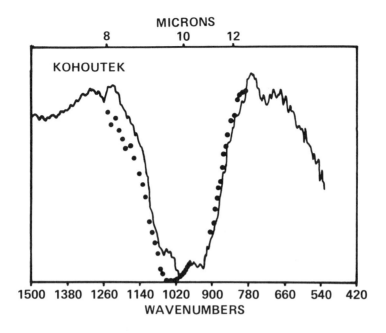

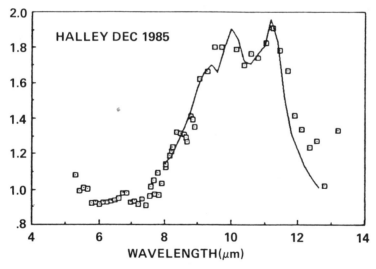

Fig. 11.1.10. Comparison of IDP spectral composites with 10 μm data from comets Kohoutek and Halley. *Top:* IDP spectral composite containing equal amounts of layer-lattice silicate and pyroxene spectra is compared to inverted comet Kohoutek emission data (cometary data from Merrill 1974). *Bottom:* Inverted IDP data are compared to the 10 μm emission from comet Halley calculated with an assumed dust temperature of 325 K. The match is provided by a spectral composite containing 56% olivine, 36% pyroxene and 8% layer-lattice silicate IDP spectra (Bregman et al. 1987).

Halley data require a substantial fraction of IDP spectra from the olivine class in order to produce a good match (the composite in Fig. 11.1.10 contains 36% pyroxene, 8% layer-lattice silicate, and 56% olivine spectral contributions). It is interesting to note that the Halley data also show the presence of a weak 6.8 µm feature, indicating that carbonates may be present in the cometary dust. Within the uncertainties, the relative contribution of layer-lattice silicates to the composite match as inferred by the 10 µm spectral fit is consistent with the strength of the 6.8 µm carbonate feature.

Unfortunately, comparisons between the spectra of IDPs and asteroids, the other major possible contributors to the collected dust, are not possible due to the lack of suitable asteroid data in the mid-infrared and appropriate IDP data in the visible and near infrared.

It is also interesting to compare the spectra of IDPs with those of interstellar dust. One such comparison is with "protostellar" objects, which consist of strong infrared sources surrounded by clouds of cold dust. Such systems are presumably undergoing processes similar to those that occurred early in the history of our own solar system. Figure 11.1.11 shows a comparison between the spectrum of the infrared object W33A and the spectrum of an IDP from the layer-lattice silicate infrared class. The general similarity of the two spectra is obvious, although differences exist. The presence of 3.0, 6.0 and 9.8 µm bands is consistent with the presence of water and layer-lattice silicates (Sandford and Walker 1985). The major differences can be explained by the presence of icy mantles on the protostellar grains. For instance, the features at 4.6 µm seen in the protostellar data are thought to be due to CO condensed in "dirty" ice (Lacy et al. 1984), a volatile material that would not have survived in the collected IDPs.

The Raman spectra of IDPs also show similarities to the spectra of certain interstellar objects. Figure 11.1.12 shows a comparison of the Raman features of an IDP with the infrared emission spectrum of the Orion nebula. The similarity between these spectra suggests that the carbonaceous component in IDPs may be related to the material responsible for the interstellar emission features (Allamandola et al. 1987). The interstellar emission features are thought by many to be due to polycyclic aromatic hydrocarbons or related molecular species (Léger and Puget 1984; Allamandola et al. 1985).

The luminescence seen in some IDPs also has astrophysical counterparts. These include several reflection nebulae (Witt et al. 1984), the "Red Rectangle," HD 44179 (Duley 1985) and the diffuse interstellar medium (van Breda and Whittet 1981). The position and width of the luminescence in IDPs is also similar to that observed in laboratory samples of hydrogenated amorphous C (Watanabe et al. 1982; Lin and Feldman 1982), a material consisting of randomly interlinked aromatic domains.

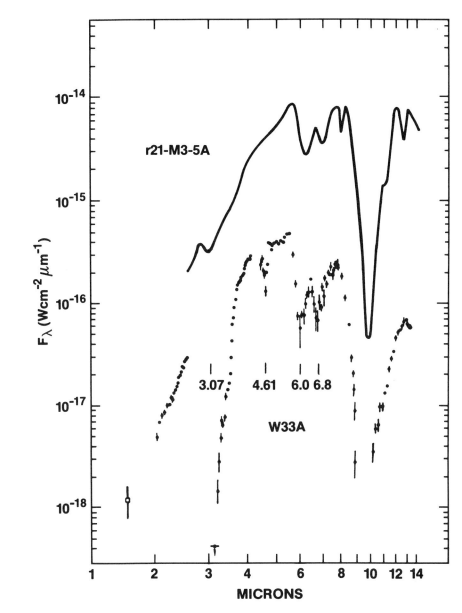

Fig. 11.1.11. A comparison between the spectrum of the protostellar object W33A (solid dots) and the interplanetary dust particle r21-M3-5A (solid line) which falls in the layer-lattice silicate infrared class. The W33A data are taken from Soifer et al. (1979). The vertical flux axis is for the W33A data. The IDP spectrum is plotted in arbitrary log units. The bands at 7.9 and 12.5 μm in the IDP spectrum are probably due to residual silicone oil.

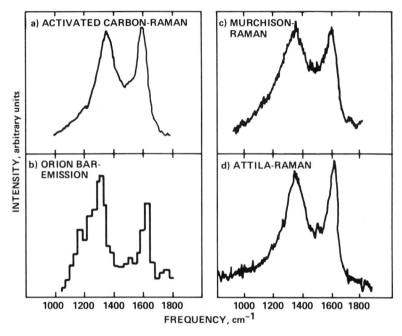

Fig. 11.1.12. A comparison of the Raman spectra of an IDP (Attila), a sample of the CM carbonaceous chondrite Murchison and activated C, with the emission spectrum of the Orion Bar nebular region.

11.1.5. ISOTOPIC MEASUREMENTS OF IDPs

The first isotopic results on IDPs were reported by Esat et al. (1979), who measured the Mg-isotopic compositions of 13 individual IDPs using standard mass-spectrometric techniques. Eight of the particles had terrestrial compositions within two parts per thousand and one showed a mass fractionation of 10^{-2} per mass unit. Although there was a hint of nonlinear isotopic effects at the level of 3 to 4 parts per thousand (now confirmed by the recent, more precise work of Esat and Taylor [1987]), the first-order result of this work is that the isotopic composition of Mg in IDPs is close to normal. The Ca isotopic composition of one particle studied was also normal within 2%.

Subsequent isotopic work on IDPs has been performed using the ion probe (Zinner et al. 1983; McKeegan et al. 1985,1987), which is capable of making measurements on dispersed fragments of individual IDPs. These investigations have shown that a substantial fraction of chondritic IDPs have large D enrichments relative to terrestrial D/H values. In five particles it has been shown that the excesses are spatially correlated with the relative abundance of C (McKeegan et al. 1985), indicating a carbonaceous carrier of the D/H anomalies. Analogous D enrichments are also seen in certain primitive meteorites and they have been attributed to the survival of organic molecules formed by ion-molecule reactions in a cold interstellar molecular cloud and/or to selective photodissociation of interstellar aromatic molecules (Allamandola et al. 1987). In either case, the presence of D-rich phases points to the presence of material that predates the formation of the solar system (see Chapter 13.2 for a more complete discussion of these points).

Ion-probe D/H measurements have now been made on 31 chondritic IDPs. A total of 13 of these have fragments with $\delta D > 100‰$ and 3 of them have parts with $\delta D \geq 2000‰$ [δD is defined as $1000 \times \{[(D/H)_{sample}/(D/H)_{smow}] - 1\}$. SMOW refers to standard mean ocean water (Hagemann et al. 1970).]. Of those which were also measured by infrared spectroscopy, 4 out of 7 identified as pyroxene types and 5 out of 12 identified as layer-lattice silicate types had substantial D enrichments. In contrast, none of the 5 particles classified as olivine types showed such effects. The results demonstrate that isotopically distinct material is a common constituent of many particles, and they are consistent with the possibility that pyroxene and layer-lattice silicate IDPs are derived from a common parent.

Although primitive material is present, it is *not* uniformly distributed. Large variations in δD are seen in different fragments of a given IDP and, in at least one, it has been shown (McKeegan et al. 1987) that a D-rich carrier is concentrated in a small spot that is at the limit of the spatial resolution of the ion probe ($\sim 1 \mu m$). A lower limit of δD of $> 9000 ‰$ is estimated for this spot, making it the most D-enriched natural material yet seen in the solar system. Its extreme enrichment and small size, suggest that it may be a preserved interstellar grain.

Apart from D, no large isotopic anomalies have been observed in chondritic IDPs. Nonetheless, small but significant differences in $\delta^{13}C$ of up to 40 ‰ between particles have been reported (McKeegan et al. 1985). In contrast to D, the measured $^{13}C/^{12}C$ ratios of a given particle were constant from one fragment to the next and C-isotopic effects appear to be decoupled from the D-enrichment effects. Ion probe measurements of Mg and Si isotopes in three chondritic IDPs gave normal values within experimental errors.

Large isotopic effects similar to those seen in primitive meteorites (see Chapter 14.3) may yet be found. Isotopically anomalous material constitutes a very small fraction of the material in meteorites, and we have not yet learned to separate the various phases in IDPs efficiently in ways analogous to those used for meteorites (e.g., etching, stepwise combustion, etc.). The collection of bigger particles using large area collectors (Zolensky 1986) should make such separations easier in the future.

Although most measurements have been made on chondritic IDPs, some isotopic data are available for other types of stratospheric dust particles. Several samples containing minerals typical of refractory meteorite inclusions were found by Zolensky (1985), who suggested that they might be extraterrestrial. Subsequent ion probe measurements of O isotopes by McKeegan et al. (1986) showed that three out of four of these particles had substantial ^{16}O enrichments and were certainly extraterrestrial. (However, it has not yet been established that these refractory grains existed as small particles in space; they could be remnants of larger objects that broke up in the atmosphere.)

Recently, McKeegan et al. (1987) have measured H isotopes in two particles of the FSN type, one of which exhibited a δD value of 173 ‰ suggestive of an extraterrestrial origin.

Magnesium and H-isotopic measurements of a number of particles in the Al' category showed some mass-fractionation effects but no large nonlinear anomalies (McKeegan 1986). Most particles in this category are probably man-made contaminants (Mackinnon et al. 1982).

11.1.6. ORIGIN OF IDPs

Interplanetary dust is responsible for zodiacal light and gegenschein, the Infrared Astronomical Satellite (IRAS) dust bands, meteors and impact pits on lunar rock surfaces. The dust has also been detected by space probes (Berg and Grün 1973; McDonnell et al. 1975), and it has been shown that it persists out to at least 18 AU (Humes 1980).

Dohnanyi (1978) calculated that the Poynting-Robertson effect and particle-particle collisions limit dust particles $\sim$ 10 μm in size to lifetimes of $\sim 10^4$ yr in the inner solar system. The density of solar-flare tracks observed in IDPs confirms this estimate. The dust cloud presently observed in the solar system is thus not primordial, but must either be a transient phenomenon or must be continuously replenished by one or more sources. Studies of surface impact craters and solar flare tracks in lunar samples show that the flux of micrometeorites has been relatively constant for the last 10^6 to 10^9 yr (Morrison and Zinner 1976; Poupeau et al. 1977), and thus the dust is being continually resupplied. Whipple (1967) calculated that a total production rate of $\sim$ 8 ton/s is required to maintain the present cloud. Although there are a number of possible sources for the interplanetary dust, the major ones, as discussed below, are comets and asteroids.

Cometary Origin

Various authors have estimated that comets produce anywhere between a few percent and essentially all of the material required to maintain the dust cloud in equilibrium (Whipple 1967; Delsemme 1976; Kresák 1980). Although there is disagreement on the magnitude of the cometary contribution, the following example indicates that comet dust must be an important constituent of the interplanetary dust complex. Comet Kohoutek was observed to lose $\sim 2 \times 10^{13}$ g of solid material inside 2 AU (Ney 1982) of which approximately one half is expected to remain within the solar system. Since the total mass of the zodiacal dust cloud is $\sim 10^{17}$ g, it would take 10,000 comet passages like that of comet Kohoutek to replenish the cloud. Therefore, the passage of one such comet every five years would provide enough dust. Indeed, some calculations suggest that even a single comet (comet Encke) situated in a unique short-period orbit, could be responsible for most of the dust presently observed (Whipple 1967). The cometary origin of interplanetary dust is further supported by in situ measurements made by spacecraft. The orbital parameters of dust near 1 AU as determined by experiments on Pioneers 8, 10 and 11 are best fit by cometary sources (Gerloff and Berg 1971; Leinert et al. 1983).

Several major meteor showers are also associated with the orbits of some well-known comets. For example, the Orionid and Aquarid meteoroid streams are associated with Halley's comet, the Taurids with comet Encke, and the Draconids with comet Giacobini-Zinner (Lovell 1954). These streams should be steadily dispersed by nongravitational effects (Dohnanyi 1970) and may be the source of sporadic meteoroids as well. Observations of meteor showers show that many of them are characterized by chondritic elemental and isotopic abundances (Herrmann et al. 1978), and that the particles have relatively low structural strengths (Verniani 1969,1973). Meteors from the Draconids are particularly weak, having strengths considerably lower than those of chondritic meteorites.

All the properties of IDPs are consistent with those inferred for cometary dust from telescopic and meteor data. The observation that many of the chondritic IDPs have high porosities and low crushing strengths supports the view that many of the collected dust particles are cometary (Bradley and Brownlee 1986). The infrared spectral comparisons discussed in the previous section demonstrate that a cometary origin of IDPs is reasonable *provided* comets produce more than one type of dust.

Comparison of IDPs with Comet Halley Data

New information about comet dust has recently been gathered by the several spacecraft that made flybys of comet Halley. These probes found that Halley was producing particles with masses at least as low as 10^{-16} g (Vaisberg et al. 1986; McDonnell et al. 1986) and was injecting approximately 3 $\times$ 10^6 g/s of dust into the interplanetary medium. If this rate were continuously maintained while Halley was in the inner solar system, on the order of 10^{14} g of material would have been deposited into the interplanetary medium during its recent passage. This is somewhat higher than the mass deposition estimated earlier for comet Kohoutek.

Comet Halley dust particles were found to have variable compositions (Kissel et al. 1986*a,b*). Many of the grains had roughly chondritic elemental abundances, but some particles were found that consisted of O, Mg, Si and Fe. These are probably individual silicate mineral grains. Other particles were rich in H, C, N and O, suggesting an organic composition. Various mixtures of the different compositional types were also seen. Unfortunately, the encounter data are not fully analyzed and it is not clear what fraction of the mass is represented by each compositional type. It should be kept in mind that the total amount of dust analyzed during the Halley flybys corresponds to the mass of a single 3 to 5 μm particle. Thus, one should not compare IDPs to the Halley dust data directly, but should instead consider the small (<1000 Å) individual components that make up IDPs.

Inasmuch as comparisons can be made at this time, the properties of IDPs are consistent with those of the dust from comet Halley (Walker 1987; Brownlee et al. 1987). The cometary grains with chondritic elemental abundances are matched by IDPs as a whole and by the small chondritic "tar balls" contained within IDPs. The silicate-only particles correspond to the individual mineral grains seen in IDPs, and the organic fraction would correspond to the carbonaceous mantles and matrix materials. Of the common minerals seen in IDPs, only magnetite appears to be missing from the Halley data. This may emerge as a critical issue in the future.

Brownlee et al. (1987) have compared a selected group of mass spectra obtained by the Vega spacecraft and primitive meteorites. They conclude that the distributions of (Fe/Fe + Mg) are better matched by the more coarse-grained, anhydrous IDPs than by either the layer-lattice silicate IDPs or the meteorites.

As discussed earlier, the infrared data for Halley dust can be fitted by combinations of IDPs in different infrared classes with the dominant contribution coming from anhydrous particles. The surface material of comet Halley was observed to be dark (albedo $\leq$ 0.04) (Keller et al. 1986) and IDPs also appear black. In analogy with the IDPs, it should be stressed that the dark appearance of comet Halley does not necessarily imply that the comet mass is dominated by carbonaceous material.

The preliminary isotopic data available from the Halley flyby are also consistent with what is known about IDPs. The isotopic compositions of C, Mg and Si in IDPs are close to terrestrial values and the same appears to be true of the Halley dust particles. The only large anomalies so far found in IDPs are enrichments of D in many of the particles. If spectra can be analyzed with adequate precision for D/H in the Halley data, this would form a crucial point of comparison.

Asteroidal Origin

Asteroids must also contribute to the interplanetary dust complex. The strongest evidence for this comes from the results of the Infrared Astronomical Satellite (IRAS) which showed bands of dust running continuously around the solar system (Low et al. 1984; Neugebauer et al. 1984). The bands have color temperatures of 165 to 200 K, which places them at the position of the main asteroid belt and suggests that the dust is derived from asteroid-asteroid collisions (Dermott et al. 1984; Sykes and Greenberg 1986). However, an asteroidal origin of the dust bands is not yet proven since they could also be due to the episodic injection of dust from short-period comets having the appropriate inclinations (Dermott et al. 1984).

While asteroids are unlikely to produce enough dust to maintain the entire interplanetary dust cloud (Dohnanyi 1976), it has been suggested that they could potentially supply about half the material required to keep the dust cloud in equilibrium (Zook and McKay 1986).

Flynn (1987a,b) has recently pointed out that orbital effects favor the survival during entry of grains derived from asteroids over those from comets. Thus, regardless of the relative populations in space, the stratospheric collections may be biased in favor of asteroidal particles, and cometary particles may be underrepresented.

Other Sources

Several additional sources of dust exist including the Sun, the planets with their rings and moons, and the general interstellar medium; however, as discussed by Sandford (1987), these are not expected to contribute substantially to the present-day interplanetary dust complex.

Possible Discrimination of Origin Using Solar-Flare Tracks

The study of solar-flare tracks may make it possible to determine the fractions of IDPs which come from asteroids and comets. The density of tracks produced within an IDP depends on its space-exposure time, its distance from the Sun, and the flux of track-producing solar-flare nuclei. All of these factors are functions of the orbit of a particle. Since the distance a solar flare ion can penetrate into a mineral grain is small ($\sim$ 100 μm), the presence of tracks in IDPs (see Fig. 11.1.1), coupled with the observation of amorphous rims produced by solar-wind exposure, demonstrates that the particles have been exposed in space as small entities (Bradley et al. 1984a; Bradley and Brownlee 1986). The evolution of the orbits of particles in the size range collected in the stratosphere is dominated by the solar gravitational field and by the Poynting-Robertson drag which causes the dust grains to spiral into the Sun in orbits that become progressively more circular (Wyatt and Whipple 1950).

It has recently been noted (Sandford 1986b) that different track-density distributions should be produced depending on whether IDPs are from comets or asteroids. For a given mass and size, particles from an asteroidal source should have a narrow range of nonzero track densities at 1 AU, while cometary particles should display a wider range of track densities that extends to lower values. This is because asteroidal particles will undergo substantial space exposure before arriving at 1 AU in nearly circular orbits. These orbits rapidly evolve from their first to last contact with the Earth's orbit, resulting in a narrow range of observed track densities. Comets, on the other hand, produce dust in highly elliptical orbits that may originally cross the Earth's

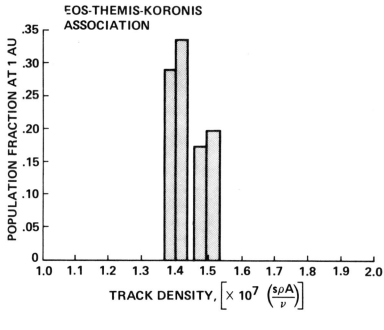
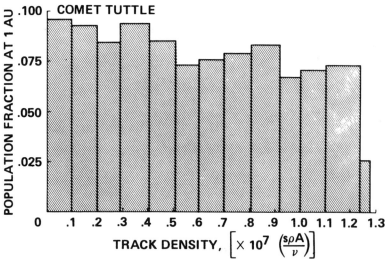

Fig. 11.1.13. A comparison of calculated track-density distributions in IDPs at 1 AU for asteroidal and cometary origins. (a) The scaled solar-flare track-density distribution expected if all the dust were derived from the Eos-Koronis-Themis asteroid associations. The scaling factors for the bottom axis are: s = particle radius, ρ = particle density, ν = fraction of incident solar light absorbed by the particle ($\nu = 1$ for a blackbody) and A = solar-flare track-production rate at 1 AU. (b) The scaled solar flare track-density distribution that would be expected if an object like comet Tuttle were the source of the IDPs. Note that comets produce track-density distributions that are significantly wider than those produced by individual asteroids. In addition, cometary objects are capable of supplying dust at 1 AU that has much lower track densities than those in dust from asteroids (figure from Sandford 1986b).

orbit. Thus, cometary particles with short space exposures and low track denisities can be collected. The higher eccentricity of cometary orbits also results in larger time scales for the evolution from first to last Earth orbit contact and thus an increased range in collectible track densities. Figure 11.1.13 shows a comparison between the expected track-density distributions of particles from these two different types of sources. The track densities observed to date are consistent with those calculated by Sandford (1986b), but insufficient numbers of particles have so far been measured to distinguish between a cometary or asteroidal origin for the majority of IDPs.

11.1.7. DISCUSSION

Returning to the questions that we posed at the beginning of this chapter, we have shown that several different types of particles collected in the stratosphere are, in fact, extraterrestrial. Different classes of IDPs exist both within the chondritic subset and the larger ensemble of collected particles. They have similarities with meteorites but also important differences. The particles are complex, unequilibrated assemblages that reflect a variety of formation processes. Some constituents appear to be the results of direct vapor-solid reactions and others give evidence for the formation of carbonaceous solids by catalytic reactions. Small spots of D-enriched material are found in IDPs, and these may be interstellar grains that survived the formation of the solar system.

The major sources of IDPs are comets and asteroids, although their relative contributions have yet to be determined. Because of their high porosities, the anhydrous IDPs are prime candidates for cometary particles. Their presence in comets is further suggested by the necessity of adding their spectral contributions to match comet Halley data. However, it is likely that the layer-lattice silicate IDPs also contain a cometary contribution since these particles are required to match the 6.8 μm feature seen in the Halley spectra. The similar D enrichments seen in some members of both IDP classes suggests a related origin. In any event, the distinction between a cometary and asteroidal origin may not be a key issue. As discussed in Chapter 2.1, some asteroids are highly evolved objects, while others are not. Regardless of their origin, many of the collected IDPs appear to be the most primitive and unaltered extraterrestrial materials that have yet been studied.

The most primitive solar-system solids imaginable would be clumps of interstellar grains that escaped processing in the solar nebula. Such a model has, in fact, been proposed for comet dust. On the basis of astronomical observations of interstellar dust and laboratory experiments, Greenberg (1982) has suggested that such dust would consist of aggregates of elongated silicate grains mantled by complex organic materials produced by photoprocessing of simple ices. Some IDPs do contain trace amounts of elongated mineral grains and flat platelets; in addition, most of the grains are imbedded in, or mantled by, carbonaceous material. However, no IDPs directly resemble Greenberg's "bird nest" model of clumped interstellar grains.

The D/H results show that both layer-lattice silicate and anhydrous IDPs contain material that can be ascribed to surviving interstellar grains. However, the anomalous grains are small and represent a tiny fraction of the total volume. Most of the material appears to have normal solar-system isotopic compositions; the simplest interpretation is that most of the solids in IDPs were formed in the solar system itself.

It must be said, however, that it might not be easy to demonstrate that an IDP consisted of a collection of interstellar grains even if, in fact, this were the case. The individual sub-μm constituents of an IDP might be isotopically distinct but the ensemble of many such subgrains might have a composition close to the solar-system average. Further advances in analytic techniques will be required to explore this issue. It is also true that pre-existing interstellar grains might not, on the average, be very different from solar-system material. Although there is strong evidence that *some* circumstellar grains with distinctive isotopic signatures have survived intact in meteorites, theoretical calculations indicate that most grains are quickly destroyed in the diffuse interstellar medium (Seab and Shull 1986; but see Chapter 13.3). Thus, the grains found in a protostellar gas-dust cloud may themselves consist of interstellar grains whose compositions have been homogenized in the interstellar medium to give compositions similar to solar-system values. In this connection, it is interesting to note that the isotopic composition of galactic cosmic rays is, with some exceptions, not strikingly different from average solar-system material (Wiedenbeck 1984).

Interplanetary dust particles are small, and this poses both experimental and philosophical problems. Great progress has been made on the experimental side where it has proven possible to make a variety of sophisticated measurements on parts of a single IDP. Serious technical challenges remain; for example, characterizing the molecular state of the light elements and measuring the absolute formation age of the particles. The remarkable progress in microanalysis in recent years leaves us sanguine about solving these and other experimental problems. The philosophic question is more difficult; perhaps the particles are simply too small to give us important insights into the large-scale questions of the origin of matter and of the solar system. Only time will tell; here too, we are optimistic.

Acknowledgments. This work was supported by grants from the National Aeronautics and Space Administration.

REFERENCES

Alexander, C., Barber, D. J., and Hutchison, R. 1986. Hydrous phases and hydrous alteration in U.O.C.'s. *Meteoritics* 21:328.
Allamandola, L. J., Tielens, A. G. G. M., and Barker, J. R. 1985. Polycyclic aromatic hydro-

carbons and the unidentified infrared emission bands: Auto exhaust along the Milky Way. *Astrophys. J.* 290:L25–L28.
Allamandola, L. J., Sandford, S. A., and Wopenka, B. 1987. Interstellar polycyclic aromatic hydrocarbons and the carbon in interplanetary dust particles and meteorites. *Science* 237: 56–59.
Anderson, R. B. 1956. Hydrocarbon synthesis, hydronation, and cyclization. In *Catalysis 4*, ed. P. H. Emmett (New York: Reinhold), p. 371.
Barber, D. J. 1985. Phyllosilicates and other layer-structured materials in stony meteorites. *Clay Minerals* 20:415–454.
Berg, O. E., and Grün, E. 1973. Evidence of hyperbolic cosmic dust particles. In *Space Research XIII*, eds. M. J. Rycroft and S. K. Runcorn (Berlin: Akademie-Verlag), pp. 1047–1055.
Bradley, J. P., and Brownlee, D. E. 1983. Microanalyses of dispersed interplanetary dust particles. In *Microbeam Analysis*—1983, ed. R. Gooley (San Francisco: San Francisco Press), pp. 187–190.
Bradley, J. P., and Brownlee, D. E. 1986. Cometary particles: Thin sectioning and electron beam analysis. *Science* 231:1542–1544.
Bradley, J. P., Brownlee, D. E., and Veblen, D. R. 1983. Pyroxene whiskers and platelets in interplanetary dust: Evidence for vapour phase growth. *Nature* 301:473–477.
Bradley, J. P., Brownlee, D. E., and Fraundorf, P. 1984*a*. Discovery of nuclear tracks in interplanetary dust. *Science* 226:1432–1434.
Bradley, J. P., Brownlee, D. E., and Fraundorf, P. 1984*b*. Carbon compounds in interplanetary dust: Evidence for formation by heterogeneous catalysis. *Science* 223:56–58.
Bradley, J., Carey, W., and Walker, R. M. 1986. Solar max impact particles: Peturbation of captured material. *Lunar Planet. Sci.* XVII:80–81 (abstract).
Bregman, J. D., Campins, H., Witteborn, F. C., Wooden, D. H., Rank, D. M., Allamandola, L. J., Cohen, M., and Tielens, A. G. G. M. 1987. Airborne and ground base spectrophotometry of Comet Halley from 5–13 micrometers. *Astron. Astrophys.*, in press.
Brownlee, D. E., 1978*a*. Microparticle studies by sampling techniques. In *Cosmic Dust*, ed. J. A. M. McDonnell (New York: Wiley), pp. 295–426.
Brownlee, D. E. 1978*b*. Interplanetary dust: Possible implications for comets and presolar interstellar grains. In *Protostars and Planets*, ed. T. Gehrels (Tucson: Univ. of Arizona Press), pp. 134–150.
Brownlee, D. E. 1985. Cosmic dust: Collection and research. *Ann. Rev. Earth Planet. Sci.* 13:147–173.
Brownlee, D. E., Ferry, G. V., and Tomandl, D. 1976*a*. Stratospheric aluminum oxide. *Science* 191:1270–1271.
Brownlee, D. E., Tomandl, D., Blanchard, M. B., Ferry, G. V., and Kyte, F. T. 1976*b*. An atlas of extraterrestrial particles collected with NASA U-2 aircraft: 1974–1976. NASA TMX-73,152.
Brownlee, D. E., Wheelock, M. M., Temple, S., Bradley, J. P., and Kissel, J. 1987. A quantitative comparison of Comet Halley and carbonaceous chondrites at the submicron level. *Lunar Planet. Sci. Conf.* XVIII:133–134 (abstract).
Bunch, T. E., and Chang, S. 1980. Carbonaceous chondrites II: Carbonaceous chondrites phyllosilicates and light element geochemistry as indicators of parent body processes and surface conditions. *Geochim. Cosmochim. Acta* 44:1543–1577.
Christoffersen, R., and Buseck, P. R. 1983. Epsilon carbide: A low-temperature component of interplanetary dust particles. *Science* 222:1327–1329.
Christoffersen, R., and Buseck, P. R. 1986*a*. Mineralogy of interplanetary dust particles from the "olivine" infrared class. *Earth Planet. Sci. Lett.* 78:53–66.
Christoffersen, R., and Buseck, P. R. 1986*b*. Refractory minerals in interplanetary dust. *Science* 234:590–592.
Christophe Michel-Levy, M., and Lautie, A. 1981. Microanalysis by Raman spectroscopy of carbon in the Tieschitz chondrite. *Nature* 292:321–322.
Cohen, R. E., Kornacki, A. S., and Wood, J. A. 1983. Mineralogy and petrology of chondrules and inclusions in the Mokoia CV3 chondrite. *Geochim. Cosmochim. Acta* 47:1739–1757.
Delsemme, A. H. 1976. Can comets be the only source of interplanetary dust? In *Interplanetary Dust and Zodiacal Light, Lecture Notes in Physics 48*, eds. H. Elsässer and H. Fechtig (Berlin: Springer-Verlag), pp. 481–484.

Dermott, S. F., Nicholson, P. D., Burns, J. A., and Houck, J. R. 1984. Origin of the solar system dust bands discovered by IRAS. *Nature* 312:505–509.
Dohnanyi, J. S. 1970. On the origin and distribution of meteoroids. *J. Geophys. Res.* 75:3468–3493.
Dohnanyi, J. S. 1976. Sources of interplanetary dust: Asteroids. In *Interplanetary Dust and Zodiacal Light, Lecture Notes in Physics 48,* eds. H. Elsässer and H. Fechtig (Berlin: Springer-Verlag), pp. 187–206.
Dohnanyi, J. S. 1978. Particle dynamics. In *Cosmic Dust,* ed. J. A. M. McDonnell (New York: John Wiley and Sons), pp. 527–605.
Duley, W. W. 1985. Evidence for hydrogenated amorphous carbon in the Red Rectangle. *Mon. Not. Roy. Astron. Soc.* 215:259–263.
Esat, T. M., and Taylor, S. R. 1987. Mg isotopic systematics of some interplanetary dust particles. *Lunar Planet. Sci.* XVIII:269–270 (abstract).
Esat, T. M., Brownlee, D. E., Papanastassiou, D. A., and Wasserburg, G. J. 1979. The Mg isotopic composition of interplanetary dust particles. *Science* 206:190–197.
Fleischer, R. L., Price, P. B., and Walker, R. M. 1975. *Nuclear Tracks in Solids* (Berkeley: Univ. of California Press).
Flynn, G. J. 1987a. Earth encounter velocities and exposure ages of IDPs from asteroidal and cometary sources. *Lunar Planet. Sci.* XVIII, 294–295.
Flynn, G. J. 1987b. Atmospheric entry heating: a criterion to distinguish between asteroidal and cometary sources of interplanetary dust. *Icarus,* in press.
Flynn, G. J., Fraundorf, P., Shirck, J., and Walker, R. M. 1978. Chemical and structural studies of "Brownlee" particles. *Proc. Lunar Planet. Sci. Conf.* 9:1187–1208.
Fraundorf, P. 1981. Interplanetary dust in the transmission electron microscope: Diverse materials from the early solar system. *Geochim. Cosmochim. Acta* 45:915–943.
Fraundorf, P., and Shirck, J. 1979. Microcharacterization of "Brownlee" particles: Features which distinguish interplanetary dust from meteorites. *Proc. Lunar Planet. Sci. Conf.* 10:951–976.
Fraundorf, P., Brownlee, D. E., and Walker, R. M. 1982a. Laboratory studies of interplanetary dust. In *Comets,* ed. L. L. Wilkening (Tucson: Univ. of Arizona Press), pp. 383–409.
Fraundorf, P., McKeegan, K. D., Sandford, S. A., Swan, P., and Walker, R. M. 1982b. An inventory of particles from stratospheric collectors: Extraterrestrial and otherwise. *Proc. Lunar Planet. Sci. Conf. 13, J. Geophys. Res.* 83:A403–A408.
Fraundorf, P., Hintz, C., Lowry, O., McKeegan, K. D., and Sandford, S. A. 1982c. Determination of the mass, surface density, and volume density of individual dust particles. *Lunar Planet. Sci.* XIII:225–226 (abstract).
Ganapathy, R., and Brownlee, D. E. 1979. Interplanetary dust: Trace element analysis of individual particles by neutron activation. *Science* 206:1075–1076.
Gerloff, U., and Berg, O. E. 1971. A model for predicting the results of in situ meteoroid experiments: Pioneer 8 and 9 results and phenomenological evidence. In *Space Research XI,* eds. K. Y. Kondratyev, M. J. Rycroft, and C. Sagan (Berlin: Akademie-Verlag), pp. 397–413.
Goldstein, J. I., Costley, J. L., Lorimer, G. W., and Reed, S. J. B. 1977. Quantitative X-ray analysis in the electron microscope. *Scanning Electron Microscopy/1977,* vol. 1, ed. O. Johari, pp. 315–324.
Greenberg, J. M. 1982. What are comets made of? In *Comets,* ed. L. L. Wilkening (Tucson: Univ. of Arizona Press), pp. 131–163.
Hagemann, R., Nief, G., and Roth, E. 1970. Absolute isotopic scale for deuterium analysis of natural waters, absolute D/H ratio for SMOW. *Tellus* 22:712–715.
Hayatsu, R., and Anders, E. 1981. Organic compounds in meteorites and their origins. In *Cosmo- and Geochemistry,* vol. 99, *Topics in Current Chemistry* ed. F. L. Bosche (Berlin: Springer-Verlag), pp. 1–39.
Herrmann, U., Eberhardt, P., Hidalgo, M. A., Kopp, E., and Smith, L. G. 1978. Metal ions and isotopes in sporadic E-layers during the Perseid meteor shower. In *Space Research XVIII,* eds. M. J. Rycroft and A. C. Strickland (New York: Pergamon Press), pp. 249–252.
Hodge, P. W. 1981. *Interplanetary Dust* (New York: Gordon and Breach Science Publ.).
Hudson, B., Flynn, G. J., Fraundorf, P., Hohenberg, C. M., and Shirck, J. 1981. Noble gases in stratospheric dust particles: Confirmation of extraterrestrial origin. *Science* 211:383–386.

Humes, D. H. 1980. Results of Pioneer 10 and 11 meteoroid experiments: Interplanetary and near-Saturn. *J. Geophys. Res.* 85:5841–5852.
Keller, H. U., Arpigny, C., Barbiere, C., Bonnet, R. M., Cazes, S., Coradini, M., Cosmovici, C. B., Delamere, W. A., Huebner, W. F., Hughes, D. W., Jamar, C., Malaise, D., Reitsema, H. J., Schmidt, H. U., Schmidt, W. K. H., Seige, P., Whipple, F. L., and Wilhelm, K. 1986. First Halley multicolor camera imaging results from Giotto. *Nature* 321:320–326.
Kessler, D. J., and Cour-Palais, B. G. 1978. Collisional frequency of artificial satellites: The creation of a debris belt. *J. Geophys. Res.* 83:2637–2646.
Kissel, J., Sagdeev, R. Z., Bertaux, J. L., Angarov, V. N., Audouze, J., Blamont, J. E., Büchler, K., Evlanov, E. N., Fechtig, H., Fomenkova, M. N., Langevin, Y., Leonas, V. B., Levasseur-Regourd, A. C., Managadze, G. G., Podkolzin, S. N., Shapiro, V. D., Tabaldyev, S. R., and Zubkov, B. V. 1986a. Composition of comet Halley dust particles from Vega observations. *Nature* 321:280–282.
Kissel, J., Brownlee, D. E., Büchler, K., Clark, B. C., Fechtig, H., Grün, E., Hornung, K., Igenbergs, E. B., Jessberger, E. K., Krueger, F. R., Kuczera, H., McDonnell, J. A. M., Morfill, G. M., Rahe, J., Schwehm, G. H., Sekanina, Z., Utterback, N. G., Völk, H. J., and Zook, H. A. 1986b. Composition of comet Halley dust particles from Giotto observations. *Nature* 321:336–337.
Kresák, L. 1980. Sources of interplanetary dust. In *Solid Particles in the Solar System, IAU Symp. 90*, eds. I. Halliday and B. McIntosh (Dordrecht: D. Reidel), pp. 211–222.
Lacy, J. H., Baas, F., Allamandola, L. J., Persson, S. E., McGregor, P. J., Lonsdale, C. J., Geballe, T. R., and van der Bult, C. E. P. 1984. 4.6 micron absorption features due to solid phase CO and cyano group molecules toward compact infrared sources. *Astrophys. J.* 276:533–543.
Léger, A., and Puget, J. L. 1984. Identification of the "unidentified" IR emission features of interstellar dust? *Astron. Astrophys.* 137:L5–L8.
Leinert, C., Röser, S., and Buitrago, J. 1983. How to maintain the spatial distribution of interplanetary dust. *Astron. Astrophys.* 118:345–357.
Lin, S., and Feldman, B. J. 1982. Sidebands in the luminescence spectra of amorphous hydrogenated carbon. *Phys. Rev. Lett.* 48:829–831.
Lovell, A. C. B. 1954. *Meteor Astronomy* (Oxford: Clarendon Press).
Low, F. J., Beintema, D. A., Gautier, T. N., Gillett, F. C., Beichman, C. A., Neugebauer, G., Young, E., Aumann, H. H., Boggess, N., Emerson, J. P., Habing, H. J., Hauser, M. G., Houck, J. R., Rowan-Robinson, M. Soifer, B. T., Walker, R. G., and Wesselius, P. R. 1984. Infrared cirrus: New components of the extended infrared emission. *Astrophys. J.* 278:L19–L22.
Mackinnon, I. D. R., and Rietmeijer, F. J. M. 1984. Bismuth in interplanetary dust. *Nature* 311:135–138.
Mackinnon, I. D. R., and Rietmeijer, F. J. M. 1987. Mineralogy of chondritic interplanetary dust particles. *Rev. Geophys.*, in press.
Mackinnon, I. D. R., McKay, D. S., Nace, G., and Isaacs, A. M. 1982. Al-prime particles in the cosmic dust collection: Debris or not debris? *Meteoritics* 17:245.
McDonnell, J. A. M., Berg, O. E., and Richardson, F. F. 1975. Spatial and time variations of the interplanetary microparticle flux analyzed from deep space probes Pioneers 8 and 9. *Planet. Space Sci.* 23:205–214.
McDonnell, J. A. M., Alexander, W. M., Burton, W. M., Bussoletti, E., Clark, D. H., Grard, R. J. L., Grün, E., Hanner, M. S., Hughes, D. W., Igenbergs, E., Kuczera, H., Lindblad, B. A., Mandeville, J.-C., Minafra, A., Schwehm, G. H., Sekanina, Z., Wallis, M. K., Zarnecki, J. C., Chakaveh, S. C., Evans, G. C., Evans, S. T., Firth, J. G., Littler, A. N., Massone, L., Olearczyk, R. E., Pankiewica, G. S., Stevenson, J. J., and Turner, R. F. 1986. Dust density and mass distribution near comet Halley from Giotto observations. *Nature* 321:338–341.
McKeegan, K. D. 1986. Hydrogen and magnesium isotopic abundances in aluminum-rich stratospheric dust particles. *Lunar Planet. Sci.* XVII:539–540 (abstract).
McKeegan, K. D., Walker, R. M., and Zinner, E. 1985. Ion microprobe isotopic measurements of individual interplanetary dust particles. *Geochim. Cosmochim. Acta* 49:1971–1987.
McKeegan, K. D., Zinner, E., and Zolensky, M. 1986. Ion probe measurements of O isotopes in refractory stratospheric dust particles: Proof of extraterrestrial origin. *Meteoritics* 21:449–450.

McKeegan, K. D., Swan, P., Walker, R. M., Wopenka, B., and Zinner, E. 1987. Hydrogen isotopic variations in interplanetary dust particles. *Lunar Planet. Sci.* XVIII:627–628 (abstract).
Merrill, K. M. 1974. 8–13 μm spectrophotometry of Comet Kohoutek. *Icarus* 23:566–567.
Morrison, D. A., and Zinner, E. 1976. Distribution and flux of micrometeoroids. *Phil. Trans. Roy. Soc. London* A285:379–384.
Nemanich, R. J., and Solin, S. A. 1979. First- and second-order Raman scattering from finite-size crystals of graphite. *Phys. Rev.* B20:392–401.
Neugebauer, G., Beichman, C. A., Soifer, B. T., Aumann, H. H., Chester, T. J., Gautier, T. N., Gillett, F. C., Hauser, M. G., Houck, J. R., Lonsdale, C. J., Low, F. J., and Young, E. T. 1984. Early results from the infrared astronomical satellite. *Science* 224:14–21.
Ney, E. P. 1982. Optical and infrared observations of bright comets in the range 0.5 μm to 20 μm. In *Comets*, ed. L. L. Wilkening (Tucson: Univ. of Arizona Press), pp. 323–340.
Poupeau, G., Rajan, R. S., Walker, R. M., and Zinner, E. 1977. The modern and ancient flux of solar wind particles, solar flare particles and micrometeoroids. In *Space Research XVII* eds. M. J. Rycroft and A. C. Strickland (New York: Pergamon), pp. 599–604.
Rajan, R. S., Brownlee, D. E., Tomandl, D., Hodge, P. W., Farrar, H., and Britten, R. A. 1977. Detection of ^{4}He in stratospheric particles gives evidence of extraterrestrial origin. *Nature* 267:133–134.
Rietmeijer, F. J. M. 1985. What predictions can be made on the nature of carbon and carbon-bearing compounds (hydrocarbons) in the interstellar medium based on studies of interplanetary dust particles? In *Interrelationship Among Circumstellar, Interstellar and Interplanetary Dust*, eds. J. A. Nuth III and R. E. Stencel, NASA CP-2403, pp. A23–A27.
Rietmeijer, F. J. M., and Mackinnon, I. D. R. 1984. Layered silicates in chondritic porous aggregate W7029*687–688. *Lunar Planet. Sci.* XV:687–688.
Rietmeijer, F. J. M., and Mackinnon, I. D. R. 1985. Layer silicates in a chondritic porous interplanetary dust particle. *Proc. Lunar Planet. Sci. Conf.*, 15, *J. Geophys. Res.* 90:D149–D155.
Rietmeijer, F. J. M., and McKay, D. S. 1985. An interplanetary dust analog to matrices of CO/CV carbonaceous chondrites and unmetamorphosed unequilibrated ordinary chondrites. *Meteoritics* 20:743–744.
Rietmeijer, F. J. M., and McKay, D. S. 1986. Fine-grained silicates in chondritic interplanetary dust particles are evidence for annealing in the early history of the solar system. *Lunar Planet. Sci.* XVII:710–711.
Rose, L. A. 1979. Laboratory simulation of infrared astrophysical features. *Astrophys. Space Sci.* 65:47–67.
Sandford, S. A. 1984. Infrared transmission spectra from 2.5 to 25 μm of various meteorite classes. *Icarus* 60:115–126.
Sandford, S. A. 1986a. Acid dissolution experiments: Carbonates and the 6.8-micrometer bands in interplanetary dust particles. *Science* 231:1540–1541.
Sandford, S. A. 1986b. Solar flare track densities in interplanetary dust particles: The determination of an asteroidal versus cometary sources of the zodiacal dust cloud. *Icarus* 68:377–394.
Sandford, S. A. 1987. The collection and analysis of extraterrestrial dust particles. *Fund. Cosmic Phys.* 12:1–73.
Sandford, S. A., and Walker, R. M. 1985. Laboratory infrared transmission spectra of individual interplanetary dust particles from 2.5 to 25 microns. *Astrophys. J.* 291:838–851.
Seab, C. G., and Shull, J. M. 1986. Shock processing of interstellar grains. In *Interrelationships Among Circumstellar, Interstellar, and Interplanetary Dust*, eds. J. A. Nuth and R. E. Stencel, NASA CP-2403, pp. 37–53.
Soifer, B. T., Puetter, R. C., Russell, R. W., Willner, S. P., Harvey, P. M., and Gillett, F. C. 1979. The 4–8 micron spectrum of the infrared source W33A. *Astrophys. J.* 232:L53–L57.
Sykes, M. V., and Greenberg, R. 1986. The formation and origin of the IRAS zodiacal dust bands as a consequence of single collisions between asteroids. *Icarus* 65:51–69.
Tomeoka, K., and Buseck, P. R. 1984. Transmission electron microscopy of the "LOW-CA" hydrated interplanetary dust particle. *Earth Planet. Sci. Lett.* 69:243–254.
Tomeoka, K., and Buseck, P. R. 1985a. Hydrated interplanetary dust particle linked with carbonaceous chondrites? *Nature* 314:338–340.

Tomeoka, K., and Buseck, P. R. 1985b. Indicators of aqueous alteration in CM carbonaceous chondrites: Microtextures of a layered mineral containing Fe, S, O and Ni. *Geochim. Cosmochim. Acta* 49:2149–2163.

Tomeoka, K., and Buseck, P. R. 1986a. A carbonate-rich, hydrated, interplanetary dust particle: Possible residue from protostellar clouds. *Science* 231:1544–1546.

Tomeoka, K., and Buseck, P. R. 1986b. Phyllosilicates in the Mokoia CV3 carbonaceous chondrite: Petrographic and transmission electron microscope observations. *Lunar Planet. Sci.* XVII:899–900.

Vaisberg, O. L., Smirnov, V. N., Gorn, L. S., Iovlev, M. V., Balikchin, M. A., Klimov, S. I., Savin, S. P., Shapiro, V. D., and Shevchenko, V. I. 1986. Dust coma structure of comet Halley from SP-1 detector measurements. *Nature* 321:274–276.

van Breda, I. G., and Whittet, D. C. B. 1981. Very broadband structure in the extinction curves of southern Milky Way stars. *Mon. Not. Roy. Astron. Soc.* 195:79–88.

Veblen, D. R., and Post, J. E. 1983. A TEM study of fibrous cuprite (chalcotrichite): Microstructures and growth mechanisms. *Amer. Min.* 68:790–803.

Verniani, F. 1969. Structure and fragmentation of meteoroids. *Space Sci. Rev.* 10:230–261.

Verniani, F. 1973. An analysis of the physical parameters of 5759 faint radio meteors. *J. Geophys. Res.* 78:8429–8462.

Walker, R. M. 1987. Are IDPs and Halley dust similar and, if so, so what? *Lunar Planet. Sci. Conf.* XVIII:1048–1049.

Watanabe, I., Hasegawa, S., and Kurata, Y. 1982. Photoluminescence of hydrogenated amorphous carbon films. *Jap. J. Appl. Phys.* 21:856–859.

Whipple, F. L. 1967. On maintaining the meteoritic complex. In *The Zodiacal Light and the Interplanetary Medium,* ed. J. L. Weinberg, NASA SP-150, pp. 409–426.

Wiedenbeck, M. E. 1984. The isotopic composition of cosmic rays. *Adv. Space Res.* 4(2-3):15–24.

Witt, A. N., Schild, R. E., and Kraiman, J. B. 1984. Photometric study of NGC 2023 in the 3500 Å to 10000 Å region: Confirmation of a near-IR emission process in reflection nebula. *Astrophys. J.* 281:708–718.

Wopenka, B. 1987. Raman observations of individual interplanetary dust particles. *Lunar Planet. Sci. Conf.* XVIII:1102–1103.

Wopenka, B., and Sandford, S. A. 1984. Laser Raman microprobe study of mineral phases in meteorites. *Meteoritics* 19:340–341.

Wyatt, S. P., and Whipple, F. L. 1950. The Poynting-Robertson effect on meteor orbits. *Astrophys. J.* 111:134–141.

Zinner, E., McKeegan, K. D., and Walker, R. M. 1983. Laboratory measurements of D/H ratios in interplanetary dust. *Nature* 305:119–121.

Zolensky, M. E. 1985. CAI's among the cosmic dust collection. *Meteoritics* 20:792–793.

Zolensky, M. E. 1986. Cosmic Dust Courier 7, Code SN2, NASA/Johnson Space Center, Houston, TX.

Zolensky, M. E., and Mackinnon, I. D. R. 1985. Accurate stratospheric particle size distributions from a flat plate collection surface. *J. Geophys. Res.* 90:5801–5808.

Zook, H. A., and McKay, D. S. 1986. On the asteroidal component of cosmic dust. *Lunar Planet. Sci. Conf.* XVII:977–978.

PART 12
Inhomogeneity of the Nebula

12.1. HETEROGENEITY IN THE NEBULA: EVIDENCE FROM STABLE ISOTOPES

MARK H. THIEMENS
University of California at San Diego

This chapter addresses the meteoritic isotopic evidence, primarily for O, which defines the existence of different nebular reservoirs. The basic physical-chemical principles of isotope effects associated with processes, such as exchange and distillation/condensation, that probably affected meteoritic components, are presented. Chemically produced mass-independent (non-Boltzmann) isotope effects are discussed as they may have been an agent in the production of some of the observed meteoritic isotopic anomalies. Various nebular reservoirs of O are defined by the isotopic compositions of bulk meteorites, and interrelations are implied by the internal meteoritic systematics, particularly for inclusions and chondrules. The mechanism by which these reservoirs were originally produced, whether by a cosmic chemical memory, late-stage injection or chemical processing, remains an open question; representative models are presented.

12.1.1. INTRODUCTION

As late as 1967, it was still maintained that the solar nebula was isotopically homogenous. As concluded by Reynolds (1967), if the noble gases are excluded, the isotopic composition of meteoritic material is essentially indistinguishable from terrestrial. (The apparently distinct noble-gas reservoirs are discussed in Chapters 7.9 and 13.1.) Homogeneity of the solar nebula was at that time entirely consistent with the widely held notion that the nebula was completely vaporized, hot and well mixed (see Chapter 6.3). It was not until 1973 that clear, unambiguous evidence for a distinct nonterrestrial isotopic

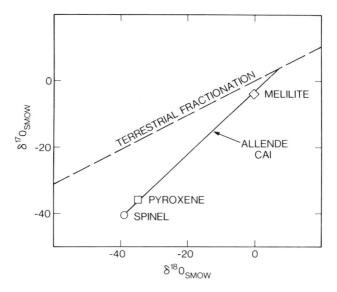

Fig. 12.1.1. The O-isotopic composition of individual minerals separated from Ca-Al-rich inclusions in the Allende CV chondrite. These data define a line with slope ∼1, in contrast to the terrestrial fractionation line with slope ∼0.5.

signature was found (Clayton et al. 1973). The O-isotopic composition of anhydrous, high-temperature minerals in the Allende meteorite exhibits, in a 3-isotope plot of $\delta^{17}O$ vs $\delta^{18}O$, a line of slope ∼1, rather than the terrestrial slope ∼0.5 (see Fig. 12.1.1), indicating the existence of a reservoir of O distinct from terrestrial. This observation led to many isotopic measurements being made, and a host of nonterrestrial isotopic compositions have subsequently been recognized in meteorites for elements such as Mg, Si, Ca, Ba, Sr, Ti, Ag, Nd and Sm (for reviews see Clayton 1978; Wasserburg et al. 1980; Podosek 1978). These isotopic anomalies possess two major types of information: the first relates to the nucleosynthetic sources of this material, and the second to processes within and structure of the early solar system. The nucleosynthetic aspects are discussed in Chapter 14.3. The second aspect forms the basis for most of this chapter.

The chapter is divided into four main parts, though they are not necessarily completely distinct. The first part (Sec. 12.1.2) is a brief description of isotope effects as they may influence isotope ratios in meteoritic material. In the second part (Sec. 12.1.3), meteoritic O isotopes are discussed in terms of the existence of presumably distinct nebular components, as evidenced primarily from the whole-rock O-isotopic systematics. The third part (Sec. 12.1.4) discusses the internal isotopic systematics (i.e., for chondrules and inclusions) and their role in placing constraints on conditions and processes in the solar nebula. The final section (Sec. 12.1.5) briefly discusses the question of the actual origin of these reservoirs.

12.1.2. CAUSES OF ISOTOPIC VARIATION

The development of the analytical capability to distinguish small isotopic variations ($< 0.1‰$) in samples of nanogram to even picogram size has provided a wealth of information on the physical, chemical and nucleosynthetic history of the solar system. Besides the possible addition of material from different nucleogenetic reservoirs, a wide variety of processes may intervene to modify meteoritic isotopic abundances. These include the following:

1. Radioactive decay, such as $^{87}Rb \rightarrow {}^{87}Sr$ or $^{26}Al \rightarrow {}^{26}Mg$, increases the abundance of the stable daughter nuclide relative to the nonradiogenic isotopes of, for example, Sr and Mg. These effects are well known and, in fact, may be used as chronometers, as discussed in other chapters. The involvement of yet another process, isotopic exchange between different reservoirs, in some cases limits the application of these chronometers.

2. Nuclear reactions, in particular cosmic-ray-induced spallation, may also produce variations in stable-isotope abundance patterns, though for major elements such as O, Si and Mg, these effects are generally unobservable. For the elements where such effects are observable, they are well understood and may also be employed as chronometers (see Chapter 4.1). Both these types of process share with nucleogenetic variations the characteristic feature that the magnitude of the isotopic variations they introduce is not a simple function of isotopic mass difference.

3. Mass-dependent isotopic fractionations can occur as a result of many physical or chemical processes, such as phase changes, diffusion, thermodynamic equilibration and chemical kinetics. The partitioning of isotopes by these processes is quite sensitive to temperature and chemistry and can therefore produce characteristic information regarding the conditions that produced the observed stable-isotope abundance pattern. Modeling efforts require *a priori* a thorough knowledge of the laws of such isotope effects.

4. Chemically produced mass-independent fractionations constitute a recently discovered class of isotope effects which are potentially relevant to meteoritic isotope data and their interpretation. With the experimental demonstration of such effects (Thiemens and Heidenreich 1983; Heidenreich and Thiemens 1983, 1985*a*), it is no longer possible to attribute an observed mass-independent signature in meteoritic data to a nuclear process strictly on the basis of its mass-independent variation.

In the most general sense, an isotope effect is defined as some physical or chemical process whose rate, magnitude or position of equilibrium is altered by isotopic substitution. For example, the vapor pressure of water (H_2O) at its boiling point (100° C) is by definition 760.00 torr (1 atm), whereas for deuterated water (D_2O) at 100° C, the vapor pressure is 721.7 torr. The total energy of a molecule or an ensemble of molecules is a sum of three energy components: translational, vibrational and rotational. However, for gas molecules, the use of the Teller-Redlich product rule permits isotopic equilibrium constants to be determined on the basis of the ratio of the vibrational frequencies of the isotopically substituted molecules (Urey 1947). The quantitative theoretical basis of isotope effects within a statistical framework was given by Urey (1947) and Bigeleisen and Mayer (1947). Relevant reviews of this subject are given by Bigeleisen (1965), Clayton (1981) and Wolfsberg (1982).

If isotope effects are attributed to molecular vibrational energies (or frequencies), then the techniques of statistical mechanics may be applied to determine quantitatively the magnitude and direction of isotopic fractionations. If we consider the vibrations of a diatomic molecule, it is known from quantum theory that the potential energy of these vibrations is a function of the intramolecular force. This force, in turn, depends upon the electronic configuration, molecular structure and the coulombic force. The molecular vibrations must, furthermore, be quantized and possess discrete states, or energy levels, with the minimum or zero point energy equal to $1/2\ h\nu$, where ν is the fundamental vibrational frequency. For isotopically substituted molecules, e.g., $^{14}N^{16}O$ and $^{14}N^{18}O$, the vibrational frequency ratio ($\nu^{16}O/\nu^{18}O$) is determined by the square root of the reduced mass (μ) ratio ($\mu^{16}O/\mu^{18}O)^{1/2}$, which, in this instance, is 0.9736 (Herzberg 1950). This factor physically represents the slower vibration of the heavier molecule $^{14}N^{18}O$. The overall consequence of an isotope effect, which arises from vibrational frequency differences, is that, owing to its higher vibrational frequency and greater average interatomic separation, the dissociation energy of the lighter isotopic species is slightly less than that of the heavier species, leading to more rapid reaction.

A major outcome of the work of Urey (1947) and Bigeleisen and Mayer (1947) is that an equilibrium constant K may be derived for a chemical reaction from molecular thermodynamic quantities. Since isotope exchange may be considered as a chemical reaction, this derivation has many applications, particularly for meteoritics and terrestrial geothermometry.

In a chemical reaction or isotope-exchange process, the equilibrium constant is a ratio of the molecular partition functions Q, or

$$K = \frac{\Pi Q_p}{\Pi Q_r} \qquad (1)$$

where, more explicitly, the partition function ratios, as given by Urey (1947), are:

$$\frac{Q_p}{Q_r} = \frac{\sigma_r}{\sigma_p}\left(\frac{M_p}{M_r}\right)^{3/2} \Sigma e^{-E_p/kT}/\Sigma e^{-E_r/kT} \qquad (2)$$

where σ is the molecular symmetry number, M is the molecular weight, E the appropriate energy of the molecule, T the temperature and k is Boltzmann's constant. The subscripts p and r refer to products and reactants, respectively.

As an example of the use of partition functions in determining the position of equilibrium in an exchange reaction, consider

$$C^{16}O + H_2{}^{18}O \rightleftarrows C^{18}O + H_2{}^{16}O. \qquad (3)$$

Using data from Urey (1947), the equilibrium constant is given at $T = 298.1$ K by

$$K = \frac{Q_p}{Q_r} = \frac{Q\left(\dfrac{C^{18}O}{C^{16}O}\right)}{Q\left(\dfrac{H_2{}^{18}O}{H_2{}^{16}O}\right)} = 1.1053/1.0667 = 1.0362 \qquad (4)$$

which means that once equilibrium is attained at 298.1 K, the measured $\delta^{18}O$ for CO will be 36.2‰ greater than for H_2O. Note that this has no relation to the rate at which equilibrium is attained, nor to the absolute values of the ratios, but only to the isotopic partitioning between the coexisting pair. If the same fractionation factors are calculated at different temperatures, it is found that, at 400, 500 and 600 K, the values are 24, 16 and 11‰, respectively. The decrease in fractionation is a consequence of Eq. (2), whereby the isotopic differences disappear at higher temperatures in the ratio of partition functions. Isotope exchange and temperature effects on chemical fractionations were both important processes in the early solar system. The principles detailed above are used in characterization of these processes from the measured isotopic distributions in meteoritic material.

A source of meteoritic isotopic fractionations, particularly in certain inclusions (see Chapter 10.3), is evaporation. The magnitude of fractionation is calculable on the basis of Rayleigh's (1896) theory of distillation. The isotopic fractionation during such an evaporation is given by

$$R/R_o = f^{(\alpha-1)} \qquad (5)$$

where f is the fraction of liquid remaining, R is the isotopic ratio of the liquid at that time, R_o was its isotopic composition at the start of evaporation, and α is the fractionation factor between vapor and liquid, i.e., $\alpha = R_v/R_l$, where the subscripts v and l stand for vapor and liquid, respectively. Equation (5), which strictly applies only when the reservoir remains well mixed, therefore relates the isotopic composition of a residue to the proportion of material that has been evaporated away. As discussed in Sec. 12.1.4, many CAIs, most dramatically those termed FUN inclusions (for the Fractionated and Unknown Nuclear effects seen in their isotopic compositions), and also chondrules, exhibit the effects of evaporation.

For meteoritic studies, measurements of isotope ratios for multi-isotope (≥ 3) elements may be particularly diagnostic of the types of processes that produced the observed isotopic signature. For small fractionations, the isotope ratios in a 3-isotope system will be an approximately linear function of the differences in the reciprocal masses of the different isotopic species (Hulston and Thode 1965). In the case of molecular O_2, with isotopic species $^{16}O^{16}O$, $^{16}O^{17}O$ and $^{16}O^{18}O$, the fractionated ratios $\delta^{18}O$ and $\delta^{17}O$ are related by:

$$\delta^{17}O/\delta^{18}O = \frac{\frac{1}{32} - \frac{1}{33}}{\frac{1}{32} - \frac{1}{34}} \quad \text{or} \quad \delta^{17}O = 0.515\ \delta^{18}O. \tag{6}$$

Any fractionation producing a 2‰ fractionation in $\delta^{18}O$ concomitantly produces approximately a 1‰ shift in $\delta^{17}O$. A plot of $\delta^{17}O$, on the y-axis, vs $\delta^{18}O$, on the x-axis, for different fractionations of a common reservoir of O produces a line of slope $\sim 1/2$ (0.515). The precise value of the coefficient varies depending upon whether the O experiencing, for example, a diffusion process is atomic (coefficient = 0.529) or in a heavy molecule such as SO_2 (0.5087) (Matsuhisa et al. 1978). It should be noted that this relation holds linearly over only a modest extent of fractionation (≤ 15.0‰ for $\delta^{18}O$), because the mass fractionation line is, in such a plot, a curve.

Provided that chemically or physically produced fractionation remains mass dependent, meteoritic nuclear components may be distinguished in a multi-isotopic (≥ 3) element by a fractionation pattern that is mass independent (see, e.g., Hulston and Thode 1965; Clayton et al. 1973). This has been a fundamental assumption widely employed in cosmochemistry. It is now known that this assumption is incorrect, as noted earlier, since the experimental demonstration made by Thiemens and Heidenreich (1983) and Heidenreich and Thiemens (1983) that it is possible to produce mass-independent isotope fractionations chemically. Those experiments, aside from requiring that a fundamental assumption be abandoned, showed that, for O, variations similar to the Allende "mixing" line could be produced by a simple chemical process. The experiments also suggest that the field of non-Boltzmann isotope effects may be of relevance to meteoritics and solar-system history, as origi-

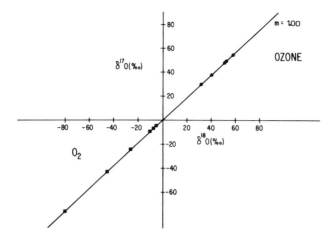

Fig. 12.1.2. The isotopic results of O_3 formation by the dissociation of O_2 at $-196°$ C. The starting composition of the O_2 is at $\delta^{18}O = \delta^{17}O = 0‰$. Chemical speciation of the two reservoirs, O_2 and O_3, is shown.

nally proposed by Arrhenius (cf. Arrhenius 1976; Arrhenius et al. 1978). The remainder of this section discusses theories and experiments involving such effects, which may have been potential fractionation processes in the early solar system.

The first experiments of Thiemens and Heidenreich employed a Tesla discharge in pure O_2 maintained at liquid nitrogen temperature. Under these conditions, O_2 molecules are dissociated by electron impact into O atoms with subsequent ozone (O_3) formation and condensation on the chilled glass walls. The O_3 and O_2 are cryogenically separated and analyzed mass spectrometrically for $\delta^{17}O$ and $\delta^{18}O$. The results are shown in Fig. 12.1.2. Ozone is enriched in ^{17}O and ^{18}O and the residual O_2 comparably depleted, with a line of slope $\cong 1.0$ defined by varying extents of reaction. It was first thought that optical self-shielding of the major isotopic species $^{16}O^{16}O$, with respect to $^{16}O^{18}O$ (abundance = 0.004081) and $^{16}O^{17}O$ (0.0007418), produces the effect (Thiemens and Heidenreich 1983; Thiemens et al. 1983a). Selective filtering of the $^{16}O_2$ lines within the Schumann-Runge bands, which are distinct from $^{16}O^{17}O$ and $^{16}O^{18}O$ lines, would result in preferential dissociation of the less abundant heteronuclear species and produce isotopically heavy O_3. In fact, large variations have been observed in the abundance ratios of CO isotopes in the outermost regions of molecular clouds, apparently due to isotope-selective photolysis of CO (Langer 1977). Since the rate of such reactions proceeds as a function of abundance rather than mass, the O_3 produced in laboratory experiments would produce a slope = 1 in a plot of $\delta^{17}O$ vs $\delta^{18}O$ as observed. Although this mechanism is, in fact, possible, kinetic limitations and isotopic exchange severely limit its effectiveness, as discussed by Kaye and Strobel (1983) and Navon and Wasserburg (1985).

The step actually responsible for the source of the experimental fractionation was demonstrated by Heidenreich and Thiemens (1985a) using CO_2 as starting material. In the dissociation of CO_2, the eventual product O_3 possessed the signature of two distinct isotopic-fractionation processes: the first mass-dependent (presumably, though not certainly, isotopic exchange between CO_2 and O) followed by a secondary mass-independent ($\delta^{17}O \cong 0.97\, \delta^{18}O$) step. Although the source of the mass-dependent step is not totally clear, the significant point is that the mass-independent fractionation does not occur in the dissociation process. A consistent mechanism has now been proposed and tested (Heidenreich and Thiemens 1985b,1986a), based on molecular symmetry. For a polyatomic molecule such as O_3, it is known that, due to its vibrational degrees of freedom, a metastable species such as O_3^* may temporarily exist with an energy in excess of its dissociation energy (cf. Herzberg 1966). These energy levels are collectively referred to as the diffuse-band system. In an atom-molecule collision (inverse predissociation), the process of producing the metastable species is rather intricate, encompassing different vibrational-rotational motions which ultimately determine the lifetime of O_3^* within the diffuse bands.

If an atom and molecule collide with energy corresponding to a level or levels within the diffuse band, there is a finite probability that the molecule may stabilize and become trapped in its potential well before it completes a vibration and redissociates. In fact, most such collisions are unsuccessful or "nonsticky." The probability of stabilization is in large part determined by the lifetime in the diffuse bands, $viz.$ the longer the metastable species lifetimes τ, the greater the probability of stabilization. The lifetime may, in turn, be a function of molecular symmetry. Ozone ($^{16}O^{16}O^{16}O$ or $^{48}O_3$) has C_{2v} symmetry and possesses half the complement of rotational lines of heteronuclear (C_s-symmetric) ozone ($^{16}O^{16}O^{17}O$, $^{16}O^{16}O^{18}O$) for rotation about its 2-fold axis (Herzberg 1950,1966). The result is that the isotopic species with C_s symmetry, due to the enhanced number of levels within the diffuse bands, have an enhanced τ, relative to that of C_{2v}-symmetric O_3 ($^{48}O_3$) and, hence, a greater probability of stabilization. The ultimate rate of O_3 formation is, in part, therefore, a function of molecular symmetry rather than of mass, and $^{49}O_3$ and $^{50}O_3$ are formed at an equal rate, greater than that of $^{48}O_3$, resulting in production of heavy O_3 with $\delta^{17}O = \delta^{18}O$, as observed. This mechanism is independent of the dissociation mechanism and, in fact, is only a feature of a gas-phase combination. It has subsequently been shown that the same effect arises in O_2 photodissociation by ultraviolet light (Thiemens and Jackson 1985,1986,1987), by 13.56 MHz electrodischarge (Thiemens et al. 1983b) and by a microwave electron impact (Bains and Thiemens 1986,1987), confirming the lack of dissociation specificity. The subject of chemically produced mass-independent isotopic fractionations is quite new, and a great deal of theoretical and experimental work needs to be done. This particular class of isotope effect is especially interesting, since it has been shown that the observed isotopic fractionations may be used as a tracer of physical-quantum chemical processes such as inverse predissociation.

With the discovery of enormous isotopic variations in meteorites for elements such as H, C and N, the possible involvement of nonequilibrium fractionation processes must be considered. Unfortunately, all of these elements suffer from the restriction that they each possess two stable isotopes, and only the magnitude and sign of fractionation may be used as an indicator. For H, ion-molecule reactions at the low temperatures characteristic of dense interstellar clouds may be relevant processes (Geiss and Reeves 1981; Robert and Epstein 1982; Kerridge 1983). The subject of isotope effects in ion-molecule reactions is discussed further in Chapter 13.2. In the context of this chapter, it must be emphasized that there are essentially no relevant measurements of these fractionations, and work in this area is needed. For D/H fractionations, the process of low-temperature quantum tunneling, which produces inverse isotope effects, can be important (Wolfsberg 1982), but the temperature dependence is inadequately characterized, especially at low temperatures where tunneling may become dominant (Weston 1975). In addition to the need for relevant measurements, the importance of low-temperature tunneling effects cannot be adequately assessed until more refined wave-mechanical calculations of the relevant processes and energy surfaces are done.

Nitrogen is another example of a 2-isotope element which exhibits extraordinarily large isotopic variations and for which it is not clear if the magnitude of the effects derives from incomplete mixing of nucleogenetically distinct components or from some nonthermally activated process (Prombo and Clayton 1985; see also Chapter 13.1). The range of $\delta^{15}N$ values observed in meteorites is from $-330‰$ to $+1033‰$ (see Chapter 13.1). Quite large enrichments (up to 423‰) of ^{15}N have been experimentally produced in NO formed by R.F. excitation at 77 K (Arrhenius et al. 1978; Manuccia and Clark 1976) and it is desirable to investigate further mechanisms of isotopic fractionation possibly appropriate to the environment of the early solar system.

If a gas molecule is vibrationally stimulated, it must eventually transfer this energy. If the gas is at room temperature and/or high pressure (e.g., 0.1 atm), the dominant transfer will be vibrational→translational ($V \rightarrow T$). If, however, the translational temperature is low and the pressure is likewise low ($\lesssim 10^{-3}$ atm), then the dominant energy-transfer mechanism is no longer $V \rightarrow T$, but becomes $V \rightarrow V$ and/or $V \rightarrow R$ (vibration→rotation). There are no strict temperature or pressure boundaries. The important criterion is rather the ratio of vibrational/translational energy or $nh\nu/kT$. As vibrational energy is increasingly stored in molecules, the energy eventually reaches the chemical activation threshold, and reaction occurs. It is well known that in the vibrational energy-transfer process the heavy isotopic species are preferentially excited, because energy may flow from the light to heavy species, but not the reverse. A complete treatment of the vibrational relaxation process in anharmonic oscillators is given by Treanor et al. (1968). The first suggestion that the $V \rightarrow V$ vibrational ladder-climbing process may lead to heavy-isotope enrichment was given by Belenov et al. (1973). Once the vibrational excitation is sufficiently high, the energy is directed into bond breakage/formation and chemical reaction may occur, such as

$$CO + CO^* \rightarrow CO_2 + C \tag{8}$$

where CO^* is the vibrationally excited species. Bergman et al. (1983) have observed $^{13}C^{16}O^*/^{12}C^{16}O^*$ enrichment of greater than 4000‰ at a rotation/translation temperature of 100°C. This heavy-isotope enrichment is subsequently passed on to the product CO_2. Rich and Bergman (1979) have observed that C_2, formed by gas phase $C + C$ reaction following the $CO + CO^* \rightarrow CO_2 + C$ reaction, possesses ^{13}C enrichments on the order of several hundred percent at room temperature. More recently, Thiemens and Meagher (1987) have shown that a mass-independent O-isotopic fractionation occurs in the $CO + CO^*$ reaction in an electrical discharge at $-196°C$. The mechanism for the fractionation is at present unknown, although under the experimental conditions it is likely to involve vibration→vibration transfer. The results are particularly interesting, since CO is a major O-bearing species in the interstellar environment and presumably in the early solar system.

In conclusion, non-Boltzmann isotope effects, particularly those which produce the $\delta^{17}O = \delta^{18}O$ fractionation, may have produced some of the observed meteoritic isotopic distributions, although at the present time this is not certain. Future experimental documentation of such effects and their parameterization will doubtless clarify their role in the early solar system.

12.1.3. ISOTOPIC COMPOSITION OF BULK METEORITES

The first use of meteoritic stable-isotope variations as a means to distinguish between the processes of nucleosynthesis and chemical or physical fractionation was made by Hulston and Thode (1965). Their study was based on the premise that a distinction is possible, based on the assumption that all physical and chemical processes produce mass-dependent fractionations, as discussed earlier. In their case, for analysis of S the relations

$$\delta^{34}S = 1.94 \, \delta^{33}S \tag{9}$$

$$\delta^{36}S = 1.89 \, \delta^{34}S \tag{10}$$

applied, and any departure would represent the involvement of a nuclear process, either nucleosynthetic or spallogenic. This premise formed a cornerstone for meteoritics and is still widely used, although, as discussed in section 12.1.2, the assumption that all nonmass-dependent variations must reflect nuclear processes is now known to be incorrect. The study of Hulston and Thode (1965) indicated that there was very little isotopic variability from one meteorite to the next, and that there was no evidence of nonmass-dependent isotopic variations attributable to nucleosynthetic processes. It is, however, likely that any original nucleosynthetic S-isotopic signature which may have been present was subsequently lost by isotopic exchange and chemical reaction. This is indeed very likely because S is both moderately volatile and possesses several available redox states, from negatively charged sulfides to positively charged sulfates.

The first non-noble-gas observation which clearly demonstrated that major isotopic inhomogeneities existed in the early solar system was made by Clayton et al. (1973). As detailed in the previous section, it had formerly been thought that all physical and chemical processes must produce mass-dependent isotopic fractionations, thus yielding a straight line with slope of approximately 0.52, the so-called terrestrial fractionation line, in a plot of $\delta^{17}O$ vs $\delta^{18}O$ (Fig. 12.1.1). Clayton et al. (1973) observed that the O-isotopic composition of anhydrous, high-temperature minerals in carbonaceous meteorites defined, in the same coordinate system, a line of slope ~ 1 (the line segment labeled Allende CAI in Fig. 12.1.1). Furthermore, it was argued that in stellar He burning or explosive C burning (see Chapter 14.2), the ^{17}O and ^{18}O abundances are considerably diminished and that of ^{16}O enhanced, suggesting that the meteoritic samples might contain an excess of material from such an astrophysical environment, compared with "normal" solar-system material. It was further suggested that an environment which produces ^{16}O should also produce large effects in Mg and Si via production of ^{24}Mg and ^{28}Si. Since that discovery, in fact, a major focus of cosmochemical research has been to determine if isotopic anomalies correlated with those in O do exist.

A striking aspect of the O-isotopic anomaly is the magnitude of the effect. Oxygen is the most abundant rock-forming element and enrichment in ^{16}O (or depletion in ^{17}O, ^{18}O) of up to 4% has now been observed. No other element exhibits such a variation in terms of the absolute number of atoms involved. Hence, the process which produced the anomaly was both major and possibly specific to O. A recent comprehensive review of meteoritic O-isotopic systematics is given by Clayton et al. (1985). As pointed out in that review and previously by Wood (1981), due to its chemical properties, O can simultaneously exist in both solid and gaseous reservoirs over a wide temperature range, since, at the temperatures at which all of the major metallic elements are condensed or combined, 80% of the O is in the gas phase, as either CO or H_2O. It is increasingly clear that different nebular isotopic reservoirs must have existed, since there are distinct bulk meteoritic O-isotopic differences, as seen in Fig. 12.1.3 (Clayton et al. 1976; Clayton et al. 1985;

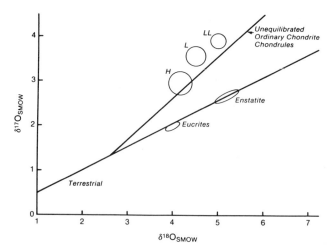

Fig. 12.1.3. The O-isotopic composition of certain bulk meteorite groups, compared with the terrestrial fractionation line. The best-fit line for separated chondrules from UOCs is also shown. All data are from the laboratory of R. N. Clayton. See also Fig. 1.1.11.

see also Fig. 1.1.11). The significance of the whole-rock isotopic differences is also discussed in Chapter 1.1. There are two aspects of meteoritic O-isotopic systematics that have major consequences. First, the observed isotopic distributions suggest that nebular heterogeneities existed on scales ranging from sub-mm to, possibly, the dimensions of the nebula itself. Secondly, there is the question as to where, when and how the isotopic anomalies were originally produced. The internal isotopic compositions, particularly for O, provide significant insight into the first aspect.

The distinct O-isotopic compositions of the different meteorite classes, as seen, for example, in the compositions of the H, L and LL chondrites, suggest separate origins, possibly in time and/or space. The enstatite chondrites' isotopic composition, shown in Fig. 12.1.3, lies (within experimental error) along the terrestrial fractionation line, with whole-rock ^{18}O values up to 1‰ higher than terrestrial ultramafic rocks (Clayton et al. 1976; Clayton et al. 1984). Internally, the EH group becomes enriched in the heavy isotopes along the terrestrial fractionation line with increasing evidence for chemical equilibration. Aubrites (enstatite achondrites) constitute the only other meteoritic group which lies along the terrestrial fractionation line, with a bulk isotopic composition close to that of EH chondrites. Clayton et al. (1984) therefore concluded that the aubrites may be related to either the EH or EL groups, or possibly both (but see Brett and Keil 1987). The colinearity of EH and EL chondrites, Moon and Earth suggests derivation from a common O reservoir, but this is presently an unresolved, though intriguing, question.

There are two other isotopically distinct igneous meteorite groups: the AMP group (eucrites, howardites, diogenites, mesosiderites and pallasites) and the SNC group (shergottites, nakhlites and chassignites (Clayton and Mayeda 1983). The eucrites define a fractionation line which is offset from the terrestrial by an ^{16}O excess of 0.4‰, while the SNC group defines a fractionation line with an ^{16}O deficit of ~ 0.6‰ relative to terrestrial. These observations suggest that the individual SNC and AMP groups are derived from separate and isotopically distinct reservoirs (see Fig. 7.8.9). These individual reservoirs might, in fact, represent two different parent bodies (Clayton and Mayeda 1983), although at present there is insufficient information regarding the spatial and temporal variations of O-isotope abundances in the early solar system to attribute a meteorite group to a specific parent body (Clayton and Mayeda 1986).

The O-containing species within iron meteorites, mainly comprising silicate inclusions, fall into several isotopically distinct groups. The classes of iron meteorites appear to be isotopically related to various stony-meteorite classes, again suggestive of derivation from a common reservoir. Group IAB and IIICD irons appear to be related to the winonaites in that they lie along a common fractionation line (δ^{17}O $= 0.52$ δ^{18}O) offset from the terrestrial line by an ^{16}O excess of 0.47‰ (Clayton et al. 1983a). Group IIE inclusions also define a fractionation line with a slope of 0.52 which is offset from terrestrial by an ^{16}O deficit of 0.61‰, as observed also for the mean value for equilibrated H-group ordinary chondrites (Fig. 12.1.3; see also Fig. 1.1.11).

The ~ 4‰ range in δ^{18}O for pyroxenes from the IIE irons and H chondrites is consistent with derivation from a common reservoir for these silicate components (Clayton et al. 1983a). The silicates in Group IVA irons similarly form a mass-fractionation line with an ^{16}O deficit of 1.2‰, which lies within the range of either the L or LL chondrites shown in Fig. 12.1.3, indicating yet another cogenetic relation between the L, LL and IVA groups. Further chemical, petrographic and isotopic analysis is needed to define the nature of these apparent relations.

Another bulk reservoir appears to be required by the ureilites, with bulk composition for δ^{18}O and δ^{17}O equal to 7.7 and 3.05‰, respectively (Clayton et al. 1976). The hydrous silicate matrix of the CM2 meteorites Murchison and Cold Bokkeveld define an apparent mass-fractionation line with an ^{16}O enrichment, relative to terrestrial, of 4‰ and ranging in δ^{18}O from 6.59 to 12.75‰ (Clayton et al. 1976,1977). A 2-stage model for the genesis of the O-isotopic composition of the major mineral phases in Murchison and other carbonaceous chondrites has been proposed (Clayton and Mayeda 1984). The question as to how these reservoirs may have been produced originally will be discussed in Sec. 12.1.6. The internal isotopic fractionation line exhibited by the matrix minerals is thought to reflect the low-temperature alteration ($\leq 20°$ C) which made the phyllosilicates and carbonates (Clayton and Mayeda 1984; see also Chapter 3.4). Interestingly, within the context of the aqueous-alteration scenario, the observed isotopic composition also places constraints on the volume fraction of water involved (>44%). The particular reservoir at $\delta^{18}O \cong \delta^{17}O \cong -40‰$, defined by the anhydrous minerals in carbonaceous chondrites, will be discussed in the context of interreservoir relations in the next section.

12.1.4. ISOTOPIC RESERVOIRS AND EXCHANGE

Examination of the internal meteoritic O-isotopic variations reveals a great deal regarding not only the isotopic composition of the different reservoirs, but their original physical state (gas or solid) and interrelationships. These issues are discussed also in Chapter 10.3 for Ca, Al-rich inclusions and in Chapter 9.1 for chondrules. As shown in Fig. 12.1.1, a unique O reservoir is apparent as defined by the point for spinel in CAIs at $\delta^{18}O \cong -40‰$, $\delta^{17}O \cong -42‰$ (cf. Clayton 1978). Detailed analyses of separated phases in CAIs from Allende indicate large ^{16}O enrichments relative to terrestrial in spinel, pyroxene and occasionally olivine, while smaller enrichments are observed in melilite, feldspathoids and grossular (Clayton et al. 1977). The observation that the most refractory phases possess the greatest ^{16}O enrichment suggests that the spinel point approximates the isotopic composition of a nebular reservoir, presumably a solid. This solid, labeled 1 in Fig. 12.1.4 apparently then underwent isotopic exchange with a nebular gas having an isotopic composition at approximately point 2' in Fig. 12.1.4. Two observa-

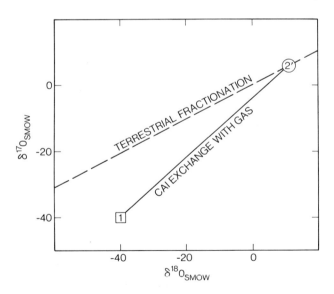

Fig. 12.1.4. Schematic representation of a solid CAI (square 1) undergoing isotopic exchange with a gas (circle 2').

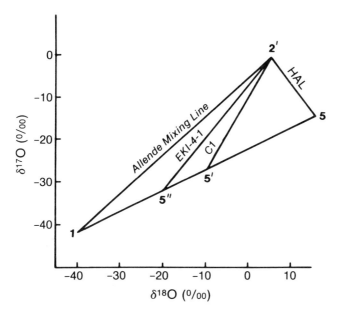

Fig. 12.1.5. Mixing lines defined by minerals separated from 3 Allende FUN inclusions, compared with that for "normal" CAIs. All inclusions appear to have exchanged with the same gas reservoir (2′). Pre-exchange compositions of FUN inclusions (5, 5′ and 5″) were apparently produced by extreme mass-fractionation of "normal" CAI material (1).

tions support this suggestion. First, the refractory spinel crystals are enveloped by melilite with $\delta^{18}O$ of $\sim +2‰$, as shown in Fig. 12.1.1, suggesting that this isotopically heavy component was a relatively late addition. Secondly, those CAIs known as FUN inclusions (see Chapter 14.3) define mixing lines between a composition close to terrestrial and a suite of components apparently derived by mass fractionation from the spinel point (1 in Figs. 12.1.4 and 12.1.5). The data suggest the following history for FUN inclusions. They originally crystallized with an isotopic composition at point 1 of Fig. 12.1.5. A secondary event occurred which resulted in mass fractionation along line segment 1-5. It has been suggested that this step may have occurred in the absence of an external reservoir, since the magnitude of the segment appears to be too large to have been simple evaporation. The fractionation for O is nearly 25‰ per mass unit. It is not clear, nor agreed upon, to what this intermediate fractionation step is due. The final step, represented by line segments 5-2′, 5′-2′ and 5″-2′, depicts exchange between solid reservoirs at 5, 5′ and 5″ with a gas reservoir near 2′, perhaps the same as terrestrial. The mineralogy is consistent with this, since the least labile hibonite component lies near the 1-5 line, while the more exchangeable rim material lies near 2′. It should be noted that the extensive effect of evaporation seen in O is also observed in Si, Mg and Ca (Clayton et al. 1985; Niederer and Papanastassiou 1984).

As discussed by Clayton et al. (1985), this extent of fractionation is well beyond that arising from isotopic exchange and requires multistage Rayleigh distillation to produce the observed range in δ^{30}Si (~ -4.00 to $+5.00$‰). Employing the Rayleigh equation (Eq. 5), it may be found that condensation from a Si gas (e.g., SiO) produces a heavy solid enriched in ^{30}Si by only about 2‰ for the first 5% of the gas condensed, which corresponds to the fraction of the cosmic abundance of Si stored in these inclusions. As discussed by Clayton et al. (1985), this is insufficient to produce the observed fractionations. Rayleigh evaporation results in ^{30}Si enrichment in the remaining few percent of an evaporated liquid (Molini-Velsko 1983). It appears then that O, Si, Mg and Ca exhibit in their mass-dependent fractionations, signs of a multistep history of evaporation. This conclusion is based on the requirement of a sufficiently large fractionation process, since isotopic exchange, as dictated by the temperature dependence of the ratio of the reduced partition functions for the relevant species, is insufficient to account for the observed isotopic fractionations in CAIs.

As discussed in the previous section, the hydrous matrix minerals of CM and CI chondrites possess O-isotopic compositions which require at least two different nebular reservoirs that interacted with one another (Clayton et al. 1976; Clayton and Mayeda 1984). In the case of Murchison, there may have been two stages of fluid-solid isotopic exchange, the first of which occurred at high temperature and generated an anhydrous-silicate line with unit slope like that shown in Fig. 12.1.1. A second episode of low-temperature aqueous alteration then produced a mass-fractionation line offset from the terrestrial by an ^{16}O-enrichment of a few per mil. For the CI hydrous matrix minerals, a similar history is suggested as for the CM group, except that the aqueous-alteration temperature was apparently 100 to 150° C, as calculated for equilibrium isotopic exchange between water and carbonate (Clayton and Mayeda 1984).

More recently, evidence from the isotopic composition of chondrules from ordinary, enstatite and carbonaceous meteorite groups has further defined the nebular reservoirs (Gooding et al. 1980,1983; Clayton et al. 1983b,1986; Clayton and Mayeda 1985). Large isotopic variations exist within chondrules separated from the unequilibrated ordinary chondrites. This variability may reflect incomplete isotopic exchange between a solid (or liquid) and a gas (Clayton et al. 1983b). In Fig. 12.1.3, the best-fit line for separated chondrules from ordinary chondrites is shown. (For a detailed description of the chondrule isotopic data, see Chapter 9.1.) The UOC data require mixing between at least two O-isotopic reservoirs. The exchange process is shown schematically in Fig. 12.1.6. Exchange is thought to have occurred between solids or liquids at point 3 and a gas near point 2. There is considerable scatter, which may indicate the presence of more than 2 isotopic

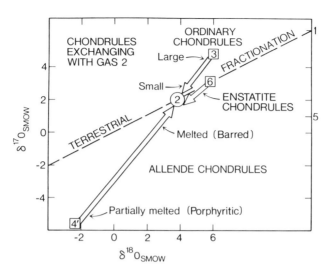

Fig. 12.1.6. Isotopic exchange between nebular gas (2) and reservoirs defined by chondrules from Allende (4′), ordinary chondrites (3) and enstatite chondrites (6). See also Fig. 9.1.9.

components or the superposition of isotopic fractionation on a 2-component mixing line (Clayton et al. 1986). There may also be intragroup isotopic variability due to fractionations during chondrule formation. In addition, as indicated on Fig. 12.1.6, there is a size-fractionation effect within the chondrules. The smaller (100–250 μm) chondrules are isotopically lighter by several per mil than larger (290–820 μm) chondrules within the same meteorite (Clayton et al. 1985). This observation is also consistent with gas-solid (liquid) exchange (with the gas ^{16}O-rich), whereby the smaller particles underwent more complete isotopic exchange. Clayton et al. (1986) suggest that heating times of minutes to hours are needed, in order to achieve measurable exchange.

The chondrules in enstatite chondrites appear to define another reservoir, point 6 in Fig. 12.1.6, which lies within ~0.1‰ of the terrestrial fractionation line (Clayton et al. 1984; Clayton and Mayeda 1985). Exchange during chondrule formation may also have caused convergence of their isotopic composition towards point 2 in Fig. 12.1.6. The heaviest chondrule has a δ^{18}O value of +5.8‰; thus the total variation is only half that observed in UOC (Clayton and Mayeda 1985). The best-fit line for chondrules in enstatite chondrites has a slope of 0.7 ± 0.1; the initial isotopic composition of the chondrule reservoir was slightly deficient in ^{16}O with respect to the terrestrial line, and had a δ^{18}O of ~+6‰ prior to exchange with the gas at point 2 of Fig. 12.1.6.

Separated chondrules from the Allende meteorite also suggest isotopic exchange, in this case between a solid at δ^{18}O ≈ −3‰ and δ^{17}O ≈ −7‰, and the same gas reservoir represented by point 2, which exchanged with the reservoirs of the UOC and enstatite-chondrite chondrules (Clayton et al. 1983b). As shown in Figs. 9.1.9 and 12.1.6 the barred chondrules, which have undergone more severe heating, appear to have undergone more extensive isotopic exchange than the porphyritic chondrules, which were not so thoroughly melted. Note also that the data shown in Fig. 12.1.6 simultaneously display four distinct reservoirs. In total, there are at least three initial reservoirs of solid, or liquid, chondrules (enstatite, UOC and CV), two apparently gaseous reservoirs which are defined by isotope exchange, in Figs. 12.1.4 and 12.1.6, as well as the spinel point, which may itself reflect exchange with a gaseous reservoir near point 1 (Clayton et al. 1985). Some of these apparently distinct reservoirs may be related to each other by mass fractionation and/or mixing.

12.1.5. NEBULAR RESERVOIRS OF ELEMENTS OTHER THAN OXYGEN

While Si possesses a modest range (5–10‰) of fractionation in many Allende inclusions (and up to 30‰ in the extreme), there are very few mass-independent effects, and essentially none observable in the bulk meteorite (cf. Molini-Velsko 1983; Clayton et al. 1985; Molini-Velsko et al. 1986). Bulk variations are quite small between the meteorite classes (1‰ per mass unit), and the average composition for each meteorite class is identical (Molini-Velsko et al. 1986). Therefore, unlike the case for O no differentiation of nebular reservoirs can be made. The close similarity of all groups suggests closed-system behavior, clearly decoupled from that of O.

Perhaps the only other element which exhibits anomalous and large isotopic effects in a variety of meteoritic classes and mineral phases is Ti. The first hint of isotopically anomalous behavior was indicated by Heydegger et al. (1979). From a statistical treatment of their combined data, a ^{50}Ti excess was suggested. However, at the precision of the measurements ($\sim \pm 1$‰), the suggested effect was within error limits. Clearly resolved excesses in ^{50}Ti were later reported at levels well outside measurement error (Niederer et al. 1980). An important feature of these measurements, aside from the requirement of multiple nucleosynthetic sources, was that the ^{50}Ti anomalies appeared to be relatively widespread compared to isotopic anomalies in other elements such as Sm (see Chapter 14.3). Further analysis of Allende (Niemeyer and Lugmair 1981) and Leoville (Niederer et al. 1981) demonstrated that the ^{50}Ti anomalies are widespread among meteorites. It is generally agreed that the ^{50}Ti excesses are derived from a nuclear equilibrium process such as neutron-rich Si burning (Niederer et al. 1980) or neutron-rich equilibrium (Niemeyer and Lugmair 1981; Papanastassiou 1986). More recent work has shown that, although ^{50}Ti excesses are ubiquitous, they do not appear to correlate with the documented ^{16}O excesses (Niemeyer and Lugmair 1984).

Using a chi-square statistical analysis of the existing meteoritic data, these authors concluded that a minimum of four distinct isotopic reservoirs of Ti are required to account for the data. The heterogeneities were interpreted as reflecting a diversity of nucleosynthetic sources for which evidence has been preserved, due to the highly refractory chemistry of Ti. The nucleosynthesis, chemistry and possible correlation with other isotopic anomalies, particularly in Ca, are discussed in Chapters 14.2 and 14.3. Regardless of their source and possible relation to other elements, it is apparent at the present time that Ti appears to be the only element, besides O, that possesses clearly defined, nebular reservoirs, of which there are at least four.

12.1.6. PRODUCTION AND SURVIVAL OF ISOTOPIC HETEROGENEITIES

Lurking behind all of the previously discussed data and concepts is the major issue of how and where were these individual isotopic reservoirs, especially of O, produced. The origin of the excess ^{16}O was first suggested by Clayton et al. (1973) to be from a nuclear process, such as stellar He burning or explosive C burning. During such processes the abundances of ^{17}O and ^{18}O are considerably diminished and that of ^{16}O enhanced, hence the presence of relict grains from such an astrophysical environment could generate the observed isotopic anomaly. It was further suggested that an environment that produced large effects in O should also have produced effects in Mg and Si, manifested by excesses in ^{24}Mg and ^{28}Si. As discussed previously, Si exhibits quite small isotopic variations in bulk and, in inclusions, fractionations up to ~ 30‰ (Molini-Velsko et al. 1986). Only a few inclusions exhibit mass-independent fractionations, and those are < 0.5‰ in size (Molini-Velsko 1983). These observations place restrictions on the nuclear model, and most certainly on the chemistry, if the O-reservoir distinction is to be maintained in bulk silicates and Si is to be homogenized.

Clayton and Mayeda (1977) and Wasserburg et al. (1977) were the first to attempt to correlate the ^{16}O anomaly to isotopic effects in another element, Mg. Those studies demonstrated that the extent of O-isotopic mass fractionation was quantitatively related to a similar mass-fractionation in Mg isotopes. However, the process(es) which originally produced the O- and Mg-isotopic compositions could not be defined. Clayton and Mayeda (1977) suggested that this particular set of observations did not require that the ^{16}O enrichment was carried by presolar grains. Alternatively, an ^{16}O-rich gaseous region in the nebula with a condensation time that was short compared to nebular mixing is consistent with the data. They further suggested that rapid injection of ^{16}O, such as from a nearby supernova, might produce the ^{16}O reservoir. Wood (1981) has also suggested a similar model whereby O-isotopic features were derived from incomplete exchange of O isotopes between the primordial gas and dust components of the presolar system. Interstellar ^{16}O-enriched grains were transported to the nebula where localized gas/dust ratio modification and exchange produced the Allende mixing line (Fig. 12.1.1) and the general meteoritic features observed in Figs. 12.1.3 and 12.1.6.

Recently, Hinton et al. (1987) have inferred, based on ion-microprobe measurements of Ti isotopes in refractory inclusions, predominantly hibonite, that at least 7% of solar system ^{50}Ti was added shortly before formation of the solar system, and that mixing of $\sim 5\%$ supernova ejecta into the presolar cloud can explain the variations observed in ^{16}O, initial ^{26}Al/^{27}Al ratio (see Chapter 15.1) and ^{50}Ti. They noted that the lack of correlation of ^{50}Ti effects with chemistry or isotopic mass fractionation is suggestive that the presolar cloud was inhomogeneous with respect to Ti, as noted previously. While the magnitude of the inferred proportion of ^{50}Ti is approximately the same as for ^{16}O, the absolute number of atoms is quite different. Future resolution of this question of origin, as discussed by Prombo and Lugmair (1986) and Zinner et al. (1986), will be achieved by further correlative isotopic studies.

A second type of nucleosynthetic model, which differs significantly from those of Clayton and Mayeda (1977), Wasserburg et al. (1977), Wood (1981) and Hinton et al. (1987), has been proposed by D. D. Clayton. Clayton has argued that the ^{16}O anomaly (as well as "anomalies" in ^{129}Xe, ^{22}Ne, ^{26}Mg and ^{50}Ti) represents cosmic chemical memories originating as condensates in extrasolar-system environments such as in the expanding shells of a supernova (SUNOCONS), and subsequently transported to the presolar environment (cf. D. D. Clayton, 1975a,b, 1979). D. Clayton emphasizes that most ^{16}O-rich meteoritic samples are also Al-rich. (For a description, see D. D. Clayton [1986].) The Al-^{16}O mixture in adiabatic stellar shell outflows may condense immediately following the burning of C and Ne. The result would be an ^{16}O-Al-rich, thermodynamically stable dust grain (such as Al_2O_3), which may be transported to the presolar nebula where its exotic chemical and isotopic memory would be retained.

The notion that the ^{16}O anomalies have a nuclear origin stems from the common assumption (see, e.g., R. N. Clayton et al. 1973) that the line with unit slope, defined by Allende inclusions (Fig. 12.1.1) cannot be chemically produced. Now that it is known that the $\delta^{17}O \cong \delta^{18}O$ fractionation may be produced by a chemical process (Thiemens and Heidenreich 1983), the possibility that some of the meteoritic O-isotopic components were produced by nonnuclear processes must be considered.

Regarding its application to meteoritic O-isotopic anomalies, there are several important features of the symmetry effect proposed by Thiemens and coworkers. First, it is not restricted to O_3; reactions such as

$$^{18}O + Si^{16}O \rightarrow Si^{16}O^{18}O \tag{11}$$

$$^{18}O + Ti^{16}O \rightarrow Ti^{16}O^{18}O \tag{12}$$

would produce the effect and not be subject to the constraint of isotopic exchange discussed by Navon and Wasserburg (1985). Furthermore, the effect is not restricted to the formation of triatomic species. In fact, as discussed by Herzberg (1950,1966), the greater the number of atoms, the higher the recombination yield. Second, the effect is not limited by a particular dissociation process, and the need for photo- or electro-dissociation in the early solar system is eliminated. Third, the effect produces both positive and negative isotopic reservoirs and, in theory, only *one* nebular reservoir would be required to account for the meteoritic O observations. In addition, for reactions such as O + SiO, the effect would be observable in O but not in Si, possibly accounting for the uniqueness of O. At the present time, there is an obvious need for more work: further clarification of the mechanism, and demonstration of the effect in other reactions such as those suggested. Such experimental work is needed before a thorough model for the early solar system may be developed. This holds true for either a nuclear or chemical model.

In regard to the observation of the effect in other elements, it is important to note that the effect must be a function of abundance, as well as symmetry (Heidenreich and Thiemens 1986b). For example, as the abundance of ^{17}O ^{18}O increases to $\sim 1\%$, contributions from species such as $^{17}O^{16}O^{17}O$, $^{18}O^{16}O^{18}O$ and $^{18}O^{18}O^{18}O$ become significant, and the ratio of the product symmetries (C_s/C_{2v}) is altered. Under these conditions, the $\delta^{17}O = \delta^{18}O$ fractionation would no longer be observed. Experiments clearly demonstrating this are important not only for mechanism confirmation but for predictions for other elements, such as S, where reactions such as S + CS $\rightarrow$ SCS are potentially important.

The recent observation by Thiemens and Meagher (1987) that mass-independent fractionations are produced in reactions of CO (one of the major O-bearing species in the astrophysical environment), demonstrates that chemical mass-independent isotopic fractionations are not restricted to O_2-O_3 chemistry. The mechanism responsible for this fractionation is unknown at the present time and clearly warrants further study. It must be concluded that the origin of the meteoritic O reservoirs, whether chemical or nuclear, is unknown.

12.1.7. SUMMARY AND CONCLUSIONS

A framework of isotope effects has been provided as an aid in interpretation of meteoritic isotope measurements. A growing body of data suggests that the study of mass-independent isotope effects is likely to be of future interest and possible application to studies of the early solar system. This represents a rather new field in cosmochemistry; it has only been a few years since the first direct experimental observation was made of a chemical mass-independent isotopic fractionation, although earlier theoretical considerations exist. Symmetry effects, photochemical isotopic self-shielding and low-temperature quantum-tunneling phenomena are areas of special interest to meteoritics and interstellar chemistry, but are fields where a paucity of relevant experiments exist.

The existing meteoritic isotopic data clearly show that the presolar nebula was inhomogeneous rather than well mixed and uniform, as was originally thought. Oxygen-isotopic measurements of the meteoritic classes show that there were several distinct isotopic reservoirs existing on a bulk meteoritic scale with apparent mixing between many of them. The sources of these reservoirs are not known. If those sources were nucleosynthetic, then the isotopic distinction must have been maintained spatially and probably temporally. The demonstration of a new isotope effect in O invalidates the original assumption that mass-independent isotopic fractionations must reflect nuclear processes. If a chemical process is responsible for the bulk meteoritic O anomalies, then it is possible that only one reservoir could have produced the observed distributions. This challenging question remains open at present. Further meteoritic isotopic measurements of the internal meteoritic components, such as chondrules and inclusions, will be of continued importance in determining the secondary mixing and reaction processes, as well as in placing constraints on the compositions of the original reservoirs.

The observation that Si-isotopic fractionations in bulk meteorites are quite small ($<4‰$) and have essentially no mass-independent signature, suggests that Si was well mixed. This remains puzzling since, besides O, Ti appears to require several distinct preserved nucleosynthetic reservoirs. Further correlative isotopic measurements with detailed textural and petrographic determinations will be of major importance in clarification of this puzzle.

In conclusion, perhaps the most exciting aspects of the studies of early solar-system history are the impact such work may have on a wide variety of fields, and that there exist an enormous number of fundamental studies to be done, both in meteoritics and chemical physics.

Acknowledgments. The author wishes to thank R. N. Clayton and I. Hutcheon for extremely thorough and helpful reviews. Thanks go also to C. Prombo for a review and many clarifying comments. Support from a grant from the National Aeronautics and Space Administration and a Camille and Henry Dreyfus Teacher-Scholar grant are acknowledged.

REFERENCES

Arrhenius, G. 1976. Chemical aspects of the formation of the solar system. In *The Origin of the Solar System,* ed. S. F. Dermott (New York: Wiley), pp. 521–581.

Arrhenius, G., Fitzgerald, R., Markus, S., and Simpson, C. 1978. Isotope fractionation under simulated space conditions. *Astrophys. Space Sci.* 55:285–297.

Bains, S. K., and Thiemens, M. H. 1986. Mass independent isotopic fractionation in the microwave region. *Lunar Planet. Sci.* XVII:20–21 (abstract).

Bains, S. K., and Thiemens, M. H. 1987. Mass independent isotopic fractionation in a microwave plasma. *J. Phys. Chem.* 91:4370–4374.

Belenov, E. M., Markin, E. P., Oraevskii, A. N., and Romanenko, V. I. 1973. Isotope separation by infrared radiation. *JETP Lett.* 18:116–117.
Bergman, R. C., Homicz, G. F., Rich, J. W., and Wolk, G. L. 1983. ^{13}C and ^{18}O isotope enrichment by vibrational energy exchange pumping of CO. *J. Chem. Phys.* 78:1281–1292.
Bigeleisen, J. 1965. Chemistry of isotopes. *Science* 147:463–471.
Bigeleisen, J., and Mayer, M. G. 1947. Calculation of equilibrium constants for isotopic exchange reactions. *J. Chem. Phys.* 15:261–267.
Brett, R., and Keil, K. 1987. Enstatite chondrites and enstatite achondrites (aubrites) were not derived from the same parent body. *Earth Planet. Sci. Lett.* 81:1–6.
Clayton, D. D. 1975a. Extinct radioactivities: Trapped residuals of presolar grains. *Astrophys. J.* 199:765–769.
Clayton, D. D. 1975b. ^{22}Na, Ne-E, extinct radioactive anomalies and unsupported ^{40}Ar. *Nature* 257:36–37.
Clayton, D. D. 1979. Supernovae and the origin of the solar system. *Space Sci. Rev.* 24:147–226.
Clayton, D. D. 1986. Isotopic anomalies and SUNOCON survival during galactic evolution. In *Cosmogonical Processes*, eds. W. D. Arnett, C. Hansen, J. Truran, and S. Tsuruta (Utrecht: VNU Science), pp. 101–122.
Clayton, R. N. 1978. Isotopic anomalies in the early solar system. *Ann. Rev. Nucl. Part. Sci.* 28:501–522.
Clayton, R. N. 1981. Isotopic thermometry. In *Thermodynamics of Minerals and Melts*, eds. R. C. Newton, A. Navrotsky, and B. J. Wood, vol. 1, *Advances in Physical Geochemistry* (New York: Springer-Verlag), pp. 85–109.
Clayton, R. N., and Mayeda, T. K. 1977. Correlated oxygen and magnesium isotope anomalies in Allende inclusions, I: Oxygen. *Geophys. Res. Lett.* 4:295–298.
Clayton, R. N., and Mayeda, T. K. 1983. Oxygen isotopes in eucrites, shergottites, nakhlites and chassignites. *Earth Planet. Sci. Lett.* 62:1–6.
Clayton, R. N., and Mayeda, T. K. 1984. The oxygen isotope record in Murchison and other carbonaceous chondrites. *Earth Planet. Sci. Lett.* 67:151–161.
Clayton, R. N., and Mayeda, T. K. 1986. Oxygen isotopes in Shergotty. *Geochim. Cosmochim. Acta* 50:979–982.
Clayton, R. N., and Mayeda, T. K. 1985. Oxygen isotopes in chondrules from enstatite chondrites: Possible identification of a major nebular reservoir. *Lunar Planet Sci.* XVI:142–143 (abstract).
Clayton, R. N., Grossman, L., and Mayeda, T. K. 1973. A component of primitive nuclear composition in carbonaceous meteorites. *Science* 182:485–488.
Clayton, R. N., Onuma, N., and Mayeda, T. K. 1976. A classification of meteorites based on oxygen isotopes. *Earth Planet. Sci. Lett.* 30:10–18.
Clayton, R. N., Onuma, N., Grossman, L., and Mayeda, T. K. 1977. Distribution of the presolar component in Allende and other carbonaceous meteorites. *Earth Planet. Sci. Lett.* 34:209–224.
Clayton, R. N., Mayeda, T. K., Olsen, E. J., and Prinz, M. 1983a. Oxygen isotope relationships in iron meteorites. *Earth Planet. Sci. Lett.* 65:229–232.
Clayton, R. N., Onuma, N., Ikeda, Y., Mayeda, T. K., Hutcheon, I. D., Olsen, E. J., and Molini-Velsko, C. 1983b. Oxygen isotopic compositions of chondrules in Allende and ordinary chondrites. In *Chondrules and Their Origins*, ed. E. A. King (Houston: Lunar and Planetary Inst.), pp. 37–43.
Clayton, R. N., Mayeda, T. K., and Rubin, A. E. 1984. Oxygen isotopic compositions of enstatite chondrites and aubrites. *Proc. Lunar Planet. Sci. Conf.* 15, *J. Geophys. Res. Suppl.* 89:C245–C249.
Clayton, R. N., Mayeda, T. K., and Molini-Velsko, C. A. 1985. Isotopic variations in solar system material: Evaporation and condensation of silicates. In *Protostars & Planets II*, eds. D. C. Black and M. S. Matthews (Tucson: Univ. of Arizona Press), pp. 755–771.
Clayton, R. N., Mayeda, T. K., and Goswami, J. N. 1986. Oxygen and silicon isotopic variations in chondrules. *Meteoritics* 21:346 (abstract).
Geiss, J., and Reeves, H. 1981. Deuterium in the solar system. *Astron. Astrophys.* 93:189–192.
Gooding, J. L., Mayeda, T. K., Clayton, R. N., Keil, K., Fukuoka, T., and Schmitt, R. A. 1980. Oxygen isotopic compositions of petrographically characterized chondrules from unequilibrated chondrites. *Meteoritics* 15:295 (abstract).

Gooding, J. L., Mayeda, T. K., Clayton, R. N., and Fukuoka, T. 1983. Oxygen isotopic heterogeneities, their petrological correlations and implications for melt origins of chondrules in unequilibrated ordinary chondrites. *Earth Planet. Sci. Lett.* 65:209–224.

Heidenreich, J. E., and Thiemens, M. H. 1983. A non-mass-dependent isotope effect in the production of ozone from molecular oxygen. *J. Chem. Phys.* 78:892–895.

Heidenreich, J. E., and Thiemens, M. H. 1985a. The non-mass-dependent oxygen isotope effect in the electrodissociation of carbon-dioxide: A step toward understanding NoMad chemistry. *Geochim. Cosmochim. Acta* 49:1303–1306.

Heidenreich, J. E., and Thiemens, M. H. 1985b. A model for the production of chemical non-mass-dependent oxygen isotope effect. *Lunar Planet. Sci.* XVI:335–336 (abstract).

Heidenreich, J. E., and Thiemens, M. H. 1986a. A non-mass-dependent oxygen isotope effect in the production of ozone from molecular oxygen: The role of molecular symmetry in isotope chemistry. *J. Chem. Phys.* 84:2129–2136.

Heidenreich, J. E., and Thiemens, M. H. 1986b. The effect of isotopic abundance of NoMaDic chemistry. *Lunar Planet. Sci.* XVII:329–330 (abstract).

Herzberg, G. 1950. *Molecular Spectra and Molecular Structure. I. Spectra of Diatomic Molecules* (New York: Van Nostrand Reinhold).

Herzberg, G. 1966. *Electronic Spectra of Polyatomic Molecules* (New York: Van Nostrand).

Heydegger, H. R., Foster, J. J., and Compston, W. 1979. Evidence of a new isotopic anomaly from titanium isotopic ratios in meteoritic materials. *Nature* 278:704–707.

Hinton, R. W., Davis, A. M., and Scatena-Wachel, D. E. 1987. Large negative ^{50}Ti anomalies in refractory inclusions from the Murchison carbonaceous chondrite: Evidence for incomplete mixing of neutron-rich supernova ejecta into the solar system. *Astrophys. J.* 313:420–428.

Hulston, J. R., and Thode, H. G. 1965. Variations in the ^{33}S, ^{34}S and ^{36}S contents of meteorites and their relation to chemical and nuclear effects. *J. Geophys. Res.* 70:3475–3484.

Kaye, J. A., and Strobel, D. F. 1983. Enhancement of heavy ozone in the earth's atmosphere? *J. Geophys. Res.* 88:8447–8452.

Kerridge, J. F. 1983. Isotopic composition of carbonaceous-chondrite kerogen: Evidence for an interstellar origin of organic matter in meteorites. *Earth Planet. Sci. Lett.* 64:186–200.

Langer, W. D. 1977. Isotopic abundance of CO in interstellar clouds. *Astrophys. J.* 212:L39–L42.

Manuccia, T. J., and Clark, M. D. 1976. Enrichment of N^{15} by chemical reactions in a flow discharge at 77° K. *Appl. Phys. Lett.* 28:372–374.

Matsuhisa, Y., Goldsmith, J. R., and Clayton, R. N. 1978. Mechanisms of hydrothermal crystallization of quartz at 250°C and 15 kilobars. *Geochim. Cosmochim. Acta* 42:173–182.

Molini-Velsko, C. A. 1983. Isotopic composition of silicon in meteorites. Ph.D. Thesis, Univ. of Chicago.

Molini-Velsko, C., Mayeda, T. K., and Clayton, R. N. 1986. Isotopic composition of silicon in meteorites. *Geochim. Cosmochim. Acta* 50:2719–2726.

Navon, O., and Wasserburg, G. J. 1985. Self-shielding in O_2—A possible explanation for oxygen isotopic anomalies in meteorites? *Earth Planet. Sci. Lett.* 73:1–16.

Niederer, F. R., and Papanastassiou, D. A. 1984. Ca isotope in refractory inclusions. *Geochim. Cosmochim. Acta* 48:1279–1293.

Niederer, F. R., Papanastassiou, D. A., and Wasserburg, G. J. 1980. Endemic isotopic anomalies in titanium. *Astrophys. J.* 240:L73–L77.

Niederer, F. R., Papanastassiou, D. A., and Wasserburg, G. J. 1981. The isotopic composition of titanium in the Allende and Leoville meteorites. *Geochim. Cosmochim. Acta* 45:1017–1031.

Niemeyer, S., and Lugmair, G. W. 1981. Ubiquitous isotopic anomalies in Ti from normal Allende inclusions. *Earth Planet. Sci. Lett.* 53:211–225.

Niemeyer, S., and Lugmair, G. W. 1984. Titanium isotopic anomalies in meteorites. *Geochim. Cosmochim. Acta* 48:1401–1416.

Papanastassiou, D. A. 1986. Cr isotopic anomalies in the Allende meteorite. *Astrophys. J.* 308:L27–L30.

Podosek, F. A. 1978. Isotopic structures in solar system materials. *Ann. Rev. Astron. Astrophys.* 16:293–334.

Prombo, C., and Clayton, R. N. 1985. A striking nitrogen isotope anomaly in the Bencubbin and Weatherford meteorites. *Science* 230:935–937.

Prombo, C., and Lugmair, G. W. 1986. Search for correlated isotope effects in Allende CAI's. *Lunar Planet. Sci.* XVII:685–686 (abstract).
Rayleigh, J. W. S. 1896. Theoretical considerations respecting the separation of gases by diffusion and similar processes. *Philosophical Magazine, 5th Series* 42:493–498.
Reynolds, J. H. 1967. Isotopic abundance anomalies in the solar system. *Ann. Rev. Nucl. Sci.* 17:253–316.
Rich, J. W., and Bergman, R. C. 1979. C_2 and CN formation by optical pumping of CO/Ar and CO/N_2/Ar mixture at room temperature. *Chem. Phys.* 44:53–64.
Robert, F., and Epstein, S. 1982. The concentration and isotopic composition of hydrogen, carbon and nitrogen in carbonaceous meteorites. *Geochim. Cosmochim. Acta* 46:81–95.
Thiemens, M. H., and Heidenreich, J. E. 1983. The mass independent fractionation of oxygen: A novel isotope effect and its possible cosmochemical implications. *Science* 219:1073–1075.
Thiemens, M. H., and Jackson, T. 1985. Production of mass independently fractionated isotopic components by ultra-violet light. *Meteoritics* 20:775–776 (abstract).
Thiemens, M. H., and Jackson, T. 1986. The effect of variable wavelength ultraviolet light on the production of oxygen isotope anomalies. *Lunar Planet. Sci.* XVII:889–890 (abstract).
Thiemens, M. H., and Jackson, T. 1987. Production of isotopically heavy ozone by ultraviolet light photolysis of O_2. *Geophys. Res. Lett.* 6:624–627.
Thiemens, M. H., and Meagher, D. 1987. Demonstration of a mass-independent isotopic fractionation in CO. *Lunar Planet. Sci.* XVII:1006–1007 (abstract).
Thiemens, M. H., Heidenreich, J. E., and Lundberg, L. 1983a. Photochemical production of isotopic anomalies in the early solar system. *Lunar Planet. Sci.* XIV:785–786 (abstract).
Thiemens, M. H., Gupta, S., and Chang, S. 1983b. The observation of mass independent fractionation of oxygen in an RF discharge. *Meteoritics* 18:408 (abstract).
Treanor, C. E., Rich, J. W., and Rehm, R. G. 1968. Vibrational relaxation of anharmonic oscillators with exchange dominated collisions. *J. Chem. Phys.* 48:1798–1807.
Urey, H. 1947. The thermodynamic properties of isotopic substances. *J. Chem. Soc. London,* pp. 562–581.
Wasserburg, G. J., Lee, T., and Papanastassiou, D. A. 1977. Correlated O and Mg isotopic anomalies in Allende inclusions: II. Magnetism. *Geophys. Res. Lett.* 4:299–302.
Wasserburg, G. J., Papanastassiou, D. A., and Lee, T. 1980. Isotopic heterogeneities in the solar system. *Early Solar System Processes and the Present Solar System,* ed. D. Lal (Amsterdam: North-Holland), pp. 144–191.
Weston, R. 1975. Quantum mechanical tunneling. In *Isotopes and Chemical Principles,* ed. P. A. Rock (Washington, DC: American Chemical Society).
Wolfsberg, M. 1982. The theoretical analysis of isotope effects. In *Stable Isotopes,* eds. H. L. Schmidt, H. Forstel, and K. Heinzinger (Amsterdam: Elsevier), pp. 3–14.
Wood, J. A. 1981. The interstellar dust as a precursor of Ca, Al-rich inclusions in carbonaceous chondrites. *Earth Planet. Sci. Lett.* 56:32–44.
Zinner, E. K., Fahey, A. J., Goswami, J. N., Ireland, T. R., and McKeegan, K. O. 1986. Large ^{48}Ca anomalies are associated with ^{50}Ti anomalies in Murchison and Murray hibonites. *Astrophys. J.* 311:L103–L107.

PART 13
Survival of Presolar Material in Meteorites

13.1. CIRCUMSTELLAR MATERIAL IN METEORITES: NOBLE GASES, CARBON AND NITROGEN

EDWARD ANDERS
University of Chicago

In addition to preserving a record of isotopically distinct reservoirs in the early solar system, some primitive meteorites contain discrete grains of presolar origin. Such grains are distinguished by the isotopically anomalous noble-gas components they contain. One such component consists of monoisotopic Ne^{22}, produced by decay of radioactive Na^{22} with a 2.6 yr half-life. Two xenon components have also been identified: one synthesized apparently in a supernova, the other probably in a red giant star. Most of the grains that carry these noble-gas components are carbonaceous and contain isotopically anomalous C, N, or both. They include diamond and silicon carbide. Two unidentified carriers of isotopically anomalous nitrogen, unaccompanied by noble gases, occur in the brecciated stony iron meteorite, Bencubbin.

13.1.1. INTRODUCTION

The solar system formed from pre-existing matter, and thus all meteoritic matter is presolar in the broadest sense of the term. However, much of this material was reprocessed in the early solar system (by vaporization, melting, mixing, isotopic exchange, chemical reactions, etc.), and became isotopically homogeneous. Such commonplace material is properly called "local," which leaves the terms "presolar" or "exotic" for material still retaining an anomalous isotopic signature that cannot be explained by solar system processes.

Most kinds of such anomalous material, specifically oxides and metals, have been diluted and assimilated by local matter, and hence no longer exist in pure form. The principal exception is carbon and some of its compounds (kerogen, carbides). Being slow to react at low temperatures, they have survived in the matrices of primitive meteorites (types 1–3, also E4) as discrete, presolar grains of anomalous isotopic composition.

The presolar carbon grains are well hidden among the 10^2 to 10^4-fold greater amounts of local C, and were found only because they are tagged with anomalous noble gases. Even though the noble gases are only $\sim 10^{-10}$ as abundant as their solid carriers, they are easier to separate, study and characterize, and thus provide a firmer taxonomic basis. Indeed, since the isotopic measurements on noble gases and C, N are usually done by different people, on different instruments and samples, the links between noble-gas and carbon components are tentative in most cases. Thus the material of this chapter is organized by noble-gas components rather than by carriers or C,N-components.

For further details and references, see the reviews by Podosek (1978), Begemann (1980), Srinivasan (1981), Lewis and Anders (1983), Pillinger (1984), Kerridge and Chang (1985) and Anders (1981,1987), as well as the papers by Tang et al. (1988) and Tang and Anders (1988a,b).

13.1.2. NOBLE GASES IN METEORITES

Among the elements in meteorites, noble gases are unique. Being volatile and unreactive, they did not fully condense even in the most primitive meteorites, and hence are present at only a small fraction of their solar abundance, from $\sim 10^{-5}$ for Xe to $\sim 10^{-9}$ for He and Ne. But this tiny amount of "primordial" noble gas is tightly bound by its "host" or "carrier" phases, and comes off only at high temperatures.

The major part of the primordial noble gas appears to be *local*, as its isotopic and elemental composition more or less resembles that of the Earth's atmosphere, and, like the latter, can be related to solar composition by plausible solar system processes such as mass fractionation. This major primordial component is called "planetary," with the understanding that this term covers a *range* of compositions rather than a single one. A minor part of the primordial gas is clearly *exotic*, however, having isotopic compositions that cannot be made by known or plausible solar system processes.

Apart from these primordial components, up to three other noble-gas components may be present.

1. *Radiogenic*, from the decay of long-lived or extinct radioactive precursors, e.g., He^4 from $U^{235,238}$ and Th^{232}, Ar^{40} from K^{40}, Xe^{129} from I^{129} ($t_{1/2} = 16 \pm 1$ Myr), and $Xe^{131-136}$ from Pu^{244} ($t_{1/2} = 80.0 \pm 0.9$ Myr).
2. *Cosmogenic* or *spallogenic*, from spallation of neighboring nuclei by cosmic rays. Spallation produces the lighter isotopes of a given element in roughly similar abundances, and the cosmogenic component therefore reveals itself by the enrichment of certain, odd or neutron-poor, isotopes that are rare in the primordial component (He^3, Ne^{21}, Kr^{78}, $Xe^{124,126}$).

3. *Solar,* from implanted solar wind or solar flare ions. The solar component is found only in those meteorites that once resided at the surface of their parent body, and hence were exposed to solar corpuscular radiation. Such meteorites usually are recognized by their brecciated texture and nuclear tracks of low-energy charged particles (see Chapter 3.5). Though the solar component is isotopically somewhat similar to the planetary component, it can be distinguished by its isotopic and elemental ratios and shallow implantation depth.

Although the *major* gas components thus are easily resolvable in a mixture, the *minor* components, often the most interesting, are hard to find unless they or their host minerals are first enriched. A useful technique, especially for reconnaissance work, is *stepped heating,* in which gas fractions released at progressively higher temperatures are analyzed separately. In a favorable case, individual gas components are released one by one, as their host minerals melt, decompose, or become permeable (Reynolds et al. 1978). Direct *separation of host minerals* by physical or chemical methods can also be used, if at least some properties of the minerals are known. But it is difficult, as the gas-bearing minerals tend to be fine grained (<1 μm) and intergrown.

13.1.3. NEON-*E*

Discovery

Until 1969, meteoritic neon was thought to be a mixture of only three components: planetary (Ne-*A*), solar (Ne-*B*), and cosmogenic (Ne-*S*). This is graphically demonstrated by a *three-isotope plot* with a common denominator (Fig. 13.1.1). On such a plot, all neon samples analyzed up to that time fell within a triangle bounded by the above three components, which implied that they were simple ternary mixtures. (It is a property of such a plot that mixtures of 2 components fall on a straight line joining these components—"mixing line"—whereas mixtures of n components fall within an n-sided polygon whose vertices are the n components.) However, when Black and Pepin (1969) analyzed six CI and CM chondrites by stepped heating, they found that the fractions between 900 and 1100°C consistently fell below the triangle, at Ne^{20}/Ne^{22} ratios down to 3.4. Evidently a new neon component was present, of very low Ne^{20}/Ne^{22} ratio (shaded region in Fig. 13.1.1). Black (1972) later named it Ne-*E*, after assigning letters *C* and *D* to two minor trapped components of less extreme composition.

Black and Pepin pointed out that a component of $20/22 < 3.4$ could hardly have been derived from solar Ne ($20/22 \approx 13$) by ordinary mass fractionation processes, such as volume diffusion or gravitational escape. A fractionation so large requires depletion by more than 10^{12}, implying an initial reservoir some 10^{10} times greater than the estimated neon content of the solar nebula.

It seemed more likely that Ne-*E* was made by nuclear processes, either

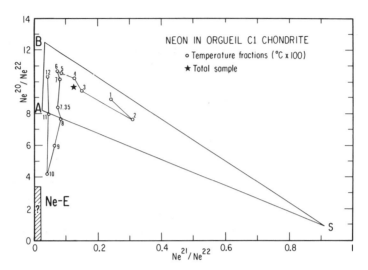

Fig. 13.1.1. Meteoritic neon is a mixture of three main components (A = planetary, B = solar, S = spallogenic), and therefore normally lies within a triangle bounded by these components. But primitive meteorites such as Orgueil also contain an apparently presolar component of very low Ne^{20}/Ne^{22} ratio (= Ne-E), first detected in the 900° and 1000° C fractions of stepped heating experiments (Black and Pepin 1969). The shaded region gives the original limits on the composition of Ne-E.

stellar nucleosynthesis under conditions favoring low Ne^{20}/Ne^{22} ratios or β^+-decay of 2.6 yr Na^{22}. The Na^{22} might be produced either by an irradiation in the early solar system (Heymann and Dziczkaniec 1976) or by stellar nucleosynthesis (Arnould and Beelen 1974; Clayton 1975; Arnould and Nørgaard 1978). Although Ne isotopes would be made at the same time, they might have become separated from Na^{22} by less efficient trapping in solid grains.

Isolation

A clean test of these two mechanisms would be provided by the isotopic composition of *pure* Ne-E and the associated complement of other noble gases. If formed from Na^{22}, Ne-E should be nearly pure Ne^{22}, but if formed directly by stellar nucleosynthesis, it should be accompanied by finite amounts of other Ne isotopes and other noble gases. For such a test it is necessary to isolate the host phase of Ne-E in pure form.

Early work on the Orgueil C1 chondrite showed that Ne-B was enriched mainly in magnetite, Ne-A in colloidal phyllosilicate (Jeffery and Anders 1970), and Ne-E in noncolloidal, low-density silicate fractions (Herzog and Anders 1974; Eberhardt 1974, 1978). The latter authors succeeded in isolating samples that had bulk Ne compositions falling below the triangle. Eberhardt and his students by ever more elaborate separations gradually lowered the 20/22 ratio to < 1. Further work (Fig. 13.1.2) showed that two distinct carriers

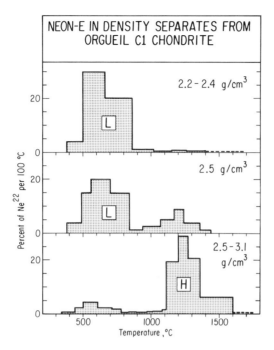

Fig. 13.1.2. Neon-E is located in two carrier phases, of low ($= L$) and high ($= H$) density and release temperature. The L-carrier is carbonaceous and has been called Cα; the H-carrier follows spinel to some extent, and appears to be silicon carbide (figure from Jungck 1982).

of Ne-E were present in both Orgueil (C1) and Murchison (C2), of densities and release temperatures < 2.3 g/cm^3, 500–700° C, and 3 to 3.5 g/cm^3, 1200–1400° C (Lewis et al. 1979c; Eberhardt et al. 1979). These two carriers were tentatively identified as a carbonaceous phase and spinel. The corresponding Ne-E components were designated Ne-$E(L)$ and Ne-$E(H)$.

Isolation of the two Ne-E components in essentially pure form was at last achieved by Jungck and Eberhardt (1979), by an elaborate density separation of the noncolloidal, nonmagnetic silicate fraction from Orgueil. Fig. 13.1.3 shows their data for the Ne-$E(L)$ carrier. Their upper limits for the composition of Ne-E are as follows:

	20/22	21/22
Ne-$E(L)$	< 0.01	< 0.001
Ne-$E(H)$	< 0.2	< 0.003.

Associated Components

No other primordial noble gases seem to be associated with Ne-E. The most stringent limits were obtained by Jungck (1982), for the L and H carriers, respectively:

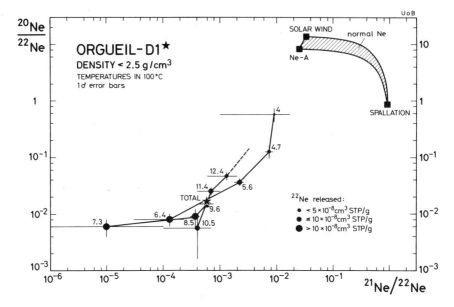

Fig. 13.1.3. The purest Ne-E to date found in a low-density fraction from Orgueil (Jungck and Eberhardt 1979). On this logarithmic plot, the triangle enclosing normal Ne is twisted into a curved shape, but the mixing line toward pure Ne^{22} still is a straight line of slope 1. The 640 to 850° C fractions are > 99% pure Ne^{22}, and even the total sample is > 98% Ne^{22}. Apparently Ne-E is monoisotopic Ne^{22} from the β^+-decay of Na^{22} ($t_{1/2} = 2.6$ yr).

$$He_p^4/Ne_E^{22} < 2.6 \text{ and } \sim 0$$
$$Ar_p^{36}/Ne_E^{22} < 0.14 \text{ and } < 0.36.$$

Thus it seems that Ne-E is simply monoisotopic Ne^{22} from the decay of Na^{22}. Moreover, it seems to be monoelemental, unaccompanied by other noble gases.

Carrier of Ne-$E(L)$

Although Ne-$E(L)$ has been isolated in > 99% purity, the corresponding carrier has not yet been enriched or characterized to a similar degree. This carrier is a fairly reactive form of carbon (noncommittally called Cα), which is destroyed by HClO$_4$ at 200° C. It seems to have a grain size of 1 to 10 μm, as shown by filtration through Nuclepore filters of known pore size (Alaerts et al. 1980). Some further clues to its nature come from *stepped combustion*, a variant of the stepped heating technique in which the sample is exposed to 5–20 torr O$_2$ at progressively higher temperatures, and the CO$_2$ formed is analyzed on a mass spectrometer (Pillinger 1984). Murchison and Murray samples enriched in Ne-$E(L)$ consistently show a component reaching

$\delta C_{PDB}^{13} \approx 340‰$ around 600 to 700° C (Swart et al. 1983; Carr et al. 1983; Tang et al. 1986), which seems to correlate with Ne-$E(L)$ and has been attributed to Cα. (Like all C,N values obtained on impure samples, this is a minimum value, as other components in the sample may have contributed.)

There are some indications that Cα is also enriched in N^{15}. Stepped combustion of two Murray samples enriched in Ne-E gave δN_{air}^{15} values up to 252‰ at $\sim$500° C, close to the combustion temperature of Cα (Tang et al. 1986). Since the samples contained large amounts of lighter N, it seems likely that the true maximum δN^{15} is even higher than 252‰. This is an interesting clue, as most stars destroy rather than produce N^{15}. However, it has not yet been proven that this component is associated with Ne-$E(L)$ or with the C component of $\delta C^{13} = 340‰$.

Carriers of Ne-$E(H)$

Although the first experiments pointed to spinel as the sole carrier, later work cast doubts on this identification, suggesting that no less than 40% of the Ne-$E(H)$ is associated with apatite, $Ca_2PO_4(OH,F)$ (Eberhardt et al. 1981; Jungck et al. 1981; Jungck 1982). This was surprising, as apatite, unlike spinel, is not a primary stellar or nebular condensate but a secondary mineral. However, when the apatite was partly dissolved in HCl, Ne-E remained in the carbonaceous residue (Jungck and Eberhardt 1985). On stepped heating, it was released largely below 1000° C, close to the range for Ne-$E(L)$. Apparently Ne-E is not present in the apatite itself, but in carbonaceous inclusions, possibly identical to Cα.

The outcome for spinel was somewhat similar. Although Ne-$E(H)$ generally follows spinel in mineral separations, it does not correlate well with spinel content or Ne_c^{21}—a good index of spinel (Ott et al. 1985; Tang et al. 1985). This suggests that Ne-$E(H)$ is located in a minor carrier, either a small part of the spinel or in some other phase tightly associated with it (Tang et al. 1985). Indeed, when spinel was dissolved in H_3PO_4 and amorphous C removed by oxidation, a 2% residue remained that contained all the Ne-$E(H)$. The carrier ("Cε") appears to be SiC (Zinner et al. 1987; Tang and Anders 1988a), which is a well-known stellar condensate at high C/O ratios (Gilman 1969).

Origin

Neon of 20/22 $\leq$ 0.01 can indeed be made by thermonuclear reactions at low densities and high temperatures (< 10 g/cm^3, $> 4 \times 10^8$ K), but always accompanied by excessively large amounts of He^4 (Arnould and Nørgaard 1978). However, most of mass 22 is made as Na^{22} under these conditions, and would readily separate from He and Ne by condensation into grains. The short half-life of Na^{22} implies explosive conditions. In principle, red giants, novae and supernovae all are possibilities, because all produce Na^{22} and eject matter under conditions where solid grains can condense within a few months, long before 2.6 yr Na^{22} has decayed. More detailed considerations favor novae, however (Clayton and Hoyle 1976; Arnould and Nørgaard 1981). The apparent N^{15} enrichment of $C\alpha$ also points to novae, as novae are thought to be the principal source of N^{15} in the universe (Trimble 1975).

A stellar origin can also account for the mineralogy of the Ne-*E* carriers. Carbon and SiC are some of the most abundant and earliest stellar condensates at high C/O ratios (Larimer and Bartholomay 1979).

13.1.4. XENON-HL

Xenon has 9 rather than 3 stable isotopes, and so the potential information content (and complexity) is correspondingly greater (Fig. 13.1.4). The dominant component in primitive meteorites is *primordial* Xe (also called *planetary* or *trapped*), differing from *solar* Xe only by a slight degree of mass fractionation. (Solar Xe itself, of course, is present in those meteorites that were exposed to the solar wind, but only in subordinate amounts.) Two *radiogenic* components are often present: Xe^{129} from the β decay of extinct 16 Myr I^{129}, and Xe^{128}, produced from iodine by cosmic-ray secondary neutrons: I^{127} (n, γ) I^{128} $(\beta^-, \nu;$ 25 m) Xe^{128}. The *spallogenic* component normally is quite small, and tends to show up only at the rare isotopes, 126 and 124. The *fissiogenic* component from Pu^{244} and U^{238} is also small.

Discovery

The first signs of another component were seen by Reynolds and Turner in 1964, when they looked for Xe^{129} in the Renazzo CR2 chondrite. They found that the 700 to 1000° C fractions were enriched in the heavy Xe isotopes 131–136 by up to 6%, compared to the remaining fractions. Similar or even greater enrichments were later found in other carbonaceous chondrites.

The rising abundances from 131 to 136 were suggestive of spontaneous fission, and this component soon became known as *CCFXe* (*c*arbonaceous *c*hondrite *f*ission).[a] But the pattern did not resemble that of any known actinide, and the amounts were too large by orders of magnitude. Thus three groups suggested almost simultaneously in 1968–1969 that a superheavy element might be responsible (Srinivasan et al. 1969; Anders and Heymann 1969; Dakowski 1969). Such elements, centered on an "island of stability" at magic numbers Z = 114 and N = 184, were expected to have relatively long half-lives, in some cases perhaps long enough to show up in the early solar system (Seaborg 1968).

An important objection to the superheavy element hypothesis was raised by Manuel et al. (1972). They pointed out that the fractions most enriched in heavy xenon isotopes also were enriched in the light, β-shielded isotopes 124,

---

[a]Other names were Xe-*X*, Xe-*R* and *DME*-Xe (demineralized, etched). We shall call it Xe-*HL*, to indicate that it is a mixture of components enriched in the light and heavy isotopes (*L*-Xe and *H*-Xe; Pepin and Phinney 1978).

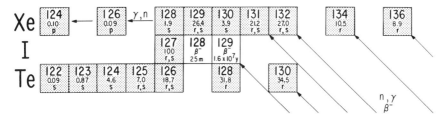

Fig. 13.1.4. Isotope chart in the region of Xe. Figures below the mass numbers indicate (terrestrial) isotopic abundances; letters refer to nucleosynthetic processes. Heavy isotopes are made by the r-process, involving buildup of very neutron-rich, short-lived progenitors by neutron capture at a rapid rate (n, γ), followed by β^--decay (diagonal arrows). Middle isotopes are made in part by a second neutron capture-β^--decay process (s-process, where s stands for slow). It is illustrated in Fig. 13.1.9. Light isotopes such as $Xe^{124,126}$ cannot be made by these processes, because they are not accessible to β^--decay, being shielded by stable isobars such as $Te^{124,126}$ that terminate the decay chain. They are made by the p-process (horizontal arrows), involving ejection of a neutron by an energetic photon (γ, n).

126 and 128 (Fig. 13.1.4) that cannot be made by fission (this trend was first noted by Reynolds and Turner [1964], but was then forgotten). Manuel et al. therefore suggested three alternative origins, of which one—nucleosynthesis in a supernova—eventually turned out to be correct. According to this hypothesis, the light and heavy Xe isotopes were made in a supernova, by p- and r-processes (p-process: proton capture or γ,n reactions; r-process: neutron capture on a rapid time scale; see Chapter 14.1). After thorough mixing, the resulting pseudocomponent then somehow found its way into the early solar system.

Isolation

Physical methods, such as density or magnetic separations, failed to enrich the carrier, but at last an unconventional chemical method succeeded. Guided by the erroneous hunch that the carrier might be pentlandite [$(Fe,Ni)_9S_8$], Lewis et al. (1975) dissolved a sample of the Allende CV3 chondrite in HF-HCl. A black residue of about 0.5% remained that was greatly enriched in noble gases but contained no pentlandite; only chromite, amorphous C, spinel and about 2% of sulfur suggestive of a minor sulfide. Yet the Xe^{136}/Xe^{132} ratio was almost unchanged, showing that the authors had got more than they had bargained for: the host phases of both Xe-*HL and* primordial Xe.

Recalling that all sulfides are soluble in oxidizing acids, Lewis et al. etched the residue under progressively more drastic conditions. He and Ne were almost unaffected, but the three heavy gases decreased by more than an order of magnitude and the 124/132, 136/132 ratios in the residue rose nearly twofold (Fig. 13.1.5). Evidently planetary Xe was being etched away while Xe-*HL* stayed behind. The separation was far better than that by stepped heating, as shown by the fact that the 136/132 ratio in some *bulk* samples was

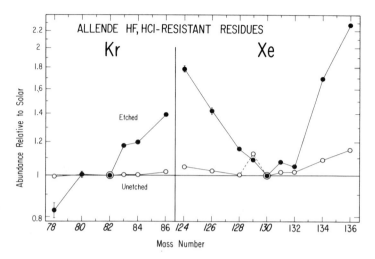

Fig. 13.1.5. Samples highly enriched in Xe-*HL* and associated Kr may be prepared by dissolving a carbonaceous chondrite in HF + HCl and etching the residue (open symbols) with an oxidant such as HNO_3, to remove the bulk of the primordial gases (Lewis et al. 1975,1979a). In such etched samples (filled symbols), Xe is enriched up to twofold in both the light, shielded isotopes (italicized) and the heavy unshielded isotopes. Krypton, too, is anomalous.

higher than that in the most enriched *temperature fractions* ever seen. The mass loss in the etch treatments leveled off at 8%, suggesting that the planetary Ar, Kr and Xe reside mainly in an oxidizable, minor phase, which was named "*Q*" for "quintessence." (Later work showed that *Q* is not a discrete phase [Ott et al. 1981,1984] but presumably a gas component adsorbed in a labyrinth of pores [Yang et al. 1982; Wacker et al. 1985; Zadnik et al. 1985; see also Chapter 7.9].)

No single process can account for the simultaneous enrichment of light and heavy xenon isotopes ($\equiv$ Xe-*L* and Xe-*H*; Fig. 13.1.5). By the supernova hypothesis, these two components must have formed in separate, neutron-poor and neutron-rich, environments. By the fission hypothesis, Xe-*H* of course formed by fission, whereas Xe-*L* formed from solar Xe by mass fractionation.

Although both hypotheses thus agreed that Xe-*H* and Xe-*L* were separate components, all attempts to separate them were unsuccessful. Virtually all meteoritic samples, no matter how treated, yielded Xe-*H* and Xe-*L* in fixed ratio. Two apparent exceptions were residues from Dimmitt (H4) and Krymka (LL3), which were ~25% enriched or ~15% depleted in Xe-*L* compared to C chondrites (Moniot 1980; Alaerts et al. 1980), but later measurements on Krymka gave at best only marginal confirmation (Schelhaas et al. 1985). Apparently Xe-*H* and Xe-*L* exist as a single pseudocomponent, Xe-*HL*, in most if not all meteorites.

A distant, shadowy relative of Xe-*L* is monoisotopic (?) Xe^{124}, which has been seen in stepped heating runs on FeS-rimmed Allende chondrules (Lewis et al. 1979d). The 550 to 650° and 1400 to 1500° C fractions had 124/132 ratios up to 60% above normal, and since no other Xe isotope was enhanced, it seems that a monoisotopic Xe^{124} component was present. Pepin and Phinney (1978) had previously seen hints of this component in an elaborate analysis of stepped heating data on bulk meteorites. Nothing more is known about this component, however.

Associated Components

Xe-*HL* is accompanied by components of the other four noble gases, which follow it in stepped heating and in mineral separations. Their elemental abundances relative to solar show a nearly linear trend of log abundance vs mass number (Lewis et al. 1975), with a slope of about 6% per mass unit (Ott et al. 1981). The Kr-component (Fig. 13.1.5) is enriched in the heavy but not the light isotopes, and hence might be regarded as Kr-*H* without Kr-*L*. The Ar component is likewise enriched in the heavy isotope: $Ar^{36}/Ar^{38} = 4.82$ compared to the atmospheric or solar ratio of 5.32 (Lewis et al. 1977).

This trend extends to Ne, which is also enriched in the heavy isotope, Ne^{22}. Surprisingly, its composition is close to that of Ne-A ("planetary" Ne), but Ne-*A* itself turns out to be complex, consisting of two subcomponents differing in etchability, release temperature, and associated Xe component (Table 13.1.1).

Ne-*A*2, the component associated with Xe-*HL*, follows it very closely through mineral separations, chemical treatments and stepped heating—"like an isotope of Xe" (J. H. Reynolds, commenting on the data of Ott et al. 1981). Presumably both gas components are located in the same carrier, and hence possibly have the same origin. Ne-*A*2 contributes some 80% of Ne-*A*, and since Ne-*A*, "planetary Ne," presumably is local, it would seem that Ne-*A*2 and hence Xe-*HL*, too, must be local. That turned out to be a false lead, however.

The He component associated with Xe-*HL* has not been well characterized. The He^3/He^4 ratio tends to rise with Xe^{136}/Xe^{132}, suggesting that the light rather than the heavy isotope is enriched. Judging from the fragmentary data, the enrichment factor is at least 1.5 relative to planetary He ($He^3/He^4 = 1.4 \times 10^{-4}$), but part of the excess He^3 may be spallogenic, from a recent or presolar exposure to cosmic rays.

Carrier

In principle, a clean test was available between the supernova and fission hypotheses. If Xe-*HL* was exotic then its carrier should also be exotic, and show isotopic anomalies. On the other hand, if the Xe had formed in the solar system, by mass fractionation and by fission of a superheavy element that had condensed in a meteoritic carrier mineral, then the carrier should show no anomalies.

TABLE 13.1.1
Planetary Neon Components in C Chondrites

Component	$\frac{Ne^{20}}{Ne^{22}}$	$\frac{Ne^{21}}{Ne^{22}}$	Etchable HNO_3?	Release T (°C)	$\frac{Xe^{136}}{Xe^{132}}$	Carrier	Ref.
Ne-A	8.2 ± 0.4	0.025 ± 0.003	—	—	—	—	a
Ne-A1	8.84 ± 0.08	0.0299 ± 0.0003	No	≤ 800	0.313	Cζ†	bj
Ne-A2	8.37 ± 0.03	0.035 ± 0.001	No	1000–1200	0.65	Cδ	bc
Q-Ne*	10.1 ± 0.2	~0.03?	Yes	≥ 1100	0.31	Q	defg
Atmosphere	9.80 ± 0.08	0.0290 ± .003	—	—	0.329	—	—
Solar wind	13.7 ± 0.3	0.0333 ± .0044	—	—	0.299	—	hi

*Black (1972) used the term Ne-C for this component in C chondrites and for an isotopically similar component in gas-rich chondrites, which seems to be derived from solar flares. As the C-chondrite component apparently is not related to solar flares but correlates with heavy planetary gases, it deserves a separate name, e.g., Q-Ne.

†Alaerts et al. (1980) suggested that the carrier of this component was a chemically resistant form of kerogen ("polymer"), but more recent work points to some form of elemental C, tentatively called Cζ (Tang and Anders 1988b).

References. a: Pepin 1968; b: Alaerts et al. 1980; c: Ott et al. 1981; d: Alaerts et al. 1979; e: Schelhaas et al. 1985; f: Smith et al. 1977; g: Black 1972; h: Geiss et al. 1972; i: Pepin and Phinney 1978; j: Tang and Anders 1988b.

As with Ne-*E*, this carrier proved as elusive as the innermost of a set of nested Russian dolls. Spinel was quickly eliminated, leaving chromite and C as the two main candidates (Lewis et al. 1975). But neither physical nor chemical methods gave *clean* separations of these two minerals, since there were at least two varieties of each that differed in chemical resistance and grain size (Lewis et al. 1977,1979*a*; Lewis and Matsuda 1980; Ott et al. 1981,1984). The Chicago group thought that fine-grained C and chromite both were carriers, but eventually it turned out that most or all of the Xe-*HL* was located in a minor part of the C, of small grain size ($\sim 20-50$ Å) but high chemical resistance (Lewis and Matsuda 1980; Swart et al. 1983; Lewis and Anders 1983; Ott et al. 1981,1984). It was named Cδ, whereas the dominant, gas-poor C was called Cγ. Subsequently Cδ was identified as diamond (Lewis et al. 1987).

The first isotopic measurement on Cδ-rich samples, by stepped combustion (Swart et al. 1983), gave δC^{13} down to $-38‰$, a bit low for meteorites but within the terrestrial range. This was another false lead, however. Nitrogen measurements by the same technique gave δN^{15} down to $-330‰$, well outside the range of solar-system processes and hence strongly suggestive of stellar nucleosynthesis (Lewis et al. 1983*a*). Thus, if both light N and Xe-*HL* are located in the same phase, as suggested by the data of Frick and Pepin (1981) and Lewis et al. (1983*a*), then Xe-*HL*, too, is nucleosynthetic. Given the wide variation in C^{12}/C^{13} in the Galaxy, from 4 to several hundred, one marvels at nature's sense of humor in choosing terrestrial type C, of ratio 93, for "packaging" highly anomalous N and Xe.

A second nail in the coffin of the fission hypothesis came from isotopic measurements of Ba, Nd and Sm in an Allende residue containing Xe-*HL* (Lewis et al. 1983*b*). If the Xe-*HL* had come from fission, then it should be accompanied by comparable amounts of isotopically anomalous Ba, Nd and Sm. No such anomalies were found, and the most stringent upper limits (for Ba^{135}) were only 0.03 the amount expected from Xe^{136}. However, now that Cδ has been identified as diamond, a loophole has become apparent in this experiment: possibly Cδ was not dissolved but was lost as a colorless colloid.

At first sight the apparent lack of Ba anomalies would seem to be a major embarrassment for both models, fission or nucleosynthesis, since any process that makes $Xe^{134,136}$ must make comparable amounts of Ba^{135}. For nucleosynthesis, chemistry offers a way out of this dilemma. As Ba, Nd and Sm have high condensation temperatures but low first ionization potentials compared to Xe, they may have separated from it by condensation or plasma processes. No such excuse is available for the fission model, as all fission fragments should have been trapped in the C carrier with nearly equal efficiency.

Origin

Heymann and Dziczkaniec (1979,1980) have shown that components resembling Xe-*HL* and its associated Kr can be produced during explosive nucleosynthesis. Using the model of Woosley and Howard (1978) for a layered presupernova of original mass 25 $M_\odot$, they found that the neutron-rich isotopes can be made in C shells at high neutron doses ($\sim 3 \times 10^{-7}$ mole s/cm³) and peak explosion temperatures $\sim 2 \times 10^9$ K, whereas the neutron-poor isotopes are made in deeper, hotter (2.7×10^9 K) zones, where photodisintegration strips neutrons from heavier isotopes. The above neutron fluences are several orders of magnitude below those for the classical r-process, but higher than those for the s-process. No attempt has yet been made to explain the presence and isotopic composition of Ar and especially He, C, N and Ne, which, being highly "combustible," require lower temperatures. But they can probably be accommodated by the model given below.

Some clues are available on how the noble gases got into the diamond. Trapping during condensation of C does not look promising; by the time the expanding gas has cooled to $\sim 1500-2000$ K where C can condense, the density is too low ($< 10^{-12}$ g/cm³) for significant trapping (Lattimer et al. 1978). Moreover, there are indications that the trapping did not involve condensation or sorption. Lewis and Anders (1981) have pointed out that Xe^{129} is not enriched relative to the neighboring isotopes (Fig. 13.1.5), although at the time of trapping it was present as an isotope of iodine rather than xenon (I^{129}; $t_{1/2} = 16$ Myr). By a similar argument, one can rule out appreciable contributions of Cl^{36} (3.01×10^5 yr) and Na^{22} (2.6 yr) to Ar^{36} and Ne^{22} Apparently the trapping was chemically nonselective, which suggests ion implantation rather than condensation or sorption.

Clayton (1981) has proposed a scenario that permits such ion implantation (Fig. 13.1.6). As the hot plasma from the supernova expands outward at $(5-10) \times 10^3$ km/s, it overtakes the dust shell that was expelled by the presupernova during its red giant or planetary nebula stage. Some of the ions come to rest within the dust grains, and thus become trapped.

A few further details can now be added to the model. The dust must have included diamond, which, by analogy to the laboratory synthesis (Roy 1987), may have condensed metastably in the expanding gas shell (as first suggested by Saslaw and Gaustad 1969). Nitrogen, which can substitute for C in the diamond lattice, would be included in the diamond. The noble-gas ions implanted at the next stage would comprise not only those formed deep within the supernova (Xe, Kr, perhaps Ar) but also some accelerated from cooler, even unprocessed regions (especially He but also Ne-Xe). Some such admixture of roughly normal gas seems to be implied by the rather subdued isotopic anomalies; even Xe-*HL* is less anomalous than expected for a mixture of pure Xe-*H* and Xe-*L,* and apparently contains a solar-like component (Pepin and Phinney 1978). In addition, the Cδ diamonds, with their enormous surface area, would adsorb normal gases during later contraction of the interstellar cloud, and thus may be the main carriers of planetary gases (Yang and Anders 1982; Zadnik et al. 1985).

There remains the question why both Cδ and average terrestrial C are

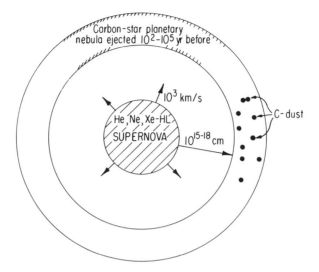

Fig. 13.1.6. A model for implanting Xe-*HL* and associated noble gases in carbon grains (Clayton 1981). Noble-gas and other ions emitted by the supernova overtake the planetary nebula shed previously, and come to rest in the dust grains of the nebula.

isotopically lighter than average present-day galactic carbon: $C^{12}/C^{13} = 90$–93 compared to 62 ± 8 (Wannier 1980). Part of the difference can be blamed on the early isolation of Cδ and solar-system C 4.5×10^9 yr ago, cutting it off from the steady supply of freshly synthesized galactic C^{13}, which causes a secular decline in C^{12}/C^{13}. But a major part of the difference must be intrinsic, possibly implying contributions from one or more massive stars. Schramm and Olive (1982) have pointed out that massive stars of ≥ 10 M$_\odot$ preferentially make the "α-particle" nuclei (= multiples of He^4) C^{12}, O^{16} and Ne^{20}, all of which seem to be enriched in the solar system. Such stars also have low C/O, and thus may be able to account for the low C/O ratio of the Sun, which has a profound effect on planetary chemistry.

Tielens et al. (1987) have proposed an alternative mechanism for the formation of Cδ: melting of carbon grains by interstellar shocks. As liquid carbon is structurally closer to diamond than to graphite, the melt would form crystalline or amorphous (= glassy) diamond on quenching. About 5% of the carbon would be transformed to diamond, of maximum grain size ~ 100 Å. This mechanism correctly predicts (better, postdicts) the size of the diamonds and the presence of amorphous diamond (Lewis et al. 1987), but it also predicts too high a graphite/diamond ratio ($\sim 20:1$), and does not readily explain either the low δC^{13} or the noble gases in the diamonds.

13.1.5. XENON-S

This component, too, was discovered by accident. Though rare (5×10^{-5} of the total Xe in the meteorite) and better hidden than Xe-*HL*, it has a

Discovery

Srinivasan and Anders (1978) were trying to characterize Xe-*HL* from the Murchison (CM2) chondrite, in the hope that it might differ from that in Allende (CV3) in its ratio of *L*-Xe to *H*-Xe. To their disappointment, the samples still contained much planetary Xe, as shown by their closeness to the planetary point on a 3-isotope diagram. Presumably the HNO_3 etch had not removed all the organic polymer, which is much more abundant in CM2s than CV3s, and is a possible carrier of planetary gas. They therefore etched the sample with two alkaline oxidizing agents, NaOCl and Na_2O_2, and examined it by stepped heating (Fig. 13.1.7).

The treatment was successful: the 1000°C fraction of the new sample, 1C10, lay well to the right of all previous Murchison samples (shaded area). However, whereas all previous samples had fallen on the mixing line between planetary Xe and Xe-*HL*, the 1100°C and succeeding fractions of the new sample veered off to the upper left. Evidently, the samples contained a new component of high release temperature and high 130/132 ratio, lying at or beyond the 1600° point on the upper left. Later work by Ott et al. (1985) and Tang and Anders (1988*a,b*) gave still more extreme compositions, with 130/132 up to 0.30.

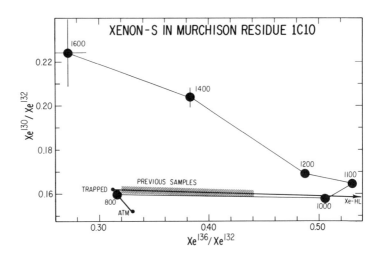

Fig. 13.1.7. All previously studied Murchison samples fall in the shaded area, on the mixing line between trapped Xe and Xe-*HL*. The first two fractions of the severely etched sample 1C10 also fall on the mixing line, but the next four fractions veer upward and to the left, indicating the presence of a new component, Xe-*S* (Srinivasan and Anders 1978).

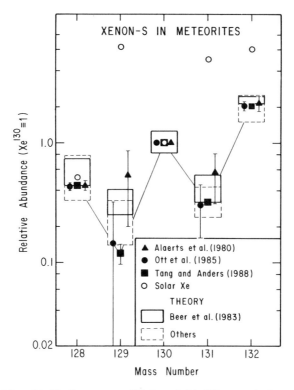

Fig. 13.1.8. Meteoritic Xe-S component (filled symbols) differs greatly from solar Xe (open circles), but agrees rather well with the theoretical composition of s-process Xe (boxes). The values of Beer et al. (1983) are systematically high, suggesting that their estimate for the normalizing isotope Xe^{130} is somewhat too low.

Origin

It was obvious on inspection that the new component was enriched mainly in the even-numbered, middle, isotopes of Xe (128, 130 and 132) but was grossly deficient in the lightest and heaviest isotopes (124, 126, 134, 136). The net pattern of the new component could then be calculated on the plausible assumption that it contained no Xe^{136} at all. This pattern, based on more recent data, is shown in Fig. 13.1.8. It looks quite different from solar Xe (triangles) but resembles the predicted pattern (boxes) for the "s-process" (neutron capture on a slow time scale), and has therefore been called Xe-S.

In this process, heavy elements are built from lighter ones by successive neutron captures alternating with β^--decays (Fig. 13.1.9). Nuclides that do not lie on this s-process path (heavy line) are destroyed but not regenerated by the neutrons, and hence are absent in s-process material (e.g., Xe isotopes 124, 126, 134 and 136 in Fig. 13.1.9).

The s-process approaches a steady state, where the abundance of each

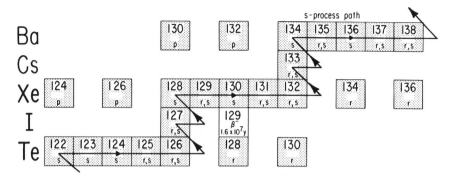

Fig. 13.1.9. The *s*-process path in the Xe region. Only the five middle isotopes of Xe are made in the *s*-process; the two lightest and two heaviest isotopes are bypassed.

nuclide N_s on the *s*-process path is inversely proportional to its neutron-capture cross section (or probability), σ (Chapter 14.1). Thus, if σ is known from laboratory measurements, then N_s can be calculated, such as the theoretical values for Xe-*S* in Fig. 13.1.8 (Clayton and Ward 1978; Beer et al. 1983).

The *s*-process occurs in red giants, stars that have exhausted hydrogen in their cores and shine by burning He^4 to C^{12} in the core along with residual H in a shell further outward. The probable site for the *s*-process is below the H-burning shell, where neutrons are produced during thermal pulses by reaction of He^4 with C^{13} or Ne^{22}, and are then captured by Fe^{56} and other heavy nuclides at $T \approx 3 \times 10^8$ K. Spectra of red giants indeed show large enrichments of elements such as Zr or Ba that are made mainly by the *s*-process. Theoretical models have given a detailed picture of these stars, agreeing well with observations (Trimble 1975; Iben 1985).

Xenon-*S* shows the characteristic signature of the *s*-process, and so must have been located in a red giant at one time. It is not so obvious, though, when it got out. Perhaps during the red-giant stage itself, as red giants are known to lose matter at high rates, up to 10^{-7} M$_\odot$/yr—or during some later stage, e.g., planetary nebula or supernova. Matter is ejected at these stages and cools by expansion and radiation until temperatures are low enough (< 2000 K) for grains to condense. Xenon and other heavy elements, once made in the *s*-process, do not change further until the supernova stage, when temperatures exceed 2×10^9 K. But lighter elements, such as C, N, O, Li, He and Ne, continue to react, and hence such elements, if associated with Xe-*S*, may eventually tell us at what stage the ejection took place.

Associated Components

Xe-*S* is accompanied by small amounts of Kr-*S*; though less conspicuous and harder to resolve, it shows the isotopic characteristics of *s*-process Kr

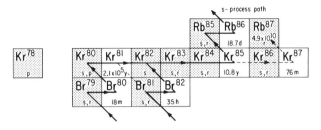

Fig. 13.1.10. The s-process path branches at Kr^{85}. If the interval between neutron captures is long compared to the 10.8 yr half-life of Kr^{85}, then Kr^{85} largely decays before capturing another neutron, thus producing little or no Kr^{86}. Consequently Kr^{86} can serve as a chronometer of the s-process.

(Srinivasan and Anders 1978; Alaerts et al. 1980; Ott et al. 1985). Of particular interest is Kr^{86}, which has been proposed as a chronometer of the s-process, because its precursor Kr^{85} has a β half-life of only 10.8 yr (Fig. 13.1.10) and hence decays before capturing another neutron if the mean neutron capture time is much longer than 10 yr (Ward et al. 1976; Walter et al. 1986). The experimental data are contradictory on this point. Matsuda et al. (1980) reported a distinct rise in Kr^{86}/Kr^{84} in the 1600°C fraction of a Murchison separate, and concluded that the mean neutron capture time of Kr^{85} was 5 to 100 yr at the s-process site(s) represented by this sample. Walter et al. (1986) have shown that these data can yield further insights into the s-process, e.g., the relative contribution of the "weak" s-process, which involved a weaker neutron flux and, in contrast to the "main" s-process, terminated below $A \approx 100$. However, Ott et al. (1985) found no enhancement of Kr^{86} in another Murchison separate, though it was more strongly enriched in Xe-S. They inferred a mean neutron capture time of 232 ± 71 yr from these data. More measurements on purer samples will be needed to settle the matter.

Nothing is known about a possible Ar component associated with Xe-S. Since very little Ar is made by the s-process, any Ar-S is probably swamped by planetary and other kinds of Ar.

Neon-$E(H)$, on the other hand, is quite prominent in all samples containing Xe-S. However, the Ne-E/Xe$_s^{130}$ ratio varies by more than 20 times, suggesting that these two components are located in different, separable grains (Alaerts et al. 1980; Tang et al. 1985).

Helium-3 also is enriched in most samples enriched in Xe-S (Alaerts et al. 1980). Only part of the He^3 can be spallogenic from recent cosmic ray exposure, but the remainder cannot be unambiguously apportioned between solar trapped, presolar trapped and presolar spallogenic components. Spallogenic Ne$_c^{21}$ is quite low, but does not help constrain the origin of the excess He^3, as Ne$_c^{21}$ cannot form from carbon.

Carrier

All these associated components could be much better characterized if the carrier of Xe-S were available in pure form. However, the carrier seems to have an abundance of only a few μg/g (Tang et al. 1985), and has been enriched to only $\sim$10% in the best samples available thus far (Tang and Anders 1988a). Even though Xe-S has been seen in $\sim$50% purity in some temperature fractions, the associated components are generally less enriched and harder to resolve, having fewer isotopes and being swamped by planetary and other components.

The carrier contains carbon and has been designated Cβ (Lewis et al. 1979c). It is quite resistant to oxidation, combusting only at 700–1000° C in 5 torr O_2 (Ott et al. 1985) and surviving treatment with boiling $HClO_4$ and other oxidants (Tang et al. 1986; Tang and Anders 1988a). Recent data suggest that Cβ is another kind of SiC, differing from the SiC that is a candidate carrier of Ne-E(H) in C-isotope composition ($\delta C^{13} \simeq 600$–$1500‰$ vs $\geq 6000‰$), and being less intimately associated with spinel (Zinner et al. 1987; Tang and Anders 1988a).

There is circumstantial evidence suggesting that Cβ (and Cα) are isotopically heavy. Swart et al. (1983) and Carr et al. (1983) analyzed by stepped combustion two Murchison separates enriched in Ne-$E(L)$ and Xe-S, and

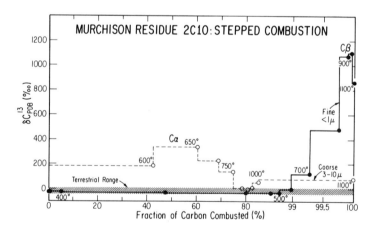

Fig. 13.1.11. Stepped combustion of Murchison separates (Swart et al. 1983; Carr et al. 1983). Most of the C in the fine sample (solid line), combusting at $\leq 600°$ C, is Cδ, with an isotopic composition within the terrestrial range (shaded). But the last $\sim$2% (shown on a 25-fold expanded scale), gets progressively heavier, reaching $1100‰$ at 1000° C. This heavy component has been tentatively assigned to Cβ, the carrier of Xe-S, though Cϵ, the carrier of Ne-E(H), may account for at least part of the peak. A third component, at 600 to 800° C, is incompletely resolved in this sample, but is seen more clearly in the coarse sample (dashed line), which is depleted in Xe-S and Xe-HL. It has been assigned to Cα, the carrier of Ne-$E(L)$.

found peaks of $\delta C^{13} \approx 340‰$ at 600 to 800° C and 1100‰ at 1000 to 1200° C (Fig. 13.1.11). They assigned these peaks to Ne-$E(L)$ and Xe-S, respectively, since they correlated roughly with the proportions of these components in the samples and with the gas-release temperatures on pyrolysis (~600 and 1400° C). The latter correlation might be questioned, as combustion and pyrolysis are manifestly different processes. But two parallel stepped combustion experiments on the same sample showed that Xe-S was released mainly at 1000° C (Ott et al. 1985), close to the 1100° C temperature where the heaviest C appeared (Yang and Epstein 1984). This match is not quite conclusive, though, as the temperature steps were rather coarse and differed somewhat in the two experiments (350, 500, 700, 1000 and 1060° in the noble-gas measurement; 350, 500, 700 and 1100° in the C measurement). Ion-microprobe data have shed further light on this question. There are *two* kinds of SiC in CM2 chondrites, one with (extrapolated) $\delta C^{13} \simeq 600–1500‰$ and $\delta N^{15} \simeq -400‰$ and the other with heavier C ($\geq 6000‰$) and heavier N (Zinner et al. 1987). It is unlikely that these two kinds of SiC were ever resolved by stepped combustion or pyrolysis and thus the peaks above 900° C presumably were mixtures of both with some lighter C. However, these SiC varieties fractionate to some extent in mineral and grain-size separations, and on that basis, the former has been tentatively assigned to Xe-S, while the latter is one of two candidates for carrier of Ne-E(H).

13.1.6. HEAVY NITROGEN

A strikingly large enrichment of N^{15} (up to 1033‰) has been found in metal and silicate from the brecciated stony irons Bencubbin and Weatherford (Prombo and Clayton 1985; Franchi et al. 1986). Two components seem to be present, differing in combustion temperature and geochemical character. Nα (900–1050° C, $\delta N^{15} = 888‰$) has abundances of 75 μg/g and 33 μg/g in metal and silicate, whereas Nβ (600–800° C, $\delta N^{15} = 754‰$) shows a preference for silicate: 5 μg/g vs 11 μg/g (Franchi et al. 1986). When enriched by HCl, HF treatment, both carriers combust at lower temperatures (300–400° and 450–600° C) but have very low C/N ratios (with $\delta C^{13} = 25‰$), and hence may not be carbonaceous. Indeed, Nα (though not Nβ) is soluble in conc HCl/HF which suggests an inorganic composition.

No noble-gas anomalies are associated with the nitrogen anomaly (Lewis 1985), and in the absence of more specific clues, it is not possible to decide whether this anomaly is nuclear or chemical. Franchi et al. (1986) suggest that the heavy N might have been produced by the hot CNO cycle in the envelope of a star evolved to the Si-burning stage (Audouze and Vauclair 1980). On the other hand, it is conceivable that the N anomaly was originally produced by ion-molecule reactions in a molecular cloud (Prombo and Clayton 1985). Either way, the similarity in δN^{15} of Nα and Nβ may imply that both carriers formed in a single source.

TABLE 13.1.2
Presolar Components in Meteorites

	Cα	Cβ (SiC)	Cδ (Diamond)	Cε (SiC)	Cθ	Nα,Nβ
Anomalous component	Ne-E(L)	Xe-S	Xe-HL	Ne-E(H)		heavy N
Associated components	—	He³,Kr-S	He,Ne,Ar,Kr	—		
Enriched in	Ne²²	Xe¹³⁰	Xe¹²⁴,¹³⁶	Ne²²		N¹⁵
Gas release T (°C)	700	1400	1000	1200		
Combustion T (°C)	600	1000	500	1000	400	300–600
δC¹³ (‰)	340	600–1500	−38	≥ 6000	−50	
δN¹⁵ (‰)	> 250	−400	−330	> 0?		1033
Abundance (µg/g)	5	3	400	2	105	< 60
Grain size (µm)	1–10	0.1	~0.005	0.1–1	0.2–2	
Probable source	nova	red giant or supernova	C: red giant Xe: supernova	nova?		red giant (Si-burning)

The presence of such a large anomaly implies that Bencubbin and Weatherford have retained some primitive features, despite their battered texture. By their O-isotope composition (Clayton and Mayeda 1978), the silicates may be related to CR2 and CM2 chondrites, and there are indications that the metal clasts may be direct nebular condensates. But otherwise these meteorites are far from pristine, having undergone severe brecciation and perhaps impact melting. This will make it difficult to trace the origin of the N anomaly from associated anomalies in other elements, although the chemical inertness of the $N\beta$ carrier offers some hope.

13.1.7. PRESOLAR C, N AND NOBLE GASES

Summary

Table 13.1.2 lists the presolar components of these elements that have been detected in meteorites. Let us briefly review their properties and origin.

$C\alpha$. Though this component seems to have been isolated only in a purity of $\leq 5\%$, its properties are fairly well known. It is a coarse-grained carbonaceous species (probably elemental) of low pyrolysis and combustion temperature and fairly low resistance to oxidation. The short, 2.6 yr, half-life of Na^{22} (parent of Ne-E) and the enrichment in N^{15} both point to a nova. An interesting but unsolved question is why the Na^{22} chose to go into C. The Ne^{22} concentration is not high (atom fraction 4×10^{-8}, from the data of Tang et al. 1985), but as the Na^{22} must have been accompanied by larger amounts of its stable isotope Na^{23}, let alone other elements, the total concentration of such trapped atoms must be much larger. Of the two principal trapping mechanisms, condensation or ion implantation, the former looks more likely, since the latter would contribute prohibitive amounts of He^4 and $Ne^{20,21}$ (Arnould and Nørgaard 1978,1981).

$C\beta$. This component seems to be silicon carbide of median grain size -0.12 μm (Tang and Anders 1988a). Its Xe component shows the signature of the s-process, and hence points to a red giant, though the ejection and grain condensation may have occurred at a later stage: planetary nebula or even supernova.

The link between Xe-S and heavy C has become somewhat tenuous, as mentioned above. Indeed, the coupling of s-process material with $C^{12}/C^{13} \approx 37$ is hard to rationalize (Schramm and Dearborn 1983). Observationally, s-process enrichments are seen both in carbon stars with $C^{12}/C^{13} \geq 100$ and in carbon-poor stars with $C^{12}/C^{13} \leq 20$, but not in stars with intermediate ratios (Schramm and Dearborn 1983). Theoretical models lead to a similar dichotomy. In view of these difficulties, Schramm and Dearborn suggest that much of $C\beta$ comes from other sources, i.e., interstellar grains that either diluted the C from the red giant or even served as the substrate in which s-

process nuclei were trapped by ion implantation. Indeed, the high extrapolated gas content (e.g., $Xe_s^{130} \approx 12 \times 10^{-8}$ ccSTP/g; Tang et al. 1985) and the high release temperature are much easier to reconcile with ion implantation than with sorption, the principal alternative (Wacker et al. 1985).

$C\delta$. The nature of this component and its link to Xe-HL have been discussed above. Its identification as diamond raises some interesting questions about the chemical state of C in interstellar dust. Saslaw and Gaustad (1969) suggested long ago that diamond (formed metastably in stellar atmospheres) might be a major constituent of interstellar dust, because diamond gives a fairly good match to the interstellar extinction curve. But this idea was not widely accepted, as the match, even when improved by adjustments in the size distribution, was not unique (Landau 1970). Moreover, the nucleation of diamond outside of its stability field was deemed improbable, as graphite forms preferentially under virtually all known conditions (Donn and Krishna Swamy 1969).

Now that $C\delta$ is known to be diamond, it seems that nature at least occasionally solves the nucleation problem in stellar atmospheres. The small grain size of the diamond, 26Å, actually suggests that the nucleation rate was high, stunting the growth of the grains. Indeed, at least in the region of the early solar system, diamond seems to have been more abundant than graphite. Nuth (1985; Chapter 13.3) has argued that crystalline graphite must be rare in the interstellar medium, since it is rare even in those C chondrites that have preserved an apparently interstellar kerogen component. This argument now is further strengthened by the preponderance of diamond over graphite, as illustrated by the C distribution for the CM2 chondrite Murray (Tang and Anders 1988a): diamond 570 μg/g, graphite <40 μg/g, amorphous C 110 μg/g, kerogen and organic compounds ~2.2 × 10^4 μg/g (of which only a small part may be presolar). Taken at face value these data suggest that in the region where the CM chondrites formed, only about 0.1% of the solar complement of C was present as elemental C; about 3% was present as solid organic matter; and 97% was present as CO and other gaseous species. Of course, these numbers only refer to the final state of the C, after reprocessing in the solar nebula. Perhaps larger amounts of elemental C were present initially, and were then reprocessed to CO and other gaseous compounds, which in turn reacted to form organic compounds and kerogen.

$C\varepsilon$. This component is SiC. It has a strong tendency to associate with spinel (presumably as inclusions or adhering grains), and thus can be partially separated from $C\beta$, which is also SiC. The $C\varepsilon$-SiC has heavy C ($\delta C \geq 6000‰$), and as it associates with spinel, it may well be the carrier of the variable C-isotope anomalies seen in Murchison spinels (-0 to $+3000‰$: Niederer et al. 1985; -0 to $7000‰$: Zinner and Epstein 1986,1987). The wide variation in bulk spinels may reflect variable proportions of isotopically light $C\theta$ (see below) and isotopically heavy $C\varepsilon$.

$C\theta$. When Murchison or Murray spinel is dissolved in H_3PO_4, up to 14% of black, amorphous carbon remains (Tang et al. 1988). It presumably comes from the numerous opaque inclusions in the spinel, 1 to 1.5 μm in size. This C component does not contain any noble gases, but its δC^{13} of -50‰ is distinctly outside the terrestrial or normal meteoritic range, and justifies assignment of a new name, $C\theta$.

$N\alpha$, $N\beta$. Nothing definite can yet be said about their carrier phases. Franchi et al. (1986) found 1.86% S in one of their residues, and therefore suggested that chromium sulfides might be the carriers. Sulfides of Cr and Ti in fact are possible stellar condensates (Clayton and Ramadurai 1977).

Other Presolar Phases. It is curious that C and SiC so predominate in Table 13.1.2, although they are only two of the major stellar condensates. One reason is their association with noble gases, which provide them with conspicuous "labels." Another is their resistance to destruction processes, such as vaporization in interstellar space and the solar nebula, or thermal metamorphism and hydrothermal alteration in the parent body. Lastly, the phases in Table 13.1.2 were all isolated by dissolving the bulk ($>99\%$) of the meteorite in HF-HCl, a treatment that would destroy most stellar condensates except carbon, carbides, and a few oxides and sulfides. Obviously, the acid treatment will have to be omitted when searching for acid-soluble phases such as silicates and metal; but without this preconcentration step, the operation becomes the proverbial search for a needle in a haystack.

Directions for Future Work

A major goal will be the isolation of all presolar solids in pure form, for intensive study by all available techniques, especially isotopic. The information return from a given presolar material rises exponentially with the number of properties that can be reliably assigned to it. In addition, there are a number of specific questions that need answering:

1. Why is $C\alpha$ so unstable, with so low a temperature of combustion and gas release (Table 13.1.2)?
2. Did $C\beta$ come from a single red giant near the forming solar system, or from general galactic background?
3. At what stage of stellar evolution was $C\beta$ ejected?
4. What is the stellar source of $C\delta$?
5. What is the origin of $C\varepsilon$?
6. How did $C\beta$ and $C\delta$ acquire their noble gases?
7. Why do $C\delta$ and $C\theta$ have C^{12}/C^{13} so close to the terrestrial value?

Evidently, meteorites contain a rich record of stellar nucleosynthesis and interstellar chemistry. The challenge to the cosmochemist is to decipher this record, thus complementing astronomical and theoretical studies by direct laboratory measurements on presolar matter.

Acknowledgments. This work was supported by a grant from the National Aeronautics and Space Administration. I am indebted to an anonymous referee for a perceptive, constructively critical review.

REFERENCES

Alaerts, L., Lewis, R. S., and Anders, E. 1979. Isotopic anomalies of noble gases in meteorites and their origins. IV. C3 (Ornans) carbonaceous chondrites. *Geochim. Cosmochim. Acta* 43:1421–1432.

Alaerts, L., Lewis, R. S., Matsuda, J., and Anders, E. 1980. Isotopic anomalies of noble gases in meteorites and their origins. VI. Presolar components in the Murchison C2 chondrite. *Geochim. Cosmochim. Acta* 44:189–209.

Anders, E. 1981. Noble gases in meteorites: Evidence for presolar matter and superheavy elements. *Proc. Roy. Soc. London* A374:207–238.

Anders, E. 1987. Local and exotic components of primitive meteorites, and their origin. *Phil. Trans. Roy. Soc. A*, in press.

Anders, E., and Heymann, D. 1969. Elements 112 to 119: Were they present in meteorites? *Science* 164:821–823.

Arnould, M., and Beelen, W. 1974. More about nucleosynthesis of the nuclei between carbon and neon. *Astron. Astrophys.* 33:215–230.

Arnould, M., and Nørgaard, H. 1978. Thermonuclear origin of Ne-E. *Astron. Astrophys.* 64:195–213.

Arnould, M., and Nørgaard, H. 1981. Isotopic anomalies in meteorites and cosmic rays, and the heavy neon puzzle. *Comments on Astrophys.* 9:145–154.

Audouze, J., and Vauclair, S. 1980. *Introduction to Nuclear Astrophysics.* (Dordrecht: D. Reidel).

Beer, H., Käppeler, F., Reffo, G., and Venturini, G. 1983. Neutron capture cross-sections of stable xenon isotopes and their application in stellar nucleosynthesis. *Astrophys. Space Sci.* 97:95–119.

Begemann, F. 1980. Isotopic anomalies in meteorites. *Rept. Prog. Phys.* 43:1309–1356.

Black, D. C. 1972. On the origins of trapped helium, neon and argon isotopic variations in meteorites. II. Carbonaceous meteorites. *Geochim. Cosmochim. Acta* 36:377–394.

Black, D. C., and Pepin, R. O. 1969. Trapped neon in meteorites. II. *Earth Planet. Sci. Lett.* 6:395–405.

Carr, R. H., Wright, I. P., Pillinger, C. T., Lewis, R. S., and Anders, E. 1983. Interstellar carbon in meteorites: Isotopic analyses using static mass spectrometry. *Meteoritics* 18:277.

Clayton, D. D. 1975. ^{22}Na, Ne-E, extinct radioactive anomalies and unsupported ^{40}Ar. *Nature* 257:36–37.

Clayton, D. D. 1981. Some key issues in isotopic anomalies: Astrophysical history and aggregation. *Proc. Lunar Planet. Sci. Conf.* 12B:1781–1802.

Clayton, D. D., and Hoyle, F. 1976. Grains of anomalous isotopic composition from novae. *Astrophys. J.* 203:490–496.

Clayton, D. D., and Ramadurai, S. 1977. On presolar meteoritic sulphides. *Nature* 265:427–428.

Clayton, D. D., and Ward, R. A. 1978. s-process studies: Xenon and krypton isotopic abundances. *Astrophys. J.* 224:1000–1006.

Clayton, R. N., and Mayeda, T. K. 1978. Genetic relations between iron and stony meteorites. *Earth Planet. Sci. Lett.* 40:168–174.

Dakowski, M. 1969. The possibility of extinct superheavy elements occurring in meteorites. *Earth Planet. Sci. Lett.* 6:152–154.

Donn, B., and Krishna Swamy, K. S. 1969. On the question of interstellar diamonds. *Nature* 224:570.
Eberhardt, P. 1974. A neon-E rich phase in the Orgueil carbonaceous chondrite. *Earth Planet. Sci. Lett.* 24:182–187.
Eberhardt, P. 1978. A neon-E rich phase in Orgueil: Results of stepwise heating experiments. *Proc. Lunar Planet. Sci. Conf.* 9:1027–1051.
Eberhardt, P., Jungck, M. H. A., Meier, F. O., and Niederer, F. 1979. Presolar grains in Orgueil: Evidence from neon-E. *Astrophys. J.* 234:L169–L171.
Eberhardt, P., Jungck, M. H. A., Meier, F. O., and Niederer, F. R. 1981. A neon-E rich phase in Orgueil: Results obtained on density separates. *Geochim. Cosmochim. Acta* 45:1515–1528.
Franchi, I. A., Wright, I. P., and Pillinger, C. T. 1986. A search for the location of isotopically heavy nitrogen in the Bencubbin meteorite. *Nature* 323:138–140.
Frick, U., and Pepin, R. O. 1981. Microanalysis of nitrogen isotope abundances: Association of nitrogen with noble gas carriers in Allende. *Earth Planet. Sci. Lett.* 56:64–81.
Geiss, J., Buehler, F., Cerutti, H., Eberhardt, P., and Filleaux, C. 1972. Solar wind composition experiment. In *Apollo 16 Prelim. Sci. Report*, NASA SP-315, Section 14, pp. 1–10.
Gilman, R. C. 1969. On the composition of circumstellar grains. *Astrophys. J.* 155:L185–L187.
Herzog, G. F., and Anders, E. 1974. Primordial noble gases in separated meteoritic minerals. II. *Earth Planet. Sci. Lett.* 24:173–181.
Heymann, D., and Dziczkaniec, M. 1976. Early irradiation of matter in the solar system: Magnesium (proton, neutron) scheme. *Science* 191:79–81.
Heymann, D., and Dziczkaniec, M. 1979. Xenon from intermediate zones of supernovae. *Proc. Lunar Planet. Sci. Conf.* 10:1943–1959.
Heymann, D., and Dziczkaniec, M. 1980. A first roadmap for kryptology. *Proc. Lunar Planet. Sci. Conf.* 11:1179–1213.
Iben, I., Jr. 1985. Nucleosynthesis in low and intermediate mass stars on the asymptotic giant branch. *Nucleosynthesis: Challenges and New Developments*, eds. W. D. Arnett and J. W. Truran (Chicago: Univ. of Chicago Press), pp. 272–291.
Jeffery, P. M., and Anders, E. 1970. Primordial noble gases in separated meteoritic minerals. I. *Geochim. Cosmochim. Acta* 34:1175–1198.
Jungck, M. H. A. 1982. *Pure ^{22}Ne in the Meteorite Orgueil: An Attempt to Characterize the Carrier Phases for Ne-E in Carbonaceous Meteorites* (München: Reinhardt).
Jungck, M. H. A., and Eberhardt, P. 1979. Neon-E in Orgueil density separates. *Meteoritics* 14:439–440.
Jungck, M. H. A., and Eberhardt, P. 1985. Ne-E in inclusions in apatite from Orgueil. *Meteoritics* 20:677.
Jungck, M. H. A., Meier, F. O., and Eberhardt, P. 1981. Apatite in Orgueil. Carrier phase for neon-E? *Meteoritics* 16:336–337.
Kerridge, J., and Chang, S. 1985. Survival of interstellar matter in meteorites: Evidence from carbonaceous material. In *Protostars & Planets II*, eds. D. C. Black and M. S. Matthews (Tucson: Univ. of Arizona Press), pp. 738–754.
Landau, R. 1970. Diamonds and the interstellar extinction curve. *Nature* 226:924.
Larimer, J. W., and Bartholomay, M. 1979. The role of carbon and oxygen in the chemistry of cosmic gases. *Geochim. Cosmochim. Acta* 43:1455–1466.
Lattimer, J. M., Schramm, D. N., and Grossman, L. 1978. Condensation in supernova ejecta and isotopic anomalies in meteorites. *Astrophys. J.* 219:230–249.
Lewis, R. S. 1985. Noble gases in the unique breccia, Bencubbin. *Meteoritics* 20:698–699.
Lewis, R. S., and Anders, E. 1981. Isotopically anomalous xenon in meteorites: A new clue to its origin. *Astrophys. J.* 247:1122–1124.
Lewis, R. S., and Anders, E. 1983. Interstellar matter in meteorites. *Sci. American* 249(2):66–77.
Lewis, R. S., and Matsuda, J. 1980. Carrier phases of CCFXe and other noble gas components in the Allende meteorite. *Meteoritics* 15:324–325.
Lewis, R. S., Srinivasan, B., and Anders, E. 1975. Host phase of a strange xenon component in Allende. *Science* 190:1251–1262.
Lewis, R. S., Gros, J., and Anders, E. 1977. Isotopic anomalies of noble gases in meteorites and their origins II. Separated minerals from Allende. *J. Geophys. Res.* 82:779–792.

Lewis, R. S., Alaerts, L., and Anders, E. 1979a. Ferrichromite: A major host phase of isotopically anomalous noble gases in primitive meteorites. *Lunar Planet. Sci.* X:725–727 (abstract).
Lewis, R. S., Alaerts, L., and Anders, E. 1979b. Isotopic anomalies in the Orgueil meteorite: Neon-E, s-process Xe, and CCFXe. *Lunar Planet. Sci.* X:728–730 (abstract).
Lewis, R. S., Alaerts, L., Matsuda, J., and Anders, E. 1979c. Stellar condensates in meteorites: Isotopic evidence from noble gases. *Astrophys. J.* 234:L165–L168.
Lewis, R. S., Hertogen, J., Alaerts, L., and Anders, E. 1979d. Isotopic anomalies in meteorites and their origins. 5. Search for fission fragment recoils in Allende sulfides. *Geochim. Cosmochim. Acta* 43:1743–1752.
Lewis, R. S., Anders, E., Wright, I. P., Norris, S. J., and Pillinger, C. T. 1983a. Isotopically anomalous nitrogen in primitive meteorites. *Nature* 305:767–771.
Lewis, R. S., Anders, E., Shimamura, T., and Lugmair, G. W. 1983b. Barium isotopes in meteorites: Evidence against an extinct superheavy element. *Science* 222:1013–1015.
Lewis, R. S., Tang, M., Wacker, J. F., Anders, E., and Steel, E. 1987. Interstellar diamonds in meteorites. *Nature* 326:160–162.
Manuel, O. K., Hennecke, E. W., and Sabu, D. D. 1972. Xenon in carbonaceous chondrites. *Nature* 240:99–101.
Matsuda, J., Lewis, R., and Anders, E. 1980. Neutron capture time scale of the s-process, estimated from s-process krypton in a meteorite. *Astrophys. J.* 237:L21–L23.
Moniot, R. K. 1980. Noble-gas-rich separates from ordinary chondrites. *Geochim. Cosmochim. Acta* 44:253–271.
Niederer, F. R., Eberhardt, P., Geiss, J., and Lewis, R. S. 1985. Carbon isotope abundances in Murchison residue 2C10c. *Meteoritics* 20:716–717.
Nuth, J. A. 1985. Meteoritic evidence that graphite is rare in the interstellar medium. *Nature* 318:166–168.
Ott, U., Mack, R., and Chang, S. 1981. Noble-gas-rich separates from the Allende meteorite. *Geochim. Cosmochim. Acta* 45:1751–1788.
Ott, U., Kronenbitter, J., Flores, J., and Chang, S. 1984. Colloidally separated samples from Allende residues: Noble gases, carbon and an ESCA-study. *Geochim. Cosmochim. Acta* 48:267–280.
Ott, U., Yang, J., and Epstein, S. 1985. s-Process Xe and Kr and Ne-E in a ^{13}C rich Murchison sample; Noble gas analysis by stepped combustion. *Meteoritics* 20:722–723.
Pepin, R. O. 1968. Neon and xenon in carbonaceous chondrites. In *Origin and Distribution of the Elements*, ed. L. H. Ahrens (Oxford: Pergamon Press), pp. 379–386.
Pepin, R. O., and Phinney, D. 1978. Components of xenon in the solar system. University of Minnesota preprint.
Pillinger, C. T. 1984. Light element stable isotopes in meteorites—from grams to picograms. *Geochim. Cosmochim. Acta* 48:2739–2766.
Pillinger, C. T. 1986. Interstellar dust components in meteorites—Implications for a comet nucleus sample return. In *Comet Nucleus Sample Return*, ESA SP-249 (Paris: Eur. Space Agency), pp. 41–45.
Podosek, F. A. 1978. Isotopic structures in solar system materials. *Ann. Rev. Astron. Astrophys.* 16:293–334.
Prombo, C. A., and Clayton, R. N. 1985. A striking nitrogen isotope anomaly in the Bencubbin and Weatherford meteorites. *Science* 230:935–937.
Reynolds, J. H., and Turner, G. 1964. Rare gases in the chondrite Renazzo. *J. Geophys. Res.* 69:3263–3281.
Reynolds, J. H., Frick, U., Neil, J. M., and Phinney, D. L. 1978. Rare-gas-rich separates from carbonaceous chondrites. *Geochim. Cosmochim. Acta* 42:1775–1797.
Roy, R. 1987. Diamonds at low pressure. *Nature* 325:17–18.
Saslaw, W. C., and Gaustad, J. E. 1969. Interstellar dust and diamonds. *Nature* 221:160–162.
Schelhaas, N., Ott, U., and Begemann, F. 1985. Trapped noble gases in some type 3 chondrites. *Meteoritics* 20:753.
Schramm, D. N., and Dearborn, D. S. P. 1983. Nucleosynthetic constraints from carbon isotope anomalies in meteorites. Preprint, EFI 83–02.
Schramm, D. N., and Olive, K. A. 1982. Chemical evolution of OB associations. *Ann. New York Acad. Sci.* 395:236–241.

Seaborg, G. T. 1968. Elements beyond 100, present status and future prospects. *Ann. Rev. Nucl. Sci.* 18:53–152.
Smith, S. P., Huneke, J. C., Rajan, R. S., and Wasserburg, G. J. 1977. Neon and argon in the Allende meteorite. *Geochim. Cosmochim. Acta* 41:627–647.
Srinivasan, B. 1981. Host phases and origin of noble gases in meteorites. *Naturwissenschaften* 68:341–353.
Srinivasan, B., and Anders, E. 1978. Noble gases in the Murchison meteorite: Possible relics of s-process nucleosynthesis. *Science* 201:51–56.
Srinivasan, B., Alexander, E. C., Jr., Manuel, O. K., and Troutner, D. E. 1969. Xenon and krypton from the spontaneous fission of californium-252. *Phys. Rev.* 179:1166–1169.
Swart, P. K., Grady, M. M., Pillinger, C. T., Lewis, R. S., and Anders, E. 1983. Interstellar carbon in meteorites. *Science* 220:406–410.
Tang, M., and Anders, E. 1988a. Isotopic anomalies of Ne, Xe, and C in meteorites. II. Interstellar diamond and SiC: Carriers of exotic noble gases. *Geochim. Cosmochim. Acta*, in press.
Tang, M., and Anders, E. 1988b. Isotopic anomalies of Ne, Xe, and C in meteorites. III. Local and exotic noble gas components and their interrelations. *Geochim. Cosmochim. Acta*, in press.
Tang, M., Lewis, R. S., and Anders, E. 1985. Anomalous noble gases and presolar carbon in Murray (C2M) chondrite. *Meteoritics* 20:712–713.
Tang, M., Lewis, R. S., Anders, E., Grady, M. M., Wright, I. P., and Pillinger, C. T. 1986. Presolar components in the Murray, C2M chondrite: Heavy carbon, nitrogen and anomalous noble gases. *Meteoritics* 21:458–459.
Tang, M., Lewis, R. S., Anders, E., Grady, M. M., Wright, I. P., and Pillinger, C. T. 1988. Isotopic anomalies of Ne, Xe, and C in meteorites. I. Separation of carriers by density and chemical resistance. *Geochim. Cosmochim. Acta*, in press.
Tielens, A. G. G. M., Seab, C. G., Hollenbach, D. J., and McKee, C. F. 1987. Shock processing of interstellar dust: Diamonds in the sky. *Astrophys. J.* 319:L109–L113.
Trimble, V. 1975. The origin and abundances of the chemical elements. *Rev. Mod. Phys.* 47:877–976.
Wacker, J. F., Zadnik, M. G., and Anders, E. 1985. Laboratory simulation of meteoritic noble gases. I. Sorption of xenon on carbon: Trapping experiments. *Geochim. Cosmochim. Acta* 49:1035–1048.
Walter, G., Beer, H., Käppeler, F., and Penzhorn, R.-D. 1986. The s-process branching at ^{85}Kr. *Astron. Astrophys.* 155:247–255.
Wannier, P. G. 1980. Nuclear abundances and evolution of the interstellar medium. *Ann. Rev. Astron. Astrophys.* 18:399–437.
Ward, R. A., Newman, M. J., and Clayton, D. D. 1976. s-Process studies: Branching and the time scale. *Astrophys. J. Suppl.* 31:33–59.
Woosley, S. E., and Howard, W. M. 1978. The p-process in supernovae. *Astrophys. J. Suppl.* 36:285–304.
Yang, J., and Anders, E. 1982. Sorption of noble gases by solids, with reference to meteorites. III. Sulfides, spinels, and other substances; on the origin of planetary gases. *Geochim. Cosmochim. Acta* 46:877–892.
Yang, J., and Epstein, S. 1984. Relic interstellar grains in Murchison meteorite. *Nature* 311:544–547.
Yang, J., Lewis, R. S., and Anders, E. 1982. Sorption of noble gases by solids, with reference to meteorites. I. Magnetite and carbon. *Geochim. Cosmochim. Acta* 46:841–860.
Zadnik, M. G., Wacker, J. F., and Lewis, R. S. 1985. Laboratory simulation of meteoritic noble gases. II. Sorption of xenon on carbon: Etching and heating experiments. *Geochim. Cosmochim. Acta* 49:1049–1059.
Zinner, E., and Epstein, S. 1986. Heavy carbon in individual oxide grains from Murchison acid residue CFOc. *Lunar Planet. Sci.* XVII:967–968 (abstract).
Zinner, E., and Epstein, S. 1987. Heavy carbon in individual oxide grains from the Murchison meteorite. *Earth Planet. Sci. Lett.* 84:359–367.
Zinner, E., Tang, M., and Anders, E. 1987. Interstellar silicon carbide and oxynitride in the Murray meteorite: Carriers of large isotopic anomalies of Si, C, N, and noble gases. *Nature*, submitted.

13.2. INTERSTELLAR CLOUD MATERIAL IN METEORITES

ERNST ZINNER
Washington University

In addition to material of apparently circumstellar origin, many primitive meteorites contain material for which an origin in interstellar clouds is indicated. Such material, apparently organic in nature, is revealed by enrichments of deuterium relative to terrestrial material. Astronomical observations have shown that simple organic molecules in dense interstellar clouds have large D/H ratios. The H-isotopic compositions of those molecules are believed to be the result of ion-molecule reactions taking place at low temperatures.

13.2.1. INTRODUCTION

All elements from C on up are produced in stars and their isotopic compositions mainly reflect the different possible nucleosynthetic reactions taking place in different sources (see Chapters 14.1 and 14.2). Physicochemical processes usually lead to mass-dependent fractionation (see Chapter 12.1). For elements with only two stable isotopes (H, C, N), nuclear effects cannot be distinguished from mass-dependent fractionation and in many cases the interpretation of anomalies in C and N as being of nucleosynthetic origin is based on the size of the effects (Swart et al. 1983; Franchi et al. 1986). The case for nucleosynthetic anomalies of C and heavier elements in carbonaceous material has been dealt with in the previous chapter (13.1).

Many primitive meteorites and interplanetary dust particles (see Chapter 11.1) exhibit large deuterium (D) excesses compared to terrestrial materials. Since the mass ratio of the isotopes D and H is much larger than for any other elements, large fractionation effects are expected for H. Both its stable isotopes were present before star formation started, primordial synthesis having occurred in the Big Bang (Reeves et al. 1973; Epstein et al. 1976). Subsequent processing in stars tends to destroy D and leads to a decrease of the D/H ratio. The model of Audouze and Tinsley (1974) predicts a decrease of a factor 3 since Galaxy formation and a factor 1.5 since solar system formation. More astronomical information exists on D/H ratios in different parts of the Galaxy and the solar system (Penzias 1980; Wannier 1980; Geiss and Reeves 1981; Owen et al. 1986) than on isotopic ratios of other elements. The observations make it plausible that the D excesses in meteorites and interplanetary dust particles are due to isotopic fractionation having occurred in interstellar clouds by low-temperature ion-molecule reactions during the synthesis of precursor molecules of organic material found in meteorites. In the following, the experimental evidence for D enrichments in meteorites and interplanetary dust particles is reviewed, the interpretation of the isotopic record is discussed and questions are raised concerning the identification of interstellar material in meteorites.

13.2.2. EXPERIMENTAL ASPECTS

Two methods of mass spectrometry have been used to measure H isotopes in meteoritic material: gas mass spectrometry (see, e.g., Pillinger 1984) and secondary ion mass spectrometry (see, e.g., McKeegan et al. 1985). By far the most data have been obtained by the first, traditional, technique.

In many (perhaps all) cases, the carrier phases of isotopic anomalies constitute only a small fraction of the meteorite. As a consequence, isotopic anomalies are frequently masked by abundant isotopically normal material. Uncovering the anomalies in such cases requires separation of the host phase by a series of mechanical, chemical and/or physical treatments. Since the chemical form of host phases is unknown, these separation steps frequently proceed on a trial and error basis (cf. Lewis and Anders 1983). For hydrogen gas mass spectrometric measurements, chemical treatments have involved dissolution of silicates with acids like HF and HCl (Smith and Rigby 1981; Robert and Epstein 1982; Kerridge 1983; Yang and Epstein 1983), the removal of soluble organic compounds with organic solvents like benzene, hexane, methanol and acetone (Becker and Epstein 1982; Yang and Epstein 1983, 1984), as well as additional removal of phases with oxidizing agents such as H_2O_2, HNO_3 and $HClO_4$ (Yang and Epstein 1983,1984). At various stages of the chemical dissolution process, residues can be divided into different size fractions. In this way, for exmaple, Yang and Epstein (1984) have separated as many as eight different residues from the meteorite Murchison.

In addition to analyzing the residues, the isotopic compositions of organic compounds extracted with organic solvents have also been measured (Briggs 1963; Becker and Epstein 1982). A different way of separating inorganic components from organic material was attempted by Kolodny et al.

(1980) who oxidized the meteorites in a plasma discharge and made measurements of the bulk and oxidized material to infer the hydrogen isotopic composition of the removed fraction. In a similar fashion, Yang and Epstein (1983) inferred the composition of phases dissolved by the HF-HCl acid treatment (Lewis et al. 1975) from measurements of meteorite bulk and residue.

In gas mass spectrometric measurements, hydrogen or H-bearing compounds are released by heating samples either in vacuum (pyrolysis) or in an oxygen atmosphere (combustion). Stepwise heating offers the possibility of further separation. The H released at different temperatures is attributed to the breakdown of different host phases. Stepwise pyrolysis also allows the separation of H_2 gas from condensible gases such as H_2O and from CH_4 (Epstein and Taylor 1970). In many instances, experimenters have also analyzed C and N isotopes along with H to look for correlated effects (Boato 1954; Robert and Epstein 1982; Kerridge 1983,1985; Yang and Epstein 1983,1984).

Various problems are inherent in the chemical separation and stepwise heating techniques. Sometimes, an isotopically anomalous phase can be only partially separated if it is chemically similar to isotopically normal material. In favorable cases, incomplete separation leads only to a dilution of the magnitude of the isotopic anomaly; however, when several isotopically distinct components are present, it may result in the incorrect identification of carrier phases. Problems associated with stepwise heating techniques have been pointed out by Halbout and Robert (1986). The thermal release patterns of different phases can overlap over wide temperature ranges or can be quite different for different types of a given mineral (e.g., different types of phyllosilicates release gases at different temperatures [McNaughton et al. 1982a]). This makes separation by stepwise heating and the unambiguous assignment of gas released in certain temperature intervals to specific carrier phases often virtually impossible, especially in view of the fact that the phases involved are sometimes only poorly characterized or not identified at all.

In correlative studies of different elements, the differing chemical and physical behavior of these elements during chemical treatment and stepwise heating can lead to incorrect associations. The thermal release patterns of H and C differ from one another but also depend on the type of sample (bulk or residue) and on the heating technique used (pyrolysis or combustion) (Kerridge 1983; Yang and Epstein 1983). Hydrogen and C that were originally present in the same phase might be released at completely different temperatures. On the other hand, the association of different carriers in a given residue may be purely accidental. For example, the heavy C found in refractory oxides (Zinner and Epstein 1987) may have nothing to do with the organic component with which the oxides were mixed in the residue (Yang and Epstein 1984). The chemical treatment thoroughly destroys any spatial relationship between the host phase and other phases that might have provided information about the incorporation of the host phase into the meteorite. Gas mass spectrometric analysis of chemically treated samples involves relatively large amounts (on the order of 10^{-6} to 10^{-5}g of analyzed H) and averages over any heterogeneous spatial distribution of isotopically anomalous components. In spite of these limitations, almost all of the information we have on D/H ratios in meteorites comes from gas mass spectrometric measurements.

A completely different approach is the measurement of H isotopes by secondary ion mass spectrometry in the ion microprobe (Hinton et al. 1983; McKeegan et al. 1985). This technique allows the measurement of subnanogram samples (total mass of analyzed H $< 10^{-11}$g). In principle, it allows also in situ measurement, i.e., analysis within the petrographic context. The spatial resolution of the ion probe of ~ 1 μm can help in locating of anomalous carrier phases if their sizes exceed this limit. On the other hand, a uniform, fine-grained distribution of anomalous and normal phases on a smaller scale cannot be resolved in the ion probe, while chemical means can result in their separation. As shown below, the ion probe can be successfully applied for measurements of heterogeneously distributed deuterium.

13.2.3. THE EXPERIMENTAL RECORD: D/H RATIOS IN METEORITES AND INTERPLANETARY DUST PARTICLES

This section contains a survey of D and H experimental data. A review of measurements of H and other light-element stable isotopes in meteorites has previously been presented by Pillinger (1984). It is customary to express the isotopic composition of H in the δD notation, where δD is the deviation of the D/H ratio from the terrestrial standard, standard mean ocean water (SMOW), in parts per thousand or per mil (‰):

$$\delta D = \left[\frac{(D/H)_{sample}}{(D/H)_{SMOW}} - 1 \right] \times 1000 \quad (1)$$

where $(D/H)_{SMOW} = 1.5576 \times 10^{-4}$ (Hoefs 1980).

Boato (1954) was the first to measure D/H ratios in primitive meteorites. Of twelve carbonaceous chondrites analyzed by combustion of bulk samples above 180°C, four, Ivuna and Orgueil (both CI), Murray (CM2) and Mokoia (CV3), gave δD values ranging from 50 to 300‰, clearly exceeding the range seen in terrestrial rocks (-200 to +20‰). Boato's findings of excesses were corroborated by Briggs (1963) who obtained similar D enrichments (84 to 275‰) in benzene-methanol extracts from Orgueil, Murray and Mokoia. The next step was taken by Robert et al. (1977,1978,1979) who found one unequilibrated chondrite, Chainpur (LL3.4), that displayed large D excesses. Water released from its chondrules showed δD ranging up to +4400‰. Subsequent measurements of Chainpur chondrules by Yang and Epstein (1983) could not duplicate these results, and the question whether the results obtained by Robert et al. (1979) can be attributed to chondrules remains open. Nevertheless, a series of experiments by McNaughton et al. (1981,1982a) confirmed the existence of D-enriched material in unequilibrated ordinary chondrites. Deuterium excesses in Chainpur bulk samples ranged to +670‰, and were even considerably higher in Bishunpur (L3.1), where δD values reached +3100‰ and Semarkona (LL3.0) with δD values of up to 4000‰. In particular, during stepwise pyrolysis, water released from Semarkona between 550 and 650°C had a δD value of 5740‰, i.e., a D/H ratio almost seven times that of SMOW.

Since the above measurements were made on chemically untreated meteorite samples, no detailed information on the chemical nature of the D carrier phase was obtained. Kolodny et al. (1980) attacked the question of carrier phases in carbonaceous chondrites by oxidizing the organic matter in an oxygen plasma and measuring the hydrogen isotopic composition of oxidized and unoxidized samples. In this way, they inferred D enrichments of up to +1600‰ in the organic matter, which was, however, contaminated by some inorganic H also removed by the plasma treatment. The presence of D excesses in organic material was confirmed by Smith and Rigby (1981) and Robert and Epstein (1982) who measured high D/H ratios in residues obtained by HF-HCl treatment of carbonaceous chondrites. Becker and Epstein (1982) found that material extracted from carbonaceous chondrites by organic solvents shows minimum enrichments of D of up to 500‰.

Most of the information on the distribution of the anomalous H in meteorites stems from a series of detailed measurements on bulk samples and a variety of acid residues by Robert and Epstein (1982), Kerridge (1983,1985) and Yang and Epstein (1983,1984,1985). These measurements clearly established the presence of D excesses in acid-insoluble organic material from a large variety of meteorites.

Data on the distribution of D in meteorites are summarized in Table 13.2.1. In many cases, only bulk measurements or measurements on residues were made; frequently the amounts of anomalous H are not given. For these reasons, Table 13.2.1 presents only results on selected meteorites. Unfortunately, there is still not enough detailed information to answer all the relevant questions concerning the extent of D excesses and the nature and distribution of carrier phases.

Apparent discrepancies in Table 13.2.1 stem from several factors. In some cases, D excesses were reported for bulk samples and in others from residues prepared in different ways. Some investigators excluded H released at the lowest temperature steps when giving total δD values. Furthermore, there are sometimes large differences between different measurements on similar types of samples from the same meteorites (e.g., the bulk measurements on Cold Bokkeveld, Semarkona, Bishunpur and Chainpur) that are due to heterogeneities in the D distribution within a given meteorite. The major observations obtained so far are given in the following subsections.

Acid-Insoluble Organic Material

In most meteorites, the dominant carrier of D excesses is an acid-resistant organic phase present in very different concentrations in CI, CM2, CR2 and unequilibrated ordinary chondrites. This material, also termed "organic polymer" or "kerogen," consists of a highly aromatic cross-linked 3-dimensional network with the approximate elemental composition $C_{100}H_{48}N_{1.8}S_2O_{12}$ (Hayatsu and Anders 1981; Robert and Epstein 1982; Kerridge 1983; Yang and Epstein 1983). Since the term "polymer" does not accurately describe this material, "kerogen" or "acid-insoluble residue" will be used henceforth. Kerridge (1983) noticed the almost identical isotopic structure of H, C and N in the kerogen obtained from Orgueil and Murray. However, the δD values in acid residues from other meteorites vary considerably. It is not yet clear how much this variability reflects the presence of carrier phases with different D enrichments or the variable dilution of a single phase with a very high D enrichment with isotopically normal kerogen. There is some evidence for the second possibility. Treatments with acids other than HF-HCl and with additional organic solvents and grain-size separation can enrich the D, as shown in the case of Murchison where Yang and Epstein (1984) obtained a separate (CFOc) with $\delta D = 2350‰$, larger than the values measured in HF-HCl Murchison residues (Robert and Epstein 1982; Yang and Epstein 1983).

If the D-rich phase is only a small fraction of the total kerogen, it is not sufficiently chemically different from the isotopically normal kerogen that it can be easily separated. Oxidizing agents such as HNO_3 and $HClO_4$ seem to dissolve preferentially the D carrier (Yang and Epstein 1984). Hayatsu et al. (1983) and Anders (1986) conjectured that cyanoacetylenic polymers, which account for only a small part of the meteoritic kerogen, are the D carriers. There exists no experimental proof whether or not this carrier hypothesis is correct, but the chemical diversity displayed by the D enrichments makes it unlikely that a single carrier, such as cyanoacetylenic polymers, is responsible.

Acid-Soluble (Organic?) Phase

In Semarkona and Bishunpur, D/H ratios measured in bulk samples are larger than those of the acid residues (McNaughton et al. 1982a; Yang and Epstein 1983). Yang and Epstein (1983) computed the δD values of the dissolved fraction in these two meteorites and Renazzo, and obtained D excesses that, in Semarkona, reached values of more than 4000‰. However, the large variations of δD for different bulk samples from Semarkona and Bishunpur (McNaughton et al. 1982a; Yang and Epstein 1983) introduce considerable uncertainties as to the δD value of the acid-soluble phase. In any event, the δD value of 5740‰ measured by McNaughton et al. (1982a) does indicate the presence of a component with $\delta D \geq 6000‰$ in Semarkona.

Yang and Epstein (1983) suggested that the carrier of the D-enrichments

TABLE 13.2.1
Summary of Distribution of D in Meteorites

Meteorite* (Class)	H† (μg/g)	δD Whole Rock (‰)	δD Residue (‰)	References
Orgueil (CI)	3620–8100	170–235	400–1080	a–g
Ivuna (CI)	6640–8160	180–300	—	b, c, h
Murchison (CM2)	7340–7880	−60 – −12	360–2350	a, c, e, g, i, j
Murray (CM2)	6410–7800	−50–100	400–1090	a–c, e–g
Cold Bokkeveld (CM2)	6420–11790	−100–525	780	a, e, h
Bells (CM2)	3770–4180	385–990	—	h
Mighei (CM2)	9030–9600	−100 – −70	500	b, c, g
Essebi (CM2)	6270	450	—	h
Kaidun (anom.)	4620	1045	—	h
Renazzo (CR2)	3880–4610	530–1010	2080–2920	a, c, e, k
Al Rais (CR2)	6700–6800	520–690	—	h, l
Felix (CO3)	240	−60	—	h
Ornans (CO3)	60	2150	—	h
ALHA 77003 (CO3)	167	60–150	—	l

Allende (CV3)	17–200	64–440	120	a, h
Mokoia (CV3)	570–620	190–300	63	b, c, g, h
Semarkona (LL3.0)	225–490	1400–4380	270–440	a, m, n
Bishunpur (L3.1)	275	590–3100	1490	a, e, l
Chainpur (LL3.4)	120	140–4400	412	a, m, o–q
Mezö-Madaras (L3.7)	—	—	770	a
Carraweena (L3.9)	154	−130	640	a
Kelly (LL4.4)	—	—	28	a
Mosquito (IDP)	up to 15000^s	130–2530	—	r
Calrissian (IDP-LLS)	—	370–2190	—	r
Butterfly (IDP-PX)	—	−320–9000	—	t

* For the infrared classification of interplanetary dust particles see Chapter 11.1.
† Weight fraction of the total meteorite.
References. a: Yang and Epstein 1983; b: Boato 1954; c: Kolodny et al. 1980; d: Robert et al. 1978; e: Robert and Epstein 1982; f: Kerridge 1983; g: Smith and Rigby 1981; h: Kerridge 1985; i: Fallick et al. 1983; j: Yang and Epstein 1984; k: Grady et al. 1983; l: McNaughton et al. 1982b; m: McNaughton et al. 1982a; n: McNaughton et al. 1981; o: Robert et al. 1977; p: Robert et al. 1979; q: Robert et al. 1983; r: McKeegan et al. 1985; s: Estimated from (r) and unpublished data; t: McKeegan et al. 1987.

in the dissolved material is an acid-soluble organic phase. This argument is based on the assumption that the phyllosilicates in Renazzo, Semarkona and Bishunpur have a δD of −110‰, the same δD as that inferred for the phyllosilicates of other meteorites (Yang and Epstein 1983). However, there is no direct experimental evidence for this assumption.

In Renazzo, the D enrichment in the kerogen is much higher than the inferred D enrichment in the acid-soluble fraction. Low-temperature exchange during aqueous alteration on the meteorite parent body (McSween 1979; Bunch and Chang 1980) could have led to isotopic redistribution of the D-rich component between different phases within each meteorite (Halbout et al. 1986a). The thermal release pattern for Renazzo does not exclude a D-rich phyllosilicate carrier (Robert and Epstein 1982). In addition, organic extracts from Renazzo are isotopically rather normal and do not exhibit the D excesses present in extracts from Orgueil, Murchison and Murray (Becker and Epstein 1982). Finally, ion probe measurements of individual 10 μm grains from the matrix of Renazzo and Semarkona (McKeegan and Zinner 1984 showed very different associations of the D. In Semarkona, the D excess is generally correlated with carbonaceous material. In Renazzo, there is no correlation between the D excess and C, but there exists a good correlation between the OH⁻ and H⁻ signals, and both are correlated with δD (Fig.

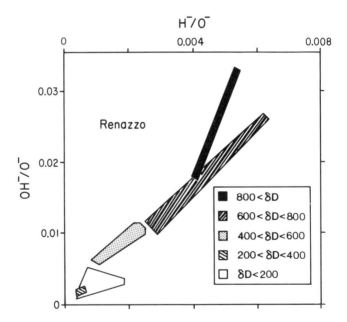

Fig. 13.2.1. The deuterium excess in individual 10μm matrix grains from Renazzo plotted as function of the negative secondary ion ratios OH⁻/O⁻ and H⁻/O⁻. Measurements were made by ion microprobe mass spectrometry. The correlation of δD with H and OH suggests that water of hydration is the carrier of excess D (figure from McKeegan and Zinner 1984).

13.2.1), indicating that the D excess in Renazzo might be associated with water of hydration in the silicates. Besides Renazzo, bulk samples of the carbonaceous chondrites Bells and Cold Bokkeveld (both CM2), Al Rais (CR2), Ornans (CO3) and Kaidun (anomalous) show δD values of greater than 500‰ (Kerridge 1985). For these meteorites the partitioning of the D excess between acid-soluble and insoluble material is still unknown.

For Semarkona and Bishunpur, the process of hydrothermal alteration at low temperature, considered for Renazzo, cannot be invoked because δD is higher in bulk samples than in the kerogen. Furthermore, type 3 ordinary chondrites do not seem to have been affected by as much aqueous alteration as the CI, CM and CR meteorites (McSween 1979). At present, no evidence exists as to the nature of the soluble D carrier in Semarkona and Bishunpur.

Many other type 3 ordinary chondrites show substantial D enrichments in whole rock samples (McNaughton et al. 1982a), but since only bulk measurements exist, the carriers of these enrichments are still not identified. McNaughton et al. (1982a) observed a dependence of the δD values measured in type 3 OCs on the petrographic subtype as defined by TL measurement (Sears et al. 1982; Chapter 1.1). This relationship is shown in Fig. 13.2.2. One glaring exception is the LL3.0 chondrite Krymka for which no D excess was found (Fallick et al. 1983). However, based on its volatility content, Krymka was reclassified as belonging to a higher subtype (Anders and Zadnik 1985)

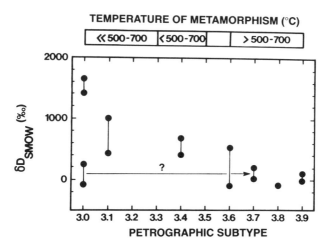

Fig. 13.2.2. Deuterium contents of combusted total H in type 3 ordinary chondrites plotted against petrographic subtype (TL after Sears et al. 1982). There is a correlation between the δD and the subtypes which experienced different temperatures of metamorphism. Krymka, a LL3.0 chondrite according to the TL classification, has small δD values and does not fit this relationship. However, a reclassification of Krymka according to volatility (Anders and Zadnik 1985) shifts Krymka to a higher subtype (arrow) in better agreement with the general trend (figure adapted from McNaughton et al. 1982a).

and thus fits much better into the relationship between δD and subtype (Fig. 13.2.2). Unfortunately, the total D excess cannot be obtained from the measurements by McNaughton et al. (1982a). In analogy with Semarkona and Bishunpur, one could expect contributions from an acid-soluble phase in these and other UOCs, but the relevant measurements still have to be made. Such a phase might also be present in the CV3 meteorite Mokoia which shows higher δD values in bulk samples (Kolodny et al. 1980) than in the acid residue (Smith and Rigby 1981).

The D-rich carrier(s) is (are) heterogeneously distributed within a given meteorite. This is shown by the data on carbonaceous chondrites where different bulk samples of Bells, Cold Bokkeveld, Renazzo and Allende show large variations in δD values (Kerridge 1985). The same is the case for the UOCs Semarkona, Bishunpur and Chainpur (see Table 13.2.1).

Deuterium-Depleted Components in Meteorites

Low D/H ratios have been found in several cases. Yang and Epstein (1983) obtained $\delta D = -490‰$ for the enstatite chondrite Abee (E4). Zinner et al. (1983) measured $\delta D = -500‰$ in one grain from the Renazzo matrix and Robert et al. (1983) observed low δD values ($-290‰$) in ordinary chondrites and postulated the existence of an end member with δD smaller than $-430‰$. So far, no low-δD carrier has been identified.

How Many H-Isotopic Components in Meteorites?

In summary, there appear to exist at least four distinct H isotopic components in meteorites: water of $\delta D = -110‰$ in phyllosilicates, a D-rich organic component in acid-insoluble residues (kerogen), an inferred D-rich acid-soluble component (of possibly organic composition), and a D-depleted component. Except for the phyllosilicates, these components are operationally defined, but, so far, are chemically not well characterized.

It is also not clear how the D-rich material extracted with solvents (Briggs 1963; Becker and Epstein 1982) is related to the kerogen and the postulated D-rich acid-soluble phase. The extractable organic phase contributes only a small part of the total D excesses in the meteorites studied. Extracts from Renazzo are isotopically relatively normal compared to a D-enriched inferred acid-soluble component. However, incomplete extraction and terrestrial contamination may play an important role in explaining these results. Extraction with organic solvents removes $\sim 5\%$ of the total C of the meteorite, while the acid treatment removes approximately half of the total C (Becker and Epstein 1982; Robert and Epstein 1982). Becker and Epstein (1982) showed that at least part of this difference is due to water-soluble organic material which is incompletely extracted due to shielding by silicates. Thus, if one could extract these compounds more completely, they could indeed carry a significant fraction of the D excesses of the total meteorites.

The first measurements on chemically well-characterized organic compounds were recently made on extracts from Murchison. Epstein et al. (1987) analyzed amino acids and monocarboxylic acids obtained from the water-soluble fraction of this meteorite. The amino acid fraction yielded a δD value of 1370‰, the carboxylic acid fraction 377‰. Also the C and N isotopic compositions were measured in these two fractions and gave $\delta^{13}C = 23.1‰$ and $\delta^{15}N = 90.0‰$ in the amino acids and $\delta^{13}C = 6.7‰$ and $\delta^{15}N = -1.3‰$ in the carboxylic acids. These measurements constitute an important first step toward the more complete chemical characterization of the D-rich carriers in meteorites.

Interplanetary Dust Particles

In addition to the classes of meteorites listed in Table 13.2.1, D excesses are also present in interplanetary dust particles. Out of 19 interplanetary dust particles of the "pyroxene" and "layer-lattice silicate" infrared classes (see Chapter 11.1) measured in the ion probe (Zinner et al. 1983; McKeegan et al. 1985,1987), roughly half showed D excesses >100‰ with δD values ranging up to 2700‰. In both classes, the D excesses are correlated with C, indicating a carbonaceous carrier. It is not yet known whether or not this carrier is acid soluble or whether both types of carriers are present. In one interplanetary dust particle, Mosquito, the carbonaceous phase makes up at least one third of the volume, and, as a consequence, the total H concentration is higher than in any of the meteorites (Table 13.2.1).

The D/H ratios in individual interplanetary dust particles vary on a μm-size scale (McKeegan et al. 1985). An extreme example is a recently analyzed interplanetary dust particle of the pyroxene class, Butterfly (McKeegan et al. 1987). Ion probe measurements of individual fragments of this particle yielded δD values between -320 and $+2700‰$. However, even within a D-rich fragment, the D/H ratio varies from spot to spot. Figure 13.2.3 shows the spatial distribution, obtained by secondary-ion imaging, of the H and the D signal inside this fragment. Most of the D is concentrated in a much smaller area than the H. The δD value of this D-rich "hot spot," which is smaller than 1 μm, is at least 9000‰. Similar measurements in meteorites are needed to establish whether the D excesses are concentrated in minor carriers, present in a small fraction of the organic material. Based on indirect arguments, Anders (1986) suggested this possibility, but experimental proof is still lacking.

Correlated Measurements of Other Isotopes

Starting with Boato (1954), C-isotopic measurements in bulk samples and residues have shown that the organic material carrying large D excesses is characterized by an essentially normal C-isotopic composition. The $\delta^{13}C$ values in residues lie within a fairly narrow range around $\delta^{13}C = -15‰$, despite a large range in corresponding δD values (Robert and Epstein 1982; Kerridge 1983; Yang and Epstein 1983,1984). Although no general correlations are seen between δD and $\delta^{13}C$, there is a suggestion of correlated effects

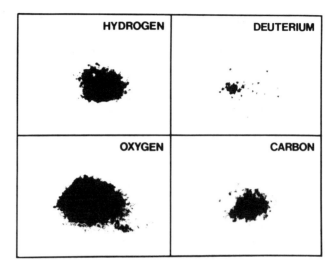

Fig. 13.2.3. Secondary negative ion images of the distribution of H, D, C and O in an individual fragment of the interplanetary dust particle Butterfly. One frame is 20 μm wide. The spatial resolution of the images is approximately 1 μm. Different elements show different spatial distributions in this fragment, but also H and D are differently distributed. The D/H ratio averaged over the whole fragment corresponds to a D excess of 2000‰. Since the D is concentrated into a much smaller area than the H, in the area of high D concentration ($\leq$ 1 μm in size) the δD value is at least 9000‰ (figure from McKeegan et al. 1987).

in a few special cases. Yang and Epstein (1983) noticed a negative correlation between the δD and δ^{13}C values measured in bulk samples of Semarkona, Bishunpur and Renazzo; Kerridge (1983) pointed to a positive correlation between these quantities in low-temperature combustion data of Orgueil and Murray kerogen.

The situation is less well defined when it comes to possible H and N-isotopic correlations because fewer data exist. Kerridge (1985) presented the most comprehensive set of data on correlated H and N measurements in bulk samples of carbonaceous chondrites. All data taken together suggest a positive correlation between δD and δ^{15}N, but this trend rests on the three meteorites, Renazzo, Al Rais and Kaidun which show large δD and δ^{15}N values. Renazzo has high δ^{15}N values (150 to 190‰) in both bulk samples and residues (Kung and Clayton 1978; Robert and Epstein 1982; Grady et al. 1983) which have very different δD values (see Table 13.2.1). On the other hand, data for CO and CV chondrites show an apparent anticorrelation.

There appears to be no simple general correlation between δD and δ^{15}N in residues (Robert and Epstein 1982; Kerridge 1983). The same is the case for organic extracts (Becker and Epstein 1982). Nitrogen-15 excesses are seen in residues from two Antarctic meteorites, 94‰ for Y-790112 (CR) and 214‰ for Y-790003 (CM), but their δD values are only 90‰ and 21‰

(Grady et al. 1983). Yang and Epstein (1984) claim to have observed a relationship between δD and $δ^{15}N$ in different Murchison residues but the data are too scant to be conclusive.

The only correlations for H and O are based on the work of Halbout et al. (1986a) who measured O isotopes in acid residues from Orgueil, Murchison, Murray and Renazzo. The authors claimed to have found a positive correlation between their $δ^{18}O$ values and δD values from previous measurements by Robert and Epstein (1982) and Kerridge (1983) of these meteorites, but the experimental evidence is inconclusive.

13.2.4. INTERSTELLAR MOLECULES IN METEORITES

Since D is synthesized in the Big Bang and subsequent processing in stars leads to its destruction, the number of mechanisms producing D enrichments relative to H is limited. Robert et al. (1979) suggested that spallation could be the cause of the D excesses found in Chainpur chondrules, but the lack of isotopic effects in other light nuclei subsequently ruled out such an irradiation origin (Birck and Allègre 1980).

Geiss and Reeves (1981) analyzed the D enrichments in primitive meteorites in the framework of D abundances in the Galaxy and the solar system. Table 13.2.2 updated from the compilation by these authors, shows a list of D abundances in the Galaxy and in different solar system objects. The D/H ratio of the interstellar H gas determined from Lyman lines has recently undergone several revisions, from 1.5×10^{-5} (Laurent et al. 1979) to 2.25×10^{-5} (Bruston et al. 1981) to 0.5×10^{-5} (Vidal-Madjar and Gry 1984). The protosolar value is derived from ^{3}He data in meteorites and the solar wind and is in good agreement with D/H ratios measured in the atmospheres of Jupiter and Saturn, especially if fractionation between methane and H_2 gas is taken into account (Owen et al. 1986). The D in the Venus atmosphere is enriched by a large factor relative to the protosolar value (McElroy et al. 1982; Donahue et al. 1982). This enrichment has been attributed to large H losses (see Chapter 7.10).

Relative to the protosolar nebula, the Earth is enriched in D by a factor of eight. While volatile loss also must be considered for the Earth, one viewpoint is that the enrichment is due to fractionation during isotopic equilibration at a temperature of 200 K between H_2 gas, and water and methane, respectively, during solar-system formation (Geiss and Reeves 1972; Hubbard and MacFarlane 1980). Figure 13.2.4 shows the thermodynamic equilibrium abundances resulting from the exchange reactions

$$H_2O + HD \rightleftarrows HDO + H_2 \text{ and } CH_4 + HD \rightleftarrows CH_3D + H_2 \quad (2)$$

TABLE 13.2.2
Observed Enrichments of Deuterium

	D/H ($\times 10^5$)	Enrichment Factor	References*
Interstellar	0.5–2		a, b
Protosolar Nebula	2±1	1	c
Earth (SMOW)	15.7	8	
Venus	1600±200	800	d, e
Jupiter (methane)	3.6±1.2	1.8	f, g
Saturn (methane)	2.0±1.5	1	h
Uranus (methane)	$9^{+9}_{-4.5}$	4.5	h
Titan (methane)	$16.5^{+16.5}_{-8.8}$	8	h
Interstellar Molecules			
HCN	80–680	340	i
HCO$^+$	40–1000	500	i
	1000–10000	5000	j
HNC	>50000	>25000	k, l
C$_3$H$_2$	2900	1500	m
Meteorites			
Carbonaceous chondrites	8–60	30	Table 13.2.1
Ordinary chondrites	8–105	50	Table 13.2.1
Interplanetary Dust	12–160	80	Table 13.2.1

*References. a: Bruston et al. 1981; b: Vidal-Madjar and Gry 1984; c: Geiss and Reeves 1981; d: McElroy et al. 1982; e: Donahue et al. 1982; f: Kunde et al. 1982; g: Knacke et al. 1982; h: Owen et al. 1986; i: Penzias 1979; j: Wootten et al. 1982; k: Snell and Wootten 1979; l: Brown and Rice 1981; m: Bell et al. 1986.

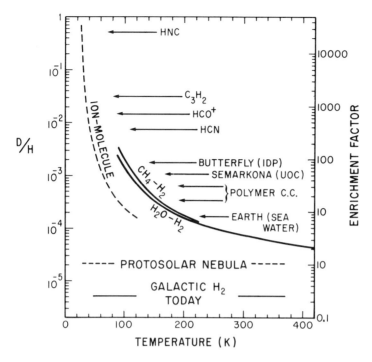

Fig. 13.2.4. Deuterium abundances in different galactic reservoirs (hydrogen gas, molecules in dense interstellar clouds), the protosolar nebula, the Earth and meteorites. Also shown are the D/H ratios expected from isotopic exchange between neutral species and from ion-molecule reactions as function of temperature (figure adapted from Geiss and Reeves 1981).

as function of temperature. Most of this equilibration probably occurred prior to the accretion of the inner planets. During accretion the Earth preferentially sampled the D-enriched water while the gas was left behind. Whether or not this is the dominant mechanism for the relative D enrichment on Earth depends on the equilibration temperature. Geiss and Reeves (1981) pointed out that different cosmothermometers indicate that equilibrium was frozen in at 360 K (Anders et al. 1973), too high a temperature for an eightfold increase in the D/H ratio.

The same authors, by examining all alternatives, concluded that the much larger D enrichments found in meteorites (Fig. 13.2.4) must have resulted from ion-molecule reactions at low temperatures under conditions characteristic of molecular clouds, a possibility previously suggested by Kolodny et al. (1980). Neutral reactions cannot be invoked because of the low isotope-exchange-reaction rates between neutral species at the low temperatures necessary to produce the observed effects. Earlier radio astronomical observations had established the existence of large D/H values in simple interstellar molecules (see Table 13.2.2).

Organic Matter in Meteorites

The conventional point of view is that most of the organic matter in meteorites was synthesized in the solar nebula from gases such as CO, H_2 and NH_3 by catalytic (Fischer-Tropsch-type) reactions on magnetite and clay minerals at temperatures of ≤ 400 K (Anders et al. 1973; Hayatsu and Anders 1981; see also Chapter 10.5). Laboratory Fischer-Tropsch synthesis can produce many of the organic compounds found in CI and CM meteorites (Hayatsu and Anders 1981). There are alternative views, though, such as late synthesis on parent bodies (Bunch and Chang 1980). However, none of these mechanisms can produce the D enrichments seen in meteorites. Kerridge (1983) considered the possibility of high ionization levels in the solar nebula because of short-lived radionuclides enabling ion-molecule reactions to take place, but expressed doubts that the solar nebula was cold enough (≤ 60 K) for substantial D fractionation. This leaves interstellar space as the most likely locale for the production of D-rich molecules.

Interstellar Synthesis of Deuterated Molecules

Since the discovery of the first interstellar molecules in 1940, interstellar chemistry has grown into a large field (Winnewisser 1981; Duley and Williams 1984). The accepted mechanism for the formation of interstellar molecules is based on ion-molecule reactions (Watson 1976) which, because of the lack of activation energy barriers, can take place rapidly at low temperatures. They can synthesize complex molecules and, most important in the context of the subject of this chapter, they lead to the deuteration of synthesized molecules. In fact, the strongest argument in favor of the ion-molecule synthesis scheme is the observation of large D enrichments of molecules in interstellar clouds (Table 13.2.2).

Ions are originally produced by cosmic rays (cr). The most important process is the ionization of molecular hydrogen (see, e.g., Duley and Williams 1984):

$$H_2 \xrightarrow{cr} \begin{cases} H + H^+ + e & 2\% \\ H_2^+ + e & 88\% \\ 2H & 10\%. \end{cases} \quad (3)$$

In dark interstellar clouds with abundant H_2 this is followed by

$$H_2^+ + H_2 \rightleftarrows H_3^+ + H. \quad (4)$$

The synthesis of H_2 takes place on the surface of interstellar dust grains which serves as catalyst:

$$H + H + \text{grain} \rightarrow H_2 + \text{grain}. \quad (5)$$

The reason for D/H fractionation is that ion-molecule exchange reactions are exothermic, favoring the formation of deuterated molecules. An example is the important reaction:

$$H_3^+ + HD \rightleftarrows H_2D^+ + H_2 + \Delta E \qquad (6)$$

which is considered to be the first stage for the enhancement of deuterated molecules (Smith 1981; Smith et al. 1982). The rate coefficients of this reaction have been measured in the laboratory, the value of $\Delta E/k$ is 140 K for ground-state species. Other deuterating reactions proceed from the above one like, for example,

$$H_2D^+ + CO \rightleftarrows DCO^+ + H_2 \qquad (7)$$

There are many different pathways even for the synthesis of a simple molecule such as DCO^+ (Adams and Smith 1985). Furthermore, competing reactions destroying the deuterated species have to be taken into account. This leads to complicated reaction networks as the basis of model calculations of the time evolution of molecule synthesis. Brown and Rice (1981), by considering 529 reactions between 108 species in their network, predicted D enrichment factors between 100 and 1500 for different molecules compared to hydrogen gas. While these numbers still fall short of the values observed in interstellar clouds (Table 13.2.2), other workers have suggested that additional reactions not considered previously contribute to D fractionation (Dalgarno and Lepp 1984).

Most of the reactions considered are assumed to occur in the gas phase. Reactions on grain surfaces also play an important role. In particular, the synthesis of H_2 from atomic H on interstellar grains has already been mentioned. It is still not clear how much grain surface reactions contribute to the deuteration of molecules. Tielens (1983) combined 1530 gas-phase, ion-molecule reactions with 440 grain surface reactions in a model for the deuteration of grain mantles. This model predicts D enrichment of up to 10,000-fold for a series of molecules such as H_2CO, NH_3, H_2O in grain mantles, but also leads to large D/H ratios of certain molecules (H_2CO) in the gas phase.

For completeness, it must be mentioned that there are several alternative models for the interstellar synthesis of simple organic molecules, all of which involve interstellar grains. Anders et al. (1974) proposed catalytic reactions on grain surfaces. Two other models involve photochemistry. Sagan and Khare (1979) produced organic solids from simple gases by ultraviolet irradiation or spark discharge (Miller-Urey reactions). They envision that this organic material synthesized in protostellar nebulae is ejected as grains into the interstellar medium and that its degradation provides a source of interstellar molecules. Finally, Greenberg and coworkers (Greenberg 1983, 1986; d'Hendecourt et al. 1986) developed a model of interstellar dust grains, based on astronomical observations and laboratory experiments, which postulates the existence of core-mantle grains. These grains undergo several cycles between dark molecular clouds and diffuse clouds. In diffuse clouds, complex molecules are produced by ultraviolet photochemical reactions of simple ices deposited as mantles on silicate grains in dense clouds. Explosive desorption returns these molecules to the gas phase. Greenberg expects most of the meteoritic organic material to be produced in this way. His views are widely accepted in the cometary community.

There are major objections to each of these models. Fischer-Tropsch-type catalytic reactions require higher temperatures than those found in interstellar clouds. There is no evidence that photochemical reactions can produce large D/H fractionations. Fractionations of N and O have been observed in discharges (Manuccia and Clark 1976; Arrhenius et al. 1978; Thiemens and Heidenreich 1983). There are no experimental data on the question whether photolysis of grain mantles can lead to deuteration. Other problems with Greenberg's photolysis model are the observations of higher D/H ratios in dense interstellar clouds than in diffuse clouds and the likelihood of destruction of grain mantles by shocks in diffuse clouds (Draine and Salpeter 1979) before extensive photochemical processing can take place. Further arguments against the Greenberg model were advanced by Anders (1986).

In spite of the objections against any of the grain-related models as the sole mechanism for the synthesis of interstellar organic material, condensation of interstellar molecules produced by gas-phase ion-molecule reactions onto dust grains and subsequent processing and further synthesis in grain mantles may play an important role in the history of D-rich interstellar material. This question will be taken up again in the last part of this chapter.

Fractionation Effects in Other Elements

Ion-molecule reactions are expected to produce also isotopic fractionation in elements other than H. Watson et al. (1976) were the first to consider C fractionation. The reaction

$$^{13}C^+ + {}^{12}CO \rightleftarrows {}^{12}C^+ + {}^{13}CO + \Delta E \tag{8}$$

has a $\Delta E/k$ of 35 K, leading to ^{13}C enrichment in CO relative to C. Models by Liszt (1978) and Langer (1977) predict an enrichment of ^{12}C in organic molecules relative to CO which is enriched in ^{13}C. However, the predicted enrichments are relatively modest compared to D enrichments.

The most comprehensive modeling effort has been made by Graedel et al. (1982) who included isotopic exchange reactions for C and O in a network of 1067 reactions. Langer et al. (1984) used this time-evolution model of dense interstellar clouds to predict C-isotopic fractionations in molecular clouds for various C species. The detailed predictions depend on several parameters such as temperature, gas density and C/O ratio. The isotopic frac-

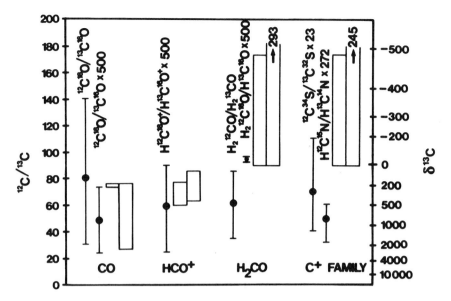

Fig. 13.2.5. $^{12}C/^{13}C$ ratios in CO, HCO$^+$, H$_2$CO and other C-bearing molecules ("C$^+$ family") predicted from ion-molecule reactions are compared with astronomical measurements in giant molecular clouds. The observed values are given by the dots and error bars. The predicted values are calculated for two different assumed "metal abundances" (abundance of all elements heavier than He) and are given by the boxes (figure adapted from Langer et al. 1984).

tionation in evolved clouds relative to the total C material at the beginning of the dense cloud stage is different for different species: CO is enhanced in ^{13}C, all other C-bearing species except HCO$^+$ are enhanced in ^{12}C, and HCO$^+$, depending on conditions, can be enhanced in either ^{12}C or ^{13}C. Figure 13.2.5 shows a comparison of predictions with observations in giant molecular clouds. The experimental data show smaller fractionations between the different species than the theoretical predictions and overlap with one another within experimental uncertainties. Recent revisions in reaction rates are believed to reduce the predicted ^{12}C enrichments in formaldehyde and "C$^+$-family" molecules (Langer et al. personal communication, 1987).

The essential conclusion is that C-fractionation effects in interstellar clouds are at most 50–100%, approximately 1000 times smaller than those seen for H. Since ^{12}C is predicted to be enhanced in most organic molecules, one might expect an anticorrelation between δD and δ^{13}C in organic matter. If the effects for both elements are diluted by the same factor, the predicted $\Delta\delta^{13}$C is $\sim -1‰$ for samples with δD = 1000‰. The meteoritic data are inconclusive. Yang and Epstein (1983) claim a negative correlation between δD and δ^{13}C for primitive meteorites (see their Fig. 12) but there is much scatter and it is not clear whether or not a correlation exists. In any event, it is likely that the C-isotopic variation seen in residues and bulk samples is dominated by other components. For example, at present it is still unknown how many different carriers of isotopically heavy C components exist and how they are distributed (Swart et al. 1983; Yang and Epstein 1984; Halbout et al. 1986b; Zinner and Epstein 1987).

Ion-molecule reactions are also expected to lead to isotopic fractionation effects for N. Adams and Smith (1981) measured the exchange reaction

$$^{14}N_2H^+ + {}^{14}N^{15}N \rightleftarrows {}^{15}N^{14}NH + {}^{14}N_2 + \Delta E \qquad (9)$$

and obtained $\Delta E/k = 9 \pm 3$ K. In equilibrium this reaction would give an enrichment factor for $^{15}N^{14}NH/^{14}N_2H$ of 1.6 at 16 K. So far, no complete isotopic model calculations including N have been performed. However, since not as many pathways are available, it is believed that isotopic fractionation effects for N should be even smaller than for C (Langer, personal communication). Astronomical observations do not indicate deviations of much more than 50% from the terrestrial $^{15}N/^{14}N$ ratio in HCN molecules (Wannier 1980). Thus N isotopic effects in meteorites are expected to be on the order of ≤ 1 ‰, much smaller than the measured variations. As for C, the N-isotopic composition in organic material is probably dominated by other components (Geiss and Bochsler 1982; Lewis et al. 1983; Franchi et al. 1986).

For oxygen, the calculations by Langer et al. (1984) indicate that isotopic fractionations by ion-molecule reactions are much smaller than for carbon, casting doubt on the interpretation of Halbout et al. (1986a), that their claimed correlation between δD and $\delta^{18}O$ may result from fractionation in interstellar molecules.

13.2.5. FROM INTERSTELLAR MOLECULES TO METEORITES

The evidence for an interstellar origin of the D-rich material in meteorites is largely circumstantial. Astronomical observations clearly show the presence of D-enriched simple molecules in dark molecular clouds, but even if we accept deuteration during ion-molecule reactions as the ultimate cause of D enrichments in meteorites, there are many uncertain steps between these two end points. As a consequence, this section consists mostly of questions to which there exist no ready answers. Figure 13.2.6 schematically depicts the journey of interstellar organic matter between molecule formation and its final destination in the meteorite.

Where did the Synthesis of the Simple Molecules Observed in Interstellar Clouds into more Complex Molecules Occur?

The length and complexity of molecules that can be obtained by gas-phase ion-molecule reactions is still unknown but there certainly exists a limit on the abundances of complex molecules (Winnewisser 1981). Most prob-

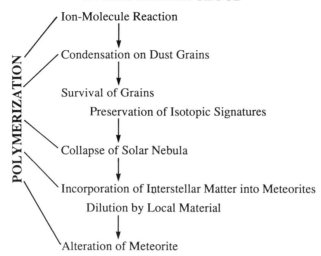

Fig. 13.2.6. Schematic diagram depicting evolutionary stages of D-rich organic matter from synthesis in interstellar molecular clouds to its presence in meteorites.

ably, grain surface reactions will play some role during subsequent synthesis (Hayatsu and Anders 1981; Tielens 1983). Many simple molecules are reactive and should polymerize if mechanisms exist that can provide the necessary energies. Photolysis, advocated by Greenberg (1983,1986), could be important even if the starting materials are not ices of CO, NH_3 and H_2O but already deuterated organic molecules (Tielens 1983).

Since photolysis is thought to occur in diffuse clouds not opaque to ultraviolet, grain survival becomes an important issue (Seab and Shull 1986). This topic is covered in Chapter 13.3. Ice mantles are effectively eroded by sputtering in shock waves (Draine et al. 1983). A way around this problem is to assume that the synthesis of deuterated molecules found in meteorites occurred in the dense molecular cloud from which the solar nebula collapsed and that dust grains were not subsequently processed in diffuse clouds. However, this scenario excludes photolysis according to the Greenberg model. An alternative possibility is that polymerization of interstellar molecules on grain surfaces occurred in the solar nebula when temperatures were high enough for catalytic reactions.

When was the Interstellar Material Mixed with Isotopically Homogenized (Local) Material?

The organic D carrier in meteorites never experienced temperatures that would have equilibrated its H-isotopic composition. During pyrolysis of acid residues, a considerable fraction of D-rich water is already released at temperatures of 300 to 350° C (Robert and Epstein 1982). The preservation of D excesses is in agreement with model temperatures during accretion onto the parent bodies of carbonaceous and unequilibrated ordinary chondrites, obtained from independent thermometers (Larimer 1978). It is most likely that mixing of organic matter from different sources took place on the meteorite parent bodies. The thermal environment on these bodies is probably consistent with the survival of this material. Primitive meteorites are known to be disequilibrium assemblages. However, these meteorites contain also high-temperature phases (e.g., chondrules and CAIs) that underwent a complex history of high-temperature processing in the solar nebula. Such a mixture is not compatible with the concept of a uniformly hot nebula (Grossman and Larimer 1974). The localized heating hypothesis (Wood 1985) for CAI formation presents a more satisfying solution to this dilemma.

To What Degree is Interstellar Material Diluted with Local Material?

The inhomogenous distribution of D provides some evidence that interstellar material is substantially diluted with material produced in the solar system. But we do not know the degree of dilution; we lack any chemical information on the D carrier in kerogen and have even less information on the soluble D carrier in UOCs.

In What Form did the D-Rich Interstellar Matter get into the Meteorites and What are the Implications of the Presence of Organic Interstellar Matter?

The assumption that deuterated interstellar molecules condensed onto grains seems reasonable. We do not know whether this assemblage survived into meteorites but since the organic matter did not lose its isotopic signature, one expects that it has not been subjected to processes which separated the organic from the inorganic material. Where is this inorganic material? It is widely believed that isotopic anomalies of elements such as O, Ca and Ti were carried into the solar system in the form of interstellar dust grains. These anomalies, however, are generally found in refractory phases and not in the matrix. The answer to this puzzle might lie in the fact that O, Ca and Ti-isotopic measurements were mostly performed on CAIs or at least in individual mineral phases that were large compared to the size of interstellar dust grains. Measurement of matrix material might average over too many thoroughly mixed individual dust grains. But why, on the other hand, were the dust grains carrying the anomalous components preserved in individual hibonites or CAIs not mixed in the same way? Another answer might be found in the degree of the dilution of the D-rich interstellar carrier. If, for example, this carrier has a D enrichment by a factor of 1000 relative to SMOW but is diluted by a factor of 100 in the meteoritic kerogen, the accompanying (presumably isotopically anomalous) interstellar inorganic matter might have easily escaped detection. If interstellar dust grains indeed consist of silicate cores with organic mantles, the acid treatment used to enrich the D carrier would destroy the grains. New ways have to be found to separate the D carrier from normal organic matter.

What is the Effect of Processing on the Meteorite Parent Body on the D Carrier(s)?

The experimental data suggest that two different carriers are responsible for the D enrichments in carbonaceous chondrites and unequilibrated ordinary chondrites. While kerogen contributes only a minor fraction of the D excess in UOCs it is the major carrier of the D excess in carbonaceous chondrites. Were there originally two distinct carriers or are they the result of processing on the parent body? Has the soluble carrier been polymerized into insoluble kerogen in the carbonaceous chondrites that experienced aqueous alteration but not, or only to a much lesser degree, in UOCs? Or, on the other hand, has it been dissolved during these alteration processes?

In conclusion, there remain, by far, more questions than there are answers. Progress has to be made on two fronts. First, new laboratory experiments must be developed to constrain the possible interstellar processes for the synthesis of deuterated organic matter. An example would be experiments that determine whether photolysis can result in D enrichment. Secondly, improved measurements must be performed on meteorites to characterize chemically the true carriers of D, possibly together with work to establish whether the D carriers are accompanied by interstellar inorganic matter. It is hoped that progress can be made with the application of microanalytical techniques which have been applied to the study of interplanetary dust particles (Chapter 11.1).

Acknowledgments. I gratefully acknowledge the role of K. McKeegan and A. Fahey in producing Fig. 13.2.3. I also thank them and E. Anders, T. Bernatowicz, G. Crozaz and R. Walker for critical reading of the manuscript. This chapter benefited from exemplary reviews by R. Becker and J. Halbout. E. Koenig and S. Moody were of great help during the preparation of the manuscript. This work was supported by the National Aeronautics and Space Administration and by the National Science Foundation.

REFERENCES

Adams, N. G., and Smith, D. 1981. $^{14}N/^{15}N$ isotope fractionation in the reaction $N_2H^+ + N_2$: Interstellar significance. *Astrophys. J.* 247:L123–L125.

Adams, N. G., and Smith, D. 1985. Laboratory studies of the reactions of HCO^+ (and DCO^+) and N_2H^+ (and N_2D^+) with D (and H) atoms: Interstellar implications. *Astrophys. J.* 294:L63–L65.

Anders, E. 1986. What can meteorites tell us about comets? In *Comet Nucleus Sample Return*, Proc. of ESA Workshop, Canterbury, UK, pp. 31–39.

Anders, E., and Zadnik, M. G. 1985. Unequilibrated ordinary chondrites: A tentative subclassification based on volatile-element content. *Geochim. Cosmochim. Acta* 49:1281–1291.
Anders, E., Hayatsu, R., and Studier, M. H. 1973. Organic compounds in meteorites. *Science* 182:781–790.
Anders, E., Hayatsu, R., and Studier, M. H. 1974. Interstellar molecules: Origin by catalytic reactions on grain surfaces? *Astrophys. J.* 192:L101–L105.
Arrhenius, G., Fitzgerald, R., Markus, S., and Simpson, C. 1978. Isotope fractionation under simulated space conditions. *Astrophys. Space Sci.* 55:285–297.
Audouze, J., and Tinsley, B. M. 1974. Galactic evolution and the formation of the light elements. *Astrophys. J.* 192:487–500.
Becker, R. H., and Epstein, S. 1982. Carbon, hydrogen and nitrogen isotopes in solvent extractable organic matter from carbonaceous chondrites. *Geochim. Cosmochim. Acta* 46:97–103.
Bell, M. B., Feldman, P. A., Matthews, H. E., and Avery, L. W. 1986. Detection of deuterated cyclopropenylidene (C_3HD) in TMC-1. *Astrophys. J.* 311:L89–L92.
Birck, J. L., and Allègre, C. J. 1980. Li^6/Li^7 variations in meteorites. *Meteoritics* 15:267.
Boato, G. 1954. The isotopic composition of hydrogen and carbon in the carbonaceous chondrites. *Geochim. Cosmochim. Acta* 6:209–220.
Briggs, M. H. 1963. Evidence for an extraterrestrial origin for some organic constituents of meteorites. *Nature* 197:1290.
Brown, R. D., and Rice, E. 1981. Interstellar deuterium chemistry. *Phil. Trans. Roy. Soc. London* A303:523–533.
Bruston, P., Audouze, J., Vidal-Madjar, A., and Laurent, C. 1981. Physical and chemical fractionation of deuterium in the interstellar medium. *Astrophys. J.* 243:161–169.
Bunch, T. E., and Chang, S. 1980. Carbonaceous chondrites II: Carbonaceous chondrites phyllosilicates and light element geochemistry as indicators of parent body processes and surface conditions. *Geochim. Cosmochim. Acta* 44:1543–1577.
Dalgarno, A., and Lepp, S. 1984. Deuterium fractionation mechanisms in interstellar clouds. *Astrophys. J.* 287:L47–L50.
d'Hendecourt, L. B., Allamandola, L. J., Grim, R. J. A., and Greenberg, J. M. 1986. Time-dependent chemistry in dense molecular clouds. *Astron. Astrophys.* 158:119–134.
Donahue, T. M., Hoffman, J. H., Hodges, R. R., Jr., and Watson, A. J. 1982. Venus was wet: A measurement of the ratio of deuterium to hydrogen. *Science* 216:630–633.
Draine, B. T., and Salpeter, E. E. 1979. Destruction mechanism for interstellar dust. *Astrophys. J.* 231:438–455.
Draine, B. T., Roberge, W. G., and Dalgarno, A. 1983. Magnetohydrodynamic shock waves in molecular clouds. *Astrophys. J.* 264:485–507.
Duley, W. W., and Williams, D. A. 1984. *Interstellar Chemistry* (London: Academic Press), 251 pp.
Epstein, R. I., Lattimer, J. M., and Schramm, D. N. 1976. The origin of deuterium. *Nature* 263:198–202.
Epstein, S., and Taylor, H. P., Jr. 1970. The concentration and isotopic composition of hydrogen, carbon and silicon in Apollo 11 lunar rocks and minerals. *Proc. Apollo 11 Lunar Sci. Conf.*, pp. 1085–1096.
Epstein, S., Krishnamurthy, R. V., Cronin, J. R., Pizzarello, S., and Yuen, G. U. 1987. Unusual stable isotope ratios in amino acid and carboxylic acid extracts from the Murchison meteorite. *Nature* 326:477–479.
Fallick, A. E., Hinton, R. W., McNaughton, N. J., and Pillinger, C. T. 1983. D/H ratios in meteorites: Some results and implications. *Annales Geophysicae* 1:129–134.
Franchi, I. A., Wright, I. P., and Pillinger, C. T. 1986. Heavy nitrogen in Bencubbin: A light-element isotopic anomaly in a stony-iron meteorite. *Nature* 323:138–140.
Geiss, J., and Bochsler, P. 1982. Nitrogen isotopes in the solar system. *Geochim. Cosmochim. Acta* 46:529–548.
Geiss, J., and Reeves, H. 1972. Cosmic and solar system abundances of D and ^3He. *Astron. Astrophys.* 18:126–132.
Geiss, J., and Reeves, H. 1981. Deuterium in the solar system. *Astron. Astrophys.* 93:189–199.
Grady, M. M., Wright, I. P., Fallick, A. E., and Pillinger, C. T. 1983. The stable isotopic composition of carbon, nitrogen and hydrogen in some Yamato meteorites. *Proc. 8th Symposium on Antarctic Meteorites, 1983*, pp. 292–305.

Graedel, T. E., Langer, W. D., and Frerking, M. A. 1982. The kinetic chemistry of dense interstellar clouds. *Astrophys. J. Suppl.* 48:321–368.
Greenberg, J. M. 1983. Interstellar dust, comets, comet dust and carbonaceous meteorites. In *Asteroids, Comets, Meteors*, eds. C. I. Lagerkvist and H. Rickman (Uppsala: Uppsala Univ. Press), pp. 259–268.
Greenberg, J. M. 1986. The role of grains in molecular chemical evolution. *Astrophys. Space Sci.* 128:17–31.
Grossman, L., and Larimer, J. W. 1974. Early chemical history of the solar system. *Rev. Geophys. Space Phys.* 12:71–101.
Halbout, J., and Robert, F. 1986. Numerical simulations of stable isotope results of progressive heating experiments. *Meteoritics* 21:384–386.
Halbout, J., Robert, F., and Javoy, M. 1986a. Oxygen and hydrogen isotope relations in water and acid residues of carbonaceous chondrites. *Geochim. Cosmochim. Acta* 50:1599–1609.
Halbout, J., Mayeda, T. K., and Clayton, R. N. 1986b. Carbon isotopes and light element abundances in carbonaceous chondrites. *Earth Planet. Sci. Lett.* 80:1–18.
Hayatsu, R., and Anders, E. 1981. Organic compounds in meteorites and their origins. In *Cosmo- and Geochemistry*, vol. 99, *Topics in Current Chemistry* (Berlin: Springer-Verlag), pp. 1–39.
Hayatsu, R., Scott, R. G., and Winans, R. E. 1983. Comparative structural study of meteoritic polymer with terrestrial geopolymers coal and kerogen. *Meteoritics* 18:310.
Hinton, R. W., Long, J. V. P., Fallick, A. E., and Pillinger, C. T. 1983. Ion microprobe measurement of D/H ratios in meteorites. *Lunar Planet. Sci.* XIV:313–314 (abstract).
Hoefs, J. 1980. *Stable Isotope Geochemistry* (Berlin: Springer-Verlag).
Hubbard, W. B., and MacFarlane, J. J. 1980. Theoretical predictions of deuterium abundances in the Jovian planets. *Icarus* 44:676–682.
Kerridge, J. F. 1983. Isotopic composition of carbonaceous-chondrite kerogen: Evidence for an interstellar origin of organic matter in meteorites. *Earth Planet. Sci. Lett.* 64:186–200.
Kerridge, J. F. 1985. Carbon, hydrogen and nitrogen in carbonaceous chondrites: Abundances and isotopic compositions in bulk samples. *Geochim. Cosmochim. Acta* 49:1707–1714.
Knacke, R. F., Kim, S. J., Ridgway, S. T., and Tokunaga, A. T. 1982. The abundances of CH_4, CH_3D, NH_3, and PH_3 in the troposphere of Jupiter derived from high-resolution 1100–1200 cm^{-1} spectra. *Astrophys. J.* 262:388–395.
Kolodny, Y., Kerridge, J. F., and Kaplan, I. R. 1980. Deuterium in carbonaceous chondrites. *Earth Planet. Sci. Lett.* 46:149–158.
Kunde, V., Hanel, R., Maguire, W., Gautier, D., Baluteau, J. P., Marten, A., Chedin, A., Husson, N., and Scott, N. 1982. The tropospheric gas composition of Jupiter's north equatorial belt (NH_3, PH_3, CH_3D, GeH_4, H_2O) and the Jovian D/H isotopic ratio. *Astrophys. J.* 263:443–467.
Kung, C. C., and Clayton, R. N. 1978. Nitrogen abundances and isotopic composition in stony meteorites. *Earth Planet. Sci. Lett.* 38:421–435.
Langer, W. D. 1977. Isotopic abundance of CO in interstellar clouds. *Astrophys. J.* 212:L39–L42.
Langer, W. D., Graedel, T. E., Frerking, M. A., and Armentrout, P. B. 1984. Carbon and oxygen isotope fractionation in dense interstellar clouds. *Astrophys. J.* 277:581–604.
Larimer, J. W. 1978. Meteorites: Relics from the early solar system. In *The Origin of the Solar System*, ed. S. F. Dermott (New York: Wiley), pp. 347–393.
Laurent, C., Vidal-Madjar, A., and York, D. G. 1979. The ratio of deuterium to hydrogen in interstellar space. IV. The lines of sight to δ, ε and ι Orionis. *Astrophys. J.* 229:923–941.
Lewis, R. S., and Anders, E. 1983. Interstellar matter in meteorites. *Sci. Amer.* 249(2):66–77.
Lewis, R. S., Srinivasan, B., and Anders, E. 1975. Host phase of a strange xenon component in Allende. *Science* 190:1251–1262.
Lewis, R. S., Anders, E., Wright, I. P., Norris, S. J., and Pillinger, C. T. 1983. Isotopically anomalous nitrogen in primitive meteorites. *Nature* 305:767–771.
Liszt, H. S. 1978. Time-dependent CO formation and fractionation. *Astrophys. J.* 222:484–490.
Manuccia, T. J., and Clark, M. D. 1976. Enrichment of N^{15} by chemical reactions in a glow discharge at 77°K. *Appl. Phys. Lett.* 28:372–374.

McElroy, M. B., Prather, M. J., and Rodriguez, J. M. 1982. Escape of hydrogen from Venus. *Science* 215:1614–1615.

McKeegan, K. D., and Zinner, E. 1984. On the distribution of excess deuterium in Renazzo and Semarkona: An ion microprobe study. *Lunar Planet. Sci.* XV:534–535 (abstract).

McKeegan, K. D., Walker, R. M., and Zinner, E. 1985. Ion microprobe isotopic measurements of individual interplanetary dust particles. *Geochim. Cosmochim. Acta* 49:1971–1987.

McKeegan, K. D., Swan, P., Walker, R. M., Wopenka, B., and Zinner, E. 1987. Hydrogen isotopic variations in interplanetary dust particles. *Lunar Planet. Sci.* XVIII:627–628.

McNaughton, N. J., Borthwick, J., Fallick, A. E., and Pillinger, C. T. 1981. Deuterium/hydrogen ratios in unequilibrated ordinary chondrites. *Nature* 294:639–641.

McNaughton, N. J., Fallick, A. E., and Pillinger, C. T. 1982a. Deuterium enrichments in type 3 ordinary chondrites. *Proc. Lunar Planet. Sci. Conf.* 13, in *J. Geophys. Res.* 87:A297–A302.

McNaughton, N. J., Hinton, R. W., Pillinger, C. T., and Fallick, A. E. 1982b. D/H ratios of some ordinary and carbonaceous chondrites. *Meteoritics* 17:252.

McSween, H. Y. 1979. Are carbonaceous chondrites primitive or processed? *Rev. Geophys. Space Sci.* 17:1059–1078.

Owen, T., Lutz, B. L., and de Bergh, C. 1986. Deuterium in the outer solar system: Evidence for two distinct reservoirs. *Nature* 320:244–246.

Penzias, A. A. 1979. Interstellar HCN, HCO$^+$, and the galactic deuterium gradient. *Astrophys. J.* 228:430–434.

Penzias, A. A. 1980. Nuclear processing and isotopes in the galaxy. *Science* 208:663–669.

Pillinger, C. T. 1984. Light element stable isotopes in meteorites: From grams to picograms. *Geochim. Cosmochim. Acta* 48:2739–2766.

Reeves, H., Audouze, J., Fowler, W. A., and Schramm, D. N. 1973. On the origin of light elements. *Astrophys. J.* 179:909–930.

Robert, F., and Epstein, S. 1982. The concentration and isotopic composition of hydrogen, carbon and nitrogen in carbonaceous meteorites. *Geochim. Cosmochim. Acta* 46:81–95.

Robert, F., Merlivat, L., and Javoy, M. 1977. Water and deuterium content of ordinary chondrites. *Meteoritics* 12:349–354.

Robert, F., Merlivat, L., and Javoy, M. 1978. Water and deuterium content in the Chainpur meteorite. *Meteoritics* 13:613–615.

Robert, F., Merlivat, L., and Javoy, M. 1979. Deuterium concentration in the early solar system: A hydrogen and oxygen isotope study. *Nature* 282:785–789.

Robert, F., Halbout, J., Javoy, M., Dimon, B., and Merlivat, L. 1983. The D/H ratio and petrological types of chondrites. *Meteoritics* 18:387–388.

Sagan, C., and Khare, B. N. 1979. Tholins: Organic chemistry of interstellar grains and gas. *Nature* 277:102–107.

Seab, C. G., and Shull, J. M. 1986. Shock processing of interstellar grains. In *Interrelationships Among Circumstellar, Interstellar, and Interplanetary Dust*, eds. J. A. Nuth and R. E. Stencel, NASA CP-2403, pp. 37–53.

Sears, D. W., Grossman, J. N., and Melcher, C. L. 1982. Chemical and physical studies of type 3 chondrites-I: Metamorphism related studies of Antarctic and other type 3 ordinary chondrites. *Geochim. Cosmochim. Acta* 46:2471–2481.

Smith, D. 1981. Laboratory studies of isotopic exchange in ion-neutral reactions: interstellar implications. *Phil. Trans. Roy. Soc. London* A303:535–542.

Smith, D., Adams, N. G., and Alge, E. 1982. Some H/D exchange reactions involved in the deuteration of interstellar molecules. *Astrophys. J.* 263:123–129.

Smith, J. W., and Rigby, D. 1981. Comments on D/H ratios in chondritic organic matter. *Earth Planet. Sci. Lett.* 54:64–66.

Snell, R. L., and Wootton, A. 1979. Observations of interstellar HNC, DNC and HN^{13}C temperature effects on deuterium fractionation. *Astrophys. J.* 228:748–754.

Swart, P. K., Grady, M. M., Pillinger, C. T., Lewis, R. S., and Anders, E. 1983. Interstellar carbon in meteorites. *Science* 220:406–410.

Thiemens, M. H., and Heidenreich, J. E., III. 1983. The mass-independent fractionation of oxygen: A novel effect and its possible cosmochemical implications. *Science* 219:1073–1075.

Tielens, A. G. G. M. 1983. Surface chemistry of deuterated molecules. *Astron. Astrophys.* 119:177–184.

Vidal-Madjar, A., and Gry, C. 1984. Deuterium, helium, and the big-bang nucleosynthesis. *Astron. Astrophys.* 138:285–289.
Wannier, P. G. 1980. Nuclear abundances and evolution of the interstellar medium. *Ann. Rev. Astron. Astrophys.* 18:399–437.
Watson, W. D. 1976. Interstellar molecule reactions. *Rev. Mod. Phys.* 48:513–552.
Watson, W. D., Anicich, V. G., and Huntress, W. T., Jr. 1976. Measurement and significance of the reaction $^{13}C^+$ + $^{12}CO \rightleftarrows$ $^{12}C^+$ + ^{13}CO for alteration of the $^{13}C/^{12}C$ ratio in interstellar molecules. *Astrophys. J.* 205:L165–L168.
Winnewisser, G. 1981. The chemistry of interstellar molecules. In *Cosmo- and Geochemistry*, vol. 99, *Topics in Current Chemistry* (Berlin: Springer-Verlag), pp. 39–71.
Wood, J. A. 1985. Meteoritic constraints on processes in the solar nebula. In *Protostars and Planets II*, eds. D. C. Black and M. S. Matthews (Tucson: Univ. of Arizona Press), pp. 687–702.
Wootten, A., Loren, R. B., and Snell, R. L. 1982. A study of DCO$^+$ emission regions in interstellar clouds. *Astrophys. J.* 255:160–175.
Yang, J., and Epstein, S. 1983. Interstellar organic matter in meteorites. *Geochim. Cosmochim. Acta* 47:2199–2216.
Yang, J., and Epstein, S. 1984. Relic interstellar grains in Murchison meteorite. *Nature* 311:544–547.
Yang, J., and Epstein, S. 1985. A search for presolar organic matter in meteorite. *Geophys. Res. Lett.* 12:73–76.
Zinner, E., and Epstein, S. 1987. Heavy carbon in individual oxide grains from the Murchison meteorite. *Earth Planet. Sci. Lett.* 84:359–368.
Zinner, E., McKeegan, K. D., and Walker, R. M. 1983. Laboratory measurements of D/H ratios in interplanetary dust. *Nature* 305:119–121.

13.3. ASTROPHYSICAL IMPLICATIONS OF PRESOLAR GRAINS

JOSEPH A. NUTH, III
NASA Goddard Space Flight Center

The widely accepted hypothesis that presolar materials survive intact in meteorites and can be studied in our laboratories is of great potential astrophysical significance. Two first-order conclusions have been drawn from the evidence supporting this hypothesis: graphitic grains are rare in the interstellar medium and grain lifetimes are considerably longer than 10^8 yr. Both conclusions are in conflict with currently accepted astrophysical models but might still be consistent with astronomical observations. Many more astrophysical revelations could be concealed within the meteorites, but until astronomers are again given some training in meteoritics, only the meteoritics community has the tools necessary to decypher the message.

13.3.1 INTRODUCTION

Most astronomers are unaware that presolar material has been identified in meteorites (see Chapters 13.1 and 13.2) and of those few who know of such developments, most are too unfamiliar with the complexities of meteoritic materials to pursue research opportunities in this field without a significant amount of basic study. A bridge between these fields is needed and it is up to the meteoritics community to build it by pursuing meteoritic studies with astrophysical goals and by publishing such results in journals read by the astrophysics community.

In what follows I will briefly describe two areas in which meteoritic data yield information about the interstellar medium. In the first case, I will show that graphite is a very rare component of the interstellar grain population (Nuth 1985), contrary to currently accepted models which predict that up to

88% of interstellar C is in the form of graphite (Mathis et al. 1977). In the second case, I will argue that grain lifetimes must be significantly longer than 100 Myr, the lifetime currently predicted by models of shock propagation in the interstellar medium (Seab and Shull 1985). Finally, I will suggest a few additional areas where meteoritic studies could be performed in order to yield astrophysical data. Throughout this chapter, the reader is urged to remain skeptical. Only recently have a large number of meteoriticists accepted the hypothesis that presolar materials can be found in primitive meteorites. The techniques by which these rare grains are isolated are still quite crude; a significant fraction of the information contained in the spatial distribution and "nearest neighbor" associations of these grains is lost. Considerable analytical advances are needed in order to rectify such problems. If such advances occur, then a new window will be opened on the early history of the solar nebula and on many presolar materials and processes.

13.3.2. PRESOLAR GRAINS AND THE INTERSTELLAR MEDIUM

Carbon Grains

In Chapter 13.1, the temperatures at which individual presolar noble-gas components are released from their carbonaceous carriers were noted. Some components are released at temperatures as low as 500 to 600° C (see Table 13.1.2). The fact that these components still exist in meteorites indicates that the individual carbonaceous grains which contain them never experienced temperatures in excess of 1000 K while dispersed in the vacuum of space. At such low temperatures, graphite is quite stable. Presolar grains not only survived transport through the interstellar medium but also survived the collapse of the solar nebula and the processing inherent in the formation of the parent bodies of carbonaceous chondrites. Such processes obviously could not have been vigorous enough to destroy all pre-existing carbonaceous grains, yet such carbonaceous grains are, thermodynamically, less stable than graphite. If graphitic interstellar grains existed at the time the nebula formed, they should have survived to at least the same degree as did the less stable carbonaceous carriers of the anomalous noble gases. Yet no evidence for pristine, presolar graphite exists in *primitive* meteorites.

It is impossible to overstate the potential influence of secondary metamorphic processes which may have altered the structure or composition of interstellar grains that had survived the nebular phase of solar-system formation. These processes are undoubtedly the source of the abundant graphite found in iron meteorites (Dodd 1981) and it has been suggested that similar processes may be the source for all meteoritic graphite, even that found in chondrites (Rietmeijer and Mackinnon 1985). Interstellar carbonaceous material has been detected in meteorites (Chapter 13.2) and is distinguished by a greatly enhanced D/H ratio, thought to have been produced by ion-molecule reactions in very cold interstellar clouds (see Chapters 13.2, 10.5; Watson

1976). Such material is not graphite, but consists in part of a macromolecular, kerogen-like substance (Hayatsu and Anders 1981). Various forms of circumstellar C have also been identified in meteorites based on isotopically distinct noble-gas components trapped in the grains (Chapter 13.1). Such primitive carbonaceous material could have been the starting point in a metamorphic process which produced the graphite seen in chondrites (Rietmeijer and Mackinnon 1985); however, the reverse process of turning graphite to a kerogen-like macromolecule is thermodynamically impossible under any scenario of meteorite formation yet postulated. Therefore, since we do not find graphite grains in primitive meteorites, yet do find both fragile, presolar, stellar noble-gas components and a relatively unstable, kerogen-like, interstellar carbonaceous precursor which could have been metamorphosed into graphite under more extreme conditions, we are forced to conclude that little, if any, graphite existed in the interstellar grain population of the giant molecular cloud from which the solar system formed.

Lewis et al. (1987) have suggested that the carrier phase which they label Cδ (see Table 13.1.2) consists in large measure of 50 Å diameter diamonds and have suggested that such diamonds may exist in the interstellar medium. This discovery has stirred up a flurry of interest on the part of the astrophysical community. Nuth (1987a,b) has shown that for such small particles the surface-free energy of the grains is a significant fraction of the total free energy of the system and that 50 Å diamonds may be thermodynamically stable with respect to graphite. Hecht (1987) has used the strength of the ultraviolet absorption edge of diamond and IUE observations in order to calculate that not more than 1% of the interstellar C can be in the form of diamond. Tielens et al. (1987) have shown that diamonds less than about 200 Å in diameter may be produced via shock processing of interstellar graphite grains. Meteoriticists are now in a position to characterize the carbonaceous component of the interstellar grain population. It would be extremely useful if the size distributions and ultraviolet-optical properties of known presolar grain components were measured so that model calculations of interstellar extinction could be constructed based on meteoritic data.

Grain Destruction and Mixing in the Interstellar Medium

Another consequence of the hypothesis that grains produced in several distinct astrophysical environments survive today in meteorites is a need to compare the hypothesis with current models of grain formation and destruction in the interstellar medium (see Seab and Shull 1985). Such models yield interstellar-grain lifetimes on the order of 10^8 yr, with shock waves as the primary agent of destruction. However, the time scale on which new grains are injected into the interstellar medium as replacements is on the order of a few times 10^9 yr. Only 1 grain in 10^5 would be expected to survive passage through the interstellar medium intact. This implies that the majority of grains now residing in the interstellar medium have each been destroyed and reformed an average of 10 times (e.g., once every 10^8 yr). For "single component" anomalies such as Ne-E (Chapter 13.1) in which only one isotope of a particular element is involved, once the carrier is destroyed and the anomalous component released, no known process could re-isolate such an isotopically pure phase from the general interstellar gas. For more complex mixtures such as Xe-HL or Xe-S where the isotopic pattern of the element is of importance, it may be possible to construct a scenario in which a somewhat diluted anomaly could form in the interstellar medium from destroyed carriers if we assume that little mixing has occurred between the stellar outflow and the average interstellar gas. Such scenarios become progressively more unlikely as grains travel farther from their stellar source and more mixing occurs; such scenarios seem ludicrous on 10^9 yr time scales.

There is good evidence that relatively efficient mixing occurs on large spatial scales in the interstellar medium. The assumption of a cosmic abundance appears to hold along any random line of sight in the Galaxy. Only a few rare stars formed in the last several billion years have anomalous chemical abundances not easily explained as a product of their own internal nuclear processing. Yet the achievement of "average" composition requires the admixture of elements and isotopes produced in a wide variety of stellar environments (see Chapters 14.1 and 14.2). Indeed, meteoritic isotopic anomalies have been found which are consistent with grain production in supernovae, novae and red giants. There is even some indication that at least two different red giants may have contributed to the solar system's Xe reservoir, since in a few carriers Kr-S is associated with Xe-S, while in other meteorites no Kr-S is found (see Chapter 13.1). Thus, these anomalies could have been produced in different stars with differing neutron fluxes or mass-loss rates. Although the universe appears homogeneous on large scales, on smaller spatial scales considerable diversity may persist.

One conclusion which can be drawn from the generally accepted interpretation of isotopic anomalies as presolar grains is that it is not unusual for individual grains to survive, relatively unchanged, from the time of their formation in particular stellar environments to the time when they are removed from the interstellar medium during the birth of new stars. Although the conclusion is at odds with currently accepted models of grain destruction, there are ways of reconciling these models with meteoritic observations. One such scenario may, for instance, involve the assumption of a much less homogeneous interstellar medium than is normally assumed, in which shocks are confined to relatively low-density "tunnels" or "wormholes" whereas the bulk of the interstellar mass is shielded in small cloudlets (Cox and Smith 1974; McKee and Ostriker 1977; McCray and Snow 1979). In such a scenario it is possible that some grains survived for billions of years while others were repeatedly destroyed and re-formed. However, this might also imply that grains from the same stellar source remained associated in clumps even during nebular collapse. This might imply that all presolar red-giant grains in a particular nebular region come from the same star whereas red-giant grains in another nebular region may have been derived from another source. Such a hypothesis could be tested once a method has been found to isolate and analyze individual presolar meteoritic grains in situ. A crude test of this hypothesis could be carried out using existing techniques on samples both from different meteorites and from different sections of the same meteorite.

13.3.3. FUTURE WORK

Meteoriticists studying the systematics of the composition and structure of presolar grains have the potential to make significant contributions to our understanding of many astrophysical processes. These studies will not be done by astrophysicists within the next decade because few astrophysicists are knowledgeable enough about meteoritics. A few of the questions which may be addressed by careful analysis of presolar grains can be found in Chapters 13.1 and 13.2. Many more have yet to be posed. In what follows I will discuss three additional areas where meteoritic studies might prove to be of considerable astrophysical significance.

1. It may be possible to determine the degree of mixing and even the turbulent mixing scale length in supernovae explosions by studies of the chemical composition of particular isotopic anomalies. As an example, both ^{48}Ca and ^{50}Ti are produced deep within the core of the explosion (see Chapter 14.2, Fig. 14.2.5). If little mixing occurs, then both ^{48}Ca and ^{50}Ti should be found as sulfides or metals whereas if more extensive mixing occurs then both could exist as oxides which would show a highly enriched ^{16}O anomaly. Of course, a careful study of the relevant chemical reaction rates, thermodynamic stabilities and nucleation rates must be made before reliable predictions of various isotopic distributions could be attempted. Much of the information needed for a preliminary analysis of this problem already exists in the literature.

2. It may be possible to determine the degree of mixing in circumstellar, interstellar and nebular environments; however, significant advances in analytical instrumentation and technique are needed in order to decypher fully the information that may be contained in primitive meteorites. What is needed is a method by which individual presolar grains can be identified in situ in the meteorite matrix. This may be possible within the next decade or so when one considers the fact that Zinner (Chapter 13.2) already discusses the isotopic variation observable within a single interplanetary dust particle. If coagulation in circumstellar environments is important *and* if grain aggregates are not efficiently destroyed in either the interstellar medium or the protosolar nebula, then one should be able to locate μm-sized regions within a matrix which are characterized by

distinct isotopic anomalies. Such regions would represent small grains which aggregated within a single circumstellar outflow and survived to be incorporated into the meteorite parent body. If additional regions of isotopically distinct material, identical to the first, were located within the same meteorite, then this would imply either that a very large dust aggregate broke up before incorporation in the parent body or that numerous grains (and probably a considerable quantity of gas) remained associated as a single parcel from the time the parcel was expelled in a circumstellar wind until it was incorporated into the meteorite. (Although this latter alternative seems unlikely, it may nevertheless be possible.) It seems more likely that if additional regions of isotopically distinct material were found, they would not be identical to one another. In this case it would be possible, by careful analysis, to set a lower limit on the number of distinct astrophysical sources required to account for the anomalies observed in the meteorite and therefore to set a lower limit on the mass fraction of presolar grains which survived nebular processing and transport through the interstellar medium.

3. There are numerous models for the formation of interstellar molecules both in dense molecular clouds (see, e.g., Leung et al. 1984; Herbst 1983) and behind shocks in such clouds (see, e.g., Mitchell 1984a,b; Tarafdar et al. 1985). In some cases isotopic fractionation mechanisms for major elements such as H/D (Dalgarno and Lepp 1984), $^{12}C/^{13}C$ and $^{16}O/^{17}O/^{18}O$ (Langer et al. 1984; Chu and Watson 1983; Penzias 1983) have been considered explicitly. There have been numerous observational studies of the isotopic composition of H, C, N and O in molecular environments (see Chapter 13.2). The object of the many models of interstellar cloud chemistry is to predict the relationships between various observable molecules as a function of the time since a cloud's internal chemical clock was started or was reset via shock processes. If a supernova blast wave actually caused the collapse of the solar nebula, then this same shock wave should have had a significant effect on the chemistry of the presolar molecular cloud. Detailed calculations of the time-dependent formation of the isotopic composition of the H, C, N, O and S-bearing condensible molecules in a dense cloud could be compared to the interstellar carbonaceous material found in meteorites (Chapter 13.2). If the isotopic relationships found in the meteoritic material are characteristic of the molecules one might expect in a young ($t \leq 10^6$ yr) molecular cloud, then additional evidence of a supernova trigger would be available. If, however, the abundance ratios are those characteristic of an old cloud ($t > 10^7$ yr), then it would appear unlikely that a supernova was responsible for the onset of collapse of the nebula.

13.3.4. SUMMARY

The particular projects outlined in Sec. 13.3.3 might never be attempted because the analytical capabilities of the meteoritics community never advance sufficiently for definitive data to be collected. This would be disappointing but acceptable. However, if such studies were not done because meteoriticists were not interested in the astrophysical implications of their work, then this would be extremely unfortunate. Meteoritics has a more direct link to astrophysics than to virtually any other scientific discipline; the solar nebula represents only one instance of star and planet formation but it is the only instance for which a detailed record of the event is potentially available for study. The meteoritic link to astrophysics is through the solar nebula, but it does not stop there. Exploration of the newly discovered scientific territory beyond the nebula holds tremendous promise. Are we up to the challenge?

REFERENCES

Chu, Y. H., and Watson, W. D. 1983. Further analysis of the possible effects of isotope-selective photodissociation on interstellar carbon monoxide. *Astrophys. J.* 267:151–155.

Cox, D. P., and Smith, B. W. 1974. Large scale effects of supernova remnants on the galaxy: Generation and maintenance of a hot network of tunnels. *Astrophys. J. Lett.* 189:L105–L108.

Dalgarno, A., and Lepp, S. 1983. Deuterium fractionation mechanisms in interstellar clouds. *Astrophys. J. Lett.* 287:L47–L50.

Dodd, R. T. 1981. *Meteorites: A Petrochemical Synthesis* (Cambridge: Cambridge Univ. Press).

Hayatsu, R., and Anders, E. 1981. Organic compounds in meteorites and their origins. *Topics Curr. Chem.* 99:1–37.

Hecht, J. H. 1987. Observational constraints on interstellar diamonds. *Nature* 328:765.

Herbst, E. 1983. Ion-molecule syntheses of interstellar molecule hydrocarbons through C_4H: Toward molecular complexity. *Astrophys. J. Suppl.* 53:41–53.

Langer, W. D., Graedel, T. E., Frerking, M.A., and Armentrout, P. B. 1984. Carbon and oxygen isotope fractionation in dense interstellar clouds. *Astrophys. J.* 277:581–604.

Leung, C. M., Herbst, E., and Heubner, W. F. 1984. Synthesis of complex molecules in dense interstellar clouds via gas-phase chemistry: A pseudo time-dependent calculation. *Astrophys. J. Suppl.* 56:231–256.

Lewis, R. S., Ming, T., Wacker, J. F., Anders, E., and Steel, E. 1987. Interstellar diamonds in meteorites. *Nature* 326:160–162.

Mathis, J., Rumpl, W., and Norsieck, K. H. 1977. The size distribution of interstellar grains. *Astrophys. J.* 217:425–433.

McCray, R., and Snow, T. 1979. The violent interstellar medium. *Ann. Rev. Astron. Astrophys.* 17:213–240.

McKee, C. F., and Ostriker, J. P. 1977. A theory of the interstellar medium: Three components regulated by supernova explosions in an inhomogeneous substrate. *Astrophys. J.* 218:148–169.

Mitchell, G. F. 1984*a*. Effects of shocks on the sulfur chemistry of a dense interstellar cloud. *Astrophys. J.* 287:665–670.

Mitchell, G. F. 1984*b*. Effects of shocks on the molecular composition of a dense interstellar cloud. *Astrophys. J. Suppl.* 54:81–101.

Nuth, J. A. 1985. Meteoritic evidence that graphite is rare in the interstellar medium. *Nature* 318:166–168.

Nuth, J. A. 1987*a*. Interstellar diamonds and the physics of small particles. *Nature* 329:589.

Nuth, J. A. 1987*b*. Are small diamonds thermodynamically stable in the interstellar medium? *Astrophys. Space Sci.*, in press.

Penzias, A. A. 1983. Isotopic fractionation and mass motion in giant molecular clouds. *Astrophys. J.* 273:195–201.
Rietmeijer, F. J. M., and Mackinnon, I. D. R. 1985. A multi-stage history for carbonaceous material in extraterrestrial chondritic porous aggregate W7029*A and a new cosmothermometer. *Lunar Planet. Sci.* 16:700–701 (abstract).
Seab, C. G., and Shull, J. M. 1985. Shock processing of interstellar grains. In *Interrelationships Among Circumstellar, Interstellar, and Interplanetary Dust,* ed. J. A. Nuth and R. E. Stencel, NASA CP-2403, pp. 37–53.
Tarafdar, S. P., Prasad, S. S., Huntress, W. T., Villere, K. R., and Black, D. C. 1985. Chemistry in dynamically evolving clouds. *Astrophys. J.* 289:220–237.
Tielens, A. G. G. M., Seab, C. G., Hollenbach, D. J., and McKee, C. F. 1987. Shock processing of interstellar dust: Diamonds in the sky. *Astrophys. J.* 319:L109–L112.
Watson, W. D. 1976. Interstellar molecule reactions. *Rev. Mod. Phys.* 48:513–522.

PART 14
Nucleosynthesis

14.1. SOLAR-SYSTEM ABUNDANCES AND PROCESSES OF NUCLEOSYNTHESIS

DOROTHY S. WOOLUM
California State University at Fullerton

This chapter provides an overview of the fundamental processes of nucleosynthesis from an historical perspective. Primordial solar-system abundances of the nuclides are inferred from meteoritic (primarily CI) and solar data, which yield abundances that show a marked similarity to those observed astronomically in a wide variety of situations. Patterns observed in these nuclide abundances are used as clues to their nuclear origin. For the very lightest nuclides, cosmological and galactic cosmic-ray spallation production dominate. For elements heavier than B, nuclear reactions (charged-particle and neutron-capture reactions, and possibly photodisintegration reactions) in stars or supernovae are required. In discussions of modern work, emphasis is placed on the synthesis of the heavy elements (i.e., heavier than Fe), where contributions from the field of meteoritics have been substantial. Elements up through Fe will be emphasized in Chapter 14.2, where the astrophysical settings for nucleosynthesis and the chemical evolution of the Galaxy will be discussed. Information based on observed isotopic anomalies in meteorites will be the subject of Chapter 14.3.

14.1.1. INTRODUCTION

The origin of the elements is the subject of this chapter. We discuss the average elemental composition of the solar system and, from this, infer the primordial solar-system abundances of the individual nuclides. Patterns in these nuclide abundances are used as clues to their origin. These patterns may, in fact, have universal (cosmic) significance. A large number of astronomical studies indicate that the solar-system elemental abundance patterns show an overall similarity to those determined in stars, stellar systems and interstellar media elsewhere in our Galaxy and in other galaxies (see, e.g., reviews by York 1982; Pagel et al. 1981; Mould 1982), in supernova ejecta and remnants (see, e.g., Trimble 1982,1983), and in the galactic cosmic rays (see, e.g., Mewalt 1983; Simpson 1983; Meyer 1985). Absolute abundances of the elements are generally similar, too, typically within a factor of 2. When abundance discrepancies are observed, they are usually understandable, for example, in terms of differences in stages of evolution or star-formation efficiency, in terms of the singular nature of the particular environment, or, for galactic cosmic rays, in terms of the effects due to propagation. Thus, these astronomical data suggest that the nucleosynthetic processes inferred from solar-system matter may in general be broadly representative of the nucleosynthetic processes of stellar systems.

From nuclear physics, we know how nuclear transmutations from one nuclear species to another can occur, via nuclear reactions and beta decays, and we use this in interpreting the patterns in the nuclide abundances. We can learn from astronomical observations about possible astrophysical settings for element production, and, in light of our nucleosynthetic theories, we can assess whether the proper environments and the necessary energies for the formation of the various nuclear species are, indeed, available in the observable universe.

In this chapter, we will outline the development of nucleosynthetic theories from an historical perspective. This will facilitate the discussion and lead, we hope, to an appreciation for the way in which advancements in science often proceed. In the final section, modern models for heavy-element ($>$ Fe) nucleosynthesis are discussed.

14.1.2. EARLY SOLAR-SYSTEM ABUNDANCE COMPILATIONS AND THE SUESS "ABUNDANCE RULES"

The Sun represents 99.9% of present solar-system matter. Therefore, a good analysis of the bulk composition of the Sun is all we need to determine the average solar-system abundances of the elements. Natural radioactive decay, nuclear burning in the Sun's interior, and cosmic-ray-induced effects must have modified the primordial abundances of the nuclides in the 4.55 Gyr lifetime of the solar system. However, these effects are quite well understood, particularly for elements heavier than Li, and they can either be quantitatively corrected for or dismissed as being negligible.

Solar elemental abundances can be determined from studies of the absorption-line spectra from the photosphere and sunspots and of the emission-line spectra from the chromosphere and corona. They can also be determined by studying the composition of the solar wind and the solar-flare particles emitted by the Sun. The photospheric abundance data are particularly interpretable because the photosphere is well mixed convectively and through vigorous turbulence, and because the physical conditions and processes are better known for the photosphere than for other solar regions. It is the photospheric data that provide, in principle, the most reliable and most accurate solar-abundance values for the chemical elements. Nevertheless, in practice, these analyses are very difficult for most elements and are impossible for some others. The primary problem, particularly for the heavier elements (atomic mass $\gtrsim 60$), is that some elements have few strong spectral lines. Some have only weak lines, and some have none at all. Other problems include poorly known atomic transition probabilities, spectral interferences, and uncertainties in the solar atmospheric structure.

Of all the materials available for chemical analyses, the chondrites (the CI chondrites, in particular) have an elemental composition closest to solar matter, except for the noble gases and the major volatiles, as first shown for the more abundant metallic elements by Russell (1929) in his pioneering study of the composition of the solar atmosphere. Therefore, chondrites can be analyzed to obtain solar-system abundances for the condensable rock-forming elements relatively readily and precisely, although it is important to recognize that small fractionations might exist and to be alert for evidence of such possibilities as more accurate solar and meteoritic data become available. That elements can be fractionated by common geochemical/cosmochemical processes has been known for a long time, and most solar-system objects do show the effects of significant elemental fractionation (chondritic fractionations are treated in Chapters 7.2 through 7.6). To determine the average content of the so-called rock-forming elements, Goldschmidt (1937,1954) combined elemental data from meteoritic silicate, metal and troilite (FeS) in the ratio 10 (silicate): 2 (metal): 1 (troilite), although other workers (see, e.g., Urey 1952) used chondrites alone, feeling that chondrites might be the appropriate mixture in themselves.

Earliest attempts to look for regularities in elemental abundances as a function of atomic number Z yielded few clues concerning the origin of the elements, although it was observed that even-Z elements are more abundant than odd-Z elements (Harkin's rule). In reviewing the pioneering study of elemental abundances by Goldschmidt (1937,1954), and using the available isotopic data for the elements, Suess (1947a,b) was the first to recognize and emphasize that there were regularities in the patterns of the nuclide abundances with increasing mass number A. Suess's insight gave credence to the idea that solar-system abundances were meaningful, and not just determined by chance. The observation that nuclide abundance regularities were functions of A, and not Z, fostered the notion that the abundances were determined by nuclear structure, not atomic structure.

Specifically, Suess noticed that the abundance of the odd-A nuclides tended to be a smooth function of mass in the high-mass region, which included the rare-earth elements, where analytical uncertainties were small and where elemental fractionations were expected to be insignificant. What was impressive was that the abundance variations of the odd-A nuclides as a function of A defined a *single, smooth* curve, with the slopes of the tie lines connecting isotopic pairs (pairs of odd-A nuclides of the same element) fitting well with the local trends of the generally smooth abundance curve of neighboring odd-A nuclides. Since isotopic patterns reflect nuclear processes, while elemental patterns can reflect the effects of chemical, as well as nuclear processes, this observation suggested that no significant chemical fractionations had occurred to alter the patterns established by nuclear processes alone.

Suess then postulated a smooth odd-A abundance curve with mass for *all* odd-A nuclides. He adjusted the Goldschmidt abundance values to achieve the postulated smoothness and conformity with the trends fixed by the odd-A isotopic pairs. The assumption here was that where there were relatively small irregularities in the abundance patterns, these were due to analytical errors and/or chemical fractionation. In making these adjustments, additional regularities appeared, particularly for the heavy even-A nuclei, although exceptions were noted for nuclei having closed neutron shells (i.e., for nuclei containing magic numbers of neutrons: $N = 50, 82, 126$).

Explicitly, as restated in English by Suess and Urey (1956), the regularities, or semi-empirical "abundance rules", given by Suess (1947a,b) were:

1. Odd mass number nuclides: the abundances of odd mass numbered nuclear species with $A > 50$ change steadily with A. When isobars occur, the sum of the abundances of the isobars must be used instead of the individual abundances.
2. Even mass number nuclides: (a) for elements with $A > 90$ the sums of the abundances of the isobars with even mass number change steadily with mass number; (b) in the regions with $A < 90$, the abundances of the nuclear species with equal numbers of excess neutrons change steadily with mass number.
3. In the region of the lighter elements with $A < 70$, the isobar with the higher excess of neutrons is the less abundant one at each mass number. For $A > 70$, the isobar with the smallest excess of neutrons is the least abundant.
4. Exceptions to these rules occur at mass numbers where the numbers of neutrons have certain values, the so-called magic numbers.

Rule (1) has proved to be the most useful rule. Figure 14.1.1 shows the log of the abundances of the odd-A nuclides determined by Suess (1947a,b) (normalized to $Si = 10^6$). For the rare cases where there are two nuclides with the same mass number (i.e., where isobars occur), the sum of the abundances are plotted. Data for isotopes of the same element are connected by tie lines.

The Suess "abundance rules" contributed to the development of the nuclear shell model, and his proposed adjustments of the elemental abundances met with impressive successes. For example, Suess (1947a,b) suggested a

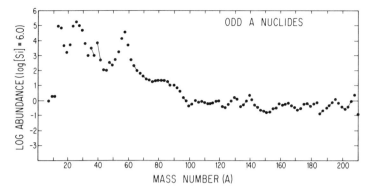

Fig. 14.1.1. The logarithm of the atomic abundances of the odd-mass number nuclides, relative to log Si = 6.0, plotted vs mass number A. The abundances are those semi-empirical abundances from Suess (1947a,b), which were obtained by adjusting the Goldschmidt (1937) abundances to obtain a smooth pattern, in conformity with trends defined by isotopic nuclides; these trends are indicated by the tie lines in the figure, which connect the pairs of isotopes from the same element (figure adapted from Suess 1987).

revision of the Re abundance by a factor of about 100. In 1949, Brown and Goldberg measured Re, and the inferred abundance was well within 50% of Suess's prediction. Suess's rules also led to the identification of carbonaceous-chondrite abundances with the average solar-system abundances of rock-forming matter, since they obeyed rule (1) of Suess better than any other available sample. Further, the abundance rules were used by Suess and Urey in 1956 (along with Goldschmidt's 1937 abundance compilation and the more recent empirical chondrite data compiled by Urey) to estimate the primordial solar-system abundances of the nuclides.

The semi-empirical abundances derived by Suess and Urey (1956) preserved the main features of the Suess abundance distribution. In interpreting their abundances, Suess and Urey clearly presaged the need to invoke several nuclear processes in the production of the elements. They discussed some of the abundance features in terms of the neutron-capture theory, and qualitatively explained the general features of the abundance peaks at nuclides with magic numbers of neutrons. With this, one of the seeds for the initial detailed models of the nucleosynthesis of the elements was sown.

14.1.3. EARLY THEORIES OF NUCLEOSYNTHESIS: INTERPRETATIONS OF THE SUESS-UREY ABUNDANCE PATTERNS

Burbidge, Burbidge, Fowler and Hoyle (1957)—reference known as B²FH—and Cameron (1957) published the first modern detailed models of the nucleosynthesis of the elements. These were based on seminal earlier studies (Hoyle 1946,1954; Cameron 1954,1955; Hoyle et al. 1956; Fowler and Greenstein 1956) and on the examination of the trends and patterns of the Suess-Urey abundance curve. B^{2}FH and Cameron (1957) underscored the evidence for the existence of several different abundance components, each having a different nuclear origin, and, therefore, requiring a different nucleosynthetic situation. This, further, argued for the need for *stellar* nucleosynthesis, where different star types and the different stages in stellar evolution could provide the variety of nucleosynthetic environments needed to explain solar-system abundances. This was a nontrivial development, given that the main competing theories at the time (reviewed by Alpher and Herman 1953) all involved some cosmic event in a primordial state of the universe. Cosmological nucleosynthesis in the early universe is now thought to contribute significantly only to the lightest nuclides (up through ^{4}He, plus ^{7}Li: see, e.g., Trimble 1975; Wagoner 1980; Truran 1984; Boesgaard and Steigman 1985; Arnould 1986).

Eight different processes of synthesis were described in B^{2}FH, but the most significant new results (particularly in light of the role meteoritics has had in the guiding of nucleosynthesis theory) included the detailed analysis of three separate processes to produce most of the heavier, naturally occurring nuclides. Because of the Coulomb-barrier problems in charged-particle reactions on heavier nuclides, neutron-capture reactions were expected to dominate. In B^{2}FH there were two main processes involving neutron capture, called the *s*-process and *r*-process, and a rarer proton-capture process named the *p*-process. (Cameron [1957] considered similar processes, but the B^{2}FH terminology has been universally adopted.)

Now, three decades later, B^{2}FH and Cameron (1957) remain classics in the field and provide the foundation for current theories of nucleosynthesis. Specifically, the status for heavy-element synthesis can be simply summarized. The *s*-process appears to be relatively well established, but the *p*-process is poorly understood. The *r*-process is better understood than the *p*-process but does not adequately explain many of the detailed features in the abundance patterns. Unlike the *s*-process, there is no well-established astrophysical site for the *r*-process. Modifications and alternate or competing processes have been proposed (see, e.g., Amiet and Zeh 1968; Suess and Zeh 1973; Marti and Zeh 1985; Kruse 1976; Woosley and Howard 1978; Blake and Schramm 1976), but the basic assumptions have not changed. Thus, it seems worthwhile to review the basic nuclear systematics and summarize the B^{2}FH modes of element synthesis, emphasizing, for present purposes, the modes of synthesis important for the heavy elements.

The *s*-process of neutron capture involves (n, γ) reactions which occur at a slow (s) rate compared to the intervening beta (β^-) decays, in which a neutron decays into a proton, plus electron, plus neutrino. The capture path of the *s*-process when plotted in the Z,A plane wanders along a trajectory with

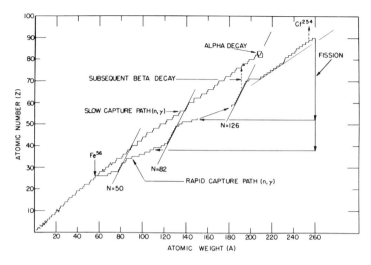

Fig. 14.1.2. The neutron-capture paths of the s-process and the r-process shown in the Z,A plane, where Z is the proton number and A is the atomic mass number. In this plane, neutron capture is represented by a horizontal step and beta decay by a vertical step. The s-process path wanders along the valley of stability in the Z,A plane and terminates in alpha decay above Bi. In the r-process, an intense neutron flux drives the matter to the neutron-rich side of the valley of stability. Once the intense neutron flux is over, the neutron-rich nuclides approach the valley of stability with beta decays, as represented by the vertical dotted line labeled "subsequent beta decay." The r-process terminates in fission at about $A = 260$, and the nuclear matter is fed back into the process. Thus, by this process nuclei up through the actinides can be produced. Lines of constant N (neutron number) for magic neutron numbers 50, 82 and 126 are shown; where a magic neutron number is obtained along the r-process path a "staircase effect" arises (see discussion in text) (figure adapted from a figure provided by W. A. Fowler which originally appeared in B²FH).

generally positive slope, along the swath occupied by stable nuclides, which we call the "valley of (nuclear) stability." This is shown schematically in Fig. 14.1.2, which is adapted from B²FH. On such a plot, neutron capture corresponds to a horizontal step, and beta decay to a vertical step. Because neutron capture is slow compared to beta decays, beta decays persistently drive the s-process trajectory back toward the valley of stability before multiple neutron captures can drive it very far to the right of the stable nuclides, i.e., to exceedingly neutron-rich unstable nuclides. The s-process finally terminates at $A = 209$ (at Pb and Bi) with alpha decay. The prevalent abundance peaks near $A = 138$, 208 and perhaps 90 (see Fig. 14.1.1) can be explained in terms of the s-process, as already recognized by Suess and Urey (1956). (These abundance peaks appear in the similar plot for even-A nuclei, too.) This is because the neutron-capture cross section for nuclei containing a magic number of neutrons ($N = 50, 82, 126$ for the above abundance peaks) is exceptionally small. If the cross section is small, neutron capture is particularly slow, providing a bottleneck in a steady neutron-capture chain. It is reasonable to expect a buildup of the abundances of the species at these points where the nuclear transmutations are impeded.

In the r-process, neutron capture occurs at a rapid (r) rate compared to the beta decays. An intense neutron flux is necessary, and the nuclear matter is driven far to the neutron-rich side of the valley of stability (i.e., far to the right of the s-process path; see Fig. 14.1.2). Following the exhaustion of the intense neutron flux, the neutron-rich capture products approach the valley of stability by beta decays. Further, because of the r-process time scales (typically, beta decay lifetimes are <1 s), r-process production can proceed through to the actinide element region, where fission cycling occurs. In addition to producing nuclides not produced in the s-process, this process also can contribute to the abundances of s-process nuclides. This is shown better in Fig. 14.1.3, which is a small region of the chart of the nuclides, where nuclides are plotted in the Z,N plane (cf. Fig. 14.1.2). The s-process nuclides are identified by the s-process path, which is indicated by the zig-zag line

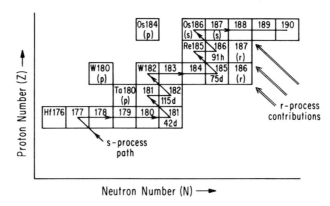

Fig. 14.1.3. A portion of the chart of the nuclides, showing the path followed by the s-process along the valley of stability. Neutron captures (horizontal steps) occur until an unstable nuclide beta decays (step up and to the left), which is usually followed by further neutron capture(s) until another beta-unstable nuclide is reached. More neutron-rich nuclides are produced in the r-process, which takes place far to the neutron-rich side of the valley of stability, but which results, in subsequent beta decays, in the production of nuclides at the neutron-rich edge of the valley of stability. For example, ^{186}W and ^{187}Re are missed by the s-process path and are produced only by the r-process. They are labeled r. Nuclides along the s-process path may be produced by both s- and r-processes (as, for example, ^{188}Os), but some stable r-process nuclides shield some of the s-process nuclides from r-contributions (e.g., ^{187}Os is produced only in the s-process because it is shielded by ^{187}Re). Rarer nuclides occur on the neutron-deficient (proton-rich) side of the valley of stability. These are labeled p. They are not reached by the s-process path, and they are shielded from r-process contributions by the stable s-process nuclides. Therefore, they require an additional process, known as the p-process, for their production.

with arrows. Stable nuclides produced primarily in the r-process are indicated with an r (e.g., ^{186}W). ^{188}Os is an example in this figure of a nuclide produced by both the r- and s-processes.

Starting with ^{56}Fe as an r-process seed nuclide, and referring to Fig. 14.1.2, we can examine the r-process in greater detail. With the intense neutron flux, neutrons are rapidly added to nuclei until the neutron binding energy (and, hence, the neutron-capture cross section) decreases sufficiently to halt neutron capture so that beta decay must occur before further neutrons can be added. A "staircase effect" in the r-process path is achieved at magic neutron numbers. When the nuclei have a magic number of neutrons, the cross section for adding another neutron is small, and a beta decay is the next step. Another neutron capture yields the magic number of neutrons, and is again followed by a beta decay. This staircase path (see Fig. 14.1.2) continues, moving closer to the valley of stability; with each step the beta-decay lifetime increases and excess abundances are established, until beta-decay rates become so slow relative to neutron-capture rates that multiple neutron captures again become favored. These effects lead to broad maxima in the abundance curve. Lines of constant (magic) N are shown in Fig. 14.1.2 and they intersect the s- and r-process paths at different points. From this view, it is clear that the r-process maxima would occur with a smaller value of A and Z than is associated with the magic peaks produced in the s-process.

In abundance plots like Fig. 14.1.1, the r-process abundance maxima would be displaced to lower A relative to the magic s-process peaks. In Fig. 14.1.1, r-process maxima do occur in the abundance curve at masses displaced to the low side of the s-process peaks. These are at $A = 130$ and 194, corresponding to $N = 82$ and 126. There is an indication of a peak at $A = 80$ ($N = 50$), but this is less clear. As evident from Fig. 14.1.2 and the preceding discussion, it is reasonable that r-process abundance maxima are much broader (involve more mass numbers) than that obtained for the s-process abundance peaks; this qualitative feature is evident in the abundance patterns.

The p-process of B^{2}FH included considerations of proton capture (p,γ) reactions and photodisintegration (γ,n) reactions (both presumably on s-process seed nuclei), which were proposed to explain the much rarer (fewer isotopes, relative to r- and s-process-produced isotopes), proton-rich heavy-element isotopes that could not be produced in the s- and r-processes (e.g., ^{184}Os, ^{180}W, ^{180}Ta, in Fig. 14.1.3). Generally, these proton-rich nuclides have lower abundances when compared to other stable nuclides.

The other B^{2}FH nucleosynthetic processes included:

1. H burning (the main energy-generating mechanism in normal stars);
2. He burning (He reactions, to produce the abundant multiple-alpha nuclei like ^{12}C and ^{16}O);
3. α-process (He reactions, using alpha particles of a different source from

that in He burning, to produce higher-mass multiple-alpha nuclei, starting with ^{20}Ne);

4. *e*-process (or equilibrium process, in which matter is raised to a sufficiently high temperature that the nuclei come into statistical equilibrium with one another; this was originally discussed by Hoyle [1954] and Hoyle et al. [1956] and was needed to produce the abundant Fe-group elements);

5. *x*-process (the unknown process needed to produce the low-abundance, light elements: Li, Be and B).

Many contributions have been made to our understanding of the nucleosynthesis of solar-system matter since 1957 (for general reviews see, e.g., Trimble 1975; Truran 1984). For example, the *x*-process is now called the *l*-process, and it is galactic cosmic-ray spallation (primarily in the interstellar medium) that is thought to produce most of the Li, B and Be nuclides, except for ^{7}Li, for which stellar and cosmological production are possible (see, e.g., Audouze and Reeves 1982; Truran 1984; Arnould 1986). In modern calculations, the dominant reactions in the so-called α-process are C-burning, Ne-burning and O-burning reactions; and the *e*-process is roughly equivalent to modern Si-burning. These latter processes (see Chapter 14.2) occur in the cores of massive stars and in explosive burning conditions in supernovae. But, before considering further the current state of theories of stellar nucleosynthesis and the contributions made to its development by meteoritic studies, we first address the current status of our determinations of primordial solar-system abundances. Then, we will focus on modern models of heavy-element synthesis, primarily involving neutron capture. Chapter 14.2 follows with emphasis on recent studies of the processes involved in the synthesis of elements up to the Fe-group elements.

14.1.4. PRIMORDIAL SOLAR-SYSTEM ABUNDANCES OF THE NUCLIDES

All stable, and many unstable, nuclei occur in nature, and their relative abundances vary by about 16 orders of magnitude. There are a total of roughly 400 naturally occurring nuclei, stable and radioactive. In the intervening years since the work by Suess and Urey (1956), numerous abundance compilations have followed. They confirm the abundance rules and their interpretation from the viewpoint of the B^{2}FH theory. Palme et al. (1981) have restated the empirical abundance rules (see, also, Amiet and Zeh 1968; Suess and Zeh 1973). The three abundance rules holding for the heavier ($A > 60$) nuclides assume smoothness with mass number for:

1. σN, the product of *s*-component abundance N times the neutron-capture cross section σ;

2. The β^- (or r)-component abundances;
3. The β^+ (or p)-component abundances.

Based on these new abundance rules, it was suggested that meteoritic data might be in error, and this has been confirmed with new analyses of Zr, Hf and Mo. These abundances have been reduced by 30% to a factor of 3 (Palme and Rammensee 1981; Palme et al. 1981).

Palme et al. (1981), Cameron (1982a) and Anders and Ebihara (1982) have published recent compilations of primordial solar-system abundances of the elements and nuclides, and throughout this book the most recent one of these, by Anders and Ebihara, has been adopted. In this chapter, the Anders-Ebihara work has been updated to include the new Y, Zr, Nb and Ta abundances from the work of Jochum et al. (1986), although the maximum correction involved (13% for Ta) is not really significant for our present purposes. As with all such compilations, the Anders and Ebihara work is based largely on direct determinations of CI elemental abundances. It uses the tabulation of isotope abundances by Lederer et al. (1978). In some cases, data for other meteorite classes were used to calculate the CI abundance, taking into account the empirically determined elemental fractionation factors, relative to CI meteorites, for elements of similar cosmochemical behavior. For some elements, solar/astronomical data were used (H, C, N, O, He and Ne) or interpolations were made (Ar, Kr, Xe, Hg).

The table in Appendix 4 is adapted from the Anders and Ebihara compilation. It tabulates the abundances of the nuclides (given as number of atoms relative to 10^6 Si atoms). For radioactive and radiogenic nuclides, the nuclide abundances are starred and are the primordial abundances (the abundances 4.55 Gyr ago), as calculated from the corresponding present-day values.

We can carry out some tests to see how closely these CI-chondrite abundances approximate primordial solar-system abundances, much as Anders and Ebihara (1982) and Anders (1971) did. First, we compare CI and solar abundances, and then, we examine abundance patterns for possible irregularities.

Figure 14.1.4 shows the ratio of solar to meteoritic elemental abundances, using the solar data from Ross and Aller (1976) and Grevesse (1984). The elements have been grouped according to their primary cosmochemical affinity, as given by Anders and Ebihara (1982). The data symbols distinguish the quality of the solar data, as judged by Grevesse (1984), with triangles denoting accurate results and ellipses denoting doubtful results; the squares denote values which should be accurate but which are either based on a limited number of spectral lines or involve atomic data that need to be checked. The uncertainties represent two standard deviations of the mean.

Aside from the well-understood solar Li depletion (solar thermonuclear reactions consume Li), relatively large discrepancies between solar and me-

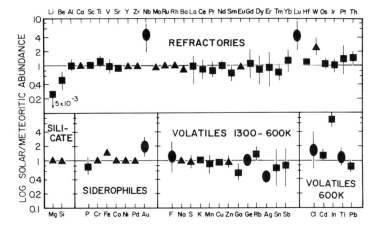

Fig. 14.1.4. The logarithm of the ratio of solar to meteoritic abundances of the elements is shown for elements where comparison is possible and where there is some estimate of uncertainties from solar studies. The elements are grouped according to their primary cosmochemical affinity (adopted from Anders and Ebihara [1982]). The solar abundances are from Grevesse (1984) and Ross and Aller (1976); the meteoritic abundances are from Anders and Ebihara (1982) and Jochum et al. (1986). As defined by Grevesse (1984), the triangles denote the data points that are believed to be accurately known; the squares are data which should be accurate, but which either involve the analysis of only one or two blended lines in the solar spectrum or need further checking when more accurate atomic data become available. The ellipses indicate doubtful results.

teoritic abundances occur only for Nb, Lu, W, Au, Ga, Ag and In. All of these have solar abundances that are judged doubtful, except for W, Ga and In, and only W is judged to be among the most accurate of the solar determinations. The discrepancies occur for elements having very different cosmochemical affinities, and there appears to be no convincing suggestion of a systematic depletion or enrichment of any cosmochemical element group in CI meteorites, relative to the Sun, as would be expected if they had suffered any significant fractionation. It has been impressive that over the years, as data of greater precision are obtained, the agreement between solar and CI abundances has generally improved. (The case of Fe was the classic example, although recent revisions of the solar Fe abundance now yield a small, but significant, discrepancy for Fe. See Grevesse [1984] for discussion of the results and the need for future work.) Further, even where discrepancies remain, it appears that they are likely due to errors in the solar data/model or in the adopted oscillator strength (which is the effective number of electrons in the atom, and is called the f-value) used in the reduction of the solar data.

Another test of the validity of the association of CI abundances with primordial solar-system abundances is to examine the abundance patterns for irregularities. The presumption is that the nucleosynthetic processes produce

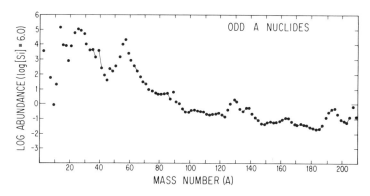

Fig. 14.1.5. The current solar-system abundances of the odd-A nuclides are plotted vs A, normalized to Si = 10^6 (data from Anders and Ebihara [1982] and Jochum et al. [1986]: cf. Fig. 14.1.1).

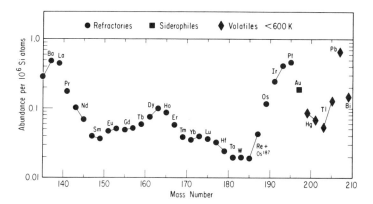

Fig. 14.1.6. A portion of Fig. 14.1.5, at an expanded scale, for the range $135 < A < 209$. The odd-mass abundances are a smooth function of mass number, and the elemental abundances conform to the trends set by the (accurately known) isotopic abundances of the elements (indicated by the tie lines). The irregularity at Sm-Eu is probably not analytical (Anders and Ebihara 1982), but may reflect nuclear processes or a modest cosmochemical fractionation (Eu is the most volatile of the rare earth elements [Boynton 1975]; figure from Anders and Ebihara 1982).

largely smooth trends in the nuclide abundances that would be destroyed if there were any significant chemical fractionation of the elements. In Fig. 14.1.5 we show a plot of the odd-A nuclide abundances from Table A1 (see Appendix), using a log scale normalized to log Si = 6.0. At this scale, general smoothness is apparent.

In their compilation of solar-system abundances, Anders and Ebihara (1982) reported that their CI abundances of odd-A nuclides between A = 65

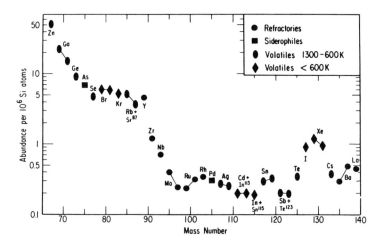

Fig. 14.1.7. A portion of Fig. 14.1.5, at an expanded scale, for the range $67 < A < 139$. The principal smoothness irregularity, apart from regions near anticipated s- or r-process abundance peaks, is the Ag-Cd-In-Sn-Sb region. Anders and Ebihara (1982) believe that analytical uncertainties in the Ag determinations are probably the explanation, but cosmochemical fractionation is a possibility. Cadmium and In are among the most volatile elements, and depletions in Cd and In in CI meteorites due to volatility are, therefore, not unreasonable (figure adapted from Anders and Ebihara 1982).

and 209 show an almost perfectly smooth trend with mass, with elemental abundances conforming well to the slopes defined by nearby odd-A isotopic abundance pairs (see Figs. 14.1.6 and 14.1.7). In the two cases where irregularities were noted by these authors (in the Sm-Eu region and in the Ag-Cd-In region), they were reported to be of the order of 15%. Nuclear factors and cosmochemical fractionation are possibilities in the Sm-Eu case. In the Ag-Cd-In case, possibilities include analytical errors in the Ag determinations (more likely) and cosmochemical fractionation (depletion of Cd and In).

In an effort to provide a test of CI abundance smoothness which is not dependent on elemental data from separate analyses and from different laboratories, using different techniques, Woolum et al. (1986) performed PIXE (or proton probe) analyses of CI meteorites. Multielement determinations with μg/g sensitivity in a broad mass range ($60 < A < 100$, corresponding to the range from Ni to Mo) were possible in a single analysis, and preliminary data indicated that there may be deviations of up to 30% in abundance smoothness (see Fig. 14.1.8, which shows recent unpublished PIXE results). The largest of the uncertainties are shown, and for all but the highest-mass elements, they primarily reflect sample variability, not statistical precision. For the highest-mass nuclides, uncertainties in the background subtraction are the most significant source of error. The data are in good agreement with the Anders-Ebihara compilation, except for the Ivuna Br analyses. (This may

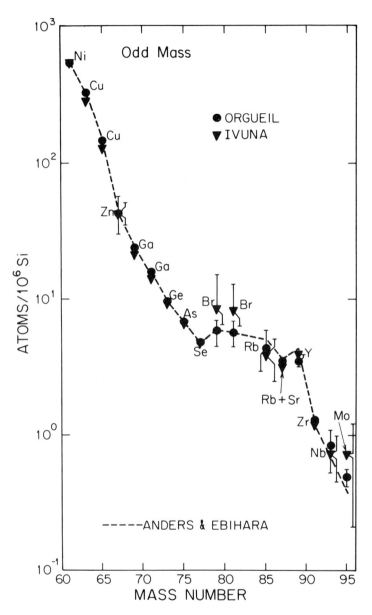

Fig. 14.1.8. Abundances of the odd-A nuclides plotted vs A, normalized to Ni (11000 μg/g; Anders and Ebihara 1982) from PIXE analyses of two CI meteorites: Orgueil and Ivuna (see Woolum et al. 1986). For comparison, abundances from Anders and Ebihara (1982) are represented by the dotted curve; the agreement is good. Irregularities in smoothness, like the Se-Br notch and the Y peak, can be understood qualitatively in terms of standard s-process neutron-capture nucleosynthesis models.

reflect contamination or sample heterogeneity. Veins produced by aqueous alteration are known to exist in the CI meteorites; see Chapter 3.4.) In particular, the Y peak and the Se-Br notch are confirmed.

Both r- and s-process contributions are expected in this mass region. A smooth abundance curve can only reasonably occur if both processes separately give smooth abundance curves. For the s-process, σN is expected to be a smooth function of mass number (new "abundance rule" (1), above; also, see discussion in Sec. 14.1.5, following), in which case, the issue of smoothness of abundance becomes a question of smoothness of cross section. The compiled cross sections (Ulrich 1982; Käppeler et al. 1982) show $\pm 30\%$ structure above $A = 70$, some of which could represent experimental uncertainties. However, the most striking feature is a sharp negative ^{89}Y anomaly, which would give theoretical justification for a sharp Y abundance peak. Theoretical s-process abundance curves are not entirely smooth in this mass region. Those calculated for core He burning (Lamb et al. 1977) show irregularities up to a factor of 2, and, in fact, reproduce the Se-Br notch obtained in the solar-system abundance curves. It is interesting that in this preliminary study, explanations for apparent nonsmooth behavior are possible using the general ideas of neutron-capture nucleosynthesis. It is somewhat ironic to think that this could ultimately provide the strongest argument for the identification of CI abundances with average solar-system abundances.

At present, the problem in interpreting irregularities in the solar-system abundance curve is that current theory for the nucleosynthesis of heavy elements allows for predictions with roughly 20% uncertainties in the best of cases; but they are often no better than order of magnitude calculations, due to uncertainties in the cross sections and in the required astrophysical parameters, like temperature. However, Marti and Zeh (1985) optimistically point to a need to improve the quality of the solar-system abundance data in the near future, because of improvements in the precision of neutron-capture cross sections that are expected in the next few years. From the work of Anders and Ebihara, we can conservatively estimate that primordial solar-system abundances are probably known to about 20%, in general (and significantly better for certain mass regions). Future improvements in this situation will require more precise empirical abundance data and more detailed assessments of how closely the CI nonvolatile-element abundances approximate those for the primordial solar system.

14.1.5. HEAVY-ELEMENT NUCLEOSYNTHESIS

Meteoritic studies have contributed most to the understanding of the nucleosynthesis of the heavier elements. The synthesis of nuclei from Ne to Fe by means of charged particle reactions is treated in Chapter 14.2. It will be seen that current theoretical models have yields which are in substantial agreement with solar-system abundances, and that these models do not involve unreasonable astrophysical settings.

Of the nucleosynthetic processes producing the heavier elements (heavier than Fe), the s-process is best understood. Clayton (1983,1968) gives a good introductory overview of the s-process. Käppeler et al. (1982), Ulrich (1982), Truran (1984) and Schatz (1986) provide useful reviews.

The basic features of the s-process were introduced in Sec. 14.1.3, above. In the standard s-process model, the equation describing the rate of change of the s-process abundance of nuclide A, $N_s(A)$, is:

$$\frac{dN_s(A)}{dt} = \lambda_n(A-1)N_s(A-1) - [\lambda_n(A) + \lambda_{\beta^-}(A)]N_s(A) \quad (1)$$

where $\lambda_n = \phi\sigma$ is the neutron-capture rate, which is proportional to the neutron flux ϕ and to the averaged neutron-capture cross section σ and where $\lambda_{\beta^-} = \ln2/T_{1/2}$ is the beta-decay rate if nuclide A is radioactive. This equation represents a system of coupled nonlinear differential equations, in the general case, where the coefficients λ are time dependent through their dependence on the temperature and on the neutron flux. These equations can be solved analytically if one makes simplifying assumptions. Standard assumptions are that (1) the temperature is a constant with time, and (2) that either, $\lambda_n \gg \lambda_{\beta^-}$, or $\lambda_{\beta^-} \gg \lambda_n$ (i.e., radioactive nuclei are either treated as stable nuclei or are neglected completely in the system of equations, with only the stable member of the isobar along the s-process path being represented). With these assumptions and a change of variable, where time t is replaced by the integrated neutron flux (called the neutron fluence) $\tau = \int \phi dt$, we have

$$\frac{dN_s(A)}{d\tau} = \sigma(A-1)N_s(A-1) - \sigma(A)N_s(A). \quad (2)$$

The general features of the solution can be inferred in the following way. It is clear that:

$$\frac{dN_s(A)}{d\tau} < 0 \quad \text{if} \quad N_s(A) > \frac{\sigma(A-1)}{\sigma(A)}N_s(A-1)$$

$$\frac{dN_s(A)}{d\tau} > 0 \quad \text{if} \quad N_s(A) < \frac{\sigma(A-1)}{\sigma(A)}N_s(A-1). \quad (3)$$

Thus, $N_s(A)$ decreases if it is too large with respect to $N_s(A-1)$ times the ratio of the cross sections; it increases if it is too small with respect to this product. With this self-adjusting feature, it is reasonable to expect that the product of the cross section and the abundance (σN) will be comparable in value for neighboring, nonmagic nuclides, and will at least be a smoothly varying function of A, in general. Detailed analyses of the s-process have focused on this property. Typically, σN values are plotted vs A, using the available solar-system abundances N and nuclear data σ for those nuclides expected to be s-process products; and theoretical s-process models are constructed to fit the data.

Deriving the solution of the above system of equations describing the s-process is beyond the scope of this text. The interested reader is referred to the references above and references therein. Briefly, Clayton et al. (1961) showed that a single irradiation of Fe-group seed nuclei was not enough to generate the observed s-nuclei abundances. (Sources of free neutrons are limited, so efficient seeds are needed for the s-process; because the Fe-group nuclei generally have high cross sections and high abundances, they are efficient seed nuclei for the s-process, compared to most of the lighter nuclides.) A distribution of neutron fluences was needed with the smaller amount of seed exposed to the larger fluences. Seeger et al. (1965) showed that an exponential distribution of exposures produced a reasonable fit to the σN curve, and they produced a simple model for galactic nuclear evolution which could produce such a distribution. Ward and Newman (1978) showed that two such distributions were required, one to explain the main distribution at large A and one to explain the nuclides in the region between the Fe peak ($A = 56$) and $A = 90$.

Figure 14.1.9 shows a recent σN plot, adapted from Käppeler et al. (1982). Only pure s-process and predominantly s-process nuclide data are shown. The trend of the data shows a "ledge-precipice" signature, being relatively flat in extended regions between precipices, which occur at magic neutron numbers. The overall character of the σN plot is predominantly determined by the cross sections, and the precipices are a reflection of the very small neutron-capture cross sections associated with nuclei having magic neutron numbers. (The σN decreases are less precipitous than the cross section decreases at magic-neutron numbers, so the abundances N show peaks at the magic-neutron nuclei, as was noted previously.) The curves in Fig. 14.1.9 show the theoretical fits to the data, assuming a steady neutron flux at $kT = 30$ KeV (Clayton 1983) and a two component exponential distribution of neutron fluences, of the form:

$$\rho(\tau) = \frac{f_1 N_{56}}{\tau_{o1}} \exp(-\tau/\tau_{o1}) + \frac{f_2 N_{56}}{\tau_{o2}} \exp(-\tau/\tau_{o2}) \quad (4)$$

where f is the fraction of the Fe-seed nuclei N_{56} that has been subjected to that component of the exponential distribution of exposures. τ_o is the mean neutron fluence.

The heavy curve in Fig. 14.1.9 corresponds to the stronger neutron irradiation, with the higher fluence ($f_2 = 0.092 \pm 0.015\%$ and $\tau_{o2} = 0.240 \pm 0.010$/mb). This component can account for virtually all s-process abun-

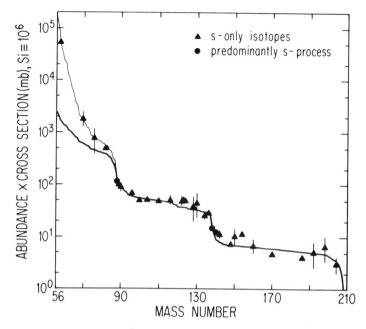

Fig. 14.1.9. The product of neutron-capture cross section times s-process abundance plotted vs mass number (figure from Käppeler et al. 1982). The abundances are taken from Cameron (1982a). Triangles denote s-only nuclides. Circles represent nuclei with closed neutron shells that are predominantly produced by the s-process. All error bars reflect only the cross section uncertainties. The solid lines correspond to the strong and weak components in the exponential neutron-flux distributions (see text).

dances with $A \gtrsim 90$. The fine curve corresponds to the weaker component ($f_1 = 2.7 \pm 0.2\%$ and $\tau_{o1} = 0.056 \pm 0.005$/mb) needed to fit the data for $A < 90$. The average number of neutrons captured per ^{56}Fe seed nucleus for the two components are 8.2 ± 0.5 and 1.1 ± 0.1 for the stronger and weaker irradiations, respectively. Further, the calculations show that the fit to the σN curve above the Fe peak does not lead to any overproduction of nuclei below the Fe group, and, in fact, the calculations indicate that s-process synthesis is practically negligible for these nuclei.

In conclusion, it appears that the s-process is well described by a simple model which is consistent across the periodic table. It is astrophysically reasonable, as well. As discussed in Chapter 14.2, steady and/or pulsed He burning can produce the necessary neutrons for the s-process and is consistent with the model. (Although, whether the α,n reaction on ^{13}C or on ^{22}Ne is the dominant neutron source is still not certain; in particular, the ^{22}Ne cross section needs to be experimentally determined.) This most likely occurs in the red-giant phase, as originally suggested by Cameron (1955). The presence of Tc (the half life of ^{99}Tc, its longest-lived isotope, is 2×10^5 yr) and the enhanced abundances in some red giants of elements that are primarily made in the s-process (e.g., Sr, Y, Zr, Ba and La) support this idea (Merrill 1952a,b).

Once the standard s-process nuclei have been isolated, the abundances due to all other processes (primarily the r-process) can be extracted. By demanding that the standard s-process component is such that the σN_s values fall on a smooth curve described by some function of A, designated $S(A)$, the remaining abundances N_r can be calculated:

$$N_r(Z,A) = N(Z,A) - N_s(Z,A) = N(Z,A) - S(A)/\sigma(Z,A). \qquad (5)$$

Käppeler et al. (1982) provide a recent compilation of such r-process residuals N_r (see also Cameron 1982b); figure 14.1.10 shows their results. From

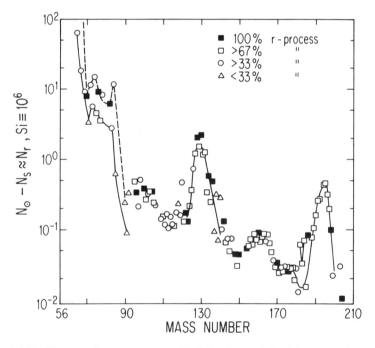

Fig. 14.1.10. The approximate r-process residual abundances derived from the difference between solar abundances (Cameron 1982a) and calculated s-process abundances, assuming σN_s smoothness. Neutron magic-number abundance maxima at mass numbers of about 130 and 195 are indicated by "eye-guide" lines and the odd-even effect below mass number 90 is indicated by the dashed line (even isotopes) and the solid line (odd isotopes) (figure from Käppeler et al. 1982).

Eq. (5), it is clear that uncertainties in the neutron capture cross sections $\sigma(Z,A)$ propagate directly and are particularly important when s-process contributions dominate the solar-system abundance, $N(Z,A)$. Thus, the calculated r-process residuals (open symbols) are plotted with different shapes, to indicate three different levels of confidence. From the figure, we note that there is a general smoothness for $A > 90$, and for odd-A nuclei (solid line) and even-A nuclei (dashed line), independently, for $A < 90$. There are pronounced maxima at $A = 130$ and 195. Further, there is good agreement between the trend defined by the solar-system abundances of the pure r-process nuclei (filled symbols) and the calculated abundances N_r, which promotes confidence in the s-process assignments and the assessment that p-process contributions to s-only nuclei are at most 10% (and thus negligible).

To be precise, we should note that N_r may not be totally attributable to an r-process, because we only subtracted an idealized, steady s-process contribution. N_r may very well contain contributions from other processes: certainly the p-process for a relatively small number of nuclides, and possibly processes such as a pulsed s-process (Truran and Iben 1977), as well. Nevertheless, a pulsed s-process cannot account for the r-process maxima (see, e.g., Käppeler et al. 1982, and references therein), and an r-process resulting in yields resembling Fig. 14.1.10 is still needed.

The general conditions required for the r-process are detailed by Norman and Schramm (1979); Fowler (1984), Truran (1984) and Schramm (1982) provide useful reviews. The classical r-process provided by B²FH and elaborated by Seeger et al. (1965) was outlined above in Sec. 14.1.3. Those original calculations assumed constant temperature and neutron density, but the required temperatures and densities were extremely high, so it was recognized early that some catastrophic dynamic event, like a supernova, was needed for the r-process. More recent work has allowed for time-varying temperature and density and nonequilibrium effects (see reviews by Schramm [1982], Truran [1984] and Fowler [1984], and references therein).

A significant problem in the theoretical modeling is that most of the nuclei involved in the calculations are so far from the valley of stability that they cannot be produced for study, and their properties (mass, beta-decay rates and neutron-capture rates) can only be estimated. Very different estimates have been used in the individual models. Nevertheless, in these studies, the gross features of the solar-system r-process abundances can be mimicked, such as the positions and relative sizes of the most prominent abundance peaks, the relative smoothness of the abundance patterns and the lack of substantial odd-even variations in the abundance patterns. Finer-scale features in the abundance patterns are not really understood and are an area of active current research (see, e.g., Marti and Suess 1986).

One feature of the r-process abundance maxima at magic neutron numbers (near $A = 130$, 195 and possibly 80: see Figs. 14.1.6–14.1.8 and 14.1.10) that provides some constraints on the r-process is the relatively narrow widths of the maxima (Schramm 1982). One expects that the solar-system r-process abundances are the result of contributions from many separate r-process events. If widely varying conditions were obtained in the different r-process events, a wide variety of r-process paths could have been achieved, with the more extreme conditions causing the more extreme deviations of the r-process path from the valley of stability (refer to Fig. 14.1.2). This would lead to a cumulative broadening of the resultant r-process path and, thus, to a broadening of the abundance peaks. That the r-process abundance peaks are relatively narrow must mean that relatively restricted conditions characterized the production of these r-process nuclides.

Nonetheless, the range of conditions allowed by the data includes the possibility of a wide variety of astrophysical sites. In fact, Blake and Schramm (1976) have shown that for a set of assumed capture cross sections and beta-decay rates, r-process nuclei can be produced by a more general neutron-capture process (called the n-process) whereby the same r-process path is obtained but extremely rapid neutron capture is not necessary; in this model, neutron-capture rates are comparable to beta-decay rates (not as in the standard r-process, where they are much greater than beta-decay rates). This lowers the required neutron densities by 5 or more orders of magnitude, extending the range of possible sites, but it still requires some catastrophic event.

Proposed astrophysical models for r-process nucleosynthesis have been grouped into two broad classes (Truran 1984): those involving the expansion and cooling of neutron-rich matter (e.g., in supernova explosions, in collapsing magnetized stars, in collisions of black holes with neutron stars) and those involving shock waves in the C and He shells of supernovae. The development of the He-core thermal-runaway r-process of Cameron et al. (1984) is an interesting possibility favored by Fowler (1984), although there is no clear astrophysical site proposed for the thermal runaway, as yet. More concerning astrophysical sites for the r-process follows in Chapter 14.2.

There has long been keen interest in the r-process, due, in part, to the fact that it is responsible for most of the nuclides involved in nucleochronology schemes. This is the subject of Chapter 15.2. No doubt, the desire to determine better the abundances of r-process nuclides used in nucleochronology and to interpret some of the isotopic anomalies observed in meteorites (the subject of Chapter 14.3) will play no small role in stimulating continued progress in our understanding of the r-process and its production sites. The recent and fortuitous occurrence of a nearby supernova (1987A) in the Large Magellanic Cloud may prove an even greater boon. But for now, we turn our attention to the remaining process by which heavy nuclei are thought to be made.

There remain some stable heavy nuclei which cannot be produced by either the s- or the r-process. This is demonstrated in Fig. 14.1.3, which shows that some, proton-rich nuclei (^{184}Os, ^{180}W, ^{180}Ta) are not reached by the s-process path, and are shielded from r-process contributions by stable s-process nuclides. Thus, yet another process (named the p-process) is required, as was introduced earlier (Sec. 14.1.3).

Recent work on the rare proton-excess nuclei (also called β^+ nuclei) indicates the possibility that these nuclei could have resulted from the destruction of pre-existing heavier nuclei by photodisintegration in a brief high-temperature event (Kruse 1976; Woosley and Howard 1978). The process producing these nuclei was called the p-process by B²FH (proton capture was a primary reaction considered), but Woosley and Howard (1978), who have explored this photodisintegration model quantitatively, have called it the γ-process. They have shown that a distribution of heavy r- and s-process seeds produces an abundance pattern very similar to that of the p-process solar-system nuclei when exposed for 1 s in an explosive event having radiation temperatures in the range 2 to 3×10^9 degrees. The transformations occur via a series of γ,n, γ,p and γ,α reactions in which photons (γ) disintegrate the seeds, producing nucleons and lighter nuclei. Unlike the p,γ reactions emphasized in earlier studies, high proton densities are not required. Unlike production models involving stable stellar evolution, this model has the advantage that the products are ejected directly into the interstellar medium without significant reprocessing. There are other models based upon explosive nucleosynthesis, some of which successfully reproduce the qualitative features of the solar-system abundances of the p-nuclei (see, e.g., Audouze and Truran 1975), but Woosley and Howard (1978) argue that these demand conditions (composition, temperature and density combinations) that may not be physically reasonable, whereas the production site in their γ-process model occurs naturally in those zones of supernovae that have experienced He and perhaps C burning prior to the explosion. On the other hand, the Woosley-Howard model seems particularly sensitive to the particular (nonsolar) choice of seed nuclei. Further, it does not seem possible to exclude the possibility that models successful in reproducing solar abundances, but not successful in other respects (e.g., in yielding astrophysically reasonable regimes) may in fact be providing valuable insights regarding actual astrophysical settings. One thing is certain, though; considerable work is necessary to flush out and refine the models for proton-excess nuclei.

In truth, additional work is required for all nucleosynthetic models, to one extent or another, even for the s-process. In this, additional/more accurate nuclear, astronomical and abundance data will be needed. More detailed theoretical studies of stellar structure and stellar evolution will be important for providing constraints on the models of nucleosynthesis.

With this brief introduction to our current understanding of the processes involved in element synthesis, we now turn to the questions of the detailed astrophysical settings for nucleosynthesis and of the chemical evolution engendered by the ancient and continuing nucleosynthesis of the elements.

Acknowledgments. The work was supported by a grant from the National Aeronautics and Space Administration. The author gratefully acknowledges discussions with and input from E. Anders, D. S. Burnett, H. Palme and H. E. Suess. Figures (for use or adaptation) were kindly provided by E. Anders, D. D. Clayton, W. A. Fowler and H. E. Suess.

Note added in proof: Marti and Suess [*Astrophys. Space Sci.* (1987), in press] have recently calculated r-process residual abundances that show systematic differences of even-A and odd-A nuclides in certain mass regions, and they propose that this odd-even effect may be due to beta-delayed neutron emission during the beta decay to the valley of stability.

REFERENCES

Alpher, R. A., and Herman, R. C. 1953. The origin and abundance distribution of the elements. *Ann. Rev. Nucl. Sci.* 2:1–40.

Amiet, J. P., and Zeh, H. D. 1968. On the origin of the heavy nuclei. *Z. Physik.* 217:485–509.

Anders, E. 1971. How well do we know "cosmic" abundances? *Geochim. Cosmochim. Acta* 35:516–522.

Anders, E., and Ebihara, M. 1982. Solar-system abundances of the elements. *Geochim. Cosmochim. Acta* 46:2363–2380.

Arnould, M. 1986. The origin of the light nuclides. *Prog. Part. Nucl. Phys.* 17:305–347.

Audouze, J., and Reeves, H. 1982. The origin of the light elements. In *Essays in Nuclear Astrophysics,* eds. C. A. Barnes, D. D. Clayton, and D. N. Schramm (Cambridge: Cambridge Univ. Press), pp. 355–373.

Audouze, J., and Truran, J. W. 1975. p-Process nucleosynthesis in postshock supernova envelope environments. *Astrophys. J.* 202:204–213.

Blake, J. B., and Schramm, D. M. 1976. A possible alternative to the r-process. *Astrophys. J.* 209:846–849.

Boesgaard, A. M., and Steigman, G. 1985. Big bang nucleosynthesis: Theories and observations. *Ann. Rev. Astron. Astrophys.* 23:319–378.

Boynton, W. V. 1975. Fractionation in the solar nebula: Condensation of yttrium and the rare earth elements. *Geochim. Cosmochim. Acta* 39:569–584.

Burbidge, E. M., Burbidge, G. R., Fowler, W. A., and Hoyle, F. 1957. Synthesis of the elements in stars. *Rev. Mod. Phys.* 29:547–650.

Cameron, A. G. W. 1954. Origin of anomalous abundances of elements in giant stars. *Phys. Rev.* 93:932 (abstract).

Cameron, A. G. W. 1955. Origin of anomalous abundances of elements in giant stars. *Astrophys. J.* 121:141–160.

Cameron, A. G. W. 1957. Stellar evolution, nuclear astrophysics and nucleogenesis. Chalk River Report, AECL (Atomic Energy of Canada, Ltd.), CRL-41.

Cameron, A. G. W. 1982a. Elemental and nuclidic abundances in the solar system. In *Essays in Nuclear Astrophysics,* eds. C. A. Barnes, D. D. Clayton, and D. N. Schramm (Cambridge: Cambridge Univ. Press), pp. 23–43.

Cameron, A. G. W. 1982b. The heavy element yields of neutron capture nucleosynthesis. *Astrophys. Space Sci.* 82:123–131.

Cameron, A. G. W., Cowan, J. J., and Truran, J. W. 1984. In *Proc. Yerkes Observatory Conf. on Challenges and New Developments in Nucleosynthesis,* ed. W. D. Arnett (Chicago: Univ. of Chicago Press).

Clayton, D. D. 1983. *Principles of Stellar Evolution and Nucleosynthesis* (Chicago: Univ. of Chicago Press). Also, New York: McGraw-Hill, 1968.

Clayton, D. D., Fowler, W. A., Hull, T. E., and Zimmerman, B. A. 1961. Neutron capture chains in heavy element synthesis. *Ann. Phys.* 12:331–408.

Fowler, W. A. 1984. The quest for the origin of the elements. *Science* 226:922–935.

Fowler, W. A., and Greenstein, J. L. 1956. Element-building reactions in stars. *Proc. Natl. Acad. Sci. U.S.* 42:173–180.

Goldschmidt, V. M. 1937. Geochem. Verteilungsgesetze der elemente IX, *Skrifter Norske Videnscaps-Akademiend, Oslo. I. Mat. Natur. Kl. No. 4.*
Goldschmidt, V. M. 1954. *Geochemistry*, ed. A. Muir (Oxford: Clarendon Press).
Grevesse, N. 1984. Accurate atomic data and solar photospheric spectroscopy. *Physica Scripta* T8:49–58.
Hoyle, F. 1946. The synthesis of the elements from hydrogen. *Mon. Not. Roy. Astron. Soc.* 106:343–389.
Hoyle, F. 1954. On nuclear reactions occurring in very hot stars I. The synthesis of elements from carbon to nickel. *Astrophys. J. Suppl.* 1:121–146.
Hoyle, F., Fowler, W. A., Burbidge, E. M., and Burbidge, G. R. 1956. Origin of the elements in stars. *Science* 124:611–613.
Jochum, K. P., Seufert, H. M., Spettel, B., and Palme, H. 1986. The solar-system abundances of Nb, Ta, and Y, and the relative abundances of refractory lithophile elements in differentiated planetary bodies. *Geochim. Cosmochim. Acta* 50:1173–1183.
Käppeler, F., Beer, H., Wisshak, K., Clayton, D. D., Macklin, R. L., and Ward, R. A. 1982. s-Process studies in the light of new experimental cross sections: Distribution of neutron fluences and r-process residuals. *Astrophys. J.* 257:821–846.
Kruse, H. G. 1976. Thermally induced particle emission as a possible formation process of β^+ nuclei. *Nuovo Cim.* 35A:211–220.
Lamb, S. A., Howard, W. M., Truran, J. W., and Iben, I., Jr. 1977. Neutron capture nucleosynthesis in the helium-burning cores of massive stars. *Astrophys. J.* 217:213–221.
Lederer, C. M., and Shirley, V. S. 1978. *Table of Isotopes* (New York: Wiley).
Marti, K., and Suess, H. E. 1986. The even-odd systematics in r-process nuclide abundances. *Meteoritics* 21:442 (abstract).
Marti, K., and Zeh, H. D. 1985. History and current understanding of the Suess abundance curve. *Meteoritics* 20:311–320.
Merrill, P. W. 1952a. Technetium in the stars. *Science* 115:484.
Merrill, P. W. 1952b. Spectroscopic observations of stars of class S. *Astrophys. J.* 116:21–26.
Mewalt, R. A. 1983. The elemental and isotopic composition of galactic cosmic ray nuclei. *Rev. Geophys. Space Phys.* 21:295–305.
Meyer, J.-P. 1985. Galactic cosmic ray composition. *Proc. 19th Internatl. Cosmic Ray Conf.* IX:141–213.
Mould, J. R. 1982. Stellar populations in the galaxy. *Ann. Rev. Astron. Astrophys.* 20:91–115.
Norman, E. B., and Schramm, D. N. 1979. On the conditions required for the r-process. *Astrophys. J.* 228:881–892.
Pagel, B. E. J., Edmunds, M. G., and Kellerman, K. I. 1981. Abundances in stellar populations and the interstellar medium in galaxies. *Ann. Rev. Astron. Astrophys.* 19:77–113.
Palme, H., and Rammensee, W. 1981. The cosmic abundance of molybdenum. *Earth Planet. Sci. Lett.* 55:356–362.
Palme, H., Suess, H. E., and Zeh, H. D. 1981. Abundances of the elements in the solar system. In *Landolt-Börnstein, Numerical Data and Functional Relationships in Science and Technology, Group VI, Vol. 2a,* ed. K.-H. Hellwege (Berlin: Springer-Verlag), pp. 257–272.
Ross, J. E., and Aller, L. H. 1976. The chemical composition of the sun. *Science* 191:1223–1229.
Russell, H. N. 1929. On the composition of the sun's atmosphere. *Astrophys. J.* 70:11–89.
Schatz, G. 1986. The s-process of stellar nucleosynthesis. *Prog. Part. Nucl. Phys.* 17:393–417.
Schramm, D. N. 1982. The r-process and nucleocosmochronology. In *Essays in Nuclear Astrophysics,* eds. C. A. Barnes, D. D. Clayton, and D. N. Schramm (Cambridge: Cambridge Univ. Press), pp. 325–353.
Seeger, P. A., Fowler, W. A., and Clayton, D. D. 1965. Nucleosynthesis of heavy elements by neutron capture. *Astrophys. J. Suppl.* 11:121–166.
Simpson, J. A. 1983. Elemental and isotopic composition of the galactic cosmic rays. *Ann. Rev. Nucl. Part. Sci.* 33:323–381.
Suess, H. E. 1947a. Über kosmische Kernhaufigkeiten. I. Mitteilung: Einige Häufigkeitsregeln und ihre Anwendung bei der Abschatzung der Häufigkeitswerte für die mittelschweren und schweren Elemente. *Z. Naturforsh.* 2a:311–321.

Suess, H. E. 1947b. Über kosmische Kernhaufigkeiten. II. Mitteilung: Einzelheiten in der Häufigkeitsverteilung der mittelschweren und schweren Kerne. *Z. Naturforsh.* 2a:604–608.

Suess, H. E. 1987. *The Chemistry of Our Solar System: An Elementary Introduction to Cosmochemistry* (New York: Wiley).

Suess, H. E., and Urey, H. C. 1956. Abundances of the elements. *Rev. Mod. Phys.* 28:53–74.

Suess, H. E., and Zeh, H. D. 1973. The abundances of the heavy elements. *Astrophys. Space Sci.* 23:173–187.

Trimble, V. 1975. The origin and abundances of the chemical elements. *Rev. Mod. Phys.* 47:877–973.

Trimble, V. 1982. Supernovae. Part I: The events. *Rev. Mod. Phys.* 54:1183–1224.

Trimble, V. 1983. Supernovae. Part II: The aftermath. *Rev. Mod. Phys.* 55:522–563.

Truran, J. W. 1984. Nucleosynthesis. *Ann. Rev. Nucl. Part. Sci.* 34:53–97.

Truran, J. W., and Iben, I., Jr. 1977. On s-process nucleosynthesis in thermally pulsing stars. *Astrophys. J.* 216:797–810.

Ulrich, R. K. 1982. The s-process. In *Essays in Nuclear Astrophysics*, eds. C. A. Barnes, D. D. Clayton, and D. N. Schramm (Cambridge: Cambridge Univ. Press), pp. 301–323.

Urey, H. C. 1952. Abundances of the elements. *Phys. Rev.* 88:248–252.

Wagoner, R. V. 1980. The early universe. In *Physical Cosmology*, eds. R. Balian, J. Audouze, and D. N. Schramm (Amsterdam: North Holland).

Ward, R. A., and Newman, M. J. 1978. s-Process studies: The effects of a pulsed neutron flux. *Astrophys. J.* 219:195–212.

Woolum, D. S., Burnett, D. S., Benjamin, T. M., Rogers, P. S. Z., Duffy, C. J., and Maggiore, C. J. 1986. Trace element contents of primitive meteorites: A test of solar system abundance smoothness. *Nucl. Instr. Meth.* B22:376–379.

Woosley, S. E., and Howard, W. M. 1978. The p-process in supernovae. *Astrophys. J. Suppl.* 36:285–305.

York, D. G. 1982. Gas in the galactic halo. *Ann. Rev. Astron. Astrophys.* 20:221–248.

14.2. STELLAR NUCLEOSYNTHESIS AND CHEMICAL EVOLUTION OF THE SOLAR NEIGHBORHOOD

DONALD D. CLAYTON
Rice University

This chapter reviews primarily the key ingredients of nucleosynthesis in stars and its role in the chemical evolution of the Milky Way Galaxy. First, the evidence for nucleosynthesis in early history is outlined, both the Big Bang and the variations in time and place of stellar abundances in the early Galaxy. The fundamentals of the theory of chemical evolution of the Galaxy are described as a quantitive framework for interpreting nucleosynthesis in stars. Two analytic models are described for clear understanding of this theory—the closed-box model, which is not observationally satisfactory, and an analytic model having continuous infall onto our Galaxy. Then, the astrophysical fabric of the yields derived from nuclear burning in stars is described. This is the major objective of this review. The ways in which this theory confronts observational data are described throughout, with special emphasis on the data derived from studies of meteorites. A table giving the origin of each isotope is included, with speculations relevant to the isotopic anomalies.

14.2.1. EARLY NUCLEOSYNTHESIS

Although it was not always so evident, it is now clear that the evolution of stars has provided the production sites for the chemical elements. Many scientists have contributed brilliantly to this conclusion, but the theory itself was largely constructed in the decade following World War II (Hoyle 1946,1954; Burbidge et al. 1957; see also Chapter 14.1). Today, observational evidence states clearly that the heavy elements were not always present in the H and He gas that is the basic atomic stuff of the universe. For one thing, galaxies apart from our own Milky Way can be seen by astronomers to be deficient in the heavy elements. In dwarf spheroidal galaxies the absorption lines of Fe are so weak that the Fe concentration can be no greater than a percent or so of its concentration in the gas of our Milky Way. Especially fascinating is the blue compact galaxy, I Zw 18, whose emission lines are similar to massive HII regions in galaxies except for the extreme weakness of the C, N and O emission lines. Their low ionic concentrations, only a few percent of the solar concentration, make this compact galaxy the most CNO deficient galaxy known (Kunth and Sargent 1983,1986). Such observations speak more eloquently than theory of places in our universe where star formation and evolution has not been efficient enough to contaminate the primordial gas with its fallout.

Early Nucleosynthesis in our Galaxy

Even more extreme evidence is found in the extreme Population II stars of our own Galaxy. These stars condensed from the primordial gases while they were in the very process of galaxy formation. As gravity collected falling primordial gas clouds, stars formed within that falling gas (Eggen et al. 1962; Larson 1976). These stars today retain spheroidal orbits. They appear to be high-velocity stars with respect to the local mean circular motion in today's galactic disk. The disk stars, such as our Sun, formed within gas that was first able to establish a gaseous galactic disk. When the shock waves of the early infalling gas were dissipated, that gas settled into a flat rotating system, but the early Population II stars retained the out-of-plane elliptical motion with which they first formed. It is in these spheroidal-orbit Population II stars that we see almost the primordial chemical composition of our Galaxy. Their photospheric absorption lines commonly reveal an abundance ratio Fe/H $\sim 10^{-2}$(Fe/H)$_\odot$. (Iron is the best tracer because of its resonance lines in the visible spectrum.) These observed abundance ratios represent the matter from which these old stars formed. They have masses slightly less than 1 M$_\odot$, because only such low-mass stars have the longevity to survive from the galaxy-formation epoch until today. The most metal-poor star so far discovered (Bessel and Norris 1984) is of this type, a low-mass red giant having the incredibly small Fe abundance Fe/H $\simeq 10^{-4.5}$(Fe/H)$_\odot$. More than 99.99% of galactic Fe has been synthesized since this old star formed, an eloquent testament to nucleosynthesis in stars. This particular star is 3000 pc away, out of the plane toward the south galactic pole, and has a radial velocity 54 km/s relative to the disk orbital velocity of the Sun.

Luck and Bond (1981) carried out a thorough spectroscopic study of 21 such Population II red giants, which are easier to study than the main sequence stars because they are so much more luminous. Their Fe/H ratios, when expressed logarithmically to base ten relative to the Sun, spanned the range [Fe/H] = -1.4 to -2.7. They showed that although other elements are deficient by similarly large factors, interesting and significant variations exist. The α-nucleus metals (Mg, Si, Ca, Ti) are less underabundant than Fe by factors of about 2 to 3, showing that the mass-weighted average of C-, O- and Si-burning phases of stars occurred somewhat earlier than the average equilibrium processes. Current interpretation is that the former are dominated by massive stars, say 15 to 50 $M_\odot$, which evolve rapidly, whereas a major fraction of the Fe emerges from longer-lived intermediate-mass stars. Similar differential-growth effects can be found in the abundances of C and N, which also experience slightly delayed growth. These differential effects can be seen in early stellar nucleosynthesis, when the time delays for stars of differing lifetimes are noticeable, but they are much harder to detect in the later disk stars, to whose initial abundances stars of larger lifetime range have been able to contribute. A very noticeable maturation effect measured also in Luck and Bond's (1981) sample is found in the abundance ratio of s-process Ba to that of Fe from which it is synthesized. In the stars of smallest Fe/H ratio, the Ba/Fe ratio is about 40 times less than the solar ratio; but by the time Fe/H has grown above 10^{-2} solar, the Ba/Fe ratio has achieved the solar value. Such data strengthen in many ways the theoretical picture of chemical evolution of galactic abundances. There is no question that nucleosynthesis in stars is responsible for almost all of our heavy elements.

Our solar system and its meteorites formed much later than these early metal-poor stars, and it formed from gas that had settled early into a disk. Apparently, this disk gas was already somewhat enriched by nucleosynthesis fallout from the halo stars, because there are very few disk population (Population I) stars having metal concentrations less than Fe/H $\simeq 0.1$ (Fe/H)$_\odot$, or [Fe/H] $\simeq -1$ in the astronomer's logarithmic notation. But their existence shows that 90%, say, of the heavy elements in our galactic disk have in fact been synthesized in that disk. Twarog (1980) and others have described the anticorrelation of (Fe/H) in dwarf stars in the solar neighborhood with the ages of those stars (the age, of course, being determined with the aid of theoretical models of how the appearance of main-sequence stars changes while they mature). His results are shown as the data points in Fig. 14.2.1, where the curve is a theoretical one discussed below. Figure 14.2.1 confirms that the oldest stars in our neighborhood of the disk are indeed the most metal-poor ones, reflecting a steady but not constant growth of metallicity with time, a growth reflecting stellar evolution and nucleosynthesis during the chemical evolution of the disk of our Galaxy.

No survey would be complete without reference to the differential nucleosynthesis with radial position in disk galaxies. This galactic abundance gradient for the nearby spiral galaxy M81 is shown in Fig. 14.2.2 in the form of the O/H ratio. It obviously decreases by a full order of magnitude between 4 kpc and 15 kpc, again demonstrating clearly the effects of ongoing nucleosynthesis. These O abundances, by the way, are of interstellar gas rather than of stars, so they measure the gas concentrations *today* in that galaxy. The O emission is from singly and doubly charged ions within HII regions: interstellar gas ionized by the ultraviolet radiation from hot young stars. The abun-

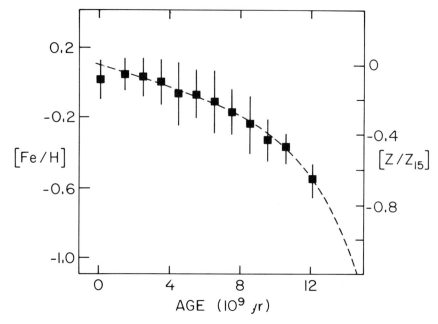

Fig. 14.2.1. The metal concentration in main-sequence dwarf stars as a function of their age is thought to measure how that concentration has increased in time in the interstellar gas from which those stars formed. [Fe/H] is the logarithm of the Fe/H abundance ratio in that star divided by its value in the Sun, and is determined from an analysis of absorption lines of Fe and H in the stellar spectra (Twarog 1980). Clearly Fe/H has increased by a factor of about 4 since the oldest of these stars formed. Right-hand ordinate $[Z/Z_{15}]$ is the logarithm of the ratio $Z(t)/Z(15\text{ Gyr})$ computed from Eq. (15) of Sec. 14.2.2 describing an analytic model of the chemical evolution of the Galaxy and is shown as the dashed curve. This particular analytic model (Clayton 1985) provides a good fit to many observable objects in the solar neighborhood.

dance gradient in our own Milky Way is very similar to this one (Shaver et al. 1983).

A theoretical framework for interpreting the chemical evolution of galaxies will be found in Sec. 14.2.2.

Nucleosynthesis in the Big Bang

The observations of the existence of old stars having heavy-element mass fraction $Z < 10^{-3} Z_\odot$, where $Z_\odot$ is about 2%, demonstrates that the universe in the form we know it today began without those heavy elements. But the earliest stars did contain He already in the concentration $Y = 0.25 \pm 0.02$ that we find today. Both the metal-poor blue compact galaxies like I Zw 18 and the metal-poor Population II stars already contain He in normal amounts (Kunth and Sargent 1983). This is to be expected, actually, because the evo-

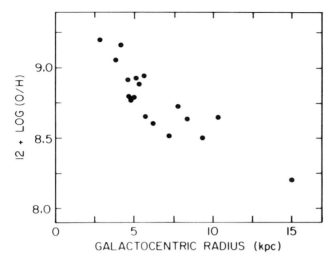

Fig. 14.2.2. The gradient in the O concentration in the interstellar medium today of the spiral galaxy M81 is measured by emission lines in HII regions of ionized interstellar gas (Garnett and Shields 1987). Very evidently the gas at increasing distance from the center of this galaxy has experienced a decreasing amount of stellar processing, providing good evidence of nucleosynthesis in stars.

lution of stars ejects no greater mass of He than of heavier elements, so far as we understand their evolutions, causing one to expect $\Delta Y \approx Z$, where Z is the heavy-element mass synthesized in stars, and ΔY is the increment to the He mass so synthesized. So it is evident that stellar nucleosynthesis cannot have produced the He-rich matter that pervades the visible universe. For a long time this was a major unexplained puzzle—the composition of the primordial gas from which galaxies formed. But because calculation of the expansion of the early universe in accord with Einstein's field equations (the Big Bang) quite naturally yields a composition that is about 25% He by mass, it is now felt by most astrophysicists that the initial composition of galaxies is explained (see, e.g., Yang et al. 1984). Indeed, the statement is much stronger: there is complete consistency between the data on the abundances of ^{1}H, ^{2}H, ^{3}He, ^{4}He and ^{7}Li and the concentrations expected from a standard Big Bang. That calculation assumes that the ratio of nucleons to baryons in the universe is equal to that observed in spread-out galaxies and in the 3K microwave radiation, namely $\eta \simeq 5 \pm 2 \times 10^{-10}$ baryon/photon. It also assumes three types of neutrinos. So impressive is the agreement of these five primordial abundances with the expectations of the simple theory that most regard it as correct. Were these five abundances coincidentally in agreement with an incorrect theory, we would be justified perhaps in reversing Einstein's own famous aphorism to read instead, "God is not subtle. He is malicious!"

In all honesty however, the situation is not all rosy. The abundance of ^{2}H, ^{3}He and ^{7}Li are very hard to determine. Each is destroyed in stars, while the latter two may also be synthesized in stars (certainly ^{3}He is). Much of the evidence from meteorites, solar wind, and interstellar emissions is subject to reasonable doubts. These prevent a definitive statement of the values of their prestellar abundances. But without a Big Bang, the rationalization of each abundance value would be very difficult. The main problem with Big Bang nucleosynthesis is its apparent exclusion of the possibility of a closed universe dominated by baryons.

In summary, however, the relationship of nucleosynthesis to the topic of meteorites and the origin of the solar system relies overwhelmingly on the elements heavier than He, and these, with the likely exception only of ^{7}Li, owe their origins to stellar evolution and ejection.

14.2.2. CHEMICAL EVOLUTION OF GALAXIES

Data of the type shown in Figs. 14.2.1 and 14.2.2 demonstrate that star formation, star evolution and partial ejection of the thermonuclearly evolved debris caused gradients in time and in space of the abundances of the heavy elements. It was in such an evolving system that the solar system was born about 4.6 Gyr ago about 10 kpc (~30,000 light-years) from the center of the Milky Way. Because star formation occurs following the hydrodynamic collapse of interstellar clouds, the newly forming stars are composed of the bulk abundances (locally) within those clouds. That is why the lower main-sequence stars, the dwarf stars having surface convection zones, reveal surface compositions equal to the bulk composition of the clouds from which they formed long ago. The planetary systems were consequences of the hydrodynamic collapses, but, because of the dynamics of planetary formation, they were not able to retain a chemically unfractionated sample of the interstellar medium, the very situation that makes it possible to obtain information on the origin of the solar system from the study of meteorites. The meteorites contain so much chemical fractionation that an incorrect theory is doomed to fail. The meteorites, at least when understood, contain not only the most accurate evidence for the relative abundances of the nonvolatile elements but also a unique record of chemical fractionation during the formation of the first solid aggregates of dust and of isotope ratios involving radioactive nuclei that independently measure the nucleosynthesis history of the matter from which the solar system formed. For these reasons and more, the chemical evolution of the Milky Way is an essential part of meteoritic theory as well.

The topic, *chemical evolution of galaxies,* is the name commonly used to describe the evolution in time and in space of the *bulk abundances* in the interstellar medium. Whether the solar system actually formed of locally well-mixed (chemically homogeneous) interstellar matter is not known. Perhaps it contained instead a disproportionate admixture of ejecta from another star or

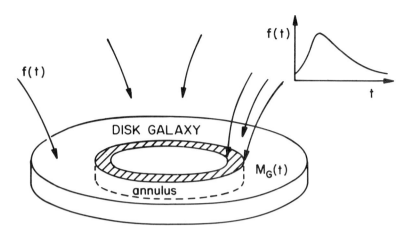

Fig. 14.2.3. The annulus containing the Sun and other stars also contains a mass $M_G(t)$ of interstellar gas that has changed with time in response to its incorporation into stars and to time dependent infall $f(t)$ of extragalactic material. The objective of chemical evolution theory is the calculation of the concentration $Z(t)$ of heavy element Z in that interstellar medium in response to these circumstances (figure from Clayton 1986a).

stars in the same mother cloud. This remains to be demonstrated. But at the very least one requires a theory for the mean evolution of a chemically homogeneous medium as a backdrop against which peculiarities can be identified, and it is this mean evolution that is the basic topic of chemical evolution. The word "chemical" is in this sense a misnomer that is astrophysically well entrenched in what should really be called "abundance evolution."

An idealized disk galaxy is shown in Fig. 14.2.3. This matter, in differential rotation about the galactic center, is not easily mixed radially because of angular momentum differences. The viscosity is not great enough for radial transport, at least efficient radial transport. The radial abundance gradient in the Galaxy, similar to that of M81 shown in Fig. 14.2.2, would not exist if matter were well mixed radially. For simplicity, one often assumes for the approximate theory that radial mixing is negligible. This reduces the problem to the chemical evolution of an annulus, specifically the one containing the Sun, as shown in Fig. 14.2.3. Within the solar annulus all matter is regarded as having the same average nucleosynthesis history, for two reasons: (1) the average history of star formation and evolution is azimuthally symmetric even though at a specific time some portion of the annulus is within spiral arms and some without; (2) mixing of gas over the annulus does occur owing to differential rotation and to the turbulent state of the interstellar medium. By the same token, the annulus is taken to be mass-conserving except for the infall $f(t)$ of new primordial matter onto the disk. This infall is envisioned as a continuation of the process that formed the disk galaxy, but it is not known if it continues today in the solar annulus.

The simplified picture of the solar neighborhood is then that interstellar matter is well mixed within the annulus as shown in Fig. 14.2.1, that stars moving on orbits in that annulus formed from that well-mixed gas and dust, and that the ejecta from those stars following their evolution is remixed into the annular medium. The quantity $M_G(t)$ represents the mass of interstellar material contained therein, and although the subscript G refers to "gas," it is especially important for meteoritics to remember that gas contains interstellar dust of all types, also well mixed. The problem is then to treat the chemical evolution of this solar neighborhood in the presence of stellar evolution and nucleosynthesis.

The physics of stellar structure, stellar evolution and nucleosynthesis is described in my textbook on the subject (Clayton 1968,1983), and has a huge subsequent literature. That text could not include a quantitative account of the nucleosynthesis yields from advanced stages of stellar evolution, however. One objective of this chapter will be a description of recent results on these nucleosynthesis processes. Before addressing those details, however, it is instructive to see how they fit into the elementary mathematical theory of chemical evolution.

The rate at which interstellar matter is turned into new stars is conventionally designated by $\psi(t)$, and the rate at which matter is ejected from aging stars by $E(t)$. The mass of interstellar material in the solar annulus is then described by the differential equation

$$\frac{dM_G}{dt} = -\psi(t) + E(t) + f(t). \qquad (1)$$

New stars are born with a distribution of masses called the initial mass function $\phi(m)$, and it is convenient to normalize that function such that $\int m\phi(m)\, dm = 1$, because then the number of new stars formed during the interval $(t, t + dt)$ and in the mass range $(m, m + dm)$ is

$$dN = \phi(m)\, \psi(t)\, dm\, dt. \qquad (2)$$

There has been much speculation about the shape of the initial mass function changing with time as galaxies evolve, but in the absence of an unambiguous demonstration that it has done so, it is common in the simplest theory to take $\phi(m)$ to be time-independent. Salpeter (1955) originally concluded from the numbers of stars observed in the solar neighborhood that $\phi(m) \propto m^{-2.35}$, a result still much used. But following the much more thorough modern analysis by Miller and Scalo (1979), Tinsley (1980) has recommended the piecewise representation

STELLAR NUCLEOSYNTHESIS AND CHEMICAL EVOLUTION

$$\begin{aligned} m\phi(m)\,\psi(t_o) &= 1.00\ m^{-0.25} & 0.4 < m < 1 \\ &= 1.00\ m^{-1.0} & 1 < m < 2 \\ &= 1.23\ m^{-1.3} & 2 < m < 10 \\ &= 12.3\ m^{-2.3} & 10 < m < 50 \end{aligned} \quad (3)$$

in units $M_\odot/pc^2 Gyr$, with the star mass m expressed in solar mass, and $\psi(t_o)$ being the total star formation rate today in the solar neighborhood. The integral over this mass function gives $\psi(t_o) = 3.0\ M_\odot/pc^2 Gyr$ for new stars between 0.4 and 50 $M_\odot$. One problem is that the full range of new stellar masses is not really well known, especially downward to 0.1 $M_\odot$. Any additional star formation below 0.4 $M_\odot$ only leaves dwarf remnants, however, so its rate, if any, affects only the gas consumption. The importance of a reliable prescription for $\phi(m)$ lies in two facts: stars of different mass require greatly different times to come to the end of their life and they eject quite different yields of heavy elements into the interstellar medium. Stars near the upper mass end live several Myr, whereas the Sun's lifetime is expected to total 10 Gyr. Stars near the upper mass end return huge yields of new elements, whereas stars like the Sun do little nucleosynthesis other than for a few special nuclei. This makes the overall time-dependent calculation numerically complicated.

The composition of the interstellar medium (ISM) as a function of time is conveniently expressed in terms of the mass fraction $Z(t)$ of element Z in that medium. Thus

$$M_{GZ}(t) = Z(t)M_G(t) \quad (4)$$

in bulk, so that the chemical state (e.g., atomic gas, ions, molecules, dust types) is not described by $Z(t)$. In principle one thinks of $Z(t)$ as the sum of the mass fractions of element Z over all of the chemical forms containing element Z in the ISM. That decomposition into chemical forms must be reckoned separately, and in it is to be found the seeds of the cosmic-chemical-memory theory of isotopic anomalies (Clayton 1978,1979,1982). That is a separate aspect of the truly chemical evolution of the Galaxy. And although that theory and its relationship to the data on the origin of the solar system comprise one of the key topics of this book, we suppress explicit treatment of it at this point by letting $Z(t)$ be the bulk mass fraction. Having done so, an expression analogous to Eq. (1) can be written for the interstellar mass balance of element Z:

$$\frac{d}{dt}(ZM_G) = -Z\psi(t) + Z_e E(t) + Z_f f(t) \quad (5)$$

where Z_e is the mean value of Z in the stellar ejecta and where Z_f is the value of Z in the newly infalling matter. One sets $Z_f = 0$ if it is truly primordial matter or $Z_f \approx 0.1 Z_\odot$ if the infalling material has been enriched by ejecta of halo stars to a value comparable to that seen in the most metal-poor disk stars. In neither case will the unknown value of Z_f be significant for the composition of the material that went into the meteorites, so we may simplify by taking $Z_f = 0$. The mean value Z_e of Z in the material ejected from stars is the difficult calculation that depends on the full power of a correct theory of star formation and evolution. At any time t, a large number of stars of differing initial mass are dying owing to their formation in differing prior epochs, and the amount of element Z in each one will differ. Fully time-dependent calculations must therefore be quite detailed.

It is at this point that one can make a very good approximation that greatly simplifies the problem. At the time of solar formation, the Galaxy was already quite mature, between 5 and 15 Gyr old. The rate of star formation $\psi(t)$ had been changing only slowly over several Gyr. The stars responsible for most of the ejecta $E(t)$ and for most of the new nucleosynthesis are rather massive, having lifetimes between a few Myr and a few hundred Myr. In that case, the rate of star death for each significant star mass m is approximately equal to the rate of star birth of the same stars. The stars recycle their matter very quickly in comparison to the galactic age. One inserts this into the differential equation by assuming that the stars die immediately after they are born, commonly called "the instantaneous recycling approximation." The total ejection rate from stars, $E(t) = R\,\psi(t)$, is then the simple product of the so-called "return fraction" R with the rate at which mass is being put into stars. When $\psi(t)$ goes into stars, immediately thereafter $R\psi(t)$ comes back out and $(1-R)\,\psi(t)$ remains either as stellar remnants (white dwarfs, pulsars etc.) or as dwarf main-sequence stars with $m < 1$ that return nothing by the time of solar formation. Thus Eq. (1) becomes

$$\frac{dM_G}{dt} = -(1-R)\,\psi(t) + f(t). \qquad (6)$$

The value of R depends strongly on the mass function $\phi(m)$. For the form of $\phi(m)$ listed earlier $R \simeq 0.5$; i.e., about half the mass that enters stars comes back out again. But if there were born a much larger number of very low-mass stars ($m < 0.4$), the return fraction would be smaller because those low masses return nothing. The rate of formation of stellar remnants is then $\dot{M}_{\text{star}} = (1-R)\psi$.

The instantaneous recycling approximation also makes Eq. (5) for the growth of mass of nucleosynthesis product more simple. The stellar ejecta $E(t)$ is then just $R\psi(t)$, the return fraction of the star-formation rate. We define the yield y_z of each heavy nucleus Z as being the ratio of new mass of Z in the ejection rate to the new mass of stellar remnants: $y_z = (Z_e - Z)\,E(t)/\dot{M}_{\text{star}} = (Z_e - Z)\,R/(1-R)$. Then inserting Eq. (6) into the expanded derivative of Eq. (5) gives the differential equation for the rate of growth of the concentration:

$$\frac{dZ}{dt} = y_z (1 - R) \frac{\psi(t)}{M_G(t)} - (Z - Z_f) \frac{f(t)}{M_G(t)}. \qquad (7)$$

This standard formulation applies to stable nucleosynthesis products; however, some of the most important meteoritic data concern the abundances of radioactive nuclei. One needs the average concentrations of U, Th and Pu, for example, in the interstellar annulus. If Z is radioactive with decay rate λ_z, an extra term, $-\lambda_z Z$, is added to Eq. (7), as is easily confirmed by adding $-\lambda_z(ZM_G)$ to the right-hand side of Eq. (5) and repeating the subsequent simplification. Their evaluation from meteoritic science is discussed in Chapter 15.2.

The rate of star formation $\psi(t)$ is a complicated and unknown function of galactic conditions, involving global properties of the Galaxy and detailed properties of star-forming clouds. However, if one concentrates on the solar annulus, as appropriate for the solar system, the radial problems of global physics (density waves, etc.) are irrelevant if constant in time. In that case, the star-formation rate in molecular clouds seems likely to be proportional to the total mass of cloud material (i.e., the total number of clouds if their mass spectrum is constant). A great mathematical simplification results from this assumption, commonly called the linear star-formation models, for then one takes the star-formation rate $\psi(t)$ to be some constant times the gas mass $M_G(t)$. Defining that constant ω according to

$$(1 - R) \psi(t) \equiv \omega M_G(t) \qquad (8)$$

the basic equations take their most transparent representation:

$$\frac{dM_G}{dt} = -\omega M_G + f \qquad (9)$$

and

$$\frac{dZ}{dt} = y_z \omega - (Z - Z_f) \frac{f}{M_G} - \lambda_z Z. \qquad (10)$$

The major task of stellar-evolution theory is then the calculation of y_z, the mass of new element Z ejected from the spectrum $\phi(m)$ of stars born per mass of stars left as permanent remnants (too small to evolve, or degenerate remnants). This means that one requires the mass ΔM_Z of Z from a star of mass m, plus the average of ΔM_Z over the mass function $\phi(m)$, normalized per unit mass of residual star.

Before discussing that problem of nucleosynthesis mechanism, however, one should see how simple solutions of Eqs. (9) and (10) behave for some interesting assumptions about the rate of infall $f(t)$.

A Closed Box

A simple example of these evolution equations is to be found in a closed box of gas turning slowly into stellar remnants. It is to be found as an example in almost all papers on chemical evolution of galaxies. If there is no infall, the total mass is constant. With $f = 0$ one sees from Eq. (9) that the mass of the interstellar medium declines exponentially: $M_G(t) = M_G(o) \exp(-\omega t)$. Because the star-formation rate is proportional to M_G in linear models, it also declines exponentially, as does, therefore, the bulk rate of nucleosynthesis. But the integral of Eq. (10) for a stable nucleus ($\lambda = 0$) is clearly $Z = y_z \omega t$, a linear growth with respect to ωt having a slope equal to the yield y_z. Even though the nucleosynthesis declines exponentially, the stable concentrations grow linearly, because the mass M_G absorbing the nucleosynthesis is also declining exponentially. Thus, the nucleosynthesis rate *per unit mass of intersteller medium* is a constant. This distinction has confused many.

The application to nuclear cosmochronology is one of the very significant connections with meteorites, which concerns itself with the concentrations of various radioactive nuclei within them (see Chapter 15.2). The closed-box solution of Eq. (10) for a radioactive concentration Z_λ in the interstellar medium is

$$Z_\lambda = y_{Z\lambda}\, \omega (1 - e^{-\lambda_z t})/\lambda_z \qquad (11)$$

again corresponding to a constant rate of production per unit mass of gas. The precise statement is that the age distribution of nuclei is flat in this simple model. The significant chronometric quantity is "the remainder," defined (Clayton 1985) as the ratio of a radioactive concentration to the value it would have had were it stable. For the closed box that remainder has the value

$$r \equiv Z_\lambda/Z = (1 - e^{-\lambda_z t})/\lambda_z t \qquad (12)$$

which is totally independent of the decay constant ω measuring the rates of decline of nucleosynthesis and of interstellar medium.

As nice as this simple model is, it does not fit the facts. The stable metallicity has not grown linearly in time, although Fig. 14.2.1 demonstrated that it has grown systematically. Nor is the observed rate of growth as great as the slope $y_z \omega$ that would describe $Z = y_z \omega t$ if one inserts calculated yields y_z from stellar structure and an observed rate of consumption ω of interstellar gas. Nor does it correctly describe the numbers of dwarf stars of differing metallicities. For these reasons it is commonly suspected that the closed-box assumption is wrong. That seems quite likely, in that infalling matter $f(t)$

associated with the formation of the Galaxy probably continued over billions of years (Larson 1976) and may still do so at a much reduced rate.

Analytic Solutions with Continuous Infall

More realistic models of chemical evolution seem to require that the infall rate have built quickly to a large value and declined to a low value today. The final annulus mass greatly exceeds its value when star formation began. In this way the ISM first increases in mass before beginning a slower decline, and the number of stars born early then does not greatly exceed those born later. The primordial infall dilutes the growth of metallicity to its lower observed value, and the cosmochronologies are affected. Clayton (1984b, 1985) showed analytic solutions of all of those quantities that have pedagogic virtues of clarity and physical reasonableness. His "standard model," which is both the simplest and a useful fit to observations, takes the infall rate to be described by a time parameter Δ and an integer k,

$$f(t) = \frac{M_G(o)k}{\Delta} \left(\frac{t + \Delta}{\Delta}\right)^{k-1} \exp(-\omega t) \tag{13}$$

where $M_G(o)$ is the initial annulus mass when star formation began. The solution of Eq. (9) yields the ISM mass

$$M_G(t) = M_G(o) \left(\frac{t + \Delta}{\Delta}\right)^{k} \exp(-\omega t). \tag{14}$$

Clayton (1985) presents the solutions of all quantities for arbitrary k, but for this pedagogic review it is adequate to look only at the result for the $k = 1$ case of exponentially declining infall rate. The stable concentration Z solving Eq. (10) is then

$$Z = \frac{y\omega t}{2} \left(\frac{t + 2\Delta}{t + \Delta}\right) \tag{15}$$

which is growing quasi-linearly for $t \gg \Delta$ when the solar system formed, but at only half the slope of the closed model. It is this result that is shown as the dashed line in Fig. 14.2.1, calculated for $\omega = 0.3/\text{Gyr}$ and $\Delta = 1$ Gyr and plotted as the logarithm of the ratio of $Z(t)$ to the value it achieves at $t = 15$ Gyr. By such comparisons one constructs credible representations of the history of matter leading up to the birth of the solar system. The use of credible (rather than arbitrary) galactic histories is crucial for meteoritic science.

For nuclear cosmochronology, the remainder expressing the ratio of Z_λ to the value it would have had were the nucleus stable becomes (for radioactive nucleus of decay rate λ_z)

$$r = \frac{2}{t^2 + 2t\Delta} \left[\frac{t + \Delta}{\lambda_z} - \frac{1}{\lambda_z^2} - e^{-\lambda_z t}\left(\frac{\Delta}{\lambda} - \frac{1}{\lambda_z^2}\right)\right]. \tag{16}$$

This exact result can be recommended as as good way to compute radioactive chronologies. As an example, consider the ratio $^{235}U(t)/^{238}U(t)$ so important to that topic. Because the remainder $r(235)$ is the fraction of all ^{235}U ever introduced into solar material that survives, one has

$$\frac{^{235}U}{^{238}U} = \frac{r(235)}{r(238)} \frac{P(235)}{P(238)} \tag{17}$$

where $P(235)/P(238)$ is their production ratio in r-process events. The value of Δ does not much matter because $t \gg \Delta$ when the solar system formed, since $\Delta \simeq 10^9$ yr and $t_\odot \approx 10^{10}$ yr. Clayton (1984b) gives examples of this application, also for $k > 1$, and notes as well that the result does not depend on the gas-consumption constant ω, as long as the initial galactic metallicity and the infall metallicity are very small, which is certainly expected to be true for the radioactive chronometers. This independence of the paramter ω is an important conclusion, because it shows that uncertainty to be irrelevant to the reliability of nuclear cosmochronology. Even the remaining weak dependence on Δ is virtually nonexistent. Thus, to escape the chronologic implications of, say, U(235)/U(238) one must advance models of chemical evolution that violate the rather plausible assumptions of this standard analytic model. Examples would be production ratios that are not constant in time, or a nucleosynthesis rate that is not a smooth curve in time (a late spike, say), or a nucleosynthesis rate that has a very different time dependence than does the star-formation rate. But in the absence of such special assumptions, the topic of nuclear cosmochronology must above all be consistent with acceptable models of the chemical evolution of the Galaxy. It is in that sense that an acceptable analytic model presents a more realistic framework for this topic than the arbitrary parameterizations that have been popular (because of their ease) in research works on nuclear cosmochronology.

The student of meteoritics and astrophysics should be aware that the great value of radioactive chronologies is that they depend upon time in an essential way, whereas many other features of chemical evolution depend upon time only implicitly. For example, one sees that in the closed-box model, the results $Z = y_z\omega t$ and $M_G = M_G(o) \exp(-\omega t)$ can be combined to give $Z = y_z \ln M_G(o)/M_G(t)$, which is normally written $Z = y_z \ln(1/\mu)$, with μ being the common notation for the fraction of total mass in the form of ISM material. Now the point is that this result is really more fundamental,

STELLAR NUCLEOSYNTHESIS AND CHEMICAL EVOLUTION   1035

because temporal history can be eliminated by considering the ratio of Eq. (10) to Eq. (9), which in the closed-box model leaves

$$\frac{dZ}{dM_G} = -\frac{y_z}{M_G}$$

showing that $Z = -y_z \ln(1/\mu)$ is *independent* of the temporal rate of star formation. Time has been physically eliminated. For radioactivity this is not so, and time remains essential. This explains why it is in chemical-evolution studies that radioactivity plays such an important role.

This standard analytic model of chemical evolution is very useful in its clear representation of the problem. Clayton (1984b, 1986a) has shown how other analytic families can be explicitly displayed for other functional forms for the infall rate $f(t)$. They can easily be turned to by the reader needing more details. The analytic growth of concentrations that depend upon an existing seed nucleus (i.e., "secondary" nucleosynthesis) was solved by Clayton and Pantelaki (1986) for the same sets of analytic representations. They also demonstrated the validity of the instantaneous recycling approximation for the solar neighborhood near the time of solar birth. One new theoretical interpretation (Clayton 1988) of the isotopic anomalies derives from the slower growth of the interstellar abundances of the secondary nucleosynthesis products in comparison with the abundances of the primary isotopes.

At this point, one sees that the yields y_z from the mass function $\phi(m)$ are needed to compute theoretically the abundances at solar birth. This is where the science of stellar nucleosynthesis enters explicitly. It bears repeating, moreover, that the foregoing theoretical results apply to a *well-mixed average* of the solar neighborhood. Special admixtures have been considered both for the isotopic-anomaly problems and for the extinct radioactivities. To be specific, many have argued that solar-system radioactivities were influenced by a final spike, a production that represents a discontinuity in the age spectrum of solar-system nuclei. Others have argued that the isotopic anomalies represent an incompleteness in the spatial mixing within the solar annulus, leading to one portion of the solar system having a different bulk isotopic composition from another portion of it. This writer is skeptical of such special admixtures, but it must be explicitly noted that they must, if they existed, be handled separately from the mean-annulus treatment described above. This mean treatment has yielded the interstellar concentrations for matter homogeneously mixed spatially and in which the nucleosynthesis rate is a smooth function of time. Special events can be added to the results of this mean treatment.

14.2.3. NUCLEAR BURNING PROCESSES AND YIELDS

The mapping of the evolution of stars and the associated nuclear evolution of their interiors has been a huge community effort, and it is by no means complete today. The remaining problems fall into primarily three classes: (1)

nuclear data and details of thermonuclear evolution; (2) detailed questions on the evolution of stars, single and binary; (3) the hydrodynamic description of the final states of stars, how a portion of the mass is ejected, and how much those ejection processes alter the composition established up to that time. The basic principles are best approached by textbook (especially Clayton 1968,1983), because the details are too numerous to be addressed in this chapter. Unfortunately there exists no self-contained account of the advanced evolution of stars to carry on beyond my textbook, so that the new student is forced to examine a huge literature showing a field in a healthy state of continuous development for the past two decades. A hand-picked bibliography of papers of special clarity and significance (in this writer's opinion) can be found in the new preface to the 1983 edition of my textbook. What I attempt in this chapter is an understandable presentation of those ideas and results that will speed up comprehension.

Although low-mass stars are more numerous in the mass spectrum $\phi(m)$ of newly born stars (Eq. 2), the bulk of heavy elements is ejected from massive stars. This is true not only because more massive stars eject more mass, but more specifically because the fraction of the star that is converted to heavy elements increases with mass. Weaver and Woosley (1980) and Woosley and Weaver (1986) find that the mass fraction of a star that is converted to elements $Z \geqq 6$ is well approximated by

$$Z_{ej} \simeq 0.5 - \frac{6.3}{M} \qquad (19)$$

where M is the mass of the star in solar units and is in the range of $15 < M < 60$ for this relation. That is, for a star of 60 $M_\odot$, fully 40% or 24 $M_\odot$ of initial H and He is converted to elements with $Z > 6$ during the thermonuclear evolution. Much of those 24 $M_\odot$ of heavy elements will be returned to the interstellar medium after the roughly 4 Myr evolution time of this star; at least it must be returned unless it is the fate of such stars to result in black holes, because only a few $M_\odot$ can be left as a neutron-star remnant if that is the fate of the cores of these stars. By contrast, only a quarter of the 15 $M_\odot$ star, about 4 $M_\odot$, is converted to heavy elements, of which perhaps 2 $M_\odot$ remains as a neutron-star remnant, leaving ejecta of perhaps 2 $M_\odot$ of heavy elements. The total yield of heavy elements is determined by an integral of the spectrum $\phi(M)$ of stellar masses,

$$\text{absolute yield} = \int [Z_{ej}(M)M\phi(M) - M_{\text{remnant}}]dM. \qquad (20)$$

This integrand has a maximum at $M = 18.5$ $M_\odot$ but a median at $M = 29$ $M_\odot$ (Weaver and Woosley 1980), arguing that a 25 $M_\odot$ star should be as typical as any of the producers of heavy elements. This same yield integrand declines steeply for $M < 10$ $M_\odot$ because those stars lock up the lion's share of their heavy element production in an eventual white-dwarf remnant. (An exception to this conclusion applies to the production of C and N by stars with $M < 10$ $M_\odot$, for reasons to be summarized below.) For the science of nucleosynthesis, therefore, it is believed that the evolution of massive stars ($M > 10$ $M_\odot$) plays the dominant role.

Some essential features of the evolution of massive stars are illustrated in Fig. 14.2.4. The cartoon half illustrates a sequence of successively smaller thermonuclear cores that are generated by the thermonuclear evolution of matter. When He first ignites by the 3α process at the center of a H-exhausted core, that core mass is labeled M_α. For a 25 $M_\odot$ star, this occurs at an age of 7 Myr, when M_α is about 37% of the mass of the star (Arnett 1972a), or about $M_\alpha \simeq 9.5$ $M_\odot$, as shown in the line drawing in Fig. 14.2.4. This core of convective He burning produces, after 0.5 Myr more, a core of C and O having a mass $M_C = 6.0$ $M_\odot$ at the time when C ignition occurs in that core, as shown by the M_C curve in Fig. 14.2.4 (Arnett 1972b). Similarly, when the

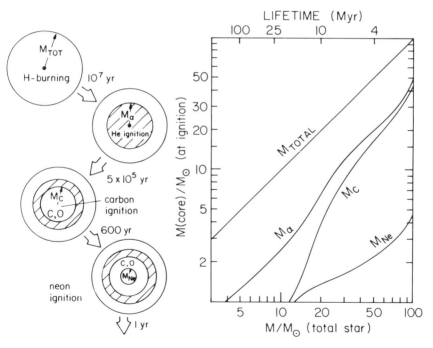

Fig. 14.2.4. The left-hand cartoon shows the evolution of the mass distribution within a star of 25 $M_\odot$. The several successively declining core masses are defined as the mass of that particular zone of fuel at the time when that fuel ignites at the stellar center (Arnett 1972a,b). The evolution develops a layer sequence of such nuclear ashes. The right-hand graph shows how those core masses depend upon the total initial mass of the star. The upper abscissa indicates the total expected lifetime of a star having that mass. This evolution leads to a complicated layered structure shown in Fig. 14.2.5 at the time of core collapse.

C-exhausted core first starts Ne melting, its core is $M_{Ne} \simeq 1.5$ M$_\odot$ for the 25 M$_\odot$ star. Such burning cores continue through phases of O burning, Si melting, core collapse and photodisintegration. A proper description of the ultimate demise of this star, long presumed to be a Type II supernova, is a matter of intense contemporary research.

Of special interest in Fig. 14.2.4 is the steep rise of the C-O core mass M_C with the mass of the initial star, because it determines the heavy-element yield. Of equal interest is the observation that the two nuclei comprising that core, ^{16}O and ^{12}C, are the two most abundant nuclei in the universe that are synthesized primarily in stars. The chemical activity of our universe, of which meteorites are such a small but highly significant portion, is dominated by the chemical properties of C and O (along with that of primordial H). The yield of C relative to O in the cores M_C of massive stars has been difficult to pin down owing to the great experimental difficulty of measuring accurately the nuclear cross section for the reaction ^{12}C$(\alpha,\gamma)^{16}$O which determines the final share of each in the total core mass M_C. That final ratio also depends upon the total mass of the star because of the increase of burning temperature with stellar mass. Using a cross section a factor $f = 3$ times larger than that recommended by Fowler's group in 1975, Arnett and Thielemann (1985) calculate that the final mass fractions in M_C fall from $Z(C):Z(O) = 49\%:48\%$ for $M_\alpha = 2$ M$_\odot$ to $Z(C):Z(O) = 20\%:76\%$ for $M_C = 16$ M$_\odot$. Varying the cross section factor f between $f = 1$ and $f = 5$, well within the current controversy over the correct value, leads to still wider ranges on the C/O ratio, especially for the more massive stars. At $M = 40$ M$_\odot$, this dependence is approximately $Z(C) \simeq 0.5/f$. This uncertainty remains the single most significant nuclear uncertainty in stellar nucleosynthesis theory, not only by controlling the yield ratio of its two most abundant products, but also by controlling the relative masses of the nucleosynthesis products to be produced from the burning of C and O themselves. Thus, displays of the relative yields of the products of C and O burning are difficult to normalize; however, the following can be said with some confidence. The cores M_C of massive stars are the major source of ^{16}O, and it is possible, though very uncertain, that they may on average produce the number ratio O/C $= 1.7$ characterizing the Sun. Most workers in this field currently suspect, however, that the average over massive stars produces something closer to O/C $\simeq 3$, in which case the remainder of the C yield is to be sought in intermediate mass stars, $2 < M < 10$, which undergo a special history described briefly below.

Another prime question for nucleosynthesis and chemical evolution is the absolute value of the O yield $y_z(O)$ for use in the structure of Sec. 14.2.2. Chiosi and Matteuchi (1984) have shown that for a wide range of reasonable assumptions about stellar evolution and the mass spectrum $\phi(M)$,

$$0.9 \times 10^{-3} < y_z(O) < 39 \times 10^{-3} \text{ grams O/grams remnant} \quad (21)$$

which is a large range of uncertainty. A simpler range of standard assumptions of many different workers suggests $y(O) \simeq 3 \times 10^{-3}$, however. Is that number at all agreeable with the observation that O is very nearly 1% of the mass in the solar neighborhood? This question is most easily answered with the aid of a standard model of chemical evolution that fits many features of the solar annulus. Such a model, one consequence of which is shown as a fit in Fig. 14.2.1, is the $k = 1$, $\omega = 0.3$/Gyr, $\Delta = 1$ Gyr standard model of Clayton (1985). Applying Eq. (15) to the element O today ($t = 15$) gives

$$Z(O) = \frac{y(O)\omega t}{2} \frac{t + 2\Delta}{t + \Delta} = 2.39 \ y_z(O) \qquad (22)$$

so that $y(O) = 3 \times 10^{-3}$ is not far from the observed $Z(O) = 0.01$ for that standard model. About the best that can be concluded from this agreement is that those numbers look reasonable, even though hardly circumscribed in a convincing way. But even this reasonableness is a triumph for the theory, for it could have been wildly in conflict with observation. As it stands, one can assert that the calculated O yield from massive stars, folded into a complete mass spectrum of star formation and infall onto the Galaxy, yields an interstellar O concentration of approximately 1% by mass. It also results in a ^{12}C abundance about 1/2 to 1/4 that of ^{16}O, by number, depending on the value of the enhancement factor f of the $^{12}C(\alpha,\gamma)$ reaction.

Two additional very abundant nuclei are synthesized during the foregoing process. During H burning the initial C and O are converted to ^{14}N, leaving it the most abundant nucleus in the core M_α, except for He itself. And during He burning, ^{14}N is converted to ^{22}Ne by two α captures.

Shell Nucleosynthesis in Massive Stars

Following the photodisintegration rearrangement of Ne (Clayton 1968, Ch.5), or Ne "melting" or "burning," the core's next major thermonuclear epoch is the fusion of O, primarily into ^{28}Si. At this time the core M_O, not shown in Fig. 14.2.4, is approximately the same mass as M_{Ne}. Most of this core, except for the very massive stars, is destined for collapse into a final dead remnant, because a central core of ^{28}Si equal to the Chandrasekhar mass, $M_{Si} = M_{Ch} = 1.4 \ M_\odot$, will definitely collapse owing to inadequate pressure support. Before that final event, however, the Si core will find itself surrounded by an onionskin series of shells of differing thermonuclear debris, each shell capped by a layer of thermonuclear burning of the previous fuel, which is gradually still increasing the mass of its ashes. But the Si core evolution, itself a photodisintegration rearrangement (Clayton 1968,Ch.7), is very rapid, so that insufficient remaining life exists for much change in mass of the overlying shells. If a core-bounce mechanism is capable of ejecting those overlying shells, their compositions will be expected to remain very nearly what they were. There is the possibility that an outward moving shock

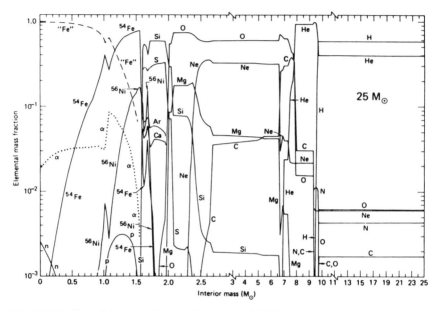

Fig. 14.2.5. The radial composition profile in a star of 25 $M_\odot$ at the time of gravitational collapse of the central 1.5 $M_\odot$ of Si, converted to Fe. Overlying that core are shells of thermonuclear debris, indicated by chemical symbol (Weaver and Woosley 1980). Isotopic information is not shown explicitly (except in the collapsing core) but in all cases except ^{22}Ne in the He shell, the elements shown are dominated by their most abundant isotope in the solar system, often exclusively that isotope. This abundance distribution will be explosively perturbed isotopically upon sudden ejection.

wave may heat those overlying layers, however, resulting in a brief burst of accelerated burning, commonly called "explosive burning," that will create most of the Fe peak nuclei and modify the yields of the rarer nuclei in the shell. Figure 14.2.5 shows the composition of the 25 $M_\odot$ presupernova star as calculated by Weaver and Woosley (1980). The radial mass scale is progressively contracted outward for easier display. Each of the shells is there evident, and one can instructively locate the radius at which each major fuel disappears, leading within to an extended shell of its ashes. The Fe core in the central 1.5 $M_\odot$ is already inwardly falling, and one sees about $\Delta M(\text{Si}) \simeq 0.4\ M_\odot$ remaining atop that collapsing core, capped in turn by an O-burning layer at the base of 6 $M_\odot$ of O. In the remainder of this section, I will attempt to illustrate the composition of those ejected layers, relying for numerical results primarily on Woosley's Saas Fee Lectures (Woosley 1986).

a. Carbon and Neon Burning. The core M_{Ne} in Fig. 14.2.4 has exhausted ^{12}C. The primary reactions occurring during C burning are listed in Table 14.2.1. Many others occur at low rates, especially neutron reactions

STELLAR NUCLEOSYNTHESIS AND CHEMICAL EVOLUTION

TABLE 14.2.1
Important Reactions in C Burning

Primary:

$^{12}C + {}^{12}C \to {}^{24}Mg^* \to {}^{20}Ne + \alpha$ (~50%)
$\to {}^{23}Na + p$ (~50%)
$\to {}^{23}Mg + n$ (~0.5%)

$^{23}Na(p,\alpha){}^{20}Ne, {}^{23}Na(p,\gamma){}^{24}Mg$

Down to 10^{-2} of the above:

$^{20}Ne(\alpha,\gamma){}^{24}Mg$ $^{23}Na(\alpha,p){}^{26}Mg(p,\gamma){}^{27}Al$

$^{20}Ne(n,\gamma){}^{21}Ne(p,\gamma){}^{22}Na(\beta^+){}^{22}Ne(\alpha,n){}^{25}Mg(n,\gamma){}^{26}Mg$

$^{21}Ne(\alpha,n){}^{24}Mg$ $^{25}Mg(p,\gamma){}^{26}Al$

Significant: all n,γ reactions

Beginning of Ne melting:

$^{20}Ne(\gamma,\alpha){}^{16}O$ $^{20}Ne(\alpha,\gamma){}^{24}Mg$

Significant: $^{24}Mg(\alpha,\gamma){}^{28}Si, {}^{25}Mg(\alpha,n){}^{28}Si, {}^{26}Mg(\alpha,n){}^{29}Si$
$^{26}Mg(\alpha,\gamma){}^{30}Si, {}^{27}Al(\alpha,p){}^{30}Si, {}^{30}Si(p,\gamma){}^{31}P$

---

with every heavy nucleus present. However, those in Table 14.2.1 indicate the major nuclear currents that establish the setting for any other trace participant. At the moment of core collapse in the 25 $M_\odot$ star shown in Fig. 14.2.5, the C has been exhausted within 6.5 $M_\odot$, whereas Ne is burning near 2.5 $M_\odot$, within which it is exhausted at that time. This 4 $M_\odot$ wide zone contains an important contribution to nucleosynthesis, being very likely the primary origin of natural ^{20}Ne, ^{23}Na, ^{24,25,26}Mg, ^{27}Al, ^{29,30}Si and ^{31}P, and perhaps also ^{21}Ne, ^{36}S and ^{40}K. Most of these C ashes are lifted intact from the supernova in the core-bounce ejection process. It is probably not to be expected that much explosive C burning occurs further out when the reflected shock wave passes through it, because that shock is probably not strong enough to raise the temperature of residual C to the required level for a quick burn. The rarity of explosive C burning is also attested to by the very low natural abundance of ^{46}Ca, which is overproduced by large factors there (Howard et al. 1972; Woosley 1986), and though a likely origin of that single interesting nucleus, explosive C burning seems otherwise not a large contributor to the natural abundances. Those nuclei resulting from it were largely contained already in the ashes of hydrostatically exhausted C. Note the change from our review of 17 yr ago (Arnett and Clayton 1970), when we felt, in momentary enthusiasm, that explosive burning was the key.

It does seem more plausible that a portion of the ^{20}Ne in the C ashes will be explosively burned during ejection, however. In Fig. 14.2.6 I compare this expectation, based on a recent survey of burning by Woosley (1986), with the

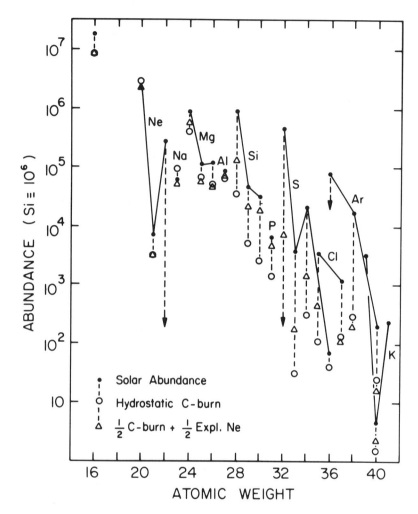

Fig. 14.2.6. The solar abundances for $16 < A < 41$ are compared to the results of hydrostatic C burning (open circles) and with an equal mix of those ashes with explosively burning Ne ejecta (triangles). The explosive Ne burning happens naturally if C-exhausted material is shock ejected. The normalization is that one gram of ejecta is compared to 162 grams of solar material, a dilution chosen to achieve agreement between solar ^{20}Ne and the ^{20}Ne content of the ejecta. Nuclei produced in this material in amounts comparable to solar are ^{20,21}Ne, ^{24,25,26}Mg, ^{27}Al, ^{29,30}Si, ^{31}P and ^{40}K. Points are plotted from tabular calculations by Woosley (1986).

solar abundances, which are shown as black dots, with isotopes of the same element connected by solid line. The calculations show specifically the abundances in ashes having a total mass 162 times smaller than the mass of solar material containing the indicated abundances. The large circles show the results of hydrostatic C exhaustion, whereas the open triangles have allowed half of the C ashes to explosively burn a portion of the Ne at temperatures $2.2 < T_9 < 2.4$. The total mass was chosen so that this final abundance of ^{20}Ne would agree with its solar abundance (i.e., 1/162 of solar material is C ashes, a not implausible amount). The conclusion for all to see is that Ne, Na, Mg, Al, P and a few other minor isotopes, owe their origin to this setting.

There is another conclusion of Fig. 14.2.6 that is of great importance for the problem of isotopic anomalies in the meteorites—namely that in the initial ejection of these nuclei into the interstellar medium, they were initially separated from other elements and other isotopes. It is up to subsequent history to mix them into the well-mixed material that most solar system samples are. Any lack of such mixing has the potential to generate isotopic anomalies. Although lack of spatial mixing in the interstellar medium (Cameron 1973) has been considered by many workers, it seems more believable to this writer that chemical fixation has preserved the anomalies. The dust that condenses during the adiabatic expansion of these supernova interiors has been named SUNOCONs in recognition of their key conceptual role in the cosmic-chemical-memory theory of the isotopic anomalies (Clayton 1978, 1982). The reader's attention is called in this regard to the isotopically pure ^{16}O which is still the most abundant nucleus in the Fig. 14.2.6 ejecta, and which bathes the most refractory element Al whose origin lies here, with such ^{16}O fugacity that an aluminous memory of excess ^{16}O is forever stamped into interstellar solids. This provides, among other things, one possible reason for finding that the most ^{16}O-rich meteoritic solids are very Al-rich.

b. Explosive O Burning. Consider again the core M_{Ne} in which C has already been exhausted by the prior hydrostatic evolution. The shock wave stimulating ejection of a portion of that core will, in its hottest portions, cause O to undergo thermonuclear fusion. The many channels of the initiating reaction

$$\begin{align} ^{16}\text{O} + {}^{16}\text{O} &\rightarrow {}^{31}\text{S} + n \sim 5\% \\ &\rightarrow {}^{30}\text{P} + d \sim 5\% \\ &\rightarrow {}^{31}\text{P} + p \sim 56\% \\ &\rightarrow {}^{28}\text{Si} + \alpha \sim 34\% \end{align} \quad (23)$$

are followed by a host of secondary reactions, leading chiefly to the nucleosynthesis of ^{28}Si, ^{32,33,34}S, ^{35,37}Cl, ^{36,38}Ar, ^{39,41}K, ^{40,42}Ca, ^{46}Ti, ^{50}Cr and significant amounts of ^{44}Ca and ^{54}Fe. Woosley (1986) has provided the latest and most accurate in a long history of calculations of explosive O burning (Woosley et al. 1973), tabulating his results for a wide range of peak temperatures before the adiabatic expansion. In Fig. 14.2.7 I have constructed the comparison of solar abundances with the yields in 0.2% as much matter from an explosive O shell at peak temperature of 3.7×10^9K ($T_9 = 3.7$). The actual stellar ejecta will average over a range of peak temperatures, some of which burn less and some more of the initial O. This particular shell has burned 80% of the initial ^{16}O, so that, after burning, the ejecta in Fig. 14.2.7 have ^{28}Si/^{16}O ~ 2.5 by number. This means that O is so depleted that Si cannot condense in its preferred SiO_2 units within silica and silicate SUNOCONs. For peak burning temperatures $T_9 < 3.5$, there would be enough O left to do so, however. Such considerations as this constitute a key interface between nucleosynthesis theory and the theory of cosmic chemical memory. Perhaps the most significant conclusion of this kind for Fig. 14.2.7, and the first historically, was the realization that the bulk of Ca- and Ti-bearing SUNOCONs were sulfidized rather than oxidized (Clayton and Ramadurai 1977).

The reader should notice the blend of Fig. 14.2.7 with Fig. 14.2.6. The explosive O ejecta are poor in the two heavy isotopes of Si and in P, but these three nuclei are contained satisfactorily in the C-burned ejecta (Fig. 14.2.6). It is of great potential significance that the nucleosynthesis of the isotopes of Si have been separated in this fashion, and the lack of huge isotopic anomalies in Si requires explanation in the cosmic chemical memory theory. The reader should also notice in Fig. 14.2.7 that the lightest isotopes of Ti, Cr and Fe are largely created in explosive O burning and are ejected as almost single isotopes of those elements. Huge SUNOCON isotope anomalies in ^{46}Ti, ^{50}Cr and ^{54}Fe were therefore among the earliest predictions of the cosmic chemical memory theory (Clayton and Ramadurai 1977; Clayton 1978), and the reason for their lack of detection in meteoritic samples is yet to be clearly detailed.

An interesting aspect of calculations of this type is that several key nuclei are synthesized as radioactive progenitors having sufficiently long half-lives that they would condense in SUNOCONs as the chemical parent and then transmute to the daughter within the grain. An outstanding example (see Fig. 14.2.7) is ^{41}K, which exists during the shell expansion as the refractory long-lived ^{41}Ca parent. The resulting chemical fractionation of ^{41}K from ^{39}K was emphasized as one of the best indicators of a cosmic chemical memory in the earliest papers on such effects (Clayton 1975,1977).

What is very evident from Figs. 14.2.6 and 14.2.7, and also from Fig. 14.2.8, is that there is undeniable correctness to the basic idea of thermonuclear synthesis. It can be no accident that the elemental and isotopic structures calculated are so like those observed in the solar abundances. Factor of 2 errors in a sawtooth pattern covering 6 powers of 10 can be regarded in this sense as small. What still is to be solved is the exact superposition of shell histories which have occurred, their ranges of peak temperatures, and so forth.

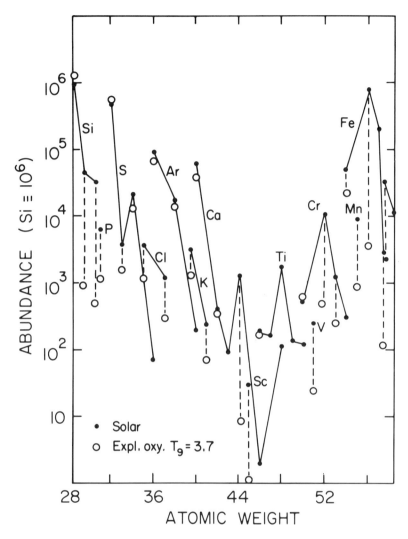

Fig. 14.2.7. Ejecta explosively burning O at initial temperature T(initial) = 3.7×10^9K during ejection are compared to solar abundances. From half to all of ^{28}Si, ^{32,33,34}S, ^{35,37}Cl, ^{36,38}Ar, ^{39,41}K, ^{40,42}Ca, ^{46}Ti, ^{50}Cr and ^{54}Fe are produceable in this way. The normalization compares one gram of this ejecta with 500 grams of solar material. This particular shell has burned 80% of its ^{16}O, leaving ^{28}Si/^{16}O ≃ 2.5 in the ejecta (figure prepared from calculated results tabulated by Woosley 1986). Large and predictable isotopic anomalies are contained in SUNOCONs condensed during the expansion and cooling of this gas.

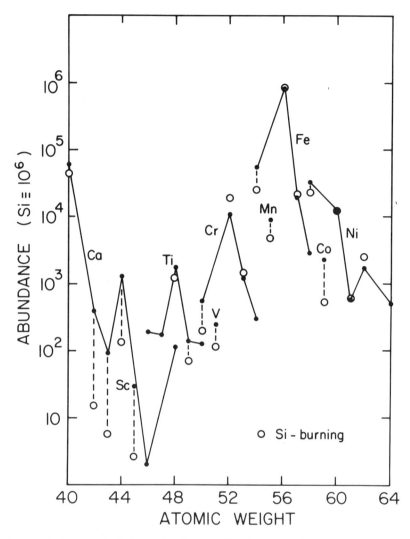

Fig. 14.2.8. Ejecta explosively burning Si (half with T(initial) = 5.0 × 10⁹K and half with T(initial) = 5.5 × 10⁹K) are compared with solar abundances after normalizing at ^{56}Fe. The nuclei ^{40}Ca, ^{48,49}Ti, ^{51}V, ^{52,53}Cr, ^{55}Mn, ^{56,57}Fe and ^{58,60,61,62}Ni and substantial fractions of ^{50}Cr, ^{54}Fe and ^{59}Co are produced satisfactorily. The higher initial temperature leads to a nuclear statistical equilibrium, or e-process, called "eq" in Table 14.2.2. Huge and predictable isotopic anomalies exist in any SUNOCONs that condense during the expansion, but O and S are not present to serve as anions (figure prepared from calculated results tabulated by Woosley 1986).

c. Explosive Si Melting. As the peak ejection temperature exceeds $T_9 = 4.5$, the very stable ^{28}Si nucleus begins to transform into the Fe-abundance peak via a process of photodisintegration rearrangement (Clayton 1968,Ch.7). In particular, this builds up the yield in the Sc-Fe region, while it transfers the excess neutrons from (primarily) ^{34}S and ^{38}Ar into the heavier metals. This is illustrated in Fig. 14.2.8, where I have taken equal parts of $T_9 = 5.0$ and $T_9 = 5.5$ peak temperatures during the explosive ejection, again based on Woosley's (1986) recent survey. The complementary nature of Figs. 14.2.8 and 14.2.7 is most satisfying. The Fe peak (actually dominated by ^{56,57}Ni) is an excellent fit to the solar abundances. This is a direct result of the nuclear statistical equilibrium (*e*-process established for $T_9 > 5$ (Clayton 1968,Ch.7). The recent detection of gamma-ray lines from the decay of radioactive ^{56}Co in supernova 1987a (Matz et al. 1988) confirms that Fe is synthesized in this way.

d. Neutron-rich Equilibria. Although Figs. 14.2.7 and 14.2.8 blend together in a very convincing way, several nuclei are conspicuous by their absence. The heaviest isotopes of Ca, Ti, Cr, Fe and Ni, namely $A = 48, 50, 54, 58$ and 64, are produced in the innermost fraction of the ejecta in nuclear statistical equilibrium, where electron capture has made the matter so neutron-rich that those nuclei dominate the isotopes of those elements (Hainebach et al. 1974; Hartmann et al. 1985). It is satisfying that large and correlated anomalies of these isotopes are found in meteoritic material (see Chapter 14.3). From the cosmic-chemical-memory point of view, this arises because the chemical environment of those isotopes differs so greatly from that of the other isotopes of those elements (Clayton 1981*a*), whereas the inhomogeneous admixture would seem to rely on solar injection of this supernova core (Hinton et al. 1987). Although some controversy has attended discussions of the origin of ^{48}Ca, I now feel confident, along with Hartmann et al. (1985), that these five heaviest isotopes are coproduced in this setting.

e. Specific Problem Nuclei of Importance. There are four nuclei in this range of intermediate masses that have not been explicitly addressed even though they are of great importance to meteoritic science. In each case the true synthesis of these natural abundances is not known even to this day. The reasons for their importance must be commented upon, even though largely self-evident to the discriminating reader.

The nucleus ^{26}Al has been detected by the presence of excess daughter ^{26}Mg in meteoritic samples and by its associated gamma-ray emission from the interstellar medium (Mahoney et al. 1984; Share et al. 1985). The latter is the first such detection, proving that nucleosynthesis of ^{26}Al continues until the present day. The sublime interplay between meteoritics, the origin of the solar system, the chemical state of the interstellar medium, and nucleosynthesis is nowhere more evident than in considerations of this nucleus. Fortunately this entire saga has recently been reviewed in depth (Clayton and Leising 1987), to which the reader is recommended to turn for details, and the meteoritic evidence is itself addressed in Chapters 15.1 and 15.2. Of relevance here is the uncertain nucleosynthesis origin of ^{26}Al, but it is almost certainly not to be found within the mantles of massive stars as described here. Before the gamma-ray observations, nucleosynthesis theory had not predicted that 4.7% of ^{26}Mg was synthesized as this parent, even though that large yield is now indicated by the facts (Clayton 1986b). Only Woosley and Weaver (1980), in their concluding remarks on novae, predicted detectable ^{26}Al gamma radiation by arguments that appear to me to have been sound. Clayton (1986b) showed also that this large yield suggests that 1 to 2% of interstellar ^{26}Mg today resides in the same fossil site within interstellar dust that the ^{26}Al previously held. Most plausible sites of origin of this nucleus are novae, hot H burning in envelopes of Wolf-Rayet or asymptotic-giant-branch stars, or explosive He burning in the cap of a Type I supernova (see below).

Two additional problem nuclei are isotopes of Ca, of great importance to meteoritics because of its highly refractory chemical nature and because of the dramatic isotopic anomalies found in that element. Curiously, ^{46}Ca, whose origin is rather uncertain, has not been singled out for isotopic variability despite being by far the rarest of its isotopes. That tiny abundance creates, of course, an experimental problem for detection of anomalies, recalling the aphorism, "Absence of evidence is not evidence of absence." Only the neutron reactions on seed nuclei in explosive C burning (Howard et al. 1972; Woosley 1986) seem to produce a big overabundance locally, a result that suggests today that very little C is ejected in a state of explosive burning. If ^{46}Ca is not highly variable, that fact needs explaining by the competing theories of isotopic anomalies. For ^{44}Ca, its second most abundant isotope, the situation is even richer and more puzzling. Strong nuclear-systematics evidence suggests that ^{44}Ca was synthesized as a radioactive progenitor, 47 yr ^{44}Ti, which led to its identification among the first targets of gamma-ray astronomy to be predicted by nucleosynthesis theory (Clayton et al. 1969). Because SUNOCON condensation traps the radioactive parent, a strong prediction (Clayton 1975) of an extinct radioactive fossil anomaly has stood before the community for more than a decade. And yet the isotopic patterns are normally reported after renormalizing the data to a standard ^{40}Ca/^{44}Ca isotopic ratio. Clearly legitimate as a convention, this procedure nonetheless hides ^{44}Ca variations in a psychological sense. Of relevance here, however, is the long-standing confusion over the nucleosynthesis of this important nucleus. In part, this is because the early work on Si burning failed to realize (see, e.g., Clayton 1968, Fig. 7–8) a freeze-out problem for ^{44}Ti that reduces its abundance during ejection. But clear underabundance of ^{44}Ti is evident now in both Figs. 14.2.7 and 14.2.8. Woosley (1986) presents calculations supporting the origin of ^{44}Ti in explosive He burning in the He cap of the exploding-white-dwarf model of Type I supernovae, where an excess of hot

α particles solves the problem. This exciting result further intensifies the isotopic-anomaly puzzle, however, because it then gives ^{44}Ca not only a radioactive progenitor, but even a different object of origin than the synthesis of ^{40}Ca. If variations do not exist, reconciling that fact with a consistent theory of the isotopic anomalies remains a target for the future.

A similar problem exists for ^{47}Ti, which is contributed to at low levels by many processes, but is not in evidence in Figs. 14.2.7 or 14.2.8 at significant levels. This nucleus does participate in endemic Ti isotopic anomalies, as described in Chapter 14.3, a rather plausible situation for a nucleus with such a puzzling origin. What is that origin? A consensus still does not exist. Neutron captures during C burning (Howard et al. 1972) was one of the better possibilities; and Woosley (1986) threw a new hat in the ring with the demonstration of high yields from the same explosive He shells that produced ^{44}Ti. Ascertaining this correlation between ^{47}Ti and ^{44}Ti nucleosynthesis will grow in theoretical importance in the future.

Much of this material is summarized in Table 14.2.2, which assigns modes of nucleosynthesis origin to each nucleus in order of importance. This table is an updated version of the table constructed by Arnett and Clayton (1970), and has been strongly influenced in particular by Woosley (1986). This table includes also the light nuclei synthesized by the main line of H burning in stars (^{13}C, ^{14}N, ^{17}O) and by their subsequent conversion during He burning in stars to ^{18}O and ^{22}Ne. The rare nuclei ^{15}N and ^{19}F apparently arise from explosive H ejecta, perhaps from the common nova. The reader is reminded that the chemical designation of the burning source has a somewhat different meaning for hydrostatic and explosive burnings. As an important prototypical example, the designation "O" refers to a large abundance within a shell or core that has already consumed its O hydrostatically before the matter is ejected from the star, whereas "ExpO" refers to a large abundance created within a shell that is explosively burning O at the time it is ejected. Designation "eq" labels matter in nuclear statistical equilibrium as it freezes out during ejection, although, as a technical point, that matter may have "excess alphas" in certain important nucleosynthesis circumstances (Woosley 1986). The neutron-rich nuclear equilibrium so important to observed isotopic anomalies is designated "n-eq." Certain question marks appear as reminders that much remains to be learned on several key nuclei.

Nucleosynthesis Related to Degenerate Cores

A major branch point in stellar evolution occurs at stellar mass near 9 $M_\odot$. Above this mass the C core M_C is sufficiently massive that it is heated by compression as it grows to the point of ignition of thermonuclear fusion reactions between C ions. This is the "massive-star" region that has been the topic of the previous section. Below 9 $M_\odot$, however, a quite different evolution ensues with great repercussions on nucleosynthesis science. The contraction of the C core is arrested by the exclusion principle forbidding electrons

TABLE 14.2.2
Primary Nucleosynthetic Sources[a]

Element	A	Source[b]
carbon	12	He
	13	H, ExpH
nitrogen	14	H
	15	ExpH
oxygen	16	He
	17	H, ExpH
	18	H, ExpH, He
fluorine	19	ExpH
neon	20	C
	21	C, ExpNe
	22	He
sodium	23	C, Ne, ExpNe
magnesium	24	Ne, ExpNe, C
	25	Ne, ExpNe, C
	26	Ne, ExpNe, C
aluminum	27	Ne, ExpNe, C
silicon	28	O, ExpO
	29	Ne, ExpNe, C
	30	Ne, ExpNe, C
phosphorus	31	Ne, ExpNe, C
sulfur	32	O, ExpO
	33	ExpO
	34	O, ExpO
	36	ExpC, Ne
chlorine	35	ExpO, ExpNe
	37	ExpO, C
argon	36	ExpO, ExpSi
	38	O, ExpO
	40	Ne, C, ?
potassium	39	ExpO
	40	He, Ne, ExpNe
	41	ExpO
calcium	40	ExpO, ExpSi
	42	ExpO
	43	ExpHe, C ?
	44	ExpHe, ExpSi ?
	46	ExpC, Ne, ExpNe
	48	n-eq.
scandium	45	Ne, ExpNe, ExpHe
titanium	46	ExpO
	47	ExpHe, ExpSi, ExpC
	48	ExpSi

TABLE 14.2.2 (continued)
Primary Nucleosynthetic Sources[a]

Element	A	Source[b]
titanium	49	ExpSi, ExpHe
	50	n-eq.
vanadium	50	ExpNe
	51	ExpSi
chromium	50	ExpO, ExpSi
	52	ExpSi
	53	ExpSi
	54	n-eq.
manganese	55	ExpSi, eq.
iron	54	ExpSi, ExpO
	56	ExpSi, eq.
	57	ExpSi, eq.
	58	n-eq., He, C
cobalt	59	eq., C, ?
nickel	58	ExpSi, eq.
	60	eq.
	61	eq., ExpNe, C, ?
	62	eq., ExpNe, O
	64	n-eq., ExpNe

[a] Table from Arnett and Clayton (1970) and Woosley (1986) and countless works in between.
[b] He, C (for example): hydrostatically burned He, or C; ExpNe (for example): explosively burning neon; eq: nuclear equilibrium; n-eq: neutron-rich equilibrium.

to occupy the same state. They are forced into high-momentum states exerting great pressure (degeneracy pressure) capable of arresting contraction (cf. Clayton 1968,1983,Ch.2). In essence a white dwarf is established within the star, with the core mass $M_C < M_{Ch} = 1.44$ $M_\odot$. An extensive literature on related astrophysical phenomena exists, but the consequences for nucleosynthesis are primarily of two types, which are summarized below without apology for a vast number of uncertainties and exciting complications that are therein suppressed.

a. Double-Shell Asymptotic-Giant-Branch Burning. A white-dwarf core provides a compact inert source of gravity that compresses the He-dominated matter above it. In a thin shell at the boundary, He is fusing to C, but rather than doing so steadily it does so in unstable bursts. Following each He-shell flash, a large subsequent convection zone appears to carry the overheated matter outward, along with expansion that extinguishes the He flash. Much higher up (in mass), atop the H-exhausted core M_α within which all of the above happens, a thin shell of H burning is fusing more He ashes via the CN cycle. This burning too occurs unsteadily, and, significantly, out of phase with the He-burning shell. The two do not burn at the same time. Rather, after the He-shell flash and its convection has subsided, the H-convection zone at the surface deepens into the star while the intershell region is recompressing, and this deepening convection reaches material that had previously been mixed outward to that point by the prior He-shell convection. This is called the "dredge up" phase because it brings to the surface matter that has been processed deep within. Following this, a quiet period of non-convective He burning increases the mass of the C-O core. At this point, the H-burning shell ignites once again and provides, for a time, enormous power of its own to generate convection and surface luminosity of this giant star, which exists on an evolutionary track in the Hertzspring-Russell diagram close to the prior giant branch and called "the asymptotic giant branch." This phenomenon has been studied with unparalleled rigor by Iben in a series of works that constitutes a major landmark in theoretical astronomy. See particularly his Darwin Lecture of the Royal Astronomical Society (Iben 1985). Figure 14.2.9, taken from that review by Iben, shows a temporal succession of structure profiles of a 7 $M_\odot$ star in that small portion of its mass just outside its C-O core of $M_C = 0.95$ $M_\odot$. In particular, the C-abundance profile shows how an enhancement of ^{12}C is produced in the He shell and convectively mixed outward to $M(r) = 0.959$ $M_\odot$ and how the surface convection zone later dips into that material and carries it on outward to the surface during the dredge up. As a result the red-giant envelope is gradually enriched in C, eventually becoming more abundant than O in the envelope. At that time the giant becomes a "carbon star," in which the C_2 molecule becomes spectroscopically observable in the stellar atmosphere. Prior to this time, atmospheric C has been locked up in the very stable CO molecule so that C_2 molecules have extremely low abundance, a situation that is suddenly and dramatically altered when C > O.

The most important nucleosynthesis consequence of all this lies in a sizeable increase of the ^{12}C yield from stars in this mass range. The need for this was already noted from the likelihood that massive stars produce much more ^{16}O than ^{12}C. As a rule of thumb, considering the present uncertainties, I would recommend regarding half of the ^{12}C yield as being from massive stars and half of it from these double-shell-source stars of intermediate mass. Perhaps it should be clarified here that the difference in He-burning product lies not in temperature or some other such variable, but rather in the fact that ^{12}C is always the overwhelming *initial* product of He fusion, and is only converted to ^{16}O later in the burning while He is being exhausted (see, e.g., Clayton 1968,Ch.5). But in the intermediate-mass stars, one is continuously mixing the first fruits of He fusion to the surface, and that means ^{12}C rather than ^{16}O.

A second nucleosynthesis consequence lies in the gradual nucleosynthesis of heavy nuclei by neutron capture (the *s*-process). Introduction and basic principles of heavy-element nucleosynthesis are described by Clayton

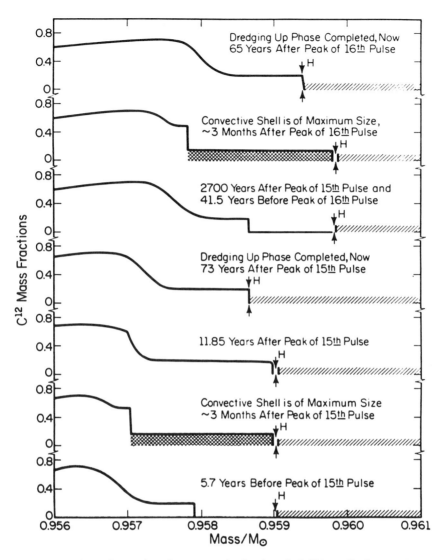

Fig. 14.2.9. The nuclear engine of an asymptotic giant branch (AGB) star lies between 0.957 $M_\odot < M(r) < 0.960$ $M_\odot$ in this example of a 7 $M_\odot$ AGB star from Iben (1985). The C mass fraction $X(C)$ is profiled over this limited interior mass coordinate in a series of seven snapshots. One sees that the C-O degenerate core reaches out to 0.957 $M_\odot$, and during the 15th He-burning shell flash convection mixes the excess ^{12}C uniformly between 0.957 and 0.959 $M_\odot$, almost to the point designated "H," where H is found. After the He-shell convection dies down, the base of the convective envelope moves inward in mass (snapshot 4), dredging up to the surface freshly synthesized ^{12}C. Notice the overlap between the inner convective shell (cross hatched) and the subsequent surface convective shell (diagonal hatched). The ^{22}Ne$(\alpha, n)^{25}$Mg neutron source operates during the convective He burning, driving Fe-peak nuclei to neutron-rich isotopes of heavier elements. The ^{13}C $(\alpha, n)^{16}$O neutron source can occur near the top of the He shell if mixing can occur between that shell and the overlying H (which at H in this figure does not show any such mixing for this particular stellar model).

(1968, 1983, Ch.7) and in Chapter 14.1. An updated evaluation based on many improved cross sections and including a decomposition of each element abundance into s-process and r-process portions has been provided by Käppeler et al. (1982) and by Mathews and Käppeler (1984). What one needs is the natural occurrence within stellar evolution of a series of reactions involving light nuclei that emit an abundant supply of free neutrons so that they in turn may be captured by the nuclei heavier than Fe, where large Coulomb barriers effectively prohibit capture of charged particles. Truran and Iben (1977) evaluated Iben's scenario (see also Ulrich 1973) for the occurrence of such reactions in the asymptotic-giant-branch (AGB) double-shell stars. The neutron source ^{22}Ne (α,n) ^{25}Mg had been suggested two decades earlier by Cameron, but without a well-tuned stellar scenario it could not produce enough captures above Fe to produce the requisite amount of Ba, for example. The AGB stars almost overcome this difficulty by continuously mixing new-source fuel for neutrons into the region where those neutrons are liberated by the above reaction. The nuclear line is direct, because the CN-cycle exhaustion of H leaves ^{14}N the most abundant nucleus except for He within the core M_α, and that ^{14}N is converted to ^{22}Ne by two successive (α,γ) reactions at temperatures lower than required for the fusion of He by the 3α process. Thus, when the 3α process does occur, transiently raising the temperature in the flashing shell, the ^{22}Ne (α,n) ^{25}Mg reaction occurs as well, liberating ultimately about 10^{-3} free neutrons per amu in the interior. The heavier nuclei compete for these in proportion to the product of their respective concentrations and their (n,γ) cross sections. But the real benefit comes from the subsequent convection, which mixes new ^{22}Ne fuel into the mass region where the next shell flash will occur. At the same time, it mixes part of the irradiated nuclei outward toward the surface while leaving some behind for the next neutron dose. The seed nuclei are subjected thereby to an exponential distribution of neutron fluences. A recent evaluation of the yield of these repeated cycles has been provided by Howard et al. (1986). The results are impressive, leading to wide acceptance of this phenomenon as central to the scheme of nucleosynthesis; however, serious quantitative puzzles remain. The yield at large atomic weight is rather too steeply declining to match solar abundances easily and is even harder pressed to account for the large Ba concentrations seen in the Ba stars, which are believed to have inherited their heavy-element-rich atmospheres from a prior AGB binary companion star (Lambert 1985). For the S-stars, which are rich in Technetium and s-process elements, an even larger problem is the normalcy of the Mg-isotopic composition in their surfaces, which can be measured with optical transitions within the MgH molecule. Solar isotope ratios, about 8:1:1, are observed there (Smith and Lambert 1986) instead of approximately 2:1:1 as required if ^{22}Ne (α,n) ^{25}Mg (n,γ) ^{26}Mg is to be the source of s-process neutrons. This seems to exclude that neutron source, but perhaps only for the low-mass AGB stars that Smith and Lambert (1986) observed. The unobserved more massive AGB stars could still be the major sources of the galactic s-process yield, and they could operate on the ^{22}Ne (α, n) source even if the low-mass stars do not.

There exists still the possibility that the ^{13}C (α, n) ^{16}O neutron source works in a similar context, but it has not yet done so in a totally natural way, i.e., without a little help from its friends. Iben and Renzini (1982) described the workings, similar in spirit but different in detail from Ulrich's (1973) reasoning. During the interpulse period a semi-convective region forms in the cool C-rich material at the top of the intershell region which enables protons from the envelope to be admixed to the intershell region where they are captured by the sequence ^{12}C (p, γ) ^{13}N $(\beta^+ \nu)$ ^{13}C. The advancing convective front generated by the next thermal pulse sweeps this matter down to hotter zones where the ^{13}C (α, n) ^{16}O reaction provides the desired free neutrons. One major difference that may yet prove to be the key diagnostic is that these neutrons are liberated and captured near $T = 1.5 \times 10^8$ K, whereas those from ^{22}Ne do so near 3×10^8 K. The major problem in this is the theoretical justification of the semi-convection. However, the giants bear ample testimony to overabundance of s-process elements in concentrations consistent with their being the major nucleosynthesis source of these heavy isotopes. So it seems still that one of these mechanisms at least works better in real stars than it does on paper.

Meteoritic science has produced totally new and unexpected evidence to support the astronomical belief that the nucleosynthesis of s-process isotopes and of r-process isotopes has happened separately, and that our solar system inherited a mixture of the two. Previous support for this view rested on theoretical decompositions of the solar abundances into s and r isotopes (see, e.g., Clayton 1968,1983,Ch.7; Käppeler et al. 1982). Chapter 14.1 addresses this decomposition of solar abundances. The heavy-element isotopic anomalies detected in some meteoritic inclusions provide new and independent evidence of the reality of this decomposition, a nontrivial point in natural philosophy. These anomalous abundances are reviewed in Chapter 14.3. Clayton (1979) discussed these decompositions to make the above point and to argue that a cosmic chemical memory of their different origins provides the basis for the observed isotopic anomalies. In his interpretation, polarizations of the average s-process and r-process yields are possible because these two heavy-element components are ejected in different chemical surroundings and therefore begin their interstellar lifetimes in different chemical interstellar forms. Gas/dust fractionation is the simplest example of s/r polarizations, and it becomes possible if these components did not condense with equal efficiency when they were ejected from stars. Mathews and Käppeler (1984) improved the nuclear details of those decompositions, which are also discussed in Chapter 14.3. Other workers, especially the discoverers of these anomalies, have envisioned instead an inhomogeneous admixture of anomalous material into solar material. In either case, one must probably take into account the likelihood that these AGB stars are the primary sources of the s isotopes.

The r-process nuclei are, by contrast, ejected explosively from a rapid dynamic event. This requirement results from the need of a high neutron density, $n_n = 10^{20} - 10^{24}/cm^3$, to move nuclei onto a path through very neutron-rich isotopes of the heavy elements. For a review of the basic physics, see Clayton (1968,1983,Ch.7) and Schramm (1982). The dynamic and nuclear problems are both so difficult that no confident knowledge of the astrophysical site has yet been established. In my own mind, the best prospect lies in re-expanding compressed and neutronized matter that is almost trapped in the neutron star during the core collapse of massive stars. Schramm (1982) reviews the possibilities. In any case, the circumstances are chemically and dynamically quite different from those of the loss of the s-process-bearing envelope of the AGB stars discussed above, a loss resulting in a planetary nebula and a remnant white dwarf. Thus a formidable mixing history before the s and r isotopes can even become homogenized is to be anticipated. One must consider both spatial homogeneity (Cameron 1973) and the diversity of chemical carriers within the interstellar medium described above. Ample opportunity therefore exists for competing theories of the origin of the isotopic anomalies in the heavy elements. Although many refinements are possible, the major competing theories are presently either the cosmic chemical memory of the ISM or the inhomogeneous admixture into the forming solar system of ejecta from a neighboring event (or two, or more).

b. Type I Supernovae. The asymptotic-giant-branch evolution discussed in the preceding section for intermediate-mass stars (3 $M_\odot < M <$ 10 $M_\odot$) is believed to leave behind a white dwarf star when the envelope is driven away in a structural instability leading to the planetary-nebula phenomenon. If the white dwarf is isolated and has mass less than the Chandrasekhar limit (1.4 $M_\odot$), it will slowly cool to a stable remnant. But white dwarfs in binary stellar systems will in many cases accrete mass from their companion star, especially when it enters its giant phases, at which time the white dwarf companion may even exist within the giant envelope for close binaries. These binaries are quite common in the sky, leading to many exciting events including, apparently, the Type I supernova. SNI are believed to result when accretion of mass by a white dwarf drives its mass above the Chandrasekhar limit, at which time the white dwarf begins to contract and cannot be halted by degenerate electrons. But, the white dwarf can explode. The fuel supply available in C burning alone is adequate to disrupt the entire white dwarf at huge kinetic energies of order 10^{51} erg. Just when C ignition occurs and just how the wave of C burning propagates through the white dwarf are contentious matters of robust current investigation. Whatever the details, the C burning will also ignite O, and the exploding dwarf will be a very rich source of nucleosynthesis between Si and Ni.

The details of this nucleosynthesis are very hard to compute because they are so dependent upon the correct description of the hydrodynamics associated with the triggering burning of C. If the burning front propagates as a supersonic shock wave from center to surface, the so-called C detonation, the entire star is converted to Fe group elements in an e-process. The ^{56,57}Ni progenitors of ^{56,57}Fe produce the Type I lightcurve by supplying radioactive power that keeps the expanding gases hot (Colgate and McKee 1969; Weaver et al. 1980). The nucleosynthesis debris is in this case very similar to that shown in Fig. 14.2.8, and is probably the major source of this nucleosynthesis pattern. If, on the other hand, the C ignites as a result of a subsonic pressure wave propagating through the star, the attendant expansion can cool and quench the burning before it is complete. This is the C-deflagration model, and its nucleosynthesis is a linear combination of Figs. 14.2.7 and 14.2.8 (Nomoto et al. 1984) and is capable of accounting for the Fe-peak elements and a portion of the Si-Ti region as well. Unfortunately the Ni/Fe ratio produced (after decays) is so large (about five times solar) and the ^{54}Fe/^{56}Fe ratio is also so large (about 2.4 times solar) that either important burning details are being overlooked or these objects have not been major contributors to the yields of galactic nucleosynthesis, despite the fact that the average observed rate of SNI is sufficiently great that their ejecta of this type should weigh in noticeably into the overall pot. This is, however, a very active area of astrophysics theory in which much progress can be hoped for in the years ahead.

One very interesting variant of this model involves the detonation of He in addition to C and O. This might come about if the C, O white dwarf carries a mantle of He at the time of detonation, or perhaps in the merger of binary white dwarfs, one of which is composed of He. It is in this setting that the explosive ejection of He carries large overabundances of ^{44}Ti and ^{47}Ti, as mentioned earlier and indicated in Table 14.2.2, leading Woosley (1986) to suggest that their origin lies here.

The SNI should be very bright sources of nuclear gamma-ray lines emitted following the decays of ^{56,57}Ni and of ^{44}Ti. This possibility was predicted on the basis of nucleosynthesis theory (Clayton et al. 1969) and remains the brightest hope for confirming continuous nucleosynthesis by this observational technique. The occurrence on 23 February 1987 of a type-II supernova in the Large Magellanic Cloud has allowed successful detection (Matz et al. 1988) of the smaller radioactive mass ejected from SN II.

c. Novae. If the matter accreted by a white dwarf has substantial H content and does not accrete too rapidly, a thermonuclear runaway occurs on the surface of the white dwarf after about 10^{-3} $M_\odot$ of H has been added. This scenario constitutes the standard thermonuclear model of the common nova explosion (Truran 1982). Accretion of envelope from a binary red-giant companion is the likely setting for this. Only 10^{-5} to 10^{-4} $M_\odot$ of this envelope is actually ejected during each (recurrent) outburst, so even with forty such events annually in the Galaxy, no great admixture of nova ejecta characterizes the average ISM. Yet for those nuclei of low relative natural abundance that have high concentration in the ejecta, the nova is important. The isotopes ^{13}C, ^{15}N, ^{17,18}O and ^{19}F are much augmented by this nucleosynthesis, and perhaps ^{7}Li as well.

Gamma-ray astronomy can also teach us much about nova nucleosynthesis (Clayton and Hoyle 1974). The e^+ annihilations in the atmosphere may be detectable for half an hour after outburst, and both ^{22}Na and ^{26}Al may be ejected in detectable concentrations. The ways in which the nuclear lifetimes couple with convection times in the envelope are crucial to the numerical yield of ^{22}Na and ^{26}Al (Woosley 1986), and the gamma radiation from ^{7}Be (Clayton 1981b) relies on a large ^{3}He concentration in the atmosphere of the red-giant companion. Of these, only ^{26}Al has been detected but it has been necessary to employ astrophysical arguments to implicate novae as the best source (Clayton 1984a).

14.2.4. DISCUSSION

The science of nucleosynthesis deals with observations in four main ways, two of which involve meteorites directly:

(1) The attempt to reproduce solar abundance ratios played a large part in the historical origins of nucleosynthesis theory, as described in Chapter 14.1. For the nonvolatile elements the meteorites have been the best source of accurate relative abundances. The theory succeeds to the extent that the solar abundances are reproduced by the appropriately weighted yields from the sites of stellar nuclear burning. To update this approach has been the primary purpose of this chapter. It involves the heroic efforts of nuclear laboratories to obtain the relevant nuclear data and the equally heroic efforts of computational astrophysicists to clarify the stellar dynamics. This is the bread and butter of nucleosynthesis.

(2) Astronomers have measured the abundances in stars of differing ages and locations, showing conclusively that their concentrations have grown over galactic history and that they have done so more efficiently near the galactic central regions than they have in regions exterior to the solar orbit. This led to the science of chemical evolution of the galaxy, placing stellar nucleosynthesis into the context of a changing galaxy, thereby involving almost all of astrophysical science. A pedagogically clear presentation of chemical evolution, with a simple but reasonable analytic model of that evolution, was the burden of Sec. 14.2.2 of this chapter. This context is needed to evaluate the meaning of radioactive nuclei in the initial solar system: for example, the ^{235}U/^{238}U ratio or the ^{244}Pu concentration in the first meteorites. The initial radioactive abundances thereby provide corroboration of the concept of continuous galactic nucleosynthesis.

(3) Astronomers can find evidence of new nucleosynthesis in the high abundance of nucleosynthesis products in supernova remnants. This long and difficult struggle has now found several remnants carrying excess abundance

in the Mg-Ni region, evidence of explosive ejections similar to those discussed in Sec. 14.2.3. One effort of this type is especially noteworthy. Gamma-ray astronomy, by detecting radioactivity in supernova remnants or in the ISM, has the capability of demonstrating that nucleosynthesis continues *today* in the Galaxy. A radioactive nucleus announces not only its presence by its emissions but also that it was recently created. This special feature has caused this writer to make a career objective of this search, and the reader is referred to Clayton (1982*b*) for a historical account of the emergence of this new astronomy of nucleosynthesis. The radioactive cobalt ejected from supernovae has recently been detected (Matz et al. 1988) for the first time. Interstellar ^{26}Al has also been detected by this technique, and already the implications for astrophysics and for meteoritics are vast and only partly perceived (Clayton and Leising 1987).

(4) The isotopic anomalies in meteorites have provided a new and independent challenge to astrophysics in general and to nucleosynthesis in particular. Many of these are reviewed in this book and have been the subject of many reviews in the literature. But for an account of the astrophysical contexts of these anomalies, the reader may still be profitably referred to my own review (Clayton 1979), which explicitly addressed this astrophysical context rather than the meteoritic one emphasized by other reviews. Indeed, continuing in a personal vein, I have perceived a new field of astronomy, called "cosmic chemical memory" (Clayton 1982*a*), arising from these small meteoritic peculiarities. Others have found equally exciting astrophysical contexts for these data, so that the story will be a good one whatever the correct interpretation.

Acknowledgments. I especially thank S. Woosley for providing me with a copy of his survey of nucleosynthesis calculations that provided the numerical data for Figs. 14.2.6, 14.2.7, 14.2.8 and, in a more general sense, for the countless energetic clarifications of this subject matter that he has made over the years. This research was supported in part by the Robert A. Welch Foundation and in part by a grant from the National Aeronautics and Space Administration.

REFERENCES

Arnett, W. D. 1972*a*. Advanced evolution of massive stars. I. Helium burning. *Astrophys. J.* 176:681–698.

Arnett, W. D. 1972*b*. Advanced evolution of massive stars. II. Carbon burning. *Astrophys. J.* 176:699–710.

Arnett, W. D., and Clayton, D. D. 1970. Explosive nucleosynthesis in stars. *Nature* 227:780–784.

Arnett, W. D., and Thielemann, F. K. 1985. Hydrostatic nucleosynthesis. I. Core helium and carbon burning. *Astrophys. J.* 295:589–603.

Bessel, M. S., and Norris, J. N. 1984. The ultra-metal-deficient (Population III ?) red giant CD-38°245. *Astrophys. J.* 285:622–636.

Burbidge, G. R., Burbidge, E. M., Fowler, W. A., and Hoyle, F. 1957. Synthesis of elements in stars. *Rev. Mod. Phys.* 29:547–650.

Cameron, A. G. W. 1973. Are large time differences in meteorite formation real? *Nature* 246:30–32.

Chiosi, C., and Matteuchi, F. 1984. Problems in the chemical evolution of galaxies. In *Stellar Nucleosynthesis,* eds. C. Chiosi and A. Renzini (Dordrecht: D. Reidel), pp. 359–380.

Clayton, D. D. 1968 and 1983. *Principles of Stellar Evolution and Nucleosynthesis* (New York: McGraw-Hill, 1968; Chicago: Univ. of Chicago Press, 1983).

Clayton, D. D. 1975. ^{22}Na, Ne-E, extinct radioactive anomalies and unsupported ^{40}Ar. *Nature* 257:36–37.

Clayton, D. D. 1977. Interstellar potassium and argon. *Earth Planet. Sci. Lett.* 36:381–390.

Clayton, D. D. 1978. Precondensed matter: Key to the early solar system. *Moon and Planets* 19:109–137.

Clayton, D. D. 1979. Supernovae and the origin of the solar system. *Space Sci. Rev.* 24:147–226.

Clayton, D. D. 1981a. Some key issues in isotopic anomalies: Astrophysical history and aggregation. *Proc. Lunar Planet. Sci.* 12B:1781–1802.

Clayton, D. D. 1981b. ^{7}Li gamma ray lines from novae. *Astrophys. J.* 244:L97–L98.

Clayton, D. D. 1982a. Cosmic chemical memory: A new astronomy. *Quar. J. Roy. Astron. Soc.* 23:174–212.

Clayton, D. D. 1982b. Cosmic radioactivity: A gamma-ray search for the origins of atomic nuclei. In *Essays in Nuclear Astrophysics,* eds. C. A. Barnes, D. D. Clayton, and D. N. Schramm (Cambridge: Cambridge Univ. Press), pp. 401–426.

Clayton, D. D. 1984a. ^{26}Al in the interstellar medium. *Astrophys. J.* 280:144–149.

Clayton, D. D. 1984b. Galactic chemical evolution and nucleocosmochronology: Standard model with terminated infall. *Astrophys. J.* 285:411–425.

Clayton, D. D. 1985. Galactic chemical evolution and nucleocosmochronology: A standard model. In *Nucleosynthesis: Challenges and New Developments,* eds. W. D. Arnett and J. W. Truran (Chicago: Univ. of Chicago Press), pp. 65–88.

Clayton, D. D., 1986a. Analytic models of the chemical evolution of galaxies. *Publ. Astron. Soc. Pacific* 98:968–972.

Clayton, D. D. 1986b. Interstellar fossil ^{26}Mg and its possible relationship to excess meteoritic ^{26}Mg. *Astrophys. J.* 310:490–498.

Clayton, D. D. 1988. New cosmic-chemical-memory mechanism for isotopic anomalies. *Lunar and Planet. Sci.* 19, in press.

Clayton, D. D., and Hoyle, F. 1974. Gamma ray lines from novae. *Astrophys. J.* 187:L101–L104.

Clayton, D. D., and Leising, M. D. 1987. ^{26}Al in the interstellar medium. *Phys. Rept.* 144:1–50.

Clayton, D. D., and Pantelaki, I. 1986. Secondary metallicity in analytic models of chemical evolution of galaxies. *Astrophys. J.* 307:441–448.

Clayton, D. D., and Ramadurai, S. 1977. On presolar meteoritic sulfides. *Nature* 265:427–428.

Clayton, D. D., Colgate, S. A., and Fishman, G. J. 1969. Gamma ray lines from young supernova remnants. *Astrophys. J.* 155:75–82.

Colgate, S. A., and McKee, C. 1969. Early supernova luminosity. *Astrophys. J.* 157:623–644.

Eggen, O., Lynden-Bell, D., and Sandage, A. R. 1962. Evidence from the motions of old stars that the galaxy collapsed. *Astrophys. J.* 136:748–766.

Garnett, D. R., and Shields, G. A. 1987. The composition gradient across M81. *Astrophys. J.* 317:82–101.

Hainebach, K. L., Clayton, D. D., Arnett, W. D., and Woosley, S. E. 1974. On the e-process: Its components and their neutron excesses. *Astrophys. J.* 193:157–168.

Hartmann, D., Woosley, S. E., and El Eid, M. F. 1985. Nucleosynthesis in neutron-rich supernova ejecta. *Astrophys. J.* 297:837–845.

Hinton, R. W., Davis, A. M., and Scatena-Wachel, D. E. 1987. Large negative ^{50}Ti anomalies in refractory inclusions from the Murchison carbonaceous chondrite—evidence for incomplete mixing of neutron-rich supernova ejecta into the solar system. *Astrophys. J.* 313:420–428.

Howard, W. M., Arnett, W. D., Clayton, D. D., and Woosley, S. E. 1972. Nucleosynthesis of rare nuclei from seed nuclei in explosive carbon burning. *Astrophys. J.* 175:201–216.
Howard, W. M., Mathews, G. J., Takahashi, K., and Ward, R. A., 1986. Parametric study of pulsed neutron source models for the s-process. *Astrophys. J.* 309:633–652.
Hoyle, F. 1946. The synthesis of the elements from hydrogen. *Mon. Not. Roy. Astron. Soc.* 106:343–383.
Hoyle, F. 1954. On nuclear reactions occurring in very hot stars I. The synthesis of elements from carbon to nickel. *Astrophys. J. Suppl.* 1:121–146.
Iben, I., Jr. 1985. The life and times of an intermediate mass star in isolation/in a close binary. *Quar. J. Roy. Astron. Soc.* 26:1–39.
Iben, I., Jr., and Renzini, A. 1982. On the formation of carbon star characteristics and the production of neutron rich isotopes in asymptotic giant stars of small core mass. *Astrophys. J.* 263:L23–L27.
Käppeler, F., Beer, H., Wisshak, K., Clayton, D. D., Macklin, R. L., and Ward, R. A. 1982. s-Process studies in light of new experimental cross sections: Distribution of neutron fluences and r-process residuals. *Astrophys. J.* 257:821–846.
Kunth, D., and Sargent, W. L. W. 1983. Spectrophotometry of 12 metal-poor galaxies: Implications for the primordial helium abundance. *Astrophys. J.* 273:81–98.
Kunth, D., and Sargent, W. L. W. 1986. I Zw 18 and the existence of very metal poor blue compact dwarf galaxies. *Astrophys. J.* 300:496–499.
Lambert, D. L. 1985. The chemical composition of cool stars: I. The barium stars. In *Cool Stars with Excesses of Heavy Elements,* eds. M. Jaschek and P. C. Keenan (Dordrecht: D. Reidel), pp. 191–223.
Larson, R. B. 1976. Models for the formation of disc galaxies. *Mon. Not. Roy. Astron. Soc.* 176:31–52.
Luck, R. E., and Bond, H. E. 1981. Extremely metal-deficient red giants. II. Chemical abundances in 21 halo giants. *Astrophys. J.* 244:919–937.
Mahoney, W. A., Ling, J. C., Wheaton, W. A., and Jacobson, A. S. 1984. HEAO3 discovery of ^{26}Al in the interstellar medium. *Astrophys. J.* 286:578–585.
Mathews, G. J., and Käppeler, F. 1984. Neutron capture nucleosynthesis of neodymium isotopes and the s-process from A = 130 to 150. *Astrophys. J.* 286:810–821.
Matz, S. M., Share, G. H., Leising, M. D., Chupp, E. L., Vestrand, W. T., Purcell, W. R., Strickman, M. S., and Reppin, C. 1988. Detection of gamma-ray line emission from SN 1987a. *Nature,* in press.
Miller, G. E., and Scalo, J. M. 1979. The initial mass function and stellar birthrate in the solar neighborhood. *Astrophys. J. Suppl.* 41:513–547.
Nomoto, K., Thielemann, F., and Yokoi, K. 1984. Accreting white dwarf models for Type I supernovae. *Astrophys. J.* 286:644–658.
Salpeter, E. E. 1955. The luminosity function and stellar evolution. *Astrophys. J.* 121:161–167.
Schramm, D. N. 1982. The r-process and nucleocosmo chronology. In *Essays in Nuclear Astrophysics,* eds. C. A. Barnes, D. D. Clayton, and D. N. Schramm (Cambridge: Cambridge Univ. Press), pp. 325–353.
Share, G. H., Kinzer, R. L., Kurfess, J. D., Forest, D. J., Chupp, E. L., and Rieger, E. 1985. Detection of galactic ^{26}Al gamma radiation by the SMM spectrometer. *Astrophys. J.* 292:L61–L65.
Shaver, P. A., McGee, R. X., Newton, L. M., Danks, A. C., and Pottasch, S. R. 1983. *Mon. Not. Roy. Astron. Soc.* 204:53–112.
Smith, V. V., and Lambert, D. L. 1986. The chemical composition of red giants. II. Helium burning and the s-process in the MS and S stars. *Astrophys. J.* 311:843–863.
Tinsley, B. M. 1980. Evolution of the stars and gas in galaxies. *Fund. Cosmic Phys.* 5:287–388.
Truran, J. W. 1982. Nuclear theory of novae. In *Essays in Nuclear Astrophysics,* eds. C. A. Barnes, D. D. Clayton, and D. N. Schramm (Cambridge: Cambridge Univ. Press), pp. 467–493.
Truran, J. W., and Iben, I., Jr. 1977. On s-process nucleosynthesis in thermally pulsating stars. *Astrophys. J.* 216:797–810.
Twarog, B. A. 1980. The chemical evolution of the solar neighborhood. II. The age-metallicity relation and the history of star formation in the galactic disk. *Astrophys. J.* 242:242–259.

Ulrich, R. K. 1973. The *s*-process in stars. In *Explosive Nucleosynthesis,* eds. D. N. Schramm and W. D. Arnett (Austin: Univ. of Texas Press), pp. 139–167.

Weaver, T. A., and Woosley, S. E. 1980. Evolution and explosion of massive stars. *Ann. New York Acad. Sci.* 336:335–357.

Weaver, T. A., Axelrod, T. S., and Woosley, S. E. 1980. Type I supernova models versus observations. In *Type I Supernovae,* ed. J. C. Wheeler (Austin: Univ. of Texas Press), pp. 113–154.

Woosley, S. E. 1986. Nucleosynthesis and stellar evolution. In *Nucleosynthesis and Chemical Evolution,* eds. B. Hauck, A. Maeder, and G. Meynet (Sauverny: Geneva Observatory), pp. 1–195.

Woosley, S. E., and Weaver, T. A. 1980. Explosive neon burning and ^{26}Al gamma ray astronomy. *Astrophys. J.* 238:1017–1025.

Woosley, S. E., and Weaver, T. A. 1986. The physics of supernova explosions. *Ann. Rev. Astron. Astrophys.* 24:205–253.

Woosley, S. E., Arnett, W. D., and Clayton, D. D. 1973. The explosive burning of oxygen and silicon. *Astrophys. J. Suppl.* 26:231–312.

Yang, J., Turner, M. S., Steigman, G., Schramm, D. N., and Olive, K. A. 1984. Primordial nucleosynthesis: A critical comparison of theory and observation. *Astrophys. J.* 281:493–511.

14.3. IMPLICATIONS OF ISOTOPIC ANOMALIES FOR NUCLEOSYNTHESIS

TYPHOON LEE
Institute of Earth Sciences, Academia Sinica

The study of isotopic anomalies found in meteorites has improved our understanding of the nucleosynthetic origins of solar-system matter. Isotopic anomalies of Sm, Nd, Ba and Sr in FUN inclusions EK1-4-1 and C1 show systematic trends that can be readily interpreted in terms of the relative excesses and deficits of material from the sources of the s-, r- and p-processes, thought to be responsible for the nucleosynthesis of heavy elements beyond the Fe peak. Meteoritic data thus provide strong empirical support for these theoretical processes. Furthermore, the independent variations of the anomalies due to these three processes require that they were produced in different sources, transported in different forms in the interstellar space, and preserved in the early solar nebula. In addition, a consistency seen in the data for the three heavier elements does not continue into Sr, suggesting that there may be two types of s-process contributing to these two mass regions. Anomalies in Ca, Ti and Cr are widespread in carbonaceous meteorites. The data imply the existence of many contributing components. Among these, the most pronounced one is rich in ^{48}Ca, ^{50}Ti and ^{54}Cr. Most likely, it came from the e-process in a neutron-rich environment, probably Si burning near the high-density core of a massive progenitor star of a supernova. Nucleosynthetic modeling of such processes has led to the conclusion that this may be the mechanism that produced many neutron-rich nuclides between 50 and 90 amu in solar-system matter.

14.3.1. INTRODUCTION

The preservation of isotopic heterogeneities in primitive meteoritic samples offers a valuable opportunity to study the nucleosynthetic history of solar-system material. That history consists of the production of nuclei inside stars, the transport of the products through interstellar space, and their incorporation into solid objects in the early solar system. The history of the quest for such heterogeneities was introduced in Chapter 12.1. The evidence for the incorporation in primitive meteorites of circumstellar and interstellar material with their distinct isotopic signatures has been discussed in Chapters 13.1 and 13.2 for elements H, C, N and the noble gases. The nuclear astrophysical production in galactic stellar sources and the solar-system abundance pattern resulting from many such contributing sources were the subjects of Chapters 14.1 and 14.2. The purpose of this chapter is primarily to examine how nucleosynthetic processes responsible for isotopic anomalies can be identified from the experimental results and what insights the interpretation of these results may provide for the nucleosynthetic processes themselves.

Signatures of distinct nucleosynthetic processes have been recognized in observed isotopic anomalies for the following 7 elements: Ca, Ti, Cr, Sr, Ba, Nd and Sm (Table 14.3.1). These are in addition to the Xe and Kr anomalies already discussed in Chapter 13.1. There are also anomalies in Mg and Si that apparently have nucleosynthetic origin but these elements have too few isotopes to permit the identification of the specific processes involved. Besides, it may be possible to generate these anomalies through nonnuclear means (Esat et al. 1986). The question of whether the O anomaly represents a nuclear effect has already been extensively discussed in Chapter 12.1.

The distribution of these anomalies among meteoritic samples is quite uneven. The Nd anomaly occurs only in Allende inclusion EK1-4-1. This sample and another Allende inclusion C1 so far also monopolize the nuclear anomalies in Si, Mg, Sr, Ba and Sm. In contrast, anomalies in Ca, Ti and Cr are more widespread. Anomalies are present in many Allende CAIs for Ca, and Cr anomalies show up in most CAIs as well as in some bulk meteorites.

TABLE 14.3.1
Isotopic Anomalies

Element	Occurrence	Effect
O	all meteorites	$\pm ^{16}O$
Mg	C1, EK1-4-1, EGG-3?	$+^{25}Mg$ or $-^{26}Mg$
Si	C1, EK1-4-1	$+^{29}Si$ or $-^{30}Si$
Ca	many CAI	$\pm ^{48}Ca$ and others
Ti	all CAI, many meteorites	$\pm ^{50}Ti$ and others
Cr	most CAI, some meteorites	$\pm ^{54}Cr$ and others
Sr	EK1-4-1 & C1	$-p$ isotope
Ba	EK1-4-1 & C1	$\pm\ r$ isotopes
Nd	EK1-4-1	$+r$ isotopes
Sm	EK1-4-1 & C1	$+r$ & p isotopes

Anomalies involving Ti are ubiquitous, appearing in all CAIs as well as in bulk samples of most carbonaceous meteorites.

It is evident that inclusions EK1-4-1 and Cl are important since not only the majority of the nuclear effects discussed in this chapter were found solely in them but also even for the Ca-Ti-Cr region, where the anomalies are more widespread, the size of the effects is much larger in them than in "normal" CAIs. These two inclusions are characterized by isotopic effects caused by nuclear processes in virtually every element analyzed so far, coupled with even more pronounced isotopic effects due to mass fractionation in O, Mg and Si. Thus, they were dubbed "FUN" (for "fractionation and unknown nuclear") and have been the subject of numerous reviews (see, e.g., Lee 1979; Begemann 1980; Wasserburg et al. 1980). Extensive efforts to search for more such samples have so far yielded no success. Worse, the distinction between FUN and non-FUN samples has become somewhat blurred because of the discovery of samples with fractionation in Mg and nuclear effects in other elements, both being intermediate in magnitude between those in "normal" and FUN inclusions (e.g., EGG-3; Esat et al. 1980). In this chapter we will reserve the FUN title for Cl and EK1-4-1 only. Furthermore, the coupling between the fractionation and the nuclear effects has recently been called into question by the discovery of samples with fractionation effects in O comparable to, but nuclear effects in Ti much smaller than, those in Cl and EK1-4-1 (e.g., EK25-S2TE; R. Clayton et al. 1984; Niederer et al. 1985; Jungck et al. 1984). Despite intensive petrographic and chemical study so far, there is no real clue as to why certain inclusions carry such strong nuclear signals. Another place where exceedingly large (up to 10%) effects are sometimes present in Ca and Ti are the tiny hibonite inclusions in carbonaceous meteorites (Lee et al. 1979; Ireland et al. 1985; Hinton et al. 1987; Fahey et al. 1985,1986; Zinner et al. 1986). The small size and refractory nature of this rare mineral are thought to be related to the presence of large effects but again the mechanism is not understood.

Obviously, we have little idea as to why some elements have anomalies while others do not. We are equally ignorant as to why anomalies appear in some samples but not in others. All we know is that in some cases various nucleosynthetic products from distinct stellar sources were able to escape complete homogenization between their ejection from stars and their incorporation into meteorites. Since it is well known that most heavy nuclei in interstellar space are locked up in dust, and that one of the first important steps toward forming meteorites in the solar nebula was dust coagulation, the transport and preservation of presolar dust were believed to be important to the anomalies from early on, essentially when they were first discovered. In principle, a lot about the transport of nucleosynthetic products through the interstellar media could eventually be learned from the study of meteoritic anomalies. A few sample questions are: What chemical form characterized the products of distinct nucleosynthetic processes while residing in dust? How did circumstellar dust form? How long can such dust survive against destructive processes such as sputtering? This approach has been tirelessly expounded by D. Clayton (1982). An obstacle to this approach is that no presolar dust grains have been identified in meteoritic samples with the possible exception of the C-rich carriers of the noble gas anomalies discussed in Chapter 13.1. For the anomalies discussed here, even the tiny hibonite grains with pronounced effects are probably not the original pristine interstellar dust but have been substantially diluted with normal solar-system material and significantly reprocessed (Fahey et al. 1986). Most of the discussions on this part of the nucleosynthetic history of solar-system material thus remain speculative at present. Therefore, in the following we concentrate on the comparison of nucleosynthetic modeling with the observed anomalies, although the transport of the nuclei through interstellar space, and the chemistry in the solar nebula must have introduced complications, especially when anomalies between different elements are to be correlated.

14.3.2. HOW TO INTERPRET THE ANOMALY DATA

In order to understand the implications of the meteoritic observations, it is instructive first to illustrate with an example the intricate interplay between the isotopic data and nucleosynthesis theory. Element Sm has been selected for this purpose because it has many isotopes from diverse origins and because the relevant nucleosynthesis theory and nuclear experimental data are in better shape than for most other elements. The isotopes of Sm are believed to have been synthesized in r, s and p processes (cf. Chapters 14.1 and 14.2 and the chart of nuclides for this mass region in Fig. 14.3.2 below). Figure 14.3.1 is a schematic plot of the four Sm isotopes which are thought to have come entirely from single processes. In addition ^{152}Sm, consisting of roughly a 25% s-process contribution and a 75% r-process contribution is included. The other two isotopes are analogous to ^{152}Sm and are omitted. In Fig. 14.3.1a, the average solar-system composition is shown as a mixture of these three processes at a proportion of 1:1:1 for the sake of simplicity. For comparison, let us now consider a hypothetical mixture which has 1% less s-contribution (Fig. 14.3.1b). Obviously all isotopes having an s contribution should show deficits as indeed is the case for ^{148,150,152}Sm. In this manner one can *empirically* identify from the data which isotopes have received contribution from a single nucleosynthetic process. Furthermore, the production ratios for that process can be inferred quantitatively by the sizes of the isotope effects. In this example, one would deduce that the s-process production ratio for ^{152}Sm/^{150}Sm is 0.25 from the fact that the ratio between their deficits is 0.25. Of course, we have no *a priori* reason not to attribute the observed deviations from the normal to an excess of r and p contributions, as only isotope ratios are measured. In that case, we would infer just as correctly that the r-process must produce ^{152}Sm/^{154}Sm at a ratio of 0.75. Naturally, in gen-

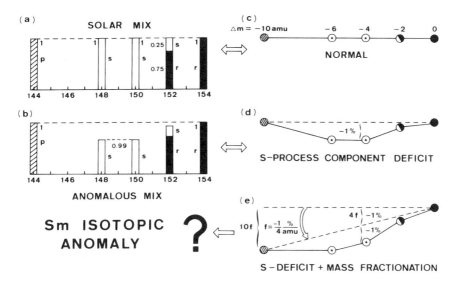

Fig. 14.3.1. Interpreting isotope anomalies using Sm as an example. (a) Four isotopes: ^{144}Sm, ^{148}Sm, ^{150}Sm and ^{154}Sm are entirely from p-, s-, s- and r- processes, respectively. ^{152}Sm, consisting of a 75% r contribution and a 25% s contribution, is used as an example of an isotope of mixed origin while the remaining Sm isotopes are omitted for clarity. Normal solar-system material is assumed to have isotopic ratios of 1 (not actually correct) for the sake of simplicity. (b) The isotopic composition of an anomalous sample with a 1% deficit of s contribution (or equivalently r and p excesses) is exaggeratedly plotted for comparison. Note that, owing to its 25% s contribution, the deficit for ^{152}Sm is 0.25%. (c) An abbreviated way of displaying the normal data, Δm is the mass difference from ^{154}Sm. (d) The abbreviated way of viewing the case given in (b). (e) The case in (d) with a $f = -0.25\%$ per amu mass-fractionation effect superimposed. Note that since the fractionation is proportional to mass difference, its effect is equivalent to rotating the pattern in (d). The usual way of treating the data is to select two normalizing isotopes (^{148}Sm and ^{150}Sm are used here), draw a straight line through them and assume the rotation between the line and the normal (horizontal line) is caused by fractionation. This is corrected by rotating the pattern in the reverse direction and reporting the deviation from the line as "nonlinear" or "nuclear" effects. This interpretation can be ambiguous since different selections of normalizing isotopes cause different "nonlinear" effects.

eral one would be dealing with isotope ratios different from unity. Since all observed deviations from the normal ratios are small, one can simply divide the observed ratios by the normal values and deal with fractional deviations away from the normal, usually in units of per mil (δ) or 0.01% (ε).

An abbreviated way to view the data is shown in Figs. 14.3.1c and 14.3.1d where the normal solar pattern is plotted as a horizontal line while the deficits in the anomalous sample now appear as dips. From this anomalous pattern, all conclusions based on Fig. 14.3.1b can be reproduced. The alternative of r and p excesses is the same as drawing the baseline through ^{148,150}Sm. However, mass-dependent isotopic fractionation processes, arising from both analytical procedures and sample history, that are superimposed on the pattern caused by nuclear effects alone can often complicate the interpretation. Small fractionation effects on an isotopic ratio are linearly dependent upon the mass difference between the two isotopes involved for the vast majority of processes (cf. Chapter 12.1). Therefore, their effect on the pattern is equivalent to rotating it. In Fig. 14.3.1e the effect of superimposing a fractionation factor f of -0.25% per amu of mass difference on the pattern in Fig. 14.3.1d is shown. Note that ^{144}Sm is fractionated by $10 \times f$ and ^{150}Sm by $4 \times f$ and so forth, hence we can no longer recognize any regular trends in the data. Based on these data alone, we may even be tempted to assume ^{144}Sm and ^{148}Sm are from the same process because of the accidental superposition of the two different effects. When feasible, a "double spike" method can sometimes be applied to remove the fractionation effects caused by the analytical procedures. The resulting ratios (often called "absolute") would still include both the nuclear effects and natural fractionation effects (see, e.g., Russell et al. 1978). Commonly, Δ is used to denote the fractional deviations in this case.

Thus, the interpretation of the data, as originally suggested by Lugmair et al. (1978; see also McCulloch and Wasserburg 1978a), usually proceeds as follows: guided by theory, one first assumes that the ratio between s-process-only isotopes, ^{148}Sm and ^{150}Sm, is affected only by mass fractionation. Then the entire pattern is corrected for this fractionation effect by rotating it in the reverse direction. Finally, the relative deviations from the normal can be quantitatively compared with the production pattern of the s-process. Note that graphically this is equivalent to drawing a straight line in parallel with the fractionation line through the two normalizing isotopes and identifying the nonlinear deviations from this linear trend as the nuclear effects. In general, interpreting anomalies through such deconvolution of two effects is subject to ambiguities, especially when there is no well-established nucleosynthetic model to guide the fractionation-correction procedure. In the simple case where the anomalies are considered as a mixture between a normal solar component (o) and an exotic component (*) the following equation holds:

$$\varepsilon_i/\varepsilon_j = (N_i^*/N_j^*)/(N_i^o/N_j^o) \qquad (1)$$

where ε_i and ε_j are the observed ε values for isotopes i and j. The right-hand side of Eq. (1) is the production ratio for the * component divided by the solar-system abundance ratio. Note that this can also be written as the ratio between enhancement factors (N^*/N^o) for i and j, which are usually the quantities reported from nucleosynthesis modeling. In this manner, we can compare the observed effects against theoretical calculations. The general case of computing nonlinear effects from theoretically derived production ratios has been discussed by Sandler et al. (1982), whereas the transformation between experimental nonlinear effects using different correction procedures can be found in McCulloch and Wasserburg (1978a).

In the preceding discussion, only isotopes of the same element are involved. In order to obtain better constraints, it is necessary to combine isotopic effects of different elements. Then one immediately encounters the difficult question of how to take into account chemical differentiation between different elements during the journey from source to sample. The critical issue is how, and with what, a nucleosynthetic component was mixed. The simplest case is of course to assume that all previous conclusions such as Eq. (1) also apply to isotopes of different elements. For the case of Eq. (1), however, this implies the validity of the following sequence of events. The two nuclides i and j were ejected from the star, transported through interstellar space and reached the solar system without chemical differentiation. They then mixed with average solar-system matter and finally the anomalous mixture solidified into the inclusion. However, if the nuclides were in presolar dust which mixed with solar-system condensates, and the mixture was then recrystallized into inclusions, it is more appropriate to compute the ratio between i and j in the presolar component using an analogous equation:

$$(N_i^*/N_j^*) = (N_i^o/N_j^o)(\varepsilon_i/\varepsilon_j) \qquad (2)$$

except that the superscript o no longer refers to the solar system but to the inclusion instead. For two refractory elements such as Ca and Ti, Eqs. (1) and (2) may not be very different. But for a refractory element and a relatively volatile element (e.g., Ti vs Cr), the two estimates can be drastically different.

14.3.3. HEAVY ELEMENTS AND THE r-, s- AND p- PROCESSES

The Sm, Nd, Ba and Sr isotopic anomalies discovered in the two Allende FUN inclusions constitute a coherent set which can be understood reasonably well in term of the r-, s- and p- processes for heavy-element nucleosynthesis. They thus substantiate those theoretical constructs and provide some additional insight. This is particularly important for the p-process which is the least understood of the three. In fact, the abundances of p-process nuclides are so low that they can only be measured using laboratory analysis of terrestrial, lunar or meteoritic rock samples and have never been measured by remote-sensing methods such as optical spectroscopy. The chart of nuclides for Sm, Nd and Ba is shown in Fig. 14.3.2 and the data for EK1-4-1 and C1 are displayed in Figs. 14.3.3a-through-d. In these figures, each isotope is labeled using its mass, and the contributing processes are identified. The solid symbols indicate the pair of isotopes which were used to remove mass-fractionation effects from the data.

For Sm, in Fig. 14.3.3a (McCulloch and Wasserburg 1978a; Lugmair et

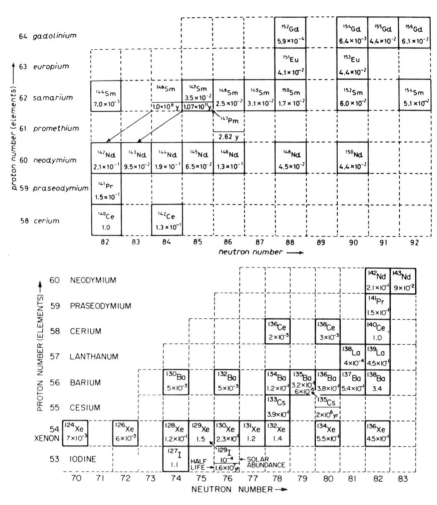

Fig. 14.3.2. The chart of nuclides for the mass region around Ba-Nd-Sm. The solar-system abundance normalized to Si = 10^6 atoms is given for each stable or long-lived isotope. Radioactive isotopes with half-lives longer than about a year are also shown with their half-lives.

al. 1978), the fractionation effect is corrected using ^{148}Sm and ^{150}Sm as normalizing isotopes, as already argued above. The resulting nonlinear effects show up as excesses in all p- and r-process isotopes for sample EK1-4-1, whereas in C1 only the p-process isotope is in excess. The EK1-4-1 data thus clearly demonstrate that s-process products are distinct from the r- and p-process products. The C1 data further imply that p products are decoupled from r products.

The data for Nd (McCulloch and Wasserburg 1978b; Lugmair et al.

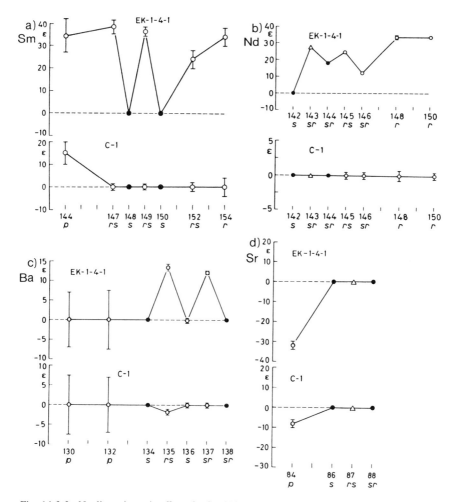

Fig. 14.3.3. Nonlinear isotopic effects for Sm-Nd-Ba-Sr found in FUN inclusions EK1-4-1 and C1. The units are fractional deviations from the normal in 0.01% (ε). Each isotope is labeled using its mass and the contributing processes are identified. The normalizing isotopes, generally s-only or s-dominated, are plotted as solid symbols. (a) Sm data show the independent variations of r, s and p components. (b) Nd data are consistent with the pattern of r excess (or s deficit) in EK1-4-1 seen in Sm. (c) Same consistency is extended to Ba except for the deficit at ^{135}Ba (however, see text). (d) Consistency does not extend to Sr data suggesting that nucleosynthesis in this mass region is different. Data for these elements strongly support the general validity of the r, s and p classifications for nuclides beyond the Fe peak (figure after McCulloch and Wasserburg 1978a,b; Papanastassiou and Wasserburg 1978).

1978), shown in Fig. 14.3.3b, suffer from the problem that there is only one s- only isotope, ^{142}Nd, so that the correction procedure used for Sm cannot be simply copied here. Luckily, Nd and Sm are neighboring elements of the rare-earth group, so that they possess very similar chemical properties. Thus the effect of chemical differentiation on their relative abundances should have been minimal, and it is anticipated that the excess for r- only ^{150}Nd should be close to that for the nearby r-only ^{154}Sm. The data were therefore corrected by assuming, somewhat arbitrarily, a ^{144}Nd excess of 18 ε because this would yield an excess of ^{150}Nd identical to that of ^{154}Sm. The resulting pattern clearly shows that the remaining Nd isotopes, which all receive r contribution, all have excesses for sample EK1-4-1. For C1, the Nd data are normal. This excess of r component relative to s component in EK1-4-1 but not in C1, is consistent with the observation for Sm, and thereby reinforces the validity of the above interpretation. Note that there is no p-only Nd isotope.

This consistent pattern is repeated in the Ba data shown in Fig. 14.3.3c (McCulloch and Wasserburg 1978b). Here the normalizing isotopes are ^{134}Ba, which is s-only, and ^{138}Ba, which is a predominantly s-process isotope because of its magic neutron number of 82. The results would have remained essentially the same, but with larger experimental errors, had the s-only ^{136}Ba been used instead of ^{138}Ba. The nonlinear effects again show up as excesses in r isotopes for EK1-4-1. The p-only Ba isotopes have very low abundances so the data are not precise enough to permit the detection of an excess of the size for the r excess. There is no Ba effect in C1 except a small deficit for ^{135}Ba which may indicate a breakdown of the consistency. However, ^{135}Ba has a relatively long-lived progenitor ^{135}Cs whose half-life is 2 Myr. Therefore, if some ^{135}Cs existed in the early solar system when C1 formed, and if C1 incorporated less than the normal amount of ^{135}Cs, then the ^{135}Ba deficit in C1 can be explained as the "hold-up" of this isotope in the form of ^{135}Cs.

The above consistent trend, however, does not extend to mass regions below 90, as can be seen in the Sr data shown in Fig. 14.3.3d (Papanastassiou and Wasserburg 1978). The data were corrected by assuming the normal value for the ratio between s-only ^{86}Sr, and ^{88}Sr, which is also predominantly an s-process isotope because of its magic neutron number of 50. Instead of excesses in r or p isotopes, the nonlinear effects show up in both EK1-4-1 and C1 as deficits in p-only ^{84}Sr. This interpretation is not unique, however. The deficits can in principle be due to excesses of the same fractional size in all the remaining three isotopes. This may sound implausible since it requires the appearance of anomalies of the right sizes in so many more isotopes. But note that, except for ^{84}Sr, the Sr isotopes are all mainly s nuclides, hence the excess of a single s component would suffice and is a plausible alternative. Even then, an s-process excess in both C1 and EK1-4-1 would still be a surprise which may be telling us something about s-process nucleosynthesis. We will come back to this point later.

The above results clearly show that the r, s and p classification is well founded based on observations of isotopic anomalies. Prior to such meteoritic studies, these nucleosynthetic processes were primarily theoretical constructs whose empirical basis consisted mainly of the regular trends of the abundance-cross-section products for the s-only nuclides in the solar system, and the spectroscopic observation in peculiar stars of enrichments of *elemental* abundances that are dominated by s-process contribution, such as Ba, Y, Zr and Sr (see, e.g., Tomkin and Lambert 1983). Only in rare cases have s-process isotopic abundances been measured in stellar sources (e.g., the Zr-band study of Zook [1978]). The direct observation of correlated *isotopic* effects which are readily interpretable, at least qualitatively, in terms of these processes is a gratifying success for the nucleosynthesis theory for elements beyond the Fe peak.

From the data on Ba, Nd and Sm, a consistent pattern of p and r excesses (or equivalently, s deficits) in EK1-4-1, and p excess (or possibly normal for Ba) in C1 seems to exist for nuclei between 130 amu and 154 amu. The contrasting pattern of p-deficits (or s excesses) in both EK1-4-1 and C1 observed for nuclei between 84 amu and the 88 amu requires that nucleosynthesis for those two mass regions is distinct in some sources. Note that it has been suggested before, based on the fit of the solar-system abundance vs cross-section curve and theoretical models of s-process, that there may be actually three s-process components coming from different astrophysical sites characterized by different temperatures and neutron densities (Beer 1986). One of these components supposedly contributes significantly only in the mass region below $A = 90$ whereas the others dominate production between $A = 90$ and $A = 200$. It will be worthwhile to investigate whether the distinct behavior of s-nuclide anomalies, if interpreted as such, in these two mass regions may be related to those proposed components.

The independent variations for the abundance of r, s and p isotopes imply that they must have been produced in different astrophysical sites. Furthermore, the mechanisms for their ejection from the source sites, for their transport in the interstellar medium, and for their incorporation into solar-system objects must have been such that they were able to retain their identity against complete mixing. These constraints have far-reaching consequences which have not yet been fully explored.

A fundamental question regarding the relationship between the anomalies and the nucleosynthetic history of solar-system material is whether the anomalies represent the input to the solar system from one or a few special source(s). The alternative is that they are residual fluctuations of the mixing process whose grand average is the normal solar-system composition, as suggested by Lugmair et al. (1978). Potentially, an observational test to distinguish between these two alternatives lies in the production ratios of the components required to explain the anomaly data. If they agree with those determined from the normal solar-system composition, it is likely that they were not from special source(s), since generally one would expect that the

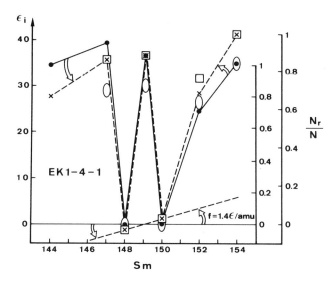

Fig. 14.3.4. Quantatitive comparison between the anomalies and the solar-system r-process production ratios. Solid dots are data from Fig. 14.3.3a. The ovals are the anomalies predicted by Eq. (1), further normalized to ^{154}Sm, using the r-process production ratios inferred from solar-system abundances. The fit is not good in spite of the general similarity with the observed pattern. However, if there is an additional mass-fractionation effect of the magnitude shown, the data would be represented by the crosses and the predicted ratios according to Eq. (1) become the squares; the fit is then much improved. Thus, within the ambiguities of the data interpretation, nonsolar production ratios are not required (figure after D. Clayton 1979).

production at individual sources would be different from that for the average of all sources. An attempt to do this for Sm is shown in Fig. 14.3.4. In that figure, the Sm data given in Fig. 14.3.3a are reproduced as solid dots. According to Eq. (1) $\varepsilon_i/\varepsilon_{154} = (N^r_i/N_i)/(N^r_{154}/N_{154})$. If we consider the solar system r-process production ratio, which can be obtained by subtracting the s contribution, then $N^r_{154}/N_{154} = 1$, $N^r_{152}/N_{152} = 0.75$, and so on. Thus, we plot the fractional contribution of average r-process to each of the isotopes (N_r/N) as ovals in the same figure. Graphically, Eq. (1) then means that the data points should coincide with the ovals. It is obvious from Fig. 14.3.4 that the pattern for the dots is similar to that of the ovals but there exist disagreements, especially at ^{147}Sm and ^{149}Sm, outside the nominal analytical errors, seemingly suggesting a nonsolar special source. However, D. Clayton (1979) has suggested that if an additional mass fractionation effect of $f = 0.014\%$ per amu is present, the true nuclear effects to be explained would correspond to the pattern of crosses in Fig. 14.3.4, which is just the pattern of dots rotated counterclockwise by the fractionation factor. In this case, the expected effect from the average r-process is given by the squares, which apparently fit the crosses much better. As discussed earlier, such an interpretation is permitted by the ambiguities. Therefore, at present, the data can be brought into agreement with the r-process production ratios inferred from the normal solar-system composition, hence do not have to imply special sources. This conclusion is generally true for the effects in the four elements discussed here.

14.3.4. IRON-GROUP ELEMENTS

The low-mass wing of the Fe abundance peak, namely Ca, Ti and Cr (cf. the chart of nuclides in Fig. 14.3.5), is another mass region where correlated isotopic anomalies have been found. Anomalies in this group are much more widespread and their impact on nucleosynthesis theory has been significant. The nucleosynthetic process responsible for the Fe peak itself from V through Ni is believed to be the e-process which takes place during Si burning in the deep interior of massive stars at a late stage of their evolution or in exploding white dwarfs (cf. Chapters 14.1 and 14.2). We will see that the largest anomalies for Ca, Ti and Cr occur at the most neutron-rich isotopes but that there is no large anomaly at the neighboring rare isotope ^{46}Ca. The favored interpretation thus invokes a neutron-rich variety of the e-process whose site is even closer than the usual e-process to the stellar core. These anomalies therefore yield valuable insight to not only the nucleosynthetic origin of solar-system matter but also late-stage stellar evolution of supernova progenitors. The origin of other isotopes in this mass region is less clear. Many nucleosynthetic processes (explosive O burning, seed nuclei capture during explosive C burning, etc.; see Chapter 14.2) have been proposed to contribute but none is

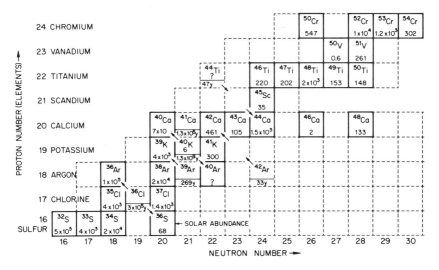

Fig. 14.3.5. The chart of nuclide for the Ca-Ti-Cr region. See Fig. 14.3.2 for explanation.

convincingly demonstrated by either solar-system abundance patterns or isotopic anomaly study. Smaller anomalies do exist in these Ca,Ti and Cr isotopes, which seem to have been the result of multiple component mixing. However, the difficulty of data interpretation without the guidance of established nucleosynthetic models so far prevents a clearer understanding.

The Ca data for the FUN inclusions EK1-4-1, C1, plus those for the hibonite inclusion HAL are summarized in Fig. 14.3.6 (Lee et al. 1978,1979). The lower panel shows the data obtained using the double-spike method and therefore reveals natural mass fractionation effects coupled with nuclear effects. For both EK1-4-1 and C1, the light Ca isotopes fall close to a linear trend whereas ^{48}Ca deviates most from this trend. Thus the linear trend is considered to be a manifestation of natural mass fractionation, and

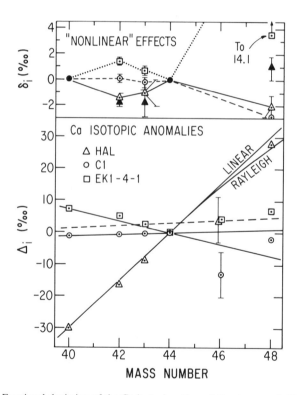

Fig. 14.3.6. Fractional deviation of the Ca isotopic ratios relative to normal. The lower panel shows the "absolute" ratios, after the instrumental fractionation is removed using the double-spike method, showing natural mass fractionation and nuclear effects. Note the linear trends for isotopes ^{40,42,43,44}Ca. The upper panel shows the "nonlinear" effects after normalizing to ^{40}Ca and ^{42}Ca. The large excess of ^{48}Ca for EK1-4-1 is off scale. Smaller effects also exist at ^{42}Ca. Note that the rare nuclide ^{46}Ca shows effects smaller than that of ^{48}Ca (Lee et al. 1978,1979; Niederer and Papanastassiou 1984).

the largest nuclear effect is assigned to ^{48}Ca. The data are normalized to ^{40}Ca/^{44}Ca because they are the two most abundant Ca isotopes and thus propagate the least amount of experimental uncertainty. The nonlinear effects obtained this way are shown in the upper panel. A substantial excess of ^{48}Ca is seen in EK1-4-1 and a significant deficit in C1. Note that there are also smaller nonlinear effects at mass 42 and perhaps also 43. In HAL there is a ^{42}Ca deficit but the fractionation effect is too large to permit definitive identification of a nonlinear effect for ^{48}Ca. In ordinary Allende inclusions, careful studies revealed that the majority seem to show small but resolvable nonlinear effects only at ^{48}Ca (Jungck et al. 1984; see also Fig. 14.3.11b). A crucial point is that no large nonlinear effect has been found in ^{46}Ca (Lee et al. 1978; Niederer and Papanastassiou 1984; Jungck et al. 1984), which accounts for only 0.003% of Ca. This is surprising since one would expect a ^{46}Ca nuclear effect about sixty times larger than that of ^{48}Ca if the anomalous component contains similar amounts of both species. Recently, both positive and negative ^{48}Ca nonlinear effects of sizes even larger than those observed in FUN samples were discovered in hibonites and associated mineral grains using the ion microprobe (Zinner et al. 1986; see also Fig. 14.3.11a). No nonlinear effects were detected for other Ca isotopes in hibonites, probably because of the lower precision of the ion microprobe.

Isotopic anomalies in Ti are quite widespread among meteoritic samples. Data for EK1-4-1, C1 and some non-FUN samples are shown in Fig. 14.3.7 (Niederer et al. 1985). The solid symbols are data corrected for instrumental fractionation using the double-spike method. It is apparent that both natural mass-fractionation effects and nuclear effects are present. However, lacking the guidance of a reliable nucleosynthetic model for the origin of the Ti isotopes, one is unsure as to which normalizing isotopes to choose for the correction of natural fractionations. One possible choice is ^{46}Ti/^{48}Ti and the resulting patterns of nonlinear effects are shown as broken lines in Fig. 14.3.7. Presented in this manner, the largest nuclear effects appear at ^{50}Ti for both FUN and non-FUN samples. The data for EGG-3 typify those for non-FUN inclusions and most Allende CAI show such nonlinear effects in ^{50}Ti (Niederer et al. 1980,1981; Niemeyer and Lugmair 1981,1984). In addition, ^{50}Ti effects have been found in bulk samples of CV, CO, CM and perhaps CI meteorites (Niemeyer 1985). Recently, the assignment of the largest nuclear effect to ^{50}Ti has been substantiated by ion-microprobe studies of Ti in hibonites from carbonaceous meteorites (Hutcheon et al. 1983; Fahey et al. 1985,1986; Ireland et al. 1985; Hinton et al. 1987). Exceedingly large excesses and deficits of ^{50}Ti have been found and the extreme examples are shown in Fig. 14.3.8. In one sample (MY-H3), the huge ^{50}Ti excess is clearly beyond the possible instrumental fractionation range. The fact that the abundances for all the lighter Ti isotopes are close to normal indicates that the ^{50}Ti assignment is likely to be correct since any other assignments would require the existence of anomalies for many other Ti isotopes in just the right propor-

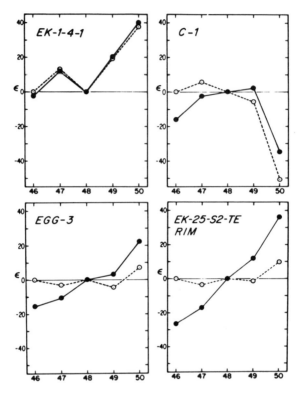

Fig. 14.3.7. Isotopic composition of Ti in EK1-4-1, C1 and non-FUN inclusions. The solid symbols give the absolute composition. The open symbols are the data normalized to ^{46,48}Ti. There are no large Ti mass-fractionation effects in EK1-4-1. For other inclusions, Ti-isotopic effects are present in all isotopes including ^{46}Ti. The largest effect is at ^{50}Ti (figure after Niederer et al. 1985).

tions by coincidence. It is still debatable if small nonlinear effects are present for lighter Ti isotopes in samples other than EK1-4-1 and C1. However, based on data for FUN samples alone and normal Ti, it is already clear that at least three nuclear components are required because, on a three-isotope plot such as Fig. 14.3.9, those data do not fall on a mixing trend but occupy a triangular area. When all the Ti data are combined, it appears that at least four components are required (Niemeyer and Lugmair 1984; Fahey et al. 1986). Since this element has only four independent ratios and one of the ratios is used to remove the fractionation effect, one thus faces the dilemma that the number of components exceeds the number of ratios.

Isotopic anomalies in Cr have been observed in non-FUN samples (Birck and Allègre 1984, 1985) as well as in the two FUN inclusions (Papanastassiou

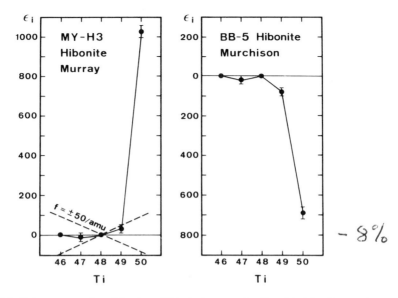

Fig. 14.3.8. Extreme examples of the large Ti-isotopic anomalies found in hibonite samples from CM meteorites. MY-H3 data are from Fahey et al. (1985), which are clearly outside the possible range of instrumental mass fractionation (± 50 ε/amu). Thus the assignment of a nuclear effect to ^{50}Ti seems to be secure. Data for BB-5 are from Hinton et al. (1987).

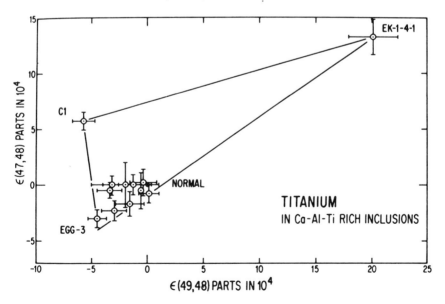

Fig. 14.3.9. Plot of $\varepsilon(^{47}\text{Ti}/^{48}\text{Ti})$ vs $\varepsilon(^{49}\text{Ti}/^{48}\text{Ti})$ for CAIs. These are the nonlinear effects using ^{46,48}Ti normalization. Since the data populate a triangular area, they are not consistent with the linear trend expected from mixing between two end members. At least three components are necessary to explain the Ti effects (figure after Niederer et al. 1981).

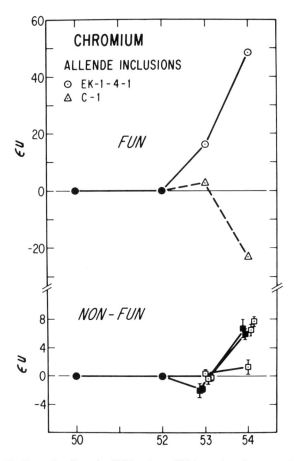

Fig. 14.3.10. Nonlinear Cr effects for FUN and non-FUN samples after normalization to ^{50}Cr/^{52}Cr. The largest effects are at mass 54. More than two components are suggested in view of the sign reversals for ^{53}Cr effects. (Figure after Papanastassiou 1986.)

1986) (Fig. 14.3.10). When ^{50}Cr/^{52}Cr is used for normalization, EK1-4-1 shows a sizeable positive ^{53}Cr effect and a large ^{54}Cr excess while C1 shows almost no effect at ^{53}Cr and a large deficit in ^{54}Cr. For some non-FUN inclusions, small ^{54}Cr excesses are seen and there seem to be some hints of ^{53}Cr deficits. Note that the ^{53}Cr effects may not be intrinsic to itself but related to the extinct nuclide ^{53}Mn discussed in Chapter 15.1.

Figure 14.3.11a,b displays the relationship between anomalies in ^{48}Ca and those in ^{50}Ti. It may be seen that the size of the anomalies increases in the order: bulk meteorites < non-FUN CAI < FUN CAI < hibonite grains; i.e., large anomalies are found in small places. However, it should be emphasized that even the hibonite grains are orders of magnitude larger than typical

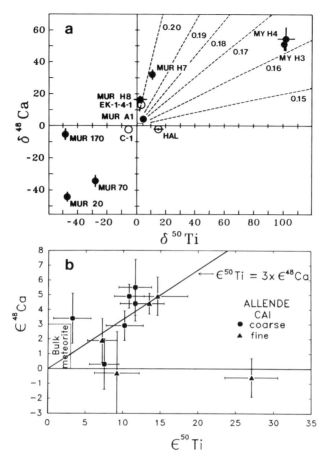

Fig. 14.3.11. Attempt to correlate nonlinear ^{48}Ca and ^{50}Ti effects. (a) Large effects for FUN samples and hibonites. The dashed lines are the expected correlation for the products from MZM e-process nucleosynthesis and the values are the maximum neutron excess η_{max} (figure after Zinner et al. 1986). No strict correlation between these two isotopic effects is apparent. Note that the small rectangle is the range covered in plot (b). (b) Small effects for non-FUN inclusions and whole meteorites (figure after Jungck et al. 1984; Niemeyer 1985). On average $\varepsilon(^{50}\text{Ti}) = 3 \times \varepsilon(^{48}\text{Ca})$ fits most of the non-FUN inclusions but many exceptions exist. Thus only a rough correlation exists between these two effects in the sense that their signs are almost always the same. The region in the lower left corner is for bulk meteorite data.

interstellar grains and after extensive studies, there is as yet no evidence that they are presolar in origin (see, e.g., Fahey et al. 1986). Bulk samples for CV, CO and CM meteorites have no well-resolved ^{48}Ca effects but show distinct ^{50}Ti effects, which could be due to the difficulty of getting high-precision data for ^{48}Ca (Niemeyer 1985). Most CAI have positive ^{48}Ca and ^{50}Ti non-

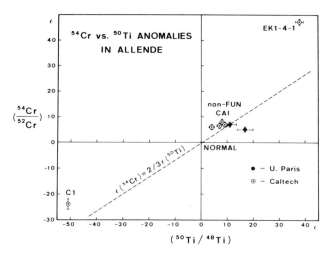

Fig. 14.3.12. Attempt to correlate nonlinear ^{54}Cr and ^{50}Ti effects. No strict correlation is observed either among non-FUN inclusions or the FUN samples. Similarly for the ^{48}Ca — ^{50}Ti case, only a rough correlation of covarying signs for the nonlinear effects exists (data from Birck and Allègre 1984,1985; Papanastassiou 1986).

linear effects (Niederer and Papanastassiou 1984; Niederer et al. 1980, 1981,1985; Niemeyer and Lugmair 1981,1984; Jungck et al. 1984) which roughly follow a correlation line with a slope δ^{50}Ti/δ^{48}Ca = 3. However, some samples do not follow this trend (Jungck et al. 1984) and one of them may even have a negative ^{48}Ca effect (Niederer and Papanastassiou 1984). The data for EK1-4-1 and HAL also do not follow this trend (Lee et al. 1978,1979; Niederer et al. 1980; Fahey et al. 1986). Among the large positive and negative effects observed in hibonites, this trend is again not rigorously followed (Zinner et al. 1986). In short, there exists only a rough correlation between ^{48}Ca and ^{50}Ti effects in the sense that their signs are almost always the same, but no strict mixing trend can be established. The situation for ^{54}Cr effects is similar. So far Cr and Ti data have been published for EK1-4-1, C1 and 6 non-FUN CAI (Birck and Allègre 1984,1985; Papanastassiou 1986). When plotted on a δ^{50}Ti vs δ^{54}Cr plot (Fig. 14.3.12), they also show the same sign but no strict correlation. For the non-FUN CAI, the average ^{54}Cr effect is roughly 2/3 of that for ^{50}Ti. From these, one has the impression that the ^{50}Ti and ^{54}Cr effects for "normal" CAIs are comparable and both are somewhat larger than the corresponding ^{48}Ca effects.

Theoretical interpretation of the Ca-Ti-Cr anomalies in terms of nucleosynthetic models has been directed towards explaining the two salient features from the above data, namely the qualitative correlation among the neutron-rich isotopes and the lack of large nuclear effects for the rare isotope ^{46}Ca. The former suggests production in a neutron-rich environment while the latter requires a physical mechanism to enhance specifically the synthesis of ^{48}Ca but not that of ^{46}Ca. In Table 14.3.2 we list 6 nucleosynthetic studies motivated by the Ca-Ti-Cr anomalies. It should be noted that all of these models are highly idealized and parametrized. Some require physical conditions that have not been demonstrated to exist in any stars. Furthermore, the calculated results are usually subject to uncertainties such as in estimated reaction cross sections. These caveats notwithstanding, the discussion below serves to illustrate the active interaction between nucleosynthesis theory and meteoritic isotope data. The first 3 processes were specifically designed to fit the data for *all* Ca and Ti isotopes in EK1-4-1. The $n\beta$-process (Sandler et al. 1982) envisages the exposure of these elements for about 1000 s to neutrons with a density of 10^{-7} mole/cc and a temperature of about 30 keV (3.5×10^8 K). Under these conditions neutron-rich isotopes are enriched at the expense of neutron-poor ones along a neutron capture path not too far from the β-stability valley, dictated by the competition of (n,γ) reactions and β decays (hence the name). A ten-fold increase of the theoretically estimated neutron-capture rates for ^{46}K and ^{49}Ca, due to possible resonances, is necessary to meet the requirement of a high ^{48}Ca/^{46}Ca production ratio.

The βn_d-process (Ziegert et al. 1985) envisages captures in a high-neutron-density environment (about 10^{-4} mole/cc) on a time scale of 1 s or less. This is followed by β decays towards the stable isobars like the classical *r*-process. The mechanism to enhance ^{48}Ca production specifically is the high β-delayed neutron-emission probability of ^{49}K. Thus, after this nuclide is built up by competition between (n) and β-decay, it will decay primarily by emitting a delayed neutron to form ^{48}Ca instead of ^{49}Ti. Both the normal ^{46}Ca/^{48}Ca ratio and the even lower ratio for EK1-4-1 can be produced by adjusting the neutron density.

The *sp*-process (Harris 1983) was motivated by the apparent fractionation trend favoring light Ca isotopes in EK1-4-1 (lower panel of Fig. 14.3.6). In this model, protons are captured slowly (compared with β decays) in a high-temperature environment ($T > 1.25 \times 10^9$ K) so that ^{40}Ca is hardly affected by (p,γ) because of the rapid reverse reaction of ^{41}Sc(γ,p). The isotopes ^{42,43,44}Ca enter into a local (p,γ) steady state similar to that involving (n,γ) in a classical *s*-process, while all three are gradually destroyed. Both ^{46}Ca and ^{48}Ca are rapidly exhausted by (p,γ), so that a pattern resembling that in the lower panel of Fig. 14.3.6 can be established for Ca. Note that this is a case of nuclear effects mimicking a seemingly linear trend which could be mistaken for a mass-fractionation effect. If the seed abundance is that of the *s*-process taking place in the core He-burning stage inside massive stars and if the proton capture rate for ^{47}Ti is adjusted slightly, the *sp*-process can also reproduce the entire Ti pattern of EK1-4-1.

The remaining 3 processes in Table 14.3.2 are the neutron-rich versions of the well-known *e*(equilibrium)-process or quasi-equilibrium process, which operate near the high-density neutronized core of massive stars (or

TABLE 14.3.2
Observed δ_i vs Theoretical Enhancement Factors (Normalized to ^{48}Ca)

		^{48}Ca	^{46}Ca	^{49}Ti	^{50}Ti	^{53}Cr[a]	^{54}Cr[a]	Astrophysical Sites
Observation:								
EK1-4-1		≡1(13.7)	0.24	0.15	0.28	0.12	0.35	
non-FUN CAI		≡1(0.3)	<3	<1	3	(0.7)	2	
Neutron-process:								
$n\beta$		≡1	0.35	0.20	0.41	0.05	0.05	hydrostatic shell He burning?
$r\beta_{nd}$ (6.5 × 10^{-5} mole/cc n exposure)		≡1	0.42	0.11	0.33	?	?	^{13}C neutron source explosive He burning? (r-process)
Proton process + large fractionation:								
sp (run #6*)		≡1	0.48	1.5	2.1	1.6	8.3	hydrostatic O burning?
n-rich equilibrium:								
quasi- (incomplete Si burnings)	$\eta = \begin{cases} 0.117 \\ 0.120 \end{cases}$	≡1 ≡1	0.9 0.1	0.14 0.02	1.7 0.02	— —	0.01 —	?
Single zone	$\eta = \begin{cases} 0.128 \\ 0.147 \end{cases}$	≡1 ≡1	0.19 0.02	0.7 0.15	14 0.64	0.08 —	8.4 0.18	supernova progenitor (core bounce of massive stars or exploding white dwarf)
Multi-zone mixing	$\eta_{max} = \begin{cases} 0.159 \\ 0.195 \end{cases}$	≡1 ≡1	0.33	0.06	3 0.4	0.03	2.5 0.3	

[a] — indicate values < 0.01; see text for references.

exploding white dwarfs). Since equilibrium conditions are achieved in these processes, their outcome does not depend on reaction rates but only on the binding energies of the nuclei. Therefore, these models are easier to calculate and the results are inherently more reliable. Because of the increase in binding energies at closed nuclear shells, the production of the doubly magic ^{48}Ca (20 protons and 28 neutrons) is highly enhanced relative to that of ^{48}Ca. The most important parameter in these processes is the neutron excess η, defined by $\eta = (N-Z)/(N+Z)$ where N and Z are the number of neutrons and protons (including both bound and free varieties) of the source zone. The η for a nuclide can be similarly defined but N and Z are now the neutron number and the atomic number, respectively, for that nuclide instead. For example, ^{48}Ca has $\eta = (28-20)/(28+20) = 0.167$. It is a distinct feature of the e-process that, in a zone characterized by a particular η, those nuclides having that value of η would be prodigiously produced. Thus, as shown in Fig. 14.3.13, the yield of ^{48}Ca is high in the zone with $\eta = 0.17$ while that for ^{50}Ti and ^{54}Cr is high in zones of lower η (0.11–0.12). Cameron (1979) was the first to realize that equilibrium is a good way to produce large ^{48}Ca/^{46}Ca ratios. He calculated the product abundances of two processes at several high η values: (1) full nuclear statistical equilibrium where the temperature is high enough that all nuclides have reached their equilibrium proportions; and (2) quasi-equilibrium burning of Si at somewhat lower temperature where equilibrium is only achieved in local mass regions but not with free protons, neutrons and alphas. Both processes can meet the requirement of high ^{48}Ca/^{46}Ca and produce Ti primarily consisting of ^{50}Ti. Thus, in principle both could be the source of ^{48}Ca and ^{50}Ti anomalies. He also investigated whether these processes could be the main supplier of solar-system ^{48}Ca. The n-rich e-process at the η values investigated was found to produce too many other n-rich nuclides relative to ^{48}Ca, hence could not be the major supplier for the bulk of solar-system ^{48}Ca. On the other hand, the n-rich Si-burning did not suffer from this shortcoming and thus was favored. Hartmann et al. (1985) carried out more extensive and accurate calculations of the n-rich e-process. They found that mixing of all zones with $\eta < \eta_{max}$ at equal weight would alleviate the problem of overproduction of other nuclides encountered in the single zone e-process described above. Therefore, it was concluded that a group of 8 rare n-rich isotopes including ^{48}Ca could have originated from such multiple zone mixing (MZM) of e-process with η_{max} around 0.1 to 0.2. The above 3 equilibrium processes do not produce n-poor isotopes at all so that the anomalies in those nuclides require additional processes. This is in contrast to the group of capture processes in the previous paragraphs which were designed to fit specifically all the effects in sample EK1-4-1. In view of the multiple-component nature of the Ca-Ti-Cr anomalies, it is not immediately clear that fitting all effects in a single sample is a valid approach.

Table 14.3.2 attempts to confront the models with the data for ^{46,48}Ca, ^{49,50}Ti and ^{53,54}Cr. Following Eq. (1), we tabulate the observed δ_i/δ^{48}Ca and

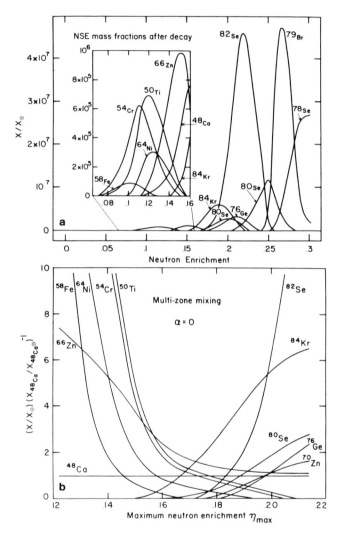

Fig. 14.3.13. Neutron-rich equilibrium nucleosynthesis (figure after Hartmann et al. 1985). (a) Enhancement factors of nuclides as a function of the neutron excess η for the source zone. Production of a nuclide is favored in the zones whose η are similar to that of the nuclide. Thus the abundances for ^{54}Cr and ^{50}Ti peak around $\eta = 0.12$ while that for ^{48}Ca peaks around $\eta = 0.16$. (b) Enhancement factors relative to that of ^{48}Ca for material from the mixing of multiple zones (at equal weights) with $\eta < \eta_{max}$ as a function of η_{max}. Note the covariation of ^{54}Cr and ^{50}Ti relative to ^{48}Ca. Anomalies in ^{64}Ni and ^{58}Fe are expected, especially for samples showing large $\varepsilon(^{50}\text{Ti})/\varepsilon(^{48}\text{Ca})$. Also note that the large enhancement of n-rich isotopes around mass 70 to 85 relative to ^{48}Ca for high η_{max} implies that stellar source zones characterized by these high η must not be ejected during supernova explosions, to avoid oversupplying them to the solar system.

the theoretical $[N_i/N(^{48}Ca)]^*/[N_i/N(^{48}Ca)]^\circ$. The first two rows list the data for EK1-4-1 and a "typical" non-FUN CAI. The data for the latter are necessarily approximate since the ^{48}Ca, ^{50}Ti and ^{54}Cr anomalies are not quantitatively correlated in them. All the theoretical models were proposed before the discovery of ^{54}Cr anomalies. Thus the Cr observation becomes a testing ground for their predictive powers. Clearly, $n\beta$ and n-Si processes failed the test and hence are no longer valid candidates, at least in their present form. The investigators of the βn_d-process did not report their predictions on Cr. They are encouraged to reexamine their models in light of the Cr data, in particular, to see if there exists yet another "accident" in detailed nuclear structures to enhance the production of the progenitors of ^{54}Cr. As mentioned above, the product abundance pattern of the sp-process, being in favor of the proton-rich isotopes, is quite different from the observed nonlinear effects. This is because the nuclear effects happen to mimic the linear trends produced usually by mass fractionation. Thus most of the nuclear effects were removed by the fractionation-correction procedure. The absolute Ca ratios (i.e., uncorrected for linear effects) indeed show this tendency of p-rich isotope enrichments. However, this is not true for the absolute Ti ratios for EK1-4-1. Therefore, although it is still possible that nature happens to superimpose a mass-fractionation effect of the right magnitude in the reverse direction, the Ti data do not support the sp-process. Also, it can be seen from Table 14.3.2 that this process does require quite different dilution factors for different elements in order to explain the data. For example, a factor of 4 between Ti and Cr is needed since the observed Ca and Ti effects are similar in size while the predicted Ti effects are much larger. This is not fatal since chemical differentiation could very well cause such elemental fractionation but such large differences do not seem to be required for the other processes in the table. We are thus left with the n-rich e-process. The single-zone version does not seem to yield comparable Ca-Ti-Cr effects for any single η in contrast to the observations. On the other hand, the MZM model naturally produces comparable effects in all three elements. Therefore, based on the above reasoning, MZM n-rich e-process seems to be the best nucleosynthetic process to explain the anomalies in the n-rich isotopes for elements Ca, Ti and Cr. Furthermore, it is also capable of producing many n-rich rare nuclides between mass 40 and 90 at their solar proportions and thus could have been the main contributor for those whose origin was not previously understood. This is an example where the desire to understand some anomalies in rare tiny meteoritic objects has resulted in a better understanding of the nucleosynthetic origin of many nuclides in bulk solar-system matter. Based on this better understanding, one can estimate how much highly neutronized material that supernovae on the average may eject into the interstellar medium without over-enriching the Galaxy with n-rich nuclides. In turn, such an estimate can be used to constrain the structure and evolution of supernova progenitors as well as the explosion and ejection mechanism (Hartmann et al. 1985). This is a beautiful case where the science of nuclear astrophysics integrates diverse phenomena together to bring us a more coherent picture of the Universe.

Acknowledgment. I thank H. F. Chiu and Y. S. Liu for assistance in manuscript preparation. E. Anders, E. Zinner and D. A. Papanastassiou kindly provided exciting preprints. A constructive and thorough review by S. Niemeyer is gratefully appreciated. This work was supported in part by a grant from the National Science Council of the Republic of China.

REFERENCES

Beer, H. 1986. S-process nucleosynthesis below A = 90. In *Nucleosynthesis and Its Implications on Nuclear and Particle Physics,* eds. J. Audouze and M. Mathieu (Dordrecht: D. Reidel), pp. 263–270.
Begemann, F. 1980. Isotopic anomalies in meteorites. *Rept. Prog. Phys.* 43:1309–1356.
Birck, J. L., and Allègre, C. J. 1984. Chromium isotopic anomalies in Allende refractory inclusions. *Geophys. Res. Lett.* 11:943–946.
Birck, J. L., and Allègre, C. J. 1985. Evidence for the presence of Mn-53 in the early solar system. *Geophys. Res. Lett.* 12:745–748.
Cameron, A. G. W. 1979. The neutron-rich silicon-burning and equilibrium processes of nucleosynthesis. *Astrophys. J.* 230:L53–L57.
Clayton, D. D. 1979. On the isotopic anomalies in samarium. *Earth Planet. Sci. Lett.* 42:7–12.
Clayton, D. D. 1982. Cosmic chemical memory: A new astronomy. *J. Roy. Astron. Soc.* 23:174–212.
Clayton, R. N., MacPherson, G. J., Hutcheon, I. D., Davis, A. M., Grossman, L., Mayeda, T. K., Molini-Velsko, C., and Allen, J. M. 1984. Two forsterite-bearing FUN inclusions in the Allende meteorite. *Geochim. Cosmochim. Acta* 48:535–548.
Esat, T. M., Papanastassiou, D. A., and Wasserburg, G. J. 1980. The initial state of Al-26 and Mg-26/Mg-24 in the early solar system. *Lunar Planet. Sci.* XI:262–264 (abstract).
Esat, T. M., Spear, R. H., and Taylor, S. R. 1986. Isotope anomalies induced in laboratory distillation. *Nature* 319:576–578.
Fahey, A. J., Goswami, J. N., McKeegan, K. D., and Zinner, E. 1985. Evidence for extreme Ti-50 enrichments in primitive meteorites. *Astrophys. J.* 296:L17–L20.
Fahey, A. J., Goswami, J. N., McKeegan, K. D., and Zinner, E. 1986. Al-26, 244-Pu, 50-Ti, REE, and trace element abundances in hibonite grains from CM and CV meteorites. *Geochim. Cosmochim. Acta* 50:445–452.
Harris, M. J. 1983. The *sp*-process and Allende isotope anomalies in calcium and titanium. *Astrophys. J.* 264:613–619.
Hartmann D., Woosley, S. E., and El Eid, M. F. 1985. Nucleosynthesis in neutron-rich supernova ejecta. *Astrophys. J.* 297:837–845.
Hinton, R. W., Davis, A. M., and Scatena-Wachel, D. E. 1987. Large negative Ti-50 anomalies in refractory inclusions from the Murchison carbonaceous chondrite—Evidence for incomplete mixing of neutron-rich supernova ejecta into the solar system. *Astrophys. J.* 313:420–428.
Hutcheon, I. D., Steele, I. M., Wachel, D. E. S., MacDougall, J. D., and Phinney, D. 1983. Extreme Mg fractionation and evidence of titanium isotopic variations in Murchison refractory inclusions. *Lunar Planet. Sci.* XIV:339–340 (abstract).
Ireland, T. R., Compston, W., and Heydegger, H. R. 1985. Titanium isotopic anomalies in hibonites from the Murchison carbonaceous chondrite. *Geochim. Cosmochim. Acta* 49:1989–1993.
Jungck, M. H. A., Shimamura, T., and Lugmair, G. W. 1984. Ca isotope variations in Allende. *Geochim. Cosmochim. Acta* 48:2651–2658.
Lee, T. 1979. New isotopic clues to solar system formation. *Rev. Geophys. Space Phys.* 17:1591–1611.

Lee, T., Papanastassiou, D. A., and Wasserburg, G. J. 1978. Calcium isotopic anomalies in the Allende meteorite. *Astrophys. J.* 220:L21–L25.
Lee, T., Russell, W. A., and Wasserburg, G. J. 1979. Calcium isotopic anomalies and the lack of aluminum-26 in unusual Allende inclusions. *Astrophys. J.* 228:L93–L98.
Lugmair, G. W., Marti, K., and Scheinin, N. B. 1978. Incomplete mixing of products from r-, p-, and s-process nucleosynthesis: Sm-Nd systematics in Allende inclusion EK-1-04-1. *Lunar Planet. Sci.* IX:672 (abstract).
McCulloch, M. T., and Wasserburg, G. J. 1978a. More anomalies from the Allende meteorite: Samarium. *Geophys. Res. Lett.* 5:599–602.
McCulloch, M. T., and Wasserburg, G. J. 1978b. Barium and neodymium isotopic anomalies in the Allende meteorite. *Astrophys. J.* 220:L15–L19.
Niederer, F. R., and Papanastassiou, D. A. 1984. Ca isotopes in refractory inclusions. *Geochim. Cosmochim. Acta* 48:1279–1293.
Niederer, F. R., Papanastassiou, D. A., and Wasserburg, G. J. 1980. Endemic isotopic anomalies in titanium. *Astrophys. J.* 240:L73–L77.
Niederer, F. R., Papanastassiou, D. A., and Wasserburg, G. J. 1981. The isotopic composition of titanium in the Allende and Leoville meteorites. *Geochim. Cosmochim. Acta* 45:1017–1031.
Niederer, F. R., Papanastassiou, D. A., and Wasserburg, G. J. 1985. Absolute isotopic abundances of Ti in meteorites. *Geochim. Cosmochim. Acta* 49:835–851.
Niemeyer, S. 1985. Systematics of Ti isotopes in carbonaceous chondrite whole rock samples. *Geophys. Res. Lett.* 12:733–736.
Niemeyer, S., and Lugmair, G. W. 1981. Ubiquitous isotopic anomalies in Ti from normal Allende inclusions. *Earth Planet. Sci. Lett.* 53:211–225.
Niemeyer, S., and Lugmair, G. W. 1984. Titanium isotopic anomalies in meteorites. *Geochim. Cosmochim. Acta* 48:1401–1416.
Papanastassiou, D. A. 1986. Cr isotopic anomalies in the Allende meteorite. *Astrophys. J.* 308:L27–L30.
Papanastassiou, D. A., and Wasserburg, G. J. 1978. Strontium isotopic anomalies in the Allende meteorite. *Geophys. Res. Lett.* 5:595–598.
Russell, W. A., Papanastassiou, D. A., and Tombrello, T. A. 1978. Ca isotope fractionation on the earth and other solar system materials. *Geochim. Cosmochim. Acta* 42:1075–1090.
Sandler, D. G., Koonin, S. E., and Fowler, W. A. 1982. Ca-Ti-Cr anomalies in an Allende inclusion and the nβ process. *Astrophys. J.* 259:908–919.
Tomkin, J., and Lambert, D. L. 1983. Heavy-element abundances in the classical barium star HR 774. *Astrophys. J.* 273:722–741.
Wasserburg, G. J., Papanastassiou, D. A., and Lee, T. 1980. Isotopic heterogeneities in the solar system. In *Early Solar System Processes and the Present Solar System,* ed. D. Lal (Amsterdam: North-Holland), pp. 144–191.
Ziegert, W., Wischer, M., Kratz, K. L., Möller, P., Krumlinde, J., Thielemann, F. K., and Hillebrandt, W. 1985. Interpretation of the solar Ca-48/Ca-46 abundance ratio and the correlated Ca-Ti isotopic anomalies in the EK-1-4-1 inclusion of the Allende meteorite. *Phys. Rev. Lett.* 55:1935–1938.
Zinner, E. K., Fahey, A. J., Goswami, J. N., Ireland, T. R., and McKeegan, K. D. 1986. Large Ca-48 anomalies are associated with Ti-50 anomalies in Murchison and Murray hibonites. *Astrophys. J.* 311:L103–L107.
Zook, A. C. 1978. A preliminary determination of the relative abundances of the isotopes of zirconium in R Cygni and V Cancri. *Astrophys. J.* 221:L113–L116.

PART 15
Extinct Radionuclides and Nucleocosmochronology

15.1. EXTINCT RADIONUCLIDES

F. A. PODOSEK AND T. D. SWINDLE
McDonnell Center for the Space Sciences

"Extinct radionuclides" are radioactive isotopes with lifetimes of the order of 10^6 to 10^8 yr, long enough to survive the interval between nucleosynthesis and the formation of solids in the solar system but short enough so that they are essentially fully decayed and now extinct in the solar system. There is good evidence for the presence of several such radionuclides in the early solar system: ^{26}Al, ^{53}Mn, ^{107}Pd, ^{129}I, ^{244}Pu, ^{146}Sm. Together with longer-lived radionuclides the abundances of these short-lived species provide significant constraints on nucleosynthetic time scales and the history of solar system materials before they became the solar system. Their rapid decay provides high-resolution chronological information concerning the formation of meteorites and the evolution of their parent bodies. The shortest-lived species is ^{26}Al, which provides the strongest constraints on the very earliest history and prehistory of the solar system; ^{26}Al may also have been sufficiently abundant to be the major heat source for meteorite parent-body metamorphism or igneous differentiation.

15.1.1. INTRODUCTION

In the various nucleosynthetic processes which have contributed to the mixture that now constitutes the solar system, the extent of production of a given nuclide depends on its own particular nuclear properties, on astrophysical conditions during the nucleosynthesis, and on the abundances of seed or target nuclides, but not on whether or not that given nuclide will ultimately decay to some other nuclide. The "products" of nucleosynthesis thus include a large number of radioactive species, with a wide range of lifetimes. At one extreme, there are many cases where the lifetime is sufficiently long that decay in the age of the Galaxy has not significantly depleted present-day abundances below initial nucleosynthetic abundances. There are also cases where the lifetimes are sufficiently short that decay below nucleosynthetic abundances has been substantial but sufficiently long that the radioisotopes are still extant and readily measurable today, e.g., ^{235}U and ^{40}K. At the other extreme, there are many cases where the lifetime is sufficiently short that decay of the species in question is essentially complete in the interval between nucleosynthesis and formation of the solar system, or in the interval between nucleosynthesis and the formation from these nucleosynthetic products of materials accessible for scientific study today.

As commonly understood, "extinct radionuclides" are those which fall between these two extremes: radioisotopes whose lifetimes are sufficiently short that they are not extant today but also sufficiently long that they survived in detectable abundance to the formation of the solar system. The term "extinct radionuclides" connotes a field of investigation as much as it does a group of nuclides. As already implied, detectability is an important parameter; there are a number of radionuclides of appropriate lifetimes which plausibly may have been present at the formation of the solar system but for which no direct evidence has yet been found. The boundaries of this field of investigation need not be nor are they sharply delineated or based on rigorous definition. For example, for the parameters described below for ^{244}Pu, the *present* abundance of ^{244}Pu in many common terrestrial rocks is of the order of one atom per kilogram, which is certainly not extinct; nevertheless, direct observation of "live" ^{244}Pu is not feasible, and for all practical purposes it is "extinct." Conversely, the field is usually understood to be restricted to "primordial" radionuclides—those which were present at the time the solar system first took identifiable form, as opposed to those which have subsequently been made within the solar system. Thus, relatively short-lived species such as ^{14}C and ^{230}Th are certainly extinct in terms of primordial abundance; they are, however, being continually produced within the solar system, ^{14}C by cosmic-ray spallation and ^{230}Th in the decay chain of ^{238}U. Useful as study of such nuclides may be, they are generally studied by different means and to different ends from "extinct radionuclides" and will not be considered further here. Restriction to "extinct primordial radionuclides" is not always a sharp distinction, however, nor an uncontested one; it can be and has been argued that at least some of the radionuclides generally considered primordial were actually produced within the solar system, e.g., by particle bombardment during an early active phase of the Sun.

Since, by definition, an extinct radionuclide cannot be observed directly today, evidence for its prior presence must be indirect. Usually, such evidence appears in the form of a variable relative abundance of the daughter isotope. Such variation is generated if decay of the parent isotope takes place in a variety of environments with different parent/daughter elemental ratios, i.e., chemical fractionation is necessary; since the extinct radionuclides generally have rather low abundances, observation is easiest in samples of very high parent/daughter elemental ratio. The daughter element must have at least two isotopes in order that variable additions of the daughter isotope be detectable; small variations are best detected in elements that have at least three isotopes, for which a specific isotope variation can be distinguished from mass-dependent isotopic fractionation. In at least one case, evidence for the existence of an extinct radionuclide can be physical: the recoiling fragments from spontaneous fission of ^{244}Pu generate "tracks" of lattice disorder which can be revealed by laboratory etching. Whether isotopic or physical, evidence for the existence of an extinct radionuclide is more convincing if evidence can also be found for direct association with the parent.

Preservation of the evidence for the prior existence of an extinct radioisotope requires some sample which is very old—having formed, in the sense of isotopic closure, at a time when the now extinct radioisotope was still extant—and which has not been so disturbed that the evidence is lost. Thus, the study of extinct radionuclides focuses primarily on analysis of meteorites, although evidence for some extinct radionuclides can also be found on the Earth and the Moon.

As will be described below, extinct radionuclides provide information in a variety of contexts: they provide chronological constraints on the formation of the solar system and planetary bodies within it, on the metamorphic history of chondrites, and on the chronology of nucleosynthesis shortly before the formation of the solar system. In all these contexts, chronologies based on short-lived radionuclides have a potentially finer resolution than those based on long-lived extant activities. One extinct radionuclide, ^{26}Al, may also figure prominently as an energy source for early planetary differentiation.

15.1.2. ROSTER

There are some instances in which the evidence for an extinct radionuclide is clear and compelling, others for which the evidence is marginal or controversial, and others for which interesting upper limits can be established. In this section we will survey candidate extinct radionuclides and describe features of special interest. A summary is given in Table 15.1.1.

Iodine-129

^{129}I (half-life 16 Myr) is detected by excesses of its β-decay daughter ^{129}Xe. It was the first extinct radionuclide to be discovered (Fig. 15.1.1) and, correspondingly, offered the first opportunity to put close constraints on the maximum *time* that could have elapsed between nucleosynthesis and the formation of solid bodies in the solar system (Reynolds 1960).

Jeffery and Reynolds (1961) performed a key experiment in irradiating the meteorite Abee with neutrons, thereby producing excess ^{128}Xe by ^{127}I (n,γ) ^{128}I $\rightarrow$ ^{128}Xe; in stepwise heating of the irradiated sample, it was found

TABLE 15.1.1
Short-lived Radionuclides in the Early Solar System

Radio-nuclide	Half-life (Myr)	Decay	Daughter	Abundance
Clear Positive Evidence				
^{146}Sm	103	α	^{142}Nd	^{146}Sm/^{144}S $\sim$ 0.005
^{244}Pu	82	α, SF	fission Xe tracks	^{244}Pu/^{238}U $\sim$ 0.004–0.007
^{129}I	16	β	^{129}Xe	^{129}I/^{127}I $\sim$ 10^{-4}
^{107}Pd	7	β	^{107}Ag	^{107}Pd/^{108}Pd $\sim$ 2×10^{-5}
^{53}Mn	3.7	β	^{53}Cr	^{53}Mn/^{55}Mn $\sim$ 4×10^{-5}
^{26}Al	0.7	β	^{26}Mg	^{26}Al/^{27}Al $\sim$ 5×10^{-5}
Restrictive Upper Limits				
^{247}Cm	16	α	^{235}U	^{247}Cm/^{235}U $<$ 0.004
^{41}Ca	0.13	β	^{41}K	^{41}Ca/^{40}Ca $<$ 10^{-8}

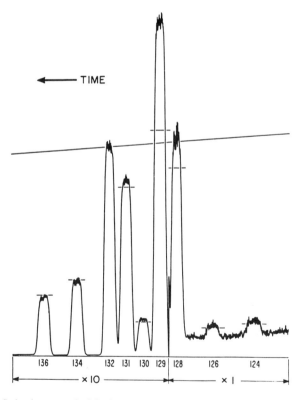

Fig. 15.1.1. Strip-chart record of the isotopic spectrum of Xe in the ordinary chondrite Richarton (from Reynolds 1960). The subhorizontal continuous line shows the overall decrease in signal intensity with time due to pumpout in the mass spectrometer; the short dashed lines at each peak indicate the expected isotopic abundance, relative to ^{132}Xe. The most prominent effect is a large fractional excess of ^{129}Xe, attributed to decay of ^{129}I. This was the first experimental evidence for the presence of a now-extinct radionuclide in the early solar system.

that, at the higher extraction temperatures, there was a close correlation between excess ^{129}Xe and excess ^{128}Xe. This "high-temperature correlation" was taken as clear evidence that both isotopic excesses originated in the same mineral sites in the meteorite (i.e., wherever the element I resides), and thus that the source of the excess ^{129}Xe was indeed an isotope of I, namely ^{129}I, and that the decay had taken place in situ, rather than in some other reservoir possibly long before the formation of Abee.

The early-solar-system abundance of ^{129}I is low, ^{129}I/^{127}I ≈ 10^{-4}. Since the daughter *element* Xe is a noble gas, however, and generally depleted in meteoritic material by several orders of magnitude relative to cosmic abundances, prominent excesses of ^{129}Xe are the rule rather than the exception. The high-temperature correlation first observed by Jeffery and Reynolds has subsequently been observed many times. Because of the near-ubiquity of excess ^{129}Xe in meteorites, and the relative ease with which it can be observed and correlated with ^{127}I, ^{129}I is the only extinct radionuclide on which a chronological methodology with an extensive data base has been founded; this methodology is discussed separately in Chapter 15.3.

Plutonium-244

^{244}Pu (half-life 82 Myr) decays primarily by α-emission, yielding ^{232}Th. Since extant Th is monisotopic, there is essentially no prospect for quantifying or even detecting ^{244}Pu via ^{232}Th excesses. There is, however, an alternative decay by spontaneous fission (branching ratio 1.25×10^{-3}), which is detectable in two ways. One way is in excesses of the heavy isotopes of Xe produced as fission fragments; as is the case with ^{129}I, this is experimentally feasible only because the daughter element Xe is a highly depleted noble gas, so that small additions of nuclear components are more readily identifiable as isotopic variations. Excesses of heavy Kr isotopes attributable to ^{244}Pu fission are also detectable. ^{244}Pu-fission Xe is detectable in bulk chondrites, but it is most readily detected and measured in samples which are relatively enriched in Pu and depleted in normal trapped Xe, such as the igneous achondrites or specific phases of chondrites such as refractory-rich inclusions in carbonaceous chondrites or phosphates in ordinary chondrites (Fig. 15.1.2). The second way in which ^{244}Pu fission is detectable is through "tracks": ~ 10 μm trails of lattice disorder in dielectric grains which are caused by passage of the fission fragments and which can be revealed by selective dissolution in laboratory etching. It is noteworthy that these fossil fission tracks constitute the only case in which detection of the prior presence of an extinct nuclide has been made by means other than isotopic excess. The only other spontaneously fissioning nuclide which can interfere with identification of ^{244}Pu fissions is ^{238}U; the fission branching ratio for ^{238}U is sufficiently low, however, that ^{244}Pu dominates both Xe and track production in samples approximating the canonical 4.5 Gyr meteorite age.

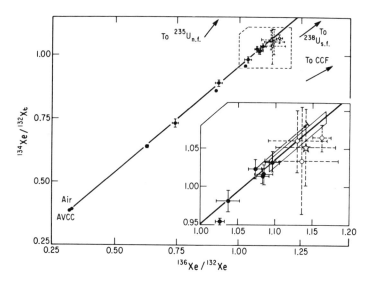

Fig. 15.1.2. Heavy-Xe-isotope correlation diagram for samples in which Xe from spontaneous fission of ^{244}Pu is prominent (from Lewis 1975). These include several meteorites and various extraction steps (labeled by extraction temperature) from St. Séverin "whitlockite". The open circles are calculated fission compositions; the agreement of compositions calculated from meteorite data with that from man-made ^{244}Pu (open star) clearly shows that the responsible nuclide is indeed ^{244}Pu.

Heavy-Xe-isotope effects plausibly attributable to ^{244}Pu were first observed in the Pasamonte achondrite (Rowe and Kuroda 1965). Heavy-Xe-isotopic effects were clearly associated with fission via correlation with tracks in chondritic phosphates by Wasserburg et al. (1969). A clear identification of ^{244}Pu as the fissioning nuclide in both types of sample was made when Alexander et al. (1971) measured the fission Xe yield spectrum of man-made ^{244}Pu and found that it agreed with the meteoritic spectrum.

A major difficulty in the study of ^{244}Pu arises from the circumstance that there is no stable or long-lived isotope of the element Pu. This does not directly hamper measurement of ^{244}Pu, but what is usually wanted is not an absolute abundance but a relative abundance; for ^{244}Pu this necessarily implies normalization to some other element, and this circumstance has impaired both the determination of the cosmic abundance of ^{244}Pu and its use as a solar-system geochronometer.

The cosmic abundance of ^{244}Pu is, in principle, best assessed by measurement of its abundance in chondrites, which nominally contain the refractory elements, including Pu, in unfractionated abundance. For this reason, the cosmic abundance of ^{244}Pu was for some time generally taken to be specified by ^{244}Pu/^{238}U $\cong 0.015$, on the basis of measurement in a single meteorite, St. Séverin, by Podosek (1970a). This is a difficult measurement in chondrites, however. By the usual stratagem of finding samples or phases in which the parent/daughter elemental ratio is chemically enhanced, there are other cases where measurement of ^{244}Pu abundance is relatively easier, but in this case application requires knowledge or assumption of the chemical behavior of Pu, generally in the form of assumption that the ratio of Pu to some other element remains "cosmic" in spite of the chemical fractionation. U and Th are demonstrably unsuitable normalization elements for Pu. Lugmair and Marti (1977) have proposed that Nd is a suitable normalization for Pu and inferred a substantially lower cosmic abundance, ^{244}Pu/^{238}U $\cong 0.004$. Hudson et al. (1982) also recommend a lower value, ^{244}Pu/^{238}U $\cong 0.007$, on the basis of further analyses of St. Séverin.

The principal application of ^{244}Pu as a meteorite geochronometer has been assessment of cooling rates via fission track densities (cf. Pellas and Storzer 1981).

^{244}Pu has a special significance in that it is the shortest-lived nuclide detected within the solar system which assuredly must have been produced in galactic nucleosynthesis prior to the formation of the solar system; its production requires rapid neutron captures and cannot, in any proposed scenario, have been produced within the solar system. Furthermore, most of the canonical abundance of ^{244}Pu was demonstrably "alive", rather than fossil, in the early solar system; this conclusion can be reached by comparing abundances of (arguably fossil) fission Xe with fission tracks in grains which must have formed in the solar system (cf. Drozd et al. 1977).

A Superheavy Element?

As was first observed by Reynolds and Turner (1964), carbonaceous chondrites characteristically display correlated heavy-isotope (especially ^{134}Xe and ^{136}Xe) compositional variations which are evident in stepwise-heating or in meteorite-to-meteorite comparisons. The prominence of heavy-isotope variations of course was suggestive of a fission component, which came to be known widely as CCF (carbonaceous chondrite fission). It was quickly recognized that CCF Xe could not be attributed to decay of ^{244}Pu; its abundance in carbonaceous chondrites is far too high, and its composition is different from that of ^{244}Pu fission Xe. CCF Xe resisted identification with any other known fissioning nuclide, and it also resisted explanation in terms of effects other than fission, such as isotopic fractionation. One of the hypotheses advanced to account for CCF was that it was indeed spontaneous fission, but not from one of the "known" nuclides but rather from an extinct superheavy element from a hypothesized "island of stability" such as ^{298}114 (cf. Anders and Heymann 1969).

The superheavy-element fission hypothesis remained in a sort of limbo for some time, lacking either proof or disproof. The case for superheavy fission was weakened when Manuel et al. (1972) pointed out that excesses of the light isotopes ^{124}Xe and ^{126}Xe, shielded in fission, correlated with the heavy isotope excesses which defined CCF. The correlation persisted in the much larger excesses found by Lewis et al. (1975) and subsequent workers.

The definitive experiment was performed by Lewis et al. (1983), who looked for excesses of isotopes of Ba and the light rare earths which should have been produced along with Xe in spontaneous fission. They found no such excesses; the strongest limit was set by ^{135}Ba, for which any excess was limited to more than an order of magnitude below the amount predicted from heavy Xe isotopes. The superheavy fission hypothesis is now defunct, and the excess heavy (plus light) Xe isotope component is now generally believed to be a stellar nucleosynthetic component which was never homogenized with other gases in the solar system (see Chapter 13.1).

Palladium-107

^{107}Pd (half-life 6.5 Myr) can be detected by excesses of its β-decay daughter ^{107}Ag. Since Ag has only two isotopes, excesses of ^{107}Ag must be relatively large (more than a few per mil) to be identified, and are thus to be expected only in materials of enhanced Pd/Ag ratio. Convincing excesses of ^{107}Ag (as high as a factor of 8) have been found in the metal phase of iron meteorites by Kelly and Wasserburg (1978) and Kaiser and Wasserburg (1983). Correlation of ^{107}Ag excess with Pd/Ag is strong evidence that the excess is indeed due to ^{107}Pd (Fig. 15.1.3).

An important feature of the ^{107}Pd observations is that they are made in differentiated planetary material (iron meteorites), and that the strong ^{107}Ag excesses could not have been generated until after the chemical differentiation

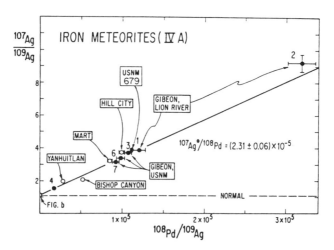

Fig. 15.1.3. Correlation of Ag composition with Pd/Ag in iron meteorites (from Chen and Wasserburg 1984). Correlation of the isotopic effect with the elemental effect in objects produced by planetary differentiation indicates not only that the effect is attributable to ^{107}Pd but also that the ^{107}Pd was "alive" rather than fossil.

enhanced the Pd/Ag ratio. ^{107}Pd is the shortest-lived radionuclide observed to have been present in such differentiated planetary material.

Samarium-146

^{146}Sm (half-life 103 Myr) can be detected by variations in the relative abundance of its α-decay daughter ^{142}Nd. In normal circumstances this requires comparison of phases of different Sm/Nd ratio which formed while ^{146}Sm was still extant. Variations in ^{142}Nd abundance indicative of the presence of ^{146}Sm in Angra dos Reis were reported by Lugmair and Marti (1977) and Jacobsen and Wasserburg (1981), although in both cases the effects were small and only marginally significant. Clear evidence of the presence of ^{146}Sm was found by Lugmair et al. (1983) in large (0.5%) excesses of ^{142}Nd, accompanied by enormous (up to 36%) excesses of ^{143}Nd (from α-decay of ^{147}Sm) in carbonaceous residues from Allende. The Nd isotopic excesses were unsupported by Sm, and were presumably generated by α-recoil into carbon coatings of Sm-bearing grains. The ratio of excess ^{142}Nd to excess ^{143}Nd allows determination of the ^{146}Sm abundance, ^{146}Sm/^{147}Sm $(0.93 \pm 0.10) \times 10^{-3}$. ^{146}Sm is of particular significance among extinct radionuclides in that it is the only short-lived p-process radionuclide observed.

Aluminum-26

^{26}Al (half-life 0.75 Myr) β-decays to ^{26}Mg, and the most reasonable place to search for evidence of its existence is thus phases of high Al/Mg ratio. After some initial claims for the discovery of primordial ^{26}Al, subsequently refuted, clear evidence for excess ^{26}Mg, as a nuclear component correlated with the Al/Mg ratio in the form of an isochron (Fig. 15.1.4), were reported by Lee et al. (1976,1977). Subsequently there have been numerous reports of excess ^{26}Mg, clearly associated with ^{26}Al in the form of isochrons, in refractory-rich minerals and polymineralic inclusions in CM2 and CV3 meteorites.

A substantial data base for different samples and from different laboratories makes a clear and cogent case for ^{26}Al. There are, however, some complexities, and the subject is a very active research area. One of the complexities is that several samples with positive evidence for ^{26}Al yield isochron slopes close to what has become a "canonical" value ^{26}Al/^{27}Al $\approx 5 \times 10^{-5}$, while other samples, ostensibly of comparable antiquity and of sufficient enrichment and range in Al/Mg, yield no positive results, i.e., "isochron" slopes consistent with ^{26}Al/^{27}Al ≈ 0. Such results are readily interpretable in terms of a few Myr of decay, although this has not been verified independently, and there is no simple explanation for an "all or none" occurrence. Subsequently, isochrons of intermediate slope lower than 5×10^{-5} have been observed, but the prevalence of the canonical value remains curious. There remain instances of minerals, notably hibonite, which have very high Al/Mg ratios and which on other grounds are expected to have formed as early or

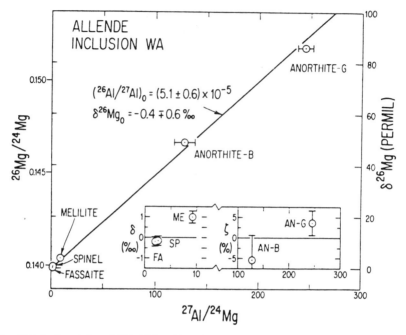

Fig. 15.1.4. Correlation of ^{26}Mg/^{24}Mg with ^{27}Al/^{24}Mg in Allende inclusion WA (figure from Lee et al. 1977). The quality of the correlation clearly indicates that the ^{26}Mg isotopic excesses are due to decay of ^{26}Al.

earlier than any other solar system objects but which nevertheless show no evidence for the presence of ^{26}Al (specifically, ^{26}Al/^{27}Al $\ll 5 \times 10^{-5}$) (cf. Lee et al. 1979; Fahey et al. 1987a); such cases have inspired the view that ^{26}Al was rather heterogeneously distributed in the early solar system. Additional complexities appear in the form of ^{26}Mg excesses which are not well correlated with Al/Mg, and apparent ^{26}Mg/^{24}Mg ratios both above and below the normal terrestrial ratio, indicative of isotopic heterogeneity in Mg; a more thorough review is given by Wasserburg and Papanastassiou (1982).

Of all the "extinct radionuclides" for whose existence in the early solar system there is some positive evidence, ^{26}Al has the shortest lifetime. Its lifetime is sufficiently short, comparable to estimated free-fall times for collapse for a presolar cloud, that it severely constrains models for the formation of the solar system. This has lent support to a number of variant or alternative views to the "orthodox" view that ^{26}Al (along with other short-lived radionuclides) was simply "at large" in the interstellar medium from which the solar system formed. These include a causal relationship between the nucleosynthetic event which made ^{26}Al and the formation of the solar system (the "supernova trigger" hypothesis; Cameron and Truran 1977), late injection of interstellar material after formation of the solar nebula, local (rather than galactic) production through particle irradiation from an early active Sun (cf. Lee 1978), and the hypothesis that ^{26}Al was never "live" at all in the early solar system, but existed only in the form of its fossil ^{26}Mg (see Sec. 15.1.3 below).

^{26}Al also assumes special importance in models for the early solar system because, as Lee et al. (1977) point out, ^{26}Al in the abundance ^{26}Al/^{27}Al ≈ 5 × 10^{-5} is quite adequate as a heat source to melt asteroidal-size bodies which accreted within a few Myr. The problem of providing a rapid heat source to produce ancient igneous meteorites is a long-standing one (Urey 1955), for which ^{26}Al is the favorite solution. The fossil-^{26}Mg viewpoint, explored in Sec. 15.1.3, would of course exclude ^{26}Al as a heat source. The consensus view at present is that ^{26}Al really was a live heat source, primarily on the basis of well-defined correlations between ^{26}Mg/^{24}Mg and ^{27}Al/^{24}Mg (Fig. 15.1.4), especially in macroscopic refractory-rich inclusions that appear to have been melted. As Wasserburg (1985) points out, however, evidence for ^{26}Al has not been found in the igneous meteorites for which it is the presumed energy source for melting, so that solidification must have been delayed by a few Myr.

Calcium-41

^{41}Ca (half-life 0.13 Myr) β-decays to ^{41}K; if it had been present in the early solar system, the most likely place to find evidence for it would be in early-formed phases rich in Ca and poor in K, i.e., in refractory inclusions in carbonaceous chondrites. Searches for the effect of excess ^{41}K have shown this to be a complicated experiment. The most definitive work is that of Hutcheon et al. (1984), who found no definite evidence for the presence of ^{41}Ca and set a strict upper limit ^{41}Ca/^{40}Ca < (8 ± 3) × 10^{-9} in Allende inclusions.

Curium-247

^{247}Cm is an important case because it has the next-shortest half-life (16 Myr) from ^{244}Pu among the transuranic nuclides which absolutely require rapid neutron addition for synthesis and thus surely require presolar stellar nucleosynthesis. It decays by α-emission to ^{235}U; evidence for its existence would thus be found in phases which formed while ^{247}Cm was still extant and which fractionated Cm relative to U. The experimental situation is complicated by the circumstance that U has only two long-lived isotopes, making it difficult to distinguish nuclear effects at ^{235}U from mass-dependent fractionations, and there have been both claims for evidence of ^{247}Cm and refutations of these claims. The presently most definitive report is that of Chen and Wasserburg (1981), who used a double-spike (^{233}U and ^{236}U) to monitor fractionation; they found no positive evidence for ^{247}Cm and, with reasonable assumptions about chemical fractionation between Cm and U, obtained an upper limit ^{247}Cm/^{235}U < 0.004 in Allende.

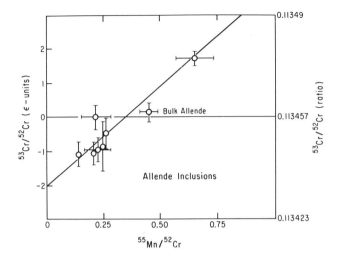

Fig. 15.1.5. Correlation of ^{53}Cr/^{52}Cr with ^{55}Mn/^{52}Cr in bulk inclusions of Allende (figure from Birck and Allègre 1985). The correlation is interpreted to indicate that variability in ^{53}Cr abundance is due to decay of ^{53}Mn. The "initial" ^{53}Cr/^{52}Cr intercept suggests that at least 0.02% (2 ε-units) of present normal ^{53}Cr was originally present as ^{53}Mn.

Manganese-53

^{53}Mn (half-life 3.7 Myr) β-decays to ^{53}Cr. Birck and Allègre (1985) have reported evidence for the existence of ^{53}Mn in Allende inclusions, both as a whole rock isochron (Fig. 15.1.5) for various inclusions, and as an internal isochron within a single inclusion. Their results correspond to ^{53}Mn/^{55}Mn $\approx$ 4 $\times$ 10^{-5} at the time of formation of the inclusions. It is noteworthy that some of the phases they examined were Cr-rich and relatively Mn-poor. In such cases the presence of ^{53}Mn is manifested as a deficiency of ^{53}Cr relative to "normal" values; the results of Birck and Allègre suggest that about 0.02% of present normal ^{53}Cr was still ^{53}Mn at the time of formation of the inclusions.

More?

The roster presented above has described cases for which there is evidence for the presence of a now-extinct radionuclide in the early solar system, as well as some interesting cases of nonoccurrence within significant upper limits. It is not, however, exhaustive, in that there are several other radionuclides with lifetimes in the range consistent with their possible presence in the early solar system, e.g., ^{60}Fe, ^{135}Cs and ^{205}Pb. It is not unreasonable to expect that significant results concerning such radionuclides may emerge from additional systematic searches, improvements in analytical technique, or identification of the right samples with the right chemistry.

15.1.3. COSMIC CHEMICAL MEMORY?

Throughout this chapter it is assumed, implicitly or explicitly, that the radionuclides under consideration were "live" in the early solar system, i.e., actually present as the indicated nuclear species. For long-lived species still extant today and thus directly measurable, such as ^{235}U or ^{40}K, this assumption hardly even needs to be stated; for short-lived species now extinct, such as ^{129}I or ^{26}Al, this assumption merits explicit consideration.

Primordial ^{129}I, for example, is now extinct, and its prior presence in meteorites is inferred only from isotopic evidence, i.e., excesses of its daughter ^{129}Xe. These isotopic excesses are, in a way, "fossils," which have been preserved until today because of the *chemical* state in which the newly produced daughter atoms of ^{129}Xe found themselves, i.e., trapped in some phase of the meteorite and unable to mix with any ambient Xe and achieve chemical equilibrium with its surroundings. The radiogenic ^{129}Xe has fossilized the ^{129}I abundance present at the time its host phase achieved this chemical state, or, in the current parlance, achieved isotopic closure. This is, of course, the basis of isotopic geochronology, but in the absence of any residual abundance of the parent radioisotope it is impossible to assign an absolute age to this event. Although this radiogenic ^{129}Xe may now be found in some specific meteorite, it is possible to argue that the decay of ^{129}I did not take place in that meteorite, but rather in some prior configuration, subsequent to which the host phase of the ^{129}Xe was included in the mechanical assembly of the meteorite in such a way that the isotopic closure was not disturbed. This would constitute a chemical memory: the isotopic effect of radiogenic ^{129}Xe "remembers" a prior history (prior to assembly of the meteorite) because of its chemical state. For the specific case here used as an example, i.e., ^{129}I, it is not only possible to make such an argument, but in certain cases it is reasonable as well. Some meteorites are clearly mechanical assemblages of constituents which have never equilibrated with each other, and in some cases different constituents of the same meteorites have different I-Xe ages (Chapter 15.3); in such cases, it is quite reasonable to argue that these different ages are chemical memories of histories prior to the final assembly of the meteorite lithology.

The concept discussed above has, of course, several parallels in terrestrial geochronology, and is not unique to meteorites. It is noteworthy that the first experimental evidence for the existence of a now-extinct radionuclide, ^{129}I (Reynolds 1960), was closely followed by concern about the extent to which it might be a chemical memory effect (see discussion in Jeffery and Reynolds 1961).

It can be argued that some isotopic effects observed in meteorites are chemical memories of histories that predate the solar system. Presolar histories are now frequently referred to as "cosmic chemical memory"; the isotopic effects themselves comprise what are generally called "isotopic anomalies." The basic ideas involved are simple: different nuclides are synthesized in different nuclear processes in different stars or even different times and places in a single star. Different nucleosynthetic products ejected from these stars can be incorporated in solid circumstellar or interstellar grains before being thoroughly mixed in the interstellar medium. Thus the interstellar medium, even if well mixed on a large scale, can preserve isotopic heterogeneity on a small scale, with gas and solid grains, and different populations of grains distinguished by chemical composition, size and structure, separately "remembering" their nucleosynthetic histories. Some of these memories can survive processing in the solar nebula, e.g., by gas-dust separation, preferential survival of refractory grains, etc., and thereby generate isotopic anomalies in meteorites.

The topic addressed in this section is the question of whether, as elsewhere assumed, the short-lived radionuclides in Table 15.1.1 really were "alive" in the early solar system or, in contrast, were present only as chemical memory fossils, i.e., whether they are isotopic anomalies. It should be made clear that the question is not whether a given isotopic effect is correctly identified as reflecting decay of its assigned parent; these associations seem quite satisfactorily made at present. The question is rather whether or not the decays took place in the solar system. A fossil interpretation for some of the nuclides in Table 15.1.1 has been advocated, notably by D. D. Clayton as part of an eloquent description of the nature and ramifications of cosmic chemical memory (cf. D. D. Clayton 1975,1982,1986; D. D. Clayton and Leising 1987). The "live" vs "fossil" question is clearly of crucial importance for the chronological interpretations which have been based on short-lived radionuclides, both for meteorite chronology (Chapter 15.3) and for presolar nucleosynthetic chronology (Chapter 15.2). An additional relevance for ^{26}Al is its potential role as a heat source for early planetary differentiation.

The question cannot be attacked in general terms. Cosmic chemical memory is clearly an important reality. Even the first isotopic anomalies to be discovered and recognized as such, for example, were immediately associated with interstellar grains (Black 1972; R. N. Clayton et al. 1973). Rather, the question has to be addressed on a case-by-case basis.

As an interesting example, it is instructive to consider the case of neon-E (Black 1972), a noble-gas component which is sufficiently close to monisotopic ^{22}Ne (Eberhardt et al. 1981) that it is generally believed to have been generated by decay of ^{22}Na (see Chapter 13.1). In principle, this situation is qualitatively similar to generation of, say, radiogenic ^{129}Xe from ^{129}I, and ^{22}Na should appear in Table 15.1.1. However, we have not otherwise discussed Ne-E in this chapter, nor is the ^{22}Na-Ne-E association generally considered along with "extinct radionuclides." The reason is simple: the half-life of ^{22}Na is so short (2.6 *years*) that it is impossible to imagine that *live* ^{22}Na in the interstellar medium survived to be *live* ^{22}Na in the solar system. Certainly no live primordial ^{22}Na existed in the solar system; it is conceivable that some ^{22}Na was produced in the solar system, but the more reasonable and prevailing view is that the ^{22}Na was produced in stellar nucleosynthesis and quickly ejected and incorporated into circumstellar grains which were poor in or devoid of the element Ne, and that subsequent decay produced the ^{22}Ne now known as Ne-E. This ^{22}Ne has been preserved as a fossil because of its chemical state (see Chapter 13.1), i.e., it is an isotope anomaly.

A similar scenario—fossil rather than live—was in fact proposed for ^{129}I by D. D. Clayton (1975). This interpretation for ^{129}I seems untenable, and it is useful to compare the evidence for live ^{129}I with the case for Ne-E. One line of evidence is that the observed correlation between radiogenic ^{129}Xe with ^{127}I (actually, ^{128}Xe produced from ^{127}I) not only establishes the parent as an isotope of I but shows that the isotopic effect is correlated with the *element* I. In some cases, notably enstatite chondrites, the radiogenic ^{129}Xe is well correlated with nearly all the I in the meteorite, and a fossil interpretation for radiogenic ^{129}Xe would imply that essentially all the normal solar system I was in unhomogenized presolar grains, which is clearly untenable. For Ne-E, in contrast, there is clearly no correlation with normal solar system Na. A second line of evidence is based on the history that radiogenic ^{129}Xe-bearing objects have experienced within the solar system, which is fully expected to have homogenized Xe and removed any pre-existing association between radiogenic ^{129}Xe and the element I; this includes not only the high metamorphic temperatures of many ordinary chondrites and the even higher temperatures which silicate inclusions in iron meteorites must have experienced, but also at least partial melting in chondrules and enstatite achondrites. Yet all of these objects contain radiogenic ^{129}Xe well correlated with the element I (Chapter 15.3), and the only plausible interpretation is that these correlations were generated by decay of live ^{129}I subsequent to these solar system histories. Clearly no such statement can be made for Ne-E, which does not even correlate with Na and whose maintenance as an isotopically distinct component has been established only in unmetamorphosed meteorites.

By the arguments above, we conclude that primordial ^{129}I was alive in the early solar system and that primordial ^{22}Na was not. Other radionuclides clearly must be evaluated on an individual basis as well. Actually, there is strong evidence that most and probably all of the radionuclides in the upper part of Table 15.1.1 were present in the early solar system in live rather than fossil form. For ^{244}Pu, the strongest evidence is the presence of isotopically distinct fission Xe in presumably melted objects such as igneous achondrites and calcium-aluminum-rich inclusions, and the production of fission tracks in the same objects. ^{107}Pd was clearly alive because its presence is detected in the *metal* phase of iron meteorites. For ^{146}Sm, the evidence is an apparent presence in the igneous meteorite Angra dos Reis and separation of its daughter ^{142}Nd by recoil in correlation with the same recoil separation of daughter ^{143}Nd from parent ^{147}Sm. The most interesting special case is ^{26}Al, discussed below; the case for ^{53}Mn is similar.

In terms of the examples considered above, the situation for ^{26}Al is clearly more nearly like that for ^{129}I than that for ^{22}Na. The quality of the correlation between excess ^{26}Mg and Al in objects such as Allende inclusion WA (Fig. 15.1.4) indicates not only that the source of the excess ^{26}Mg is the element Al, i.e. decay of ^{26}Al, but also that the daughter ^{26}Mg has remained associated with parent element Al since the decay occurred. It should be noted that the daughter ^{26}Mg is correlated with all of the Al in WA, not just a small fraction of it. As stressed by Lee et al. (1977), the history of the macroscopic object WA has involved very high temperatures and at least partial if not total melting. This history, which must have occurred within the solar system, would be expected to destroy any prior association of radiogenic ^{26}Mg with the element Al and homogenize Mg isotopic composition. The conclusion from this line of argument is that the observed correlation (Fig. 15.1.4) must have been generated by ^{26}Al after this high-temperature solar system history and thus that the ^{26}Al was live.

It should be noted that presently available experimental data are consistent with the live ^{26}Al interpretation and do not demand or suggest the alternative fossil interpretation. Arguments for a fossil interpretation are circumstantial and originate in astrophysics rather than geochemistry. One such argument is that in the *present* local interstellar medium ^{26}Al/^{27}Al $\approx 10^{-5}$; live ^{26}Al in the early solar system with ^{26}Al/^{27}Al $\approx 5 \times 10^{-5}$ thus requires that at least locally the interstellar cloud region from which the Sun formed 4.6 Gyr ago was at least five-fold richer in ^{26}Al. Another argument is that the lifetime of ^{26}Al is so short that it severely limits the length of time available from synthesis of ^{26}Al to isotopic closure of solar-system solids; the lifetime is comparable to free-fall times for reasonable molecular cloud densities and substantially shorter than the lifetime of the cloud itself.

As most recently espoused by Clayton and Leising (1987), the fossil interpretation is based on the proposition that the production ratio of ^{26}Al, relative to ^{27}Al, is relatively high, and that the element Al is very refractory, specifically more refractory than the element Mg. Stellar ejecta bearing recently synthesized ^{26}Al are condensed into refractory solids in which Al is enriched relative to Mg. Subsequent decay of ^{26}Al in such grains thus generates an isotopic excess of ^{26}Mg, e.g., of the order of 10% in Fig. 15.1.6a. When this interstellar dust enters the solar system, at some later time long after the ^{26}Al decay, it can be processed ("distilled") at very high temperatures, losing Mg in preference to Al, and thereby increasing the Al/Mg ratio but maintaining the same ^{26}Mg isotopic excess in residual Mg (Fig. 15.1.6b). Subsequently, this material can take up isotopically normal solar Mg in variable amounts. If this uptaken Mg does not homogenize with the residual Mg from stage (b), a mixing line will be generated (Fig. 15.1.6) which has the appearance of an isochron, even though no live ^{26}Al was present during these solar system events.

The key phase of this proposed scenario is the uptake of normal Mg without isotopic homogenization with the residual Mg which "remembers"

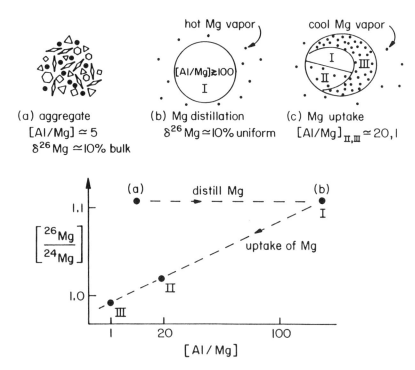

Fig. 15.1.6. Cartoon illustrating the proposed scenario whereby apparent Al-Mg isochrons might be attributed to "cosmic chemical memory" rather than live ^{26}Al (figure from Clayton and Leising 1987). Live ^{26}Al trapped in Al-rich interstellar grains decays and produces excess ^{26}Mg which is preserved in these grains throughout their history in the interstellar medium. When they enter the solar system they can be "distilled" with loss of most of the Mg but preserving the same isotopic excess of ^{26}Mg. Subsequent uptake of normal Mg, which is not allowed to mix with the residual Mg, may generate a mixing line which mimics an isochron.

its prior association with Al in the form of excess ^{26}Mg, since isotopic homogenization of the Mg would produce an isochron of zero slope. Evaluation of this possibility is thus a matter of kinetics. Most investigators consider that it is not possible to introduce the normal Mg without mixing it with the residual Mg enriched in ^{26}Mg, especially in samples like WA (Fig. 15.1.4) for which the data represent well-defined macroscopic mineral phases which must have formed in place (Lee et al. 1977). Most investigators thus consider that ^{26}Al was indeed live rather than fossil.

Additional arguments for live ^{26}Al have been advanced by Fahey et al. (1987b). One argument concerns isotopic fractionation. It would be expected that in Mg loss by distillation (Fig. 15.1.6b) the residual Mg should be fractionated. Enrichment of heavy Mg isotopes is indeed frequently observed in coarse-grained inclusions which show evidence for ^{26}Al, but in such cases

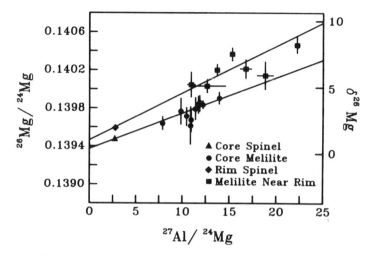

Fig. 15.1.7. Correlation of $^{26}Mg/^{24}Mg$ with $^{27}Al/^{24}Mg$ for ion probe analyses in an inclusion from Efremovka (figure from Fahey et al. 1987b). Data from the core of the inclusion and data from the rim separately define correlations with different slopes and different intercepts. The slope of the rim data correlation indicates $^{26}Al/^{27}Al \approx 4.9 \times 10^{-5}$; the slope of the core data correlation indicates $^{26}Al/^{27}Al \approx 3.7 \times 10^{-5}$.

both the low Al/Mg and the high Al/Mg data show the Mg fractionation. In the fossil model this would require that the uptaken normal Mg (Fig. 15.1.6c) either be fractionated in the vapor phase or become fractionated during uptake, in either case coincidentally to the same degree as the residual Mg. Such a coincidence would be highly unlikely unless the residual Mg equilibrates with the uptaken Mg, but this would then defeat the argument by destroying the correlation between excess ^{26}Mg and Al. A second argument is based on results for a rimmed inclusion from Efremovka shown in Fig. 15.1.7. Fahey et al. present evidence that this rim was formed by condensation from vapor; since the rim material also shows the Al-correlated excess ^{26}Mg, ^{26}Al *must* have been live, because the fossil correlation could not have been preserved in the vapor phase.

15.1.4. PLANETARY OCCURRENCES

Evidence of extinct radionuclides shows up in major planetary bodies— the Earth, the Moon and Mars, as well as in meteorites. This is neither surprising nor extraordinary. As might be expected, however, occurrences are restricted to cases that do not require large parent-daughter chemical enhancements beyond those of ordinary bulk meteoritic material, i.e., cases where the daughter is a noble gas. In practice, evidence for extinct radionuclides in large planetary bodies is restricted to ^{129}I and ^{244}Pu.

In the Earth's atmosphere, the abundance of ^{129}Xe is about 7% higher than would be expected on the basis of models for primordial composition, presumably reflecting the radiogenic daughter of ^{129}I. There is also a contribution from spontaneous fission of ^{244}Pu which amounts to about 4% of atmospheric ^{136}Xe plus commensurate amounts of the other fission isotopes; but identifying the exact amount of fission from ^{244}Pu is considerably more difficult and ambiguous than identifying how much ^{129}Xe is attributable to ^{129}I (Pepin and Phinney 1979; Ozima and Podosek 1983). The excesses of radiogenic ^{129}Xe and ^{244}Pu-fission Xe in the Earth's atmosphere are noteworthy not because they are present at all but because they are not larger than they actually are. The usual interpretation of these smaller-than-expected excesses is that the Earth as a whole did not become closed to Xe loss until substantially later, of the order of 10^8 yr later, than the time of formation of meteorites (cf. Bernatowicz and Podosek 1978; Ozima and Podosek 1983); this is about the same time scale projected for accretion of the terrestrial planets on the basis of orbital dynamics; see Chapter 6.4.

There are also observations of some terrestrial samples which contain excesses of ^{129}Xe relative to atmospheric Xe, notably some midocean ridge basalts (Staudacher and Allègre 1982), and well gases in New Mexico (Smith and Reynolds 1981). In the common interpretation that these excesses represent decay of primordial ^{129}I, such observations are indeed remarkable, since they imply current degassing of reservoirs which have been isotopically isolated from the principal source(s) of the atmosphere since very early in the Earth's history, before decay of ^{129}I was complete. These excesses of ^{129}Xe, relative to atmospheric composition, are further remarkable in that they are *not* accompanied by the excesses of fission Xe (from ^{244}Pu) which would be expected on the basis of such early isotopic isolation. The absence of excess fission Xe accompanying excess ^{129}Xe requires either special constraints on the formation of the Earth or revision of the usually perceived relationship between terrestrial and meteoritic Xe (cf. Ozima and Podosek 1983; Ozima et al, 1985; Staudacher 1987) or that identification of the ^{129}Xe excesses with *primordial* ^{129}I is incorrect.

Excess ^{129}Xe, again presumably from primordial ^{129}I, is also prominent in the atmosphere of Mars (Owen et al. 1977), where ^{129}Xe/^{132}Xe ≈ 2.5. This corresponds to between one and two orders of magnitude greater ratio of radiogenic ^{129}Xe to primordial Xe than in the Earth's atmosphere, and can be interpreted as a reflection of a different overall volatile budget of Mars and/or a different degassing history (cf. Anders and Owen 1977; see also Chapter 7.10).

The Moon also exhibits evidence of ^{244}Pu and ^{129}I, in the form of excesses of daughter Xe in surface-correlated gases in old lunar highland soils and breccias (cf. Drozd et al. 1976; Swindle et al. 1986). The effect is believed to arise via degassing of daughter Xe from the lunar interior into a transient atmosphere, from which it can be implanted into regolith grain surfaces by interaction with the solar wind and electromagnetic fields.

REFERENCES

Alexander, E. C., Lewis, R. S., Reynolds, J. H., and Michel, M. 1971. Plutonium-244: Confirmation as an extinct radioactivity. *Science* 172:837–840.

Anders, E., and Heymann, D. 1969. Elements 112 to 119: Were they present in meteorites? *Science* 164:821–823.

Anders, E., and Owen, T. 1977. Mars and Earth: Origin and abundance of volatiles. *Science* 198:453–465.

Bernatowicz, T. J., and Podosek, F. A. 1978. Nuclear components in the atmosphere. In *Terrestrial Rare Gases*, eds. E. C. Alexander and M. Ozima (Tokyo: Japan Scientific Society Press), pp. 99–135.

Birck, J.-L., and Allègre, C. J. 1985. Evidence for the presence of ^{53}Mn in the early solar system. *Geophys. Res. Lett.* 12:745–748.

Black, D. C. 1972. On the origins of trapped helium, neon and argon isotopic variation in meteorites—II. Carbonaceous meteorites. *Geochim. Cosmochim. Acta* 36:377–394.

Cameron, A. G. W., and Truran, J. W. 1977. The supernova trigger for formation of the solar system. *Icarus* 30:447–461.

Chen, J. H., and Wasserburg, G. J. 1981. The isotopic composition of uranium and lead in Allende inclusions and meteorite phosphates. *Earth Planet. Sci. Lett.* 52:1–15.

Clayton, D. D. 1975. Extinct radioactivities: Trapped residuals of presolar grains. *Astrophys. J.* 199:765–769.

Clayton, D. D. 1982. Cosmic chemical memory: A new astronomy. *J. Roy. Astron. Soc.* 23:174–212.

Clayton, D. D. 1986. Interstellar fossil ^{26}Mg and its possible relationship to excess meteoritic ^{26}Mg. *Astrophys. J.* 310:490–498.

Clayton, D. D., and Leising, M. D. 1987. ^{26}Al in the interstellar medium. *Physics Rept.* 144:1–50.

Clayton, R. N., Grossman, L., and Mayeda, T. K. 1973. A component of primitive nuclear composition in carbonaceous meteorites. *Science* 182:485–488.

Drozd, R. J., Kennedy, B. M., Morgan, C. J., Podosek, F. A., and Taylor, G. J. 1976. The excess fission Xe problem in lunar samples. *Proc. Lunar Sci. Conf.* 7:599–623.

Drozd, R. J., Morgan, C. J., Podosek, F. A., Poupeau, G., Shirck, J. R., and Taylor, G. J. 1977. Plutonium-244 in the early solar system? *Astrophys. J.* 212:567–580.

Eberhardt, P., Jungck, M. H. A., Meier, F. O., and Niederer, F. R. 1981. A neon-E rich phase in Orgueil: Results obtained on density separates. *Geochim. Cosmochim. Acta.* 45:1515–1528.

Fahey, A., Goswami, J. N., McKeegan, K. D., and Zinner, E. 1987a. ^{26}Al, ^{244}Pu, ^{50}Ti, REE and trace element abundances in hibonite grains from CM and CV meteorites. *Geochim. Cosmochim. Acta* 51:329–350.

Fahey, A. J., Zinner, E., and Crozaz, G. 1987b. Microdistributions of Mg isotopes and REE abundances in a Type A CAI from Efremovka: Constraints on rim formation. *Geochim. Cosmochim. Acta*, submitted.

Hudson, G. B., Kennedy, B. M., Podosek, F. A., and Hohenberg, C. M. 1982. The early solar system abundances of ^{244}Pu as inferred from the St. Séverin chondrite. *J. Geophys. Res.*, submitted.

Hutcheon, I. D., Armstrong, J. T., and Wasserburg, G. J. 1984. Excess ^{41}K in Allende CAI: Confirmation of a hint. *Lunar Planet. Sci.* XV:387 (abstract).

Jacobsen, S. B., and Wasserburg, G. J. 1981. Sm-Nd isotopic systematics in Angra dos Reis. *Lunar Planet. Sci.* XII:500–502 (abstract).

Jeffery, P. M., and Reynolds, J. H. 1961. Origin of excess Xe^{129} in stone meteorites. *J. Geophys. Res.* 66:3582–3583.

Kaiser, T., and Wasserburg, G. J. 1983. The isotopic composition and concentration of Ag in iron meteorites and the origin of exotic silver. *Geochim. Cosmochim. Acta* 47:43–58.

Kelly, W. R., and Wasserburg, G. J. 1978. Evidence for the existence of ^{107}Pd in the early solar system. *Geophys. Res. Lett.* 5:1079–1082.

Lee, T. D. 1978. A local proton irradiation model for isotopic anomalies in the solar system. *Astrophys. J.* 224:217–226.

Lee, T., Papanastassiou, D. A., and Wasserburg, G. J. 1976. Demonstration of ^{26}Mg excess in Allende and evidence for ^{26}Al. *Geophys. Res. Lett.* 3:109–112.
Lee, T., Papanastassiou, D. A., and Wasserburg, G. J. 1977. ^{26}Al in the early solar system: Fossil or fuel? *Astrophys. J. Lett.* 211:L107–L110.
Lee, T., Russell, W. A., and Wasserburg, G. J. 1979. Calcium isotopic anomalies and the lack of aluminum-26 in an unusual Allende inclusion. *Astrophys. J. Lett.* 228:L93–L98.
Lewis, R. S. 1975. Rare gases in separated whitlockite from the St. Séverin chondrite: Xenon and krypton from fission of extinct ^{244}Pu. *Geochim. Cosmochim. Acta* 39:417–432.
Lewis, R. S., Srinivasan, B., and Anders, E. 1975. Host phase of a strange xenon component in Allende. *Science* 190:1251–1262.
Lewis, R. S., Anders, E., Shimamura, T., and Lugmair, G. W. 1983. Barium isotopes in Allende meteorite: Evidence against an extinct superheavy element. *Science* 222:1013–1015.
Lugmair, G. W., and Marti, K. 1977. Sm-Nd-Pu timepieces in the Angra dos Reis meteorite. *Earth Planet. Sci. Lett.* 35:273–284.
Lugmair, G. W., Shimamura, T., Lewis, R. S., and Anders, E. 1983. Samarium-146 in the early solar system: Evidence from neodymium in the Allende meteorite. *Science* 222:1015–1018.
Manuel, O. K., Hennecke, E. W., and Sabu, D. D. 1972. Xenon in carbonaceous chondrites. *Nature* 240:99–101.
Owen, T., Biemann, K., Rushneck, D. R., Biller, J. E., Howarth, D. W., and Lafleur, A. L. 1977. The composition of the atmosphere at the surface of Mars. *J. Geophys. Res.* 82:4635–4639.
Ozima, M., and Podosek, F. A. 1983. *Noble Gas Geochemistry* (Cambridge: Cambridge Univ. Press).
Ozima, M., Podosek, F. A., and Igarashi, G. 1985. Terrestrial xenon isotope constraints on the early history of the earth. *Nature* 315:471–474.
Pellas, P., and Storzer, D. 1981. ^{244}Pu fission track thermometry and its application to stony meteorites, *Proc. Roy. Soc. London* A374:253–270.
Pepin, R. O., and Phinney, D. 1979. Components of xenon in the solar system. *Moon and Planets*, submitted.
Podosek, F. A. 1970. The abundance of ^{244}Pu in the early solar system. *Earth Planet Sci. Lett.* 8:183–187.
Reynolds, J. H. 1960. Determination of the age of the elements. *Phys. Rev. Lett.* 4:8.
Reynolds, J. H., and Turner, G. 1964. Rare gases in the chondrite Renazzo. *J. Geophys. Res.* 69:3263–3281.
Rowe, M. W., and Kuroda, P. K. 1965. Fissiogenic xenon from the Pasamonte meteorite. *J. Geophys. Res.* 70:709–714.
Smith, S. P., and Reynolds, J. H. 1981. Excess ^{129}Xe in a terrestrial sample in a pristine system. *Earth Planet. Sci. Lett.* 54:236–238.
Staudacher, T. 1987. Origin of Harding County CO_2 well gases. *Nature*, in press.
Staudacher, T., and Allègre, C. J. 1982. Terrestrial xenology. *Earth Planet. Sci. Lett.* 60:389–406.
Swindle, T. D., Caffee, M. W., Hohenberg, C. M., and Taylor, S. R. 1986. I-Pu-Xe dating and the relative ages of the Earth and Moon. In *Origin of the Moon,* eds. W. K. Hartmann, R. J. Phillips, and G. J. Taylor (Houston: Lunar and Planetary Inst.) pp. 331–357.
Urey, H. C. 1955. The cosmic abundances of potassium, uranium and thorium and the heat balances of the Earth, the Moon, and Mars. *Proc. Natl. Acad. Sci. U. S.* 41:127–144.
Wasserburg, G. J. 1985. Short-lived nuclei in the early solar system. In *Protostars & Planets II,* eds. D. C. Black and M. S. Matthews (Tucson: Univ. of Arizona Press), pp. 703–754.
Wasserburg, G. J., and Papanastassiou, D. A. 1982. Some short-lived nuclides in the early solar system—A connection with the placental ISM. In *Essays in Nuclear Astrophysics,* eds. C. A. Barnes, D. D. Clayton and D. N. Schramm (Cambridge: Cambridge Univ. Press), pp. 77–140.
Wasserburg, G. J., Huneke, J. C., and Burnett, D. S. 1969. Correlation between fission tracks and fission-type xenon in meteoritic whitlockite. *J. Geophys. Res.* 74:4221–4232.

15.2. NUCLEOCOSMOCHRONOLOGY

F. A. PODOSEK AND T. D. SWINDLE
McDonnell Center for the Space Sciences

The relative abundances of radioactive nuclides can be used to determine the time scale for nucleosynthesis of those nuclides and, by extension, of many other nuclides in the solar system. Such time scales can also yield estimates for the age of the universe, and information about the time scale for protosolar collapse.

15.2.1. INTRODUCTION

As is described in Chapters 14.1 and 14.2, the elements heavier than He were produced primarily in stellar (rather than Big Bang) nucleosynthesis and are present in the interstellar medium because some fraction of stellar production is ejected. Time thus appears in the equations describing the composition of the interstellar medium, i.e., nucleosynthesis has a history, elucidation of which is the subject of this section. This history is different in different places and is not the same for all elements. Attention here will be limited to the nucleosynthetic history of the particular patch of the interstellar medium from which our solar system formed about 4.6 Gyr ago.

Time appears in nucleosynthetic history equations in two ways: one is as the parameter describing the dynamics of stellar formation and evolution, infall of primordial material into the galactic disk, etc.; the second way is as absolute clock time describing the decay of radioactive nuclides (we will assume that, at least in the interstellar medium, decay rates are absolutely constant). The scope of this chapter will be limited to the usual understanding of the term "nucleocosmochronology" as establishing constraints on nucleosynthetic history through radioactive nuclides; it should be noted, however, that further powerful constraints follow from considerations of stellar and galactic evolution (Chapter 14.2) and that the radioactive species to be considered are rather minor ingredients in the cosmic soup.

In general terms, the rate of production of some nuclear species i, as a mass fraction of the interstellar material from which the solar system eventually formed, can be rendered as $p_i(t)$, explicitly a function of time and net of losses to astration. Without loss of generality, nucleocosmochronological time is conventionally divided into three periods (Fig. 15.2.1):

1. A period $0 \leq t \leq T$ of active nucleosynthesis, during which $p_i(t) \geq 0$;
2. A free-decay period Δ, during which all $p_i(t) = 0$ and at the end of which the solar system "forms";

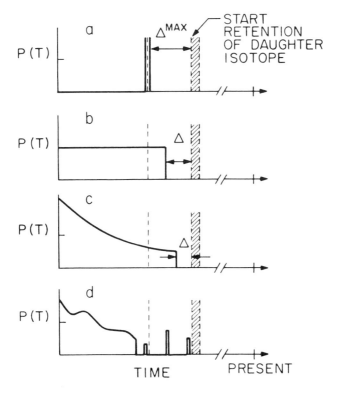

Fig. 15.2.1. Illustration of various models of nucleosynthetic production histories (figure from Wasserburg and Papanastassiou 1982). The ordinate is a generalized nucleosynthetic production rate, the abscissa is time, and the shaded region is the interval over which solar-system solids may have become closed to radioactive parent-daughter systems of interest. Case (a) is essentially instantaneous synthesis of all heavy elements; case (b) is constant production until some cut-off time at which the solar-system-to-be is isolated from further nucleosynthetic input; case (c) is a smoothly decreasing production function; case (d) is a generally decreasing but irregular production with one or more late single episodes of nucleosynthesis separately identified.

3. The ~ 4.6 Gyr age of the solar system.

The time $t = 0$ is the beginning of nucleosynthesis (the Big Bang) and the "age of the universe" is $T + \Delta + 4.6$ Gyr. The concept of a free-decay interval Δ was introduced because, as described later, the concentrations of short-lived radionuclides in meteorites are lower than expected for average interstellar medium at the time of formation of the solar system.

It is conventional, and for short-lived radionuclides now extinct, necessary, to consider radionuclide abundances at time $T + \Delta$, the time of formation of the solar system. Since the abundances of extinct radionuclides in particular must be measured in some physical object (meteorite), formation of the solar system as used here necessarily means formation in the geochemical sense of isotopic closure. Thus, precursor events such as collapse of the protosolar cloud, generation of a nebula, etc., occur within the period Δ. This introduces some ambiguity, because different radionuclide abundances are generally measured in different physical objects and thus do not necessarily refer to exactly the same time. This ambiguity, on a scale of the order of a few (?) Myr, marks the transition from presolar history to planetary history, but within the scope of this chapter will not cause significant difficulty.

With these definitions the solution to the growth equation for a radioactive species i, with decay constant λ_i, is straightforward: at time $T + \Delta$ the abundance A_i is

$$A_i = e^{-\lambda_i \Delta} \int_0^T p_i(t) e^{-\lambda_i(T-t)} dt. \qquad (1)$$

This equation also applies to a stable species as the special case $\lambda_i = 0$, and it can be evaluated by brute force for any chosen $p_i(t)$. To avoid difficulties associated with absolute abundances A_i or absolute production rates p_i, it is customary to consider any given species i in conjunction with some other species j in terms of the ratio $R_{ij} = A_i/A_j$; this is especially useful if j is chosen so that a plausible case can be made that i and j are produced in the same nucleosynthetic process so that $p_i/p_j = P_{ij}$ = constant.

This subject is amenable to and has indeed received substantial formal mathematical treatment (cf. Wasserburg et al. 1969; Hohenberg 1969; Schramm and Wasserburg 1970; Schramm 1973; Symbalisty and Schramm 1981; D. D. Clayton 1984). Any measured ratio R_{ij}, together with its corresponding P_{ij}, provides a constraint on nucleosynthetic history, e.g., determination of Δ or T in conjunction with some model for $p_i(t)$. In principle, each R_{ij} provides an independent constraint; with enough data, then, and some parameterized form of the $p_i(t)$, one could determine nucleosynthetic history to an arbitrary level of detail. In practice, the point of diminishing returns is reached very quickly, and there is little profit in trying to pursue the rigorous mathematics very far. Certainly the number of different radioisotopes from which an R_{ij} could be constructed is not very large, nor is experimental determination perfectly precise. But the limit is more quickly reached because of uncertainty in theoretical prediction of the P_{ij}, and especially because of uncertainty in the assumption that different nucleosynthetic processes will have the same time dependence in $p_i(t)$. Most of the illumination comes from consideration of simple special cases, which is the approach that will be taken below.

15.2.2. LONG-LIVED RADIONUCLIDES

There are a few long-lived radionuclides which are particularly relevant to the overall time scale for nucleosynthesis—roughly speaking, to the problem of the age of the universe, or at least the age of the elements. These radionuclides are measurably extant today, in terrestrial as well as meteoritic materials, and the subject is a venerable one.

As a simple special case, we can consider the extreme case that nucleosynthesis occurred essentially instantaneously, i.e., $T \rightarrow 0$. From Eq. (1) we then obtain

$$A_i = P_i e^{-\lambda_i \Delta} \quad R_{ij} = P_{ij} e^{-(\lambda_i - \lambda_j)\Delta} \tag{2}$$

where P_i is the total nucleosynthetic production of species i. This formulation will be approximately valid whenever production is rapid compared to decay, i.e., $T \ll 1/\lambda_i$. It has been clear for some time that instantaneous nucleosynthesis is unrealistically simple, but it is an interesting limit, since it provides the *maximum* value of Δ that is consistent with a given isotope pair i, j, and also the *minimum* value of $T + \Delta$, i.e., a minimum age of the universe.

Application of an instantaneous nucleosynthesis model to ^{232}Th, ^{235}U and ^{238}U gives results included in Table 15.2.1. The strongest limit is given by the pair ^{235}U and ^{238}U, for which $\Delta_{max} = 1.9$ Gyr. This gives the first-order result that the solar system could not have been formed until at least ~ 2 Gyr in the history of the universe had elapsed; qualitatively, at least this much time is

TABLE 15.2.1
Nucleocosmochronology Data[a]

R_{ij} at $T+\Delta$	P_{ij}	Δ_{max} (Gyr)
^{235}U/^{238}U $= 0.31$	$1.5^{+0.5}_{-1.0}$ 0.6	$1.9^{+0.5}_{-0.7}$
^{232}Th/^{238}U $= 2.5$	$1.9^{+0.2}_{-0.4}$	$2.6^{+3.9}_{-1.7}$
^{187}Re/^{187}Os		5.9 ± 1.8
^{244}Pu/^{238}U $= 0.004 - 0.007$	$0.9^{+0.1}_{-0.2}$	$0.59 - 0.65$
^{129}I/^{127}I $= 1 \times 10^{-4}$	$1.5^{+1.4}_{-0.5}$	0.24

[a] Data as given by Schramm (1982) except for abundance of ^{244}Pu.

needed to allow for ^{235}U decay. A somewhat larger Δ_{max} follows from comparison of the pair ^{238}U and ^{232}Th (Table 15.2.1), but this has a larger uncertainty because it is more sensitive to uncertainty in the production ratio.

Another simple nucleosynthetic model which is interesting to consider as a potential limiting case is that of continuous production: $p_i(t) = p_i =$ constant. In this case Eq. (1) becomes

$$A_i = p_i e^{-\lambda_i \Delta} \frac{1}{\lambda_i} (1 - e^{-\lambda_i T}) . \qquad (3)$$

If relative production rates P_{ij} are presumed known, this model contains two unknowns, Δ and T, and could, in principle, be solved by two pairs, such as ^{235}U and ^{238}U plus ^{238}U and ^{232}Th. In practice, however, sensitivity to uncertainty in the production ratios is too great to permit useful simultaneous determination of Δ and T.

It is clearly impossible to set a rigorous upper limit to T, since one could always hypothesize a model in which a small amount of synthesis occurred an arbitrarily long time ago. Schramm and Wasserburg (1970) stressed that a more useful parameter is a weighted *mean* age for nucleosynthesis; they also showed that for sufficiently long-lived radionuclides, $\lambda_i T \ll 1$, this mean age could be determined from the R_{ij} and P_{ij} independently of any specific model for the time dependence in the $p_i(t)$. The mean age is given by $\Delta_{max} - \Delta$, and in the limit that the real Δ is small compared to Δ_{max}, the mean age is simply Δ_{max}.

Besides the pair ^{238}U and ^{232}Th, the only other pair which is suitable for this kind of analysis is ^{187}Re and ^{187}Os. Uncertainties in the relative production rate as well as in the half-life of ^{187}Re have made application of this pair difficult. Schramm (1982), using the most recently available data for this pair, cites a Δ_{max} of 4.1 Gyr as the lower limit which is consistent with both the ^{238}U-^{232}Th and ^{187}Re-^{187}Os pairs. Added to the 4.6 Gyr age of the solar system, the corresponding minimum age of the universe is 8.7 Gyr.

For a nucleosynthetic model in which production is sharply peaked at the beginning, the total age T will approach the mean age (Δ_{max} for long-lived activities). For a model which is symmetric about the mean (including continuous synthesis; Eq. 3), the total age T will be twice the mean age. Plausible nucleosynthetic histories would be expected to lie somewhere between these extremes, i.e., plausible total ages T would be expected to be between Δ_{max} and $2 \Delta_{max}$. Most current nucleosynthetic models evidently approach continuous synthesis more closely than sudden synthesis; with the values above, the total age of the universe thus approaches a minimum 12.8 Gyr.

It should be noted that all of the nuclides discussed above are produced in the *r*-process, so that the chronology based on them is really an *r*-process chronology. It has not been possible to base similar calculations on long-lived radioisotopes other than those made in the *r*-process. Kazanas et al (1978) have attempted to constrain the "age of the universe" on the basis of the nucleocosmochronology considerations outlined above and also including considerations of Big Bang nucleosynthesis and stellar and galactic evolution; they infer a rather narrow range of 13.5 to 15.5 Gyr for the actual age of the universe.

There are other radionuclides, sufficiently long-lived that they are still extant, which in principle could be applied to provide further constraints on nucleocosmochronological models but which in practice have not been invoked. A notable example is ^{40}K, which presumably has not been exploited because its production rate (and even its production mechanism) is so poorly known. There are many natural radionuclides whose lifetimes are comparable to or greater than the age of the universe; these have not been exploited in nucleocosmochronology because decay must proceed far enough so that abundance at $T + \Delta$ is significantly lower than production in order that they be useful in this context.

15.2.3. SHORT-LIVED RADIONUCLIDES

Short-lived radionuclides (specifically $\lambda_i T \gg 1$), now extinct but present in the early solar system, also play an important role in nucleocosmochronology, especially in constraining the free-decay interval Δ. It is again useful to consider the two extreme cases of sudden and continuous nucleosynthesis. For sudden synthesis, Eq. (2) remains valid; for continuous synthesis, Eq. (3) also remains valid, but for $\lambda_i T \gg 1$ simplifies to the approximate form

$$A_i = p_i \tau_i e^{-\lambda_i \Delta} = P_i \frac{\tau_i}{T} e^{-\lambda_i \Delta} \qquad (4)$$

where $\tau_i = 1/\lambda_i$ is the mean life. Equation (4) indicates that for continuous synthesis, at time T the abundance of species i is equal to the amount synthesized during its last mean life, and is reduced from total production P_i by the factor $\tau_i/T \ll 1$.

As described in Chapter 15.1.2, the first short-lived radionuclide discovered to have existed in the early solar system was ^{129}I (Reynolds 1960). Its short half-life (17 Myr), coupled with an abundance approximately specified by ^{129}I/^{127}I $\approx 10^{-4}$, set a strong upper limit $\Delta_{max} \cong 2 \times 10^8$ yr for the free decay interval Δ. It was also clear that the existence of ^{129}I at this level is incompatible with sudden synthesis of U and Th; with any realistic T of the order of a few Gyr, and any reasonable approach to uniform synthesis, the free decay interval Δ for ^{129}I is reduced to the neighborhood of ~ 100 Myr.

This was an important and exciting result. At the time, meteorites were known to be older than any terrestrial rocks, and were clearly bunched around an upper limit age around 4.5 Gyr, but it still might have been supposed that the solar system had been in existence for some considerable time before the formation of meteorites. By requiring a relatively short time between the last addition of stellar nucleosynthesis and formation of meteorites, the presence of ^{129}I set an upper limit to the age of the solar system and made it clear that meteorites were formed very early in the history of the solar system.

Applying similar considerations to ^{244}Pu, the second short-lived radionuclide to be discovered, will obviously give a Δ which scales by the $\sim 10^8$ yr lifetime of ^{244}Pu, and since this is of the same order of magnitude as the Δ inferred from ^{129}I, these results are qualitatively consistent. In detail, however, it is not easy to account for the abundances of ^{129}I and ^{244}Pu simultaneously in simple nucleosynthetic models in which production is reasonably uniform and/or peaked toward early times.

For a uniform synthesis model with total time T of several Gyr, the ^{244}Pu/^{238}U ratio in average interstellar medium at the time of solar system formation would be expected to be about 0.025 to 0.030, and less if production peaked early in galactic history. Early observations suggested that ^{244}Pu/^{238}U in the solar system was rather high, around 0.03. This clearly corresponds to a short Δ and is incompatible with the ~ 100 Myr Δ based on ^{129}I. Several resolutions of this apparent discrepancy, some of which are described below, are possible. One which has received considerable attention, especially before radionuclides with even shorter lifetimes than ^{129}I were discovered, is to postulate a nucleosynthetic production function $p_i(t)$ with a peak around $t \approx T$, i.e., a "last-minute spike" (cf. Hohenberg 1969; Wasserburg et al. 1969). Introduction of a new free parameter, essentially the strength of the last-minute spike, permits solution for both ^{129}I and ^{244}Pu, consistent with the scale for T derived from longer-lived radionuclides, with the general result of a longer Δ, approaching 200 Myr, with the terminal spike accounting for much or most of the ^{129}I and ^{244}Pu and even a significant fraction of total nucleosynthesis.

It has become seemingly evident that the solar ratio ^{244}Pu/^{238}U was actually substantially lower than the values that prompted the last-minute spike. Ratios in the general neighborhood of 0.01 are reasonably consistent with the ^{129}I abundance and quasi-uniform nucleosynthesis, without the need for the terminal spike, but if the true abundance of ^{244}Pu is near the lower end of the presently likely range (Chapter 15.1.2), the Δ for ^{244}Pu becomes too long to be consistent with the Δ for ^{129}I, assuming coproduction at roughly uniform rates. As noted, there are other possible resolutions of such discrepancies, and there has not been much recent formal tinkering with general galactic nucleosynthetic production functions.

It should be noted that both ^{129}I and ^{244}Pu, like the long-lived radioisotopes of U, Th and Re which are used to constrain the overall time scale for galactic nucleosynthesis, are r-process nuclides. Strictly speaking, then, all these isotopes describe an r-process chronology. In this context it is particularly interesting to consider ^{146}Sm, the only short-lived p-process radionuclide detected in the early solar system. Lugmair et al. (1983) point out that the ratio of ^{146}Sm to ^{144}Sm, another pure p-process isotope, is ^{146}Sm/^{144}Sm ≈ 0.005, and that this is consistent with a production rate of ~0.5 and continuous synthesis (Eq. 4) followed by a free-decay period Δ of around 150 Myr; this is thus quite consistent with the picture of r-process time scales which emerge from ^{129}I and ^{244}Pu.

The astrophysical interpretations of a free-decay period Δ of the order of 10^8 yr remain unclear. Shortly after the discovery of short-lived radionuclides, attention was seemingly focused on the shortness of Δ, as establishing an upper limit on the age of the solar system and formation of meteorites very early in solar system history. More recently, attention has shifted toward an appreciation of how long a 10^8 yr Δ is, substantially longer than time scales inferred from even shorter-lived radionuclides (see below) and also time scales anticipated for protosolar collapse or the lifetime of the solar nebula. One possible view is that this is the approximate time scale for passage of a galactic spiral density wave in the vicinity of the Sun, so that the Sun may have formed during passage of a density wave through a region in which the last major nucleosynthetic production had been during or shortly after passage of the prior wave. Another view, suggested by D. D. Clayton (1983), is that the implicit assumption of a uniform interstellar medium is unwarrantedly simplistic and that an apparent 10^8 yr time scale is essentially a reflection of the time required for transport from major nucleosynthetic sites into the large molecular cloud in which the Sun presumably formed.

15.2.4. VERY SHORT-LIVED RADIONUCLIDES

The nucleosynthetic chronologies considered in the previous two sections involve nucleosynthetic production over a period T of several Gyr and a free decay interval Δ of about 100 Myr, or at least several tens of Myr. The free decay interval Δ is constrained primarily by the abundances of short-lived ^{129}I and ^{244}Pu.

Subsequent to the discovery of ^{129}I and ^{244}Pu, and chronologies based on them, other extinct radionuclides have been discovered, some with shorter half-lives than ^{129}I. Even without formal equations, it is clear that the inferred abundances of these very short-lived species are incompatible with the time scales discussed above, and demand a much shorter Δ. The extreme case is ^{26}Al, of course; there is simply no way that ^{26}Al in the interstellar medium could have survived more than a few Myr of free decay before being incorporated in meteoritic minerals. Even for ^{107}Pd, however, the 6.5 Myr half-life is sufficiently short that its abundance (Table 15.1.1) is incompatible with the free decay time scales based on ^{129}I and ^{244}Pu. Some modification of nucleosynthetic models is clearly required; a number of alternative viewpoints are possible, but a clear consensus has not yet emerged.

One possible resolution of these discrepancies is to postulate that different radionuclide abundances refer to significantly different times; in particular, the longer decay time is demanded by ^{129}I and ^{244}Pu, both of whose abundances are based on retention of the noble gas Xe, and it might be supposed that this did not happen until several to many tens of Myr after the isotopic closures which record the abundances of ^{27}Al, ^{107}Pd and ^{53}Mn. It would be difficult to exclude this possibility rigorously, but it nevertheless seems an unattractive possibility which has not won much favor. There are classes of objects such as refractory-rich inclusions in carbonaceous chondrites, chondrules or igneous meteorites, which have approximately the canonical abundances of ^{129}I and ^{244}Pu and for which long delays before Xe closure seem quite unlikely. Refractory inclusions and chondrules, for example, are believed to have formed as hot small dispersed objects and then cooled quickly; it would be difficult to keep them hot for so long a time in the solar nebula, or in parts of parent bodies which have not been metamorphosed. In many cases such long delays are also not evident in absolute chronologies based on extant radionuclides, notably including K-Ar (^{40}Ar-^{39}Ar) chronologies for which closure for daughter ^{40}Ar should occur at lower temperatures and later times than closure for Xe.

A second possible resolution is to postulate that some of the radionuclides under consideration were never actually present in the solar system, i.e., they were fossils rather than live species. Their measured "abundances" would then yield information about the chemistry of the interstellar medium, but would impose no constraints on nucleosynthetic chronologies. This viewpoint is discussed in Chapter 15.1.3, and we will here assume the validity of the prevalent view that the radionuclides in question, including ^{26}Al, were indeed present in live rather than fossil form in the early solar system.

A third possible resolution, which in one variant or another has won the greatest support, is simply to allow the synthesis of the various short-lived radionuclides to be decoupled, since the synthesis occurs in different astrophysical environments. Decoupling of the nucleosynthetic production functions is particularly reasonable as shorter lifetimes are considered, since injection of nucleosynthetic products into the interstellar medium is ultimately episodic rather than continuous, and it can be expected that the granularity of nucleosynthesis would become progressively more important and evident as progressively shorter times before formation of the solar system are considered.

Thus, it has already been noted that the long-lived nuclides considered in Sec. 15.2.2, along with ^{129}I and ^{244}Pu in Sec. 15.2.3, are all r-process nuclides. It is not unreasonable to suppose that a formation interval Δ of several tens of Myr does indeed characterize the last addition of r-process nuclides to the interstellar medium from which the solar system formed, but that other nuclides made in other stars by other nucleosynthetic processes were added more recently. These would include ^{26}Al, ^{53}Mn and also ^{107}Pd, which can be produced in the s-process as well as the r-process. In such a case there would be no single formation interval Δ; different nucleosynthetic processes, perhaps individual radionuclides, would have their own time scales.

One variant of this viewpoint is that some short-lived radionuclides might have been produced locally, within the solar system, by particle irradiation from an energetically active early Sun (cf. Lee 1978; Wasserburg and Papanastassiou 1982; Wasserburg 1985). Local production of ^{26}Al, in particular, would circumvent the astrophysical constraints involved in getting ^{26}Al synthesized in other stars into the solar system so quickly. Irradiation of the whole solar nebula to produce the canonical level of ^{26}Al can be excluded (Lee 1978) but the irradiation might have been more localized. At this point, the possibility of producing some short-lived radionuclides within the solar system is more a suggestion than a quantitative model and is quite difficult to assess. While ^{26}Al could be produced by protons, it is unclear whether species requiring neutrons, e.g., ^{107}Pd, could also be produced, nor is it clear whether the Sun would have produced the required proton irradiation.

Another variant, expressed by Wasserburg and Papanastassiou (1982), is based on the observation that ^{26}Al, ^{107}Pd and ^{129}I (and now ^{53}Mn) all have early solar system abundances roughly of the order of 10^{-4} times production. If each of these nuclides had independent histories including different free decay intervals Δ, arriving at roughly the same level would be coincidental. An alternative view would be that most of the observed abundances of these nuclides were added to solar system material together in a single batch shortly before formation of solar system solids (the time scale would be limited to no more than a few Myr by ^{26}Al). This would correspond to adding of the order of 10^{-4} to 10^{-3} solar masses of fresh material. It is suggestive that this is about the same order of magnitude as many isotopic anomalies.

These two variants are not necessarily mutually exclusive; synthesis of small amounts of radionuclides within the solar system would provide a natural explanation for "batch" production. It is unclear, however, whether proton irradiation synthesis would produce these diverse nuclides in proportions similar to those expected for the usual stellar interior syntheses, and the same reservations as noted above about local synthesis apply.

If a freshly synthesized batch of material were indeed added to solar system material, whether very shortly before formation of the solar nebula or even within the solar nebula, in such quantity as to produce normalized abundance around 10^{-4} for several short-lived radionuclides, then whether or not this fresh batch contained any ^{244}Pu or ^{146}Sm at all, it would be unlikely to be the major source of these radionuclides, since their normalized abundances are much higher than 10^{-4} (Table 15.1.1). It would also not have added enough ^{247}Cm to exceed the present upper limit to its abundance (Table 15.1.1).

If freshly synthesized material added to the solar system mix shortly (less than a few Myr) before formation actually was the dominant source of ^{129}I, then production of ^{129}I would be decoupled from production of ^{244}Pu, and chronologies based on their assumed coproduction (cf. Sec. 15.2.3) would be invalid. Actually, however, this would not make much of a qualitative difference. In a continuous synthesis model the formation interval Δ based on ^{244}Pu/^{238}U $\approx$ 0.007 is close to a half-life of ^{244}Pu, around 90 Myr; with allowance for uncertainty in the production ratio and especially the abundance of ^{244}Pu, Δ could be somewhat lower or significantly higher. It is not reasonable to postulate that this level of ^{244}Pu was added in a single late event, no more than a few Myr before solar system formation, since in this case we would expect comparable abundance of ^{247}Cm, which is not observed. A free decay interval of a few half-lives of ^{247}Cm, i.e., a few tens of Myr, is required. As long as ^{247}Cm is coproduced with ^{244}Pu, this order of free decay time seems required for at least the trans-bismuth elements even if not for the r-process in general. The abundance of the p-process nuclide ^{146}Sm, even if it does not require this order of time scale, is certainly consistent with it.

15.2.5. HETEROGENEITY?

In one fashion or another, the principal reason for interest in short-lived radionuclides is usually chronology, and in general terms the shorter the lifetime the finer the chronological resolution. One aspect of chronological information provided by short-lived radionuclides concerns the history of solar-system materials before they were organized to form the solar system, which thus far has been the principal focus of this chapter. The complementary aspect is chronological information about events within the early solar system, which will be considered in Chapter 15.3.

An experimental measurement can provide a value for the abundance of a now-extinct radionuclide, and more specifically its ratio to another isotope, e.g., ^{26}Al/^{27}Al or ^{129}I/^{127}I. This does not provide any information directly on the absolute age of isotopic closure. It *can* provide information for relative chronologies. If two samples have measured ratios R_1 and R_2 then

$$\frac{R_2}{R_1} = e^{-\lambda \delta t} \tag{5}$$

where δt is the difference in formation times of the two samples. The high value of λ, in comparison with the decay constants for the extant radionuclides on which absolute chronologies are based, is what permits a fine resolution for the relative chronology δt.

This application requires the assumption that at any given time the ratio R is the same in all materials considered, and does not change except by decay. Avoidance of chemical fractionation effects is the reason why normalization of the radionuclide to another isotope of the same element is used where possible, e.g., ^{26}Al to ^{27}Al, or ^{129}I to ^{127}I. It is possible to suppose, however, that the isotope ratio in question may not have been uniform, e.g., that at a given time the ratio of ^{26}Al to ^{27}Al may not have been the same in now-accessible samples because these two isotopes have separate nucleosynthetic histories and were never homogenized with each other. In such a case, ^{26}Al could not be used as a chronometer. Measurements of ^{26}Al/^{27}Al would then provide information about the distribution of ^{26}Al, which is itself a useful objective. Studies of the *distribution* of an extinct radionuclide, however, require an assumed or measured independent chronology in order to separate variations due to heterogeneity from variations due to radioactive decay.

It is now widely believed that ^{26}Al was indeed heterogeneously distributed in the early solar system, i.e., that ^{26}Al/^{27}Al at any given time was not everywhere the same (cf. Wasserburg and Papanastassiou 1982). This follows from observations that some refractory-rich samples show evidence for ^{26}Al and others have no or much less ^{26}Al, but are not thought to have formed so much later that this should be attributable to decay (Chapter 15.1.2). This view seems reasonable, but in the absence of an independent chronology it is difficult to be conclusive. Recently, Fahey et al. (1987) have presented direct evidence for heterogeneous distribution of ^{26}Al. As seen in Fig. 15.1.7, the core and rim of an Efremovka inclusion define separate isochrons with different ^{26}Al/^{27}Al, with the rim having the higher value. Since the rim must have formed after the core, it must have formed in a different region of the nebula, which was characterized by a higher ^{26}Al/^{27}Al than wherever the core of the inclusion formed.

A heterogeneous distribution of ^{26}Al is not unreasonable, since it must have been added very shortly before solar system formation or possibly synthesized within the solar system. The hypothesized association of several short-lived radionuclides (Sec. 15.2.4) suggests, but does not establish, that other extinct radionuclides, e.g., ^{129}I, may also have been inhomogeneously distributed, whether because of recent additions or local production. It remains unclear whether this is actually so, since the association is only hypothesized, and even if it is valid, each radionuclide would have to be evaluated separately because each has its own chemistry as well as its own nucleosynthetic history. The special case for ^{129}I is discussed in Chapter 15.3.

REFERENCES

Clayton, D. D. 1983. Extinct radioactivities: A three-phase mixing model. *Astrophys. J.* 268:381–384.

Clayton, D. D. 1984. Galactic chemical evolution and nucleocosmochronology: Standard model with terminated infall. *Astrophys. J.* 285:411–425.

Fahey, A. J., Zinner, E., and Crozaz, G. 1987. Microdistributions of Mg isotopes and REE abundances in a Type A CAI from Efremovka: Constraints on rim formation. *Geochim. Cosmochim. Acta,* in press.

Hohenberg, C. M. 1969. Radioisotopes and the history of nucleosynthesis in the galaxy. *Science* 166:212–215.

Kazanas, D., Schramm, D. N., and Hainebach, K. 1978. A consistent age for the universe. *Nature* 274:672–673.

Lee, T. D. 1978. A local proton irradiation model for isotopic anomalies in the solar system. *Astrophys. J.* 224:217–226.

Lugmair, G. W., Shimamura, T., Lewis, R. S., and Anders, E. 1983. Samarium-146 in the early solar system: Evidence from neodymium in the Allende meteorite. *Science* 222:1015–1018.

Reynolds, J. H. 1960. Determination of the age of the elements. *Phys. Rev. Lett.* 4:8.

Schramm, D. N. 1973. Nucleo-cosmochronology. *Space Sci. Rev.* 15:51–67.

Schramm, D. N. 1982. The r-process and nucleocosmochronology. In *Essays in Nuclear Astrophysics*, eds. C. A. Barnes, D. D. Clayton and D. N. Schramm (Cambridge: Cambridge Univ. Press), pp. 325–353.

Schramm, D. N., and Wasserburg, G. J. 1970. Nucleochronologies and the mean age of the elements. *Astrophys. J.* 162:57–69.

Symbalisty, E. M. D., and Schramm, D. N. 1981. Nucleocosmochronology. *Rept. Prog. Phys.* 44:293–328.

Wasserburg, G. J. 1985. Short-lived nuclei in the early solar system. In *Protostars & Planets II*, eds. D. C. Black and M. S. Matthews (Tucson: Univ. of Arizona Press), pp. 703–754.

Wasserburg, G. J., and Papanastassiou, D. A. 1982. Some short-lived nuclides in the early solar system—A connection with the placental ISM. In *Essays in Nuclear Astrophysics*, eds. C. A. Barnes, D. D. Clayton and D. N. Schramm (Cambridge: Cambridge Univ. Press), pp. 77–140.

Wasserburg, G. J., Schramm, D. N., and Huneke, J. C. 1969. Nuclear chronologies for the galaxy. *Astrophys. J.* 157:L91–L96.

15.3. IODINE-XENON DATING

T. D. SWINDLE AND F. A. PODOSEK
McDonnell Center for the Space Sciences

The most readily and widely studied of the extinct radionuclides in meteorites is ^{129}I, and there is an extensive data base for meteorite chronology based on this isotope, but also significant uncertainty about how to interpret many of the data. If the data are interpreted as a straightforward chronology, a time span is inferred for most meteorite classes that appears too long for the events being dated to have taken place in the nebula.

15.3.1. PRINCIPLES

The only extinct radionuclide for which a chronology has been proposed and extensively tested is ^{129}I. Despite the availability of data from more than 75 meteorites, the question of whether the I-Xe system is a chronometer and, if so, what it is dating have not been conclusively answered.

Jeffery and Reynolds' (1961) experiment to demonstrate that excess ^{129}Xe ($^{129}Xe^*$) was the result of in situ decay of ^{129}I pioneered the technique still used in most I-Xe studies: neutron irradiation followed by analysis of Xe released in stepwise heating. Typically, the ratio of ^{129}Xe to some Xe isotope not produced in the irradiation, such as ^{130}Xe or ^{132}Xe, is plotted against the ratio of ^{128}Xe to that same isotope (see Fig. 15.3.1). If the excess ^{128}Xe and ^{129}Xe are derived from I that had a uniform isotopic composition, then the data points will define a straight line, with the slope proportional to the ratio of ^{129}I to ^{127}I at the last time Xe isotopes were in equilibrium. The trapped $^{129}Xe/^{130}Xe$ (or $^{129}Xe/^{132}Xe$) ratio is determined by the intercept of the correlation line with an assumed $^{128}Xe/^{130}Xe$ (or $^{128}Xe/^{132}Xe$) value, usually the

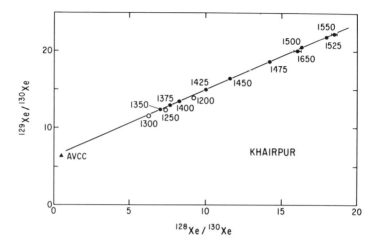

Fig. 15.3.1. Correlation of ^{129}Xe (from decay of ^{129}I) with ^{128}Xe (produced from ^{127}I by neutron irradiation) for the enstatite chondrite Khairpur (figure from Kennedy 1981). The nine highest-temperature extractions (solid points) define a single line while lower-temperature extractions (open points and other points not plotted) fall to the right of the line. The slope of the line is proportional to the ^{129}I/^{127}I ratio at the time of Xe isotopic closure. The intercept of the correlation line with a vertical line representing the ^{128}Xe/^{130}Xe ratio of the average carbonaceous chondrite value (AVCC, Eugster et al. 1967) gives the ^{129}Xe/^{130}Xe ratio of the trapped Xe at the time of isotopic closure.

carbonaceous chondrite average value (AVCC) determined by Eugster et al. (1967).

Hohenberg et al. (1967) analyzed the data from 10 chondrites (assuming a single trapped Xe composition), and found that the initial ^{129}I/^{127}I ratios varied by < 10%. This "sharp isochronism" corresponds to a time difference of <2 Myr, assuming variations in initial I isotopic composition are solely the result of decay of ^{129}I. Podosek (1970) showed that when corrections are made for spallation and fission contributions to the Xe spectrum, and when the isotopic composition of the trapped Xe is not assumed to be known, some of these meteorites do have measurably different initial I compositions, corresponding to age differences as large as 4 or 5 Myr. Since then, the spread in initial I compositions has expanded. The range of reported values is now about a factor of 20 (corresponding to more than 60 Myr). Furthermore, the assumption of homogeneous initial I composition from one object to another has been questioned (Jordan et al. 1980; Crabb et al. 1982).

Early I-Xe results were most frequently discussed in terms of formation intervals (Chapter 15.2.3), assuming various models of I nucleosynthesis (Reynolds 1960; Turner 1965). After the realization that there was little difference in initial I compositions from one meteorite to another (Hohenberg et al. 1967), results usually have been discussed in terms of ages relative to other meteorites. In particular, many studies have given ages relative to Bjurböle, a chondrite in the original "sharp isochronism" suite, because Bjurböle's I-Xe system gives reproducible results (Hohenberg and Kennedy 1981). The meteorite St. Séverin has also been used for this purpose but, unfortunately, it is apparently inhomogeneous (Rison and Zaikowski 1980).

For the remainder of this chapter, the initial I isotopic composition will be discussed either in terms of relative ages or in terms of $\mathbf{R_0}$, the ratio of ^{129}I to ^{127}I at the time of Xe isotopic closure. Note that a higher $\mathbf{R_0}$ corresponds to an earlier age.

15.3.2. CHRONOLOGY BY ^{129}I?

For the I-Xe system to work as a chronometer, the radiogenic Xe we measure today must have gotten into meteorites from in situ decay of ^{129}I, and this ^{129}I must have been homogeneously mixed with ^{127}I throughout the solar system. Whether those two requirements are likely to have been met depends to a large extent on when and where the ^{129}I was made. Thus, it is important to consider the source of ^{129}I. Approaching the problem from the other direction, we can also look for correlations that might indicate that the system is (or is not) a clock.

Sources of ^{129}I

A common assumption is that ^{129}I was produced in r-process nucleosynthesis in galactic sources (Reynolds 1963). It may have been produced continuously (Wasserburg et al. 1960), in a late supernova explosion (Cameron and Truran 1977), or through some combination of those two (Cameron 1962). If it is produced by galactic sources, the ^{129}I/^{127}I ratio at the time of the formation of the solar system can be used to constrain models of the history of nucleosynthesis of the material in the solar system (see Section 15.2.3).

Alternatively, the production site for ^{129}I might have been within the solar system. Fowler et al. (1961) proposed that ^{129}I (and other short-lived radionuclides) might have been produced in an early irradiation of icy planetesimals by an active young Sun. While this specific model is now defunct (cf. Reynolds 1967), variant local irradiation models for production of ^{129}I, along with ^{26}Al and ^{107}Pd, can still be entertained (Section 15.2.4). If ^{129}I was indeed produced locally, then it seems likely that some or all of the variations in $\mathbf{R_0}$ are the result of incomplete mixing of this freshly synthesized material.

Nonchronological Models of I Isotopic Variations

Even if ^{129}I was not produced locally, there is not universal agreement that I was isotopically well mixed, or even that the ^{129}I was ever alive in the early solar system. D. D. Clayton and others have proposed a number of mechanisms that might produce variations in R_0 that have no chronological significance.

The most extreme proposal (Clayton 1975) was that there never was any live ^{129}I in the early solar system, that fossil ^{129}Xe* came into the solar system in interstellar grains and apparent isochrons just represent mixing between these interstellar grains and ^{129}Xe*-free solar-system material (Section 15.1.3). The strongest evidence against this theory comes from samples where most of the ^{129}Xe* and iodine are correlated (Jordan et al. 1980; Niemeyer 1979; Wasserburg and Huneke 1979). In these samples, if interstellar grains contain most or all of the ^{129}Xe*, then all or most of the I must be in the same interstellar grains to preserve the correlation. For samples that are almost certainly not interstellar (e.g., the sodalite-rich Allende chondrule studied by Wasserburg and Huneke [1979]), the fossil grain theory breaks down.

In another model that might not require live ^{129}I, Huneke (1976) suggested that apparent isochrons could be produced by diffusive mixing during the stepwise heating experiment itself. However, this could only work if the correlated Xe is only a small fraction of the total gas (Drozd and Podosek 1977). While that is sometimes the case, there are samples, such as those listed above, where the ^{129}Xe*/^{128}Xe* ratio is constant throughout virtually all the heatings.

A more plausible variation of the fossil-grain theory stipulates that the live ^{129}I in the solar system arrived in interstellar grains, and was never thoroughly mixed with the I (in gas form) already in the nebula. In this scenario, variations in R_0 are largely the result of variations in the abundance of interstellar grains at different times and places in the solar nebula (Crabb et al. 1982; Clayton 1983). Crabb et al. noted that this theory would be consistent with their observation that R_0s among carbonaceous chondrites seem to correlate inversely with I, Br and Cd content. They argued that as time progressed in the solar nebula, the ratio of (high ^{129}I/^{127}I, ^{127}I-poor) dust to (low ^{129}I/^{127}I, ^{127}I-rich) gas decreased. Thus, the carbonaceous chondrites with lower R_0s did form later, but Crabb et al. estimate a total time span of less than 1 Myr (based on the expected duration of the nebula) rather than the 10 Myr that would be required if R_0 were truly a chronometer.

One drawback with this theory is that it does not address the common trend of increasing trapped ^{129}Xe/^{132}Xe ratio with decreasing R_0. This trend, which will be discussed in more detail later, is difficult to produce in a gas-dust mixing model. In fact, Clayton (1983) pointed out that some ^{129}I-rich interstellar grains might have rather large ^{129}Xe/^{132}Xe ratios, particularly if they formed shortly after nucleosynthesis (with an ^{129}I/^{127}I ratio of order unity) and were closed systems for millions of years. If this were the case, one would expect no correlation between R_0 and ^{129}Xe/^{132}Xe (which is what Crabb et al. found for the carbonaceous chondrites), or else a positive correlation. But an inverse trend is more commonly seen. Another problem is that the inverse correlation of R_0 with I content is not seen in the data on enstatite chondrites (Kennedy 1981), iron meteorite silicates (Niemeyer 1979) or Chainpur chondrules (Swindle 1986), although such a correlation does exist for the L chondrites of Shukolyukov et al. (1986).

Another variation on the chemical memory theme would have the ^{129}I enriched in the gas (Clayton 1980). If this were the case, the ^{129}I would eventually end up concentrated on the smaller grains (which have higher surface-to-volume ratios). Clayton (1980) suggested that this might lead to systematic variations among meteorite classes that formed in different parts of the solar system and to higher R_0s in (fine-grained) matrix samples than in other materials. The predicted systematic variations, if they occur, are masked by other effects, because variations within types are larger than variations among types. Also, although an Allende matrix sample yielded a higher R_0 than many Allende chondrules and inclusions (Swindle et al. 1983b), a sample of Bjurböle matrix (Caffee et al. 1982b) and two samples of Chainpur matrix (Podosek 1970; Swindle 1986) fell well within the range of values determined for chondrules from the same meteorites. Again, this scenario would not seem to explain variations of R_0 with ^{129}Xe/^{132}Xe.

Iodine isotopic variations are larger than those in almost any other element (Shima 1986), but R_0 does not seem to correlate with isotopic compositions of other elements where anomalies are seen (Kennedy 1981). Some of the most dramatic isotopic anomalies are in FUN inclusions (Chapter 14.3), but in the two FUN inclusions for which R_0 has been determined, EK 1-4-1 (Papanastassiou et al. 1978) and EGG-3 (Swindle et al. 1987), the I isotopic composition is indistinguishable from that in other Allende objects, although the uncertainties could conceal variations of 10 to 50%.

Even if the I-Xe system is not usually controlled by isotopic inhomogeneity, it is possible that there was some inhomogeneity present. In three chondrules out of more than 40 studied, high-temperature extractions were observed with apparent R_0s up to 2% lower than those of lower-temperature extractions from the same chondrule (Caffee et al. 1982b; Swindle et al. 1983b). These could be explained either as inhomogeneity or as artifacts produced by mixing, as suggested by Kuroda (1976) or Rison and Zaikowski (1980). Iodine isotopic inhomogeneity at the percent level would be consistent with variations seen in other elements (Shima 1986), but would introduce an uncertainty of only a million years or less into I-Xe ages, which have a span of more than 50 Myr.

Correlations of I-Xe Ages with Other Chronometers

The question of whether variations in R_0 have chronometric significance is clearly still open to debate. The most conclusive proof that the I-Xe system works as a clock would be a demonstration that I-Xe ages correlate with ages determined using some other chronometer. However, no absolute chronometer has a comparable precision. In two studies where I-Xe and ^{40}Ar-^{39}Ar ages were compared, there was not even a qualitative correlation (Hohenberg et al. 1981; Kennedy 1981). This was interpreted as evidence that the I-Xe system is dating higher-temperature events, or is less affected by post-formational metamorphism.

It is difficult to correlate I-Xe ages with the Rb-Sr (or initial Sr) or U-Pb ages discussed in Chapter 5.2. In part, this is because the conditions that lead to precise U-Pb or initial Sr ages (high U/Pb or Sr/Rb ratios) often occur in I-poor meteorites (such as the eucrites and Angra dos Reis) for which the I-Xe system is difficult to interpret, while the meteorites for which I-Xe ages can be determined usually do not give precise enough U-Pb or Rb-Sr ages to make comparisons useful. The discovery of large variations in R_0 among Chainpur chondrules has led to the hope that such a comparison might now be feasible (Swindle et al. 1986). However, since it is not known what event the I-Xe system is dating, even assuming that it is a clock, I-Xe ages might not correlate with ages determined by other chronometers.

Correlations of I-Xe Ages with $^{129}Xe/^{132}Xe$ Ratios

In the absence of a direct comparison with an absolute chronometer, the relationship between R_0 and the trapped Xe isotopic composition may provide the key to understanding the I-Xe system. The $^{129}Xe/^{132}Xe$ ratio in the trapped component, which can be found on a three-isotope plot (Fig. 15.3.1), should evolve with time, just as the initial Sr isotopic composition does (Chapter 5.1). Two samples that evolved in the same system, but underwent Xe isotopic closure at different times, should have different trapped $^{129}Xe/^{132}Xe$ ratios.

Several workers have noted a trend of higher $^{129}Xe/^{132}Xe$ ratios with later apparent I-Xe ages in studies of related samples (Podosek 1970; Jordan et al. 1980; Kennedy 1981; Shukolyukov et al. 1986; Swindle 1986). In general, nothing more quantitative than a trend should be expected, since different meteorites presumably evolved in environments with different I/Xe ratios. However, within the enstatite chondrites, Kennedy (1981) found that the trapped Xe isotopic compositions are consistent with closed-system evolution within each of the two types of enstatite chondrites, and open-system evolution from one type to the other (Fig. 15.3.2). Since it is difficult to generate the observed trends through simple gas-dust mixing, their existence argues that the I-Xe system is dating something.

The lowest trapped $^{129}Xe/^{132}Xe$ ratios are, surprisingly, lower than the value of about 1.0 inferred for the solar wind from analyses of lunar breccias, by as much as 3σ (Neimeyer 1979; Kennedy 1981; Jordan et al. 1980). It is difficult to explain the presence of trapped $^{129}Xe/^{132}Xe$ values lower than the solar-wind value (Podosek 1970; Drozd and Podosek 1976). Since the Sun presumably has an I/Xe elemental ratio on the order of unity, if it started with an $^{129}I/^{127}I$ ratio of about 10^{-4}, the decay of ^{129}I would not cause a perceptible change in the $^{129}Xe/^{132}Xe$ ratio. Lunar breccias contain surface-correlated

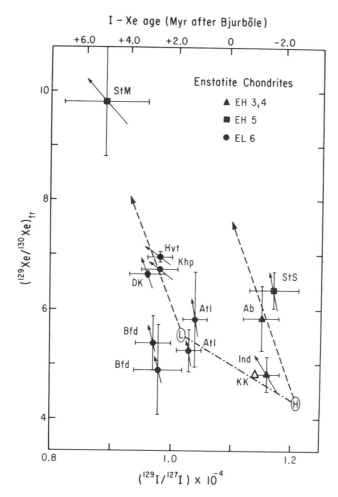

Fig. 15.3.2. Plot of evolution of enstatite chondrite $^{129}I/^{127}I$ and (trapped) $^{129}Xe/^{130}Xe$ (figure from Kennedy 1981). Note from top axis that time moves from right to left. The solid arrows correspond to the trajectory that would be followed in closed system evolution, based on the high-temperature $^{127}I/^{130}Xe$ ratios. Ovals represent the initial $^{129}I/^{127}I$ and $^{129}Xe/^{130}Xe$ ratios inferred for the EL and EH chondrites from apparent "whole rock" isochrons. Dashed and dashed-dot lines represent Kennedy's proposed evolutionary trajectory, in which meteorites within each type are related by closed system evolution, while the two groups are related to one another by open system evolution (with a lower $^{127}I/^{130}Xe$ ratio).

radiogenic Xe, so determinations of the solar-wind ^{129}Xe/^{132}Xe ratio from lunar samples may be too high (Basford et al. 1973), but the effect is probably small enough that the true solar-wind value is no less than about 1.0. Thus if some meteorites contain trapped Xe with a ^{129}Xe/^{132}Xe ratio much less than this, the explanation is probably Xe isotopic inhomogeneity in the early solar system. Drozd and Podosek (1976) suggested that low trapped ^{129}Xe/^{132}Xe ratios might be found if much of the ^{129}Xe in the solar system came in as dust and, before this dust vaporized and contributed its ^{129}Xe to the gas, some meteorite parent bodies formed in regions depleted in this dust. Clayton (1977) suggested that such effects ("ghosts"), which might be found in other elements as well, are evidence for I isotopic variations being the result of incomplete mixing of interstellar material.

Alternatively, the lowest apparent ^{129}Xe/^{132}Xe ratios could be artifacts, perhaps involving thermal disturbance of the I-Xe system (Podosek 1970; Rison and Zaikowski 1980). Many of the most extreme values are poorly defined, so this sort of disturbance might be more common than has been previously recognized. Whether all the apparent trapped ^{129}Xe/^{132}Xe ratios that are less than 1.0 can be explained this way is another question. If so, it may mean that the I-Xe system, already described as "puzzling" (Jordan et al. 1980), may be even more difficult to interpret correctly than is currently believed.

Correlations with Chemical Properties

Kennedy (1981) found that EH chondrites (generally E4 and E5) had $\mathbf{R_0}$s about 15% higher (4 Myr earlier) than ELs (Fig. 15.3.3). Furthermore, the trapped Xe compositions were consistent with a chronological interpretation (Fig. 15.3.2). Kennedy (1981) argued that while it was not possible strictly to rule out isotopic heterogeneity as the cause of these effects, it seemed unlikely.

Niemeyer (1979) found that $\mathbf{R_0}$ in silicate inclusions from IAB iron meteorites correlated with Ga, Ge and Ni contents of the metal. He also argued that it would be easier to envision these correlations occurring if the variations in $\mathbf{R_0}$ are due to decay of ^{129}I than if they are just the result of isotopic inhomogeneity.

As discussed above, Crabb et al. (1982) also found correlations of $\mathbf{R_0}$ with chemical properties in CO3 and CV3 carbonaceous chondrites. However, they concluded that the variations in $\mathbf{R_0}$ could not be true ages, since they were too large to represent nebular time scales and CO3 and CV3 meteorites have undergone little thermal processing.

The situation for Chainpur chondrules is similar to that of the carbonaceous chondrites, because these chondrules show an even larger variation in $\mathbf{R_0}$ and also come from a reputedly primitive meteorite. In the Chainpur chondrules, $\mathbf{R_0}$ does seem to correlate positively with refractory lithophile abundance (Swindle 1986) which would seem to be consistent with a mixing model

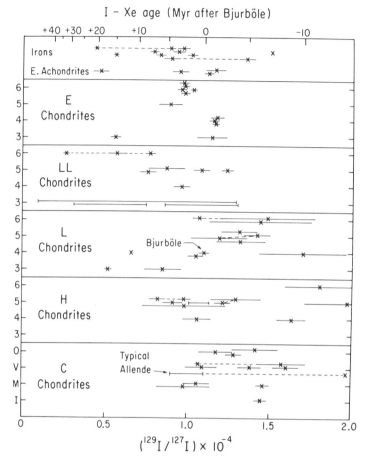

Fig. 15.3.3. Compilation of $^{129}I/^{127}I$ ratios (or apparent I-Xe ages) from Appendix 5. Time moves from right to left. For meteorites where different analyses (or analyses of different components) yield different results, the range is plotted (or distinct values are connected with a dashed line). For Allende, only a typical range and the most extreme value are plotted.

in which a refractory, ^{129}I-rich material mixed with a halogen-rich, ^{129}I-poor material. However, the inverse correlation with halogen abundance that led Crabb et al. to argue for a mixing model interpretation is not present in the Chainpur chondrules. The situation for the enstatite chondrites is even worse for such a model, since R_0 correlates positively with I abundance and negatively with refractory lithophile abundance. Swindle (1986) suggested that the correlations of R_0 with chemistry in Chainpur chondrules are related to the ease with which Xe can become re-equilibrated, since K-Ar ages, which presumably reflect partial loss of radiogenic Ar, tend to correlate with the same parameters (but not with $\mathbf{R_0}$).

15.3.3. RESETTING THE CLOCK

If the I-Xe system is a clock, to make sense of it we must know what event or events can reset it. A number of factors bear on this question.

Where is the ^{129}Xe*?

Since I is a trace element, and a very volatile one at that, little is known of its geochemical properties (Reed 1971). Numerous studies have attacked, or at least discussed, the problem of the location of radiogenic ^{129}Xe, but the carrier is known with confidence only in rare cases.

In the Allende meteorite, much of the ^{129}Xe* is contained in sodalite ($Na_8(Al_6Si_6O_{24})Cl_2$), where I presumably substitutes for Cl. Sodalite was first suggested because of its melting temperature (Zaikowski 1979) and solubility in HNO_3 (Zaikowski 1980). More recently, Kirschbaum (1985a,b) found that ^{129}Xe* and Cl are correlated in Allende inclusions.

The other case where the location of the ^{129}Xe* is known with some certainty is in type E6 (enstatite) chondrites. There, the carrier phase has been identified as enstatite on the basis of its density and solubility in acids (Crabb and Anders 1982).

Kirsten et al. (1978) argued that phosphates contain I in some ordinary chondrites, since they found that a phosphate separate from Shaw had more than 10 times as high a concentration of ^{129}Xe* as a bulk sample. However, the phosphate separates that they analyzed from other ordinary chondrites had no more ^{129}Xe* than bulk samples (cf. Jordan et al. 1980), and even in Shaw, mass balance calculations show that most of the I is not in phosphates.

Another logical place to look for ^{129}Xe* is in sulfides. Goles and Anders (1962) concluded that I is somewhat chalcophile, and Lewis and Anders (1975) and Srinivasan et al. (1978) suggested that lower-temperature ^{129}Xe* release peaks in Orgueil "magnetite" and Abee, respectively, might be from sulfides. However, the results of several I-Xe studies suggest that sulfides are usually depleted in ^{129}Xe*. Jeffery and Reynolds (1961), in their pioneering work on Abee, argued from the lack of correlation of ^{129}Xe* or ^{128}Xe* with (presumably Te-derived) ^{131}Xe* that if Te was concentrated in the sulfides, the high-temperature I was not. More direct evidence against a sulfide carrier phase was found in later studies. Merrihue (1966), Reynolds et al. (1980) and Niemeyer (1979) found lower concentrations of ^{129}Xe* in sulfides than in silicates in mineral separates from Bruderheim, Allende and IAB iron meteorites, respectively. Kerridge et al. (1979a) and Murty and Marti (1987) found no evidence of any ^{129}Xe* in high-purity sulfide samples from Orgueil and

Cape York. Also, Wacker and Marti (1983) found that the ^{129}Xe* content of Abee was not affected when sulfides were chemically removed.

Reynolds et al. (1980) concluded that the ^{129}Xe* resides in a minor phase spottily included in bulk phases, and that the trends from one mineral to another result from different affinities for this minor phase. This is compatible with the proposal of Wood (1967), who suggested that the high-temperature I-Xe correlation is established not by diffusion from mineral grains but by the rupturing of "cages." These "cages" would be sites within minerals from which neither I nor Xe could escape until melting occurred. Thus the I-Xe system would be a high-temperature clock.

Two experiments by the Berkeley group seem to support the "cage" hypothesis. High-temperature correlation lines in Abee and Allende were unaffected by pre-irradiation heating to 900°C and 1200°C, respectively, despite a significant loss of Xe (Hohenberg and Reynolds 1969; Rison and Zaikowski 1980), although an intermediate-temperature isochron in the latter sample was severely disturbed. On the other hand, Srinivasan et al. (1978) argued that the stability of the correlation line in Hohenberg and Reynolds' Abee sample was simply the result of a fortuitous choice of preheating temperatures that melted one carrier mineral completely but left another unaffected.

The idea of "cages" originally arose because the I-Xe system seems to be remarkably stable against thermal resetting, even though I and Xe are both extremely volatile. For example, meteorite cooling rates would seem to require tens of millions of years differences in closure times, if closure occurs at a few hundred degrees centigrade. Wood's (1967) "cages" might provide a mechanism for Xe isotopic equilibration to cease at higher temperatures than Fe and Mg diffusion within minerals. The I-Xe system also has been suggested to be more stable than the Rb-Sr or K-Ar systems. Swindle et al. (1983a) pointed out that matrix samples give I-Xe ages that are similar to those of chondrules, while matrix ages in the Rb-Sr or ^{40}Ar-^{39}Ar system tend to be hundreds of millions of years younger than chondrule ages. Furthermore, they noted that the Rb-Sr and ^{40}Ar-^{39}Ar systems of Bjurböle are disturbed, yet the I-Xe system is so well behaved it can be used as a standard.

One way to quantify whether the I-Xe system is a high-temperature system would be to find the closure temperature for the Xe responsible for the correlation. Closure temperatures calculated or estimated from ^{129}Xe* release curves range from 400°C to 1000°C (Lewis and Anders 1975; Hohenberg et al. 1981; Hudson 1981; Kennedy 1981; Caffee et al. 1982b). One type of calculation assumes diffusion from spherical grains and some specific cooling rate (Hudson 1981), while another type assumes that Xe isotopic closure occurred when a presumed carrier phase condensed. The first type requires the assumption that release of ^{129}Xe* is controlled by diffusion (not "cages"), while the second assumes a knowledge of the carrier phase (which is generally not known). In any case, calculated Xe closure temperatures are always higher than those calculated for the ^{40}Ar-^{39}Ar system in the same samples, since the correlated Xe usually does not begin to appear until more than 90% of the ^{40}Ar has been released.

In summary, the site of the I responsible for the high-temperature I-Xe correlation is still not known in most cases. This makes it extremely difficult to determine what events reset the I-Xe clock simply by calculating the thermal history of those events. A further complication is the possibility that some nonthermal mechanisms can disturb or reset the I-Xe system.

Mechanisms for Resetting the I-Xe System

To reset the I-Xe system, it is necessary to completely equilibrate the Xe isotopes, since the parameter determined in I-Xe studies is the I isotopic composition at the last time Xe isotopes were in equilibrium. In most I-Xe studies, it has been implicitly or explicitly assumed that the I-Xe system has only been reset by formation of objects or by later thermal events. Obviously, if a sample is heated sufficiently, the Xe isotopes will be equilibrated. What is less obvious is how hot the sample has to be (see above). The Xe isotopes should be in equilibrium in material that has been molten (e.g., to form chondrules) or material that has condensed from a gas, although how long they will remain in equilibrium will depend on the closure temperature which, of course, is unknown. Metamorphism should also be capable of resetting the I-Xe system if it is sufficiently intense. Again, it is not clear how much heat is sufficient, but most estimates of peak metamorphic temperature in equilibrated chondrites (Dodd 1981; see also Chapter 3.3) are higher than most of the estimates of Xe closure temperatures cited above.

Many meteorites were presumably subjected to shock events during or after accretion (see Chapter 3.6). Artificial shock has been shown to be capable of disturbing the I-Xe system today, reducing the range in both ^{129}Xe/^{132}Xe and I/Xe ratios (Fig. 15.3.4). Therefore, it is worth considering whether natural shock could have completely reset the I-Xe clock. This would be the case if equilibration of Xe isotopes could be complete without complete homogenization of I with Xe. Although this seems reasonable, the only shock data available (Caffee et al. 1982*a*) are inconclusive. Then, the question is whether the shocks experienced are sufficiently intense to reset the I-Xe system. Some Chainpur chondrules have apparently experienced shocks of 50 to 70 kbar (Ruzicka 1986), and since shock effects were present in the shocked Bjurböle samples at 70 kbar (Caffee et al. 1982*a*), Swindle (1986) suggested that a 60 Myr spread in I-Xe ages might be a spread in shock ages. They noted that R_0 is positively correlated with the spread in I/Xe ratios in both the Chainpur chondrules and Crabb et al.'s (1982) carbonaceous chondrites, another set of samples where it is difficult to attribute the spread in R_0 to thermal effects.

Fluid alteration is another process whose effects on the I-Xe system have only recently been considered. In Allende, where there are signs of alteration

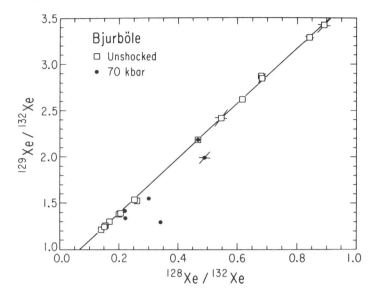

Fig. 15.3.4. Demonstration of the effects of shock on the I-Xe system (data from Caffee et al. 1982a,b). Open symbols are from an unshocked Bjurböle sample, while filled symbols are from a sample artifically shocked to 70 kbar. Both samples were irradiated, receiving the same neutron fluence. Note that the points for the shocked sample do not fall on the same correlation line, show more scatter about any line that could be fit to them (i.e., they would define an apparent correlation line of lower quality), and show less spread in both ratios, suggesting isotopic homogenization of Xe and homogenization of I with Xe.

involving a fluid (Meeker et al. 1983), most objects have multiple isochrons. These properties might be related, particularly since Allende may not have been reheated strongly enough to reset the I-Xe clock. In Semarkona chondrules, preliminary results (Swindle and Grossman 1987) suggest that the I-Xe system might be dating aqueous alteration, particularly if I is concentrated in the glassy mesostasis. Two radial pyroxene chondrules, where the higher surface-to-volume ratio of the glass could make it more susceptible to such alteration, have I-Xe ages 10 Myr later than three porphyritic chondrules.

All this suggests that while the I-Xe system may be harder to reset thermally than might be expected from the volatility of I and Xe, it might be easier to reset by processes such as fluid alteration or shock than has been considered previously.

Disturbances of the I-Xe System

The I-Xe system can be disturbed, but not completely reset by heating (Rison and Zaikowski 1980) or shock (Caffee et al. 1982a). In both cases, the apparent isochrons of samples that had been artificially disturbed are of lower quality than those of the original sample (see Fig. 15.3.4). This suggests that low-quality isochrons should be viewed with suspicion. Also, both effects apparently are capable of producing artificially old ages as artifacts.

Iodine contamination may also affect the I-Xe system, even at high-temperature sites. The problem seems to be particularly severe for Antarctic meteorites, in which most of the I is probably terrestrial contamination (Dreibus and Wänke 1983). The two Antarctic meteorites on which I-Xe studies have been performed both give late apparent ages (Honda et al. 1983). This means that the large Antarctic meteorite collection may not be usable for I-Xe studies.

Another aspect of the I-Xe system that can cause difficulties in interpretation is the fact that the ratio of radiogenic ^{129}Xe to I-derived ^{128}Xe is usually lower in low-temperature extractions. Jeffery and Reynolds (1961), in the first analysis of Xe from a neutron-irradiated meteorite, discussed this effect. However, they said, "It is not necessary to propose different sites for I and ^{129}Xe to explain this effect, since any of three plausible mechanisms could be responsible: prior diffusion of ^{129}Xe over the long lifetime of the meteorite, slight surface contamination of the specimen by terrestrial iodine, or recoil effects in the (n,γ) reaction." Diffusive loss, though difficult to demonstrate experimentally, is still considered a possibility, and I contamination has been demonstrated in one sample (M. W. Caffee, personal communication). However, the recoil energy is small enough (< 220 eV) that recoil is not likely to affect the correlation except in extremely fine-grained material (Drozd and Podosek 1977). Whether or not the cause of the excess ^{128}Xe at low temperatures is known, it is sometimes difficult to tell when all the uncorrelated ^{128}Xe has been removed.

15.3.4. IMPLICATIONS OF I-XENON RESULTS

Having considered the many ways in which the I-Xe system can be disturbed, including the uncertainty in how to interpret the low-temperature excess ^{128}Xe*, it might seem that one would never find a high-quality isochron. This is not the case.

In a number of studies, isochrons have been found which involve several well-determined isotopic ratios. In many of these, the low-temperature points are clearly resolved or absent. It is the existence of these high-quality isochrons that leads to the interpretation of lower-quality data as isochrons, although the latter should be interpreted with caution.

We have tabulated the results of I-Xe studies on samples from 79 meteorites, recalculating R_0s with respect to a common set of standards ($R_0 = 1.095 \times 10^{-4}$ for Bjurböle, $R_0 = 0.58 \times 10^{-4}$ for whitlockite-free St. Séverin, etc.), using a half-life of 15.7 Myr for ^{129}I (Emery et al. 1972). These results are given in Appendix 5 and Fig. 15.3.3. In the following subsections, we will discuss the implications of the I-Xe ages that have been obtained. We will explicitly assume that the I-Xe system can be used as a chronometer (i.e., that "I-Xe ages" are really ages), although we recognize that this assumption is unproven, as discussed above.

The Span of I-Xe Ages

I-Xe ages span 10 Myr or more for most meteorite classes. As Crabb et al. (1982) have pointed out, this means that I-Xe ages, if they are ages, probably are not dating any nebular process, since the solar nebula probably lasted no more than about 1 Myr. In the case of the Chainpur chondrules (Swindle et al. 1987b), this means that either (1) chondrule formation was not a nebular process (but see Chapters 9.3 and 9.4), or (2) I-Xe ages of individual chondrules must be dating some event other than chondrule formation.

This span is consistent with estimates of the duration of metamorphism and/or accretion, since absolute-age chronometers such as Rb-Sr and ^{40}Ar-^{39}Ar, with uncertainties of 10 Myr or more, generally see only hints of age differences among most chondrites (Chapter 5.2). However, it is not straightforward to explain I-Xe results as metamorphic ages. If chondrites were metamorphosed in a parent body with an onion-shell structure, where the more equilibrated (higher petrographic type) chondrites were closer to the center, a correlation between I-Xe age and petrographic type would be expected, with I-Xe ages of type 6 (the most equilibrated) chondrites having later ages. As can be seen from Fig. 15.3.3, no such correlation is found.

I-Xe ages may be consistent with other models of meteorite metamorphism, even though not with the simplest onion-shell model. Because metallographic cooling rates do not correlate with metamorphic type either, Scott and Rajan (1981) proposed that either metamorphism occurred in smaller planetesimals prior to parent-body accretion or that parent bodies were broken up and reassembled during metamorphism. Then, no correlation between petrographic type and I-Xe age would be expected. However, I-Xe ages might be expected to correlate with metallographic cooling rates. This correlation is not present either, although unequilibrated ordinary chondrites, which tend to have lower $\mathbf{R_0}$s, also tend to have lower metallographic cooling rates (Wood 1979). In fact, Chainpur, with the lowest metallographic cooling rate in the tabulation of Wood (1979) contains chondrules with the lowest $\mathbf{R_0}$s. This would be consistent with a scenario in which equilibrated chondrites were metamorphosed in smaller planetesimals (and thus I-Xe ages, cooling rates and type are unrelated). The thermal histories of the unequilibrated chondrites might have been distinct from those of the equilibrated chondrites, leading to the unequilibrated chondrites' later I-Xe ages and lower metallographic cooling rates.

The fact that many of the lowest $\mathbf{R_0}$s (latest I-Xe ages) for ordinary chondrites are found in type 3 material suggests that I-Xe ages might have been established during accretion of meteorite parent bodies or by surface processes, if the type 3 material accreted last and was nearest the surface.

This might explain the large ranges seen among Chainpur chondrules, Semarkona chondrules and carbonaceous chondrites, none of which have been extensively reheated. The span in ages of equilibrated meteorites (typically 10–20 Myr) could reflect either the duration of metamorphism or nonthermal post-metamorphic effects, if the metamorphism was more rapid than that.

Alternatively, the lower R_0s in unequilibrated chondrites might reflect a greater sensitivity of their I-Xe systems to low-temperature thermal events. Swindle et al. (1987b) suggested that this might happen if metamorphism created stable "caging" sites that could not be disturbed at low temperatures, but pointed out that the release patterns for ^{129}Xe* seem similar for chondrules from the equilibrated meteorite Bjurböle and the unequilibrated meteorite Chainpur.

Specific Cases

In some systematic studies of related samples, the I-Xe ages obtained have been incorporated in specific models. For example, Kennedy (1981) pointed out that the fact that EL chondrites have later I-Xe ages than EH chondrites is consistent with models where the EL chondrites condensed and accreted later than EH chondrites (Baedecker and Wasson 1975), rather than earlier (see, e.g., Takahashi et al. 1978). Wasson et al. (1980) used the correlation between metal content and I-Xe ages of silicate inclusions in IAB iron meteorites to argue for a formation model involving impact melting. Also, Swindle et al. (1987) have argued that the spread in apparent ages seen in many Allende samples suggests that the alteration process responsible was planetary, not nebular.

The Oldest Apparent I-Xe Ages

Apparent I-Xe ages are all relative ages, so it is important to have some absolute reference point. Although Bjurböle is an excellent monitor in terms of reproducibility (Hohenberg and Kennedy 1981), it does not have the oldest I-Xe age, and it is not clear exactly what event(s) established the Xe isotopic closure in Bjurböle.

With this in mind, Herzog et al. (1973) and Lewis and Anders (1975) analyzed magnetite separates from CI and CM meteorites, which have nearly solar elemental abundances (Chapter 1.1) and thus appear to have undergone very little secondary processing (although the magnetite itself may be secondary; Kerridge et al. 1979b). Samples from Orgueil and Murchison gave ages of about 7 Myr before Bjurböle, the earliest ages seen at that time. The Murchison magnetite separate, in particular, was extremely radiogenic (^{129}Xe/^{132}Xe ratios greater than 100 in several extractions), so it gave a precise R_0 of $(1.456 \pm 0.006) \times 10^{-4}$. Since these precise, early ages came from meteorites that are chemically primitive, Herzog et al. (1973) and Lewis and Anders (1975) argued that this R_0 characterized the condensation of the solar nebula, a suitable time zero.

Later I-Xe studies on other samples have resulted in reported R_0s significantly higher than those in the Murchison magnetite, corresponding to ages before the proposed time zero. However, reanalysis of some of these "older" samples have resulted in lower R_0s than originally reported, and none of the isochrons involved are as well defined as those of the Murchison and Orgueil magnetites.

An old age reported for Arapahoe (Drozd and Podosek 1976) was apparently a shock artifact (Caffee et al. 1982a), while old ages reported for troilite from Mundrabilla (Niemeyer 1979) and some inclusions from Allende (Zaikowski 1979) were calculated using an incorrect assumption about a calibration monitor's R_0 (Rison and Zaikowski 1980). Similarly, Fireman et al. (1970) calculated total I-Xe ages using ^{128}Xe produced by neutron captures in space as the I tracer and ^{60}Co as the fluence monitor, and found that one sample of two Allende chondrules seemed to have an R_0 at least as high as the magnetites. However, that technique has never been calibrated experimentally, and the ^{129}Xe/^{128}Xe ratio that gave the high R_0 is comparable to the ^{129}Xe/^{128}Xe ratio in other samples that give later apparent I-Xe ages when analyzed in a conventional I-Xe experiment (Swindle et al. 1987). Kirschbaum (1985b) found an apparent R_0 twice as high as Bjurböle in a laser extraction of gas from an Allende inclusion that had been preheated before being irradiated. This is the highest R_0 reported, but it is from a single extraction, and the effects of the preheating are not known, although it would increase the apparent R_0 only if I were preferentially lost relative to Xe in the preheating.

Jordan et al. (1980) determined apparent R_0s up to 2.4σ higher than the magnetites for Menow, Nadiabondi and Kernouve, despite rather large uncertainties. The data would give significantly later ages (by 2σ to 5σ, so that none are more than 1σ older than the magnetites) if plotted using ^{130}Xe instead of ^{132}Xe as the reference isotope, suggesting a complicated isotopic system. Although these meteorites were selected for the absence of shock or weathering effects, they have been metamorphosed, so Rison and Zaikowski (1980) suggested that the high apparent R_0s and large uncertainties might be caused by reheating effects.

The best candidate for a sample with a higher R_0 than the Murchison magnetite is Vigarano, which Crabb et al. (1982) found to be 2.0 ± 1.0 Myr older than the magnetite. This isochron is reasonably well defined (seven points) and reheating effects are unlikely. Even so, it is only 2σ (about 10%) away from Murchison magnetite. Furthermore, Vigarano, like Murchison, is a carbonaceous chondrite, presumably a relatively primitive object. Thus, it seems plausible that Vigarano may be dating the condensation of the nebula. The Murchison and Orgueil magnetite results could then be interpreted as (1) dating aqueous activity on parent bodies, since Kerridge et al. (1979b)

argued that the magnetites were formed by such activity, rather than condensation, (2) dating a late stage in a condensation process that took 1 to 2 Myr, or (3) indicating inhomogeneity at the 10% level in the I isotopes.

Thus, although several samples have yielded more primitive initial iodine compositions than Murchison magnetite, when reproducibility and quality of the isochrons are considered, all except Vigarano are unconvincing. However, if all the questionable ages are accepted at face value, the oldest age then comes from an Allende inclusion (Kirschbaum 1985b), again a plausible candidate for a primitive age (see Chapter 5.2). The prevalence of ages 10 to 20 Myr later, even in carbonaceous chondrites, might then be attributed to resetting of the I-Xe clocks by one or more of the alteration processes that apparently affected most, if not all, meteorites (Chapters 3.1–3.6).

REFERENCES

Baedecker, P. A., and Wasson, J. T. 1975. Elemental fractionations among enstatite chondrites. *Geochim. Cosmochim. Acta.* 39:735–765.

Basford, J. R., Dragon, J. C., Pepin, R. O., Coscio, M. R., Jr., and Murthy, V. R. 1973. Krypton and xenon in lunar fines. *Proc. Lunar Sci. Conf.* 4:1915–1955.

Caffee, M. W., Hohenberg, C. M., Horz, F., Hudson, B., Kennedy, B. M., Podosek, F. A., and Swindle, T. D. 1982a. Shock disturbance of the I-Xe system. *Proc. Lunar Planet. Sci. Conf.* 13, *J. Geophys. Res. Suppl.* 87:A318–A330.

Caffee, M. W., Hohenberg, C. M., Swindle, T. D., and Hudson, B. 1982b. I-Xe ages of individual Bjurböle chondrules. *Proc. Lunar Planet. Sci. Conf.* 13, *J. Geophys. Res. Suppl.* 87:A303–A317.

Cameron, A. G. W. 1962. The formation of the sun and planets. *Icarus* 1:13–69.

Cameron, A. G. W., and Truran, J. W. 1977. The supernova trigger for formation of the solar system. *Icarus* 30:447–461.

Clayton, D. D. 1975. Extinct radioactivities: Trapped residuals of presolar grains. *Astrophys. J.* 199:765–769.

Clayton, D. D. 1977. Cosmoradiogenic ghosts and the origin of Ca-Al-rich inclusions. *Earth Planet. Sci. Lett.* 35:398–410.

Clayton, D. D. 1980. Chemical and isotopic fractionation by grain size separates. *Earth Planet. Sci. Lett.* 47:199–210.

Clayton, D. D. 1983. Chemical state of pre-solar matter. In *Chondrules and Their Origins,* ed. E. A. King (Houston: Lunar and Planetary Institute), pp. 26–36.

Crabb, J., and Anders, E. 1982. On the siting of noble gases in E-chondrites. *Geochim. Cosmochim. Acta.* 46:2351–2361.

Crabb, J., Lewis, R. S., and Anders, E. 1982. Extinct ^{129}I in C3 chondrites. *Geochim. Cosmochim. Acta.* 46:2511–2526.

Dodd, R. T. 1981. *Meteorites: A Petrologic-Chemical Synthesis* (Cambridge: Cambridge University Press).

Dreibus, G., and Wänke, H. 1983. Halogens in Antarctic meteorites. *Meteoritics* 18:291–292.

Drozd, R. J., and Podosek, F. A. 1976. Primordial ^{129}Xe in meteorites. *Earth Planet. Sci. Lett.* 31:15–30.

Drozd, R. J., and Podosek, F. A. 1977. Systematics of iodine-xenon dating. *Geochem. J.* 11:231–237.

Emery, J. F., Reynolds, S. A., Wyatt, E. I., and Gleason, G. I. 1972. Half-lives of radionuclides—IV. *Nucl. Sci. Eng.* 48:319–323.

Eugster, O., Eberhardt, P., and Geiss, J. 1967. Krypton and xenon isotopic composition in three carbonaceous chondrites. *Earth Planet. Sci. Lett.* 3:249–257.

Fireman, E. L., DeFelice, J., and Norton, E. 1970. Ages of the Allende meteorite. *Geochim. Cosmochim. Acta* 34:873–881.

Fowler, W. A., Greenstein, J. L., and Hoyle, F. 1961. Nucleosynthesis during the early history of the solar system. *Geophys. J.* 6:148–220.
Goles, G. G., and Anders, E. 1962. Abundances of iodine, tellurium and uranium in meteorites. *Geochim. Cosmochim. Acta* 26:723–737.
Herzog, G. F., Anders, E., Alexander, E. C., Jr., Davis, P. K., and Lewis, R. S. 1973. Iodine-129/xenon-129 age of magnetite from the Orgueil meteorite. *Science* 180:489–491.
Hohenberg, C. M., and Kennedy, B. M. 1981. I-Xe dating: Intercomparisons of neutron irradiations and reproducibility of the Bjurböle standard. *Geochim. Cosmochim. Acta* 45:251–256.
Hohenberg, C. M., and Reynolds, J. H. 1969. Preservation of the iodine-xenon record in meteorites. *J. Geophys. Res.* 74:6679–6683.
Hohenberg, C. M., Podosek, F. A., and Reynolds, J. H. 1967. Xenon-iodine dating: Sharp isochronism in chondrites. *Science* 156:233–236.
Hohenberg, C. M., Hudson, B., Kennedy, B. M., and Podosek, F. A. 1981. Noble gas retention chronologies for the St. Séverin meteorite. *Geochim. Cosmochim. Acta* 45:535–546.
Honda, M., Bernatowicz, T. J., and Podosek, F. A. 1983. ^{129}Xe-^{128}Xe and ^{40}Ar-^{39}Ar chronology of two Antarctic enstatite meteorites. *Mem. Natl. Inst. Polar Res., Spec. Issue* 30:275–291.
Hudson, G. B. 1981. Noble gas retention chronologies in the St. Séverin meteorite. Ph.D. thesis, Washington Univ., St. Louis.
Huneke, J. C. 1976. Diffusion artifacts in dating by stepwise thermal release of rare gases. *Earth Planet. Sci. Lett.* 28:407–417.
Jeffery, P. M., and Reynolds, J. H. 1961. Origin of excess Xe^{129} in stone meteorites. *J. Geophys. Res.* 66:3582–3583.
Jordan, J., Kirsten, T., and Richter, H. 1980. ^{129}I/^{127}I: A puzzling early solar system chronometer. *Z. Naturforsch.* 35a:145–170.
Kennedy, B. M. 1981. Potassium-argon and iodine-xenon gas retention ages of enstatite chondrite meteorites. Ph.D. thesis, Washington Univ., St. Louis.
Kerridge, J. F., Macdougall, J. D., and Marti, K. 1979a. Clues to the origin of sulfide materials in CI chondrites. *Earth Planet. Sci. Lett.* 43:359–367.
Kerridge, J. F., Mackay, A. L., and Boynton, W. V. 1979b. Magnetite in CI carbonaceous meteorites: Origin by aqueous activity on a planetesimal surface. *Science* 205:395–397.
Kirschbaum, C. 1985a. Radiogenic ^{129}Xe$_r$ in individual minerals in Allende as studied with a laser microprobe. *Lunar Planet. Sci.* XVI:439–440 (abstract).
Kirschbaum, C. 1985b. Iodine and chlorine in two fine-grained inclusions from the Allende meteorite. *Meteoritics* 20:683–684.
Kirsten, T., Jordan, J., Richter, H., Pellas, P., and Storzer, D. 1978. Plutonium and uranium distribution patterns in phosphates from ten ordinary chondrites. U.S.G.S. Open-File Report 78-701, pp. 215–219.
Kuroda, P. K. 1976. Xenon-iodine dating: Primordial xenon in meteorites. *Geochem. J.* 10:65–70.
Lewis, R. S., and Anders, E. 1975. Condensation time of the solar nebula from extinct ^{129}I in primitive meteorites. *Proc. Natl. Acad. Sci. U.S.A.* 72:268–273.
Meeker, G. P., Wasserburg, G. J., and Armstrong, J. T. 1983. Replacement textures in CAI and implications regarding planetary metamorphism. *Geochim. Cosmochim. Acta* 47:707–721.
Merrihue, C. 1966. Xenon and krypton in the Bruderheim meteorite. *J. Geophys. Res.* 71:263–313.
Murty, S. V. S., and Marti, K. 1987. Nucleogenic noble gas components in the Cape York iron meteorite. *Geochim. Cosmochim. Acta* 51:163–172.
Niemeyer, S. 1979. I-Xe dating of silicate and troilite from IAB iron meteorites. *Geochim. Cosmochim. Acta* 43:843–860.
Papanastassiou, D. A., Huneke, J. C., Esat, T. M., and Wasserburg, G. J. 1978. Pandora's Box of the nuclides. *Lunar Planet. Sci.* IX:859–861 (abstract).
Podosek, F. A. 1970. Dating of meteorites by the high-temperature release of iodine-correlated Xe. *Geochim. Cosmochim. Acta* 34:341–365.
Reed, G. W. 1971. Iodine. In *Handbook of Elemental Abundances in Meteorites*, ed. B. Mason (New York: Gordon and Breach), pp. 401–406.
Reynolds, J. H. 1960. Determination of the age of the elements. *Phys. Rev. Lett.* 4:8.
Reynolds, J. H. 1963. Xenology. *J. Geophys. Res.* 68:2939–2956.

Reynolds, J. H. 1967. Isotopic abundance anomalies in the solar system. *Ann. Rev. Nucl. Phys.* 17:253–316.

Reynolds, J. H., Lumpkin, G. R., and Jeffery, P. M. 1980. Search for ^{129}Xe in mineral grains from Allende inclusions: An exercise in miniaturized rare gas analysis. *Z. Naturforsch.* 35a:257–266.

Rison, W., and Zaikowski, A. 1980. Proportional retention of I and radiogenic ^{129}Xe in preheated Allende. *Proc. Lunar Planet. Sci. Conf.* 11:977–994.

Ruzicka, A. 1986. Deformation histories of chondrules in the Chainpur chondrite. *Meteoritics* 21:499–500.

Scott, E. R. D., and Rajan, R. S. 1981. Metallic minerals, thermal histories and parent bodies of some xenolithic, ordinary chondrite meteorites. *Geochim. Cosmochim. Acta* 45:53–67.

Shima, M. 1986. A summary of extremes of isotopic variations in extra-terrestrial materials. *Geochim. Cosmochim. Acta* 50:577–584.

Shukolyukov, Yu. A., Vu Minh, D., Kolesov, G. M., Fugzan, M. M., and Ivanova, M. A. 1986. ^{129}I/^{129}Xe data on the relative interval of formation for some type L chondrites. *Geochem. Int.* 23(2):130–144.

Srinivasan, B., Lewis, R. S., and Anders, E. 1978. Noble gases in the Allende and Abee meteorites and a gas-rich mineral fraction: Investigation by stepwise heating. *Geochim. Cosmochim. Acta* 42:183–198.

Swindle, T. D. 1986. Iodine-xenon and other noble-gas studies of individual chondrules from the Chainpur meteorite. Ph.D. thesis, Washington Univ., St. Louis.

Swindle, T. D., and Grossman, J. N. 1987. I-Xe studies of Semarkona chondrules: Dating alteration. *Lunar Planet. Sci.* XVIII:982–983 (abstract).

Swindle, T. D., Caffee, M. W., and Hohenberg, C. M. 1983a. Radiometric ages of chondrules. In *Chondrules and their Origins*, ed. E. A. King (Houston: Lunar and Planetary Inst.), pp. 246–261.

Swindle, T. D., Caffee, M. W., Hohenberg, C. M., and Lindstrom, M. M. 1983b. I-Xe studies of individual Allende chondrules. *Geochim. Cosmochim. Acta.* 47:2157–2177.

Swindle, T. D., Caffee, M. W., and Hohenberg, C. M. 1987. I-Xe studies of Allende inclusions: EGGs and the Pink Angel. *Geochim. Cosmochim. Acta*, submitted.

Takahashi, H., Gros, J., Higuchi, H., Morgan, J. W., and Anders, E. 1978. Volatile elements in chondrites: Metamorphism or nebular fractionation. *Geochim. Cosmochim. Acta.* 42:1859–1869.

Turner, G. 1965. Extinct iodine 129 and trace elements in chondrites. *J. Geophys. Res.* 70:5433–5445.

Wacker, J. F., and Marti, K. 1983. Noble gas components in clasts and separates of the Abee meteorite. *Earth Planet. Sci. Lett.* 62:147–158.

Wasserburg, G. J., and Huneke, J. C. 1979. I-Xe dating of I-bearing phases in Allende. *Lunar Planet. Sci.* X:1307–1309 (abstract).

Wasserburg, G. J., Fowler, W. A., and Hoyle, F. 1960. Duration of nucleosynthesis. *Phys. Rev. Lett.* 4:112–114.

Wasson, J. T., Willis, J., Wai, C. M., and Kracher, A. 1980. Origin of iron meteorite groups IAB and IIICD. *Z. Naturforsch.* 35a:781–795.

Wood, J. A. 1967. Criticism of paper by H. E. Suess and H. Wänke, "Metamorphosis and equilibration in chondrites." *J. Geophys. Res.* 72:6379–6383.

Wood, J. A. 1979. Review of the metallographic cooling rates of meteorites and a new model for the planetesimals in which they formed. In *Asteroids*, ed. T. Gehrels (Tucson: Univ. of Arizona Press), pp. 849–891.

Zaikowski, A. 1979. I-Xe chronology of Allende inclusions. *Lunar Planet. Sci.* X:1392–1394 (abstract).

Zaikowski, A. 1980. I-Xe dating of Allende inclusions: Antiquity and fine structure. *Earth Planet. Sci. Lett.* 47:211–222.

PART 16
Summary

16.1. BOUNDARY CONDITIONS FOR THE ORIGIN OF THE SOLAR SYSTEM

JOHN F. KERRIDGE
University of California at Los Angeles

and

EDWARD ANDERS
University of Chicago

From the meteoritic results presented in the preceding chapters, we identify a set of boundary conditions that can be applied, with variable degrees of confidence, as constraints to models that attempt to describe the origin and early evolution of the solar system. Those constraints bear upon the time scale, physical and chemical environment, and processes that characterized the early solar system.

16.1.1. INTRODUCTION

The most fundamental contributions of meteoritics to science in general are almost certainly those that have shaped our knowledge of the origin and distribution of the elements, and the age of the Galaxy and the solar system. Progress in those areas has been described in Parts 5, 14 and 15. Of more specialized interest, though no less basic importance, are the insights that meteorites have given us about conditions and processes in the early solar system, and such insights have formed the major thrust of this book. From the preceding chapters, it will have been apparent that, although many different aspects of meteoritics bear upon the earliest history of the solar system, these aspects have rarely been the primary goal of meteorite studies. It is therefore not surprising that the record resulting from these studies is quite patchy; some things we know very well, others remain enigmatic. Another consequence is that it is quite difficult to summarize in a concise fashion everything that meteorites have taught us about the early solar system. Nonetheless, in this chapter we list a set of boundary conditions on the environment of the early solar system that can presently be derived from the study of meteorites. Unless otherwise stated, they apply to the formation location of the chondrites, generally taken to be in the range 2 to 4 AU.

16.1.2. TIME SCALES FOR NEBULAR EVOLUTION

Although interpretation of extinct-radionuclide data in terms of early solar-system processes is subject to uncertainties, such as the extent to which differences in isotopic ratios at closure reflect inhomogeneities (e.g., dilution of "live" nuclides with "dead" material) rather than temporal evolution, model-dependent ages can still be calculated and used, at least tentatively, as boundary conditions. With that *caveat,* solid objects apparently formed in the solar system, presumably by reprocessing of interstellar grains, within about 10^8 yr of the last r- and p-process productions. Timing of the last s-process production is less certain but may have been about 10^7 yr before solid-object formation. Some freshly synthesized (nova-produced?) material was incorporated into solids within about 10^6 yr of production. *Rapid accretion* of solids in the solar nebula is suggested by the survival of apparently primordial compositional trends in chondrites that might otherwise have been smeared out. If it is *assumed* that initial solid formation in the solar system is dated by the oldest extant CAIs, then several different processes took place within about 10^7 yr of condensation. Those processes include *formation of chondrules* in the nebula and a number of secondary processes in asteroids: *core formation,* production of basalts during *igneous differentiation* and carbonate mineralization during *aqueous activity.* Iodine-Xe data imply a duration for chondrule formation of the order of 10^7 yr, but this span is considered implausibly large and may reflect the influence of asteroidal metamorphism or initial isotopic inhomogeneity. *Chondrite metamorphism* is believed, on other grounds, to have spanned about 10^8 yr, which probably reflects the size of the parent bodies, not the time scale of the heat source (which may have been very brief). *Final compaction* of carbonaceous chondrites is tentatively dated at within about 3×10^8 yr of solid formation, presumably reflecting post-accretional brecciation processes. Until some of the uncertainties mentioned above are resolved, i.e., until less ambiguous chronometers are available, it may be true, though regrettable, that astronomical observations can supply tighter constraints than meteoritic data on some aspects of early solar-system history.

16.1.3. TEMPERATURE HISTORY OF THE NEBULA

At the chondrite formation location, temperatures have been estimated from: (1) refractory-element fractionations: $T_{max} \simeq 2000$ K; (2) CAI formation (multiple heating events?): $T_{max} = 1800$ to 2000 K, $dT/dt \simeq -0.5$ to $-50°$ C/hr; (3) chondrule formation (flash heating, multiple events?): $T_{max} \simeq 1850$ K, $dT/dt \simeq -100$ to $-2000°$ C/hr, $T_{ambient} < 650$ K. Spatial extent of these heating events is not known directly, though short time scales imply limited linear dimensions; duration of CAI/chondrule-formation epochs is poorly constrained, but no chondrules are known that are younger than ~ 4.4 Gyr. During accretion of chondrite parent asteroids, $T_{ambient} = 300$ to 500 K (for CI to ordinary chondrites, respectively).

16.1.4. PRESSURE HISTORY OF THE NEBULA

Meteorites contain few robust clues to the pressures that prevailed in the solar nebula. The apparent particulate density during chondrule formation suggests a pressure around 10^{-4} atm, but this may well have been enhanced by local concentration of dust relative to gas. *Minimum* estimates based on reconstruction of the mass of the nebula yield 10^{-7} to 10^{-5} atm for the chondrite formation location. Abundances of highly volatile elements in ordinary chondrites, whose parent asteroids apparently accreted between the formation temperatures of FeS, on the one hand, and Fe_3O_4, on the other, imply pressures in the range 10^{-6} to 10^{-3} atm.

16.1.5. OXYGEN FUGACITY IN THE NEBULA

Occurrences in primitive meteorites of (1) Fe-bearing silicates, (2) refractory inclusions depleted in Ce, (3) refractory metal grains depleted in Mo and W, (4) scheelite and V-rich magnetite in Fremdlinge, and (5) magnetite and awaruite in Allende chondrules, all point towards O fugacities higher by factors of up to 10^4 than that characteristic of a gas of solar composition. This is commonly attributed to *reheating* of regions of the nebula *locally enriched in dust* (O-rich) relative to gas (H-rich), e.g., by settling of the former towards the nebular midplane. Such regions would also have been characterized by relatively high pressures that would have had the effect of raising condensation temperatures and possibly stabilizing the formation of liquid condensates. (Note that some of the Fe-bearing silicates, also the scheelite and the magnetites, could result from secondary alteration.) By contrast, locally reducing conditions (O fugacities about an order of magnitude lower than solar) are implied by the chemistry and mineralogy of the enstatite chondrites.

16.1.6. COMPOSITIONS OF SOLID COMPONENTS IN THE NEBULA

Elemental fractionation patterns in bulk chondrites require a minimum of 7 *chemically distinct* components that could be separated either from each other or from nebular gas. These have been identified as: (1) early (i.e., refractory) condensate; (2) remelted silicate, i.e., in chondrules; (3) unremelted silicate; (4) remelted metal; (5) unremelted metal; (6) sulfide; and (7) volatiles. Note that the exact compositions inferred for these components are based on assumptions of chemical equilibrium during condensation.

Compositions of individual chondrules can also be interpreted in terms of 7 components. These are: (1) olivine (refractory-rich); (2) pyroxene (Fe-rich, refractory-poor); (3) metal; (4) refractory-rich metal; (5) sulfide; (6) alkalies; and (7) zinc. Agreement between the two sets of components is only partial, suggesting that the story is not yet complete or definitive. The individual components broadly resemble the predictions of equilibrium condensation but show some differences in detail. Titanium-isotopic data require a minimum of four *isotopically distinct* components, presumably solid.

16.1.7. COMPOSITIONS OF GASEOUS RESERVOIRS

Oxygen-isotopic data for chondrules and inclusions can be modeled using three *isotopically distinct* components, one apparently solid and two gaseous. These span a range in $^{18}O/^{16}O$ of at least 5%. The presence of more components is not excluded and may actually be suggested by bulk-meteorite data. Similar gas-phase isotopic inhomogeneity is reflected in enrichment of inclusion rims in ^{26}Al with respect to cores. Isotopically distinct reservoirs are suggested also by data for H and N, spanning ranges of factors of >4 and >2, respectively, but the extent to which these represent gaseous, as distinct from solid, reservoirs is not clear. The existence of *chemically distinct* gaseous reservoirs is suggested by inter- and intra-chondrite variations, of up to about 10^5, in O fugacity, though these may reflect local reheating following dust/gas fractionation (see above). Note that such fractionations would have led to redistribution of O isotopes between solid and gas, complicating interpretation of the meteoritic data in terms of exchange between discrete reservoirs.

16.1.8. PROCESSES IN THE NEBULA

The best evidence in chondrites for *condensation* of solids from gas are the Group II rare-earth-element patterns in some CAIs, whiskers of minerals such as wollastonite within vugs, also in some CAIs, and enstatite whiskers occasionally observed in chondritic IDPs. The process of *evaporation* is strongly suggested by mass-dependent isotopic fractionations observed in Si,

Mg, Ca and other elements in CAIs. There is growing evidence that chondritic material was subjected to repeated cycles of heating and cooling: the nebula was apparently a violently energetic environment.

Solid/gas fractionation is apparently recorded in the bulk-chondrite patterns both for refractory elements (variations of a factor of about 2) and for volatile elements (variations of about 10^3). *Solid/solid fractionation,* in the form of separation of metal from silicate in the nebula, was apparently responsible for the variations of about a factor of 2 in siderophile-element contents among chondrite groups. A significant fraction of the organic molecules in carbonaceous chondrites seem to have resulted from *mineral-catalyzed reactions.* The initial noble-gas inventories of the terrestrial planets were probably trapped by *adsorption* on carbon grains, possibly in surficial micropores, or by *implantation* at low energies.

16.1.9. GRAIN SIZE DISTRIBUTION IN THE NEBULA

The size distribution of solid particles in the early solar nebula presumably changed with time, first through *condensation,* later through *agglomeration.* Any primitive material surviving from that epoch could, in principle, yield information about those distributions, which are needed as input parameters for many nebular models. *If* matrix in type-3 chondrites represents pristine nebular material, grain sizes in the range 0.05 to 10 μm are indicated. Relict grains in chondrules extend that range to about 100 μm, but both populations may have experienced selection effects and for neither population is an origin by nebular condensation rigorously established. If some CAIs represent aggregates of minimally altered nebular condensates, grain sizes in the range 1 to 10 μm are indicated, with evidence for agglomeration to at least cm size. The presence of sub-μm grains is implied by observations of IDPs, but discriminating between nebular and presolar grains is not straightforward in such cases.

16.1.10. MIXING OF MATERIAL BETWEEN NEBULAR REGIONS

The presence of xenoliths in many primitive chondrites attests to *radial transport of material* within the solar system after planetesimal accretion. Presumably, this mainly reflects mixing within the asteroid belt as it is at present, characterized by approach velocities of the order of 5 km/s. Mixing of material between different zones in the nebula prior to accretion, when velocities were presumably much lower, may be recognized, in principle, by distributions of apparently primitive components such as chondrules or CAIs. The marked differences between CAIs in different chondrite types suggest that any mixing was limited in extent, though the evidence from chondrule populations is less clearcut. The spatial scale of such transport is presently poorly constrained (e.g., obviously extrasolar material is rare in primitive chondrites).

16.1.11. MAGNETIC FIELDS IN THE NEBULA

Natural remanent magnetization of components in carbonaceous chondrites yields estimates for *paleofield intensities* in the range 10 to 100 μT. These fields apparently predate accretion of the carbonaceous planetesimals but are otherwise not temporally constrained. Also, the mineral carriers of the remanent magnetization are not all well characterized, adding further uncertainty to interpretation of the paleofields.

16.1.12. SUMMARY

Clearly, the data presented above vary considerably in precision, reliability and applicability to early solar-system issues. Given the presently high level of interest in meteorite research, and assuming that a reasonable funding level for such research can be achieved, it is safe to predict that our constraints on conditions and processes in the early solar system will become ever more rigorous. In the next chapter, we explore specific lines of inquiry that we believe could contribute to that increase in understanding. We note also that the usefulness of meteoritic data for constraining theory is a strong function of the quality of communication between the meteorite community and the appropriate theoreticians.

16.2. FUTURE DIRECTIONS IN METEORITE RESEARCH

E. ANDERS
University of Chicago
and
J. F. KERRIDGE
University of California at Los Angeles

The study of meteorites has already provided us with important constraints on several key aspects of the origin and early evolution of the solar system. These include: the thermal evolution of the solar nebula; the chronology of cloud collapse, nebula formation, grain nucleation and growth, planetesimal accretion, and formation and development of asteroids; the nature and properties of grains in the solar nebula, and; the extent to which the chemistry of nebular phases can be modeled within an equilibrium framework. None of these issues are rigorously resolved at this time, however, and much work remains to be done. In this chapter, we discuss several lines of meteorite research that we believe could shed additional light on the earliest history of the solar system. These lines of inquiry range from the parochial (e.g., how large were the chondrite parent asteroids) to the interdisciplinary (e.g., what do meteorites tell us about the planets), and from the technical (e.g., to what extent are chondrules depleted in volatile elements) to the fundamental (e.g., what was the thermal history of the nebula). It is inherent in the nature of scientific inquiry that no such account of future directions can hope to be comprehensive.

16.2.1. CLASSIFICATION

The Van Schmus and Wood (1967) chemical-petrographic classification scheme for chondrites has proven to be of inestimable value and it seems safe to assume that any future developments will involve modification, rather than replacement, of that scheme. Such modification should probably focus initially on petrographic type 3, where two alteration sequences meet: thermal metamorphism, mainly of ordinary chondrites (types 3 to 6) and aqueous alteration of carbonaceous chondrites (types 3 to 1).

With the recent development, and general acceptance, of subdivision of "ordinary type 3" by means of thermoluminescence sensitivity (Sears et al. 1980), recognition of metamorphic effects in CO and CV chondrites (McSween 1977a,b) and discovery of (quantitatively minor) aqueous-alteration effects in some type-3 ordinary chondrites (Hutchison et al. 1987; see also Chapters 3.4 and 10.2), modest overhaul of the function and criteria of type 3 seems appropriate. Another aspect of such a project would be the search for chondrites that have escaped both the Scylla of aqueous alteration and the Charybdis of thermal metamorphism. Such meteorites would be the most promising places to search for truly pristine material, particularly for possible interstellar oxide grains that would be very vulnerable to secondary alteration.

More general recognition of the importance of the different chemical, as distinct from petrographic, groupings among carbonaceous chondrites is desirable. Furthermore, the possible existence of genetically significant chemical groups in addition to those presently recognized should be more closely investigated. Also, if, as seems plausible, the CM2 chondrites of ~ 0.3 Myr exposure age were all derived from a relatively restricted region of their parent body, the degree of their homogeneity as a subgroup would be indicative of the scale of variability within a single C-type asteroid.

Within the classification framework, it may also be useful to devise a scheme that includes the presence of xenoliths showing different degrees of equilibration within brecciated chondrites.

By its nature, classification tends to stress the differences among chondrites, but it should be remembered that chondrites as a whole are remarkably similar to each other, compared, for example, to terrestrial rocks. In fact, a continuum of properties may be traced through the different chondrite groups, suggesting that they are all derived from rather similar parent objects occupying a fairly restricted region within the solar system. There is now general agreement that that region is the asteroid belt: the task is therefore to locate the different chondrite formation locations within the belt.

16.2.2. ASTEROIDS AND METEORITES

Although identification of a specific asteroid as the parent body for each known meteorite may not be an attainable goal, recent progress in this area suggests that more refined spectral comparisons of asteroids with meteorites will succeed in identifying likely parent bodies for many of the major meteorite types, including that long-sought goal, the sources of the ordinary chondrites. Similarly, time may tell whether nature has an efficient mechanism for transporting basaltic achondrites from Vesta, or whether another, but presumably disrupted, basaltic asteroid is lurking in a more favorable source region. Such spectral identifications, even if coupled with dynamically plausible transport mechanisms, will probably not be universally accepted without ground-truth calibration of the spectral data. Analysis of one or more carefully chosen asteroids, either *in situ* or by sample return, would do much to dispel such uncertainty. Meanwhile, further experimental studies of the effects of the space environment on the optical properties of mineral grains are very desirable: small Apollo-Amor objects come much closer to matching ordinary-chondrite spectra than do large main-belt asteroids, suggesting that regolith production on the latter may be affecting their spectral signature, leading to the apparent mismatch between S asteroids and ordinary chondrites.

Another topic worth exploring is the relationship, if any, between E asteroids, aubrites and enstatite chondrites. The latter two apparently required a domain of locally high C/O somewhere in the early solar system, and the asteroids tell us where at least some of the former inhabitants of that domain now reside. As one of the rare cases where a clue remains to the original formation location, this association certainly deserves a closer look than it has so far received.

An issue that needs to be addressed is whether or not the meteorites that fell over the past $\sim 3 \times 10^5$ yr, sampled by the Antarctic finds, represent a population distinctly different from that sampled by recent falls (see Chapter 7.6; Dennison et al. 1986; Wetherill 1986). The apparent differences in volatile-trace-element contents, if real, ought to be accompanied by other systematic differences detectable by a thorough survey.

An important question, alluded to in various chapters but not addressed directly in this book, concerns the evidence within meteorites on the sizes of their parent bodies. All available indicators of this parameter need further improvement. Cooling rates based on Pu^{244} tracks give rather consistent and plausible results for H and LL chondrites (Pellas 1986; see also Pellas and Storzer 1981; Taylor et al. 1987), but with occasional exceptions. Doubling the data base should make matters decidedly better or worse. In contrast, the metallographic method (see, e.g., Willis and Goldstein 1981) suffers not from a lack of data, but from uncertainties in the method itself. Cooling rates have recently risen by 10 to 100 times as a result of new Ni diffusion rates and other changes, and it is therefore necessary to re-examine the foundations of the method (Goldstein 1986) and to calibrate it against the Pu^{244} method. A particularly important experiment would be to determine Pu^{244}-based cooling rates of several iron meteorites with normal Widmanstätten patterns, as in the pioneering study of Toluca by Fleischer et al. (1968).

With certain assumptions about closure temperatures, differences in apparent age of a single object based on different radiometric schemes can be used to infer a cooling rate. Thus, a comparison of Sm-Nd, Pb-Pb, Rb-Sr and Ar^{40}-Ar^{39} ages yields cooling rates of about 3 K/Myr for St. Séverin (LL6) and 5 K/Myr for Guareña (H6), suggesting a larger size (130 to 150 km) for the LL asteroid than for the H asteroid (80 to 100 km) (Lipschutz et al. 1988). Such work is very promising, but requires very careful sample selection and coordinated studies by several laboratories. It would also be worthwhile to explore Wlotzka's (1987) novel approach to cooling-rate determination, based on the distribution of Fe and Mg between chromite and olivine.

For chondrites, it may be possible to use porosity as a measure of overburden pressure and hence of size. Porosity is a function of pressure, temperature and time, and if the last two (as well as the relative burial depth) are independently estimated, then the pressure can be determined.

Looming over these relatively detailed issues is the more general question of the origin of the asteroid belt itself. Presumably, detailed observations of meteorites associated with specific asteroids, or narrow regions within the asteroid belt, will ultimately constitute the strongest available tests of models of the origin and evolution of the belt. That nexus between observation and theory will likely involve the fine-scale chronology of meteorites and increasingly sophisticated models for accretion within the asteroid belt.

16.2.3. SECONDARY PROCESSING

Igneous Processes

The most basic issue concerning the igneous differentiation of certain of the asteroids sampled by meteorites is: what heat source caused it? Was live Al^{26} widespread throughout the inner solar system? Did a T-Tauri solar wind generate heating currents in planetesimals? Did impacts deposit enough heat in accreting objects to produce widespread melting in at least some of them? Whether these questions can be answered definitively through study of the products of the differentiation is problematical. Evidence for an Al-correlated Mg^{26} excess in a differentiated meteorite would be a decisive clue. A complementary approach would involve a systematic study of the distribution of $(Al^{26}/Al^{27})_0$ among CAIs: is that distribution totally haphazard, or can a gross pattern be discerned? Possibly such information could afford a rough idea of the distribution of Al^{26} in the early solar system (see also Sec. 16.2.15). Failing such direct evidence, future study of the igneous meteorites themselves can be of value by constraining the nature and precise chronology of differentiation events, and the size and composition of the parent bodies. In particular, extremely precise dating might establish the relative ages of common eucrites, noncumulate eucrites and diogenites. The ages of the ureilites are still virtually unconstrained, and previous models for their origin as igneous cumulates on a single large body have been called into question by newly discovered O-isotopic heterogeneities.

Igneous meteorites tend to belong to fractionation sequences, the trends of which can only be established by studying numerous samples from each sequence. Relationships among cumulate, noncumulate and polymict eucrites, diogenites and several other types of meteorites that may also be from the same parent asteroid (most notably the howardites, but possibly also the main-group pallasites, IIIAB iron meteorites and mesosiderites) can be clarified by petrologic studies of additional samples, such as individual clasts from polymict eucrites and howardites. Polymict ureilites also contain a great diversity of clasts that are still little studied. By clarifying how those lithologies may be related to one another, and most importantly to monomict ureilites, we should gain valuable new insights into the origin of ureilites.

Antarctica is particularly useful as a collecting area for igneous stony meteorites, which in less-icy surroundings are often difficult to distinguish from typical Earth rocks. Recent years have seen the discovery in Antarctica of only the second angrite (LEW86010), the second brachinite (ALH84025) and the third polymict eucrite (EETA83309). Discoveries such as these can be expected to help clarify the petrogenesis of igneous meteorites in general. The uniquely Martian(?)-gas-rich SNC meteorite EETA79001 is another example of the key finds that are coming from Antarctica.

Experimental petrology can also be applied further to constrain models of igneous differentiation. For example, more precise constraints on the effects of melt MgO/FeO ratio and pressure on the olivine-pyroxene liquidus boundary could be used to test whether the main-group eucrites could have formed by fractional crystallization, or, as argued by Stolper (1977), their distinctive compositions could only have been derived by partial melting without subsequent fractional crystallization.

Thermal Metamorphism

As in the case of igneous processes, the ultimate issue under the heading of thermal metamorphism is the cause of the metamorphic heating. Again, it seems unlikely that study of the products of metamorphism will by itself identify that heat source, though the details, such as chronology, depth of burial, etc., may well serve as useful constraints on models of planetesimal heating.

Of more immediate concern, however, is the question of defining the thermal history of brecciated chondrites of low petrographic type. As noted in Chapters 3.5 and 10.2, many, and perhaps most, type-3 ordinary chondrites contain lithic clasts of higher petrographic type. While in most cases such clasts can be easily recognized, and avoided if inappropriate for that particular study, their presence raises the larger issue of the extent to which the seemingly regular type-3 host material is strictly uniform, i.e. has experienced a common thermal history. This question takes on added significance in light of the prevailing view that such meteorites represent our best sources of nebular material that may have escaped parent-body processing. It is probably desirable for studies aimed at such primitive material to be placed in a well-defined petrographic context. For example, one sample of Semarkona matrix analyzed for its volatile trace elements may not have had the same thermal history as another one analyzed for noble gases. The same concern applies also, of course, to the effects of shock and aqueous alteration. Failing adequate petrographic control, it may be desirable to expand analytical strategies to include measurement of intrameteorite variability as well as average values.

Aqueous Alteration

The main objective here is the refinement of quantitative models defining the conditions of alteration, e.g., temperature, pH, Eh and duration of the active stage. This will require more O-isotopic studies on carefully separated fractions of minerals produced by the aqueous activity (see, e.g., Clayton and Mayeda 1984) and more detailed studies of the stability regimes of minerals such as tochilinite. Experimental studies may also indicate whether the observed elemental partitioning among secondary minerals, e.g., Fe and Mn distribution between carbonates and phyllosilicates, might be used to quantify certain parameters of the alteration environment.

At a more descriptive level, there is probably still more to be learned concerning the extent to which aqueous alteration has operated on ordinary chondrites, following the pioneering study of Hutchison et al. (1987).

Regolith Breccias

Thanks to the Apollo Program, we now know a great deal about the processes involved in breccia formation. This is not to suggest that there is nothing left to be learned, but perhaps a more appropriate near-term goal is the application of breccia characteristics to clarify other issues, e.g., the thermal history of primitive chondrites, as discussed earlier in this section, and the nature of the accretionary processes, and/or the post-accretion regolith environment, of the parent asteroids. In addition, a thorough inventory of the distinguishable components in primitive meteoritic breccias might help test models of transport and mixing of materials in the early solar system.

16.2.4. IRRADIATION EFFECTS

Limits upon a general irradiation of *all* meteoritic matter by the Sun have become ever more restrictive over the years. Such an irradiation is, in fact, inherently implausible as it could begin only after dissipation of the nebula. At that time, dust and small bodies would be lost along with the gas, leaving only the bigger bodies, whose radii would be large compared to the range of solar corpuscular radiation. However, at least the surfaces of those bodies would be irradiated, thus acquiring a record of early solar corpuscular radiation, perhaps all the way back to the T-Tauri stage. Stellar theory and astronomical observations of young solar-mass stars suggest that the Sun probably went through such a stage of greatly enhanced activity, but it would be desirable to document it from the meteoritic record. In particular, such a direct record might shed light on the question of the observed interstar variations in T-Tauri intensity: are such differences a function of stellar age alone, or do they also reflect differences in rotation rate?

Unfortunately, application of the meteoritic record to this question would not be straightforward. Because that record yields a measure of the fluence, but not the flux, experienced by certain meteoritic components, it is necessary to infer the duration of exposure independently if the flux is to be estimated. For that reason, even if a particular meteoritic component was exposed to the peak of the T-Tauri radiation, it may not be feasible to recognize that fact, nor to estimate the actual flux. Furthermore, any T-Tauri record could have been erased during subsequent processing of the meteoritic material. Finally, it would take very special circumstances to preserve a regolith sample from that epoch, given the high erosion rates of asteroids. However, it still seems desirable to study the meteoritic record for evidence of possible changes in solar activity through time. Recognition of a period of enhanced activity would be significant even if it did not extend to a T-Tauri stage.

The best indication of such an irradiation would involve tracks from solar energetic particles (SEP), implanted noble gases from the solar wind and SEP, and spallation products (Ne^{21}, Ar^{38}) from the most energetic particles. All approaches other than solar-wind analysis require grains larger than about 10 μm, so that only chondrules, inclusions or relatively large individual mineral grains are suitable, not matrix. No record has yet been found for noble gases implanted during a period of enhanced activity, but suggestive evidence has been found for "excess" spallogenic nuclides in some populations of SEP-irradiated grains in certain gas-rich meteorites (Chapter 4.1). This excess may be due to SEP-induced spallation caused by an active early Sun (Caffee et al. 1987), though spallation by galactic cosmic rays, at present-day intensity, during a "pre-irradiation" stage (in a regolith, in transit between two asteroids, or both) cannot be ruled out (Pedroni et al. 1988).

Future work should focus on discriminating between these alternatives. It may be advantageous to study meteorites that have very short cosmic-ray exposure ages, so that heavily irradiated grains will stand out. In addition, for gas-rich carbonaceous chondrites, the irradiated grains appear to be typical of the general population, suggesting that careful reconstruction of the evolutionary history of such meteorites might permit the irradiation epoch(s) to be effectively constrained. An important part of this task will involve the measurement of compaction ages for these meteorites (Chapter 5.4) to determine whether any observed irradiation phenomena were acquired 4.5 Gyr ago. If successful, the study would also address the debate over whether carbonaceous chondrites acquired their irradiation record while in small, i.e., roughly m-sized, bodies during accretion of the parent asteroids (see, e.g., Goswami and Lal 1979), or in a regolith setting.

Further characterization of the isotopic composition of N and noble gases in ancient solar wind (post-T Tauri) is needed. Again, retentive mineral grains in different types must be studied. So far, olivines and pyroxenes have been studied; these are convenient minerals because the presence of flare tracks allows picking of irradiated grains. The results have to be synthesized with the lunar data to yield a consistent picture of the evolutionary history of wind and flare compositions. At present there is a curious discrepancy between the lunar regolith, which apparently records an ancient solar wind highly depleted in N^{15} (e.g., Kerridge 1980), and gas-rich meteorites which reveal no trace of such light N, though in the latter case, indigenous N tends to obscure any possible solar-wind signature.

16.2.5. SOLAR-SYSTEM CHRONOLOGY

The contribution of meteorite chronology to our detailed understanding of the early solar system has already been of profound importance. A major emphasis in future work will undoubtedly continue to be the fine-scale chronology of nebular and immediately post-nebular events. A start has already been made in this area, e.g. Fig. 5.2.3., but much more remains to be done. We need to know the extent to which the different radiometric systems are affected by isotopic inhomogeneity, what event is being dated in each case, and what the former host phases of extinct I^{129} are (see also Sec. 16.2.15 and Chapter 15.3).

In addition, significant developments in methodology, discussed in Chapter 5.1, will permit many ages to be determined with much greater precision and accuracy then heretofore. Redetermination of many of the "benchmark" meteorite ages (e.g., Allende CAIs, Angra dos Reis, eucrites and eucritic clasts) therefore seems worthwhile. A desirable goal, even though it may seldom prove feasible, would be to try to address the question of uncertainty in the decay constants by performing age determinations using more than one pair of isotopes. The problems with this approach are that each isotopic system has its own characteristic closure temperature, so that differently determined "ages" do not necessarily refer to the same event, and that many of the available chronometers lack the appropriate range to monitor the early solar system with adequate precision. As a specific example of an outstanding issue in this general area, it is particularly important to resolve the present uncertainty surrounding the decay constant of Rb^{87} (Chapter 5.2).

16.2.6. THE EARLY SOLAR SYSTEM

Progress in this area is what this book is all about. Increments of knowledge derived from the study of meteorites should lead to corresponding increments of reality in models of the earliest solar system. Increasing refinement of those models should lead, in turn, to increasingly stringent observational or experimental tests involving meteorites or other solar-system objects. It is not within the purview of this book to address the details of nebular or stellar-collapse models, or how those details might be further refined. However, a general comment worth making is that there is a fundamental difference in scale between the predictions of current nebula models and the results of meteorite analyses. This leads to a situation in which many of the meteoritic data can be accommodated by the models without really testing them. In essence, the data become lost in the "noise." Because the nature of the meteoritic data is unlikely to change significantly in scale, we must look mainly to the theoreticians for progress in this area. However, the meteorite community can aid such progress by making their literature more accessible to the theoretician. It is important that a selection process be brought to bear on that dense and technically complex literature in order to extract and draw attention to those facts that are essential to the topic, thereby generating constraints that are relevant, understandable and, of course, right.

In the interim, there are several issues where models and observations can fruitfully interact, e.g., the thermal history of the nebula, the mixing of material between different nebular regions and the chronology of the nebula and its solid contents. Two questions, in particular, that require both theoretical and experimental treatment have arisen at various points in the book. Were the high-temperature events recorded in the primitive chondrites local or nebula-wide? To what extent did dust/gas fractionation lead to local variations in O fugacity?

Clear-cut observational responses to these questions are not immediately available; the route to such insights probably lies through the gradual accumulation of information along the lines summarized in the other sections of this chapter. However, the ultimate value of such information will depend upon the extent to which it can be related to specific times and locations within the early solar system. In principle, the time parameter can be supplied by radiochronology, but defining location accurately is more difficult. High-resolution spectral identification of parent asteroids will improve this situation, particularly in light of the apparent primordial zoning of the asteroid belt (Gradie and Tedesco 1982; see also Chapter 2.1), but will not rigorously constrain the formation location of the components of primitive meteorites. For that, some measure of heliocentric distance embedded in the component itself seems to be required. In principle, solar irradiation could form the basis of such a measure (see, e.g., Anders 1978) but currently lacks the appropriate spatial resolution. In the interim, the temptation to use model-dependent parameters, e.g., redox state, prevalence of high-temperature phases, as proxies for heliocentric distance should probably be avoided.

16.2.7. ELEMENTAL COMPOSITION OF CHONDRITES

Refractory Lithophiles

These elements provide important constraints on the temperature history of the solar nebula. Some further clues may come from Group-II CAIs, which show paradoxical features. They are depleted in super-refractory elements (which suggests condensation from a gas depleted in these elements), yet they show isotopic anomalies (which suggest survival of presolar solids). Several explanations may be invoked, such as rapid cooling, mixture of materials from several reservoirs, etc. An answer may emerge from coordinated petrographic, chemical, and isotopic studies of Group-II inclusions, complemented by laboratory studies of thermodynamic properties, reaction rates, and exchange rates.

None of the hypothetical components involved in the refractory-element fractionation have been located thus far, though the amoeboid olivine inclusions in CV chondrites have most of the expected features. These "lost" components may have found their way into planets, other asteroids, or comets. They would manifest themselves by high Mg/Si and U, and—in undifferentiated bodies—by an assemblage of forsteritic olivine and Ca, Al-rich minerals. Perhaps the high Mg/Si and U of the Earth reflects the presence of such "lost" components. One should certainly continue to look for such enrichments (and for the above mineral assemblage) in all new meteorite classes, especially those from Antarctica, although the negative results thus far seem to suggest that the "lost" components did not stay in the asteroid belt.

Another approach is to look for small variations in refractory-element content as a function of petrographic type. Such variations would be expected if accretion began before the lost component was completely removed. Some variations between types 3 and 4 through 6 have been seen for REE (Nakamura 1974) and refractory siderophiles (Rambaldi et al. 1979; Morgan et al. 1985), but they have not yet been observed within types 4 to 6. The effects to be looked for are quite small ($\leq 10\%$), requiring large samples and elements that can be measured with high precision.

Lastly, one should look for the lost component in (cometary) IDPs. Both the presence of Ca-rich phases and an overall enrichment in Ca, Al would be positive indications, provided one can rule out an asteroidal origin for the IDPs in question. The chemical and textural criteria that have been used for this purpose are not quite conclusive, and it would therefore be desirable to collect IDPs during meteor showers, which are assuredly of cometary origin. Additional, if less accurate, data can be obtained from optical spectra of shower meteors and measurements of metal ions in the ionosphere. Some meteor spectra show no Na lines and hence may consist of CAI-type material (Wilkening 1975) or of the lost component, but as these were sporadic rather than shower meteors, they may have been asteroidal rather than cometary.

Siderophiles

The major problems in this field are similar to those in the preceding section, i.e.:

1. The abundance pattern of the "lost" components involved in the metal-silicate fractionation of the chondrite groups;

2. The fate of these components;
3. Variations in metal content and composition with petrographic type.

Problems (1) and (3) both require multi-element analyses to better than 3 to 5%. Isotope dilution is the ideal method, but as it is hard to apply to some of the elements in question, it would have to be supplemented by high-precision RNAA—preferably by the metal-extraction technique of Rammensee and Palme (1982). Samples should be large enough to make sampling error negligible, which may require > 10 g for some metal-poor meteorites (Keil 1962).

H-chondrites actually seem to show systematic variations with petrographic type in the proportions of Os, Ir, Re, Ni, Au, and Pd, at the 5% level (Morgan et al. 1985). However, this study involved 0.1 to 0.2 g samples of only 10 meteorites, and should be repeated on larger samples of a larger suite of meteorites.

Problem (2) is less tractable. A lost metal component could still be recognized in undifferentiated meteorites, where it would manifest itself by enrichment of siderophiles in a characteristic pattern, but not in differentiated asteroids or planets, where neither the magnitude of the enrichment nor the pattern is recognizable.

Moderately Volatile Elements

Condensation temperatures need to be more accurately determined, especially for alkalis and halogens. The present order (Table 7.5.1) is Rb $>$ K $>$ Na, but the abundance pattern in chondrites shows just the reverse trend: Rb $<$ K $<$ Na. For some minor and trace elements—especially incompatibles—the nature of the carrier phase is not known, and even when it is, the activity coefficients often are poorly known.

It is also important to determine the abundance of moderately volatile elements in chondrules, to see whether they were volatilized during chondrule formation. Only purely lithophile elements count, as siderophiles and chalcophiles seem to have been lost mechanically with metal and troilite. The six principal candidates are B, Mn, Na, K, Rb, and F. Of course, only elements depleted in the bulk meteorite are suitable for this test. This leaves all six for C-chondrites but only B, Rb and F for ordinary or E chondrites. Two of these diagnostic elements—Na and Mn—have been measured in C chondrites, and indeed show depletions by 0.15 to 0.75 times (Schmitt et al. 1965). But no comparable data on the right elements exist for ordinary or E chondrites; the elements measured (Mn, Na, K) are not appreciably depleted in bulk meteorites of these classes and hence it is a foregone conclusion that they show no depletion in chondrules. It would be interesting to combine the Rb measurements with isotopic measurements of Sr, to check the reality of any apparent depletion. Micro-mapping of B also is a very promising technique (Shaw et al. 1986).

Highly Labile Elements

A major question that must be addressed in future studies is the siting and chemical state of the labile elements. Volatilization at 400 to 500° C presumably implies desorption from surfaces, whereas higher temperatures may imply release from grain interiors by diffusion, melting, or chemical reactions. Only when the actual siting is known will it be possible to calculate precise nebular condensation curves for comparison with the observed abundance data. A promising start has been made with micro-mapping of Bi and Pb (Woolum and Burnett 1981). But until complete information on siting becomes available, the range of possible condensation curves remains much wider than indicated in Figs. 7.6.4, 7.6.9 and 7.6.10, precluding a choice between condensation and metamorphism (Hertogen et al. 1983).

A related question is the original, premetamorphic volatile-element content of chondrites of low petrographic type. For C3s the abundance pattern is flat, suggesting survival with little change, but for type-3 ordinary chondrites, there are large variations, with many instances of "superenrichments" above CI-chondrite ratios (Fig. 7.6.6). Possible reasons are enrichments by left-behind volatiles during late stages of condensation ("mysterite"; such material has actually been seen as discrete clasts) or metamorphic losses/enrichments. These possibilities can be studied experimentally; as a first step, the available data should be examined in the light of the subclassifications of type-3 ordinary chondrites (Chapter 1.1).

The same question must of course be asked for the higher petrographic types. Here it may be harder to get a clear answer. Still, at least partial answers may emerge from a coordinated study of trace-element siting, abundance, and release pattern as a function of cooling rate, permeability, and peak metamorphic temperature.

Oxidation State in Chondrites

The most important question in this field is the redox state and composition of the solar nebula, in those regions where the meteorites formed. The answers provided by CAIs or chondrules must be treated with caution, as they formed by brief, small-scale events and (judging from the inconsistent fO_2s of different minerals in a single CAI) generally did not fully equilibrate with the nebula. Presumably their redox state reflects that of its parent material to a considerable extent. It may be necessary to focus on this parent material—matrix, relict crystals in chondrules, the "chondrule precursors" of Grossman and Wasson (1983), etc.—for further clues to the redox state of the nebula. This is a formidable task, as these materials themselves may be mixtures, representing different formation conditions and degrees of equilibration.

The E chondrites are particularly vexing, as their extreme state of reduction requires a separate, highly reducing, domain in the solar nebula. Indeed, no fewer than three such domains seem to be required by the present distribution of E asteroids (which have a similarly reduced, Fe^{2+}-poor, mineralogy): 434 Hungaria (1.94 AU), 44 Nysa (2.42 AU), and 64 Angelina (2.68 AU). There is a hint that the reducing conditions developed gradually; relict grains in chondrules from Qingzhen (E3) have higher Fe^{2+} than do the chondrules themselves. This trend, as well as the multiplicity of reducing domains, is easier to reconcile with local gas-dust fractionations than with pockets of H_2O-poor gas. But it remains to be seen whether gas-dust fractionation can explain all the evidence. In particular, a satisfactory model must provide a self-consistent O-isotope balance among all chondrite classes as well as nebular dust.

A very interesting lead is the variation in fayalite content among H chondrites (Fig. 7.7.2). As this may reflect a similar variation in depth within the parent body, it would be interesting to look for correlations with other properties that are related to depth: cooling rate, chemical composition (including all classes of trace elements), gas-retention age, cosmic-ray exposure age, etc.

Planetary Compositions

Further advances in this field depend on progress in two areas.

Fractionation Processes in the Solar Nebula. We must understand the composition of all meteorite classes, the processes that produced them, and the variation of these processes with heliocentric distance and with time. This includes not only the chondrites but also other classes such as eucrites—close analogues of the Moon that are consistently neglected in attempts to make the Moon by *ad hoc* processes, such as fission or giant impacts. Moreover, we must find out which properties of planetary matter were inherited from its presolar stage and which were acquired by "reprocessing" in the solar nebula. This includes not only *isotopic* but also *chemical* traits, such as Fe^{2+} content, organic compounds, "poorly characterized phases," etc.

Composition of Planets. Traditionally, density has been the principal clue, but it is highly ambiguous and must be supplemented by other constraints. Much of the work thus far has been based on the assumption of a uniform, chondritic composition, but as seismic data become available from planetary landers, it may be possible to invert the problem, inferring first the mineralogy and then the composition. Further insights may come from heat flow data (which indicate K, U, Th contents) and geochemical data, especially key element ratios (K/U, Fe/Mn), FeO content of surface rocks, etc. As cosmochemically similar elements fractionate by similar factors, it is possible to estimate an idealized bulk composition of a planet from a small number of element ratios (Morgan and Anders 1980).

Planetary Atmospheres

A key point made by Hunten et al. is that the noble gases in planets and meteorites were "derived from the solar composition . . . through different degrees of processing by the same kinds of fractionating mechanisms." The principal if not dominant mechanism is "adsorption of nebular gas on dust." Thus the trapping of planetary gases by meteorites is one of the principal clues to the formation of planetary atmospheres.

One problem awaiting solution is the characterization of "Q," the carrier of heavy planetary noble gases. It appears to be not a discrete phase but a set of adsorption sites accessible to oxidants but slow to adsorb or desorb noble gases. Several of its properties match those of the "labyrinth of pores" in carbon-black particles (Wacker et al. 1985; Zadnik et al. 1985), and the latter authors specifically suggested that Q might be the pore network in $C\delta$, the fine-grained (25 Å) carrier of the anomalous component Xe-HL. However, now that $C\delta$ has been shown to be mainly diamond (Lewis et al. 1987), the problem must be re-examined from the ground up.

First, it is necessary to check, by the most exquisite separation methods, whether Xe-HL and planetary Xe are indeed located in the same phase, and if so, whether this phase is crystalline diamond or some other phase (e.g., amorphous diamond) accompanying it. Next, the trapping of noble gases by the proper type of diamond should be studied. First results show surprisingly large uptake of Ar in diamond (2 ccSTP/g atm; Fukunaga et al. 1987), but as the gas was partly ionized by microwaves, the true equilibrium solubility is likely to be lower. Similarly, thus far there are no suitable data on the *adsorption* of noble gases by diamond, especially amorphous diamond. To be applicable to meteorites, the diamonds must match meteoritic diamonds in surface properties (i.e., no exposure to O), radiation damage, etc. Moreover, as the diamonds are presolar, the gases may have been trapped in a molecular cloud rather than the solar nebula (Yang and Anders 1982; Huss and Alexander 1987). Thus the experiments will have to be designed so as to permit extrapolation to the low temperatures and pressures of molecular clouds.

A touchstone of any model will be its ability to account for the very high gas contents of chondrites. Even the most recent studies (Zadnik et al. 1985) have failed by 10^6-fold to account for the Xe contents of meteoritic carrier phases, thus being forced to invoke special factors such as higher concentrations of active sites or larger internal surface area. With diamonds there are additional possibilities, such as the larger negative heat of sorption.

A third major topic is the distribution of noble gases within planets. Distribution coefficients (melt/solid) and solubilities (gas/melt) should be measured, to permit modeling of planetary degassing and the evolution of the various reservoirs in the mantle, crust, and atmosphere.

A fourth topic is the abundance and origin of Xe^{129}, C, N, and H on Mars, Earth, and Venus. Radiogenic Xe^{129}, in particular, provides important constraints on meteoritic sources of planetary noble gases, and the extent and timing of degassing and loss of primary and secondary atmospheres. But the isotopic and elemental ratios of C, N, and H can also provide information on sources and loss processes.

16.2.8. MAGNETIC STUDIES OF METEORITES

This field has great potential, but requires some fundamental work on the carriers of natural magnetism in meteorites. There are several interesting questions that may be resolved by magnetic studies.

1. Paleomagnetic field in the solar nebula; its strength, origin, and duration.
2. Origin of the NRM of ureilites and other achondrites: solar-nebula or parent-body field.
3. Origin of NRM in C chondrites; possible presolar components.
4. Chondrite parent bodies: one or two generations.
5. Identification of chondrites shock-heated to $>550°$ C and then cooled rapidly.
6. Magnetite in type 3 chondrites: formed in the nebula or *in situ*, by hydrothermal alteration.

16.2.9. CHONDRULES

The recent explosion in information about the composition and petrology of chondrules has helped to eliminate many of their proposed origins, but we are still left without a single thoroughly satisfactory model for their formation. Clearly, such a model has to constitute a major goal of meteoritics, both for its own sake and also for the insight it will afford us into the energetic and thermal regimes of the solar nebula. The following lines of inquiry seem potentially fruitful in this regard, though it is likely that they would operate by progressively tightening constraints rather than by yielding a revelation of the unique solution.

A comprehensive search should be carried out for systematic differences (chemical, isotopic, chronological and petrographic) between the chondrule populations present in the different chondrite groups. This would yield information on the spatial, and possibly temporal, distribution not only of the chondrules themselves but also of their precursors. More extensive study would also presumably serve to define those precursors more precisely in both chemical and mineralogical terms. Their distribution among chondrites is related to the scale of chemical processing in the solar nebula, though our inability to assign a heliocentric distance to each recognizable component thus far precludes actual chemical mapping of the nebula.

An issue of importance not only to the question of chondrule formation but also to that of fractionation of moderately volatile elements is the extent to which chondrules are depleted in such elements. Because of the depletion of chondrules in siderophile and chalcophile elements, diagnostic elements must be lithophiles, e.g., Na, K and Rb (see also Sec. 16.2.7). From the work of Tsuchiyama et al. (1981), Na loss during chondrule formation is a function of cooling rates and O fugacity. The same parameters would be expected to influence the oxidation state of Fe during chondrule formation. Although studies to date have revealed no relationship between oxidation state and Na content (Chapter 9.1), it might nonetheless be instructive to investigate this issue more closely. Such a relationship might involve, for example, Fa zoning in olivine or reduction of fayalitic olivine to "dusty" relict grains, and would shed light on the nature of the gaseous environment in which chondrules were formed. That information might in turn serve to constrain models of chondrule formation.

The Fe-rich relict olivine grains are also of interest in their own right, as a convincing explanation for their origin is lacking. Strong arguments were made in Chapter 9.1 against their origin in an earlier generation of chondrules. However, production of a relatively large, euhedral grain of fayalitic olivine within a nebular context is not straightforward (Kerridge 1977; see, however, Anders 1986). (The same issue confronts the study of matrix [Chapter 10.2] and isolated olivine grains [Chapter 10.4].) A plausible interpretation is condensation from a gas with a substantially greater fO_2 than the canonical nebular value, presumably brought about by local enhancement of the dust-to-gas ratio (Wood 1988). It would be instructive to model the minor-, and possibly trace-, element chemistry of fayalitic olivines so produced and to test such models against observed compositions of relict grains. If the match is no better than that between relict and extant chondrule olivines, it would be desirable to re-examine the plausibility of a chondrule origin for at least the Fe-rich relict olivines. Identification of an earlier generation of chondrules no longer represented as such in chondrites could provide important new constraints on chondrule origin.

Even if the relict grains were not formed originally in chondrules, observations of chondrules within chondrules and of apparently sintered rims on many chondrules suggest that chondrule material experienced more than a single heating episode. Quantifying the number, duration and severity of those episodes would further constrain mechanisms of chondrule formation. Similarly, the apparent absence of any compound objects containing both a mafic-silicate chondrule and a CAI is presumably explicable in terms of episodic heating in the nebula and could therefore help to define such episodes. In fact, the relationship between CAIs and chondrules needs further clarification with respect to time and place of formation and heat sources. Also, is there any evidence for a dependence of intensity of chondrule heating on heliocentric distance (assuming that chondrite bulk differences reflect, at least in part, differences in heliocentric distance)? Severity of heating might be indicated by loss of volatiles, enrichment of refractories, scarcity of relict grains, increased sphericity, and abundance of chondrules.

Many recent results of experimental chondrule petrology reveal the influence of terrestrial gravity on the crystallizing droplets. This suggests that experimentation in a zero- or micro-gravity environment would likely be beneficial, by better reproducing the actual chondrule-formation conditions. Although a truly zero-gravity environment would be hard to achieve, the microgravity facility planned for the Space Station might be appropriate for such a study.

Further work would also be desirable on the partitioning of minor and trace elements between minerals and melt, paying particular attention to the development of zoning patterns. The data would not only help constrain the chondrule-forming environment and the cooling rate of newly formed chondrules, but would also clarify the origin of relict grains, as discussed above, and of isolated olivine grains (see Chapter 10.4).

The observational and experimental studies proposed here would certainly be useful, but the most pressing need is for construction of a model that rigorously defines both the energy source and the mechanism for converting that energy into chondrules. The ball would seem to be in the modelers' court at this time.

16.2.10. PRIMITIVE MATERIAL SURVIVING IN CHONDRITES

It is now clear that interpreting the properties of individual lithic constituents of chondrites in terms of conditions and processes in the solar nebula is much less straightforward than might be inferred from the minimally fractionated bulk compositions of the chondrites themselves. This does not lessen the appeal of studying such properties, but means that in practice, we must try to discern a nebular signature that is not only commonly overprinted by the effects of secondary processing but also significantly more complex than predicted by current cosmochemical models. It follows that much further work is needed in observational, experimental and theoretical areas.

Matrix

Controversy permeates every aspect of the study of matrix material in chondrites, from definition of matrix to theories of its origin. The former issue, at least, should yield to a systematic survey of the nonchondrule/inclusion/aggregate material in chondrites of low petrographic type. Fayalitic olivine is a prevalent, and commonly dominant, phase in such matrices. To what extent can empirical relationships be established among these different olivines; e.g., do all CV-matrix olivines constitute a single population and if so, how is this population related to olivines in OC matrix? Furthermore, existence of a relationship between matrix olivines and those in chondrules and aggregates is still debated and more detailed comparisons are warranted. Even in cases where alteration may have obscured some chemical or isotopic signatures that might have established petrogenetic connections, others may have survived.

A related question is the relative equilibration rate of major cations and O isotopes in minerals such as olivine. This question is prompted by the observation that chondrules and rims are out of chemical equilibrium with each other, but appear to have related O-isotopic compositions (Clayton et al. 1987; see also Chapter 10.2). An experimental approach to this question seems feasible, though difficult. Besides being informative about the origin and evolution of matrix, it would also help constrain the chondrule-formation conditions if, as seems likely, rim formation was related to chondrule formation.

Despite the many uncertainties that currently surround the study of chondrite matrix, it seems likely that this is the material that contains the best record of early solar-system processes that occurred at moderate temperatures, i.e., between the nominal condensation temperatures of S and H_2O. Consequently, even though that record has been overprinted in most, if not all, cases, it would be interesting to trace a number of properties across a suite of type-3 chondrites, perhaps by means of a multidisciplinary consortium.

Refractory Inclusions

Although the study of refractory inclusions now involves most chondrite types, the overwhelming majority of the data still come from just two meteorites: Allende (CV3) and Murchison (CM2). An immediate need is therefore to broaden the data base to make it more representative of the chondrite population. In particular, systematic studies are needed of CO, ordinary and enstatite chondrites. Such studies should combine petrographic observations, O-isotopic analyses (probably by ion-microprobe in view of the small size of most non-CV inclusions) and major-element compositions. (Even for CV inclusions, bulk major-element data are less than copious: the extent to which CAIs correspond to the putative refractory-lithophile component is still not clear [see Chapter 7.3].)

Besides broadening the data base, it is desirable to search for further anomalous CAIs, whose extreme compositions can serve as sensitive tests of models for CAI production, e.g., the ultra-refractory inclusions from Ornans described by Palme et al. (1982) and Davis (1984).

Experimental petrology has already been of value in constraining the origin of many CAIs and more studies of that kind are needed, particularly for compositions corresponding to types other than CV. This is an area where active collaboration between meteoriticists and experimental petrologists would be beneficial. A major goal of such studies is an answer to the question: did a particular inclusion form as an equilibrium assemblage, or is it an aggregate of unequilibrated material? A corollary to that question is: within a single inclusion, do different elemental and isotopic systems reflect the same state of equilibration, and if not, how do they differ? We also need to understand better how processes such as vaporization, metamorphism and alteration might have affected inclusions (see Chapter 10.3).

A specific area where an experimental approach has already made valuable progress is in the study of Fremdlinge. The work of Blum et al. (1988) on the Fe-Ni-Ru-Pt system suggests strongly that most Fremdlinge could have formed by low-temperature re-equilibration within CAIs, and that the details of Ru-Os exsolution in Fremdlinge could yield cooling-rate information for those CAIs. Furthermore, an origin by condensation alone seems unlikely because the selective condensation of $\sim 1\ \mu m$ grains of two or three rare metals (e.g., Ir, Ru) would have required temperatures constant to about $10°\ C$ for about 10^3 yr (Palme and Wlotzka 1976). However, although at least some features of Fremdlinge can apparently be produced by parent-body processes (see, e.g., Armstrong et al. 1985), many aspects of the chemistry of Fremdlinge are still not well understood, such as their contents of volatile elements (e.g., Ge, Sn, As), the occurrence of minerals such as molybdenite, MoS_2, and the occurrence of sharp contacts between Pt-metal Fremdlinge and Ni-Fe which would have been destroyed by diffusion in very short times (e.g., 1 day at 900 K; Arrhenius and Raub 1978).

It is noteworthy that, despite the prevalence of isotopic anomalies in refractory inclusions, no surviving presolar grains have been identified *in situ* in such inclusions. Search for such material will presumably continue, with the ion microprobe being the obvious technique of choice, but it is, of course, quite plausible that the thermal and chemical history of such inclusions could have altered the putative presolar grains beyond recognition, particularly if they consisted of reactive compounds such as oxides (Zinner et al. 1987).

Individual Mineral Grains

The fundamental problem here, as with matrix, is to distinguish between grains derived by comminution of larger entities, i.e., chondrules, inclusions and aggregates, and material that was formed by nucleation and growth of grains dispersed in the nebula, i.e., condensation. The key diagnostic criterion is likely to remain minor-, and possibly trace-, element chemistry, though this has the drawback that it restricts the usefulness of such grains for deriving fresh insights into chemical fractionation patterns in the solar nebula. At least in the case of silicates, the presence or absence of an irradiation record in a particular grain population can help constrain its possible origin, so that compositional studies should be accompanied by track studies whenever possible. Metal grains of putatively nebular origin represent an even less tractable problem. Data from the relatively few studies conducted so far are ambiguous and identification of "primitive" criteria seems problematical. A crucial issue is the extent to which metal grains presently observed in chondrites were produced or reworked by the chondrule-forming process.

Given a population of meteoritic grains that apparently had independent existence in the nebula, the first question is whether they were formed there or whether they were of presolar origin, possibly modified during residence in the nebula. Note that although observations of interstellar, and cometary, grains are generally taken to imply sub-μm sizes, larger grains could be optically inconspicuous and therefore might be overlooked. Criteria for discriminating between these two possible origins could presumably be based on the presence or absence of detectable isotopic anomalies. However, these might be "reset" by thermal processing in the nebula. Assuming, nonetheless, that a population of preserved nebular condensate grains can be identified, and that the effects of any secondary alteration are either trivial or amenable to correction, what can we learn about the nebula?

Two questions stand out: first, were the initial condensates stoichiometric mineral grains or amorphous substances with ill-defined compositions? Second, under what circumstances, if any, did FeO enter silicate structures in the solar nebula? This question was discussed earlier in connection with relict grains in chondrules. If a nebular origin could be assigned with some certainty to a population of Fe-bearing silicates that were not connected in any way with chondrule formation, the properties of those grains would probably shed light on how such silicates formed in the nebula. This issue has recently come into sharper focus with the detection of appreciable numbers of Fe- and Si-bearing dust grains emanating from Halley's comet. Recent interpretations of the Vega 1 spacecraft data suggest that in addition to chondritic particles, Fe-bearing monomineralic olivine and pyroxene grains are present (Jessberger et al. 1988). Further detailed comparisons of the distribution of these compositions with those for analogous chondritic populations (see, e.g., Brownlee et al. 1987) would be desirable.

Although the origin of Fe-bearing silicates has been mentioned several times in this chapter already, it is an important issue that deserves emphasis. To recapitulate, the various competing hypotheses (low-temperature reactions in solar gas; condensation/reaction in nonsolar gas; chondrule crystallization; metasomatic alteration) should all be refined to yield specific mineral-chemical, petrographic and other predictions that can then be matched against the meteoritic record in a more rigorous fashion than has been employed to date.

Besides being of fundamental significance in their own right, answers to these two questions can provide clues as to whether nebular chemistry can be reconstructed using equilibrium considerations. This issue, in turn, influences the extent to which planetary compositional components can be predicted *ab initio*. An additional question, upon which observations of pristine metal grains, particularly of their size distribution, would bear, is the opacity of the solar nebula during the earliest stages of planetesimal accumulation.

Organic Matter

The major goal is to determine the *sources* of the organic compounds and the *processes* that made them. This will require sharply focused studies of compound classes that are diagnostic of particular formation processes. But in addition, it is also desirable to make a systematic, comprehensive survey of all compound classes by the most modern techniques. Many of them have not been studied since the early 1970s. The following sources and processes must be considered.

1. Molecular Cloud
   a. Ion-molecule reactions in the gas phase
   b. Radiation chemistry in the solid phase (grain mantles)
2. Stellar Atmospheres
3. Solar Nebula
   a. Equilibrium reactions
   b. Surface catalysis (Fischer-Tropsch-Type reactions)
   c. Spontaneous, uncatalyzed but kinetically controlled reactions
   d. Radiation chemistry (Miller-Urey reactions)
   e. Reprocessing by heat, reactions with the gas, isotopic exchange
4. Meteorite Parent Body
   a. Reactions with liquid water
   b. Reprocessing by heat; isotopic exchange

Amino Acids. A strong case has been made for formation of amino acids by the Strecker cyanohydrin reaction in the aqueous phase of the meteorite parent body. Of the reactants required (NH_3, H_2O, HCN, and an aldehyde), the first two are assumed to come from the solar nebula, and the last two, from the Miller-Urey reaction. However, three additional sources must be considered: ion-molecule reactions, interstellar grain mantles, and FTT reactions. Whatever the source, Comet Halley contains both in substantial amounts ($HCN/H_2O \approx 10^{-3}$; $HCHO/H_2O \approx 10^{-2}$; Schloerb et al. 1986; Combes et al. 1986).

The D enrichment of amino acids (up to 2.5 times SMOW) clearly implies that the reactants included a contribution from interstellar molecules, but as this enrichment is some 3 to 5 orders of magnitude smaller than in molecules such as HCN or HCHO, the interstellar contribution may not have been large. It would be interesting to determine the isotopic composition of H, C, and N in several individual amino acids, to see whether they are consistent with a single set of reactants in the Strecker synthesis. The water phase in the CM2 parent body had fairly normal δD, and hence some other molecule(s) must have been the principal carrier(s) of the D anomaly.

Hydrocarbons and Other Aliphatic Compounds. The key point is the predominance of straight chains, which implies formation by a selective process such as surface catalysis (FTT synthesis), rather than by a random (Miller-Urey) synthesis. Thus far, such straight-chain dominance has been reported for alkanes from C_8 upward and for hydroxyalkanoic acids, but not for amino, alkanoic, and alkandioic acids, as well as amines.

However, the data for the last 4 classes end at C_5 to C_8, just where the straight-chain dominance of alkanes begins. There is reason to believe that considerable isomerization and aromatization has taken place below C_9: C_4 is present as butene, not butane; C_6 is present as benzene, not hexane; C_7 is present as toluene or heptene, not heptane; C_8 is present mainly as xylene, etc. Thus the question of synthesis mechanism cannot be decided for any family until the data have been extended well above C_8, and to the full range of possible alteration products below C_8. Isotopic measurements of C and H on individual compounds would also be of interest, as the expected fractionations differ greatly among the mechanisms: negligible for Miller-Urey, moderate for Fischer-Tropsch, and large (especially for H) for ion-molecule reactions.

Kerogen. The largest and most daunting task is to establish the origins (sic) of the kerogen. As with amino acids, the D enrichment of up to 3-fold shows convincingly that ion-molecule reactions played a role, but since this enrichment is only 10^{-2} to 10^{-4} that of interstellar molecules and is not uniform, the kerogen cannot simply be polymerized interstellar molecules. Either the entire kerogen formed at higher T, where D/H fractionation is smaller, or only a minor part of it comes from interstellar molecules.

The structure of the kerogen at first seems to tell a very different story. It consists of aromatic ring systems bridged by $-CH_2-$ or $-CH_2-O-CH_2-$ groups, strikingly similar to coal. This has been taken to imply similar formation conditions: aromatization of organic matter at a few hundred degrees C. Such conditions are attainable in the solar nebula but not in molecular clouds, and hence at least the aromatization—if not the formation of the parent organic matter—seems to have taken place in the solar nebula.

However, a third source of the parent material must now be considered: stellar atmospheres. There are indications that circumstellar dust includes minute graphite grains, of ~ 90 atoms (Sellgren 1984). As such grains would acquire H atoms at their unsatisfied valence positions along the periphery, they would resemble polynuclear aromatic hydrocarbons (PAH). Indeed, numerous emission bands resembling those of PAH have been found in interstellar infrared spectra (Léger and Puget 1984; Allamandola et al. 1985). Thus aromatic structures apparently *can* form in interstellar space, and hence it is conceivable that the meteoritic kerogen is largely or entirely presolar. Isotopic measurements should be made on free aromatic hydrocarbons as well as aromatic fragments released by degradation of the kerogen.

There is further evidence pointing to a contribution from stellar atmospheres. On heating to 250 to 330° C, kerogen releases C_n and C_mN chains (n

= 2 to 10; m = 3 to 6), which, according to the low release temperatures, can be only weakly bound to the kerogen (Hayatsu et al. 1980). These fragments resemble two prominent families of interstellar molecules, polyacetylenes and polycyanoacetylenes. They are difficult to make by conventional processes, but form readily on vaporization of graphite; consequently, it has been proposed that these molecules form in stellar atmospheres (Kroto et al. 1986). Again, isotopic studies of these acetylenic fragments would be of interest.

A fourth process is radiation chemistry in the solid phase. It may alter and polymerize the icy mantles of dust grains in a molecular cloud and analogous condensed volatiles in the solar nebula. Being the least selective of the processes considered, it should be recognizable by a broad distribution of compounds and isomers, and an absence of aromatic rings and other, more ordered structures, as well as substantial isotope fractionations.

Evidently, given this multitude of possible sources, it will be necessary to tease apart and study the kerogen (chemically and isotopically) by the most exquisite techniques.

16.2.11. MICROMETEORITES

The principal goal of micrometeorite research is, like that of meteorite research, the exploration of conditions and processes in the early solar system, and therefore relies similarly on the identification of primitive traits or pristine entities. Consequently, its progress tends to mirror that of meteorite research, and many of the lines of inquiry described elsewhere in this chapter apply also to micrometeorite research. Specific to that area, however, are the following two topics.

First, which micrometeorites come from comets and which from asteroids? Besides the obvious significance of this question for reconstruction of each micrometeorite's history, its resolution is needed if the "primitive" information derived from the particles is to be related to a specific location within the early solar system. The comet-vs-asteroid issue has been controversial but, at least for a substantial fraction of IDPs, may be resolvable in the future using the approach of Sandford (1986), based on the different track distributions predicted for each source region (see Chapter 11.1). An additional line of inquiry that would address this issue would be to schedule some of the stratospheric collections to sample IDPs associated with known meteor showers and hence with specific comets.

Second, the number of micrometeorites collected, and hence the size of the largest particles available for study, are direct functions of the area-time product of the collection process. The advent of larger-area collectors will therefore lead to larger micrometeorites becoming available. This will permit analysis for less-abundant species and/or broader-based multidisciplinary (i.e., consortium) approaches to the study of individual particles of particular interest. The staple analytical techniques in this field will presumably continue to be the ion-microprobe and various kinds of electron-optical devices.

16.2.12. HETEROGENEITY IN THE NEBULA

In a broad sense, the fact of isotopic inhomogeneity in the solar nebula has been established qualitatively, suggesting that the next step should be quantification of the phenomenon, i.e., establishment of the spatial scale, composition and temporal evolution of the individual reservoirs. This is an important and technically nontrivial task, probably requiring ground-truth measurements on specific asteroids, comets and major planets, but begs the question of how those reservoirs were originally produced. Here the issue is clearly the discrimination between nucleogenetic isotope effects and those produced by mass-independent fractionation. At present, the two elements that yield the most broad-based data that bear upon this issue are Ti and O. For Ti, the pattern of anomalies clearly indicates a nucleogenetic origin. For that reason, it is highly unlikely that O could be devoid of nucleogenetic anomalies: the issue becomes whether unmixed nuclear components are necessary *and* sufficient to explain the full range of meteoritic observations. The arguments advanced so far, mostly of a plausibility nature, have not led to a clean resolution of this issue. What seems to be needed here is an approach based on formulation and testing of rigorous predictions based on the competing viewpoints, though it must be admitted that the apparent lack of any meteoritic property correlating reliably with the O^{16} anomaly represents a difficulty for this approach.

The relatively new field of mass-independent fractionation needs a good deal more experimental and theoretical work, as discussed in Chapter 12.1. In particular, the astrophysical environments must be defined in which such fractionations might occur and be preserved for incorporation into the early solar system.

Returning to the more general question of nebular inhomogeneity, there is good evidence that the isotopes of H and N were distributed inhomogeneously in the early solar system, though whether discrete gaseous reservoirs were involved, or gas-solid differences, or simply variability among solids is not known. Meteoritic data that bear upon the primordial distribution of N and H isotopes were discussed in Chapters 13.1 and 13.2 from a somewhat different perspective. Both these elements suffer from their vulnerability to subsequent mass-dependent fractionation, but the observed effects greatly exceed those likely to have been generated by plausible early-solar-system processes. In addition to the meteorite data, nebula-wide inhomogeneity is implied for H by data for the major planets (see, e.g., Owen et al. 1986; Geiss and Reeves 1981), and for N by solar data (see, e.g., Geiss and Bochsler 1983; Kerridge 1982). Recent data for Mars (for H and N) and Venus (for H) reveal striking departures from the terrestrial values (see Chapter 7.10), but the effects of exospheric processes have obscured the initial compositions. The primordial distribution of these isotopes should be greatly clarified by analyses of cometary nuclei and the major planets, particularly the analysis of the Jovian atmosphere by Galileo.

16.2.13. SURVIVAL OF PRESOLAR MATERIAL IN METEORITES

Circumstellar Material in Meteorites

The first major goal is isolation and characterization of all presolar solids in *pure* form. Past work has relied mainly on destructive chemical techniques, thus favoring chemically resistant phases such as carbon, spinel, and silicon carbide. However, there is little doubt that these phases originally were accompanied by other solids, such as silicates, nickel-iron, sulfides, and nitrides. Some of these phases would be destroyed by parent-body processes such as hydrothermal alteration or thermal metamorphism, but others may have survived. Perhaps these solids can be separated from their solar-system counterparts by nondestructive techniques, based on grain size, density, or magnetism.

A second goal is the isotopic characterization of each presolar solid. Ideally all 62 polyisotopic elements should be measured, but most will be too rare for present or forseeable techniques. Such measurements will be easier to interpret if the samples are "pure-bred" rather than mixtures, and for this reason progress in the field will depend mainly on development of better methods for the separation, handling, and analysis of sub-μm-sized grains. Ion-probe analysis of SiC has indeed shown large, variable anomalies of Si, C, and N (Zinner et al. 1987), as well as noble gases (Tang and Anders 1988). However, the small size of the SiC grains (~ 0.1 μm) has made it necessary to measure clusters of tens or hundreds of grains, with consequent averaging of the anomalies. It would be very desirable to improve analytical sensitivity to the point where individual grains can be measured.

A third goal is the search for genetic relationships among individual, well-defined types of interstellar grains, e.g., diamond, spinel, silicate. This task is similar to that for meteorites, and in principle, the same markers could be used: redox state, isotopic anomalies of O or other elements, cosmic-ray age, primordial noble gases, etc. However, it will be enormously difficult to use most of these markers for μg amounts of nm-sized grains. An important first step will be to decide whether a mineralogically well-defined type of grain comes from one or many stars. One possible approach would be to look for isotopic and chemical variations with size and density.

Interstellar Cloud Material in Meteorites

The principal problem in this field is characterization of the interstellar components in meteoritic organic matter. This topic has already been reviewed in Sec. 16.2.10, but from the broader perspective of the origins of all organic matter, not just the interstellar components. The task is truly daunting, as some IDPs show coexistence of several isotopically distinct components on a ~ 10-μm scale (Fig. 13.2.4). Ideally, a complete organic and isotopic analysis should be conducted on such particles, μm by μm. But since at least the organic analysis cannot be done in sufficient detail on so small a scale, these measurements will have to be complemented by studies of bulk meteorites, where the large sample size permits separation and characterization of individual components.

Special efforts should be made to find actual interstellar molecules, of molecular weight < 100. Although most of them are likely to have been incorporated in larger molecules by secondary processes, some may be trapped in kerogen or otherwise loosely bound, and may be released by grinding, mild heating, etc. Methane, CO, as well as acetylenic chains to C_{10} and C_7N have already been detected by such techniques, but have not yet been isotopically analyzed. Naturally, all separations will have to be done under the gentlest possible conditions, to minimize secondary reactions or isotopic exchange. Patient souls may prefer to wait until a sample of cometary ices has become available.

Astrophysical Implications

As described in more detail in Sec. 13.3.3, there are several areas in which the study of presolar grains surviving in meteorites can be focused in order to address specific astrophysical issues. These include: degree and scale of mixing within supernovae, inferred from intercorrelations and chemical affinities of certain isotopic anomalies; lifetimes of grains in the interstellar environment; time-dependent production of molecules in interstellar clouds. Additional examples, focusing more on the production of anomalies than on their survival, are given in Sec. 16.2.14.

16.2.14. NUCLEOSYNTHESIS

The circumstances that have led to this being a major area of achievement for meteorite research will presumably continue to apply: the fact that CI chondrites are only a step or two away from solar composition, and the precision with which they, and other meteoritic samples, such as the FUN inclusions, can be analyzed still make them the tests of choice for most nucleosynthetic models. Despite the impressive quality of present tabulations of solar-system abundances, further refinement of such abundances is not only feasible (see, e.g., Woolum et al. 1986) but desirable in light of anticipated improvements in neutron-capture cross sections (Marti and Zeh 1985).

Many details concerning both the r and p processes remain to be worked out, with the ultimate objective of narrowly defining the astrophysical environments in which they take place. Progress in this area, at least that based on observation rather than theory, will presumably continue to depend on availability of suitable samples, in particular of further FUN inclusions (Chapter 14.3). In this connection, the discovery by Brigham et al. (1988) that FUN inclusions are more than three times as abundant in purple, coarse-grained, spinel-rich CAIs as they are in the general population of coarse-grained inclusions, is of considerable interest. The purple inclusions contain a high abundance of Na- and Fe-rich secondary phases and exhibit textures that are consistent with a molten origin. Apparently the spinel is remarkably resistant to mineralogical alteration and isotopic exchange.

An issue that deserves more attention is the extent to which meteoritic data are in accord with the predictions of specific theories, such as nucleosynthetic models (recently summarized in Mathews 1988), or cosmochemical models, which also take into account correlations caused by volatility (see, e.g., Clayton 1982). For example, the latter theory predicts an association between O^{16} and Al in interstellar grains. Although a hint of such an association may be found in the meteorite record, it is far from a robust correlation, possibly because of dilution of the O^{16} carrier by mundane material. Identification of that carrier is, of course, a long-sought goal of meteorite isotopy, but even partial isolation might serve to test the theory. It should be noted, in general, that in a nucleogenetic mixture such as the protosolar system, any individual theory might be "right" for some isotopic phenomena and "wrong" for others. Thus, several features of the Fe-peak anomalies seem attributable to cosmic chemical memory (Niemeyer and Lugmair 1984) whereas applicability of that theory to the evidence for extinct nuclides is still controversial.

16.2.16. NUCLEOCOSMOCHRONOLOGY

Extinct Radionuclides

Perhaps the most urgent problem in this area is one alluded to earlier in this chapter in connection with possible asteroidal heat sources, namely determination of the abundance and distribution of Al^{26} in the early solar system. Within the context of our present state of knowledge there are two aspects to this question: what is the significance of the variations in Mg^{26*}/Mg^{24} observed in apparently identical CAIs, and was the Mg^{26*} brought into the solar system as Al^{26} or as a fossil anomaly? The answer to the former is presumably isotopic heterogeneity analogous to that observed for several other elements; but in this case, how much is due to spatial heterogeneity and how much is temporal? Existing evidence, especially the mineral isochrons and occasional higher Al^{26}/Al^{27} in inclusion rims *vis à vis* cores, strongly supports the "live" interpretation for the latter (Chapter 15.1; see, however, Clayton and Leising 1987), but further work seems desirable. Thus, a search for evidence of nuclides coproduced with Al^{26} in the cosmic-chemical-memory model would be worthwhile, and more-detailed mapping of the distribution of Al^{26} gamma-ray activity in the Galaxy would indicate whether local concentrations capable of yielding the inferred meteoritic values exist today. Evidence that Al^{26} was live in the early solar system would not, of course, rule out the possible existence also of a fossil Mg^{26} signal.

Whether the observed variability among CAI values reflects temporal variations or spatial inhomogeneities could in principle be resolved by a correlated search for the daughter products of another appropriate radionuclide. The logical candidate would be Ca^{41} but since no convincing evidence has yet been found for its existence in the early solar system (Chapter 15.1), the feasibility of a test based upon it is problematical.

The search for a K^{41} anomaly indicative of *in situ* decay of Ca^{41} should continue, regardless of its possible supporting role in the Al^{26} story. Similarly, evidence for the decay products of extinct nuclides such as Cm^{247}, Fe^{60}, Cs^{135} and Pb^{205} will continue to be sought in appropriate samples. Plausible host materials for these nuclides are: CAIs for Cm^{247} (although studies of such inclusions have so far yielded only upper limits); CI magnetite for Fe^{60}; fine-grained, alkali-bearing CAIs for Cs^{135}; and perhaps S-bearing Fremdlinge for Pb^{205}.

Although data are still scanty, there is evidence that the distribution of radiogenic Cr^{53} from the decay of Mn^{53}, with a half-life of 3.7 Myr, might be used for fine-scale chronology of events in the early solar system (Birck and Allègre 1987). For all studies of extinct nuclides, however, the possibility of variation brought about by initial inhomogeneity must be borne in mind.

Nucleocosmochronology

Several interesting questions emerge from the discussion in Chapter 15.2, but few of them seem to be amenable to attack through meteorite analysis. If indeed the 100-Myr free-decay interval inferred from some of the extinct-radionuclide data is real, it is obviously of importance to ascertain its astrophysical significance. However, it is difficult to see how meteoritic material could have retained a sufficiently explicit record of the pertinent presolar history to address that issue. Perhaps more accessible to a meteorite-based study would be a rigorous test of the concept that different, apparently discrepant, radionuclides are dating the same event in the early solar system. This question is part of the more general issue of the fine-scale chronology of the early solar system, an area in which continuing progress can be expected, in step with ongoing improvements in methodology (see Sec. 16.2.5 and Chapter 5.1).

Iodine-Xenon Dating

From its inception, the technique of I-Xe dating has been one of the most promising tools for the study of the earliest solar system but for several reasons application of this approach is not straightforward. First, the technique has the potential to resolve events of the order of 1 Myr apart, which is probably an appropriate scale for the study of early planetesimal and planetary development, but may be too coarse for study of nebular events (see Chapters 6.3 and 6.4). Second, accuracy of the technique is limited by the extent to which initial isotopic homogeneity can be assumed. This is presently uncertain; i.e., much of the spread of ages in Fig. 15.3.3 may be due to initial isotopic inhomogeneity. Third, one needs to know the host phases of the Xe^{129}, their closure temperatures and the physical meaning of such closures (see also Chapter 15.3).

Given resolution of these issues and availability of suitable samples, the following processes could in principle be addressed, provided that they took place within the time span covered by I^{129}: condensation of volatile-rich phases, either locally or on a nebular scale; chemical alteration of CAIs; chondrule formation; thermal metamorphism; and aqueous mineralization. In some cases, not only the timing but also the duration of the event could be determined. In the case of the chronology of thermal metamorphism, it would be of particular interest to search for a relationship between I-Xe "age" and parent-body structure, which could bear upon the debate between the "onion-shell" and "rubble-pile" models for chondrite parent bodies (Chapter 3.3; see also Miyamoto et al. 1981; Scott and Rajan 1981).

16.2.16. CONCLUDING REMARKS

Those who have contributed to this book, and their colleagues in the field, if they agree on nothing else, share a common belief that the study of meteorites and the early solar system represents a source of extraordinary intellectual excitement. We hope that this excitement has communicated itself in the pages of this book, and that younger readers, in particular, will see in this chapter, and throughout the book, a challenge to which they simply must respond.

Acknowledgments. We thank the following for their assistance in preparing this chapter: C. A. Brigham, D. D. Clayton, D. Lal, G. J. MacPherson, J. A. Nuth, A. E. Rubin and P. H. Warren.

REFERENCES

Allamandola, L. J., Tielens, A. G. G. M., and Barker, J. R. 1985. Polycyclic aromatic hydrocarbons and the unidentified infrared emission bands: Auto exhaust along the Milky Way! *Astrophys. J.* 290:L25–L28.

Anders, E. 1978. Most stony meteorites come from the asteroid belt. In *Asteroids: An Exploration Assessment*, eds. D. Morrison and W. C. Wells, NASA CP-2053, pp. 57–75.

Anders, E. 1986. What can meteorites tell us about comets? In *Comet Nucleus Sample Return*. Proc. ESA Workshop, Canterbury, pp. 1–9.

Armstrong, J. T., El Goresy, A., and Wasserburg, G. J. 1985. Willy: A prize noble Ur-

Fremdling—Its history and implications for the formation of Fremdlinge and CAI. *Geochim. Cosmochim. Acta* 49:1001–1022.

Arrhenius, G., and Raub, C. J. 1978. Thermal history of primordial metal grains. *J. Less-Common Metals* 62:417–430.

Birck, J. L., and Allègre, C. J. 1987. ^{53}Cr isotopic anomalies related to ^{53}Mn decay in old solar system matter. *Meteoritics* 22:325–326 (abstract).

Blum, J. D., Wasserburg, G. J., Hutcheon, I. D., Beckett, J. R., and Stolper, E. M. 1988. "Domestic" origin of opaque assemblages in refractory inclusions in meteorites. *Nature* 331:405–409.

Brigham, C. A., Hutcheon, I. D., Papanastassiou, D. A., and Wasserburg, G. J. 1988. Isotopic heterogeneity and correlated isotope fractionation in purple FUN inclusions. *Lunar Planet. Sci.* XIX:132–133 (abstract).

Brownlee, D. E., Wheelock, M. M., Temple, S., Bradley, J. P., and Kissel, J. 1987. A quantitative comparison of Comet Halley and carbonaceous chondrites at the submicron level. *Lunar Planet. Sci.* XVIII:133–134 (abstract).

Caffee, M. W., Hohenberg, C. M., Swindle, T. D., and Goswami, J. N. 1987. Evidence in meteorites for an active early sun. *Astrophys. J.* 313:L31–L35.

Clayton, D. D. 1982. Cosmic chemical memory: A new astronomy? *J. Roy. Astron. Soc.* 23:174–212.

Clayton, D. D., and Leising, M. D. 1987. ^{26}Al in the interstellar medium. *Phys. Rept.* 144:1–50.

Clayton, R. N., and Mayeda, T. K. 1984. The oxygen isotope record in Murchison and other carbonaceous chondrites. *Earth Planet. Sci. Lett.* 67:151–161.

Clayton, R. N., Mayeda, T. K., Rubin, A. E., and Wasson, J. T. 1987. Oxygen isotopes in Allende chondrules and coarse-grained rims. *Lunar Planet. Sci.* XVIII:187–188 (abstract).

Combes, M., Moroz, V., Crifo, J. F., Bibring, J. P., Coron, N., Crovisier, J., Encrenaz, T., Sanko, N., Grigoriev, A., Bockelée-Morvan, D., Gispert, R., Emerich, C., Lamarre, J. M., Rocard, F., Krasnopolsky, V., and Owen T. 1986. Detection of parent molecules in Comet Halley from the IKS-Vega experiment. *Proc. 20th ESLAB Symp. on Exploration of Halley's Comet,* Heidelberg, 27–31 Oct. 1986, ESA-SP-250 1:353–358.

Davis, A. M. 1984. A scandalously refractory inclusion in Ornans. *Meteoritics* 19:214 (abstract).

Dennison, J. E., Lingner, D. W., and Lipschutz, M. E. 1986. Antarctic and non-Antarctic meteorites: Different populations. *Nature* 319:390–393.

Fleischer, R. L., Price, P. B., and Walker, R. M. 1968. Identification of Pu^{244} fission tracks and the cooling of the parent body of the Toluca meteorite. *Geochim. Cosmochim. Acta* 32:21–31.

Fukunaga, K., Matsuda, J. I., Nagao, K., Miyamoto, M., and Ito, K. 1987. Noble gas enrichment in vapour growth diamonds and the origin of diamonds in ureilites. *Nature,* 328:141–143.

Geiss, J., and Bochsler, P. 1983. Nitrogen isotopes in the solar system. *Geochim. Cosmochim. Acta* 46:529–548.

Geiss, J., and Reeves, H. 1981. Deuterium in the solar system. *Astron. Astrophys.* 93:189–199.

Goldstein, J. I. 1986. Metallographic cooling rates—Recent advances. *Meteoritics* 21:370–371.

Goswami, J. N., and Lal, D. 1979. Formation of the parent bodies of the carbonaceous chondrites. *Icarus* 40:510–521.

Gradie, J., and Tedesco, E. F. 1982. Compositional structure of the asteroid belt. *Science* 216:1405–1407.

Grossman, J. N., and Wasson, J. T. 1983. Refractory precursor components of Semarkona chondrules and the fractionation of refractory elements among chondrites. *Geochim. Cosmochim. Acta* 47:759–771.

Hayatsu, R., Scott, R. G., Studier, M. H., Lewis, R. S., and Anders, E. 1980. Carbynes in meteorites: Detection, low-temperature origin, and implications for interstellar molecules. *Science* 209:1515–1518.

Hertogen, J., Janssens, M.-J., Takahashi, H., Morgan, J. W., and Anders, E. 1983. Enstatite chondrites: Trace element clues to their origin. *Geochim. Cosmochim. Acta* 47:2241–2255.

Huss, G. R., and Alexander, E. C., Jr. 1987. On the pre-solar origin of the "normal planetary" noble gas component in meteorites. *Proc. Lunar Planet. Sci. Conf.* 17, *J. Geophys. Res. Suppl.* 92:E710–E716.

Hutchison, R., Alexander, C. M. O., and Barber, D. J. 1987. The Semarkona meteorite: First recorded occurrence of smectite in an ordinary chondrite. *Geochim. Cosmochim. Acta* 51:1875–1882.

Jessberger, E. K., Christoforidis, A., and Kissel, J. 1988. Aspects of the major element composition of Halley's dust. *Nature,* in press.

Keil, K. 1962. On the phase composition of meteorites. *J. Geophys. Res.* 67:4055–4061.

Kerridge, J. F. 1977. Iron: Whence it came, where it went. *Space Sci. Rev.* 20:3–68.

Kerridge, J. F. 1980. Secular variations in composition of the solar wind: Evidence and causes. In *Proc. Conf. Ancient Sun,* eds. R. O. Pepin, J. A. Eddy, and R. B. Merrill (New York: Pergamon Press), pp. 475—489.

Kerridge, J. F. 1982. Whence so much ^{15}N? *Nature* 295:308.

Kroto, H. W., Heath, J. R., O'Brien, S. C., Curl, R. F., and Smalley, R. E. 1987. Long carbon chain molecules in circumstellar shells. *Astrophys. J.* 314:352–355.

Léger, A., and Puget, J. L. 1984. Identification of the "unidentified" IR emission features of interstellar dust? *Astron. Astrophys.* 137:L5–L8.

Lewis, R. S., Tang, M., Wacker, J. F., Anders, E., and Steel, E. 1987. Interstellar diamonds in meteorites. *Nature* 326:160–162.

Lipschutz, M. E., Gaffey, M. J., and Pellas, P. 1988. Meteoritic evidence on asteroid parent bodies. In *Asteroids II,* eds. R. P. Binzel, T. Gehrels, and M. S. Matthews (Tucson: Univ. of Arizona Press) in press.

Marti, K., and Zeh, H. D. 1985. History and current understanding of the Suess abundance curve. *Meteoritics* 20:311–320.

Mathews, G. J., ed. 1988. *The Origin and Distribution of the Elements* (Singapore: World Scientific Publ.), in press.

McSween, H. Y. 1977a. Carbonaceous chondrites of the Ornans type: A metamorphic sequence. *Geochim. Cosmochim. Acta* 41:477–491.

McSween, H. Y. 1977b. Petrographic variations among carbonaceous chondrites of the Vigarano type. *Geochim. Cosmochim. Acta* 41:1777–1790.

Miyamoto, M., Fujii, N., and Takeda, H. 1981. Ordinary chondrite parent body: An internal heating model. *Proc. Lunar Planet. Sci. Conf.* 12:1145–1152.

Morgan, J. W., and Anders, E. 1980. Chemical composition of the Earth, Venus, and Mercury. *Proc. Natl. Acad. Sci. USA* 77:6973–6977.

Morgan, J. W., Janssens, M.-J., Takahashi, H., Hertogen, J., and Anders, E. 1985. H-chondrites: Trace element clues to their origin. *Geochim. Cosmochim. Acta* 49:247–259.

Nakamura, N. 1974. Determination of REE, Ba, Fe, Mg, Na and K in carbonaceous and ordinary chondrites. *Geochim. Cosmochim. Acta* 38:757–775.

Niemeyer, S., and Lugmair, G. W. 1984. Titanium isotopic anomalies in meteorites. *Geochim. Cosmochim. Acta* 48:1401–1416.

Owen, T., Lutz, B. L., and de Bergh, C. 1986. Deuterium in the outer solar system: Evidence for two distinct reservoirs. *Nature* 320:244–246.

Palme, H., and Wlotzka, F. 1976. A metal particle from a Ca,Al-rich inclusion from the meteorite Allende, and the condensation of refractory siderophile elements. *Earth Planet. Sci. Lett.* 33:45–60.

Palme, H., Wlotzka, F., Nagel, K., and El Goresy, A. 1982. An ultra-refractory inclusion from the Ornans carbonaceous chondrite. *Earth Planet. Sci. Lett.* 61:1–12.

Pedroni, A., Baur, H., Wieler, R., and Signer, P. 1988. T-Tauri irradiation of Kapoeta grains? *Lunar Planet. Sci.* XIX:913–914 (abstract).

Pellas, P. 1986. Onion-shell structure of the H-asteroid: A confirmation. *Meteoritics* 21:482 (abstract).

Pellas, P., and Storzer, D. 1981. ^{244}Pu fission track thermometry and its application to stony meteorites. *Proc. Roy. Soc. London* A374:253–270.

Rambaldi, E. R., Wänke, H., and Larimer, J. W. 1979. Interelement refractory siderophile fractionation in ordinary chondrites. *Proc. Lunar Planet. Sci. Conf.* 10:997–1010.

Rammensee, W., and Palme, H. 1982. Metal-silicate extraction technique for the analysis of geological and meteoritical samples. *J. Radiochem.* 71:401–418.

Sandford, S. A. 1986. Solar flare track densities in interplanetary dust particles: The determination of asteroidal versus cometary sources of the zodiacal dust cloud. *Icarus* 68:377–394.

Schloerb, F. P., Kinzel, W. M., Swade, D. A., and Irvine, W. M. 1986. HCN production from Comet Halley. *Proc. 20th ESLAB Symp. on Exploration of Halley's Comet*, Heidelberg, 27–31 Oct., ESA SP-250, 1:577–581.

Schmitt, R. A., Smith, R. H., and Goles, G. G. 1965. Abundances of Na, Sc, Cr, Mn, Fe, Co, and Cu in 218 individual meteoritic chondrules via activation analysis, 1. *J. Geophys. Res.* 70:2419–2444.

Scott, E. R. D., and Rajan, S. 1981. Metallic minerals, thermal histories, and parent bodies of some xenolithic, ordinary chondrites. *Geochim. Cosmochim. Acta* 45:53–67.

Sears, D. W. G., Grossman, J. N., Melcher, C. L., Ross, L. M., and Mills, A. A. 1980. Measuring metamorphic history of unequilibrated ordinary chondrites. *Nature* 287:791–795.

Sellgren, K. 1984. The near-infrared continuum emission of visual reflection nebulae. *Astrophys. J.* 277:623–633.

Shaw, D. M., Higgins, M. D., and Truscott, M. G. 1986. Status of alpha track imaging of chondrites for B/Li distributions. *Meteoritics* 21:510 (abstract).

Stolper, E. 1977. Experimental petrology of eucritic meteorites. *Geochim. Cosmochim. Acta* 41:587–611.

Tang, M., and Anders, E. 1988. Isotopic anomalies of Ne, Xe, and C in meteorites. II. Interstellar diamond and SiC: Carriers of exotic noble gases. *Geochim. Cosmochim. Acta*, in press.

Taylor, G. J., Maggiore, P., Scott, E. R. D., Rubin, A. E., and Keil, K. 1987. Original structures, and fragmentation and reassembly histories of asteroids: Evidence from meteorites. *Icarus* 69:1–13.

Tsuchiyama, A., Nagahara, H., and Kushiro, I. 1981. Volatilization of sodium from silicate melt spheres and its application to formation of chondrules. *Geochim. Cosmochim. Acta* 45:1357–1367.

VanSchmus, W. R., and Wood, J. A. 1967. A chemical-petrologic classification for the chondritic meteorites. *Geochim. Cosmochim. Acta* 31:747–765.

Wacker, J. F., Zadnik, M. G., and Anders, E. 1985. Laboratory simulation of meteoritic noble gases. I. Sorption of xenon on carbon: Trapping experiments. *Geochim. Cosmochim. Acta* 49:1035–1048.

Wetherill, G. W. 1986. Unexpected Antarctic chemistry. *Nature* 319:357–358.

Wilkening, L. L. 1975. High temperature condensates among meteors? *Nature* 258:689–690.

Willis, J., and Goldstein, J. I. 1981. A revision of the metallographic cooling rate curves for chondrites. *Proc. Lunar Sci. Conf.* 12:1135–1143.

Wlotzka, F. 1987. Equilibration temperatures and cooling rates of chondrites: A new approach. *Meteoritics* 22:529–531.

Wood, J. A. 1988. Chondritic meteorites and the solar nebula. *Ann. Rev. Earth Planet. Sci.* 16, in press.

Woolum, D. S., and Burnett, D. S. 1981. Metal and Bi/Pb microdistribution studies of an L3 chondrite: Their implications for a meteorite parent body. *Geochim. Cosmochim. Acta* 45:1619–1632.

Woolum, D. S., Burnett, D. S., Benjamin, T. M., Rogers, P. S. Z., Duffy, C. J., and Maggiore, C. J. 1986. Trace element contents of primitive meteorites: A test of solar system abundance smoothness. *Nucl. Instr. Meth.* B22:376–379.

Yang, J., and Anders, E. 1982. Sorption of noble gases by solids, with reference to meteorites. III. Sulfides, spinels, and other substances; on the origin of planetary gases. *Geochim. Cosmochim. Acta* 46:877–892.

Zadnik, M. G., Wacker, J. F., and Lewis, R. S. 1985. Laboratory simulation of meteoritic noble gases. II. Sorption of xenon on carbon: Etching and heating experiments. *Geochim. Cosmochim. Acta* 49:1049–1059.

Zinner, E., Tang, M., and Anders, E. 1987. Large isotopic anomalies of Si, C, N and noble gases in interstellar silicon carbide from the Murray meteorite. *Nature* 330:730–732.

APPENDIX 1:
MINERAL NAMES

MINERAL NAMES

Mineral	Formula	Mineral	Formula
Åkermanite	$Ca_2MgSi_2O_7$	Hematite	Se_2O_3
Alabandite	$(Mn,Fe)S$	Hercynite	$(Fe,Mg)Al_2O_4$
Albite	$NaAlSi_3O_8$	Hibonite	$CaAl_{12}O_{19}$
Andradite	$Ca_3Fe_2Si_3O_{12}$	Ilmenite	$FeTiO_3$
Anorthite	$CaAl_2Si_2O_8$	Kaersutite	$Ca_2(Na,K)(Mg,Fe)_4TiSi_6\text{-}$
Apatite	$Ca_3(PO_4)_2$		$Al_2O_{22}F_2$
Aragonite	$CaCO_3$	Kamacite	$\alpha\text{-}(Fe,Ni)$
Armalcolite	$FeMgTi_2O_5$	Krinovite	$NaMg_2CrSi_3O_{10}$
Augite	$Mg(Fe,Ca)Si_2O_6$	Lawrencite	$(Fe,Ni)Cl_2$
Awaruite	Ni_3Fe	Lonsdaleite	C
Baddeleyite	ZrO_2	Mackinawite	FeS_{1-x}
Barringerite	$(Fe,Ni)_2P$	Maghemite	Se_2O_3
Bassanite	$CaSO_4 \cdot 1/2H_2O$	Magnesiochromite	$MgCr_2O_4$
Bloedite	$Na_2Mg(SO_4)_2 \cdot 4H_2O$	Magnesite	$(Mg,Fe)CO_3$
Brezinaite	Cr_3S_4	Magnetite	Fe_3O_4
Brianite	$CaNa_2Mg(PO_2)$	Majorite	$Mg_3(MgSi)Si_3O_{12}$
Buchwaldite	$NaCaPO_4$	Marcasite	FeS_2
Calcite	$CaCO_3$	Melilite solid solution	
Carlsbergite	CrN	åkermanite (Ak)	$Ca_2MgSi_2O_7$
Caswellsilverite	$NaCrS_2$	gehlenite (Ge)	$Ca_2Al_2SiO_7$
Chalcopyrite	$CuFeS_2$	Merrihueite	$(K,Na)_2Fe_5Si_{12}O_{30}$
Chamosite	$Fe_6Mg_3[(Si_4O_{10})(OH)_8]_2$	Merrillite	$Ca_9MgH(PO_4)_7$
Chaoite	C	Mica	$(K,Na,Ca)_2Al_4[Si_6Al_2O_{70}]\text{-}$
Clinopyroxene	$(Ca,Mg,Fe)SiO_3$		$(OH,F)_4$
Chlorapatite	$Ca_5(PO_4)_3Cl$	Molybdenite	MoS_2
Chromite	$FeCr_2O_4$	Monticellite	$Ca(Mg,Fe)SiO_4$
Cohenite	$(Fe,Ni)_3C$	Montmorillonite	$Al_4(Si,Al)_8O_{20}(OH)_4Mg_6\text{-}$
Copper	Cu		$(Si,Al)_8O_{20}(OH)_4$
Cordierite	$Mg_2Al_4Si_5O_{18}$	Nepheline	$NaAlSiO_4$
Corundum	Al_2O_3	Niningerite	$(Mg,Fe)S$
Cristobalite	SiO_2	Oldhamite	CaS
Cronstedtite	$(Mg,Fe)_2Al_3Si_5AlO_{18}$	Olivine	$(Mg,Fe)_2SiO_4$
Cubanite	$CuFe_2S_3$	Olivine solid solution	
Daubreelite	$FeCr_2S_4$	fayalite (Fa)	Fe_2SiO_4
Diamond	C	forsterite (Fo)	Mg_2SiO_4
Diopside	$CaMgSi_2O_6$	Orthoclase	$KAlSi_3O_8$
Djerfisherite	$K_3CuFe_{12}S_{14}$	Orthopyroxene	$(Mg,Fe)SiO_3$
Dolomite	$CaMg(CO_3)_2$	Osbornite	TiN
Enstatite	$MgSiO_3$	Panethite	$(Ca,Na)_2(Mg,Fe)_2(PO_4)_2$
Epsomite	$MgSO_4 \cdot 7H_2O$	Pentlandite	$(Fe,Ni)_9S_8$
Farringtonite	$Mg_3(PO_4)_2$	Perovskite	$CaTiO_3$
Fassaite	$Ca(Mg,Ti,Al)(Al,Si)_2O_6$	Perryite	$(Ni,Fe)_5(Si,P)_2$
Fayalite	Fa_2SiO_4	Pigeonite	$(Fe,Mg,Ca)SiO_3$
Feldspar solid solution		Plagioclase	
albite (Ab)	$NaAlSi_3O_8$	albite	$NaAlSi_3O_8$
anorthite (An)	$CaAl_2Si_2O_8$	anorthite	$CaAl_2Si_2O_8$
orthoclase (Or)	$KAlSi_3O_8$	Portlandite	$Ca(OH)_2$
Ferrosilite	$FeSiO_3$	Potash feldspar	$(K,Na)AlSi_3O_8$
Forsterite	Mg_2SiO_4	Pyrite	FeS_2
Gehlenite	$Ca_2Al_2SiO_7$	Pyrope	$Mg_3Al_2(SiO_4)_3$
Gentnerite	$Cu_8Fe_3Cr_{11}S_{18}$	Pyroxene solid solution	
Graftonite	$(Fe,Mn)_3(PO_4)_2$	enstatite (En)	$MgSiO_3$
Graphite	C	ferrosilite (Fs)	$FeSiO_3$
Greigite	Fe_3S_4	wollastonite (Wo)	$CaSiO_3$
Grossular	$Ca_3Al_2Si_3O_{12}$	Pyrrhotite	$Fe_{1-x}S$
Gypsum	$CaSO_4 \cdot 2H_2O$	Quartz	SiO_2
Haxonite	$Fe_{23}C_6$	Rhönite	$Ca_4(Mg,Al,Ti)_{12}(Si,Al)_{12}O_{40}$
Heazlewoodite	Ni_3S_2	Richterite	$Na_2CaMg_5Si_8O_{22}F_2$
Hedenbergite	$CaFeSi_2O_6$	Ringwoodite	$(Mg,Fe)_2SiO_4$
Heideite	$(Fe,Cr)_{1+x}(Ti,Fe)_2S_4$	Roaldite	$(Fe,Ni)_4N$

MINERAL NAMES continued

Mineral	Formula	Mineral	Formula
Roedderite	$(K,Na)_2Mg_5Si_{12}O_{30}$	Stanfieldite	$Ca_4(Mg,Fe)_5(PO_4)_6$
Rutile	TiO_2	Suessite	Fe_3Si
Sanidine	$KAlSi_3O_8$	Sulfur	S
Sarcopside	$(Fe,Mn)_3(PO_4)_2$	Taenite	$\gamma\text{-}(Fe,Ni)$
Scheelite	$CaWO_4$	Tetrataenite	$FeNi$
Schöllhornite	$Na_{0.3}(H_2O)[CrS_2]$	Thorianite	ThO_2
Schreibersite	$(Fe,Ni)_3P$	Tridymite	SiO_2
Serpentine (or chlorite)	$(Mg,Fe)_6Si_4O_{10}(OH)_8$	Troilite	FeS
Sinoite	Si_2N_2O	Ureyite	$NaCrSi_2O_6$
Smythite	Fe_9S_{11}	V-rich magnetite	$(Fe,Mg)(Al,V)_2O_4$
Sodalite	$Na_8Al_6Si_6O_{24}Cl_2$	Valleriite	$CuFeS_2$
Sphalerite	$(Zn,Fe)S$	Vaterite	$CaCO_3$
Spinel	$MgAl_2O_4$	Whewellite	$CaC_2O_4 \cdot H_2O$
Spinel Solid Solution		Wollastonite	$CaSiO_3$
spinel	$MgAl_2O_4$	Yagiite	$(K,Na)_2(Mg,Al)_5(Si,Al)_{12}O_{30}$
hercynite	$FeAl_2O_4$	Zircon	$ZrSiO_4$
chromite	$FeCr_2O_4$		
magnesiochromite	$MgCr_2O_4$		
V-rich magnetite	$(Fe,Mg)(Al,V)_2O_4$		

APPENDIX 2:
IMPACT BRECCIAS

IMPACT BRECCIAS[a]

	Equilibrium Shock Pressure Regime (GPa)	Post-Shock Temperature (°C)	Post-Depositional Equilibrium Temperature (°C)	Petrographic Characteristics	Geological Setting after Shock and Brecciation
Shocked meteorites or shocked clasts of meteorites:	uniform in the range of	uniform in the range of	variable depending on geological setting	(1) *low and medium shock*: primary rock texture largely preserved but shock effects in constituent minerals (deformations, mineral glasses, high pressure phases); shock veins and melt pockets may be present (2) *strong shock*: whole rock melt with glassy or devitrified texture and droplet or irregular shape: quench texture in metal	(1) as lithic clasts and melt particles within surface regolith (2) as lithic clasts or melt particles within polymict breccias of impact crater fills and ejects blankets (subsurface or megaregolith) (3) part of central crater uplift (4) components of an impact-fragmented, reaccreted parent body
(a) Chondrites	~ 5 to 100	300 to 2000			
(b) Achondrites	~ 10 to 100	100 to 2000			
(c) Iron meteorites	~ 1 to 150	0 to 2000			
Monomict breccias	uniform < 10 to 20	uniform < 100 to 200	variable depending on geological setting	cataclastic texture of a preexisting uniform lithology (igneous or metamorphic)	(1) as clasts within polymict breccias formed in impact craters (2) crater basement (3) component of fragmented and reaccreted parent body
Dimict breccias[b]	bimodal < 10 to 20 and > ~ 10 or > ~ 80	bimodal < 100 to 200 > ~ 100 or > ~ 1500	variable depending on geological setting	one lithology is a cataclastic rock (monomict breccia), the other is either a polymict fragmental breccia or an impact melt breccia with intrusive texture	dike breccias within crater basement or allochthonous megablocks of impact craters

Polymict breccias				Matrix	Occurrence
(a) Regolith breccias[c]	variable <~100	variable <~2000	variable	lithic and mineral clasts and melt particles set in a clastic matrix with intergranular matrix glass	near surface regolith of parent body
(b) Fragmental breccias[c]		<300 to 1000	<ca 500	lithic and mineral clasts in a fragmental matrix of rock debris	(1) breccias lens of crater cavity and ejecta blanket of impact craters; (2) dike breccias in basement of craters; (3) part of an impact-fragmented and reaccreted parent body
(c) Impact melt breccias			<ca 2000	(1) lithic and mineral clasts in a fine-grained crystalline, igneous-textured matrix; (2) lithic and mineral clasts in a coherent glassy or (partially) devitrified matrix	(1) melt sheets on top of breccia lens in the cavity or on top of near-rim ejecta of impact craters; (2) dike breccias in basement of craters; (3) component of an impact-fragmented and reaccreted parent body
(d) Granulitic breccias[d]			variable, with post-depositional thermal annealing at ~800 to 1000	lithic and mineral clasts in a crystalline metamorphic (granoblastic to poikiloblastic) matrix	(1) inclusions in impact-melt sheets; (2) thermally annealed breccia formation within parent body

[a] Pressure-temperature, textural and geological characteristics of shocked meteorites and meteoritic breccias (modified after Stöffler et al. 1980; Keil 1982; Taylor 1982); post-shock temperature relative to 0°C ambient temperature.
[b] For meteorites only dimict breccias with one lithology representing an impact melt are known (Taylor 1982).
[c] Include also "genomict breccias" as defined by Wasson (1974): breccia consisting of fragments of the same chondritic group (e.g., LL) but of different petrographic type (degree of recrystallization).
[d] In principle, all other dimict and polymict breccias may be the precursor lithologies of these breccias.

APPENDIX 3:
CHONDRITE BULK ELEMENTAL ANALYSES

CHONDRITE BULK ELEMENTAL ANALYSES

	CI	CM	CO	CV	H	L	LL	EH	EL
Na mg/g	4.95 (4.4–6.2)	4.10 (3.8–4.5)	4.10 (3.8–4.5)	3.25 (3.1–3.7)	6.35 (5.8–7.0)	7.00 (6.3–7.8)	6.98 (5.9–7.9)	6.80 (5.4–7.4)	5.80 (5.4–5.9)
Mg mg/g	96 (86–100)	117 (110–130)	145 (140–150)	145 (135–152)	140 (130–150)	147 (140–160)	148 (140–160)	106 (96–110)	141 (130–150)
Al mg/g	8.60 (7.9–9.6)	11.8 (10–13)	14.3 (12–16)	17.5 (16–21)	11.3 (10–13)	12.2 (11–14)	11.9 (11–13)	8.10 (7.7–8.6)	10.5 (10–12)
Si mg/g	103 (94–110)	129 (120–135)	159 (155–165)	156 (150–160)	169 (160–180)	187 (180–195)	190 (180–195)	167 (155–175)	186 (175–205)
P μg/g	800 (760–800)	900 (740–1000)	1080 (1000–1200)	990 (930–1100)	1080 (1000–1200)	950 (830–1100)	850 (800–900)	2000 (1900–2100)	1170 (1000–1300)
S mg/g	59.2 (50–67)	33.0 (28–37)	20.0 (18–24)	22.3 (20–25)	20.0 (13–26)	22.2 (17–28)	23.4 (11–55)	57.7 (55–62)	33.2 (26–45)
K μg/g	530 (370–650)	400 (340–490)	345 (310–380)	310 (265–340)	780 (610–890)	825 (630–920)	793 (610–920)	800 (430–970)	735 (600–845)
Ca mg/g	9.40 (8.7–10)	12.7 (11–15)	15.8 (15–17)	19.0 (17–23)	12.5 (11–15)	13.1 (11–16)	12.8 (11–15)	8.50 (7.5–10)	10.1 (8.5–12)
Sc μg/g	5.82 (5.6–6.5)	8.15 (7.2–9.3)	9.60 (8.9–10)	11.3 (11–12)	7.95 (7.3–8.3)	8.60 (7.7–9.7)	8.35 (7.4–9.3)	5.70 (5.1–5.9)	7.40 (6.8–7.8)
Ti μg/g	430 (400–450)	580 (500–700)	780 (600–900)	980 (850–1100)	600 (580–620)	630 (600–670)	620 (600–640)	450 (430–470)	580 (430–700)
V μg/g	56 (50–59)	75 (67–82)	92 (89–97)	96 (93–100)	74 (60–80)	77 (65–84)	75 (64–81)	54 (51–59)	60 (52–68)
Cr mg/g	2.60 (2.5–2.7)	3.05 (2.6–3.5)	3.55 (3.2–3.8)	3.60 (3.4–3.8)	3.66 (3.1–4.2)	3.88 (3.5–4.2)	3.74 (3.3–4.0)	3.15 (2.9–3.7)	3.05 (2.6–3.4)
Mn mg/g	1.94 (1.9–2.0)	1.70 (1.6–1.8)	1.65 (1.6–1.7)	1.45 (1.4–1.5)	2.32 (2.1–2.5)	2.57 (2.4–2.8)	2.62 (2.5–2.7)	2.20 (1.8–2.3)	1.63 (1.2–2.2)
Fe mg/g	182 (180–190)	210 (200–220)	248 (240–250)	235 (230–240)	275 (250–330)	215 (190–230)	185 (160–200)	290 (270–300)	220 (200–250)
Co μg/g	510 (500–520)	575 (550–600)	688 (680–700)	655 (640–670)	810 (760–890)	590 (490–670)	490 (380–570)	840 (750–900)	670 (560–810)
Ni mg/g	10.4 (9.8–11)	12.0 (11–13)	14.0 (13–15)	13.4 (12–14)	16.0 (14–18)	12.0 (10–14)	10.2 (8.0–12)	17.5 (16–19)	13.0 (11–15)
Rb μg/g	2.10 (1.5–2.6)	1.70 (1.2–2.9)	1.45 (1.1–1.9)	1.25 (1.0–2.0)	2.90 (2.0–3.5)	3.10 (1.9–4.1)	3.10 (1.0–5.0)	2.60 (2.1–2.7)	2.50 (1.5–3.0)
Sr μg/g	7.70 (6.9–9.3)	10.1 (8.6–12)	12.7 (12–14)	15.3 (14–17)	10.0 (9.3–11)	11.1 (10–12)	11.1 (10–12)	7.20 (6.5–7.6)	8.2 (7.5–8.5)
La ng/g	232 (220–250)	317 (280–340)	387 (370–400)	486 (470–500)	295 (260–330)	310 (260–350)	315 (270–360)	235 (220–260)	190 (170–220)
Ir ng/g	450 (430–480)	595 (550–610)	735 (700–760)	765 (750–790)	770 (700–830)	490 (430–560)	360 (250–470)	565 (540–620)	525 (420–620)
U ng/g	8.5 (7.3–9.0)	13 (12–14)	15 (12–18)	17 (13–19)	13 (10–16)	14 (10–16)	14 (11–16)	9 (7–10)	8 (7–9)

APPENDIX 4:
SOLAR SYSTEM ABUNDANCES
OF THE NUCLIDES

SOLAR-SYSTEM ABUNDANCES OF THE NUCLIDES[a]

Element	A	Atom Percent	Process[b]	Abundance[c] at ./10⁶ Si
1 H	1	−100		2.72×10^{10}
	2	0.002	U	5.4×10^{5}
2 He	3	0.0142	U, H	3.10×10^{5}
	4	−100	U, H	2.18×10^{9}
3 Li	6	7.5	X	4.48
	7	92.5	X, H, U	55.22
4 Be	9	100	X	0.78
5 B	10	19.8	X	4.8
	11	80.2	X	19.2
6 C	12	98.89	He	1.20×10^{7}
	13	1.11	H	1.34×10^{5}
7 N	14	99.634	H	2.47×10^{6}
	15	0.366	H	9.08×10^{3}
8 O	16	99.76	He	2.01×10^{7}
	17	0.038	H	7.64×10^{3}
	18	0.204	He, N	4.10×10^{4}
9 F	19	100	N	843
10 Ne	20	92.99	Ex	3.25×10^{6}
	21	0.226	He, N	7.91×10^{3}
	22	6.79	He, N	2.38×10^{5}
11 Na	23	100	Ex	5.70×10^{4}
12 Mg	24	78.99	Ex	8.49×10^{5}
	25	10.00	Ex	1.07×10^{5}
	26	11.01	Ex	1.18×10^{5}
13 Al	27	100	Ex	8.49×10^{4}
14 Si	28	92.23	Ex	9.22×10^{5}
	29	4.67	Ex	4.67×10^{4}
	30	3.10	Ex	3.10×10^{4}
15 P	31	100	Ex	1.04×10^{4}

SOLAR-SYSTEM ABUNDANCES OF THE NUCLIDES[a] *continued*

Element	A	Atom Percent	Process[b]	Abundance[c] at./10^6 Si
16 S	32	95.02	Ex	4.89×10^5
	33	0.75	Ex	3860
	34	4.21	Ex	2.17×10^4
	36	0.017	Ex	88
17 Cl	35	75.77	Ex	3970
	37	24.23	Ex	1270
18 Ar	36	84.2	Ex	8.76×10^4
	38	15.8	Ex	1.64×10^4
	40	6.1×10^{-4}	Ex	0.55
	40			*0.02*
19 K	39	93.258	Ex	3516
	40	0.01167	Ex	0.440
	40			*5.48*
	41	6.730	Ex	253.7
20 Ca	40	96.94	Ex	5.92×10^4
	42	0.647	Ex, HeS	395
	43	0.135	Ex, HeS	82.5
	44	2.09	Ex, HeS	1277
	46	0.0035	Ex	2.14
	48	0.187	Ex	114
21 Sc	45	100	Ex, E	33.8
22 Ti	46	8.2	E	197
	47	7.4	E	178
	48	73.7	E	1769
	49	5.4	E	130
	50	5.2	E, Ex	125
23 V	50	0.250	E	0.74
	51	99.750	E	294
24 Cr	50	4.35	E	583
	52	83.79	E	1.12×10^4
	53	9.50	E	1280
	54	2.36	E	316
25 Mn	55	100	E	9510
26 Fe	54	5.8	E	5.22×10^4
	56	91.8	E	8.26×10^5
	57	2.15	E	1.94×10^4
	58	0.29	E	2.61×10^3
27 Co	59	100	E	2250
28 Ni	58	68.3	E	3.37×10^4
	60	26.1	E	1.29×10^4
	61	1.13	E	557
	62	3.59	E	1770
	64	0.91	E	449
29 Cu	63	69.1	E	356
	65	30.8	E	158

SOLAR-SYSTEM ABUNDANCES OF THE NUCLIDES[a] *continued*

Element	A	Atom Percent	Process[b]	Abundance[c] at./10^6 Si
30 Zn	64	48.6	E	612
	66	27.9	E	352
	67	4.10	E, HeS	51.7
	68	18.8	E, HeS	234
	70	0.62	E, HeS	7.81
31 Ga	69	60.1	E, HeS	22.7
	71	39.9	E, HeS	15.1
32 Ge	70	20.5	E, HeS	24.2
	72	27.4	E, HeS	32.3
	73	7.8	E, HeS	9.20
	74	36.5	E, HeS	43.1
	76	7.8	E, HeS	9.20
33 As	75	100	HeS, ?	6.79
34 Se	74	0.87	P	0.54
	76	9.0	HeS	5.59
	77	7.6	HeS, ?	4.72
	78	23.5	HeS, ?	14.6
	80	49.8	HeS, ?	30.9
	82	9.2	?	5.71
35 Br	79	50.69	HeS, ?	5.98
	81	49.31	HeS, ?	5.82
36 Kr	78	0.339	P	0.154
	80	2.22	HeS, P	1.01
	82	11.45	HeS	5.19
	83	11.47	HeS, ?	5.20
	84	57.11	HeS, ?	25.9
	86	17.42	R	7.89
37 Rb	85	72.17	S, R	5.12
	87	27.83	R	1.97
	87			*2.10*
38 Sr	84	0.56	P	0.132
	86	9.82	S	2.34
	87	7.41	S	1.76
	87			*1.63*
	88	82.22	S, R	19.57
39 Y	89	100	S, R	4.64
40 Zr	90	51.5	S, R	5.75
	91	11.2	S, R	1.25
	92	17.1	S, R	1.91
	94	17.4	S, R	1.94
	96	2.80	R	0.313
41 Nb	93	100	S, R	0.696
42 Mo	92	14.8	P	0.373
	94	9.3	P	0.234

SOLAR-SYSTEM ABUNDANCES OF THE NUCLIDES[a] *continued*

Element	A	Atom Percent	Process[b]	Abundance[c] at./10^6 Si
	95	15.9	S, R	0.401
	96	16.7	S	0.421
	97	9.6	S, R	0.242
	98	24.1	S, R	0.607
	100	9.6	R	0.242
44 Ru	96	5.5	P	0.102
	98	1.86	P	0.0346
	99	12.7	S, R	0.236
	100	12.6	S	0.234
	101	17.0	S, R	0.316
	102	31.6	S, R	0.588
	104	18.7	R	0.348
45 Rh	103	100	S, R	0.344
46 Pd	102	1.0	P	0.0139
	104	11.0	S	0.153
	105	22.2	S, R	0.309
	106	27.3	S, R	0.379
	108	26.7	S, R	0.371
	110	11.8	R	0.164
47 Ag	107	51.83	S, R	0.274
	109	48.17	S, R	0.255
48 Cd	106	1.25	P	0.0199
	108	0.89	P	0.0142
	110	12.5	S	0.199
	111	12.8	S, R	0.204
	112	24.1	S, R	0.383
	113	12.2	S, R	0.194
	114	28.7	S, R	0.456
	116	7.5	R	0.119
49 In	113	4.3	P, S, R	0.0079
	115	95.7	S, R	0.176
50 Sn	112	1.01	P	0.0386
	114	0.67	P	0.0256
	115	0.38	P, S, R	0.0145
	116	14.8	S	0.565
	117	7.75	S, R	0.296
	118	24.3	S, R	0.929
	119	8.6	S, R	0.329
	120	32.4	S, R	1.24
	122	4.56	R	0.174
	124	5.64	R	0.215
51 Sb	121	57.3	S, R	0.202
	123	42.7	S, R	0.150
52 Te	120	0.091	P	0.0045
	122	2.5	S	0.123

SOLAR-SYSTEM ABUNDANCES OF THE NUCLIDES[a] continued

Element	A	Atom Percent	Process[b]	Abundance[c] at ./10⁶ Si
	123	0.89	S	0.044
	124	4.6	S	0.226
	125	7.0	S, R	0.344
	126	18.7	S, R	0.918
	128	31.7	R	1.56
	130	34.5	R	1.69
53 I	127	100	S, R	0.90
54 Xe	124	0.114	P	0.00496
	126	0.111	P	0.00483
	128	2.16	S	0.0939
	129	27.60	S, R	1.20
	130	4.34	S	0.189
	131	21.64	S, R	0.941
	132	26.53	S, R	1.15
	134	9.69	R	0.421
	136	7.82	R	0.34
55 Cs	133	100	S, R	0.372
56 Ba	130	0.106	P	0.00462
	132	0.101	P	0.00440
	134	2.42	S	0.106
	135	6.59	S, R	0.287
	136	7.85	S	0.342
	137	11.2	S, R	0.488
	138	71.7	S, R	3.13
57 La	138	0.089	P	4.0×10^{-4}
	139	99.911	S, R	0.448
58 Ce	136	0.190	P	0.0022
	138	0.254	P	0.0029
	140	88.5	S, R	1.026
	142	11.1	R	0.129
59 Pr	141	100	S, R	0.174
60 Nd	142	27.2	S	0.227
	143	12.2	S, R	0.102
	143			*0.101*
	144	23.8	S, R	0.199
	145	8.3	S, R	0.0694
	146	17.2	S, R	0.144
	148	5.7	R	0.0477
	150	5.6	R	0.0468
62 Sm	144	3.1	P	0.0081
	147	15.1	S, R	0.0394
	147			*0.0406*
	148	11.3	S	0.0295
	149	13.9	S, R	0.0363
	150	7.4	S	0.0193

SOLAR-SYSTEM ABUNDANCES OF THE NUCLIDES[a] *continued*

Element	A	Atom Percent	Process[b]	Abundance[c] at ./10⁶ Si
	152	26.6	R	0.0694
	154	22.6	R	0.0589
63 Eu	151	47.9	S, R	0.0466
	153	52.1	S, R	0.0506
64 Gd	152	0.20	P	0.00066
	154	2.1	S	0.00695
	155	14.8	S, R	0.0490
	156	20.6	S, R	0.0682
	157	15.7	S, R	0.0520
	158	24.8	S, R	0.0821
	160	21.8	R	0.0722
65 Tb	159	100	S, R	0.0589
66 Dy	156	0.057	P	0.000227
	158	0.100	P	0.000398
	160	2.3	S	0.00915
	161	19.0	S, R	0.0756
	162	25.5	S, R	0.101
	163	24.9	S, R	0.0991
	164	28.18	S, R	0.112
67 Ho	165	100	S, R	0.0875
68 Er	162	0.14	P	0.000354
	164	1.56	P, S	0.00395
	166	33.4	S, R	0.0845
	167	22.9	S, R	0.0579
	168	27.1	S, R	0.0686
	170	14.9	R	0.0377
69 Tm	169	100	S, R	0.0386
70 Yb	168	0.135	P	0.000328
	170	3.1	S	0.00753
	171	14.4	S, R	0.0350
	172	21.9	S, R	0.0532
	173	16.2	S, R	0.0394
	174	31.6	S, R	0.0768
	176	12.6	R	0.0306
71 Lu	175	97.39	S, R	0.0359
	176	2.61	S	0.000964
	176			*0.00106*
72 Hf	174	0.16	P	0.00028
	176	5.2	S	0.0092
	176			*0.00902*
	177	18.6	S, R	0.0327
	178	27.1	S, R	0.0477
	179	13.7	S, R	0.0241
	180	35.2	S, R	0.0620

SOLAR-SYSTEM ABUNDANCES OF THE NUCLIDES[a] continued

Element	A	Atom Percent	Process[b]	Abundance[c] at ./10⁶ Si
73 Ta	180	0.0123	P, S, R	2.78×10^{-6}
	181	99.9877	S, R	0.020
74 W	180	0.13	P	0.000178
	182	26.3	S, R	0.0360
	183	14.3	S, R	0.0196
	184	30.7	S, R	0.0421
	186	28.6	R	0.0392
75 Re	185	37.40	S, R	0.0190
	187	62.60	S, R	0.0317
	187			*0.0343*
76 Os	184	0.018	P	0.000129
	186	1.60	S	0.0115
	187	1.60	S	0.0115
	187			*0.0089*
	188	13.3	S, R	0.0954
	189	16.1	S, R	0.115
	190	26.4	S, R	0.189
	192	41.0	R	0.294
77 Ir	191	37.3	S, R	0.246
	193	62.7	S, R	0.414
78 Pt	190	0.013	P	0.000178
	192	0.78	S	0.0107
	194	32.9	S, R	0.451
	195	33.8	S, R	0.463
	196	25.3	S, R	0.347
	198	7.2	R	0.0986
79 Au	197	100	S, R	0.186
80 Hg	196	0.15	P	0.00078
	198	10.0	S	0.052
	199	16.8	S, R	0.0874
	200	23.1	S, R	0.120
	201	13.2	S, R	0.0686
	202	29.8	S, R	0.155
	204	6.9	R	0.0359
81 Tl	203	29.5	S, R	0.0542
	205	70.5	S, R	0.130
82 Pb	204	1.94	S	0.0612
	206	19.12	S, R	0.603
	206			*0.594*
	207	20.62	S, R	0.650
	207			*0.644*
	208	58.31	S, R	1.838
	208			*1.830*
83 Bi	209	100	S, R	0.144

SOLAR-SYSTEM ABUNDANCES OF THE NUCLIDES[a] continued

Element	A	Atom Percent	Process[b]	Abundance[c] at ./10⁶ Si
90 Th	232	100	RA	0.0335
	232			*0.0420*
92 U	235	0.720	RA	6.49×10^{-5}
	235			*5.73×10^{-3}*
	238	99.275	RA	0.00894
	238			*0.0181*

[a] Slightly modified from Anders and Ebihara (1982).
[b] U = cosmological nucleosynthesis
    H = hydrogen burning
    N = hot hydrogen burning
   He = helium burning
   Ex = explosive nucleosynthesis
    E = nuclear statistical equilibrium
    S = s-process
  HeS = helium-burning s-process
    R = r-process
   RA = r-process producing actinides
    P = p-process
    X = cosmic-ray spallation
[c] Italicized values refer to abundances 4.55 AE ago.

REFERENCE

Anders, E., and Ebihara, M. 1982. Solar-system abundances of the elements. *Geochim. Cosmochim. Acta* 46:2363–2380.

APPENDIX 5:
IODINE-XENON AGES

IODINE-XENON AGES

Sample	Class	Mon.[a]	$^{129}Xe/^{132}Xe_t$[b]	R_o[c]	ΔT^d	Ref.[a]
Carbonaceous Chondrites						
Orgueil magnetite	CI	StS		1.40(13)	-5.6 ± 2.1	1
		KI	1.02*	1.45(1)(4)	-6.4 ± 0.6	2
Cold Bokkeveld	CM	KI		<0.89	>4.7	2
Mighei	CM	KI	**		**	2
Murchison magnetite	CM	KI	1.02*	1.465(6)(39)	-6.6 ± 0.6	2
Murray H$_2$O$_2$ etch	CM	Bjb,3	1.026(10)	0.94(8)(8)	3.5 ± 1.9	4
Renazzo	CM	Bjb,6	0.99(2)	1.02(16)	1.6 ± 3.6	5,6
Felix	CO3	MM	1.018(12)	1.41(13)(14)	-5.7 ± 2.2	7
Lancé	CO3	MM	1.028(12)	1.17(10)(10)	-1.5 ± 1.9	7
Ornans	CO3	MM	1.041(10)	1.28(4)(5)	-3.5 ± 0.9	7
Warrenton	CO3	MM	**	**	**	7
Grosnaja	CV3	MM	1.012(15)	1.57(14)(15)	-8.2 ± 2.2	7
		Bjb	0.81*	1.07(2)	0.5 ± 0.4	8
Kaba	CV3	MM	1.010(17)	1.09(10)(10)	0.1 ± 2.1	7
Mokoia	CV3	MM	0.988(16)	1.38(6)(7)	-5.2 ± 1.1	7
Vigarano	CV3	MM	0.996(9)	1.60(7)(8)	-8.6 ± 1.1	7
Karoonda	CV5	KI,6	0.86(5)	1.29(1)(3)	-3.7 ± 0.5	6
		KI,9		1.25(10)	-3.0 ± 1.8	9
magnetite		KI	1.30(10)*	1.35(3)(4)	-4.7 ± 0.7	2
Allende (CV3)						
Bulk (IT)		StS,10		0.90(1)(2)	4.4 ± 0.5	10
(HT)				1.04(5)(5)	1.2 ± 1.1	10

Sample	Type				Ref
White inclusions		Kl,9			
(800–1000)			1.22(6)(6)	−2.4 ± 1.1	9
(1100–1300)			0.98(1)(2)	2.5 ± 0.5	9
(1100–1300)			1.04(1)(2)	1.2 ± 0.4	9
(1400–1700)			1.14(7)(8)	−0.9 ± 1.6	9
3509 rim		Abee			
			1.17	−1.5	11
center			1.15	−1.1	11
T-1 (IT)		StS,10	1.28(40)	0.94(1)(2)	3.5 ± 0.5	12
(HT)			0.72(8)	1.11(4)(4)	−0.3 ± 0.8	12
T-3 (IT)		StS,9	1.02(4)	0.97(1)(2)	2.7 ± 0.5	12
(HT)			0.94(9)	1.15(7)(7)	−1.1 ± 1.4	12
EGG-1		Bjb,13		1.042(4)(28)	1.1 ± 0.6	14
EGG-2		Bjb,13		0.97(4)(5)	2.7 ± 0.9	14
EGG-3		Bjb,13		0.82–1.34 (m)	−4.6 − +6.6 (m)	14
EGG-4		Bjb,13		0.89(38)	4.7 ± 10.3	14
Pink Angel		Bjb,13		0.86–1.10 (m)	0 − +6 (m)	14
rim (IT)		Bjb,13		0.93–1.10 (m)	0 − +4 (m)	14
(HT)		Bjb,13		1.11(4)(5)	−0.3 ± 1.0	14
Chondrules		Co		>0.42–>1.39	<−5.4–<+21.7	15
Chondrules		Bjb,13		1.013–1.155	−1.2–+1.8	16
Matrix		Bjb,13	1.01	1.172(2)(33)	−1.5 ± 0.6	16
Preheated inclusion		Bjb		2.0	−14	41

H Chondrites

Sample	Type				Ref
Beaver Creek	H4	Bjb	1.12(4)	1.06(8)	0.7 ± 1.7	17
Menow	H4	Bjb	0.95(2)	1.63(9)	−9.0 ± 1.3	17
Allegan chondrules	H5	Kl,6	1.05(2)	1.21(4)(5)	−2.3 ± 0.9	6
matrix		Kl,6	1.00(5)	1.29(15)(15)	−3.7 ± 2.6	6
Ambapur Nagla	H5	Bjb	1.09(2)	0.91(6)	4.2 ± 1.5	17
Nadiabondi	H5	Bjb	0.92(3)	1.98(27)	−13.4 ± 3.1	17

IODINE-XENON AGES *continued*

Sample	Class	Mon.[a]	$^{129}Xe/^{132}Xe_t$[b]	R_o[c]	ΔT[d]	Ref.[a]
Pantar light	H5	Bjb,5	1.06(1)	0.82(4)(5)	6.6±1.4	18,6
dark	H5	Bjb,5	0.95(1)	0.98(3)(4)	2.5±0.9	18,6
Plainview PV1	H5	Bjb,13		1.01–1.13 (m)	−0.9−+2.0 (m)	19
Richardton	H5	KI,20	1.06(8)	0.98(25)	2.5±5.9	20,6
Kernouve	H6	Bjb	1.07(4)	1.80(21)	−11.3±2.7	17
L Chondrites						
ALHA 77011	L3	Bjb,13	1.025(11)	0.52(1)(2)	16.9±0.9	19
Rakity	L3	Bjb	1.02*	0.85(11)	5.7±2.9	8
McKinney	L4	Bjb,22a		0.66	11.5	23
Nikolskoe	L4	Bjb	0.83*	1.70(27)	−10.0±3.6	8
Saratov	L4	Bjb	1.07*	1.05(4)	1.0±0.9	8
Arapahoe	L5	Bjb,3	0.56(4)	1.62(8)	−8.9±1.1	3
		Bjb,22a			System disturbed by shock	23
Ausson	L5	Bjb	1.01(7)	1.32(15)	−4.2±2.6	17
Barwell	L5	Bjb,41		1.19(17)	−1.9±3.3	42
Inclusion	H(?)	Bjb,41		1.42(8)	−5.9±1.2	42
Khmelevka	L5	Bjb	0.98*	1.31(11)	−4.1±1.9	8
Tsarev	L5	Bjb	0.99*	1.09(53)	0±12	8
Bruderheim	L6	KI,20	0.84(9)	1.48(29)	−6.8±4.5	24,6
chondrule		Bjb,18	1.00(1)	1.07(2)(4)	0.5±0.8	24,6
Peetz	L6	Bjb	1.02(6)	1.44(31)	−6.2±5.0	17
Stavropol	L6	Bjb		**	**	8
Bjurböle	L4		1.02–1.07		≡0.0	17
				1.095(29)		21
			1.045(9)		≡0.0	13
chondrules		Bjb,13		1.07–1.15	−1.1−+0.5	13

LL Chondrites

Chainpur matrix	LL3	Bjb,6	1.01(1)	0.97(1)(2)	2.7 ± 0.5	6
chondrules		Bjb,6	1.04(1)	0.71(1)(3)	9.8 ± 1.0	6
individual chondrules		Bjb,25	0.92–1.07	0.15–1.29	−3.7–+57	25
Manych	LL3			0.31–0.75	8.5–28.6	26
Semarkona						
individual chondrules	LL3	Bjb,25	0.98–1.06	0.86–1.30	−3.9–+5.5	40
Soko-Banja	LL4	Bjb	1.023(13)*	0.96(4)(5)	3.0 ± 1.2	27
Alta'ameen	LL5	Bjb	1.029(16)*	1.08(3)(5)	0.2 ± 1.0	27
Guidder	LL5	Bjb	1.11(3)*	0.87(11)(11)	5.2 ± 2.9	27
Olivenza	LL5	Bjb	1.15(3)*	0.76(4)(5)	8.2 ± 1.5	27
Tuxtuac	LL5	Bjb	1.036(10)*	1.23(1)(4)	−2.6 ± 0.7	27
St. Séverin	LL6	Bjb,6	1.07(2)*	0.81(2)(5)	6.8 ± 1.4	6,28
light		Bjb,22a	1.08(2)*	0.78(2)(3)	7.7 ± 0.9	28
		Bjb,22b	1.04(3)*	0.76(2)(3)	8.3 ± 0.9	28
dark		Bjb,22a	1.22(4)*	0.31(10)(10)	28.6 ± 7.6	28
		Bjb,22b	1.27(2)*	0.26(2)(2)	32.6 ± 1.7	28
whitlockite removed		Kl,10		0.57(6)	14.8 ± 2.4	10

Enstatite Chondrites

Kota-Kota	EH3	Bjb,22b	≡0.78	1.14(9)(m)	−0.9 ± 1.8	22
Y-6901	EH3	Bjb	0.926(12)*	0.56(3)	15.2 ± 1.2	29
Abee	EH4	Kl,20	0.95(10)*	1.15(3)	−1.1 ± 0.6	31,6,22
Indarch	EH4	Kl,20	0.78(5)*	1.16(2)	−1.3 ± 0.4	30,6,22
St. Sauveur	EH4	Bjb,22b	1.03(5)*	1.17(2)(4)	−1.5 ± 0.8	22
St. Mark's	EH5	Kl,20	1.59(16)*	0.89(7)	4.7 ± 1.8	30,6,22
Atlanta-A	EL6	Bjb,3	0.85(6)*	1.03(1)(2)	1.4 ± 0.4	22
B		Bjb,3	0.95(14)*	1.04(1)(2)	1.2 ± 0.4	22

IODINE-XENON AGES continued

Sample	Class	Mon.[a]	$^{129}Xe/^{132}Xe_i$[b]	R_o[c]	ΔT[d]	Ref.[a]
Blithfield	EL6	Bjb,22a	0.88(8)*	0.97(2)(3)	2.7±0.7	22
		Bjb,22b	0.79(14)*	0.98(3)(4)	2.5±0.9	22
Daniel's Kuil	EL6	Bjb,22b	1.080(8)*	0.96(2)(3)	3.0±0.7	22
Hvittis	EL6	Bjb,3	1.132(16)*	0.98(1)(2)	2.5±0.5	22
Khairpur	EL6	Bjb,22b	1.094(5)*	0.98(1)(3)	2.5±0.7	22
Enstatite Achondrites (Aubrites)						
ALHA 7813		Bjb	1.165(10)*	0.00021(2)	194±2	29
Bishopville		KI,6	0.92(4)	0.47(5)(5)	19.2±2.3	6,32
Happy Canyon	Anomalous	Bjb,22a	0.71(14)	1.16(5)(6)	−1.3±1.2	22
Peña Blanca Spring		KI,6	0.94(3)	0.95(4)(5)	3.2±1.2	6,32
Shallowater		Bjb	1.02(1)	1.12(1)(3)	−0.5±0.6	33,6
Iron Meteorite Silicates						
Copiapo	IAB	Bjb,34	0.79(8)	0.83(2)(3)	6.3±0.8	34
El Taco (Campo del Cielo)	IAB	KI,6	1.17(5)	0.944(4)(4)	3.5±1.0	6
Landes	IAB	Bjb,34	3.50(16)	0.79(2)(3)	7.4±0.9	34
Pitts (unetched IT)	IAB	Bjb,34	1.09(1)	0.89(1)(3)	4.7±0.8	34
(etched IT)		Bjb,34	1.06(3)	0.97(2)(4)	2.7±0.9	34
troilite (LT)		Bjb,34	1.01(1)	0.44(1)(2)	20.7±0.8	34
Toluca	IAB	NaI	≡0.98	1.5	−7	35
Woodbine	IAB	Bjb,34	1.23(3)	1.02(1)(3)	1.6±0.7	34
Mundrabilla	IAB Anom.	Bjb,34	1.17(14)	0.90(2)(3)	4.4±0.8	34
troilite		Bjb,34	0.97(1)	1.35(4)(5)	−4.7±0.8	34
Netschaëvo	IIE Anom.	Bjb,34	**	**	**	36
Weekeroo Station	IIE	Bjb,34	0.84(5)	0.56(1)(2)	15.2±0.8	36

Other Meteorites

ALHA 77005	Shergottite	Bjb,22b	($>99.9\%$ ^{128}Xe*)	**	37
Brachina	Anom. Achondrite		Correlated (no monitor)		38
Eagle Station silicate	Pallasite	Bjb,34		**	39
Enon silicate	Anom. Meso.	Bjb,34	<1	0.92(15)(15)	39
Lafayette	Nakhlite	Bjb	3.9 ± 3.7	**	33,6,32
Petersburg	Eucrite	Bjb,6	>68	**	32
			>26		

[a] Monitors: Bjb = Bjurböle (Hohenberg and Kennedy 1981); StS = St. Séverin (see Rison and Zaikowski 1980); MM = Murchison magnetite; Co = ^{60}Co used to monitor production of neutrons during space irradiation. Where a number is given, the irradiation is discussed in that reference. All samples with the same number are in the same irradiation.

[b] Trapped ^{129}Xe/^{132}Xe ratio. Where original analysis was normalized to ^{130}Xe, trapped ^{129}Xe/^{130}Xe (and error) are divided by 6.15 (Eugster et al. 1967) and noted by an asterisk.

[c] Ratio of ^{129}I to ^{127}I (in 10^{-4}) at time of Xe isotopic closure. The first number in parentheses is "internal error," uncertainty to be used in comparisons with samples in same irradiation. Second number is "external error," which includes uncertainty in slope of irradiation monitor and, if meteoritic monitor was used, uncertainty in determination of monitor. Double asterisk indicates a lack of ^{129}Xe*. Model ages are denoted by (m). Where no uncertainty is given, none was stated in the original reference.

[d] Apparent time of isotopic closure, in Myr after Bjurböle, assuming a half-life of 15.7 Myr for ^{129}I. Error is mean of (asymmetric) "external errors" (see note c).

[e] Where two references are given, the data are given in the first reference, and a reanalysis (the results of which are listed in the table) is given in the second. 1: Herzog et al. 1973; 2: Lewis and Anders 1975; 3: Drozd and Podosek 1976; 4: Niemeyer and Zaikowski 1980; 5: Reynolds and Turner 1964; 6: Podosek 1970; 7: Crabb et al. 1982; 8: Shukolyukov et al. 1986; 9: Podosek and Lewis 1972; 10: Rison and Zaikowski 1980; 11: Wasserburg and Huneke 1979; 12: Zaikowski 1980; 13: Caffee et al. 1982b (range of values for individual chondrules); 14: Swindle et al. 1987a; 15: Fireman et al. 1970; 16: Swindle et al. 1983 (range of values for chondrule plateaus); 17: Jordan et al. 1980; 18: Turner 1965; 19: Caffee et al. 1988; 20: Reynolds 1963; 21: Hohenberg and Kennedy 1981; 22: Kennedy et al. 1988 (a = SLC4, b = SLC5); 23: Caffee et al. 1982a; 24: Merrihue 1966; 25: Swindle et al. 1986a (range of results for individual chondrules); 26: Podosek and Hohenberg 1970; 27: Bernatowicz et al. 1988; 28: Hohenberg et al. 1981, and Hudson 1981; 29: Honda et al. 1983; 30: Hohenberg et al. 1967; 31: Jeffery and Reynolds 1961; 32: Podosek 1972; 33: Hohenberg 1967,1968; 34: Niemeyer 1979 (recomputed using his Bjurböle monitor); 35: Alexander et al. 1969,1970; 36: Niemeyer 1980; 37: Swindle et al. 1986; 38: Bogard et al. 1983; 39: Niemeyer 1983; 40: Swindle and Grossman 1987; 41: Kirschbaum 1985; 42: Kirschbaum 1986.

REFERENCES

Alexander, E. C., Srinivasan, B., and Manuel, O. K. 1969. I-Xe dating of silicates from Toluca iron. *Earth Planet. Sci. Lett.* 6:355–358.

Alexander, E. C., Srinivasan, B., and Manuel, O. K. 1970. Erratum. *Earth Planet. Sci. Lett.* 8:188.

Bernatowicz, T. J., Podosek, F. A., Swindle, T. D., and Honda, M. 1988. Iodine-xenon studies of LL-chondrites. *Geochim. Cosmochim. Acta* 52:1113–1121.

IODINE-XENON AGES continued

Bogard, D. D., Nyquist, L. E., Johnson, P., Wooden, J., and Bansal, B. 1983. Chronology of Brachina. *Meteoritics* 18:269–270.

Caffee, M. W., Hohenberg, C. M., Hörz, F., Hudson, B., Kennedy, B. M., Podosek, F. A., and Swindle, T. D. 1982a. Shock disturbance of the I-Xe system. *Proc. Lunar Planet. Sci. Conf.* 13, *J. Geophys. Res. Suppl.* 87:A318–A330.

Caffee, M. W., Hohenberg, C. M., Swindle, T. D., and Hudson, B. 1982b. I-Xe ages of individual Bjurböle chondrules. *Proc. Lunar Planet. Sci. Conf.* 13, *J. Geophys. Res. Suppl.* A303–A317.

Caffee, M. W., Hohenberg, C. M., Swindle, T. D., and Hudson, B. 1987. Noble gases from graphite-magnetite inclusions in ordinary chondrites. *Geochim. Cosmochim. Acta*, submitted.

Crabb, J., Lewis, R. S., and Anders, E. 1982. Extinct ^{129}I in C3 chondrites. *Geochim. Cosmochim. Acta* 46:2511–2526.

Drozd, R. J., and Podosek, F. A. 1976. Primordial ^{129}Xe in meteorites. *Earth Planet. Sci. Lett.* 31:15–30.

Fireman, E. L., DeFelice, J., and Norton, E. 1970. Ages of the Allende meteorite. *Geochim. Cosmochim. Acta* 34:873–881.

Herzog, G. F., Anders, E., Alexander, E. C., Jr., Davis, P. K., and Lewis, R. S. 1973. Iodine-129/xenon-129 age of magnetite from the Orgueil meteorite. *Science* 180:489–491.

Hohenberg, C. M. 1967. I-Xe dating of the Shallowater achondrite. *Earth Planet. Sci. Lett.* 3:357–362.

Hohenberg, C. M. 1968. Studies in I-Xe and Pu-Xe dating of pile-irradiated achondrites. PH.D. thesis, Univ. of California, Berkeley.

Hohenberg, C. M. 1969. Radioisotopes and the history of nucleosynthesis in the galaxy. *Science* 166:212–215.

Hohenberg, C. M., and Kennedy, B. M. 1981. I-Xe dating: Intercomparisons of neutron irradiations and reproducibility of the Bjurböle standard. *Geochim. Cosmochim. Acta* 45:251–256.

Hohenberg, C. M., Hudson, B., Kennedy, B. M., and Podosek, F. A. 1981. Noble gas retention chronologies for the St. Séverin meteorite. *Geochim. Cosmochim. Acta* 45:535–546.

Honda, M., Bernatowicz, T. J., and Podosek, F. A. 1983. ^{129}Xe–^{128}Xe and ^{40}Ar–^{39}Ar chronology of two Antarctic enstatite meteorites. *Mem. Natl. Inst. Polar Res., Special Issue* 30:275–291.

Hudson, G. B. 1981. Noble gas retention chronologies in the St. Séverin meteorite. Ph.D. thesis, Washington Univ., St. Louis.

Jeffery, P. M., and Reynolds, J. H. 1961. Origin of excess Xe^{129} in stone meteorites. *J. Geophys. Res.* 66:3582–3583.

Jordan, J., Kirsten, T., and Richter, H. 1980. ^{129}I/^{127}I: A puzzling early solar system chronometer. *Z. Naturforsch.* 35a:145–170.

Kennedy, B. M., Hudson, B., Hohenberg, C. M., and Podosek, F. A. 1988. ^{129}I/^{127}I variations among enstatite chondrites. *Geochim. Cosmochim. Acta* 52:101–111.

Kirschbaum, C. 1985. Iodine and chlorine in two fine-grained inclusions from the Allende meteorite. *Meteoritics* 20:683–684.

Kirschbaum, C. 1986. Iodine-xenon and ^{40}Ar–^{39}Ar dating of an unusual inclusion from the Barwell meteorite. *Meteoritics* 21:414–415.

Lewis, R. S., Anders, E., Shimamura, T., and Lugmair, G. W. 1983. Barium isotopes in Allende meteorite: Evidence against an extinct super-heavy element. *Science* 222:1013–1015.

Merrihue, C. 1966. Xenon and krypton in the Bruderheim meteorite. *J. Geophys. Res.* 71:263–313.

Niemeyer, S. 1979. I-Xe dating of silicate and troilite from IAB iron meteorites. *Geochim. Cosmochim. Acta* 43:843–860.

Niemeyer, S. 1980. I-Xe and ^{40}Ar–^{39}Ar dating of silicate from Weekeroo Station and Netschaevo IIE iron meteorites. *Geochim. Cosmochim. Acta* 44:33–44.

Niemeyer, S. 1983. I-Xe and ^{40}Ar-^{39}Ar analyses of silicate from the Eagle Station pallasite and the anomalous iron meteorite Enon. *Geochim. Cosmochim. Acta* 47:1007–1012.

Niemeyer, S., and Zaikowski, A. 1980. I-Xe age and trapped Xe components of the Murray (C-2) chondrite. *Earth Planet. Sci. Lett.* 48:335–347.

Podosek, F. A. 1970. Dating of meteorites by the high-temperature release of iodine-correlated Xe. *Geochim. Cosmochim. Acta* 34:341–365.

Podosek, F. A. 1972. Gas retention chronology of Petersburg and other meteorites. *Geochim. Cosmochim. Acta* 36:755–772.

Podosek, F. A., and Hohenberg, C. M. 1970. I-Xe dating: Evidence for cold assembly of an unequilibrated chondrite. *Earth Planet. Sci. Lett.* 8:443–447.

Podosek, F. A., and Lewis, R. S. 1972. ^{129}I and ^{244}Pu abundances in white inclusions of the Allende meteorite. *Earth Planet. Sci. Lett.* 15:101–109.

Reynolds, J. H. 1963. Xenology. *J. Geophys. Res.* 68:2939–2956.

Rison, W., and Zaikowski, A. 1980. Rare gases in the chondrite Renazzo. *J. Geophys. Res.* 69:3263–3281.

Shukolyukov, Yu. A., Vu Minh, D., Kolesov, G. M., Fugzan, M. M., and Ivanova, M. A. 1986. ^{129}I/^{129}Xe data on the relative interval of formation for some type L chondrites. *Geochem. Int.* 23(2):130–144.

Swindle, T. D., and Grossman, J. N. 1987. I-Xe studies of Semarkona chondrules: Dating alteration. *Lunar Planet. Sci.* XVIII:982–983 (abstract).

Swindle, T. D., Caffee, M. W., Hohenberg, C. M., and Lindstrom, M M. 1983. I-Xe studies of individual Allende chondrules. *Geochim. Cosmochim. Acta,* 47:2157–2177.

Swindle, T. D., Caffee, M. W., and Hohenberg, C. M. 1986a. I-Xe and ^{40}Ar-^{39}Ar ages of Chainpur chondrules. *Lunar Planet. Sci.* XVII:857–858 (abstract).

Swindle, T. D., Caffee, M. W., and Hohenberg, C. M. 1986b. Xenon and other noble gases in shergottites. *Geochim. Cosmochim. Acta* 50:1001–1019.

Swindle, T. D., Caffee, M. W., and Hohenberg, C. M. 1988. I-Xe studies of Allende inclusions: EGGs and the Pink Angel. *Geochim. Cosmochim. Acta,* in press.

Swindle, T. D., Caffee, M. W., Hohenberg, C. M., Lindstrom, M. M., and Taylor, G. J. 1987. Iodine-xenon and other studies of individual Chainpur chondrules. *Geochim. Cosmochim. Acta,* submitted.

Turner, G. 1965. Extinct iodine 129 and trace elements in chondrites. *J. Geophys. Res.* 70:5433–5445.

Wasserburg, G. J., and Huneke, J. C. 1979. I-Xe dating of I-bearing phases in Allende. *Lunar Planet. Sci.* X:1307–1309 (abstract).

Zaikowski, A. 1980. I-Xe dating of Allende inclusions: Antiquity and fine structure. *Earth Planet. Sci. Lett.* 47:211–222.

GLOSSARY*

Compiled by Melanie Magisos

ablation	removal of material by attrition, e.g., by passage through the atmosphere where a major process is evaporation.
absorption edge	a wavelength at which there is an abrupt change in absorption.
accretion	growth by assimilation of material from the outside, e.g., formation of planets and planetesimals by the accumulation of smaller objects in the primordial nebula.
achondrite	meteorite of nonsolar composition, also known as differentiated stony meteorite.
acicular	said of a crystal that is needlelike in form.
activation energy	minimum quantity of energy needed to initiate a process such as a chemical reaction.
activity	a quantity used in physical-chemical calculations that plays the same role in nonideal solutions that concentration plays in ideal solutions. The *activity coefficient* is the ratio of the activity of a species to its concentration in a nonideal solution.
ADOR	initial $^{87}Sr/^{86}Sr$ ratio for the Angra dos Reis achondrite, 0.69883 ± 0.00002.
adsorption	adherence of gas molecules to solid surfaces with which they are in contact.

*We have used some definitions from *Glossary of Astronomy and Astrophysics* by J. Hopkins (by permission of the University of Chicago Press, copyright 1980 by the University of Chicago), from *Astrophysical Quantities* by C. W. Allen (London: Athlone Press, 1973), and from *Glossary of Geology,* edited by M. Gary, R. McAfee, and C. L. Wolff (Washington, D.C.: American Geological Institute, 1972). We also acknowledge definitions and helpful comments from various chapter authors, especially R. T. Dodd, D. M. Hunten, and D. W. G. Sears.

AEM	analytical (scanning) electron microscope: An electron microscope equipped with a facility to permit elemental analysis.
agglutinate	a lithology, characteristic of planetary surfaces, consisting of regolith particles bonded together with impact-generated glass.
albedo	the fraction of light incident on a surface which is reflected away from the surface.
aliphatic	a type of organic compound characterized by an open-chain structure, as distinct from a cyclic (=aromatic) structure.
alkane	one of a series of saturated, open-chain organic compounds, i.e., containing no double or triple bonds.
alkene	one of a series of open-chain organic compounds containing one or more double bonds.
alkyne	one of a series of open-chain organic compounds containing a triple bond.
ALL	lowest initial $^{87}Sr/^{86}Sr$ ratio measured in a CAI in the Allende carbonaceous chondrite, 0.69877 ± 0.00002.
alteration	a change in mineralogy due to chemical reactions between solids and fluids at subsolidus temperatures.
ambipolar diffusion	the relative motion of the charged components of a gas and the neutral components under the action of electromagnetic and electrostatic forces. Since the charged particles (electrons or ions) are subject to direct interaction with the electromagnetic field, they can move relative to the neutrals, which are then dragged along by the collisions with the ions or electrons. Such diffusion, constrained by charge neutrality, is given the general name "ambipolar diffusion."
amide	a nitrogenous organic compound, generally containing the $CO.NH_2$ radical.

GLOSSARY

amine	a derivative of ammonia in which hydrocarbon radicals have replaced one or more of the H atoms.
amino acid	one of a group of nitrogenous organic compounds that serve as the structural units of proteins.
Amor asteroids	asteroids having perihelion distance 1.017 AU $< q <$ 1.3 AU.
amphoterites	obsolete name for LL chondrites.
amu	atomic mass unit; $= 1/16$ of the mass of an O atom; also called u.
angular momentum	the angular momentum of a system about a specified origin is the sum over all the particles in the system (or an integral over the different elements of the system if it is continuous) of the vector products of the radius vector joining each particle to the origin and the momentum of the particle. For a closed system it is conserved by virtue of the isotropy of space.
anorthosite	an igneous rock made up almost entirely of plagioclase feldspar.
aphelion	that point in the orbit of a body, gravitationally bound to the Sun, at which the body is farthest from the Sun. Also used more casually for nonheliocentric orbits, though terms such as apogee, apoapse and apocenter are preferred.
Apollo asteroids	asteroids having semimajor axis $a > 1.0$ AU, and perihelion distance $q < 1.017$ AU.
aromatic	one of a series of organic compounds characterized by a cyclic ring structure.
asteroid	one of a number of objects ranging in size from sub-km to about 1000 km, most of which lie between the orbits of Mars and Jupiter.
ataxites	iron meteorites with little or no structure visible to the naked eye, containing >16 wt% Ni.

atmophile	one of the geochemical classes of elements. Atmophiles are those elements which tend to be present in planetary bodies as gases, e.g., the inert gases and nitrogen.
AU	astronomical unit. The mean distance of the Earth from the Sun, equal to 1.5×10^{13} cm.
aubrite	alternate name for enstatite achondrite. A differentiated stone meteorite consisting predominantly of enstatite with very low Fe content, highly reduced, possibly related to enstatite chondrites.
axiolitic	a lithology containing a spherulitic structure (i.e., one containing radiating fibrous or acicular crystals) developed perpendicular to an axis.
BABI	acronym for basaltic achondrite best initial. It refers to a ratio value for initial $^{86}Sr/^{87}Sr$ that characterizes those achondrites, i.e., 0.69898.
baryon	a nucleon or any elementary particle that can be transformed into a nucleon and some number of mesons and lighter particles.
basalt	a dark, fine-grained, mafic igneous rock composed primarily of plagioclase and pyroxene.
basaltic achondrite	collective name for eucrites and howardites which superficially resemble terrestrial basalts or their fragmentation products.
beta decay	radioactive decay taking place by emission of an electron or positron.
body-centered cubic	a type of crystal structure in which each atom has eight nearest neighbors.
breccia	rock composed of fragments derived from previous generations of rocks, cemented together to form a new lithology. *See also* genomict, monomict, polymict and *Chapter 3.5*.
brecciation	breakage of a rock into smaller fragments.

CAI	calcium, aluminum-rich inclusions. Inclusions rich in these elements are abundant in CV and, to a lesser degree CM, chondrites.
carbonaceous chondrite	a chondritic meteorite, generally containing more than about 0.2 wt% C. Most such chondrites are highly oxidized and have nearly solar composition for all but the most volatile elements.
carbon star	a rather loosely defined category of red-giant star whose spectra show strong bands of C_2, CN or other C compounds.
cataclastic	a type of structure produced in a rock by severe deformation resulting in fracturing and rotation of mineral grains.
cathodo-luminescence	light induced in a crystal by the action of an electron beam.
CCFXe	acronym for carbonaceous chondrite fission xenon. A component of Xe in chondrites, which was thought to be due to fission of an unknown nuclide, now believed to represent Xe from a number of different zones in a supernova.
chalcophile	one of the geochemical classes of elements which tend to be present in the sulfide phase, e.g., S, Se, Cd, Zn.
Chandrasekhar limit	a limiting mass for white dwarfs. If the mass of the star exceeds this critical mass (1.44 $M_\odot$ for the expected mean molecular weight of 2), the load of the overlying layers will be so great that degeneracy pressure will be unable to support it, and no configuration will be stable.
charged-particle tracks	lines of radiation damage in minerals caused by the passage of energetic ionizing radiation.
chassignite	a very rare type of achondrite (only one known, Chassigny) consisting of olivine with minor amounts of pyroxene, plagioclase, chromite and sulfide.

chemical remanent magnetization	a remanent magnetization acquired by a rock specimen when the constituent magnetic mineral was formed by a chemical reaction in the presence of a magnetic field.
chirality	the handedness of an asymmetric molecule.
chondrite	originally defined as a meteorite that contained chondrules; now also implies a chemical composition, for all but the most volatile elements, that is not far removed from that of the Sun.
chondrite, equilibrated	chondrite that has closely approached or reached internal equilibrium, presumably as a result of thermal metamorphism, so that individual grains of the same mineral have similar compositions.
chondrule	approximately spherical assemblages, characteristic of most chondrites, that existed independently prior to incorporation in the meteorite and that shows evidence for partial or complete melting.
chromosphere	a normally transparent region lying between the photosphere and the corona of a main-sequence star such as the Sun, and of intermediate temperature.
CHUR	chondritic uniform reservoir. Derived from the averaged Sm/Nd ratio and present ^{143}Nd/^{144}Nd ratio for chondrites. From these data the isotopic composition of Nd in the reservoir at any time in the past can be calculated.
clast	an individual lithic unit, e.g., mineral grain or rock fragment, produced by disintegration of a larger rock unit.
clathrate	a structure formed by the systematic inclusion of certain molecules in cavities within a crystal lattice.
clinopyroxene	a mineral of the pyroxene group that crystallizes in the monoclinic system.
closure	the start of retention within a substance, e.g., a mineral grain, of the stable products of radioactive decay.
column density	number of atoms or molecules per cm^2 in the line of sight.

commensu- rability	a condition in which orbital frequencies are in a ratio of small integers.
concordia diagram	a diagram in which a curve depicting the locus of points for which ^{207}Pb/^{235}U age equals ^{206}Pb/^{238}U age for a specified age range. This permits identification of disturbed ages, and the nature of the disturbance in some cases. *See Chapter 5.1.*
condensation	transformation from the gaseous to a solid or liquid phase. In the context of this book, it is generally taken to refer to the formation of solid grains from nebular gas.
corona	the greatly extended and very hot outermost region of a main-sequence star such as the Sun, sometimes used for a similar region of a planetary atmosphere.
cosmic-ray exposure age	the period of time during which a meteorite was exposed to cosmic radiation, commonly the time between its final reduction in size by impact and its arrival on Earth. More generally, it is the time spent within a few m of the space environment. Nuclear reactions between the radiation and nuclides in the meteorite produce new nuclides, or associated phenomena such as tracks, whose abundances can be used to estimate the exposure age.
cosmogenic	that produced by interaction with cosmic radiation, e.g., ^{21}Ne is a cosmogenic nuclide, produced by spallation reactions.
cosmo- thermometry	term usually reserved for determining the last temperature of equilibration between solids and solar nebula gas, although it is sometimes used for any cosmically interesting temperature determination (*see* palaeothermometry).
cotectic	a situation in which two or more phases crystallize simultaneously during cooling of a single liquid.
Coulomb barrier	the repulsive force between charged particles of the same sign that acts to prevent bombardment by charged particles from penetrating the nucleus.

crypto-crystalline	a rock texture in which the individual crystallites are too small to be distinguished by conventional optical microscopy.
cumulate	a plutonic igneous rock composed chiefly of crystals that accumulated by sinking or floating from a magma.
Curie temperature	the temperature marking the transition between ferromagnetism and paramagnetism, or between the ferroelectric phase and the paraelectric phase.
daughter nuclide	a nuclide produced by decay of a radioactive parent. The daughter nuclide may be stable or radioactive.
decay constant	the decay constant multiplied by the number of atoms of a radioactive nuclide equals the instantaneous decay rate. It is also equal to the natural log of 2 divided by the half-life. Units are in reciprocal time.
dendrite	a crystal that has grown by growth along certain preferred directions resulting in a pattern resembling the twigs and branches of a tree.
deuterium	heavy stable H isotope that has a mass of 2 amu.
devitrification	crystallization of a glass in the solid state.
diagenetic	that pertaining to the chemical and physical changes brought about in a sediment after it is buried.
diaplectic glass	a disordered, glass-like substance formed by shock metamorphism of a mineral or mineral assemblage without melting.
diastereomers	stereoisomers that are not mirror image molecules. Diastereomers, unlike enantiomers, have different physical and chemical characteristics.
diogenite	class of achondritic meteorite consisting essentially of magnesian orthopyroxene.
double-spike technique	a procedure in mass spectrometry in which known amounts of two isotopes are added to the sample to permit precise correction for the mass fractionation introduced by the analysis itself.

EDS	acronym for energy dispersive system. An analytical device generally attached to an *electron microprobe* or a *SEM* for gathering qualitative or semiquantitative compositions.
Eh	oxidation potential; the difference in potential between an atom or ion and the state in which an electron has been removed to an infinite distance from that atom or ion.
ejecta	materials ejected either from a crater by the action of volcanism or a meteoroid impact, or from a stellar object, such as a supernova, by shock waves.
electron microprobe	an analytical instrument that utilizes a finely focused beam of electrons to excite characteristic X-rays in the sample. These are then analyzed using either a crystal spectrometer or a solid-state, energy-dispersive detector.
enantiomers	nonsuperimposable mirror image molecules.
endogenic	originating within a planetary or planetesimal object.
endothermic reaction	a reaction that consumes energy.
enstatite chondrite	collective name for the EH and EL classes of chondritic meteorite, highly reduced chondrites with Mg/Si around 0.83.
E-process	equilibrium process. A complex set of nucleosynthetic reactions in which photodisintegration of previously synthesized nuclides leads to a population of nuclides that approaches local statistical equilibrium. Believed to be largely responsible for the so-called iron peak.
Epstein drag	a law describing the drag of a medium (e.g., a gas) on an object moving through it. The Epstein drag law applies when the object is smaller than the mean free path of the constituent particles (or molecules) of the medium.
eucrite	class of achondritic meteorite consisting of Ca-pyroxene and plagioclase.

euhedral	well formed; used of a mineral crystal that is completely or mainly bounded by its own regularly developed crystal faces.
europium anomaly	a situation in which the element europium is fractionated relative to the other rare-earth elements. It is generally caused by its divalent nature in reducing environments, causing it to partition preferentially into minerals in which the other rare earths are incompatible, most notably plagioclase.
eutectic	the lowest temperature at which an alloy can be wholly or partly liquid. Crystallization under these conditions frequently leads to a petrologically distinctive texture.
exogenic	originating externally to a planetary or planetesimal object.
exothermic reaction	a reaction that liberates heat energy.
exposure age	*see* cosmic-ray exposure age.
exsolution	formation of a second crystalline phase by decomposition of a primary crystalline phase at subsolidus temperature.
extinct nuclides	radioactive nuclides with short *half-lives* (compared with the age of the solar system) which were present when a meteorite or meteoritic component formed but which have now decayed below detection limits. Their one-time presence in the material is indicated by their decay products.
face-centered cubic	a type of crystal structure in which each atom has twelve nearest neighbors.
fall	a meteorite that was seen to fall. Such meteorites are usually recovered soon after fall and are relatively free of terrestrial contamination and weathering effects.
find	a meteorite that was not seen to fall but was found and recognized subsequently.

GLOSSARY

Fischer-Tropsch synthesis	production of organic molecules by the hydrogenation of carbon monoxide in the presence of a suitable catalyst.
fission	a radioactive decay process in which a heavy nucleus fragments into two or more pieces of roughly equal mass. The process can be spontaneous or induced by particle bombardment.
fractionation	the physical separation of one phase, element or isotope from another.
fugacity	a measure of the chemical potential of a gaseous species; it is the equivalent for a nonideal gas of the partial pressure of an ideal gas.
FUN	fractionation and unknown nuclear effects.
FU Orionis	a newly formed star, probably a pre-main-sequence star (F5-G3 Ib) presently near the top of its Hayashi track, about 500 pc distant. In 1936 it suddenly appeared in the middle of a dark cloud and brightened by 6 magnitudes in the photographic (blue) band. Since 1960 it has slowly faded. It has developed a reflection nebula, and is a strong infrared emitter.
gardening	reworking and overturning of a *regolith*, principally by micrometeoroid bombardment.
gas-retention age	the age of a meteorite as calculated from the abundance of gaseous daughter products.
gegenschein	a relatively bright region in the sky at 180° from the Sun (i.e., the antisolar point). It is caused by scattering of sunlight from interplanetary dust particles.
genomict	a breccia in which the clasts have the same class but different petrographic properties.
geobarometry	determination of the pressure at which an assemblage (mineral pairs, minerals and a presumed gas or liquid, or different phases of a given mineral) was formed.
granulite	a rock consisting of interlocking grains of roughly uniform size.

graphite surface	surface produced on a phase diagram by plotting the O fugacity in equilibrium with an equilibrium mixture of C, CO and CO_2.
half-life	the length of time required for half of the atoms in a given sample of a radioactive nuclide to decay.
Hayashi stage	a hypothetical, high-luminosity stage early in the life of a star.
Herbig-Haro objects	semi-stellar, emission-line nebulae which are produced by shock waves in the supersonic outflow of material from young stars; also referred to as Herbig-Haro nebulae.
H II region	region of ionized hydrogen in interstellar space. H II regions occur near stars with high luminosities and high surface temperatures. Ionized hydrogen, having no electrons, does not produce spectral lines; however, occasionally a free electron will be captured by a free proton and the resulting radiation can be studied optically. H II regions have lifetimes of only a few Myr.
hopper crystal	a crystal in which the faces have grown more at the edges than at the center.
howardite	polymict brecciated achondrite consisting predominantly of lithic units similar to eucrites and diogenites, though more extreme compositions are also found.
HR diagram (Hertzsprung-Russell diagram)	in present usage, a plot of bolometric magnitude against effective temperature for a population of stars. Related plots are the color-magnitude plot (absolute or apparent visual magnitude against color index) and the spectrum-magnitude plot (visual magnitude versus spectral type, the original form of the HR diagram).
HRTEM	high resolution transmission electron microscopy.
Hugoniot curve	locus of pressure-volume-energy states achieved by shock compression; frequently represented in the pressure-volume plane.
ID	acronym for isotope-dilution mass spectroscopy. An

	analytical technique for determining trace elements in meteorites, especially useful for *REE*.
IDP	interplanetary dust particle, also known as a micrometeoroid or, after entry into the Earth's atmosphere, a *micrometeorite*.
igneous	a term used to describe the melting and subsequent solidification of a rock.
imidazole	a heterocyclic organic compound containing two N atoms within a five-membered ring.
impactite	glassy rocks associated with craters, made by the fusion of local, i.e., target, rock by the heat of impact.
INAA	acronym for instrumental neutron-activation analysis. An analytical technique for determining major, minor and trace elements in meteorites.
inclusions	aggregates of mineral grains that existed independently prior to incorporation in the meteorite. *See* CAI.
incompatible element	an element that tends to be excluded from a growing crystal (and, hence, concentrated in the residual magmatic liquid), commonly because its ionic radius is too large to fit comfortably in the crystal lattice.
ion microprobe	an analytical instrument in which a finely focused beam of ions ionizes atoms in the sample and ejects them into a mass spectrometer where they may be analyzed.
ion-molecule reaction	a reaction between an ionized atom or molecular fragment and a neutral atom or molecule. In the interior of interstellar molecular clouds, the ion is initially produced by interaction of a galactic cosmic ray with a neutral atom or molecule.
IRAS	Infrared Astronomical Satellite
iron-wüstite buffer curve	the curve produced on a phase diagram by plotting the O fugacity in equilibrium with metallic iron and wüstite $Fe_{(1-x)}O$ in any proportions as a function of temperature.

isochemical	without change in bulk chemical composition.
isochron	strict meaning: equal age. As used in this book it refers to a line of equal age for a sample suite when the radiogenic daughter nuclide is plotted against the radioactive parent nuclide. The two nuclides are usually normalized to a nonradiogenic nuclide.
isomer	one of a number of molecules that all have the same elemental composition but which differ from each other in structure.
Jeans escape	escape of the fastest atoms in a Maxwell-Boltzmann distribution when their speed exceeds the escape velocity.
kamacite	Fe,Ni alloy of 7 wt% Ni or less with the body-centered-cubic structure. It occurs as large plates or single crystals in iron meteorites, abundant grains in chondrites and rare grains in most achondrites.
Kepler velocity	the orbital velocity of a gravitationally bound object around the central object, i.e., the velocity that leads to a centrifugal force exactly balancing the gravitational attraction between the two objects.
kerogen	insoluble macromolecular organic matter, operationally defined as the organic residue left after acid demineralization of a rock.
kinetic isotope effects	changing of an element's isotopic proportions as a result of differing reaction rates for each isotope.
KREEP	a basaltic rock found on the Moon, characterized by its high contents of K, rare-earth elements and P.
Le Chatelier's principle	if a system in equilibrium is altered by an outside force, the system will tend to minimize the effect of the force by changes in itself.
lightcurve	brightness values plotted as a function of time.
light-dark structure	the structure displayed by gas-rich regolith breccias in which light clasts are located in dark matrix.

GLOSSARY

liquidus	the line or surface in a phase diagram above which the system is completely liquid.
lithology	the physical character of a rock.
lithophile	one of the geochemical classes of elements. Lithophile elements are those which tend to concentrate in the silicate phase, e.g., Si, Mg, Ca, Al, Na, K and rare-earth elements.
lithostatic pressure	pressure due to the weight of overlying rock.
Lyman lines	spectroscopic features due to H atoms.
$M_\odot$	solar mass ($= 1.989 \times 10^{33}$ g).
$M_\oplus$	Earth mass ($= 5.976 \times 10^{27}$ g).
mafic	term used to describe a silicate mineral whose cations are predominantly Mg and/or Fe. It is also used for rocks made up principally of such minerals.
magma ocean	a hypothetical stage in the evolution of a planetary object during which virtually the entire surface of the object is covered with molten lava.
magmatic	that associated with molten silicate material.
martensite	distorted body-centered cubic structure formed in Fe, Ni-alloys, originally in the gamma field, as they cool rapidly.
maskelynite	glass formed from plagioclase feldspar by disordering during intense shock.
mass fractionation	a process which causes the proportions of the isotopes or elements to change in a manner which is dependent on the differences in their mass. Most chemical and physical processes are of this type. Thus the $^{18}O/^{16}O$ ratio changes twice as much, for a given physical or chemical process, as the $^{17}O/^{16}O$ ratio. Consequently, samples which were originally similar in their proportion of O isotopes spread along a line with slope 0.5 (the mass

fractionation line) on a plot of $^{17}O/^{16}O$ against $^{18}O/^{16}O$ when the samples have experienced the usual chemical and physical processes.

mass spectrometer
an analytical instrument in which the sample is converted into a beam of ions which can be separated from each other on the basis of their mass-to-charge ratio, generally by a magnetic or electrostatic field, so that the relative proportions of entities of different mass, commonly isotopes, can be determined.

matrix
the fine-grained material that occupies the space in a rock, such as a meteorite, between the larger, well-characterized components such as chondrules, inclusions, etc. In practice, the precise definition is controversial (*see Chapter 10.2*).

mesosiderite
class of stony-iron meteorite consisting of subequal proportions of silicate material (related to eucrites and diogenites) and Fe-Ni metal.

mesostasis
interstitial, generally fine-grained material occupying the space between larger mineral grains in an igneous rock; generally, therefore, the last material to solidify from a melt.

metamorphism
solid-state modification of a rock, e.g., recrystallization, caused by elevated temperature (and possibly pressure).

metastable
said of an energy state, or the material in that state, that is characterized by a potential-energy minimum that is not, however, the ground state of the system.

meteor
the light phenomenon produced by a meteoroid experiencing frictional heating when entering a planetary atmosphere; also used for the glowing *meteoroid* itself. If particularly large, it is described as a fireball.

meteorite
a natural object of extraterrestrial origin that survives passage through the atmosphere.

meteoroid
a natural small (sub-km) object in an independent orbit in the solar system.

micrometeorite	a small extraterrestrial particle that has survived entry into the Earth's atmosphere. The actual size is not rigorously constrained but is operationally defined by the collection procedure because small particles are more abundant than large ones. In practice, the micrometeorites being studied in the laboratory after collection in the stratosphere are rarely as large as 50 μm.
Miller-Urey synthesis	formation of organic molecules by the passage of an electric discharge, or energetic radiation, through a mixture of methane and ammonia over refluxing water. It is often used for analogous experiments.
model age	a radiometric age determination based on parameters at least one of which is assumed rather than measured. (The assumed value is generally the initial concentration of the daughter nuclide.) Strictly, it might be argued that even mineral isochrons are model ages in that their interpretation rests upon assumptions such as common initial concentrations, but in most cases the quality of the isochron itself can be taken as an adequate test of those assumptions.
monoclinic	a crystal system characterized by either a single twofold symmetry axis, a single symmetry plane or a combination of the two.
Monte Carlo technique	a computational procedure in which random numbers are used to approximate the solution to otherwise intractable mathematical or physical problems.
monomict	description of a breccia in which the matrix and clasts are of the same class and type.
nakhlite	a rare type of achondritic meteorite consisting of calcic pyroxene (augite) and olivine.
natural remanent magnetization	a remanent magnetization (whose origin is not specified) in a rock specimen.
Neumann lines	lines in kamacite, visible upon etching a polished face with mild acid, caused by mechanical twinning following mild shock.

neutron star	a star whose core is composed primarily of neutrons, as is expected to occur when the mean density is in the range $10^{13} - 10^{15}$ g/cm^3. Under current theories pulsars are thought to be rotating magnetic neutron stars. A neutron star would probably be only 10 to 15 km in diameter with a magnetic field of about 10^{12} gauss, a density of 10^{13} to 10^{15} g/cm^3 (compared with a *white dwarf's* maximum density of about 10^8 g/cm^3, and a central temperature of about 10^9 K, and thus it would be both bluer and dimmer than a white dwarf.
noble gases	the gases He, Ar, Kr, Ne, Xe, Rn which rarely undergo chemical reactions, also known as inert gases and rare gases.
norite	a type of igneous rock containing plagioclase and in which the pyroxene is mainly of the orthorhombic form, rather than monoclinic.
nova	a star that exhibits a sudden surge of energy, temporarily increasing its luminosity by as much as 17 magnitudes or more (although 12 to 14 magnitudes is typical). Novae are old disk-population stars, and are all close binaries with one component, a main sequence star filling its Roche lobe, and the other component a *white dwarf*. Unlike *supernovae,* novae retain their stellar form and most of their substance after the outburst. Since 1925, novae have been given variable-star designations.
obliquity	the angle between an object's axis of rotation and the pole of its orbit.
Occam's razor	an axiom enunciated by William of Ockham (?–1348) of which one translation is: It is vain to do with more what may be done with less.
octahedrite	a type of iron meteorite consisting of a mixture of taenite and kamacite in a characteristic *Widmanstätten pattern,* revealed by mild etching, that reflects the octahedral symmetry of the taenite.
olefin	alternate term for alkene.

onion-shell	a hypothetical chondrite parent body in which petrographic types are arranged concentrically.
Oort cloud	a spherical cloud of comets having semimajor axes > 20,000 AU found by J. H. Oort in his empirical study of the orbits of long-period comets. Comets in this shell can be sufficiently perturbed by passing stars or giant molecular clouds so that a fraction of them acquire orbits that take them within the orbits of Jupiter and Saturn.
ophitic	a texture found in many igneous rocks, characterized by laths of plagioclase partially or completely included with crystals of pyroxene.
ordinary chondrite	collective name for the most common variety of chondritic meteorite, subdivided into H, L and LL groups on the basis of Fe content and distribution.
orthopyroxene	a mineral of the pyroxene group that crystallizes in the *orthorhombic* form.
orthorhombic	a crystal system characterized by three mutually perpendicular two-fold symmetry axes.
oxidation	the process of adding O to, or removing H from, an element (or of increasing the element's valence, i.e., oxidation state).
oxygen fugacity	a function expressing the molar free energy of O in a manner analogous to the way pressure measures free energy of an ideal gas. In practice, O fugacity is equivalent to the partial pressure of O.
paired falls	meteorites specimens originally recovered some distance apart and hence given separate names, but later recognized as fragments of a single parent mass, on the basis of classification, cosmic-ray or gas-retention age, texture, or other diagnostic features.
palaeo-thermometry	the determination of the temperature at which processes have occurred. The method usually involves elemental or isotopic distribution between mineral pairs, or temperature-dependent phase transformations, and the tem-

	perature obtained strictly refers to the temperature at which the system was last able to equilibrate.
pallasite	class of stony-iron meteorites in which the Fe-Ni metal forms a continuous framework enclosing nodules of the silicate olivine.
parent body	the object on or in which a given meteorite or class of meteorites was located prior to ejection as $\sim$ m-sized objects.
parent nuclide	a radioactive nuclide, e.g., ^{238}U.
parsec	measure of distance used in astronomy. 1 parsec is the distance at which 1 AU subtends an angle of 1 arcsecond, equivalent to 3.26 light years.
pc	abbreviation for parsec.
PCP	poorly characterized phase, the term initially applied to an ill-defined mineral found in certain carbonaceous chondrites. It is now known to be either tochilinite or a tochilinite-phyllosilicate intergrowth.
PDB	Pee Dee belemnite, fossil carbonate from the Cretaceous Pee Dee formation in South Carolina. The isotopic composition of its C is used as the international standard for reporting $^{13}C/^{12}C$ ratios.
peridotite	a coarse-grained, olivine-rich plutonic rock.
perihelion	the point in the orbit of an object gravitationally bound to the Sun at which it is nearest to the Sun.
peritectic	a situation in which a phase crystallizes from a mixture of a liquid plus a previously crystallized solid phase, with the newly forming solid growing at the expense of the earlier one.
phenocryst	a relatively large, and therefore conspicuous, grain in a porphyritic rock.
photolysis	chemical decomposition brought about by the action of light.

GLOSSARY

photosphere	the normally visible region of a main-sequence star such as the Sun. Most solar spectroscopic abundance data have been obtained for this region.
phyllosilicate	one of a family of silicate minerals characterized by a structure that consists of sheets or layers, invariably hydrated.
planetary nebula	an expanding (typical expansion rate, 30 to 50 km/s) envelope of rarefied ionized gas surrounding a hot *white dwarf*. The envelope receives ultraviolet radiation from the central star and re-emits it as visible light by the process of fluorescence. The central star may in some cases be as hot as 10^5 K. During the core contraction that terminates the red-giant stage, a shell of material is ejected at a velocity so high that it becomes separated from the core. Planetary nebulae represent a cosmically evanescent stage (about a 16×10^3 yr duration) between the main sequence and the white dwarf state for most modestly massive (<4 $M_\odot$) stars.
planetesimal	small rocky or icy body formed in the primordial solar nebula.
plasmasphere	a region in the upper atmosphere of a planet in which most of the constituents are ionized.
plessite	a fine-grained intergrowth of kamacite and taenite.
pmd	percent mean deviation, a statistical term commonly used as a measure of how far the silicate minerals in a chondrite are from chemical equilibrium.
poikilitic	a rock texture in which many small euhdral mineral grains are contained within a single, generally anhedral, mineral grain.
poikiloblastic	a rock texture in which relatively large recrystallized grains surround relicts of the original minerals.
polymer	a large molecule made by *polymerization*.
polymerization	joining molecules together to form a larger, macromolecular complex. Originally the joined units (mono-

	mers) were identical to each other; now the term commonly includes heteromolecular complexes.
polymict	description of a breccia in which the clasts and/or matrix have differing composition.
Populations I and II	two classes of stars introduced by W. Baade in 1944. In general, Population I (now sometimes called arm population) are young stars with relatively high abundances of metals, and are found in the disk of a galaxy, especially the spiral arms, in dense regions of interstellar gas. Population II (now sometimes called halo population) are older stars with low abundances of metals, and are typically found in the nuclear bulge of a galaxy or in globular clusters. The Sun is a rather old Population I star; old Population I is sometimes called the disk population.
porphyritic	a texture found in many igneous rocks, characterized by relatively large crystals, known as phenocrysts, set in a fine-grained or glassy matrix.
Poynting-Robertson effect	an effect of radiation on a small particle orbiting the Sun that causes it to spiral slowly toward the Sun. It occurs because the orbiting particle absorbs energy and momentum streaming radially outward from the Sun, but reradiates energy isotropically in its own frame of reference.
ppm	parts per million, generally by weight, replaced in most chapters of this book by the term $\mu g/g$.
p-process	the name of the hypothetical nucleosynthetic process thought to be responsible for the synthesis of the rare heavy proton-rich nuclei which are bypassed by the r- and s-*processes*. It is manifestly less efficient (and therefore rarer) than the s- or r-process because the protons must overcome the *Coulomb barrier,* and may in fact work as a secondary process on the r- and s-process nuclei. It seems to involve primarily (p, gamma) reactions at masses above cerium (where neutron separation energies are low). The p-process is assumed to occur in supernova envelopes at a temperature $\lesssim 10^9$ K and at densities $\lesssim 10^4$ g/cm^3.

GLOSSARY

Prairie Network	a network of 16 cameras that provided photometric and astrometric data for bright meteors, ranging in mass from a few g up to hundreds of kg. Because of their spectacular appearance, these objects are termed "fireballs." This network, operated by the Smithsonian Astrophysical Observatory, surveyed the fireball flux over an area of 10^6 km^2 in the central U.S. for a period of about 10 yr. Several thousand fireballs were photographed, and reduced data are available for about 335 objects.
purine	a heterocyclic compound containing fused *pyrimidine* and imidazole rings.
pyrimidine	a heterocyclic organic compound containing two N atoms within the six-membered ring.
pyrolysis	heating and decomposition under controlled conditions, typically in a vacuum or an inert atmosphere.
quantum tunelling	penetration by a particle into a potential energy region which is classically forbidden.
racemic	that which consists of equal molar proportions of both left- and right-handed optically active forms.
radiogenic	that which is made by radioactive decay.
radiogenic nuclide	a nuclide produced by decay of a radioactive parent nuclide, e.g., ^{206}Pb produced from the decay of ^{238}U.
radionuclide	a nuclide that is unstable against radioactive decay.
Raman spectroscopy	analysis of the characteristic spectrum produced when monochromatic light is scattered by a transparent substance.
reaction cross section	proportionality constant which relates the abundance of a target nucleus to the rate at which a given nuclear reaction occurs.
red giant	a late-type (K or M) high-luminosity (brighter than $M_v = 0$) star of very large radius that occupies the upper-right portion of the *HR diagram*. Red giants are

post-main-sequence stars that have exhausted the nuclear fuel in their cores, and whose luminosity is supported by energy production in a H-burning shell. Within the lifetime of the Galaxy, only main-sequence stars of type F and earlier have had time to evolve to the red-giant phase (or beyond). The red-giant phase corresponds to the establishment of a deep convective envelope. Red giants in a globular cluster are about 3 times more luminous than RR Lyrae stars in the same cluster. Red supergiants have a maximum luminosity of $M_v \sim -8$.

reduction process involving the addition of H or removal of O or decreasing the valence of an element, the opposite of *oxidation*.

REE acronym for rare-earth elements, the lanthanide series in the periodic table.

refractory term describing the high-temperature stability of an element or phase, the opposite of *volatile*.

regolith as used in this book, the fragmented layer found on the surface of many planetary or subplanetary objects. It is created from the local competent lithologies by *meteoroid* impact and subsequently comminuted and turned over by such impacts.

resonance a dynamical configuration of bodies in which Newtonian gravitation maintains *commensurability*.

RNAA acronym for radiochemical neutron-activation analysis, an analytical technique for determining trace and ultra-trace elements in meteorites.

r-process the capture of neutrons on a very rapid time scale (i.e., one in which a nucleus can absorb neutrons in rapid succession, so that regions of great nuclear instability are bridged), a theory advanced to account for the existence of all elements heavier than Bi as well as the neutron-rich isotopes heavier than Fe. The essential feature of the r-process is the production and consumption of great numbers of neutrons in a very short time (< 100 s). The presumed source for such a large flux of neutrons is a

GLOSSARY

	supernova, at the boundary between the neutron star and the ejected material.
rubble-pile	a hypothetical chondrite parent body in which petrographic types are arranged in random fashion.
saturation magnetization	a magnetization observed when magnetic moments in all magnetic domains are aligned in the same direction under a strong magnetic field.
secondary ion mass spectrometry	alternate name for analysis by means of the *ion microprobe*.
SEM	scanning electron microscope.
shergottite	a rare type of meteorite, consisting of pyroxene (pigeonite) and maskelynite.
shock	momentary excursion to high pressures (and temperatures) caused by impact.
shock lithification	welding of *regolith* by shock-induced intergranular melting.
shock metamorphism	alteration of a rock by shock-induced mechanical deformation or phase transformation above or below the solidus.
shock wave (or shock front)	discontinuity in temperature and pressure propagating in a solid, liquid or gas with supersonic velocity, caused by impact or explosion.
siderophile	one of the geochemical classes of elements. Siderophile elements are those which tend to go into the metal phase, e.g., Ni, Co, Au, As, Ge, Ga, Ir, Os, Re.
SMOW	acronym for standard mean ocean water, the reference standard for the measurement of O and H isotopes.
solar flare	sudden outburst of energy on the solar *photosphere*.
solar nebula	a disk of gas and dust around the proto-Sun.

solar wind	expansion of the solar corona to form a stream of ions away from the Sun.
solid solution	a substance in which two or more components, such as the atoms of two or more elements, are randomly mixed on so fine a scale that the resulting solid is homogeneous.
solidus	the line or surface in a phase diagram below which the system is completely solid.
solvus	the curved line in the phase diagram of a binary system separating a field of homogeneous solid solution from a field of two or more phases.
spallogenic	that which is produced by the violent partial disintegration of an atomic nucleus; in a meteoritic context, this is generally due to cosmic-ray bombardment.
spectroscopic binaries (SB)	stars whose binary nature can be detected from the periodic Doppler shifts of their spectra, owing to their varying velocities in the line of sight. Spectroscope binaries are typically of spectral type B, with nearly circular orbits whereas long-period M-type spectroscopic binaries have highly eccentric orbits.
s-process	a process in which heavy, stable, neutron-rich nuclei are synthesized from iron-peak elements by successive captures of free neutrons in a weak neutron flux, so that there is time for beta decay before another neutron is captured (cf., *r-process*). This is a process of nucleosynthesis that is believed to take place in intershell regions during the *red-giant* phase of evolution, at densities up to 10^5 g/cm^3 and temperatures of about 3×10^8 K (neutron densities assumed are 10^{10}/cm^3).
sputtering	expulsion of atoms or ions from a solid, caused by impact of energetic particles.
STEM	scanning transmission electron microscopy.
Stokes drag	viscous drag law stating that the force which retards a body moving through a fluid is directly proportional to the velocity, the radius of the body, and the viscosity of

	the fluid. Stokes drag applies for low Reynolds number and if the mean free path of fluid molecules is small compared with the body's size (*see* Epstein drag).
Strecker-cyanohydrin synthesis	the synthesis of amino acids from aldehydes, HCN and NH_3 in the presence of an aqueous fluid.
suevite	a fragmental breccia containing melt particles of the same composition formed in a single impact event.
supernovae, Type I	supernovae which have a spectrum characterized by detailed structure without a well-defined continuum, and appear to be deficient in hydrogen. They occur in the halo and in old disk populations (stellar *Population II*) in galaxies.
supernovae, Type II	supernovae which have a spectrum characterized by a continuum with superimposed hydrogen lines and some metal lines. They occur in the arms of spiral galaxies, implying that their progenitors are massive stars (stellar *Population I*).
Surveyor glass	a glass lens filter from the Surveyor III camera that was returned to Earth by the Apollo 12 astronauts after it had experienced 2.5 yr of exposure to the space environment while on the lunar surface.
taenite	iron with a face-centered cubic structure that is stable at high temperatures and/or when alloyed with a suitable proportion of a face-centered metal such as Ni.
TEM	transmission electron microscope.
terrestrial age	the period of time since the fall of a meteorite.
Teller-Redlich product rule	approximation technique which permits calculation of reduced partition function ratios from vibrational frequencies.
thermo-luminescence (TL)	an emission of light brought about by heating certain natural materials.

thermoremanent magnetization	a remanent magnetization acquired by a rock specimen when it was cooled in a magnetic field.
T Tauri stars	young, late-type stars that are precursors to solar-mass stars characterized by emission line spectra, infrared excesses and irregular variability. The prototype for this class of stars is T Tau.
turbostratic	partially ordered polymer-like structure.
ultramafic	an igneous rock consisting predominantly of mafic silicate minerals.
unit of magnetic field	$1\ T = 10^4$ Oe. Geomagnetic field is about 0.5 Oe on the Earth's surface.
unit of magnetization	AM^2/KG = emu/G.
ureilite	class of carbon-rich achondritic meteorite in which the silicates consist of olivine and pigeonite.
van der Waals attraction	the relatively weak attraction forces operative between neutral atoms and molecules.
vesicle	bubble-shaped cavity in a mineral or rock, generally produced by expansion of gas in a magma.
volatile	an element that condenses from a gas or evaporates from a solid at a relatively low temperature.
vug	a *vesicle*.
white dwarf	a star of high surface temperature (typically on the order of 3×10^4 K), low luminosity, and high density (10^5 to 10^8 g/cm^3) with roughly the mass of the Sun (average mass about 0.7 $M_\odot$) and the radius of the Earth, which has exhausted most or all of its nuclear fuel, believed to be a star in its final stage of evolution. The typical escape velocity from a white dwarf is 3000 km/s. The minimum possible pulsation period for a static white dwarf under general relativity is 1 s. DA white dwarfs are H rich; DB white dwarfs are He rich; DC are pure continuum; DF are Ca rich; DP are magnetic stars (some have magnetic